CLINICAL APPLICATIONS FOR MUSCLE AND MUSCULOCUTANEOUS FLAPS

CLINICAL APPLICATIONS FOR MUSCLE AND MUSCULOCUTANEOUS FLAPS

STEPHEN J. MATHES, M.D.

Associate Professor, Department of Surgery,
Division of Plastic and Reconstructive Surgery,
University of California at San Francisco,
San Francisco, California

FOAD NAHAI, M.D.

Assistant Professor, Department of Surgery,
Division of Plastic and Reconstructive Surgery;
Director of Microvascular Surgery,
Emory University School of Medicine,
Atlanta, Georgia

with 13 contributors

with 1,053 *illustrations*
including 41 *in color*

The C. V. Mosby Company

ST. LOUIS • TORONTO • LONDON 1982

MOSBY

A TRADITION OF PUBLISHING EXCELLENCE

Editor: Karen Berger
Editing supervisor: Elaine Steinborn
Manuscript editor: Ivie Lewellen Davis
Design: Susan Trail
Production: Carol O'Leary

Printed in the United States of America

The C.V. Mosby Company
11830 Westline Industrial Drive, St. Louis, Missouri 63141

Library of Congress Cataloging in Publication Data

Mathes, Stephen J.
Clinical applications for muscle and musculocutaneous flaps.

Bibliography: p.
Includes index.
1. Muscles—Transplantation. 2. Skin—Transplantation. 3. Flaps (Surgery) I. Nahai, Foad, 1943- . [DNLM: 1. Surgical flaps. WE 500 M427c] RD125.M37 617'.95 81-18913
ISBN 0-8016-3164-5 AACR2

C/CB/B 9 8 7 6 5 4 3 2 1 01/C/064

CONTRIBUTORS

BERNARD S. ALPERT, M.D.
Assistant Professor, Department of Surgery, Division of Plastic and Reconstructive Surgery, University of California at San Francisco, San Francisco, California

PHILLIP G. ARNOLD, M.D.
Assistant Professor of Plastic Surgery, Department of Plastic and Reconstructive Surgery, Mayo Clinic, Mayo Graduate School of Medicine, Rochester, Minnesota

DAVID K. BONE, M.D.
Assistant Professor, Department of Surgery, Division of Cardiothoracic Surgery, Emory University School of Medicine, Atlanta, Georgia

JOHN BOSTWICK III, M.D.
Associate Professor, Department of Surgery, Division of Plastic and Reconstructive Surgery, Emory University School of Medicine, Atlanta, Georgia

JOHN J. COLEMAN III, M.D.
Assistant Professor, Department of Surgery, Emory University School of Medicine, Atlanta, Georgia

JACK FISHER, M.D.
Assistant Professor of Plastic Surgery, Department of Plastic and Reconstructive Surgery, Mayo Clinic, Mayo Graduate School of Medicine, Rochester, Minnesota

T. RODERICK HESTER, M.D.
Assistant Professor, Department of Surgery, Division of Plastic and Reconstructive Surgery, Emory University School of Medicine, Atlanta, Georgia

H. LOUIS HILL, Jr., M.D.
Plastic and Reconstructive Surgeon, Tallahassee Regional Medical Center, Capital Medical Center, Tallahassee Plastic Surgery Clinic, Tallahassee, Florida

MAURICE J. JURKIEWICZ, M.D.
Professor and Chief, Department of Surgery, Division of Plastic and Reconstructive Surgery, Emory University School of Medicine, Atlanta, Georgia

STEPHEN J. MATHES, M.D.
Associate Professor, Department of Surgery, Division of Plastic and Reconstructive Surgery, University of California at San Francisco, San Francisco, California

FOAD NAHAI, M.D.
Assistant Professor, Department of Surgery, Division of Plastic and Reconstructive Surgery; Director of Microvascular Surgery, Emory University School of Medicine, Atlanta, Georgia

MICHAEL SCHEFLAN, M.D.
Assistant Professor of Plastic Surgery, Department of Surgery, Division of Plastic Surgery, The Medical College of Virginia, Richmond, Virginia

WILLIAM J. SCHNEIDER, M.D.
Plastic and Reconstructive Surgeon, East Tennessee Baptist Hospital, East Tennessee Children's Hospital, Fort Saunders Presbyterian Hospital, Knoxville, Tennessee

LUIS O. VASCONEZ, M.D.
Professor and Chief, Department of Surgery, Division of Plastic and Reconstructive Surgery, University of California at San Francisco, San Francisco, California

ROBERT L. WALTON, M.D.
Assistant Professor, Department of Surgery, Division of Plastic and Reconstructive Surgery, University of California at San Francisco, San Francisco General Hospital, San Francisco, California

To

Maurice J. Jurkiewicz

The great possession of any university is its great names . . . Not its wealth nor the number of its schools, not the students who throng its halls—but the men who have trodden in its service the thorny road through toil.

Osler

FOREWORD

Longevity in the academic life occasionally rewards one with an opportunity to observe the progression of people or programs from simple beginnings to great heights of excellence. This book, *Clinical Applications for Muscle and Musculocutaneous Flaps*, seems to invoke such a feeling.

In early 1971 there was no plastic and reconstructive training program at Emory University. Later that year, Dr. Maurice J. Jurkiewicz joined our faculty and less than a decade later his program was renowned throughout the world. The academic progeny of "Dr. J." are now scattered from coast to coast. Drs. Mathes and Nahai, members of this group, also trained in General Surgery at Emory, where they compiled enviable records. Their professional success has been accompanied by graciousness and good humor, engaging qualities in those achieving distinction at an early age.

The broad principles espoused in this book and the wide-ranging applicability of these reconstructive techniques are eloquent testimony to the authors' commitment to the total care of the surgical patient. Certainly this work should mute the commonly heard theme that plastic and reconstructive surgery serves only the wealthy and vain. There is great satisfaction in having played even a small part in the events that led to this outstanding surgical contribution.

W. Dean Warren, M.D.

Whitehead Professor and Chairman, Department of Surgery
Emory University School of Medicine
Atlanta, Georgia

PREFACE

"What we need to see now are a few harpsichords, rather than so many logs--recognizable, new, artistic, and fully acceptable noses, cheeks, chins, necks, legs, and arms rather than indistinguishable globs and blobs of transported tissue in those areas.

The older meaning of plastic is molder or sculptor, and this is the central reason for our existence as a specialty."*

The principle of transposition of the muscle and musculocutaneous flap has progressed from the initial phase of amazement to its present phase of refinement. Fortunately, this phase of refinement in reconstructive surgery is a never-ending process and as such is not complete. However, progress in the clinical application of the muscle and musculocutaneous flap is extensive and includes modifications in flap design and techniques for flap transposition and now transplantation. McDowell further states in the editorial cited above, "But there is a difference between a hod carrier and a sculptor, and the time has come to produce more finished works of art--something that will not only function like a leg (or whatever), but that will look like a leg (or whatever)." In the hope of demonstrating when and how the clinical application of muscle and musculocutaneous flaps will provide the "harpsichord" and not the "log" and the "finished result" and not the "blobby flap," this text is written.

Unlike the traditional tubed skin flap, transposition of the muscle and musculocutaneous flap is accomplished in one operation. When secondary revisions are necessary to achieve the optimal result, the advantage of the single operation is lost. With the rapid inflation of medical costs, the reconstructive surgeon is challenged to shorten the reconstructive process while providing the optimal result. The clinical application for muscle and musculocutaneous flap is presented for each body region, using the unique vascular anatomy for single-staged reconstruction and emphasizing flap refinements to achieve form and function.

The vascular anatomy of muscle and musculocutaneous flaps is the basis of its wide clinical application. The first part of the text presents the principles of muscle transposition. Emphasis is placed on vascular anatomy and its application to achieve the optimal result when the muscle or musculocutaneous flap is used in reconstructive surgery.

The muscle and musculocutaneous flap is but one method in the reconstructive ladder that ranges from the skin graft to the more complex methods of distant flap transfer or transplantation. The second part of this text consists of the application of muscle and musculocutaneous flaps for each specific region. Within each section, the systematic approach to flap selection initially relates the role of specific muscle and musculocutaneous flaps to the other methods in this reconstructive ladder for specific defects in the region. When the muscle or musculocutaneous flap is appropriate for a specific defect, the chapters dealing with the reconstruction of each region contain specific information regarding design and techniques

*From McDowell, F.: Plast. Reconstr. Surg. **64**:249, 1979.

for use of muscle and musculocutaneous flaps to achieve defect reconstruction. Finally, specific complications inherent to each flap are presented so that potential causes of flap failure are avoided by the reconstructive surgeon.

Microsurgery has released the muscle and musculocutaneous flap from a specific arc of rotation allowing use of certain flaps for any body region. Despite the increased complexity of composite tissue transplantation, this method of reconstruction is already standard for certain problems encountered by the reconstructive surgeon. In some instances composite tissue transplantation provides a solution for a difficult wound where none was formerly available. The third part of this text stresses the utilization of flap transplantation where other methods commonly fail to provide the optimal result.

In the final and fourth part of this text, the immediate and long-term evaluations of the muscle and musculocutaneous flap are examined. The current period of refinement as related to use of the muscle and musculocutaneous flap is filled with rapid changes in flap selection and design. The basis of these changes is presented with the expectation that future refinements are forthcoming and will further enable the reconstructive surgeon to uniformly achieve the optimal result.

ACKNOWLEDGMENTS

It is both the unique privilege and responsibility of the plastic surgeon to provide the best solution for the wound wherever it may occur. At this time, the muscle and musculocutaneous flap has become a safe method to achieve this goal. The evaluation and treatment of the problem wound require a thorough understanding of the broad principles of surgery. Acquisition of this knowledge was possible for us under the leadership of W. Dean Warren, Chairman, Department of Surgery, at Emory University in Atlanta. The advances in our knowledge of the muscle and musculocutaneous flap have occurred under the guidance of Maurice Jurkiewicz. These accomplishments in the field of reconstructive surgery reflect the dedication and enthusiasm of Maurice Jurkiewicz as a teacher of surgery, first to the medical student and then to the house officer. This unique teacher has directly touched each contributor to this text.

Our contributors have special interest in reconstructive surgery. Each is knowledgeable in its entire field and has expanded the clinical applications of muscle and musculocutaneous flaps. Although each contributor may have a chapter on one specific region of interest, all have written in our surgical literature on their approach and solution to the reconstruction of problem wounds in all regions. Each has provided encouragement to us throughout our interest in reconstructive surgery and specifically during the preparation of this text. Each chapter presents the personal experience of its authors in the management of problems specific to a region. For these chapters we thank Bud Alpert for his contributions to the chapters on head, neck, pressure sore, and lower extremity complications and his chapter on assessment of muscle and musculocutaneous flap viability; P. G. Arnold for his chapter on chest wall reconstruction; David Bone for his contributions to the chapter on reconstruction of the sternum; John Bostwick for his chapter on breast reconstruction and his contributions to the chapter on posterior trunk reconstruction; Jack Coleman for his contributions to the chapter on long-term follow-up of muscle and musculocutaneous flaps; Jack Fisher for his contributions to the chapter on reconstruction of the sternum; Rod Hester for his contributions to the chapters on head, neck, groin, perineum, and pressure sore reconstruction; H. Lewis Hill, Jr., for his contributions to the chapter on groin, perineum, and pressure sore reconstruction; Maurice Jurkiewicz for his contributions to the chapter on head and neck reconstruction; Michael Scheflan for his contributions to the chapters on posterior trunk, abdomen, and foot reconstruction; William Schneider for his contributions to the chapter on pressure sore reconstruction; Luis Vasconez for his contributions to the chapters on head and neck, pressure sore, and lower extremity complications and lower extremity reconstruction; and Robert Walton for his chapter on the inferior gluteal thigh flap.

Without the support of Paul Ebert, Professor and Chairman, Department of Surgery, University of California, and Luis Vasconez, Professor of Surgery, and Chief, Division of Plastic Surgery, this text could not have been completed. The Department of Anatomy generously supplied their staff and the required cadaver specimens. All dissections for this text were performed in the anatomy laboratories of University of California, School of Medicine, San Francisco. A special note of appreciation to Jerry D. Ritchey and Joe Hobson for their help in the anatomy laboratories. Also, a special note of appreciation to Harry K. Genant, Professor of Radiology, and Jim Stoughton for their assistance in the radiologic evaluation of muscle vascular anatomy.

The process of completing a text places a hardship on everyone associated with its contributors. Without the support of our families and those of our contributors, this text would not be possible. Luis Vasconez, Robert Walton, and Bud Alpert in San Francisco and John Bostwick, Rod Hester, and Jack Coleman in Atlanta have provided both encouragement and care of our patients during our frequent absences for text preparation. Our nurses, especially Marie Hugenberg in San Francisco and Kay Jones in Atlanta, and our secretaries, Kathy Garcia and Frances Torney of San Francisco and Lissa DeProdocini of Atlanta, have endured our frequent schedule changes while supporting our efforts to complete this book.

We especially thank Karen Berger, our editor, who has patiently guided us over the past 4 years during the preparation of this text. Throughout this period Ms. Berger and her staff have enthusiastically worked alongside us to complete this text.

We thank Vicki Moses Friedman of St. Louis, our primary medical illustrator, Patsy Bryan of Atlanta, and Rick Soloway of San Francisco, our primary photographer, for their excellent work and willingness to adjust to our irregular schedules.

Stephen J. Mathes
Foad Nahai

CONTENTS

CLINICAL APPLICATIONS FOR MUSCLE AND MUSCULOCUTANEOUS FLAPS

I

PRINCIPLES

1

SELECTION OF THE MUSCLE AND MUSCULOCUTANEOUS FLAP IN RECONSTRUCTIVE SURGERY

Stephen J. Mathes ■ Foad Nahai

Techniques available to the reconstructive surgeon for the management of a surgical defect or chronic wound are considered as a reconstructive ladder—progressing from simple to more complex procedures.

Reconstructive ladder

Complex	Distant flap
↑	↑
	Local flap
	↑
	Skin grafts
	↑
Simple	Direct closure

Selection of a technique is based on its ability to satisfy the particular reconstructive requirements of the defect. Success in reconstructive surgery requires viable coverage and restoration of form, contour, and function.

COVERAGE

The majority of surgical defects or chronic wounds maintain local capillary circulation within both the wound base and adjacent skin. Thus direct closure, skin grafts, or local flaps provide coverage.

Progression up the reconstructive ladder becomes necessary when local wound circulation is impaired (e.g., radiation vasculitis). Local random flaps generally have vascular compromise by the same mechanism that caused the nearby wound. Therefore the flap with an intact vascular pedicle distant to the defect is selected for coverage.

Axial and muscle flaps have specific vascular pedicles. When these pedicles are distant from the wound or surgical defect, the flap is considered for use in defect coverage. Unfortunately, few axial pattern flaps have been identified, including temporal, deltopectoral, groin, and dorsalis pedis. Muscles are available in most body regions. Each muscle has potential as a muscle or musculocutaneous flap. Approximately 37 muscles and 20 musculocutaneous flaps have been described (Table 1). Selection of a particular axial flap or muscle is based on its ability to provide safe coverage because of a vascular pedicle distant to the wound.

FORM AND CONTOUR

Although coverage is essential, selection of techniques from the reconstructive ladder is also based on restoration of form and contour. The technique that best restores form and contour is selected from the methods with the ability to provide safe coverage.

The skin graft provides ideal coverage for the shallow wound. Local tissue rearrangement maintains form and contour in the deep small defect. The defect with missing muscle and skin (e.g., extirpative head and neck surgery and mastectomy) may benefit from a muscle or musculocutaneous flap to improve contour as well as coverage.

In certain regions, when local tissue is unavailable for coverage (e.g., lower leg and scalp), transposition of the muscle only with skin grafts may be superior to the bulk associated with the musculocutaneous flap. The recent expansion in available techniques in the reconstructive ladder provides surgeons with adequate choice in flap selection to avoid bulky flaps for thin defects despite the complexity of specific wounds.

FUNCTION

Functional requirements at the site of reconstruction vary, depending both on location and sensibility. In regions where trauma is frequent and sensibility lost, flap coverage is desirable. However, skin grafts are used in areas of minimal trauma or where sensibility exists.

Emphasis on function at the site of reconstruction is secondary only to the necessity for successful wound or defect coverage. In acute injury safe coverage must not be jeopardized to achieve function. Skin grafts are readily removed later and replaced by the selected flap to provide durable coverage after resolution of the acute injury phase (e.g., burns and lower extremity trauma).

SELECTION OF A SPECIFIC MUSCLE OR MUSCULOCUTANEOUS FLAP

When transposition of a muscle or musculocutaneous flap is planned to accomplish a specific reconstruction, selection of the muscle has been initially based on its potential to provide coverage. The reliability for successful coverage has earned the muscle and musculocutaneous flaps a position in the armamentarium of the reconstructive surgeon.

With continuing anatomic study and clinical experience, refinements in muscle selection, flap design, and techniques of elevation have occurred. Transposition of muscle and musculocutaneous flaps now provides reliable coverage and form, contour, and function preservation. These refinements have resulted in greater use of a smaller number of muscles in the practice of reconstructive surgery. These muscles include the following in the order of frequency:

Latissimus dorsi
Pectoralis major
Tensor fascia lata (TFL)
Gluteus maximus
Gracilis
Gastrocnemius
Soleus
Rectus abdominis
Trapezius
Platysma

The frequency in which a particular muscle flap is used reflects not only the proximity of the muscle to common wounds (e.g., mastectomy defect, head and neck defect, and pressure sores) but also the specific features of muscle anatomy.

Requisites for muscle flaps

Requisites for muscle selection for use as a muscle or musculocutaneous flap are presented as criteria for safe and successful transposition. The emphasis on function preservation is now greatly expanded, since this refinement is applicable only to specific muscles. These guidelines allow selection of the muscle best suited to accomplish reconstruction in all body regions.

Size of muscle belly

The size of the muscle belly distal to the point of rotation must be adequate to cover the defect. Only with microvascular transplantation can the entire muscle belly be used for defect coverage. A part of the muscle length is also used to reach the defect after transposition. Use of muscles immediately adjacent to the defect often avoids objectionable muscle bulk between the point of flap rotation and the defect. Although upper and lower motor neuron diseases are associated with fatty changes within the muscle, the potential for defect coverage are not altered.

Size of cutaneous territory

The skin territory of superficial muscles is based on musculocutaneous perforators. Since the pattern of musculocutaneous perforators varies with each muscle, a knowledge of their location affects flap design and usefulness. A separate chapter is included in this text to present the anatomy of the musculocutaneous perforators and its clinical significance.

Location of vascular pedicles

The entire basis for the success of the muscle flap is based on a pattern of circulation that is consistent in location and resistant to the effects of radiation therapy and superficial trauma. This robust circulation allows safe transposition of both muscle and overlying skin with dimensions far exceeding the traditional random flap. However, the suitability of each muscle for both transposition or transplantation by microvascular surgery is different. This difference is based on patterns of circulation and is determined by a study of location and size of vascular pedicles to muscle. The number and locations of vascular pedicles critical to muscle survival are variable. The relationship of the pedicle(s) to the muscle belly determines the potential arc of rotation and resultant clinical usefulness. A separate chapter is devoted to anatomic studies that define the patterns of circulation to muscle and their clinical significance.

Accessibility

The majority of muscles useful as transposition flaps have a superficial location. With the knowledge of muscle origin and insertion in relationship to external landmarks, flap design and elevation are not complicated. The ease of dissection for all muscles is presented in Table 1.

Preservation of function

If muscle origin or insertion or motor nerve is divided during flap elevation, then the muscle will no longer serve its original function. This may well result in functional disability or alteration of form and contour. In selecting a particular muscle for reconstructive purposes, it is therefore essential to consider the potential functional loss.

The use of some muscles will not lead to significant functional impairment or contour deformity (e.g., soleus), whereas in other muscles the use of the entire muscle will lead to significant functional loss (e.g., gluteus maximus muscle in an ambulatory patient). Similarly, the elevation of a pectoralis major muscle may lead to loss of form in the anterior axillary fold. Thus options are available to minimize or eliminate these losses (i.e., the preservation of function through synergistic muscles or surgical techniques for function preservation).

Synergistic muscles. When a muscle from a group of synergistic muscles is transposed as a muscle flap, the functional loss is compensated by the remaining synergistic muscles. The gracilis, an adductor of the thigh, is expendable, because the remaining more powerful adductor muscles will preserve function. If one muscle of the hamstring group is used, function is preserved by the remaining two muscles.

Function-preserving techniques. Through the exact definition of the blood supply to muscle, two techniques have evolved for transposition of a part of a muscle while preserving function and form. The first technique is applicable to the muscles of the extremities with long tendons, and the second is more suited to the broad, flat muscles of the trunk.

For lower extremity coverage the application of one belly of the gastrocnemius or the soleus will not result in significant impairment of ankle plantar flexion. However, if the flexor digitorum longus, tibialis anterior, or the long extensor of the toes is divided for transposition as a flap, functional deficits are easily detected (e.g., tibialis anterior foot drop). For coverage of small defects careful dissection of the distal muscle from the tendon allows transposition of the muscle without disruption of the muscle tendon unit. Latex injections of muscle before and after flap elevation using this technique demonstrate vascularity of both muscle and tendon. This technique will only allow elevation of a small flap limited in length to 7 to 8 cm (Fig. 1-1).

The transposition of the pectoralis major and latissimus dorsi muscle flaps leads to the loss of the anterior and posterior axillary folds, respectively. Loss of the posterior axillary fold is minimized by the presence of the teres major muscle. However, if the entire pectoralis major muscle is elevated and the insertion divided, then a significant deficit of the anterior axillary fold results. Through a precise knowledge of the blood supply, intramuscular vascular patterns, and motor innervation, it is possible to use both these muscles as flaps and still preserve function and form.

Half of the latissimus dorsi muscle may be elevated and transposed anteriorly onto the chest wall. During the dissection for flap elevation, the lower origin is divided. The insertion and motor nerve are not disturbed, thus preserving form and function (Fig. 1-2).

The use of the distally based latissimus dorsi muscle flap also serves to preserve function. When the latissimus dorsi muscle is based on the thoracodorsal pedicle and advanced posteriorly for coverage of the back, the entire origin is divided. However, the insertion and motor nerve are undisturbed. The flap is then sutured either to the opposite latissimus dorsi muscle if bilateral flaps are used or to the opposite edge of the defect if a unilateral flap is used. In either case the origin of the muscle is reestablished. Function and form are preserved.

The pectoralis major has a dominant thoracoacromial pedicle, entering the muscle several centimeters from the insertion, and segmental secondary pedicles, branches of the internal mammary artery, entering the muscle medially close to its origin. The motor nerves, the medial and lateral pectoral nerves, enter the muscle with the thoracoacromial pedicle. When the pectoralis major muscle is transposed medially, based on the distal segmental vascular pedicles for coverage of sternal defects, the anterior axil-

Text continued on p. 14.

A

B

C

D

E

F

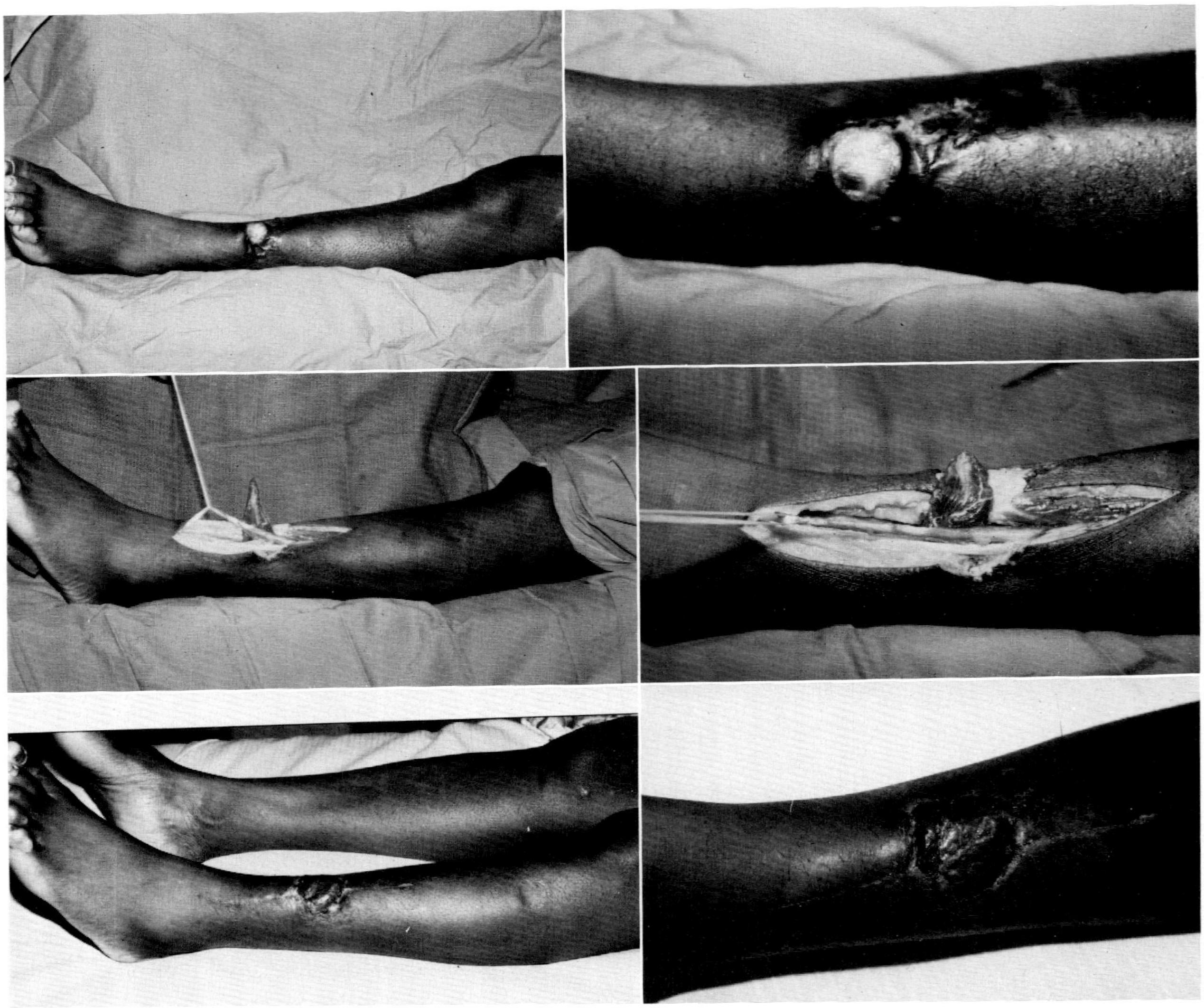

Fig. 1-1. Extensor hallucis longus muscle and skin graft for coverage of small defect of lower third of tibia with function preservation.

A and **B,** Small traumatic anterior tibial defect of lower third of leg.

C, Extensor hallucis separated from tendon. NOTE: Great toe extension with retraction of preserved tendon.

D, Muscle transposed into defect and skin grafted.

E, Postoperative lateral view demonstrates full extension of great toe.

F, Stable wound coverage provided by muscle flap and skin graft.

A

B

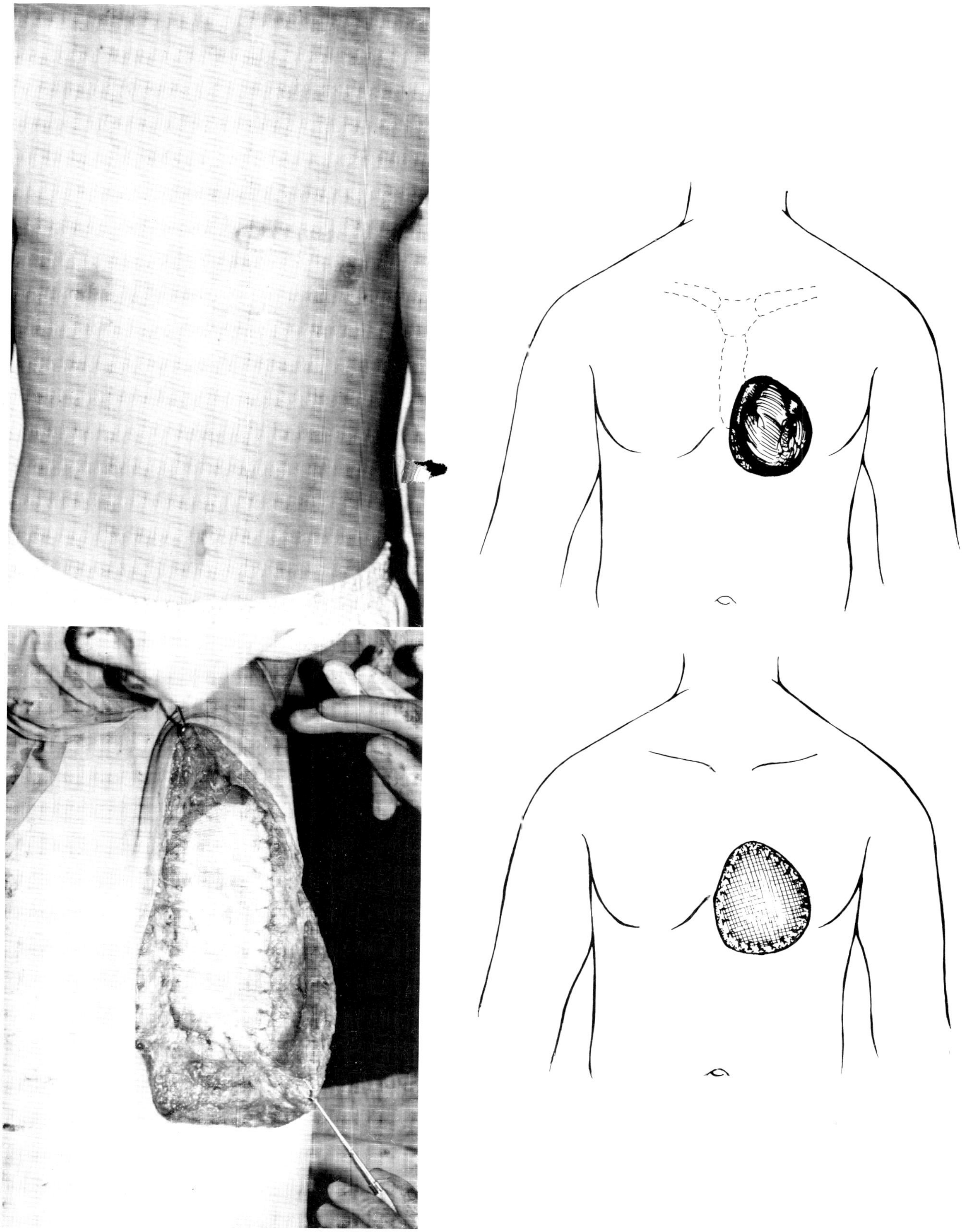

C

D

Fig. 1-2. Reconstruction of anterior chest wall with synthetic mesh and split latissimus dorsi muscle with preservation of muscle function and form.

A and **B,** Fifteen-year-old boy after resection of ribs and sternum for chondroma has thin skin coverage directly over pericardium.

C and **D,** Ribs and sternal defect reconstructed with synthetic mesh.

Continued.

E

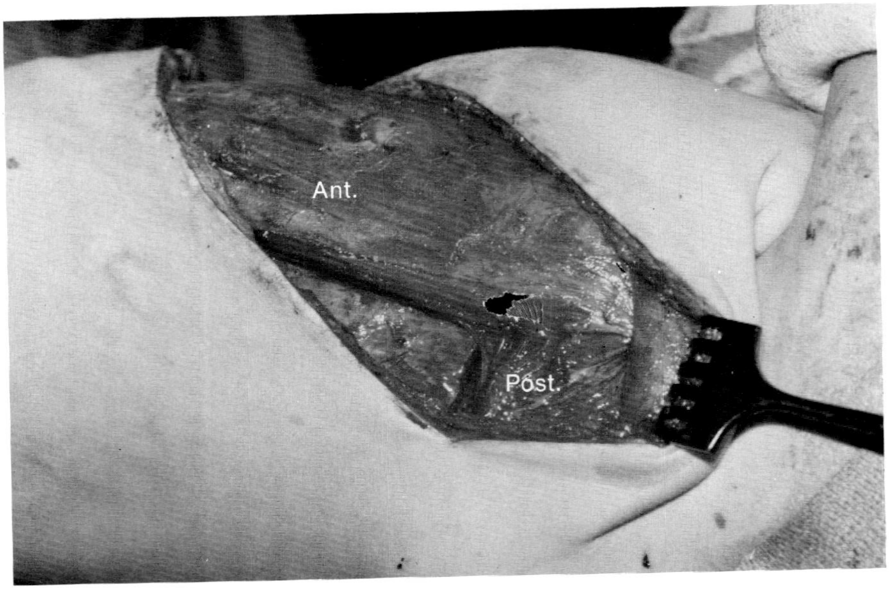

F

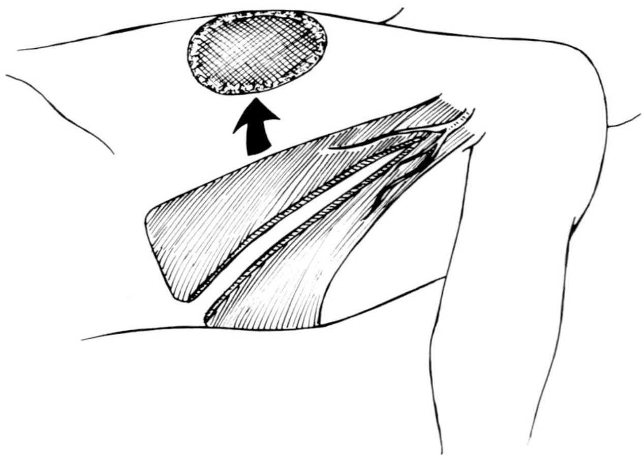

Fig. 1-2, cont'd
E and **F**, Anterior half of latissimus dorsi is elevated, *(ant)* preserving vessels and motor nerve to posterior half of muscle *(post)*.

G H I J

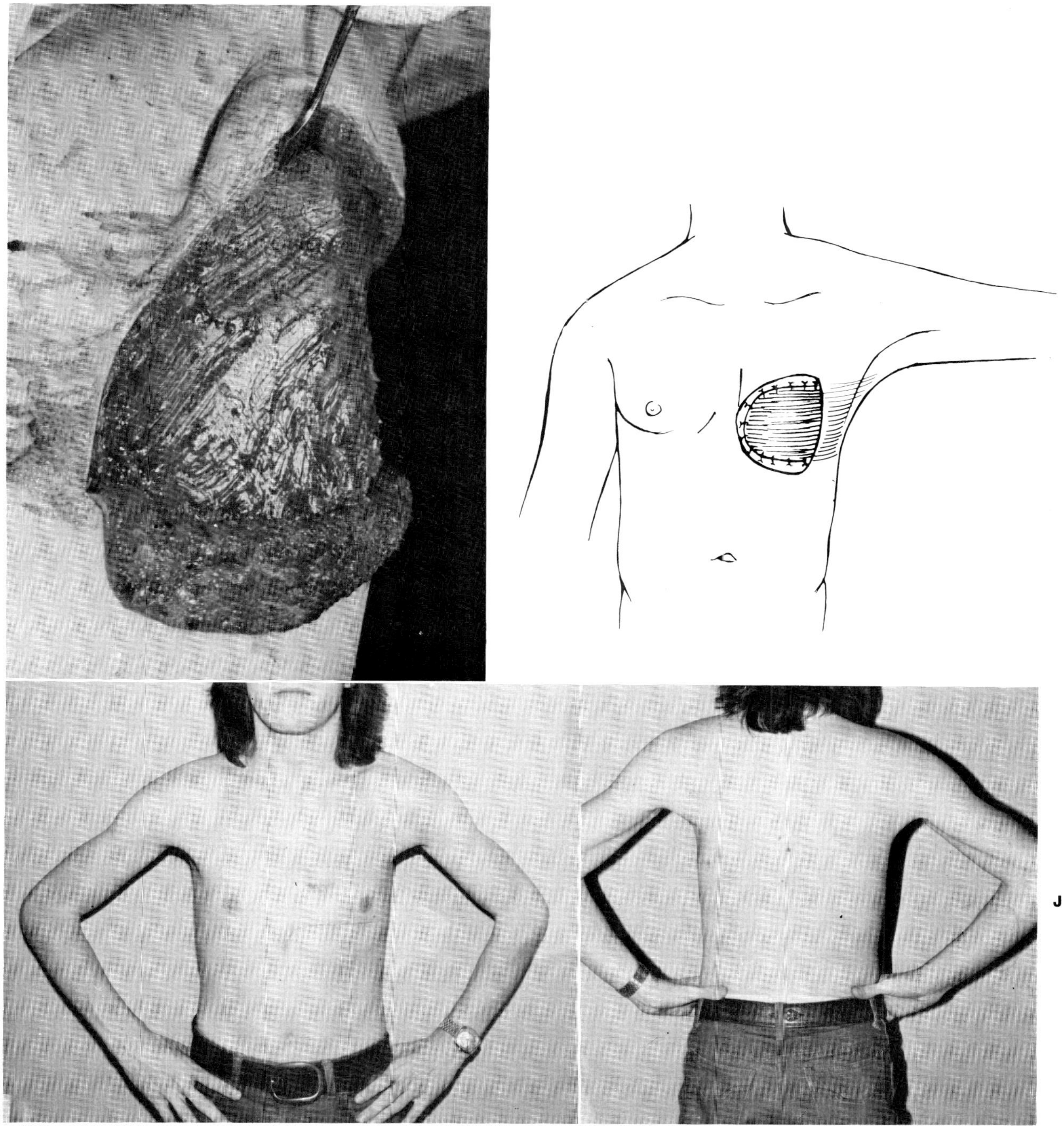

Fig. 1-2, cont'd
G, Latissimus dorsi transposed anteriorly.
H, Muscle covers mesh.
I, Three-month postoperative view.
J, Preservation of left posterior axillary fold and muscle function.

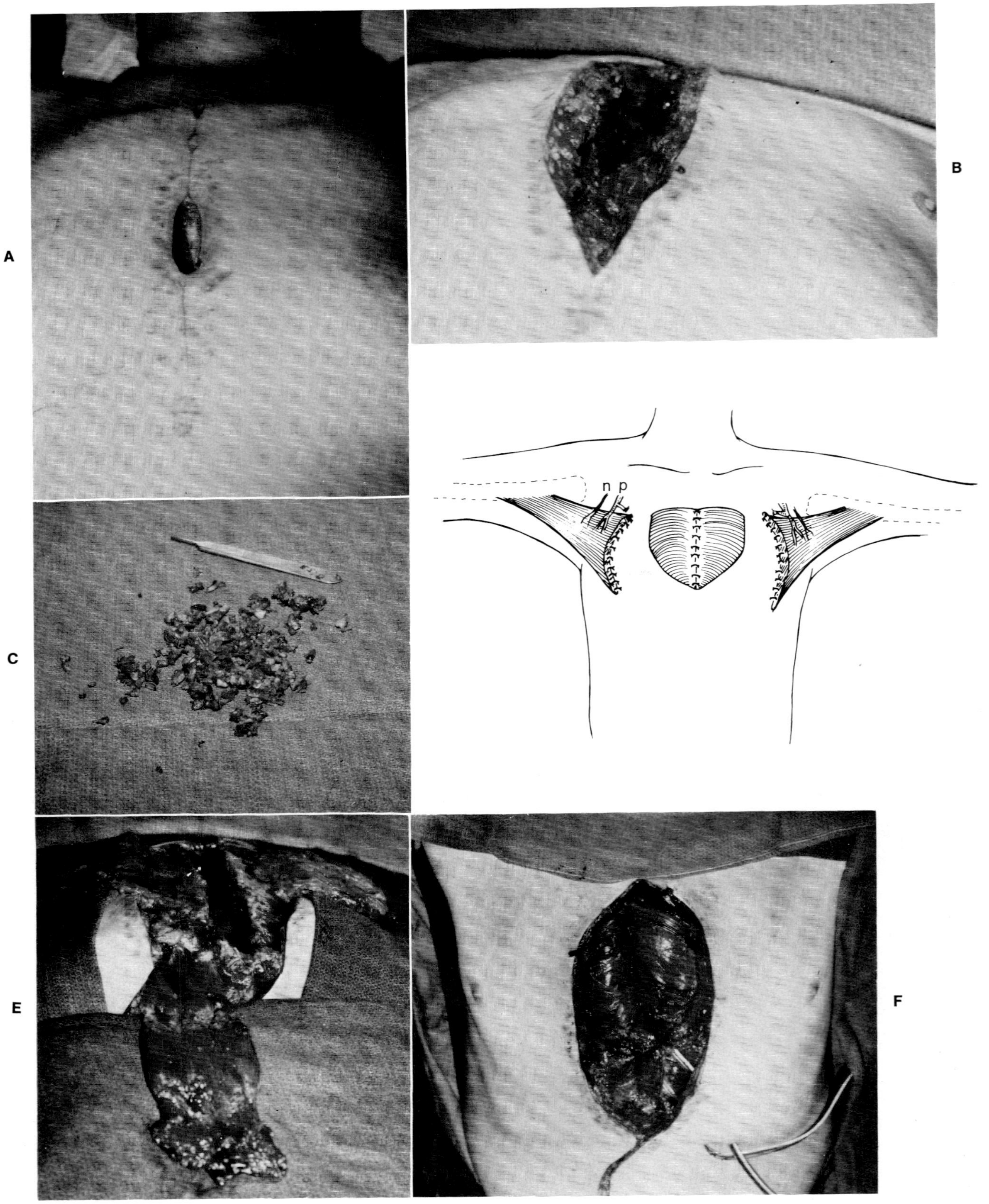

Fig. 1-3. For legend see opposite page.

G

H

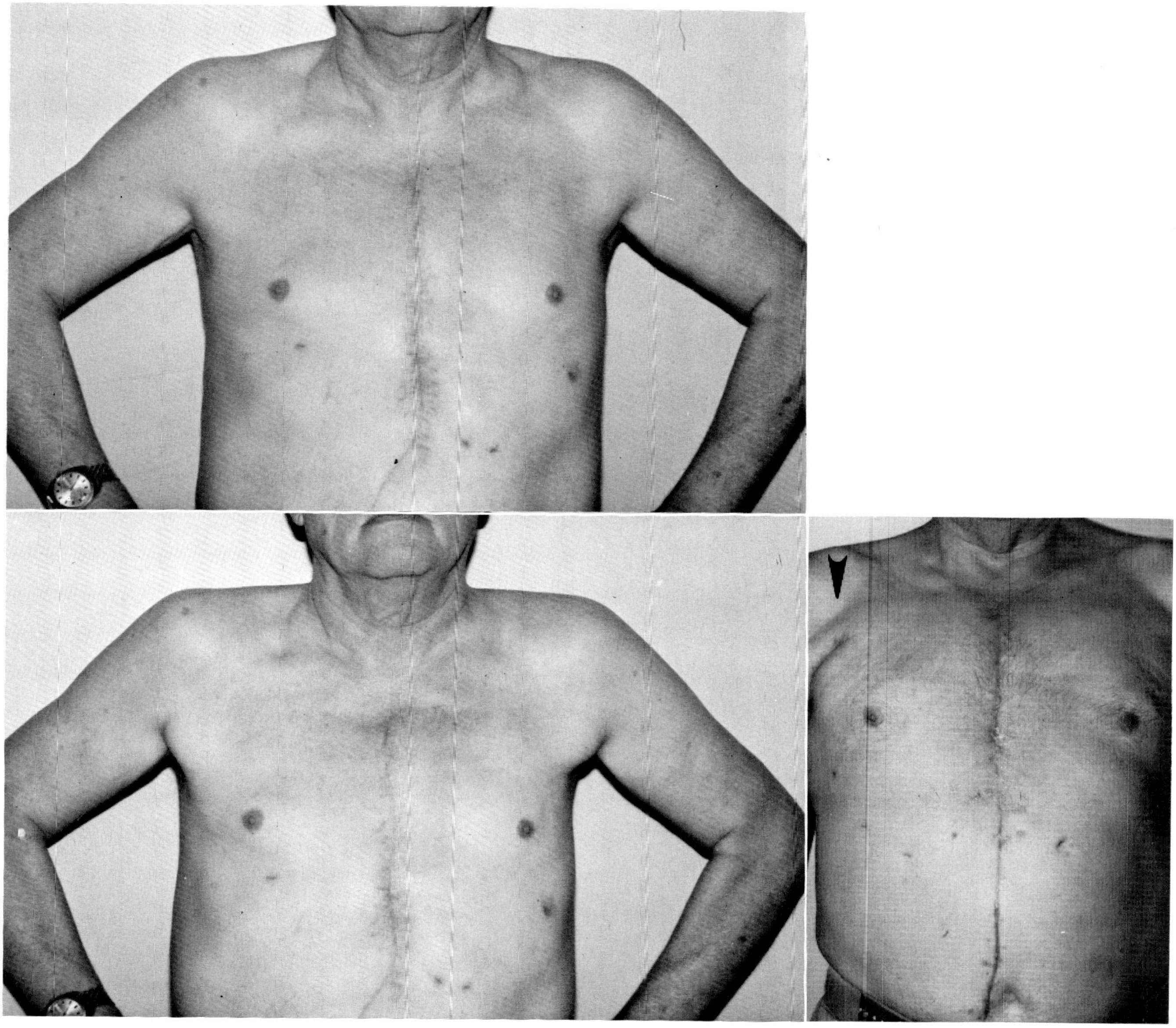

I

Fig. 1-3. Coverage of infected median sternotomy wound with bilateral pectoralis major flaps with preservation of anterior axillary fold and muscle function.

A, Fifty-six-year-old man with infected median sternotomy wound.

B, Wound debrided.

C, Resected necrotic sternum.

D, Medial pectoral nerve *(n)* and lateral branch of thoracoacromial vascular pedicle *(p)* preserved to lateral pectoralis major muscle fibers. Medial two thirds of muscle transposed into sternal defect.

E, Bilateral medially based pectoralis major muscle flaps and right rectus abdominis muscle elevated.

F, Flaps transposed into sternum.

G and **H,** Patient several months after operation at rest, and contraction of pectoralis major muscle with preservation of anterior axillary fold.

I, Anterior view of another patient without function preservation techniques demonstrates absent anterior axillary fold *(arrow)*.

TABLE 1 Specifications for muscle and musculocutaneous flaps

Muscle	Size—width × length (cm)	Skin territory	Sensory innervation—nerve and dermatome	Pattern of blood supply (type)
Abductor digiti minimi (hand)	1 × 5	Same as muscle	None	I
Abductor digiti quinti (foot)	2 × 10	Same as muscle	Lateral plantar (S1, S2)	II
Abductor hallucis	2 × 10	Same as muscle	Medial plantar (S1, S2)	II
Biceps brachii	6 × 23	Same as muscle	Medial brachial cutaneous (C8, T1)	III
Biceps femoris	8 × 35	10 × 40	Posterior femoral cutaneous (S1-S3)	II
Brachial radialis	5 × 22	Same as muscle	Lateral antebrachial cutaneous (C5, C6)	II
Dorsal interossei	1 × 5	None	None	I
Extensor digitorum longus	4 × 35	None	None	IV
Extensor digitorum brevis	4 × 8	Same as muscle	Deep peroneal nerve (L5, S1)	I
Extensor hallucis longus	3 × 24	None	None	IV
External oblique	28 × 15	Same as muscle	Intercostal, segmental (T5-T12)	IV
Flexor carpi ulnaris	5 × 20	Same as muscle	Medial antebrachial cutaneous (C8, T1)	II
Flexor digitorum brevis	5 × 12	Same as muscle	Medial (L4, L5) and lateral plantar	II
Flexor digitorum longus	4 × 35	None	None	IV
Flexor hallucis longus	3 × 24	None	None	IV
Gastrocnemius (medial)	**8 × 15**	**10 × 23**	**None**	**I**
Gastrocnemius (lateral)	6 × 12	10 × 23	None	I
Gluteal thigh flap	12 × 12	10 × 35	Posterior femoral cutaneous (S1-S3)	Axial
Gluteus maximus	**24 × 24**	**24 × 24**	**Segmental sacral (S1-S5)**	**III**
Gracilis	**6 × 24**	**8 × 15**	**Anterior femoral cutaneous (L2, L3)**	**II**
Latissimus dorsi	**25 × 35**	**30 × 40**	**Segmental intercostal (T2-T6)**	**V**
Pectoralis major	**15 × 23**	**20 × 28**	**None**	**V**
Peroneus brevis	3 × 15	Same as muscle	Superficial peroneal nerve (L5, S1)	II
Peroneus longus	3 × 20	Same as muscle	Superficial peroneal nerve (L5, S1)	II

Accessibility	Function			Osseous territory	Clinical application
		Preserved			
	Lost	Technique	Synergists		
Excellent	X			None	Hand
Excellent	X			None	Foot
Excellent	X			None	Foot
Excellent		X	X	None	Arm
Excellent			X	None	Ischium
Excellent	X			None	Arm
Excellent	X			None	Hand
Poor		X	X	None	Lower leg
Excellent			X	None	Foot
Poor	X	X		Fibula	Lower leg
Excellent			X	None	Abdomen
Excellent		X	X	None	Arm
Fair			X	None	Foot
Fair		X	X	None	Lower leg
Poor	X	X		None	Lower leg
Excellent		**X**	**X**	**None**	**Knee, leg**
Excellent		X	X	None	Knee, leg
Excellent		X		None	Pressure sore, perineum
Excellent		**X**		**None**	**Pressure sore, perineum, back**
Excellent			**X**	**None**	**Pressure sore, perineum**
Excellent		**X**	**X**	**None**	**Breast, chest, back**
Excellent		**X**	**X**	**Fifth and sixth ribs**	**Head, neck, chest**
Fair		X	X	None	Lower leg
Fair		X	X	None	Lower leg

Continued.

TABLE 1 Specifications for muscle and musculocutaneous flaps—cont'd

Muscle	Size—width × length (cm)	Skin territory	Sensory innervation—nerve and dermatome	Pattern of blood supply (type)
Platysma	**8 × 12**	**10 × 15**	**Superficial peroneal nerve (L5, S1)**	**II**
Rectus abdominis	**6 × 25**	**21 × 14**	**None**	**III**
Rectus femoris	7 × 32	12 × 32	Anterior femoral cutaneous (L2, L3)	I
Sartorius	4 × 40	5 × 7	None	IV
Semimembranosus	6 × 24	None	None	II
Semitendinosus	4 × 24	None	None	II
Serratus anterior	12 × 10	12 × 6	Segmental intercostal (T2-T4)	III
Soleus	**8 × 28**	**5 × 15**	**None**	**II**
Sternocleidomastoid	5 × 18	6 × 20	None	II
Temporalis	6 × 10	None	None	III
Tensor fascia lata	**5 × 15**	**8 × 40**	**Lateral femoral cutaneous (L2, L3)**	**I**
Tibialis anterior	4 × 25	4 × 12	None	IV
Trapezius	**34 × 18**	**20 × 8**	**None**	**II**
Vastus medialis	8 × 24	None	None	II
Vastus lateralis	10 × 26	10 × 15	None	II

lary fold may be preserved. The lower origin of the pectoralis major muscle from the external oblique fascia and lateral sixth rib is not disturbed, and the muscle is divided 1 or 2 cm medially to this border of the pectoralis major. The medial pectoral nerve and lateral branches of the thoracoacromial artery and vein are also preserved. The medial pectoralis major muscle fibers are folded into the sternal defect. The distal pectoralis major retracts toward the axilla. However, if this part of the muscle that is vascularized and innervated is then sutured to the pectoralis minor under tension, some function and form of the anterior axillary fold are preserved (Fig. 1-3).

Muscle transplantation. The improved reliability of microvascular techniques for free transplantation of muscle and musculocutaneous flaps has further enabled reconstructive surgeons to preserve function. Lower extremity defects are commonly associated with injury to entire muscle groups. Transposition of remaining muscles may further impair lower extremity function. However, the free transplantation of the gracilis muscle for small defects or the latissimus dorsi musculocutaneous flap for large defects often eliminates the necessity to use local flap techniques. The same function-preserving technique described for transposition of the latissimus dorsi muscle is also applied to elevation of the latissimus dorsi free flap. Only the anterior branches of the thoracodorsal artery and nerve are incorporated into the portion of muscle and skin elevated as the donor flap for transplantation to the lower extremity. The posterior muscle is not disturbed, with vascular pedicle, nerve, origin, and insertion intact.

Proximity of motor nerve entry to muscle origin

The motor nerve generally enters the proximal muscle close to the origin. The motor nerve is often closely associated with the dominant vascular pedicle to muscle. Muscle denervation noted on preoperative evaluation may indicate an associated injury to the vascular pedicle. The location and preservation of the motor nerve are essential in free functional muscle transplantation. Specific efforts to divide the motor nerve will not always significantly reduce muscle bulk in the denervated muscle, and the required

Accessibility	Function			Osseous territory	Clinical application
	Lost	Preserved			
		Technique	Synergists		
Excellent	**X**			**None**	**Head, neck**
Excellent			**X**	**None**	**Abdomen, chest, breast**
Excellent	X			None	Abdomen, pressure sore
Excellent			X	Iliac crest	Groin, leg
Fair			X	None	Pressure sore
Fair			X	None	Pressure sore
Fair	X	X		None	Breast
Fair		**X**	**X**	**None**	**Lower leg**
Excellent			X	Clavicle	Head, neck
Fair			X	None	Head, neck
Excellent			**X**	**Iliac crest**	**Abdomen, groin, perineum, pressure sore**
Excellent	X	X		None	Lower leg
Excellent	**X**	**X**		**Scapula**	**Head, neck, back**
Poor			X	None	Leg
Fair			X	None	Leg

proximal dissection subjects the vascular pedicle to injury.

Sensory innervation

The preservation of sensory nerves to the cutaneous territory of superficial muscles allows transposition of a sensory flap. The restoration of sensibility to pressure regions is performed whenever possible. Use of this refinement in musculocutaneous flap design has been expanded by use of nerve grafts between intact sensory nerves above the level of spinal cord injuries to the sensory nerve of the flap in staged procedures. Inclusion of a sensory nerve to the cutaneous territory of muscle allows microvascular transplantation of a sensory flap. The sensory nerve to the cutaneous territories of all superficial muscles is presented in Table 1.

Osseous territory

Vascular connections between muscle fibers and bone at either the origin or insertion allow elevation of bone with the flap. These vascular connections to the periosteum are tenuous, and careful elevation is required for transposition or microvascular transplantation of an osseous muscle or musculocutaneous flap. This muscle flap design may be applied to mandible reconstruction, incorporating rib with pectoralis major and microvascular flap transplantation for reconstruction of long bone injuries.

CONCLUSION

The transplantation of muscle and musculocutaneous flaps is now an established technique in reconstructive surgery. Progressive refinements in muscle selection and design are responsible for their consistent reliability in providing coverage and restoring form, contour, and function. Through knowledge of the requisites of muscle for use as a transposition flap, the muscle flap has been integrated into the reconstructive ladder and selected when required for optimal results in reconstructive surgery.

2

VASCULAR ANATOMY OF MUSCLE: CLASSIFICATION AND APPLICATION

Stephen J. Mathes ■ Foad Nahai

I. Head and neck
   A. Platysma
   B. Sternocleidomastoid
   C. Temporalis
II. Trunk
   A. Anterior
      1. Pectoralis major
      2. Rectus abdominis
   B. Posterior
      1. Latissimus dorsi
      2. Trapezius
III. Anterior thigh
   A. Gracilis
   B. Rectus femoris
   C. Sartorius
   D. Tensor fascia lata (TFL)
   E. Vastus lateralis
IV. Posterior thigh
   A. Biceps femoris
   B. Gluteus maximus
   C. Semimembranosus
   D. Semitendinosus
V. Medial leg
   A. Flexor digitorum longus
   B. Gastrocnemius
   C. Soleus
VI. Lateral leg
   A. Extensor digitorum longus
   B. Extensor hallucis longus
   C. Peroneus longus
   D. Tibialis anterior
VII. Foot
   A. Abductor digiti minimi
   B. Abductor hallucis
   C. Flexor digitorum brevis

The circulation of muscle is based on the vascular pedicle(s) that enter(s) the muscle belly between its origin and insertion. The pedicle(s) consist(s) of an artery, generally a branch of the major artery to the specific anatomic region of the muscle, and paired venae comitantes that drain into a corresponding major regional vein. The paired venae comitantes commonly join as a single vein immediately before entering the major vein. When more than one vascular pedicle enters a muscle, the larger pedicle generally enters that portion of the muscle located at the proximal end of an extremity or the midline area of the trunk. The size and location of the larger vascular pedicle(s) are fairly constant in the human muscle. The number and size of the small vascular pedicle(s) are subject to variability. When a muscle receives two large vascular pedicles from several regional sources, each pedicle is usually equal in size and constant in location. Each human muscle has a fairly constant configuration in regard to its source of blood supply.

The use of a muscle as a flap requires movement of a part or all of a muscle by surgical manipulation. This surgical manipulation of the muscle must not interrupt the circulation of the muscle. Therefore a precise knowledge of the location of the vascular pedicle(s) to the muscle is required. When a muscle has more than one vascular pedicle, the relative importance of each vascular pedicle in regard to muscle circulation is necessary. When division of a pedicle generally results in muscle avascular necrosis, this pedicle is defined as a *major vascular pedicle*. Minor pedicles represent smaller vascular attachments to muscle. When divided the muscle circulation is sustained by the larger dominant vascular pedicle(s). Use of muscle as a flap is generally based on a single vascular pedicle or its dominant vascular pedicle(s).

Each muscle has small vascular attachments from adjacent structures: bone, tendon, muscle, and skin. Selective division of the major vascular pedicle(s) without surgical manipulation of the muscle will not routinely result in muscle necrosis because of sustained circulation through minor vascular pedicles and these potential vascular communications to adjacent structures. Furthermore certain muscles can survive surgical manipulation and division of the major vascular pedicle when minor pedicles are preserved. However, this circulation via the minor pedicle(s) is not always satisfactory, and muscle avascular necrosis often results after such surgical manipulation. A muscle used as a flap based only on circulation from minor vascular pedicles is defined as a *distally based flap*.

The vascular anatomy of human muscles has been studied by regional vascular injections with colored barium-latex solution. Subsequent dissection of each muscle allows location and measurement of the vascular pedicle(s) to muscle. Measurements include the location of the point of entry of the vascular pedicle into the muscle, the regional source of the vascular pedicle, and the diameter and length of each specific pedicle. When a muscle has a single vascular pedicle, its potential as a flap is established by these anatomic data. Use of this muscle as a flap by detachment of either its origin, insertion, or both is safe as long as this single vascular pedicle is maintained.

When the muscle has more than one vascular pedicle, the relative importance of each vascular pedicle to muscle circulation must be determined. Selective injections of a barium-latex solution into the vascular pedicles allow study of the relative filling of the internal muscle vasculature (Fig. 2-1). The larger vascular pedicle that successfully fills the internal muscle vasculature on selective angiogram is considered a *dominant vascular pedicle*.

Anatomic studies are performed in an adynamic state and do not fully define the relative importance of each vascular pedicle to muscle circulation. Actual use of each muscle as a flap confirms the importance of each pedicle in maintaining adequate circulation for muscle survival. No vascular pedicle is described as a dominant vascular pedicle unless the pedicle has sustained muscle circulation after surgical manipulation of the muscle for use as a flap in reconstructive surgery.

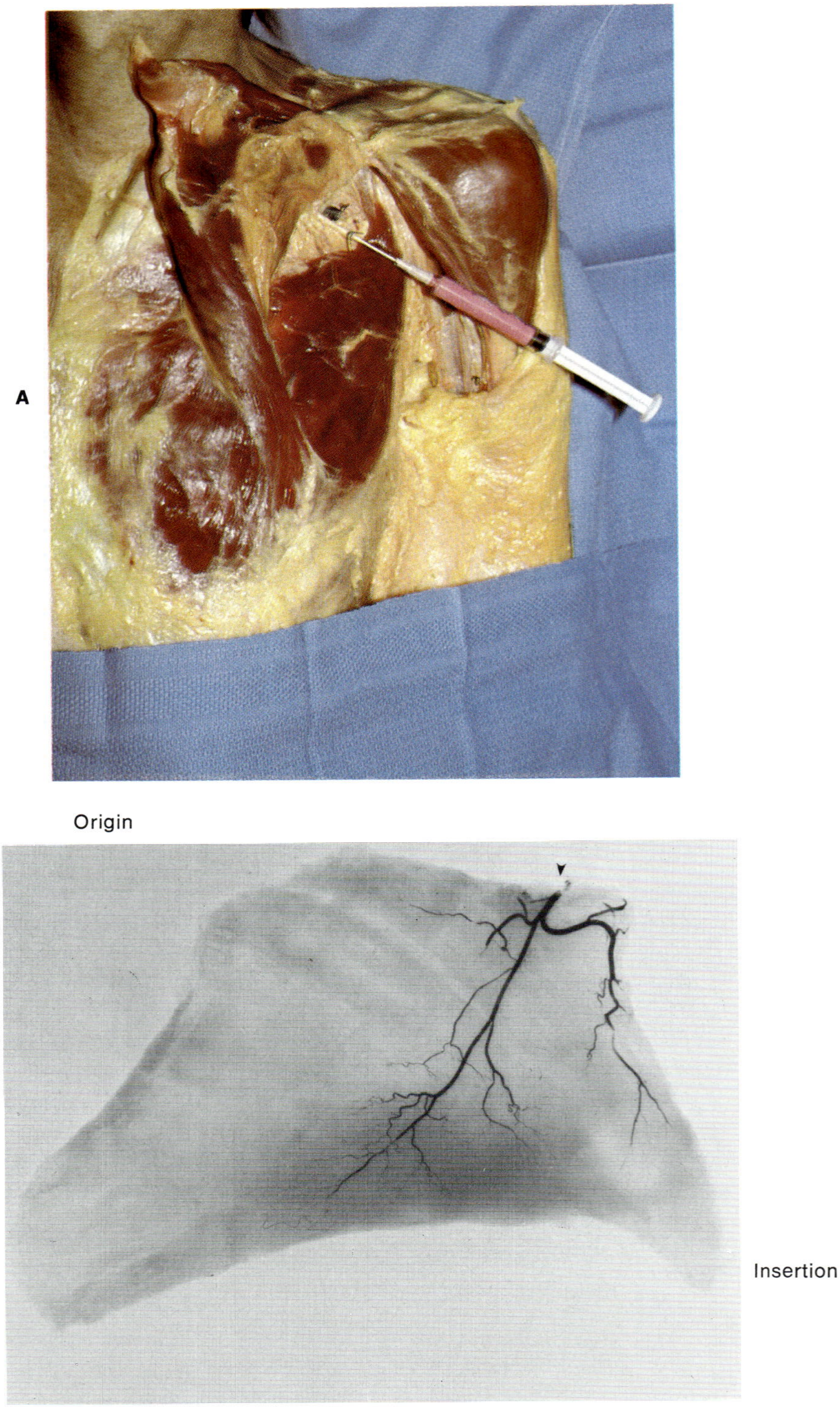

Fig. 2-1. Vascular anatomy of pectoralis major.
A, Selective barium-latex injection in thoracoacromial artery.
B, Selective thoracoacromial angiogram.

(From Mathes, S.J., and Nahai, F.: Classification of the vascular anatomy of muscles: experimental and clinical correlation, Plast. Reconstr. Surg. **67**[2]:177, 1981.)

C

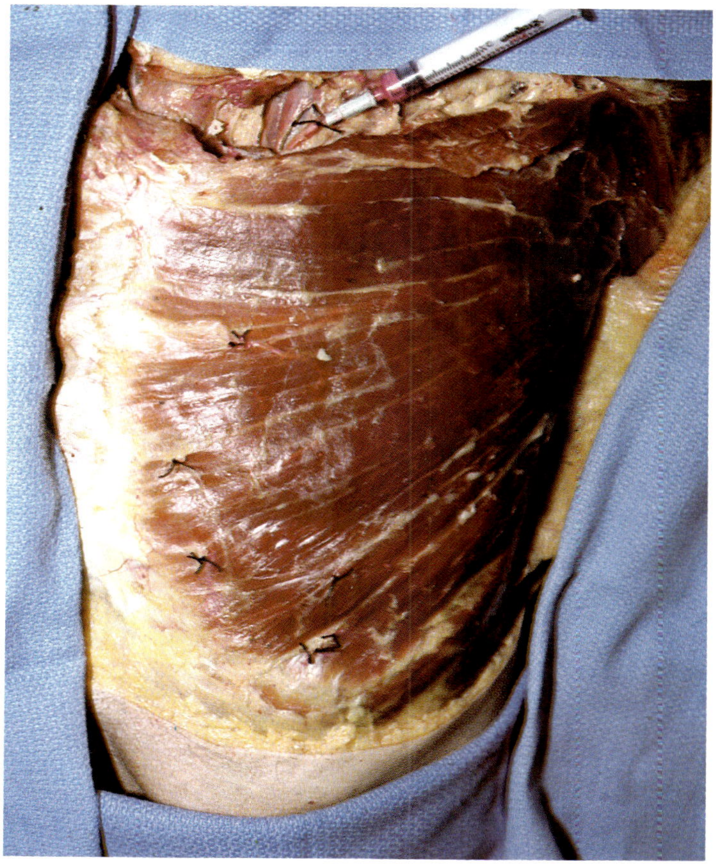

Origin

D

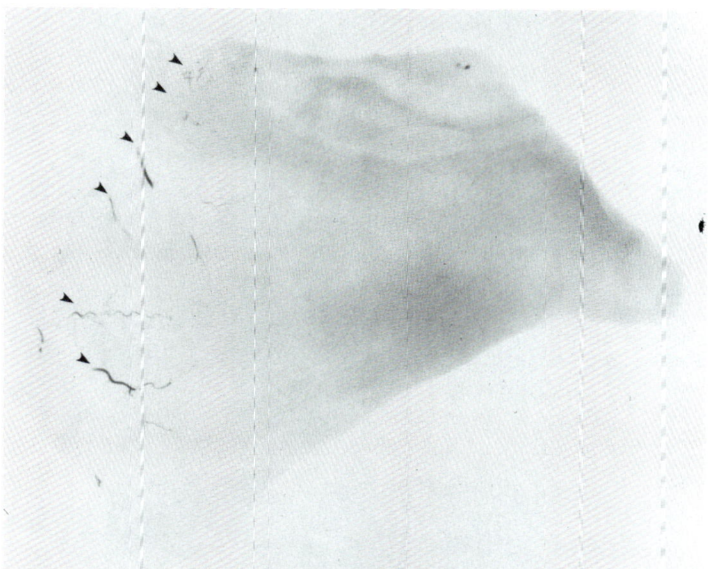

Insertion

Fig. 2-1, cont'd
C, Selective barium-latex injection in internal mammary artery. Ligatures denote site of musculocutaneous perforating vessels.
D, Selective internal mammary angiogram. Arrows denote secondary segmental pedicles, which are basis for reverse transposition of muscle.

CLASSIFICATION

Review of the vascular anatomy of human muscle obtained by cadaver studies and 10 years of clinical experience in use of muscle as a flap reveals five patterns of muscle circulation (Fig. 2-2). The potential use of each muscle as a flap is based on its individual vascular anatomy. However, generalizations regarding muscle vascular anatomy are possible. These patterns or types of circulation are presented to assist in the selection of the muscle for use as a flap with emphasis on survival of muscle after the required surgical manipulation of its vascular anatomy.

Type I: One vascular pedicle (Fig. 2-3)

Anatomic studies reveal a single vascular pedicle to these muscles. Muscles with this vascular pattern include the following:

Medial gastrocnemius
Lateral gastrocnemius
Rectus femoris
Tensor fascia lata

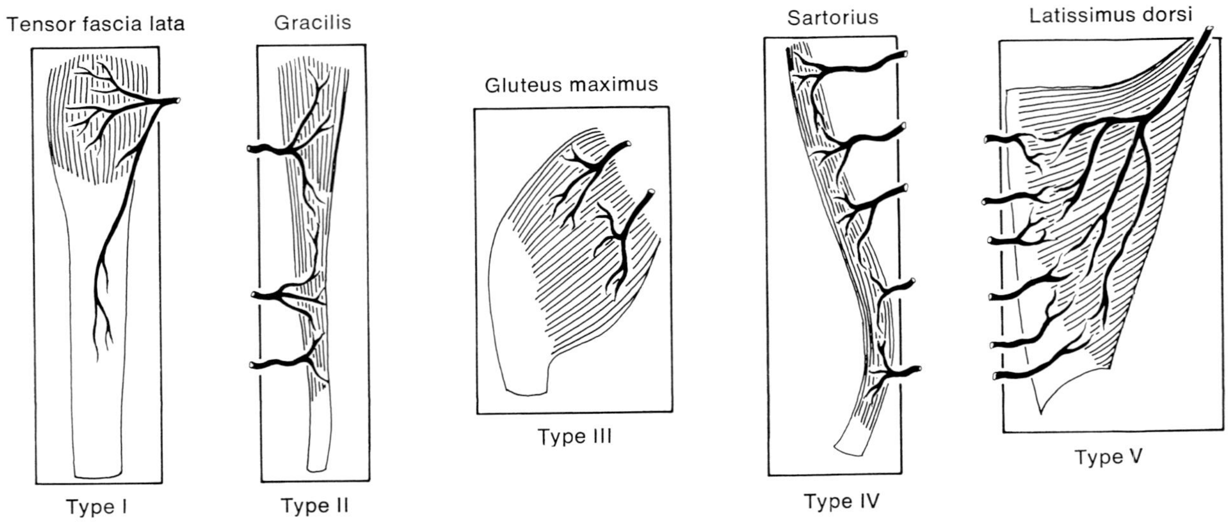

Fig. 2-2. Patterns of vascular anatomy of muscle: *type I,* One vascular pedicle; *type II,* dominant pedicle(s) and minor pedicle(s); *type III,* two dominant pedicles; *type IV,* segmental vascular pedicles; *type V,* one dominant pedicle and secondary segmental pedicles.

(From Mathes, S.J., and Nahai, F.: Classification of the vascular anatomy of muscles: experimental and clinical correlation, Plast. Reconstr. Surg. **67**[2]:177, 1981.)

Origin

A

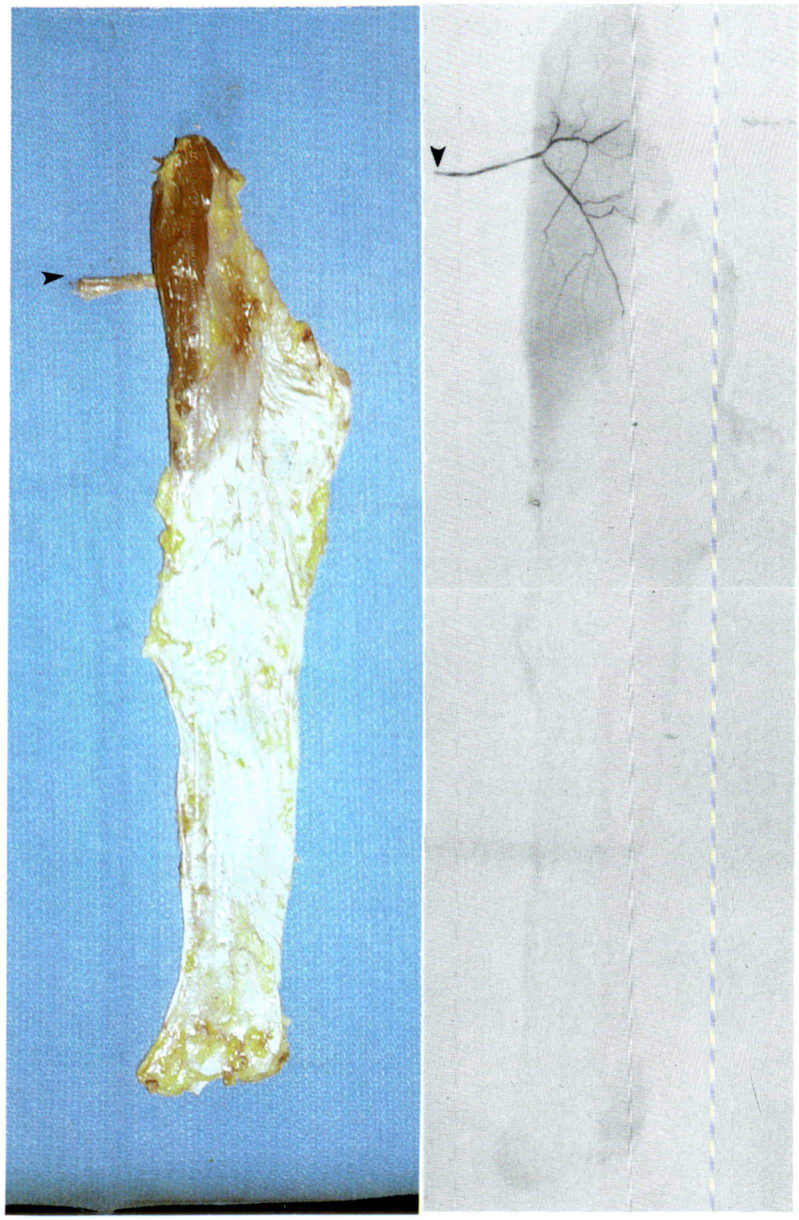

B

Insertion

Fig. 2-3. Patterns of vascular anatomy of muscle.

A, Tensor fascia lata *(type I)*. Arrow denotes vascular pedicle.

B, Angiogram of single pedicle-terminal branch of the lateral femoral circumflex artery.

(From Mathes, S.J., and Nahai, F.: Classification of the vascular anatomy of muscles: experimental and clinical correlation, Plast. Reconstr. Surg. **67**[2]:177, 1981.)

Type II: Dominant vascular pedicle(s) and minor vascular pedicle(s) (Fig. 2-4)

Anatomic studies reveal both large and small vascular pedicles to these muscles. The larger dominant vascular pedicles will sustain circulation to these muscles after their elevation as a flap when the smaller minor pedicles are divided. This is the most common pattern of circulation observed in the human muscle. Muscles with this vascular pattern include the following:

Abductor digiti minimi
Abductor hallucis
Biceps femoris
Flexor digitorum brevis
Gracilis
Peroneus longus
Peroneus brevis
Platysma
Semitendinosus
Soleus
Sternocleidomastoid
Trapezius
Vastus lateralis

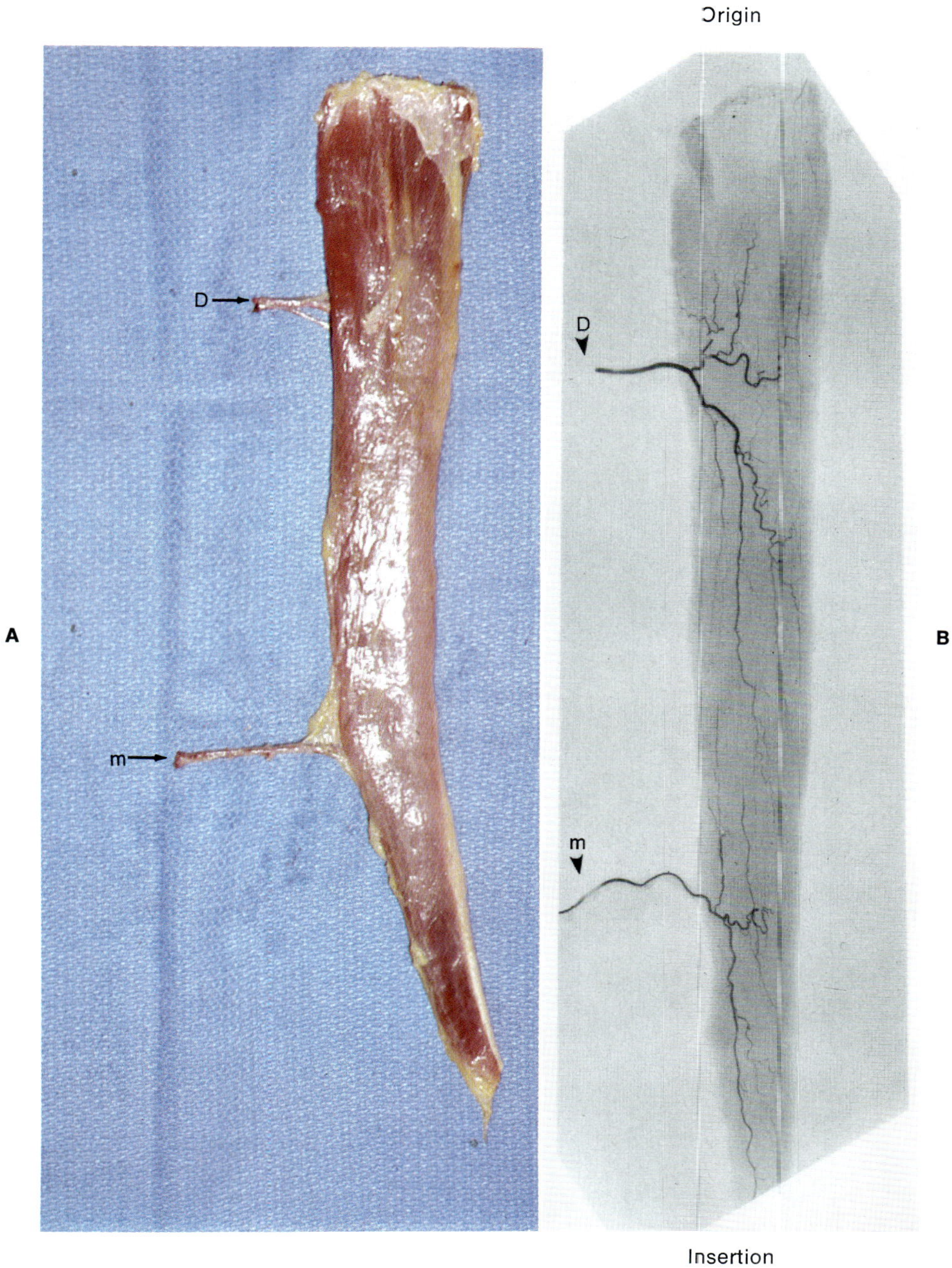

Fig. 2-4

A, Gracilis *(type II)*. *D*, Dominant pedicle; *m*, minor pedicle.

B, Angiogram of dominant pedicle (*D*, branch of medial femoral circumflex artery) and minor pedicle (*m*, branch of superficial femoral artery).

(From Mathes, S.J., and Nahai, F.: Classification of the vascular anatomy of muscles: experimental and clinical correlation, Plast. Reconstr. Surg. **67**[2]:177, 1981.)

Type III: Two dominant pedicles (Fig. 2-5)

Anatomic studies reveal two large vascular pedicles to these muscles. These pedicles have either a separate regional source of circulation or the vascular pedicles are located on opposite sides of the muscle. Since regional and selective angiograms demonstrate equal filling of the arterial vessels within the muscle by these two pedicles, division of one pedicle during flap elevation may result in loss of muscle within its vascular distribution. Muscles with this vascular pattern include the following:

Gluteus maximus
Rectus abdominis
Serratus anterior
Semimembranosus
Temporalis

Type IV: Segmental vascular pedicles (Fig. 2-6)

Anatomic studies reveal multiple vascular pedicles entering the course of the muscle belly. Each pedicle provides circulation to a portion of the muscle. Division of more than two or three of these pedicles during elevation as a flap results in distal muscle necrosis. Muscles with this vascular pattern include the following:

Extensor digitorum longus
Extensor hallucis longus
Flexor digitorum longus
Flexor hallucis longus
Sartorius
Tibialis anterior

A

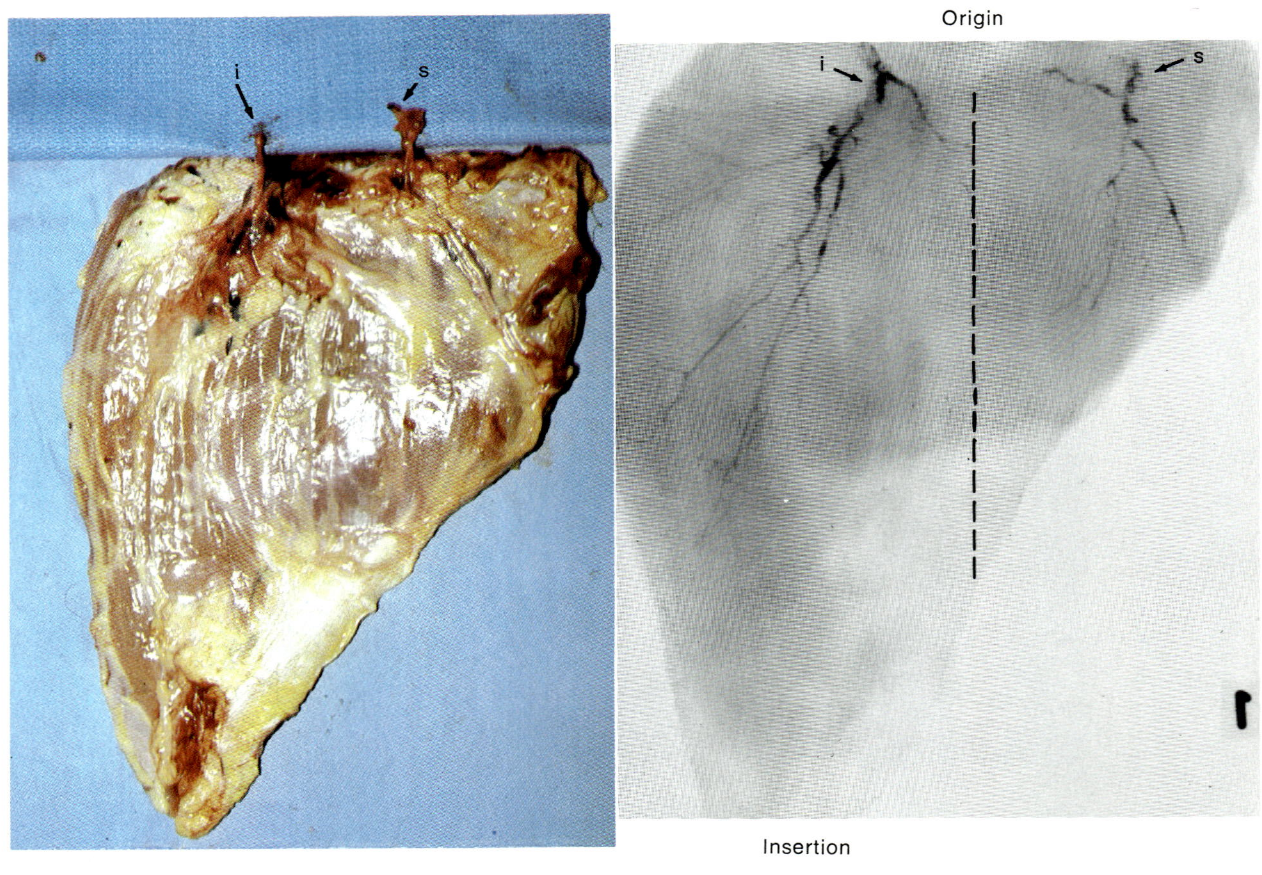

B

Fig. 2-5

A, Gluteus maximus *(type III)*. *s,* Superior gluteal artery; *i,* inferior gluteal artery.

B, Angiogram of dominant pedicles. *s,* Superior gluteal artery, *i,* inferior gluteal artery. Dotted line denotes site of division of muscle for transposition of only one half of muscle.

(From Mathes, S.J., and Nahai, F.: Classification of the vascular anatomy of muscles: experimental and clinical correlation, Plast. Reconstr. Surg. **67**[2]:177, 1981.)

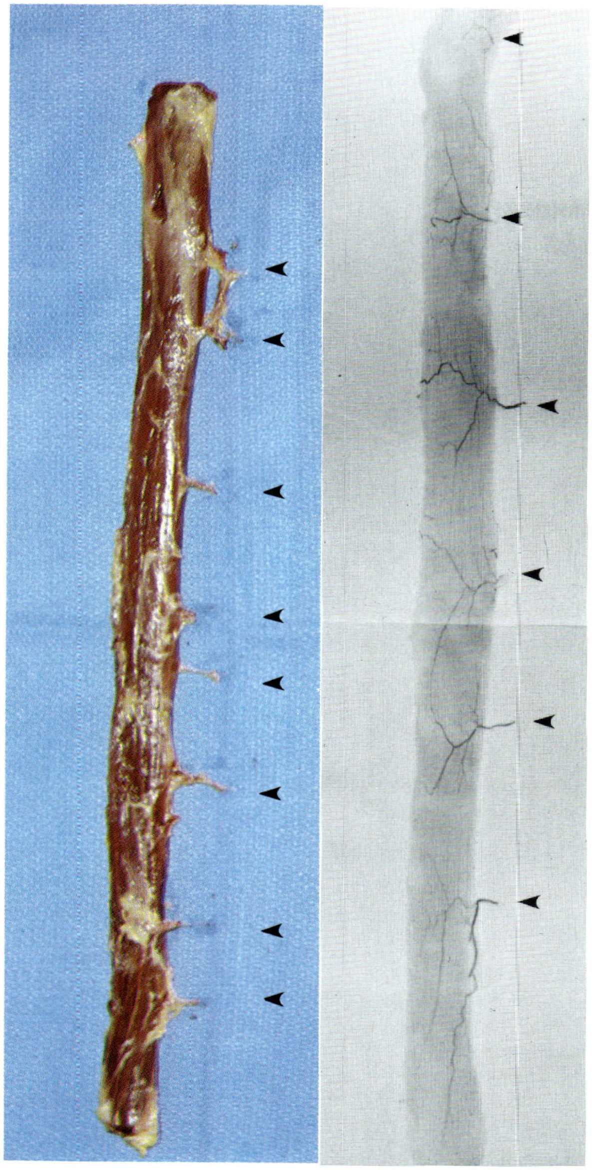

Fig. 2-6. Patterns of vascular anatomy of muscle.
A, Sartorius muscle *(type IV)*. Arrows denote segmental pedicles.
B, Angiogram of femoral artery. Arrows denote segmental pedicles.

(From Mathes, S.J., and Nahai, F.: Classification of the vascular anatomy of muscles: experimental and clinical correlation, Plast. Reconstr. Surg. **67**[2]:177, 1981.)

Type V: One dominant vascular pedicle and secondary segmental vascular pedicles (Fig. 2-7)

Anatomic studies reveal a large vascular pedicle within the shoulder girdle to these muscles. Selective angiograms demonstrate filling of the internal vasculature of the muscle via this pedicle. However, the segmental pedicles entering these muscles at the midline of the trunk also provide a significant source of circulation to these muscles. The internal vasculature of these muscles is also filled on angiograms via these secondary segmental pedicles. These muscles may be elevated as a flap on either vascular system. Muscles with this vascular pattern include the following:

Pectoralis major
Latissimus dorsi

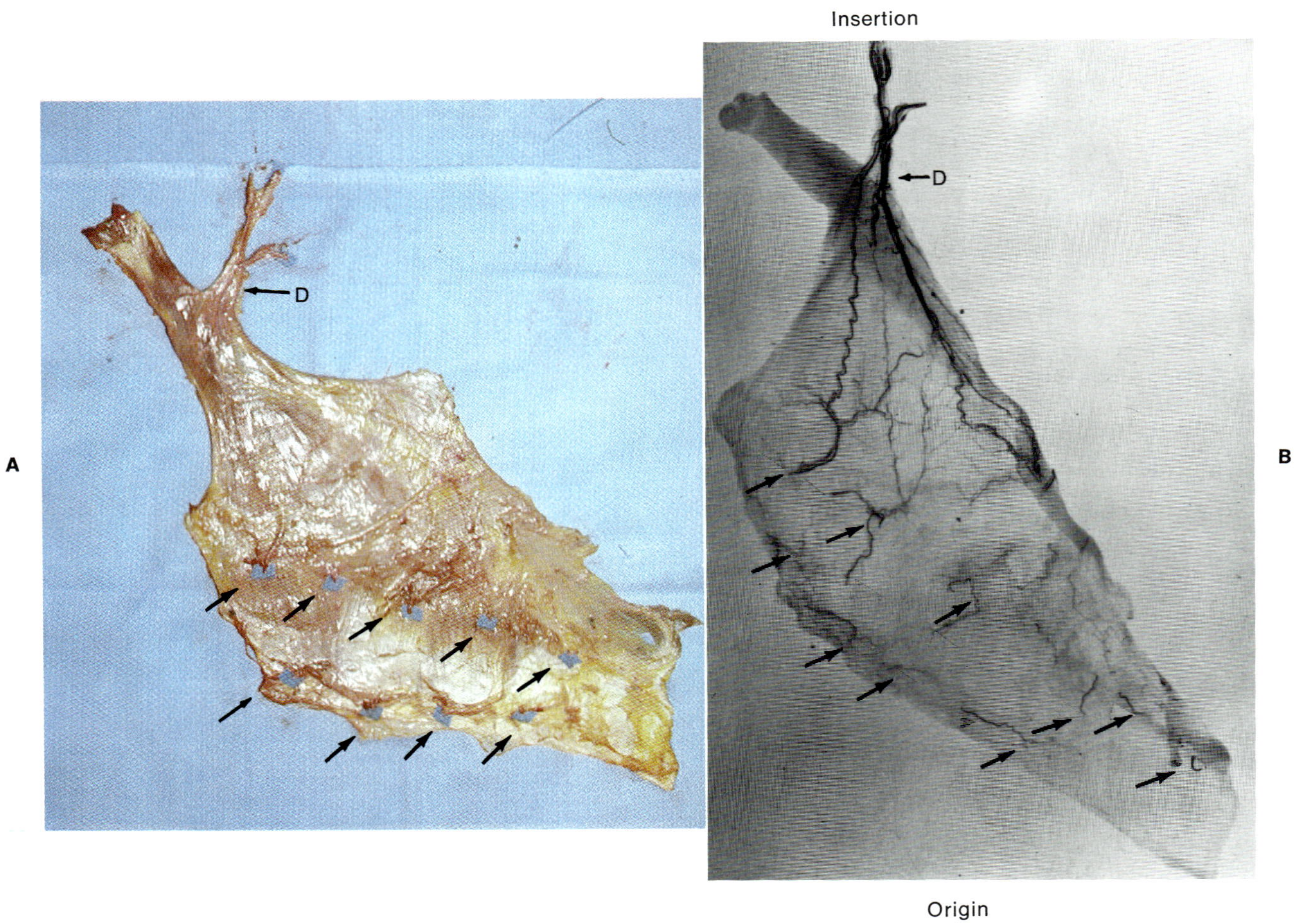

Fig. 2-7. Patterns of vascular anatomy of muscle.

A, Latissimus dorsi *(type V). D,* Thoracodorsal artery; small arrows denote secondary segmental pedicles.

B, Angiogram of dominant pedicle. *D,* Thoracodorsal artery. Arrows on outer row indicate branches of thoracic and lumbar arteries; arrows on inner row indicate branches of intercostal arteries.

(From Mathes, S.J., and Nahai, F.: Classification of the vascular anatomy of muscles: experimental and clinical correlation, Plast. Reconstr. Surg. **67**[2]:177, 1981.)

CLINICAL CORRELATION: GENERAL CONSIDERATIONS

The reconstructive surgeon selecting a muscle for use as a flap must first consider the reliability of its survival. However, other factors in selection include the resultant form and function both at the site of reconstruction and in the donor region. Refinements in the design and technique of muscle flap elevation are based on manipulation of the vascular anatomy of muscle. The patterns of circulation to muscle determine the potential of each muscle to satisfy specific reconstructive requirements.

Prediction of arc of rotation

The extent of elevation of muscle from its normal anatomic position without devascularization and its subsequent ability to reach adjacent defects determine the arc of rotation. The point of rotation is determined by the site of entrance of the dominant or major vascular pedicle into the muscle. Only the muscle distal to the point of rotation is actually used as a transposition flap. The point of rotation for muscles of types I, II, III and V generally is located at one end or the proximal one third of the muscle. However, the type IV muscle has a very limited arc of rotation, since only the proximal or distal third of the muscle is safely elevated because of the segmental vascular pattern.

In type III muscle in which the two dominant vascular pedicles are located at opposite ends of the muscle, the arc of rotation is decreased, since the whole muscle may not survive transposition on one dominant vascular pedicle.

The type V muscle has two arcs of rotation (Fig. 2-8). The entire muscle is safely elevated on the major vascular pedicle, entering adjacent to the muscle insertion in the shoulder girdle. Division of this dominant vascular pedicle allows safe transposition (reverse arc of rotation) on the secondary segmental pedicles entering the muscle adjacent to its origin in the midline of the trunk.

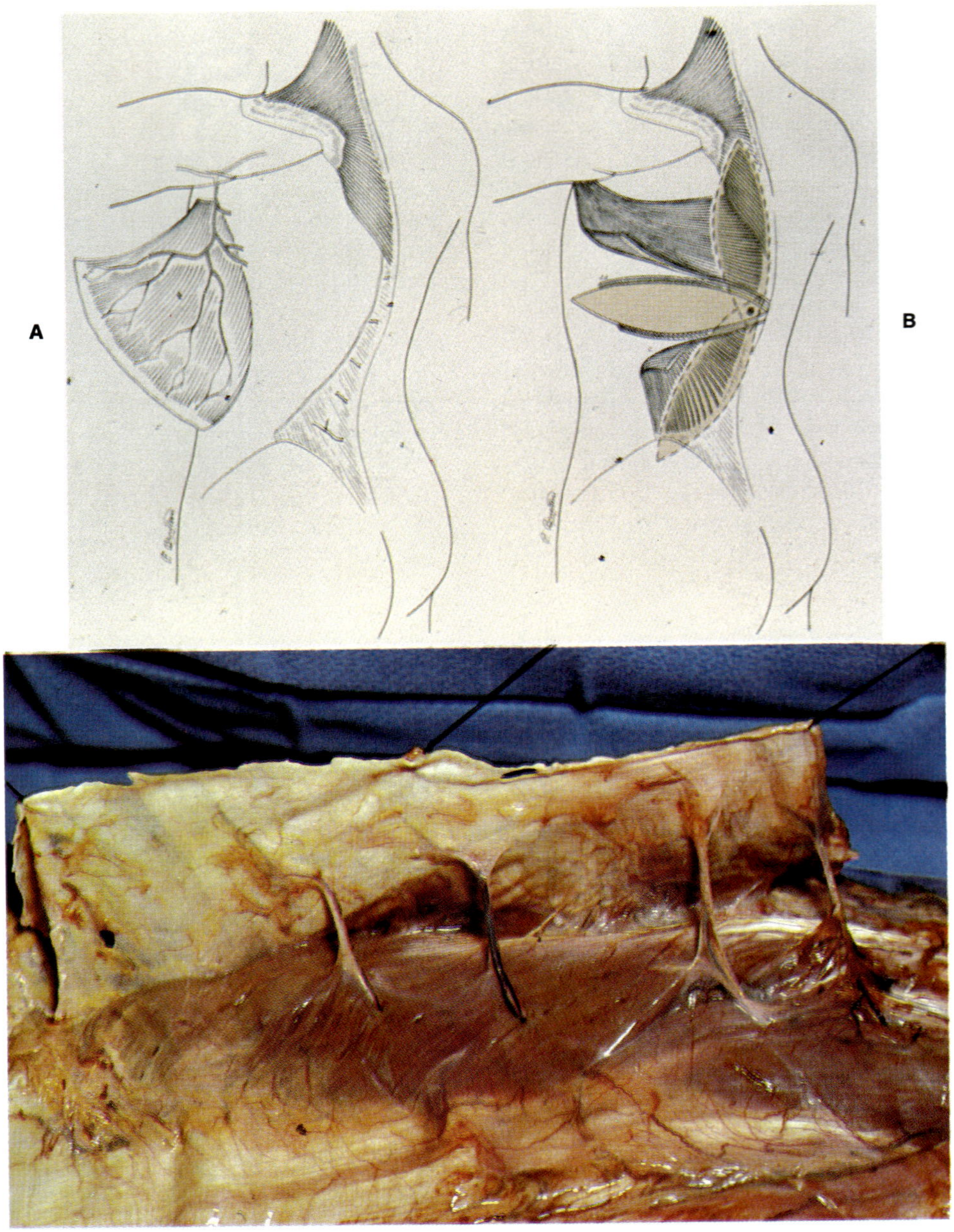

Fig. 2-8. Prediction of arc of rotation.

A, Latissimus dorsi (*type V*). Arc of rotation based on dominant vascular pedicle (thoracodorsal artery).

B, Arc of rotation based on secondary segmental pedicles.

C, Elevation of vertebral origin of latissimus dorsi muscles reveals secondary segmental pedicles.

(From Mathes, S.J., and Nahai, F.: Classification of the vascular anatomy of muscles: experimental and clinical correlation, Plast. Reconstr. Surg. **67**[2]:177, 1981.)

D E

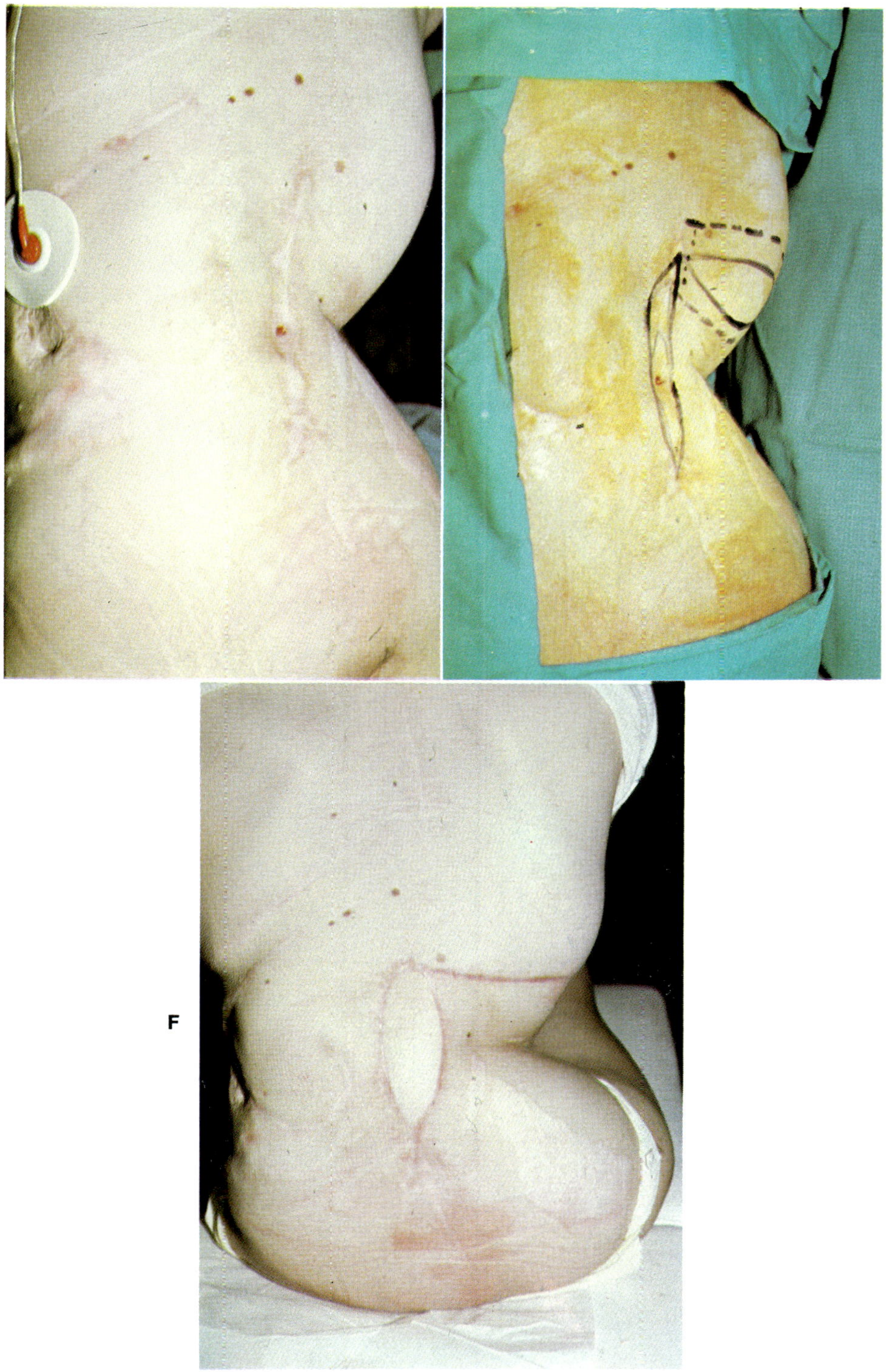

F

Fig. 2-8, cont'd
D, Child with exposed Harrington rod.
E, Outline of reverse latissimus dorsi musculocutaneous flap.
F, Postoperative view reveals stable wound coverage provided by reverse latissimus dorsi musculocutaneous flap.

Prediction of skin territory

Use of superficial muscles as a flap may include overlying skin based on musculocutaneous perforating vessels. However, the design of the skin overlying the muscle as a musculocutaneous flap is determined by the pattern of circulation to the muscle. Each pedicle to muscle contributes to the skin circulation via the musculocutaneous perforating vessels. Muscles with types II (Fig. 2-9) and IV patterns of circulation require division of minor or segmental pedicles for flap elevation. The skin that is located over the distal muscle does not always survive when these pedicles are divided during flap elevation.

The circulation to skin overlying type III muscles is always adversely affected after division of one of the major vascular pedicles.

The skin distant to segmental secondary vascular pedicles in type V muscles is rarely included with the reverse flap transposition, since the flap is generally used as a turnover flap (i.e., pectoralis major) or segmental transposition (i.e., latissimus dorsi). However, the entire skin overlying type V muscles is reliable based on the dominant vascular pedicle.

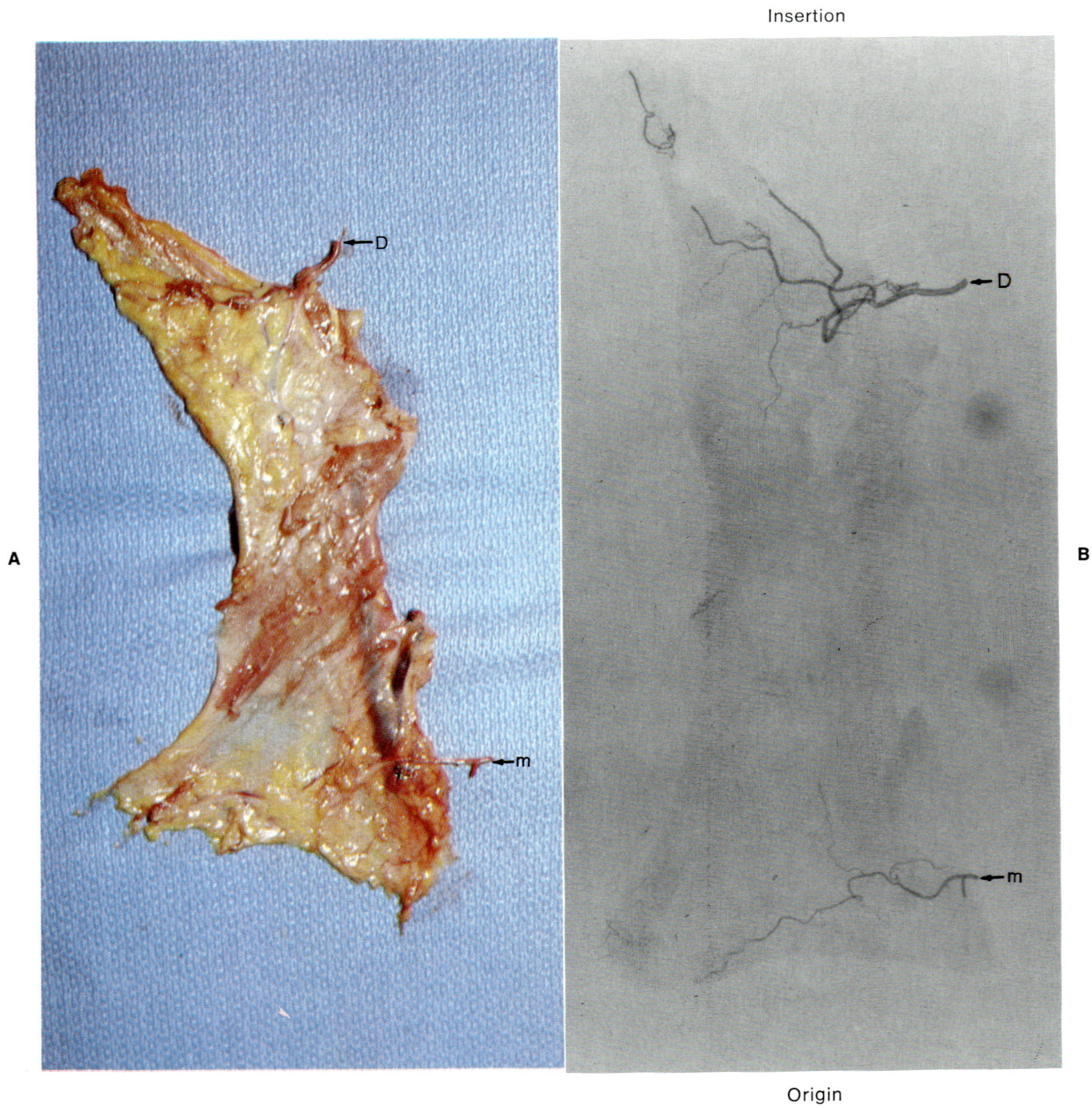

Fig. 2-9. For legend see opposite page.

C

D

E

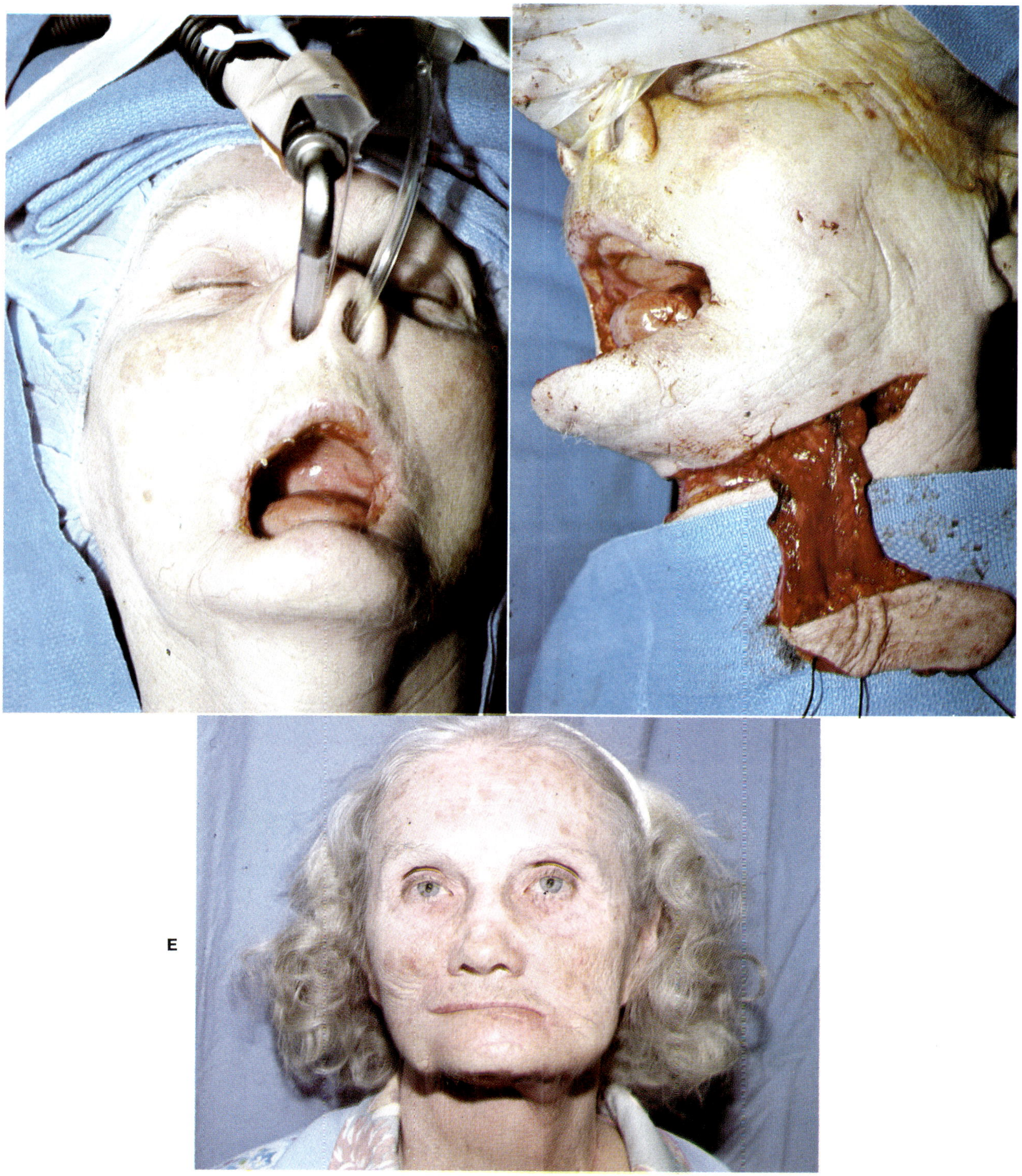

Fig. 2-9. Prediction of skin territory.

A, Platysma *(type II)*. *D,* Dominant pedicle; *m,* minor pedicle.

B, Angiogram of platysma. *D,* Branch of the facial artery; *m,* branch of the transverse cervical artery.

C, Absence of lower lip and buccal sulcus after chemosurgery.

D, Platysma muscle with distal skin island.

E, Six-month postoperative view demonstrates reconstruction of buccal sulcus and lower lip with platysma musculocutaneous flap.

(From Mathes, S.J., and Nahai, F.: Classification of the vascular anatomy of muscles: experimental and clinical correlation, Plast. Reconstr. Surg. **67**[2]:177, 1981.)

Segmental transposition

Elevation of a portion of a muscle for use as a flap while maintaining the continuity of the muscle origin, insertion, vascular pedicles, and innervation is possible. The type III muscle (Fig. 2-10) is ideally suited, since one half of the muscle is elevated on its dominant vascular pedicle, while the continuity of the other half with its vascular pedicle, nerve, origin, and insertion are maintained. Furthermore a knowledge of the intrinsic circulation of muscle allows elevation of a part of the muscle based on a selected branch of the dominant or major vascular pedicle. The continuity of the major pedicle and nerve with the remaining muscle is preserved. This technique is applicable to muscles with types II, IV, and V patterns of circulation.

A B C

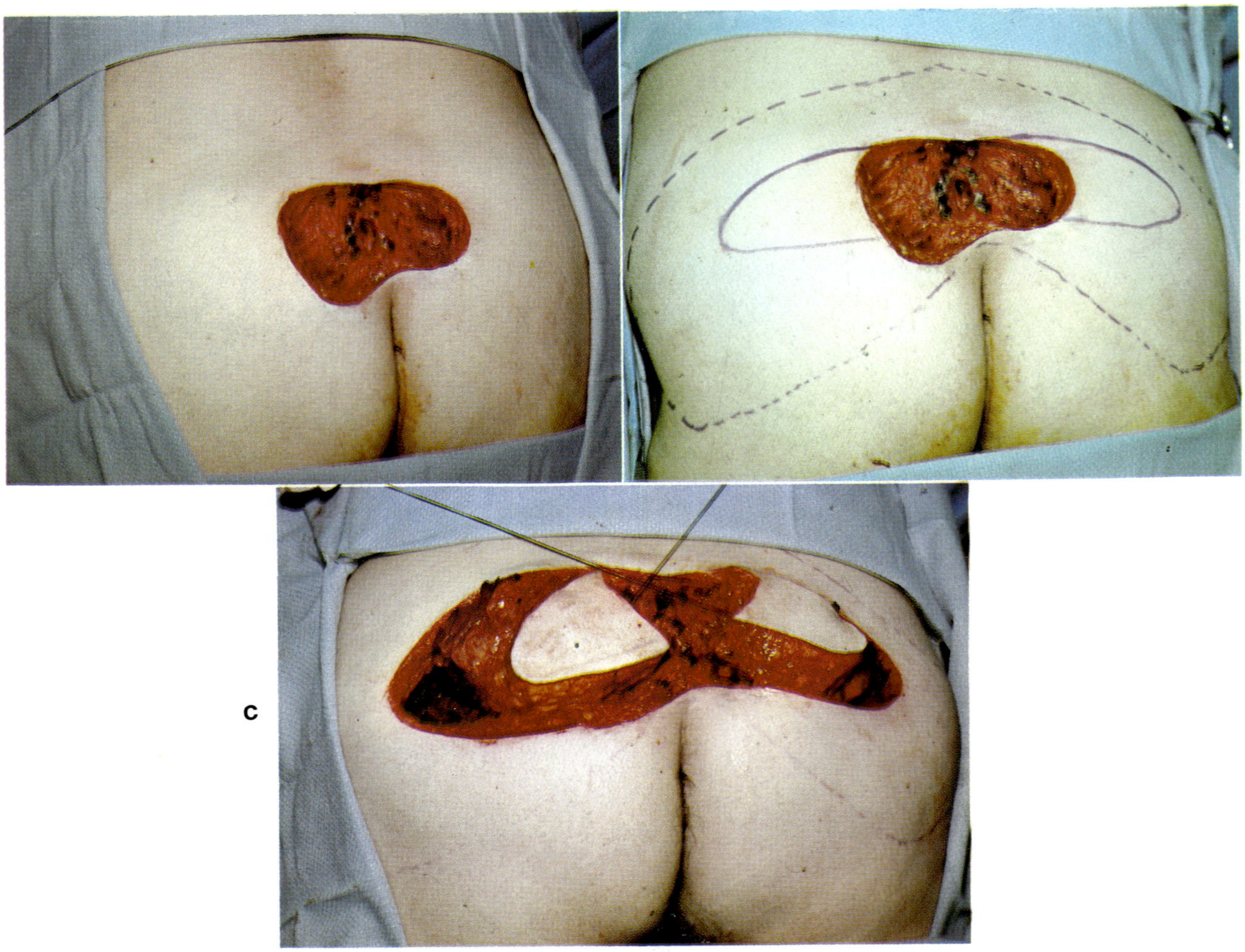

Fig. 2-10. Segmental transposition.

A, Gluteus maximus *(type III)*. Defect after excision of chronic wound of sacrum with exposed sacral bone devoid of periosteum.

B, Design of skin island located over superior aspect of each gluteus maximus muscle.

C, Origin and insertion of superior half of each gluteus maximus muscle are divided. Both musculocutaneous flaps are advanced to midline of sacrum, providing both muscle and skin coverage.

(**B** and **C** from Perry, S., and Mathes, S.J.: Bilateral gluteus maximus myocutaneous advancement flaps: sacral coverage for ambulatory patients, Ann. Plast. Surg. [In press.])

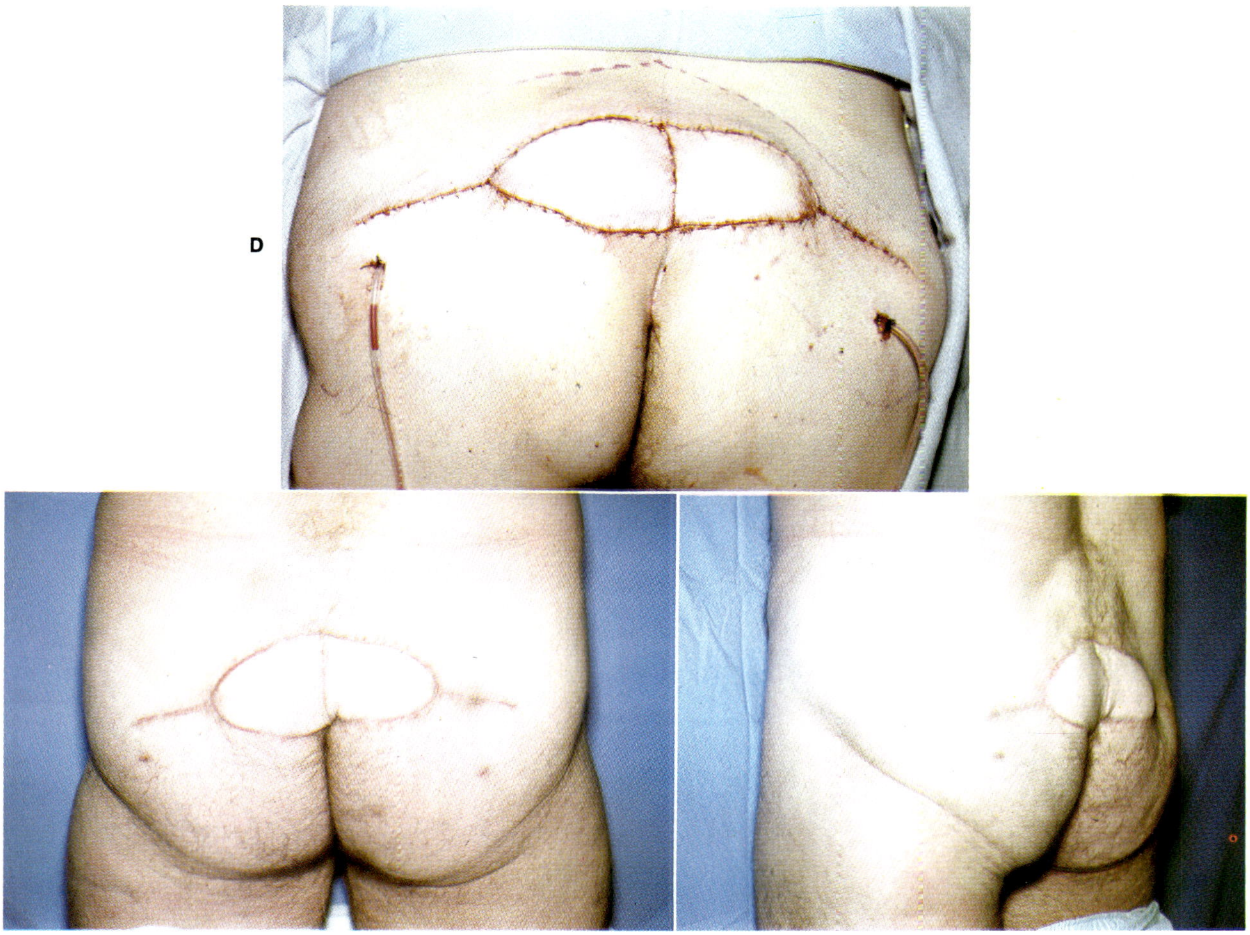

Fig. 2-10, cont'd

D, Coverage of sacrum includes superior gluteus maximus muscles and overlying skin island. NOTE: Origins of superior gluteus maximus muscles are approximated in midline. Closure of donor defect is possible with V-Y advancement of musculocutaneous flaps to midline.

E, Stable coverage provided by bilateral superior gluteal musculocutaneous flaps. Inferior halves of gluteal muscles remain intact for function preservation.

F, Direct closure of donor defect possible as result of V-Y closure. Muscle closure in midline provides soft tissue coverage of sacrum without contour deformity.

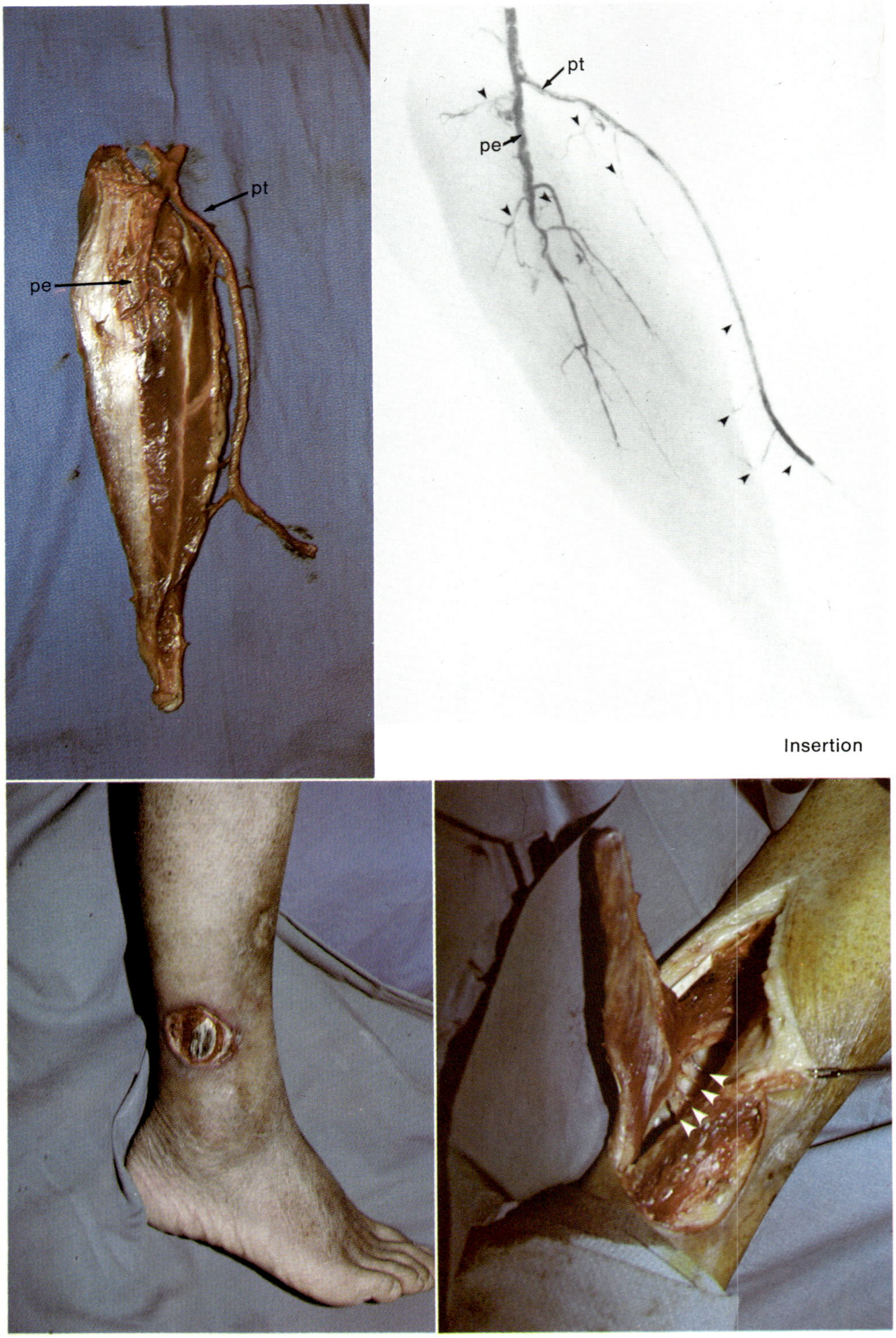

Fig. 2-11. Design of distally based flaps.

A, Soleus *(type II)*. Posterior tibial and peroneal arteries.

B, Angiograms of popliteal artery. *pe,* Peroneal artery, *pt,* posterior tibial artery. NOTE: Distal minor pedicles denoted by arrows on inferior portion of muscle.

C, Chronic wound of distal third of leg with exposed fibula and absent peroneal musculature.

D, Distal soleus muscle elevated from lateral approach. Arrows denote medially located minor pedicles to distal muscle.

(From Mathes, S.J., and Nahai, F.: Classification of the vascular anatomy of muscles: experimental and clinical correlation, Plast. Reconstr. Surg. **67**[2]:177, 1981.)

Design of distally based flaps

Muscles with type II circulation have minor vascular pedicles that are usually divided during standard flap elevation. Division of the dominant vascular pedicle and transposition of the muscle distally based on the minor vascular pedicle have been successfully performed. However, the location and size of minor pedicles are variable, and the use of the distally based flap is not reliable. Prior delay of the muscle either by selective division of the dominant vascular pedicle or use of a biochemical delay (i.e., isoxsuprine) may improve the chances of muscle survival. The transposition of only a portion of the muscle for small defects based on the minor pedicles of type II muscle is a reliable technique (Fig. 2-11).

E

F

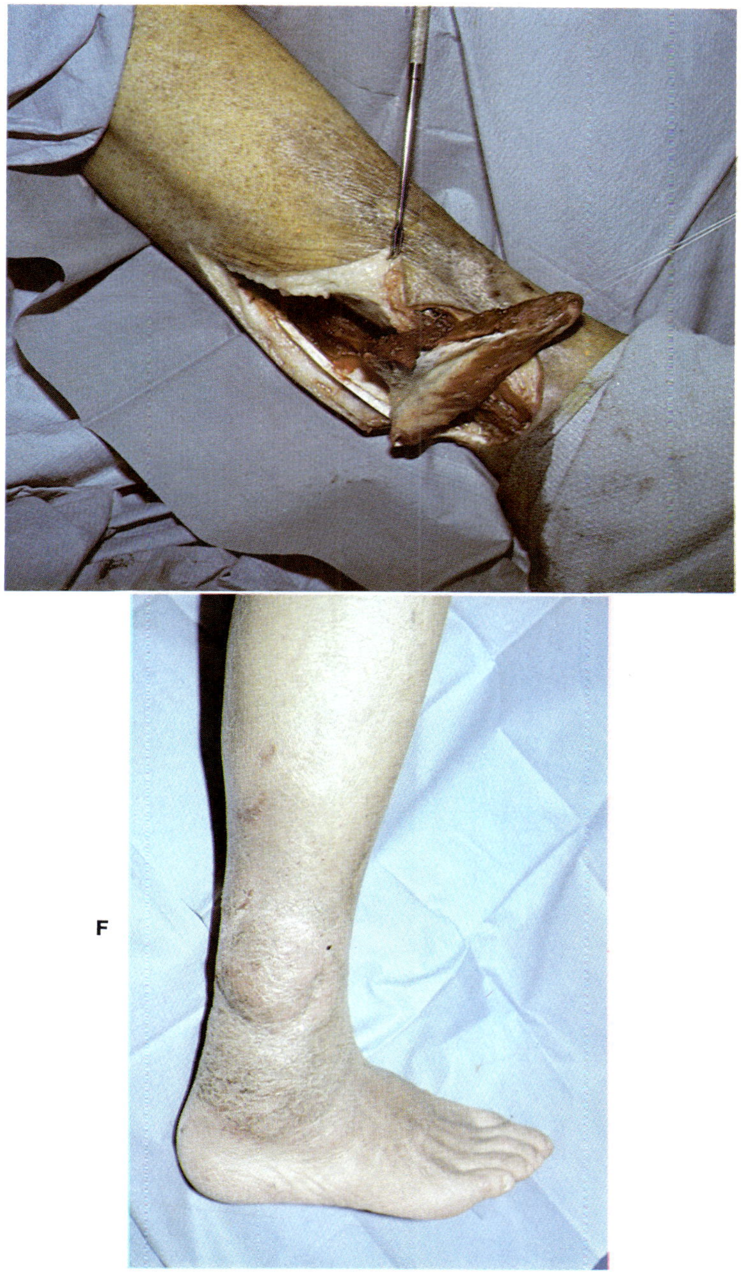

Fig. 2-11, cont'd

E, Arc of rotation of distal portion of soleus with medial muscle continuity maintained.

F, Skin graft is applied to exposed muscle belly.

G, Nine-month postoperative view reveals stable wound coverage.

Potential for microvascular composite tissue transplantation

Complete elevation of a muscle and transplantation to a distant site require restoration of circulation by microvascular techniques. Muscles with types I and V circulation patterns are ideally suited for this technique, since the single vascular pedicle is generally large and the muscle with its overlying skin reliable. The majority of type II muscles are also safely transplanted based on the dominant vascular pedicle. However, those muscles with two or more dominant vascular pedicles are not usually suitable because of the necessity for multiple microvascular repairs to reestablish circulation at the site of flap transplantation.

Vascularized bone

Muscles with all five types of circulation have vascular communication between muscle fibers and the periosteum of bone at the site of origin or insertion. However, the incorporation of vascularized bone with the transposition flap is impossible when the vascular attachments to bone are located beyond the point of flap rotation. However, vascularized bone is useful in muscles suitable for microvascular transplantation or in muscles suitable for transposition when the vascular attachments to bone are distal to the point of rotation (i.e., pectoralis major muscle).

CLINICAL CORRELATION: SPECIFIC MUSCLES

The vascular anatomy of each muscle useful as either a transposition or transplantation flap is presented. This vascular anatomy is correlated with specific characteristics of the muscle to assist the reconstructive surgeon in selection of the muscle flap best suited for a particular reconstructive requirement. These data are based on anatomic studies and correlated with clinical experience. The potential variability of muscle vascular pedicles is emphasized especially in regard to minor pedicles. Furthermore prior trauma or vascular disease will alter the vascular anatomy of muscles.

PLATYSMA

Vascular anatomy (Figs. 2-12 and 2-13)

Pattern of circulation: Type II

Dominant vascular pedicle (D)

Branch of submental artery and venae comitantes

REGIONAL SOURCE: Facial artery and vein

SIZE: Length: 3 cm
Diameter: 0.8 mm

LOCATION: Adjacent to muscle insertion beneath mandible

Minor vascular pedicle (m)

Branch of suprasternal artery and venae comitantes

REGIONAL SOURCE: Suprascapular artery and vein

SIZE: Length: 2 cm
Diameter: 0.5 mm

LOCATION: Adjacent to muscle origin above clavicle

Clinical application

ARC OF ROTATION: The entire muscle may be safely elevated on the major vascular pedicle. However, transposition of muscle alone is only useful for coverage of tracheostomy defects or other midline coverage problems.

SKIN TERRITORY: The anterior medial skin of the neck may be elevated with underlying platysmal muscle based on musculocutaneous perforating vessels. A skin island may be designed over the inferior muscle for transposition of the musculocutaneous flap to the inferior face and oral cavity for defect coverage. Skin islands designed over the middle of the muscle may be transposed inferiorly with underlying platysmal muscle for midline neck defects. This transposition is based on the minor vascular pedicle.

SEGMENTAL TRANSPOSITION: Since the platysma is small, partial muscle transposition has no clinical application.

DESIGN OF DISTALLY BASED FLAPS: The platysma and associated skin island located over the middle of the muscle belly may be transposed to the midline neck based on the minor pedicle.

POTENTIAL FOR MICROVASCULAR TRANSPLANTATION: The platysma with an overlying small skin island may be transplanted by microvascular techniques based on the facial artery and vein. However, the donor site in the neck and proximity of this flap dissection to the mandibular branch of the facial nerve may preclude its use.

VASCULARIZED BONE: No significant vascular attachments of the platysma to the clavicle periosteum are noted.

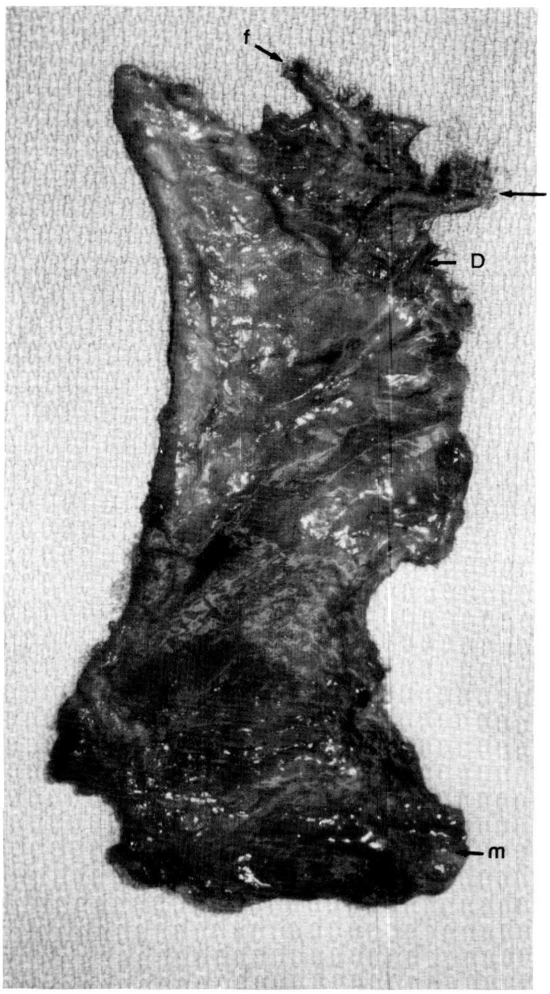

Fig. 2-12. Dominant vascular pedicle: *D*, Submental branch of facial artery *(f)*. Minor vascular pedicle: *m*, Branch of suprascapular artery.

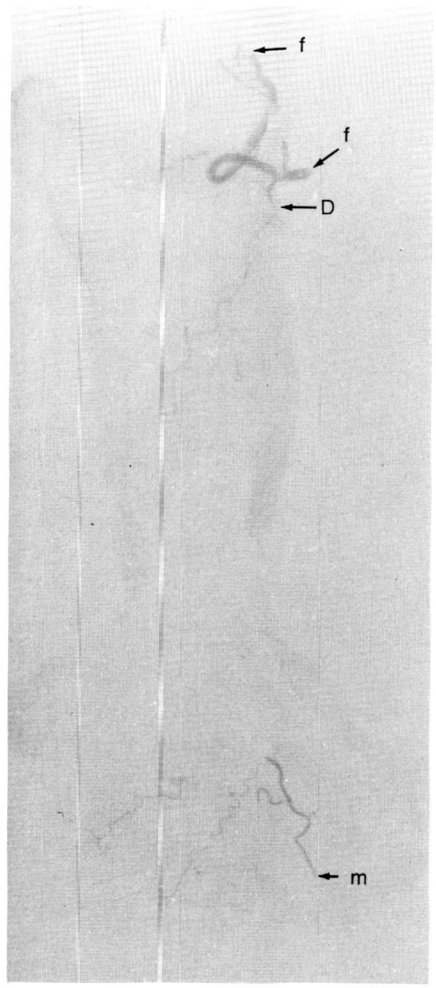

Fig. 2-13. Dominant vascular pedicle: *D*, Submental branch of facial artery *(f)*. Minor vascular pedicle: *m*, Branch of suprascapular artery.

STERNOCLEIDOMASTOID

Vascular anatomy (Figs. 2-14 and 2-15)

Pattern of circulation: Type II

Dominant vascular pedicle (D)
Branch of occipital artery and vein
REGIONAL SOURCE: External carotid artery and vein
SIZE: Length: 3 cm
Diameter: 1 mm
LOCATION: Superior third of muscle adjacent to hypoglossal (XII) nerve

Minor vascular pedicle (m)

1. Branch of posterior auricular artery and vein
   REGIONAL SOURCE: External carotid artery and vein
   SIZE: Length: 2 cm
   Diameter: 0.6 mm
   LOCATION: Adjacent to muscle insertion at mastoid process
2. Branch of superior thyroid artery and vein
   REGIONAL SOURCE: External carotid artery and vein
   SIZE: Length: 2 cm
   Diameter: 0.6 mm
   LOCATION: Distal one third of muscle
3. Branch of suprascapular artery and vein
   REGIONAL SOURCE: Thyrocervical trunk
   SIZE: Length: 2 cm
   Diameter: 0.5 mm
   LOCATION: Adjacent to muscle insertion at clavicle

Clinical application

ARC OF ROTATION: The entire muscle may be elevated on the dominant vascular pedicle (D). Transposition of the distal two thirds of the muscle may be performed with safety for coverage of defects of the anterior and posterior neck and lower face.

SKIN TERRITORY: The skin located over the lateral anterior neck receives musculocutaneous perforators from the sternocleidomastoid. Inferior neck skin may be designed as a skin island with muscle transposition with coverage of intraoral and lower facial defects.

SEGMENTAL TRANSPOSITION: Since the dominant pedicle is located on the superior third of the muscle belly, complete release of the origin is required to obtain adequate length to reach facial defects. The sternal head may be transposed for anterior midline neck defects with preservation of continuity of the clavicular head.

DESIGN OF DISTALLY BASED FLAPS: The superior muscle may be transposed inferiorly based on the distal minor pedicles.

POTENTIAL FOR MICROVASCULAR TRANSPLANTATION: This musculocutaneous unit is not useful as a flap for transplantation because of multiple vascular pedicles and its strategic location within the neck.

VASCULARIZED BONE: Vascular communications between the muscle origin and periosteum of the medial clavicle are noted. This bone may be transposed with the muscle to oral defects requiring mandibular reconstruction. However, adequate lymphadenectomy associated with extirpative surgery may require division of the proximal muscle vascular pedicles.

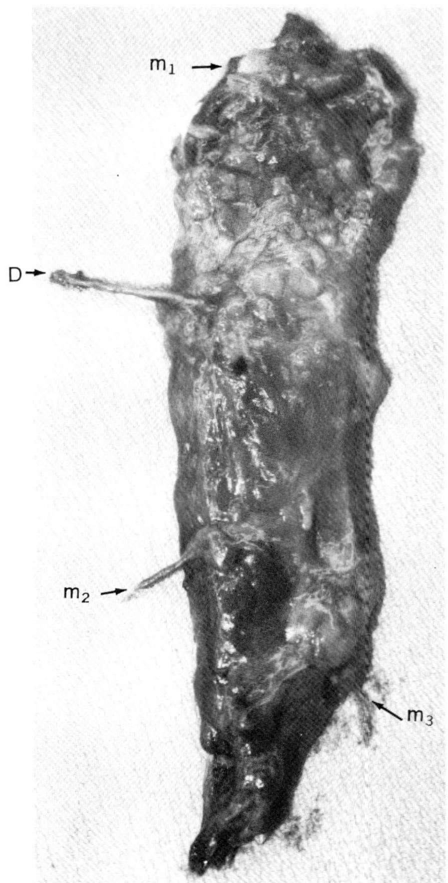

Fig. 2-14. Dominant vascular pedicle: *D*, Branch of occipital artery. Minor vascular pedicles: m_1, Branch of posterior auricular artery; m_2, branch of superior thyroid artery; m_3, branch of suprascapular artery.

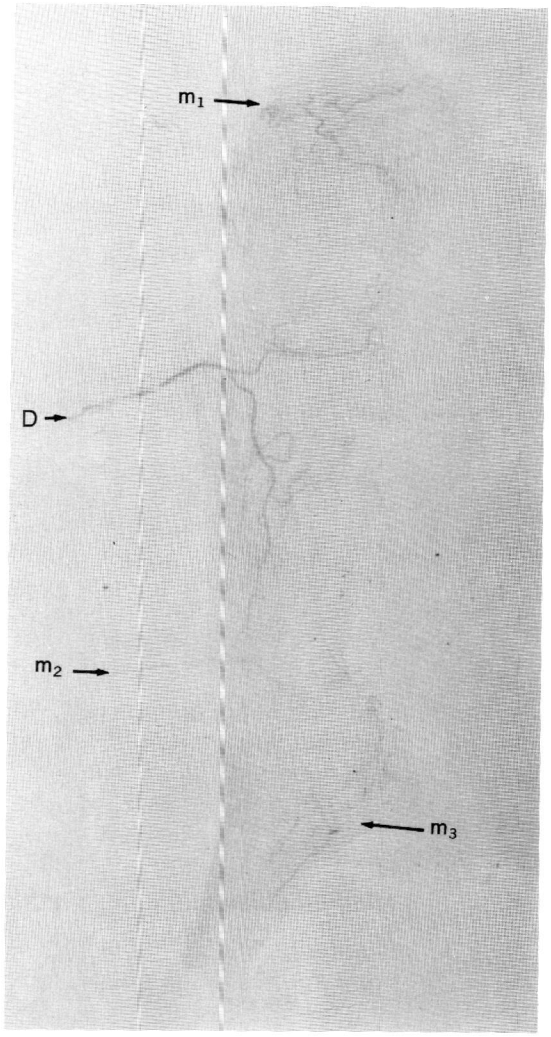

Fig. 2-15. Dominant vascular pedicle: *D*, Branch of occipital artery. Minor vascular pedicles; m_1, Branch of posterior auricular artery; m_2, branch of superior thyroid artery; m_3, branch of suprascapular artery.

TEMPORALIS

Vascular anatomy (Figs. 2-16 and 2-17)

Pattern of circulation: Type III

Dominant vascular pedicle (D)

1. Anterior deep temporal artery and venae comitantes
   REGIONAL SOURCE: Maxillary artery and vein
   SIZE: Length: 2 cm
   Diameter: 1 mm
   LOCATION: Beneath muscle adjacent to its insertion
2. Posterior deep temporal artery and venae comitantes
   REGIONAL SOURCE: Maxillary artery and vein
   SIZE: Length: 2 cm
   Diameter: 1 mm
   LOCATION: Beneath muscle adjacent to its insertion

Minor vascular pedicle (m)

Two to four branches of middle temporal artery and venae comitantes
REGIONAL SOURCE: Superficial temporal artery and vein
SIZE: Length: 1 cm
Diameter: 0.5 mm
LOCATION: Middle third of muscle along its superficial surface

Clinical application

ARC OF ROTATION: The entire muscle may be elevated on the dominant vascular pedicles. Transposition as a turnover flap is useful for coverage of the orbit, superior maxillae, and ear.

SKIN TERRITORY: Musculocutaneous perforating vessels are present between the muscle and scalp. However, since this skin is hair bearing, transposition of the scalp with muscle is rarely indicated.

SEGMENTAL TRANSPOSITION: Since the muscle has two dominant vascular pedicles, the muscle may be split, leaving the posterior half intact with its origin and insertion. Generally, the entire origin of the muscle is divided, and then the muscle is split for defect coverage (e.g., anterior half on supraorbital region and posterior half on the infraorbital region).

DESIGN OF DISTALLY BASED FLAPS: The posterior muscle may be elevated on the minor vascular pedicles. However, the superficial temporal artery is generally divided in superior facial defects where flap coverage is required.

POTENTIAL FOR MICROVASCULAR TRANSPLANTATION: The temporalis may be transplanted to the middle or lower face based on the maxillary artery and vein for use as function muscle in facial palsy. There is potential for injury to the upper division of the facial nerve during elevation of this muscle for transplantation.

VASCULARIZED BONE: Vascular communications to the periosteum of the temporal bone and the muscle origin are noted. However, transposition of vascularized temporal bone with muscle only has clinical application for local defects in cranial facial surgery.

Origin

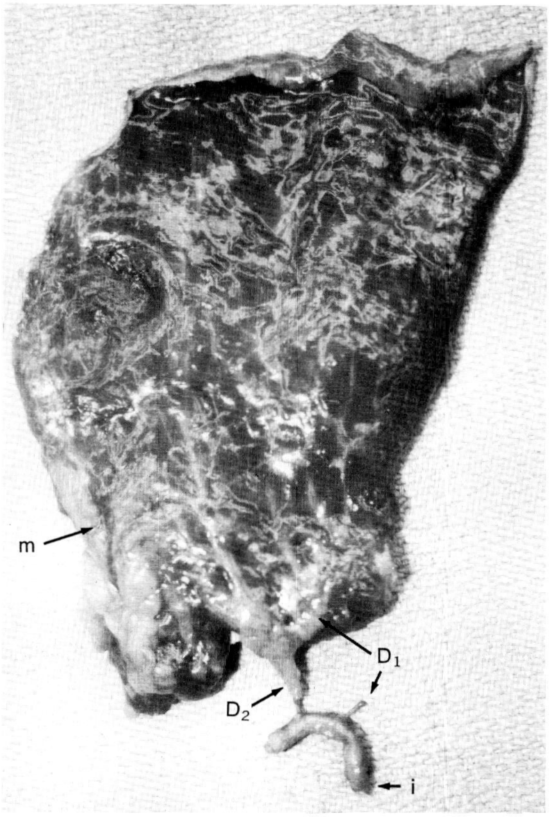

Insertion

Fig. 2-16. Dominant vascular pedicle: D_1, Anterior deep temporal artery; D_2, Posterior deep temporal artery. Minor vascular pedicles; *m*, Branch of middle temporal artery.

Origin

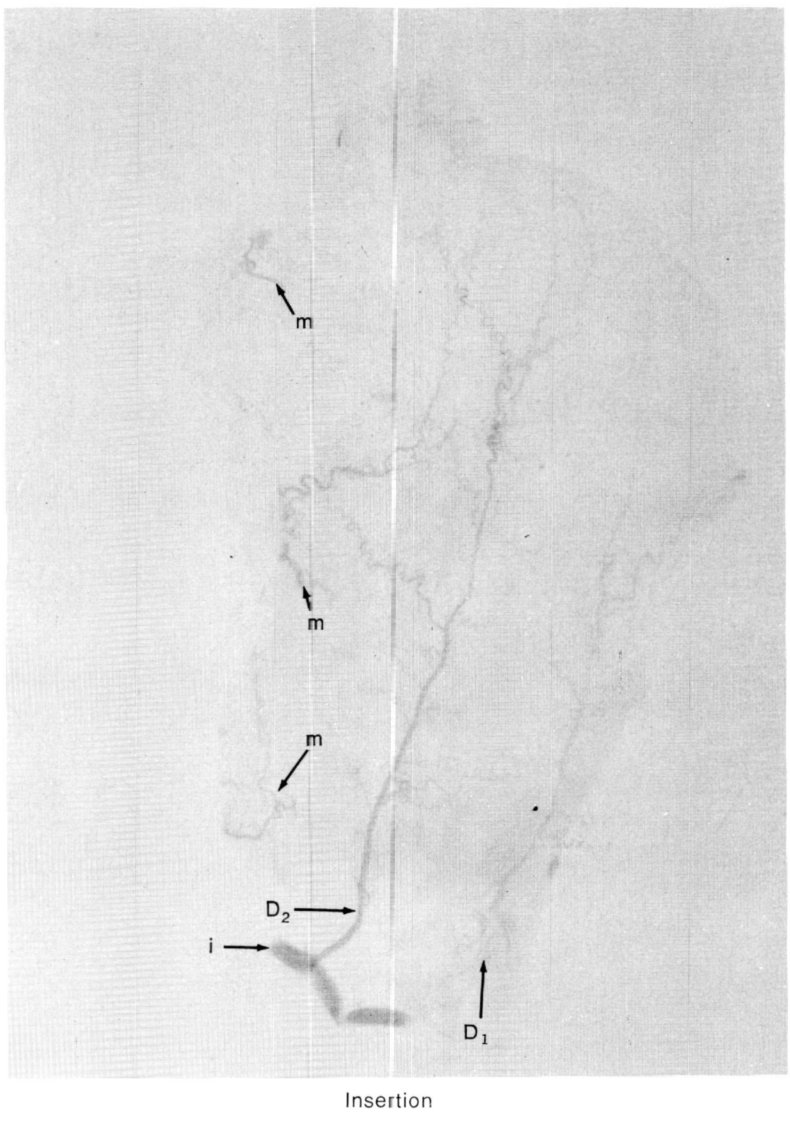

Insertion

Fig. 2-17. Dominant vascular pedicle: D_1, Anterior deep temporal artery; D_2, Posterior deep temporal artery. Minor vascular pedicles; *m*, Branches of middle temporal artery.

PECTORALIS MAJOR

Vascular anatomy (Figs. 2-18 and 2-19)

Pattern of circulation: Type V

Dominant vascular pedicle (D)

Pectoral branch of thoracoacromial artery and venae comitantes

REGIONAL SOURCE: Subclavian artery and vein

SIZE: Length: 4 cm
Diameter: 2.5 mm

LOCATION: Beneath clavicle at midsuperior edge of muscle

Segmental secondary vascular pedicles (s)—Perforating arterial branches at first through sixth intercostal spaces and venae comitantes

REGIONAL SOURCE: Internal mammary artery and vein

SIZE: Length: 2 to 3 cm
Diameter: 0.5 to 0.8 mm

LOCATION: Beneath muscle origin at sternum

Clinical application

ARC OF ROTATION: The entire muscle will survive elevation on either vascular system: The single dominant vascular pedicle or the secondary segmental pedicles. Transposition of the muscle based on the dominant vascular pedicle provides coverage for defects of the head, neck, and anterior chest. Transposition of the muscle as a turnover flap is based on the secondary segmental pedicles and is primarily useful for coverage of defects of the sternum.

SKIN TERRITORY: The skin of the anterior chest receives multiple musculocutaneous perforators from the pectoralis major muscle. All or part of this skin may be included with the pectoralis major when used as a muscle flap. Central perforating vessels also contribute to the circulation of the breast. Design of the skin island on the musculocutaneous flap is generally placed immediately below or medial to the breast in the female patient.

SEGMENTAL TRANSPOSITION: Release of the muscle origin from both the sternum and rib and insertion from the humerus are required for transposition of the muscle based on its major vascular pedicle for coverage of head and neck defects. An adequate muscle flap for sternal coverage is provided by transposition of the medial two thirds of the muscle based on the secondary segmental perforating vessels. The medial pectoral nerve and vessels are preserved with the remaining muscle fibers that are attached to the lateral chest wall to establish a new site of muscle origin. These fibers maintain the anterior axillary fold.

The musculocutaneous perforating vessels from the secondary segmental muscle pedicles provide circulation to the skin extending from the midline anteriorly to the anterior axillary line posteriorly. This skin, the deltopectoral flap, survives elevation without disruption of the pectoralis major and will cover head and neck defects.

DESIGN OF DISTALLY BASED FLAPS: Distally based flaps are not applicable, because the pectoralis major has no minor vascular pedicles.

POTENTIAL FOR MICROVASCULAR TRANSPLANTATION: The muscle is suitable for transplantation based on its dominant vascular pedicle for both coverage and functional use.

VASCULARIZED BONE: Vascular communication between the fibers of the muscle origin and the periosteum of the fifth and sixth ribs allows transposition of pectoralis major with the vascularized rib. The vascularized rib reaches the intraoral defect and is useful for mandible replacement.

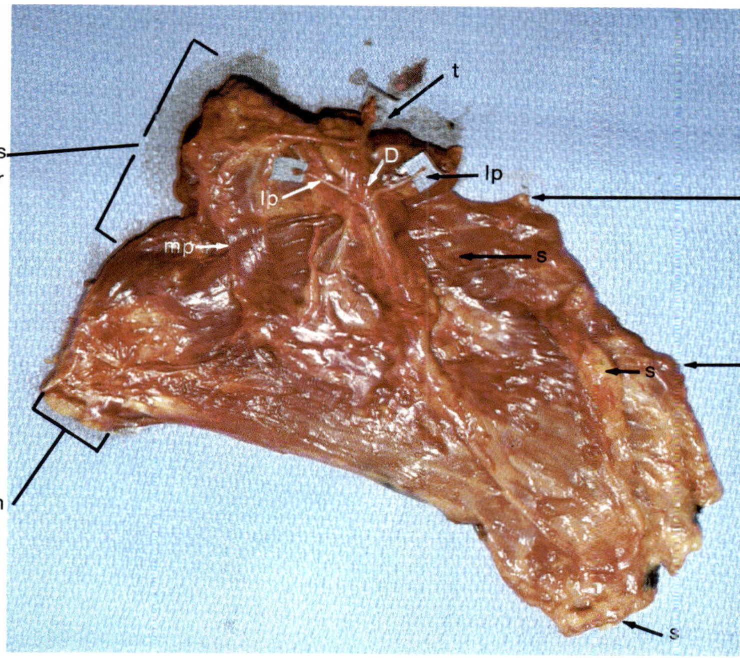

Fig. 2-18. Dominant vascular pedicle: *D*, Pectoral branch of thoracoacromial artery (*t*). Secondary segmental pedicles: *s*, Lateral pectoral nerve (*lp*): medial pectoral nerve (*mp*).

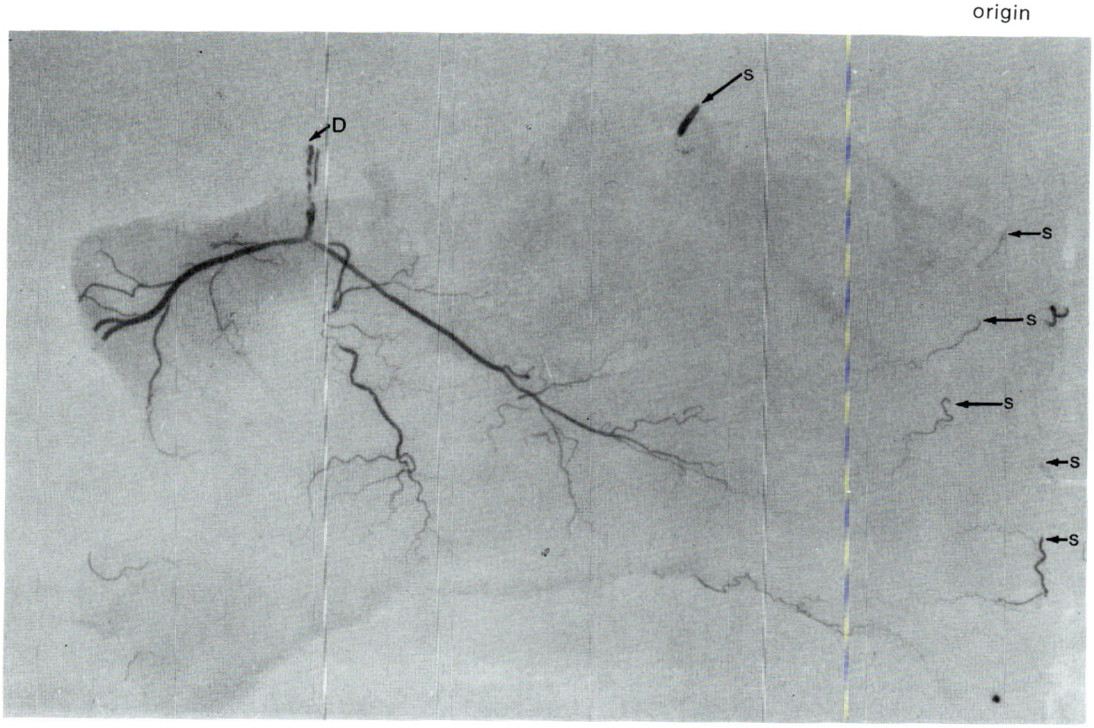

Fig. 2-19. Dominant vascular pedicle: *D*, Pectoral branch of thoracoacromial artery. *s*, Secondary segmental pedicles.

RECTUS ABDOMINIS

Vascular anatomy (Figs. 2-20 and 2-21)

Pattern of circulation: Type III

Dominant vascular pedicle (D)

1. Superior epigastric artery and vein
   REGIONAL SOURCE: Internal mammary artery and vein
   SIZE: Length: 2 cm
   Diameter: 2.5 mm
   LOCATION: Beneath muscle insertion at costal margin
2. Inferior epigastric artery and vein
   REGIONAL SOURCE: External iliac artery and vein
   SIZE: Length: 5 cm
   Diameter: 2.5 mm
   LOCATION: Beneath muscle origin at groin

Clinical application

ARC OF ROTATION: The rectus abdominis may be elevated on either major vascular pedicle. Transposition of the inferior muscle will cover the thorax, breast, and superior abdominal wall and may be used in reconstruction of these regions. Transposition of the superior muscle inferiorly based will cover the abdominal wall groin and may be used to reconstruct lower abdominal wall.

SKIN TERRITORY: The anterior midabdominal wall skin receives musculocutaneous perforating vessels from the rectus abdominis. This skin overlying either the superior or inferior half of the muscle may be included with the muscle as a transposition flap. The entire skin overlying the muscle will generally survive transposition as a musculocutaneous flap based on either the superior or inferior dominant vascular pedicle.

SEGMENTAL TRANSPOSITION: The muscle must be divided either near its insertion or origin to achieve an adequate arc of rotation to cover most defects. However, the musculocutaneous perforating vessels provide cutaneous circulation to the lateral abdominal wall. The thoracoepigastric flap is based on these perforating musculocutaneous vessels and may be elevated from the anterior axillary line to the rectus sheath without interrupting muscle continuity. This cutaneous flap will cover thoracic and abdominal wall defects.

DESIGN OF DISTALLY BASED FLAPS: Distally based flaps are not relevant to the rectus abdominis.

POTENTIAL FOR MICROVASCULAR TRANSPLANTATION: The rectus abdominis may be transplanted based on either dominant vascular pedicle and its associated venae comitantes. However, the donor defect may cause donor site disability, and for this reason the flap is not recommended for microvascular transplantation.

VASCULARIZED BONE: Although vascular communications between the muscle insertion and the perichondrium of the fifth, sixth, and seventh ribs are noted, donor deformity generally precludes transposition of the rectus abdominis with vascularized rib.

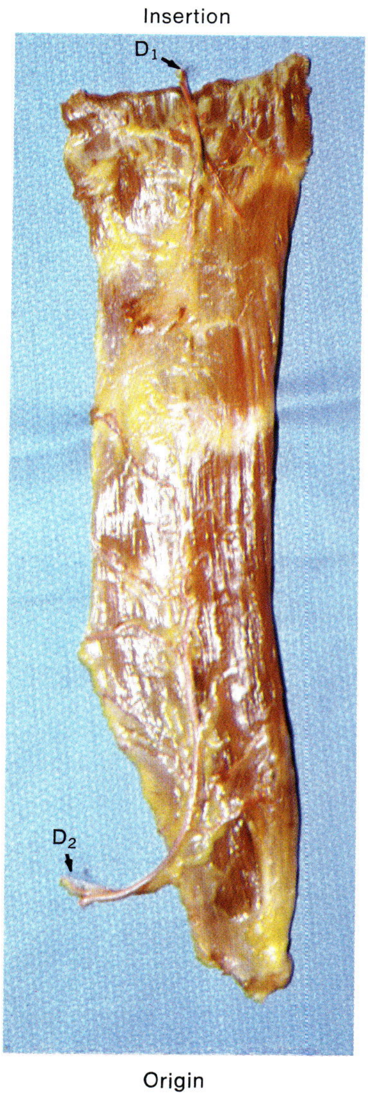

Fig. 2-20. Dominant vascular pedicles: D_1, Superior epigastric artery; D_2, inferior epigastric artery.

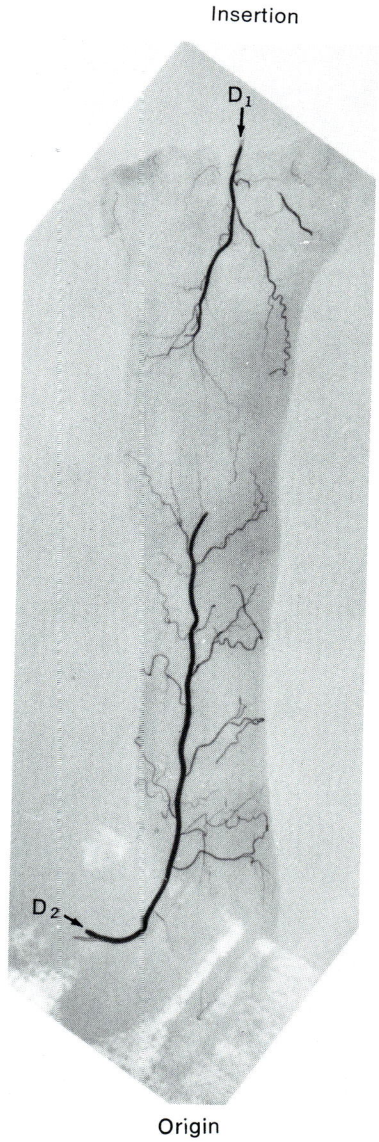

Fig. 2-21. Dominant vascular pedicles: D_1, Superior epigastric artery; D_2, inferior epigastric artery.

LATISSIMUS DORSI

Vascular anatomy (Figs. 2-22 and 2-23)

Pattern of circulation: Type V

Dominant vascular pedicle (D)

Thoracodorsal artery and venae comitantes

REGIONAL SOURCE: Subscapular artery and vein

SIZE: Length: 8 cm

Diameter: 2.5 mm

LOCATION: Adjacent to muscle insertion in posterior axillae

Secondary segmental pedicles (s)—Four to six perforating arterial branches and venae comitantes

REGIONAL SOURCE: Posterior intercostal artery and vein

SIZE: Length: 2 to 3 cm

Diameter: 0.6 mm

LOCATION: Lateral to lumbosacral fascia

Insertion

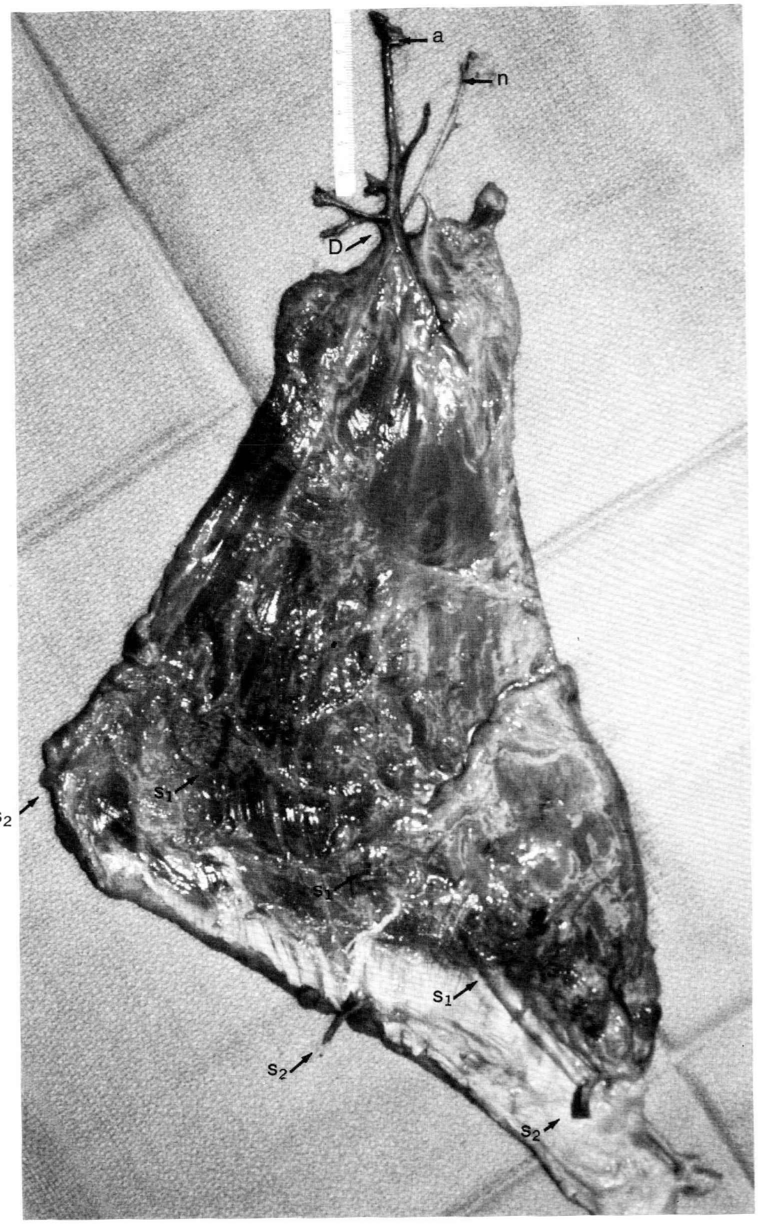

Origin

Fig. 2-22. Dominant vascular pedicle: *D*, Thoracodorsal artery (branch of subscapular artery [*a*]). Secondary segmental pedicles: s_1, Perforating branch of intercostal artery; s_2, perforating branch of lumbar artery. *n*, Thoracodorsal nerve.

Four to six perforating branches and venae comitantes

REGIONAL SOURCE: Lumbar artery and vein

SIZE: Length: 1 to 2 cm
Diameter: 0.5 mm

LOCATION: Adjacent to site of muscle origin into lumbar vertebrae

Clinical application

ARC OF ROTATION: The entire muscle will survive elevation based on either the dominant vascular pedicle or the secondary segmental perforating vessels. Transposition of the muscle based on the dominant vascular pedicle will cover defects in the head, neck, and anterior and posterior chest. Anterior transposition is useful for head, neck, and breast reconstruction. Transposition of the muscle based on the secondary segmental pedicles is primarily useful for posterior midline coverage over the vertebral column.

SKIN TERRITORY: The skin of the back is well supplied with musculocutaneous perforators from the latissimus dorsi muscle. The entire skin or a portion of skin may be transposed with this muscle based on either vascular system. Transposition of the musculocutaneous flap anteriorly based on the dominant pedicle is most useful for breast reconstruction. A portion of the skin territory is also generally included with the muscle for posterior trunk coverage based on the secondary segmental pedicles.

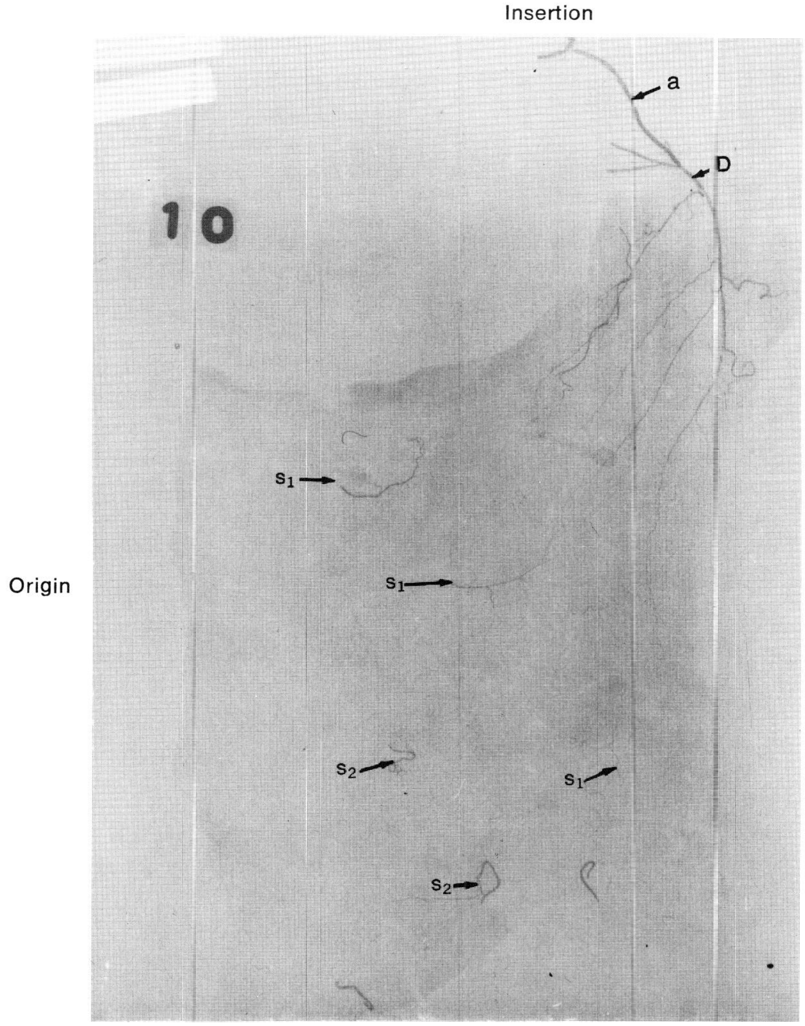

Fig. 2-23. Dominant vascular pedicle: *D*, Thoracodorsal artery (branch of subscapular artery [*a*]). Secondary segmental pedicles: s_1, Perforating branch of intercostal artery; s_2, perforating branch of lumbar artery.

SEGMENTAL TRANSPOSITION (Figs. 2-24 and 2-25): Division of the muscle origin is required to establish a wide arc of rotation for anterior defects of the head, neck, and anterior chest. However, the anterior half of the muscle will survive transposition based on the anterior branch of the thoracodorsal artery. The posterior muscle retains origin, insertions, motor nerve, and vascular continuity. Transposition of the anterior muscle covers defects of the anterior chest, breast, and axillae.

After release of the entire origin of the muscle, it will extend across the midline, providing coverage and establishing a new site of origin at the site of fixation. This advancement of the muscle origin will cover congenital posterior midline defects in children (e.g., myelomeningocele) or acquired wounds on the midline of the back.

Complete release of the muscle insertion is not required for muscle transposition based on the secondary segmental perforators. The musculocutaneous perforators derived from the secondary segmental pedicles, especially the lumbar vessels, will vascularize the back skin across the midline, extending to the posterior axillary line. Elevation of this skin as a flap, the transverse back flap, will cover the midline defects. The muscle is not elevated and retains normal function.

DESIGN OF DISTALLY BASED FLAP: Distally based flaps are not applicable to the latissimus dorsi.

POTENTIAL FOR MICROVASCULAR TRANSPLANTATION: The latissimus dorsi muscle is ideally suited for transplantation based on the subscapular vessels or on the anterior branch of the thoracodorsal vessels preserving the function of the posterior fibers of the muscle. The muscle transplantation may include skin and is suitable for both coverage and functional use in reconstructive surgery.

VASCULARIZED BONE: Vascular communication is not adequate between the fascia of origin or insertion for elevation of vascularized bone with the latissimus dorsi.

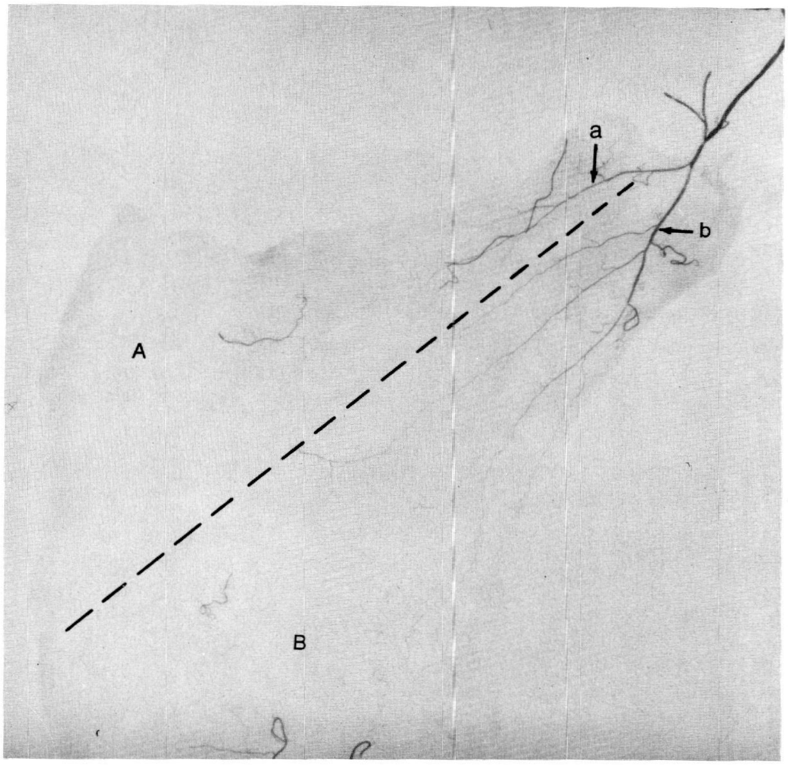

Fig. 2-24. Segmental transposition. Muscle *A* is vascularized by posterior branch of thoracodorsal artery *(a)*. Muscle *B* transposition flap is based on anterior branch of thoracodorsal artery *(b)*.

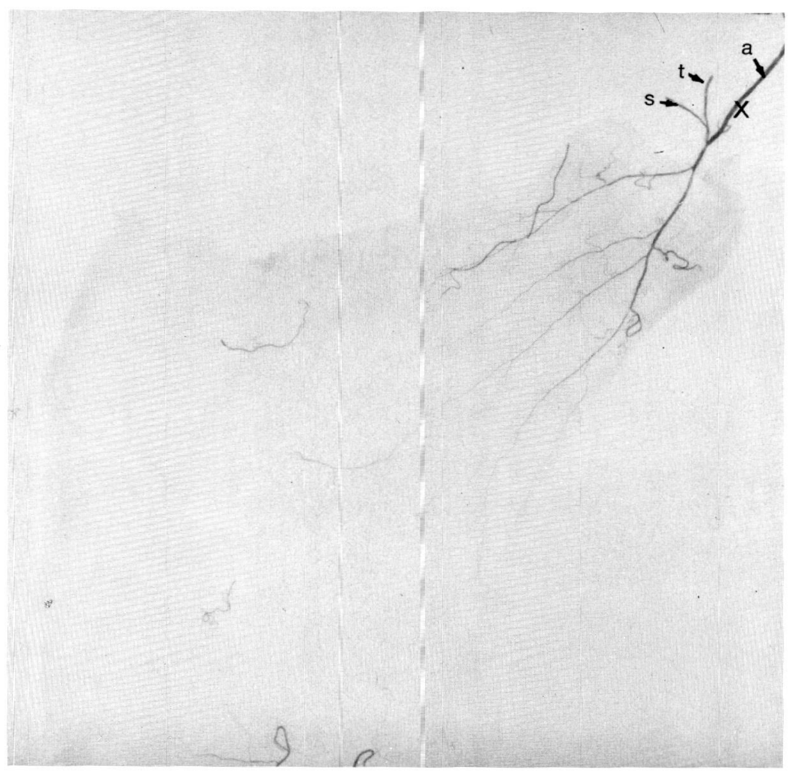

Fig. 2-25. Muscle transposition based on retrograde vascular filling through branches *(t* and *s)*. *x*, Prior ligation of subscapular artery *(a)*; *t*, branch to teres major; *s*, branch to serratus anterior.

TRAPEZIUS

Vascular anatomy (Figs. 2-26 and 2-27)

Pattern of circulation: Type II

Dominant vascular pedicle (D)

Transverse cervical artery and vein

REGIONAL SOURCE: Thyrocervical trunk or subclavian artery and vein

SIZE: Length: 4 cm
Diameter: 1.8 mm

LOCATION: Posterior inferior neck and superior anterior margin of muscle

Minor vascular pedicles (m)

1. Branch of occipital artery and vein

   REGIONAL SOURCE: External carotid artery and vein

   SIZE: Length: 3 cm
   Diameter: 1 mm

   LOCATION: Adjacent to muscle origin at external occipital protuberance

2. Perforating posterior intercostal arteries and vein

   REGIONAL SOURCE: Descending aorta and vena cava

   SIZE: Length: 1 to 2 cm
   Diameter: 0.5 mm

   LOCATION: Vascular pedicles enter the muscle along the posterior midline adjacent to the cervical and thoracic vertebral body

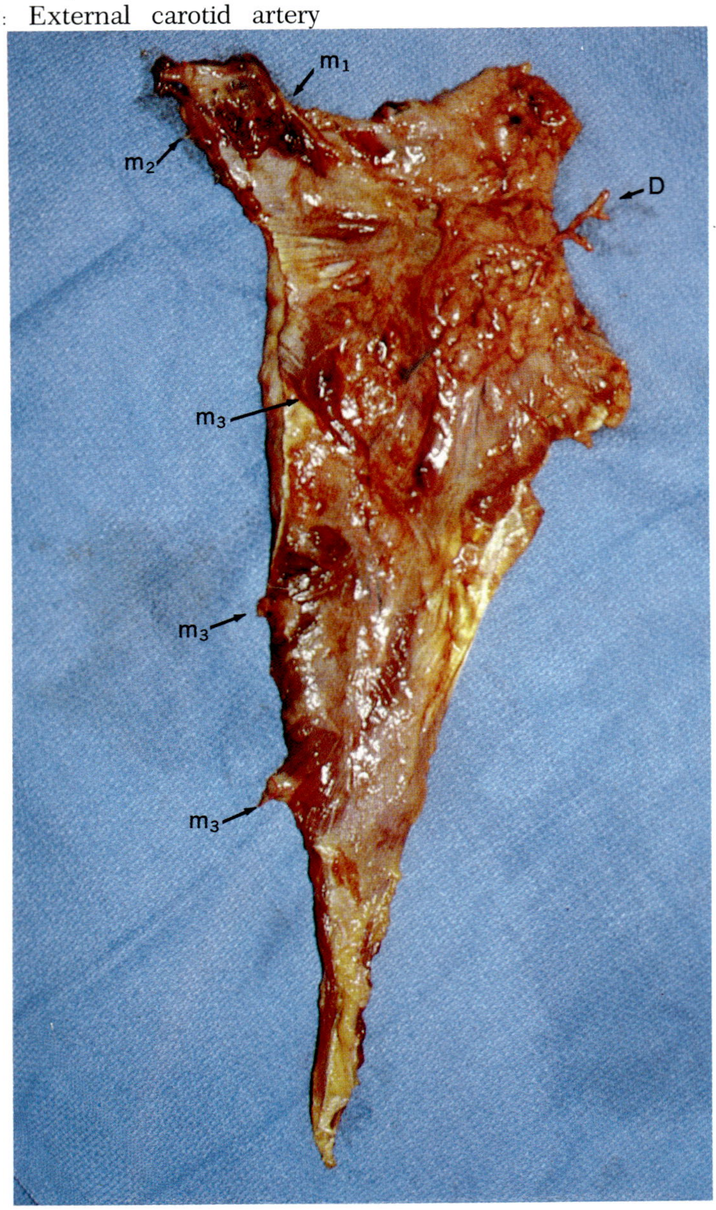

Fig. 2-26. Dominant pedicle: *D*, Transverse cervical artery. Minor pedicles: m_1, Branch of occipital artery; m_2, branch of thoracic artery; m_3, branch of posterior intercostal artery.

Clinical application

ARC OF ROTATION: The trapezius may be completely elevated safely based on its circulation from the dominant vascular pedicle. Elevation of the entire muscle for coverage or reconstruction is rarely indicated except when a very wide arc of rotation is required to reach skull defects. Prior division of the regional source of this muscle's dominant circulation, the transverse cervical artery, may significantly impair muscle circulation and preclude its use as a transposition flap.

SKIN TERRITORY: The trapezius has musculocutaneous perforating vessels to the posterior neck, shoulder, and superior back. Any portion of this skin may be elevated with this muscle during muscle transposition. The back skin is preferable, since the posterior neck skin may border a neck dissection defect.

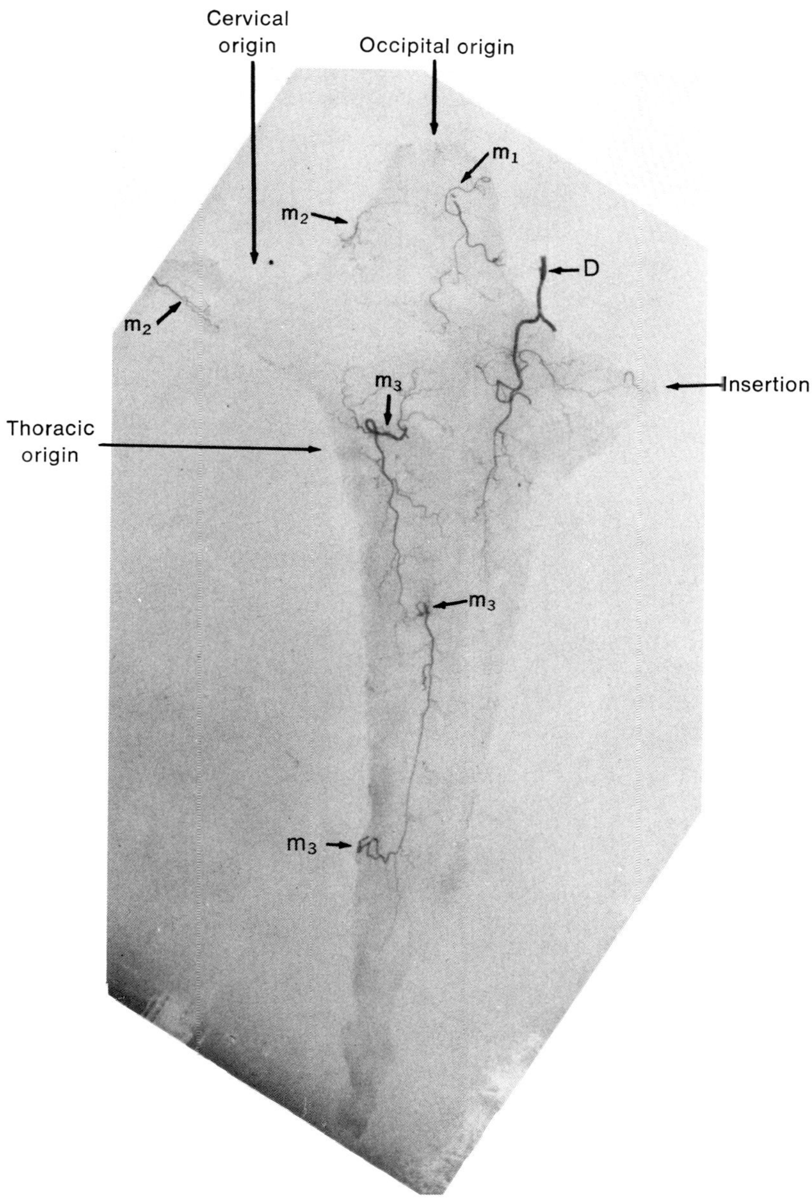

Fig. 2-27. Dominant pedicle: *D*, Transverse cervical artery. Minor pedicles: m_1, Branch of occipital artery; m_2, branch of thoracic artery; m_3, branch of posterior intercostal artery.

SEGMENTAL TRANSPOSITION: The anterior trapezius with overlying shoulder and neck skin may be safely elevated based on the anterior branches of the dominant vascular pedicle (Fig. 2-28). Division of the fibers of insertion to the lateral third of the clavicle and acromion is required for flap elevation. Although attachments of muscle to the spine of the scapula are preserved, the function of shoulder elevation is impaired after flap transposition. This flap provides coverage of anterior neck and lower facial and oral defects.

The posterior trapezius with overlying superior back skin may be safely elevated based on the descending branch of the dominant vascular pedicle (Fig. 2-28). Flap elevation only requires release of distal fibers of insertion to the spine and inferior fibers of origin to the thoracic vertebral column. Continuity of the anterior trapezius between the neck and shoulder is preserved to maintain important muscle function. This flap provides reliable coverage for anterior and posterior neck, lower and middle face, and superior back defects.

DESIGN OF DISTALLY BASED FLAPS (Fig. 2-29): The anterior trapezius and its overlying skin may be elevated based on the minor pedicle from the occipital artery. However, the skin is unreliable beyond the lateral third of the clavicle, and prior flap delay is recommended.

POTENTIAL FOR MICROVASCULAR TRANSPLANTATION: Because of the location of the trapezius, transplantation to distant defects is rarely indicated despite the presence of a single dominant vascular pedicle.

VASCULARIZED BONE: Vascular attachments between the muscle insertion and periosteum of the lateral third of the clavicle and the spine of the scapula are present. The lateral clavicle may be included with transposition of the anterior trapezius. The spine of the scapula may be included with transposition of the posterior trapezius. This vascularized bone may be used in conjunction with flap coverage for mandibular reconstruction.

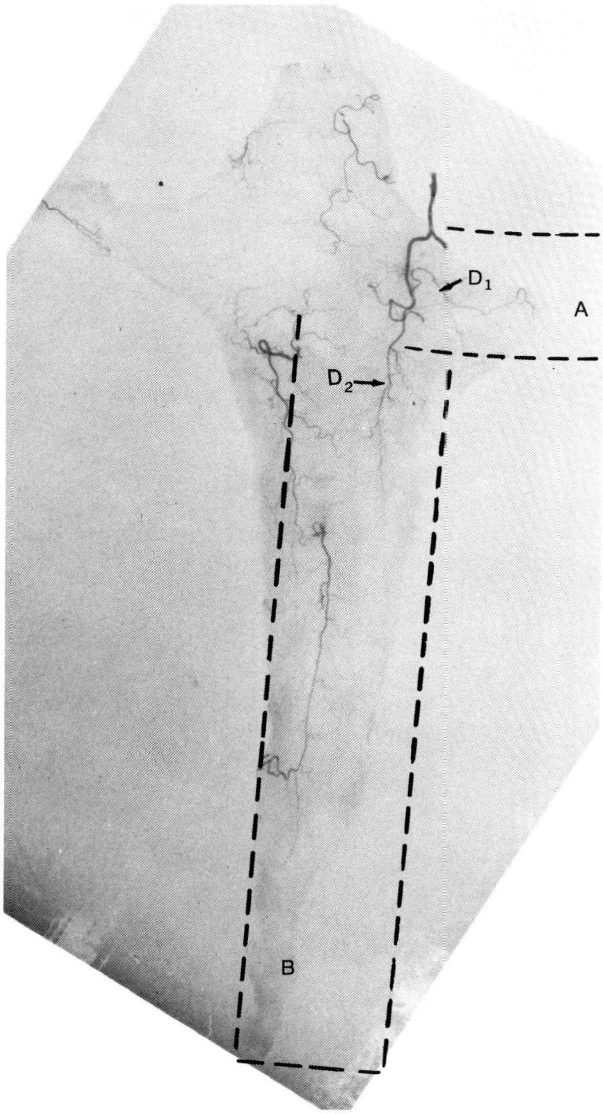

Fig. 2-28. Segmental transposition. *A*, Anterior trapezius flap (cervicohumeral flap); *B*, posterior trapezius flap. Transverse cervical artery: D_1, Anterior branch; D_2, descending branch.

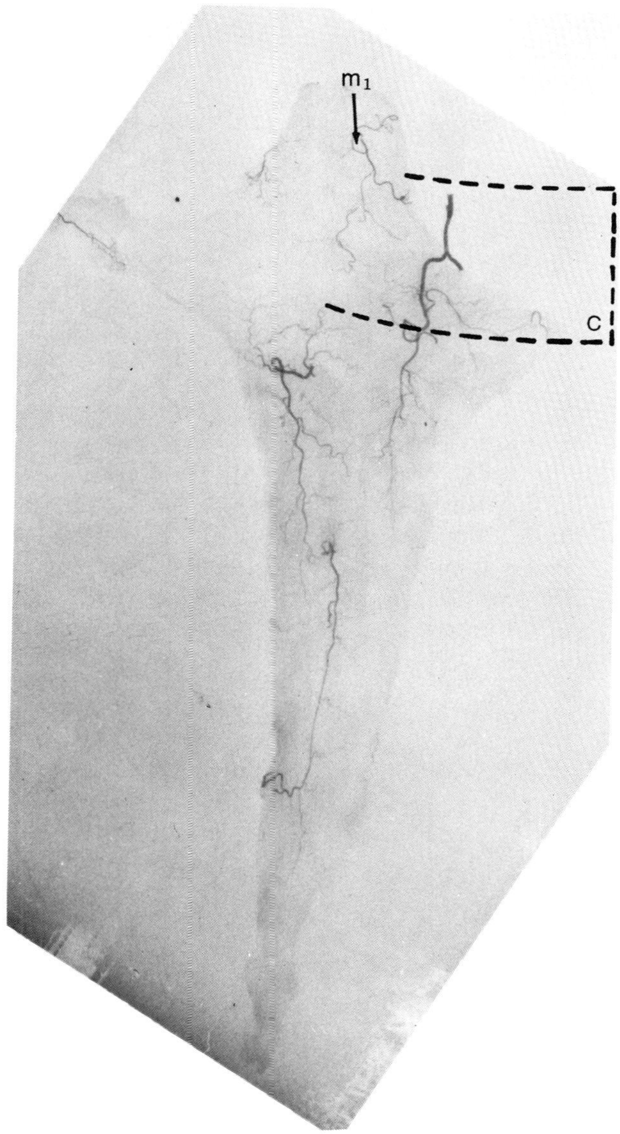

Fig. 2-29. Distally based flap. *C*, Occipital flap based on occipital artery pedicle (m_1).

GRACILIS

Vascular anatomy (Figs. 2-30 and 2-31)

Pattern of circulation: Type II

Dominant vascular pedicle (D)

Medial femoral circumflex artery and venae comitantes

REGIONAL SOURCE: Profunda femoris artery and vein

SIZE: Length: 6 cm
Diameter: 1.6 mm

LOCATION: Superior third of muscle belly

Minor vascular pedicle(s) (m)

One or two branches of the superficial femoral artery and venae comitantes

REGIONAL SOURCE: Superficial femoral artery and vein

SIZE: Length: 2 cm
Diameter: 0.5 mm

LOCATION: Inferior half of muscle

Clinical application

ARC OF ROTATION: Elevation of the entire muscle is reliable based on its dominant vascular pedicle. Transposition of this muscle provides coverage of peroneal and pelvic defect and penile reconstruction. Division of the minor pedicle(s) does not adversely affect muscle survival after transposition.

SKIN TERRITORY: The medial thigh skin is not entirely vascularized by musculocutaneous perforating vessels from the gracilis. The skin over the distal muscle when designed as a skin island based on muscle circulation is not reliable with division of the minor pedicle(s). Transposition of the gracilis with a superior skin island is a reliable flap for vaginal reconstruction and coverage of pressure sores, groin, and perineum.

SEGMENTAL TRANSPOSITION: Complete division of the muscle insertion is required for muscle transposition to reach pelvic defects. However, the medial thigh skin may be elevated as a skin flap if both proximal musculocutaneous perforators and superior skin continuity are maintained.

DESIGN OF DISTALLY BASED FLAPS: In musculocutaneous flaps based on the minor pedicle(s), distally based flaps are possible. However, a strategic delay is required 2 weeks before flap transposition. This is accomplished by selective division of the dominant vascular pedicle.

POTENTIAL FOR MICROVASCULAR TRANSPLANTATION: The muscle or musculocutaneous unit may be transplanted by microvascular techniques based on the dominant vascular pedicle for coverage and functional muscle transplantation.

VASCULARIZED BONE: There are minimal vascular communications between the muscle origin and pubic symphysis. This muscle is not useful as an osseous muscle flap for transplantation.

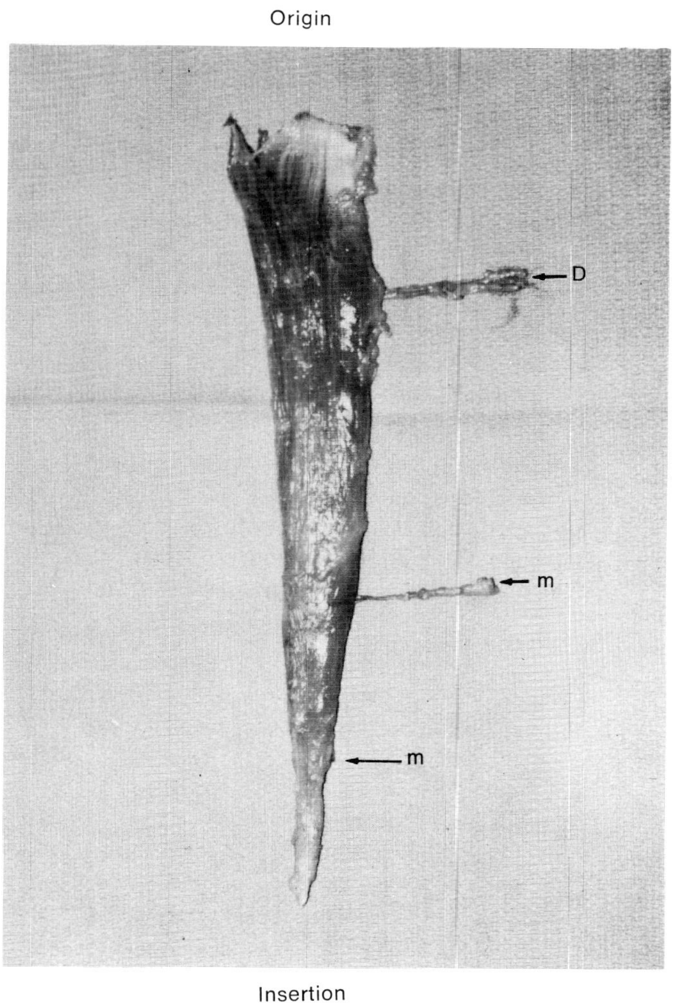

Fig. 2-30. Dominant pedicle: *D,* Medial circumflex femoral artery. Minor pedicle: *m,* Branch of superficial femoral artery.

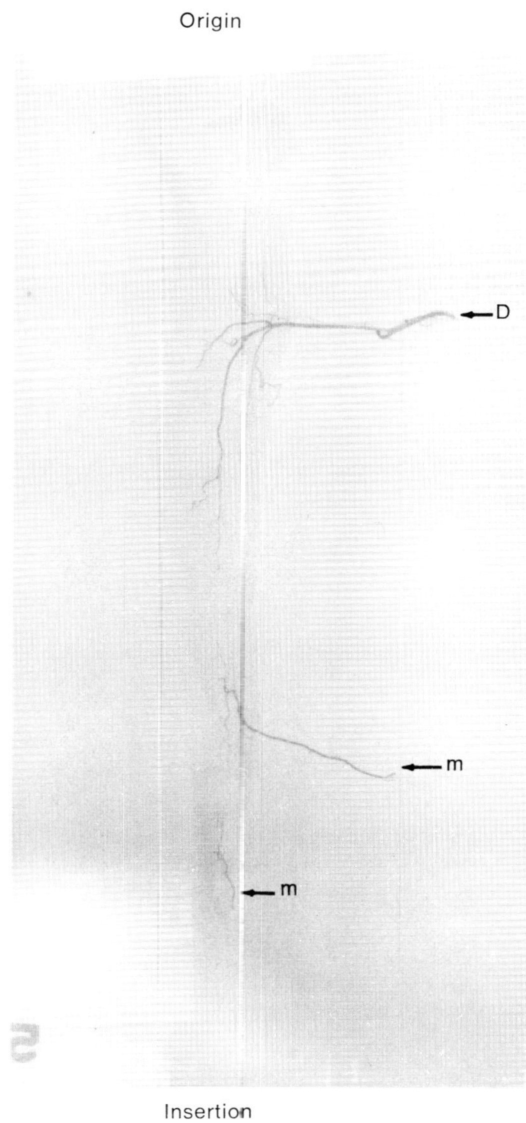

Fig. 2-31. Dominant pedicle: *D,* Medial circumflex femoral artery. Minor pedicle: *m,* Branch of superficial femoral artery.

RECTUS FEMORIS

Vascular anatomy (Figs. 2-32 and 2-33)

Pattern of circulation: Type I

Dominant vascular pedicle (D)

Lateral circumflex femoral artery and venae comitantes (NOTE: Muscle may receive minor pedicle from femoral artery.)

REGIONAL SOURCE: Profunda femoris artery and vein

SIZE: Length: 4 cm
Diameter: 2 mm

LOCATION: Superior one third of muscle

Clinical application

ARC OF ROTATION: Elevation of the entire muscle based on its single proximal vascular pedicle is possible. Transposition of this muscle provides coverage of the groin, peroneum, pressure sores, and anterior abdominal wall defects.

SKIN TERRITORY: The entire anterior thigh skin may be transposed based on vascular communications with the rectus femoris. The presence of numerous musculocutaneous perforators between the distal muscle and the overlying thigh skin allows transposition of a distal skin island for coverage of defects of both the groin and inferior abdominal wall.

SEGMENTAL TRANSPOSITION: Transposition of the rectus femoris requires complete division of its insertion for adequate length to reach defects where muscle coverage is generally required. The distal muscle may be transposed with maintenance of tendon continuity for defects located in the midthigh.

DESIGN OF DISTALLY BASED FLAPS: Because of the absence of inferior minor pedicles, a distally based flap is not possible.

POTENTIAL FOR MICROVASCULAR TRANSPLANTATION: The rectus femoris has potential usefulness for microvascular transplantation based on its single vascular pedicle for defects requiring both coverage and functional muscle transplantation.

VASCULARIZED BONE: Vascular communications between the muscle origin and periosteum of the inferior iliac spine are present, allowing the potential for transplanting bone with muscle.

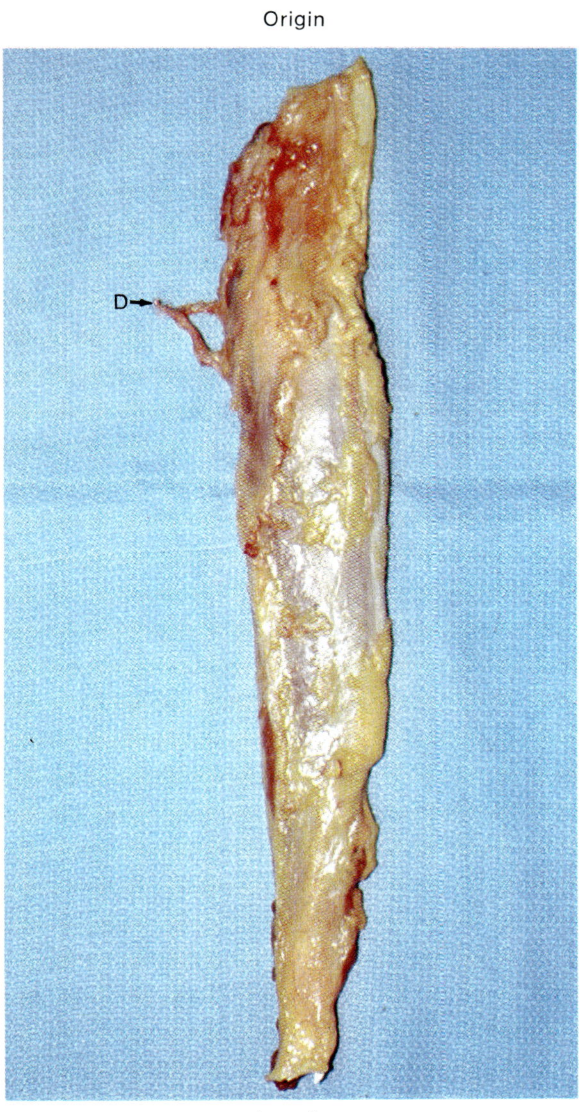

Fig. 2-32. Branch of lateral circumflex femoral artery (*D*).

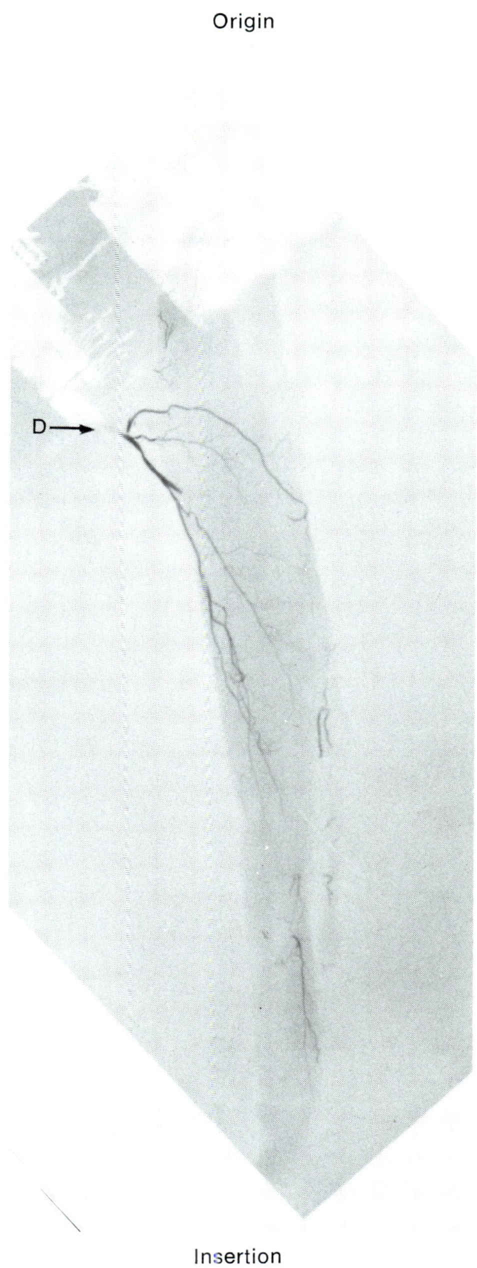

Fig. 2-33. Branch of lateral circumflex femoral artery (*D*).

SARTORIUS

Vascular anatomy (Figs. 2-34 and 2-35)

Pattern of circulation: Type IV

Segmental vascular pedicles

Six to ten segmental arterial branches from the superficial femoral artery and venae comitantes

REGIONAL SOURCE: Superficial femoral artery and vein

SIZE: Length: 2 to 3 cm
Diameter: 0.6 to 0.8 mm

LOCATION: Deep surface of entire muscle belly

Clinical application

ARC OF ROTATION: The superior or inferior third of the sartorius may be elevated with safety. Transposition will cover small defects of the groin (e.g., femoral vessels after radical groin dissection) or knee. However, division of more than three segmental pedicles may result in distal muscle necrosis.

SKIN TERRITORY: There are occasionally musculocutaneous perforating vessels between the muscle and lateral superior and medial inferior thigh skin. Skin over the superior and inferior muscle may be included with the muscle as a transposition flap. However, since this muscle will only reach adjacent defects, use of this muscle as a musculocutaneous flap is rarely indicated.

SEGMENTAL TRANSPOSITION: Either the entire origin or insertion must be divided to obtain a short arc of rotation for local defects.

DESIGN OF DISTALLY BASED FLAPS: Distally based flaps are not applicable to the sartorius.

POTENTIAL FOR MICROVASCULAR TRANSPLANTATION: Only a small portion of the sartorius could be transplanted based on a single segmental vascular pedicle.

VASCULARIZED BONE: Although vascular communications between muscle origin and insertion to the iliac spine and medial tibia, respectively, are noted, there is no clinical application for this muscle flap with vascularized bone.

Origin

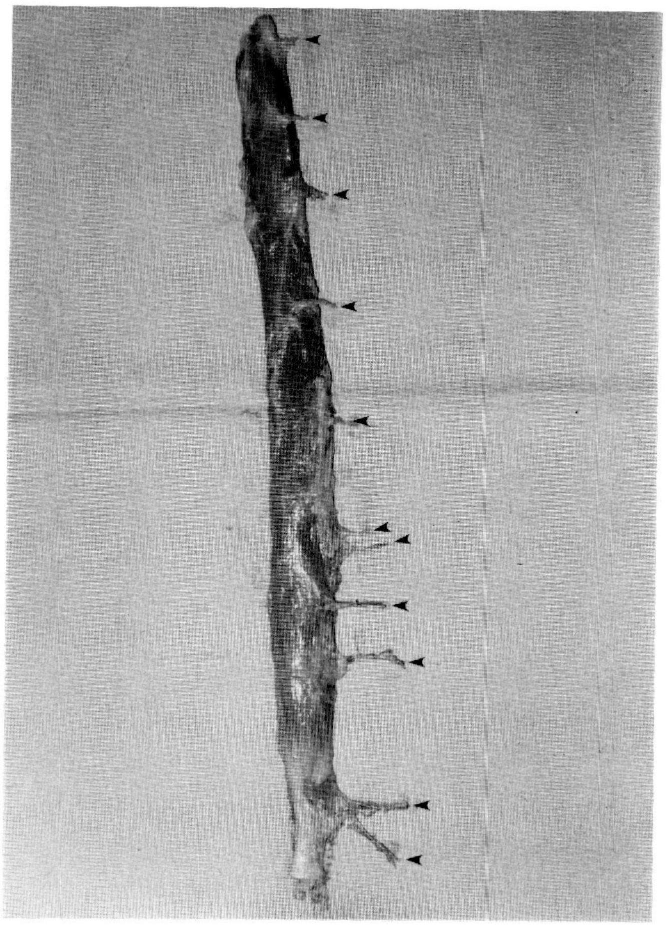

Insertion

Fig. 2-34. Arrows indicate segmental vascular pedicles.

Origin

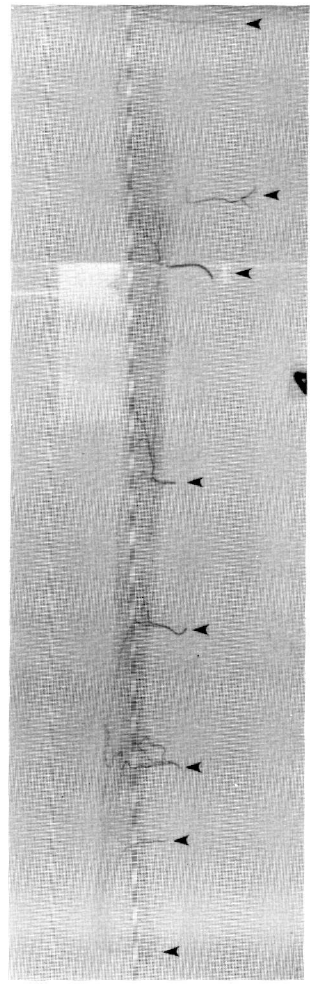

Insertion

Fig. 2-35. Arrows indicate segmental vascular pedicles.

TENSOR FASCIA LATA

Vascular anatomy (Figs. 2-36 and 2-37)

Pattern of circulation: Type I

Dominant vascular pedicle (D)

Terminal branch lateral circumflex femoral artery and venae comitantes

REGIONAL SOURCE: Profunda femoris artery and vein

SIZE: Length: 8 cm
Diameter: 2 mm

LOCATION: 10 cm inferior to the anterior superior iliac crest

Clinical application

ARC OF ROTATION: The fascia lata, the long fascial insertion of the small TFL muscle, is vascularized by the single muscle vascular pedicle. Transposition of this muscle with its fascial insertion will provide coverage for inferior abdominal wall reconstruction.

SKIN TERRITORY: The entire lateral thigh skin is vascularized by perforating vessels from the TFL muscle, which extends inferiorly between the fascia lata and subcutaneous tissue. Transposition of muscle, fascia lata, and overlying skin will provide coverage of the abdomen, groin, perineum, and pelvic pressure sores.

SEGMENTAL TRANSPOSITION: Complete division of insertion is required for a useful arc of rotation. However, distal skin may be elevated for midthigh defects based on proximal musculocutaneous perforating vessels extending from the muscle inferiorly. However, elevation of the skin distal to the site of entry of the vascular pedicle into muscle will divide musculocutaneous perforating vessels and result in unreliable skin circulation.

DESIGN OF DISTALLY BASED FLAPS: Because of the absence of inferiorly located minor pedicles, the distally based flap is not possible for the TFL.

POTENTIAL FOR MICROVASCULAR TRANSPLANTATION: TFL fascial or musculocutaneous unit is ideally suited for microvascular transplantation based on its single vascular pedicle for both coverage and functional muscle.

VASCULARIZED BONE: Vascular communications between the muscle origin and periosteum of the anterior iliac crest allow transplantation of this unit with bone.

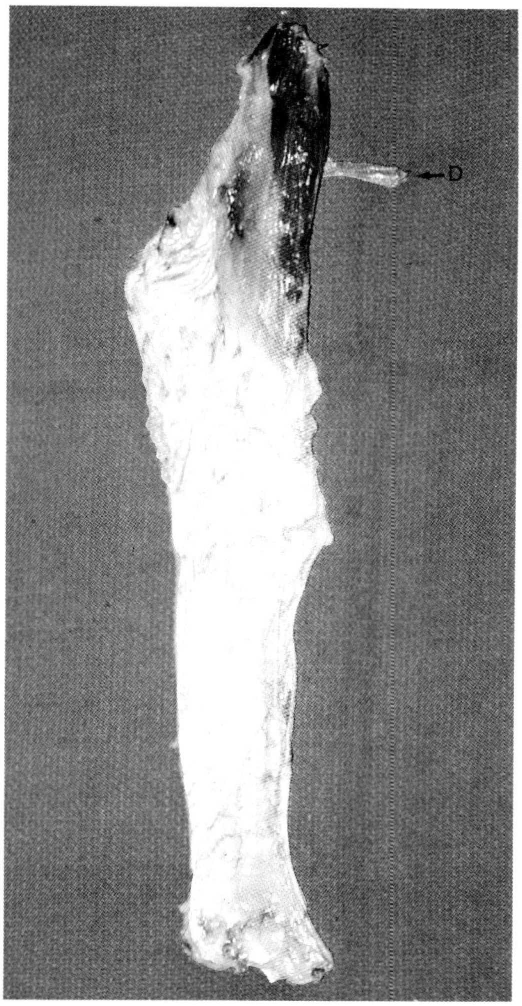

Fig. 2-36. Terminal branch of lateral circumflex femoral artery (*D*).

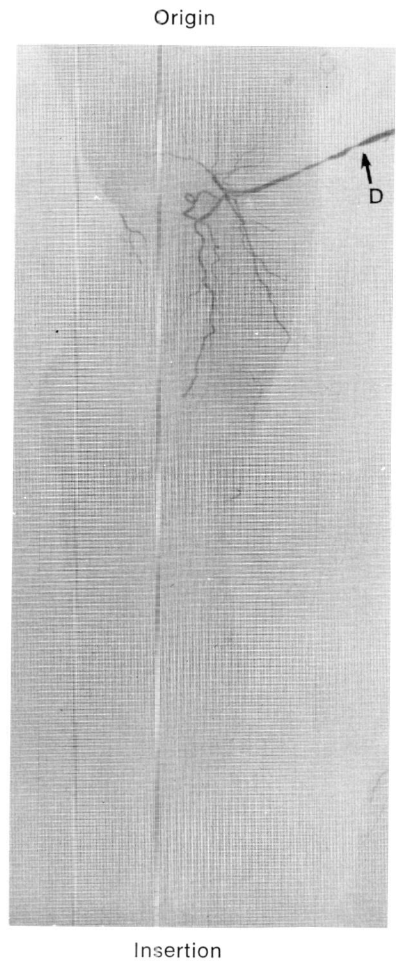

Fig. 2-37. Terminal branch of lateral circumflex femoral artery (*D*).

VASTUS LATERALIS

Vascular anatomy (Figs. 2-38 and 2-39)

Pattern of circulation: Type II

Dominant vascular pedicle (D)

Descending branch of lateral circumflex femoral artery and venae comitantes

REGIONAL SOURCE: Profunda femoris artery and vein

SIZE: Length: 4 cm
Diameter: 1 mm

LOCATION: Superior third of the muscle belly

Minor vascular pedicle(s) (m)

Intermuscular branches from superficial femoral artery and venae comitantes

REGIONAL SOURCE: Superficial femoral artery and vein

SIZE: Length: 1 to 2 cm
Diameter: 0.4 mm

LOCATION: Medial deep aspect of the muscle along the inferior half

Clinical application

ARC OF ROTATION: The entire vastus lateralis may be safely elevated on its dominant vascular pedicle. Transposition of this muscle is useful for coverage of trochanter, ischium, groin, and acetabular fossa and for reconstruction of the abdominal wall.

SKIN TERRITORY: Although located beneath the TFL unit, the vastus lateralis does contribute musculocutaneous perforating vessels to skin. Significant musculocutaneous perforators are noted to the overlying skin if the vascular pedicle to the TFL is injured. In this instance the vastus lateralis should be transposed with overlying fascia lata and skin.

SEGMENTAL TRANSPOSITION: Complete division of the insertion of the vastus lateralis is required to obtain an adequate arc of rotation for common pelvic defects.

DESIGN OF DISTALLY BASED FLAPS: The small minor pedicles will not support the entire vastus lateralis if the dominant pedicle is divided. Also, the location of minor pedicles along the middle of the muscle belly precludes any useful arc of rotation.

POTENTIAL FOR MICROVASCULAR TRANSPLANTATION: The vastus lateralis may be transplanted on its dominant vascular pedicle and may be useful either for coverage or for functional muscle transplantation.

VASCULARIZED BONE: Although the vastus lateralis has numerous vascular communications with bone periosteum, transplantation of this muscle with a portion of femur is not useful because of potential donor site disability.

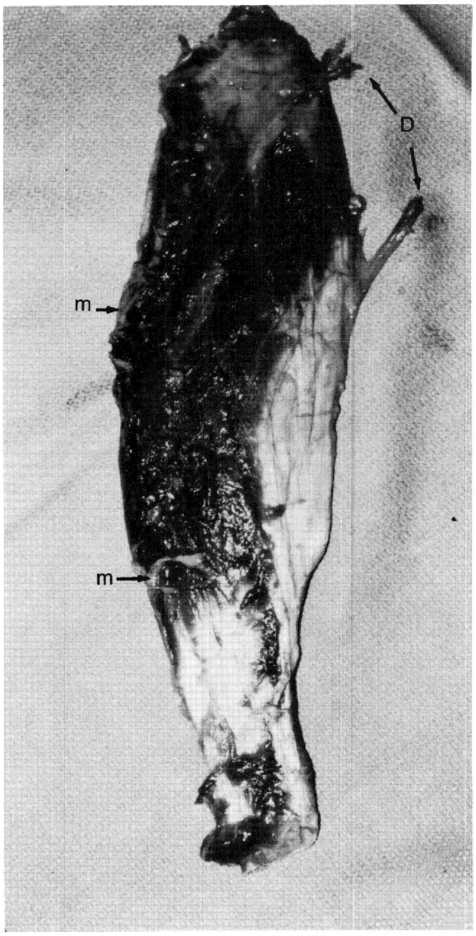

Fig. 2-38. Dominant vascular pedicle: *D*, Descending branch of lateral femoral circumflex artery. Minor pedicle: *m*, Branches of profunda femoris artery.

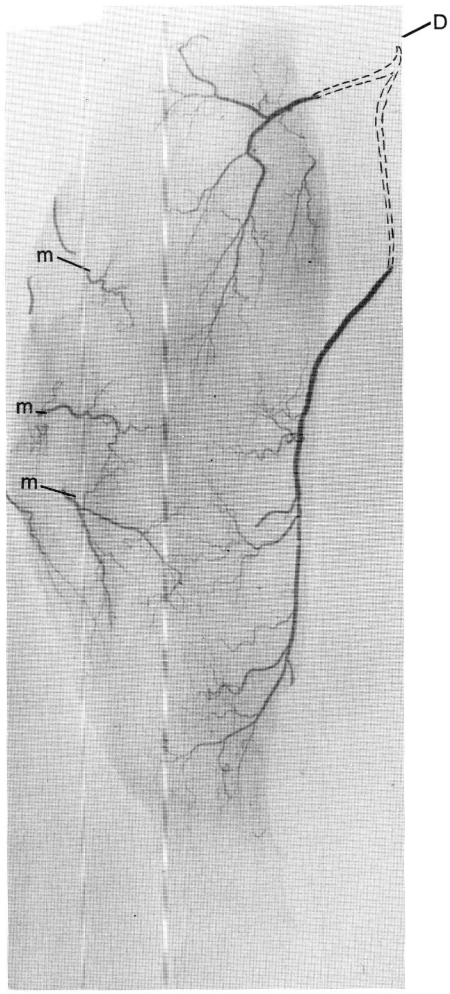

Fig. 2-39. Dominant vascular pedicle: *D*, Descending branch of lateral femoral circumflex artery. Minor pedicle: *m*, Branches of profunda femoris artery.

BICEPS FEMORIS

Vascular anatomy (Figs. 2-40 and 2-41)

Pattern of circulation: Type II

Dominant vascular pedicles (D)

Three superior muscular branches of the profunda femoris artery and venae comitantes

REGIONAL SOURCE: Profunda femoris artery and vein

SIZE: Length: 3 cm

Diameter: 1 mm

LOCATION: Superior one third of muscle

Minor vascular pedicles (m)

1. Two to three muscular branches of the profunda femoris artery and venae comitantes

   REGIONAL SOURCE: Profunda femoris artery

   SIZE: Length: 2 cm

   Diameter: 0.8 mm

   LOCATION: Inferior one third of muscle

2. Two muscular branches of popliteal artery and venae comitantes

   REGIONAL SOURCE: Popliteal artery and vein

   SIZE: Length: 3 cm

   Diameter: 0.8 mm

   LOCATION: Inferior one third of muscle

Clinical application

ARC OF ROTATION: Complete elevation of the biceps femoris is possible when the superior three dominant vascular pedicles are preserved. However, the location of these pedicles prevents an adequate arc of rotation as a transposition flap to reach most defects. The muscle is advanced superiorly after release of the origin and insertion for coverage of ischial defects.

SKIN TERRITORY: The posterior lateral thigh skin receives musculocutaneous perforating vessels from the biceps femoris. The entire posterior

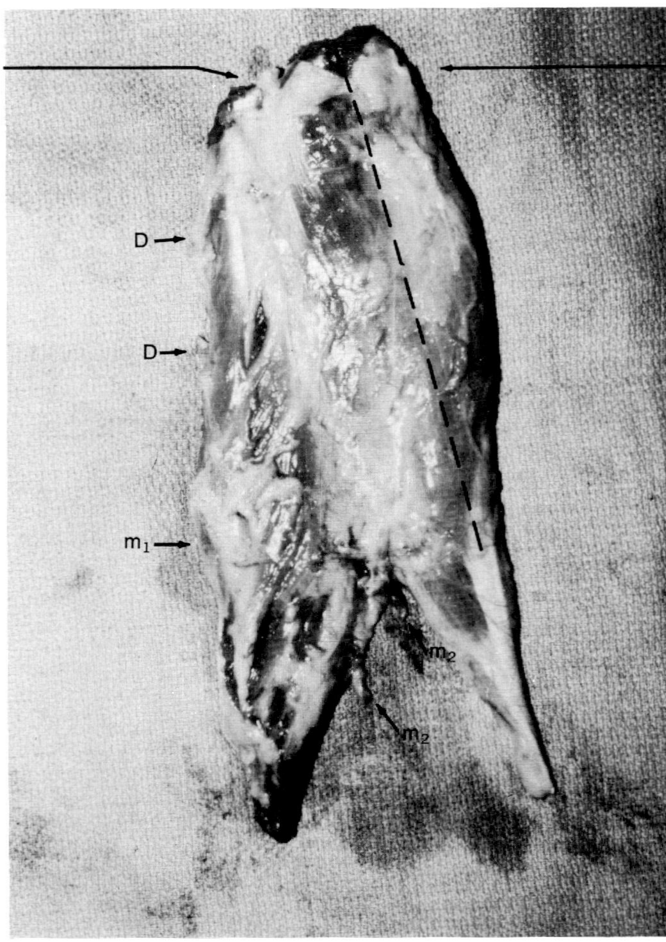

Fig. 2-40. Dominant pedicle: *D*, Branch of profunda femoris artery. Minor pedicles: m_1, Branches of profunda femoris artery; m_2, branches of popliteal artery.

thigh skin may be elevated with this muscle. Elevation of the semitendinosus muscle is often included with the posterior thigh skin and biceps femoris muscle for advancement into ischial defects.

SEGMENTAL TRANSPOSITION: Because of the short arc of rotation resulting from the location of dominant vascular pedicles in the middle of the muscle belly, a complete release of origin and insertion is required for its use as an advancement flap.

DESIGN OF DISTALLY BASED FLAPS: The entire muscle may be shifted inferiorly with preservation of the inferiorly located minor pedicles.

POTENTIAL FOR MICROVASCULAR TRANSPLANTATION: Because of the presence of multiple dominant vascular pedicles, the biceps femoris is not useful for transplantation.

VASCULARIZED BONE: The muscle has fascial attachments to bone, both at its origin and insertion. Therefore elevation of vascularized bone with this muscle is not possible.

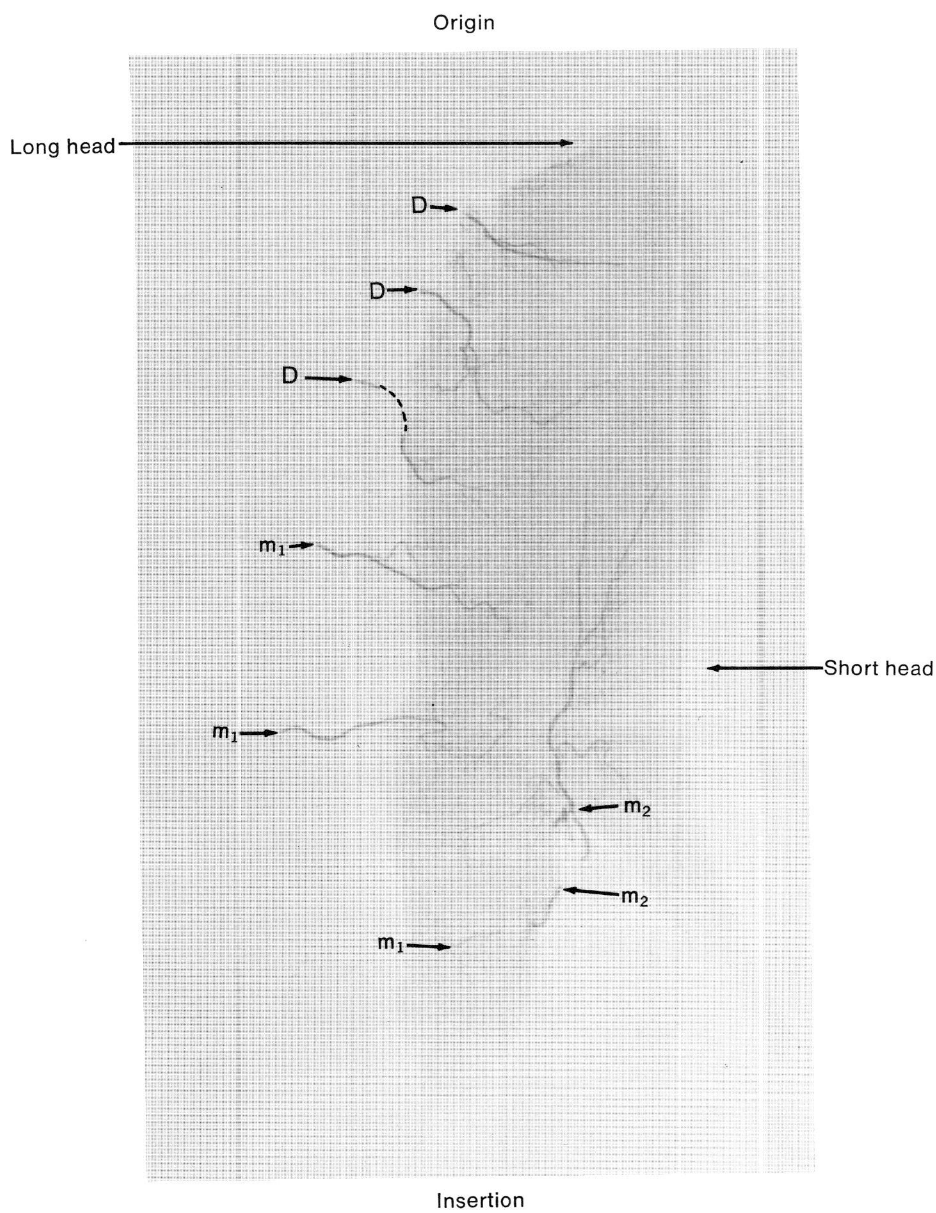

Fig. 2-41. Dominant pedicle: *D*, Branch of profunda femoris artery. Minor pedicles: m_1, Branches of profunda femoris artery; m_2, branches of popliteal artery.

GLUTEUS MAXIMUS

Vascular anatomy (Figs. 2-42 and 2-43)

Pattern of circulation: Type III

Dominant vascular pedicle (D)

1. Superior gluteal artery and venae comitantes
   REGIONAL SOURCE: Hypogastric artery and vein
   SIZE: Length: 3 cm
   Diameter: 2.5 mm
   LOCATION: Deep to muscle origin
2. Inferior gluteal artery (d) and venae comitantes
   REGIONAL SOURCE: Hypogastric artery and vein
   SIZE: Length: 3 cm
   Diameter: 2.5 mm
   LOCATION: Deep to muscle origin

Minor vascular pedicle (m)

Two to three intermuscular branches of medial circumflex femoral artery and venae comitantes
REGIONAL SOURCE: Medial circumflex femoral artery and veins
SIZE: Length: 1 cm
Diameter: 0.6 mm
LOCATION: Beneath inferior muscle insertion

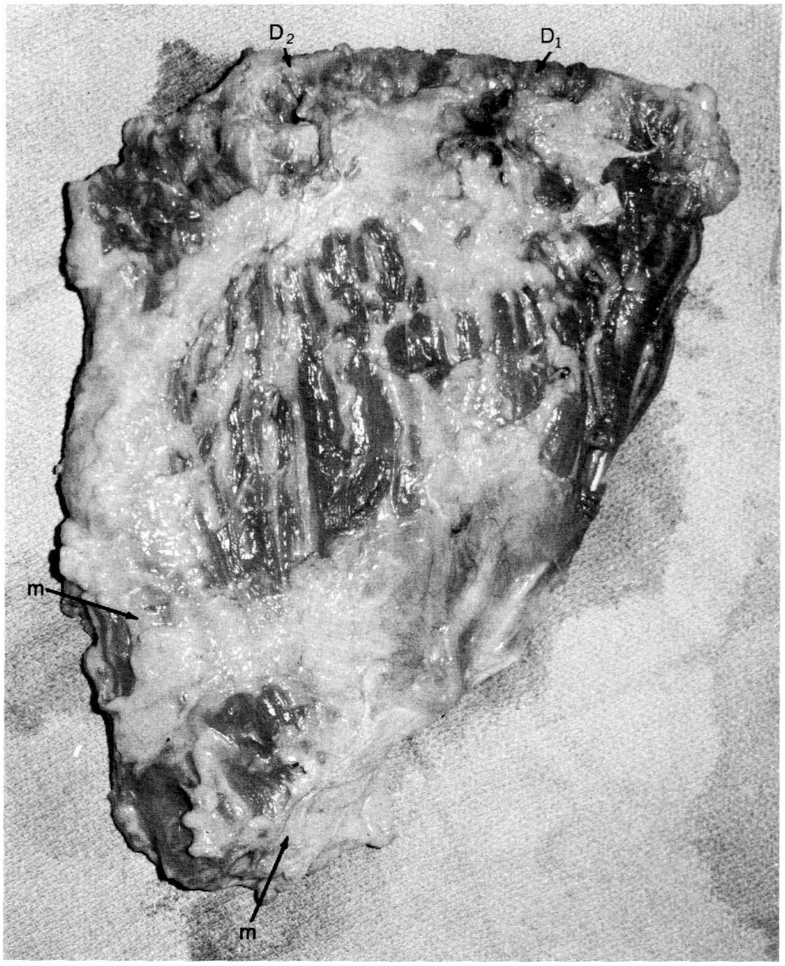

Fig. 2-42. Dominant vascular pedicles: D_1, Superior gluteal artery; D_2, inferior gluteal artery. Minor vascular pedicle: *m*, Intermuscular branch of medial femoral circumflex artery.

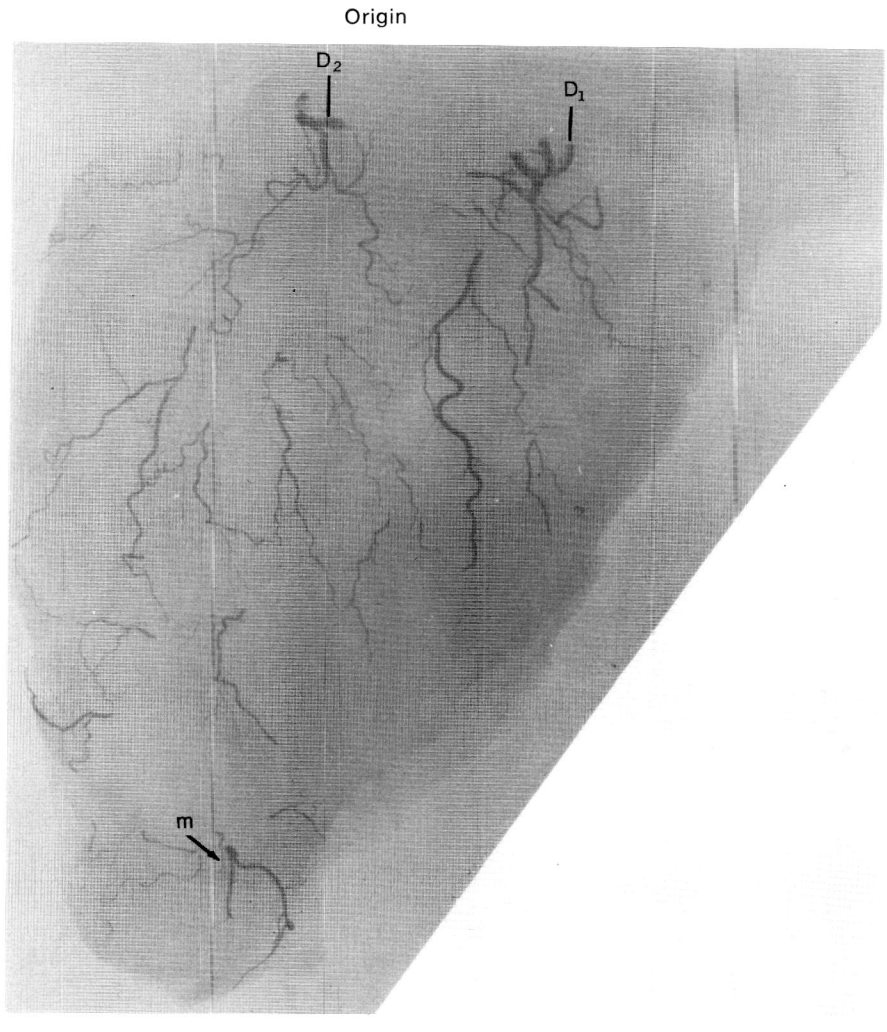

Fig. 2-43. Dominant vascular pedicles: D_1, Superior gluteal artery; D_2, inferior gluteal artery. Minor vascular pedicle: *m*, Intermuscular branch of medial femoral circumflex artery.

Clinical application

ARC OF ROTATION: The entire muscle may be elevated based on both dominant vascular pedicles. Transposition of the muscle as a turnover flap will cover sacral defects. This flap design is now rarely used, since the entire muscle is used for coverage (see segmental transposition).

SKIN TERRITORY: The skin of the entire buttock is vascularized by musculocutaneous perforating vessels from the gluteus maximus. A skin island may be designed over either the superior or inferior muscle belly.

The skin of the posterior thigh is vascularized by a descending branch of the inferior gluteal artery. This skin may be elevated as a skin-fascial flap based on this axial vascular pedicle from the inferior gluteal artery (Fig. 2-44). Transposition of this posterior thigh skin will cover defects of the ischium, trochanter, sacrum, and perineum.

SEGMENTAL TRANSPOSITION: The gluteus maximus may be split with release of the insertion of either the superior or inferior half of the muscle based on its vascular pedicle. The remaining half of the muscle is left completely intact. Transposition of the superior gluteus maximus muscle will cover defects of the sacrum. Transposition of the inferior gluteus maximus muscle will cover defects of the peroneum and ischium.

DESIGN OF DISTALLY BASED FLAPS: Release of the inferior half of the origin of the inferior gluteal muscle and division of its dominant vascular pedicle allows transposition of muscle to the hip based on minor vascular pedicles. However, the donor defect may result in further disability. The gluteal thigh flap is preferable for hip coverage.

POTENTIAL FOR MICROVASCULAR TRANSPLANTATION: Either half of the muscle may be transplanted based on its vascular pedicle. Transplantation of the superior gluteus maximus muscle is not useful for reconstructive surgery because of the muscle's short vascular pedicle and the potential for donor site disability; the inferior muscle has been transplanted for breast reconstruction.

VASCULARIZED BONE: Vascular communications between the muscle origin and periosteum of the sacrum are not useful for flap design because of potential donor site disability.

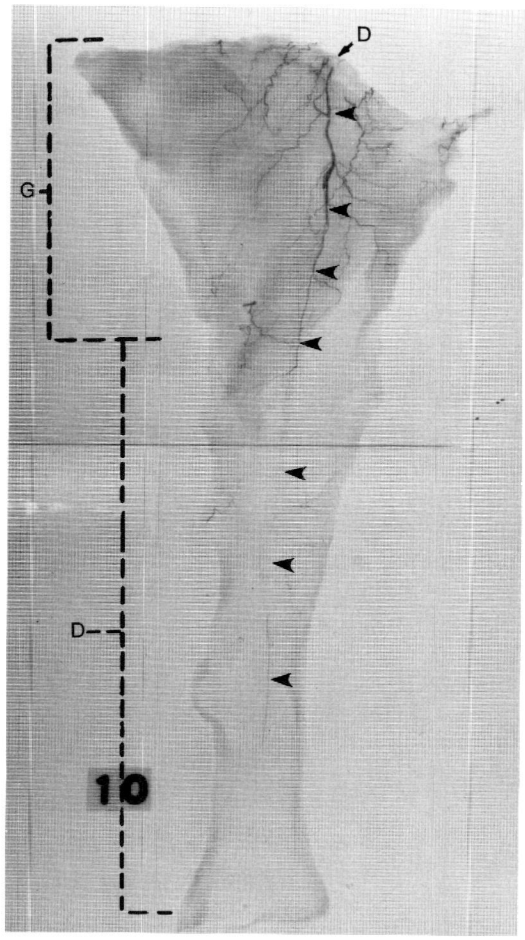

Fig. 2-44. Gluteal thigh flap. *G*, Inferior gluteus maximus; *P*, posterior thigh skin; *D*, descending branch of inferior gluteal artery. Arrows denote vessel course.

SEMIMEMBRANOSUS

Vascular anatomy (Figs. 2-45 and 2-46)

Pattern of circulation: Type III

Dominant vascular pedicles (D)

1. Two branches from profunda femoris artery and venae comitantes
   REGIONAL SOURCE: Profunda femoris artery and vein
   SIZE: Length: 2 cm
   Diameter: 0.6 mm
   LOCATION: Superior one half of muscle
2. Branch of superficial femoral artery and venae comitantes
   REGIONAL SOURCE: Superficial femoral artery and vein
   SIZE: Length: 3 cm
   Diameter: 0.6 mm
   LOCATION: Distal one third of muscle

Clinical application

ARC OF ROTATION: The semimembranosus has two independent regional sources of circulation. Each contributes significantly to the vascular supply of the muscle. Division of the inferior dominant pedicle allows elevation of the distal one half to two thirds of the muscle. Transposition of the muscle is possible for coverage of the trochanter, ischium, sacrum, and perineum. Division of the superior dominant vascular pedicles allows elevation of the superior two thirds of the muscle. Transposition of the muscle is possible for coverage of the distal thigh and knee.

SKIN TERRITORY: The semimembranosus is located deep in the posterior medial thigh. It does not have a useful skin territory for coverage as a transposition flap. However, the muscle may be elevated with the gracilis as a combined musculocutaneous flap when muscle bulk is desirable in the reconstructed defect.

SEGMENTAL TRANSPOSITION: Complete division of the insertion or origin is required for a useful arc of rotation.

DESIGN OF DISTALLY BASED FLAPS: With a type III circulation pattern, an inferior arc of rotation is possible based on the inferiorly located dominant pedicle. However, the arc of rotation does not reach the inferior aspect of the knee, so its use is rarely practical for coverage of problem wounds.

POTENTIAL FOR MICROVASCULAR TRANSPLANTATION: Because of multiple dominant vascular pedicles, the muscle is not useful for microvascular transplantation.

VASCULARIZED BONE: A fascial origin and insertion eliminate any potential for transplantation of vascularized bone.

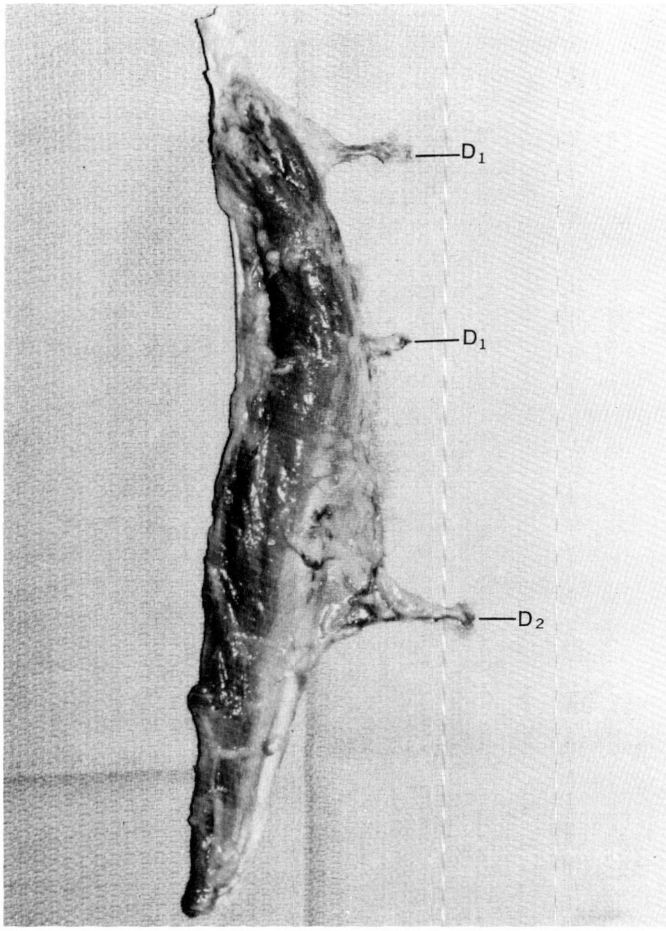

Fig. 2-45. Dominant vascular pedicles: D_1, Branch of profunda femoris artery; D_2, branch of superficial femoral artery.

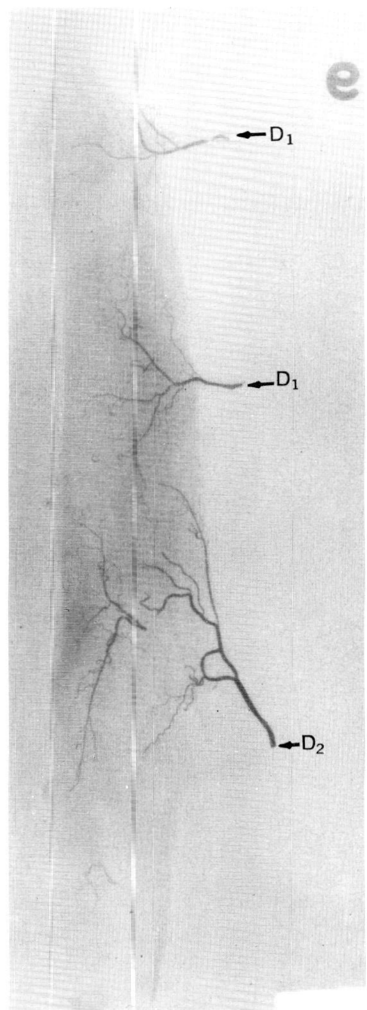

Fig. 2-46. Dominant vascular pedicles: D_1, Branch of profunda femoris artery; D_2, branch of superficial femoral artery.

SEMITENDINOSUS

Vascular anatomy (Figs. 2-47 and 2-48)

Pattern of circulation: Type II

Dominant vascular pedicles (D)

Proximal two muscular branches from profunda femoris artery and venae comitantes

REGIONAL SOURCE: Profunda femoris artery and vein

SIZE: Length: 4 cm
Diameter: 0.8 mm

LOCATION: Proximal one third of muscle

Minor vascular pedicles (m)

One or two vascular pedicles from superficial femoral artery and venae comitantes

REGIONAL SOURCE: Superficial femoral artery and vein

SIZE: Length: 2 cm
Diameter: 0.5 mm

LOCATION: Middle one third of muscle

Clinical application

ARC OF ROTATION: The distal two thirds of the semitendinosus may be elevated with safety based on the superiorly located dominant vascular pedicles. Transposition of this muscle is only occasionally useful as an alternative flap for coverage of the ischium, buttocks, and perineum. This muscle is located on the medial posterior thigh and provides musculocutaneous perforating vessels to this aspect of the thigh. However, this muscle is rarely elevated primarily as a musculocutaneous flap for coverage of pelvic defects. The muscle is elevated in conjunction with the biceps femoris as a musculocutaneous flap for advancement into ischial defects.

SEGMENTAL TRANSPOSITION: Complete division of the muscle insertion is required for an adequate arc of rotation.

DESIGN OF DISTALLY BASED FLAPS: Because of the location of the minor vascular pedicles in the middle third of the muscle, a distally based flap has a short arc of rotation and is rarely useful for coverage of inferior thigh defects.

POTENTIAL FOR MICROVASCULAR TRANSPLANTATION: Because of the multiple dominant vascular pedicles, the semitendinosus is not useful for microvascular transplantation.

VASCULARIZED BONE: A fascial origin and insertion eliminate any potential for transplantation of vascularized bone.

Origin

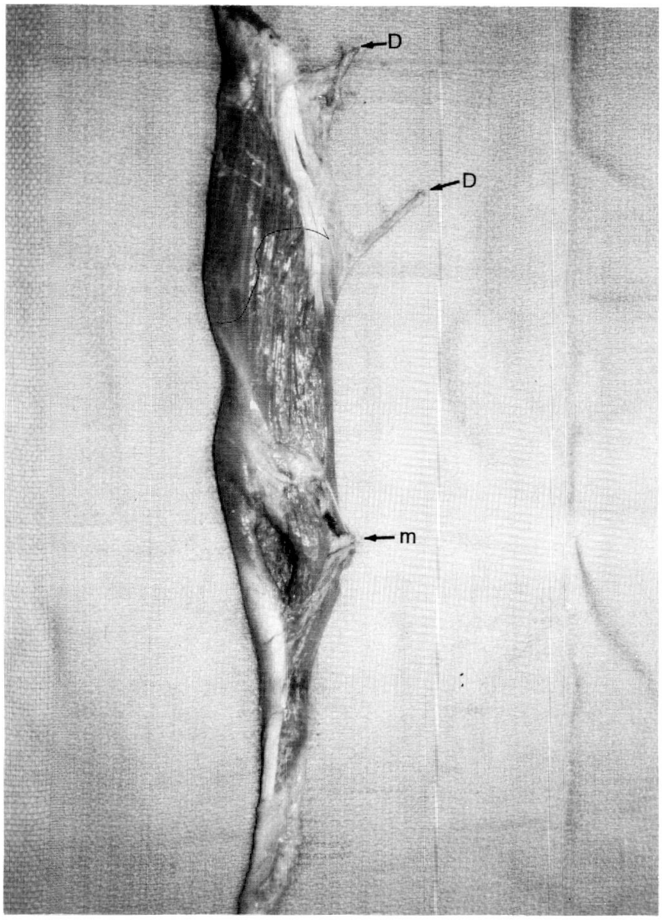

Insertion

Fig. 2-47. Dominant vascular pedicle: *D,* Branch of profunda femoris artery. *m,* Branch of profunda femoris artery.

Origin

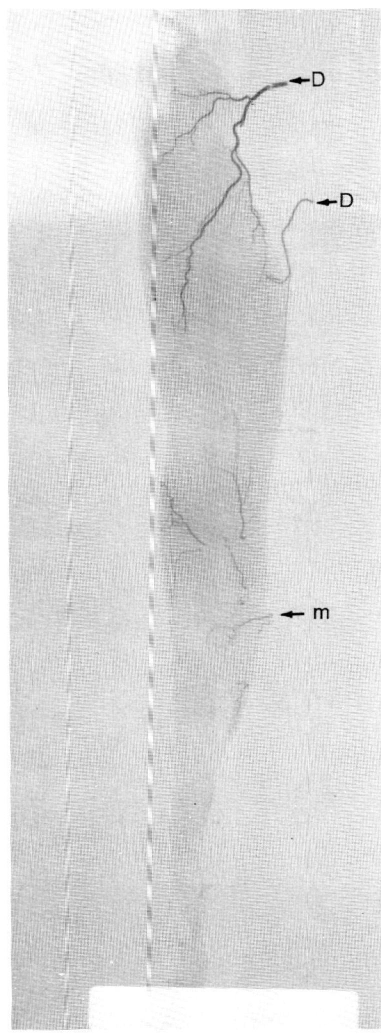

Insertion

Fig. 2-48. Dominant vascular pedicle: *D,* Branch of profunda femoris artery. *m,* Branch of profunda femoris artery.

FLEXOR DIGITORUM LONGUS

Vascular anatomy (Figs. 2-49 and 2-50)

Pattern of circulation: Type IV

Minor vascular pedicle (m)

Ten to twelve segmental muscular arterial branches and venae comitantes

REGIONAL SOURCE: Posterior tibial artery and veins

SIZE: Length: 1 to 2 cm
Diameter: 0.6 to 0.8 mm

LOCATION: Entire lateral aspect of muscle belly

Clinical application

ARC OF ROTATION: The superior or inferior half of the muscle may be elevated with safety. However, division of more than four segmental vascular pedicles may jeopardize distal muscle flap circulation. Transposition of the inferior muscle will cover small defects of the middle third of the lower leg. This muscle used in conjunction with the soleus as transposition flaps will cover large defects of the middle third of the lower leg. Transposition of the superior muscle is rarely indicated because of the associated difficult dissection and availability of other more useful muscles (e.g., soleus and gastrocnemius).

SKIN TERRITORY: This deep muscle does not contribute musculocutaneous perforators to the skin of the medial lower leg.

SEGMENTAL TRANSPOSITION: The distal half of the muscle will separate from the tendon for use as a transposition flap. The tendon of insertion and associated vascular communications are left in continuity with the superior muscle belly.

DESIGN OF DISTALLY BASED FLAPS: Distally based flaps are not applicable to the flexor digitorum longus.

POTENTIAL FOR MICROVASCULAR TRANSPLANTATION: Transplantation of the flexor digitorum longus requires use of the posterior tibial artery and vein with its vascular branches to this muscle. Use of this muscle is rarely indicated because of possible impairment of muscle circulation to the distal leg.

VASCULARIZED BONE: The vascular communications between muscle origin and tibia are noted. However, excision of bone with this flap may result in donor site disability.

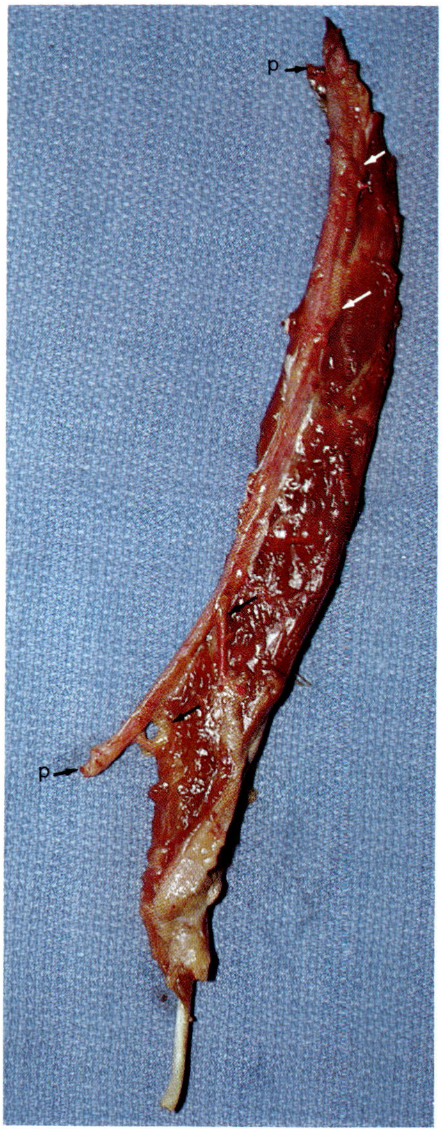

Fig. 2-49. Segmental vascular pedicles. Arrows indicate branches of posterior tibial artery *(p)*.

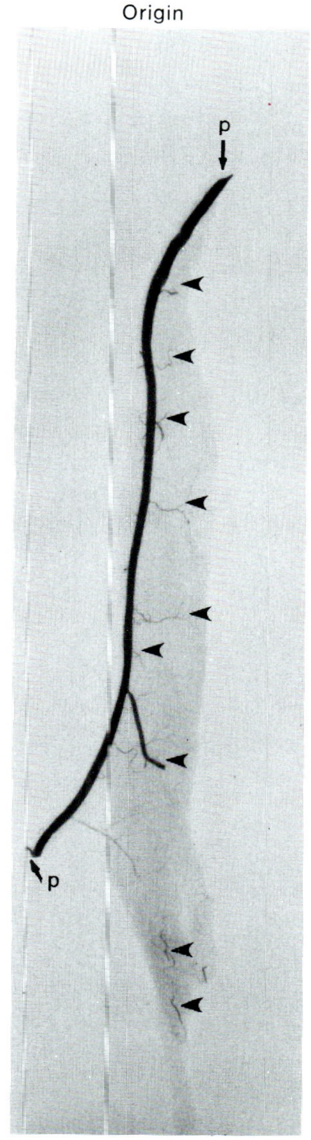

Fig. 2-50. Segmental vascular pedicles. Arrows indicate branches of posterior tibial artery *(p)*.

GASTROCNEMIUS

Vascular anatomy (Figs. 2-51 and 2-52)

Pattern of circulation: Type I

Dominant vascular pedicle (D)—medial gastrocnemius
- Medial sural artery and venae comitantes
- REGIONAL SOURCE: Popliteal artery
- SIZE: Length: 4 cm
  Diameter: 2 mm
- LOCATION: Popliteal fossa

Dominant vascular pedicle (D)—lateral gastrocnemius
- Lateral sural artery and venae comitantes
- REGIONAL SOURCE: Popliteal artery
- SIZE: Length: 4 cm
  Diameter: 2 mm
- LOCATION: Popliteal fossa

Clinical application

ARC OF ROTATION: The entire medial gastrocnemius or lateral gastrocnemius may be elevated on its respective single vascular pedicle. Transposition of either head of this muscle is useful for coverage of defects of the knee and superior third of the leg.

SKIN TERRITORY: The medial and lateral skin of the posterior lower leg receives musculocutaneous perforators from the proximal medial and lateral heads of the gastrocnemius muscle. Based on these musculocutaneous perforators, either the medial or lateral skin may be elevated safely with its respective underlying muscle. Distal medial or lateral leg skin will survive elevation as a skin-fascial flap based on proximal musculocutaneous perforating vessels.

SEGMENTAL TRANSPOSITION: Release of the posterior midline raphe allows transposition of either the medial or lateral gastrocnemius muscle. Each muscle head is considered independently with its separate vascular pedicle.

DESIGN OF DISTALLY BASED FLAPS: The absence of inferiorly located minor pedicles precludes the use of an inferiorly based muscle flap.

POTENTIAL FOR MICROVASCULAR TRANSPLANTATION: The gastrocnemius has potential usefulness for microvascular transplantation based on the dominant pedicle for a functional muscle transplantation.

VASCULARIZED BONE: Although vascular communications are noted between the distal femur and the heads of the gastrocnemius, use of this bone may result in impairment of knee function.

Origin

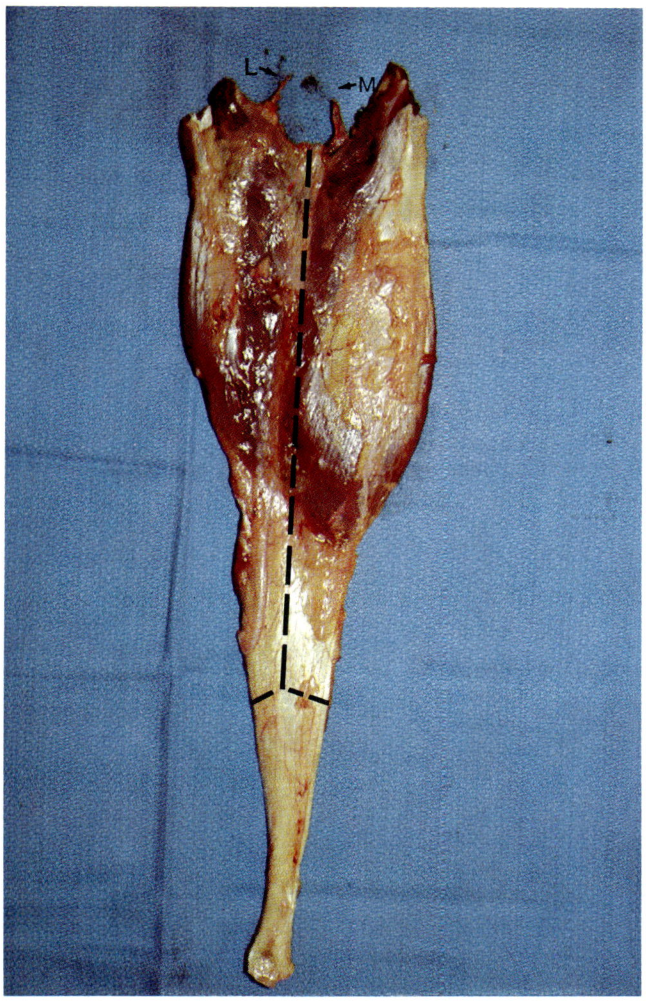

Insertion

Fig. 2-51. Lateral gastrocnemius: *L,* Sural artery. Medial gastrocnemius: *M,* Sural artery.

Origin

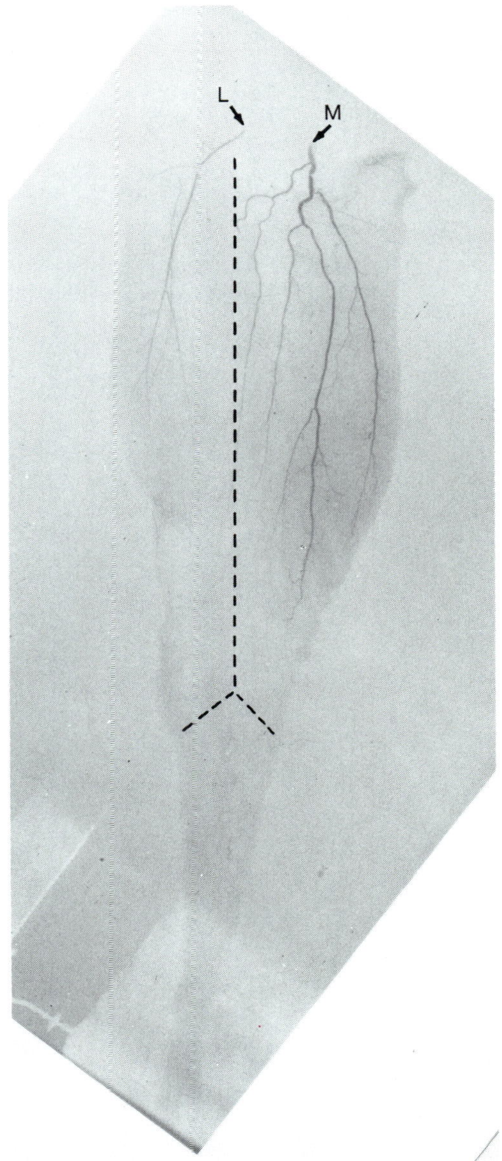

Insertion

Fig. 2-52. Lateral gastrocnemius: *L,* Sural artery. Medial gastrocnemius: *M,* Sural artery.

SOLEUS

Vascular anatomy (Figs. 2-53 and 2-54)

Pattern of circulation: Type II

Dominant vascular pedicles (D)

1. Muscular branches of popliteal artery
   REGIONAL SOURCE: Popliteal artery
   SIZE: Length: 0.5 to 1 cm
   Diameter: 0.6 mm
   LOCATION: Superior one third of muscle
2. Proximal two branches posterior tibial artery and venae comitantes
   REGIONAL SOURCE: Posterior tibial artery and vein
   SIZE: Length: 1 to 2 cm
   Diameter: 0.8 mm
   LOCATION: Superior one third of muscle
3. Proximal two branches of peroneal artery and venae comitantes
   REGIONAL SOURCE: Peroneal artery and vein
   SIZE: Length: 1 to 2 cm
   Diameter: 0.8 mm
   LOCATION: Superior one third of muscle

Minor vascular pedicles (m)

Distal three branches of posterior artery and venae comitantes
REGIONAL SOURCE: Posterior tibial artery and veins
SIZE: Length: 0.5 to 1 cm
Diameter: 0.5 mm
LOCATION: Inferior two thirds of muscle along medial border

Clinical application

ARC OF ROTATION: Elevation of the distal two thirds of the soleus is reliable based on its dominant vascular pedicles. Transposition of this muscle

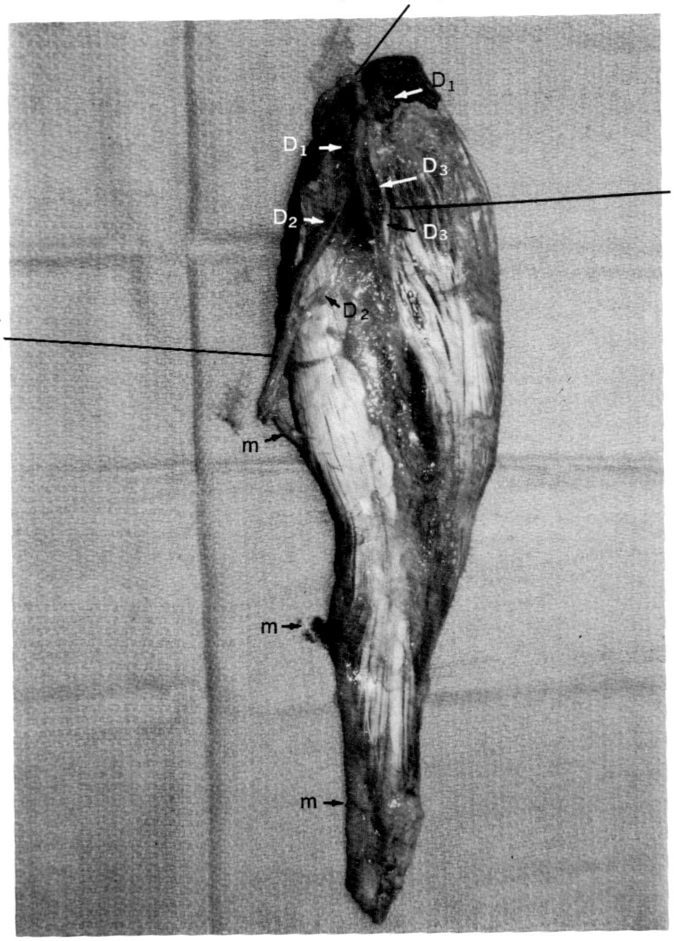

Fig. 2-53. Dominant vascular pedicles: D_1, Branch of popliteal artery; D_2, branch of posterior tibial artery; D_3, branch of peroneal artery. Minor pedicle: *m*, Branches of posterior tibial artery.

provides coverage of middle third leg defect. Transposition of the soleus muscle is generally possible based only on proximal dominant vascular pedicles either from the posterior tibial or peroneal regional sources.

SKIN TERRITORY: The circulation to the skin of the medial and lateral leg is primarily derived from musculocutaneous perforators from the gastrocnemius muscle. The soleus muscle is only useful as a muscle transposition flap. However, the distal medial and lateral muscle supplies musculocutaneous perforating vessels to adjacent skin.

SEGMENTAL TRANSPOSITION: Middle third leg defects generally require mobilization of the entire distal aspect of the muscle with division of the insertion for adequate coverage. Transposition of the medial or lateral half of the muscle allows preservation of muscle function.

DESIGN OF DISTALLY BASED FLAPS: Transposition of the soleus inferiorly for lower leg defects is possible based on minor pedicles from the posterior tibial artery and veins. However, proximal muscle will not always survive on the circulation via minor pedicles. Transposition of the lateral distal muscle based on the medially located inferior pedicles is possible for small lateral inferior leg defects. This technique allows preservation of medial muscle continuity between origin and insertion.

POTENTIAL FOR MICROVASCULAR TRANSPLANTATION: Because of multiple dominant pedicles, transposition of this entire muscle is not possible without impairment of leg circulation and is therefore not indicated.

VASCULARIZED BONE: Vascular communications between lateral muscle and the periosteum of the fibula are present. The lateral soleus with its major vascular pedicles from the peroneal artery may be included with microvascular fibular transplantation. This muscle may then be used to add coverage to the site of inset of the vascularized bone graft.

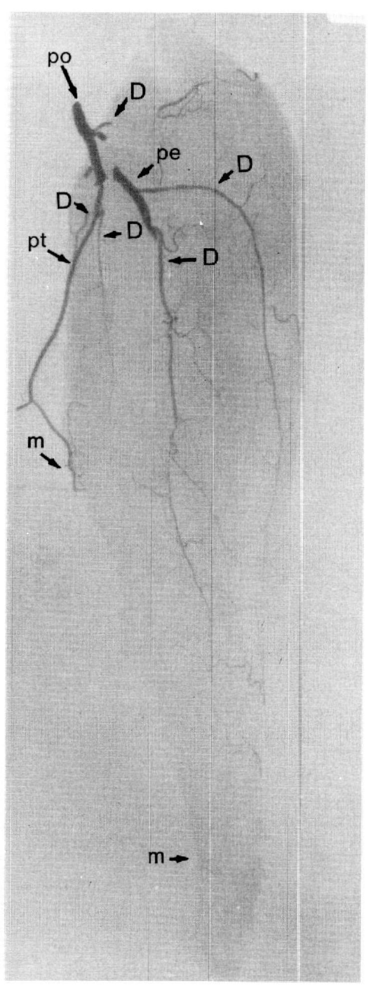

Fig. 2-54. *po,* Popliteal artery; *pe,* peroneal artery; *pt,* posterior tibial artery; *D,* dominant vascular pedicle; *m,* minor vascular pedicle.

EXTENSOR DIGITORUM LONGUS

Vascular anatomy (Figs. 2-55 and 2-56)

Pattern of circulation: Type IV

Segmental vascular pedicles

Eight to ten segmental muscular arterial branches and venae comitantes

REGIONAL SOURCE: Anterior tibial artery and veins

SIZE: Length: 1 to 2 cm
Diameter: 0.6 to 0.8 mm

LOCATION: Medial aspect of entire muscle belly

Clinical application

ARC OF ROTATION: The superior or inferior half of the extensor digitorum longus may be elevated with safety. Transposition of the inferior half of the muscle will cover defects of the middle third of the lower lateral leg. However, this portion of muscle is rarely used for coverage. The transposition of the superior half of the muscle will cover defects in the superior part of the distal third of the lower leg. Elevation of the superior two thirds of the muscle requires division of six to eight pedicles with potential loss of the distal aspect of the muscle. When the muscle survives this transposition, coverage of the distal third of the lower leg is possible.

SKIN TERRITORY: The extensor digitorum longus muscle is only useful as a muscle transposition flap.

SEGMENTAL TRANSPOSITION: The inferior half of the muscle may be separated from the tendon and transposed medially for coverage of small defects. The continuity of tendon and associated vascular communications is maintained with the superior muscle belly.

DESIGN OF DISTALLY BASED FLAPS: Distally based flaps are not applicable to the extensor digitorum longus.

POTENTIAL FOR MICROVASCULAR TRANSPLANTATION: Transplantation of the extensor digitorum longus requires use of the anterior tibial artery with vascular communications to the muscle. This dissection may result in impairment of lower extremity circulation and is rarely indicated.

VASCULARIZED BONE: Vascular communications between muscle origin and periosteum of the tibular and medial condyle of the tibia have no clinical application because of potential donor site difficulty.

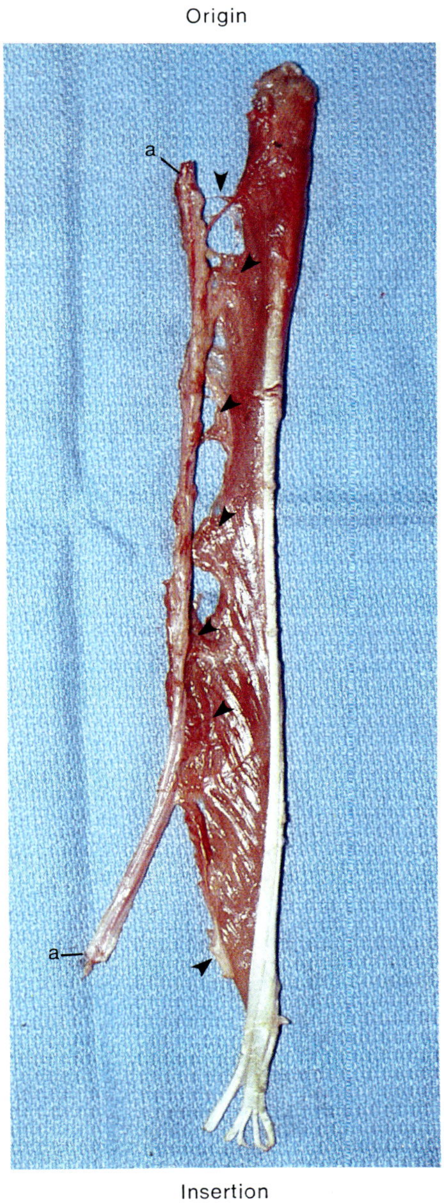

Fig. 2-55. Segmental pedicles (branches of anterior tibial artery [*a*]).

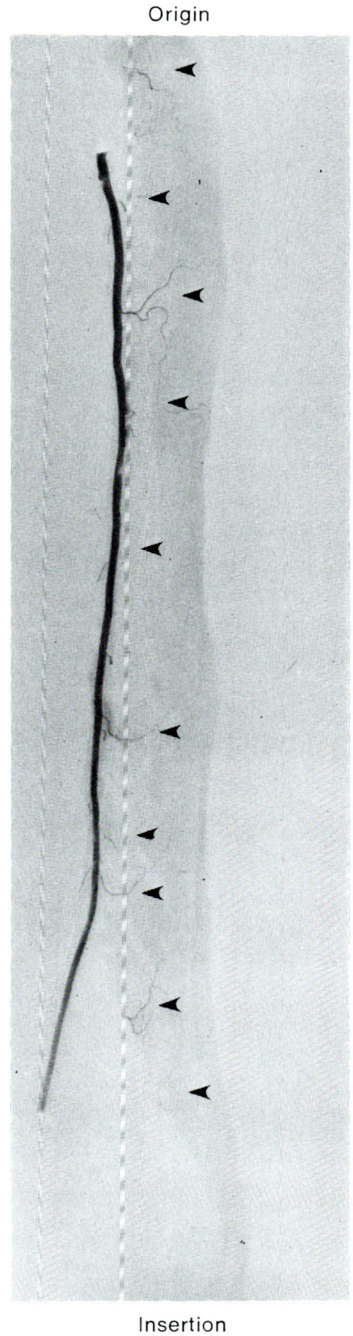

Fig. 2-56. Segmental pedicles (branches of anterior tibial artery [*a*]).

EXTENSOR HALLUCIS LONGUS

Vascular anatomy (Figs. 2-57 and 2-58)

Pattern of circulation: Type IV

Minor vascular pedicle (m)

Six to eight segmental arterial muscular branches and venae comitantes

REGIONAL SOURCE: Anterior tibial artery and veins

SIZE: Length: 1 to 2 cm
Diameter: 0.5 to 0.8 mm

LOCATION: Along entire muscle belly

Clinical application

ARC OF ROTATION: The distal or proximal half of the extensor hallucis longus may be elevated with safety. Division of more than three segmental vascular pedicles may result in distal flap necrosis. The inferior muscle is very narrow and is not useful to cover lower third leg defects. Transposition of the proximal muscle based on inferior segmental vascular pedicles will cover defects in the distal third of the leg. Usually this muscle is transposed inferiorly in conjunction with the extensor digitorum longus to provide adequate flap to cover the defect.

SKIN TERRITORY: The extensor hallucis longus is only useful as a muscle transposition flap.

SEGMENTAL TRANSPOSITION: The inferior half of the muscle may be separated from the tendon and transposed medially for coverage of small defects.

DESIGN OF DISTALLY BASED FLAP: Distally based flaps are not applicable to the extensor hallucis longus.

POTENTIAL FOR MICROVASCULAR TRANSPLANTATION: Transplantation of the extensor hallucis longus would require use of the anterior tibial artery and vein with its vascular branches to this muscle. Thus use of this muscle for transplantation for functional reconstruction is rarely indicated.

VASCULARIZED BONE: Although there are vascular communications between origin and periosteum of the fibula, clinical use of vascularized bone with the extensor hallucis longus is rarely practical.

Origin

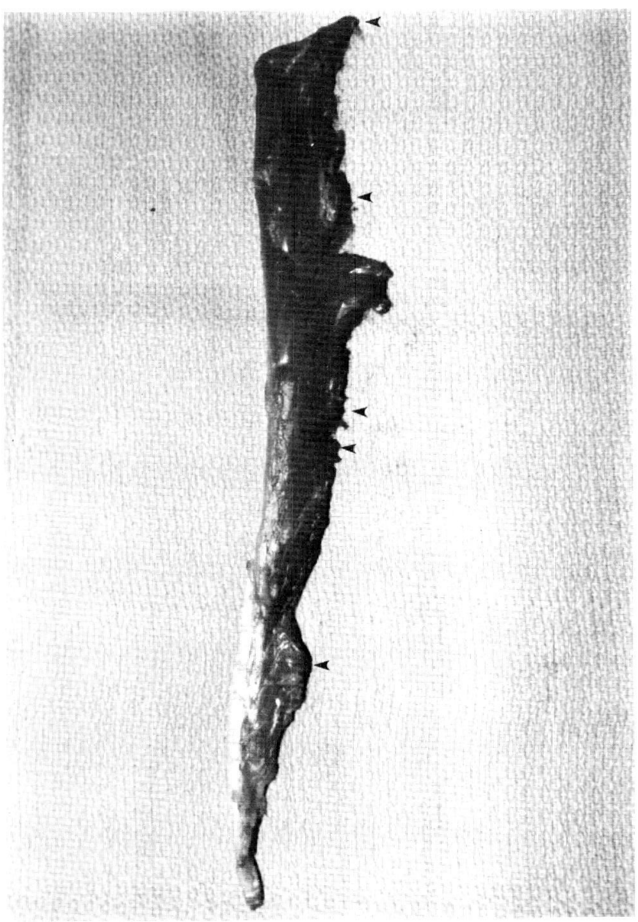

Insertion

Fig. 2-57. Segmental vascular pedicles *(arrows)*.

Origin

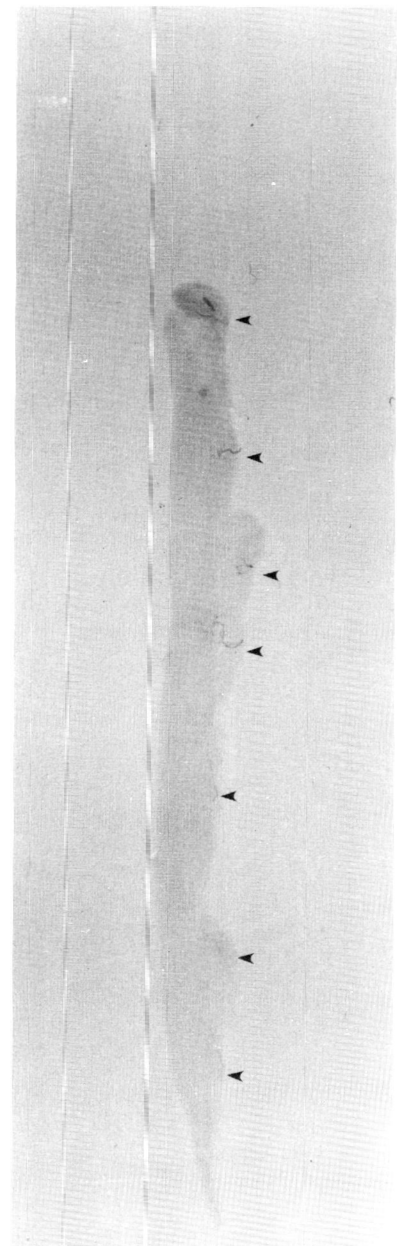

Insertion

Fig. 2-58. Segmental vascular pedicles *(arrows)*.

PERONEUS LONGUS

Vascular anatomy (Figs. 2-59 and 2-60)

Pattern of circulation: Type II

Dominant vascular pedicle (D)

Muscular branch of peroneal artery and venae comitantes

REGIONAL SOURCE: Peroneal artery and veins

SIZE: Length: 3 cm
Diameter: 1 mm

LOCATION: Superior one third of muscle

Minor vascular pedicle (m)

Muscular branch of anterior tibial artery and venae comitantes

REGIONAL SOURCE: Anterior tibial artery and vein

SIZE: Length: 2 cm
Diameter: 0.5 mm

LOCATION: Distal two thirds of muscle

Clinical application

ARC OF ROTATION: Elevation of the peroneus longus is possible based on the proximal dominant vascular pedicle. Transposition of the distal two thirds of this muscle provides partial coverage of defects located on the middle third of the lateral leg.

SKIN TERRITORY: Only a small segment of lateral leg skin may be transposed with the peroneus longus. Since the donor defect results in exposed fibula, a musculocutaneous flap is not recommended.

SEGMENTAL TRANSPOSITION: The insertion of the peroneus longus must be completely divided to achieve an adequate length for useful transposition.

DESIGN OF DISTALLY BASED FLAPS: The superior location of the minor pedicle does not allow an adequate inferior arc of rotation for a useful distally based flap.

POTENTIAL FOR MICROVASCULAR TRANSPLANTATION: The peroneus longus has potential usefulness for microvascular transplantation based on its dominant vascular pedicle for functional muscle transplantation.

VASCULARIZED BONE: Vascular communications to the fibular periosteum allow inclusion of fibula with transposition or transplantation flap. (Inclusion of the peroneal nutrient artery to the fibula is routinely performed with the peroneus longus for muscle-vascularized bone transplantation.)

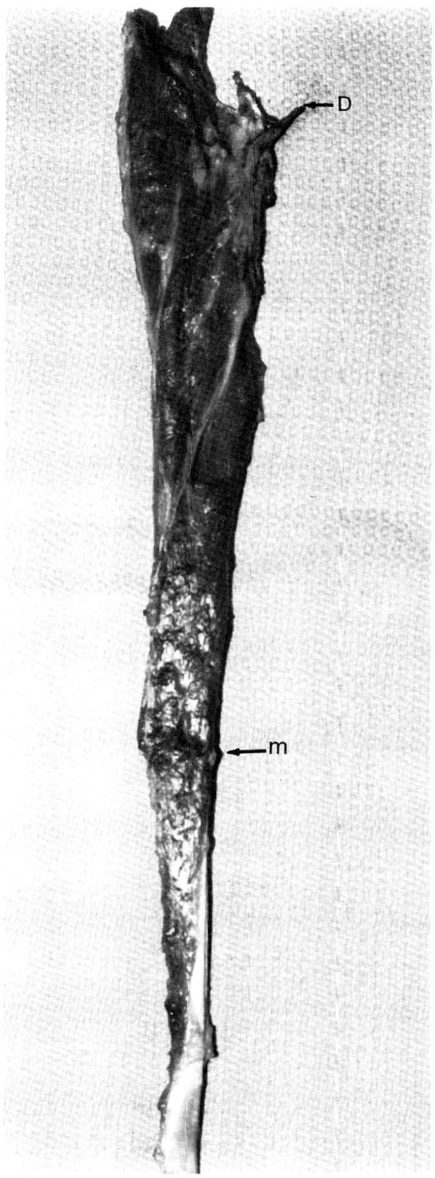

Fig. 2-59. Dominant vascular pedicle: *D*, Branch of peroneal artery. Minor vascular pedicle: *m*, Branch of anterior tibial artery.

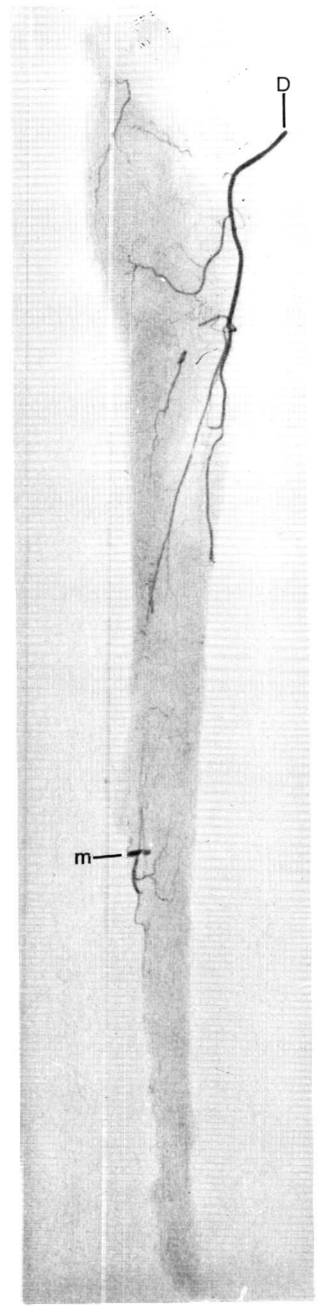

Fig. 2-60. Dominant vascular pedicle: *D*, Branch of peroneal artery. Minor vascular pedicle: *m*, Branch of anterior tibial artery.

TIBIALIS ANTERIOR

Vascular anatomy (Figs. 2-61 and 2-62)

Pattern of circulation: Type IV

Minor vascular pedicle (m)

Eight to twelve muscular arterial branches of anterior tibial artery and venae comitantes

REGIONAL SOURCE: Anterior tibial artery and veins

SIZE: Length: 1 to 2 cm
Diameter: 0.6 to 0.8 mm

LOCATION: Lateral deep aspect of muscle

Clinical application

ARC OF ROTATION: The inferior one half or superior one third of the muscle may be elevated with safety. After division of five or six vascular pedicles, the distal muscle may not survive as a transposition flap. Transposition of the inferior muscle will cover small defects of the middle and proximal thirds of the anterior lower leg. Elevation of the muscle origin requires a difficult dissection and is not useful as a transposition flap for proximal tibia coverage.

SKIN TERRITORY: The anterior lateral skin of the lower leg receives musculocutaneous perforating vessels from the tibialis anterior muscle. Transposition of this muscle with its skin is likely to result in an unacceptable donor defect.

SEGMENTAL TRANSPOSITION: The muscle is separated from the tendon of insertion in the distal third of the leg. The tibialis anterior is transposed to cover small defects of the lower leg. Both vascular connections and tendon continuity are maintained with muscle belly.

DESIGN OF DISTALLY BASED FLAPS: Distally based flaps are not applicable to the tibialis anterior.

POTENTIAL FOR MICROVASCULAR TRANSPLANTATION: The tibialis anterior has no clinical application for microvascular transplantation because of multiple segmental vascular pedicles.

VASCULARIZED BONE: The vascular communications between the tibial condyle and muscle origin have no clinical application because of potential for donor site disability.

Origin

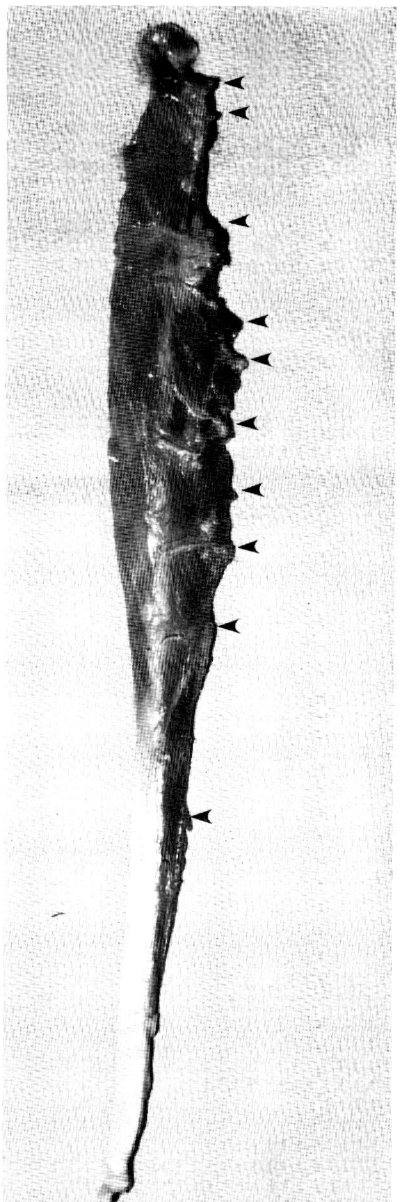

Insertion

Fig. 2-61. Segmental vascular pedicles *(arrows)*.

Origin

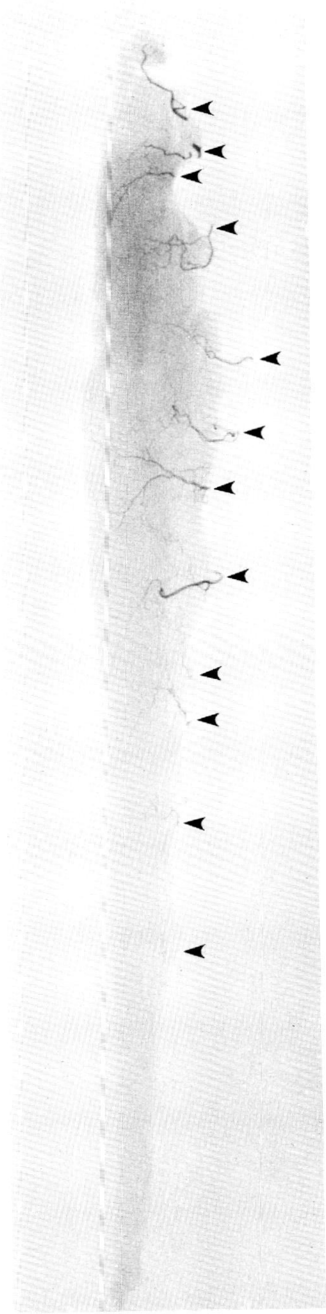

Insertion

Fig. 2-62. Segmental vascular pedicles *(arrows)*.

ABDUCTOR DIGITI MINIMI

Vascular anatomy (Figs. 2-63 and 2-64)

Pattern of circulation: Type II

Dominant vascular pedicles (D)

Two muscular branches of the lateral plantar artery and venae comitantes

REGIONAL SOURCE: Posterior tibial artery and vein

SIZE: Length: 2 cm
Diameter: 0.6 mm

LOCATION: Superior one third of muscle

Minor vascular pedicles (m)

Two to three muscular branches of the lateral plantar artery and venae comitantes

REGIONAL SOURCE: Posterior tibial artery and vein

SIZE: Length: 1 cm
Diameter: 0.5 mm

LOCATION: Inferior two thirds of muscle

Clinical application

ARC OF ROTATION: The inferior two thirds of the abductor digiti minimi may be elevated based on the dominant vascular pedicles. Since the muscle is very small, transposition of the inferior muscle will only cover small adjacent defects. The entire muscle may be elevated on the lateral plantar artery and vein. Transposition of the entire muscle will cover defects adjacent to the lateral malleolus.

SKIN TERRITORY: The lateral plantar skin of the foot receives musculocutaneous perforating vessels from the abductor digiti minimi. This skin may be elevated with the muscle as a transposition flap.

SEGMENTAL TRANSPOSITION: Release of muscle insertion is required to achieve adequate arc of rotation to reach posterior foot defects.

DESIGN OF DISTALLY BASED FLAPS: With release of muscle origin and division of the proximal lateral plantar artery, the muscle may be transposed inferiorly for small plantar defects.

POTENTIAL FOR MICROVASCULAR TRANSPLANTATION: The abductor digiti minimi may be transplanted based on the lateral plantar artery and veins.

VASCULARIZED BONE: Vascular communications between the muscle origin and periosteum of the calcaneus are noted. However, excision of calcaneus with this muscle may result in donor site disability.

Origin

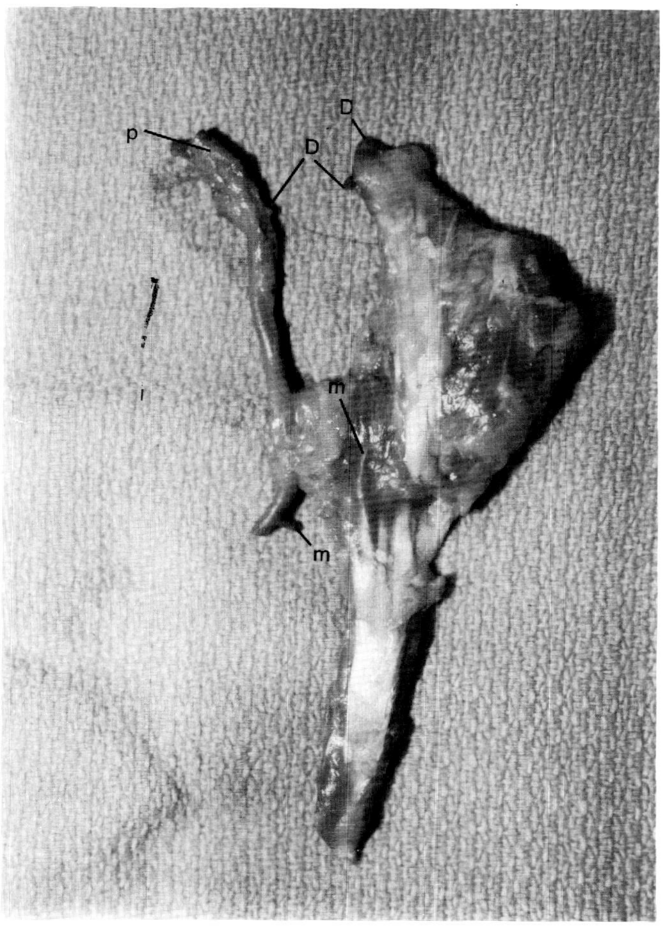

Insertion

Fig. 2-63. Dominant vascular pedicle: *D*, Branch of lateral plantar artery (*p*). Minor vascular pedicle: *m*, Branch of lateral plantar artery (*p*).

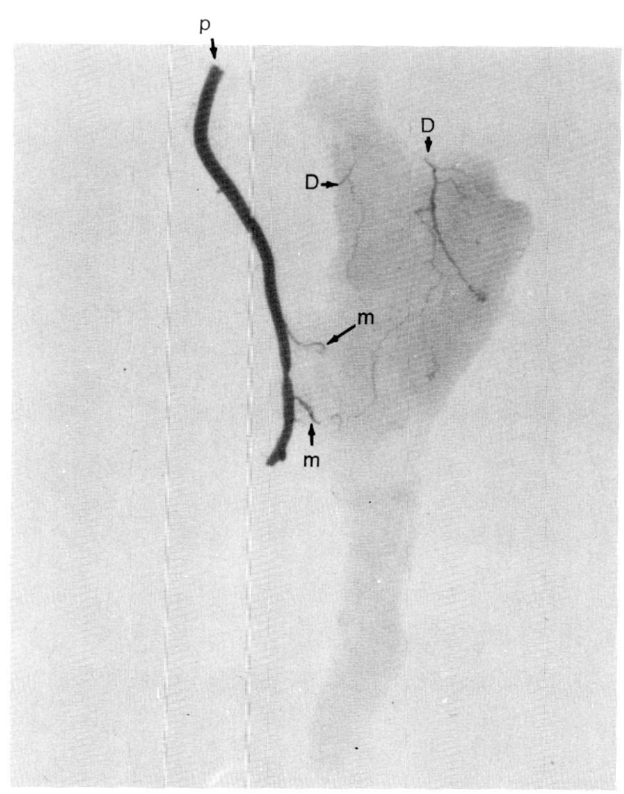

Fig. 2-64. Dominant vascular pedicle: *D*, Branch of lateral plantar artery (*p*). Minor vascular pedicle: *m*, Branches of lateral plantar artery (*p*).

ABDUCTOR HALLUCIS

Vascular anatomy (Figs. 2-65 and 2-66)

Pattern of circulation: Type II

Dominant vascular pedicles (D)

Two branches of the medial plantar artery and venae comitantes

REGIONAL SOURCE: Posterior tibial artery and vein

SIZE: Length: 2 cm
Diameter: 0.6 mm

LOCATION: Superior one third of muscle

Minor vascular pedicles (m)

Two to three branches of medial plantar artery and venae comitantes

REGIONAL SOURCE: Posterior tibial artery and vein

SIZE: Length: 1 to 2 cm
Diameter: 0.5 mm

LOCATION: Inferior two thirds of muscle

Clinical application

ARC OF ROTATION: Elevation of the inferior two thirds of the abductor hallucis is possible based on the dominant vascular pedicles. Transposition of this portion of the muscle covers defects over the medial malleolus. The entire muscle may be elevated based on the medial plantar artery and vein. Arc of rotation is further increased by division of the lateral plantar artery proximally at its junction with the medial plantar artery. Transposition of the entire muscle will cover defects over the medial malleolus and heel. However, this dissection may adversely affect foot circulation.

SKIN TERRITORY: The medial plantar skin receives musculocutaneous perforating vessels from the muscle. This skin may be included with this muscle as a transposition flap.

SEGMENTAL TRANSPOSITION: The insertion of the abductor hallucis must be divided to achieve an adequate arc of rotation for posterior foot defects. However, the skin overlying the muscle may be advanced distally for defects over the first metatarsal phalangeal joint by release of only the plantar fascia.

DESIGN OF DISTALLY BASED FLAPS: The division of the muscle origin and proximal medial plantar artery and veins allows distal transposition of the abductor hallucis.

POTENTIAL FOR MICROVASCULAR TRANSPLANTATION: The abductor hallucis may be transplanted based on the medial plantar artery and vein.

VASCULARIZED BONE: Vascular communications between the muscle's origin and the periosteum of the calcaneus are noted. However, the donor defect may result in donor site disability.

Origin

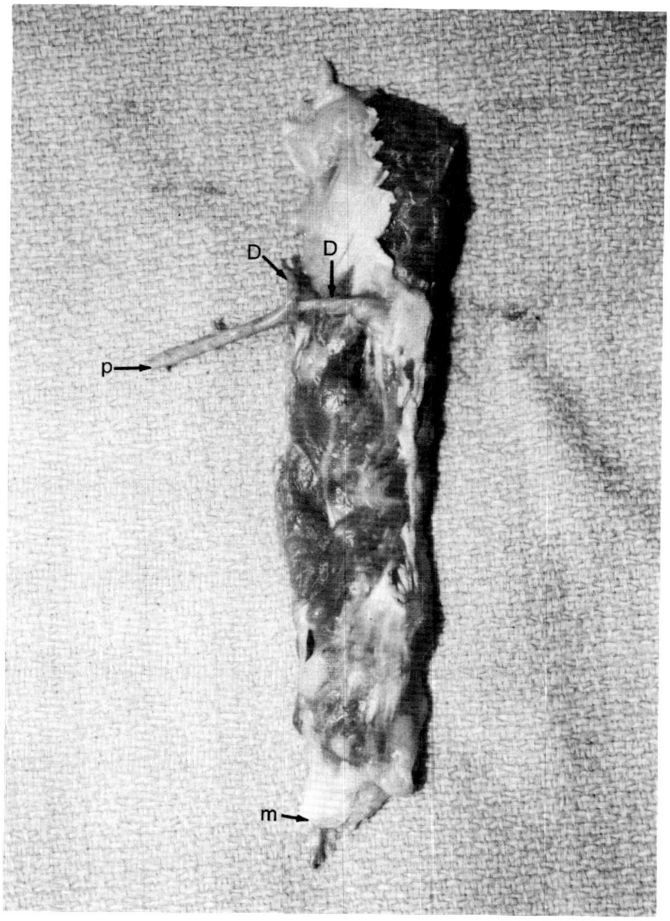

Insertion

Fig. 2-65. Dominant vascular pedicle: *D,* Branches of medial plantar artery *(p).* Minor vascular pedicle: *m,* Branch of medial plantar artery *(p).*

Origin

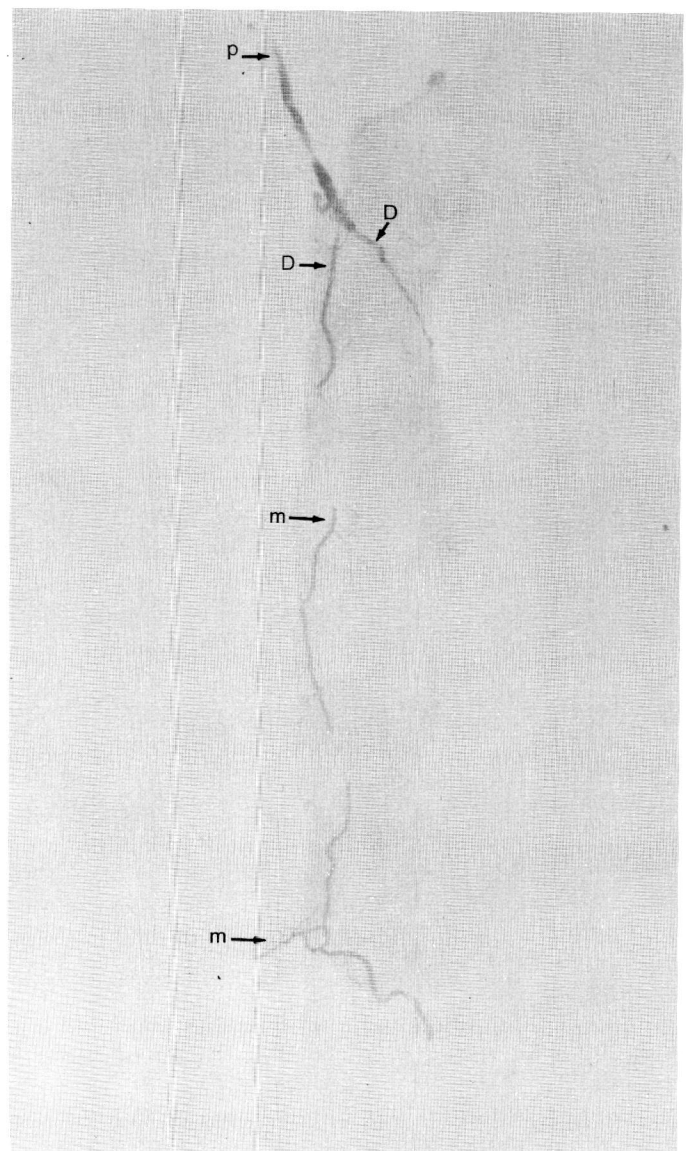

Insertion

Fig. 2-66. Dominant vascular pedicle: *D,* Branches of medial plantar artery *(p).* Minor vascular pedicle: *m,* Branches of medial plantar artery *(p).*

FLEXOR DIGITORUM BREVIS

Vascular anatomy (Figs. 2-67 to 2-69)

Pattern of circulation: Type II

Dominant vascular pedicles (D)

1. Proximal two branches of medial plantar artery and venae comitantes
   REGIONAL SOURCE: Posterior tibial artery and vein
   SIZE: Length: 1 to 2 cm
   Diameter: 0.6 mm
   LOCATION: Medial proximal muscle adjacent to origin at calcaneus
2. Proximal branch of lateral plantar artery and venae comitantes
   REGIONAL SOURCE: Posterior tibial artery and veins
   SIZE: Length: 1 to 2 cm
   Diameter: 0.6 mm
   LOCATION: Lateral superior one third of muscle

Minor vascular pedicles (m)

1. Two to three muscular branches of medial plantar artery and venae comitantes
   REGIONAL SOURCE: Posterior tibial artery and vein
   SIZE: Length: 1 cm
   Diameter: 0.5 mm
   LOCATION: Medial aspect of muscle belly
2. Two muscular branches of lateral plantar artery and venae comitantes
   REGIONAL SOURCE: Posterior tibial artery and vein
   SIZE: Length: 1 cm
   Diameter: 0.5 mm
   LOCATION: Lateral aspect of muscle belly

Clinical application

ARC OF ROTATION: Elevation of the inferior two thirds of the flexor digitorum brevis is possible based on its dominant vascular pedicles. Transposition of the muscle as a turnover flap allows coverage of the calcaneus. The entire muscle may be elevated based on the lateral plantar artery. With division of the distal lateral plantar artery and muscle origin and insertion, the muscle will reach the medial ankle. However, foot circulation may be adversely impaired by this extensive dissection.

Fig. 2-67. *pt,* Posterior tibial artery; *lp,* lateral plantar artery; *mp,* medial plantar artery; *a,* abductor hallucis; *b,* flexor digitorum brevis; *c,* abductor digiti minimi.

SKIN TERRITORY: The central plantar skin receives musculocutaneous perforating vessels from the flexor digitorum brevis. This skin may be included with the muscle as a transposition flap if both origin and insertion are divided to achieve an adequate arc of rotation to cover heel and medial ankle defects.

SEGMENTAL TRANSPOSITION: Complete division of the insertion is required to use the flexor digitorum brevis as a turnover flap. However, the posterior plantar skin may be advanced posteriorly based on axial cutaneous and musculocutaneous perforating vessels from the medial plantar and posterior tibial vessels, with division of only the plantar fascia overlying this muscle.

DESIGN OF DISTALLY BASED FLAPS: The muscle may be transposed over the distal posterior foot by release of the muscle origin and division of the medial plantar artery and vein at its junction with the lateral plantar vessel.

POTENTIAL FOR MICROVASCULAR TRANSPLANTATION: The muscle may be transplanted by microvascular techniques based on the lateral plantar artery and vein. However, excision of this specialized skin from the foot is rarely indicated.

VASCULARIZED BONE: Vascular communications between the muscle origin and the periosteum of the calcaneus are noted. Removal of this bone with this muscle may result in donor site disability.

Origin

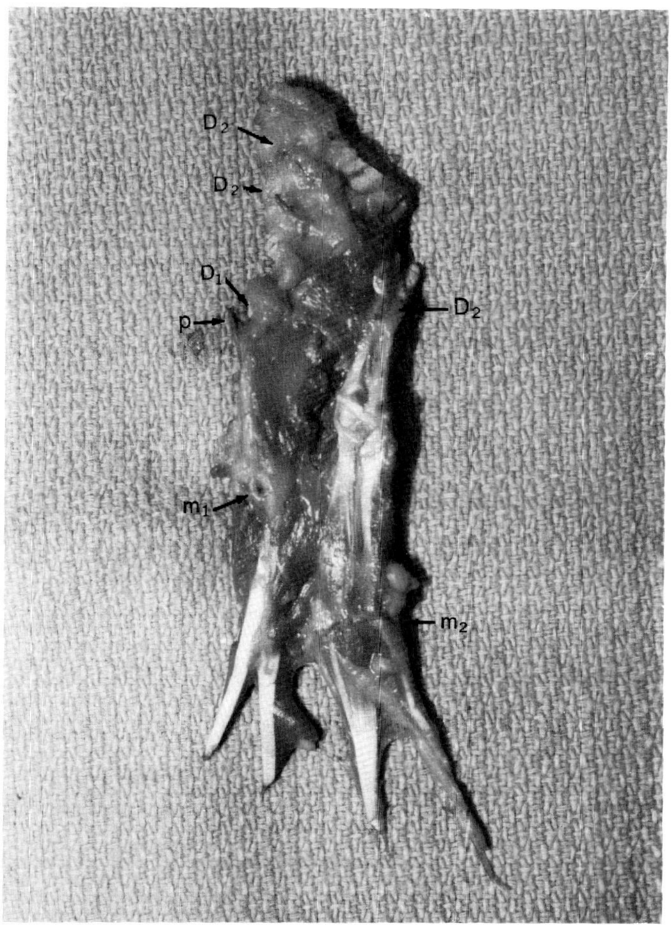

Insertion

Fig. 2-68. Dominant vascular pedicles: D_1, Branches of medial plantar artery *(p)*; D_2, branch of lateral plantar artery. Minor vascular pedicles: m_1, Branches of medial plantar artery; m_2, branches of lateral plantar artery.

Origin

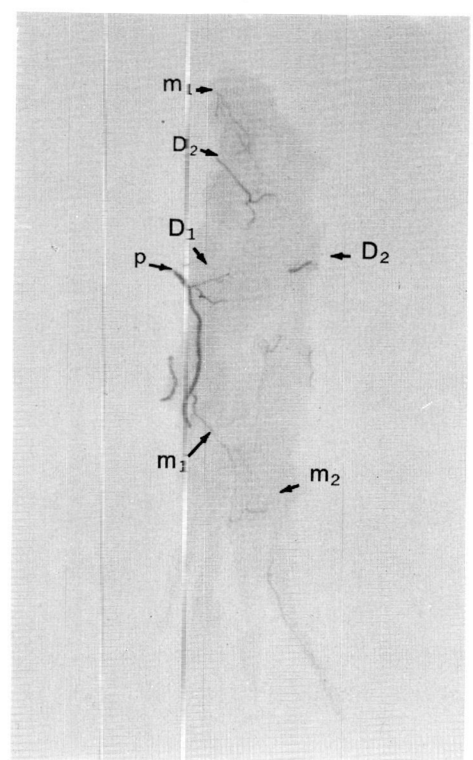

Insertion

Fig. 2-69. Dominant vascular pedicles: D_1, Branches of medial plantar artery *(p)*; D_2, branch of lateral plantar artery. Minor vascular pedicles: m_1, Branches of medial plantar artery; m_2, branches of lateral plantar artery.

ANNOTATED BIBLIOGRAPHY

1. Campbell, J., and Pennefather, C.: An investigation into the blood supply of muscles with special reference to war surgery, Lancet **1**:294, 1919.

   A description of the vascular anatomy of muscles is based on radiographic study of human muscles to assist surgeons in the management of war injuries. A knowledge of the distribution and arrangement of the arterial supply in muscles provides necessary information concerning the potential for survival after débridement of injured muscles.

2. Clark, W.E.L., and Blomfield, L.B.: The efficiency of intramuscular anastomoses, with observations on the regeneration of devascularized muscle, J. Anat. **79**:15, 1945.

   An experimental study of the vascularization of muscular tissue in rabbits is based on interruption of external and internal vascular anatomy. The variability in efficiency of intramuscular anastomosis is demonstrated based on the pattern of external and internal vascular anatomy. The practical importance of a knowledge of intramuscular vascular patterns in human muscles is emphasized.

3. Mathes, S., McCraw, J., and Vasconez, L.: Muscle transposition flaps for coverage of lower extremity defects: anatomical considerations, Surg. Clin. North Am. **54**:1337, 1974.

   The principles of muscle flap transposition are presented, emphasizing the vascular anatomy of the muscle as the major factor determining both arc of rotation and successful transposition of muscles for lower leg coverage. The use of the latex injection technique is presented for study of vascular anatomy of muscles.

4. Mathes, S., and Nahai, F.: Clinical atlas of muscle and musculocutaneous flaps, St. Louis, 1979, The C.V. Mosby Co.

   This text includes five sections that concentrate on the pertinent anatomy and techniques for use of muscle flaps. The importance of an accurate knowledge of muscle vascular anatomy is emphasized.

3

MUSCULOCUTANEOUS PERFORATORS

Foad Nahai ■ Stephen J. Mathes

The vascular supply of the skin is based on direct cutaneous arteries and musculocutaneous perforators. Musculocutaneous flaps and random flaps are based on musculocutaneous arteries, whereas axial flaps are based on the direct cutaneous arteries. Direct cutaneous vessels predominate in the face and around the limb girdles, joints, digits, and genitalia. The rest of the skin is vascularized through the underlying muscles.

Through dissection of injected cadaver specimens, the precise blood supply to muscle has been described. The same techniques have been applied to demonstrate the musculocutaneous perforators. The number, size, and location of the perforators are now known, and these data, when correlated with clinical experience, allow accurate prediction of cutaneous territories of superficial muscles. Each superficial muscle will supply the skin lying directly over it, and this skin territory may be safely extended 3 to 4 cm beyond the borders of the underlying muscle. The

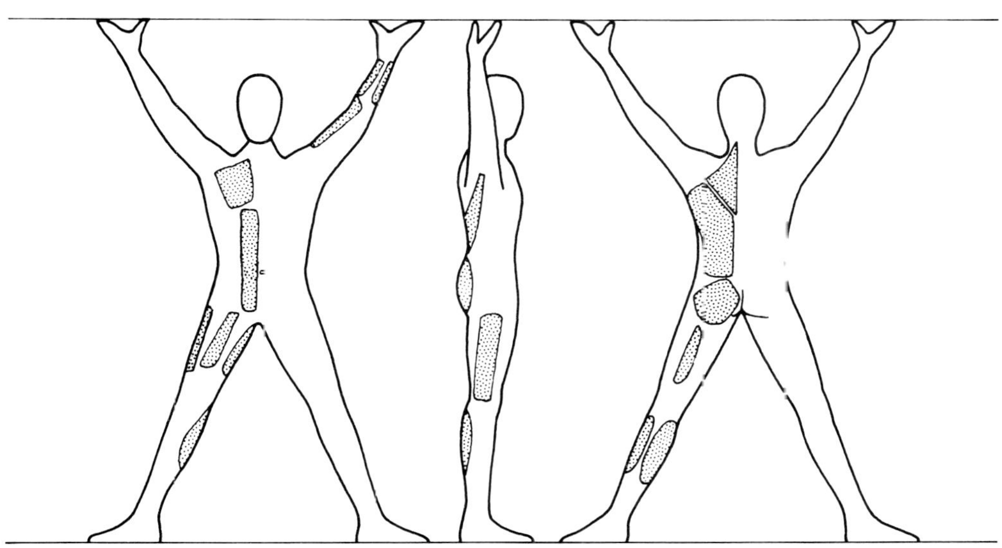

extent of this skin territory is not affected by previous incisions in the overlying skin, provided the skin and muscle are not separated at the time of elevation.

The intramuscular arteries run parallel to the muscle fibers and send perforating branches between muscle fibers into the overlying skin (Fig. 3-1). In general, these perforators are found only between muscle fibers and the overlying skin. There are no perforators between tendon or fascia into the overlying skin. The TFL and gastrocnemius are excellent examples. These muscles carry the skin over the fascia lata and Achilles tendon, respectively, through perforating vessels that enter the skin directly over the muscle fibers and not through the fascia lata or Achilles tendon (Fig. 3-2). Similarly, musculocutaneous perforators, nutrient vessels into bone from muscle origin or insertion, are only present when actual muscle fibers come into contact with bone, but not through tendon or fascial connections.

The actual pattern of blood supply to the underly-

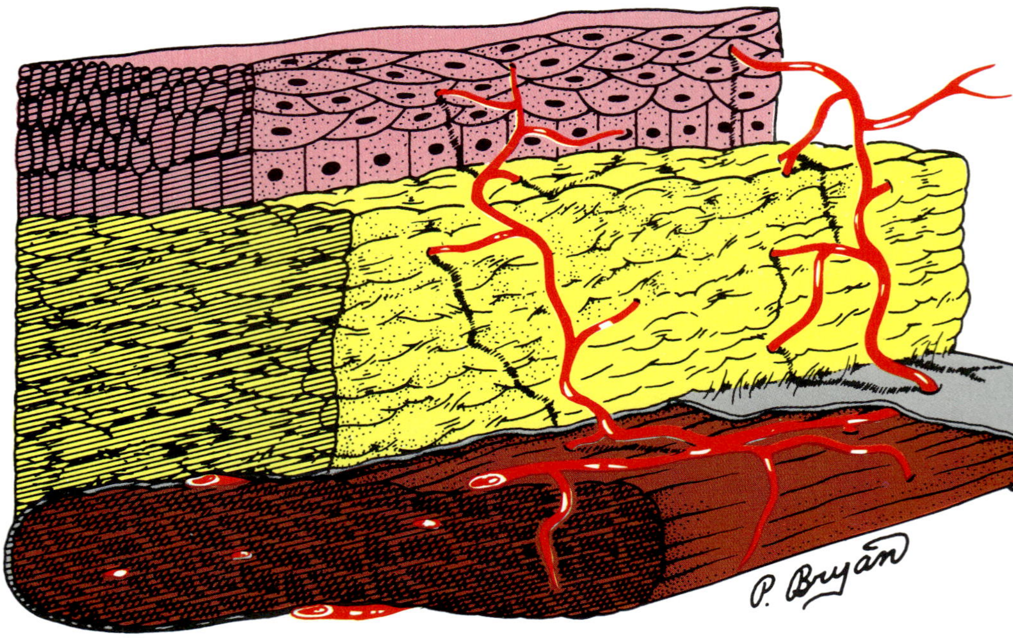

Fig. 3-1. Intramuscular arteries and perforating branches into overlying skin.

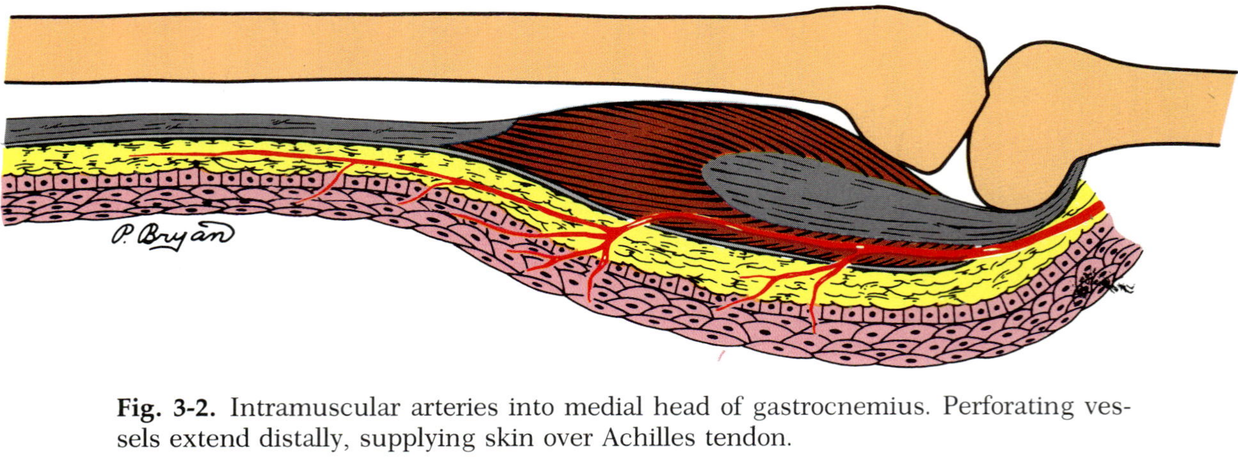

Fig. 3-2. Intramuscular arteries into medial head of gastrocnemius. Perforating vessels extend distally, supplying skin over Achilles tendon.

ing muscle must be taken into account when discussing the overlying skin island. In general, the type I blood supply will support all the overlying skin of a muscle unit. With the type II blood supply the proximal area of skin or that area over the dominant pedicle is far more reliable than the skin territory over the distal pedicles. With the type III blood supply each half of the muscle can be elevated separately with the overlying skin. The type IV has limited usefulness as a musculocutaneous flap, because each of the segmented pedicles carries a segmental area of the overlying skin. The type V is most useful, because skin islands can be based on both the proximal dominant and the secondary segmental blood supply to the muscle. There is considerable overlap in the skin territory of muscle flaps so that skin directly over one muscle may be elevated on the perforators through an adjacent muscle. For example, the skin over the lateral portion of the rectus femoris may be elevated on the TFL, and the skin of the lateral abdominal wall may be elevated either on the underlying external oblique or the rectus abdominis muscles.

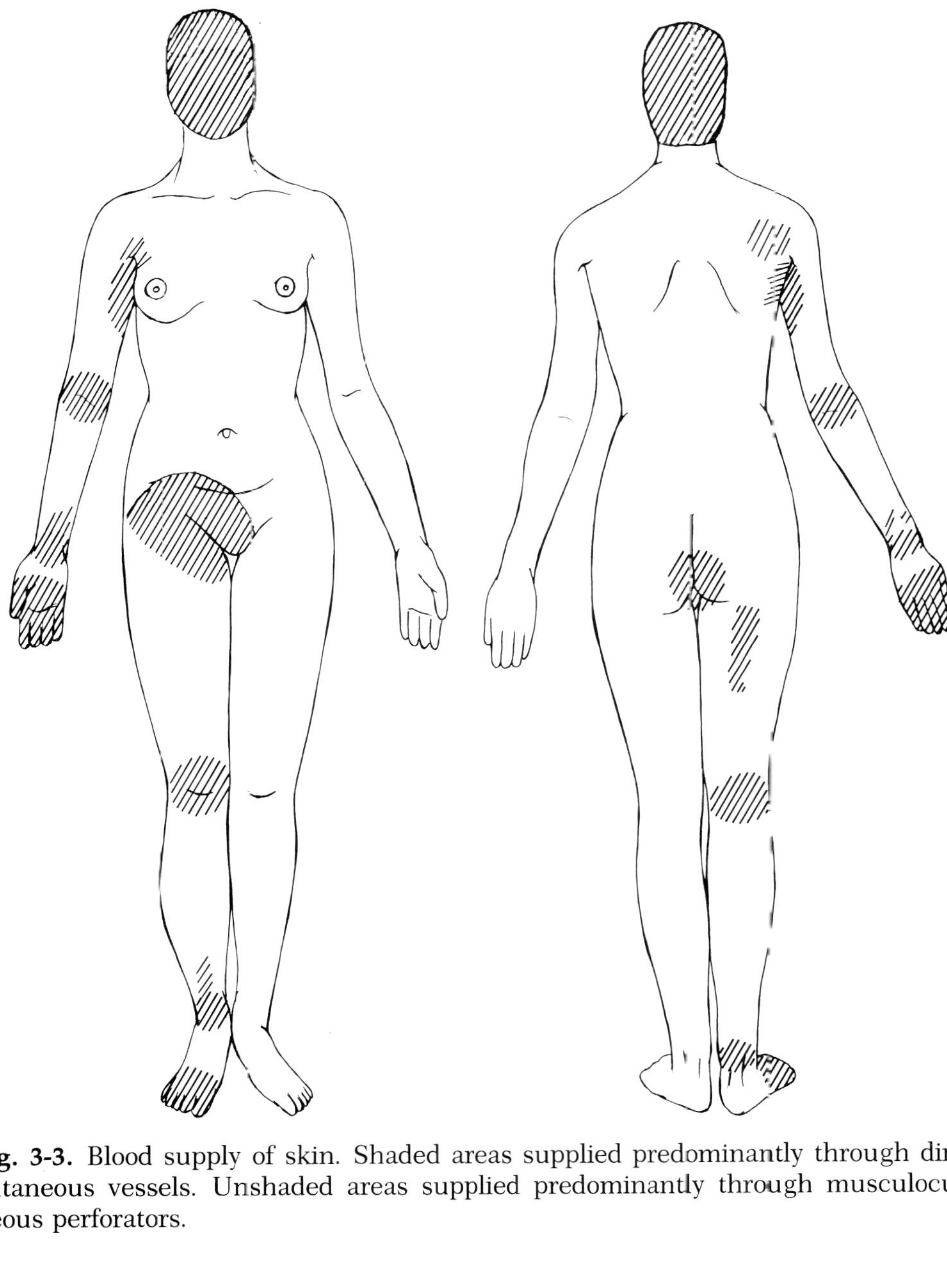

Fig. 3-3. Blood supply of skin. Shaded areas supplied predominantly through direct cutaneous vessels. Unshaded areas supplied predominantly through musculocutaneous perforators.

Similarly, there is considerable overlap in the blood supply of the skin through musculocutaneous perforators and direct cutaneous muscles. Therefore the same island of skin may be elevated in some areas either on the musculocutaneous perforators or direct cutaneous vessels. For example, the skin of the lower abdomen between the umbilicus and the groin may be elevated on the superficial inferior epigastric artery as an axial flap or on the underlying rectus abdominis muscle as a musculocutaneous flap. In designing and elevating flaps, these factors should be taken into account (Fig. 3-3).

OVERVIEW OF CUTANEOUS BLOOD SUPPLY BY REGIONS

Head and neck

Scalp. The scalp is well vascularized through direct cutaneous vessels that are branches of the supratrochlear, superficial temporal, posterior auricular, and occipital arteries.

Face. The face is vascularized through direct cutaneous arteries and musculocutaneous vessels through the facial muscles. Direct cutaneous branches of the facial, superficial temporal, and supratrochlear arteries dominate in this region.

Neck. The blood supply to the skin of the neck is predominantly through the musculocutaneous arteries. Around the ear direct cutaneous vessel branches of the occipital and posterior auricular arteries supply the upper lateral neck skin. Anteriorly the musculocutaneous perforators are based on the platysma, laterally on the sternocleidomastoid, and posteriorly on the trapezius. The supraclavicular skin is supplied by perforators through the trapezius and direct cutaneous vessels, branches of the transverse cervical artery.

Anterior trunk

The anterior trunk, extending from the clavicle down to the inguinal ligament, is supplied predominantly through musculocutaneous perforators. A continuous row of paramedian musculocutaneous perforators is seen extending from the first intercostal space down to the symphysis pubis. There are segmental vessels through the pectoralis major superiorly and the rectus abdominis inferiorly. These are relatively large perforators measuring up to 0.7 mm in diameter and emerge through the muscle 2 to 3 cm laterally to the midline. These vessels are based on the deep epigastric arcade, which is formed by the anastomosis between the internal mammary artery and its continuation as the superior epigastric artery and the deep inferior epigastric artery.

Laterally the skin is supplied superiorly by perforators through the pectoralis major and serratus anterior and a direct cutaneous vessel, a branch of the lateral thoracic artery. Below the costal margin segmental perforating branches of the intercostal arteries through the external oblique muscle supply the overlying skin.

In the lower abdomen the superficial inferior epigastric artery, a 1 to 1.5 mm branch of the femoral artery, directly supplies the skin of the lower abdomen up to the umbilicus. The superficial circumflex iliac artery, another 1 to 1.5 mm branch of the femoral artery, supplies the skin overlying the inguinal ligament and iliac crest.

Posterior trunk

The posterior trunk extends from the neck down to the ischium. The blood supply to this region is predominantly through musculocutaneous perforators. A row of segmental perforators 2 to 3 cm from the midline extends from the level of the first thoracic vertebra down to the upper border of the gluteus maximus muscle. These are segmental perforating branches of the intercostal and lumbar arteries through the trapezius and latissimus dorsi muscles. Laterally musculocutaneous perforators are seen through the trapezius, latissimus dorsi, external oblique, and gluteus maximus muscles. Direct cutaneous branches of the subscapular artery are distributed to the skin around the shoulder posteriorly and laterally.

Lower extremity

Anterior thigh. The anterior thigh is supplied through direct cutaneous arteries and musculocutaneous perforators. Proximally in the inguinal region and distally around the knee direct cutaneous vessels predominate. The superficial iliac circumflex, superficial external pudendal, and branches of the deep external pudendal arteries are distributed to the skin around the hip and genitalia, whereas branches of the geniculate arteries are distributed to the skin around the knee. Musculocutaneous perforators through the TFL, sartorius, rectus femoris, vastus medialis, and gracilis are distributed to the skin between the hip and knee anteriorly.

Posterior thigh. The posterior thigh is supplied by direct cutaneous arteries proximally around the hip and distally around the knee region. In between there are perforators through the hamstring muscles predominantly through the biceps femoris with a few perforators through the semimembranosus and semitendinosus. The inferior gluteal artery continues distal to the lower border of the inferior gluteus maximus muscle becomes a direct cutaneous vessel to the posterior thigh and is the basis of the inferior gluteal thigh flap.

Leg. The skin of the knee region is supplied through direct cutaneous branches of the anastomotic network of geniculate vessels. The skin of the ankle region is similarly supplied through an anastomotic network of vessels between the calcaneal and malleolar vessels. The area of skin between the knee and the ankle is supplied predominantly by musculocutaneous perforators, anteriorly through the tibialis anterior and peroneus longus and posteriorly through the lateral and medial gastrocnemius muscle.

Foot. The heel area is supplied through direct cutaneous branches of the calcaneal vessels, and the toes are supplied through the digital vessels. The dorsum of the foot is supplied by the dorsalis pedis artery. The rest of the sole of the foot is supplied through the calcaneal vessels and musculocutaneous perforators through the abductor hallucis, flexor digitorum brevis, and abductor digiti minimi muscles.

Upper extremity

The vascular supply to the shoulder girdle region is through direct cutaneous vessels that are branches of the subscapular, acromiothoracic trunk, and lateral thoracic arteries. Around the elbow there is a network of anastomotic vessels between branches of the brachial artery and branches of the radial and ulna arteries. Direct cutaneous branches from these vessels supply the skin around the elbow. The area of skin between the shoulder and elbow is predominantly supplied by musculocutaneous arteries through the biceps anteriorly and triceps posteriorly. The skin between the elbow and hand is similarly supplied by direct vessels and perforators through the underlying muscles. Both the dorsum of the hand and the digits are supplied through direct vessels. On the palmar surface of the hand perforating vessels are seen between the thenar and hypothenar muscles and the overlying skin.

SPECIFIC MUSCULOCUTANEOUS PERFORATORS

The perforating musculocutaneous vessels for each of the flaps will now be described. This description is based on clinical experience and dissections of injected cadaver specimens. The skin overlying the injected muscle has been dissected off, and the site of each musculocutaneous perforator has been labeled with a small, black head.

In addition to musculocutaneous perforating vessels the sensory innervation of the skin territory of most of the flaps is described. Some musculocutaneous flaps may be elevated and transposed as neurosensory flaps (e.g., TFL and rectus femoris). The skin of these flaps is usually innervated by a single nerve, which lies close to the major vascular pedicle of the flap. Preservation of the continuity of this sensory nerve to the flap allows flap transposition as a neurosensory flap. However, in other flaps (e.g., latissimus dorsi and pectoralis major) the skin territory is innervated segmentally, and the sensory nerves do not enter the skin territory in proximity to the dominant vascular pedicle; therefore the sensory nerves are divided to allow the normal arc of rotation of the flap. These flaps have little potential for transposition as neurosensory flaps.

Platysma (Fig. 3-4)

The total skin territory of the platysma located between the midline and the medial border of the sternocleidomastoid laterally and the mandible superiorly to the clavicle inferiorly measures approximately 10 × 15 cm. The perforating vessels are small, less than 0.5 mm in diameter, and are distributed evenly throughout this area.

Clinically, small 4 × 6 cm islands of skin are elevated with the platysma, and this may be safely designed anywhere on the muscle, provided the island is large enough to include one or two of the perforating vessels.

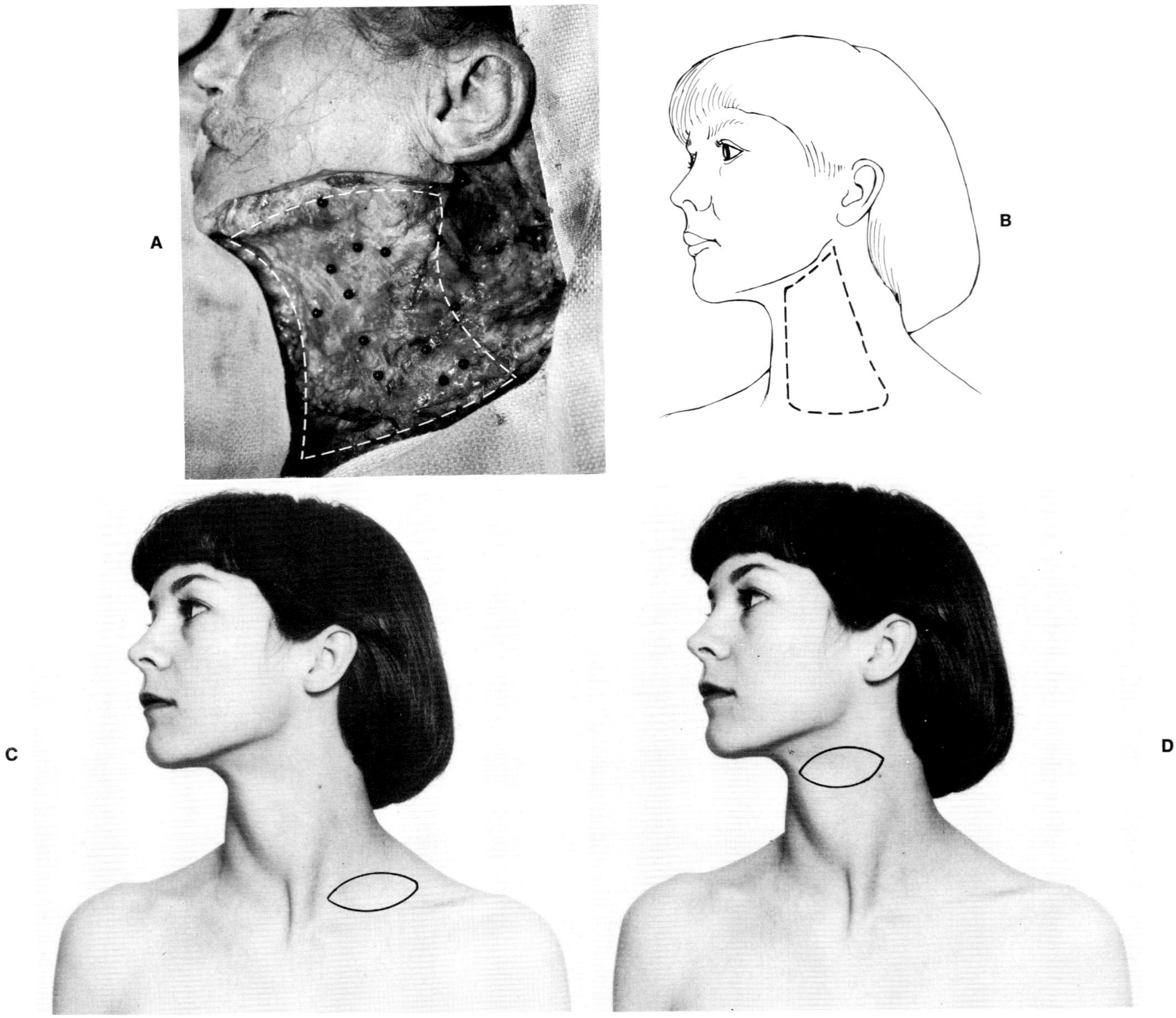

Fig. 3-4
A, Perforating musculocutaneous arteries through platysma into overlying skin.
B, Outline of platysma muscle.
C, Design of skin island for standard platysma musculocutaneous flap.
D, Design of skin island for distally based platysma flap.

Sternocleidomastoid (Fig. 3-5)

The total skin territory of the sternocleidomastoid measures 6 × 20 cm. This includes the skin overlying the muscle and extending 2 to 3 cm beyond the margins of the muscle. The perforating vessels are small, 0.5 mm in diameter, and are distributed throughout this area. However, there is a greater number of perforators proximally over the upper third of the muscle. These perforators are generally larger in size than the distal perforators. Based on these perforators the distal skin territory can be separated from the underlying muscle and transposed as a skin flap.

For clinical purposes the skin island can be based over the entire muscle, the proximal portion, or the distal portion. Distally the platysma is elevated with the sternocleidomastoid as the perforators pass through the platysma to the overlying skin.

A B C D

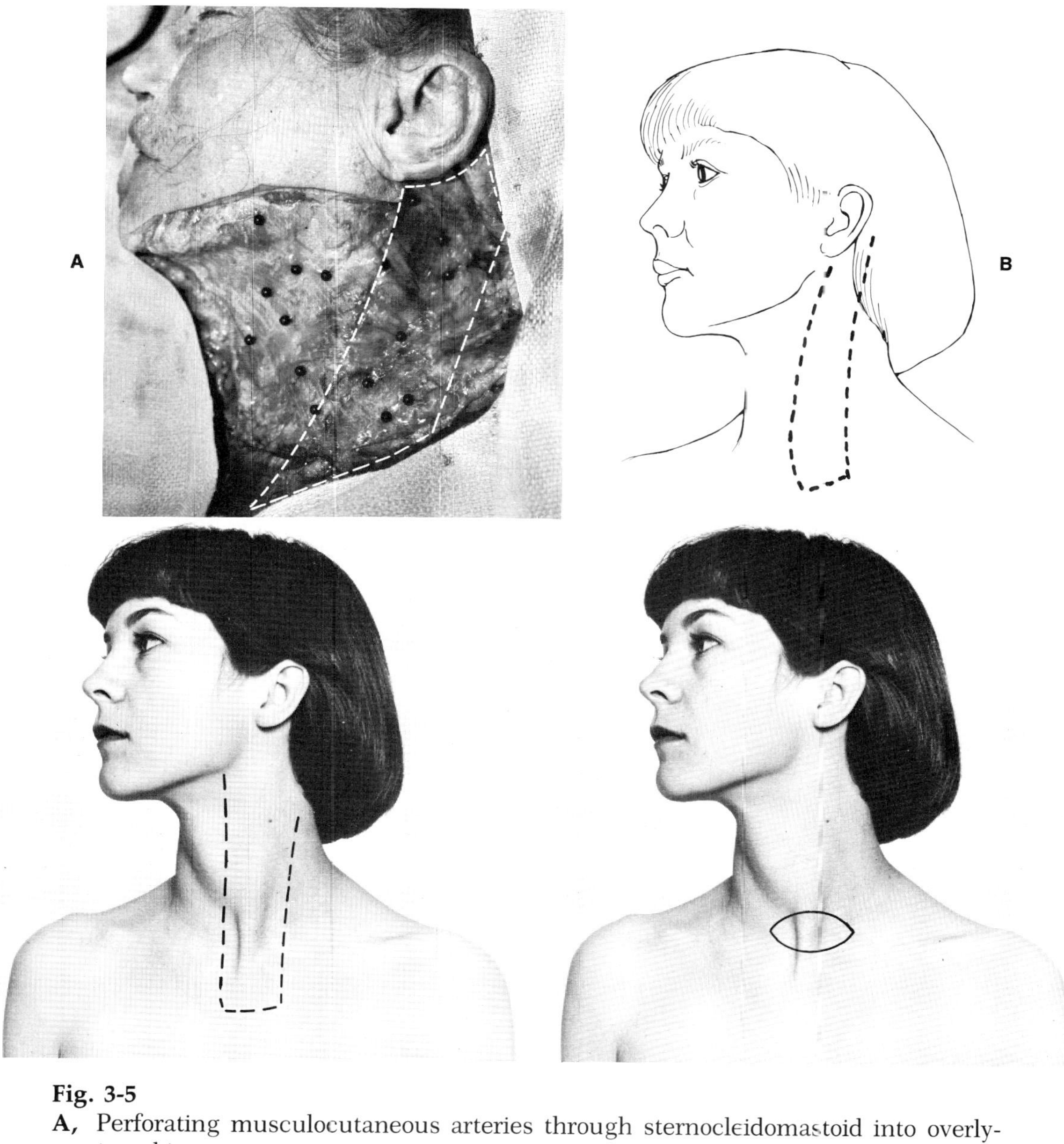

Fig. 3-5
A, Perforating musculocutaneous arteries through sternocleidomastoid into overlying skin.
B, Outline of sternocleidomastoid muscle.
C, Design of standard musculocutaneous flap.
D, Design of skin island for sternocleidomastoid musculocutaneous flap.

Pectoralis major (Fig. 3-6)

The total skin territory of the pectoralis major is 20 × 28 cm. This includes all the overlying skin and 3 to 4 cm beyond the muscle origin. The pectoralis major has a type V blood supply and a constant set of perforating arteries, measuring in size up to 1 mm in diameter, based on the secondary segmental pedicles. These are seen medially, 2 to 3 cm from the midline. There is a perforating artery in each intercostal space. The vessel in the third interspace is usually larger than the rest. The deltopectoral flap is based on these perforating musculocutaneous arteries. The remainder of the perforating vessels are smaller in size and are distributed over the entire muscle with more vessels over the lower part of the muscle than the upper portion.

For clinical purposes the entire skin territory is rarely used. Islands of skin of various sizes are based on the muscle. Particularly reliable is the distal island of skin over the inframammary area, because the perforators are more numerous in this area. This island of skin also has a much more acceptable donor defect. The skin territory is innervated by segmental intercostal nerves T2 to T6 and superiorly by the cervical nerves. However, this flap is not elevated as a neurosensory flap, because the nerves must be divided for flap elevation.

A

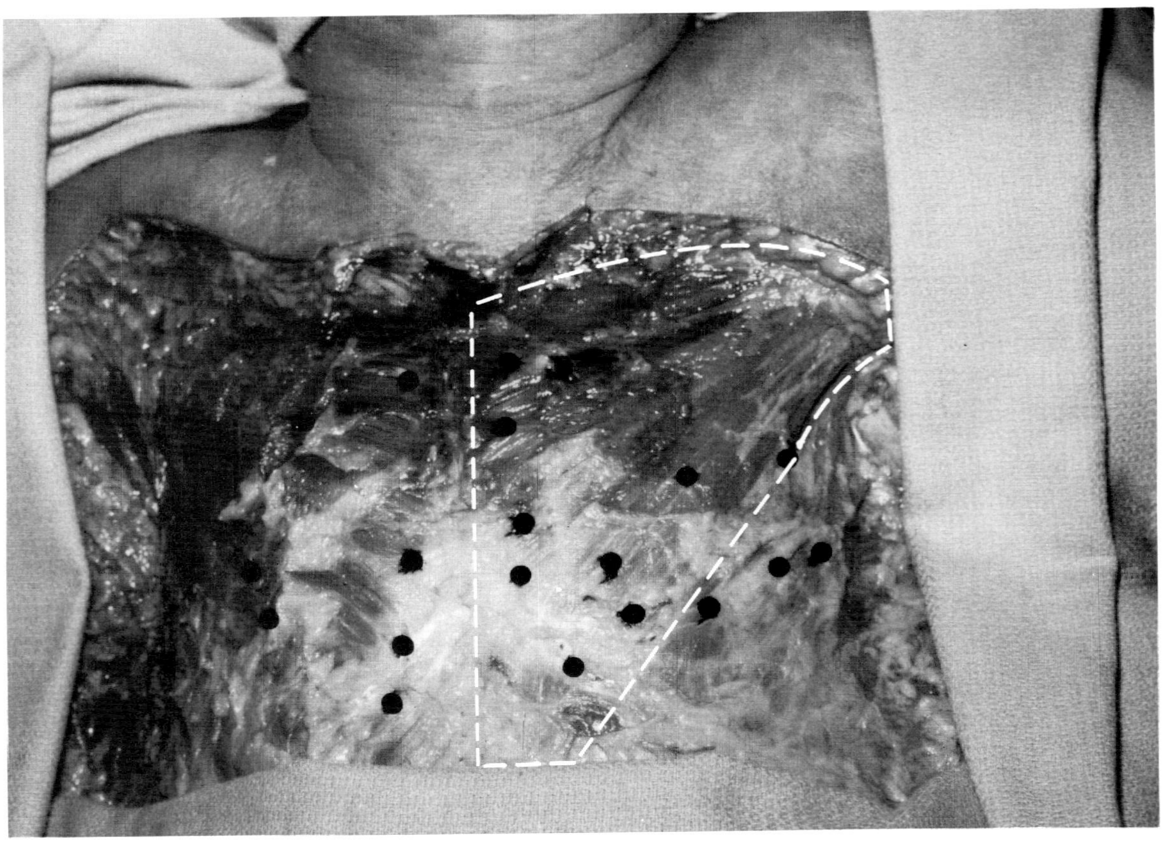

Fig. 3-6
A, Perforating musculocutaneous arteries through pectoralis major into overlying skin.

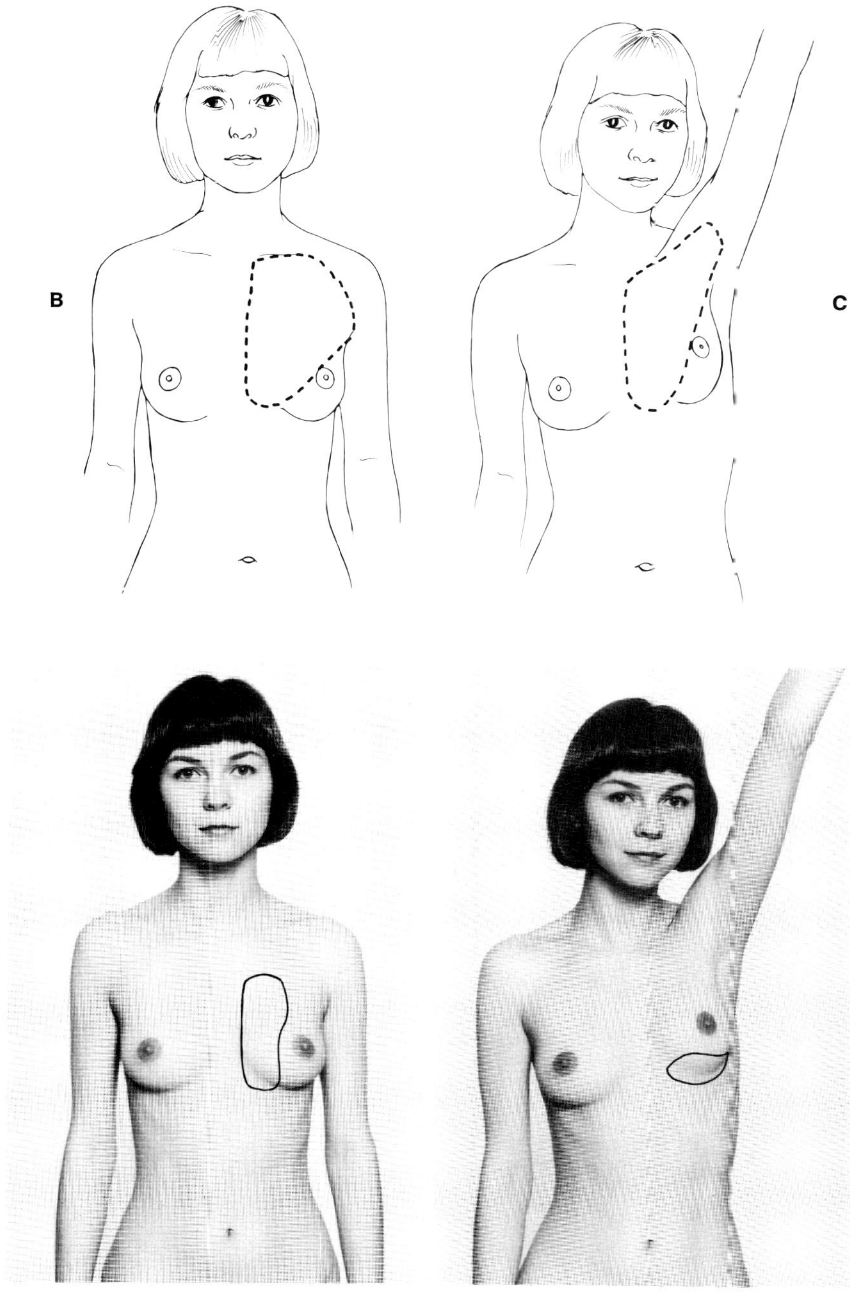

Fig. 3-6, cont'd
B and **C,** Outline of pectoralis major muscle.
D, Design of skin island. Standard pectoralis major musculocutaneous flap transposition for neck and oral cavity coverage.
E, Design of skin island pectoralis major musculocutaneous flap for intraoral coverage. NOTE: Donor defect closure located at inframammary line.

Trapezius (Fig. 3-7)

The skin territory of the trapezius measures 34 × 18 cm and includes all the skin overlying the muscle and several centimeters beyond, superiorly along the lateral border across to the shoulder and inferiorly below the lower border of the muscle. The musculocutaneous perforating arteries measure between 0.5 and 1 mm in diameter and are distributed all over the muscle. There is, however, a constant row of perforators medially, 2 to 3 cm from the midline. These vessels are usually the larger perforators.

The clinically useful islands of skin, based on the trapezius, are the anterior flap across the shoulder and the posterior island, 8 × 20 cm, based on the lower part of the muscle.

The skin territory of the trapezius is innervated anteriorly by the cervical nerves and posteriorly by the intercostal nerves. The flap is not a neurosensory flap, because the nerves must be divided for flap elevation; therefore this flap has little potential as a neurosensory flap.

A

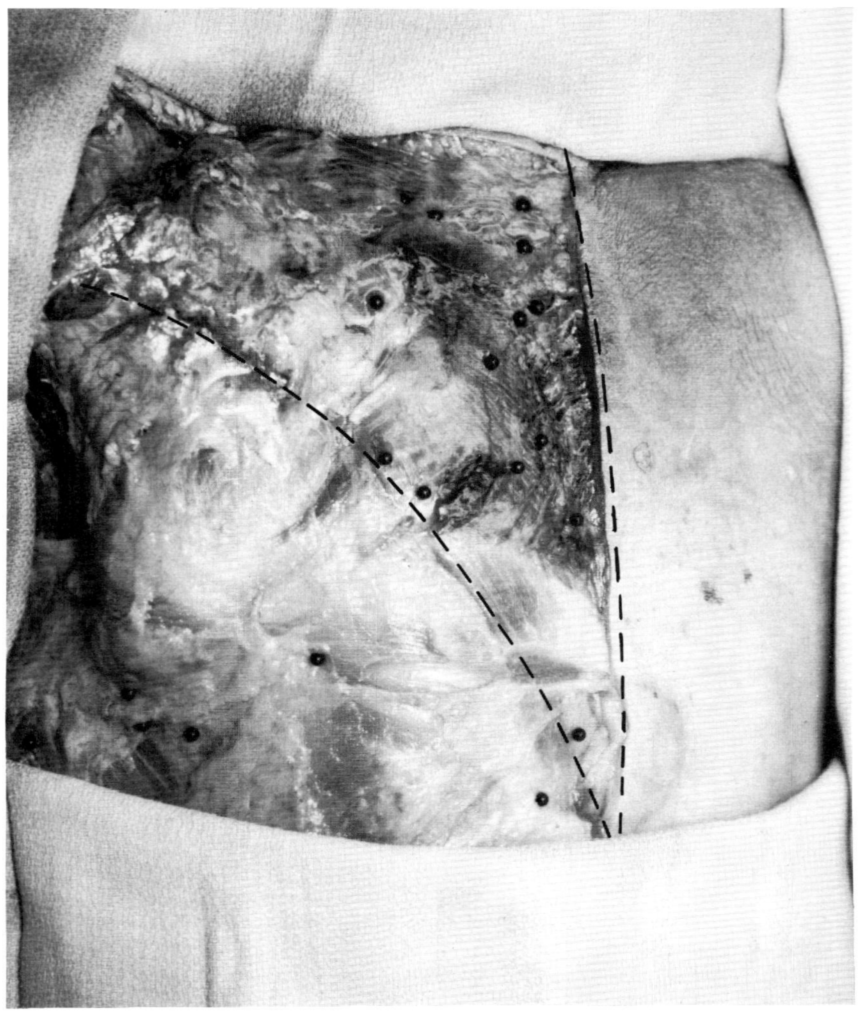

Fig. 3-7
A, Perforating musculocutaneous arteries through trapezius into overlying skin.

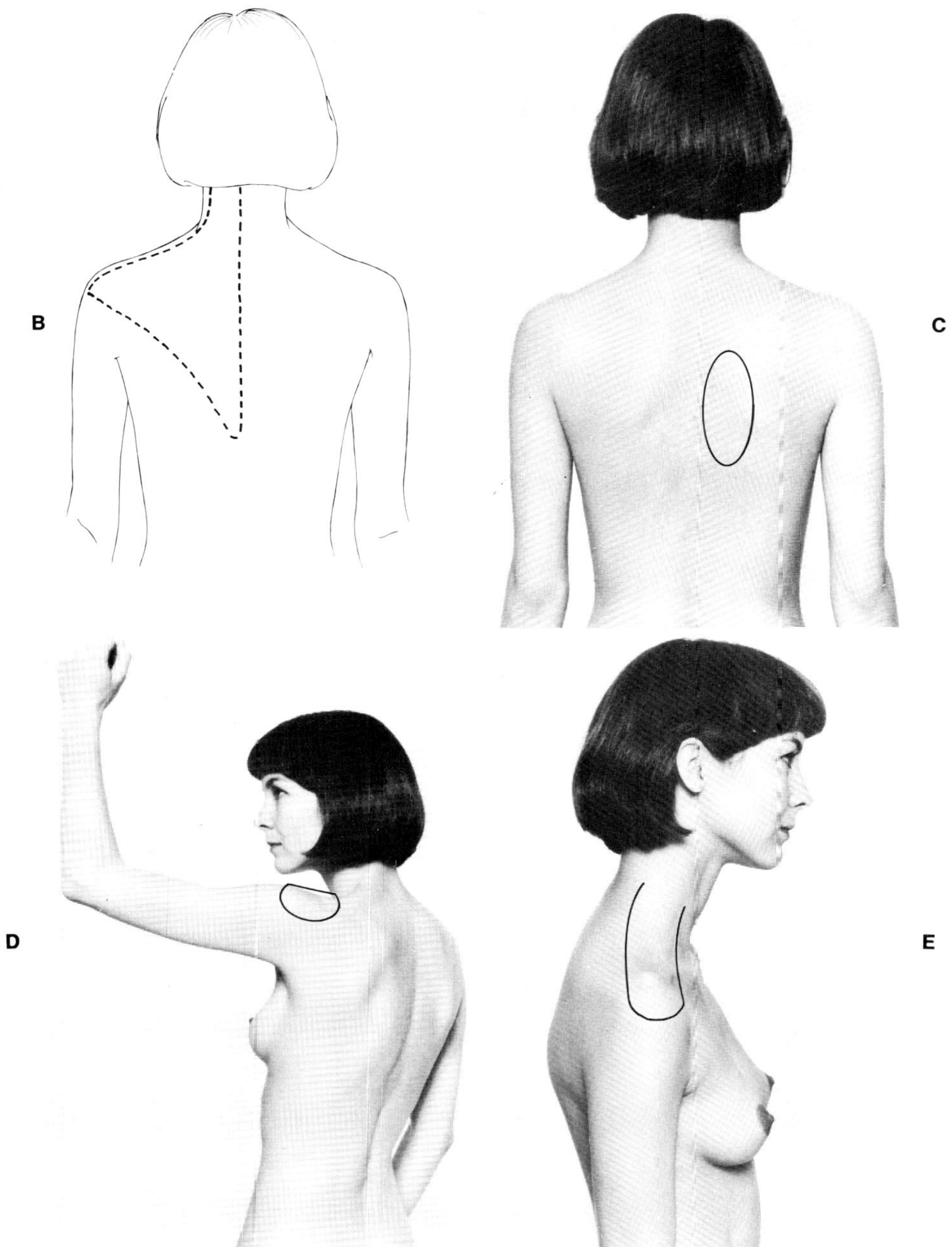

Fig. 3-7, cont'd
B, Outline of trapezius muscle.
C, Design of skin island for posterior trapezius musculocutaneous flap.
D, Design of skin island for anterior trapezius musculocutaneous flap.
E, Design of distally based trapezius musculocutaneous flap.

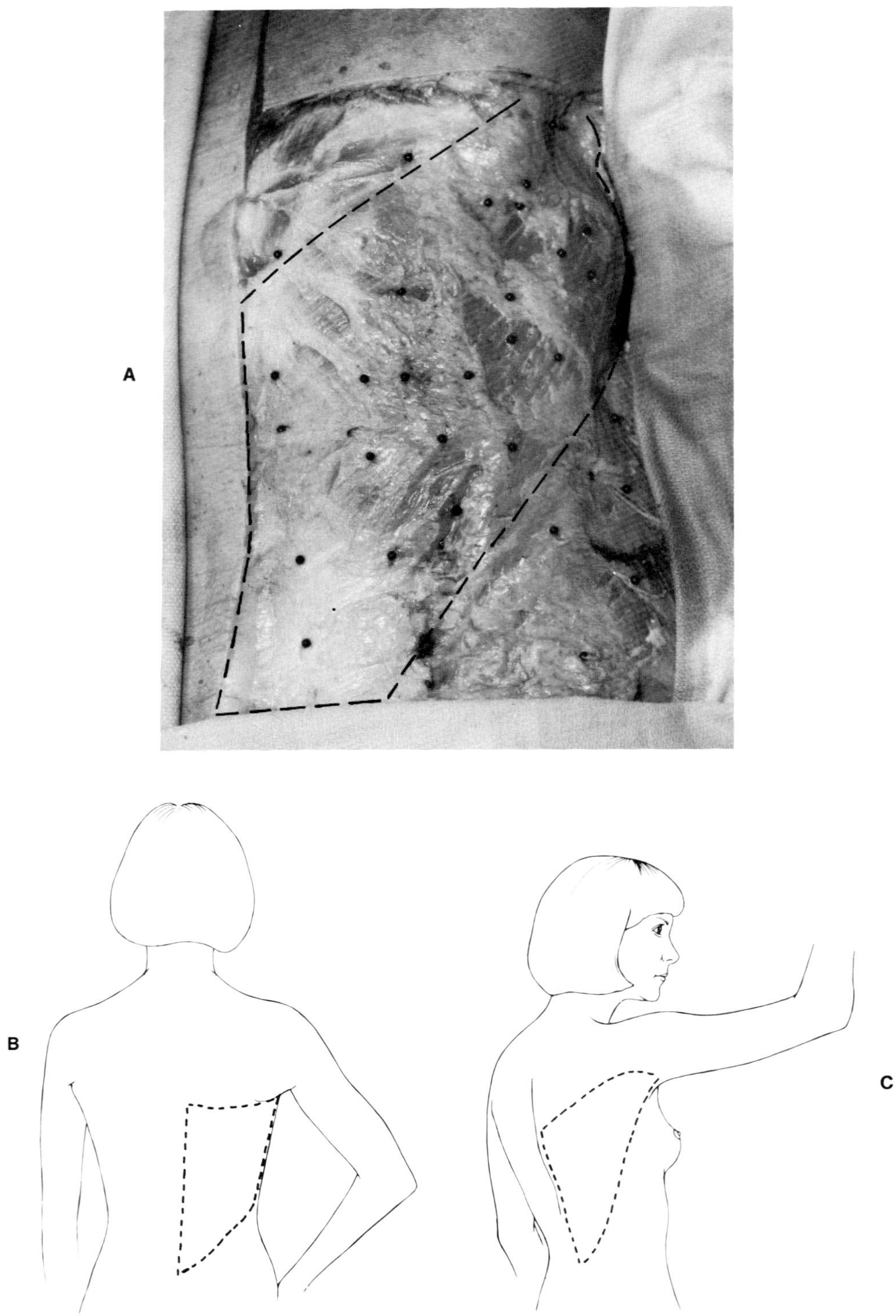

Fig. 3-8
A, Perforating musculocutaneous arteries through latissimus dorsi into overlying skin.
B and **C,** Outline of latissimus dorsi.

Latissimus dorsi (Fig. 3-8)

The skin territory of the latissimus dorsi measures approximately 30 × 40 cm and includes all the skin overlying the muscle and 4 to 5 cm beyond the margins of the muscle. There are two rows of perforators posteriorly approximately 2 to 3 cm and 4 to 5 cm from the midline. These perforators are larger, measuring up to 1 mm in diameter. This muscle has a type V blood supply, and the most medial row of perforators posteriorly is based on the segmental secondary pedicles of the muscle with the perforator corresponding to each vertebral level. The rest of the perforators are distributed all over the muscle with a constant row of perforators along the anterior free border of the muscle.

Clinically skin islands may be designed anywhere on the latissimus dorsi, depending on the reconstructive needs.

The skin territory of the latissimus dorsi is innervated segmentally by branches of the intercostal nerve T2 to T6, which have to be divided for flap elevation. Therefore the flap has little potential as a neurosensory flap.

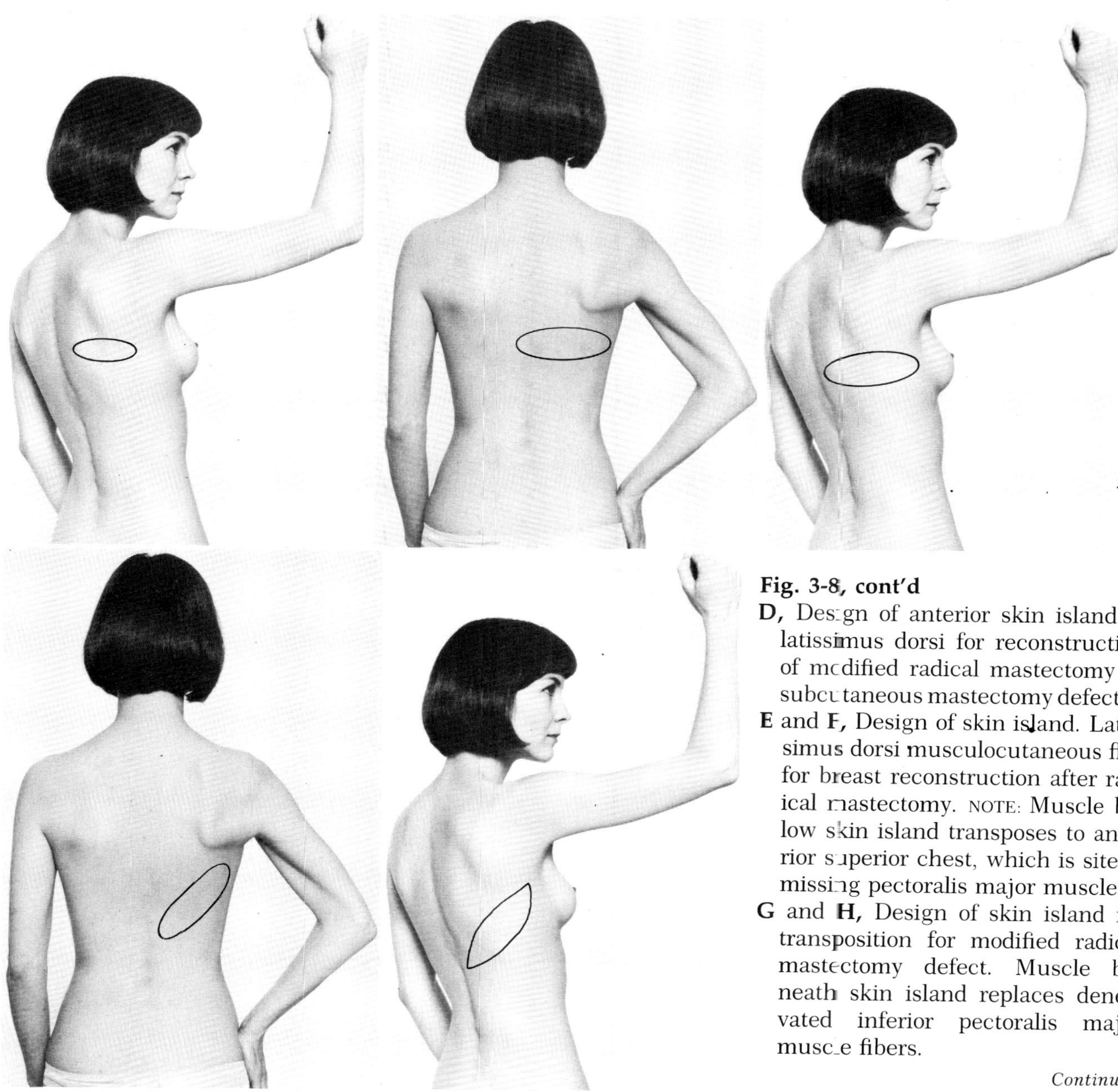

D E F G H

Fig. 3-8, cont'd
D, Design of anterior skin island of latissimus dorsi for reconstruction of modified radical mastectomy or subcutaneous mastectomy defects.
E and **F,** Design of skin island. Latissimus dorsi musculocutaneous flap for breast reconstruction after radical mastectomy. NOTE: Muscle below skin island transposes to anterior superior chest, which is site of missing pectoralis major muscle.
G and **H,** Design of skin island for transposition for modified radical mastectomy defect. Muscle beneath skin island replaces denervated inferior pectoralis major muscle fibers.

Continued.

Rectus abdominis (Fig. 3-9)

The skin territory of the rectus abdominis measures approximately 20 × 20 cm. This muscle supplies not only the entire skin over the muscle but also the skin of the abdominal wall as far as the anterior axillary line (thoracoepigastric flap).

The musculocutaneous perforating arteries are distributed over the muscle, but a continuous row of large, 1 mm in diameter, perforators is located medially 2 to 3 cm from the midline. These perforators are the basis of the thoracoepigastric flap and part of a continuous row of perforators through the pectoralis major and rectus abdominis, extending from the clavicle to the pubis (Fig. 3-10).

Clinically the useful skin island over the rectus abdominis measures 10 × 14 cm. This skin island is located at any level of muscle and may include a random extension either across the midline or laterally, depending on reconstructive requirements. This muscle has a type III blood supply with an artery superiorly and inferiorly. If the flap is based on the superior epigastric pedicle, the island is placed vertically over the mid muscle or transversely over either the inferior or superior muscle. If it is based on the inferior epigastric pedicle, the island is generally placed over the inferior two thirds of the muscle.

The skin territory of the rectus abdominis is innervated segmentally by branches of the intercostal nerves T6 to T12, which are divided for flap elevation. Therefore the flap has little potential as a neurosensory flap, although this same area of skin based on the intercostal arteries is elevated as a neurosensory intercostal flap.

A

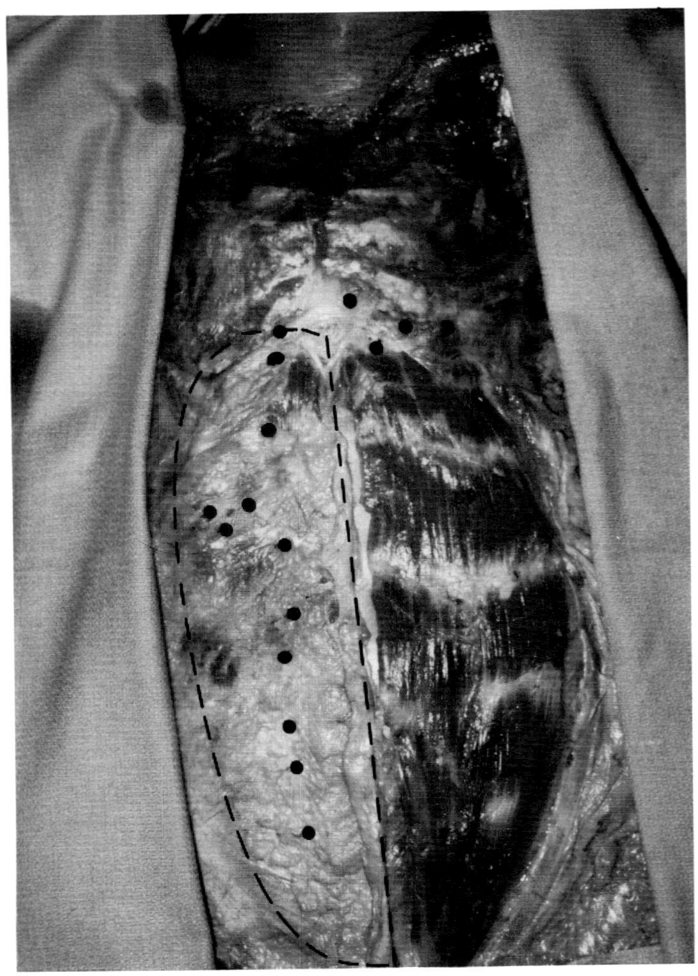

Fig. 3-9
A, Perforating musculocutaneous arteries through rectus abdominis into overlying skin.

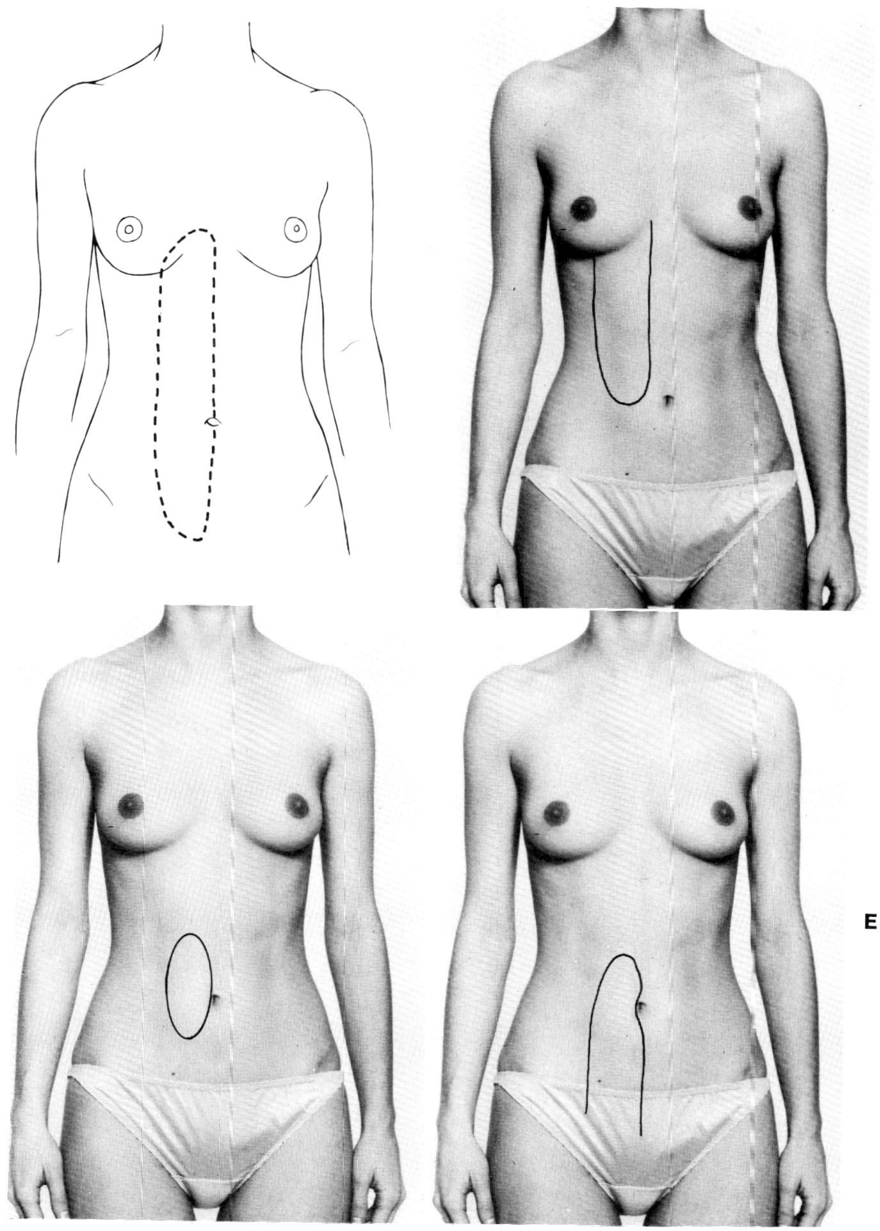

Fig. 3-9, cont'd
B, Outline of rectus abdominis.
C, Design of standard superior-based rectus abdominis musculocutaneous flap.
D, Design of skin island rectus abdominis musculocutaneous flap.
E, Design of standard inferior-based rectus abdominis musculocutaneous flap.

Continued.

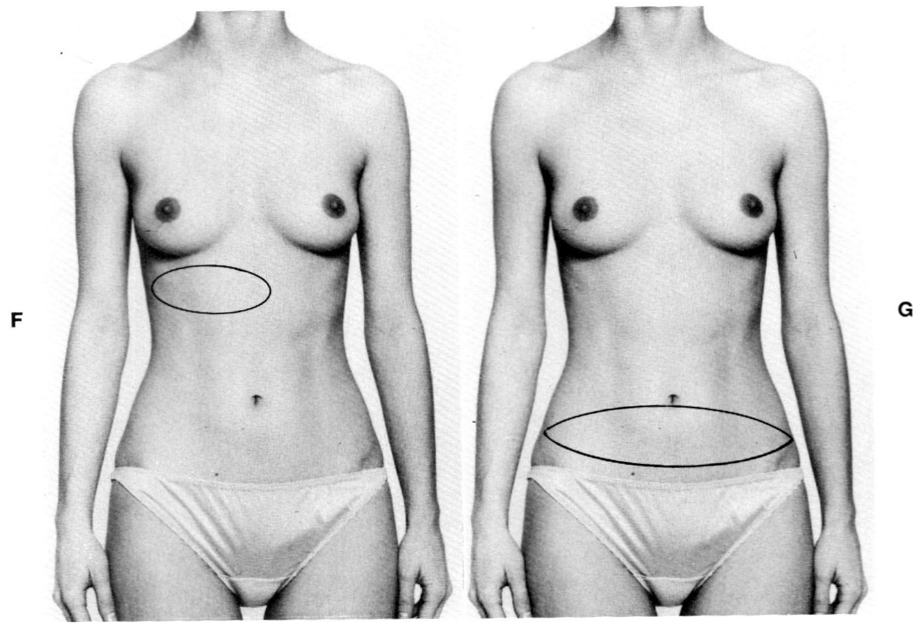

Fig. 3-9, cont'd
F, Design of upper transverse island rectus abdominis musculocutaneous flap.
G, Design of lower transverse island rectus abdominis musculocutaneous flap.

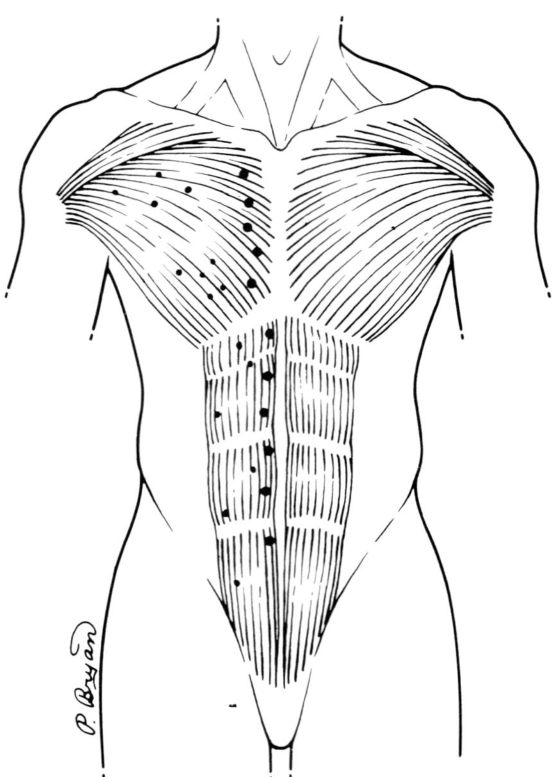

Fig. 3-10. Musculocutaneous perforators through pectoralis major and rectus abdominis into skin of anterior trunk.

Thoracoepigastric flap (Fig. 3-11)

The thoracoepigastric flap is a skin flap measuring 8 × 15 cm, based on the medial row of perforating musculocutaneous arteries through the rectus abdominis. The base of the flap is 2 to 3 cm from the midline, and it extends laterally to the anterior axillary line.

Sensory nerves to this flap are segmental and therefore divided during flap elevation. Therefore the flap has little potential as a neurosensory flap.

A

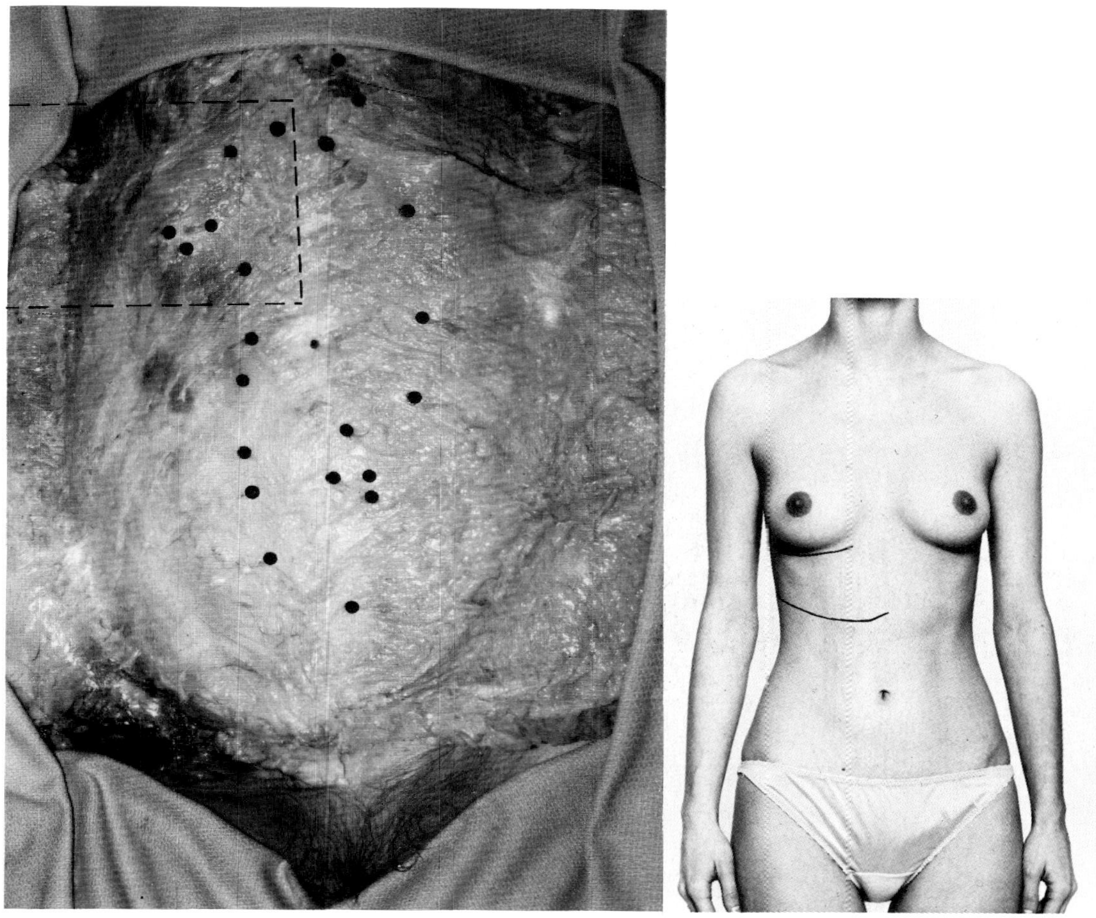

B

Fig. 3-11
A, Perforating musculocutaneous arteries through rectus abdominis into overlying skin (thoracoepigastric flap).
B, Outline of thoracoepigastric flap.

External oblique (Fig. 3-12)

The skin territory of the external oblique includes the overlying skin and 2 to 3 cm beyond the muscle margin. The area measures approximately 28 × 15 cm and includes the skin between the anterolateral border of the latissimus dorsi laterally and the rectus abdominis medially.

The musculocutaneous perforators are small, 0.5 mm in diameter, and are distributed evenly over the muscle.

The useful skin territory of the external oblique includes all the overlying skin that is usually transposed as a unit with the muscle for abdominal wall coverage.

The skin overlying the external oblique is innervated segmentally by branches of the intercostal nerves T5 to T12. Because this flap is usually advanced superiorly, sensory innervation of the skin area is preserved.

A

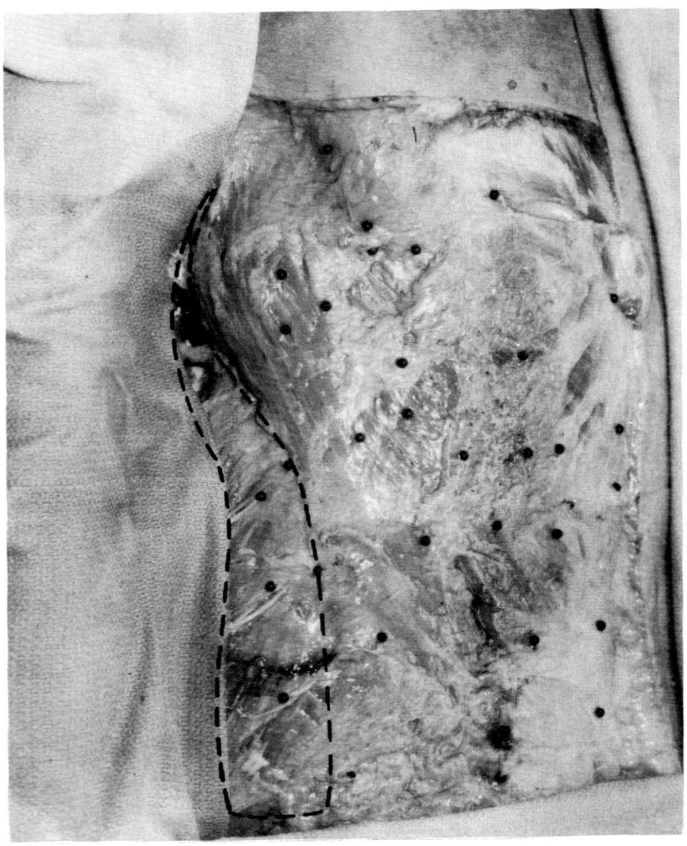

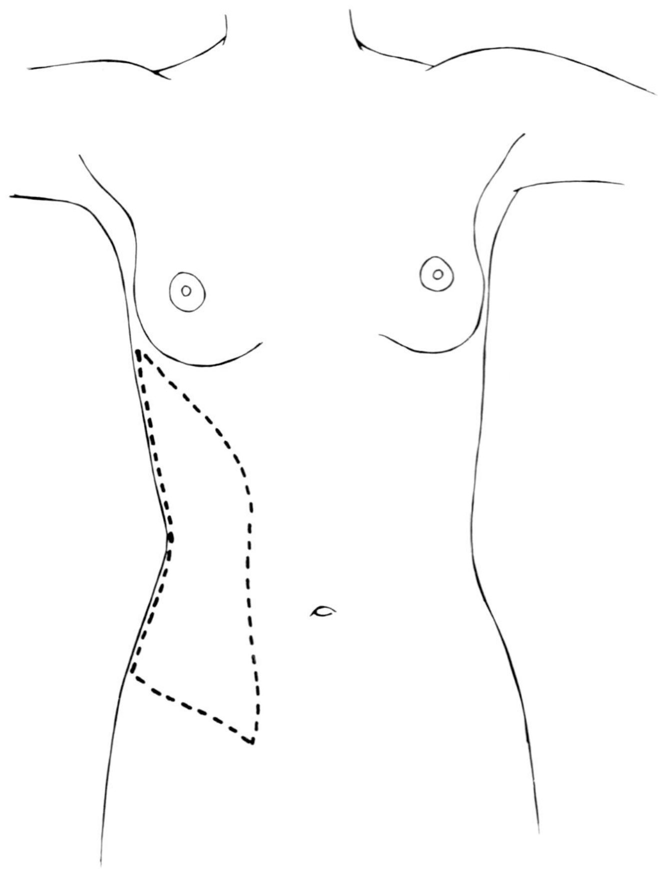

Fig. 3-12
A, Perforating musculocutaneous arteries through external oblique into overlying skin.
B, Outline of external oblique.

C

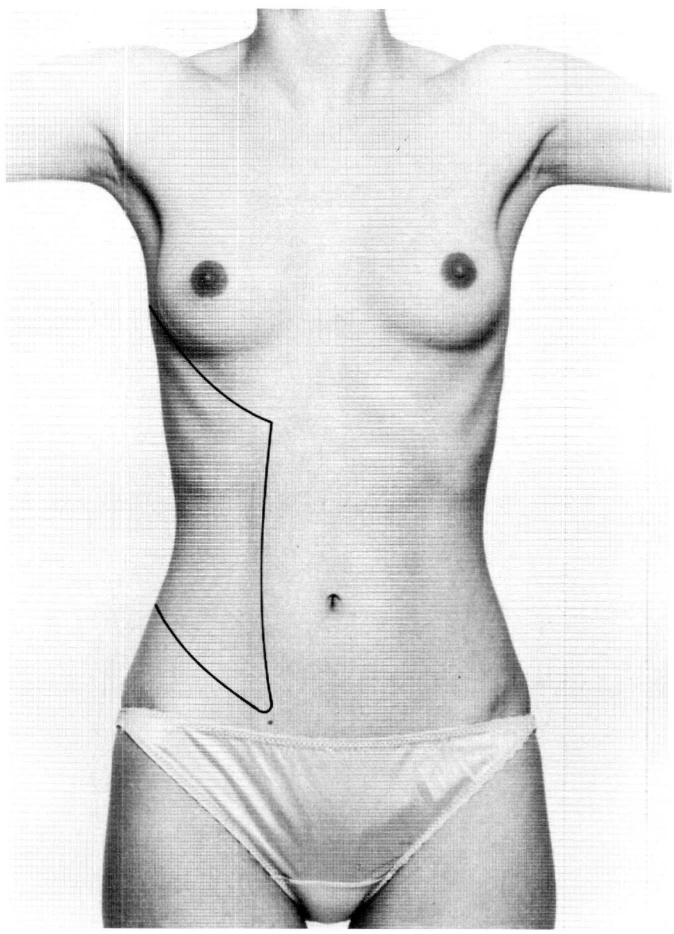

Fig. 3-12, cont'd
C, Standard design of external oblique musculocutaneous flap.

Gluteus maximus (Fig. 3-13)

The skin territory of the gluteus maximus measures approximately 24 × 24 cm and includes all the skin overlying the muscle and 2 to 3 cm beyond. Because this muscle has a type III blood supply, each half of the muscle will support half of the skin territory.

The musculocutaneous perforators are distributed evenly over the muscle into the skin territory. The perforators measure between 0.5 and 1 mm in diameter. The continuous line of medial perforators, 2 to 3 cm from the midline through the trapezius and latissimus dorsi is not seen through the gluteus maximus (Fig. 3-16).

Clinically useful skin islands may be designed anywhere on the muscle to suit the reconstructive needs.

A 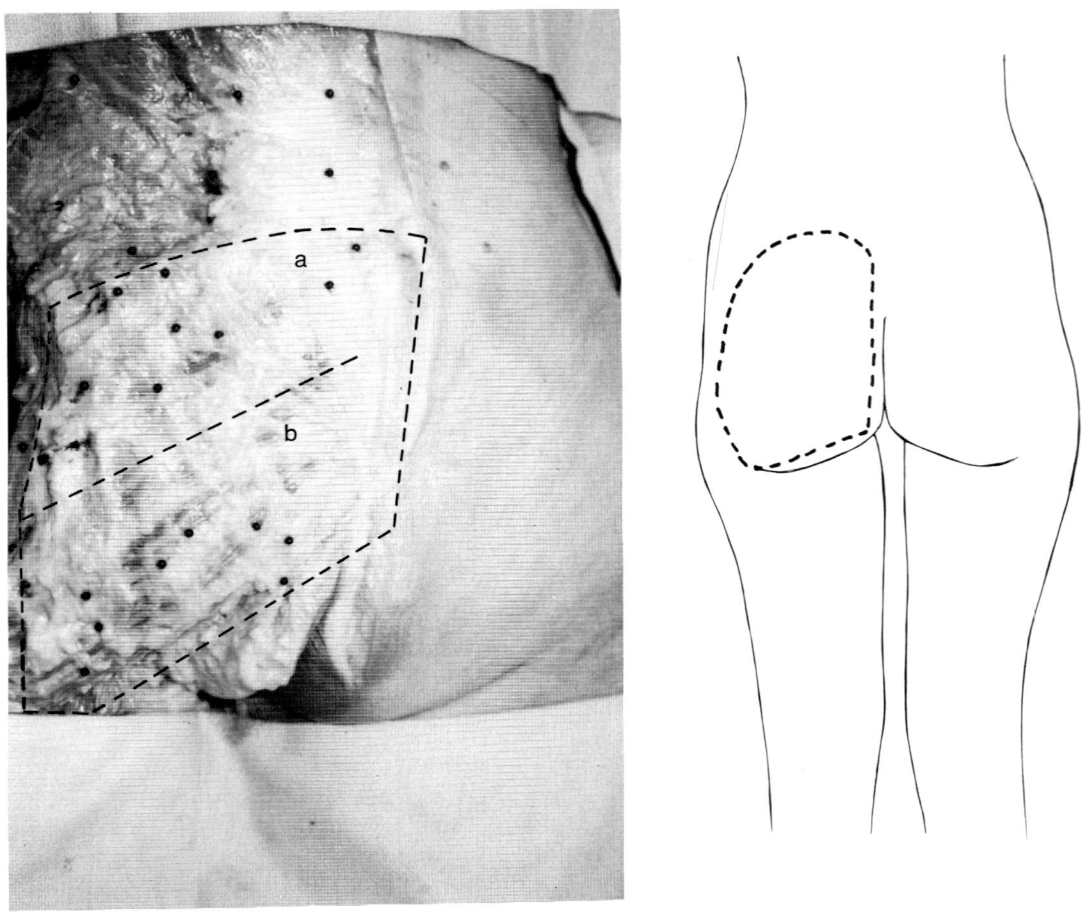 B

Fig. 3-13

A, Perforating musculocutaneous arteries through gluteus maximus into overlying skin. *a*, Superior gluteus maximus skin territory; *b*, inferior gluteus maximus skin territory.

B, Outline of gluteus maximus.

C 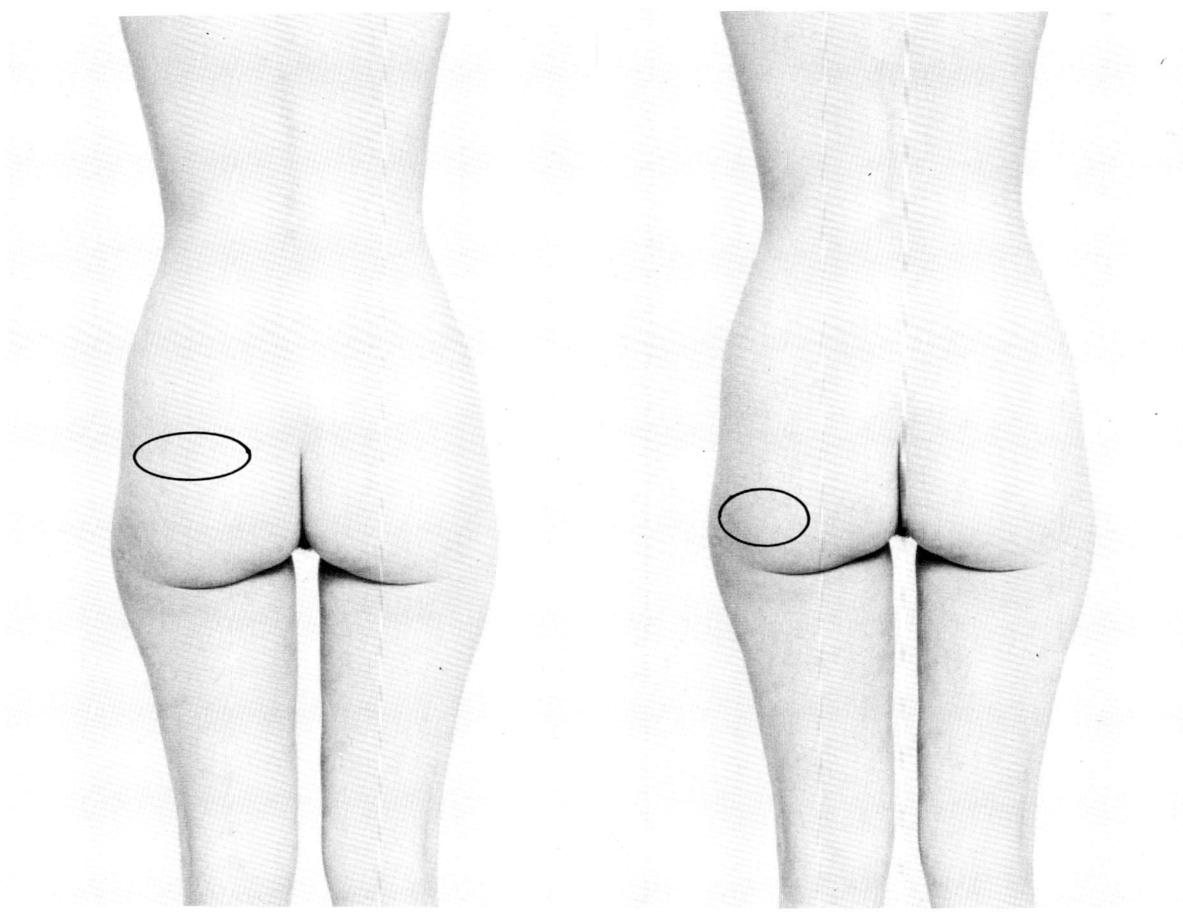 D

Fig. 3-13, cont'd
C, Design of skin island for transposition of superior gluteus maximus musculocutaneous flap for sacral coverage.
D, Design of skin island for transposition of inferior gluteus maximus musculocutaneous flap for ischial coverage.

Inferior gluteal thigh flap (Fig. 3-14)

The inferior gluteal thigh flap is designed on the posterior thigh and may measure 10 × 35 cm. It is based not on musculocutaneous perforators but on the continuation of the inferior gluteal artery. The flap may be raised with or without the inferior gluteus maximus muscle. The continuation of the inferior gluteal artery is a direct cutaneous vessel to this flap. There are some musculocutaneous perforators through the hamstrings into this skin territory, and the same flap may be elevated based on either the underlying hamstrings or the inferior gluteal artery.

The gluteal thigh flap is innervated by the posterior femoral cutaneous nerve (S1 to S2), which emerge below the gluteus maximus. This is a neurosensory flap, and the nerve is easily preserved during flap elevation.

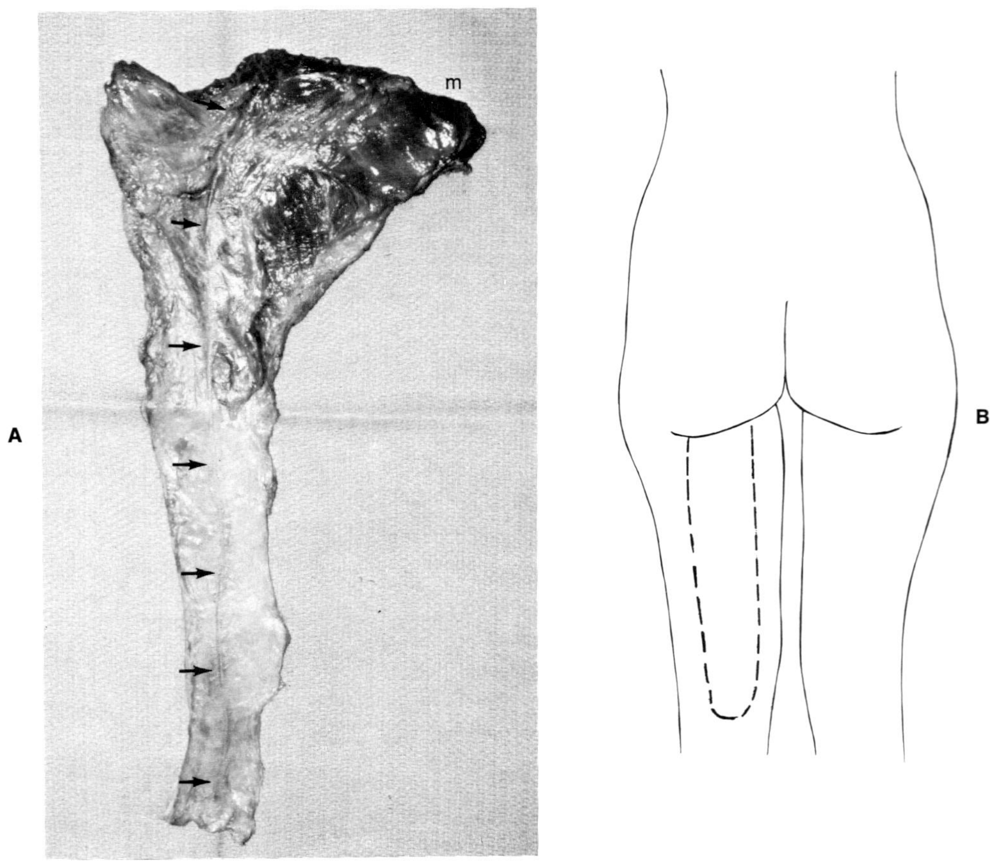

Fig. 3-14

A, Descending branch of inferior gluteal artery extends into posterior thigh skin. Arrows denote course of this axially directed branch of muscle pedicle. *m*, Inferior half of gluteus maximus muscle.

B, Outline of gluteal thigh flap.

C

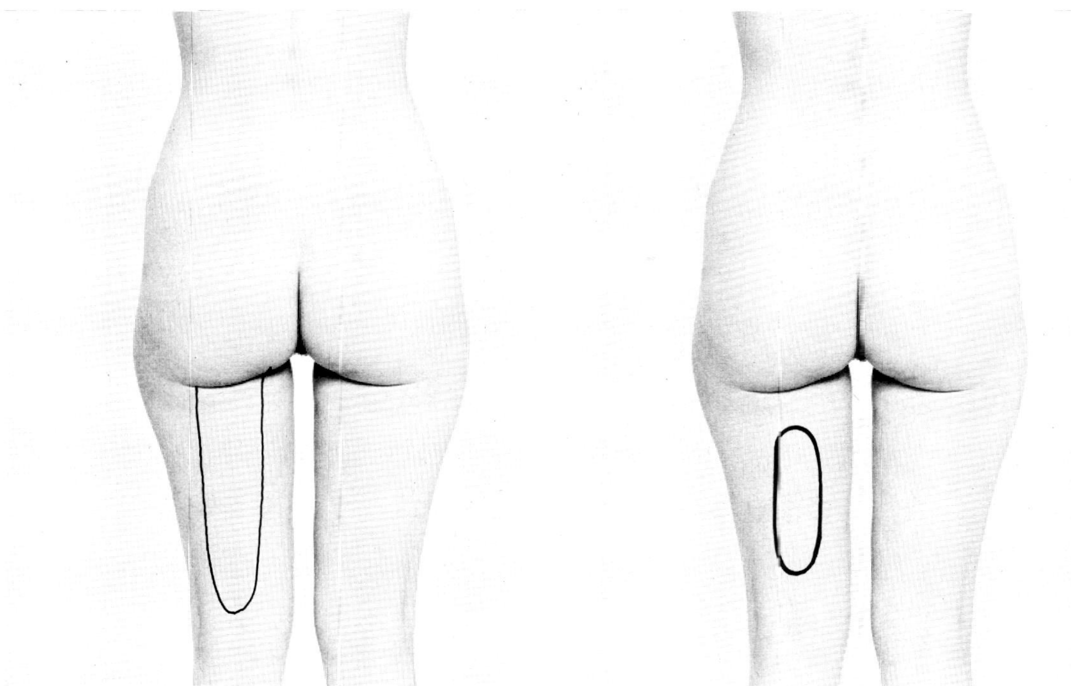

D

Fig. 3-14, cont'd
C, Design of standard gluteal thigh flap.
D, Design of skin island for gluteal thigh flap.

Transverse back flap (Fig. 3-15)

The transverse back flap is based on the row of medial musculocutaneous perforators on the contralateral side. The flap has to be based above the gluteus maximus, because the paramedian perforators are not present over the gluteus maximus (Fig. 3-16). The flap measures 8 × 15 cm and is based 2 cm laterally to the midline. It extends to the opposite anterior axillary line.

A

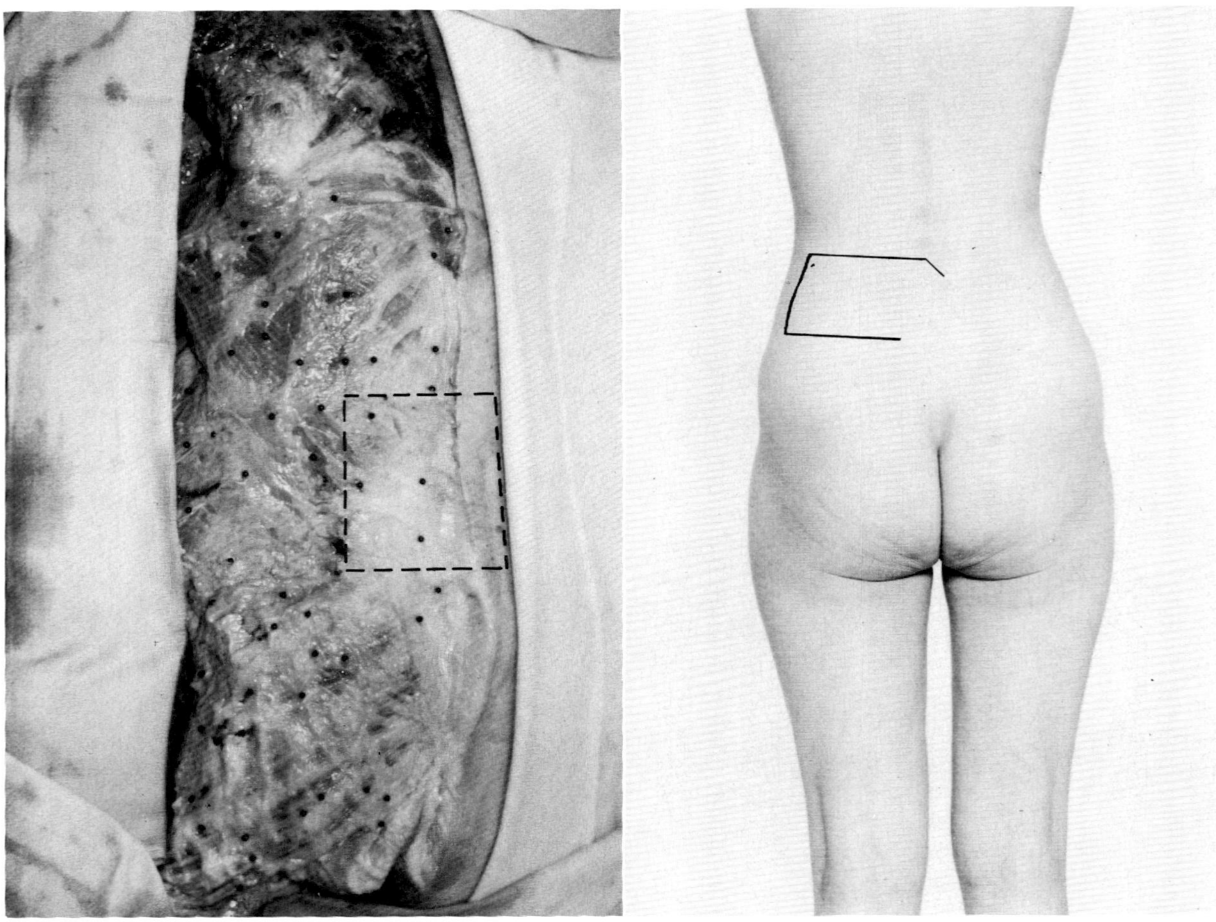

B

Fig. 3-15

A, Perforating branches of lumbar vessels through paraspinous muscles are vascular basis of transverse back flap.

B, Design of transverse back flap for sacral coverage.

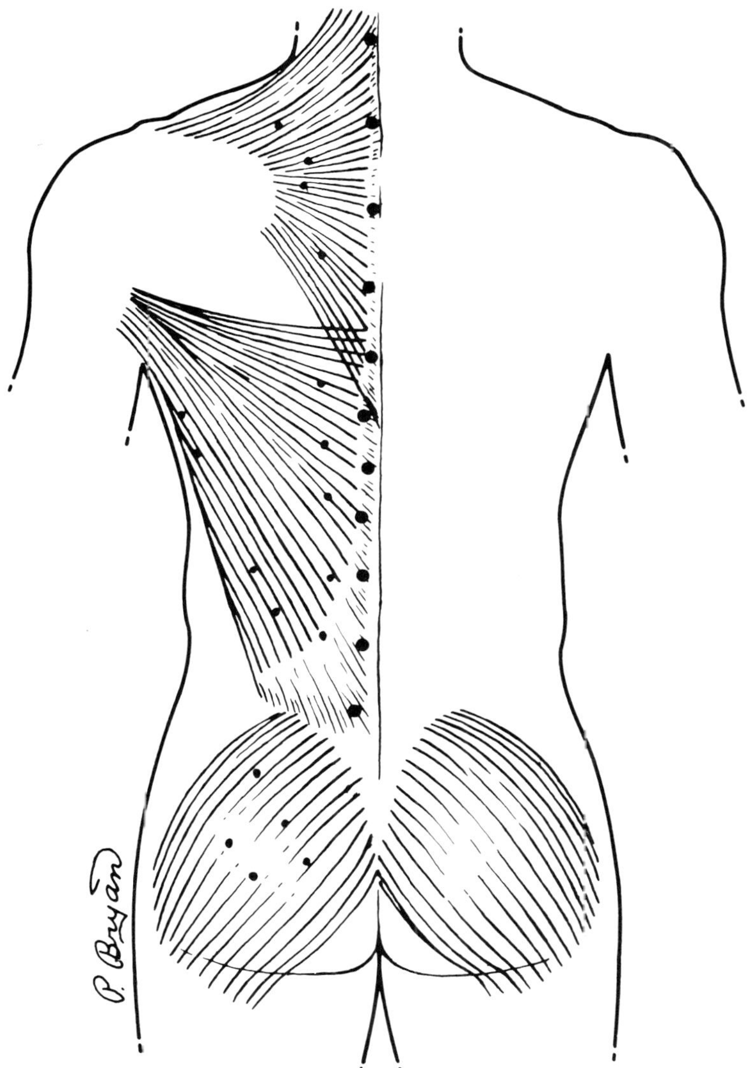

Fig. 3-16. Paramedian perforating arteries through trapezius and latissimus dorsi into skin of posterior trunk.

Sartorius (Fig. 3-17)

The total skin territory of the sartorius measures 4 × 40 cm and includes all the skin overlying the muscle. The small musculocutaneous perforators, less than 0.5 mm in diameter, are distributed throughout the muscle to the skin territory. Because this muscle has a type IV blood supply, the clinically useful island of skin that can be elevated with the muscle is limited to a 5 × 7 cm island based either near the origin of the muscle or distally near its insertion.

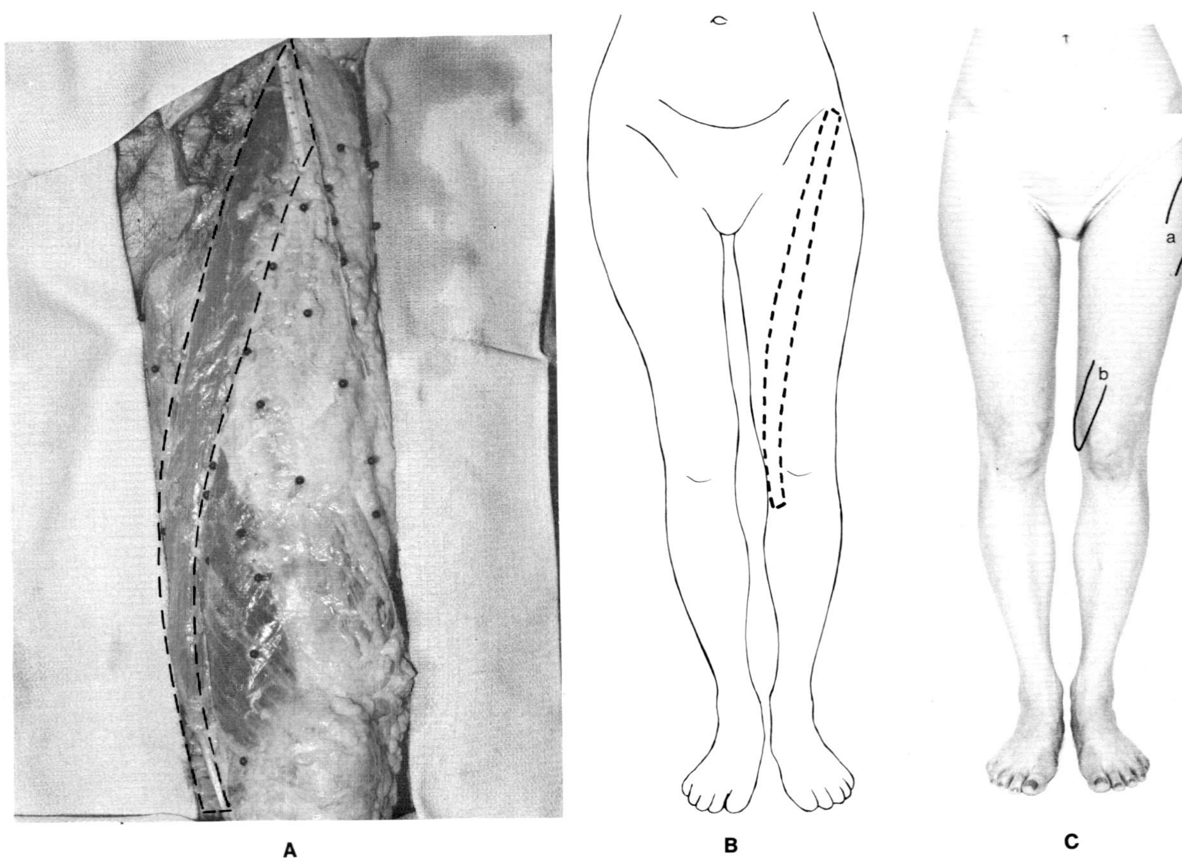

Fig. 3-17
A, Perforating musculocutaneous arteries through sartorius into overlying skin.
B, Outline of sartorius muscle.
C, Design of superior sartorius musculocutaneous transposition flap (*a*); Design of inferior-based sartorius musculocutaneous flap (*b*).

Gracilis (Fig. 3-18)

The total skin territory of the gracilis measures 6 × 24 cm and includes all the skin over the muscle. Perforating musculocutaneous arteries are small, less than 0.5 mm in diameter, and are distributed mostly over the proximal portion of the muscle. The clinically reliable island of skin based on the gracilis measures 8 × 15 cm and is usually designed over the proximal two thirds of the muscle. This muscle has a type II blood supply with the dominant pedicle located proximally. This makes the proximal island of skin more reliable than the distal skin. The distal skin is also supplied by direct cutaneous vessels.

The skin territory of the gracilis is innervated by branches of the anterior cutaneous nerve of the thigh (L2 and L3). However, this flap has little potential as a neurosensory flap, because the nerves are divided during flap elevation to permit the usual arc of rotation.

A

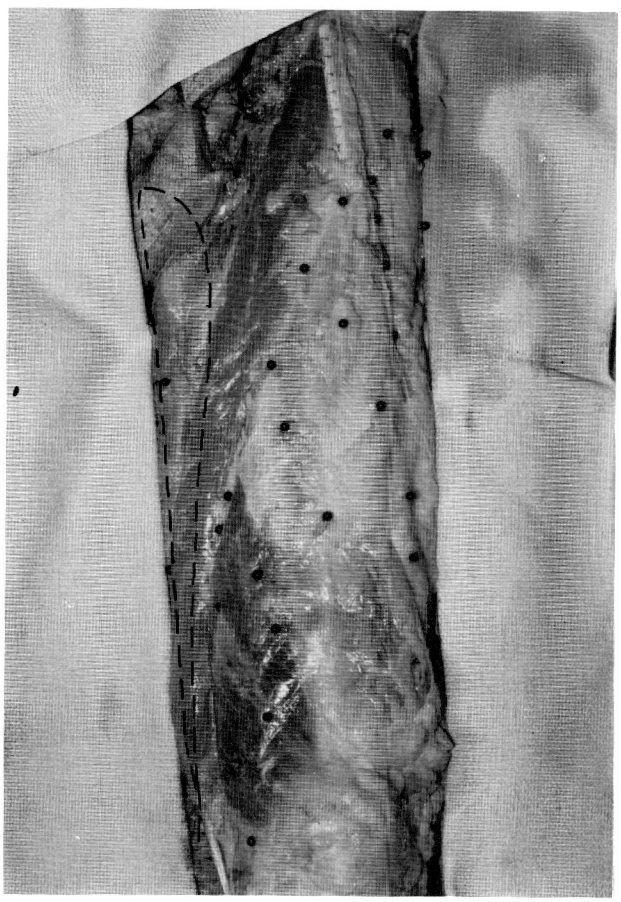

B

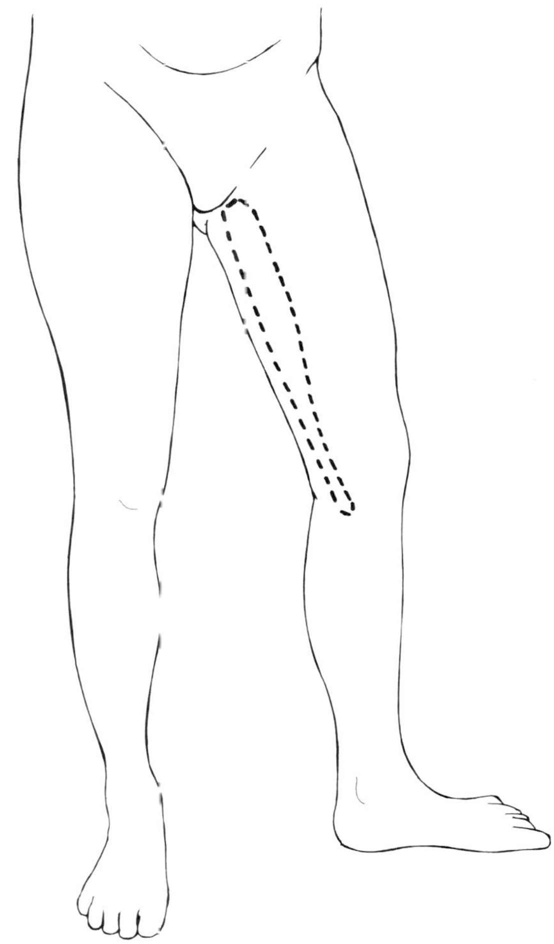

Fig. 3-18
A, Perforating musculocutaneous arteries through gracilis into overlying skin.
B, Outline of gracilis muscle.

Continued.

C

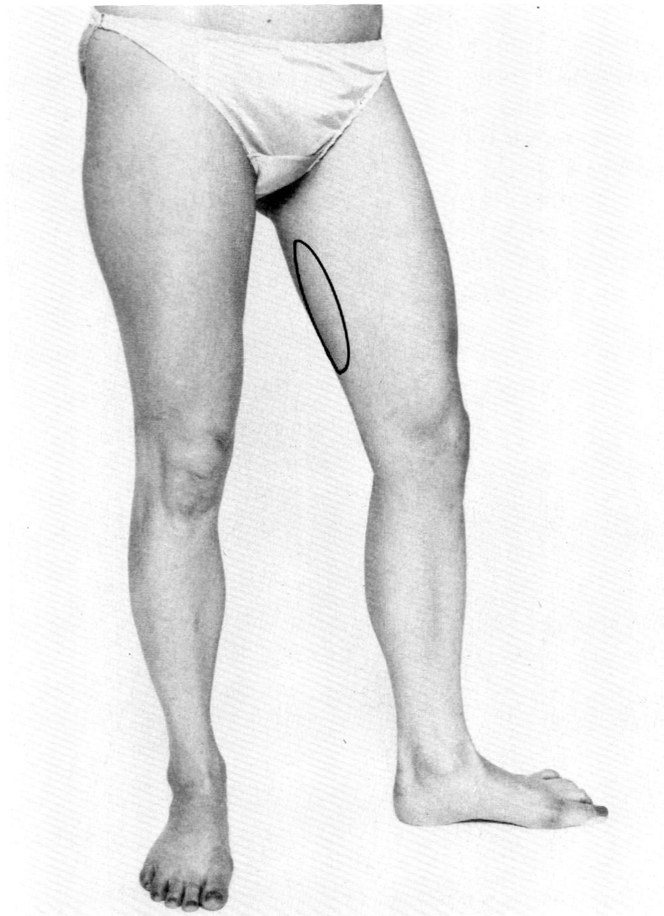

D

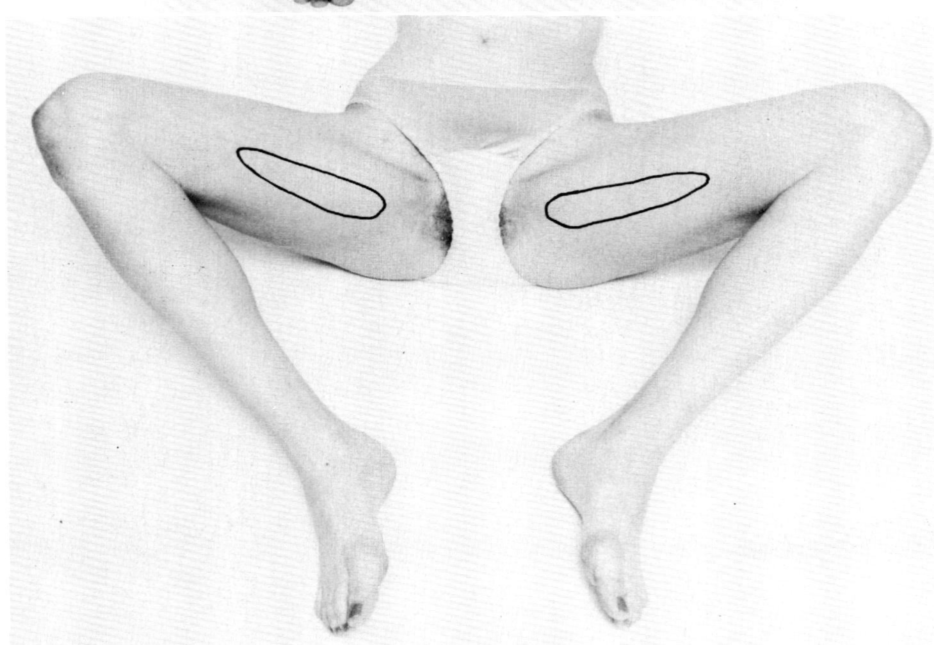

Fig. 3-18, cont'd

C, Design of skin island for gracilis musculocutaneous flap.

D, Design of skin island for gracilis musculocutaneous flap for vaginal reconstruction. NOTE: Skin island below adductor longus muscle.

Rectus femoris (Fig. 3-19)

The total skin territory of the rectus femoris measures 12 × 32 cm and includes the skin overlying the muscle and 3 to 4 cm beyond. The musculocutaneous perforators measure 0.5 to 1 mm in diameter and are distributed evenly throughout the muscle to the skin. The useful skin islands, based on the rectus femoris, may include all the overlying skin or part of it, depending on the reconstructive needs.

It should be noted that the sartorius crosses the upper part of the rectus femoris. However, the overlying skin may be safely elevated with the rectus femoris.

The rectus femoris skin territory is innervated by the femoral cutaneous nerve of the thigh (L2 and L3). This is a neurosensory flap, because the nerve can be isolated and easily elevated with the flap.

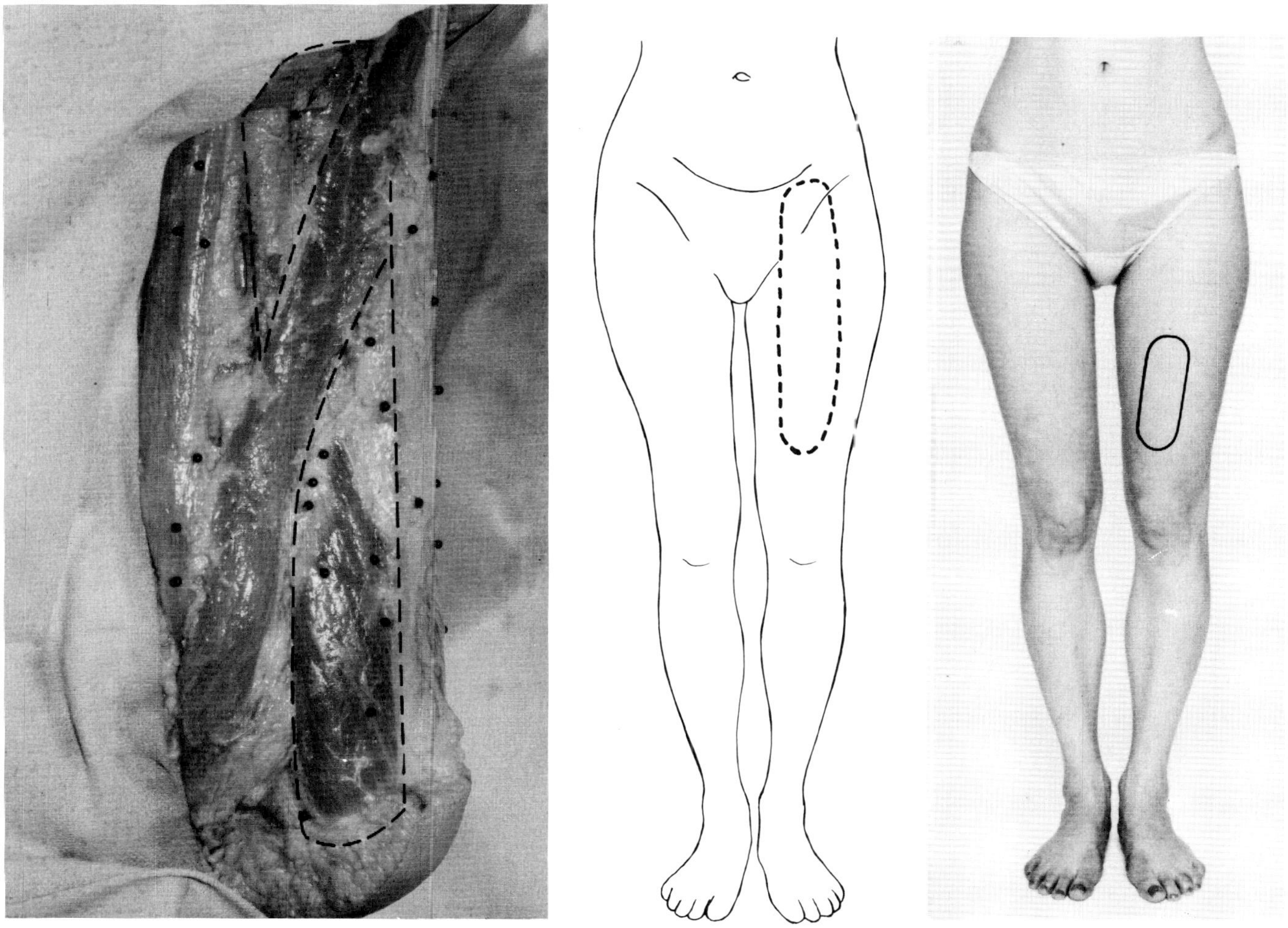

A B C

Fig. 3-19
A, Perforating musculocutaneous arteries through rectus femoris.
B, Outline of rectus femoris muscle.
C, Design of skin island for rectus femoris musculocutaneous flap

Tensor fascia lata (Fig. 3-20)

The skin territory of the TFL measures 12 × 40 cm and includes not only the skin overlying the relatively small muscle but also all of the skin of the anterolateral thigh over the fascia lata to within 6 or 8 cm of the knee.

The musculocutaneous perforators are large, 1 mm or more in diameter, and are distributed to the skin directly over the muscle. The perforators are located over the muscle proximally and support the entire skin over the fascia lata. There are small perforators from the underlying vastus lateralis through the fascia lata into the overlying skin in the distal thigh. During flap elevation the fascia lata is elevated with the overlying skin to protect the distal course of the musculocutaneous perforating vessels.

The clinically useful skin islands on the TFL can be made any length down to within 6 or 8 cm of the knee and any width up to 15 cm. The donor area can be closed directly if the width is 8 cm or less. Beyond that a skin graft may be necessary.

The skin of the TFL flap is innervated by two nerves. Superiorly below the iliac crest a small area is innervated by a lateral branch of the twelfth thoracic nerve. The rest of the skin is innervated by the lateral femoral cutaneous nerve of the thigh. Both nerves are easily dissected and elevated with the flap as a neurosensory flap.

A

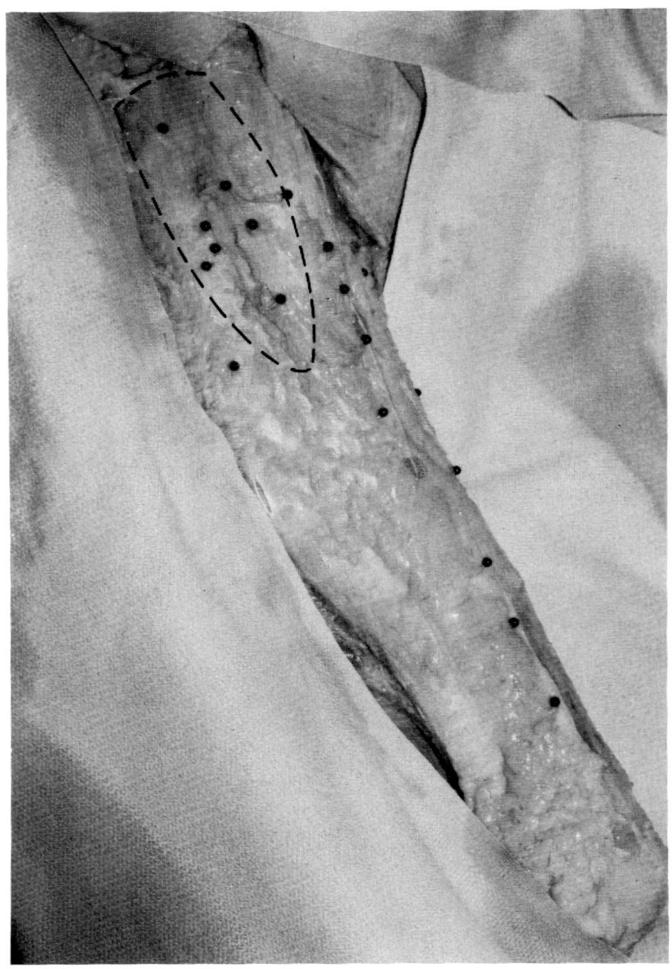

Fig. 3-20
A, Perforating musculocutaneous arteries through TFL into overlying skin.

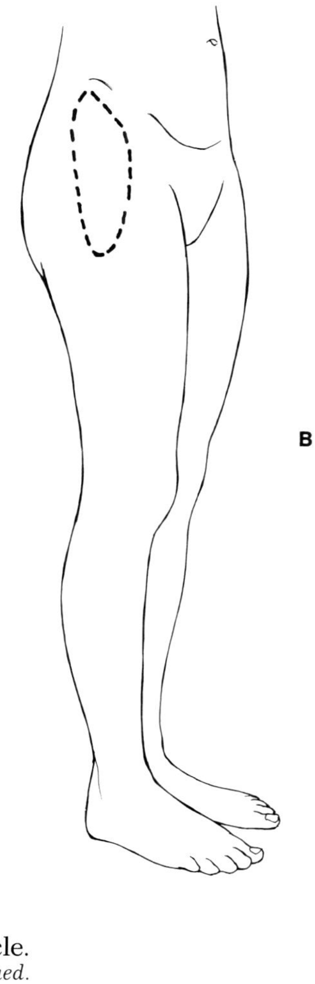

Fig. 3-20, cont'd
B, Outline of TFL muscle.
Continued.

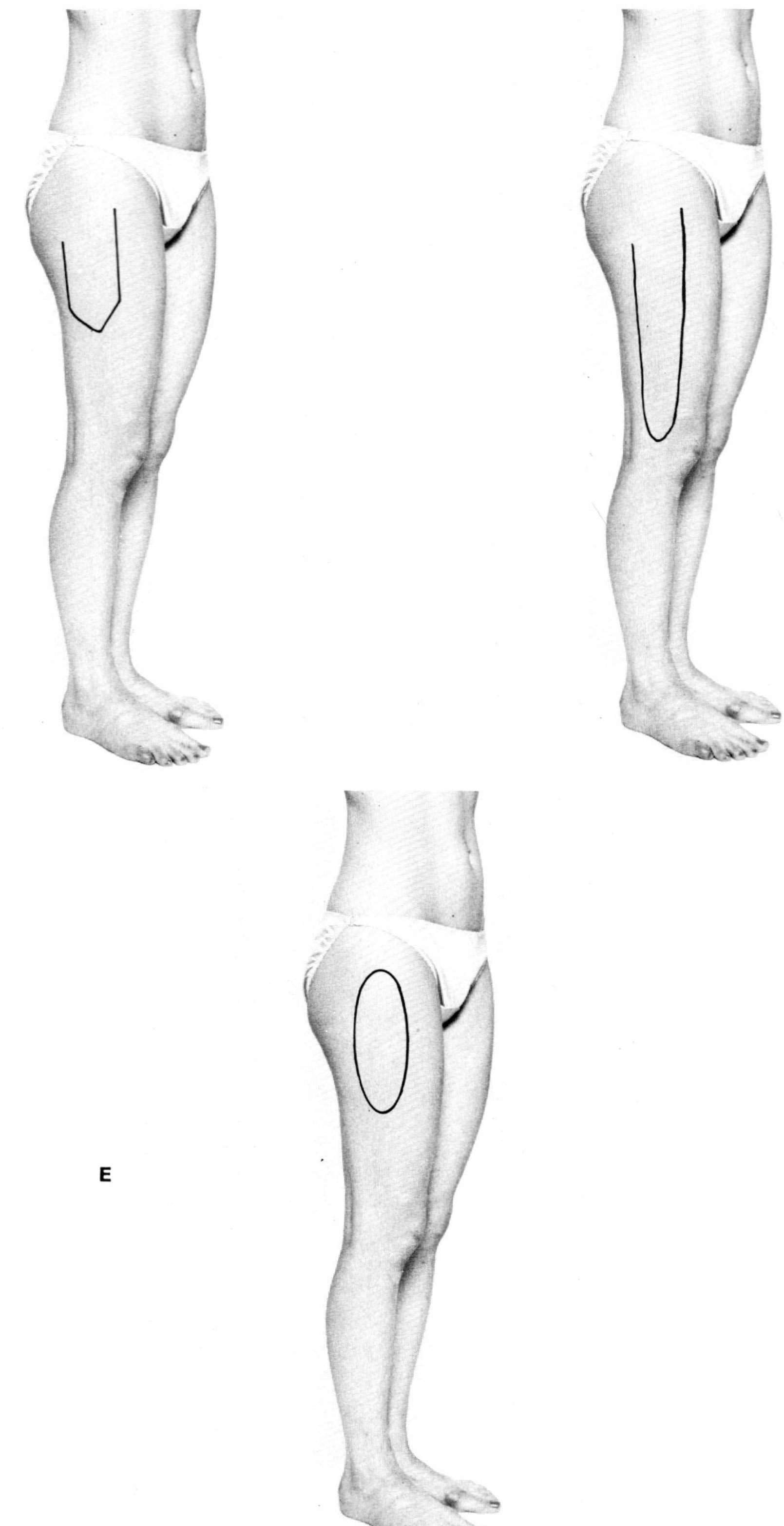

C

D

E

Fig. 3-20, cont'd
C, Design of standard TFL musculocutaneous transposition flap.
D, Design of extended TFL musculocutaneous transposition flap.
E, Design of TFL island musculocutaneous flap.

Biceps femoris (Fig. 3-21)

The skin territory of the biceps femoris measures 10 × 40 cm and includes not only the skin over the muscle but also most of the skin of the posterior thigh. The musculocutaneous perforators measure approximately 0.5 to 1 mm in diameter. These are evenly distributed over the muscle into the overlying skin. The skin territory of the biceps femoris is also supplied by a direct cutaneous branch of the inferior gluteal artery (that is, the gluteal thigh flap).

For clinical purposes the entire hamstring group with the overlying skin is elevated for closure of pressure sores. Singly as a musculocutaneous unit, the biceps must be elevated with a smaller island of overlying skin.

The skin territory of the biceps femoris is innervated by the posterior femoral cutaneous nerve of the thigh (S1 to S3). The nerve may be isolated and elevated with the flap as a neurosensory flap.

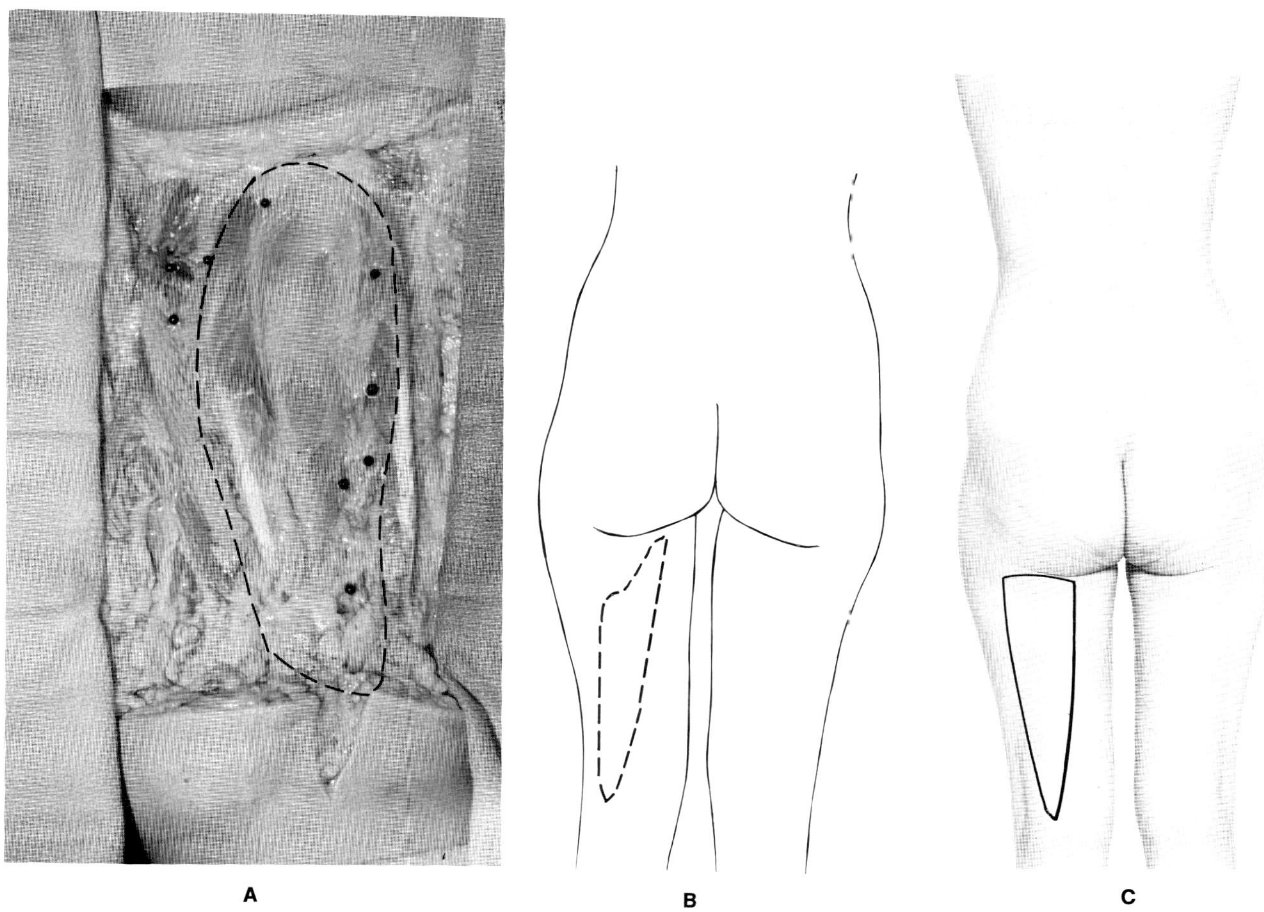

Fig. 3-21

A, Perforating musculocutaneous arteries through biceps femoris into overlying skin.

B, Outline of biceps femoris muscle.

C, Design of skin island for biceps femoris musculocutaneous advancement flap for ischial coverage.

Continued.

Gastrocnemius (Fig. 3-22)

The skin territory of each head of the gastrocnemius measures 10 × 23 cm. This includes not only the skin over the muscle but also the skin over the Achilles tendon to within 2 to 3 cm of the malleoli. The musculocutaneous arteries are rather large, 1 to 1.5 mm in diameter, and perforate through the proximal muscle into the overlying skin. There are no perforators through the Achilles tendon into the overlying skin.

The distal skin territory over the Achilles is vascularized through the proximal perforators. The distal skin can therefore be elevated up to the lower third of the muscle and transposed as a skin fascial flap without elevating the Achilles tendon or the underlying muscle.

Clinically useful islands of skin may be based either on the muscular part of the gastrocnemius or extended distally toward the malleoli. However, the gastrocnemius musculocutaneous flap is rarely recommended because of the significant donor deformity.

A

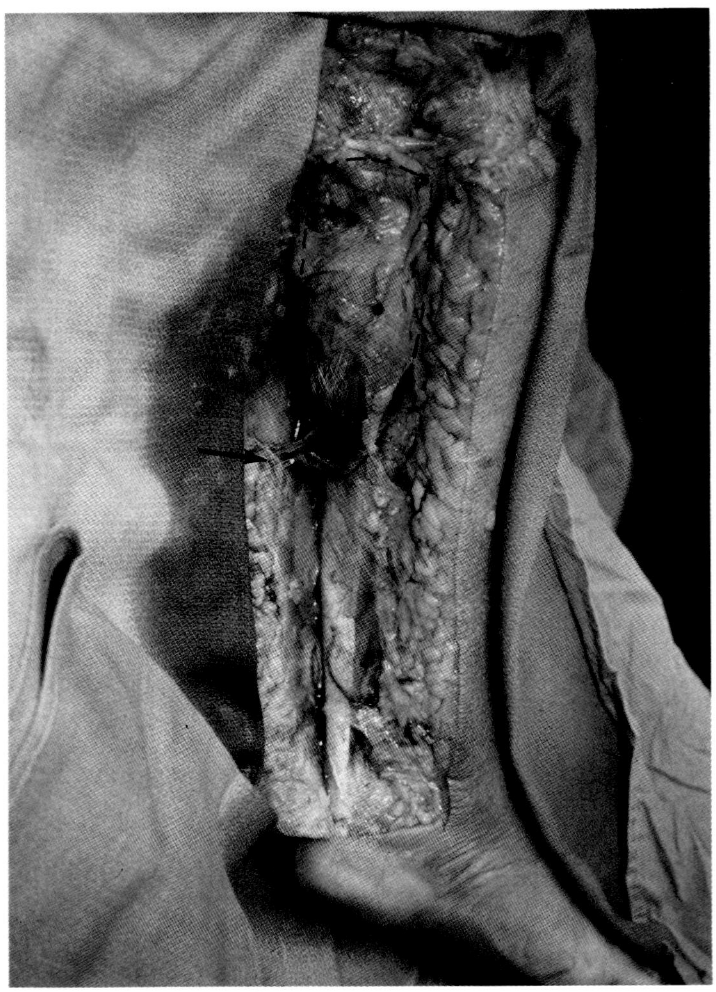

Fig. 3-22
A, Perforating musculocutaneous arteries through medial gastrocnemius into overlying skin *(arrow)*.

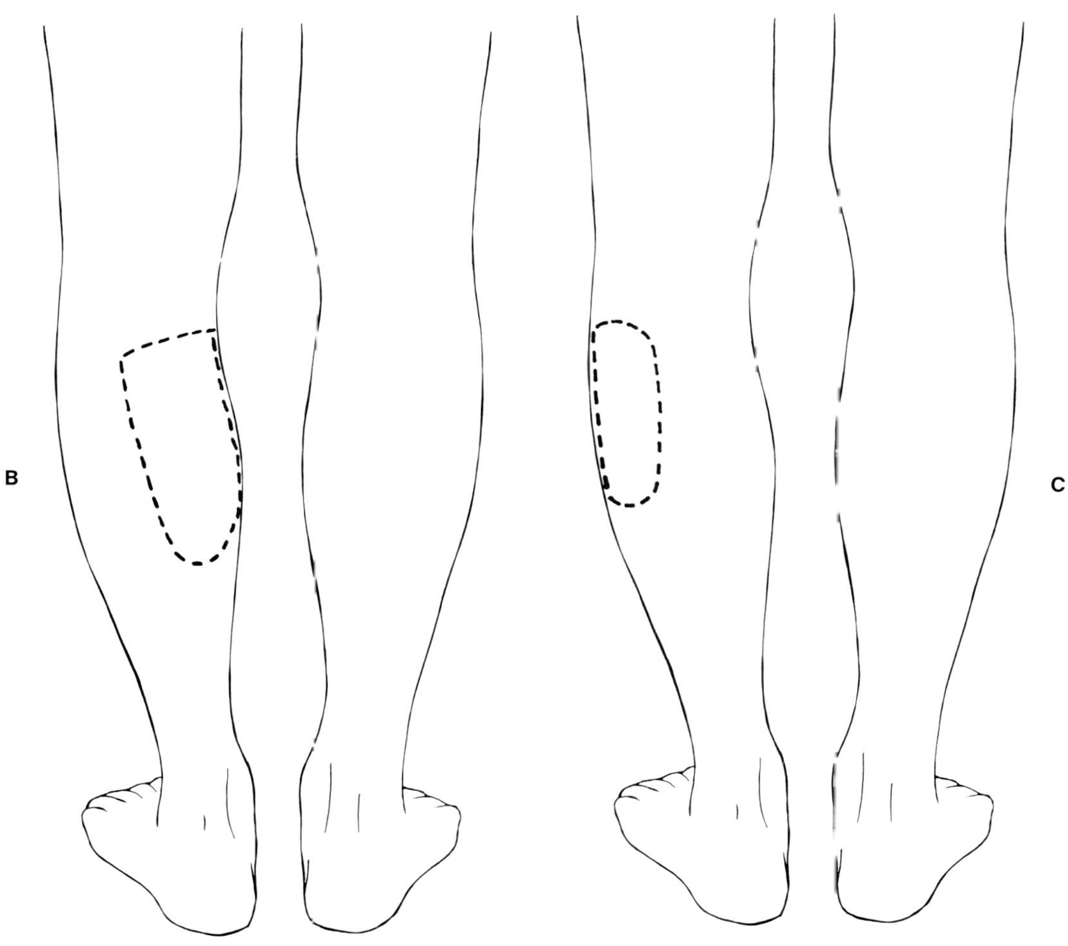

Fig. 3-22, cont'd
B, Outline of medial gastrocnemius muscle.
C, Outline of lateral gastrocnemius muscle.

Continued.

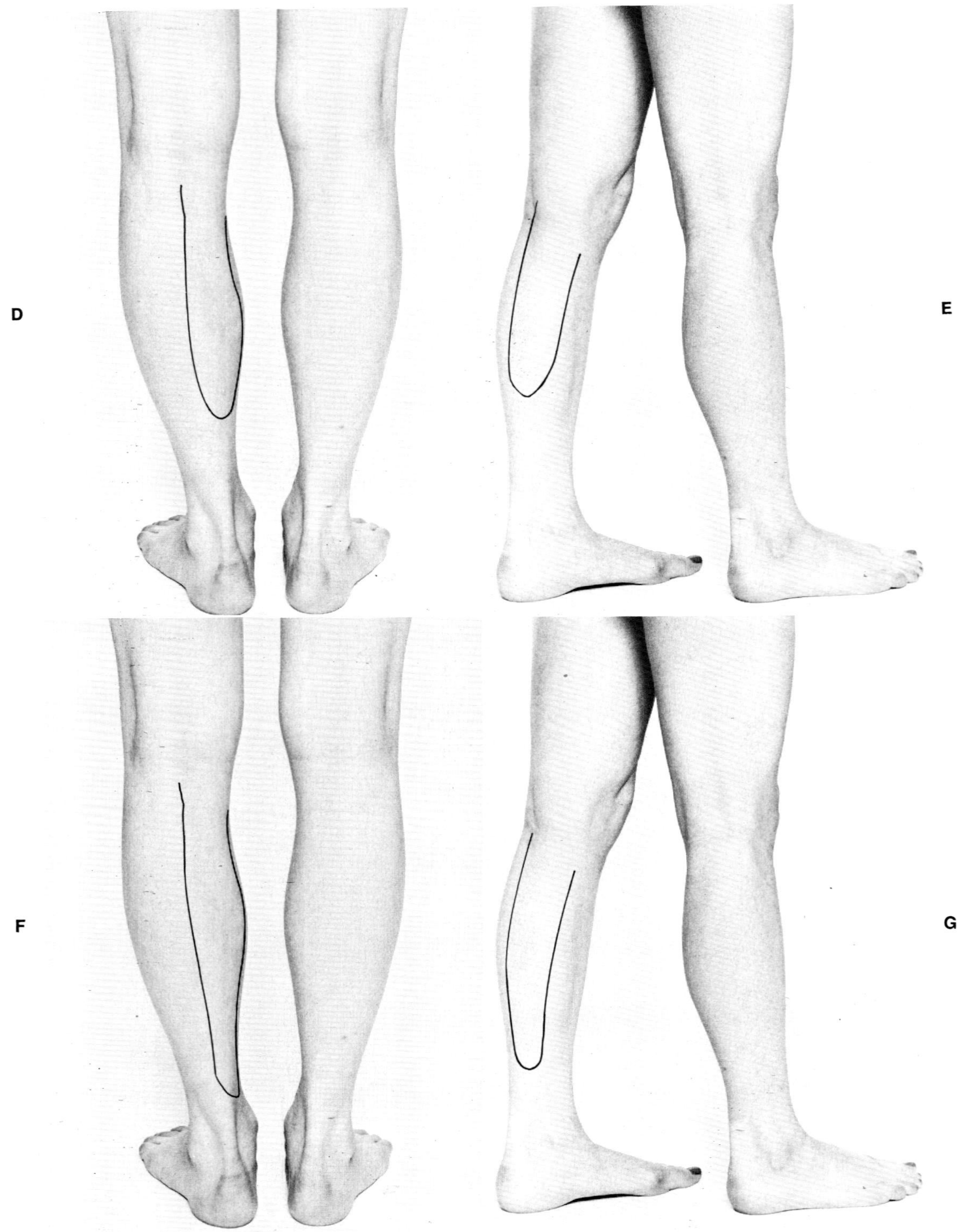

Fig. 3-22, cont'd
D and **E,** Design of standard medial gastrocnemius musculocutaneous flap.
F and **G,** Design of extended medial gastrocnemius musculocutaneous flap.

H

I

J

K

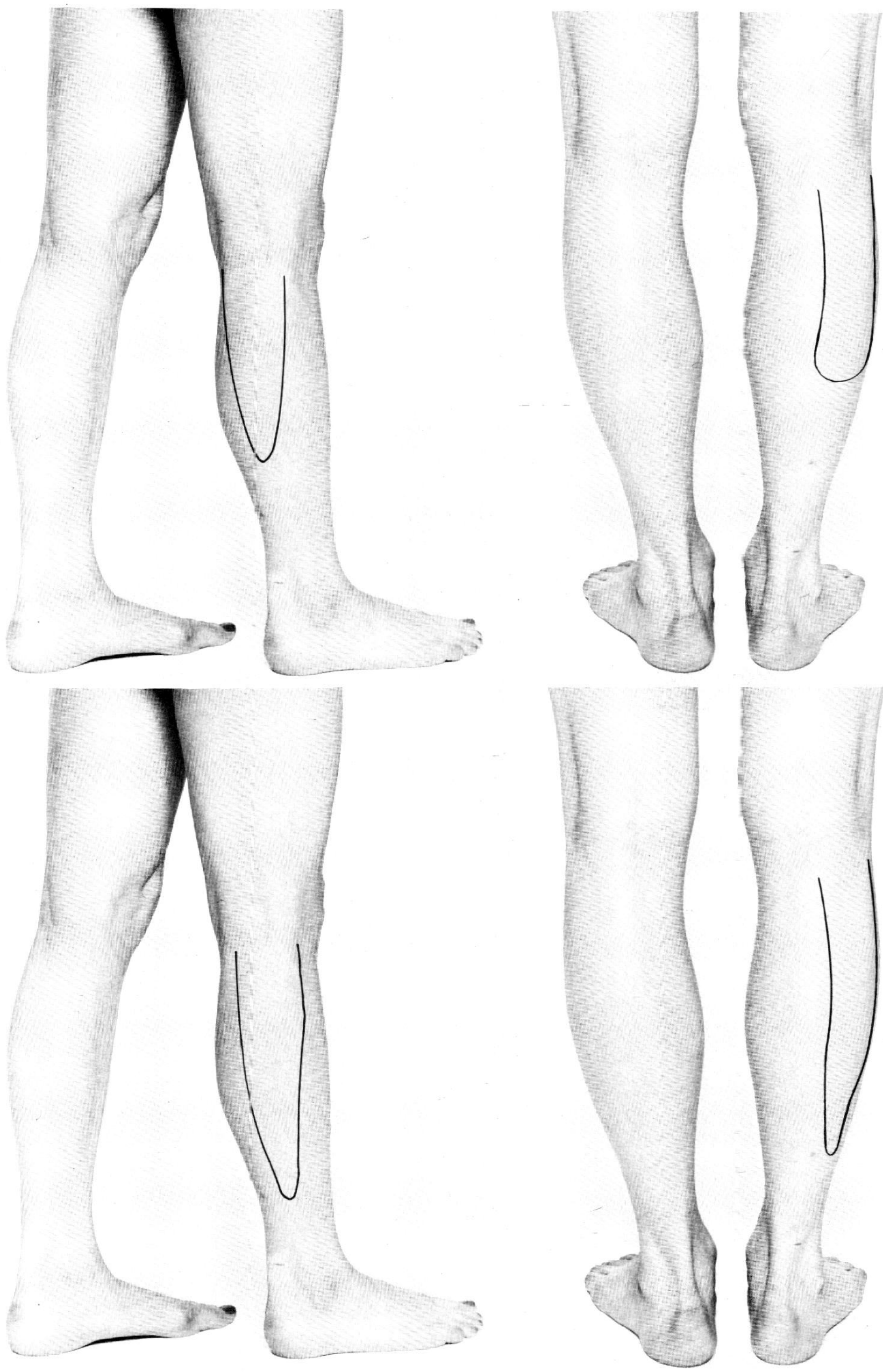

Fig. 3-22, cont'd
H and **I,** Design of standard lateral gastrocnemius musculocutaneous flap.
J and **K,** Design of extended lateral gastrocnemius musculocutaneous flap.

Tibialis anterior (Fig. 3-23)

The skin territory of the tibialis anterior is approximately 4 × 12 cm and includes its overlying skin. The musculocutaneous perforators are small, 0.5 mm in diameter, and are distributed evenly over the muscle into the overlying skin. The skin of the anterolateral leg is also supplied through small perforators from the peroneal muscles.

The tibialis anterior is not recommended as a musculocutaneous flap, because this would leave an unacceptable donor deformity.

A

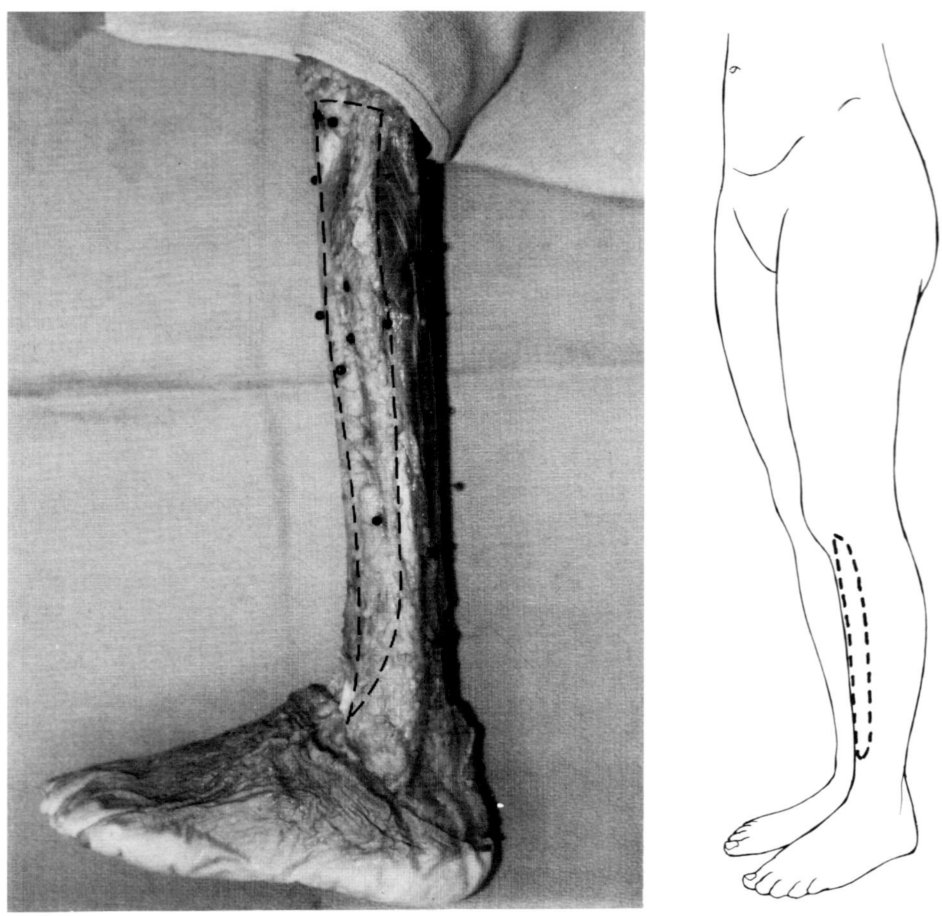

B

Fig. 3-23
A, Perforating musculocutaneous arteries through tibialis anterior into overlying skin.
B, Outline of tibialis anterior muscle.

Flexor digitorum brevis (Fig. 3-24)

The skin overlying the flexor digitorum brevis is innervated by the medial and lateral plantar nerves (L4,5 and S1,2). Branches of these nerves may be dissected and elevated with the flap as a neurosensory flap.

A

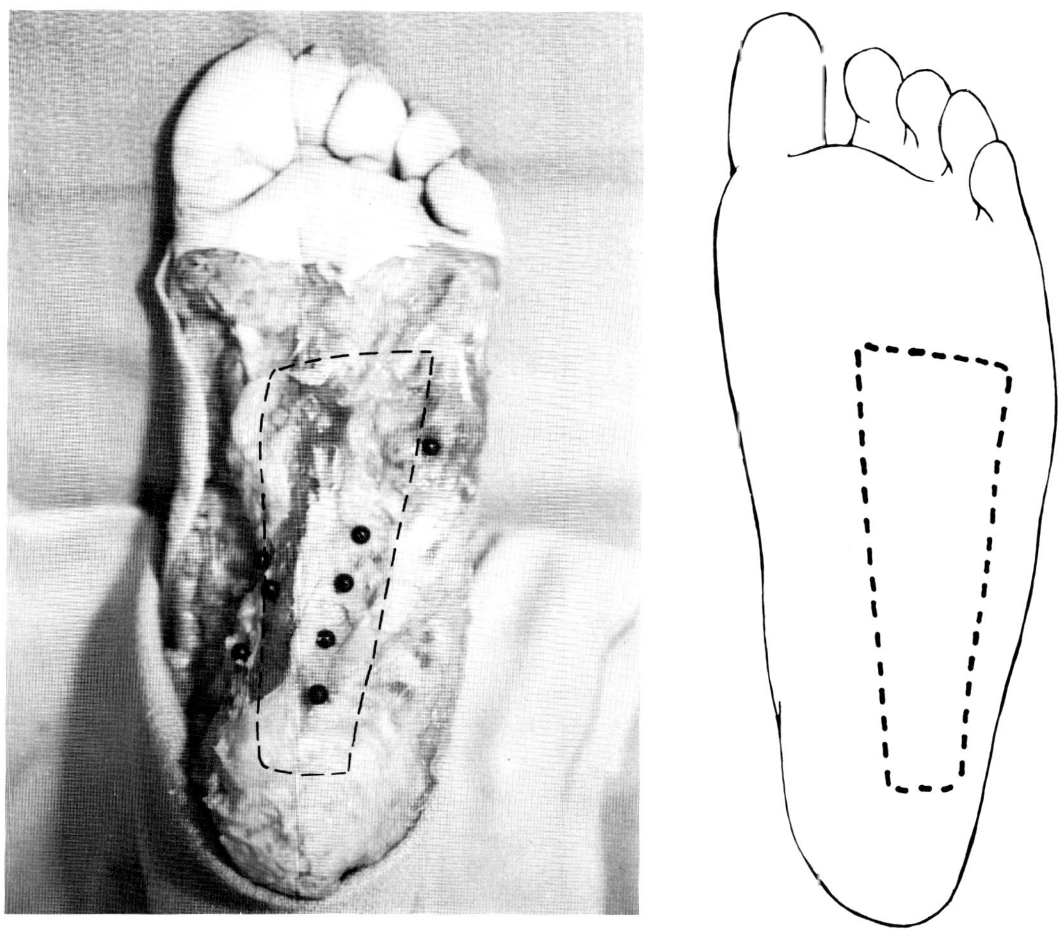

B

Fig. 3-24
A, Perforating musculocutaneous arteries through flexor digitorum brevis into overlying skin.
B, Outline of flexor digitorum brevis muscle.

Abductor digiti minimi (Fig. 3-25)

The skin of the sole of the foot is supplied by direct cutaneous vessels and perforators from the abductor digiti minimi muscle. The perforating vessels are small, 0.2 mm or less in diameter, and are evenly distributed to the overlying skin.

Clinically a small skin island may be based on this muscle.

The skin overlying the abductor digiti minimi is innervated by the sural nerve (S1,2), which may be elevated with the flap as a neurosensory flap.

Abductor hallucis (Fig. 3-26)

The abductor hallucis supplies a small area of skin along the medial border of the foot. The perforating vessels are small, measuring 0.5 mm or less in diameter.

Clinically a small island of skin may be elevated with the muscle.

The skin overlying the abductor hallucis is innervated by branches of the medial plantar nerve (L4,5), which can be isolated and elevated with the flap as a neurosensory flap.

A

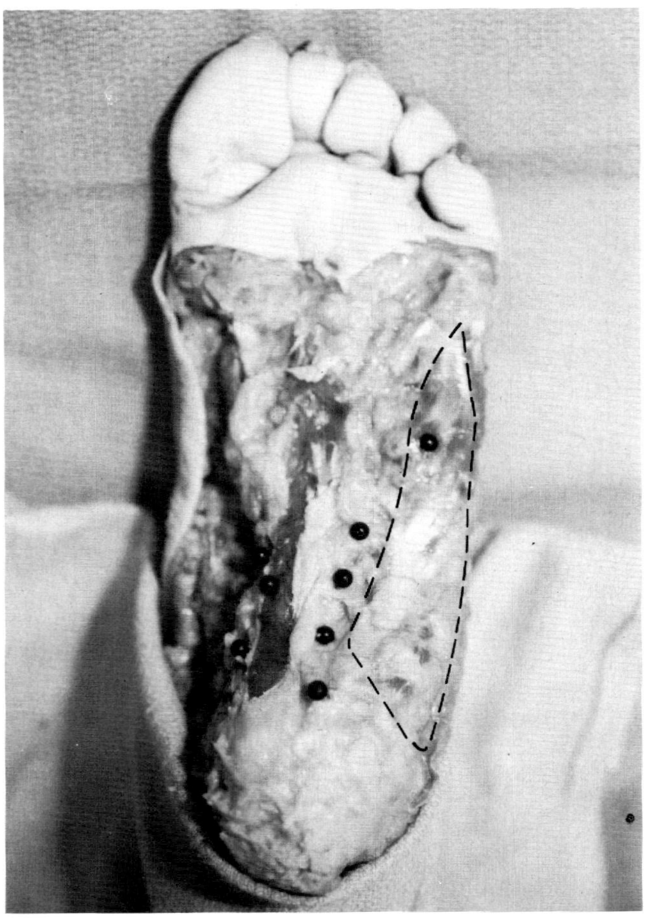

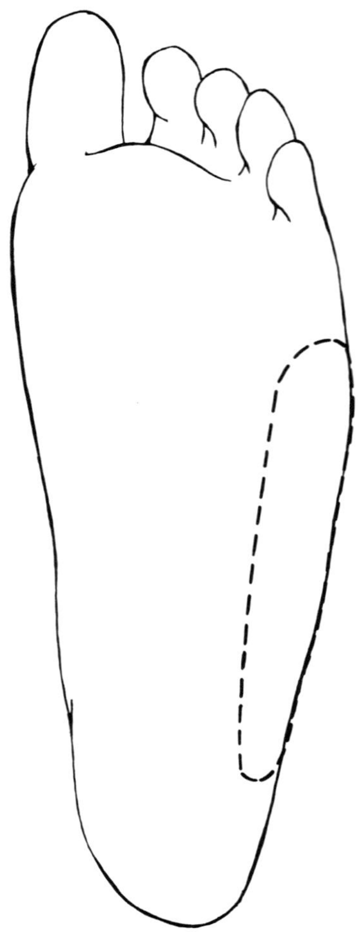

B

Fig. 3-25

A, Perforating musculocutaneous arteries through abductor digiti minimi into overlying skin.

B, Outline of abductor digiti minimi muscle.

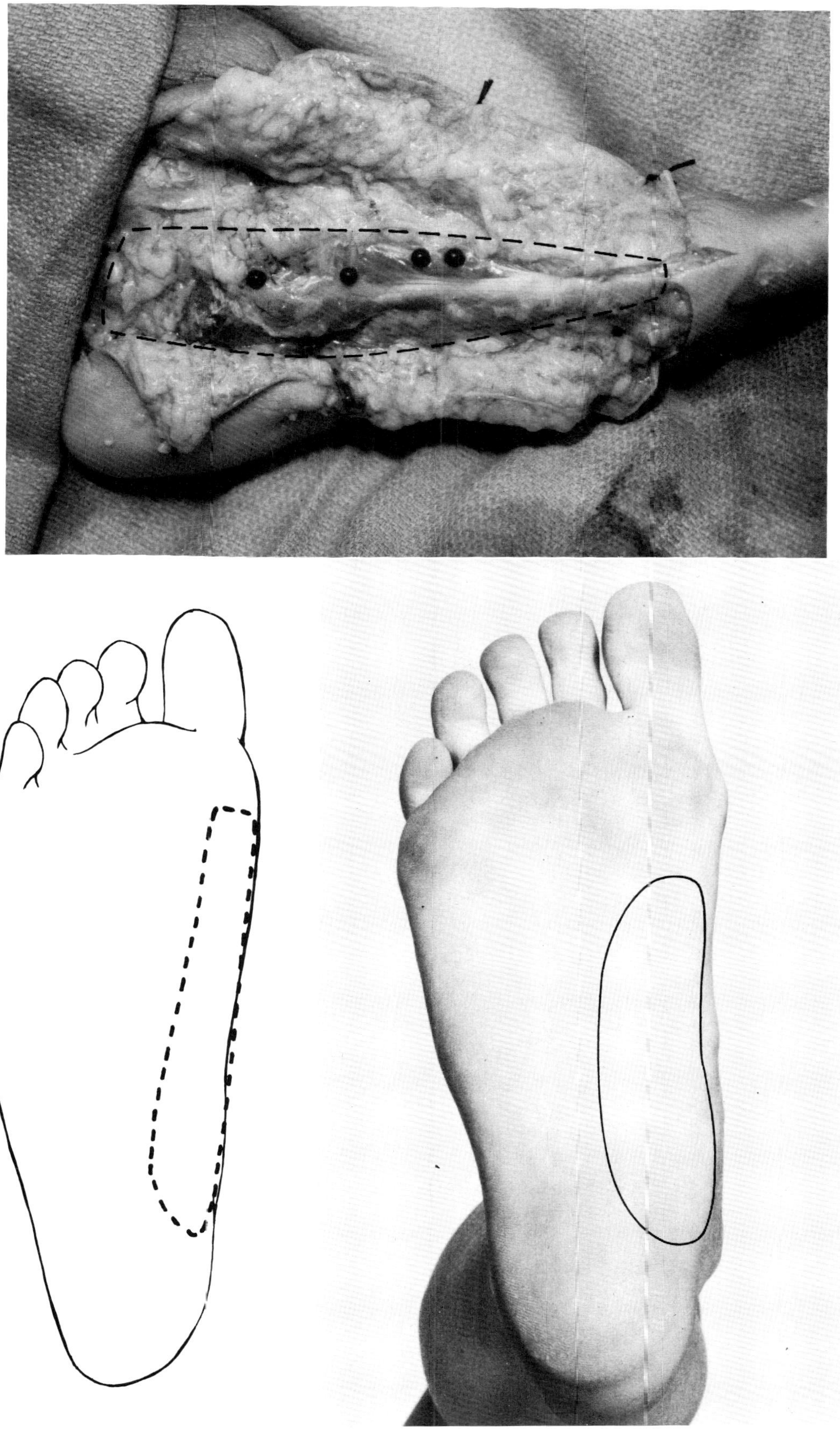

A

B

C

Fig. 3-26

A, Perforating musculocutaneous arteries through abductor hallucis into overlying skin.

B, Outline of abductor hallucis muscle.

C, Design of skin island for abductor hallucis musculocutanecus transposition or distal advancement flap.

Conclusion

The knowledge of the size, number, and location of musculocutaneous perforating arteries has clinical significance in that flaps may be elevated with greater reliability and flap design may be altered to meet various reconstructive needs. Certain skin flaps based on musculocutaneous perforating arteries may be elevated without elevating the underlying muscle (e.g., thoracoepigastric flap). The underlying muscle and overlying skin may be separated so that the muscle can be placed into an infected cavity and the skin used for coverage. The skin island or the entire musculocutaneous unit may be split so that one unit will cover two adjoining defects. Anatomic variations and previous ligation or damage to the vascular pedicle of the underlying muscle may alter the pattern of blood supply to the overlying skin. This should always be taken into account before flap selection and design.

MUSCLE AND SKIN GRAFT OR MUSCULOCUTANEOUS FLAP

Superficial muscles may be elevated as a muscle flap and skin grafted or elevated with the overlying skin as a musculocutaneous flap. The reconstructive surgeon therefore has a choice. One method may be more suitable than the other for a particular defect. A selection is therefore made on the basis of the reconstructive needs of the defect and aesthetic and functional considerations of each method (Table 2). The goal is to restore form and function with minimal donor deformity.

Durability

Muscle makes an excellent recipient site for skin graft. This well-vascularized bed accounts for the durability of the grafted skin. However, the skin associated with the musculocutaneous flap retains functional adnexal glands and hair growth. When the musculocutaneous flap is inset into a defect where its thickness does not result in excessive projection, the durability of this skin is excellent.

Aesthetic considerations

The flap. The skin island of a musculocutaneous flap may be preferable to skin graft over muscle. However, if the subcutaneous tissue is bulky, then skin-grafted muscle is preferable to a bulky flap. In general, the muscle flap is only preferred for shallow defects, and the musculocutaneous flap is preferred for deeper defects.

The donor defect. The donor defect of a muscle flap is always closed directly. However, the donor site of musculocutaneous flaps often requires skin grafting. The donor defect of the gastrocnemius musculocutaneous flap is unacceptable to many patients. It is preferable to transfer the muscle only, and cover it with a skin graft.

Flap atrophy. Denervated and tenotomized muscle will atrophy. However, the degree of atrophy cannot be predicted and depends somewhat on tension within the muscle fibers. However, muscle will atrophy, but subcutaneous fat will not.

Functional considerations

Pliability. Muscle alone without the overlying skin is more pliable and can be transposed as a turnover flap or placed into bone to obliterate osteomyelitis cavities. The epimysium may be incised and the muscle expanded. Skin cannot be stretched. Under certain circumstances a combination flap may be ideal. The overlying island of skin on a musculocu-

TABLE 2 Aesthetic and functional considerations of muscle flap and skin graft and musculocutaneous flap

Considerations	Type	
	Muscle flap and skin graft	Musculocutaneous flap
Aesthetic		
Flap	Good to fair	Excellent to fair
Donor defect	Excellent	Excellent to poor
Flap atrophy	Muscle will atrophy	Subcutaneous fat will not atrophy
Functional		
Sensibility	No	Yes
Pliability	Fair	Excellent
Size	Limited to size of muscle	May be extended beyond muscle

taneous flap may be separated in part from the underlying muscle. The muscle is transposed into a cavity, and the skin island provides coverage over it. A precise knowledge of the location of the perforators is required to avoid interruption of skin circulation with this maneuver.

Sensibility. Certain musculocutaneous flaps may be transposed with intact cutaneous sensory nerves as sensory flaps. Sensibility in these flaps is excellent. Otherwise sensibility in musculocutaneous flaps and muscle flaps with skin graft is similar in that some pressure stimuli may be transmitted through the underlying muscle if it remains innervated.

Size. The dimensions of a muscle flap are limited to the size of the muscle. However, muscle will contract when origin or insertion is cut for elevation as a flap. For maximum coverage, therefore, the muscle has to be stretched to its original size then sutured into the defect. This stretching of the muscle is facilitated by incising the epimysium. In fact these incisions allow expansion of the muscle.

The skin elevated with muscle as a musculocutaneous flap may be considerably greater in dimensions than the underlying muscle (e.g., TFL and gastrocnemius). Skin islands based on the periphery of the muscle with extension beyond the muscle may be elevated for a greater arch of rotation. However, once a musculocutaneous flap is elevated, the dimensions of the skin remain constant and cannot be extended by stretching or any other means.

ANNOTATED BIBLIOGRAPHY

1. Manchot, C.: Die hautarterien des menshlichen körpes, Leipzig, 1889, F.C.W. Vogel.

   A classic treatise on the blood supply to the skin is presented. The diagrammatic division of the skin into territories supplied by specific vessels, some of which are clearly musculocutaneous perforators, resembles the skin territory of several of the musculocutaneous flaps.

2. Mathes, S.J., and Nahai, F.: Clinical atlas of muscle and musculocutaneous flaps, St. Louis, 1979, The C.V. Mosby Co.

   A description of anatomy, vascular supply, arc of rotation, and clinical application of muscle and musculocutaneous flaps is presented. The skin territory of the musculocutaneous flaps is outlined, and the musculocutaneous perforating arteries of several flaps are demonstrated.

3. McCraw, J.B., Dibbell, D.G., and Carraway, J.H.: Clinical definition of independent myocutaneous vascular territories, Plast. Reconstr. Surg. **60:**341, 1977.

   A number of musculocutaneous flaps are described for the first time. This paper inludes a description of 13 musculocutaneous flaps. The size of each unit and its vascular basis are described.

4. Orticochea, M.: The musculocutaneous flap method: an immediate and heroic substitute for the method of delay, Br. J. Plast. Surg. **25:**106, 1972.

   The first description of the gracilis musculocutaneous flap is presented. This was the first publication of the "musculocutaneous flap era," recognizing the dependence of skin blood supply on the underlying muscle.

4

EXPERIMENTAL CONSIDERATIONS

Foad Nahai ▪ Stephen J. Mathes

The rapid advancement of the application of muscle and musculocutaneous flaps for reconstruction of defects in all regions has been based mainly on cadaver dissections and clinical experience. Experimental work, however, has also played a role in this development.

Since 1972 the dog has been used as an experimental model in the laboratories at Emory University, Atlanta, Washington University, St. Louis, and the University of California, San Francisco, to study muscle and musculocutaneous flaps. Initially the work was directed toward dissection of muscle and musculocutaneous flaps in dogs to define the vascular anatomy and potential for transposition as muscle flap or transfer as a free flap. These early experiments showed that the gracilis, rectus abdominis, and biceps femoris muscles of the dog can be transposed locally as muscle or musculocutaneous flaps or transplanted as free flaps by microsurgical technique (Fig. 4-1). It was also shown that the sartorius with a type IV pattern of blood supply will not survive as a local transposition flap when more than two of the segmental pedicles are divided (Fig. 4-2).

ANATOMIC STUDIES

Barium-latex-injected dog cadavers were dissected in a search for a musculocutaneous flap model with a pattern of blood supply similar to humans. The gracilis was chosen as a suitable model and has been our experimental flap to evaluate the musculocutaneous flap.

The dog gracilis is located on the medial thigh and measures approximately 5 × 8 cm. It has a muscular origin and a tendinous insertion. Although relatively shorter than the gracilis in humans, it resembles it in form. The pattern of blood supply of the dog gracilis is type II with a dominant proximal pedicle 1 mm in diameter and single distant minor pedicle 0.5 mm in diameter. Both are branches of the saphenous artery of the dog, which is equivalent to the superficial femoral artery in man (Fig. 4-3). The venous drainage accompanies the arteries, and the muscle is innervated by the obturator nerve (Fig. 4-4).

STUDIES OF INTRAMUSCULAR VESSELS

Dog gracilis muscles injected with a barium-latex mixture were dissected under the microscope. The intramuscular course of the major and minor vascular pedicles was dissected (Figs. 4-5 and 4-6). The vessels were found to run parallel to the muscle fibers and, in between the muscle fibers, send small perforating vessels to supply the overlying skin. These dissections and the radiographs demonstrated that the dominant pedicle was distributed to the upper two thirds of the muscle and the minor pedicle to the lower one third. Furthermore, anatomic continuity was demonstrated between the two systems. These findings were confirmed through the injection of red latex into the dominant pedicle and blue latex into the minor pedicle. Again, it was shown that the red latex through the dominant pedicle was distributed to the upper two thirds of the muscle and the

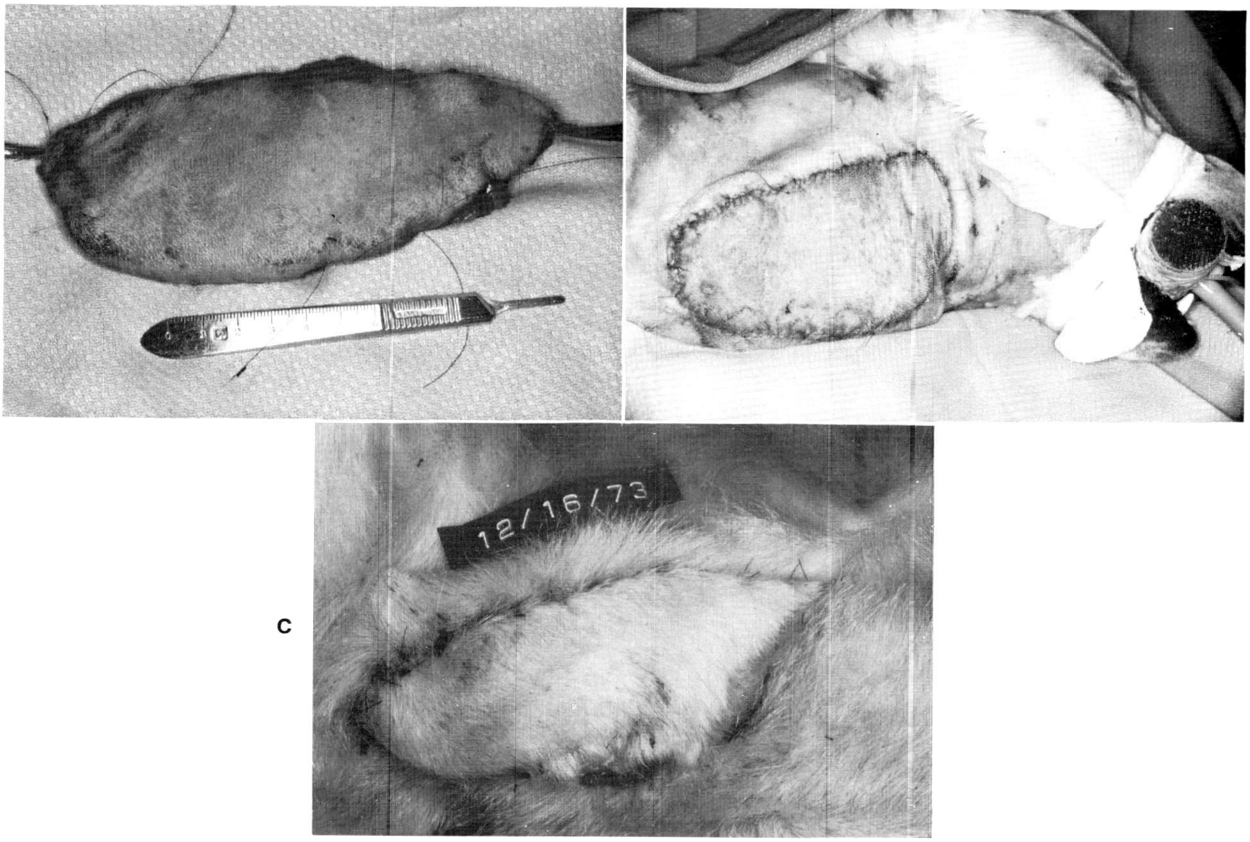

Fig. 4-1. Transplantation of biceps femoris musculocutaneous flap by microvascular techniques.

A, Biceps femoris muscle with overlying cutaneous territory for microvascular transplantation.

B, Postoperative view of musculocutaneous flap immediately after successful revascularization.

C, Three-month postoperative view demonstrates survival of both muscle and overlying skin.

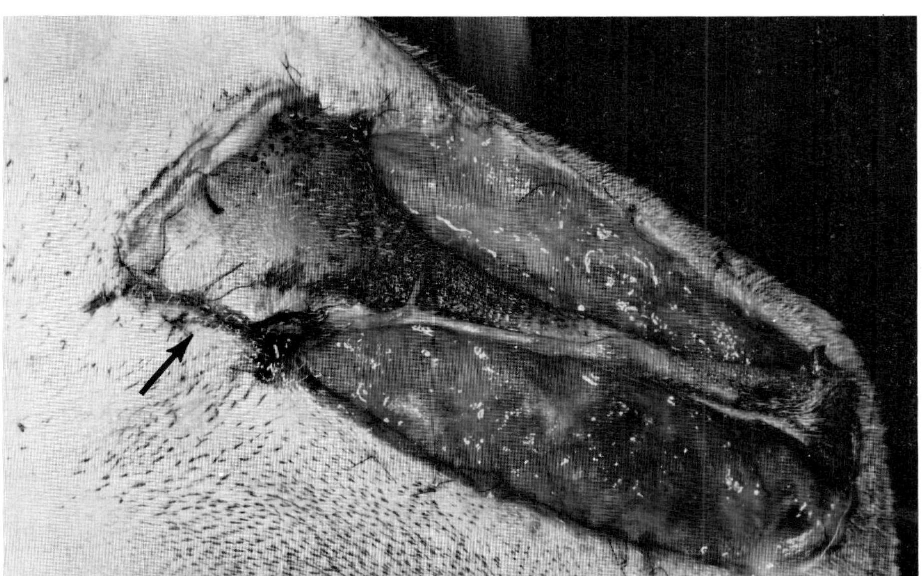

Fig. 4-2. Necrotic sartorius musculocutaneous flap in dog. Flap elevated on one proximal pedicle only. NOTE: Survival of proximal flap *(arrow)*.

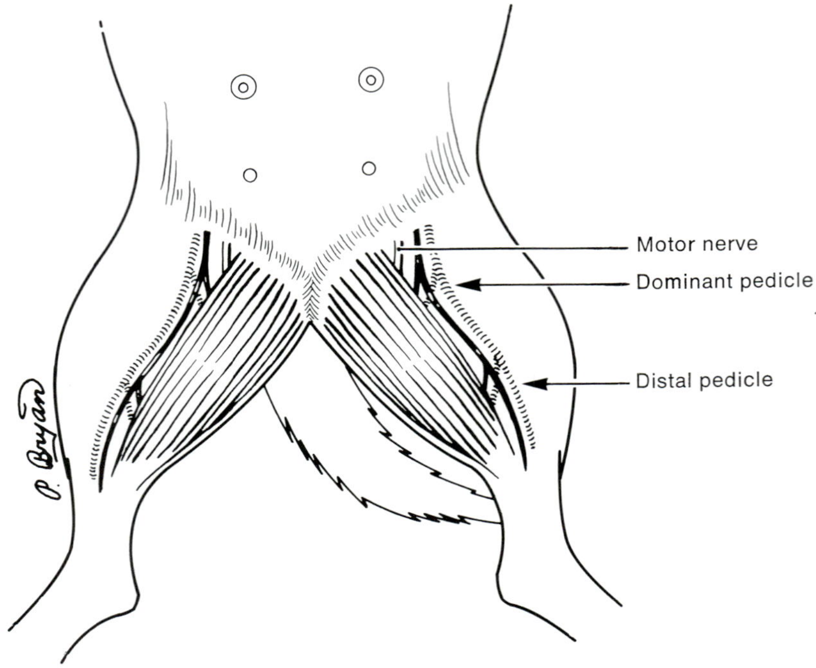

Fig. 4-3. Anatomy of dog gracilis muscle.

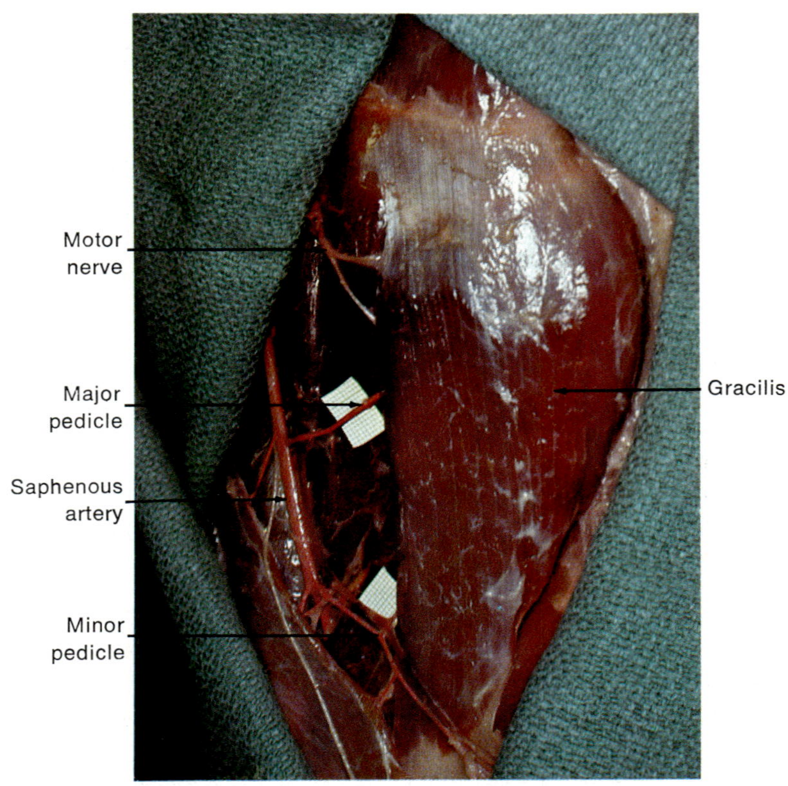

Fig. 4-4. Motor nerve and blood supply of dog gracilis muscle.

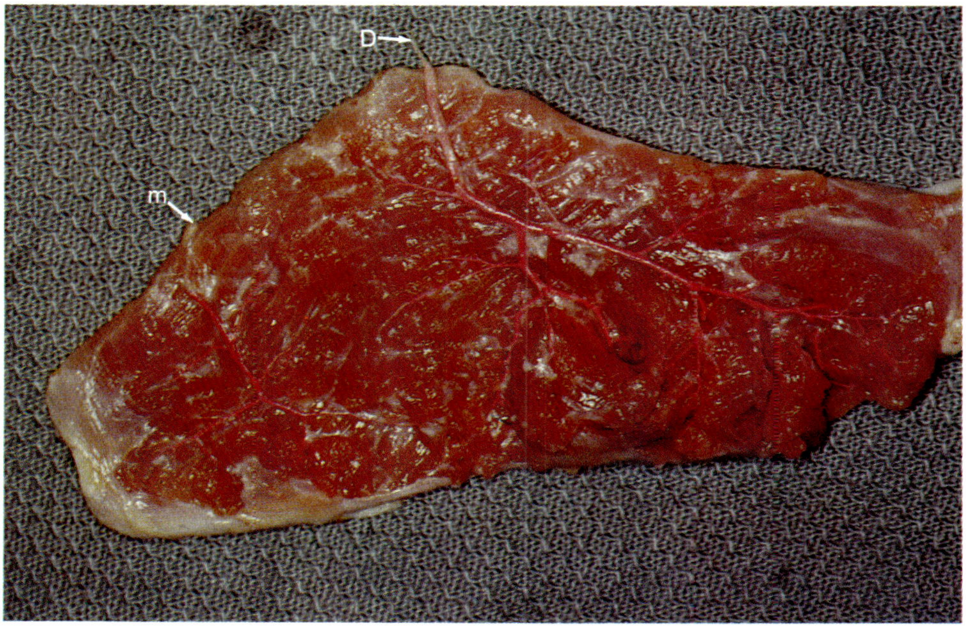

Fig. 4-5. Dissection of barium-latex–injected gracilis demonstrates intramuscular course of major and minor pedicles. *D*, Major pedicle; *m*, minor pedicle.

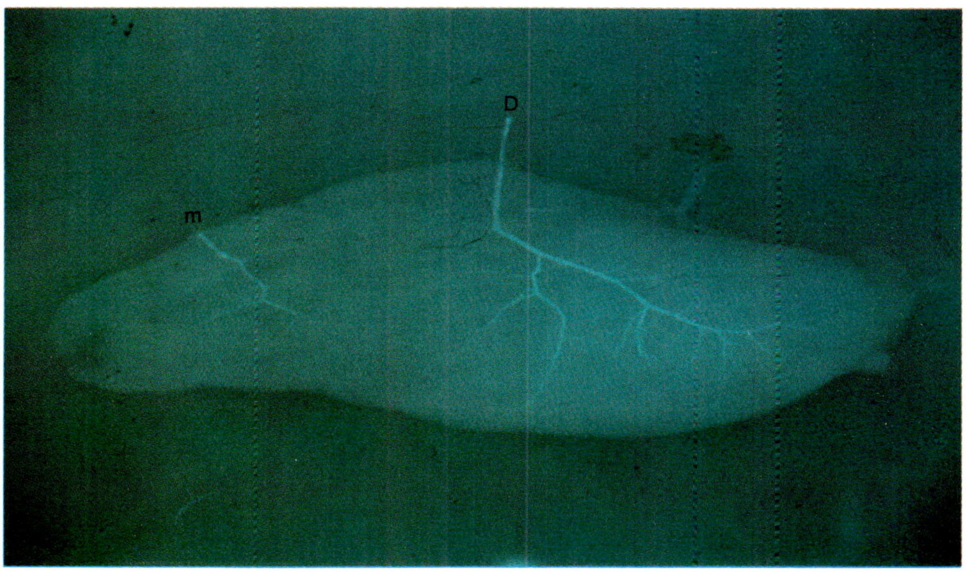

Fig. 4-6. Radiograph of muscle in Fig. 4-5. *D*, Major pedicle; *m*, minor pedicle.

blue latex through the minor pedicle to the lower one third (Fig. 4-7, *A*). Microscopic dissection of these two-color–injected muscles demonstrated several vessels with both blue and red latex in the lumen, thus confirming once again that anatomic continuity existed between the two pedicles (Fig. 4-7, *B*).

However, anatomic continuity demonstrated with latex injection under pressure does not necessarily imply physiologic continuity. Therefore cineradiography was performed under normal physiologic conditions. The minor pedicle of the muscle was clamped temporarily and the muscle perfused through the major pedicle. The dye was distributed to the entire muscle, first into the upper two thirds, then slowly into the lower one third. Conversely, with the major pedicle clamped and the minor pedicle open, dye was seen predominantly in the distal one third of the muscle and only a minimal portion of the upper two thirds.

This cineradiography study demonstrated that the

A

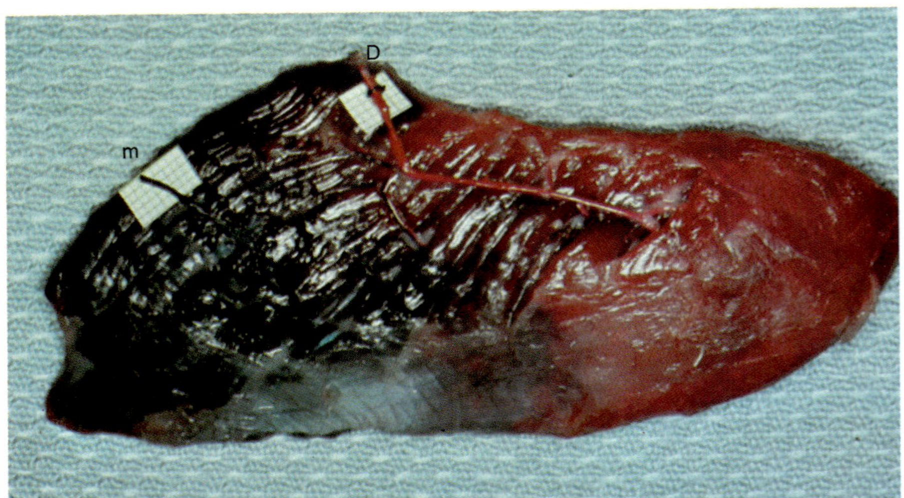

B

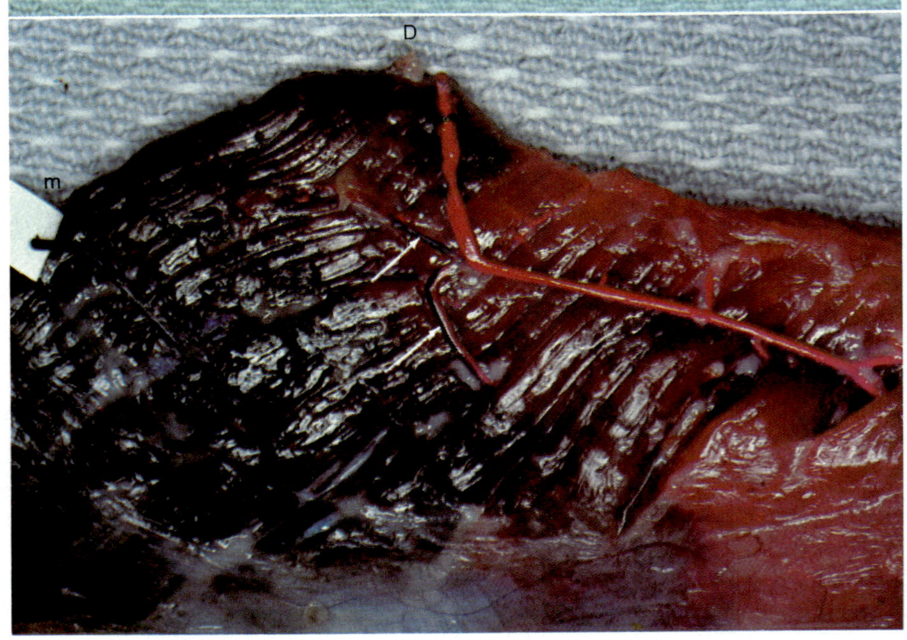

Fig. 4-7

A, Simultaneous injection of red latex through major pedicle and blue latex through minor pedicle. NOTE: Distribution of red to proximal two thirds of muscle and blue to distal one third. *D*, Major pedicle; *m*, minor pedicle.

B, Arrow indicates blue and red latex seen in lumen of vessel.

dominant pedicle would support the entire muscle. Although anatomic continuity was present, the flow through the distal pedicle was limited to the lower one third. Further experiments were then designed and undertaken to manipulate the distal blood supply so that the entire muscle could be based on the minor pedicle.

DELAY EXPERIMENTS
Rationale for delay

The arc of rotation of a muscle flap is limited by the size of the muscle and the location of the pedicle on which it is based. However, if through a delay procedure the entire muscle were to survive on the distal pedicle, then a second or distal arc of rotation would be available; for example, the distally based gracilis will cover the knee, and the distally based soleus will cover the lower one third of the leg (i.e., areas that the normal arc of rotation of these flaps will not cover). This therefore would increase the clinical application of the flap.

Experiments were designed to study the effects of vascular delay and nerve section on muscle flaps' survival based on the distal nondominant pedicle.

Vascular delay

The dog gracilis model was used to study vascular delay. On day 1 a skin incision was made on each side over the dominant pedicle to the muscle. On the right or experimental side the dominant vascular pedicle was identified, dissected free from the surrounding tissues, divided, and ligated; on the left or control side the pedicle was identified and preserved (Fig. 4-8, *A*). The skin incisions were then closed. On day 21 each flap was elevated on the minor distal pedicle (Fig. 4-8, *B*). The motor nerve was cut at this time, and the origin and insertion of the muscle were divided. The island musculocutaneous flap was then sutured into its original site. On day 24 the flaps were evaluated clinically and with fluorescein to assess the extent of muscle and skin survival. Cadaver injection studies and histologic examinations of the muscles were performed to confirm the clinical findings (Fig. 4-9).

This experiment was repeated in another group of animals after only a 14-day delay.

Results. On the experimental or delayed side 99% of the muscle survived, whereas on the control side only 60% of the muscle survived. Similar results were observed after 14 and 21 days of delay. Survival

A

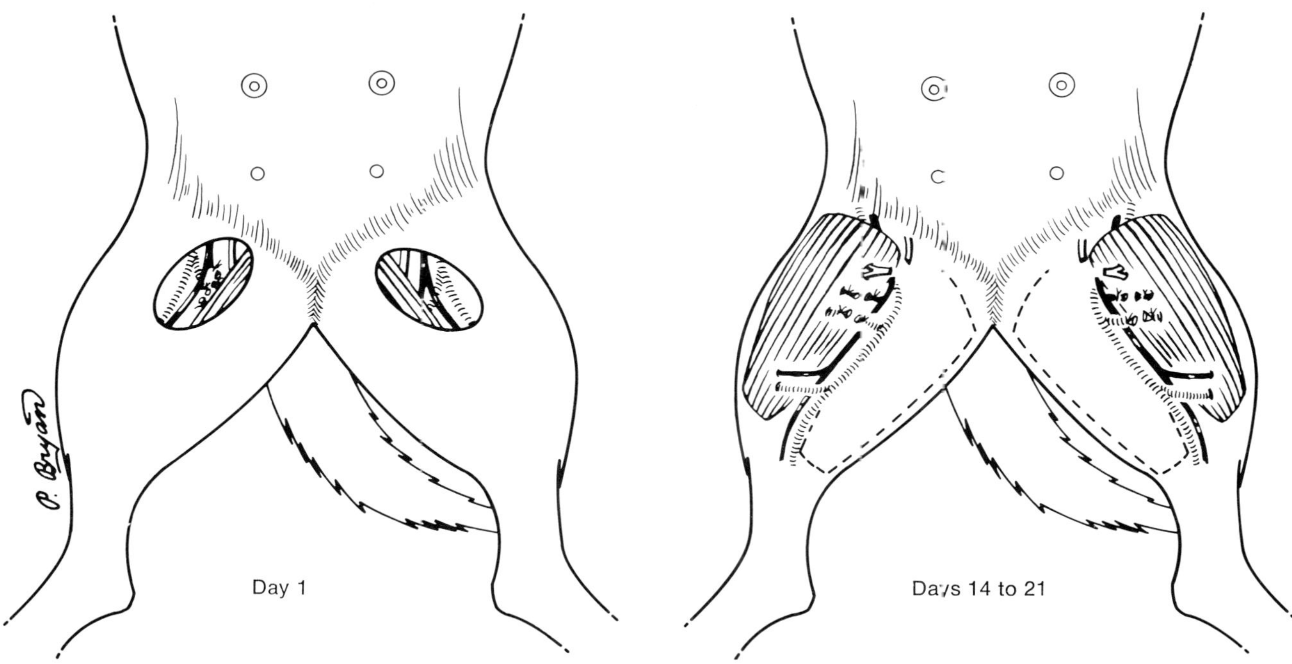

B

Fig. 4-8

A, Delay of experimental gracilis musculocutaneous flap. Major pedicle on right side ligated; sham incision performed on left side.

B, Bilateral gracilis musculocutaneous island flaps elevated on minor pedicle 14 to 21 days after flap delay.

A

B

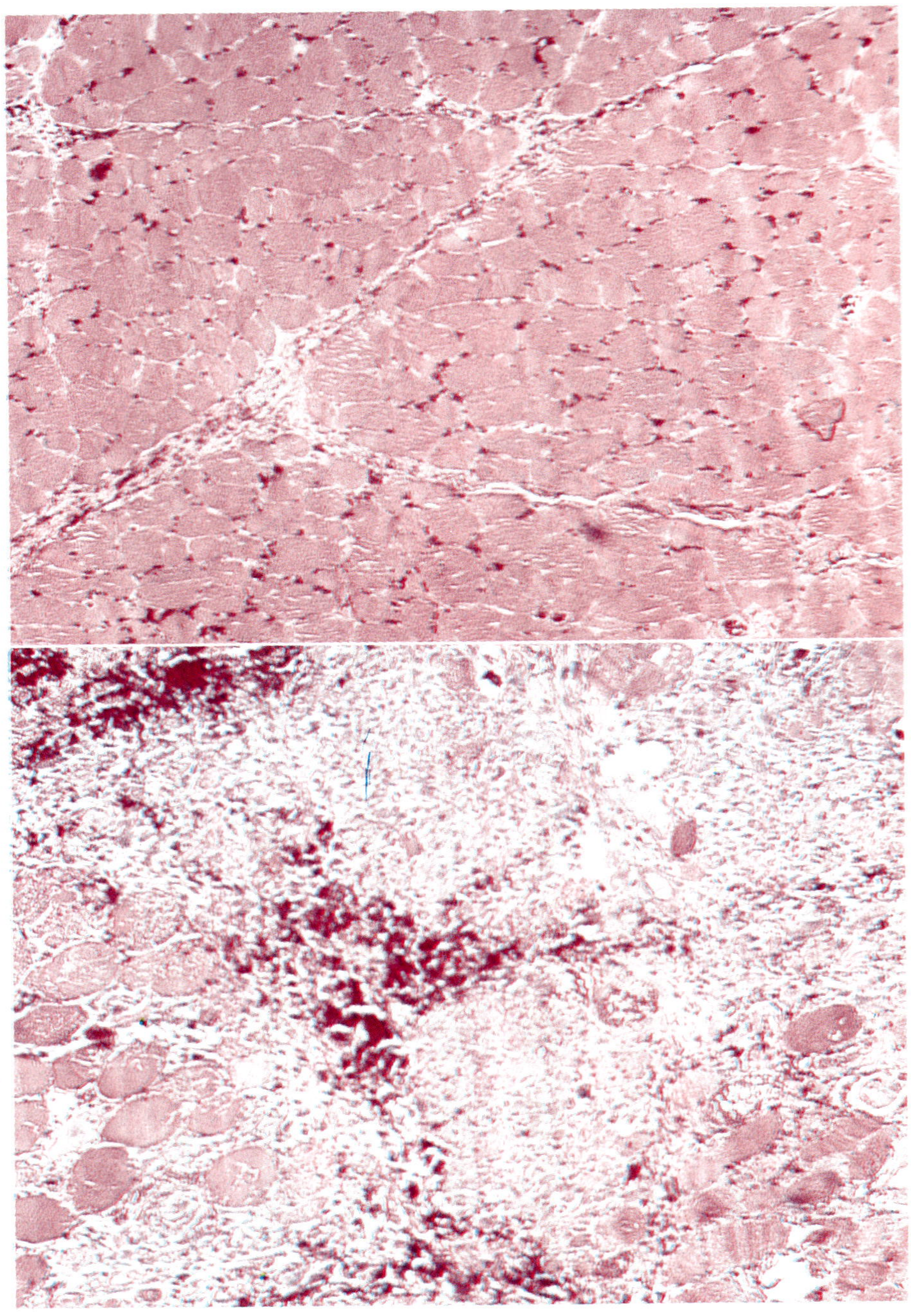

Fig. 4-9

A, Hematoxylin-eosin (H & E)–stained microscope section of proximal portion of muscle from experimental delayed side 3 days after flap elevation on distal pedicle. NOTE: Normal appearance of muscle.

B, Similar H&E-stained section taken from same location on control side. NOTE: Muscle necrosis.

of the overlying island of skin was not as predictable. On the delayed side approximately 60% of the skin island survived, and only 25% on the control side of the skin island survived.

Discussion. Blood flow through muscle is regulated by the smooth muscle in the vessel walls. According to Poiseuille's law, flow varies to the fourth power of the radius of the vessel. Thus a small change in vessel size leads to a significant change in flow. Therefore changes in smooth muscle tone within the vessels control flow into the muscle. This smooth muscle tone is controlled by neurogenic stimuli and local metabolites. The neurogenic stimuli may constrict or dilate the vessels. Alpha-adrenergic stimulation constricts, while beta-adrenergic stimulation dilates. The local metabolites vasodilate. Although abbreviated and simplified, this description of muscle blood flow accounts for the experimental findings.

The ligation of the major pedicle on day 1 on the experimental side does not lead to muscle necrosis, because in addition to the distal pedicles, smaller vessels through the origin are present to maintain muscle viability. However, this does result in muscle ischemia, which we believe is the stimulus for increased flow through the minor pedicle that will eventually support the entire muscle. More studies, including the use of flow meters, to accurately measure flow in the minor pedicle before and after vascular delay are needed to confirm these observations. The barium-latex injection studies after delay do indeed show filling of the entire muscle through the minor pedicle after the delay.

Vascular delay of the gracilis and soleus muscles has been clinically performed. The distally based delayed gracilis flap has been successfully transposed for coverage around the knee, and the distally based delayed soleus flap has provided coverage of the lower one third of the leg. However, with the selection of more appropriate alternative flaps and the availability of microvascular tissue transplantation, delayed muscle flaps have had very limited clinical applications.

Nerve delay

Hudlická has demonstrated that after denervation there is an increase in blood flow into muscle that persists for 20 days. In addition to this increase in blood flow, complicated changes at the biochemical level also take place. An experiment similar to the

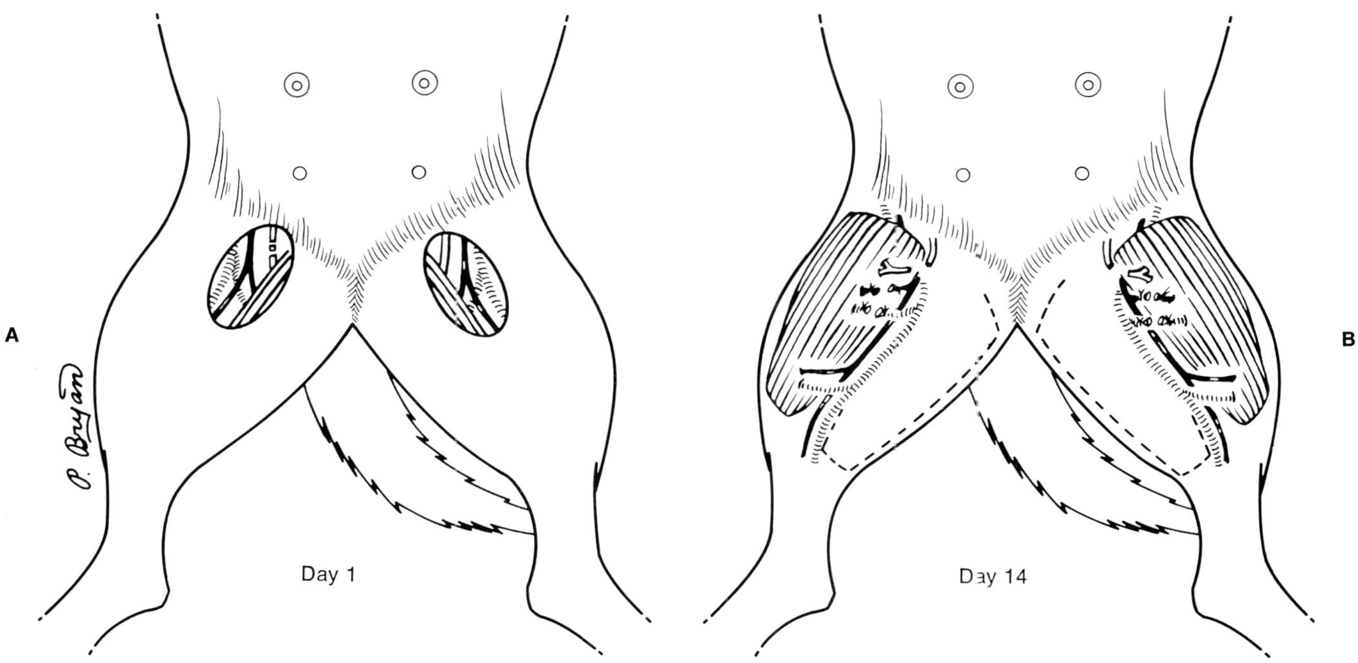

Fig. 4-10
A, Incision made over motor nerve to gracilis muscle on each side. Segment of nerve (5 mm) is excised on right side.
B, Bilateral gracilis island musculocutaneous flaps are elevated on distal minor pedicle 14 days after motor nerve section on right side.

vascular delay study was performed in our laboratory by Dr. Richard Hagerty to investigate the effects of motor nerve section on survival of gracilis flaps based on the distal pedicle.

Method. On day 1 an incision was made on each side over the proximal portion of the gracilis muscle. Through this incision the motor nerve, a branch of the obturator nerve, was identified. On the right or experimental side a 5-mm segment of the nerve was resected (Fig. 4-10, *A*). On the left or controlled side the nerve remained intact, and the skin incision was closed. On day 14 both flaps were elevated as island musculocutaneous flaps based on the distal minor pedicle (Fig. 4-10, *B*). The motor nerve on the left side was divided at this time. On day 17 the flaps were evaluated as in the vascular delay experiment.

Results. Although there were some individual variations, there was no significant difference in flap muscle or skin survival between the experimental and controlled side. Therefore in this experimental study, prior denervation of the muscle did not affect muscle flap survival based on the distal minor pedicles.

VASODILATOR DRUGS

Neurogenic stimuli affect smooth muscle tone of the vessels within skeletal muscle. Alpha-adrenergic stimuli constrict, while beta-adrenergic stimuli dilate. The drug isoxsuprine* is a beta-adrenergic stimulator, producing prolonged vasodilation. This beta-receptor effect is directed toward the precapillary sphincter.

Finseth and co-workers in a series of articles first demonstrated that this prolonged smooth muscle relaxation in the vessel wall would prevent necrosis in an experimental neurovascular island skin flap in the rat. The effect of isoxsuprine was then evaluated on island muscle and musculocutaneous flaps.

In a well-controlled experiment on the vastus medialis island muscle flap in the rabbit, they showed that pretreatment with isoxsuprine injected intraperitoneally (1 mg/kg/day) for 2 weeks before and 5 days after flap elevation led to significant increase in muscle survival as compared to untreated controls. Finseth and co-workers then studied the effects of pretreatment with isoxsuprine on island musculocutaneous flaps in the pig. The gracilis musculocutaneous unit was chosen as the experimental flap. The isoxsurpine-treated pigs were given 1 mg/kg of isoxsuprine intramuscularly for 2 weeks before and 6 days after flap elevation. Again, they showed greater survival of muscle and overlying skin in the isoxsuprine-treated group.

Cherry studied the effects of isoxsuprine on muscle and skin flap survival in the pig. Using each animal as its own control, he first raised a sartorius muscle flap and a neurovascular island skin flap on one side. Survival of the muscle flap and skin flap was then noted. Pretreatment with isoxsuprine, 15 mg/kg/day by mouth, was initiated 2 weeks later and continued for an additional 14 days. At the end of this period, the sartorius muscle flap and neurovascular island flap were elevated on the opposite side. Cherry demonstrated a significant increase in the surviving length of the sartorius flap when compared to the untreated muscle flap in the same pig. However, he was unable to show any significant difference in survival length of the neurovascular island skin flap.

Discussion. The studies of Finseth and co-workers and Cherry clearly demonstrate the effects of isoxsuprine on increased muscle flap survival in the pretreated animals. Finseth has also reported two patients in whom musculocutaneous flaps were salvaged by intraoperative and postoperative injections of isoxsuprine.

We have initiated a study on the effect of isoxsuprine on the previously described gracilis musculocutaneous flap model in the dog. Early data show increased muscle survival after isoxsuprine pretreatment when the flap is elevated on the distal minor pedicle.

Further experimental and clinical evaluation of the effects of isoxsuprine are indicated to define its role in clinical practice. Isoxsuprine may well prove to be an excellent tool for the salvage of failing flaps or as a method for increasing muscle survival on a given vascular pedicle.

Three methods for increasing muscle flap survival on a given pedicle have been investigated. These include strategic vascular delay by ligation of the dominant pedicle, nerve section, and isoxsuprine pretreatment. Our data clearly show the effectiveness of a vascular delay through the ligation of a dominant pedicle. Sections of the motor nerve in our experiments did not show any such effect, although recent studies with a different experimental model by Heckler have demonstrated improved flap survival after muscle denervation. The work of Finseth and co-workers and Cherry have demonstrated the effects of isoxsuprine. Vascular delay requires previous surgical intervention, whereas the use of isoxsuprine would obviate the need for a previous procedure.

*Vasodilan, Mead Johnson Labs, 2404 Pennsylvania St., Evansville, IN 47721.

REVASCULARIZATION OF MUSCULOCUTANEOUS FLAPS

Revascularization of island musculocutaneous flaps has been investigated in our laboratories.

Method. The previously described gracilis musculocutaneous island flap model in the dog was studied. On day 1 an island musculocutaneous flap was elevated on the dominant pedicle. The distal vascular pedicle, motor nerve, and muscle origin and insertion were divided. The flap was then sutured back into its original location. In a series of flaps on days 3 through 10 the vascular pedicle was exposed and either the artery or vein was ligated. Flap survival was then evaluated 2 days later.

Results

Arterial ligation. Although isolated flap survival was seen after arterial ligation as early as day 4, all flaps survived after arterial ligation on day 7.

Venous ligation. Although a smaller number of flaps was studied after venous ligation, it appeared that venous ligation was less well tolerated than arterial ligation. All flaps survived after venous ligation after day 8.

A barium-latex injection was then used to study the surviving flaps. The barium-latex mixture was injected into the femoral artery of the dog, and the flaps were then dissected and studied radiographically. These studies clearly showed that the experimental flap had been revascularized through the underlying muscles (Fig. 4-11).

Discussion. Further work is in progress to evaluate the effect of simultaneous ligation of the artery and vein. These studies indicate that in the flap model, revascularization through the surrounding tissues is adequate to maintain viability of the experimental flap after 8 days. It should be noted that in this study, the flaps were sutured back into normal, well-vascularized tissue. Clinically muscle flaps are transposed into avascular or irradiated tissues; therefore pedicle division should be delayed for at least 2 weeks.

FURTHER INVESTIGATION INTO MUSCLE FLAPS

Work is in progress in our laboratories to further investigate muscle and musculocutaneous flaps.

Infection. It has been a clinical impression that muscle placed directly into a contaminated wound resists and in fact eradicates infection. It is placed in osteomyelitis cavities to eradicate chronic infection. The thesis that muscle is highly vascularized and should therefore resist or eradicate infection is being evaluated experimentally. Early data confirm this thesis.

Muscle bulk. As more and more attention is focused on restoring form and function with muscle and musculocutaneous flaps, the predictability of muscle atrophy or preservation of muscle bulk is essential. Studies are in progress to evaluate atrophy of muscle bulk after denervation and flap transposition. The role of muscle tension in the flap on muscle atrophy and loss of bulk is also under study.

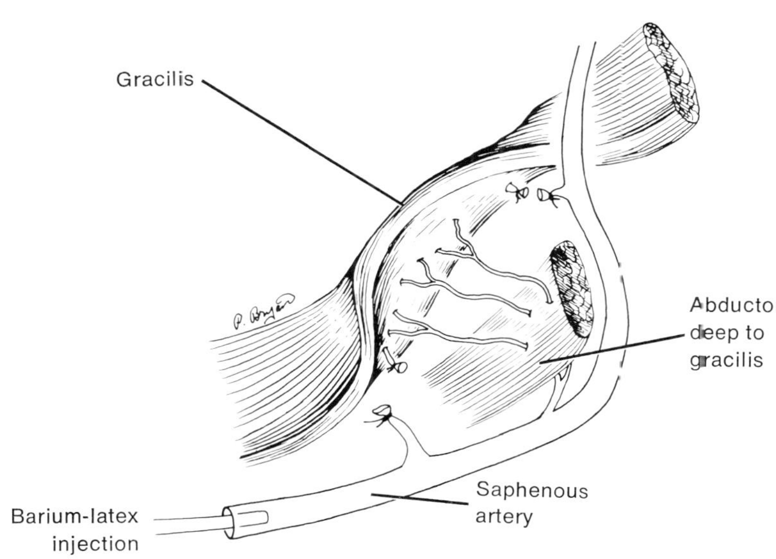

Fig. 4-11. Revascularization of gracilis muscle through underlying abductor muscles.

Lymphatics in muscle and musculocutaneous flaps. After transposition as island musculocutaneous flaps, the lymphatic drainage of the skin island is interrupted. The clearance of india ink from the skin island is markedly altered, and our initial data indicate that in time the lymphatics drain through the underlying muscle.

Cadaver injection studies. Cadaver injection studies have been used to study the blood supply to muscle and skin. Red latex and blue latex are mixed in equal amounts with barium paste to make a radiopaque mixture. This mixture is then injected slowly into the specimen. The latex solidifies in the acid pH of the cadaver tissues. For best results injections are made in fresh cadavers. Although it is still possible to inject preserved specimens, the results are not always optimal. The injections are performed under minimal pressure. The application of excessive pressure or the injection of large volumes of the mixture will force the mixture into the tissues and beyond the territory of the vessel under investigation. In study of the extremities, for example, injection of 40 to 50 ml of the mixture is sufficient.

Evaluation of skin territories through the injection of dye into underlying muscles or direct cutaneous vessels has been performed. However, a note of caution is necessary in the technique of injection, since the use of large volumes of the mixture will increase the area of distribution and hence the territory of the vessel under study. Cadaver injection studies merely demonstrate anatomic continuity and do not necessarily represent physiologic dependence of a muscle or an area of skin on a given vessel. Clinical correlation of these findings is essential before conclusions can be drawn.

Our studies have indicated that injection of the pedicle to an underlying muscle will certainly distribute the material into the overlying skin and several centimeters beyond. It is of interest that in certain areas the same area of skin can be filled with contrast material either through direct cutaneous vessels or through the underlying muscles. An excellent example of this is the skin of the anterior abdominal wall. The injections through the deep inferior epigastric artery through the rectus abdominis will demonstrate the skin overlying the rectus abdominis, stretching laterally to the anterior axillary line and across the midline to the opposite side. The same area of skin can be demonstrated through direct injection of the superficial inferior epigastric artery, which is a direct cutaneous vessel.

Cadaver injection studies have been most useful in evaluation of the blood supply to the skin. Using these techniques, it has clearly been demonstrated that direct cutaneous vessels and musculocutaneous perforators supply the skin. The areas of overlap and the areas of greater dependence on one system or the other have been demonstrated. These factors have also been confirmed through clinical correlation.

ANNOTATED BIBLIOGRAPHY

1. Campbell, J., and Pennefather, C.: An investigation into the blood supply of muscle with special reference to war surgery, Lancet **1**:294, 1919.

   An experimental study in rabbits designed to assist trauma surgeons with débridement of muscle after gunshot wounds is presented. Campbell and Pennefather describe the vascular anatomy of several muscles. Despite the anastomotic connections between vascular pedicles within muscle, they stress that division of certain dominant pedicles generally resulted in muscle necrosis.

2. Cherry, G.W.: The differing effects of isoxsuprine on muscle and skin flap survival in the pig, Plast. Reconstr. Surg. **64**:670, 1979.

   Isoxsuprine is shown to effect the survival of the sartorius muscle flap in the pig but not the survival of neurovascular skin island flaps.

3. Clark, W.E., and Blomfield, L.B.: The efficiency of intramuscular anastomoses with observations on the regeneration of devascularized muscle, J. Anat. **79**:15, 1945.

   The tibialis anterior, semimembranosus, and gracilis muscles of the rabbit were studied. Bromphenol blue dye was used to assess muscle viability after experimental ligation of different vascular pedicles in each muscle. Clark and Blomfield show that the ensuing pattern of muscle necrosis depends on the efficiency of the intramuscular anastomoses.

4. Finseth, F.: Clinical salvage of three failing skin flaps by treatment with a vasodilator drug, Plast. Reconstr. Surg. **63**:304, 1979.

   One axial and two musculocutaneous flaps that at the time of operation were judged by the fluorescein test as nonviable were salvaged with the administration of isoxsuprine.

5. Finseth, F., and Adelbert, M.G.: Prevention of skin flap necrosis by a course of treatment with vasodilator drugs, Plast. Reconstr. Surg. **61**:738, 1978.

   Isoxsuprine is shown to prevent necrosis in neurovascular island skin flaps in rats.

6. Finseth, F., and Zimmerman, J.: Prevention of necrosis in island myocutaneous flaps in the pig by treatment with isoxsuprine, Plast. Reconstr. Surg. **64**:536, 1979.

   Isoxsuprine is shown to increase survival of muscle and skin in the experimental gracilis island musculocutaneous flap in pigs.

7. Finseth, F., Zimmerman J., and Liggins, D.: Prevention of muscle necrosis in an experimental neurovascular island muscle flap by a vasodilator drug—isoxsuprine, Plast. Reconstr. Surg. **63:**774, 1979.

Isoxsuprine is shown to prevent muscle necrosis in the experimental vastus medialis island muscle flap model in rabbits.

8. Hudlická, O.: Muscle blood flow (its relation to muscle metabolism and function), Amsterdam, 1973, Swets en Zeitlinger, B.V.

An excellent monograph on the circulation of muscle and its experimental evaluation physiologically is presented. All aspects of experimental muscle circulation are described. This book is good basic reading for anyone planning research into muscle flap circulation.

9. Hudlická, O., and Renkin, E.: Blood flow and blood tissue diffusion of 86 Rb in denervated and tenotomized muscles undergoing atrophy, Microvasc. Res. **1:**147, 1968.

An increase is shown in blood flow both per 100 gm weight and for the whole muscle for 20 days after denervation.

10. Mathes, S.J., Nahai F., and Vasconez, L.O.: Myocutaneous free flap transfer: anatomical and experimental consideration, Plast. Reconstr. Surg. **62:**162, 1978.

The concept of delaying muscle and musculocutaneous flaps by ligating vascular pedicles is explored, and initial data are presented.

11. Tamai S., and co-workers: Free muscle transplants in dogs with microsurgical neurovascular anastomoses, Plast. Reconstr. Surg. **46:**219, 1970

First report of free muscle transplantation by microvascular techniques is presented. The rectus femoris in dogs is transferred. The artery, vein, and motor nerve are sutured and functional recovery is evaluated 3 to 9 months later.

12. Terzis, J.K., and co-workers: Recovery of function in free muscle transplants using microneurovascular anastomoses, J. Hand Surg. **3:**37, 1978.

Experimental work on the transfer of the rectus femoris muscle in the rabbit is presented. After free neurovascular transfer, maximum working capacity of the muscle is only one fourth of normal.

5

AN ANALYSIS OF CAUSES OF FLAP FAILURE

Stephen J. Mathes ▪ Foad Nahai

GENERAL PRINCIPLES

The complications associated with muscle flaps stem from the same causes as those with other procedures: errors in judgment, errors in technique, and concurrent illness. The best method of avoiding complications is prevention.

The inadequate preparation of the patient represents an error in judgment that may lead to failure. Proper preparation includes the assessment of the patient's overall status and correction of metabolic abnormalities. Adequate nutritional support, hydration, and maintenance of blood volume are essential.

Flap selection

The flap of choice and alternative flaps should be determined preoperatively. The final decision is not made until the full extent of the defect has been determined. The entire reconstructive armamentarium should be considered: primary closure, skin graft, random skin flaps, muscle flaps, musculocutaneous flaps, and composite tissue transplantation.

Previous experience and knowledge of all of these modalities are essential so that the reconstructive surgeon can select the method that is best suited to the defect requiring reconstruction.

Specifically for the application of muscle and musculocutaneous flaps the surgeon should be familiar with the following three basic principles of safe muscle transposition:

1. Knowledge of the anatomy of the muscle
2. A precise knowledge of the number and exact location of the vascular pedicles to the flap
3. The limits of the arc of rotation of the flap

A working knowledge of the general anatomy of the operative areas as well as the precise anatomy of a given flap is essential. The reconstructive surgeon must be familiar with the number of alternative flaps. Unexpected anatomic variability, previous damage to a specific vascular pedicle, and an unexpectedly large defect as a result of débridement or dissection may force the surgeon to use alternative flaps. The well-prepared surgeon can proceed with one of the alternatives without risking flap failure.

GENERAL CONSIDERATIONS

Complications are those that are general and likely to be seen with any muscle flap and those that are specific and only seen with certain flaps. The general complications include (1) planning errors; (2) technical, interoperative errors; and (3) postoperative care errors, all of which will contribute to flap failure.

Planning errors

Inadequate flap design. Final flap selection and design should not be undertaken until the exact extent of the defect is known. The resection of the tumor or the débridement of the infected wound may result in a defect larger than anticipated. If the flap is designed and elevated before the full extent of the defect has been determined, the flap will be too small.

Inadequate vasculature. The known vascular pedicles to a given flap may have been divided or damaged by previous surgery, trauma, or underlying disease. Selective angiography may be indicated in patients with extensive traumatic defects or atherosclerotic disease of the lower extremity before flap selection.

Prior division of a muscle generally results in permanent interruption of circulation of the muscle distal to the injury site from proximally located vascular pedicles. However, prior incisions or random cutaneous flap elevation superficial to muscle only temporarily (3 weeks) eliminates vascular connections between muscle and skin.

Obese patients. The arc of rotation of the muscle flap is limited by the location of the vascular pedicle and length of the flap. In thin patients a flap will reach much further. The extended TFL flap, for example, in a thin patient may well reach the contralateral ischium, whereas in an obese patient it may not even reach the ipsilateral ischium. These factors have to be taken into account in preoperative planning. Furthermore the orientation of the overlying skin on the muscle is less precise in obese patients.

Donor defect. The donor defect should influence flap selection. A donor defect over a pressure area in a paraplegic patient will lead to further complications and should always be avoided. The donor defect of the musculocutaneous flap in the lower extremity (e.g., gastrocnemius) results in both an unsightly donor defect and occasionally an area subject to wound breakdown with minimal trauma.

Flap selection. In general, the muscles with types I, III, and V patterns of blood supply have been the most reliable and types II and IV the least reliable. These factors concerning muscle circulation should be taken into account when selecting flaps for a given defect.

Function. The functional deficit from the use of the flap must be considered in flap selection. The gluteus maximus flap, an excellent flap for reconstruction in the paraplegic patient, may well lead to functional disability in an ambulatory patient, especially when the entire muscle is used. Whenever possible, function preservation techniques should be employed so that the patient may have the benefits of the particular muscle flap without loss of function.

Technical errors

Flap design. Flap design requires a precise knowledge of the topographic anatomy of the muscle. Failure to align the skin territory correctly over the muscle will result in necrosis of the skin flap when designed as an island. Obesity, inexperience in flap design, or narrow muscles are common causes of inaccurate flap design (Fig. 5-1). This error is avoided by the following:

1. Review of flap anatomy
2. Accurate flap design
3. Conversion of the skin territory to an island only after the muscle belly borders are fully identified

Muscle choice. Muscle choice is based on the suitability of muscle to cover the defect. Elevation of an adjacent muscle may result in a muscle that fails to provide vascularized cover for the defect because of the following:

1. Inadequate size
2. Poor vascular pattern

If the muscle fails to cover the defect, or if ischemia is noted by failure of the distal portion of the flap to bleed or fluoresce (e.g., with use of intravenous fluorescein [1 gm injection]), then the muscle should

Fig. 5-1. Same size skin island is based on two different muscles to demonstrate some problems involved in flap design. Muscle *A* is narrow with thick layer of subcutaneous fat. Muscle *B* is flat with thin subcutaneous layer. *A* is typical of limb muscles and *B* typical of trunk muscles.

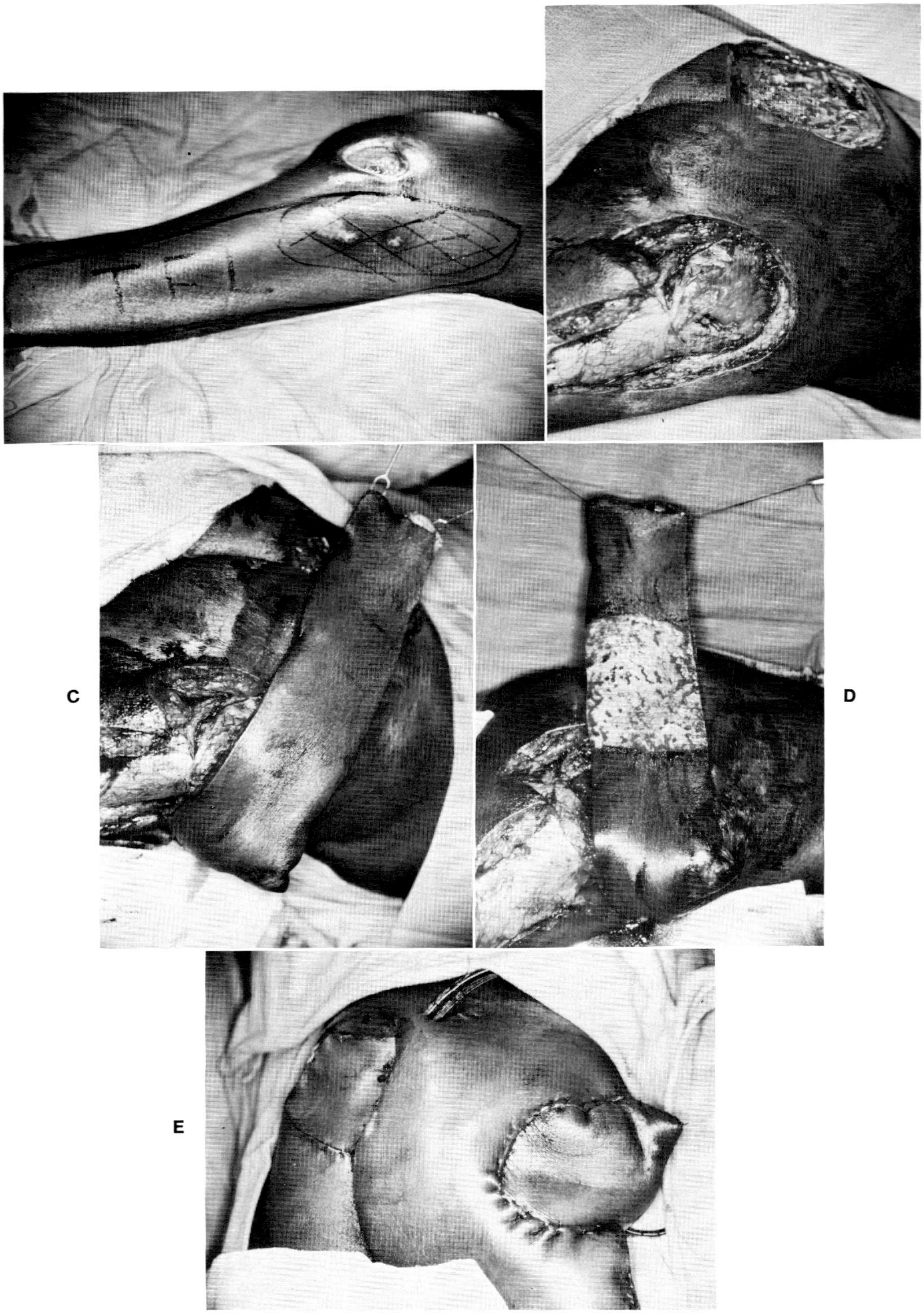

Fig. 5-2. Tight tunnel leading to distal flap necrosis.

A, Patient with combined trochanteric and sacral pressure sores. Extended TFL flap is designed.

B, Both pressure sores excised. NOTE: Intact bridge of skin between two defects.

C, Extended TFL flap elevated will cover trochanteric area and reach sacrum.

D, Central portion of flap that will be in tunnel is deepithelialized.

E, Flap sutured into sacral and trochanteric defect. Tunnel was inadequate in size and flap distal to tunnel was lost.

not be used. Alternative muscle flaps or the initially planned muscle should be identified and elevated. This error is avoided by the following:

1. Review of flap anatomy
2. Knowledge of the muscle characteristics, especially its relationship to adjacent muscles near its insertion or origin
3. Use of the tourniquet during identification and elevation of muscle flaps in the lower extremity

Pedicle division. Division of the dominant vascular pedicle results in muscle flap failure. In type II muscles the surgeon must routinely divide the minor pedicle(s). However, the dominant pedicle should be identified before division of the most proximal minor pedicle if this minor pedicle seems abnormally close to the point of flap rotation. This may be the major pedicle located more inferiorly or distally along the course of the muscle belly. If major pedicle division is suspected, flap circulation is best evaluated by examination for distal muscle bleeding and use of intravenous fluorescein (1 gm injection). If flap ischemia is confirmed, then the major pedicle should be identified. If it is divided, one may repair the pedicle by microvascular techniques when feasible. If the pedicle is intact, spasm may be suspected. In this instance topical vasodilators (e.g., Marcaine [0.5%] or Papaverine) may be applied to the pedicle and the flap observed. Failure of flap perfusion after a 30-minute period of observation may indicate pedicle injury. However, prolonged spasm has been observed, and the flap is best inset and observed for 24 hours.

This grave technical error of flap pedicle injury or division must be avoided in flap elevation. This area can be avoided by the following:

1. Precise knowledge of muscle vascular anatomy
2. Avoidance of blunt dissection when in the proximity of the dominant vascular pedicle
3. Elevation of a muscle flap to the level of entrance of the dominant vascular pedicle is not necessary in every reconstructive application of a particular muscle

Tunnel. The tunnel is the most common cause of muscle and musculocutaneous flap failure. The tunnel should be avoided. Any bridge of skin separating the muscle or musculocutaneous flap from its site of inset should be divided. A tunnel may be required in certain instances, including the following:

1. Head and neck reconstruction
2. Breast reconstruction
3. Chest reconstruction

Since muscle swelling is common in the early postoperative period, the tunnel must be twice the size of the traversing flap (Fig. 5-2).

Tension. When the flap does not reach the defect, the dominant vascular pedicle is generally the last limiting factor preventing a wider arc of rotation. Release of tissue around the pedicle generally increases flap length. However, the resultant tension on the vascular pedicle may prevent flap perfusion and lead to flap failure. When this complication is noted, tension is decreased from the flap, and a second flap is elevated to complete defect coverage.

Damage to adjacent structures. Elevation of a muscle flap necessitates dissection close to important vessels and nerve structures. Damage to these structures may result in serious sequelae. Severe functional defects after injury to the underlying nerves during flap elevation are avoided by knowledge of vascular and neural structures in proximity to the flap.

Functional loss. Complete division of origin or insertion or interruption of motor nerve eliminates muscle function. Since use of muscle as a transposition flap generally involves one or both of these maneuvers, remaining synergistic muscles must be preserved to maintain function. The function of the muscle used for coverage can be maintained by the following:

1. Use of one muscle from a group of synergistic muscles
2. Use of function preservation techniques in muscle flap design (i.e., using only a portion of muscle, leaving muscle origin, insertion, and innervation intact)
3. Use of microvascular composite tissue transplantation when use of local muscle flap transposition techniques results in unacceptable functional limitations for the patient

Hemostasis. Bleeding either in the donor defect or the recipient site may result in flap failure. Increased tension in the flap is an immediate problem often because of hematoma formation. Later, the hematoma is a site for infection. Hematoma is avoided by the following:

1. Careful hemostasis
2. Knowledge of location of the minor pedicles to the muscle and careful ligation, as required for flap elevation

Inadequate débridement. Successful coverage of the difficult wound requires preliminary adequate wound débridement. The second most common technical cause of muscle flap failure is residual nonviable bone or soft tissue. The problem is avoided by the following:

1. Aggressive wound débridement
2. Delay in flap design and elevation until wound

débridement is complete, which avoids failure to debride a wound adequately because of inadequate flap size

Postoperative errors

A successful operation and a viable flap may be lost through postoperative errors. These include the following:

Positioning. Correct positioning of the patient is essential postoperatively to prevent pressure on the flap or its pedicle. In general, the head and neck, trunk, and extremities are elevated postoperatively.

Dressings. Tight dressings, especially around the pedicle of a flap, may lead to compromise of flap circulation.

Suction drains. Proper functioning of drains is essential postoperatively. A drain that is not performing is more a potential hazard than a benefit.

Antibiotics. Perioperative antibiotics are recommended if flaps are transposed into infected or contaminated defects. Antibiotics are started preoperatively and continued for 1 to 3 days postoperatively. If wound drainage becomes purulent, it is cultured and appropriate antibiotic therapy instituted.

Mobilization. After reconstructive procedures on the trunk or head and neck, patients are mobilized early in the postoperative period. However, dependency is avoided for 2 to 3 weeks after lower extremity reconstruction and then permitted gradually with support stockings or dressings. Weight bearing is avoided in paraplegic patients for at least 3 weeks after pressure sore reconstruction.

Rehabilitation. Proper postoperative rehabilitation is essential for lower extremity reconstruction. Pressure sore coverage in a paraplegic patient is only one step in the rehabilitation process, and unless postoperative rehabilitation is instituted, the patient will have recurrence.

II

APPLICATIONS

6

A SYSTEMATIC APPROACH TO FLAP SELECTION

Stephen J. Mathes ■ Foad Nahai

ANTERIOR SCALP (Figs. 6-1 to 6-2)

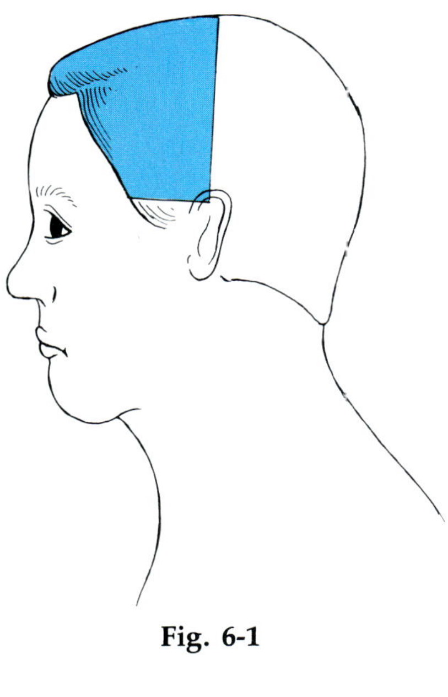

Fig. 6-1

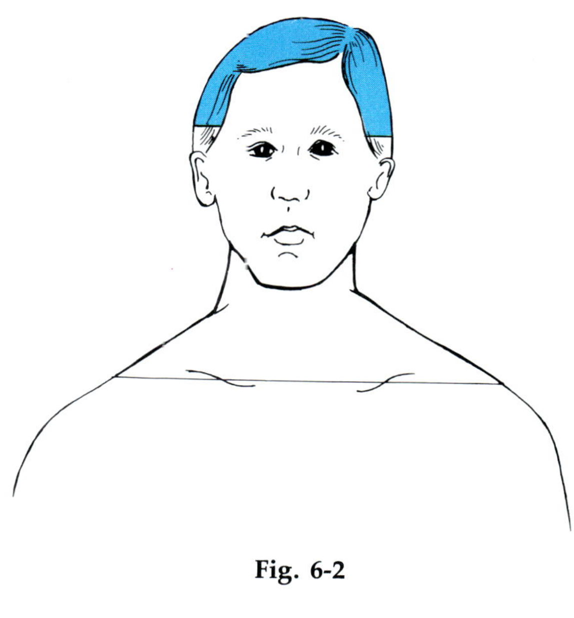

Fig. 6-2

Skin grafts

1. Skin grafts are method of choice for temporary coverage of acute injury
2. Skin grafts cannot be used on exposed bone unless outer table is excised (flap coverage is preferred; see local flaps)
3. Skin grafts are alternative method for coverage of dura (Fig. 6-3)

Local flaps (scalp flaps)

1. Local flaps are method of choice for coverage of small to moderate size defects
2. Local flaps may be designed to provide hair coverage
3. Orticochea technique allows expansion of scalp for greater coverage (Fig. 6-4)
4. Radiation therapy may contraindicate use of local tissue

Distant flaps (Fig. 6-5)

1. Major head and neck flaps (pectoralis major, latissimus dorsi, and trapezius) will not routinely reach this area
2. Free tissue transplantation is method of choice when the following conditions exist:
   a. Defect too large for local tissue (greater than one third of scalp)
   b. Radiation therapy or vascular injury precludes use of local tissue
   c. Adequate receptor vessels are a prerequisite for this method
3. Choice of free flaps in order of preference
   a. Latissimus dorsi
      (1) Musculocutaneous flap for moderate defect
      (2) Muscle only and skin graft for large defects (Fig. 6-6)
      (3) May include ribs for skull reconstruction
   b. Omentum and skin graft
   c. Tensor fascia lata musculocutaneous flap (makes an excellent dura substitute)

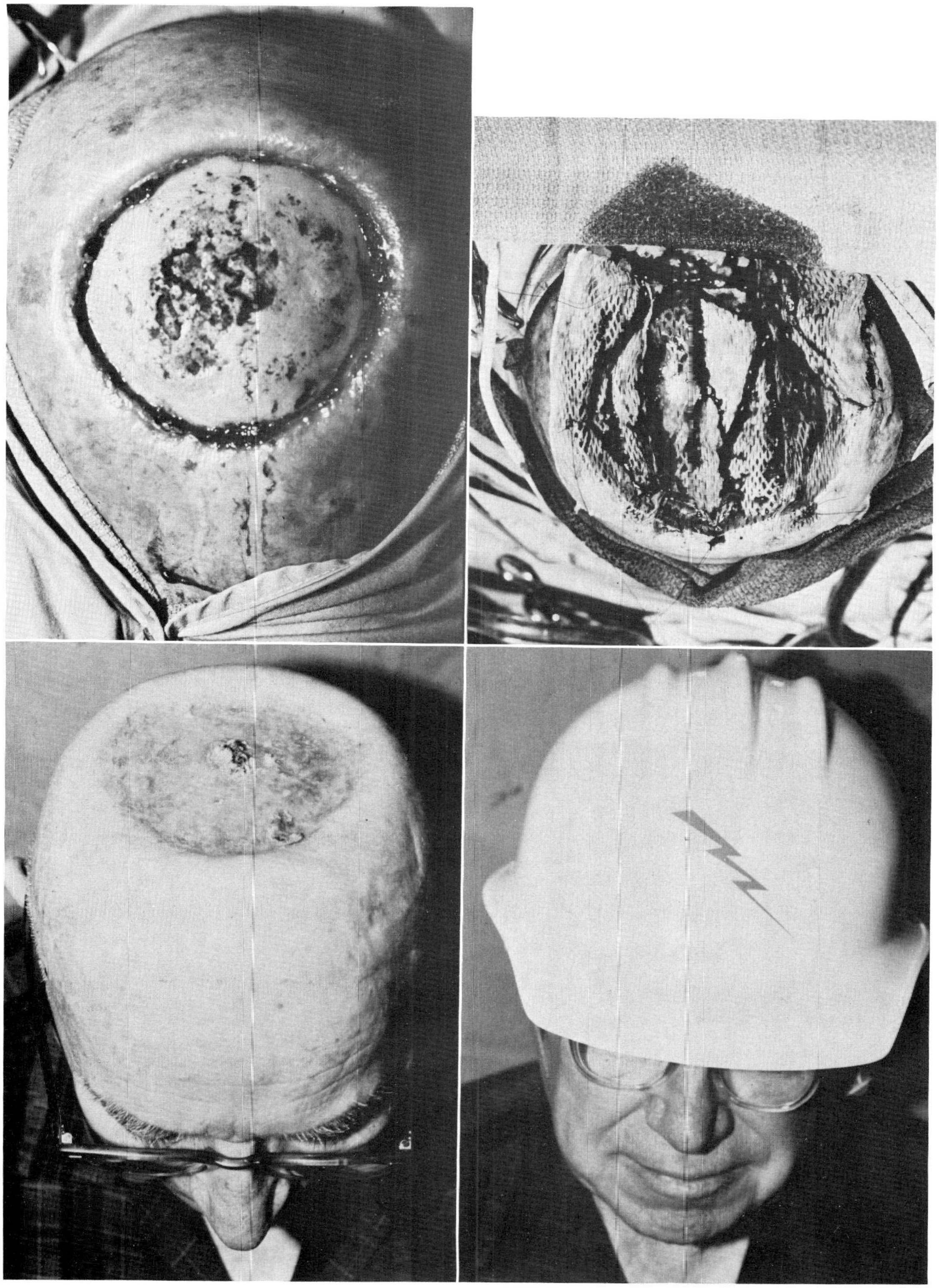

A B C D

Fig. 6-3. Skin grafts.

A, Exposed skull with recurrent fibrosarcoma after chemosurgery.

B, After craniectomy with removal of nonviable skull with tumor invasion, split-thickness skin grafts are applied to dura.

C, Coverage provided by skin grafts. This elderly patient is currently undergoing chemotherapy for tumor recurrence.

D, Hard hat required for protective covering for skull defect.

A

B

C

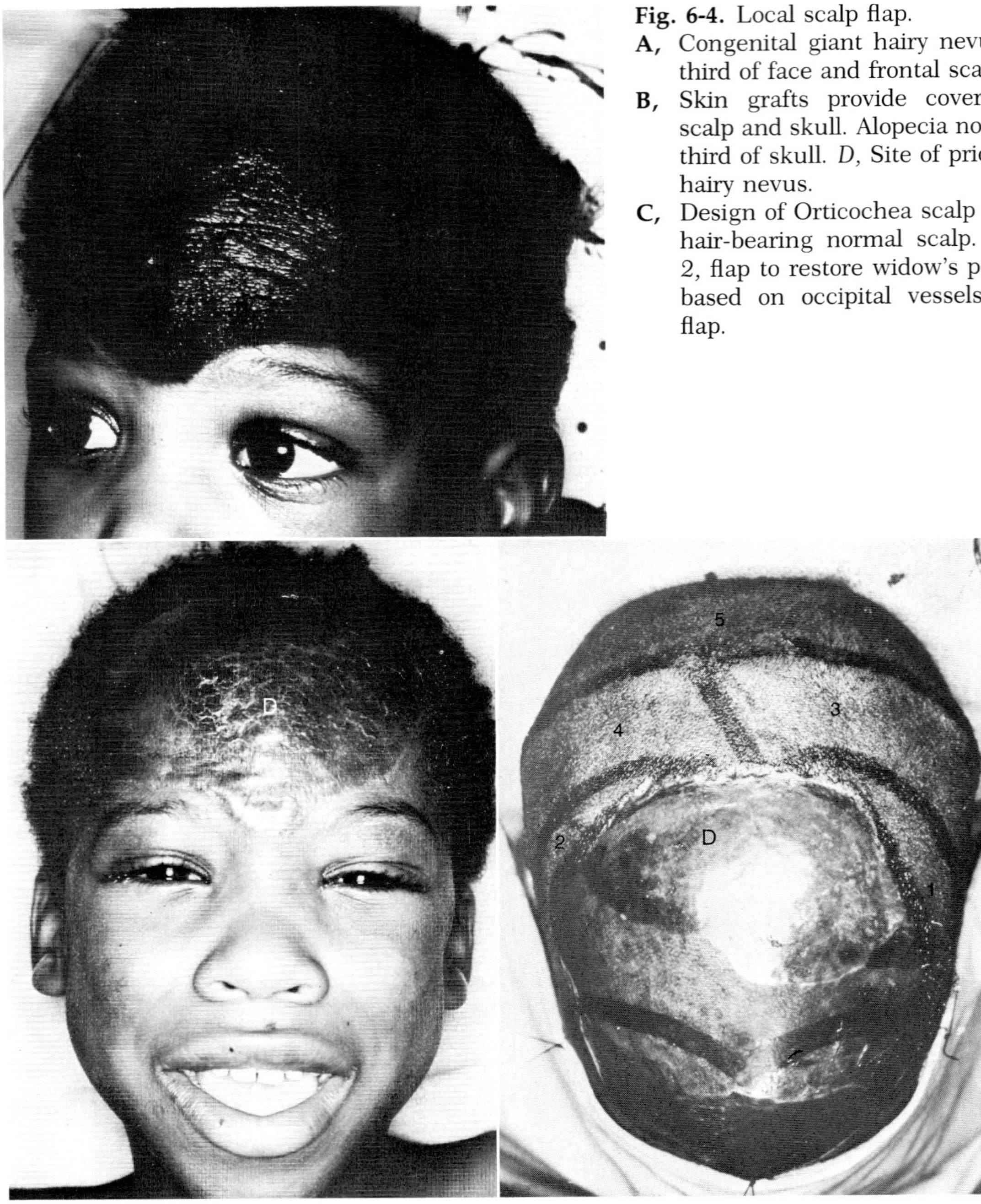

Fig. 6-4. Local scalp flap.

A, Congenital giant hairy nevus involving upper third of face and frontal scalp.

B, Skin grafts provide coverage over anterior scalp and skull. Alopecia noted on anterior one third of skull. *D,* Site of prior excision of giant hairy nevus.

C, Design of Orticochea scalp flaps incorporating hair-bearing normal scalp. *D,* Defect; *1* and *2,* flap to restore widow's peak; *3* and *4,* flaps based on occipital vessels; *5,* advancement flap.

D

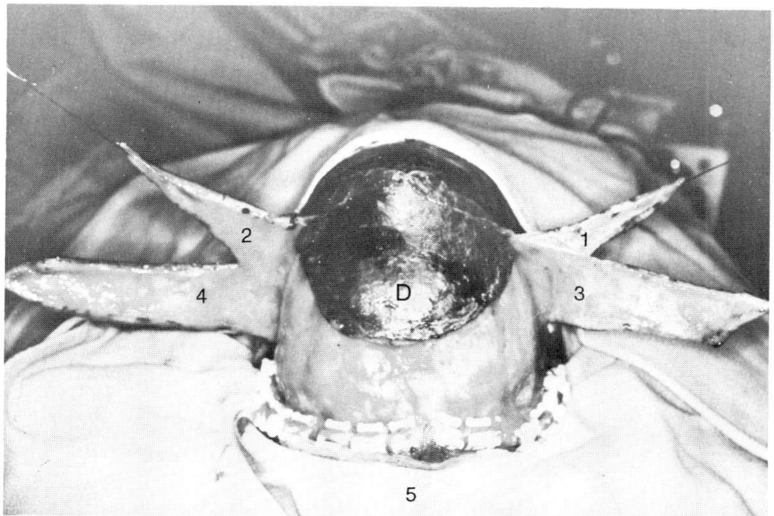

E

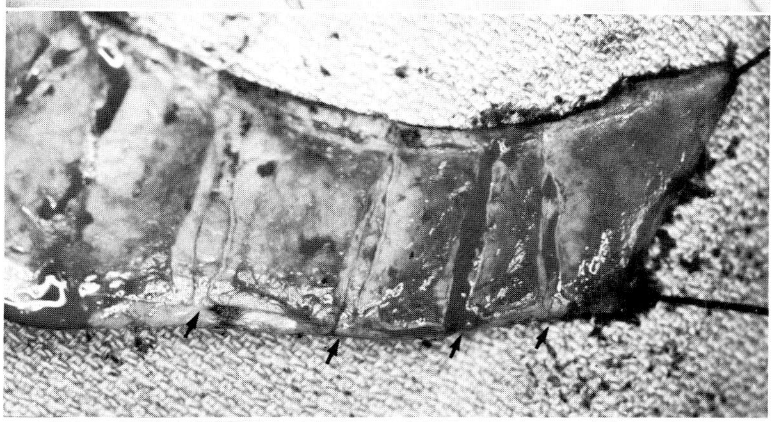

F

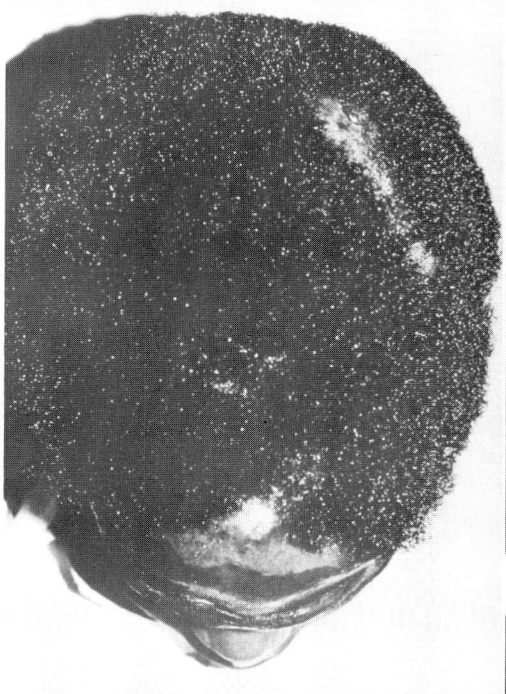

G

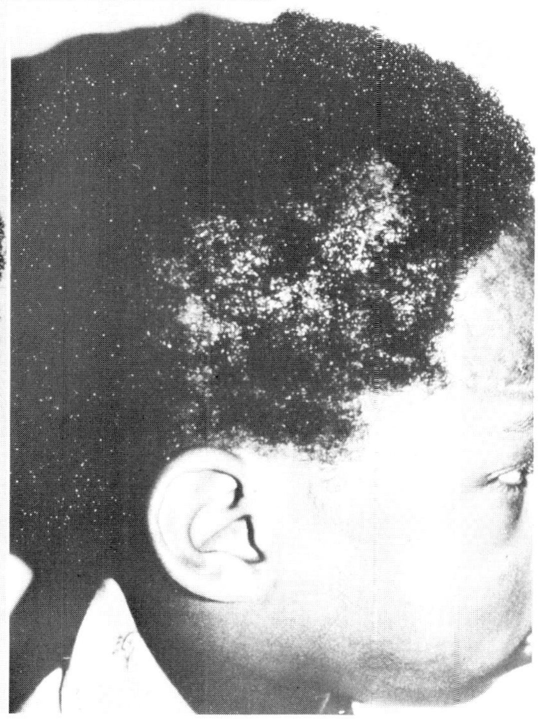

Fig. 6-4, cont'd
D, Scalp flaps elevated between galea and periosteum based on superficial temporal and occipital pedicles. *D,* Site of prior excision of giant hairy nevus.
E, Multiple incisions through galea allow expansion of scalp flaps without injury to axial vessels coursing superficially to galea. Arrows denote sites of incisions into galea. NOTE: Flap expansion at each incision site.
F, Postoperative view demonstrates complete coverage of scalp with hair-bearing skin.
G, Postoperative lateral view demonstrates normal frontal hair line without donor defect.

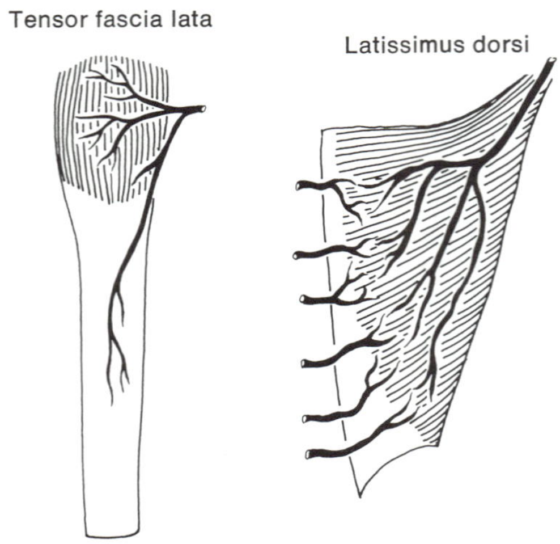

Fig. 6-5

(From Mathes, S.J., and Nahai, F.: Plast. Reconstr. Surg. **67**[2]:178, 1981.)

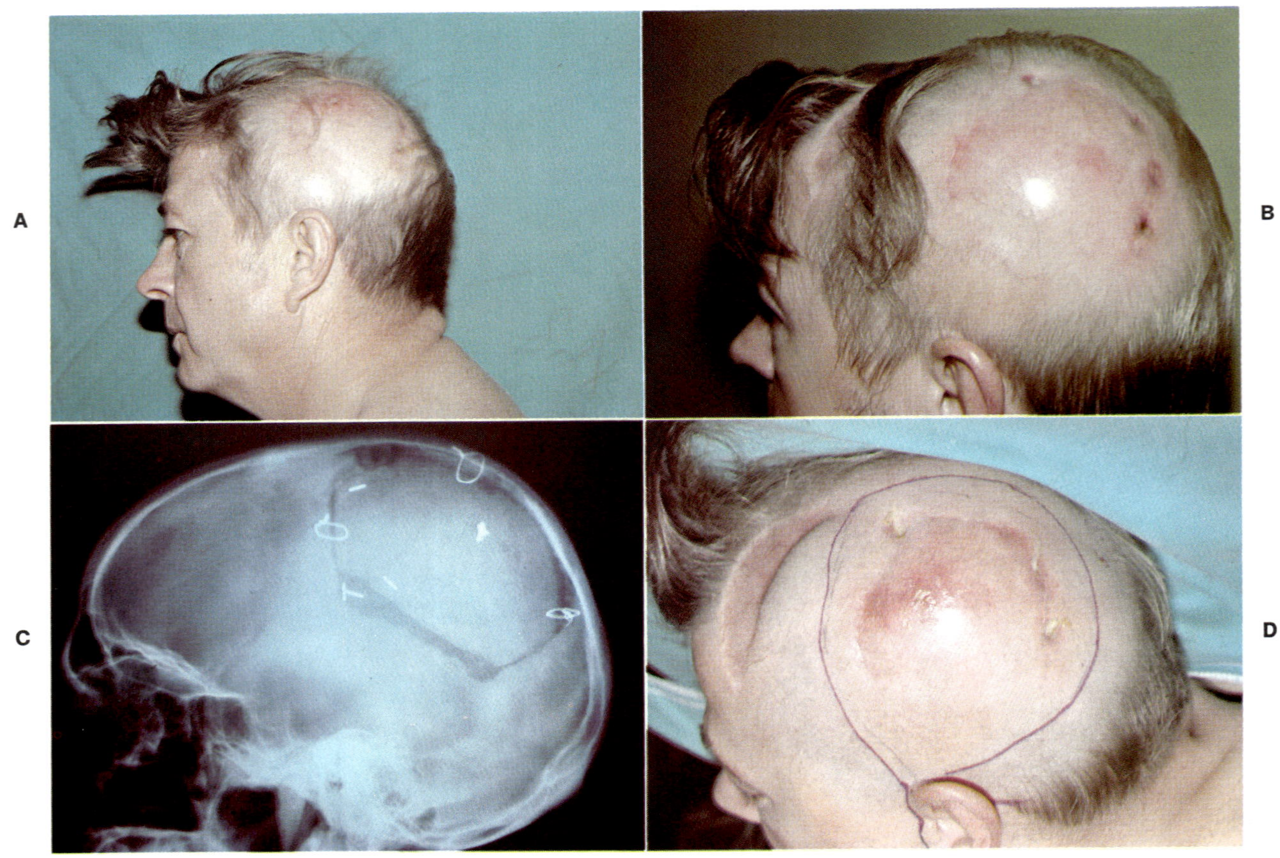

Fig. 6-6. Latissimus dorsi muscle flap transplantation.

A and **B,** Fifty-four-year-old man after resection of intracerebral tumor and irradiation. Radiation necrosis of scalp with infected draining sinuses. Previous local flap had failed.

C, Radiograph of skull.

D, Area of excision outlined.

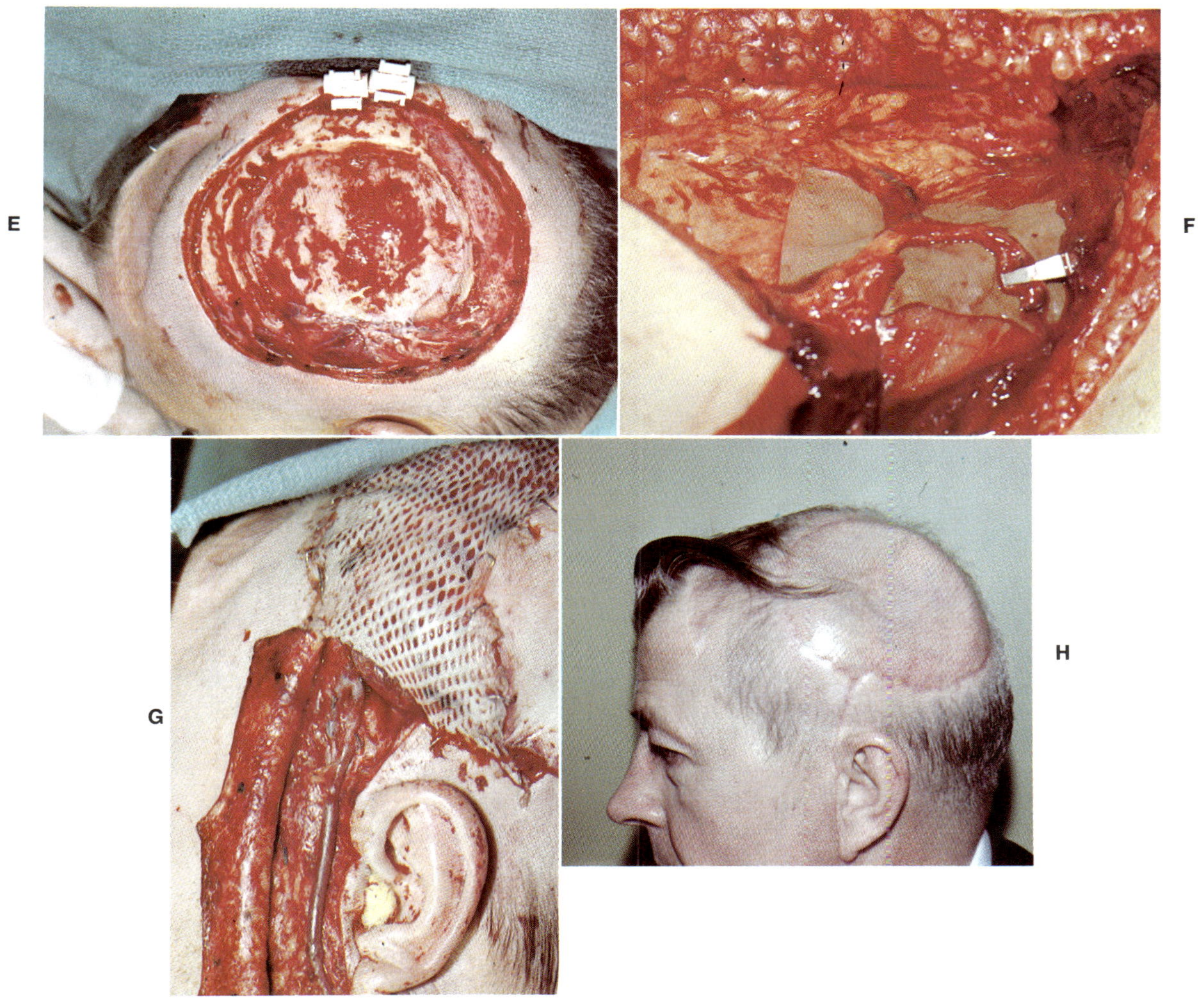

Fig. 6-6, cont'd
E, Scalp and necrotic bone resected and dura exposed.
F, Latissimus dorsi muscle flap elevated. Thoracodorsal artery and vein isolated. Venous branch from serratus anterior dissected (clipped) and used as second vein.
G, Interposition vein graft required between thoracodorsal vein and recipient neck vein. Skin graft placed directly over muscle.
H, Healed scalp several months after operation.

POSTERIOR SCALP (Figs. 6-7 to 6-8)

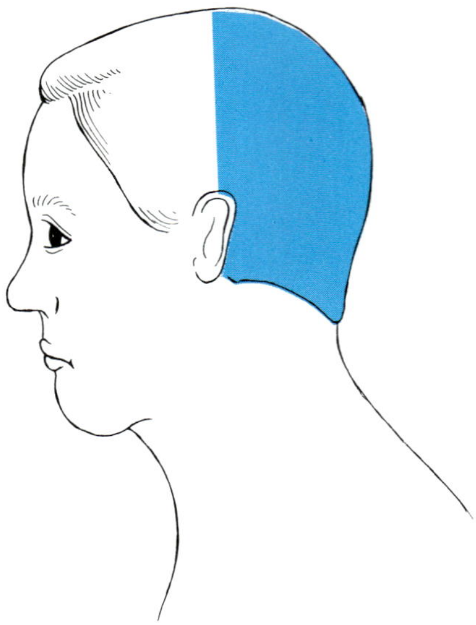

Fig. 6-7

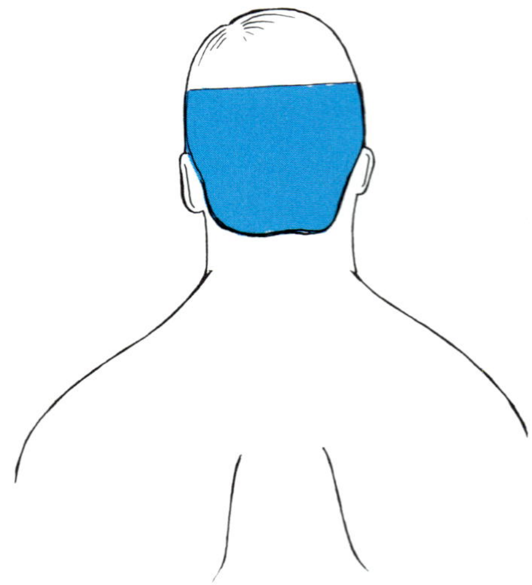

Fig. 6-8

Skin grafts

See anterior scalp

Local flaps

1. See anterior scalp
2. Local flaps are method of choice if available

Distant flaps (Fig. 6-9)

1. Trapezius musculocutaneous flap is distant flap of choice (Fig. 6-10)
2. Latissimus dorsi will reach occipital scalp
3. When distant pedicled flaps are not available, free tissue transplantation is indicated (see anterior scalp)

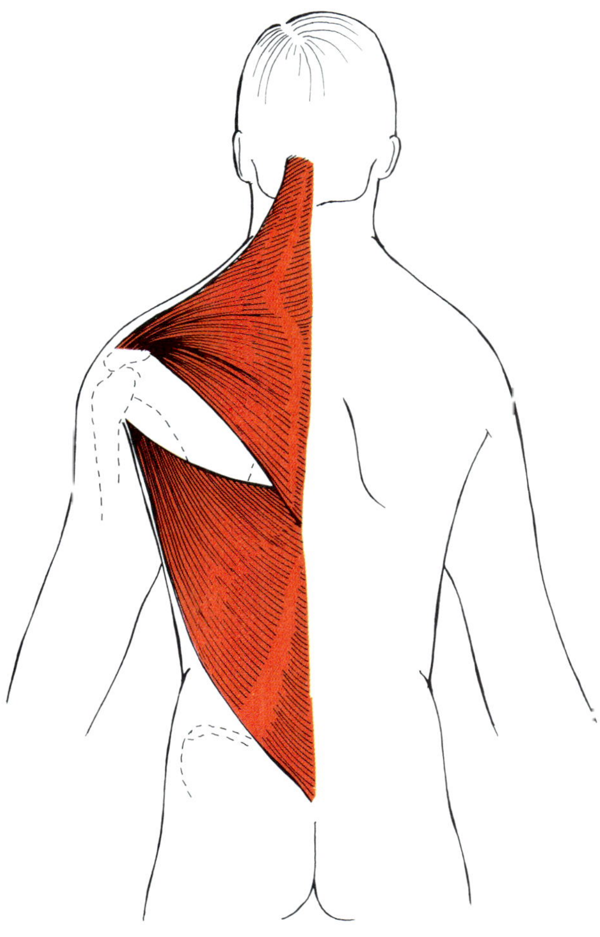

Trapezius
Latissimus dorsi

Fig. 6-9

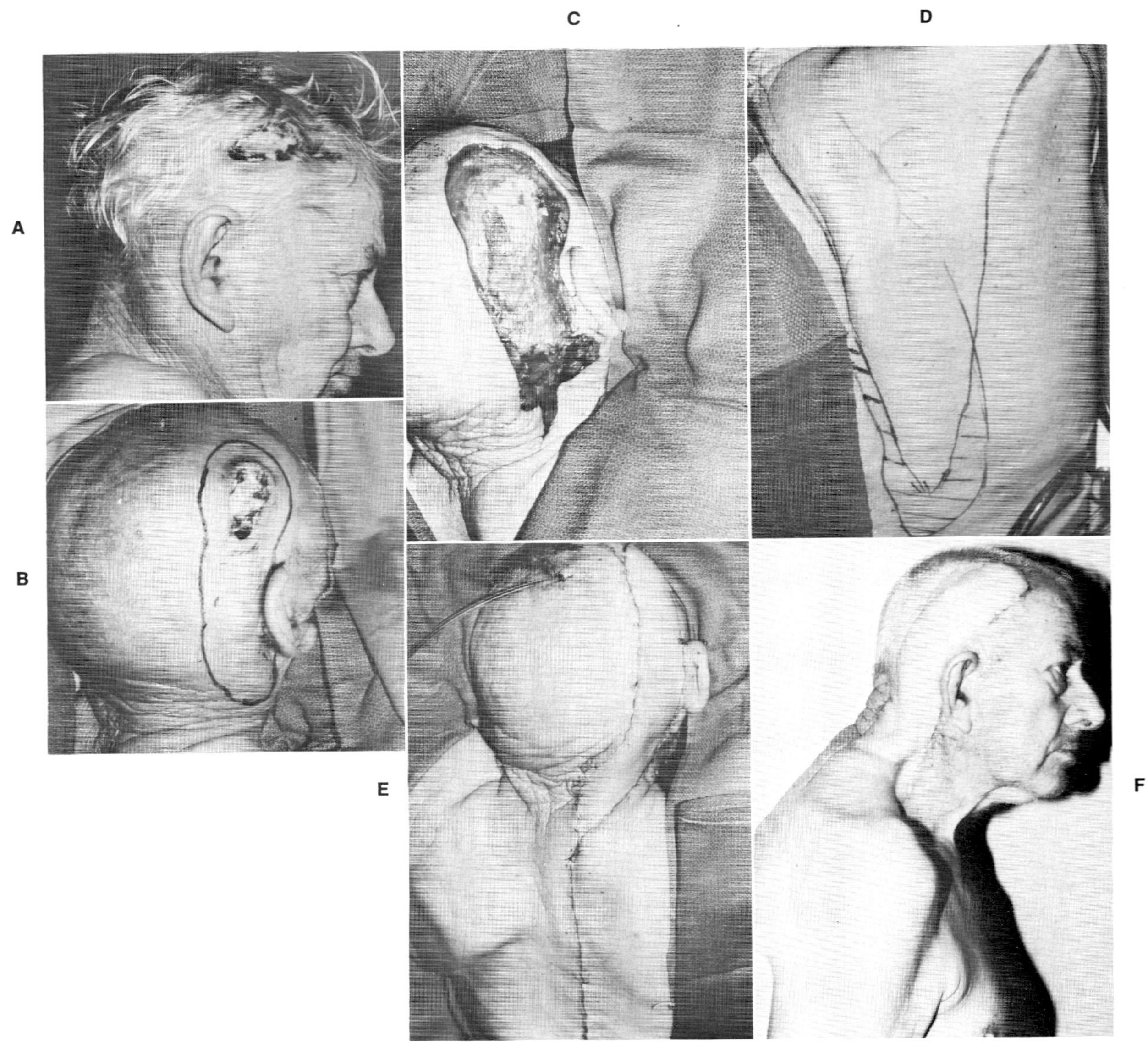

Fig. 6-10. Posterior trapezius musculocutaneous flap for coverage of neck and occipitotemporal scalp.

A, Seventy-year-old man with recurrent squamous cell carcinoma of scalp after radiation therapy.

B, Large positive posterior cervical lymph nodes and in-transit metastases are included in resection.

C, Defect after excision of tumor and posterior neck dissection.

D, Posterior trapezius flap designed.

E and **F,** Flap transposed to cover neck dissection and scalp.

UPPER FACE (Fig. 6-11)

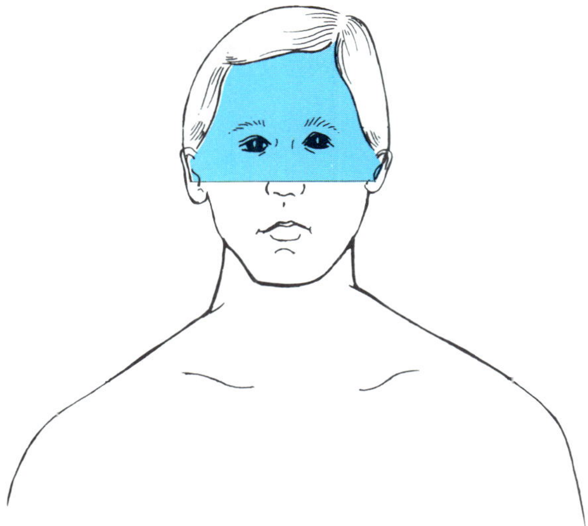

Fig. 6-11

Skin grafts

Split-skin and full-thickness grafts are method of choice

Local flaps (Fig. 6-12)

Temporalis is flap of choice for orbital and maxillary defects (Fig. 6-14)

Distant flaps (Figs. 6-12 and 6-13)

1. Sternocleidomastoid will reach this area
2. Trapezius is alternative flap for lateral defects (Fig. 6-15)
3. Free tissue transplantation is procedure of choice for massive defects involving sinuses (e.g., frontal, ethmoid, and maxillary)
4. Choice of free flaps
   a. Omentum
   b. Latissimus dorsi musculocutaneous
   c. Gracilis muscle with skin graft

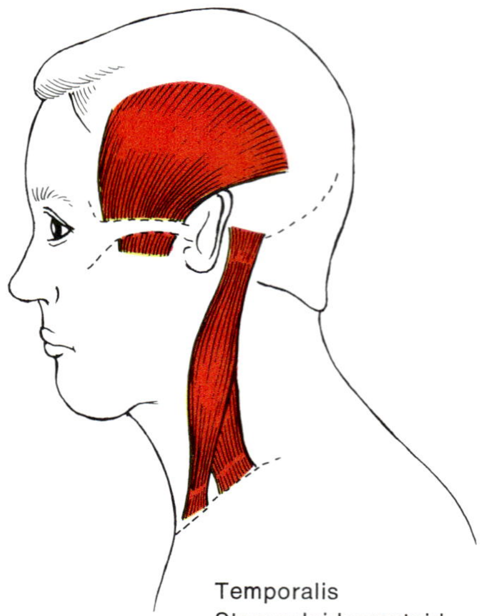

Temporalis
Sternocleidomastoid

Fig. 6-12

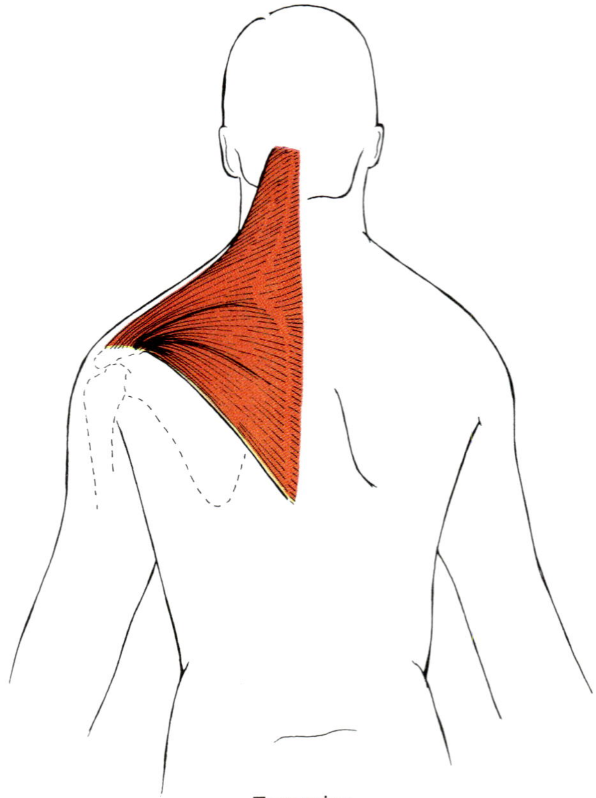

Trapezius

Fig. 6-13

A

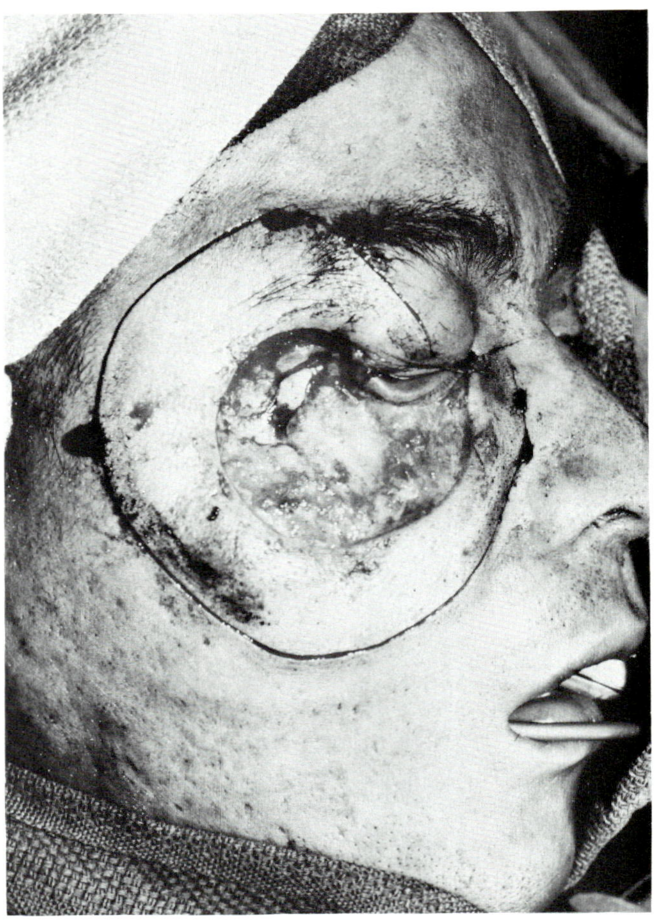

Fig. 6-14. Temporalis muscle flap for upper face coverage.

A, Young male patient has advanced squamous cell carcinoma involving maxillary skin with bone invasion.

B and **C,** Resection includes periorbital skin, zygomatic arch, superior maxillary bone, and orbital floor.

D, Temporal muscle flap is transposed into defect. Muscle is split, and upper portion of muscle used to cover upper lid defect. Lower muscle is used to cover maxillary sinus and provide support for orbital floor. Skin grafts are placed over muscle.

E, Postoperative view demonstrates stable wound coverage with use of temporalis muscle flap and skin grafts. Vision is preserved.

(Cases similar to **A** and **B** shown in Mathes, S.J., and Vasconez, L.O.: Head, neck and truncal reconstruction with the musculo-cutaneous flap: anatomical and clinical considerations, Transactions of the VII International Congress of Plastic and Reconstructive Surgery, 1979, Sao Paulo, Brazil, 1980, Cartgraf, pp. 178-182.)

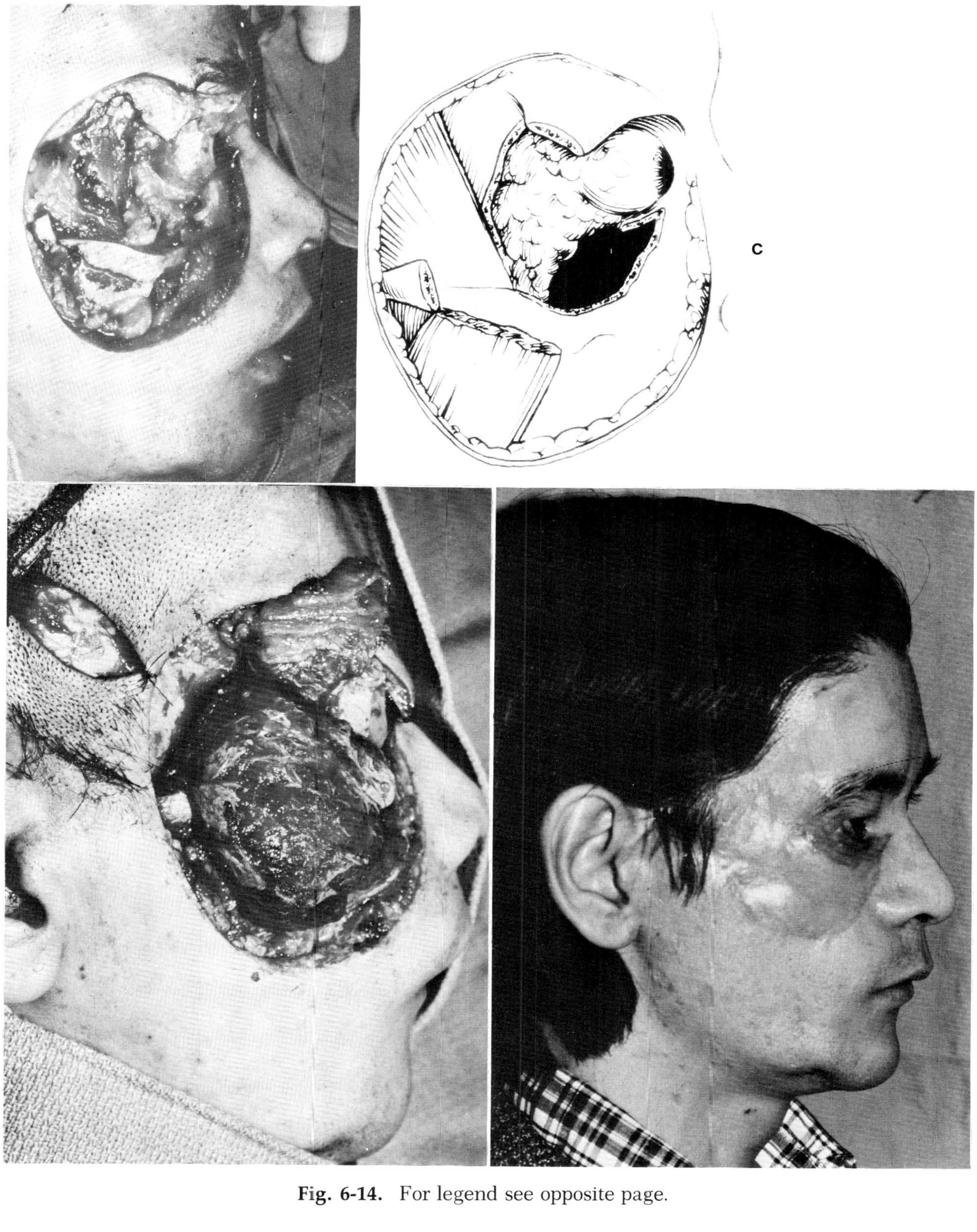

Fig. 6-14. For legend see opposite page.

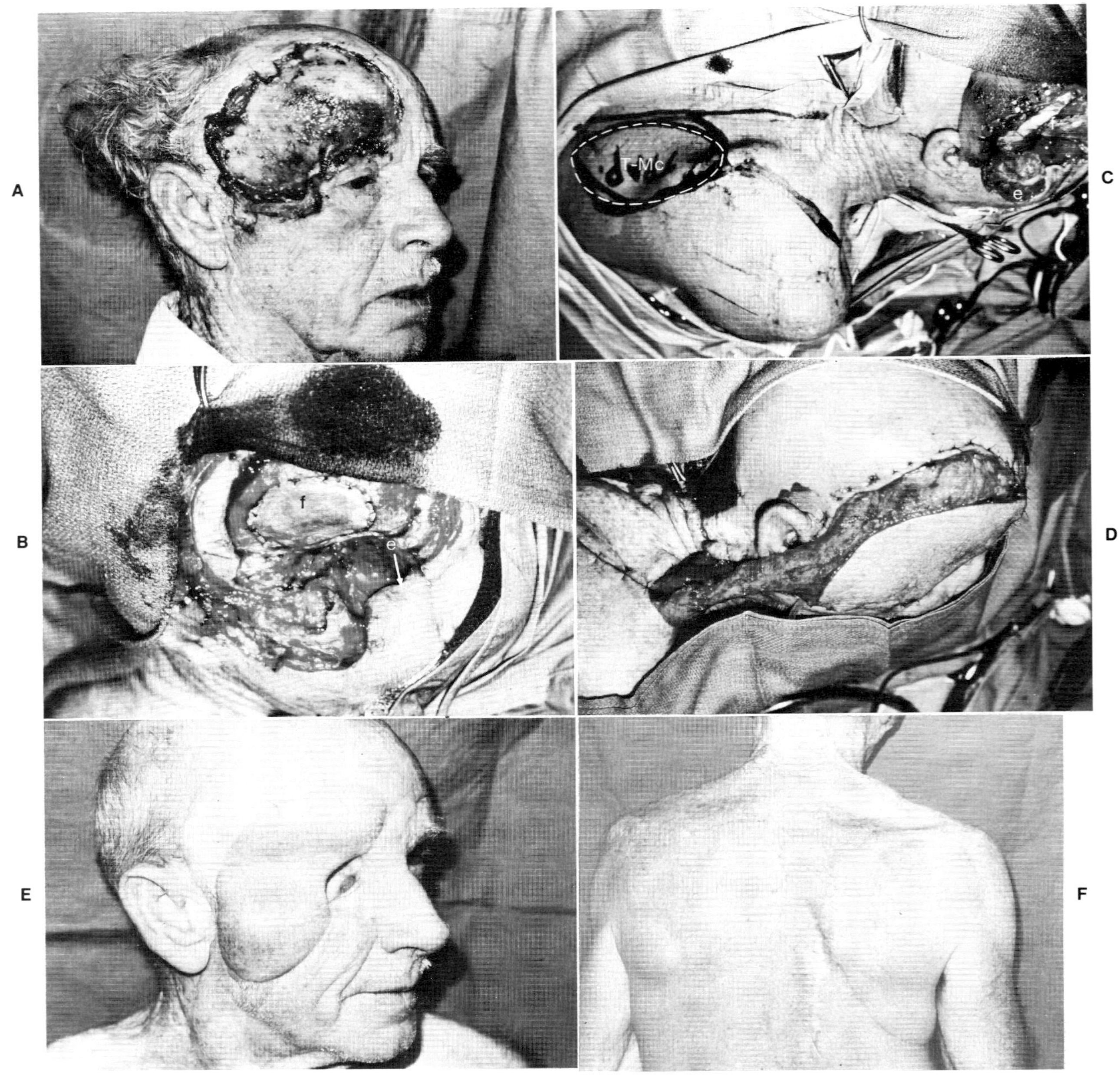

Fig. 6-15. Posterior trapezius musculocutaneous flap.

A, Patient has advanced squamous cell carcinoma involving upper and lateral midface. Patient has previously received both chemosurgery and radiation therapy.

B, Frontal-temporal craniectomy reveals bony invasion with dural involvement. Dural continuity restored with fascia lata graft (*f*). Orbital exenteration also performed because of tumor invasion into orbital cavity (*e*).

C, Design of posterior trapezius musculocutaneous flap for coverage of lateral, upper, and midface defect. *T-Mc,* Design of posterior trapezius musculocutaneous flaps; *e,* orbital exenteration defect.

D, Trapezius musculocutaneous flap covers defect. Carrier muscle is skin grafted.

E, Carrier muscle is divided and flap inset 6 weeks after extirpative procedure.

F, Postoperative view demonstrates direct closure of donor defect. Preservation of anterior trapezius fibers and eleventh cranial nerve preserves function and avoids drooping of right shoulder.

LOWER FACE (Fig. 6-16)

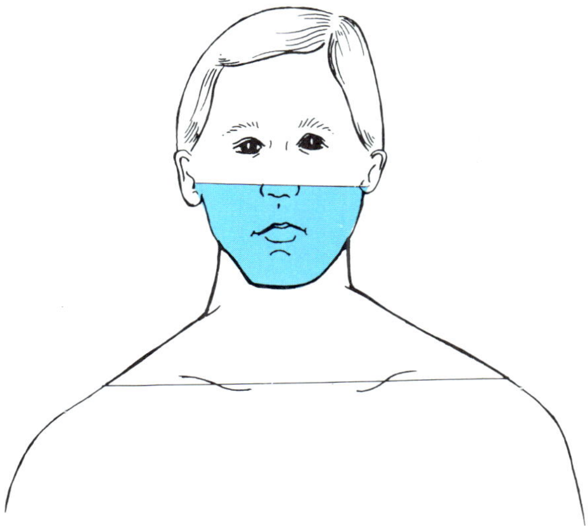

Fig. 6-16

Skin grafts

Skin grafts are method of choice for small, shallow defects

Local flaps

Local axial flaps (e.g., nasolabial and cervical flaps) are method of choice for small to moderate defects

Distant flaps (Figs. 6-17 and 6-18)

1. Large defects after extirpation surgery for tumors require distant tissue for coverage
2. Platysma musculocutaneous flap is flap of choice when local skin flaps are not available for small defects (Fig. 6-19)
3. Sternocleidomastoid muscle or musculocutaneous flap will cover this area
4. Pectoralis major musculocutaneous flap is flap of choice for composite defects of lower face (Fig. 6-20)
5. Trapezius musculocutaneous flap is preferred for coverage of the ear and parotid regions (Fig. 6-21)
6. Deltopectoral axial flap and latissimus dorsi musculocutaneous flap are alternatives to 4 and 5
7. Free tissue transplantation is considered when preceding methods are not available

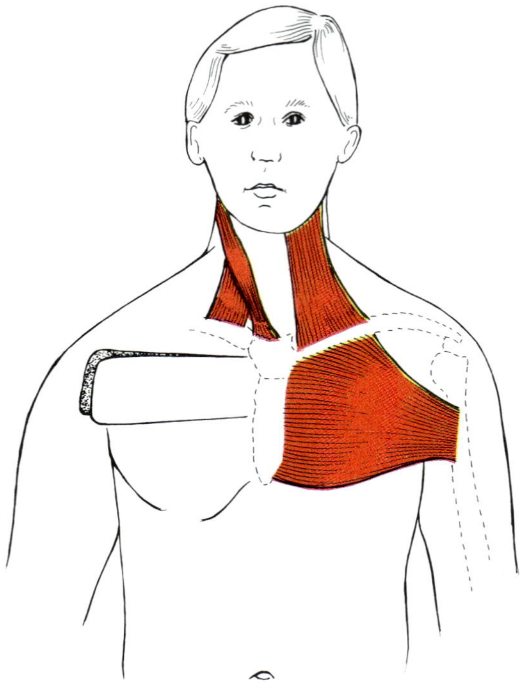

Platysma
Sternocleidomastoid
Pectoralis major
Deltopectoral flap

Fig. 6-17

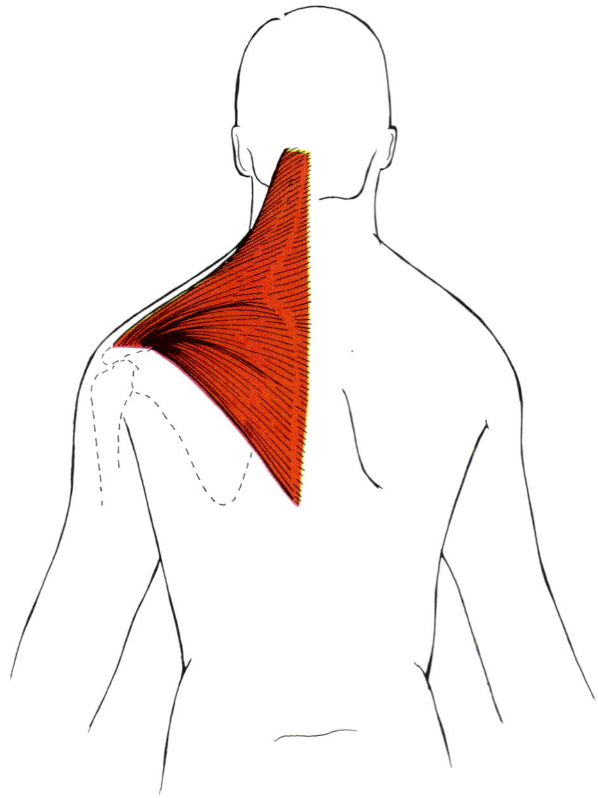

Trapezius

Fig. 6-18

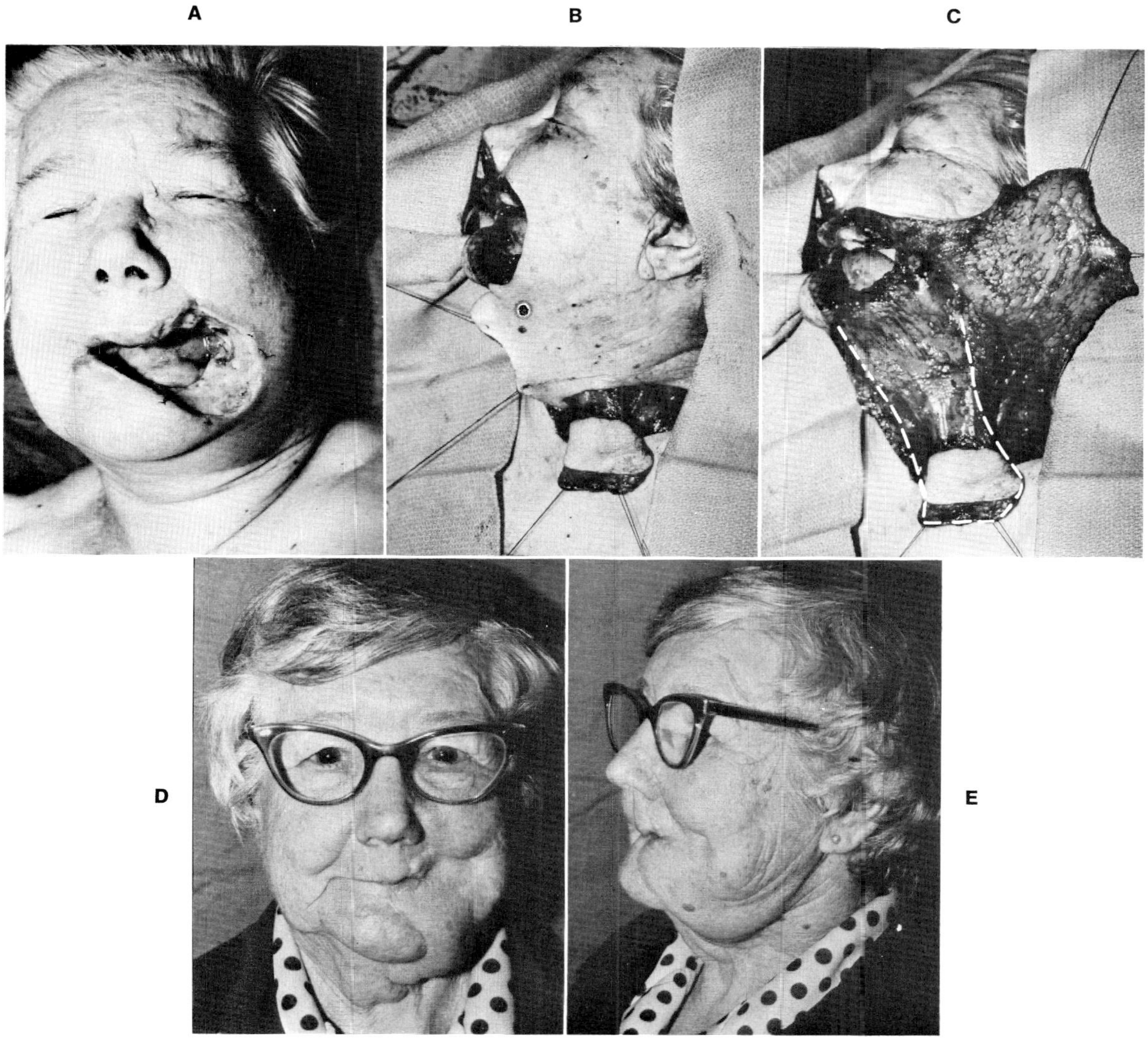

Fig. 6-19. Platysma musculocutaneous flap.
A, Buccal mucosa and lower face defect after chemosurgery for squamous cell carcinoma of oral commissure.
B, Wound reexcision performed. Left platysma musculocutaneous flap elevated for transposition to oral cavity for reconstruction of buccal sulcus.
C, Anterior neck flap elevated for lower face skin coverage Platysma musculocutaneous flap is ready for inset into buccal sulcus.
D, One-year postoperative examination demonstrates adequate oral lining provided by platysma skin island with no oral incontinence.
E, Lateral view demonstrates no donor site deformity. Adequate skin coverage provided by anterior neck flap.

A

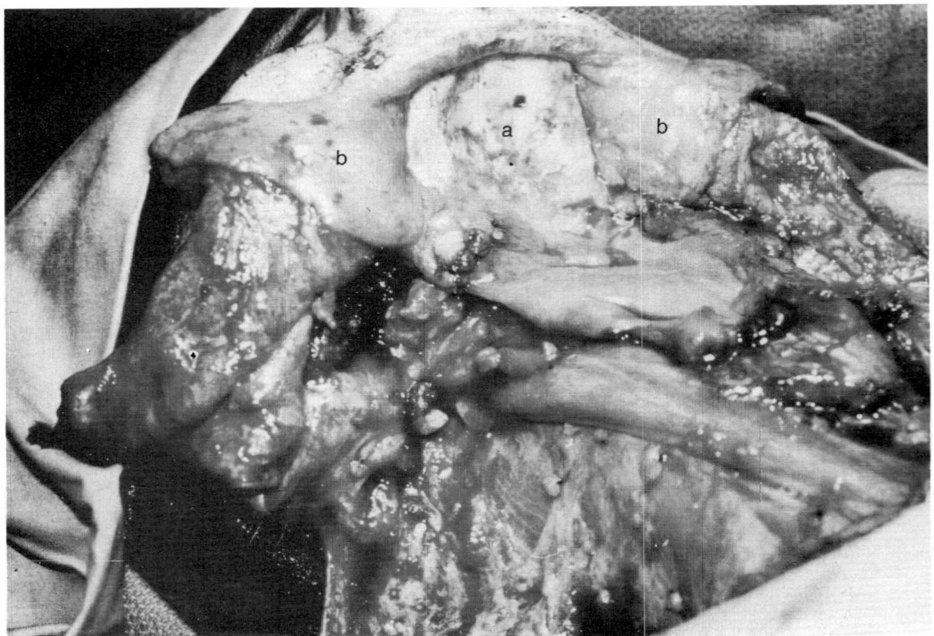

B

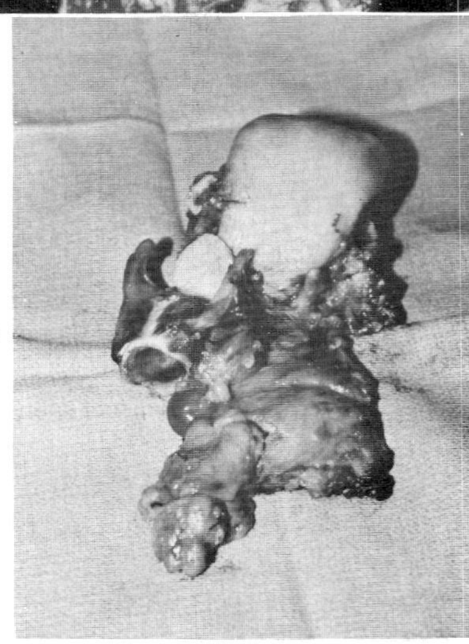

Fig. 6-20. Pectoralis major musculocutaneous flap for closure of chin defect and reconstruction of floor of mouth.

A, Resultant defect after resection of advanced floor of mouth squamous cell carcinoma invading mandible and chin. *a,* palate; *b,* buccal mucosa.

B, Resected specimen.

C, Pectoralis major flap musculocutaneous elevated.

D, Flap passed beneath neck skin. Skin island folded on itself so part will reconstruct chin and part will reconstruct floor of mouth.

E, Portion of skin island beneath lip is deepithelialized *(arrow)*.

F, Three-month postoperative view. Skin island visualized on chin and floor of mouth.

G, Donor area closed with thoracoepigastric flap.

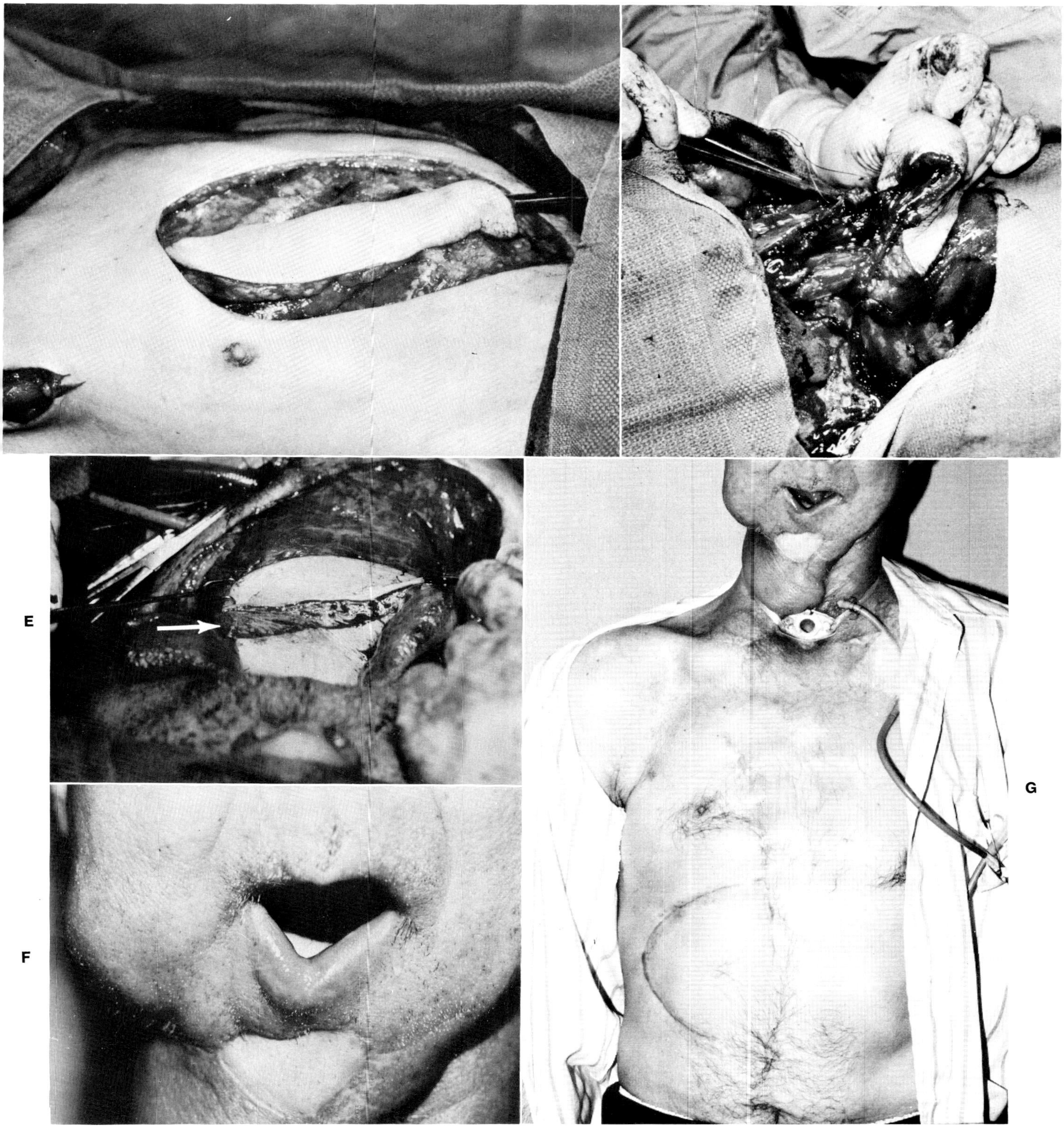

D

E

F

G

Fig. 6-20. For legend see opposite page.

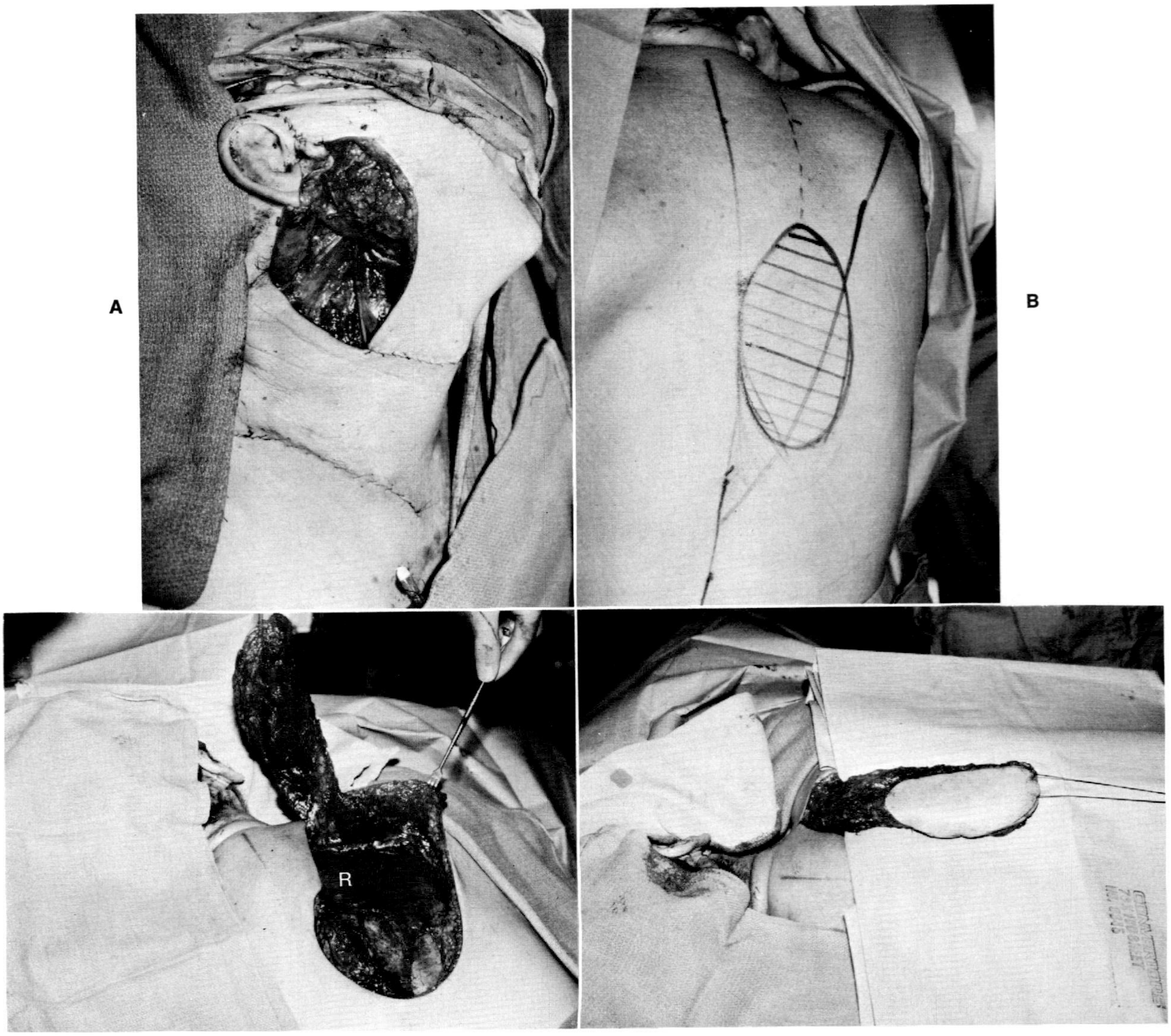

Fig. 6-21. Posterior trapezius musculocutaneous flap.

A, Defect after resection of preauricular level IV melanoma with associated neck lymphadenectomy. Coverage required for exposed branches of facial nerve and reconstruction of large skin defect.

B, Design of posterior trapezius musculocutaneous flap.

C, Posterior view of elevated posterior trapezius musculocutaneous flap. NOTE: plane of dissection between trapezius fibers and rhomboid muscles (*R*).

D, Demonstration of relationship of posterior skin island with underlying trapezius muscle.

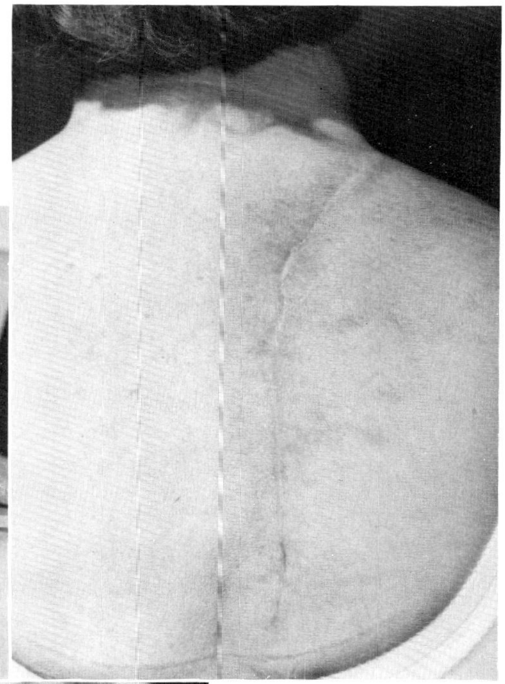

F

E

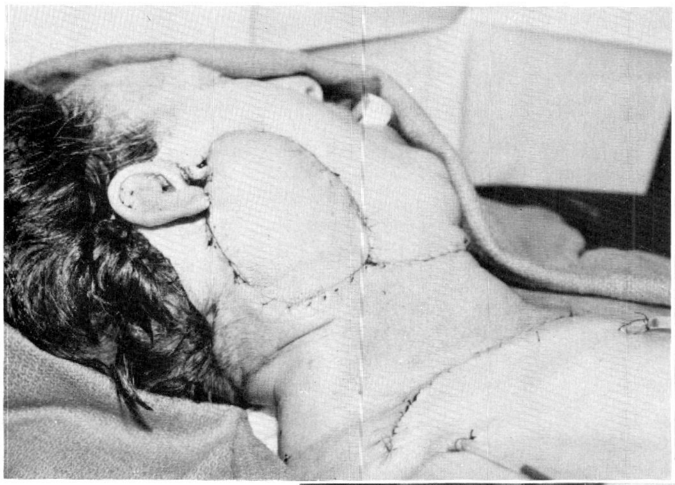

G

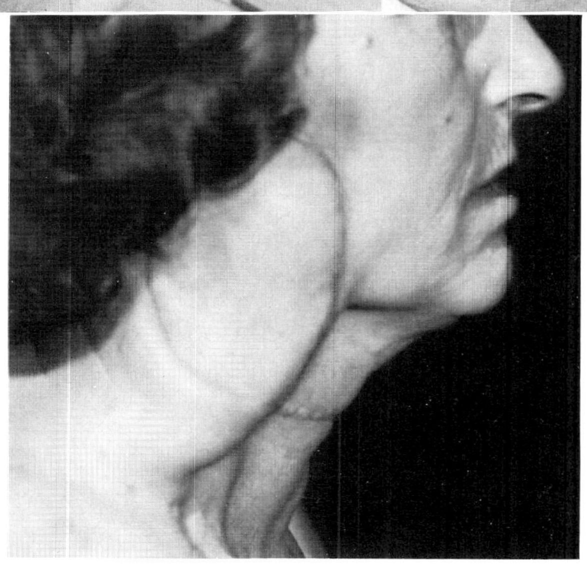

Fig. 6-21, cont'd

E, Posterior trapezius musculocutaneous flap tunneled beneath McFee incisions. Muscle will simulate absent sternocleidomastoid muscle.

F, Donor defect closed directly. NOTE: Anterior trapezius muscle intact with function preservation, avoiding shoulder droop.

G, Three-month postoperative view demonstrates stable coverage provided by posterior trapezius musculocutaneous flap. Flap debulking required in future procedure.

ORAL CAVITY

Skin grafts

Skin and mucosal grafts are method of choice for small defects with appropriate recipient site

Local flaps

Axial skin flaps (e.g., nasolabial and tongue flaps) are useful for small defects

Distant flaps (Figs. 6-22 and 6-23)

1. Platysma musculocutaneous flap is flap of choice for small defects when local flaps not available (Fig. 6-24)
2. Pectoralis major musculocutaneous flap is flap of choice for extensive defects, particularly with mandibular resection (Figs. 6-25 and 6-26)
3. Posterior trapezius and latissimus dorsi musculocutaneous flaps and deltopectoral axial flap are alternative flaps
4. Free tissue transplantation
    a. Alternative approach for intraoral reconstruction of extensive defects
    b. Procedure of choice for immediate reconstruction of mandible and soft tissues
    c. Suitable receptor vessels are a prerequisite
    d. Choice of free flap
        (1) Deep circumflex iliac flap (iliac bone and overlying skin based on deep circumflex iliac artery)
        (2) Dorsalis pedis flap

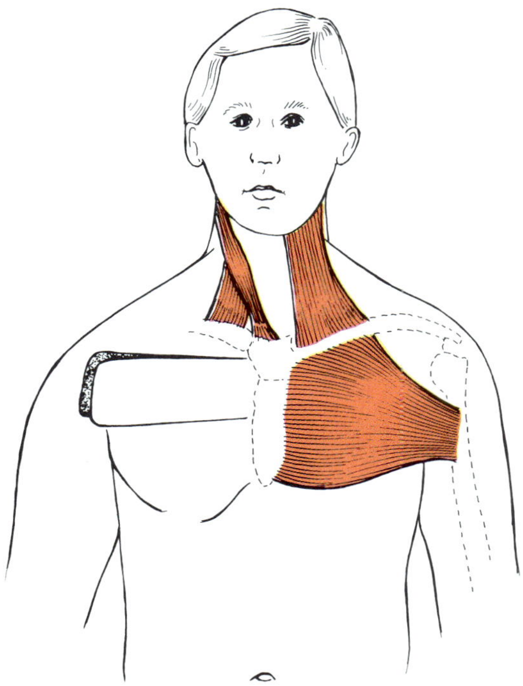

Fig. 6-22

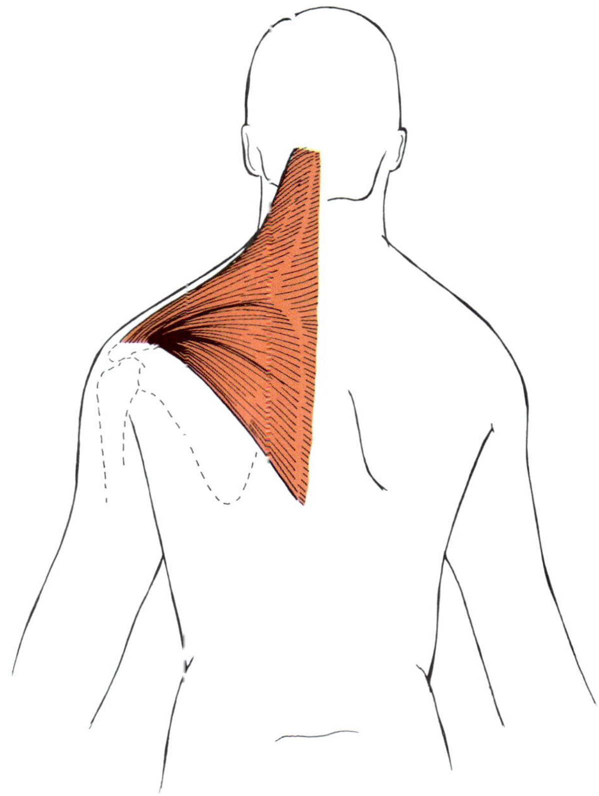

Fig. 6-23

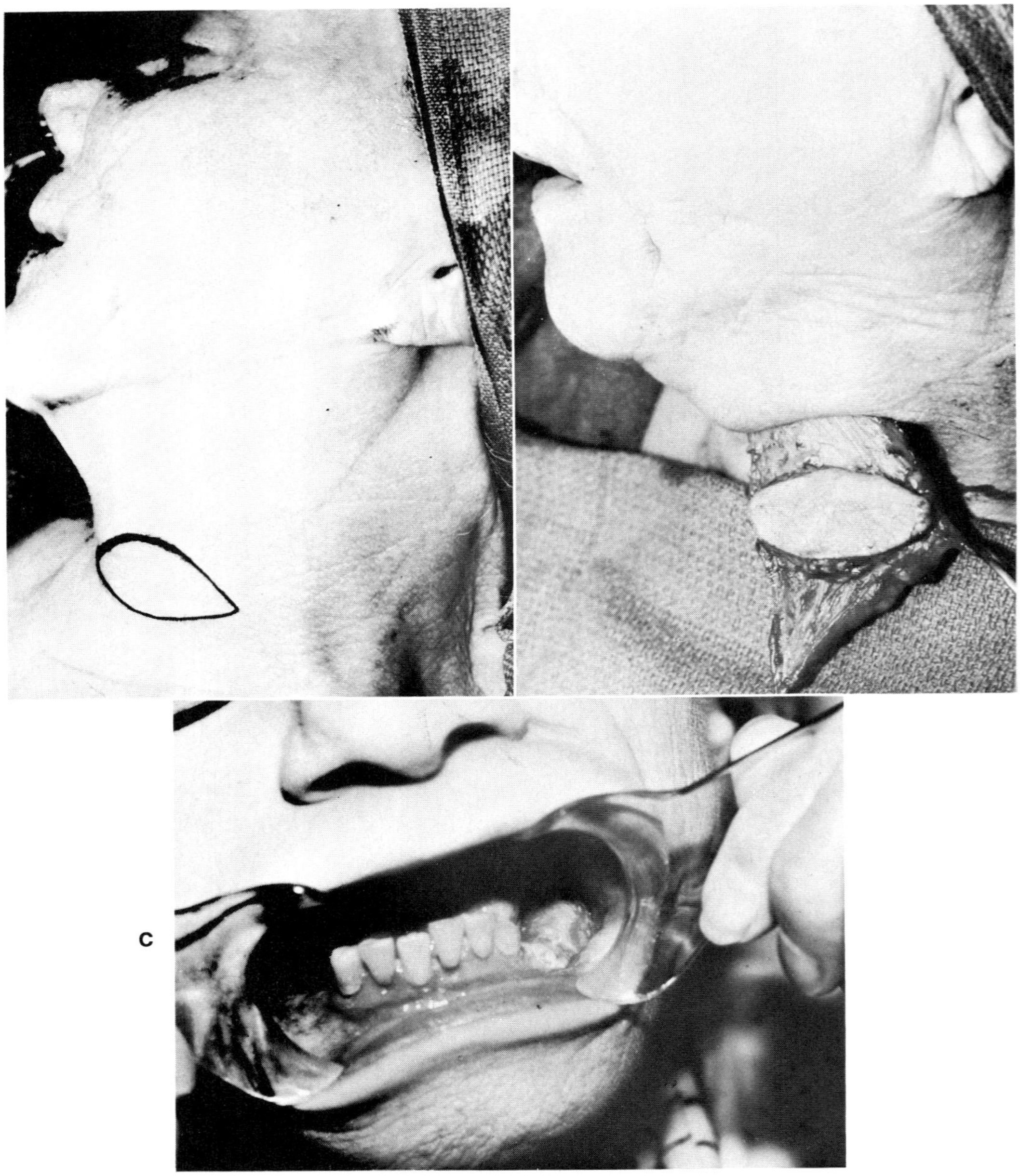

Fig. 6-24. Platysma musculocutaneous island flap for coverage of left mandible. Sixty-eight-year-old woman developed osteoradionecrosis of left mandible after radiation for tongue carcinoma. Previous nasolabial flap for coverage of mandible had failed.

A, Platysma skin island outlined.
B, Flap elevated.
C, Stable coverage of mandible provided by platysma musculocutaneous flap.

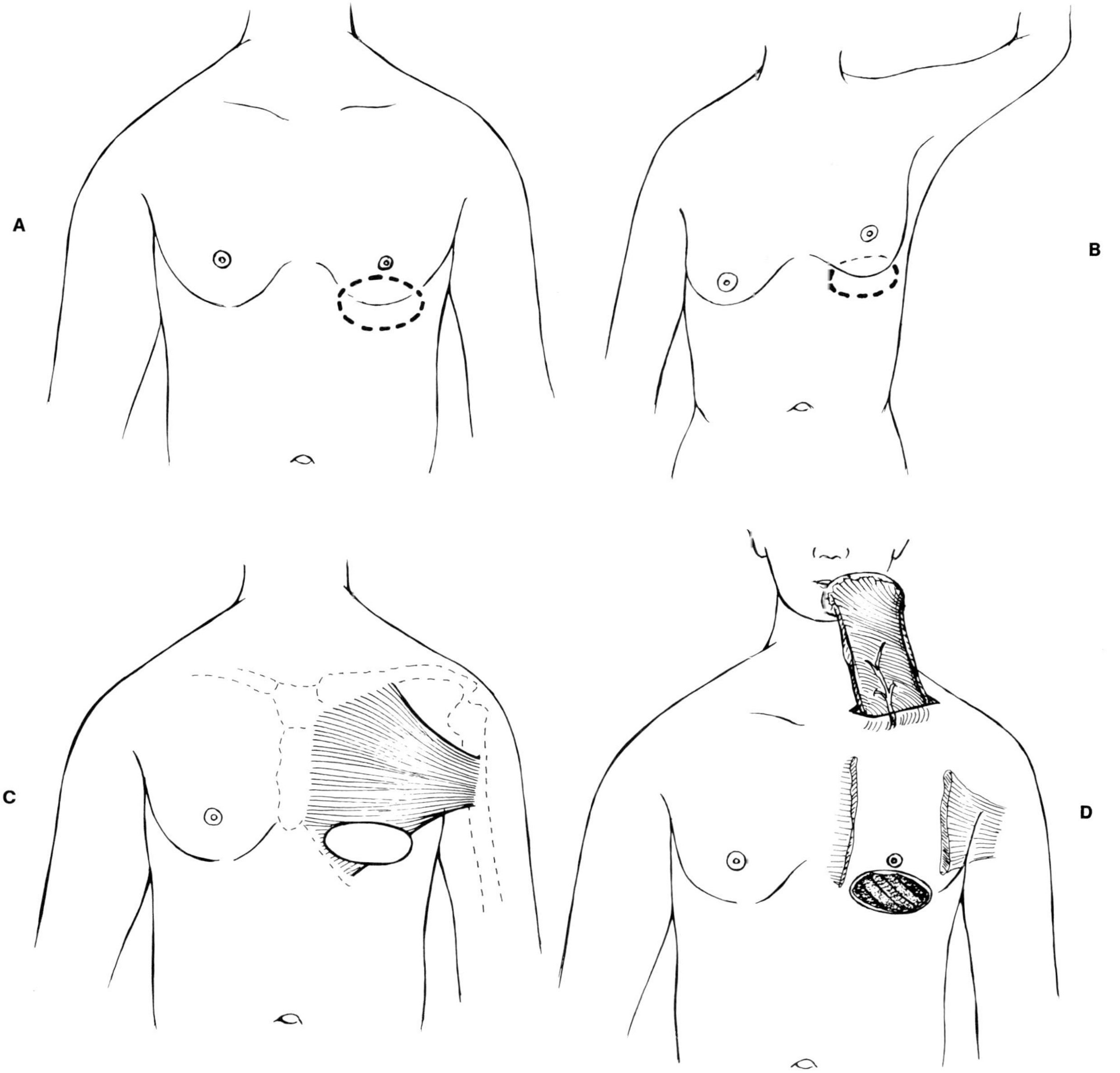

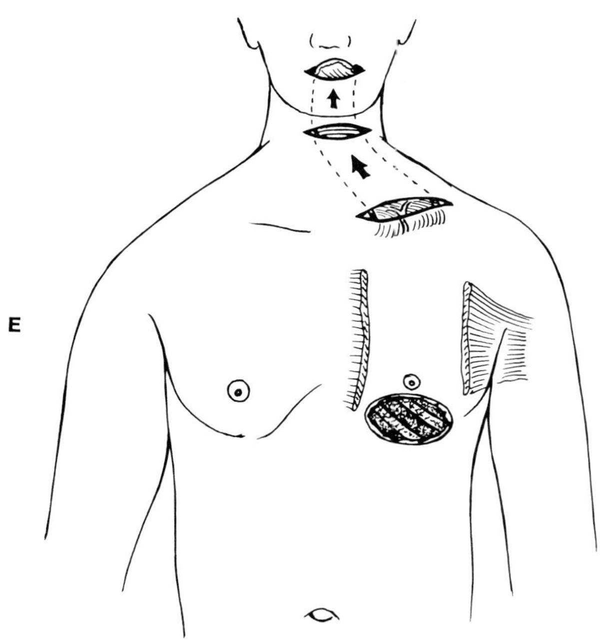

Fig. 6-25

A, Design of skin island for pectoralis major musculocutaneous flap for head and neck reconstruction.

B, Skin island is located in region of inframammary line to avoid breast deformity in female patient.

C, Skin island is located over inferior margin of pectoralis major muscle for use in oral cavity coverage.

D, Release of pectoralis major origin and insertion allows transposition of musculocutaneous flap to head and neck region.

E, The distal flap with skin island is tunneled beneath neck skin into oral cavity. Donor defect is closed directly.

A B C D

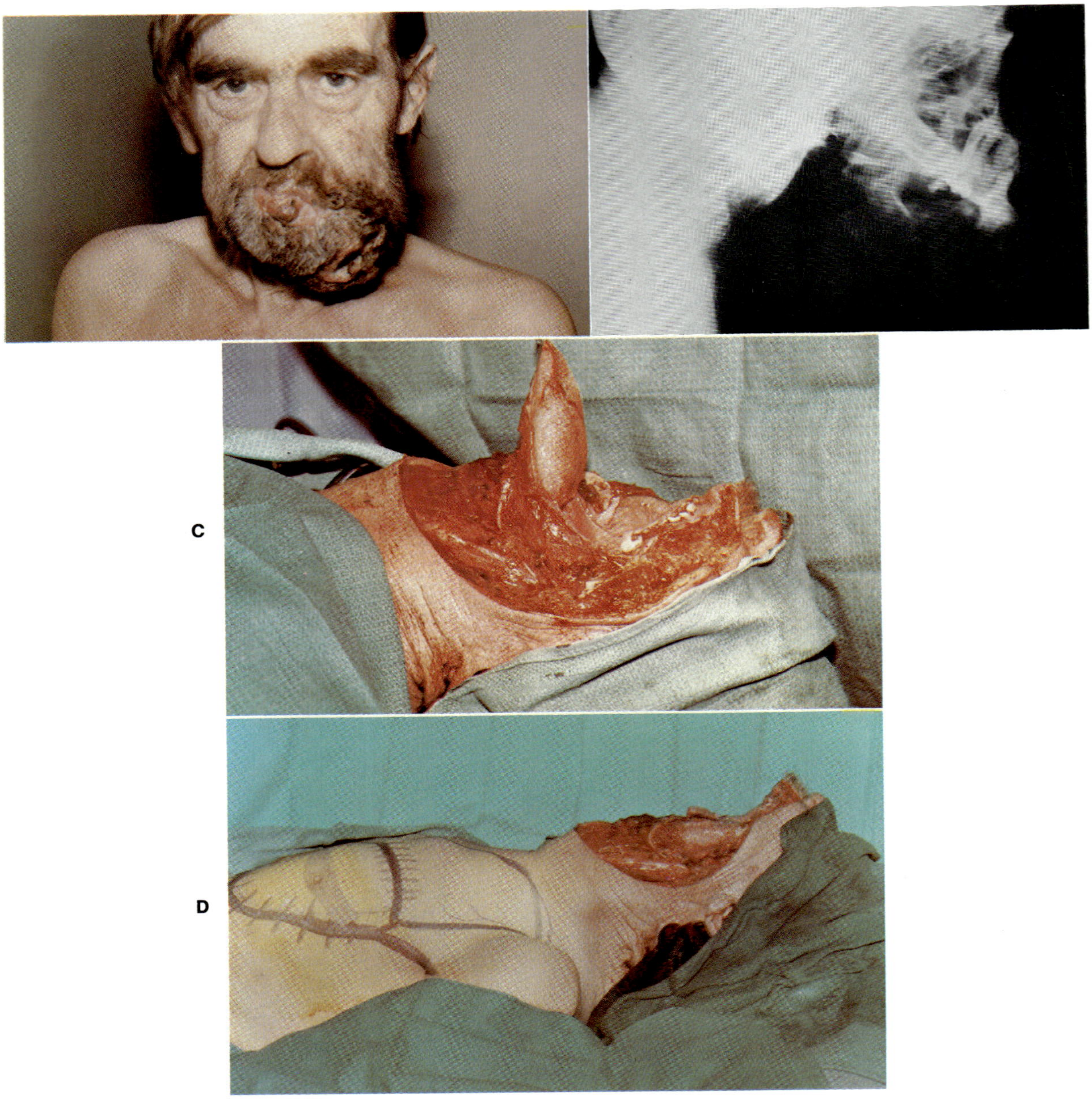

Fig. 6-26. Reconstruction of mandible, intraoral cavity, chin, and neck with pectoralis major musculocutaneous flap, including a portion of vascularized rib.

A Sixty-five-year-old man with advanced squamous cell carcinoma of lip.

B, Invasion of left hemimandible is seen on radiograph.

C, Resultant defect after resection of tumor, which included left hemimandibulectomy, left radical neck dissection, and right suprahyoid dissection.

D, Large pectoralis musculocutaneous flap with vascularized rib is designed.

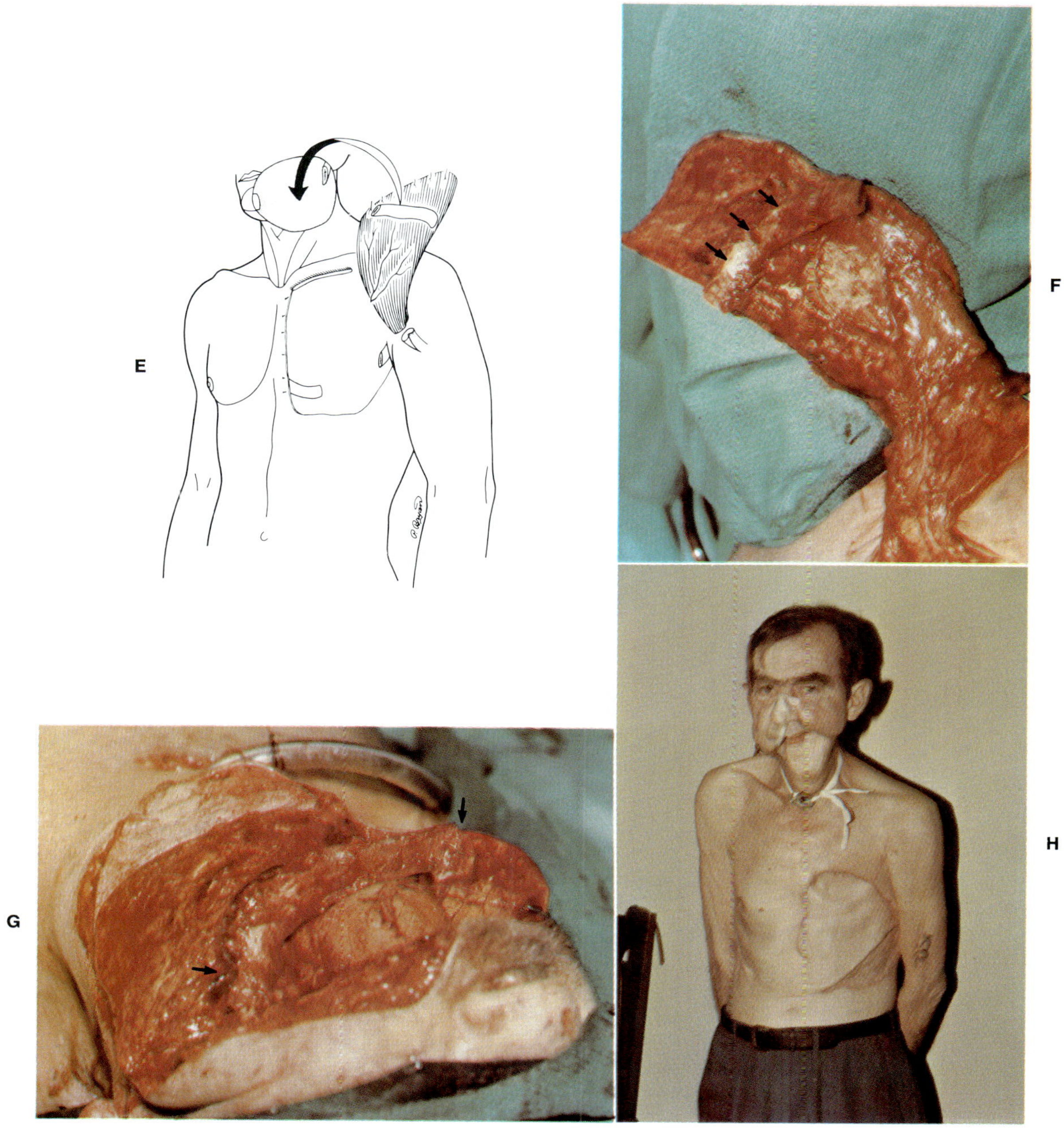

Fig. 6-26, cont'd
E and **F**, Flap elevated. Arrows point to rib.
G, Rib is wired to mandible *(arrow)*, and skin and muscle are folded over rib and sutured to oral mucosa.
H, Postoperative result. Donor area closed with thoracoepigastric flap. (Nipple was excised at later date.)

ANTERIOR NECK (Fig. 6-27)

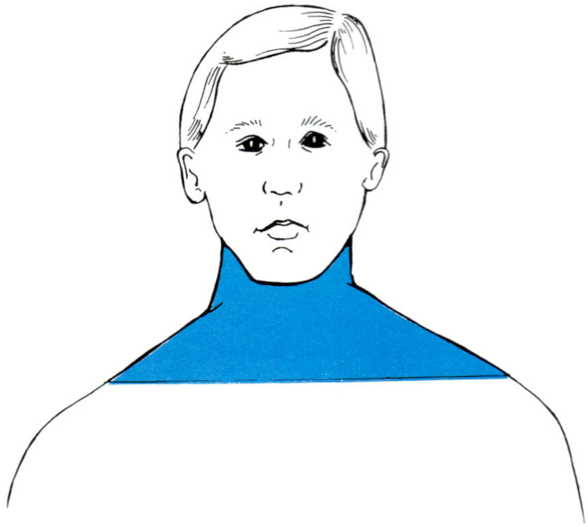

Fig. 6-27

Skin grafts

Skin grafts are method of choice when major structures are not exposed

Local flaps (Fig. 6-28)

1. Platysma musculocutaneous flap is flap of choice for small midline defects (Fig. 6-30)
2. Sternocleidomastoid muscle and musculocutaneous flaps are alternative local flaps

Distant flaps (Figs. 6-28 and 6-29)

1. Anterior trapezius musculocutaneous flap is flap of choice for extensive defects of the neck
2. Latissimus dorsi musculocutaneous flap will reach anterior neck (Fig. 6-31)
3. Pectoralis major musculocutaneous flap is alternative flap (Fig. 6-32)

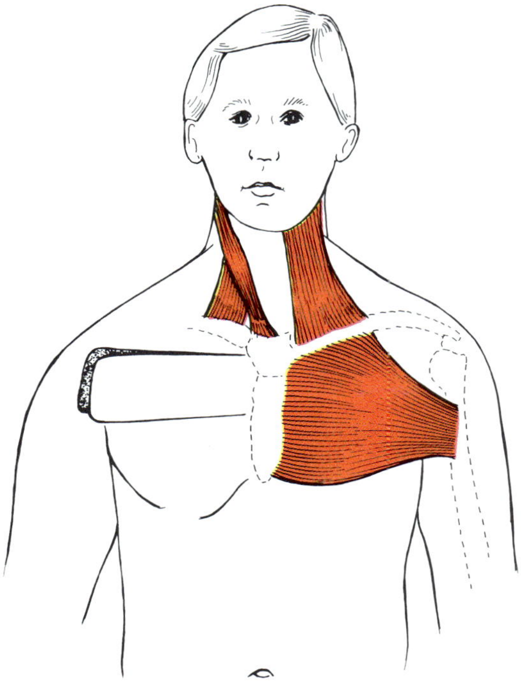

Platysma
Sternocleidomastoid
Pectoralis major
Deltopectoral flap

Fig. 6-28

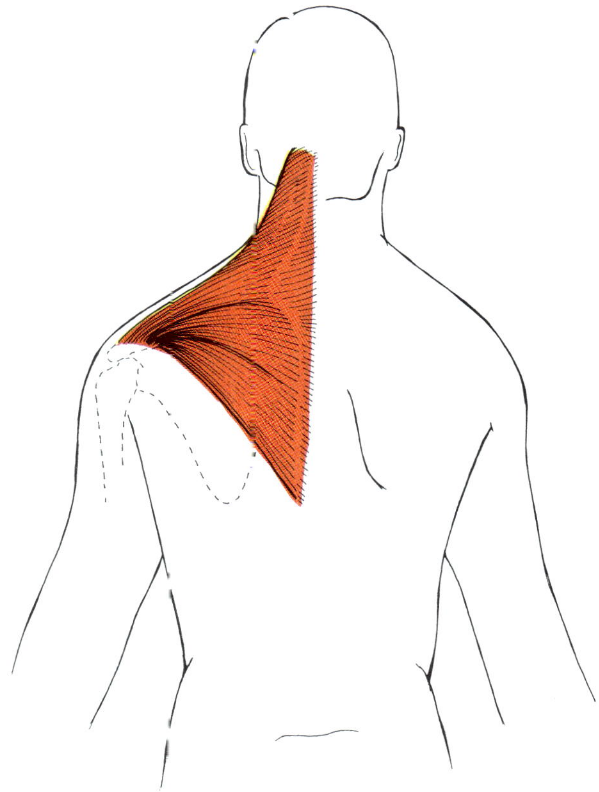

Trapezius
Latissimus dorsi

Fig. 6-29

A

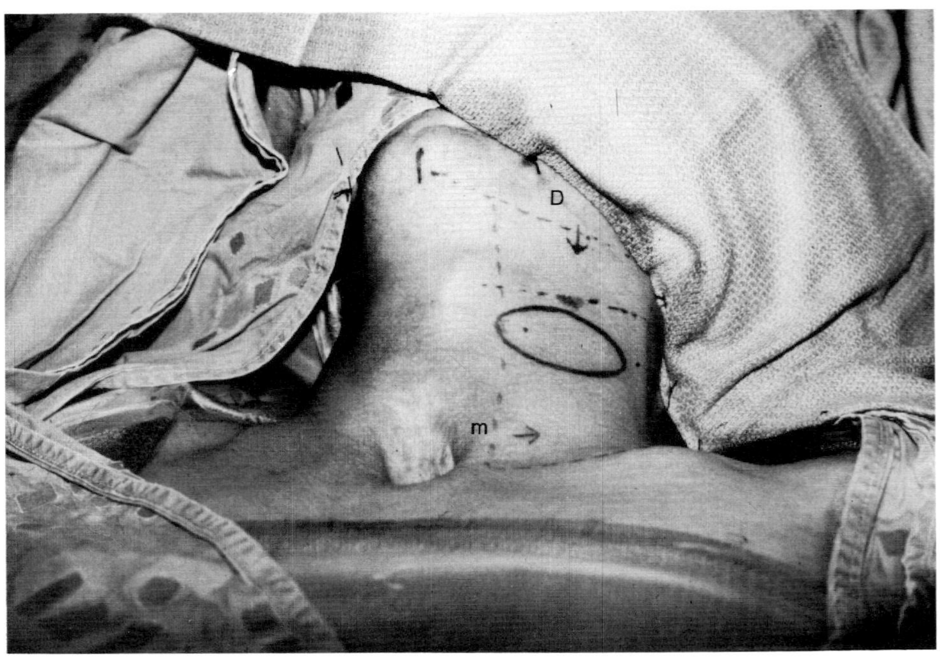

Fig. 6-30. Platysma musculocutaneous flap.

A, Adult singer notes unstable skin at site of long-term tracheostomy defect. Design of platysma musculocutaneous flap demonstrated based on inferior minor pedicle via transverse cervical artery. *D,* Site of entry of dominant pedicle (branch of facial artery); *m,* site of entry of minor pedicle (branch of transverse cervical artery).

B C

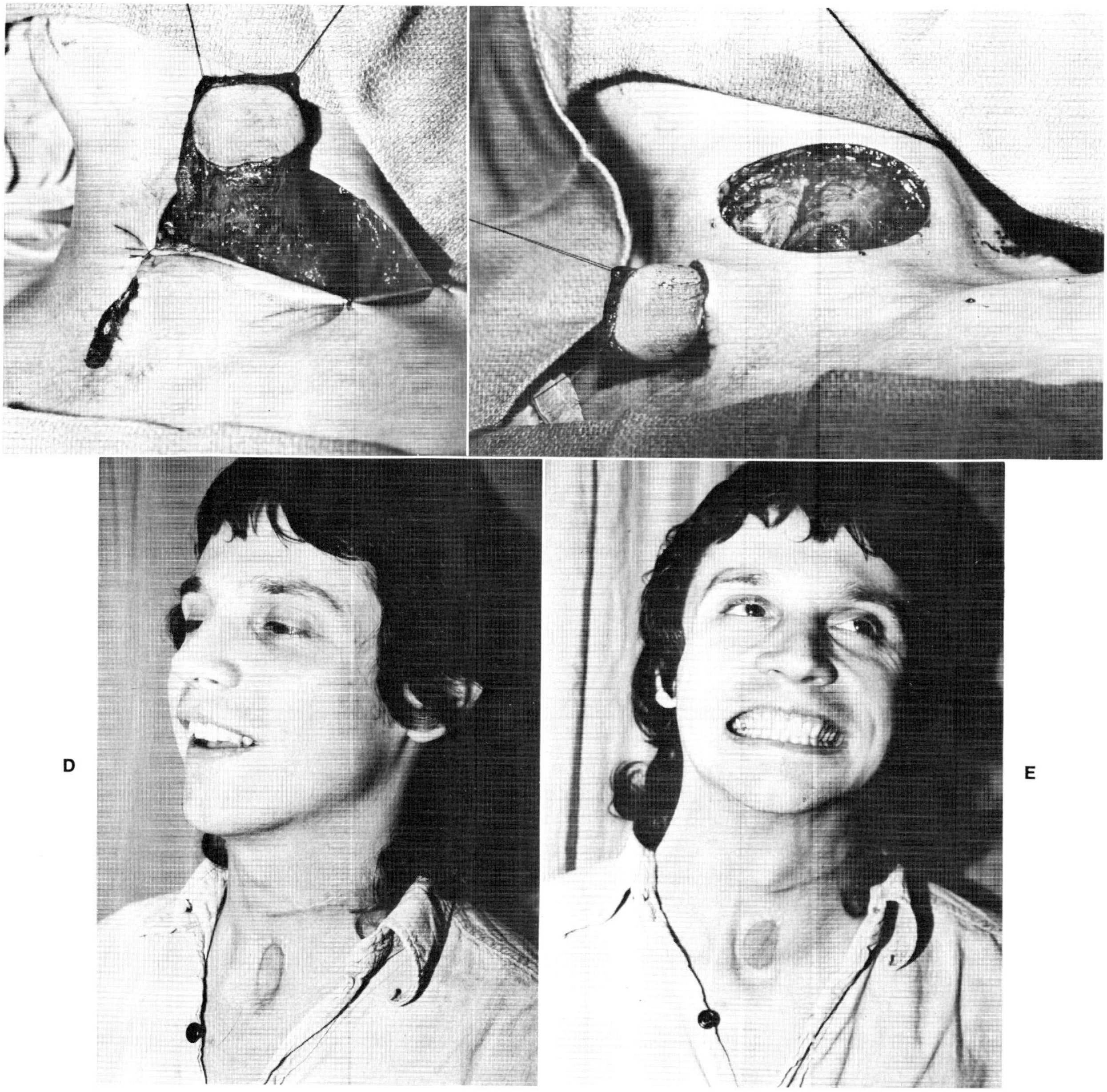

D E

Fig. 6-30, cont'd

B, Inferiorly based platysma musculocutaneous flap. NOTE: Design of skin island beneath left mandible.

C, Platysma musculocutaneous flap tunneled beneath intact anterior neck skin to tracheostomy defect.

D, Platysma muscle used for coverage for long-term tracheostomy defect.

E, Absence of platysmal contractions in left neck after transposition as flap to midline.

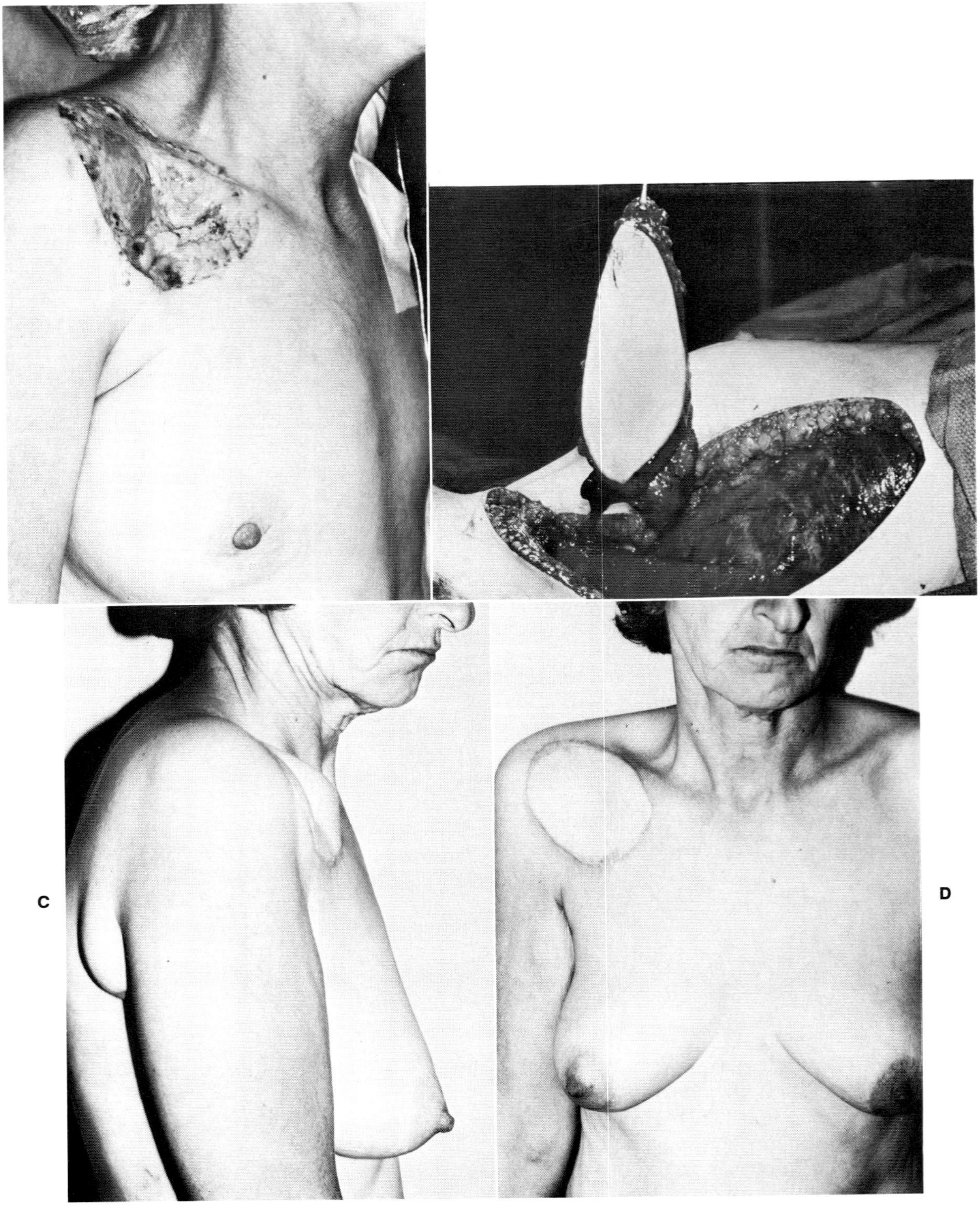

A

B

C

D

Fig. 6-31. Latissimus dorsi musculocutaneous flap for coverage of lower neck and chest. Forty-five-year-old woman after chemosurgical excision of basal cell carcinoma over right clavicle.

A, Right clavicle is exposed.

B, Musculocutaneous flap is elevated and pedicle dissected. Flap tunneled deep to pectoralis major into neck.

C and **D,** Stable coverage of lower neck and chest defect provided by latissimus dorsi musculocutaneous flap.

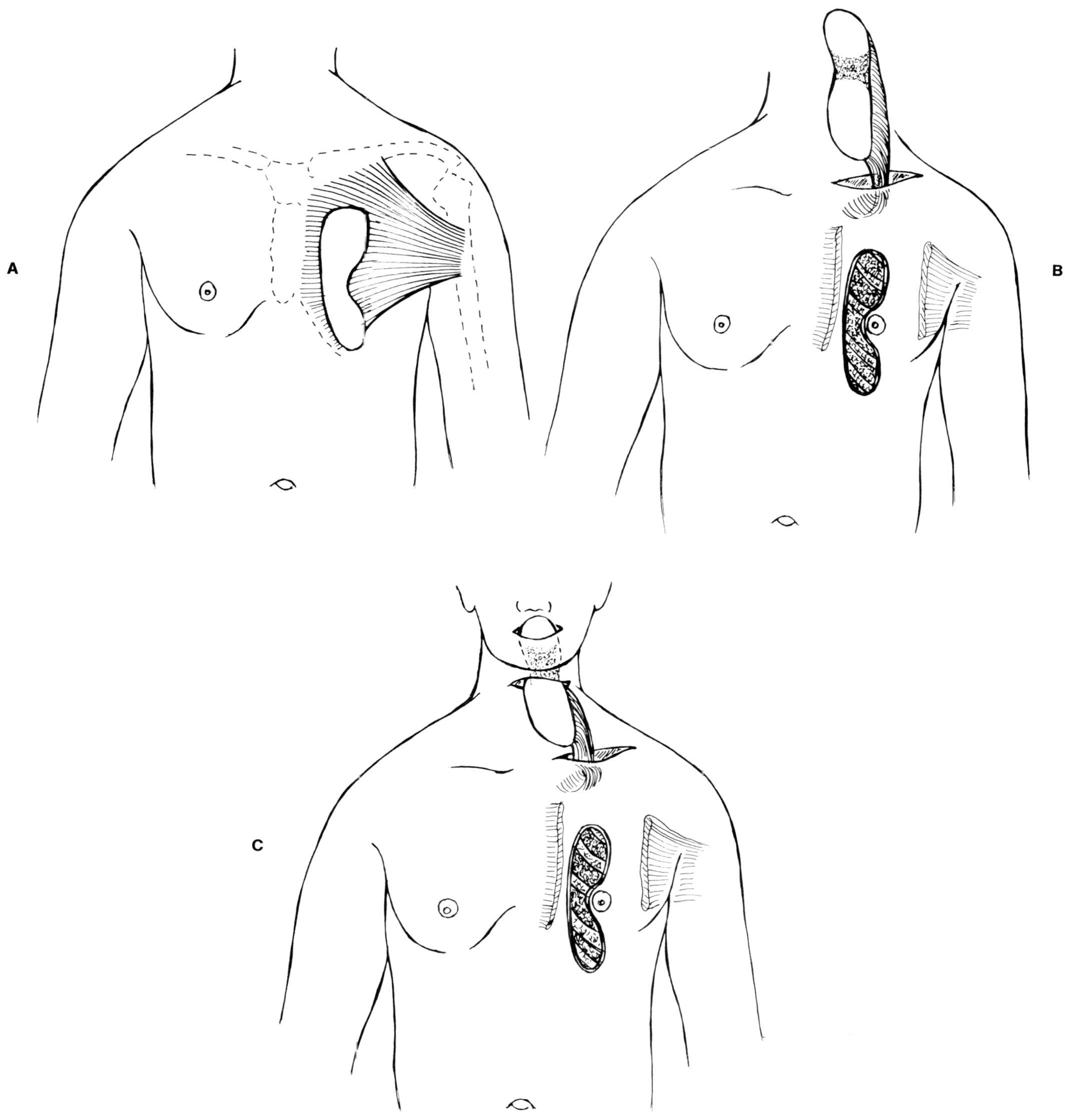

Fig. 6-32

A, Design of skin island on pectoralis major musculocutaneous flap for combined coverage of intraoral and neck defects.

B, Double skin island is established by excision of skin of midportion of flap.

C, Distal skin island is used for coverage of oral cavity defect; proximal skin island is used for coverage of neck defect. Donor defect is closed directly.

POSTERIOR NECK (Fig. 6-33)

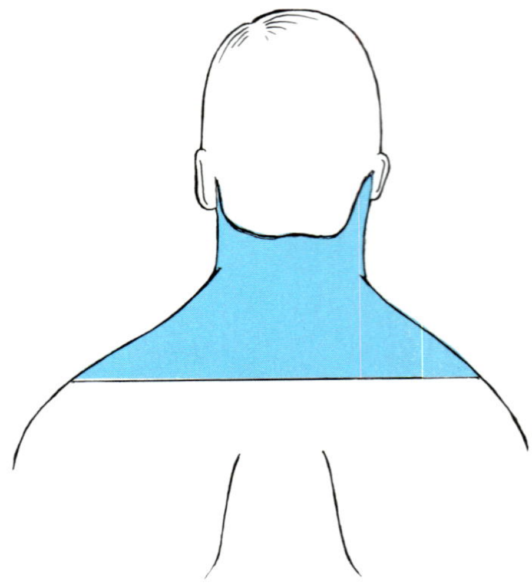

Fig. 6-33

Skin grafts

Skin grafts are method of choice when vertebral column is not exposed

Local flaps

Local flaps are not applicable in reconstruction of defects of the posterior neck

Distant flaps (Fig. 6-34)

1. Posterior trapezius muscle or musculocutaneous flap is flap of choice for extensive defects (Fig. 6-35)
2. Latissimus dorsi is alternative flap

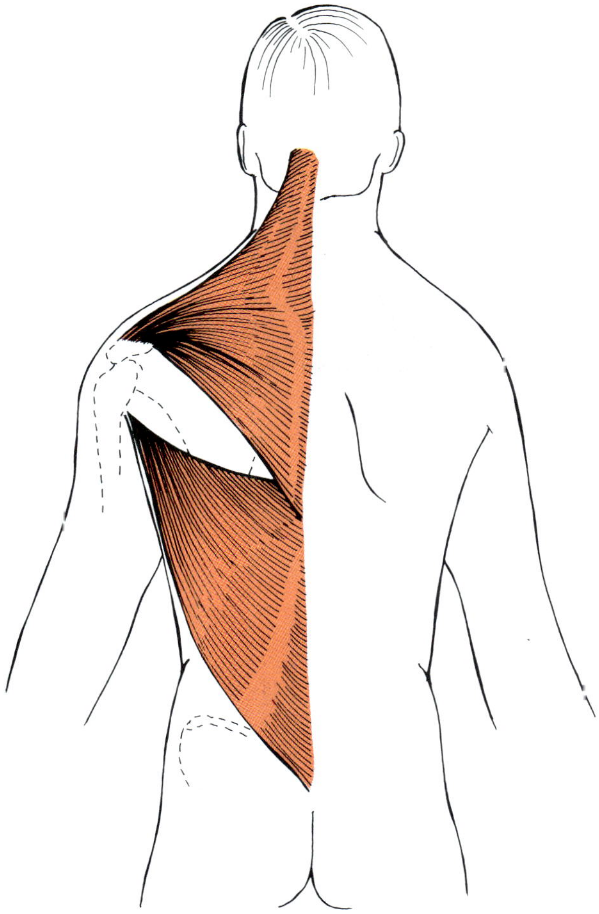

Trapezius
Latissimus dorsi

Fig. 6-34

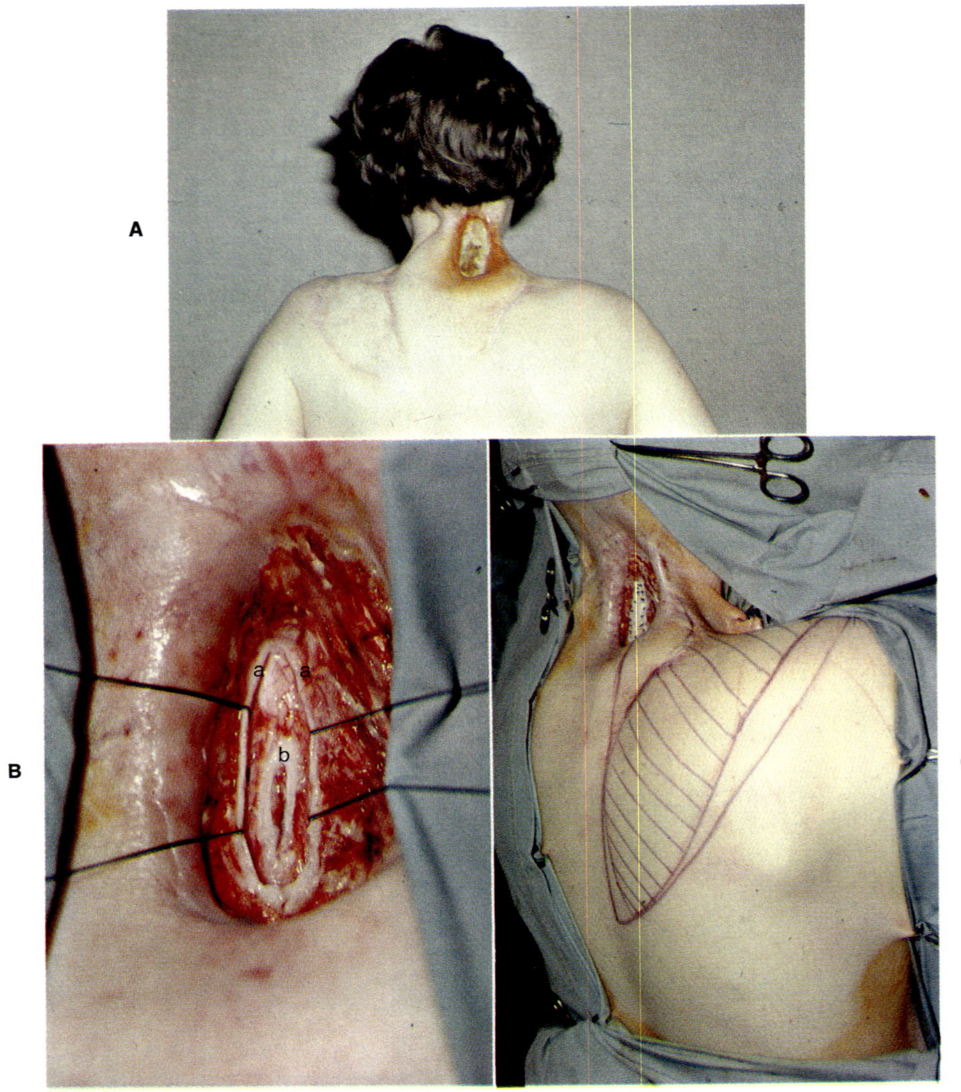

Fig. 6-35. Posterior trapezius musculocutaneous flap.

A, Defect is secondary to osteoradionecrosis of cervical vertebral column. NOTE: Local random flaps from left shoulder have failed to provide stable coverage.

B, Resection of nonviable tissue includes arachnoid. NOTE: Underlying radiation necrosis of posterior commissure of cervical cord. *a,* Arachnoid; *b,* posterior commissure of cervical cord.

C, Fascia lata graft used to restore arachnoid continuity. Design of posterior trapezius musculocutaneous flap.

(Cases similar to **A** and **C** shown in Mathes, S.J., and Vasconez, L.O.: Head, neck and truncal reconstruction with the musculo-cutaneous flap: anatomical and clinical considerations, Transactions of the VII International Congress of Plastic and Reconstructive Surgery, 1979, Sao Paulo, Brazil, 1980, Cartgraf, pp. 178-182.)

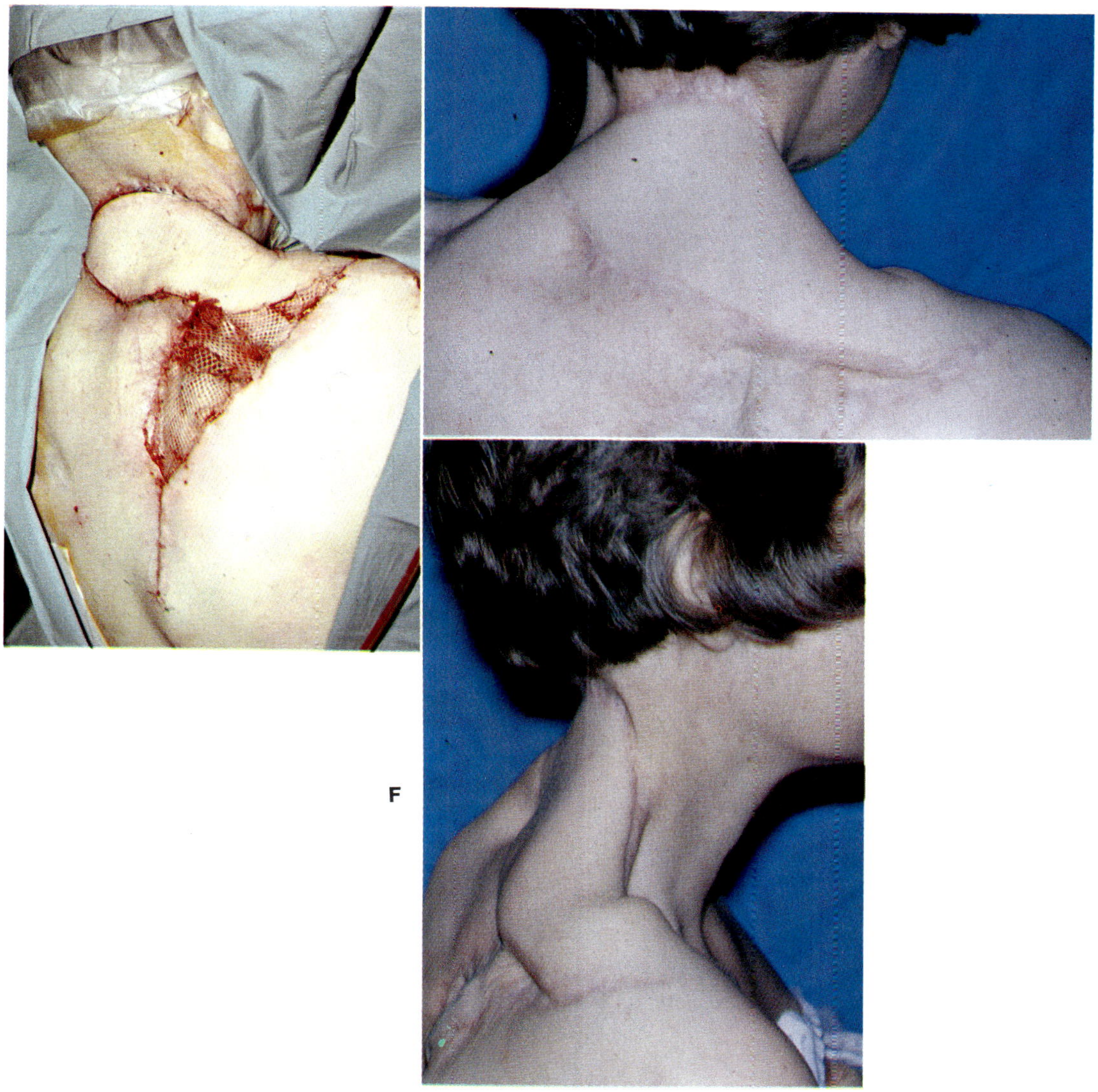

Fig. 6-35, cont'd

D, Posterior trapezius musculocutaneous flap inset into posterior neck defect. Muscle provides coverage of fascia lata graft. Donor defect skin grafted.

E, Three-year postoperative posterior view demonstrates stable coverage for osteoradionecrosis of cervical vertebral column.

F, Lateral view demonstrates flap inset. Preservation of anterior fibers of trapezius muscle provides function preservation and avoids shoulder droop.

7

RECONSTRUCTION

Maurice J. Jurkiewicz ■ T. Roderick Hester

Understandable is the reluctance of the surgeon to ablate those features that distinguish the uniqueness of a person and at the same time destroy function in a radical effort to extirpate aggressive cancer of the head and neck, particularly when complete eradication of the malignancy so often cannot be ensured. With the advent of the notion of immediate reconstruction by Edgerton in 1951 and by McGregor, some of the aversion by surgeons for this necessary mutilation was removed. Planned reconstruction could be offered to assuage the terrible grinding debility of oral incontinence and resultant social ostracism. The procedures then available and in use often required multiple stages; nonetheless, the concept was a real advance.

In 1963 Bakamjian published a technique for primary reconstruction of the palate after radical maxillectomy for cancer. He used a modification of a "compound neck pedicle" described by Owens in 1955. Owens based the flap on the mastoid area of the neck and incorporated the sternomastoid muscle not only "to offer additional tissue for increased blood supply but also to increase thickness and the possibility of subsequent transmitted animation through nerve supply and neurotization." With characteristic insight Bakamjian advised caution at the anatomic level of the posterior belly of the digastric muscle "to preserve the functional branches of the occipital artery and the sternomastoid tributaries to the common facial and external jugular veins." Unlike Owens, the incisions for the flap as described by Bakamjian are approximately 2 cm parallel to and beyond the borders of the muscle as well as a few centimeters beyond the clavicle. Clearly Bakamjian recognized the critical importance of the muscle in the blood supply of the skin. Without the muscle the flap had to be delayed. This technique of Bakamjian provided functional closure of a massive oral defect in one stage. The subsequent description of the deltopectoral flap, a brillant innovation that provided renewed interest in the treatment of head and neck cancer, obscured the hidden principle in the earlier work by Bakamjian and Owens (i.e., the blood supply of the skin comes from the muscle).

The deltopectoral flap, based on perforators of the internal mammary artery, was immediately recognized as a great advance in the treatment of patients with head and neck cancer. For the past 15 years it has been the flap that all head and neck surgeons have used to achieve primary closure after radical extirpation of cancer of the head and neck.

The deltopectoral flap as described by Bakamjian in 1965 is based primarily on perforators from the internal mammary artery via the second, third, and fourth intercostal spaces. The flap extends from the clavicle to a point just above the nipple and can be taken safely without delay to the anterior axillary line. The flap should be delayed if it is to be carried lateral to the anterior axillary line. Delay may be necessary, since the blood supply to the skin from

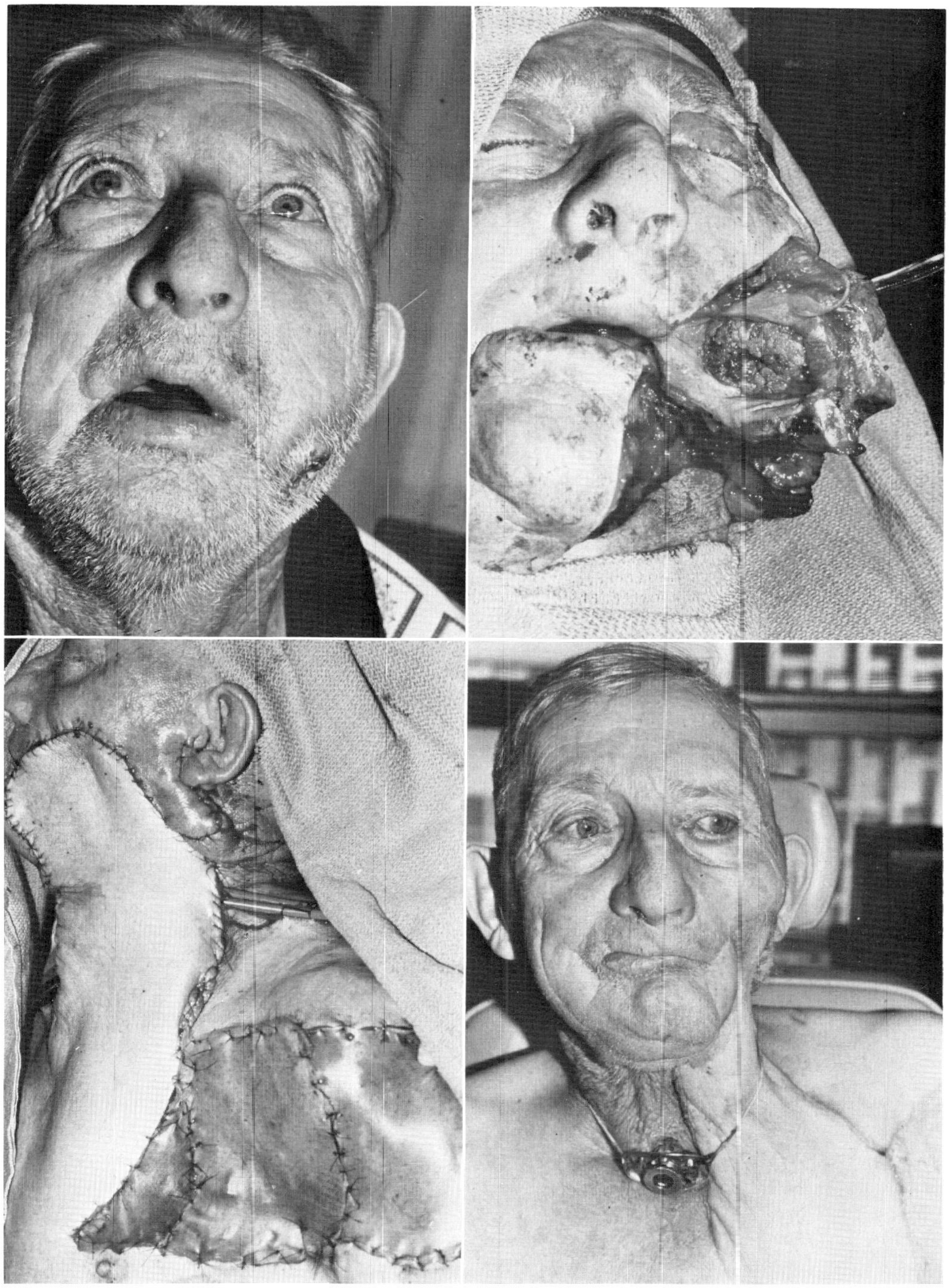

A B C D

Fig. 7-1. Deltopectoral flap.

A, Eighty-one-year-old man with carcinoma of left buccal mucosa invading cheek skin.

B, Composite resection of cheek skin, mandible, and oral mucosa was performed.

C, Defect was reconstructed with deltopectoral flap. Cheek skin was replaced by flap, and deep surface was grafted with split skin for oral lining.

D, One-year postoperative view demonstrates stable coverage.

over the shoulder comes from many perforating vessels through the deltoid muscle.

The deltopectoral flap is still a very useful flap in reconstructive surgery (Fig. 7-1). It is relatively thin, pliable, and hairless, which are characteristics that make it useful for external lining. The limitations of the deltopectoral flap are the following:

1. The arc of rotation is resticted without the shoulder extension.
2. When this flap is used for intraoral reconstruction, a temporary orocutaneous fistula is designed. Several additional procedures, including division of the pedicle and inset of the flap, are necessary.
3. There is an objectionable donor site on the anterior chest and shoulder that may be difficult for women to accept.

Many of these problems can be ameliorated or eliminated by the use of muscle or musculocutaneous flaps.

STERNOCLEIDOMASTOID FLAP

The advances in reconstructive surgery initiated by the musculocutaneous flap concept prompted the inevitable rediscovery of the sternocleidomastoid compound flap. Successive steps in the evolution of the flap were reported by Owens, Bakamjian, and Little-

A

B

C

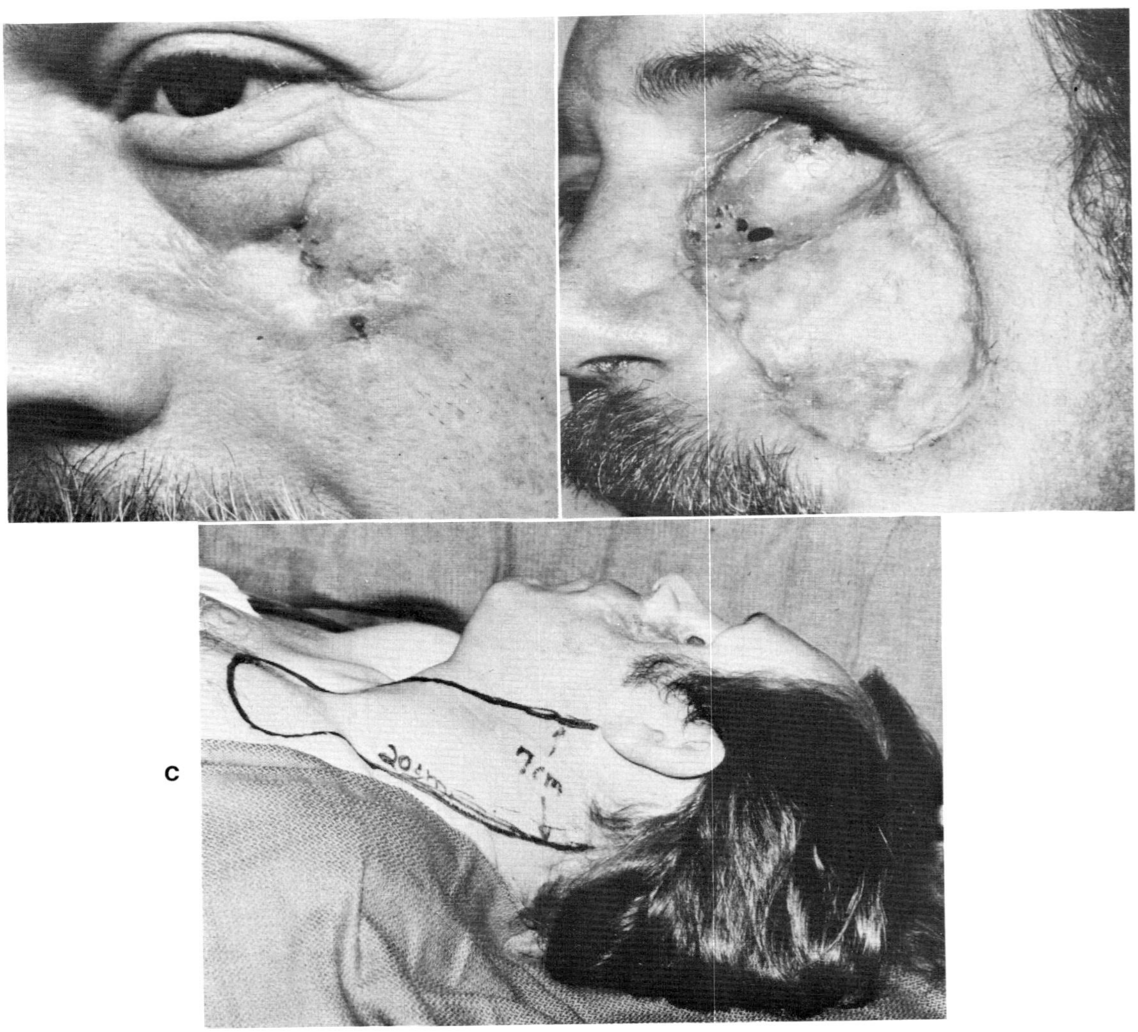

Fig. 7-2. Sternocleidomastoid flap.
A, Forty-eight-year-old man with recurrent squamous cell carcinoma of skin invading orbit.
B, Defect after orbital exenteration, partial maxillectomy, and skin graft.
C, Sternocleidomastoid flap is outlined with skin extending below clavicle.

wood. Recently the vascular supply of the sternocleidomastoid muscle and its cutaneous territory based either proximally or distally has been described.

The sternocleidomastoid flap has been useful for intraoral and pharyngeal lining and has also been used as a carrier of the clavicle as a potentially viable bone graft for mandibular reconstruction. The muscle has a double origin for the manubrium and clavicle and inserts into the mastoid process and superior nuchal line of the occipital bone. Although the muscle has a segmental pattern of blood supply (type II), a branch of the occipital artery entering the upper one third of the muscle appears to be the dominant pedicle. This muscle flap can also be based on a branch of the thyrocervical trunk that enters the middle one third of the muscle. A third pedicle, a small branch of the transverse cervical artery, enters the muscle near its insertion.

The flap is elevated through incisions parallel to the anterior and posterior borders of the muscle. The skin can be safely extended to a point 2 to 3 cm below the clavicle. The skin over the distal one half of the muscle can be elevated as an island based on the muscle and used for intraoral or pharyngeal reconstruction as a single-stage procedure. This flap may be used to reconstruct the orbit, in which instance the muscle is temporarily exteriorized and skin grafted (Fig. 7-2).

PECTORALIS MAJOR FLAP

The pectoralis major musculocutaneous flap appears to have supplanted the deltopectoral flap as the primary flap for reconstructive procedures in the head and neck region. In a case report Brown describes the use of bilateral island flaps of pectoralis major muscle to provide primary closure of a huge defect of the mediastinum and lower neck with exposure of the heart and great vessels that resulted from the extirpation of extensive cancer. Ariyan subsequently describes the use of the pectoralis major muscle with its associated skin territory for head and neck reconstruction. Recent reports have established the reliability of this muscle unit as a reconstructive modality for head and neck surgery.

The pectoralis major muscle has a clavicular and sternal origin with slips from the first seven ribs. Its insertion is in the bicipital groove of the humerus

D

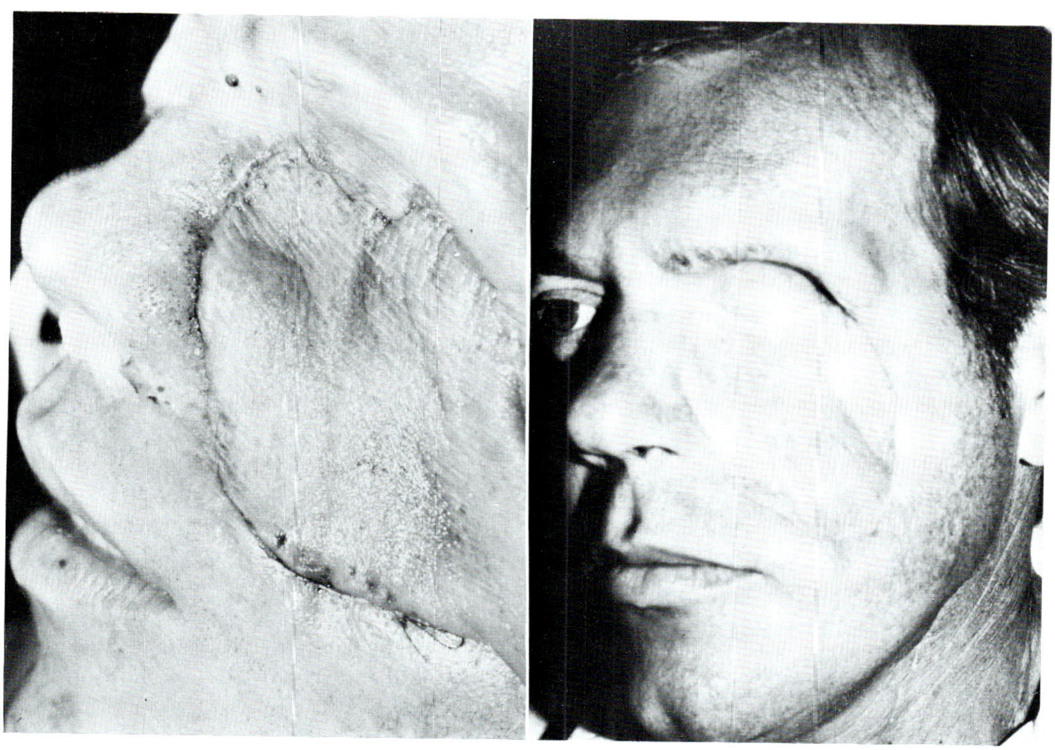

E

Fig. 7-2, cont'd
D, Flap inset.
E, Three-year postoperative view demonstrates stable coverage.

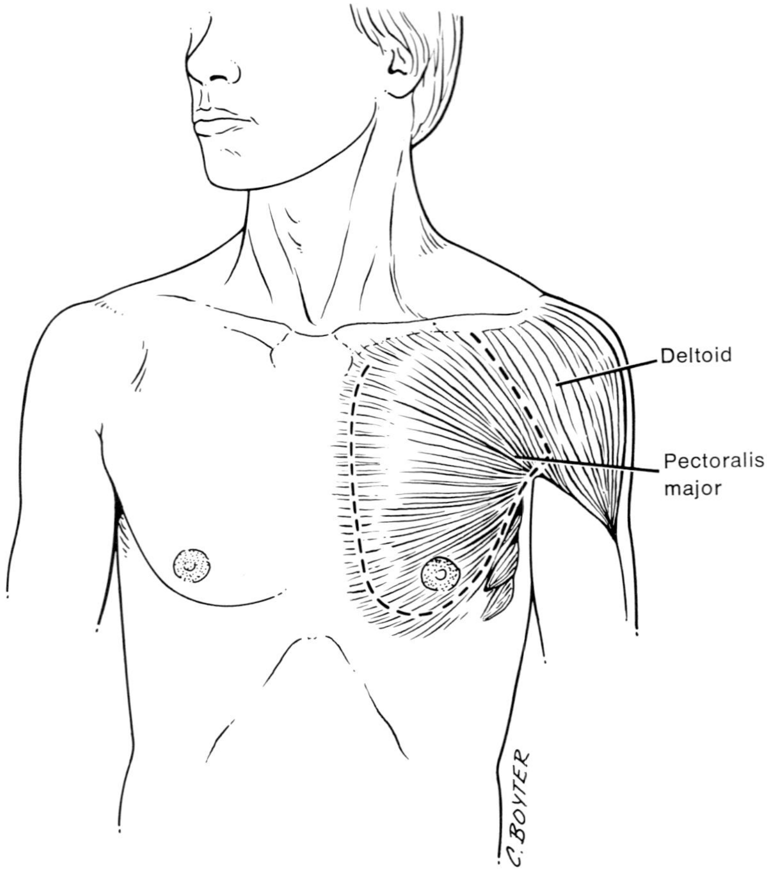

Fig. 7-3. Anatomy of pectoralis major. Dotted line represents possible skin incisions for elevation of flap.

(Fig. 7-3). The dominant pedicle is the thoracoacromial artery (Fig. 7-4). A system of secondary segmental perforators from the internal mammary artery also supplies the muscle medially (type V). The loss of adduction and medial rotation of the arm after use of this muscle is well tolerated by the patient with minimal clinical disability.

Based on the thoracoacromial artery, the pectoralis major has a wide arc of rotation. The wide arc allows the use of this muscle or musculocutaneous unit to reconstruct a variety of head and neck defects. The useful skin territory is located between the nipple and sternum over the fifth to eighth ribs (Fig. 7-5). It should be remembered that any skin taken more distally is raised as a random portion of the flap and is potentially at risk.

A variation of the pectoralis major flap that places the skin portion of the unit more distally in the inframammary crease has been described by Ariyan and Magee. This inferior extension of the skin territory (skin paddle) increases the arc of rotation (Fig. 7-6). According to Magee, the anatomic basis of this flap is an arborization of small vessels that extend over the rectus fascia and supply skin distal to the pectoralis major (Fig. 7-7). By raising the flap in a plane deep to the rectus fascia, these vessels are protected, allowing the more distal placement of the skin island. The concept is tenuous, and this random portion of skin (skin paddle) may be unreliable.

The flap is most dependable when the skin island is raised with at least a portion of the skin over the muscle itself. In elevation of the flap it is always necessary to divide the sternal and humeral attachments. The vascular pedicle, the thoracoacromial artery, is visualized as the flap is elevated superiorly and is preserved as further elevation is accomplished. In patients who have a radical neck dissection, it is easy to transpose the pedicle subcutaneously through the neck. If a neck dissection has not been performed, it is occasionally more appropriate to leave the muscle external to the skin. This allows easier later examination of the neck for possible nodal metastases. If a

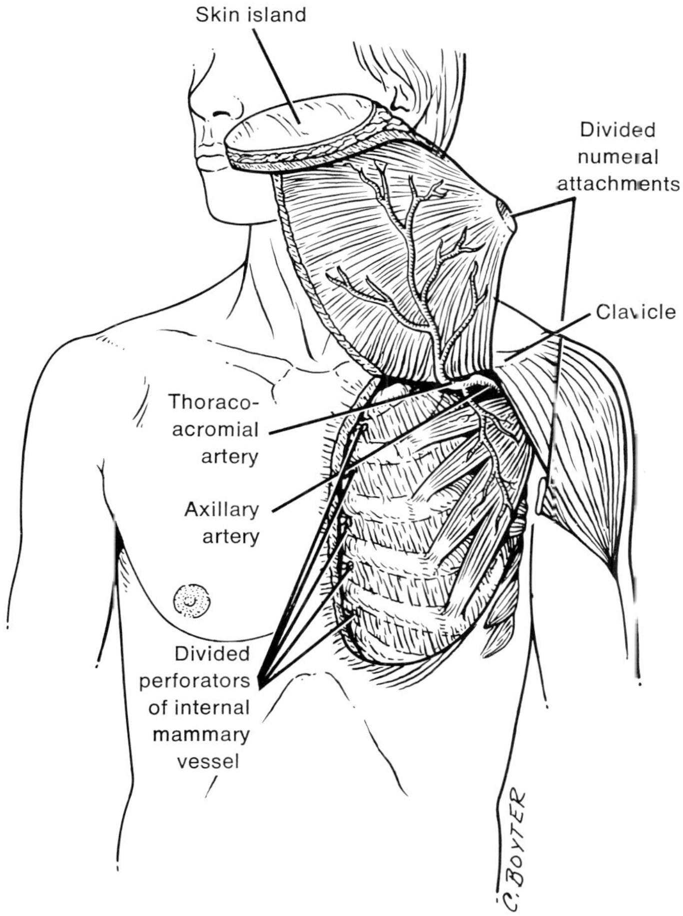

Fig. 7-4. Blood supply of pectoralis major flap.

muscle with its associated vascular pedicle is transposed externally to skin to reach the head and neck region, the division and revision of the muscle can be done safely in 4 weeks.

Variations in the design of the pectoralis major musculocutaneous flap allow its use in the following situations:

1. External resurfacing of skin of neck and face (Fig. 7-8)
2. Coverage of orbital defects (Fig. 7-9)
3. Intraoral and pharyngeal lining (Fig. 7-10)
4. Carrier for rib and skin in mandibular reconstruction (Fig. 7-11)
5. Reconstruction of the esophagus

Flap elevation

The pectoralis major musculocutaneous flap is elevated in the following steps:

1. A skin island of appropriate size is designed generally as an ellipse with a transverse axis over the fifth rib between the nipple and sternum or in the inframammary fold (Fig. 7-5, *B*).
2. A skin island is incised to the anterior surface of the pectoralis major muscle. The most inferior attachments of the muscle are divided from the sixth, seventh, and eighth ribs, allowing cephalad elevation in the retropectoral area toward the fifth and sixth intercostal space. If a rib is to be included (Fig. 7-11, *E*) (usually the fifth rib), dissection proceeds through the fifth to sixth intercostal space. Here the intercostal muscles are divided and the pleura gently dissected away from the undersurface of the fifth rib. Care is taken not to avulse attachments of the pectoralis major to the fifth rib. The rib is then divided with rib cutters medially and laterally, taking enough bone to fill the mandibular defect requiring reconstruction. The dissection is then performed superiorly through the fourth and fifth intercostal space. The plane of

Text continued on p. 208.

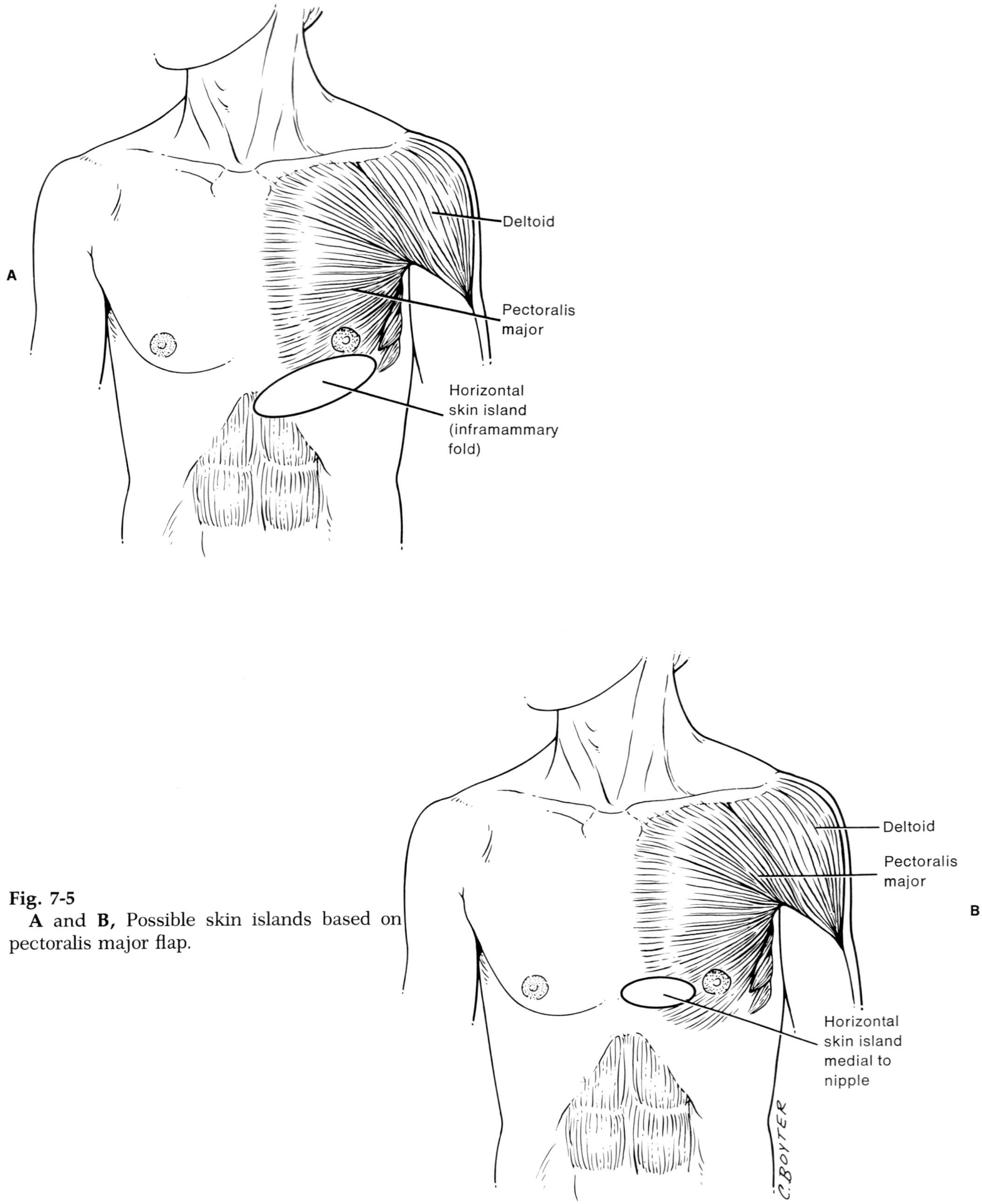

Fig. 7-5
A and **B,** Possible skin islands based on pectoralis major flap.

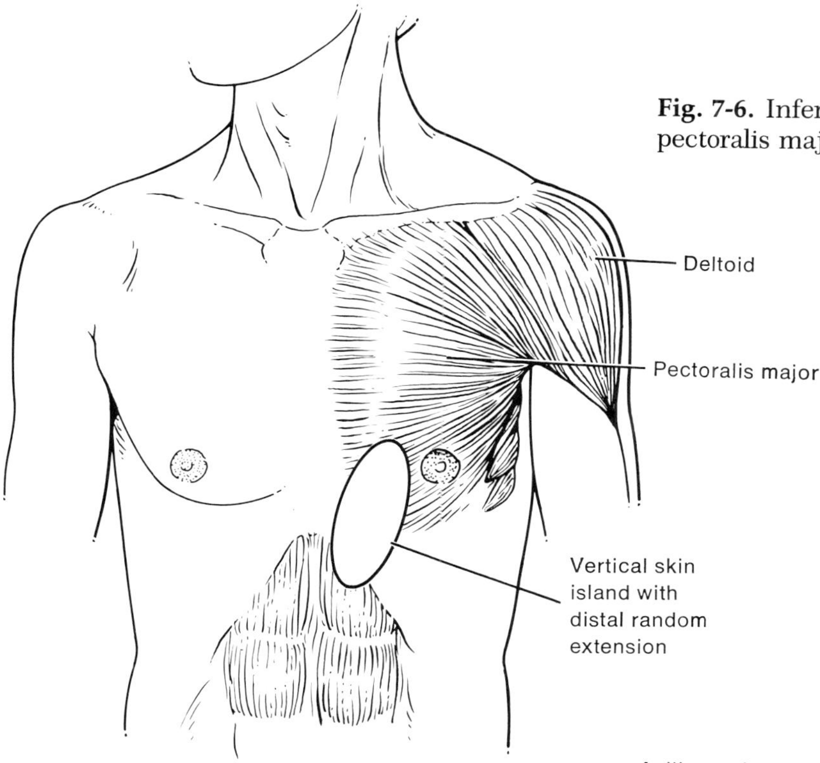

Fig. 7-6. Inferior extension of skin territory of pectoralis major (pectoralis paddle).

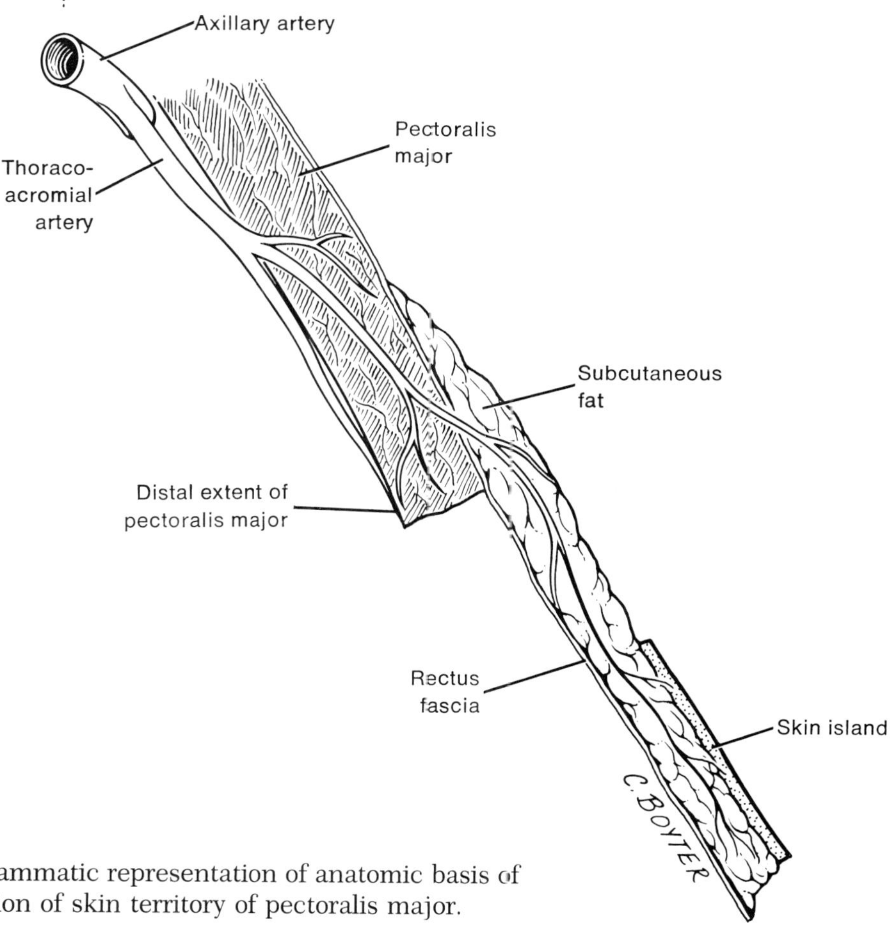

Fig. 7-7. Diagrammatic representation of anatomic basis of inferior extension of skin territory of pectoralis major.

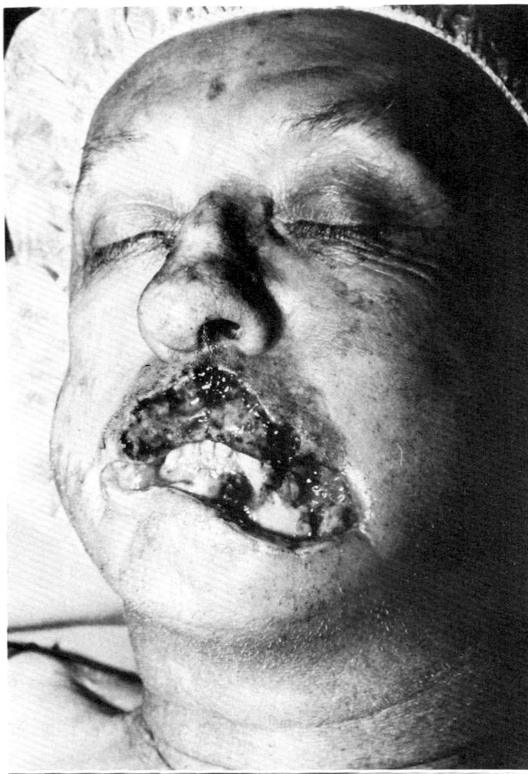

A

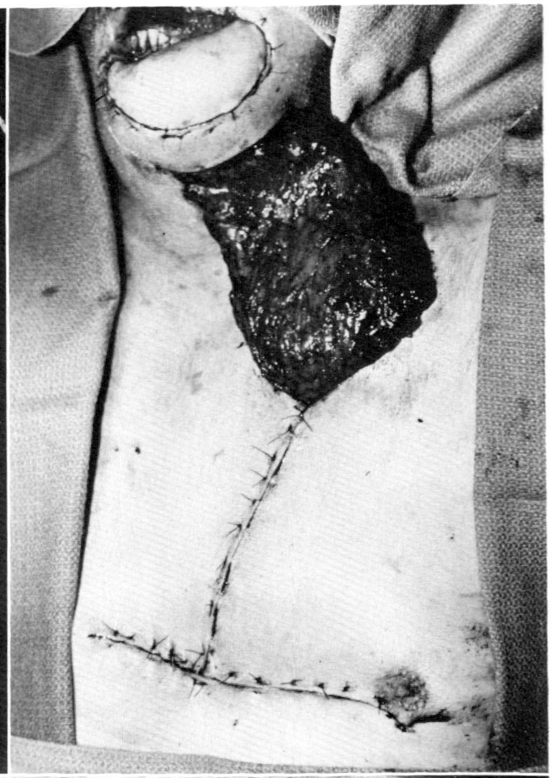

B

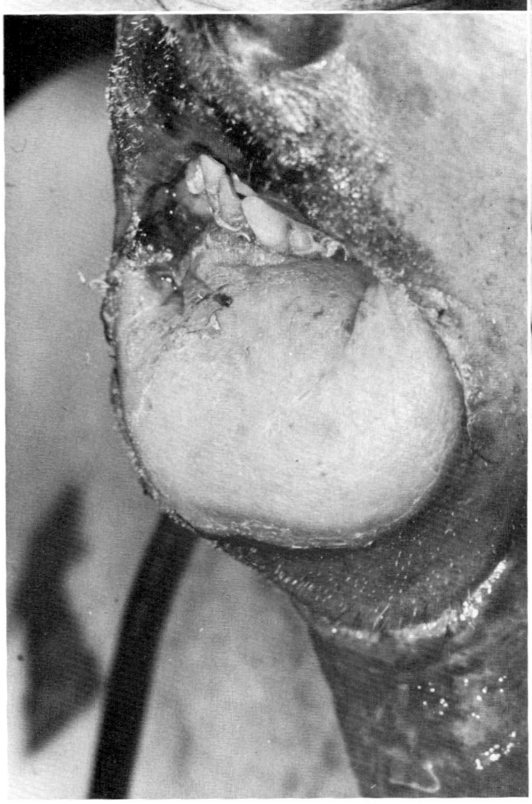

C

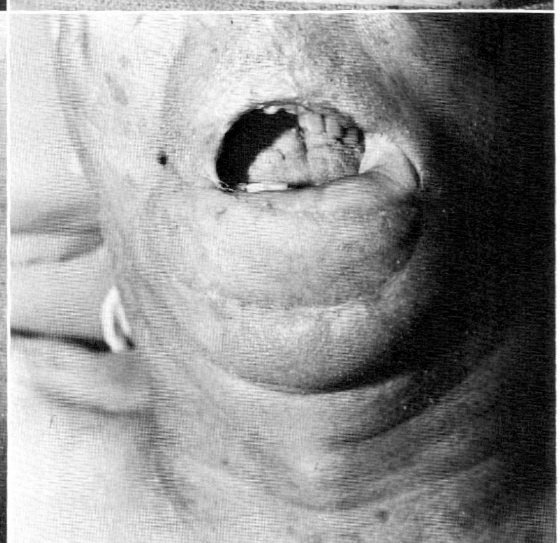

D

Fig. 7-8. Sixty-eight-year-old man who underwent chemosurgery for removal of extensive basal cell carcinoma of upper and lower lips.

A, Upper lip excision included only external skin; however, lower lip was full thickness.

B, Skin island based on pectoralis major muscle transposed into lower lip area as first stage of reconstruction of inner lining of lower lip. Standard elliptic skin island was used, and donor defect was closed directly.

C, Skin well healed 1 month after operation when muscle pedicle was divided.

D, Eight-month postoperative view after transfer of two inferiorly based nasolabial flaps to provide lower facial coverage.

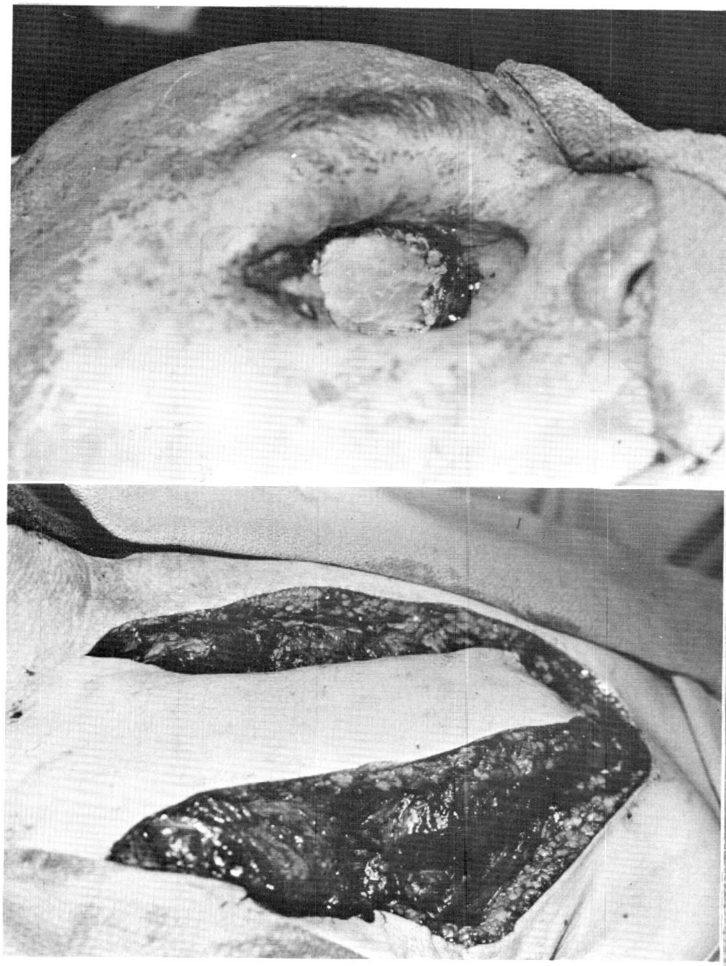

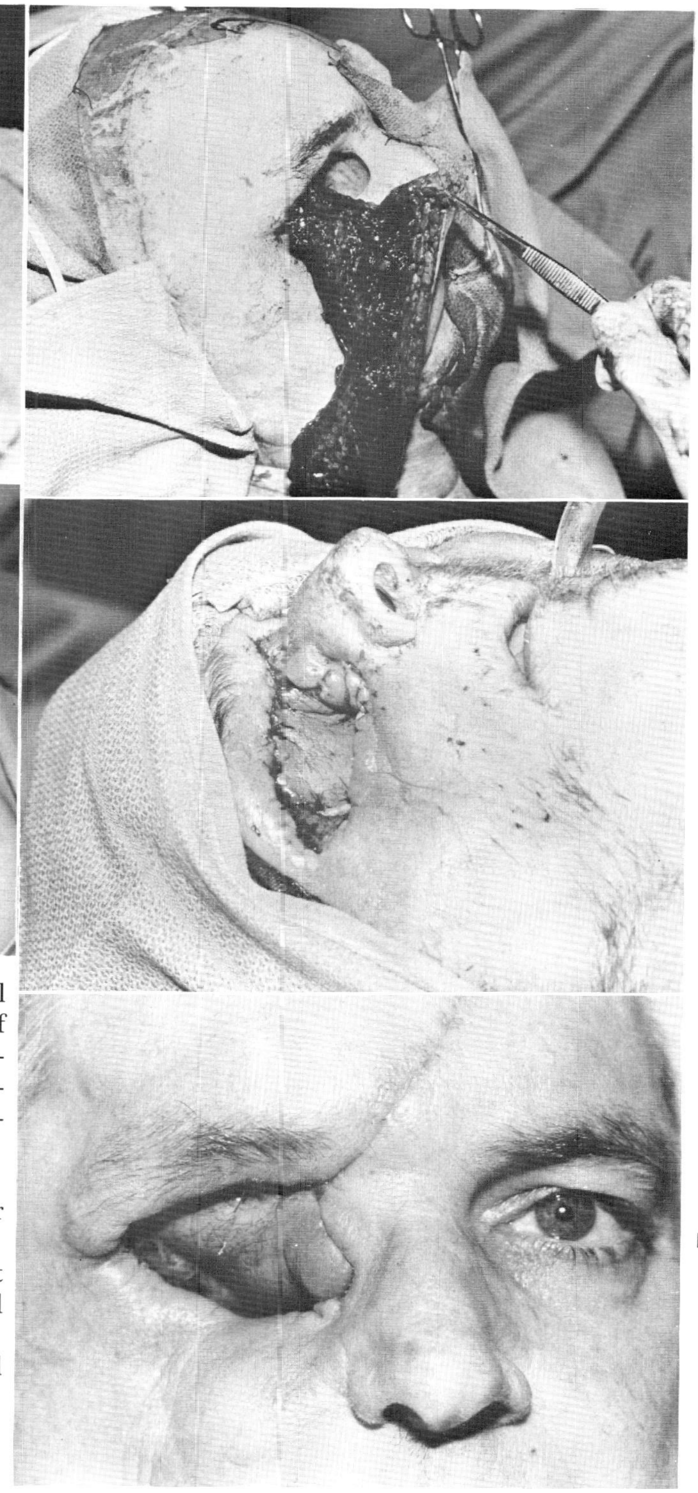

Fig. 7-9. Fifty-five-year-old man who underwent total maxillectomy and orbital exenteration for carcinoma of maxillary sinus. His postoperative course was complicated by development of dural fistula secondary to erosion by prosthetic implant. Closure of defect with pectoralis major flap was planned.

A, Fascia lata patch sutured into dural defect.

B, Pectoralis major flap dissected and prepared for transposition.

C, Transposition of muscular portion of flap into orbit followed by insetting of skin along superior orbital rim.

D, Patient 3 weeks later at time of division of flap and insetting of skin over dural patch.

E, Well-healed wound several months after operation.

A

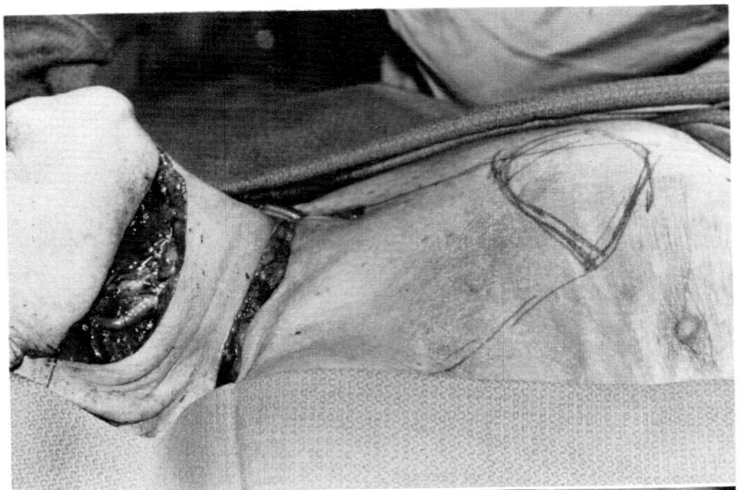

B

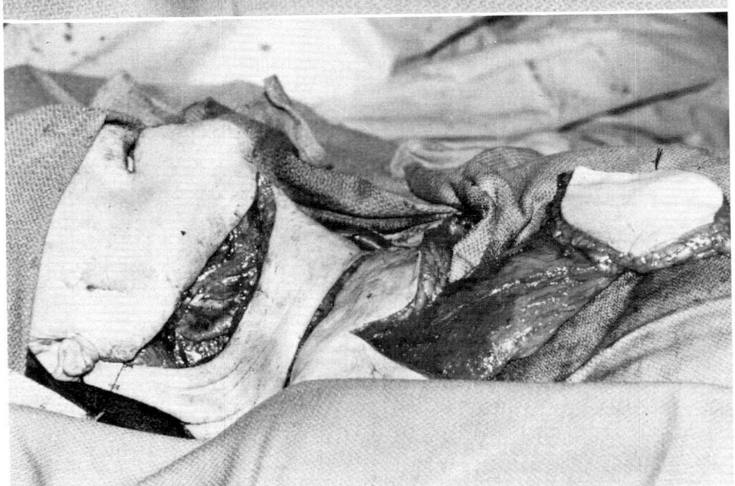

C

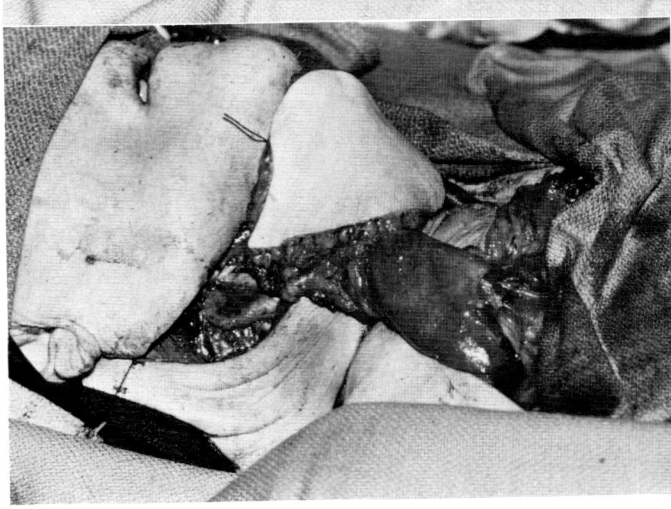

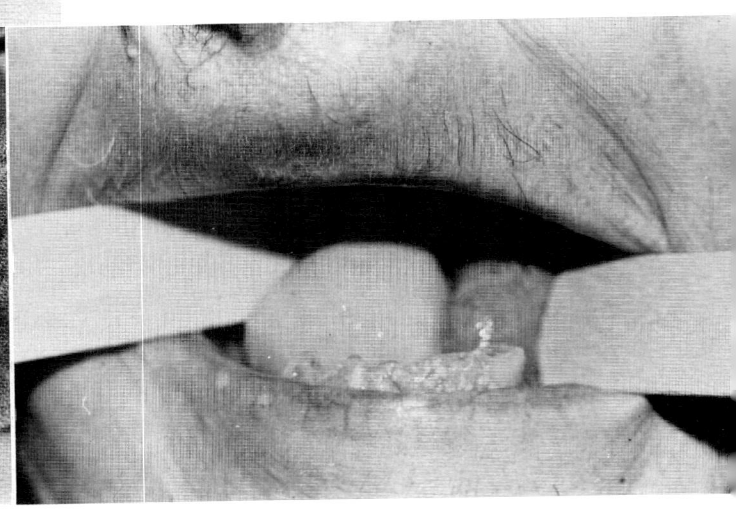

Fig. 7-10. Seventy-four-year-old woman after excision of a T3 lesion of floor of mouth and right radical neck dissection. Mandible was left intact.

A, Outline of pectoralis skin island for reconstruction of floor of mouth.

B, Skin island with pectoralis major muscle elevated after division of humeral and sternal attachments of muscle.

C, Rotation of unit into floor of mouth.

D, Skin island well healed in the floor of mouth several months after operation.

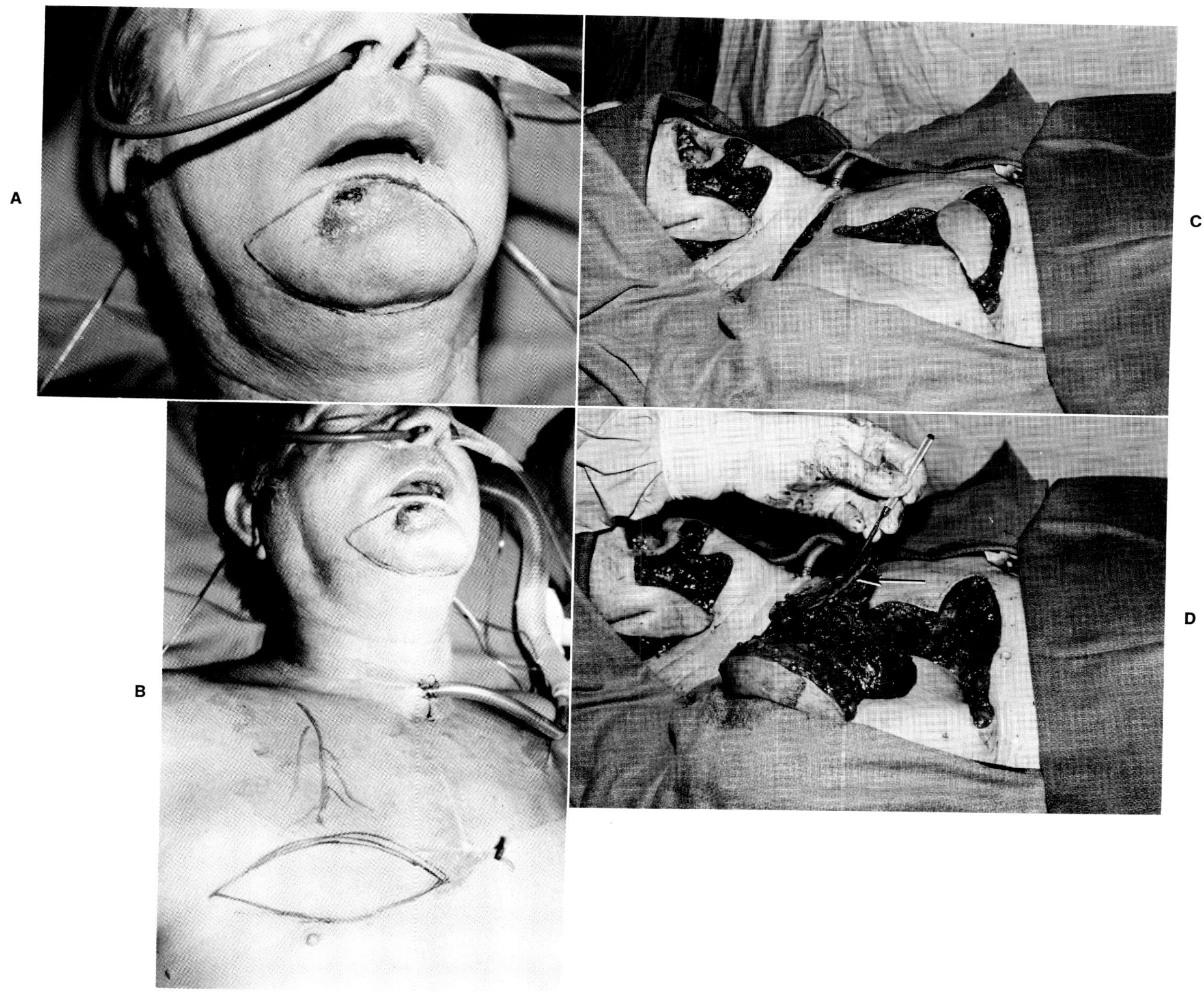

Fig. 7-11. Reconstruction of mandible with pectoralis flap, including rib.
A, Sixty-one-year-old man with recurrent squamous cell carcinoma of buccal mucosa invading symphysis of mandible and extending into cheek skin.
B, Extent of chin skin excision outlined. Pectoralis skin island outlined.
C, Resection completed. Pectoralis flap elevated.
D, Elevated flap. Arrow indicates rib.

Continued.

E

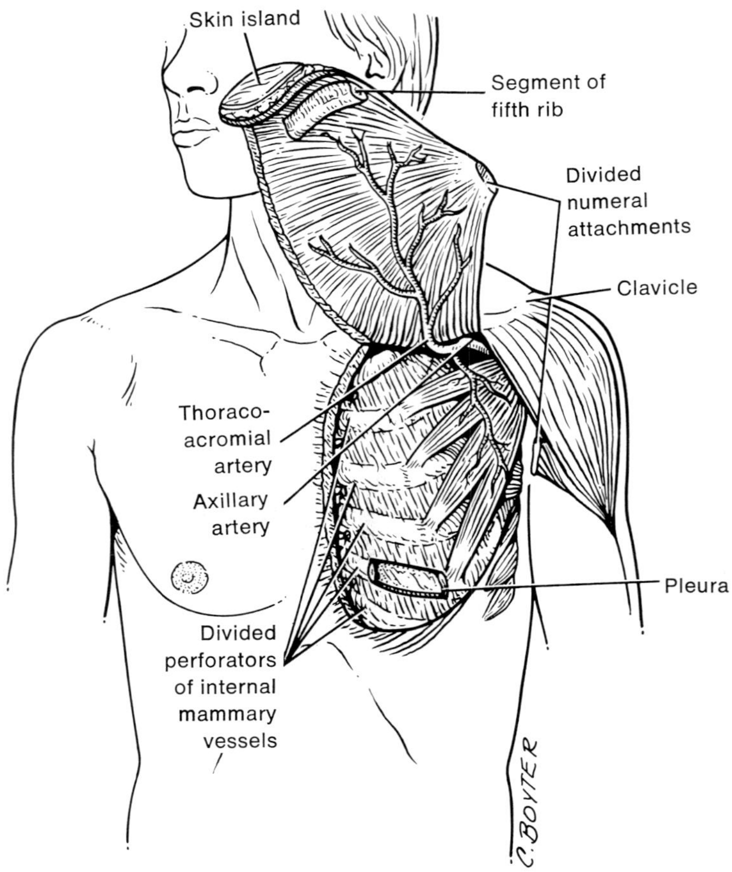

Fig. 7-11, cont'd
E, Diagrammatic representation of operative procedure.

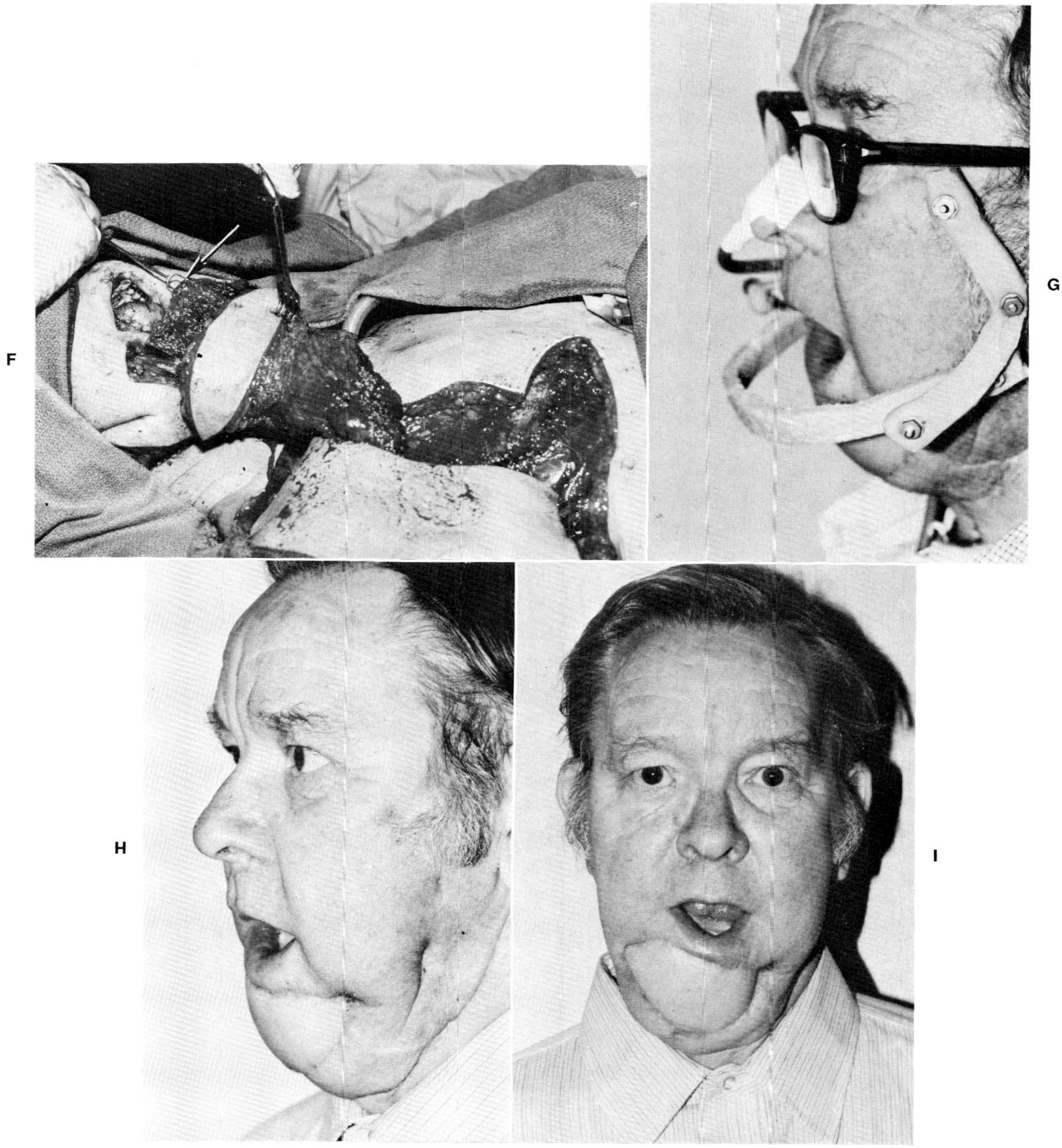

Fig. 7-11, cont'd
F, Flap transposed into position. Arrow indicates rib.
G, Early postoperative result shows patient in external fixation device.
H and **I,** Late postoperative results.

dissection now returns into the retropectoral space. Thus the segment of rib, with the pectoralis major periosteal attachments, is released from its chest wall attachments. Although the idea of a pectoralis major composite flap for mandible reconstruction is intriguing, a word of caution is in order. Postoperative fixation, preferably external (Fig. 7-11, *G*), is mandatory to allow adequate time for dense fibrous union to occur. Too early movement at the site of rib-mandible wiring can result in insufficient union and infection. There is also risk of pleural tear at the time of rib elevation, and two of our patients have required postoperative pleural drainage. Finally, it should be noted that flap bulk and inadequate creation of an inferior buccal sulcus can make fitting of postoperative dentures impossible.

3. The humeral and sternal attachments of the muscle are divided, allowing transposition into the neck.
4. The clavicular attachments are divided as required for adequate arc of rotation to reach the defect.

The use of the pectoralis major muscle with a distal skin island provides excellent mandibular coverage, especially when used for intraoral lining, and represents the most common use of this flap. The common problem of tongue fixation so often seen when tongue flaps are used to restore intraoral lining is avoided. Furthermore, the muscle portion of the flap provides coverage for the associated radical neck dissection.

The second most common use of the pectoralis major musculocutaneous flap in head and neck reconstructive surgery is as a modality for external resurfacing of skin of the neck and face. When used for this purpose, the muscle with its overlying skin island is transposed exteriorly to the neck. In this instance one does not violate an undissected neck. The external muscle and its associated pedicle are excised as a second stage.

When the pectoralis major musculocutaneous unit is used for coverage of the superior face, as for orbital coverage, the skin island may be designed vertically over the anterior surface of the pectoralis major muscle. This design is better suited for external facial coverage.

The use of the pectoralis major musculocutaneous unit for esophageal reconstruction or as a carrier for skin and bone for mandibular reconstruction represents an alternative modality for use in reconstructive surgery. Microsurgical composite tissue transplantation must be considered as the procedure of choice for immediate reconstruction of these defects.

LATISSIMUS DORSI FLAP

Although not as convenient as the pectoralis major flap, the latissimus dorsi flap may have a less significant donor defect, especially in women. The use of this flap for various problems in head and neck reconstruction has been described.

The latissimus dorsi muscle originates from the lower six thoracic vertebrae and inserts into the intertubercular groove of the humerus. This muscle receives its blood supply from the thoracodorsal artery in the shoulder girdle region and segmental secondary perforators in the posterior midline (type V). Its wide anterior superior arc of rotation based on the thoracodorsal artery allows use of this muscle for head and neck reconstruction.

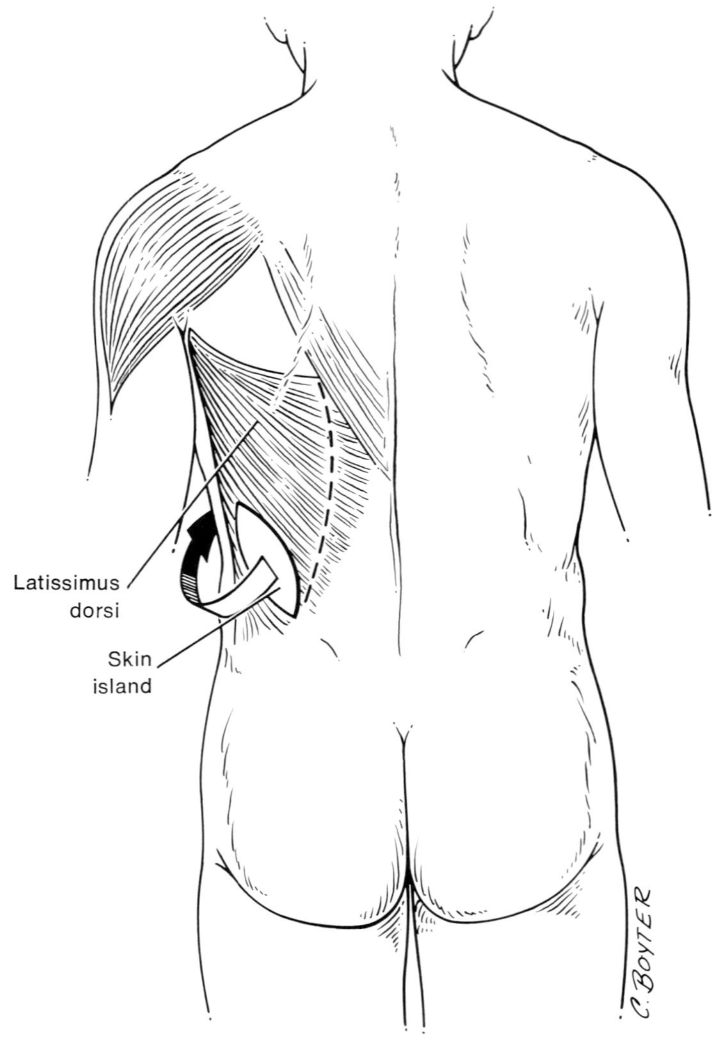

Fig. 7-12. Latissimus dorsi skin island for head and neck reconstruction.

Flap elevation

A skin island of appropriate size to cover the defect is designed over the inferior one third of the muscle, usually over the tenth, eleventh, and twelfth ribs (Fig. 7-12). Based on the numerous musculocutaneous perforators, the size and shape of this skin island can vary depending on reconstructive requirements, providing the upper part of the island is over the muscle itself. If needed, a random distal skin extension can be used to increase its arc of rotation.

The skin island is then incised to muscle, and the remaining skin elevated off the muscle medially and laterally. The medial, lateral, and inferior attachments of the muscle are divided and the muscle dissected proximally. In most instances the pedicle is mobilized and the humeral insertion of the muscle divided for easier transposition and to increase the arc of rotation. The unit is transposed externally to the pectoralis major muscle or, if necessary, through a tunnel in the retropectoral space (Fig. 7-13). This unit has its greatest application in head and neck reconstruction surgery for defects of the posterior neck and shoulder. This musculocutaneous unit can also be used for intraoral and pharyngeal reconstruction. Although in many instances the latissimus dorsi flap is not as convenient as the pectoralis major in reconstructive head and neck surgery, it is a very useful unit, and the donor site is acceptable.

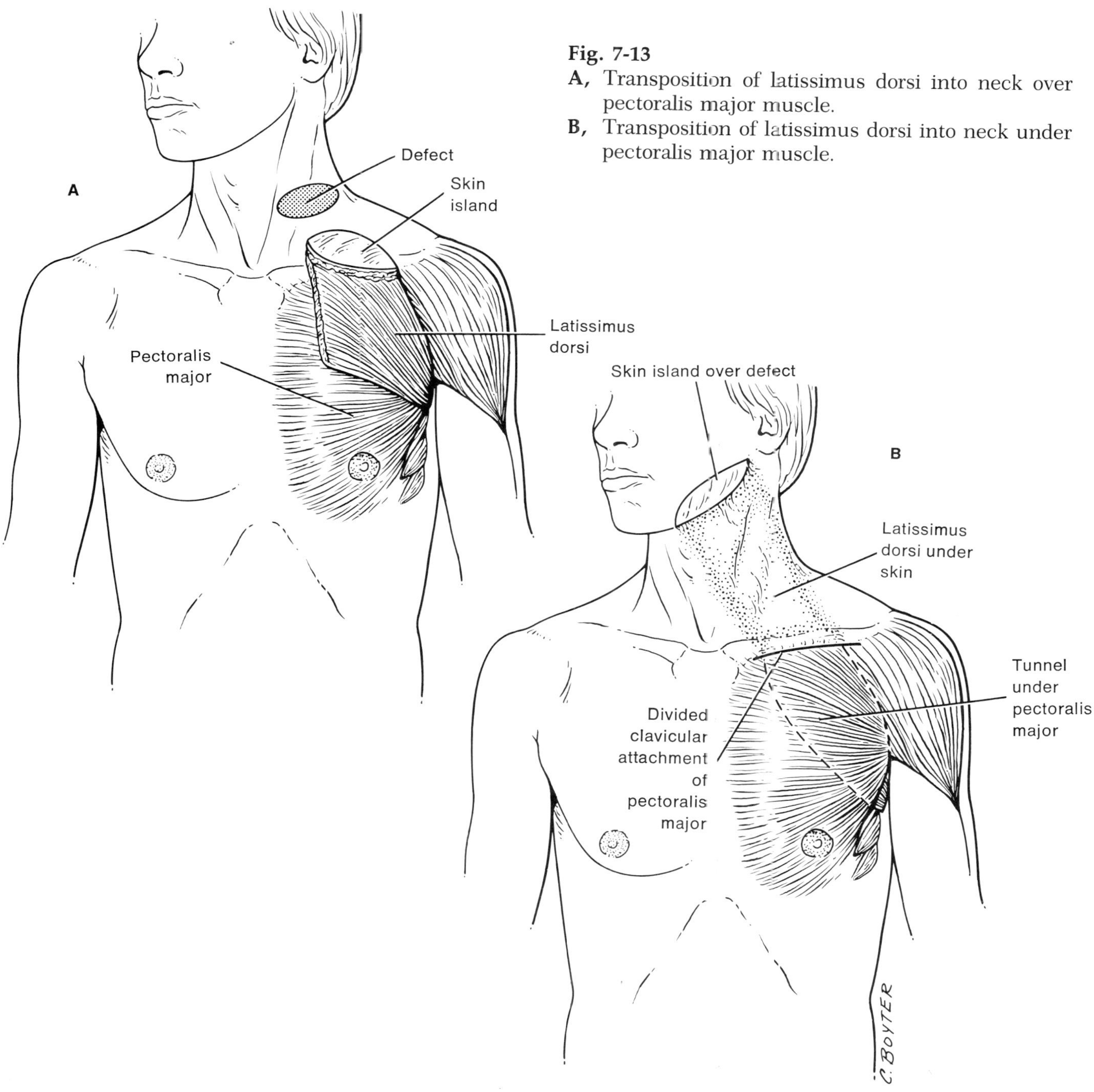

Fig. 7-13
A, Transposition of latissimus dorsi into neck over pectoralis major muscle.
B, Transposition of latissimus dorsi into neck under pectoralis major muscle.

TRAPEZIUS MUSCULOCUTANEOUS FLAP

The trapezius musculocutaneous flap has a definite role in head and neck reconstruction. Although it is less widely used than the pectoralis major musculocutaneous flap, its superior location on the upper posterior trunk and neck with its impressive anterior arc of rotation makes its potential use in this area obvious. The muscle originates from the occipital bone and from the lateral processes of the seventh cervical and all thoracic vertebrae. The medial and inferior fibers of the muscle can be used for transposition without significant clinical loss of function. The primary blood supply originates from the transverse cervical artery, a branch of the subclavian artery (type II). The artery enters the muscle at the base of the neck and divides into ascending and descending branches. Ascending branches may occasionally arise as a separate branch of thyrocervical trunk. The transverse cervical artery courses between the sternocleidomastoid muscle and scalenus anterior muscle and enters the anterior margin of the muscle. The descending branch of the pedicle extends inferiorly along the deep surface of the muscle between the vertebral column and scapula. Its skin territory based on musculocutaneous perforators extends between the spine and scapula and is the basis for a vertically designed, superiorly based flap. This muscle with overlying skin may extend 15 cm below the inferior border of the scapula. This musculocutaneous unit can also be designed in a lateral position on the superior edge of the muscle parallel to the clavicle. This modification in design is based as a skin territory with underlying muscle on the ascending branch of the transverse cervical artery.

The vertically oriented posterior flap has been easier to transpose, and the donor defect is more acceptable. This flap is most useful in instances where other flaps, such as deltopectoral and pectoralis major muscle flaps, have failed to provide adequate coverage. For this reason the posterior trapezius musculocutaneous unit has primarily been used as a secondary flap in reconstructive head and neck surgery (Fig. 7-14).

Flap elevation

The skin island is located on the inferior aspect of the trapezius muscle. It is designed between the vertebral column and the scapula with its vertical axis extending between the midscapula and the inferior origin of the muscle. The skin territory may be extended inferiorly over the territory of the latissimus dorsi muscle as a random flap extension when necessary to achieve an extended arc of rotation for this muscle unit. The skin is incised to the posterior surface of the trapezius muscle. The medial muscle fibers of origin are divided, and the flap is elevated toward the base of the neck. At the level of the tip of the scapula, care must be taken to separate the anterior surface of the trapezius muscle from the rhomboid muscles. At this level the fibers of insertion into the scapula are divided. Further proximal flap dissection will depend on the required arc of rotation. If the flap is designed for reconstruction of defects of the anterior face, then the flap is elevated to the level of the base of the neck. This requires further division of fibers of insertion into the acromioclavicular joint. When possible, the fibers between the base of the skull and the acromioclavicular joint are left intact, thus preserving the trapezius muscle function. It must be cautioned that knowledge of the status of the transverse cervical artery is required before this flap is elevated for use in head and neck reconstructive surgery. When there is doubt in regard to possible division of this vascular pedicle associated with a prior radical neck dissection, then angiography must confirm the presence of this artery before flap elevation is undertaken. The donor defect is generally closed directly.

Fig. 7-14. Sixty-eight-year-old woman who underwent total laryngectomy and partial pharyngectomy followed by 6000 R of radiation for carcinoma of larynx 5 years earlier developed pharyngocutaneous fistula that has been present for 5 years. Two previous attempts at closure with pectoralis major flaps have failed.

A, Fistula in right neck.
B, Skin flaps elevated around fistula to be turned in for mucosal lining.
C, Vertically oriented posterior trapezius musculocutaneous flap outlined.
D, Flap transposed into area of defect.
E, Wound is well healed several months later. Patient was able to eat 2 weeks after operation. *T,* Tracheostomy site.

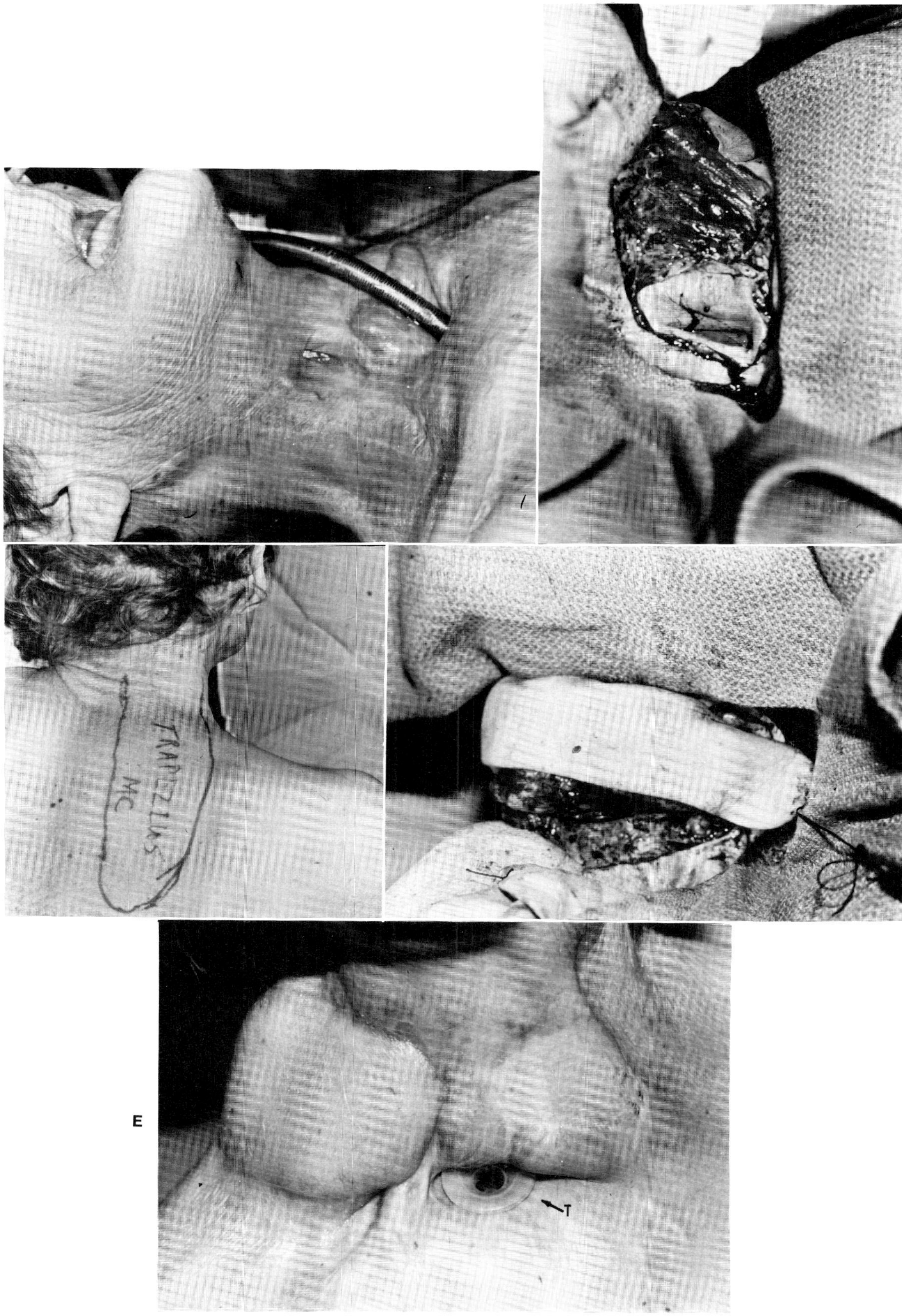

Fig. 7-14. For legend see opposite page.

Summary

The use of muscle and musculocutaneous flaps has simplified and enhanced the capability of surgeons to reconstruct defects of the head and neck. By incorporation of the skin territory with its underlying muscle, the surgeon can complete the reconstruction of these patients in one stage. The pectoralis major muscle is now used more widely than the deltopectoral axial flap in intraoral and pharyngeal reconstruction, because the stage of planned oral cutaneous fistula can be avoided, and the flap can be raised without delay with great reliability. The latissimus dorsi and trapezius muscles are also useful flaps and find their greatest applicability in second and third reconstructive failures.

MICROSURGICAL COMPOSITE TISSUE TRANSPLANTATION

Although this chapter primarily covers the use of transposition of musculocutaneous flaps, no discussion of current head and neck reconstructive techniques would be complete without the mention of the role of microsurgical composite tissue transplantation. Two specific procedures should be mentioned that show promise of becoming standard methods of reconstruction. Very often the use of microsurgical composite tissue transplantation is associated with the simultaneous use of muscle and musculocutaneous flap transposition for head and neck reconstructive surgery.

MANDIBULAR RECONSTRUCTION

Mandibular reconstruction continues to be a challenge whether after trauma or the surgical extirpation of tumors. The use of vascularized rib and clavicle transposed with muscle shows some promise but may be associated with significant failure and complication rates. Recent reports describe successful transplantation of vascularized iliac bone based on the deep circumflex iliac artery for mandibular reconstruction. The advantage of this method appears to be the use of iliac crest, which has a better shape and consistency, for mandibular reconstruction than does rib. The fact that the tissue is well vascularized both by periosteal and endosteal circulation improves the ability of this transplanted tissue to resist infection and radiation. In many instances the properly tailored iliac bone is inset more easily than a segment of rib transposed in association with a muscle flap. Preliminary data indicate rapid bony union using this approach of microvascular surgery to accomplish mandibular reconstruction.

HYPOPHARYNX AND CERVICAL-ESOPHAGEAL RECONSTRUCTION

Hypopharynx and cervical-esophageal reconstruction after tumor extirpation represents a difficult reconstructive problem not easily solved by standard flap transposition techniques. Although both the deltopectoral flap and the pectoralis major musculocutaneous flap have been successfully used to reconstitute the continuity of the upper gastrointestinal tract, recent experience with the use of free jejunal transplantation now represents the treatment of choice for this reconstructive problem (Fig. 7-15). In an experience with 30 free jejunal transplantations, only one immediate technical failure and one late failure related to infection have been noted. The use of free jejunal transplantation is often associated with the use of either the axial deltopectoral flap or musculocutaneous flap transposition from simultaneous soft tissue and skin coverage.

Technical technique

The following are steps in performing free jejunal transplantation:

1. Tumor extirpation and neck dissection are completed. During this time a branch of the external carotid artery and a branch of the external jugular vein are marked and preserved as recipient vessels.
2. A mini-laparotomy is performed, and a segment of jejunum is removed. The segment with a prominent artery and vein is identified and the approximate jejunum excised. Jejunal continuity is reestablished and the abdomen closed.
3. The donor vessels to the jejunum are prepared for microvascular anastomoses. An end-to-end microanastomosis is generally performed between the artery and vein of the jejunum and recipient vessels previously preserved in the neck.
4. Skin closure is completed, or when required, pectoralis major musculocutaneous or deltopectoral flap transposition completes a soft tissue and skin coverage.

The use of jejunal transplantation is indicated in the following head and neck reconstructive situations:

1. Massive resection of laryngopharynx when resection may extend into the oropharynx and involve a large portion of the cervical esophagus
2. Patients with radiation failure when the laryngopharyngectomy is required
3. Patients for secondary reconstruction of the hypopharynx or cervical esophagus in whom

A B C D

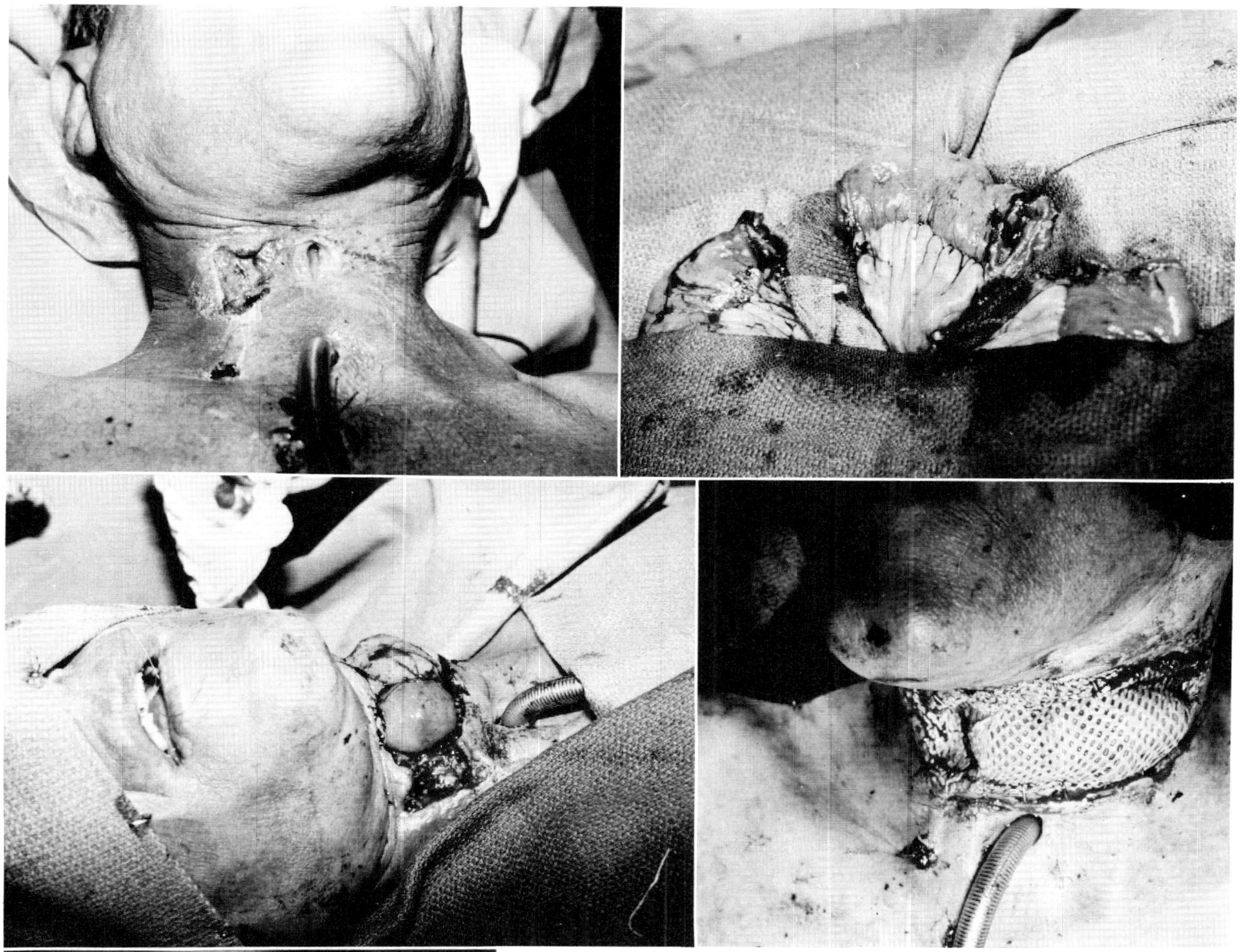

E

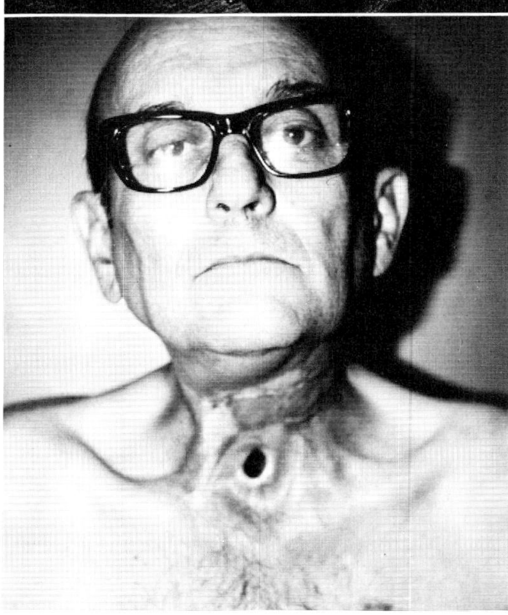

Fig. 7-15. Reconstruction of esophagus with free jejunal transplantation.

A, Sixty-two-year-old man who 5 years previously had undergone laryngectomy and radiation for carcinoma of larynx. Spontaneous pharyngocutaneous fistula persists despite attempt at closure with right sternocleidomastoid flap.

B, Isolated jejunal segment.

C, Revascularized jejunum is sutured into esophageal defect.

D, Radiated skin flaps would not close directly, therefore meshed split graft is applied directly over mesentary and bowel.

E, Three-month postoperative view shows fistula closed and skin grafts healed. Patient swallowing with no difficulties.

Continued.

F 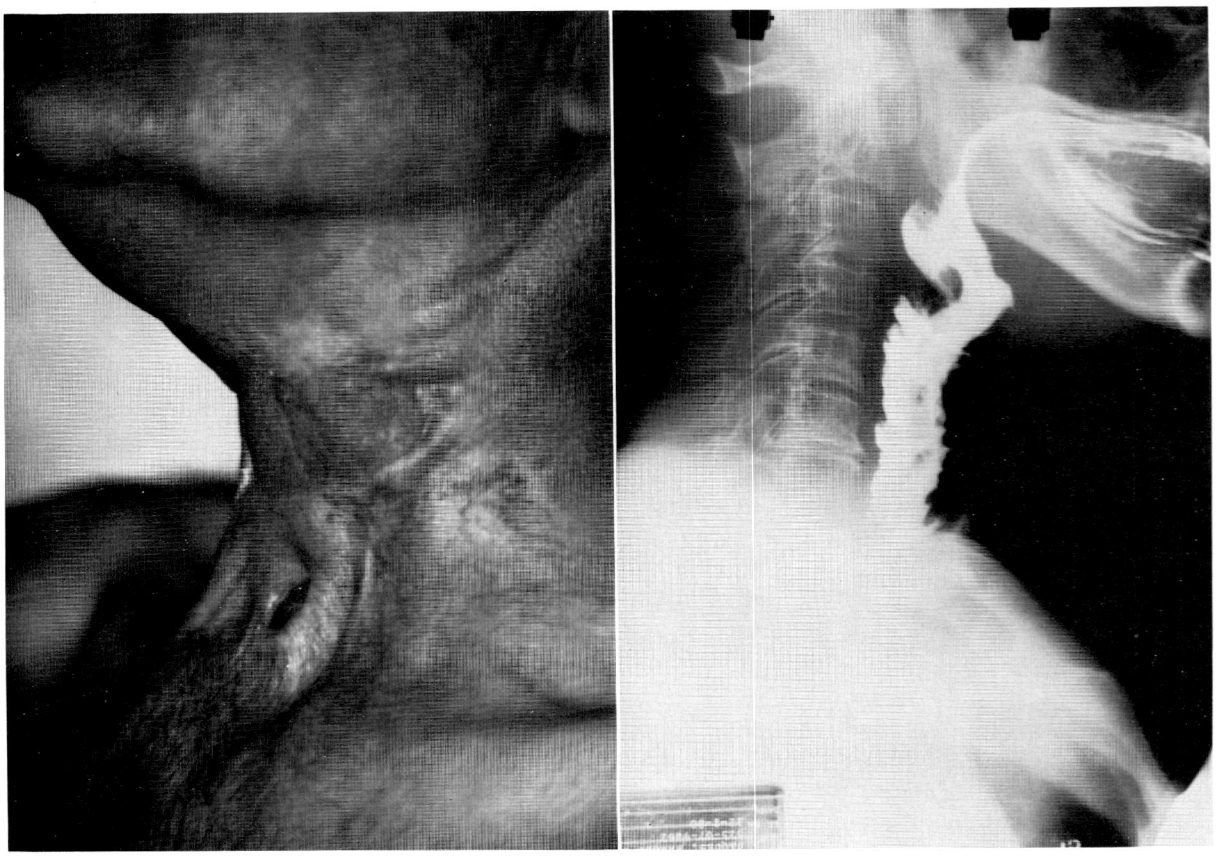 G

Fig. 7-15, cont'd
F, Close-up shows stable skin graft coverage of jejunum.
G, Barium swallow 1 year after esophageal reconstruction.

other methods have failed because of flap necrosis or radiation

4. Patients in whom pharyngeal closure has resulted in a stricture that is not responsive to dilation
5. Patients with loss of large amounts of intraoral or pharyngeal lining

Conclusion

The use of axial and musculocutaneous flaps and microsurgical composite tissue transplantation has simplified and greatly improved reconstructive efforts in a wide variety of head and neck problems. These procedures are basic; refinements are now required to simplify the reconstructive effort. Closure of a massive compound wound is now possible in one stage.

ANNOTATED BIBLIOGRAPHY

1. Ariyan, S.: One stage reconstruction for defects of the mouth using a sternomastoid myocutaneous flap, Plast. Reconstr. Surg. **63:**618, 1979.

   Using a cadaver and radiographic techniques, the blood supply to the sternomastoid muscle is described. Based on this knowledge, the author repairs defects in the mouth with the pedicle based superiorly (occipital) or in the neck with the pedicle based inferiorly (thyrocervical trunk).

2. Ariyan, S.: The pectoralis major myocutaneous flap: a versatile flap for reconstruction in head and neck, Plast. Reconstr. Surg. **63:**73, 1979.

   In a landmark paper Ariyan describes the pectoralis myocutaneous flap — one that has become in an incredibly short period the primary method of reconstruction in head and neck cancer patients.

3. Ariyan, S.: The pectoralis major, sternomastoid, and other musculocutaneous flaps for head and neck reconstruction, Clin. Plast. Surg. **7:**89, 1980.

   The extensive experience by the author in head and neck reconstruction is presented in this article. Fully described in detail are his contributions with the pectoralis major and sternomastoid musculocutaneous flaps.

4. Ariyan, S., and Cuono, C.B.: Use of the pectoralis major myocutaneous flap for reconstruction of large cervical, facial, or cranial defects, Am. J. Surg. **140:**503, 1980.

   Experience in the reconstruction of craniofacial defects in 12 patients is described, including orbit, temporal bone resection, and face and neck.

5. Baek, S., and co-workers: The pectoralis major myocutaneous island flap for reconstruction of the head and neck, Head Neck Surg. **1:**293, 1979.

   The authors' experience in 26 cases confirms the great reliability and utility of this flap. Anatomy, operative technique, and four illustrative cases are described in some detail.

6. Bakamjian, V.Y.: A technique for primary reconstruction of the palate after radical maxillectomy for cancer, Plast. Reconstr. Surg. **31:**103, 1963.

   In this paper Bakamjian documents in detail the use of the sternocleidomastoid musculocutaneous flap as a safe, reliable method of primary palatal reconstruction after its extirpation for cancer. Clearly he recognizes the importance of the underlying muscle as the major source of blood supply to the overlying skin and thus the principle of the musculocutaneous flap.

7. Bakamjian, V.Y.: A two stage method for pharyngoesophageal reconstruction with a pectoral skin flap, Plast. Reconstr. Surg. **36:**173, 1965.

   More so than any other single article, this one profoundly influenced the surgical treatment of head and neck cancer. It documents for the first time the deltopectoral flap—the most versatile and useful technique in primary reconstruction until recent times.

8. Bertolti, J.A.: Trapezius-musculocutaneous island flap in the repair of major head and neck cancer, Plast. Reconst. Surg. **65:**16, 1980.

   A variant of an island trapezius compound flap is described by the author and attributed to Demergasso, who reported it originally in 1977. It is based on the transverse cervical vessels or when these are not present on the posterior scapular artery. The experience by Demergasso in "over 100 cases" and Bertolti in 22 cases would indicate that the flap is very reliable.

9. Brown, R.G., Fleming, W.H., and Jurkiewicz, M.J.: An island flap of the pectoralis major muscle, Br. J. Plast. Surg. **30:**161, 1977.

   The remarkable arc of rotation of the pectoralis major muscle is documented in this case report of a patient with extensive cancer of the neck and mediastinum.

10. Cuono, C.B., and Ariyan, S.: Immediate reconstruction of a composite mandibular defect with a regional osteomusculocutaneous flap, Plast. Reconst. Surg. **65:**477, 1980.

    A logical extension of any musculocutaneous flap is the inclusion of underlying bone to provide skeletal support when repair calls for it. By labeling techniques the authors establish that viability can be maintained throughout the transfer on periosteal blood supply alone.

11. Demergasso, F., and Piazza, M.: Trapezius myocutaneous flap in reconstructive surgery head and neck cancer: an original technique, Am. J. Surg. **138:**533, 1979.

A detailed description of the trapezius island flap, including skin based on the transverse scapular vessels, is presented by the author who first used the method in 1975. Included is a description of the use of the acromion and scapula in the flap to reconstruct the mandible.

12. Edgerton, M.T., Jr.: Replacement of lining to oral cavity following surgery, Cancer **4:**110, 1951.

Profoundly important to all serious students of head and neck surgery, this article details techniques to achieve functional closure of the oral cavity at the time of the major tumor resection. It is perhaps the first article in modern times to suggest that primary reconstruction not only was possible but also was of distinct benefit to the patient who undergoes surgical treatment for head and neck cancer. Moreover, Edgerton presents an effective argument to counter the then-prevailing notion that the immediate repair might mask early local recurrence.

13. Futrell, J.W., and co-workers: Platysma myocutaneous flap for intraoral reconstruction, Am. J. Surg. **136:**504, 1978.

The authors present their experience with 14 patients, using a skin island in the lower neck based on the platysma muscle to reconstruct oral continuity. The article is well illustrated.

14. Gibson, T., and Robinson, D.W.: The mammary artery pectoral flaps of Jacques Joseph, Br. J. Plast. Surg. **29:**370, 1976.

Gibson calls attention to the description in "Nasenplastick" by Joseph (1931) of a flap remarkably similar to the modern deltopectoral flap. Bakamjian became aware of Joseph's description after he had reported on his own extensive clinical experience with the deltopectoral flap.

15. Hill, H.L., and Brown, R.G.: The sternocleidomastoid flap to restore facial contour in mandibular reconstruction, Br. J. Plast. Surg. **31:**143, 1978.

When not only bone but also soft tissue are missing in the mandibular region, bone grafting can be augmented by a transposition flap of the sternomastoid to provide cheek bulk.

16. Holmes, A.D. and Marshall, K.A.: Uses of the temporalis muscle flap in blanking out orbits, Plast. Reconstr. Surg. **63:**337, 1979.

The authors present an excellent description of the anatomy of the temporalis muscle and the capability of splitting the muscle sagittally and coronally to reconstruct the orbit.

17. Jabaley, M.E., Clement, R.L., and Orcutt, T.W.: Myocutaneous flaps in lip reconstruction, Plast. Reconstr. Surg. **59:**680, 1977.

Further experience with the innervated orbicularis oris myocutaneous flap of Karapandzic is described in detail.

18. Jabaley, M.E., and co-workers: Sternocleidomastoid regional flaps: a new look at an old concept, Br. J. Plast. Surg. **32:**106, 1979.

This is a comprehensive account of the blood supply of the sternocleidomastoid muscle and overlying skin.

19. Jabaley, M.E., Orcutt, T.W., and Clement, R.L.: Applications of the Karapandzic principle of lip reconstruction after excision of lip cancer, Am. J. Surg. **133:**529, 1976.

This is a well-illustrated article that confirms the work of Karapandzic, who described the method in 1974. Basically the method is to advance a functional orbicularis oris musculocutaneous unit.

20. Kirshna, B.V., and Green, M.F.: Extended role of latissimus dorsi myocutaneous flap in reconstruction of the neck, Br. J. Plast. Surg. **33:**233, 1980.

A massive defect of the lower face and neck is covered by a large latissimus dorsi myocutaneous flap complemented by a deltopectoral flap for lining.

21. Littlewood, M.: Compound skin and sternomastoid flaps for repair in extensive carcinoma of the head and neck, Br. J. Plast. Surg. **20:**403, 1967.

This work further extends the use of the sternocleidomastoid muscle flap in head and neck reconstruction and builds on the work of Bakamjian.

22. Magee, W.P., Jr., and co-workers: Pectoralis "paddle" myocutaneous flaps: the workhorse of head and neck reconstruction, Am. J. Surg. **140:**507, 1980.

A random extension of the skin in a pectoralis major compound flap is described. This one and several others have appeared as refinements on the original description of the flap so as to give a less obtrusive donor site scar.

23. Maruyama, U., Nakajima, H., and Fujino, T.: A dynamic reconstruction of a facial defect with pectoralis major myocutaneous flap, Br. J. Plast. Surg. **33:**145, 1980.

A patient with an avulsion injury of the face was treated by a transposition flap of the pectoralis major with the pectoral nerve anastomosed to a branch of the facial nerve. A subsequent operation was performed in an attempt to reanimate the face. The result is an indeterminate one.

24. Mathes, S.J., and Vasconez, L.O.: The cervicohumeral flap, Plast. Reconstr. Surg. **61:**7, 1978.

The blood supply to the skin of the lateral neck, shoulder, and arm is documented by cadaver studies. An extended flap of skin of the shoulder and arm can be raised if a portion of the trapezius is included. Blood supply comes from the transverse cervical artery.

25. Matulic, Z., and co-workers: Tongue reconstruction by means of the sternocleidomastoid muscle and a forehead flap, Br. J. Plast. Surg. **31:**147, 1978.

The authors present a technique to reconstruct the tongue by an innervated transposition flap of the sternomastoid muscle. Epithelial lining is provided by a forehead flap. Three patients so reconstructed are reported.

26. McCraw, J.B., Magee, W.P., Jr., and Kalwaic, H.: Uses of the trapezius and sternomastoid myocutaneous flaps in head and neck surgery, Plast. Reconstr. Surg. **63:**49, 1979.

    An excellent discussion of the uses of both of these myocutaneous flaps in head and neck reconstruction is presented in illustrative cases.

27. McGregor, I.A.: The temporal flap in intra-oral cancer: its use in repairing the post-excisional defect, Br. J. Plast. Surg. **16:**318, 1963.

    Ian McGregor in his wry, inimitable way sums up the work as follows: "The technique which has been presented extends the scope of surgery at least a little and as such merits consideration." What he describes in fact in exquisite detail is a technique that paved the way for the modern surgical approach to head and neck cancer, i.e., primary reconstruction.

28. Owens, N.: A compound neck pedicle designed for the repair of massive facial defects: formation, developments and application, Plast. Reconst. Surg. **15:**369, 1955.

    This article represents the first description of the sternocleidomastoid compound flap.

29. Parkash, S., Ramakrishnan, K., and Ananthakrishnan, N.: Sternomastoid based island flap for lining after resection of oral carcinomas, Br. J. Plast. Surg. **33:**115, 1980.

    Nineteen cases wherein the sternomastoid myocutaneous flap was used to reconstruct oral lining form the basis of this report. It provides additional confirmation of the flap's reliability.

30. Quillen, C.G.: Latissimus dorsi myocutaneous flaps in head and neck reconstruction, Plast. Reconstr. Surg. **63:**664, 1979.

    The author presents his experience with the flap in 10 patients. Details of the operative procedure are provided.

31. Robson, M.C., and co-workers: The undelayed Mutter flap in head and neck reconstruction, Am. J. Surg. **132:**472, 1976.

    The Mutter flap is an axial flap based on the descending branch of the occipital artery. Used as described and illustrated by the authors, it has a high degree of reliability. Although it has limited usefulness, the surgeon should be familiar with it.

32. Siemssen, S.O., Kirkby, B., and O'Connor, T.P.F.: Immediate reconstruction of a resected segment of the lower jaw using a compound flap of clavicle and sternomastoid muscle, Plast. Reconstr. Surg. **61:**724, 1978.

    Experience with 8 patients forms the basis for this report, wherein the jaw is reconstituted using clavicle with the clavicular portion of the muscle only as a carrier. In the last 12 patients, in whom full thickness of the clavicle was used, satisfactory results have been achieved without significant complication.

33. Theogaraj, S.D., and co-workers: The pectoralis major musculocutaneous island flap in single-stage reconstruction of the pharyngoesophageal region, Plast. Reconst. Surg. **65:**267, 1980.

    This article explores the use of the pectoralis major compound flap in the reconstruction of defects involving partial circumference of the pharynx and cervical esophagus.

34. Withers, E.H., and co-workers: Pectoralis major musculocutaneous flap: a new flap in head and neck reconstruction, Am. J. Surg. **138:**537, 1979.

    Well-illustrated cases are described to confirm the work of Ariyan.

35. Wustrack, K.O., and Silsby, J.T.: Reconstruction of incompetent oral commissures with dermal-muscle flaps from the lips, Plast. Reconstr. Surg. **62:**118, 1978.

    The fact that drooling can occur in the patient with a reconstructed lip is well known to every head and neck surgeon. The authors present an interesting method of achieving oral continence after overzealous commissurotomy.

8

COMPLICATIONS

Bernard S. Alpert ■ Stephen J. Mathes ■ Luis O. Vasconez

HEAD AND NECK

The complications associated with extirpative surgery and irradiation in the head and neck region have not been eliminated by muscle flap reconstructions. Orocutaneous fistula, abscess, and carotid rupture may result from flap failure.

PECTORALIS MAJOR

Planning errors

Prior radiation therapy or operative injury is unlikely to have affected the blood supply to the pectoralis major. However, the patient's blood pressure should be checked in both arms to rule out any proximal vascular obstruction resulting from atherosclerotic disease or thoracic outlet syndrome, which may affect the patency of the thoracoacromial artery.

Careful attention must be given to the size and location of the skin island for the pectoralis major musculocutaneous flap (Fig. 8-1). In women the island is best placed in the inframammary area to avoid significant donor defect with breast and nipple-areolar distortion. In obese patients or in women with large, ptotic breasts, the flap may be too bulky for intraoral use.

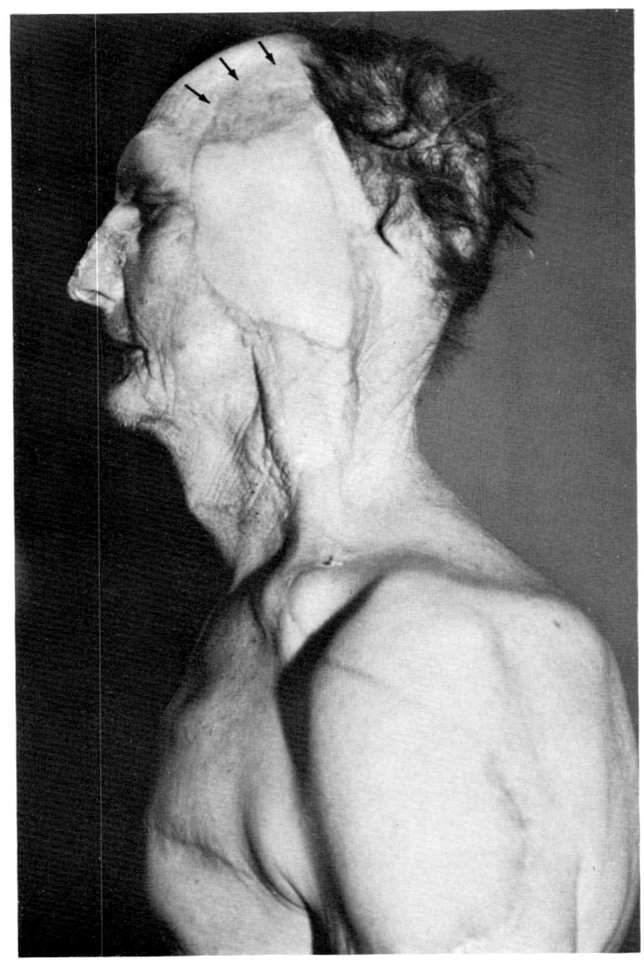

Fig. 8-1. Pectoralis major musculocutaneous flap will not reach skull defects. NOTE: Loss of superior aspect of cutaneous island of flap *(arrows)*.

The resultant donor defect must be taken into account in planning the pectoralis major musculocutaneous flap (Fig. 8-2). Loss of the anterior axillary fold may be avoided by leaving the lateral fibers of the muscle and the medial pectoral nerve intact. Problems with donor site healing over the ribs or costal cartilages may occur. For a large island of skin, a flap from the abdomen may be required for closure of the secondary defect. If rib is included, adequate skin or muscle flap coverage of the rib defect over the pleura is essential.

A

B

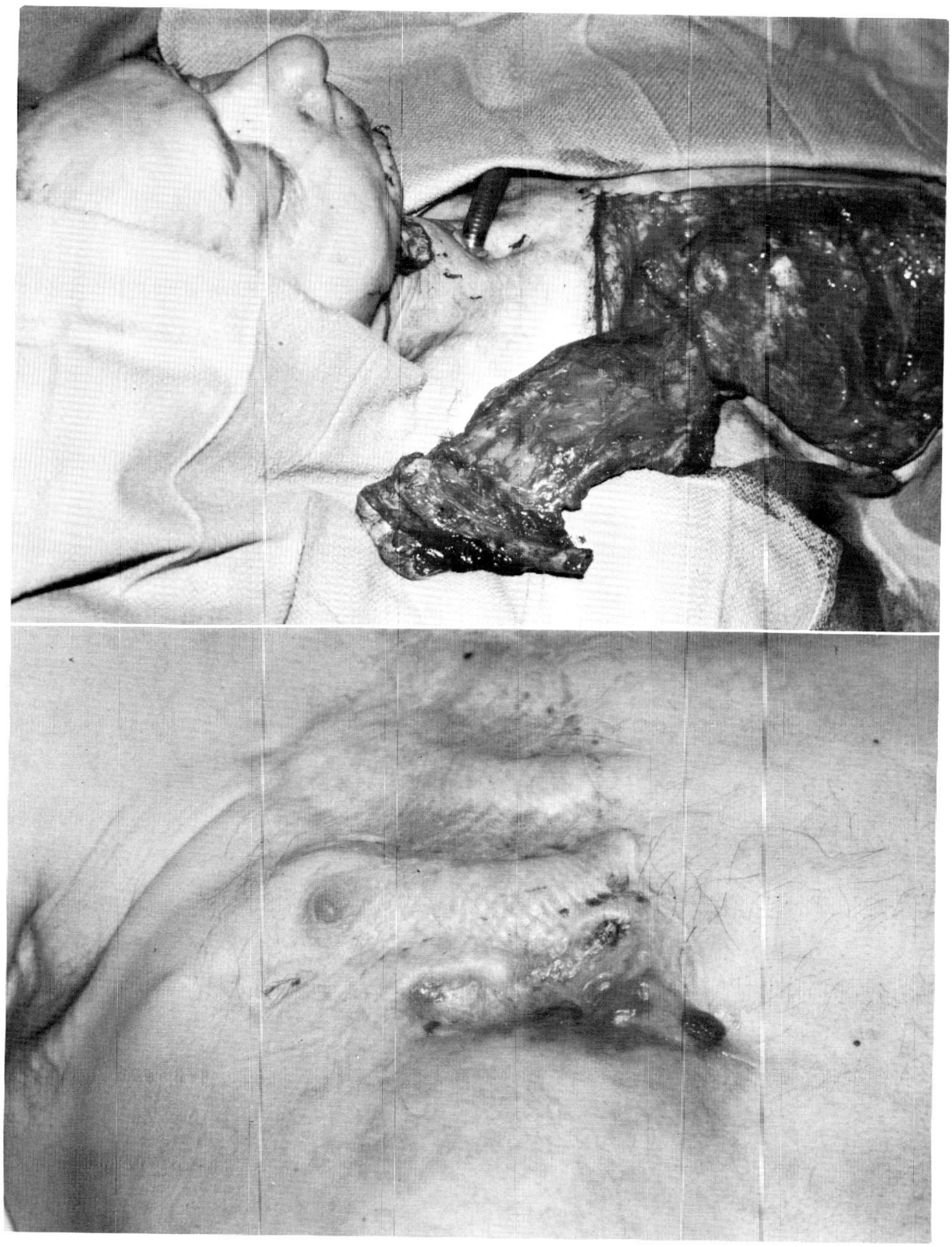

Fig. 8-2
A, Vascularized rib with pectoralis major musculocutaneous flap.
B, Failure of donor site to heal after pectoralis major musculocutaneous flap with vascularized rib. Local flap required for donor site closure.

A 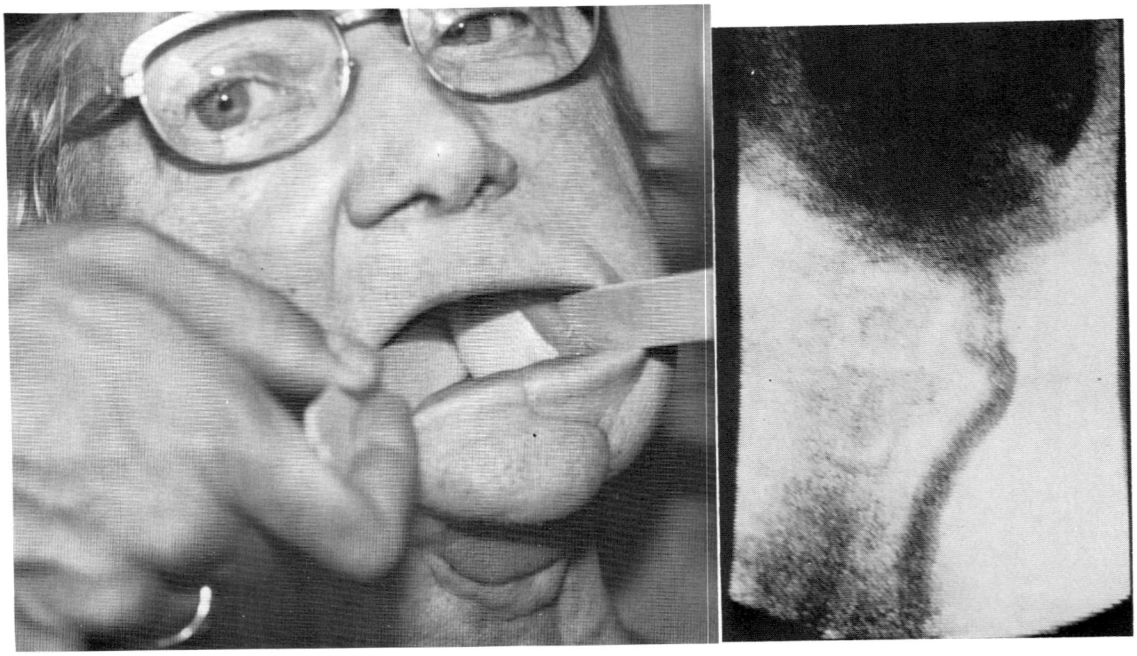 B

Fig. 8-3
A, Site of inset of pectoralis major musculocutaneous flap.
B, Esophagram reveals traction deformity of midcervical esophagus caused by adhesions with pectoralis major musculocutaneous flap used for head and neck reconstruction. Release of these muscular adhesions relieved dysphagia.

Technical errors

The most common technical errors with the pectoralis flap involve injury to the vascular pedicle. It may be damaged by direct trauma or by excessive twisting or tension. The thoracoacromial pedicle enters the muscle on its deep surface 2 cm below the clavicle at the anterior border of the pectoralis minor muscle. By dissecting from the distal border to the proximal border and following the anterior border of the pectoralis minor, the pedicle is easily identified and preserved. All proximal dissection must be done under direct vision, or the pedicle may be damaged.

The pectoralis major muscle, with distal skin over the rectus fascia, is based on perforators from the distal pectoralis muscle that must be preserved for skin island survival. Suturing the dermis to the rectus fascia is a safeguard to prevent disruptive sheering forces. Ribs based on the pectoralis major muscle will survive as vascularized bone only if the periosteal vessels between muscle and rib are preserved.

Tunneling, twisting, and tension may all contribute to flap failure with the pectoralis major. The tunnel must be wide enough so that there is absolutely no pressure on the pedicle. For intraoral reconstruction, in which the pedicle is rotated 180 degrees, further twisting of the muscle or a spiral rotation may compromise flap survival. Tension on the pedicle is avoided by correct preoperative design and skin island placement so that the skin will reach the defect without traction on the muscle. If this is not possible, careful mobilization of the pedicle, preferably under magnification, will allow an extra 3 to 4 cm of reach. If the skin island will not reach the defect because of inadequate flap length, it is best to remove the skin island and use the muscle with a skin graft.

When the pectoralis major muscle is used as a carrier for the skin island external to the neck skin, muscle division must be delayed, depending on the status of the extirpative wound. Muscle division and flap inset are performed 3 to 4 weeks after extirpation for uncomplicated wounds or longer, 6 to 8 weeks, for irradiated wounds. If the muscle bulk of the transposed muscle within the neck is excessive, or if muscle contractions are objectionable, the muscle may be divided or debulked, leaving the vascular pedicle intact at similar postoperative intervals (Fig. 8-3).

Postoperative errors

Appropriate positioning will facilitate venous drainage and avoid vascular constriction. Constrictive neck dressings are avoided.

TRAPEZIUS

Planning errors

The most common error in planning the trapezius musculocutaneous flap relates to its blood supply. The dominant vascular pedicle, the transverse cervical artery, is located in the neck and may have been divided during prior surgery. If prior pedicle division may have occurred, and alternative flaps are not available, a preoperative angiogram should be obtained.

If the anterior trapezius is used, the flap is based on its segmental secondary blood supply from the occipital artery. The skin portion is less reliable, and flap delay may be considered. Elevation of the anterior trapezius muscle results in functional loss.

Technical errors

The underlying rhomboid muscles are intimately associated with the trapezius muscles. Dissection is performed from the distal border to the proximal border to avoid inadvertent elevation of both muscles. Dissection below the rhomboid muscles will lead under the scapula. Since elevation of the posterior trapezius flap allows function preservation, the branches of the eleventh nerve to the intact anterior fibers must be preserved.

If the trapezius is used as a carrier for a skin island external to the neck, flap division must be delayed, depending on the status of the extirpative wound. Muscle division and flap inset are performed 3 to 4 weeks after extirpation for uncomplicated wounds and longer, 6 to 8 weeks, for irradiated wounds.

Postoperative errors

Appropriate positioning will facilitate venous drainage and avoid vascular constriction. Constrictive neck dressings are avoided.

TEMPORALIS

Planning errors

The proximal bulk of the temporalis muscle may produce a bulge if it is transposed over the zygomatic arch. Distally it is thin, and this differential in thickness must be considered preoperatively. The donor defect following transposition may leave a marked depression in the temporal fossa that is noticeable in the bald patient. This deformity is corrected with a Silastic implant.

Technical errors

The temporal and zygomatic branches of the facial nerve are close to the temporalis muscle near the zygomatic arch, and injury to these nerves must be avoided. The internal maxillary artery is deep to the arch at the level of the sigmoid notch and could be injured during dissection in this region.

Postoperative errors

Pressure dressings are avoided to prevent flap compression.

STERNOCLEIDOMASTOID

Planning errors

Extirpative procedures within the neck may require removal of either the sternocleidomastoid muscle or its vascular pedicles. Therefore previous neck surgery or concurrent lymphadenectomy may preclude the use of this flap. If the width of the skin island extends beyond the margins of the muscle, neck closure is compromised, leaving vital structures exposed.

Technical errors

Major vascular structures are located beneath the sternocleidomastoid muscle, including the carotid artery and internal jugular vein. Both the spinal accessory and hypoglossal nerves are located in proximity to the superior muscle belly.

Postoperative errors

Appropriate positioning will facilitate venous drainage and avoid vascular constriction. Constrictive neck dressings are avoided.

PLATYSMA

Planning errors

The arc of rotation of the platysma muscle is limited to the lower lip and buccal sulcus. The size of the skin island should allow direct closure of the donor defect.

Technical errors

The platysma is a very thin muscle and may actually be missed in the dissection. Meticulous technique, precise hemostasis, and identification of the muscle distal to the skin island are critical technical points for successful flap elevation. Excessive twisting, turning, or tension will lead to flap failure.

Postoperative errors

Excessive pressure or constriction within the tunnel or tight external dressings will lead to flap failure.

9

A SYSTEMATIC APPROACH TO FLAP SELECTION

Stephen J. Mathes ■ Foad Nahai

ANTERIOR CHEST (Fig. 9-1)

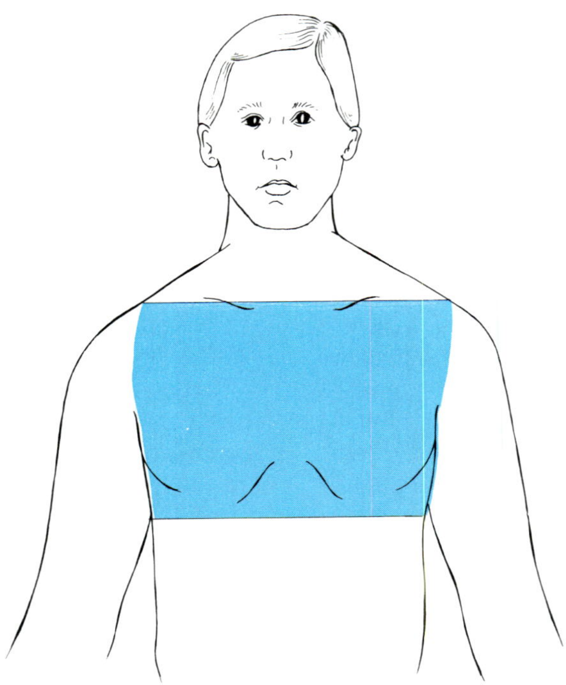

Fig. 9-1

Skin grafts

Skin grafts are method of choice for coverage of soft tissue defects where vital structures are not exposed

Local flaps (Fig. 9-2)

Pectoralis major is flap of choice for mediastinal coverage

Distant flaps (Figs. 9-2 and 9-3)

1. Latissimus dorsi musculocutaneous flap is flap of choice for anterior chest wall defects (Figs. 9-4 and 9-5)
2. Thoracoepigastric axial flap is alternative flap
3. Rectus abdominis muscle or musculocutaneous flap is flap of choice for the inferior mediastinum
4. Omentum is alternative flap
5. Composite tissue transplantation rarely indicated

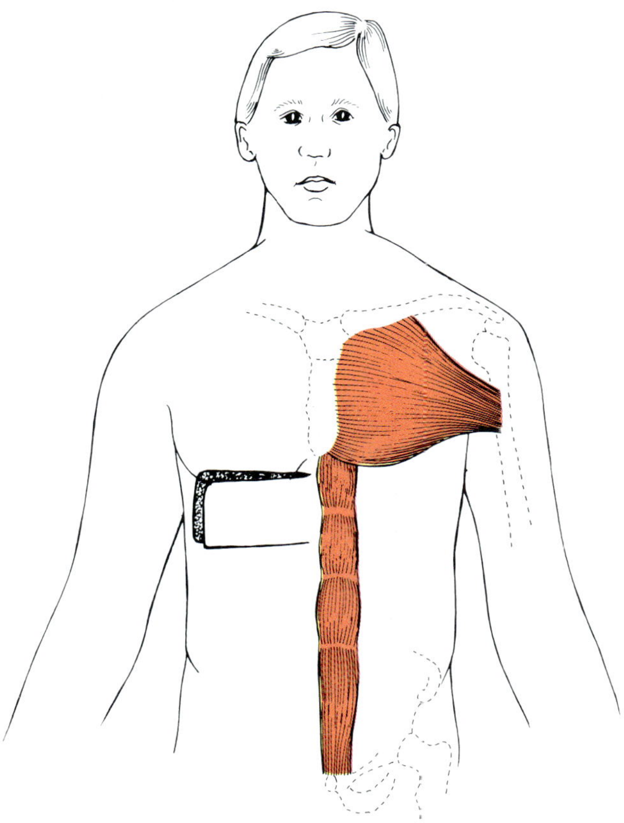

Pectoralis major
Rectus abdominis
Thoracic epigastric flap

Fig. 9-2

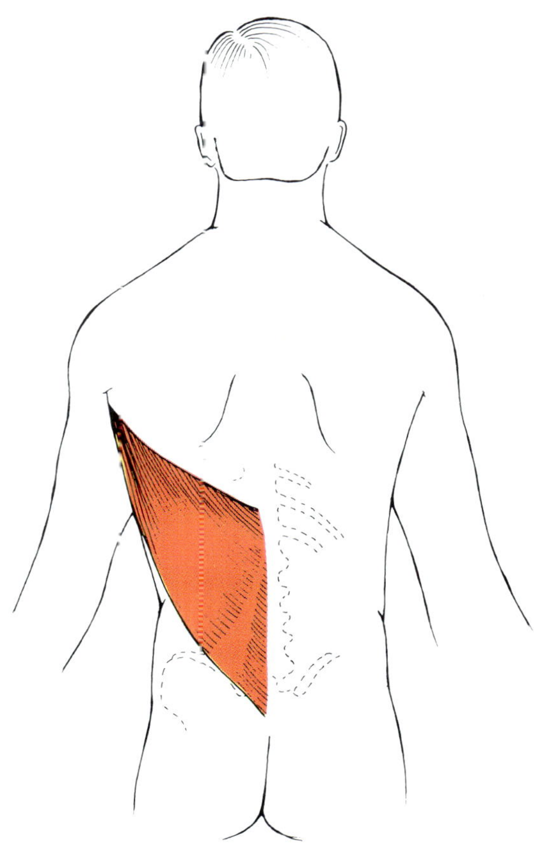

Latissimus dorsi

Fig. 9-3

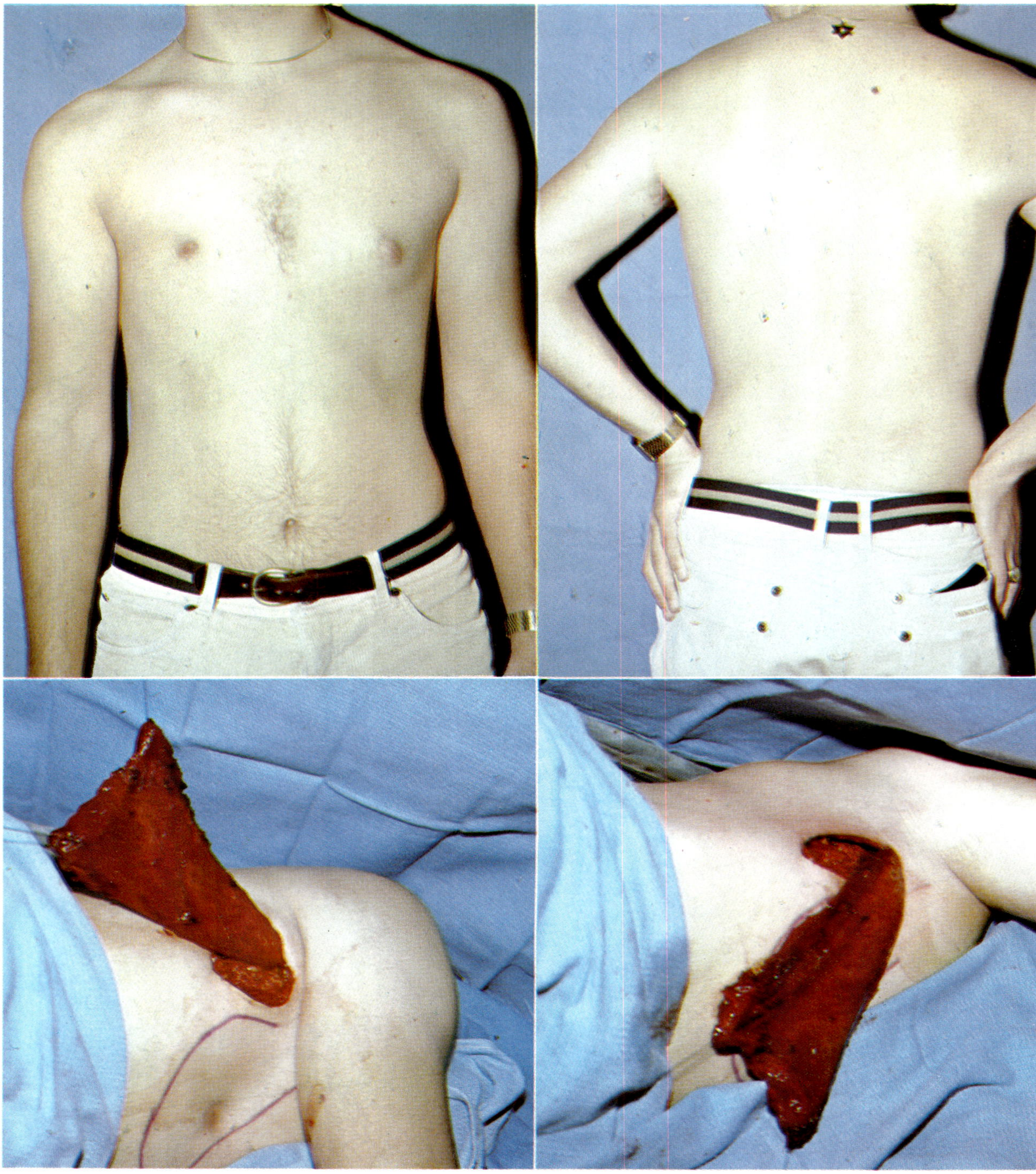

A B

C D

Fig. 9-4. Latissimus dorsi muscle flap for reconstruction of Poland's syndrome.

A, Preoperative view of patient with congenital absence of inferior two thirds of pectoralis major muscle. This patient lacks anterior axillary fold and normal right chest projection.

B, Preoperative view demonstrates normal right latissimus dorsi muscle. No hand anomalies noted.

C, Anterior half of latissimus dorsi muscle elevated through midaxillary incision.

D, Arc of rotation of latissimus dorsi muscle.

E F G H

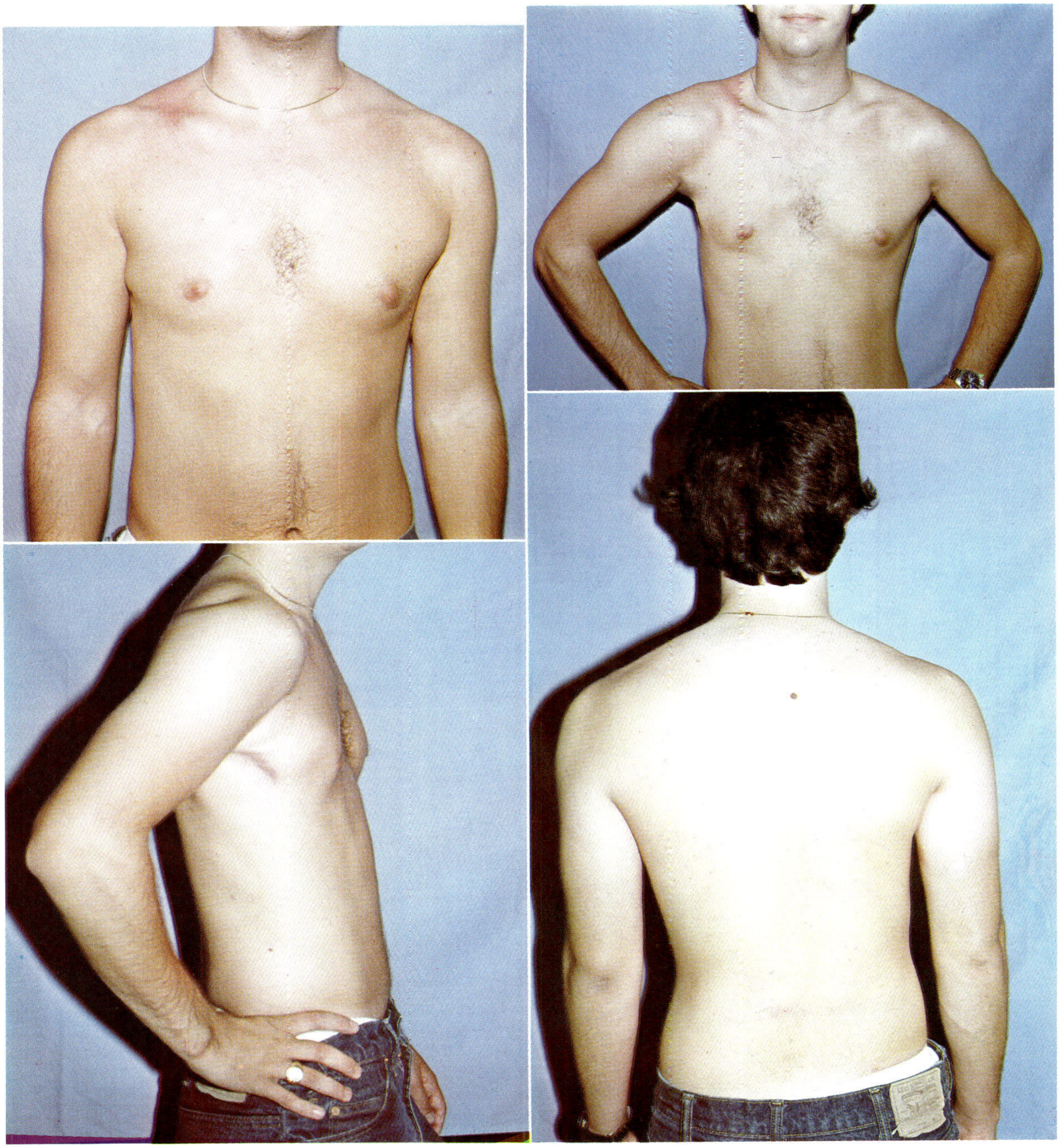

Fig. 9-4, cont'd
E, Anterior postoperative view 1 year after chest wall reconstruction.
F, Active contraction of anterior transposed latissimus dorsi muscle noted.
G, Postoperative oblique view of reconstructed anterior chest.
H, Preserved innervated posterior fibers of latissimus dorsi muscle preserve function.

A B C D

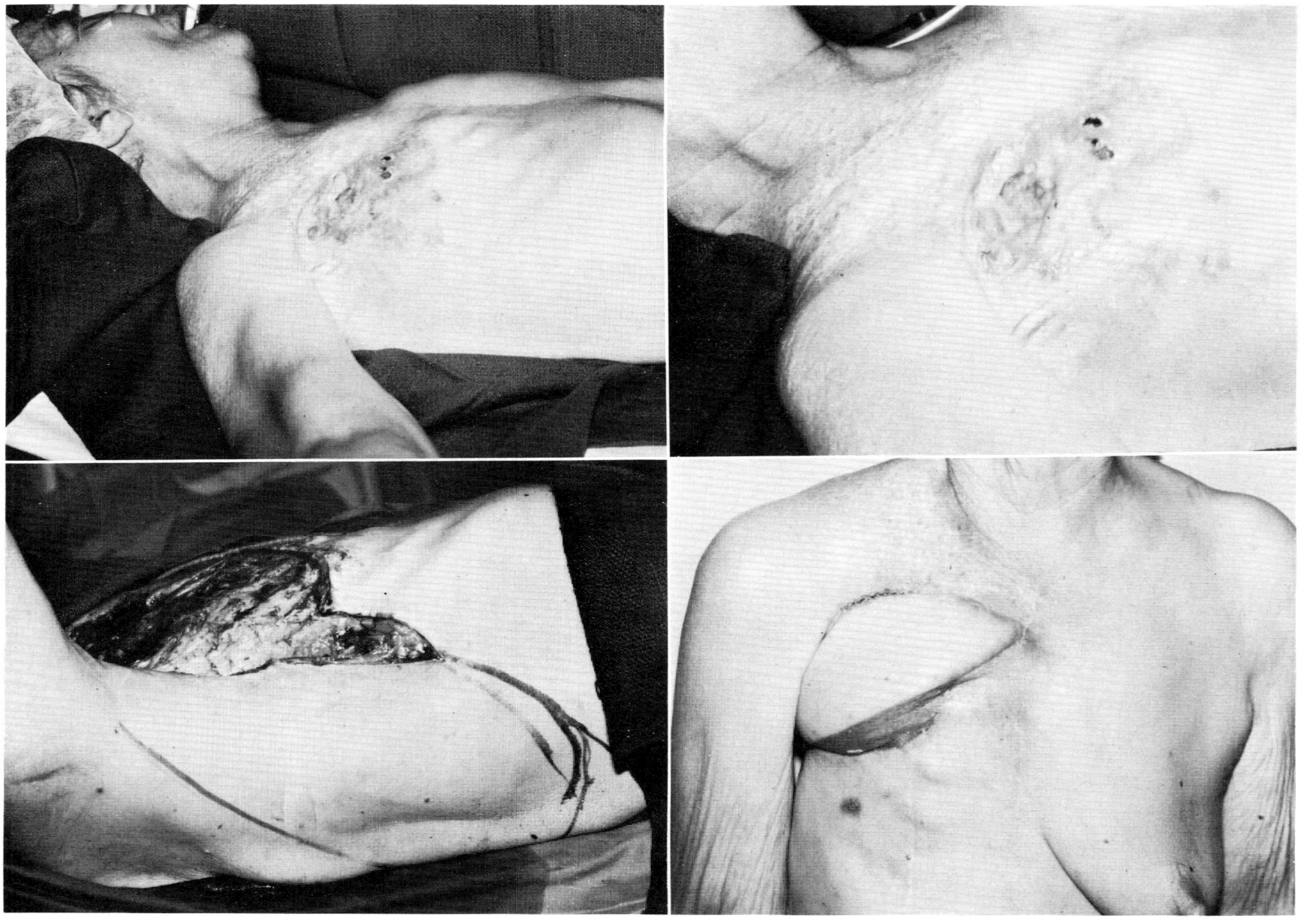

Fig. 9-5. Latissimus dorsi musculocutaneous flap. Radiation necrosis of chest wall after radical mastectomy and radiation for breast cancer.

A and **B**, Sixty-seven-year-old diabetic patient with radiation necrosis and resultant chest wall defect.

C, Denervated latissimus dorsi flap with prior ligation of thoracodorsal artery. Flap elevated based on vascular connection between latissimus dorsi and serratus anterior muscle (branch of thoracodorsal artery to serratus anterior).

D, Flap coverage of radiation ulcer.

STERNUM (Fig. 9-6, *A*)

Skin grafts

Skin grafts are not applicable in reconstruction of the sternum

Local flaps (Fig. 9-6, *B*)

Pectoralis major muscle flap is flap of choice for superior three fourths of exposed anterior mediastinum (Figs. 9-7, 9-8, 9-11, and 9-12)

Distant flaps (Fig. 9-6, *B*)

1. Rectus abdominis muscle flap is flap of choice for inferior fourth of exposed anterior mediastinum (Figs. 9-9 to 9-12)
2. Latissimus dorsi musculocutaneous flap is alternative flap for upper mediastinum
3. Omentum is alternative flap

A

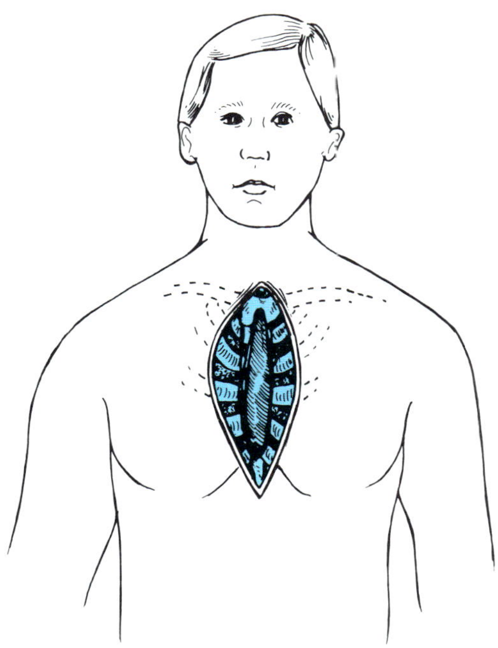

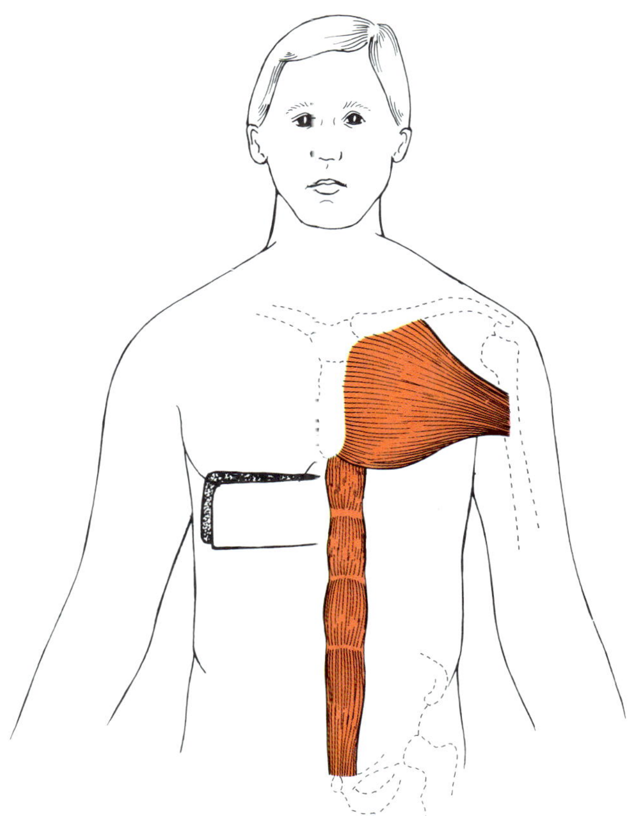

B

Pectoralis major
Rectus abdominis
Thoracoepigastric flap

Fig. 9-6

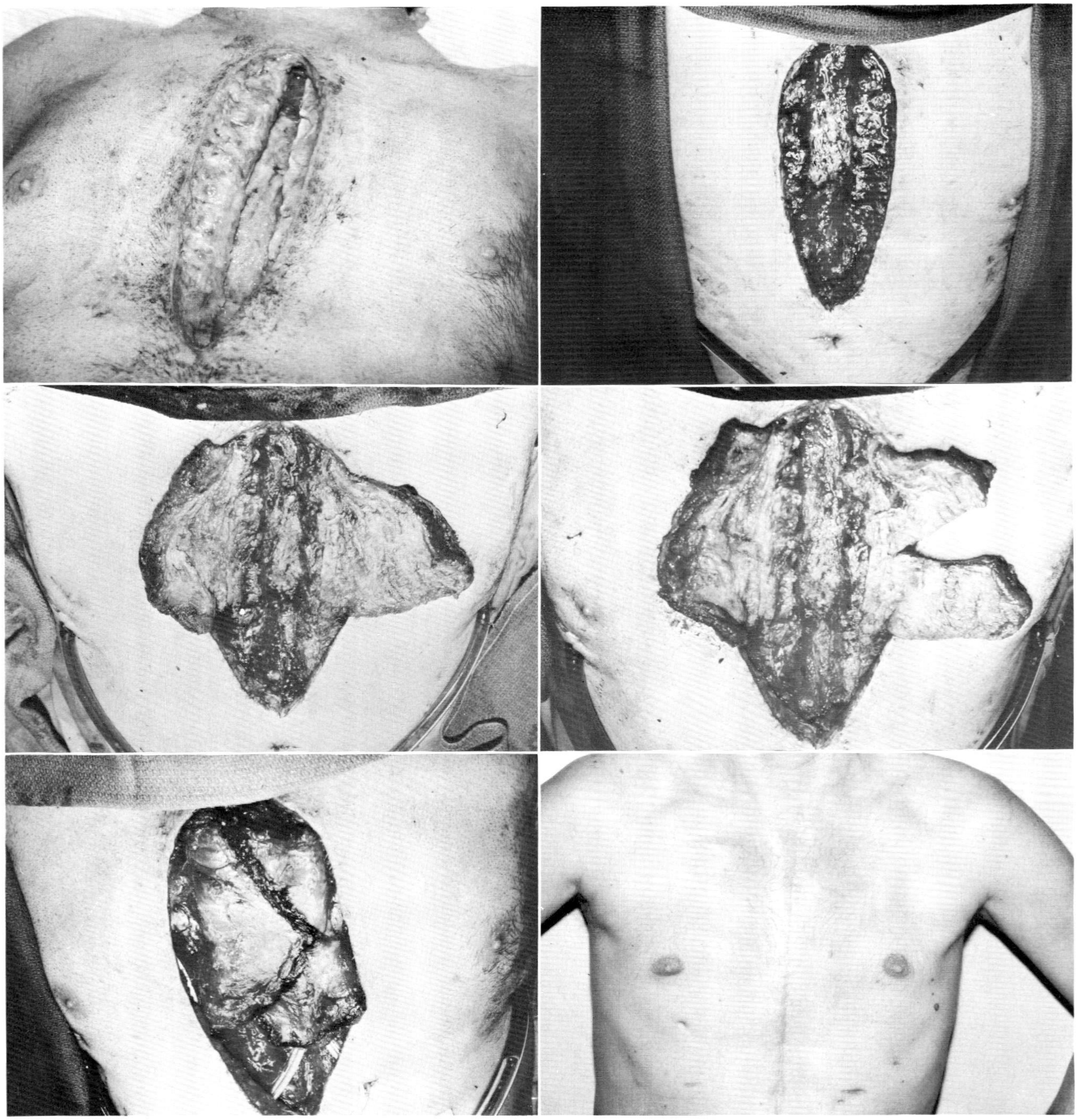

Fig. 9-7. Pectoralis major muscle flap for closure of sternal defect.
A, Infected mediastinal wound.
B, Wound debrided.
C, Bilateral pectoralis major flaps elevated on internal mammary artery.
D, Pectoralis major split in two transversely.
E, Flap sutured.
F, Wound closed directly. NOTE: Preservation of anterior axillary fold.

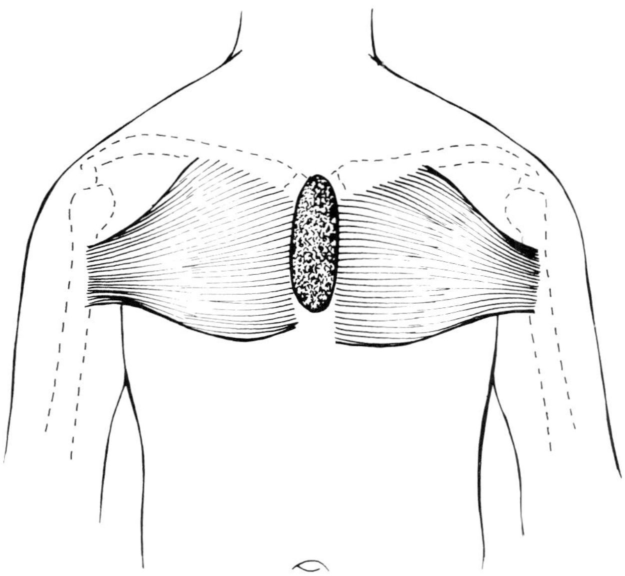

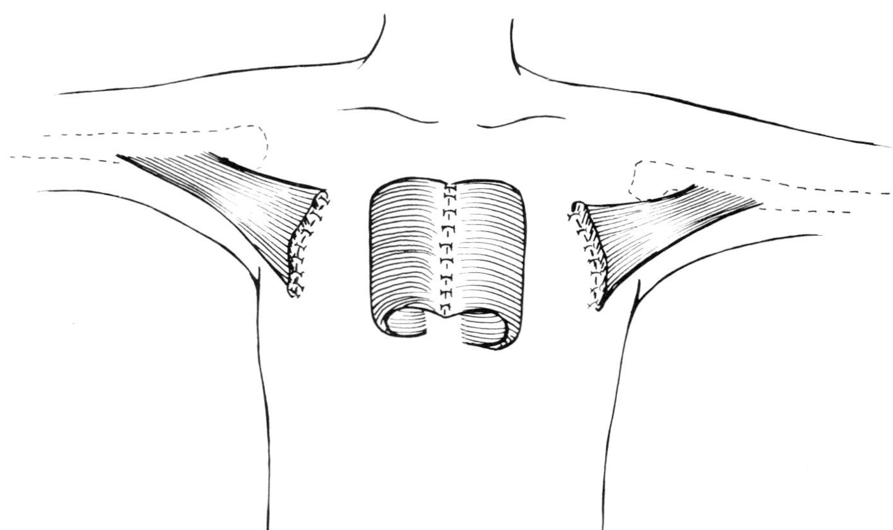

Fig. 9-8
A, Mediastinal wound dehiscence with exposure of superior two thirds of anterior mediastinum.
B, Transposition of bilateral pectoralis major muscle as turnover flaps provide coverage.

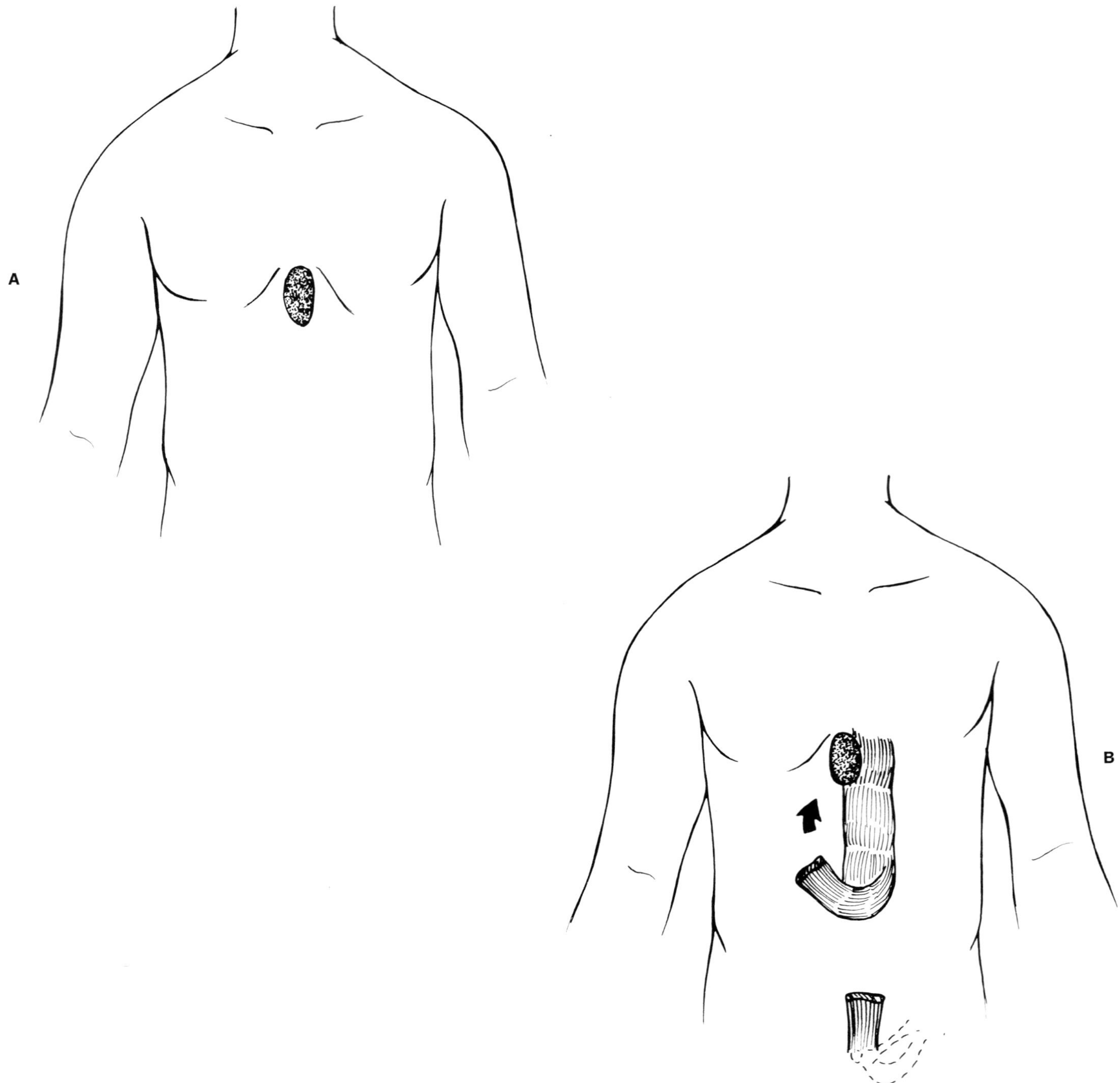

Fig. 9-9
A, Mediastinal dehiscence with exposure of inferior one third of anterior mediastinum.
B, Transposition of superior-based rectus abdominis muscle flap provides coverage.

C

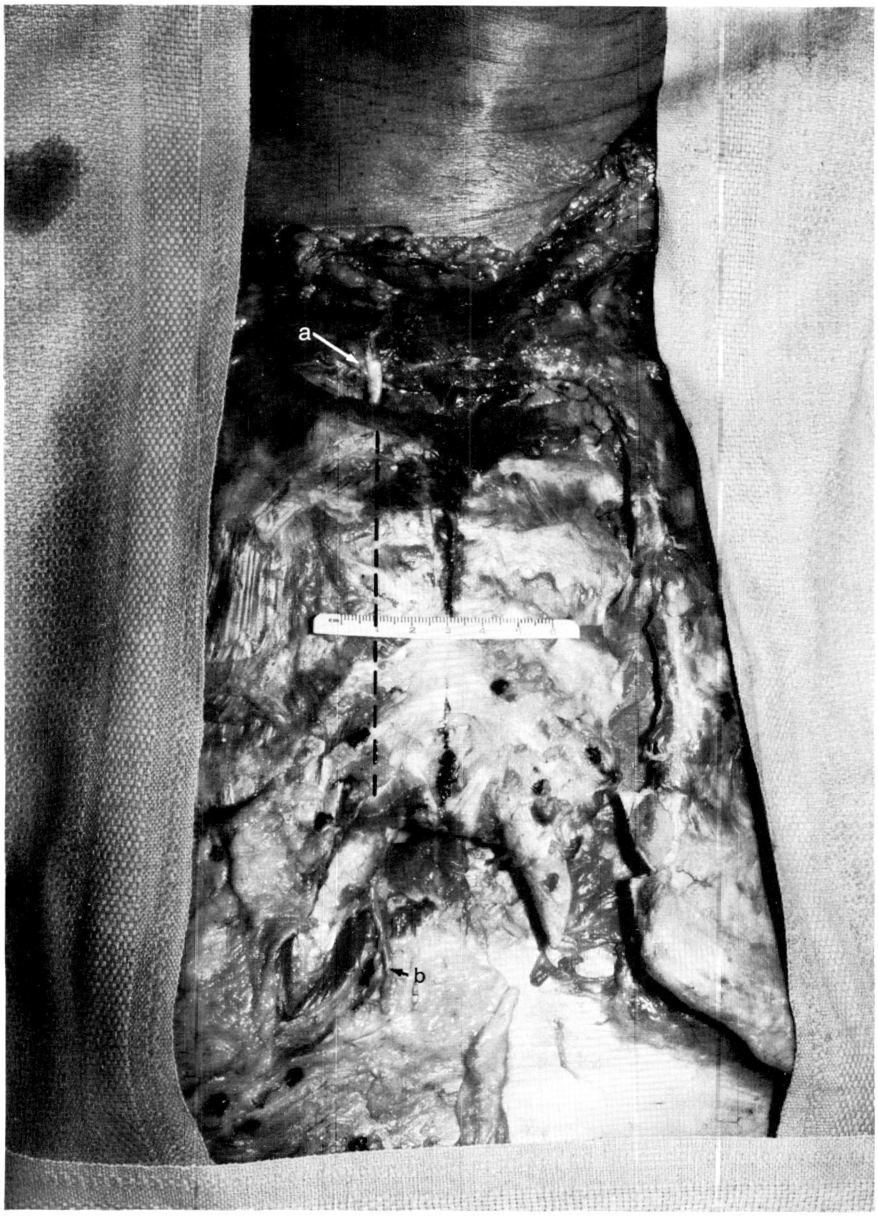

Fig. 9-9, cont'd

C, Position of internal mammary artery in relationship to sternum is demonstrated. NOTE: Injury to internal mammary artery precludes use of rectus abdominis superior–based flap. *a,* Internal mammary artery; *b,* superior epigastric artery.

A B

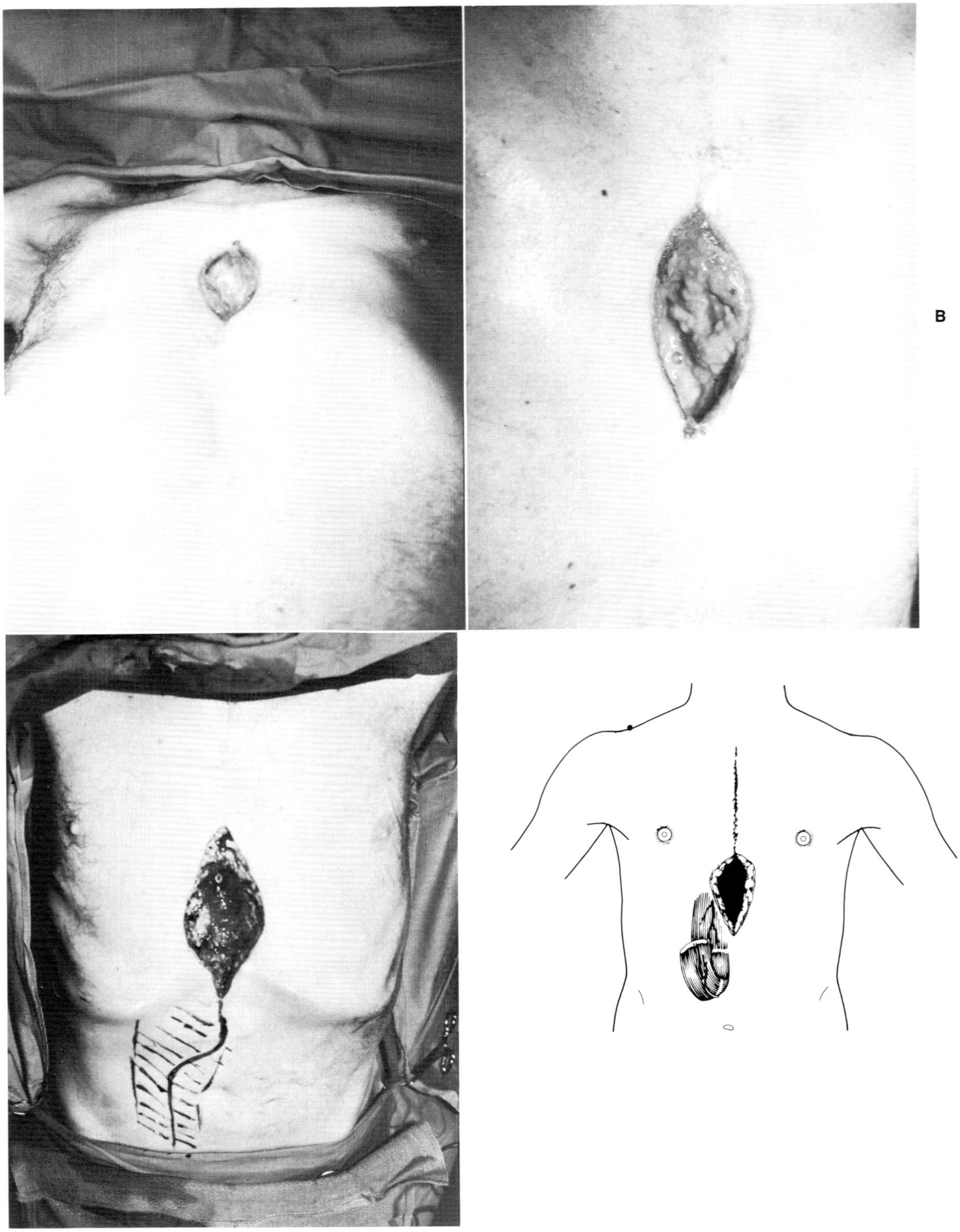

C D

Fig. 9-10. Rectus abdominis muscle flap. Sternal dehiscence following open heart surgery with infected lower sternum.
A and **B,** Open wound.
C, Débridement includes infected sternum.
D and **E,** Flap elevated.

E

F

G

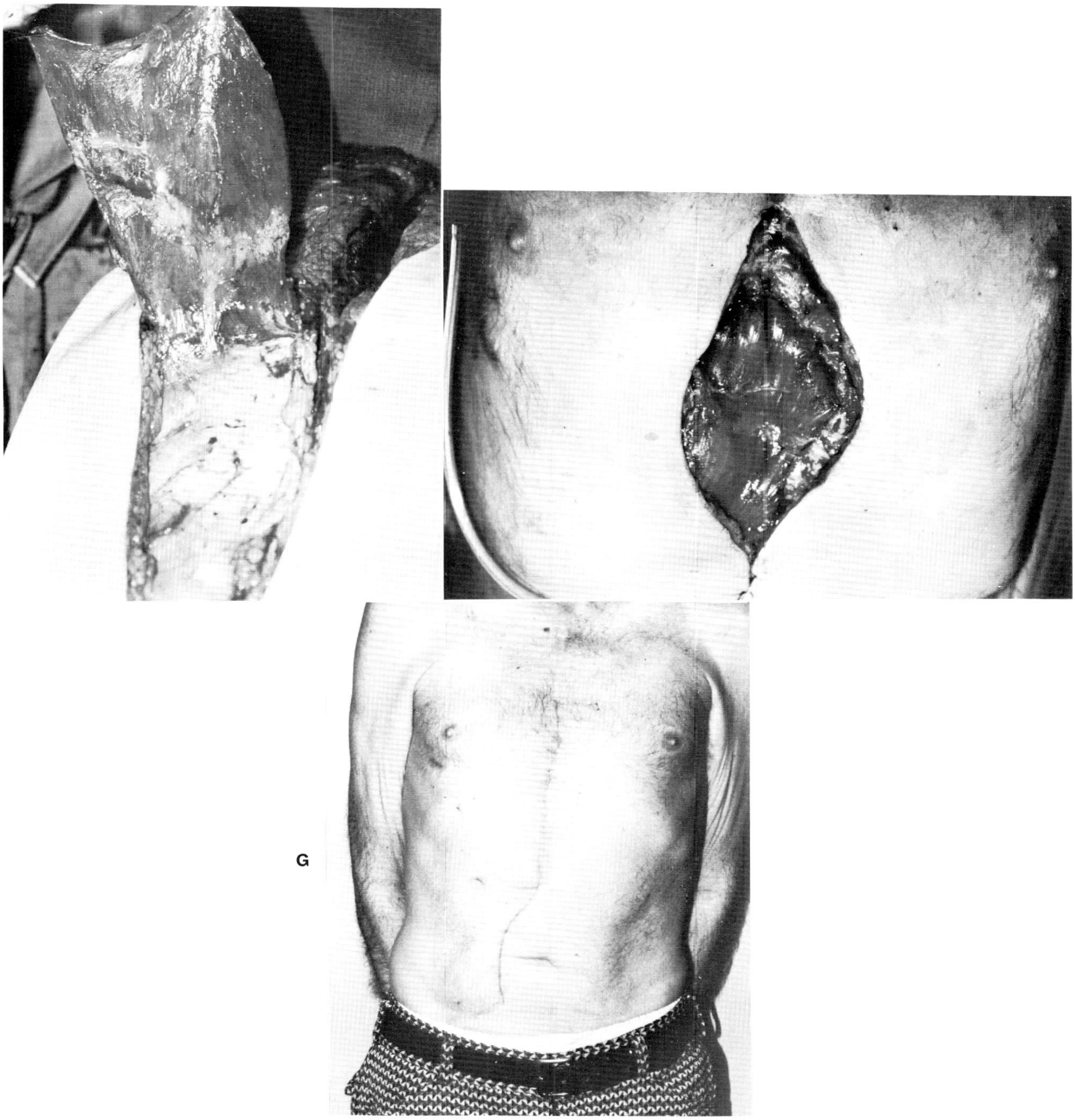

Fig. 9-10, cont'd
F, Flap sutured into defect.
G, Skin closed directly over flap. Six-month postoperative view of patient.

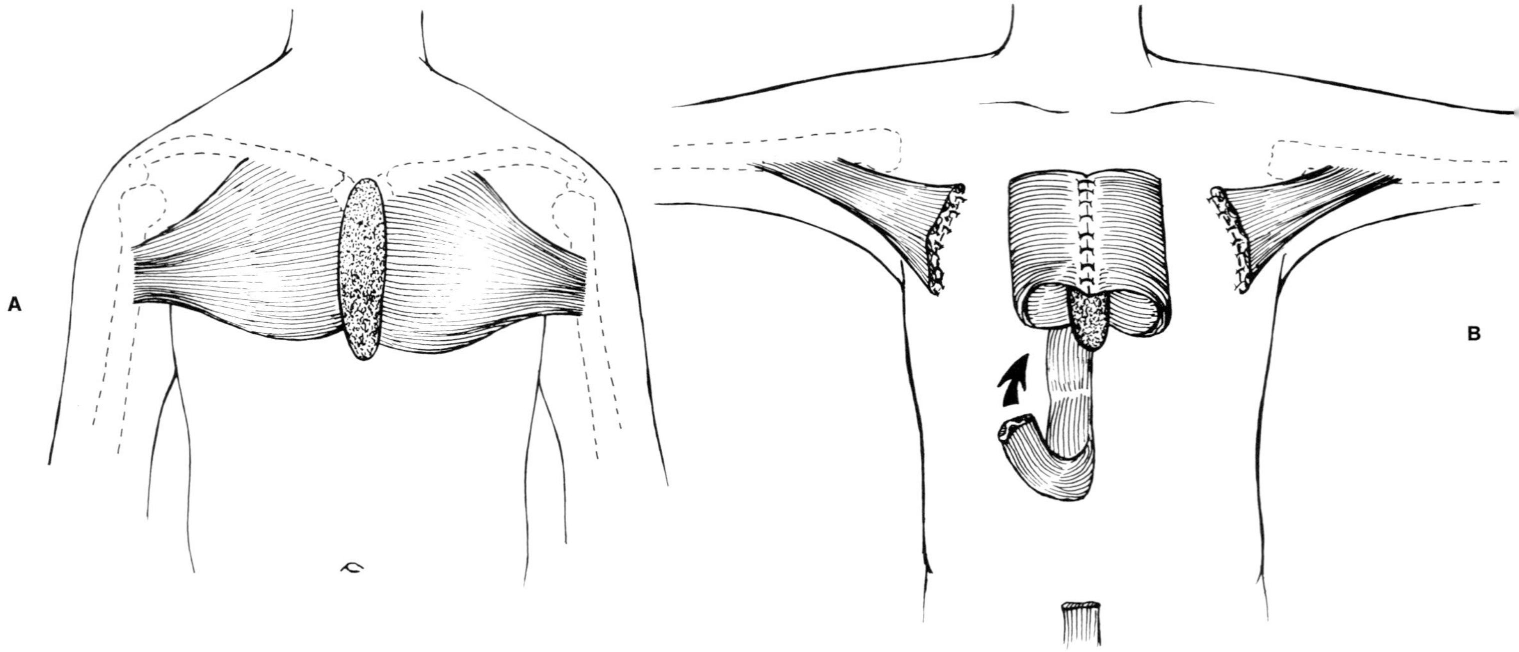

Fig. 9-11. Pectoralis major and rectus abdominis muscle flaps.
A, Mediastinal wound dehiscence with exposure of entire anterior mediastinum.
B, Transposition of both bilateral pectoralis major muscle flaps and rectus abdominis is necessary to provide coverage in one operation.

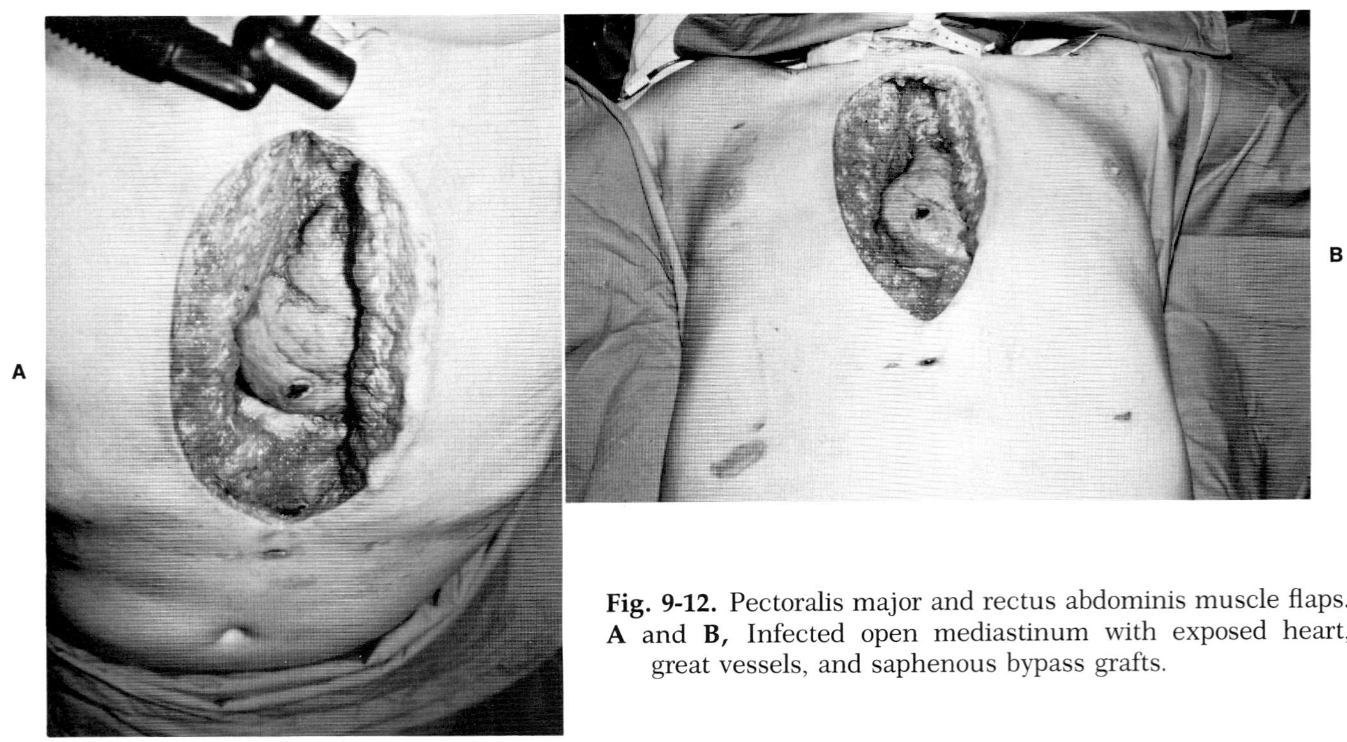

Fig. 9-12. Pectoralis major and rectus abdominis muscle flaps.
A and **B,** Infected open mediastinum with exposed heart, great vessels, and saphenous bypass grafts.

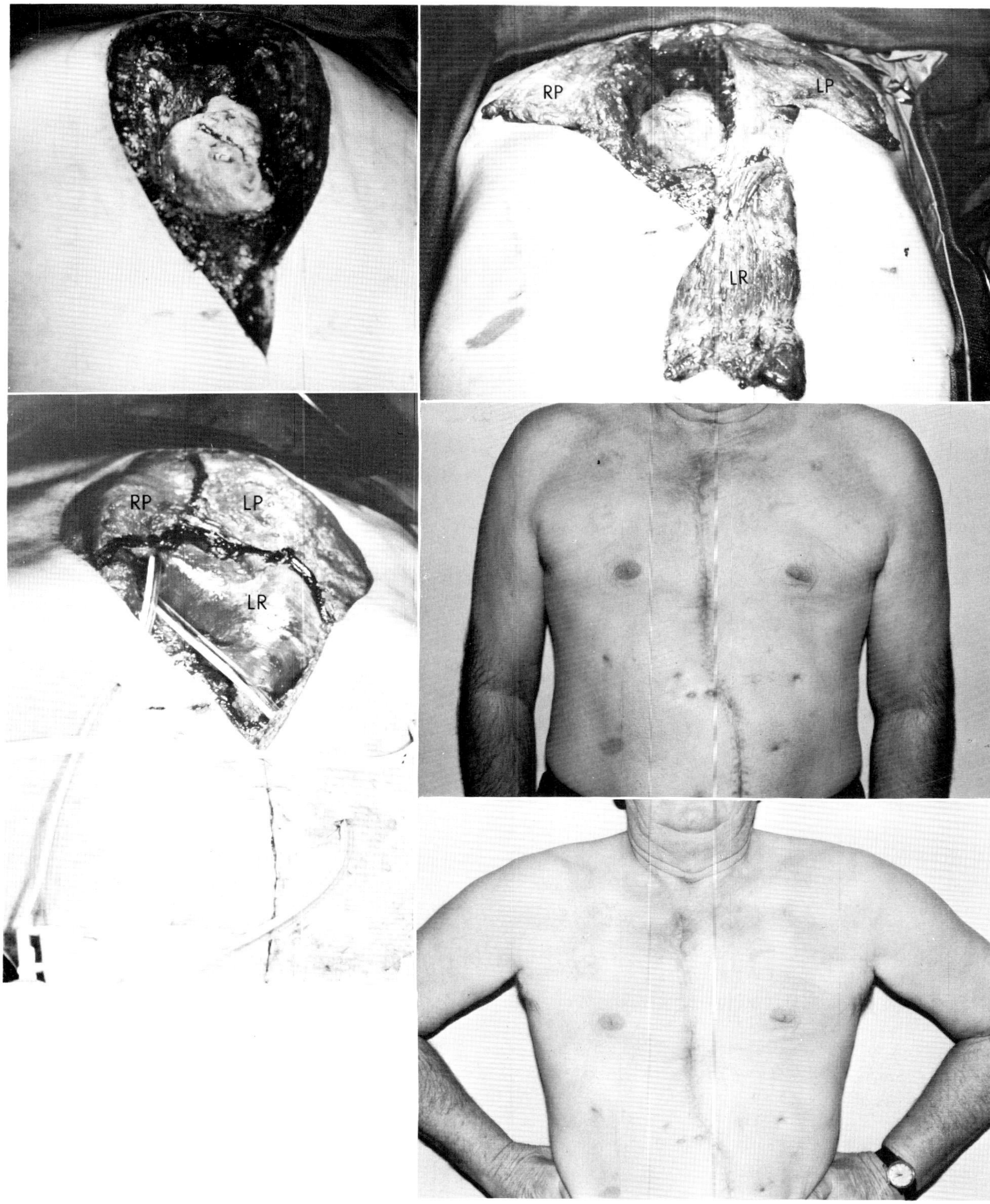

C D E F G

Fig. 9-12, cont'd

C, Wound debrided.

D, Bilateral medially based pectoralis major flaps based on segmental perforators have been elevated. Left rectus abdominis flap elevated. *RP,* Right pectoralis major muscle; *LP,* left pectoralis major muscle; *LR,* left rectus abdominis muscle.

E, Flap sutured to obliterate defect. *RP,* Right pectoralis major muscle; *LP,* left pectoralis major muscle; *LR,* left rectus abdominis muscle.

F, Postoperative view of skin closed directly over muscle.

G, Preservation of anterior axillary folds with intact lateral pectoral muscle fibers innervated by medial pectoral nerve.

10

RECONSTRUCTION OF THE CHEST WALL

Phillip G. Arnold

Chest wall defects have long challenged the creativity of reconstructive surgeons. The choice of procedures for reconstruction of the chest wall depends on four main considerations: the size, location, and thickness of the defect and the condition of the surrounding tissues. A well-vascularized piece of tissue must be used in all cases, especially if the surrounding tissues have been irradiated and are relatively ischemic secondary to the associated obliterative endarteritis that accompanies this form of therapy.

Large random flaps, tubed flaps, and the transposed greater omentum have been used successfully in chest reconstruction. The omentum has been a remarkably dependable tissue to use in radiation injury, particularly if a full-thickness defect is not present. It brings in a new, nonradiated blood supply that readily attaches itself to the excised wound and will accept and support an overlying skin graft.

Muscle and musculocutaneous flaps have also provided exceedingly useful methods of chest wall reconstruction. Muscles tend to maintain a rigidity if not denervated. They can also be voluntarily contracted to increase this rigidity. They also readily nourish underlying bone grafts to produce a more structurally sound reconstruction.

Some of the larger and more useful muscles for chest wall reconstruction include the latissimus dorsi, pectoralis major, trapezius, and external oblique. These may be used as muscle flaps alone or as musculocutaneous flaps, depending on the size and location as well as the depth of the defect. The two most frequently used muscles are the latissimus dorsi and pectoralis major.

LATISSIMUS DORSI

The latissimus dorsi is a muscle that is particularly suited for use in reconstruction of large midline, both front and back, defects and breasts after mastectomy.

History

D'Este in 1912 described Tansini's method of transposing a superiorly based flap of latissimus dorsi and overlying skin for immediate reconstruction of the mastectomy defect. Hutchins in 1939 also reported using the latissimus dorsi muscle to close the mastectomy defect, hoping to prevent edema of the ipsilateral upper limb and to stimulate the excised pectoralis major muscle. Campbell in 1950 reported using this muscle to reconstruct several large thoracic defects and very clearly pointed out its major blood supply and axis of rotation. He used the muscle for full-thickness chest wall defects, nourishing a fascia lata graft beneath it and a skin graft on its surface, as did Davis in 1949.

Desprez closed large meningomyeloceles of the

thoracolumbar region with bipedicle flaps augmented by the trapezius and latissimus dorsi muscles.

Mendelson and Masson used a large latissimus dorsi musculocutaneous flap in the treatment of severe radionecrosis of the shoulder with excellent success. Then the blood supply to the muscle was further elucidated.

Bostwick and co-workers reported 60 examples of utilization of the latissimus dorsi muscle as a functional muscle transfer (to restore elbow flexion) for arm and shoulder coverage and breast reconstruction and as a free flap.

Anatomy

This large muscle and its overlying skin, as well as the skin just anterior to the muscles, are supplied by the two terminal branches of the subscapular vessels (Fig. 10-1). The thoracodorsal vessels enter the muscle on its deep surface approximately 10 cm from its insertion into the humerus. There is also a large vascular branch going to the serratus anterior muscle that may be divided when moving the latissimus dorsi muscle. The blood vessels actually going directly to the muscle fibers are within the muscle itself and are well protected, except in the upper axilla just as the thoracodorsal artery enters the muscle. This makes elevation and mobilization of this structure a relatively safe maneuver. The latissimus dorsi muscle with its overlying skin may be transposed anteriorly across the midline of the chest or posteriorly or superiorly over the shoulder, depending on where the defect is located (Fig. 10-2).

Operative procedure

Flap elevation is usually started at the anterior border of the muscle. The presence of a functional, normal-size muscle before the operation has generally been all that is required to ensure a "safe" flap. The

A

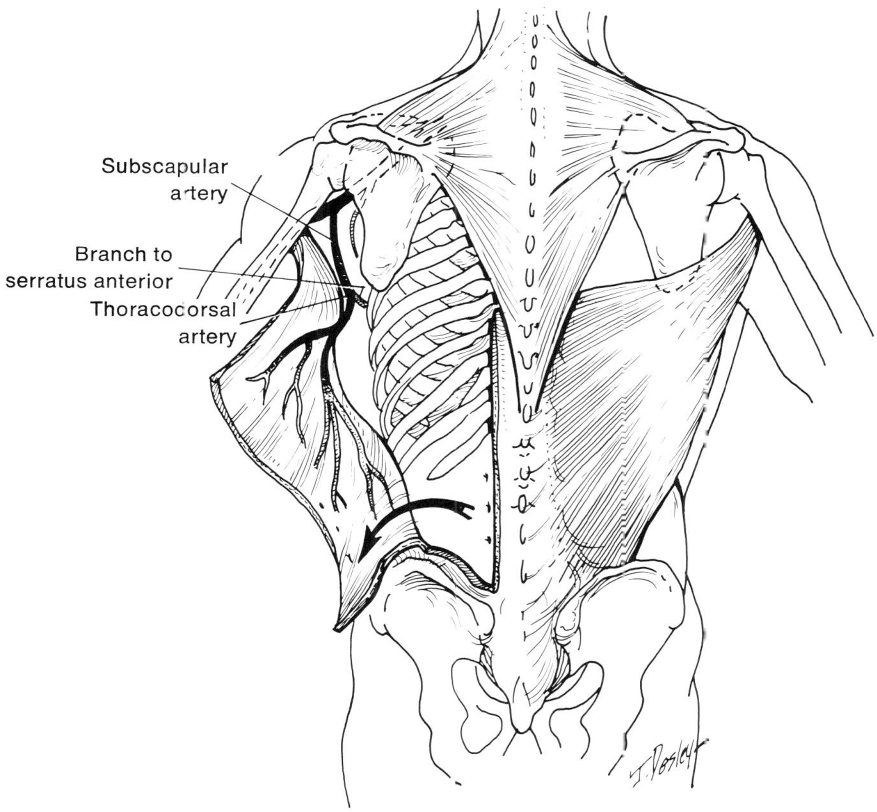

Continued.

Fig. 10-1

A to **C,** Anterior and posterior views of vascular anatomy of latissimus dorsi muscle based on thoracodorsal neurovascular leash. NOTE: Distal branches of thoracodorsal artery supply serratus anterior muscle. Cadaver dissection of same.

(**B** and **C** from Mendelson, B.C., and Masson, J.K.: Treatment of chronic radiation injury over the shoulder with a latissimus dorsi myocutaneous flap, Plast. Reconstr. Surg. **60**[5]:681, 1977.)

B

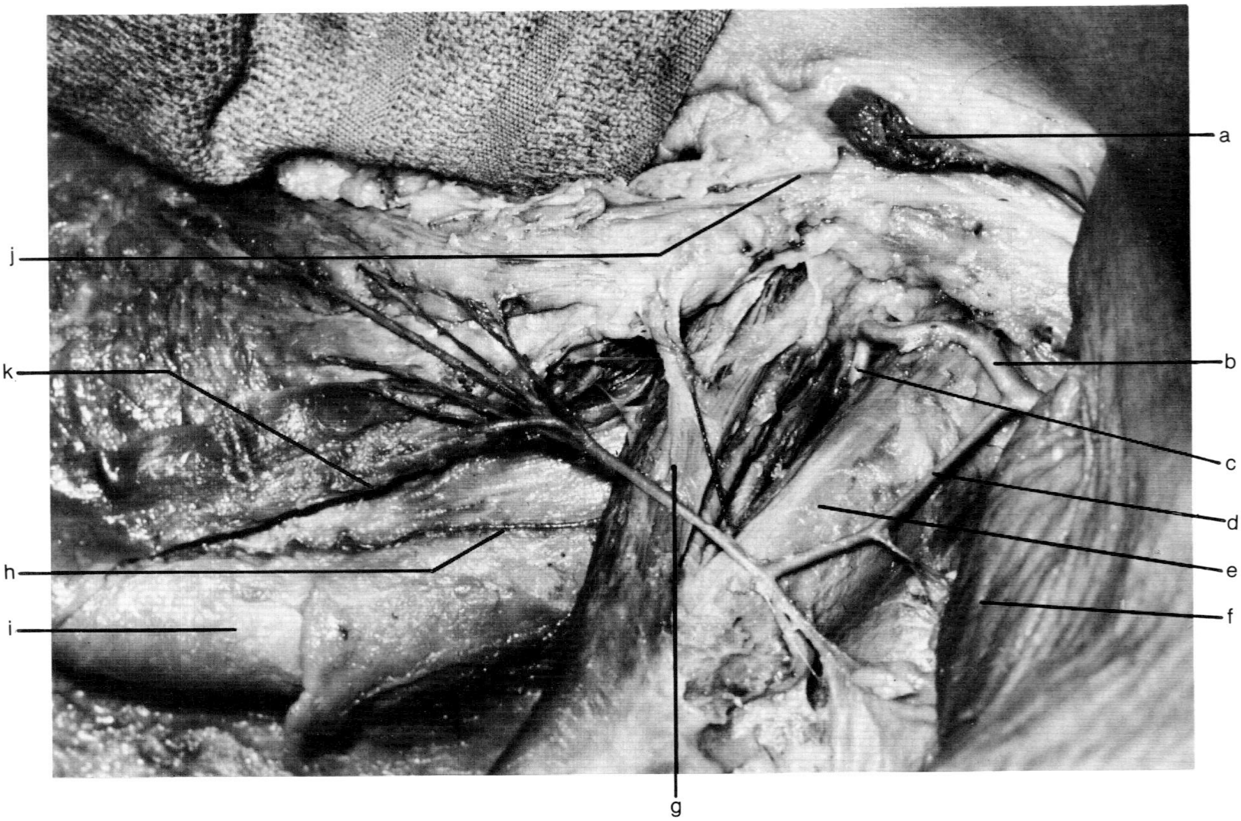

C

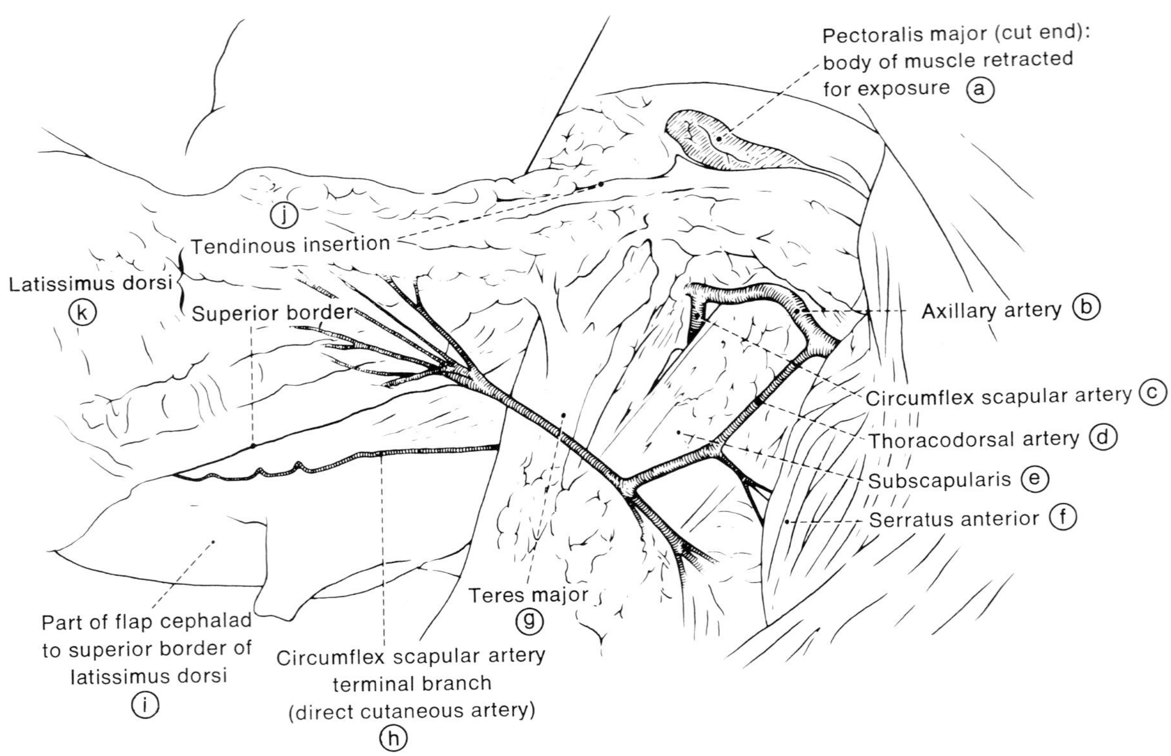

Fig. 10-1. For legend see p. 237.

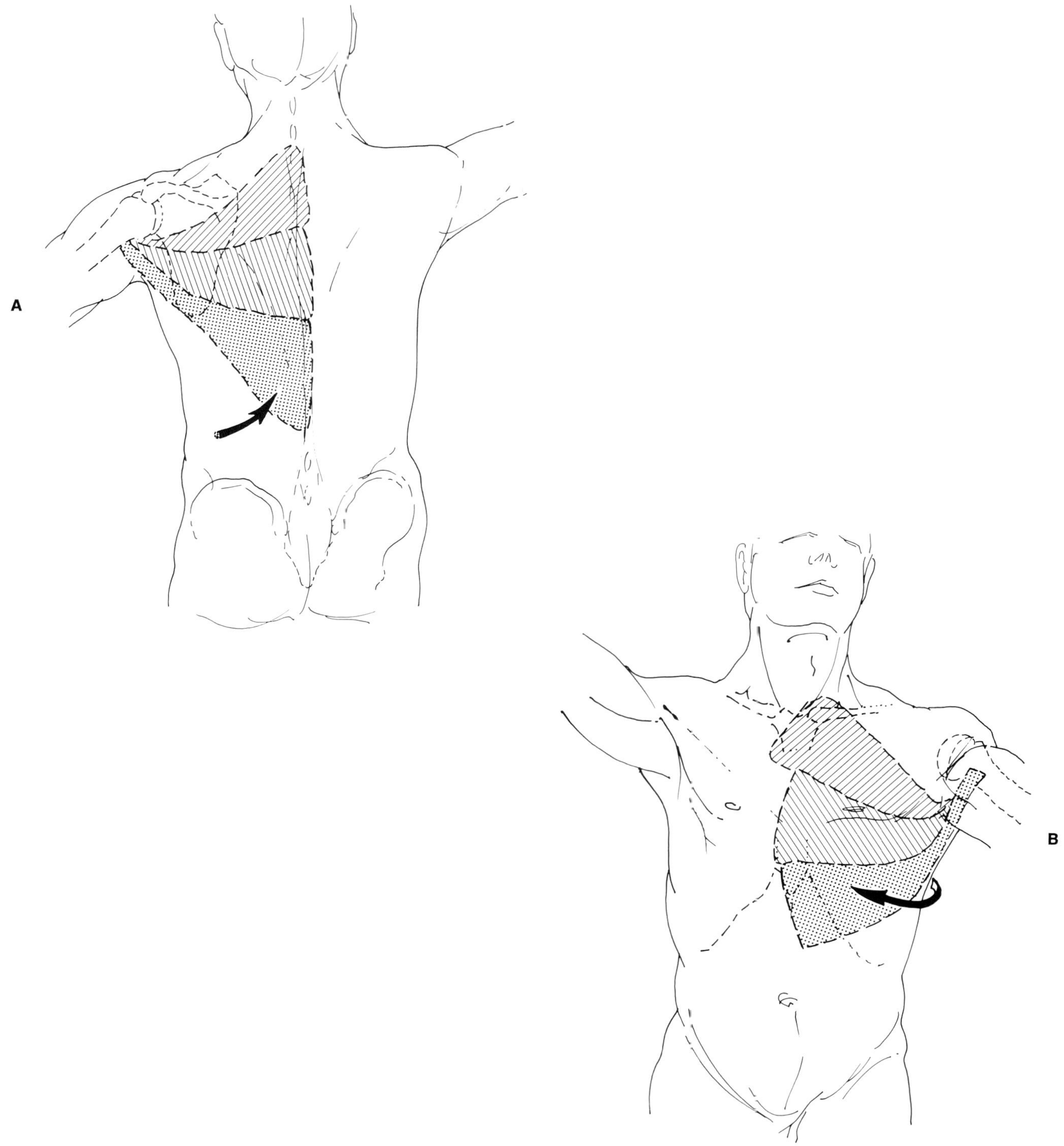

Fig. 10-2
A and **B,** Arc of rotation of latissimus dorsi muscle or musculocutaneous unit for anterior and posterior thoracic defects.

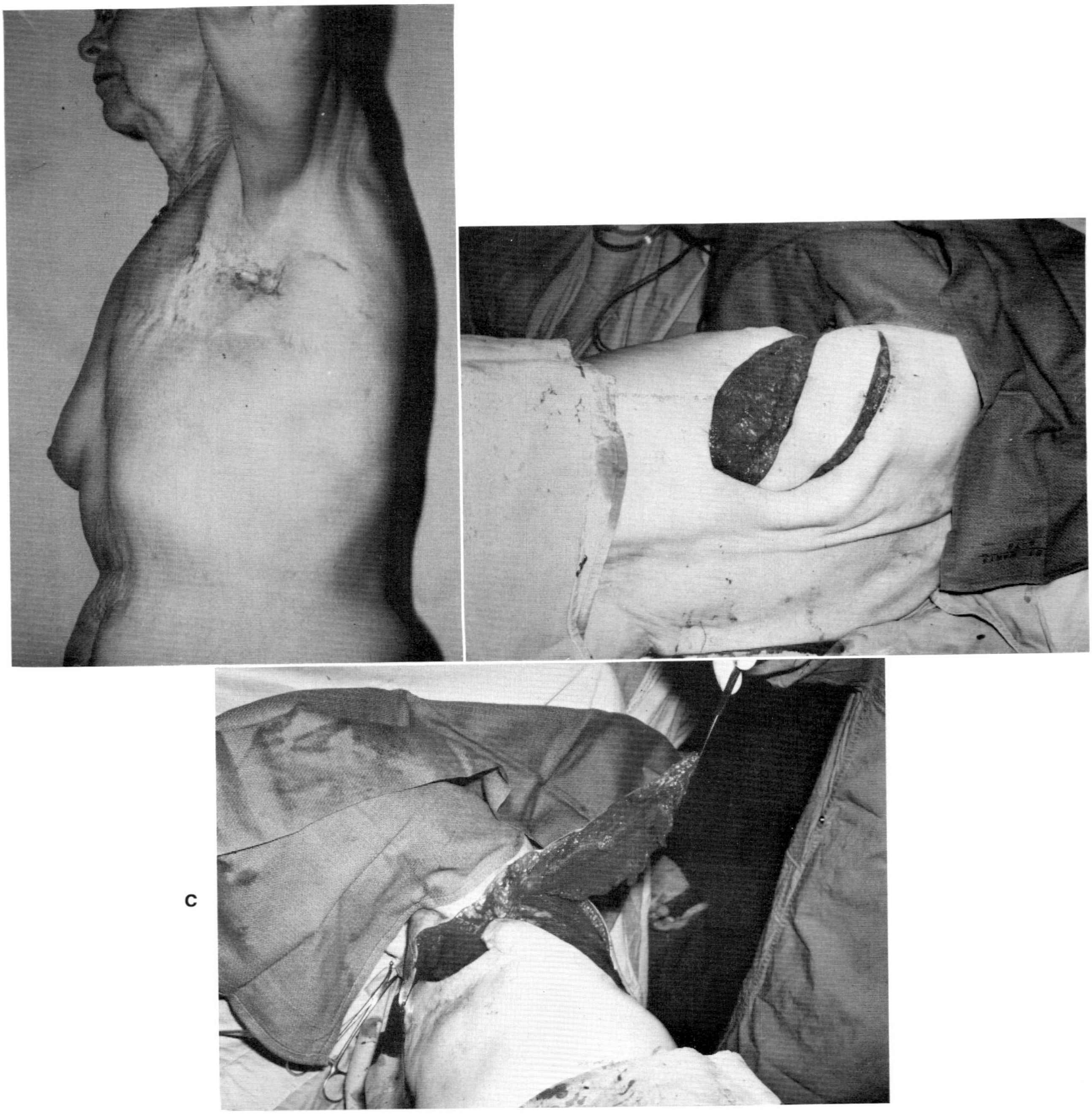

A

B

C

Fig. 10-3

A, Seventy-year-old woman with radiation ulcer of left axilla and anterior chest present for several months.

B, Latissimus dorsi musculocutaneous flap elevated.

C, Latissimus dorsi musculocutaneous flap elevated and ready for transposition.

entire muscle may then be elevated with blunt dissection over to the lateral border of the paraspinous muscles through which two rows of perforating vessels enter the muscle. The latissimus dorsi muscle has been transposed on these vessels (Chapters 10 to 17).

The elevated muscle (with or without overlying skin) may then be transposed to the anterior chest wall and the donor site closed (Figs. 10-3 and 10-4). If the defect is on the shoulder, the same flap may be rotated 180° to cover this area.

Large, posterior midline defects may be readily covered with bilateral latissimus dorsi skin-muscle flaps, using the technique described by McCraw in closing meningomyeloceles in infants. Using this principle, for even larger defects, a large bipedicled flap of each half of the chest and back may be elevated and advanced to the midline. The donor areas from the axillary line are skin grafted, if necessary. These problems may arise in patients with spinal cord injuries from trauma or tumor surgery with or without subsequent irradiation (Fig. 10-5). This procedure has the added advantage of using the large muscle and its overlying skin by only changing the origin of the muscle and thus creating no loss in function. This is an important consideration in the paraplegic patient who needs this muscle for turning and transfering (Chapter 12).

Breast reconstruction after mastectomy is a particularly suitable situation to use this musculocutaneous unit, as described by Schneider and beautifully demonstrated by Bostwick and co-workers (Chapter 14).

Precautions

If a previous thoracotomy has been done, the viability of the distal portion of the muscle is questionable. The overlying skin of this portion is even more tenuous, and at this time I would hesitate to use this without some secondary procedure available. Fluorescein, of course, answers this question at the operating table, in my experience.

D 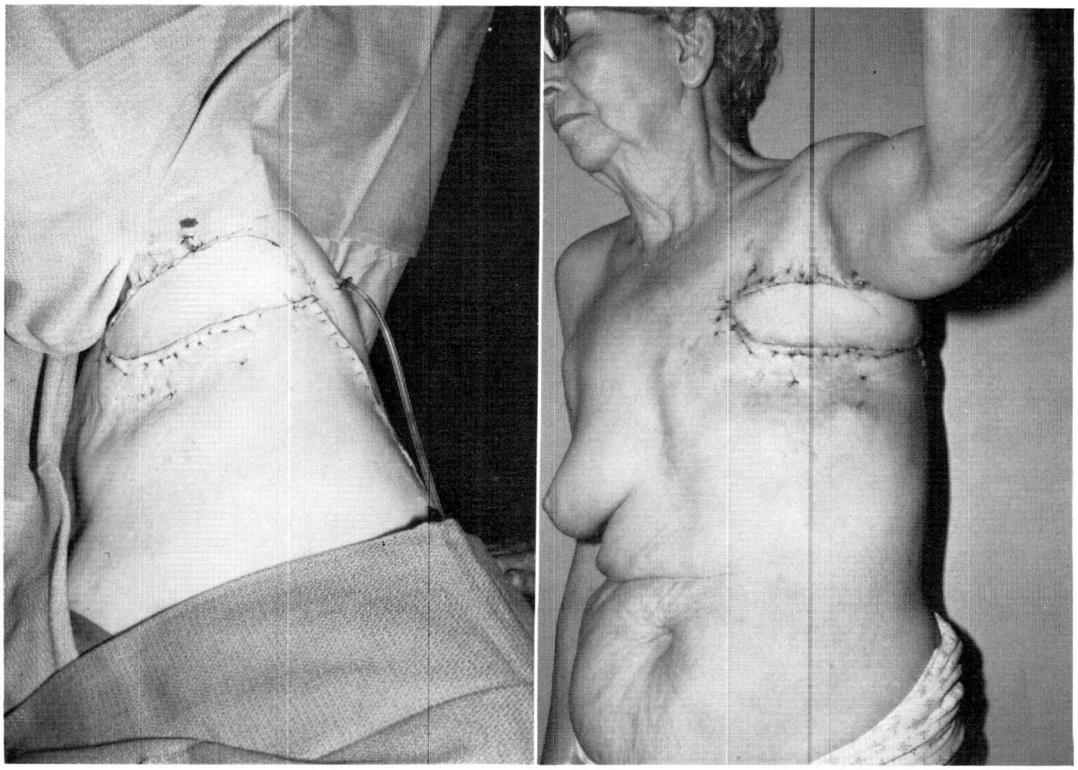 E

Fig. 10-3, cont'd
D, Latissimus dorsi musculocutaneous flap transposed and inset into defect following excision of radiated area.
E, Appearance 3 weeks after operation.

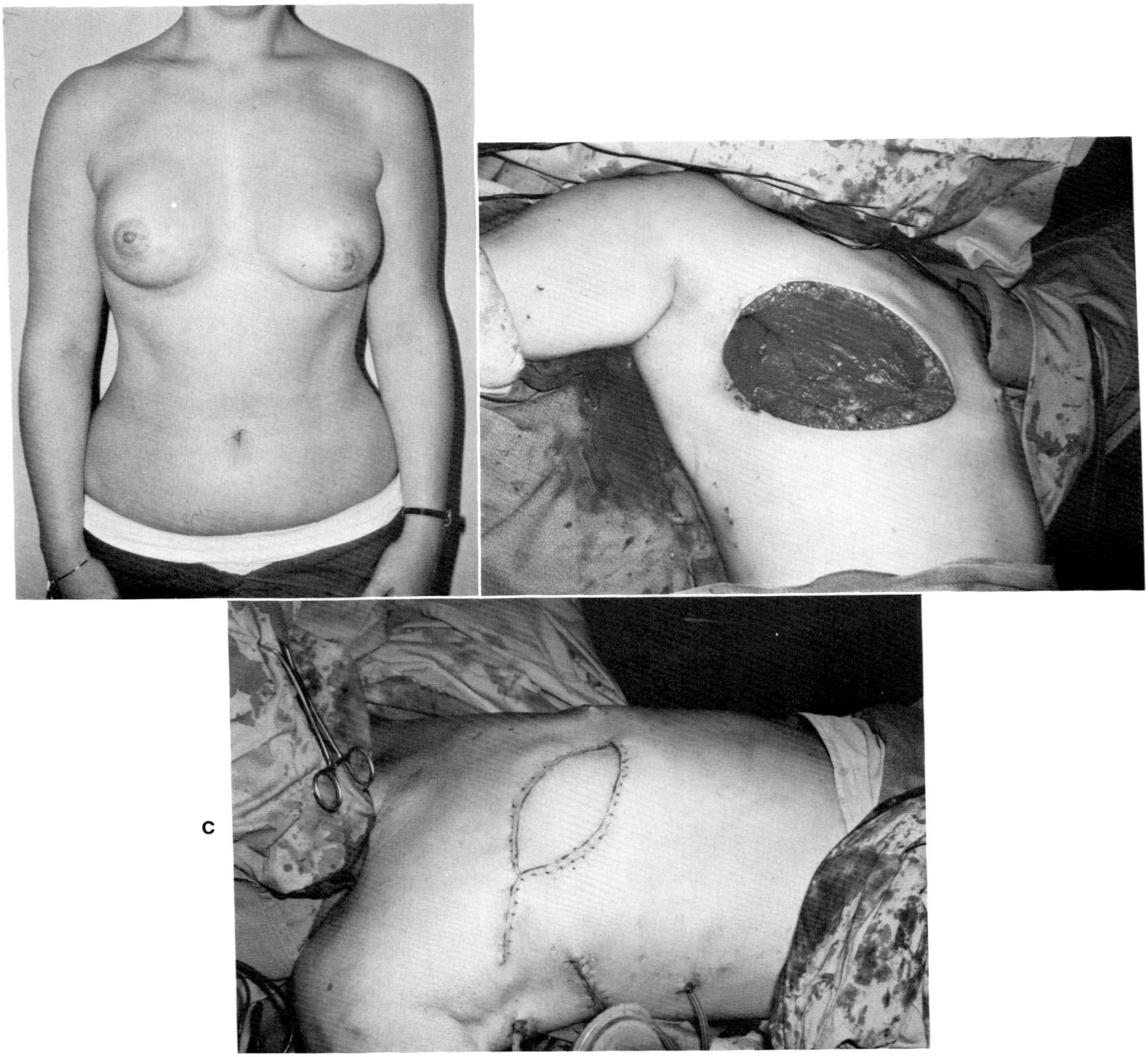

A

B

C

Fig. 10-4

A, Young woman with bilateral breast carcinoma with large amount of skin involvement over right breast. Treatment plan included mastectomies with postoperative irradiation.

B, Right mastectomy defect.

C, Defect closed with latissimus dorsi musculocutaneous flap and radiation started 3 weeks after operation.

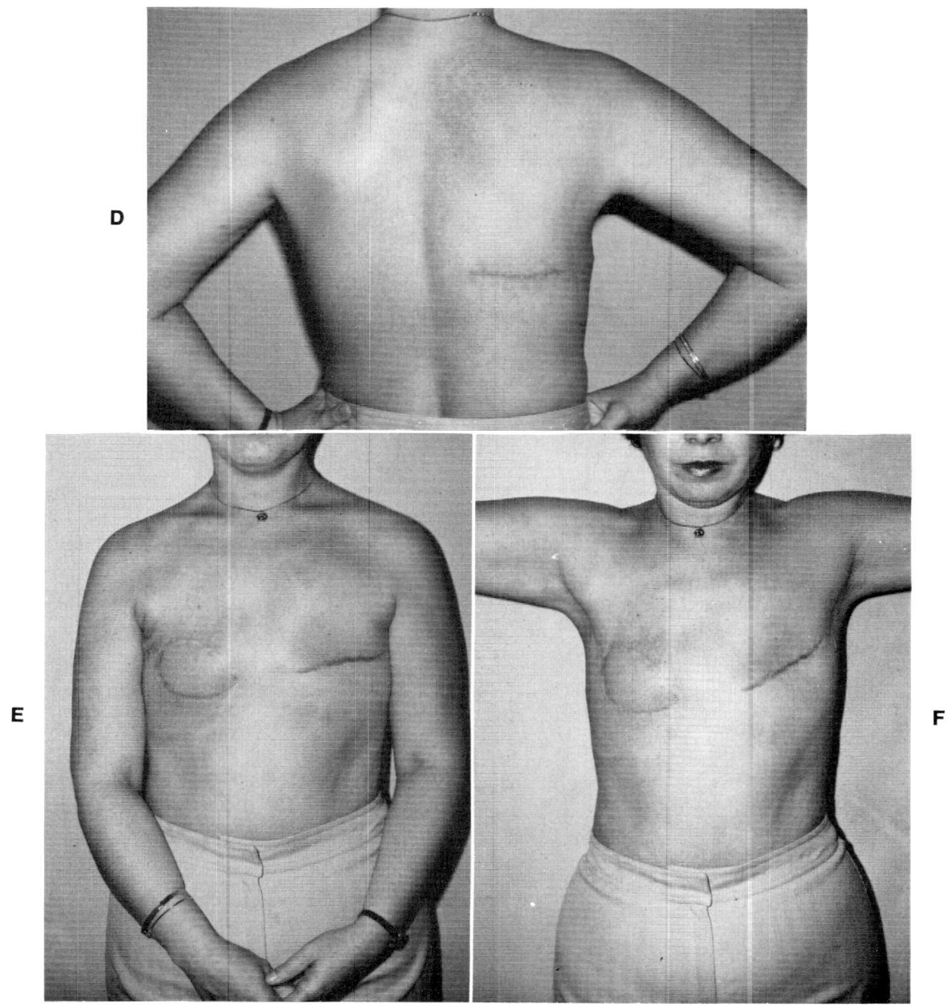

Fig. 10-4, cont'd

D, Donor defect following transposition of latissimus dorsi musculocutaneous flap defect.

E and **F,** Appearance of wound 3 months after radiation therapy with no limitation of motion.

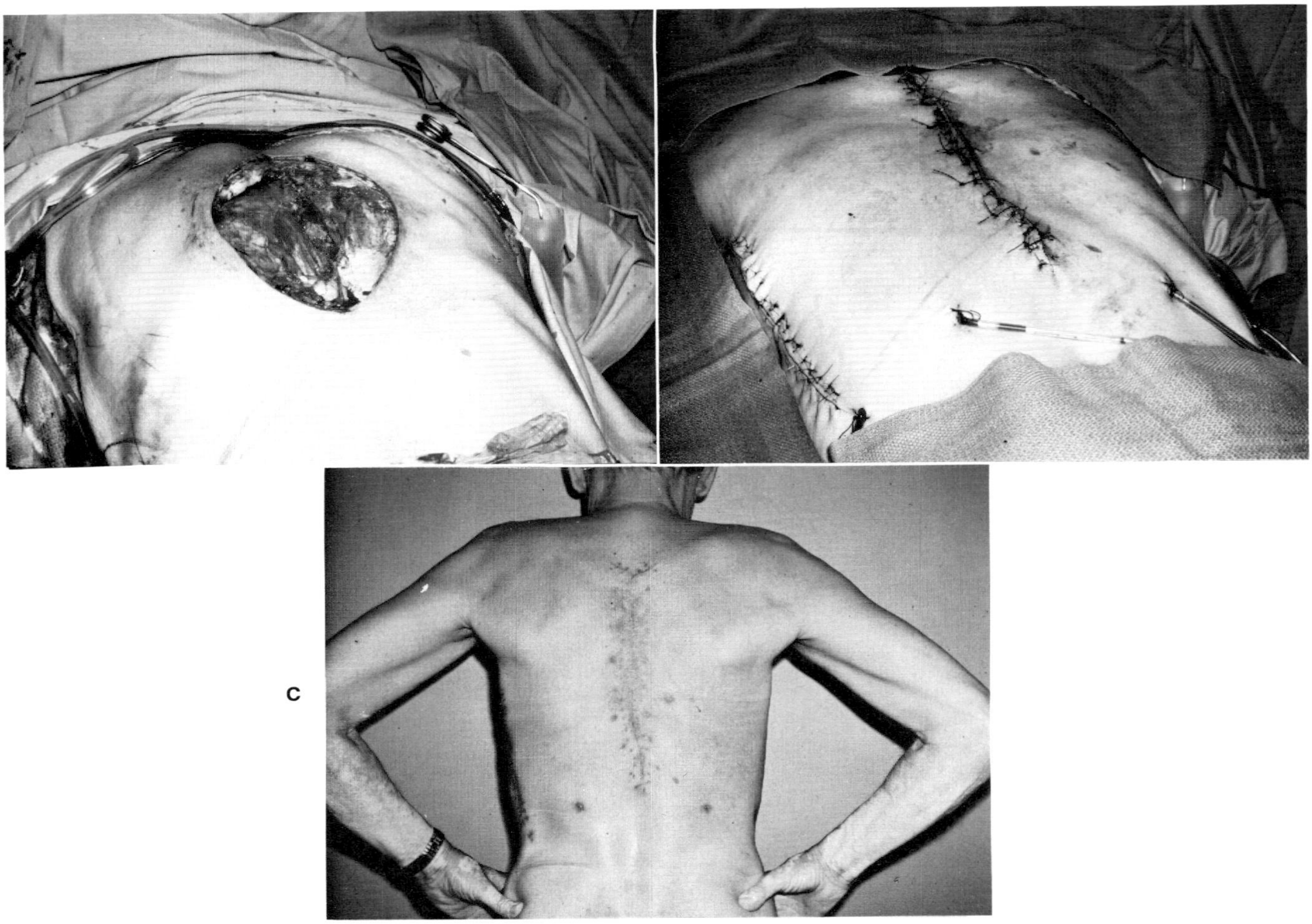

Fig. 10-5

A, Elderly man after irradiation of thoracic spine for metastatic tumor. Present wound resulted after removal of sequestrum of osteomyelitis and radiation-damaged skin.

B, Bilateral latissimus dorsi musculocutaneous flaps are elevated, and muscle-to-muscle closure is accomplished over spine as separate layer. In this case all of wound was closed directly.

C, Several months after closure with healed wound and bilaterally functioning latissimus dorsi muscles.

PECTORALIS MAJOR

History

The transposed pectoralis major muscle for closure of a chest wall defect has been described by Pickrell. Sisson uses this muscle based on its medial blood supply by placing it in the mediastinum after extended radical laryngectomy to prevent erosion of the great vessels. In 1968 Hueston and McConchie used a musculocutaneous flap of the pectoralis major muscle to successfully close a manubrial defect. In 1977 Brown and co-workers used bilateral pectoralis major muscle flaps with split-thickness skin grafts to close an anterior chest wall defect and demonstrated the excellent mobility of these muscles if the humeral attachment is divided. We have more recently used this muscle with autogenous bone grafts and skin grafts to close and maintain chest wall stability in large anterior chest wall defects. Ariyan beautifully demonstrates the versatility of this muscle as a musculocutaneous flap for reconstruction in the head and neck.

Anatomy

The pectoralis major muscle receives its primary blood supply from the thoracoacromial vessels that enter the deep surface about two thirds of the way along the clavicle. The medial and lateral pectoral nerves enter at this same level (Figs. 10-6 to 10-8). There are perforators from the internal mammary vessels that supply the muscle medially and have

A

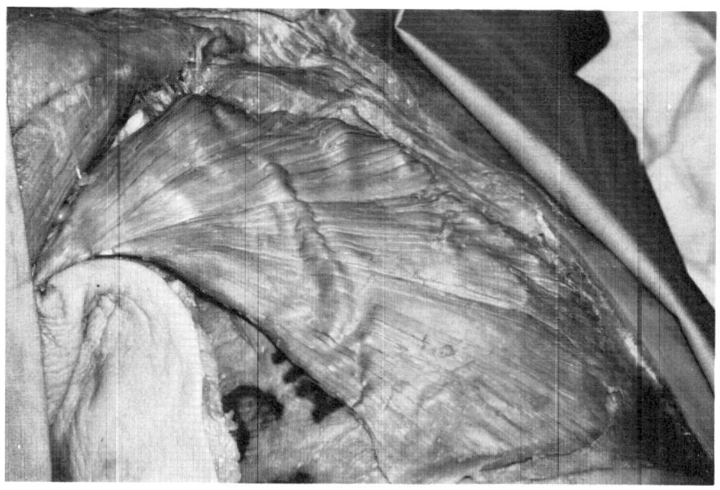

B

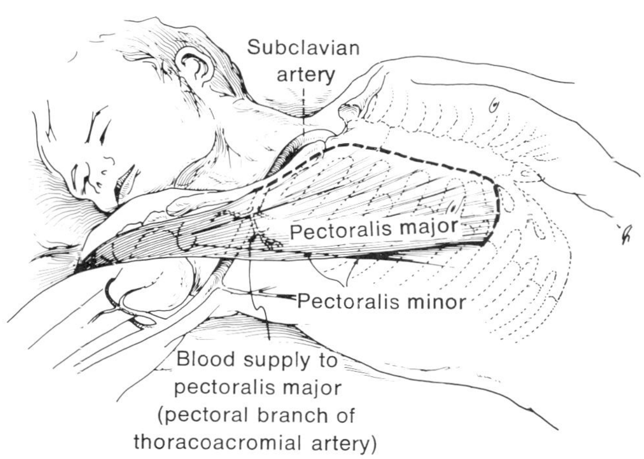

Fig. 10-6
A and **B,** Cadaver dissection of pectoralis major muscle.

(From Arnold, P.G., and Pairolero, P.C.: Use of the pectoralis major muscle to repair defects of the anterior chest wall, Plast. Reconstr. Surg. **63**[2]:205, 1979.)

A

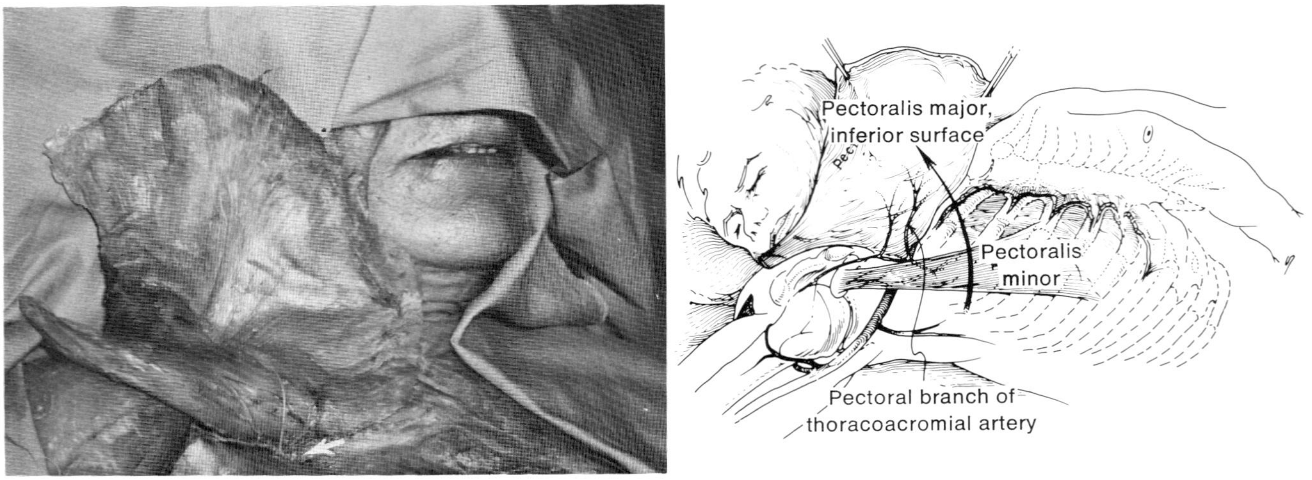

B

Fig. 10-7
A and **B,** Pectoralis major muscle reflected, demonstrating large thoracoacromial vascular leash that is its dominant blood supply.

(From Arnold, P.G., and Pairolero, P.C.: Use of the pectoralis major muscle to repair defects of the anterior chest wall, Plast. Reconstr. Surg. **63**[2]:205, 1979.)

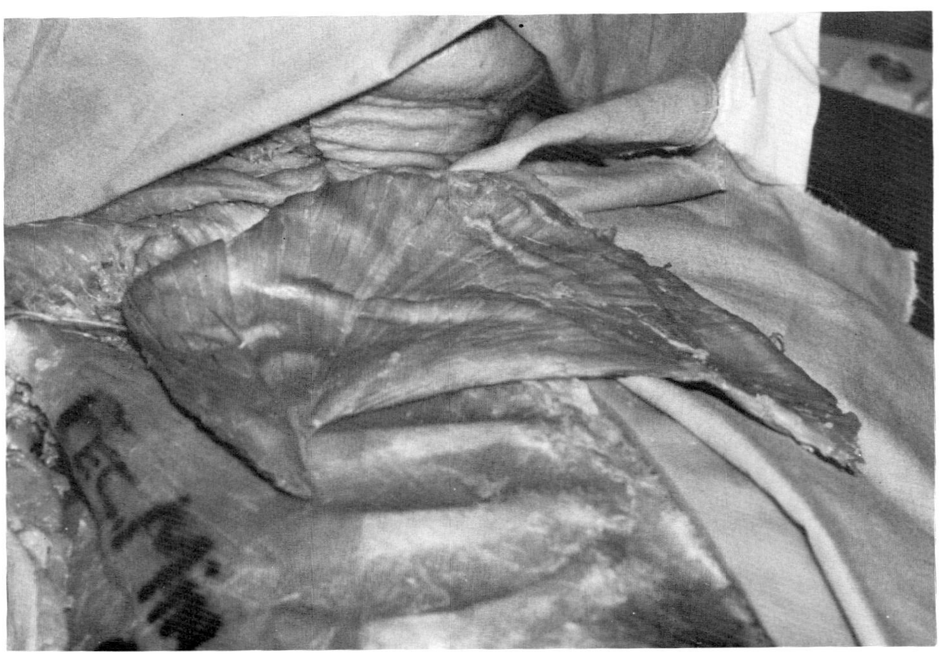

Fig. 10-8. Humeral attachment of pectoralis major muscle is divided, demonstrating ease with which it may be rotated and transposed beyond midline of upper thorax.

(From Arnold, P.G., and Pairolero, P.C.: Use of the pectoralis major muscle to repair defects of the anterior chest wall, Plast. Reconstr. Surg. **63**[2]:205, 1979.)

been the basis of some of the earlier uses of this muscle. This muscle has been used to move skin and soft tissue to the head and neck region based on its thoracoacromial blood supply. This discussion will deal only with chest wall repair using this muscle. It is most useful for the upper midline defects that have presented difficult closure problems in the past with tumor resections, radiation injury, and median sternotomy wound problems. It has been equally useful for persistent empyema and bronchopleural fistulae in the upper lobes.

Operative procedure

The defect to be closed may be created by the excision of infected or irradiated tissues or the result of an en bloc through-and-through resection of a por-

A

B

C

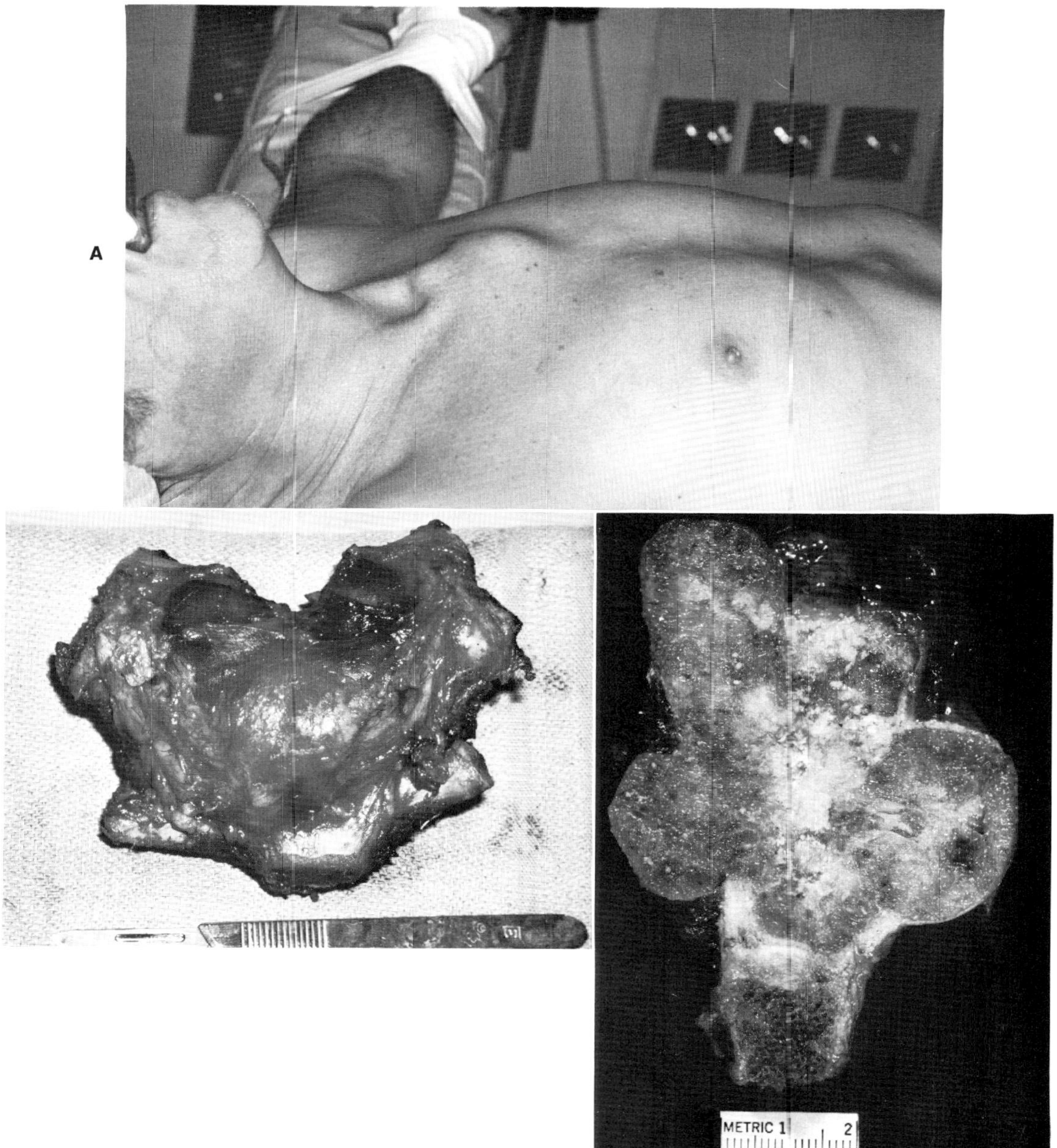

Fig. 10-9

A to **C,** Forty-five-year-old man with enlarging upper sternal and manubrial mass noted for past 1 year. Anterior and cross-sectional views of specimen, which included clavicles, manubrium, and upper one fourth of ribs bilaterally with diagnosis of grade 2 chondrosarcoma with histologically free margins.

(**A** to **C, F** to **H,** and **J** to **L** from Arnold, P.G., and Pairolero, P.C.: Mayo Clin. Proc. **53:**54, 1978; **E** and **I** from Arnold, P.G., and Pairolero, P.C.: Use of the pectoralis major muscle to repair defects of the anterior chest wall, Plast. Reconstr. Surg. **63**[2]:205, 1979.) *Continued.*

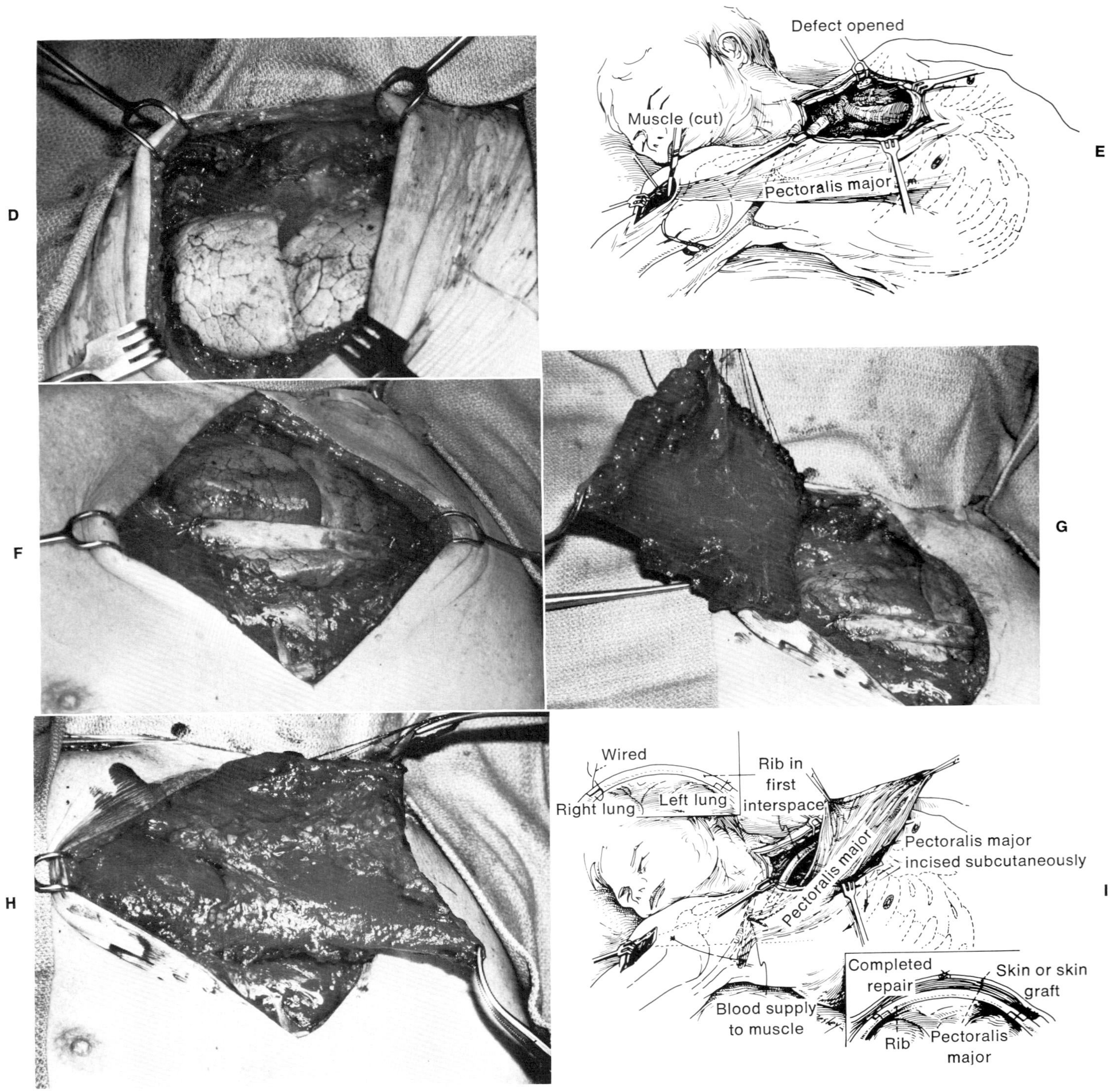

Fig. 10-9, cont'd

D and **E,** Thoracic defect following excision of tumor demonstrating mediastinal structures and the right and left lung, which are expanded.

F, Anterior bony thorax is reconstructed with autogenous rib grafts secured by a rabbet joint.

G, Right pectoralis major muscle (patient was left-handed) mobilized with humeral attachment divided.

H to **J,** Muscle transposed and sutured into position with direct skin closure.

tion of the chest wall containing a neoplasm. Once the histologic examination of the margins is complete and satisfactory, the situation may be assessed as to the feasibility of using one or both pectoralis muscles. If the defect is small enough to be closed with one muscle, the nondominant muscle should be chosen.

If the full-thickness defect includes the upper one half or more of the sternum, the anterior bony thorax should be reconstructed before soft tissue closure (Fig. 10-9, *A* to *E*). This can be done with autogenous rib grafts (usually removed through a separate thoracotomy incision) (Fig. 10-9, *F*). These can then be placed across the bony defect and the ends of the graft and recipient ribs tailored to form a marrow-to-marrow rabbet joint and secured with stainless steel wire (Fig. 10-9, *G*).

The actual dissection of the muscle is usually begun medially through the defect. The muscle is dissected free of the overlying skin or breast tissue and the underlying chest wall. The dissection is carried laterally until the lateral border of the pectoralis major is identified, and a second small incision is made in the anterior axillary fold. The pectoral fascia is dissected free of the pectoralis minor and the anterior chest wall by blunt dissection. The clavicular attachments are divided, taking care not to injure the thoracoacromial vessels and pectoral nerves. The humeral attachment is divided last through a separate incision over the inferior shoulder, since it maintains tension that facilitates the previous dissection. The muscle is now virtually an island based on the thoracoacromial neurovascular bundle and may be rotated and advanced (counterclockwise on the right, clockwise on the left) medially to span the defect.

The transposed muscle may be draped over the rib grafts or over the debrided chest wall wound and sutured to stable structures around the defect (Fig. 10-9, *H* to *L*). Care should be taken to preserve the anterior pectoral fascia during the initial dissection so that a strong piece of fascia will be present to suture at this point rather than only "red" muscle that will not hold sutures as well. Appropriate thoracostomy tubes are placed before final closure of the muscle.

If the anterior chest skin and soft tissue are not widely resected, they may be closed by direct suture over the transposed muscle and bone graft. If there is a cutaneous defect, then the muscle may be skin grafted with no loss in anterior chest wall stability, provided the bony thorax has been reconstructed and stabilized (Fig. 10-10). Meshed skin graft (1.5 to 1)

J

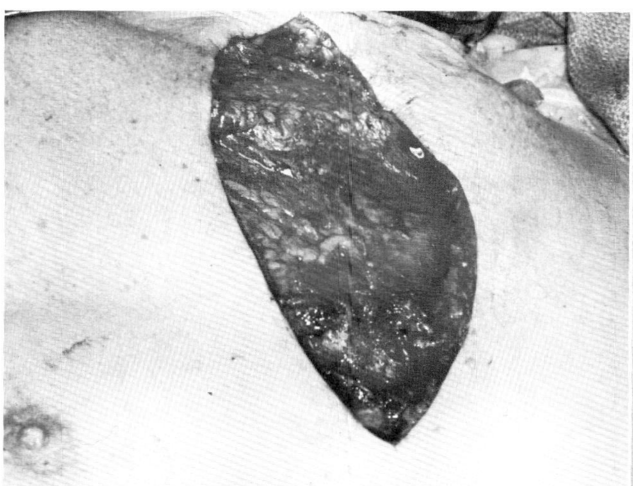

K

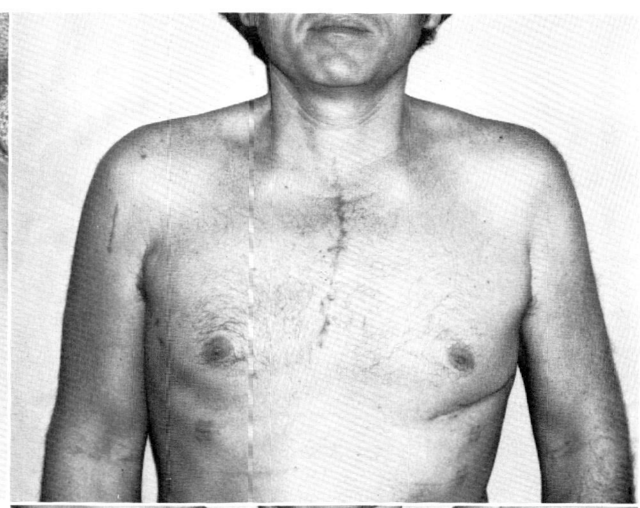

L

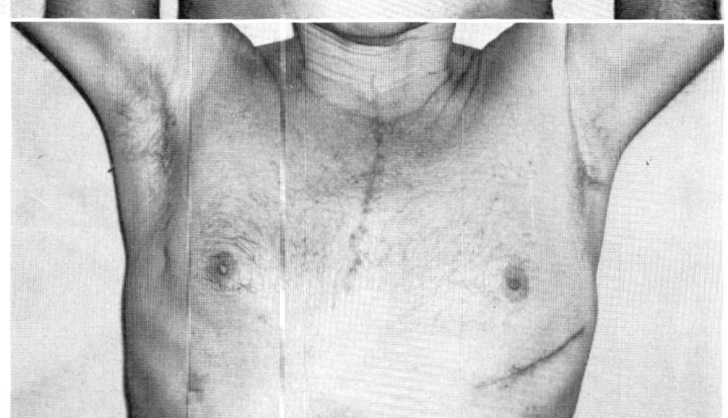

Fig. 10-9, cont'd

K and **L,** Appearance of patient 7 months after operation. He had resumed his normal physical activities of daily running and swimming.

A

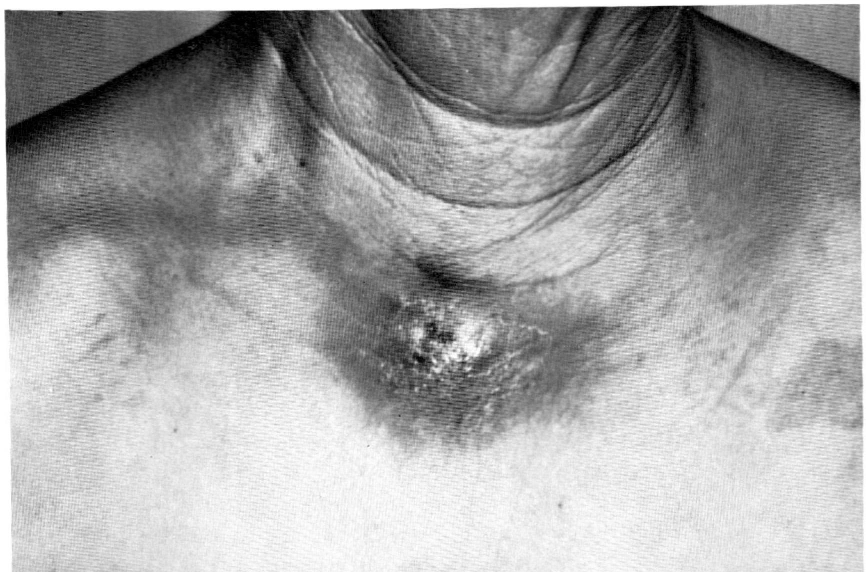

B

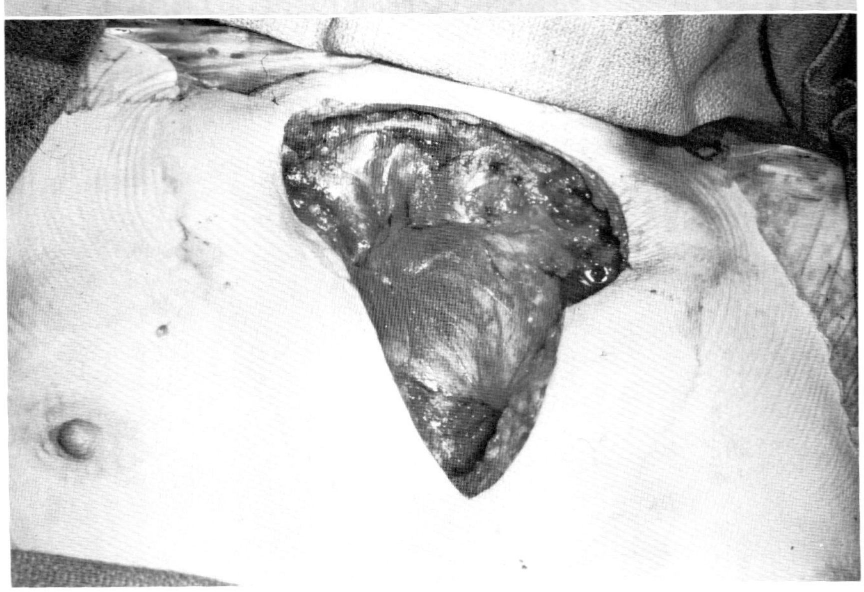

Fig. 10-10

A, Sixty-year-old woman 2 years after excision of papillary carcinoma of thyroid with reexcision in manubrial area followed by 6000 R with recurrence of her tumor in irradiated field.

B, Upper one half of sternum, clavicles, and upper five ribs excised with heart and great vessels at base of defect.

(**A** and **C** to **E** from Arnold, P.G., and Pairolero, P.C.: Use of the pectoralis major muscle to repair defects of the anterior chest wall, Plast. Reconstr. Surg. **63**[2]:205, 1979.)

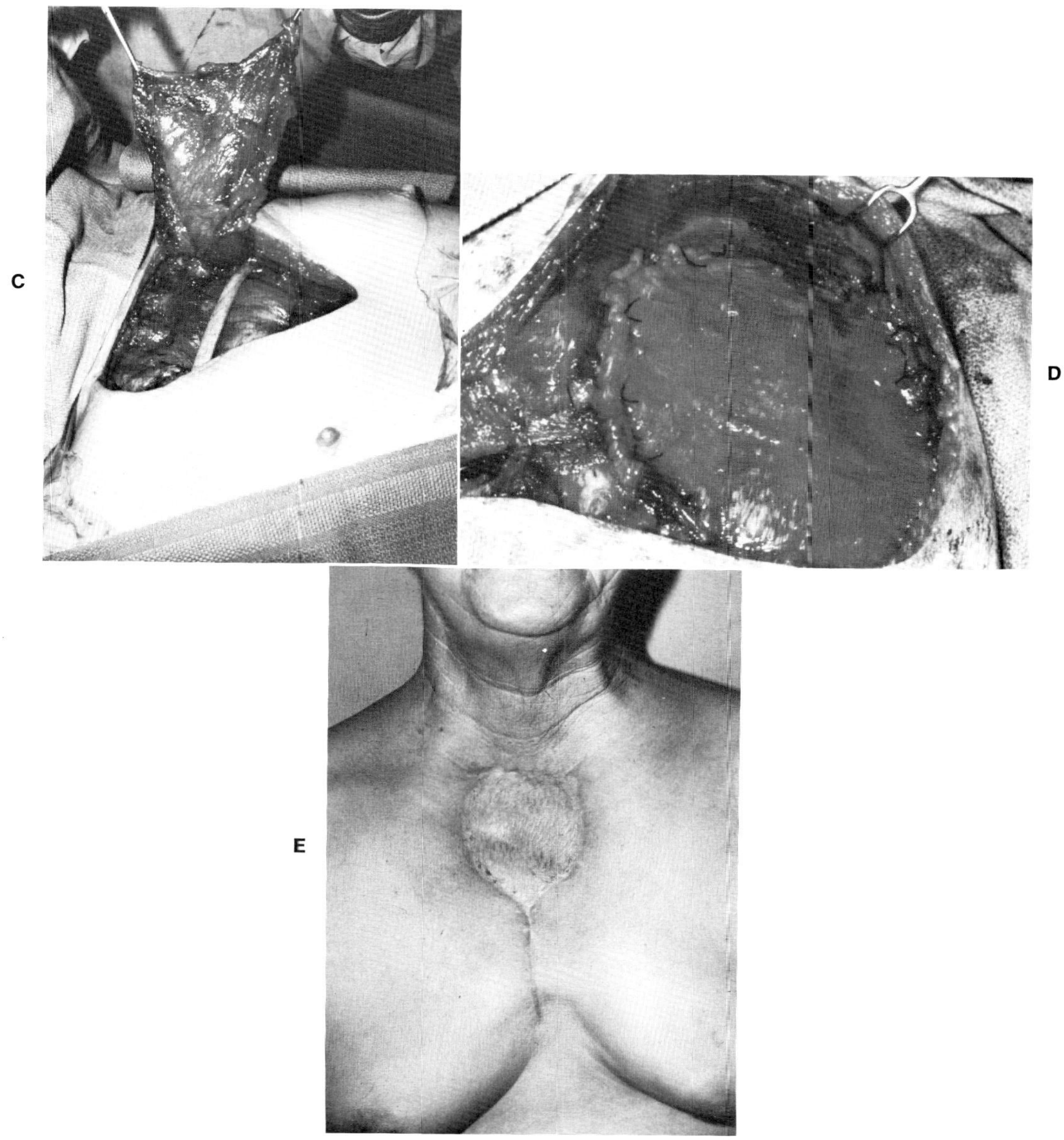

Fig. 10-10, cont'd

C, Left pectoralis major muscle elevated on its thoracoacromial leash with humeral attachment divided and autogenous rib graft placed across thoracic defect.

D, Pectoralis major muscle inset, closing bilateral chest defect and covering mediastinal structures, and autogenous grafts with a vascularized flap.

E, Appearance of chest wall 5 months after operation.

is placed directly over the muscle. Generally speaking, the patient may be extubated in the operating room or certainly by the following morning with no difficulty. The muscle may be used in the same manner to attend to problems from infected median sternotomy wounds (Chapter 11) (Figs. 10-11 and 10-12).

The muscle may be easily used as a carrier for a large area of skin of the anterior chest wall. This is particularly helpful in head and neck defects. However, a skin graft placed over the transposed muscle seems acceptable for midline defects and avoids significant distortion of the breasts. In very large defects that cannot be closed with bilateral pectoralis major muscle flaps, musculocutaneous flaps can be used for closure.

Precautions

Care should be taken in dividing the humeral attachment of the pectoralis major muscle. The brachial plexus, vessels, and cephalic vein should be avoided, and, in my opinion, this is accomplished more safely through a separate incision in the upper arm.

A

B

C

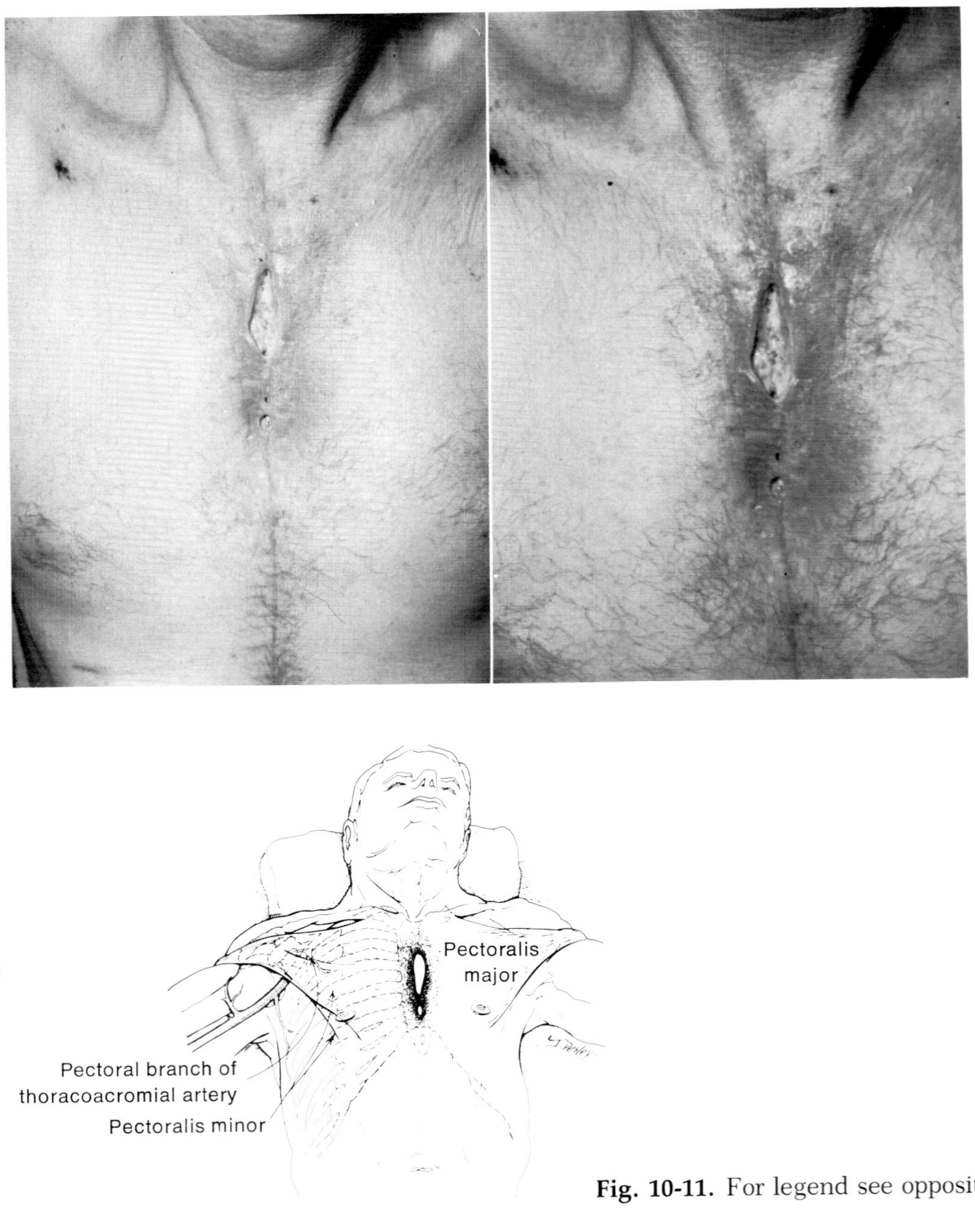

Fig. 10-11. For legend see opposite page.

D

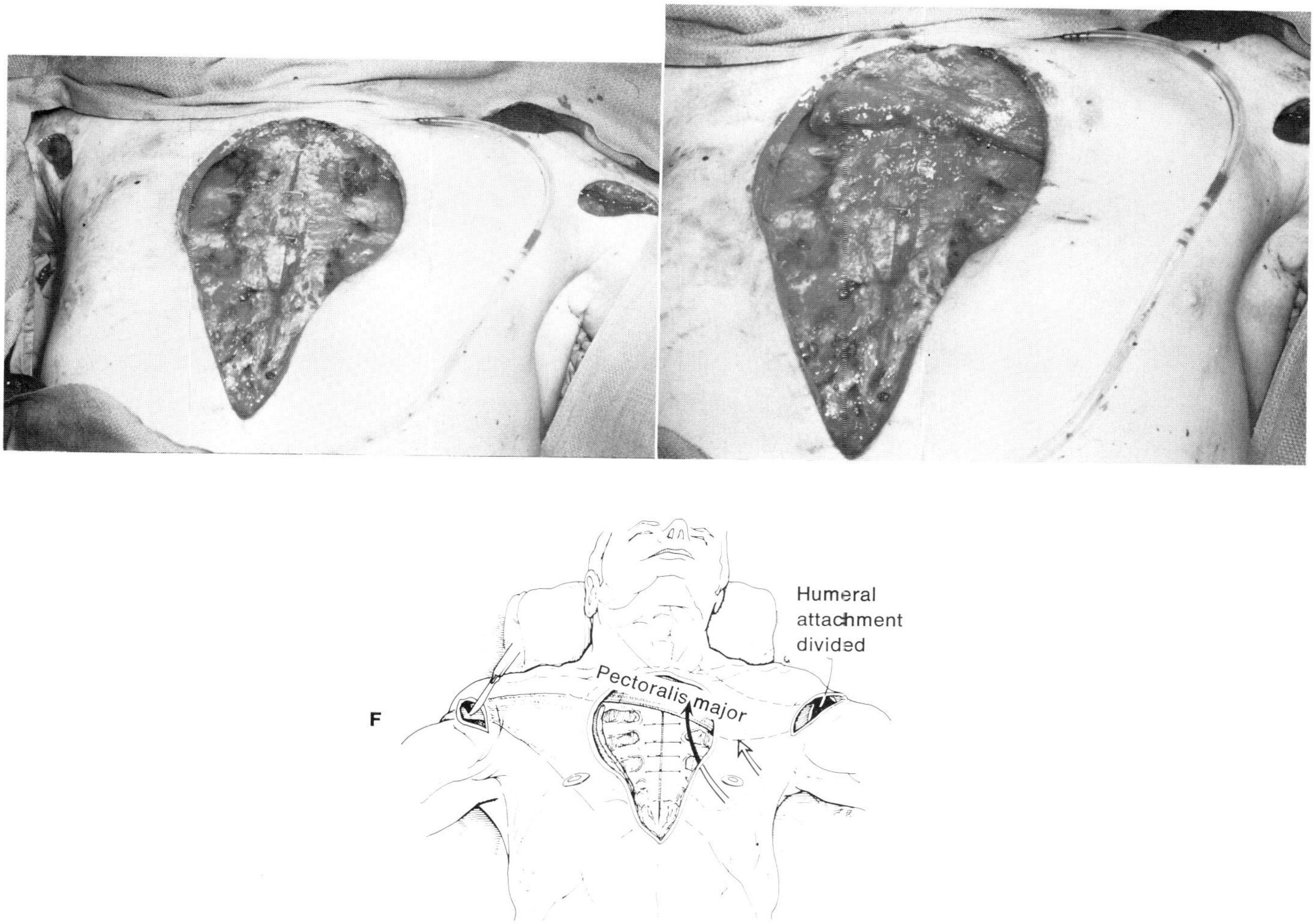

Fig. 10-11

A to **C**, Fifty-year-old man with positive mediastinoscopy for squamous cell carcinoma with no known primary some 2 years after irradiation. Wound illustrated resulted from median sternotomy done under emergency conditions for pericardial tamponade secondary to pericardiocentesis for further investigation. Wound was several months old with persistent drainage and no healing.

D, Wound was debrided widely as were edges of sternum that were reapproximated with large wire sutures. Silicone drainage catheter was placed in retrosternal position and brought out through separate wound superiorly.

E and **F**, Left pectoralis major muscle was mobilized, elevated, transposed, and only partially closes defect.

(From Arnold, P.G., and Pairolero, P.C.: Use of the pectoralis major muscle to repair defects of the anterior chest wall, Plast. Reconstr. Surg. **63**[2]:205, 1979.)

Continued.

G

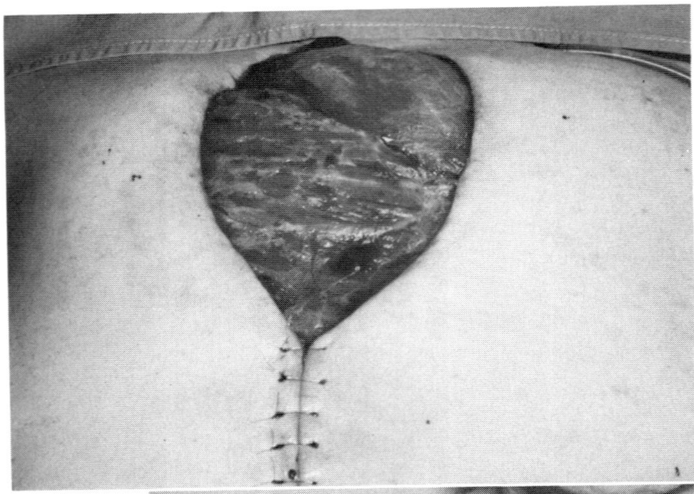

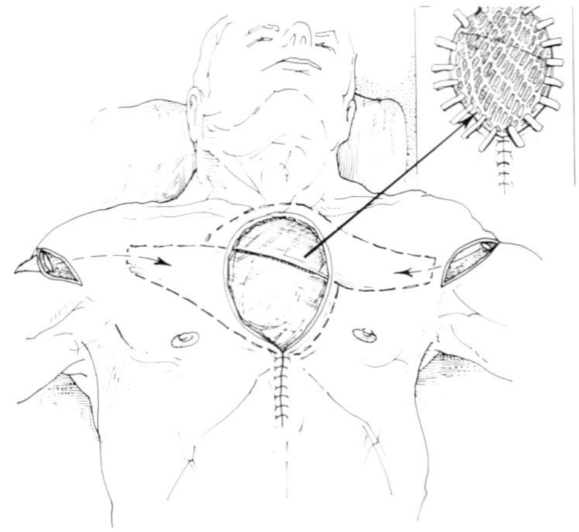

H

I

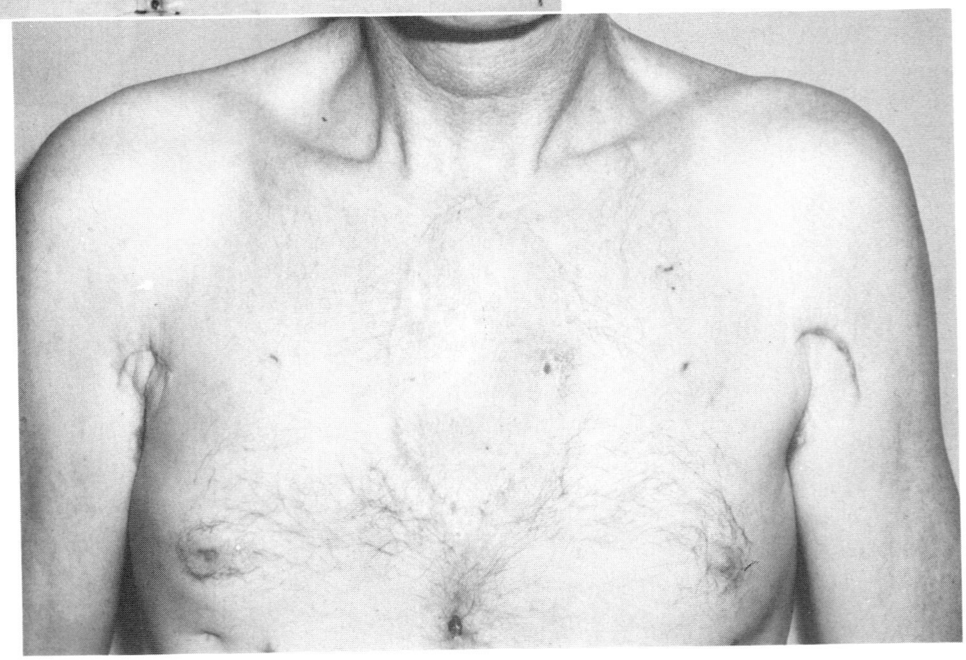

Fig. 10-11, cont'd

G and **H,** Defect closed by elevation of remaining pectoralis major muscle obtaining good closure and appropriate bed for skin graft as well as coverage for underlying irradiated sternum.

I, Appearance of wound 4 months after bilateral pectoralis major muscle flaps with skin grafts.

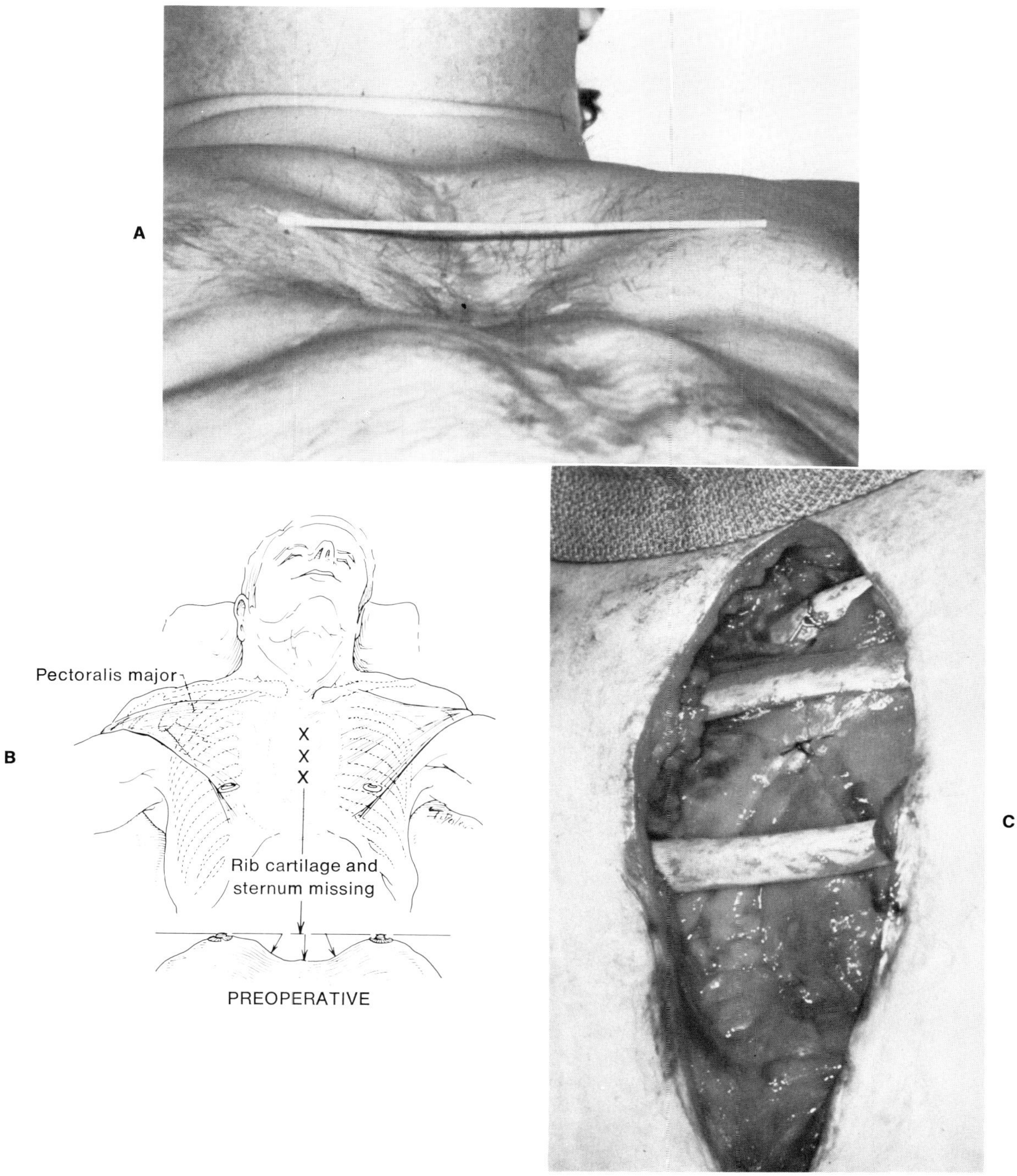

Fig. 10-12

A and **B,** Thirty-nine-year-old man after coronary bypass surgery with postoperative infection requiring 12 subsequent drainages and excisions, resulting in total sternectomy and excision of costochondral cartilages. Anterior chest was entirely flail with no skeletal structures anteriorly and only scarred and epithelialized tissue over heart and great vessels.

C, Scarred area was excised and three rib grafts placed across to reconstruct anterior bony thorax.

(From Arnold, P.G., and Pairolero, P.C.: Use of the pectoralis major muscle to repair defects of the anterior chest wall, Plast. Reconstr. Surg. **63**[2]:205, 1979.)

Continued.

D

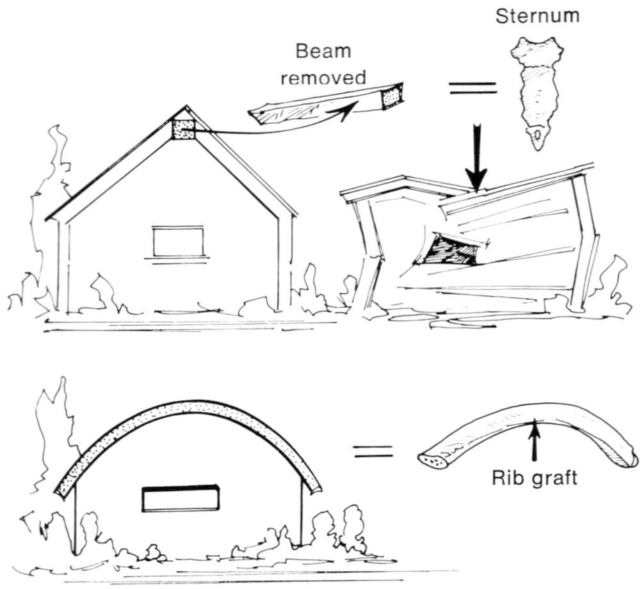

E

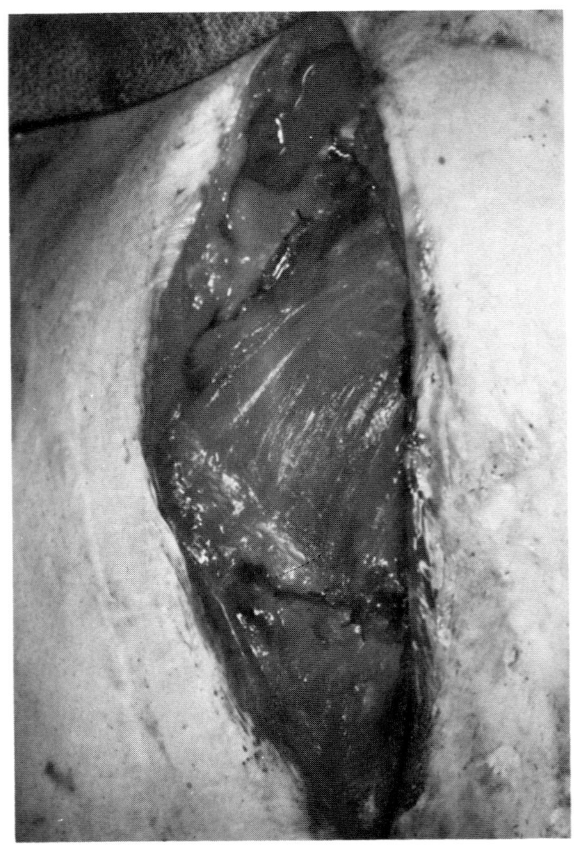

Fig. 10-12, cont'd

D, Basic plan here is to convert median tie beam-type of structure to that of Quonset hut.

E, Left pectoralis major muscle elevated, transposed, and sutured to opposite pectoralis major muscle for anterior chest wall stability and coverage for bone grafts.

F

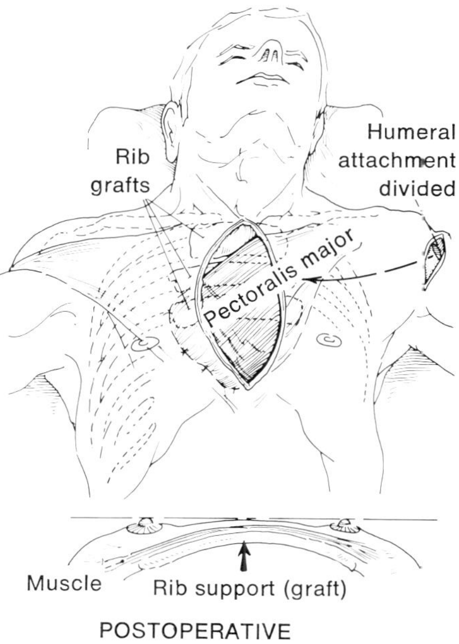

G

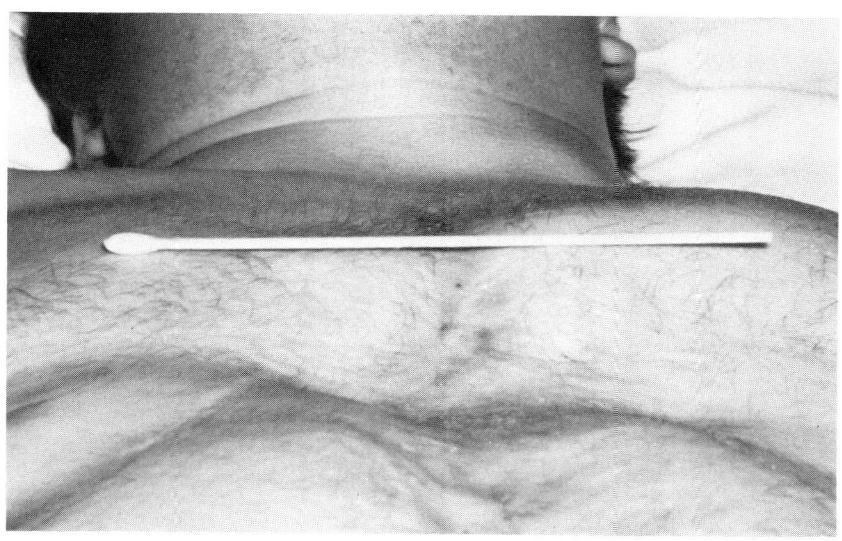

Fig. 10-12, cont'd

F, Skin edges were closed by direct suture.

G, Appearance of anterior chest wall 7 months after operation with reconstruction of bony thorax and protection to heart and great vessels.

TRAPEZIUS

History

As early as 1889 Manchot described numerous vascular territories, including the posterior neck and upper posterior back. These included territories of the occipital, scapular, and suprascapular arteries. Almost 50 years earlier Mutter had used a flap from the nape of the neck to relieve some burn scar contractures. In 1957 Zovickian[4] repaired pharyngeal fistulas with a similar flap. In 1979 McCraw and coworkers reported using the upper portion of the trapezius muscle to augment a standard Mutter flap,

A

B

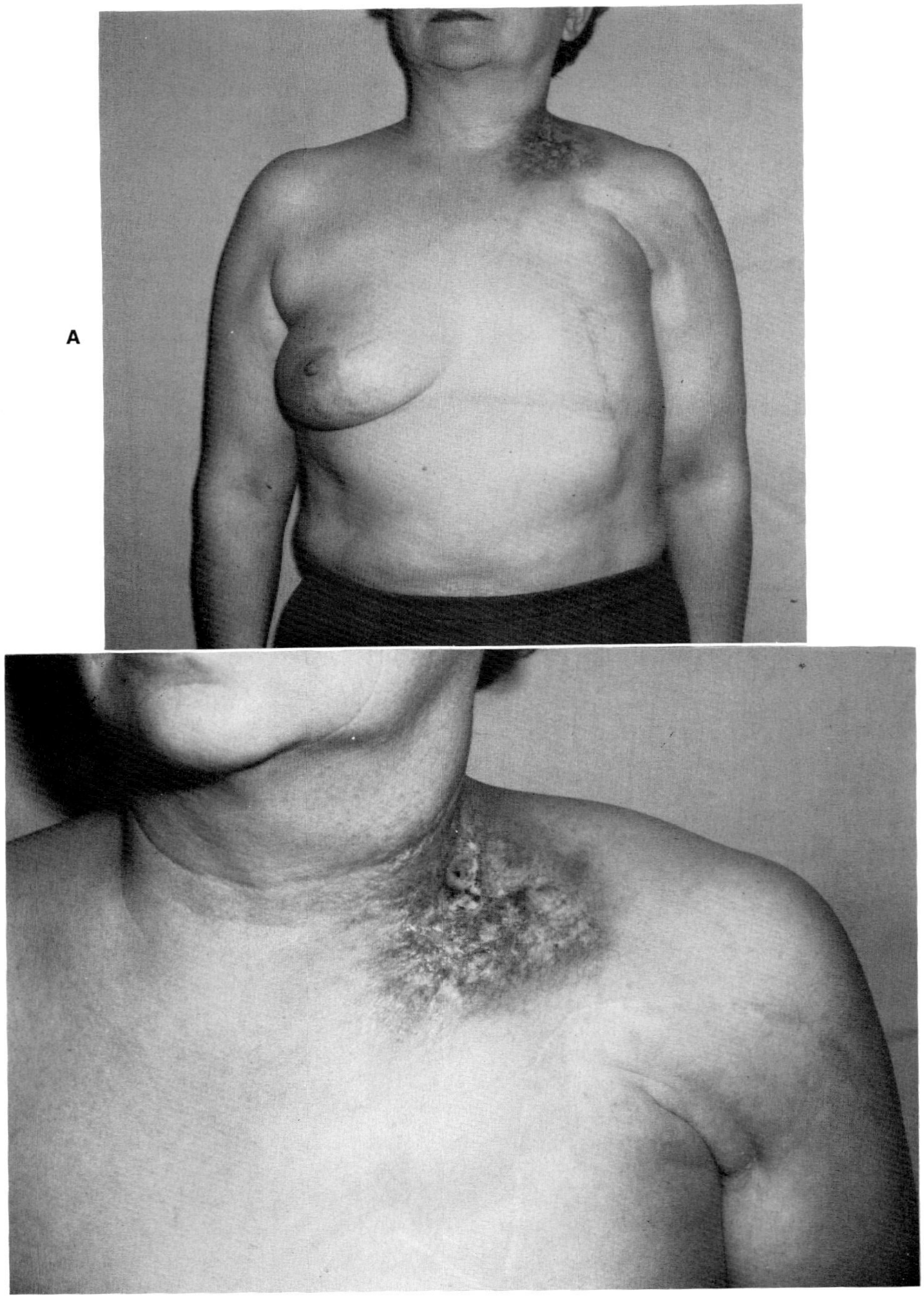

Fig. 10-13

A and **B,** Fifty-year-old woman after radical mastectomy and radiation with 18-month history of persistent radiation ulcer at lower neck and upper chest. Pectoralis major muscle is not present.

thus creating a trapezius musculocutaneous flap. This was used primarily for head and neck reconstruction after tumor resections but can be used to close upper anterior and posterior chest defects.

Anatomy

The muscular component of this musculocutaneous flap is the upper portion of the trapezius muscle. The muscle is present under the proximal one half of the flap. The distal one half is skin, subcutaneous tissue, and fascia (Fig. 10-13). It may be from 6 to 10 cm wide and 18 to 30 cm long, which is basically a 3:1 ratio.

Operative procedure

The flap may be outlined and elevated, beginning distally with skin and deep fascia. The suprascapular artery and branches of the transverse scapular artery may be encountered and divided during this elevation. A large branch of the eleventh nerve to the trapezius enters the muscle laterally and superiorly. Again, when the dissection does not have to be carried proximally to this level as just mentioned, there is less danger of injury to this nerve. One of the major contributing arteries to this flap is the occipital artery, which is also very proximal and in a safe location.

The amount of proximal dissection is related to where the axis of rotation of that particular flap must be for adequate transposition. If the arc of the flap must be increased, the flap may be delayed (Fig. 10-14, *A* to *D*). This would seem reasonable if the measurements exceeded the 3:1 ratio or if there were other unfavorable local circumstances such as pre-

A

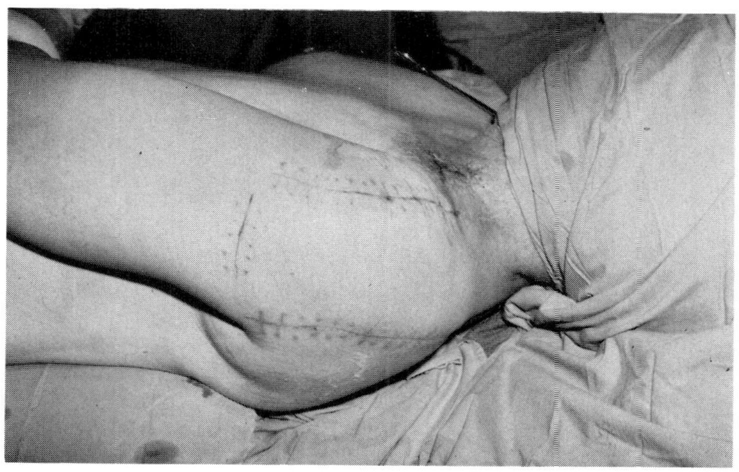

B

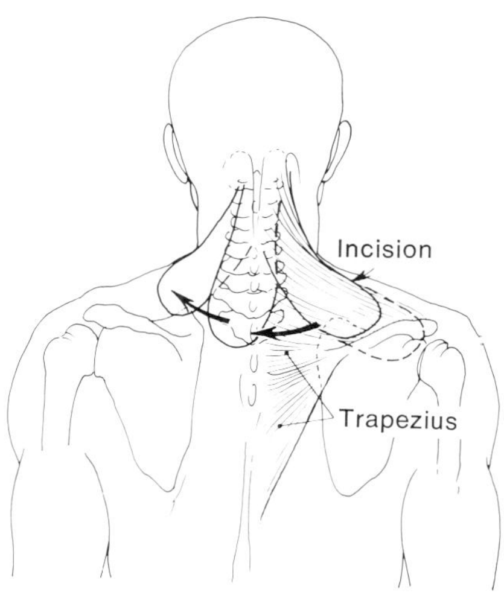

Fig. 10-14

A, "Extended" trapezius musculocutaneous flap is outlined and delayed.

B, Posterior view of trapezius muscle and general outline of skin it will "carry" without delay.

Continued.

C

D

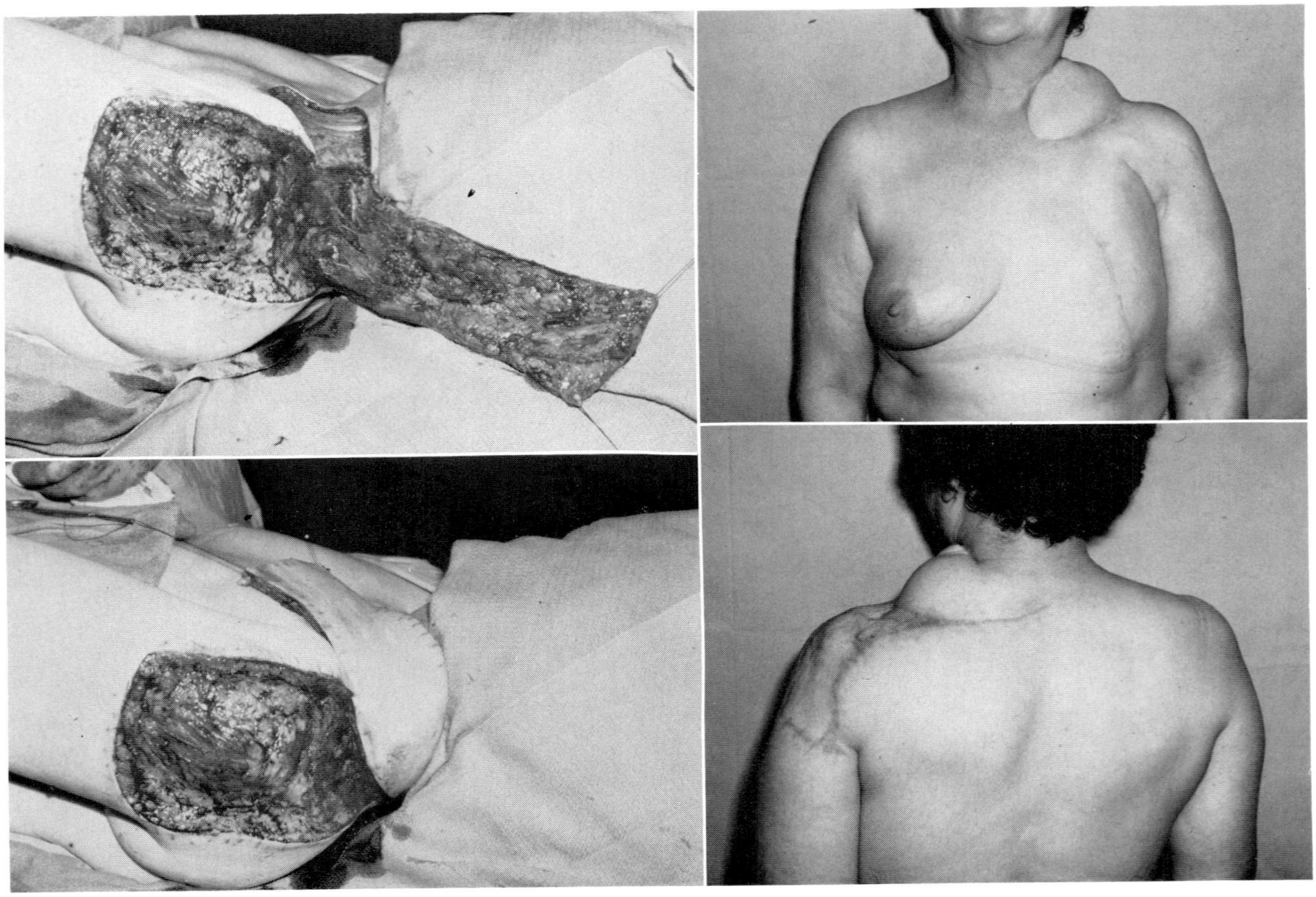

E

F

Fig. 10-14, cont'd
C and **D,** Flap is elevated and viability confirmed with fluorescein. Defect is excised and flap rotated into defect. Donor site is partially closed and then skin grafted.
E and **F,** Anterior and posterior view of healed wound 6 months after operation.

vious radiation or dissection. In reference to the chest wall, this flap would seem most useful in covering the contralateral shoulder or the ipsilateral upper chest. The donor site must frequently be grafted (Fig. 10-14, *E* and *F*).

Precautions

When elevating the flap in a distal-to-proximal fashion, care should be taken to include the upper portion of the trapezius muscle with the flap.

EXTERNAL OBLIQUE

History

In 1922 Farr reported using randomly based portions of upper abdomen and lower chest musculofascial units to close abdominal hernias. He comments that "it would seem desirable to use, whenever possible, a transplant of fascia or fascia and muscle combined where circulation is not entirely cut off." In 1953 Lesnick used the external oblique based medially to close a lower abdominal wall defect following excision of a large abdominal wall malignancy that was unresponsive to radiotherapy. Hershey in 1964 reported using a musculocutaneous flap to close a large upper abdominal–lower thoracic defect following excision of a large hemangiopericytoma.

Anatomy

The external oblique muscle is the largest and most superficial of the four muscles of the anterior abdominal wall. It originates from the inferior outer surface of the fifth to twelfth ribs. The origin from the fifth to ninth ribs interdigitates with the serratus

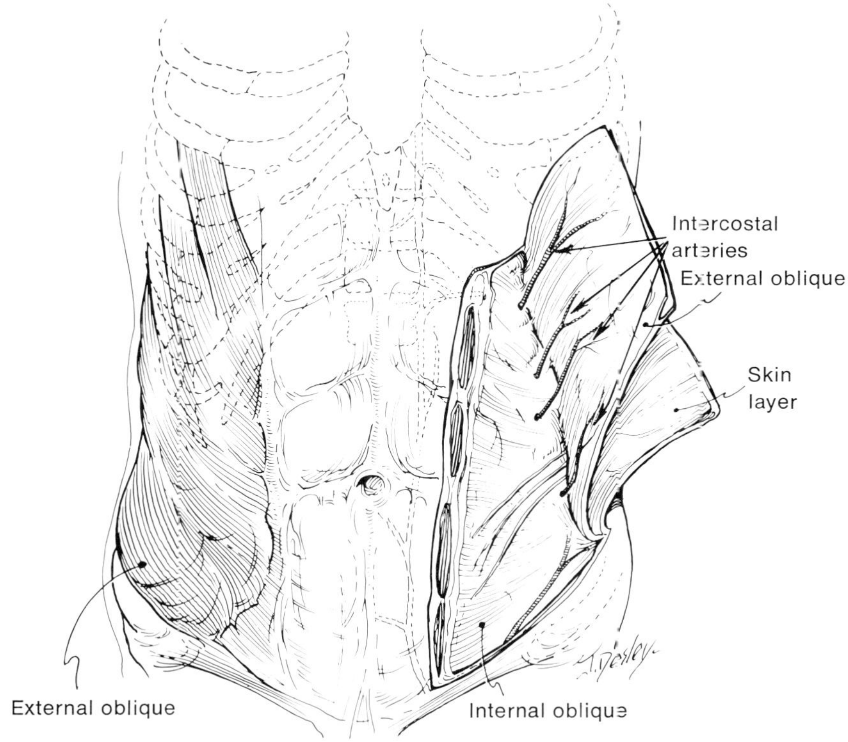

Fig. 10-15. External oblique muscle in its normal anatomic position on right and elevated on left showing intercostal blood supply.

(From Hodgkinson, D, and Arnold, P.G.: Br. J. Plast. Surg. **33**:216, 1980.)

anterior and, from the tenth to twelfth ribs, with the latissimus dorsi. Thus the origin slopes caudally and dorsally. The muscle inserts into the anterior half of the outer lip of the iliac crest and becomes aponeurotic as it passes forward to the linea alba, thus forming a portion of the inguinal ligament. The cranial portion of the fibers blends with the inferior fibers of the pectoralis major. The vascular supply of this muscle is from branches of the intercostal and lumbar arteries. The intercostal vessels run between the transversus abdominis and the internal oblique (Fig. 10-15). They give off branches to the external oblique that pierce the internal oblique as the vessel runs forward to finally terminate with epigastric vessels in the rectus sheath.

Cadaver dissections show that these vessels vary considerably in size and are not constant. They appear to be larger the higher up the abdominal wall and more posteriorly. Perforating vessels traverse the external oblique to go to the skin and subcutaneous fat. The lateral cutaneous branches of the intercostal vessels arise from the costal segment and go directly to the overlying skin. These also may be inconsistent in size and position but generally do travel forward to also terminate with the epigastric system.

The muscle alone or the muscle and skin in tandem can be transposed safely as long as a wide base is maintained.

Operative procedure

The chest wall defect to be closed with this muscle should be in the lower aspect of the chest (Fig. 10-16, *A* to *D*). Exposure of the muscle is through a vertical midline abdominal incision. The external oblique is then elevated by incising the lateral aspect of the rectus sheath where it is firmly attached. This gives a strong edge with which to maintain tension of the muscle during dissection and an excellent "cuff" for suturing the transposed muscle into the chest wall defect after the muscle is transposed. The rectus sheath is simply closed by direct suture after the external oblique is elevated. The muscle may be easily elevated posteriorly to the posterior axillary line where perforating vessels will be found entering the muscle. The inferior attachments of the muscle to the inguinal ligament and iliac bone can be divided

Text continued on p. 266.

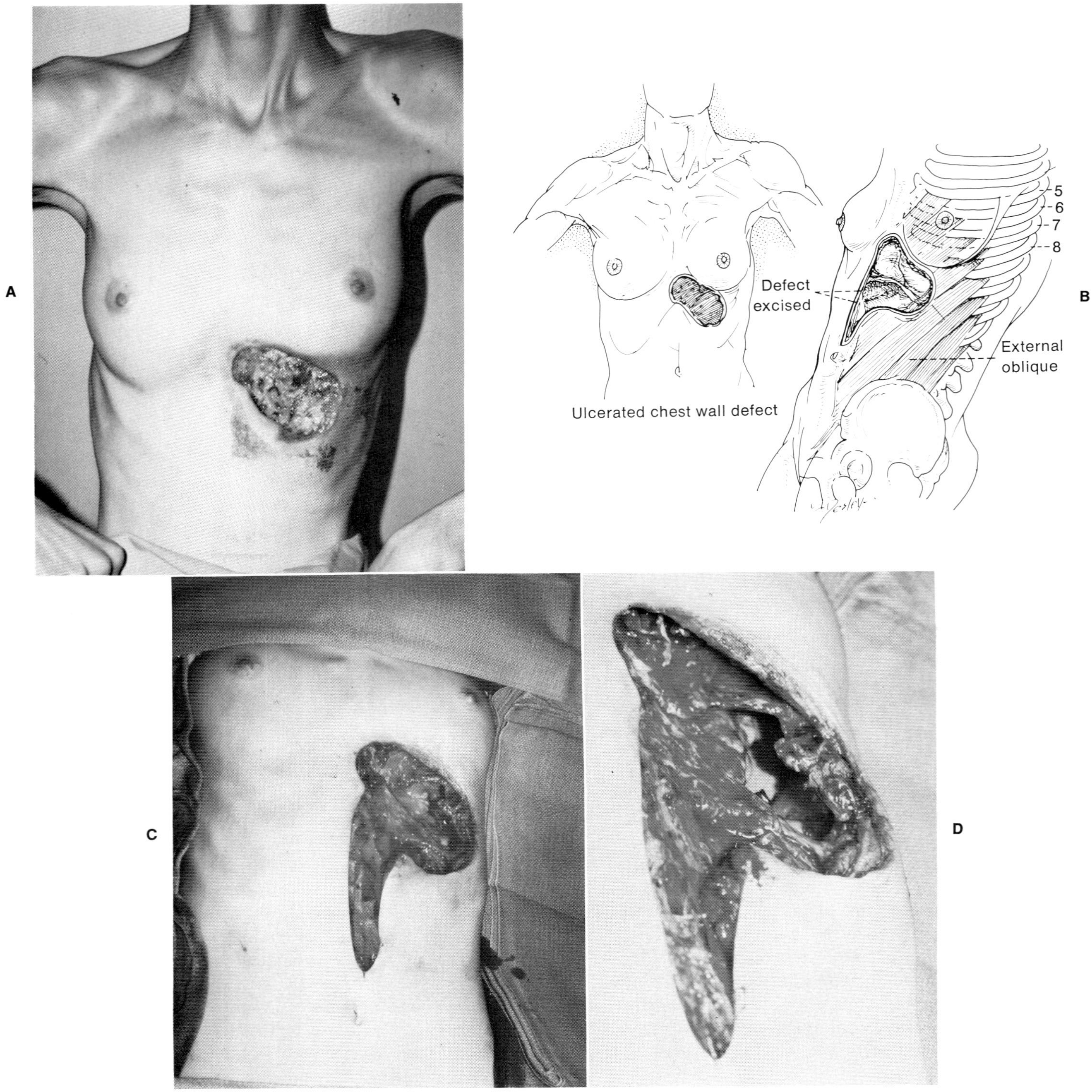

Fig. 10-16

A and **B,** Chest wall defect in 35-year-old woman after radiation therapy for Hodgkin's lymphoma. Ulcerated defect has been present over 2 years, with no other clinical, chemical, or radiologic evidence of her disease.

C, Defect after excision of infected irradiated wound, including lower chest wall (full thickness), anterior portion left hemidiaphragm, and upper abdomen.

D, Close-up of defect demonstrating comunication with left chest as well as blood supply to external oblique muscle.

(From Hodgkinson, D. J., and Arnold, P.G.: Br. J. Plast. Surg. **33:**216, 1980.)

E

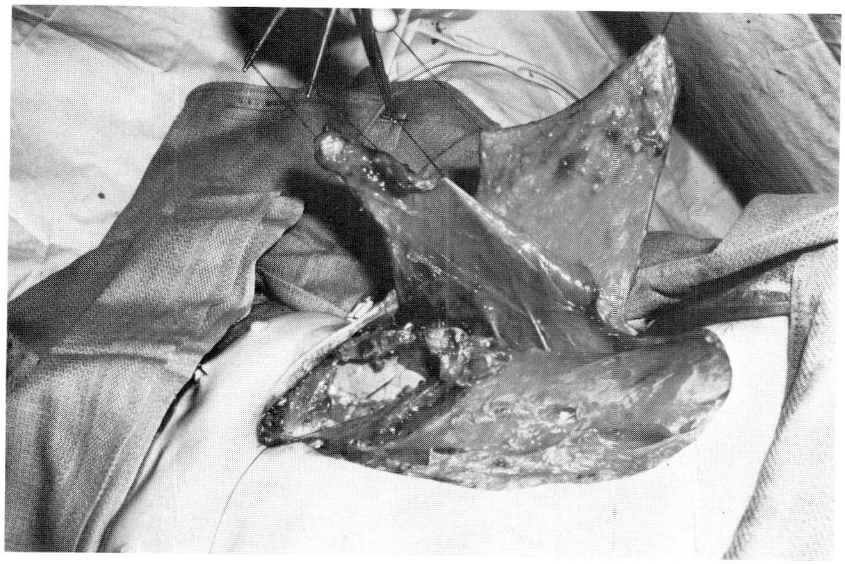

F

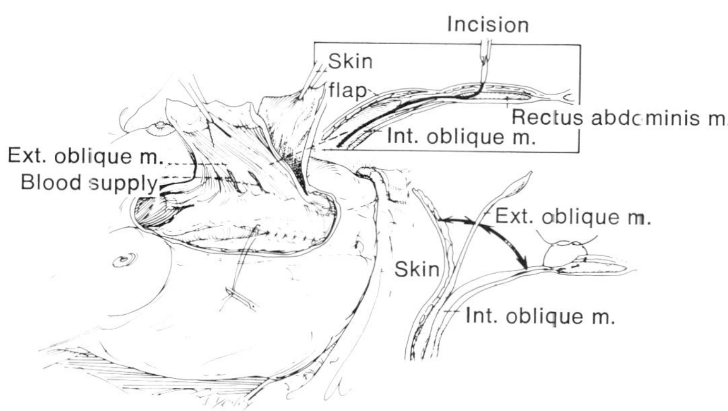

Fig. 10-16, cont'd
E and **F,** Muscle and skin elevated as separate layer, since muscle is to be transposed to higher level on chest wall than skin. Portion of lateral rectus fascia is left attached to muscle tc be transposed, which is finer piece of tissue to suture than muscle alone. (It is preferable to move these tissues together as musculocutaneous unit unless their separation is necessary to obtain proper wound closure.)

Continued.

G

H

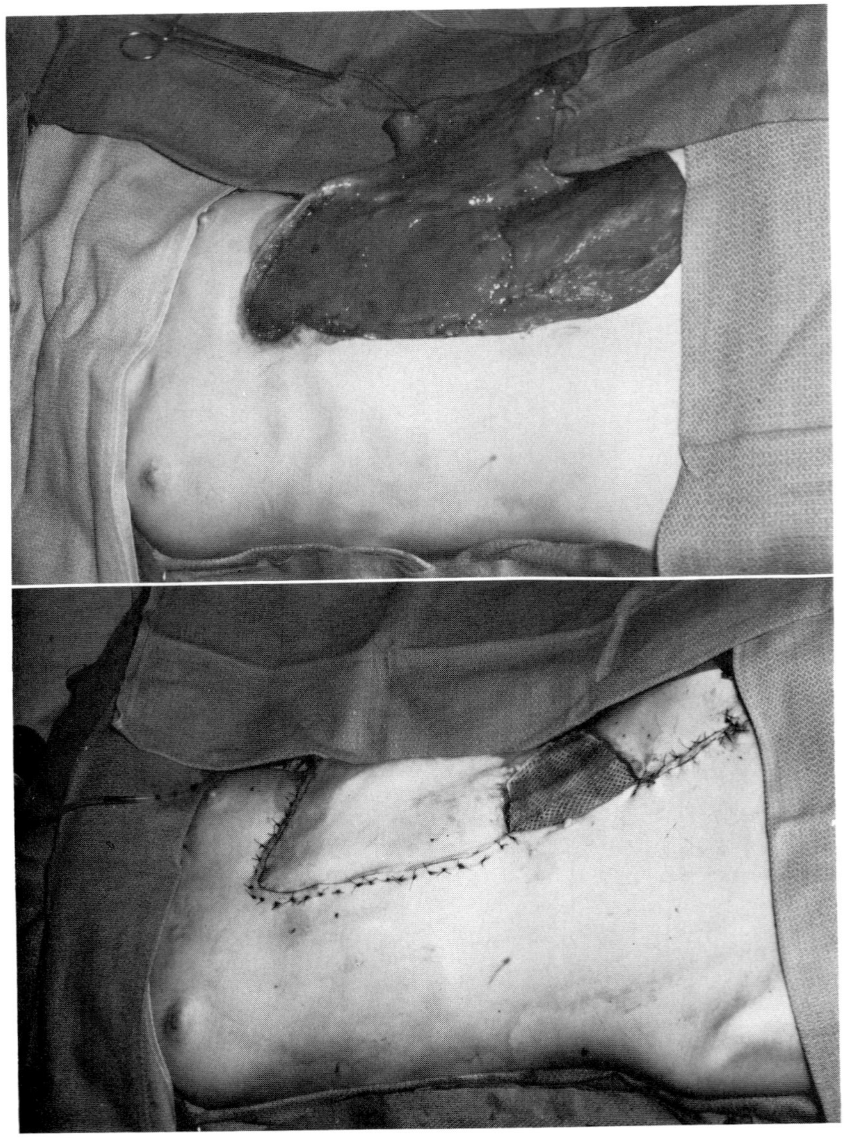

Fig. 10-16, cont'd

G, Muscle is transposed into thoracic defect beneath edge of skin and skin flap ready to be transposed.

H and **I,** Donor site in groin and lower abdomen may be closed by direct suture or split skin graft as situation dictates.

J, Three-month postoperative view.

I

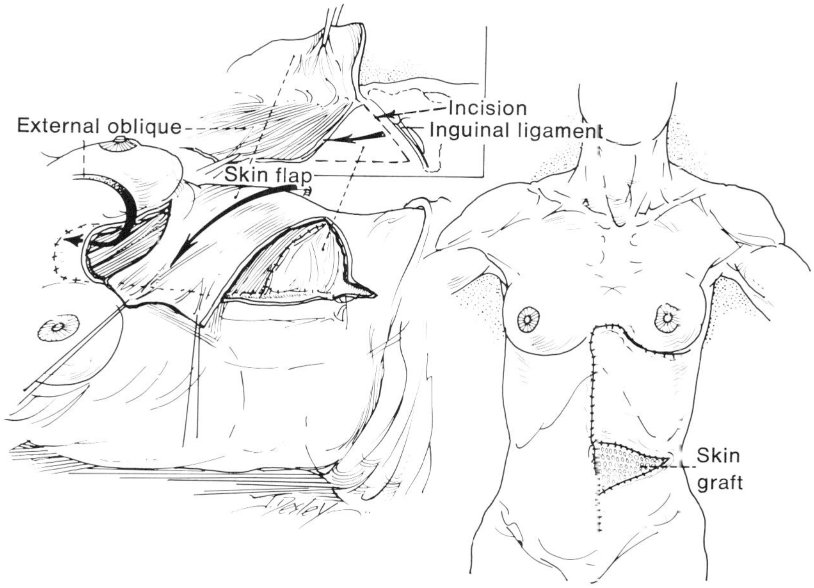

J

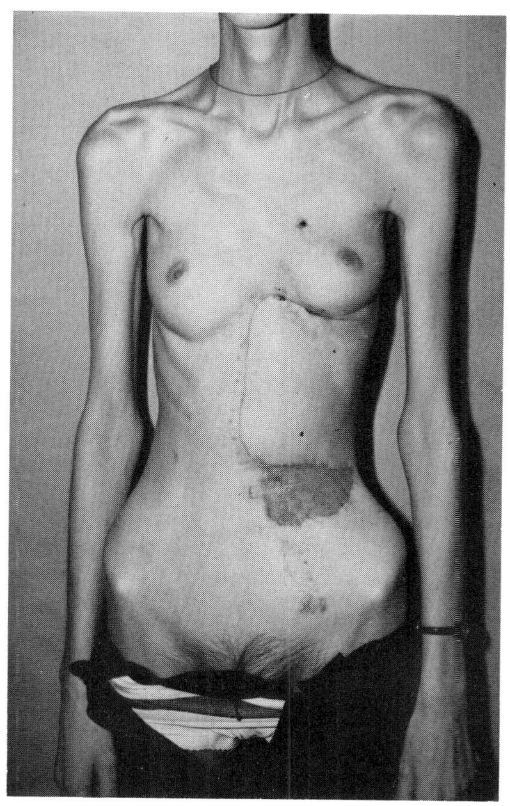

Fig. 10-16, cont'd. For legend see opposite page.

under direct vision, leaving a large posteriorly based flap. The overlying abdominal wall skin may be moved with the muscle as a musculocutaneous flap or separately as a skin flap or may simply be replaced in its normal position (Fig. 10-16, *E* and *F*). If the chest wall defect is larger than the overlying defect (Fig. 10-16, *C* and *D*), it may be advantageous to separate the skin and muscle, since they will be transposed to different levels (Fig. 10-16, *G*). The donor site may be partially or totally closed by suture and split skin grafts, depending on the presence and laxity of the lower abdominal and groin skin (Fig. 10-16, *H* to *J*). Suction catheters are used under the proximal portion of the muscle. The average patient may be extubated in the operating room or the following morning usually with no respiratory difficulty.

Precautions

Preservation of the internal oblique and transversus abdominis muscles is required to avoid hernia formation at donor site.

SUMMARY

No single method is the obvious solution for difficult problems in chest wall reconstruction. The reconstructive surgeon should have muscles high on his list of alternatives when planning these projects.

The disability following utilization of these muscles, with or without the overlying skin, has been small and seems to be well worth the "trade" in virtually every case. We have used the latissimus dorsi muscle as a musculocutaneous unit even in the absence of the pectoralis major muscle on the ipsilateral side without significant disability. This seems unusual but most fortunate. On two occasions I have mobilized and transposed the latissimus dorsi musculocutaneous flap on a muscle that was denervated. The muscle was atrophic, but the vessels were present, and the patients had no problems with flap viability. This has been confirmed by others.

The greater omentum has been very useful in chest wall problems, particularly in radiation sores and in partial-thickness defects. It does not provide the structural stability that muscles do and would not be my first choice if appropriate muscles were available. It may be most helpful in "salvage" situations where muscle flaps or laterally based flaps have been used and failed for some reason.

The latissimus dorsi and pectoralis major muscles are two excellent sources of large portions of tissues both with and without their overlying skin. The trapezius and external oblique muscles may be most helpful for smaller but special situations and would not be used as frequently as either the pectoralis major or latissimus dorsi muscles in chest wall reconstruction.

ANNOTATED BIBLIOGRAPHY

1. Ariyan, S.: The pectoralis major myocutaneous flap, Plast. Reconstr. Surg. **63**(1):73, 1979.

   The vascular supply and distribution of the pectoralis major muscle are described, and clinical cases of its usefulness in reconstruction of large head and neck defects are beautifully demonstrated.

2. Arnold, P.G., and Pairolero, P.C.: Chondrosarcoma of the manubrium—resection and reconstruction with pectoralis major muscle, Mayo Clin. Proc. **53:**54, 1978.

   A case is reported involving a primary chondrosarcoma of the manubrium with immediate reconstruction at the time of surgical extirpation, using autogenous rib grafts and transposed pectoralis major muscle.

3. Arnold, P.G., and Pairolero, P.C.: Use of pectoralis major muscle to repair defects of the anterior chest wall, Plast. Reconstr. Surg. **63**(2):205, 1979.

   The anatomy and arc of rotation of the pectoralis major muscle are discussed. The technical details of transposing the muscle alone to anterior midline defects with or without autogenous bone grafts are described. Several cases are illustrated.

4. Aston, S., and Pickrell, K.: Chest wall reconstruction plastic surgery, Philadelphia, 1977, W.B. Saunders Co.

   Random flaps, arterialized flaps, and some muscle flaps are discussed with reference to their usefulness in chest wall reconstruction.

5. Baek, S.M., and co-workers: The pectoralis major myocutaneous island flap for reconstruction of the head and neck, Head Neck Surg. **1**(4):293, 1979.

   The anatomy, access, and arc of rotation of the pectoralis major musculocutaneous flap are once again discussed with reference to their usefulness in the head and neck.

6. Bostwick, J., III, and co-workers: Sixty latissimus dorsi flaps, Plast. Reconstr. Surg. **63**(1):31, 1979.

   This is a very comprehensive article including utilization of this muscle as a muscle only as well as a skin-muscle unit. Transposition of the muscle to the upper extremity for cover as well as functional rehabilitation is also discussed. Free flap transposition to the lower extemity is described, and examples are shown. Breast reconstruction is discussed and demonstrated.

7. Bostwick, J., III, and co-workers: The reverse latissimus dorsi muscle and musculocutaneous flap: anatomical and clinical considerations, Plast. Reconstr. Surg. **65**(4):395, 1980.

A demonstration of transposition of this muscle based on the paraspinous perforating vessels is presented. Three clinical cases are described along with a brief discussion.

8. Brown, R.G., Vasconez, L.O., and Jurkiewicz, M.J.: An island flap of the pectoralis major muscle, Br. J. Plast. Surg. **30:**161, 1977.

The original paper describing transposition of both pectoralis muscles to close a large midline defect secondary to tumor removal is presented. The advantages of dividing the humeral attachment and maintaining the blood supply are clearly illustrated and well discussed.

9. Campbell, D.A.: Reconstruction of the anterior thoracic wall, J. Thorac. Surg. **19:**456, 1950.

This article is one of the early, clear descriptions of the utilization of the latissimus dorsi muscle for full-thickness chest wall defects. The method uses a fascia lata graft, mobilization and transposition of the latissimus dorsi muscle, and finally, a split-thickness skin graft.

10. Davis, H.H., Tollman, J.P., and Brush, J.H.: Huge chondrosarcoma rib, Surgery **26:**699, 1949.

A case report of a large chondrosarcoma involving the left fifth through seventh ribs, which was excised and repaired by a fascia lata graft and transposition of the latissimus dorsi muscle to maintain chest wall stability, is presented.

11. Desprez, J.D., Kiehn, C.L., and Eckstein, W.: Closure of large meningomyelocele defects by composite skin-muscle flaps, Plast. Reconstr. Surg. **47:**234, 1971.

Large thoracolumbar and lumbosacral defects are closed by mobilizing well-vascularized muscle flaps. Six cases are presented and discussed.

12. D'Este, S.: La technique de l'amputation de la mamelle pour carcinoma mammaire, Rev. Chir. Orthop. (Paris) **45:**164, 1912.

This article amplifies the Tanzini method of reconstruction of the chest wall at the time of radical mastectomy. This includes a superiorly based flap of latissimus dorsi muscle and overlying skin based in the axilla. This was transposed anteriorly at the time of mastectomy.

13. Farr, R.E.: Closure of large hernial defects of the upper abdomen, Surg. Gynecol. Obstet. **34:**264, 1922.

One of the classical considerations of abdominal wall repairs is presented with emphasis on maintaining the innervation and blood supply of muscle and muscle fascia units for abdominal wall reconstruction is discussed.

14. Hershey, F.B., and Butcher, H.R., Jr.: Repair of defects after partial resection of the abdominal wall, Ann. Surg. **107:**586, 1964.

Large defects of the upper abdominal wall and lower chest are closed, using a "musculofascial-cutaneous flap." Emphasis is placed on the blood supply of the external oblique muscle and the necessity of maintaining it. Tensor fascia lata musculofascial flaps are also used for lower abdominal closure.

15. Hodgkinson, D.J., and Arnold, P.G.: Chest wall reconstruction using the external oblique muscle, Br. J. Plast. Surg. **33:**216, 1980.

The anatomy and application of the external oblique musculocutaneous flap are described. This unit is used to accomplish reconstruction of a difficult abdominal wall defect.

16. Hueston, J T., McConchie, I.H.: A compound pectoral flap, Aust. NZ J. Surg. **38:**61, 1968.

One of the early papers using the pectoralis major muscle as well as its overlying skin, which was delayed and transposed, to close a large chest wall defect is presented.

17. Hutchins, E.H.: A method for prevention of elephantiasis, Surg. Gynecol. Obstet. **69:**795, 1939.

Hutchins discusses the problem of lymphedema of the upper extremity after mastectomy. He describes transposing the latissimus dorsi muscle into this defect to relieve or prevent the occurrence of lymphedema. He discusses the nerve and blood supply to the muscle and demonstrates nicely its transposition anteriorly. He also mentions that the anterior axillary web may be reconstructed by detaching the latissimus dorsi at its insertion and reinserting it into the coracoid process.

18. Jurkiewicz, M.J., and Arnold, P.G.: The omentum: an account of its use in reconstruction of the chest wall, Ann. Surg. **185**(5):548, 1977.

The repair of defects to the chest wall, particularly with those people afflicted by osteoradionecrosis, by using the transposed greater omentum based on either the right or left gastroepiploic vessels is discussed. Full-thickness as well as partial-thickness chest wall defects are closed, using this structure. Ten examples are given.

19. Lesnick, G.J., and Davids, A.M.: Repair of surgical abdominal wall defect with a pedicled musculofascial flap, Ann. Surg. **137:**569, 1953.

The external oblique muscle is transposed based on its medial attachments for closure of a midline defect. The utilization of fascial attachments to the lateral edge of the rectus sheath is emphasized.

20. McCraw, J.B., Phenix, J.O., and Baker, S.W.: Repair of major defects of the chest wall and spine with the latissimus dorsi myocutaneous flap, Plast. Reconstr. Surg. **62**(2):197, 1978.

Chest wall defects as well as midline back defects in the case of myelomeningoceles are closed by a transposition of the latissimus dorsi muscle. The vascular anatomy and cutaneous territories as well as primary and secondary blood supplies are discussed. Its usefulness for closure of myelomeningoceles is emphasized.

21. McCraw, J.B., Magie, W.P., Jr., and Kalwaic, H.: Use of the trapezius and sternomastoid myocutaneous flaps in head and neck reconstruction, Plast. Reconstr. Surg. **63**(1):49, 1979.

The usefulness of the trapezius as well as the sternomastoid myocutaneous flaps is discussed. Numerous case examples are cited and described. The technical details of flap elevation, considerations of delays for extended flaps, and clinical applications are nicely described.

22. Manchot, C.: Die hautarterien des menschlichen körpes, Leipzig, 1889, F.C.W. Vogel.

This book very clearly outlines a great number of cutaneous vascular territories. The territories supplied by the occipital, suprascapular, and scapular arteries are specifically illustrated.

23. Mendelson, B.C., and Masson, J.K.: Treatment of chronic radiation injury over the shoulder with a latissimus dorsi myocutaneous flap, Plast. Reconstr. Surg. **60**(5):681, 1977.

Chronic radiation injuries of the shoulder involving the skin and underlying shoulder joint are closed with large transposition of large latissimus dorsi musculocutaneous units. In one particular case the skin-muscle unit has been separated at a previous procedure but goes on to survive in its entirety.

24. Mutter, T.D.: Cases of deformity from burns relieved by operation, Am. J. Med. Sci. **4**:66, 1842.

This article presents a classic description of the nape of the neck flap that is transposed to correct contractures in the neck area after large burns.

25. Pairolero, P.C., and Arnold, P.G.: Bronchopleural fistula: treatment with pectoralis major muscle transposition, J. Thorac. Cardiovasc. Surg. **79**(1):142, 1980.

A bronchopleural fistula in the face of an old empyema is repaired via an axillary thoracotomy. The pectoralis major muscle from that same side is used to bolster the closure of the fistula and fill up the dead space from the empyema. The anatomy and technique are discussed.

26. Schneider, W.J., Hill, H.L., and Brown, R.G.: Latissimus dorsi myocutaneous flap used for breast reconstruction, Br. J. Plast. Surg. **30**:277, 1977.

One of the very early reports of a one-stage total reconstruction of the radical mastectomy defect presents the use of the transposed latissimus dorsi skin-muscle unit over an implant with direct closure of the donor site and concomitant tailoring of the remaining breast.

27. Sisson, G.A., Straehley, C.J., Jr., and Johnson, N.E.: Mediastinal dissection for recurrent cancer after laryngectomy, Laryngoscope **72**:1064, 1962.

The pectoralis major muscle is transposed on its perforating blood supply from the internal mammary vessels and turned back on itself to cover the great vessels after an extended neck dissection and mediastinal dissection. This muscle flap protected these vital exposed structures.

28. Zovickian, A.: Pharyngeal fistulas: repair and prevention using occiput based shoulder flaps, Plast. Reconstr. Surg. **19**:355, 1957.

An extension of the nape of the neck concept to delay large flaps that can be brought around for repair of pharyngofistulas is presented.

11

RECONSTRUCTION OF THE STERNUM

Jack Fisher ■ David K. Bone ■ Foad Nahai

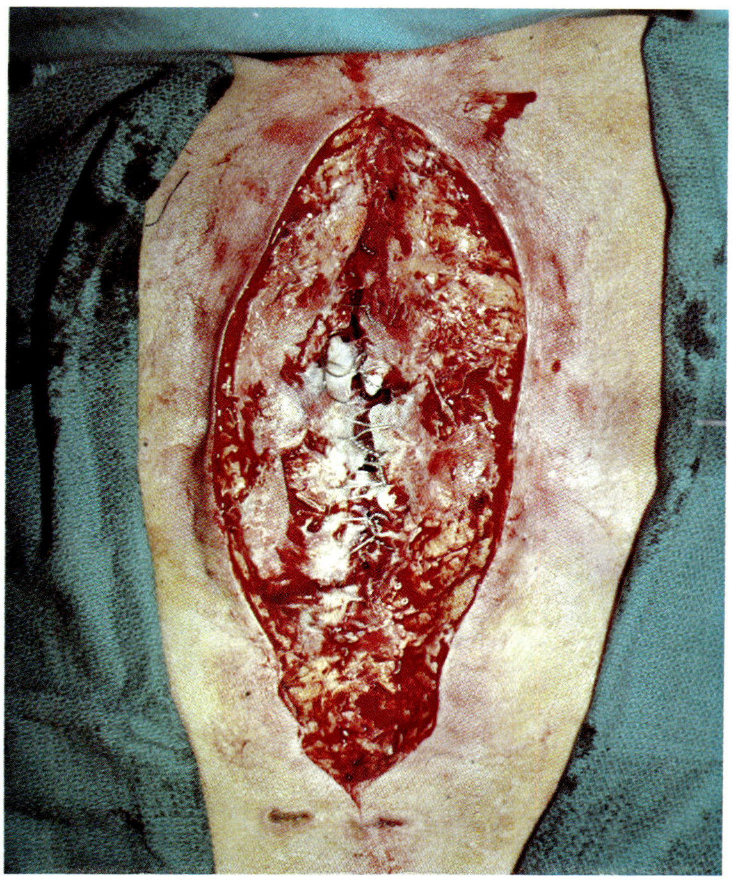

Fig. 11-1. Extensive sternal wound infection with associated mediastinitis. Reconstructive requirements are débridement and vascularized flap coverage.

INFECTED MEDIAN STERNOTOMY

Median sternotomy has become the incision of choice for most intracardiac procedures since its introduction in 1957. It is a simple and rapid technique that provides excellent exposure. Complications associated with median sternotomy are relatively infrequent, varying from 0.5% to 5.0%. When complications do occur, there is significant morbidity and mortality. Complications associated with median sternotomy include sternal dehiscence, mediastinitis, infection of cardiotomy incisions, infection in implanted prosthetic material, and chronic infection of the sternum and costal cartilages. The sternal dehiscence is usually associated with an infective process, and once infection is established in the sternum, it may be difficult to eradicate.

The bacteria associated with infection of the median sternotomy vary. A significant number are related to gram-negative organisms. Gram-positive organisms have also been cultured from these wounds, including *Staphylococcus epidermidis*. Anaerobic infection of a sternal wound has also been reported. Occasionally negative cultures are reported, and this may be related more to poor culture technique than absence of bacteria.

Several factors associated with an increased risk of infection after median sternotomy have been identified. These factors include prolonged operative time, postoperative hemorrhage requiring reexploration, prolonged mechanical ventilation, and chronic lung disease.

The diagnosis of sternal dehiscence and associated mediastinal infection is based on the presence of fever, wound erythema with drainage, and sternal instability (Fig. 11-1). Typically the onset of this complication is seen 7 to 21 days after operation. Late presentations may occur with chronic forms of osteomyelitis and chondritis.

METHODS OF TREATMENT

The open technique

The open technique consists of débridement of the infected wound and sternum, followed by open packing and delayed secondary healing. Although this is a simple technique, it has numerous disadvantages. This method is associated with high mortality, requiring prolonged hospitalization and recovery period. Since the mediastinum remains open for an extended period, this technique is frequently associated with complications such as graft or myocardial hemorrhage, pyogenic graft thrombosis, respiratory embarrassment with prolonged ventilation, and tracheostomy. The open mediastinum appears to be less tolerated when compared to other areas of the body with open, granulating wounds.

The closed technique

The closed technique for management of infected median sternotomy wounds consists of débridement, primary closure, and continuous antibiotic irrigation via catheters. The components of this method as defined by Spencer are the following:

1. Exploration of the entire sternal incision once the diagnosis is made
2. Débridement of the sternum and mediastinum
3. Insertion of irrigation and drainage catheters
4. Reapproximation of the sternal edges with wire
5. Antibiotic irrigations for 7 to 14 days
6. The use of appropriate systemic antibiotics

This technique has been modified somewhat with the use of povidone-iodine (0.5%), which has the advantage of being both bactericidal and fungicidal. It also does not have the severe ototoxicity and nephrotoxicity associated with the classical neomycin polymyxin irrigation solutions.

This method is effective in the majority of cases with limited sternal or mediastinal involvement. However, this method is less effective with advanced or extensive infection, and it is in these cases that muscle flap coverage may be necessary.

Muscle flap coverage technique

The muscle flap coverage method is discussed in detail later in this chapter and only indications for its use will be briefly mentioned here.

This technique is appropriate for patients recalcitrant to débridement and closure with continuous antibiotic irrigation. It is also appropriate as initial treatment in patients with significant involvement requiring extensive débridement of the sternum and adjacent costal cartilages. Primary closure is impossible in these cases when removal of all or part of the sternum is required.

Omental flap coverage technique

For completeness, a discussion of the methods of treatment of sternal dehiscence and infection must mention the use of the omentum. The sternal incision is extended over the abdomen, and the greater omentum is identified. It is dissected free of the stomach and brought up into the chest with its blood supply based on either gastroepiploic vessel. This technique brings healthy, vascular tissue into the chest. Its major disadvantage is the possible contamination of the abdomen from the infected median sternotomy wound and the added risk of a laparotomy.

Indications for muscle flap coverage

There are two primary indications for the use of muscle flaps for median sternotomy dehiscence with associated mediastinitis. In the first group are patients in whom conventional treatment, consisting of débridement and closure with antibiotic irrigations, has failed. These patients are at great risk and require aggressive therapy for salvage. In the second group are patients with significant sternal infection requiring extensive débridement of sternum and costal cartilages because of the extent of the infective process (Fig. 11-2, *A*). In these cases rewiring of the sternum is impossible, and significant dead space results after débridement (Fig. 11-2, *C*).

Muscle flaps bring healthy, vascular tissue into a heavily contaminated area and obliterate a potentially large dead space (Fig. 11-2, *C* to *E*). The sternum and adjacent costal cartilages have a relatively poor blood supply, enhancing their susceptibility to infection. Even after aggressive débridement and copious irrigation, the wound remains contaminated. Transposing healthy muscle with its rich blood supply appears to dramatically improve the ability of this wound to heal.

Selection of appropriate muscle flaps

The pectoralis major muscles, with a rectus abdominis muscle when necessary, provide excellent coverage of wounds after débridement for sternal infection and dehiscence.

The blood supply of the pectoralis major muscle consists of a dominant pedicle from the thoraco-

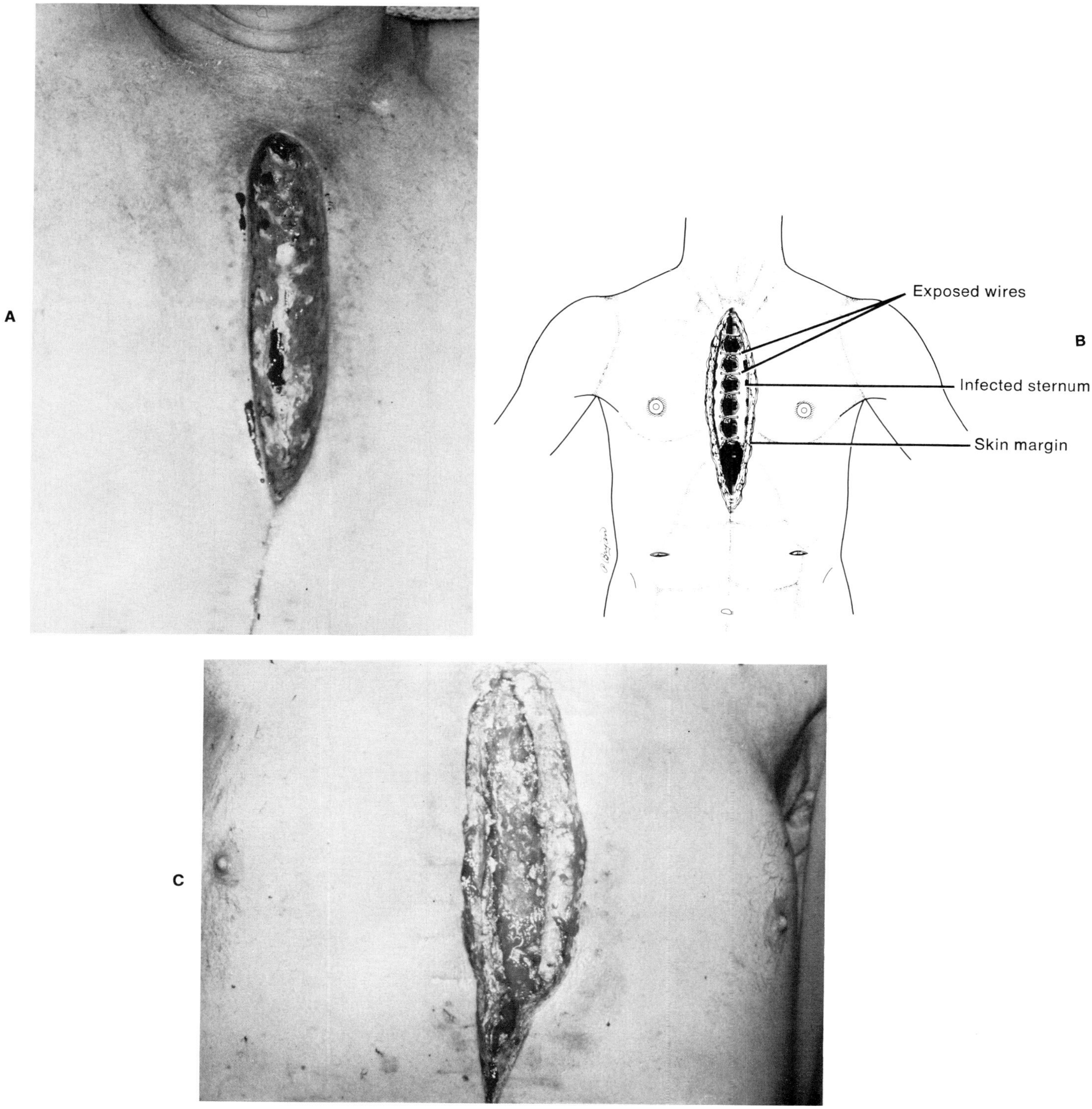

Fig. 11-2

A, Forty-nine-year-old man 6 weeks after coronary artery bypass procedure with open infected median sternotomy incisions. NOTE: Multiple exposed sternal wires.

B, Diagram of infected median sternotomy wound with exposed sternal wires before débridement.

C, Sternal wound immediately after débridement. All sternal wires and nonviable sternal edges have been removed.

Continued.

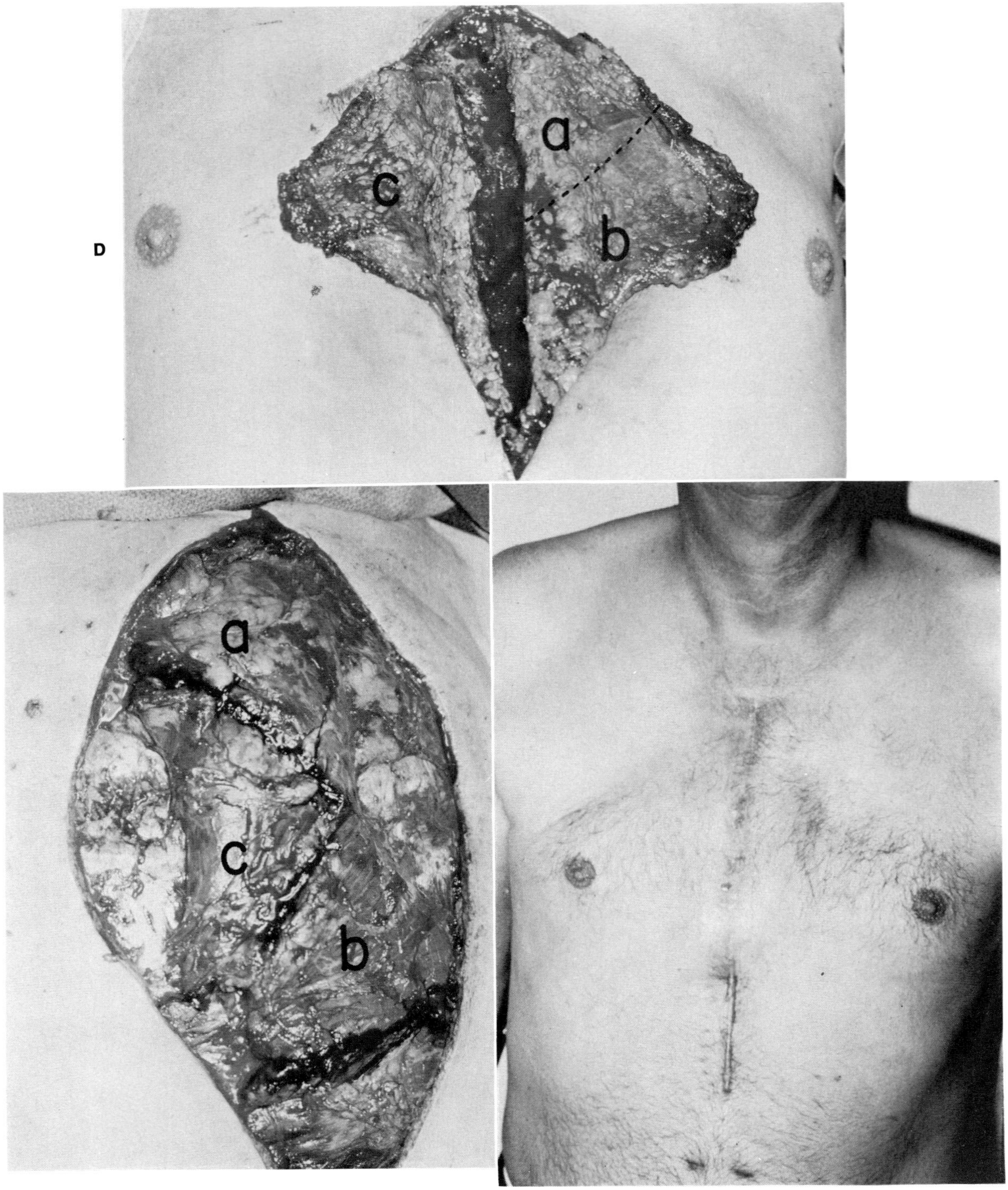

D

E

F

Fig. 11-2, cont'd

D, Bilateral pectoralis major muscle flaps based on perforating vessels from internal mammary artery. Left pectoralis major muscle is divided into two separate flaps (*a* and *b*), allowing superior half to be placed above and inferior half below contralateral muscle (*c*).

E, Both pectoralis major muscles folded on themselves into sternal defect. Left pectoralis major has been divided and placed above (*a*) and below (*b*) right pectoralis major muscle flap (*c*). Lower portion of defect is covered with rectus abdominis muscle based on the superior epigastric artery. Rectus abdominis muscle was divided at level of umbilicus and removed from rectus sheath and folded on itself.

F, Patient 8 months after closure.

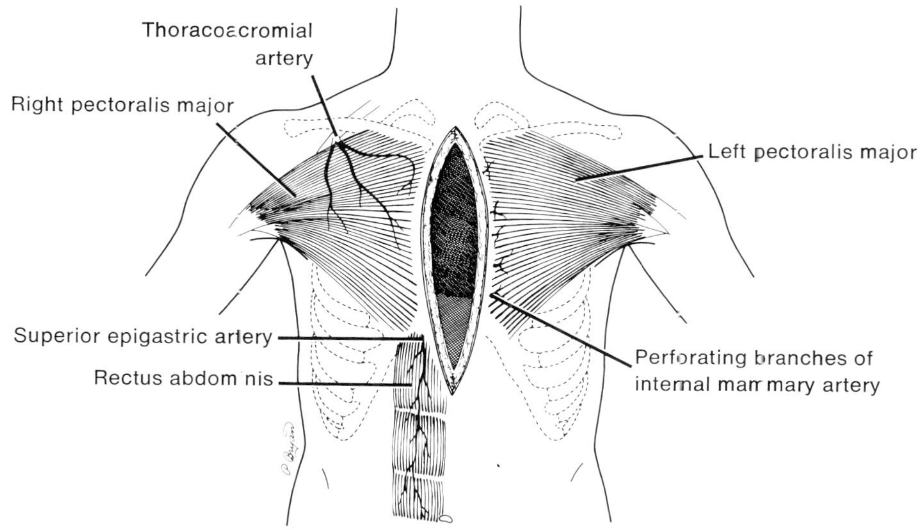

Fig. 11-3. Diagram of anatomy and blood supply of pectoralis major and rectus abdominis muscles.

acromial artery laterally and multiple perforating branches of the internal mammary artery medially (type V) (Fig. 11-3). The insertion of the muscle is in the bicipital groove and the origin from the clavicle, sternum, and first seven ribs. The pectoralis major muscle may be transposed medially based on either of its vascular supplies. To transpose the muscle based on the thoracoacromial vessel, both the origin and insertion are divided. This requires dividing the multiple perforating branches of the internal mammary artery that enter the muscle medially along its origin. Once the muscle is completely mobilized, it can be advanced medially into the sternal defect.

The other method, using the pectoralis major muscle based on its medial blood supply from perforating vessels of the internal mammary artery, appears better suited for closure of sternal defects. Here, the dominant vascular pedicle arising from the thoracoacromial artery is ligated after the muscle insertion in the humerus has been divided. The muscle is then dissected off the chest wall and folded on itself into the sternal defect (Fig. 11-4). The muscle based on its medial blood supply allows for greater fill of the sternal defect while requiring less dissection (Fig. 11-5). In small defects of the sternum, a single pectoralis major muscle or a portion of one may suffice. Usually, however, both pectoralis major muscles are required.

Often, after adequate débridement a portion of the defect remains inferiorly that cannot be covered by the pectoralis major muscles. Here, the rectus abdominis muscle, based on the superior epigastric artery, is dissected out of the rectus sheath and folded on itself. This muscle is divided at or just below the umbilicus and is transposed into the lower portion of the sternal defect.

Using both pectoralis major muscles and a single rectus abdominis muscle, it is possible to cover the entire mediastinum (Fig. 11-6).

Technical considerations

Fever, wound drainage, and sternal instability indicate possible diagnosis of sternal dehiscence and mediastinal infection. Early aggressive therapy is required. When the infective process is localized to the skin and subcutaneous tissue, opening of the wound superficially with drainage may be successful. However, when the infection involves the sternum and mediastinum, the entire wound must be opened. In many patients débridement with closure and continuous antibiotic therapy will suffice. However, there are patients who will require more aggressive therapy, and it is for these patients that muscle flap coverage is presented.

Technique

Successful coverage of a median sternotomy infection, with or without associated mediastinitis, requires thorough and aggressive débridement (Figs. 11-2, *C* and 11-6, *B*). The first step consists of excision of the wound margins down to the level of the sternum. Next the sternal wires, which are usually

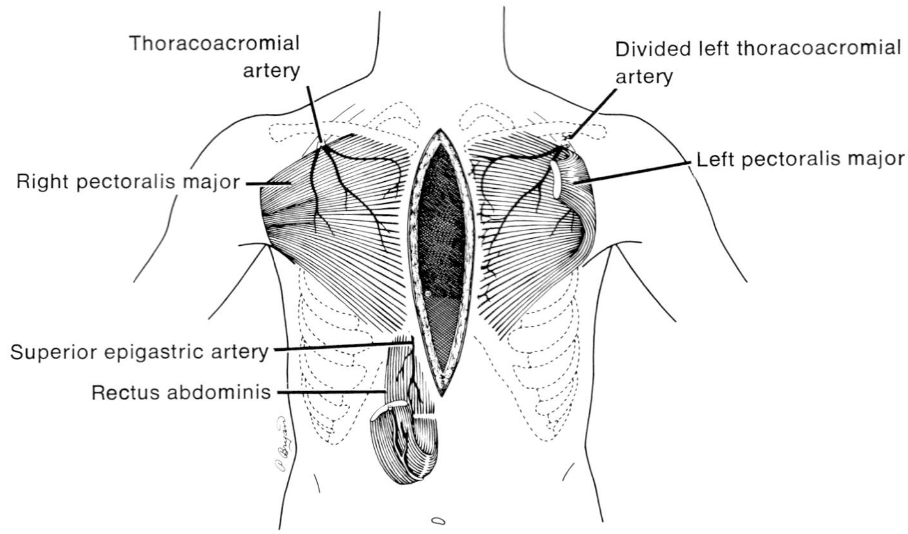

Fig. 11-4. Transposition of muscle groups into sternal defect. Left pectoralis major muscle is based on perforating branches of internal mammary artery. Its dominant pedicle, *thoracoacromial artery,* has been divided. Muscle is folded over on itself after insertion is divided. Right pectoralis major muscle is based on its thoracoacromial vascular pedicle, and insertion in humerus and origin from ribs have been divided. In both methods access to pectoralis major can be obtained by elevating skin flaps off prepectoral fascia. In obese patients counterincisions in anterior axillary fold facilitate division of insertion. Rectus abdominis muscle is divided below umbilicus after rectus sheath is opened. It is removed from rectus sheath to level of costal margin based on superior epigastric artery.

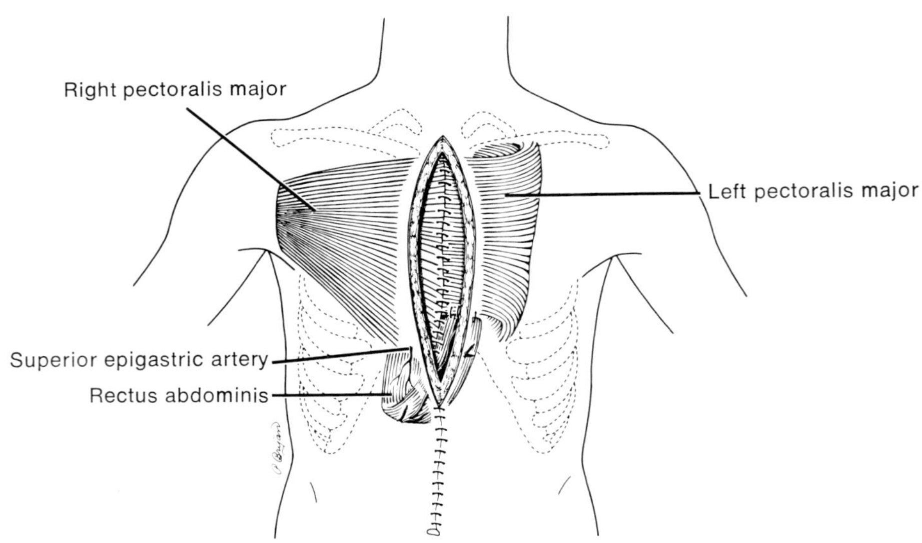

Fig. 11-5. Transposition of left pectoralis major muscle based on internal mammary artery perforators. Right pectoralis major muscle based on its thoracoacromial blood supply. Rectus abdominis muscle based on superior epigastric vasculature.

loose or free, are removed in all areas of involvement. At times only a limited number of sternal wires require removal. Typically, to eradicate the extensive disease, all the sternal wires must be removed. Once the wires are removed, débridement with rongeurs is performed until healthy, bleeding sternal edges are seen. It may be necessary to extend the débridement into adjacent costal cartilages. This type of radical débridement is possible, since the transposition of muscle flaps into these massive defects will provide adequate coverage.

Once débridement of the sternum is complete, attention is directed to the base of the wound. In this defect the heart and great vessels are the base. Therefore little, if any, débridement is undertaken over these structures. The primary objective is to open safely any loculated collections of fluid in this area. A curette is used to gently remove the granulation tissue from the base. Once adequate débridement is completed, there are two possible approaches.

If extensive collections of purulent material have not been encountered, immediate closure with muscle flaps can be performed. If extensive substernal loculations of purulent material are found, a delay of several days with frequent dressing changes at the bedside may be preferred before muscle flap coverage.

With the decision to reconstruct with muscle flaps made, the operation is begun by elevating skin flaps off the prepectoral fascia. The elevation of skin flaps begins along the excised margin of the wound and proceeds from the medial border to the lateral border. While the assistant lifts the skin flap with hooks, the surgeon elevates the skin, staying just superficial to the pectoralis major fascia with the knife or cautery. The skin flap is elevated to the lateral edge of the pectoralis major muscle. As the elevation of the skin flap proceeds from the medial border to the lateral border, visualization may become a problem. We have found the use of the lighted mammary retractor helpful in this part of the dissection. If both pectoralis major muscles are required, which is the usual case, the surgeon changes sides and begins elevation of the opposite skin flap in the manner already discussed.

Early in the course of these procedures counterincisions were made in the anterior axillary fold to provide access for dividing the muscle insertion. This method has been modified, and the lateral margin of the muscle near its insertion is divided, using the lighted mammary retractor. With the retractor elevating the skin flap, the muscle is divided as near the insertion as practical with the cautery or heavy scissors.

Initially in the majority of patients the pectoralis major muscle was based on its blood supply from the thoracoacromial artery. In these patients both the insertion and origin were divided, and the muscle was transposed medially attached only by its vascular pedicle. The technique now used consists of dividing the pectoralis major muscle near its insertion and then dissecting it off the chest wall from the lateral border to the medial border. As the muscle is folded onto itself and elevated, the dominant vascular pedicle, the thoracoacromial artery, comes into view just medial to the pectoralis minor muscle. This vessel is divided, and medial dissection of the pectoralis major muscle continues, freeing it from the chest wall. Proceeding with caution, the dissection continues until perforating vessels from the internal mammary artery are seen entering the deep surface of the muscle. At this point the muscle readily folds into the sternal defect. This procedure is performed on the contralateral pectoralis major muscle in the same manner. We have found that using this muscle based on its medial blood supply from internal mammary artery perforators provides a better arc of rotation for filling these sternal defects. In using the pectoralis major muscle based on its medial blood supply one must be sure the internal mammary artery has not been used either as an internal mammary artery bypass or encompassed previously in parasternal wire placement. If the internal mammary has been used or damaged, then the muscle flap must be based on the thoracoacromial vessel.

The two transposed pectoralis major muscles are sutured together in the midline with absorbable suture. Several sutures are placed between the muscle and defect border to secure the muscle in the sternal wound. It may be necessary to divide one pectoralis major muscles, allowing it to interdigitate above and below with the contralateral muscle, thus providing greater surface area coverage (Fig. 11-2, *D*).

In most patients requiring radical débridement of the sternum, the portion of the wound at or below the level of the xiphoid process cannot be covered by the pectoralis major muscles. This lower portion of the sternal defect can be adequately covered with rectus abdominis muscle. The sternal incision is extended to a level below the umbilicus. A skin flap is elevated off the anterior rectus sheath dissecting from the medial border to the lateral border until the lateral border of the rectus abdominis muscle is reached. Next the anterior rectus sheath is opened longitudinally over the muscle to the level of the um-

A

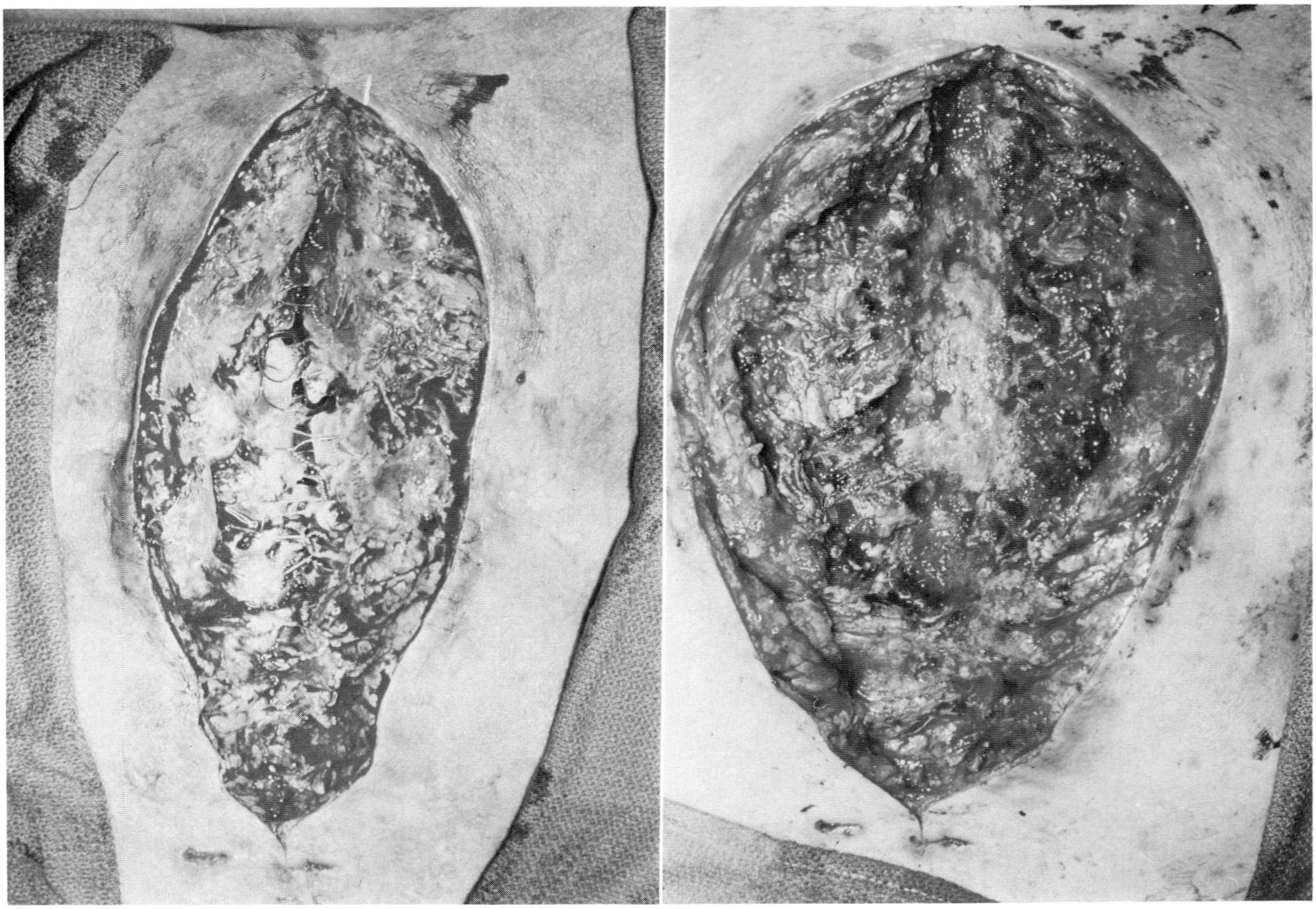

B

Fig. 11-6

A, Sixty-two-year-old woman 4 weeks after coronary artery bypass procedure with extensive sternal infection and associated mediastinitis. Patient was running a septic course at time of débridement and required mechanical ventilation.

B, Large defect after débridgement of entire sternum and adjacent costal cartilages.

C, Rectus abdominis muscle elevated out of rectus sheath. Muscle was divided 2 cm below umbilicus and elevated up to level of costal margin.

D, Pectoralis major muscles (*a* and *b*) are bilaterally folded on themselves and sutured together in midline. They are based on perforating vessels arising medially from internal mammary artery. Lower portion of wound is covered with rectus abdominis muscle (*c*) folded on itself based on superior epigastric vessels. Skin was closed primarily over muscle flaps.

E, Both pectoralis major muscles transposed medially based on perforating vessels from internal mammary artery. Rectus abdominis muscle covers lower portion of defect based on superior epigastric vasculature. Using both pectoralis major muscles in this manner transposes greatest amount of muscle into defect.

C

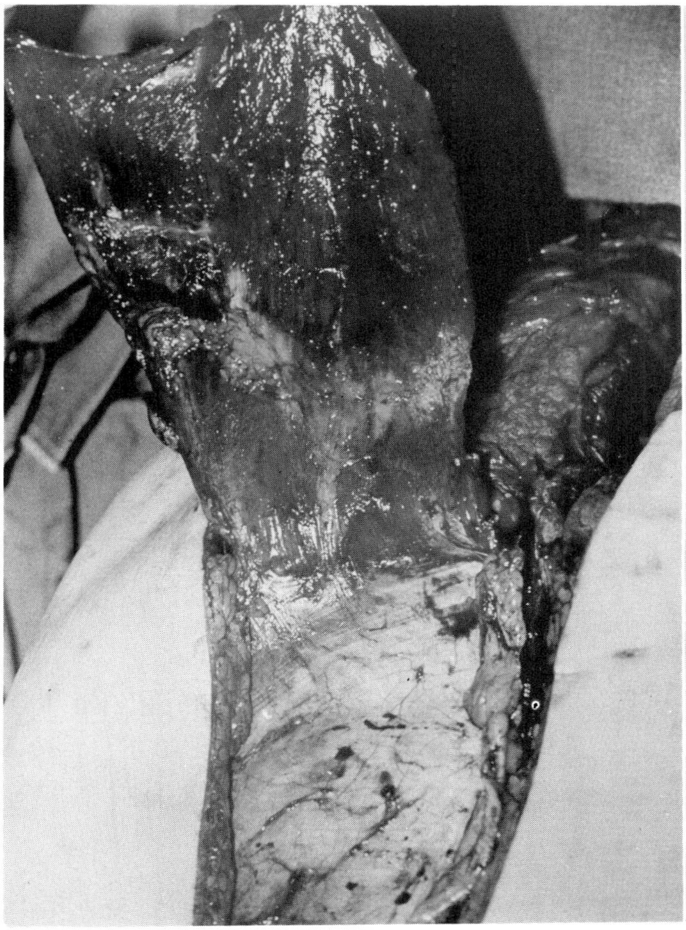

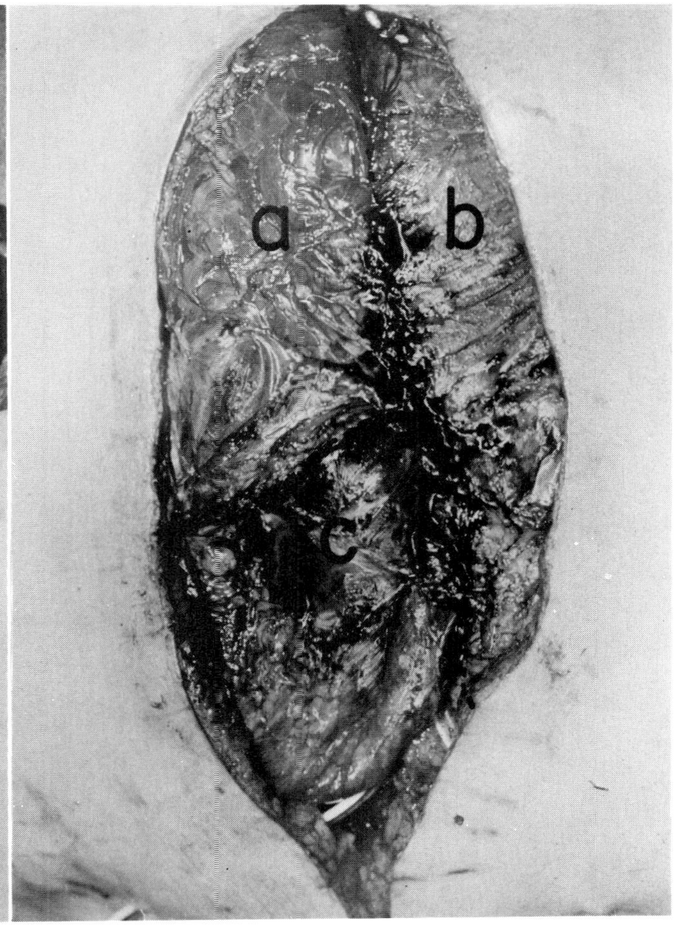

D

E

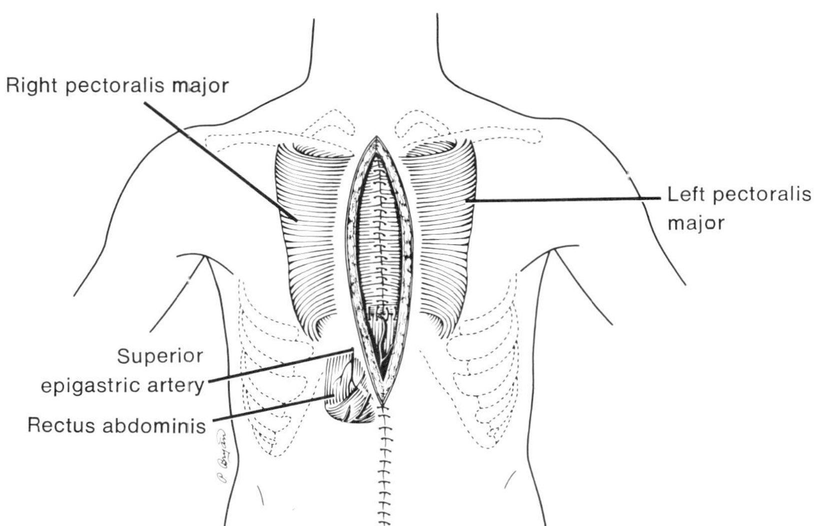

Fig. 11-6, cont'd. For legend see opposite page.

bilicus. The muscle is dissected free from the posterior sheath and divided below the level of the umbilicus. With traction on the muscle pulling it cephalad, it is dissected out of the rectus sheath to the costal margin (Fig. 11-6, *C*). The vascular supply, the superior epigastric vessel, is seen along the deep surface of the muscle. It must be remembered that the superior epigastric artery is a continuation of the internal mammary vessel, and one should be aware of a previous internal mammary artery bypass in which the internal mammary has been used. Once the costal margin is reached, the dissection is completed, and the muscle is folded on itself into the lower portion of the sternal defect (Fig. 11-6, *D*). The rectus

A

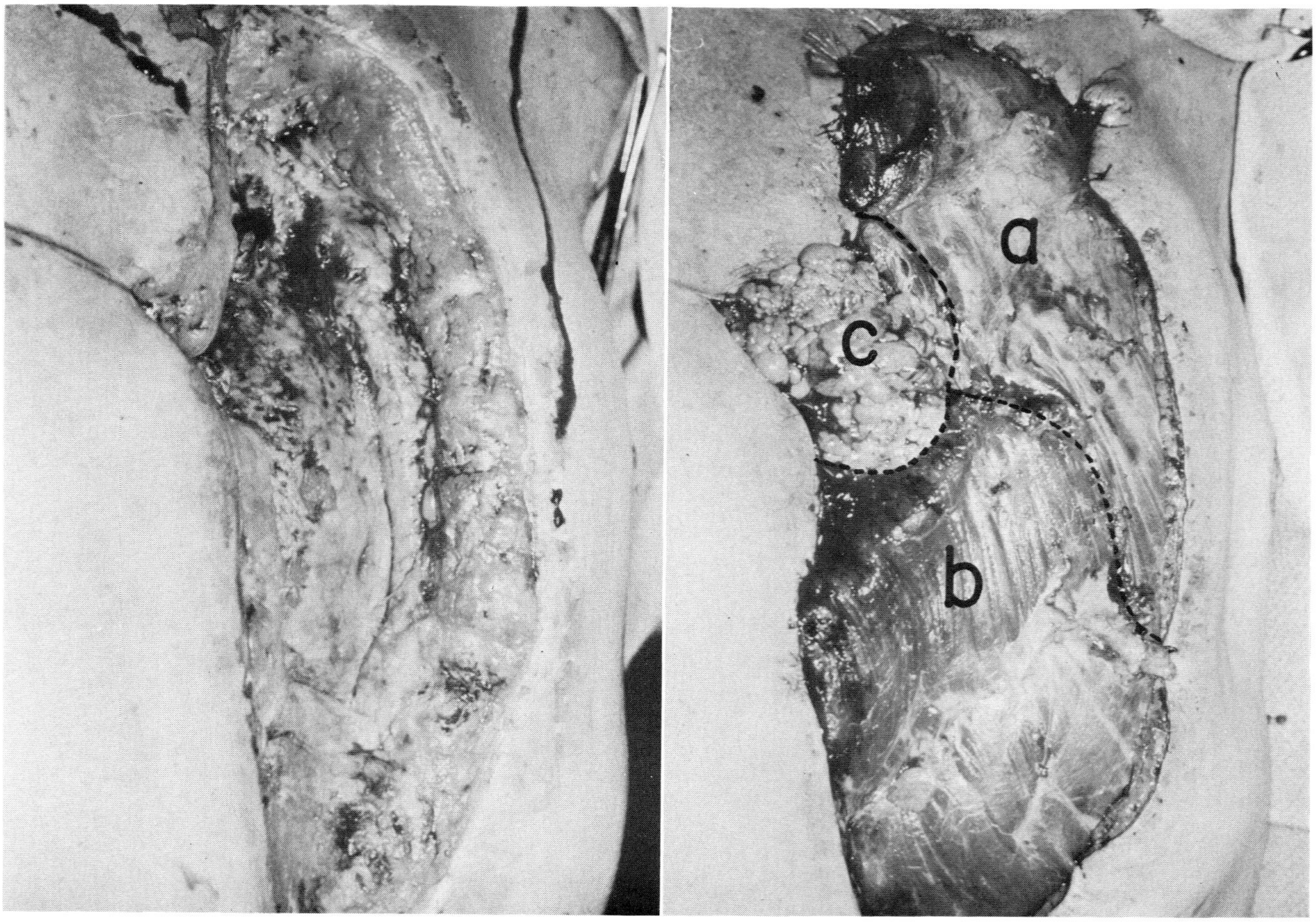

B

Fig. 11-7

A, Thirty-four-year-old man 1 month after five-vessel coronary artery bypass. Patient developed sternal infection with dehiscence and on seventh postoperative day underwent sternal débridement with closure over irrigation catheters with continuous Betadine irrigation. Cultures grew *Staphylococcus epidermidis*. Patient also developed *Candida* septicemia, and by third postoperative week sternum was again unstable, and patient required extensive débridement. Sternal wound was packed open. On twenty-eighth postoperative day patient bled from right ventricle and one of coronary artery bypass grafts. He was taken immediately to operating room where hemorrhage was controlled.

B, Massive defect was closed with combination of flaps including greater omentum (*c*), rectus abdominis muscle (*b*), and left pectoralis major muscle (*a*). Because of previous procedures right pectoralis major muscle was not available. Muscles and omentum were covered with split-thickness skin graft.

abdominis muscle is sutured to the pectoralis major muscles with absorbable suture. The anterior rectus sheath is then closed in an appropriate fashion.

After the muscles are sutured in position, large-bore drainage catheters are inserted into the mediastinum and subcutaneously. The skin is closed primarily in the majority of cases (Fig. 11-2, *F*). However, if inadequate skin is present, the muscles can be covered with split-thickness skin grafts (Fig. 11-7).

After the operation the patient is maintained on appropriate antibiotic coverage with the drainage catheters connected to continuous suction. The subcutaneous drain is removed, usually by the second

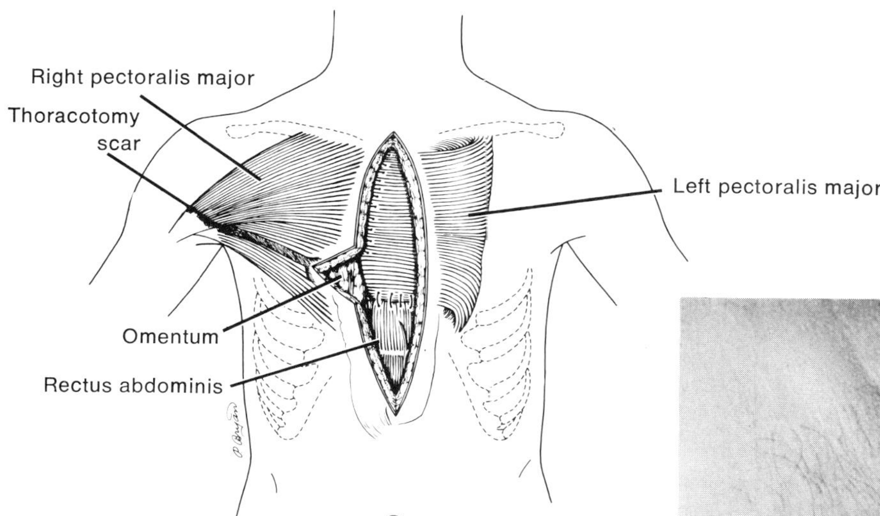

C

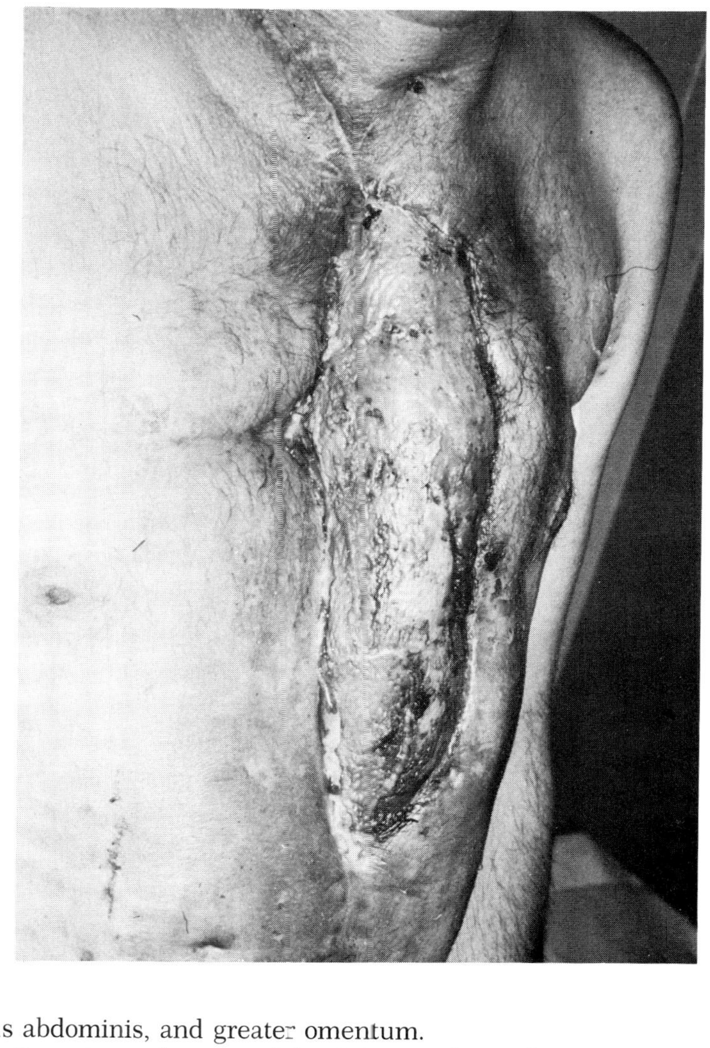

D

Fig. 11-7, cont'd
C, Closure with pectoralis major, rectus abdominis, and greater omentum.
D, Patient after closure with left pectoralis major and rectus abdominis muscles and greater omentum covered with split skin graft.

postoperative day, while the catheters beneath the muscle flaps remain until the third or fourth postoperative day. Depending on the patients' status, he or she is mobilized from bed as early as tolerated.

RESULTS

The incidence of significant infection involving the median sternotomy incision is fortunately low. However, infections involving this incision can result in excessive morbidity and mortality if not treated aggressively and successfully. In most patients adequate débridement followed by closed continuous antibiotic irrigation can produce a successful outcome. Previously those patients not responding to the closed irrigation technique were managed by open packing with its associated high morbidity and prolonged hospitalization. In these unfortunate patients, who were recalcitrant to conservative therapy or had disease too extensive for closure, we have had excellent results using the described muscle flap procedure.

In over 35 patients morbidity and mortality have been low. Specifically one patient died as a result of mediastinal infection despite adequate débridement and muscle flap closure. This death was attributed to multiple system failure. In two patients significant donor site hematomas were encountered, requiring immediate exploration and evacuation. In both patients the bleeding was in the pectoralis major donor site.

Following débridement and closure with muscle flaps there is an immediate improvement in the patient's overall status. Specifically pulmonary function was dramatically improved, and patients previously on mechanical ventilation were extubated early in the postoperative course. This is attributed to increased stability of the chest wall.

The shortened period of hospitalization is another factor associated with the muscle flap technique. Patients treated previously by more conservative methods averaged 2 months or more of hospitalization after the diagnosis of sternal dehiscence. In over 35 patients in whom muscle flap coverage was performed, the average hospitalization from time of procedure to discharge was approximately 10 days.

The technique of muscle flap closure should be available to all surgeons involved with the devastating complications of mediastinal wound dehiscence and infection after open heart surgery. The type V pattern of blood supply to the pectoralis major enables the reconstructive surgeon to design flaps to suit the reconstructive need of a given defect (Fig. 11-8).

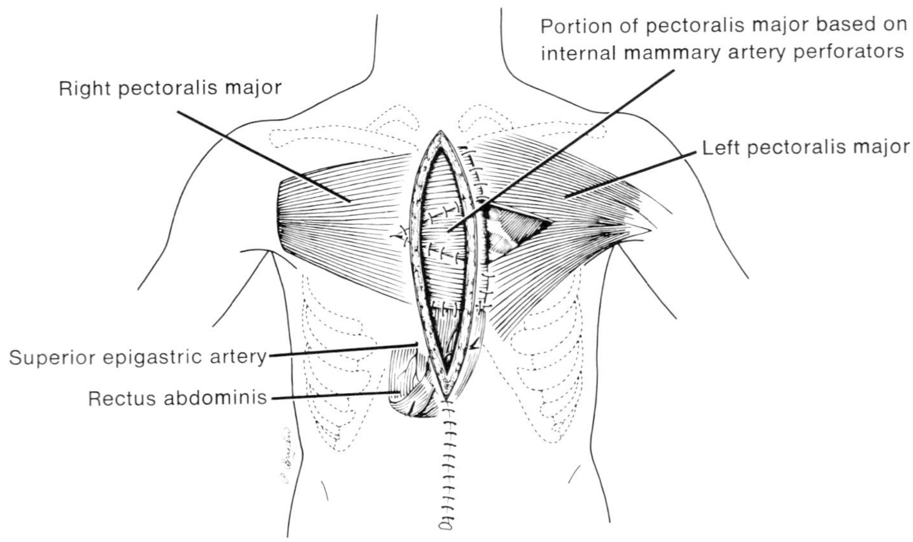

Fig. 11-8. Multiple possible variations in reconstruction of sternal defect. Small portion of left pectoralis major muscle is transposed on its internal mammary perforating vessels and interdigitated with contralateral pectoralis major muscle based on its thoracoacromial vessel. Rectus abdominis flap is used to cover inferior part of defect.

ANNOTATED BIBLIOGRAPHY

1. Bryant, L.R., Spencer, F.C., and Trinkle, J.K.: Treatment of median sternotomy infection by mediastinal irrigation with an antibiotic solution, Ann. Surg. **169**(6):914, 1969.

   Report of five patients with median sternotomy infections who were treated with débridement and primary closure followed by continuous antibiotic irrigation is discussed. Four of the five patients survived. A protocol for managing these patients is presented.

2. Cerat, G.A., McHenry, M.C., and Loop, F.D.: Median sternotomy wound infection and anterior mediastinitis caused by *Bacteroides fragilis,* Chest **69:**2, 1976.

   First report of Bacteroides fragilis causing a median sternotomy infection and mediastinitis is presented.

3. Culliford, A.T., and co-workers: Sternal and costochondral infections following open-heart surgery, J. Thorac. Cardiovasc. Surg. **72**(5):714, 1976.

   This article is a review of a large series of median sternotomy infections. The associated predisposing factors such as prolonged perfusion time and postoperative bleeding are presented. Most cases are associated with staphylococcal infections.

4. Engleman, R.M., and co-workers: Mediastinitis following open heart surgery, Arch. Surg. **107:**772,1973.

   A series of 15 patients treated with mediastinal irrigations with neomycin sulfate-bacitracin is presented. Ten patients survive. Associated factors and management are discussed.

5. Grmoljez, P.F., and co-workers: Major complications of median sternotomy, Am. J. Surg. **130:**679, 1975.

   Fever, pain and purulent drainage are the primary signs of median sternotomy infection. As in other series Staphylococcus aureus is the most common organism. These patients are treated by the open technique.

6. Jimenez-Martinez, M., and co-workers: Anterior mediastinitis as a complication of median sternotomy incisions: diagnostic and surgical considerations, Surgery **67:**929, 1970.

   Closed irrigation with antibiotic solutions is discussed. Early surgical intervention leads to an excellent salvage rate.

7. Julian, O.C., and co-workers: The median sternal incision in intracardiac surgery with extracorporeal circulation, Surgery **42:**753, 1957.

   Early report of the use of the median sternotomy in cardiac surgery is presented.

8. Jurkiewicz, M.J., and Arnold, P.G.: The omentum: an account of its use in reconstruction of the chest wall, Ann. Surg. **185:**548, 1977.

   This article presents the usefulness of omentum in difficult reconstructive problems of the chest wall.

9. Jurkiewicz, M.J., and co-workers: Infected median sternotomy wound: successful treatment by muscle flap, Ann. Surg. **191:**738, 1980.

   This article presents the use of muscle flaps for the treatment of patients with infected median sternotomies recalcitrant to débridement and closed irrigations. The pectoralis major and rectus abdominis muscles are used with excellent results. Both increased survival and decreased length of hospitalizations are reported.

10. Lee, A.G., Schimert, G., and Shatkin, S.: Total excision of the sternum and thoracic pedicle transposition of the greater omentum: useful stratagems in managing severe mediastinal infection following open heart surgery, Surgery **80:**433, 1976.

    Omentum based on the gastroepiploic vessels is used to fill extensive mediastinal defects.

11. MacManus, Q., and Okies, J.E.: Mediastinal wound infection and aortocoronary graft patency, Am. J. Surg. **132:**558, 1976.

    This article is a review of graft patency in patients after mediastinal wound infections.

12. Sanfelippo, P.M., and Danielson, G.K.: Complications associated with sternotomy, J. Thorac. Cardiovasc. Surg. **63**(3):419, 1972.

    Further data on the problems associated with median sternotomy are presented.

13. Thurer, R.J., and co-workers: The management of mediastinal infection following cardiac surgery, J. Thorac. Cardiovasc. Surg. **68**(6):962, 1974.

    This article is an important series in which povidone-iodine is used for closed sternal wound irrigation. Povidone-iodine is both bactericidal and fungicidal with low toxicity. Excellent salvage in this patient group is possible.

14. Wray, T.M., Bryant, R.E., and Killen, D.A.: Sternal osteomyelitis and costochondritis after median sternotomy, J. Thorac Cardiovasc. Surg. **65:**227, 1973.

    Series of patients with median sternotomy infections in which Pseudomonas aeruginosa is involved in four of the five patients are presented. A review of the literature is included.

12

COMPLICATIONS

Foad Nahai ■ Stephen J. Mathes

ANTERIOR CHEST

Muscle flaps have facilitated the reconstruction of difficult chest wounds. Complications specific to this area may occur.

PECTORALIS MAJOR

Planning errors

Patients undergoing chest wall reconstruction may have underlying respiratory or cardiac insufficiency. Preoperative evaluation and stabilization of these problems are essential. Cardiopulmonary bypass should be available for reconstruction of mediastinal defects. Primary bone grafting after débridement of infected sternal wounds should be avoided.

The pectoralis major may be based either on its dominant vascular pedicle from the thoracoacromial axis or on its secondary segmental blood supply for the internal mammary artery. Based on its secondary segmental blood supply from the internal mammary artery, the flap will cover a great surface area of the mediastinum. However, damage to internal mammary vessels by previous surgery or the costochondral infection may necessitate transposition of the muscle based on the thoracoacromial artery. Alternative flaps are generally required for closure of inferior defects of the sternum (e.g., rectus abdominis and omentum).

Technical errors

The thoracoacromial pedicle may be damaged in mobilizing the clavicular portion of the pectoralis major. This portion of the dissection is performed under direct vision. Inadequate débridement of infected cartilage will lead to recurrent infection. If the pleural space is entered, closed-tube thoracostomy is required. Injury to the heart, great vessels, and recent coronary vein grafts must be avoided. A significant blood loss is associated with wound débridement and flap elevation. Adequate volume replacement and massive intravascular fluid shifts must be avoided in those patients with prior cardiac compromise.

Postoperative errors

Appropriate cardiopulmonary support is essential. Mechanical respiratory support is not generally necessary postoperatively because of improved chest wall stability. Use of soft Silastic suction drains within the mediastinum is recommended to avoid great vessel or myocardial wall erosion. Postoperative antibiotic coverage is necessary to avoid mediastinal abscess formation.

RECTUS ABDOMINIS

Planning errors

Patency of the internal mammary artery is essential for use of the rectus abdominis in chest reconstruction. If muscle is elevated below the semicircular line of Douglas, there is the potential for hernia formation.

Technical errors

The superior epigastric artery enters the rectus abdominis muscle beneath the costal margin. Blunt dissection is avoided during separation of the superior muscle from the posterior rectus sheath to avoid injury to the vascular pedicle. The anterior rectus sheath should always be preserved and reapproximated to avoid the potential formation of an abdominal hernia. The posterior rectus sheath is never elevated with this flap.

Postoperative errors

Suction drainage within the rectus sheath is necessary.

LATISSIMUS DORSI

Planning errors

The latissimus dorsi will reach the anterior chest, but it is the most distal portion of the flap that will be used for mediastinal coverage. It should never be used if there is a previous thoracotomy scar, because the distal muscle will not survive. Similarly, if the muscle has been previously denervated, its bulk may not be adequate for chest wall reconstruction.

Technical errors

The thoracodorsal vascular pedicle may be damaged during proximal dissection if not performed under direct vision. The latissimus dorsi muscle must be separated from the underlying serratus anterior muscle during flap elevation.

Postoperative errors

Suction drainage is essential for donor defect.

13

A SYSTEMATIC APPROACH TO FLAP SELECTION

Stephen J. Mathes ■ Foad Nahai

TOTAL (SUBCUTANEOUS) MASTECTOMY (Fig. 13-1)

Skin grafts

Skin grafts are not applicable in reconstruction of the breast after total mastectomy

Local flaps

1. Pectoralis major muscle—the submuscular placement of the implant under this muscle is method of choice (Figs. 13-2 and 13-3)
2. Serratus anterior—same as above

Distant flaps (Figs. 13-2 to 13-4)

1. Latissimus dorsi musculocutaneous flap is flap of choice for salvage of the complicated total mastectomy wound (Fig. 13-4)
2. Rectus abdominis musculocutaneous flap is alternative flap
3. Thoracoepigastric flap is alternative flap

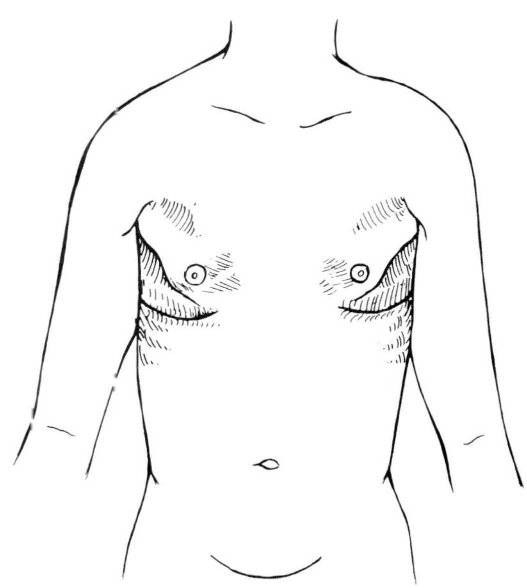

Fig. 13-1. Total (subcutaneous) mastectomy defect.

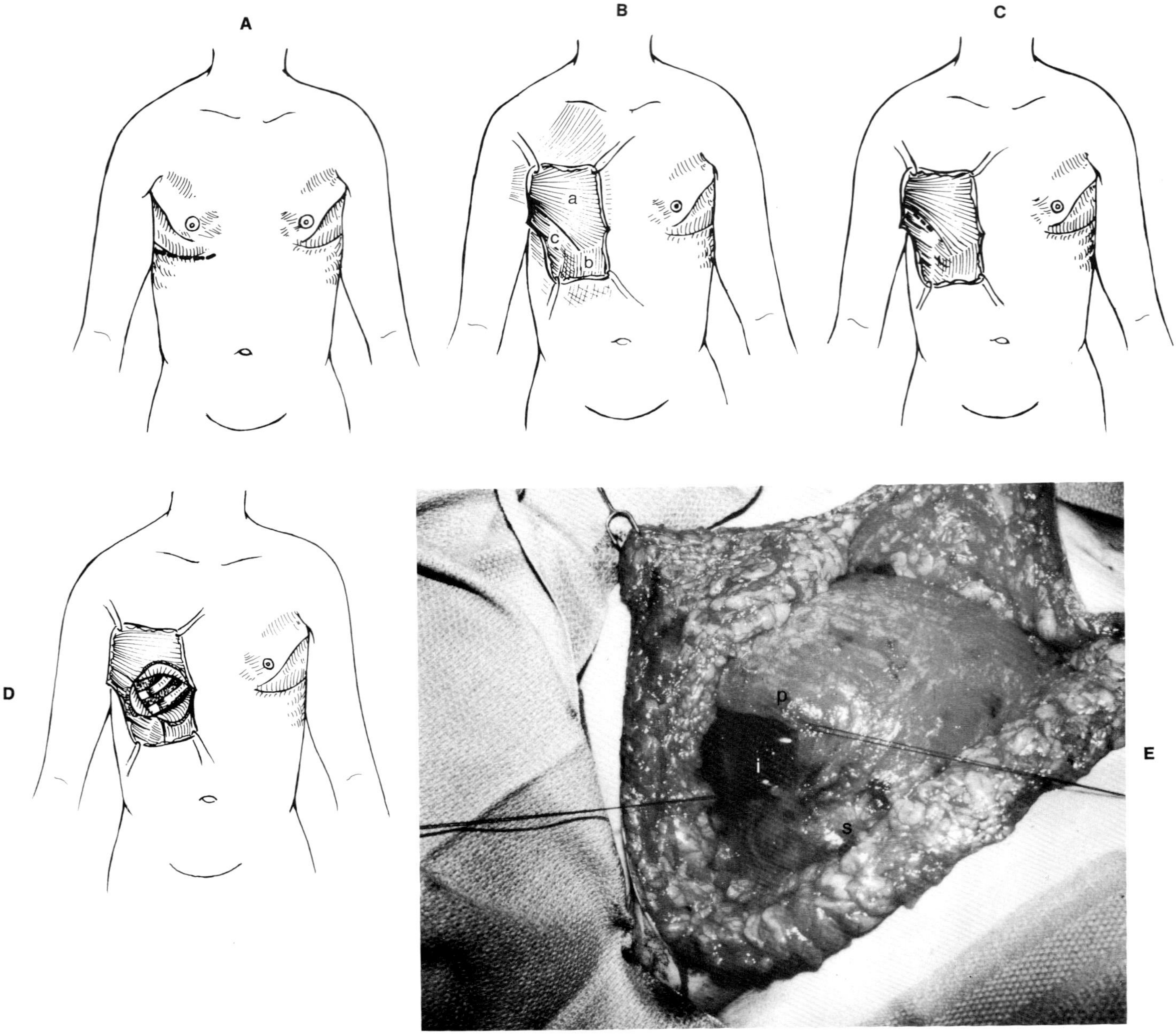

Fig. 13-2

A, Postoperative defect following bilateral total subcutaneous mastectomy. Immediate breast reconstruction performed through inframammary incision.

B, Silicone breast implant coverage provided by pectoralis major *(a)*, rectus abdominis *(b)*, and serratus anterior *(c)* muscles.

C, Serratus anterior and pectoralis major muscle fibers split near axilla for exposure of submuscular space.

D and **E,** Submuscular pocket elevated. Fibers of origin of pectoralis major from fifth and sixth ribs divided. *p,* Pectoralis major; *s,* serratus anterior; *i,* implant.

F

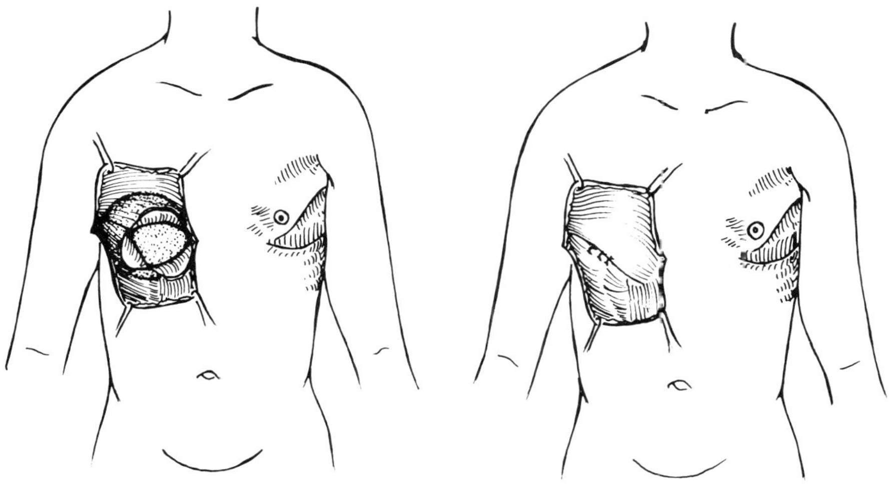

G

H

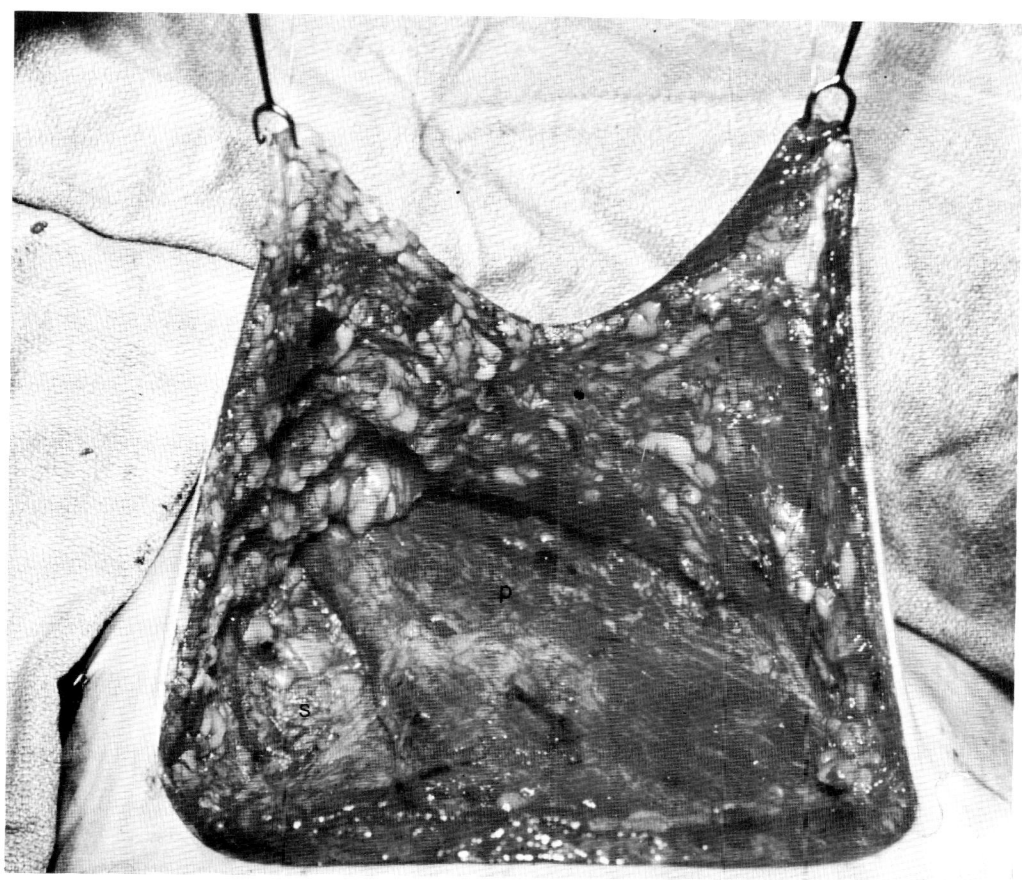

Fig. 13-2, cont'd

F, Silicone implant inserted beneath muscles with adequate pocket to ensure normal breast position.

G and **H,** Muscle closed over implant and preserved breast skin closed directly. *p,* Pectoralis major; *s,* serratus anterior.

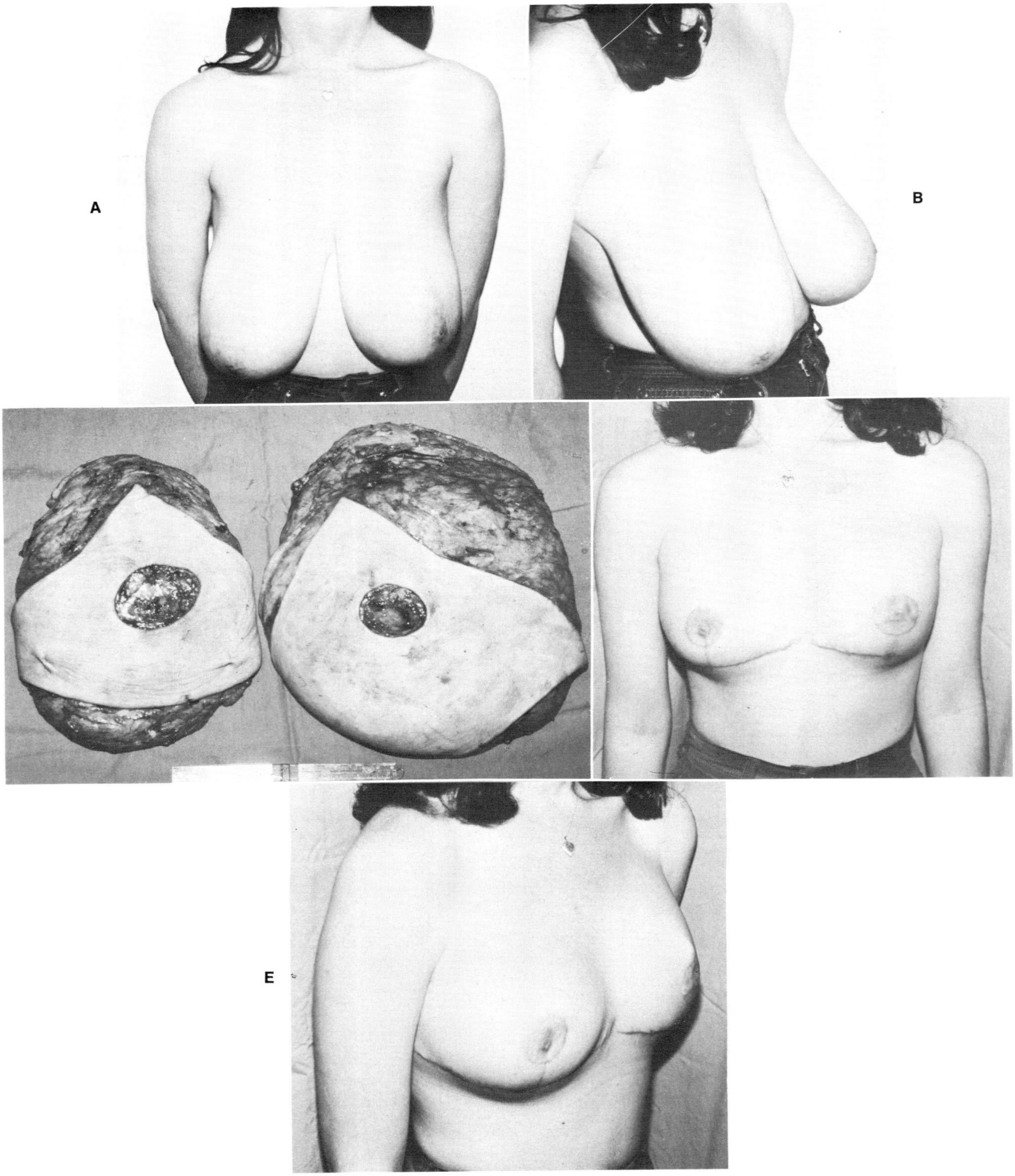

Fig. 13-3. Pectoralis major and serratus anterior muscle flaps.

A, Bilateral macromastia. Breast growth controlled only by estrogen antagonists.

B, Oblique view of bilateral breast macromastia.

C, Specimens from bilateral subcutaneous mastectomies.

D, Immediate breast reconstruction after bilateral total mastectomy incorporating placement of silicone implants beneath pectoralis major, serratus anterior, and anterior rectus sheath.

E, Six-month postoperative oblique view of bilateral breast reconstruction after subcutaneous mastectomy with nipple-areola grafts.

A

B

C

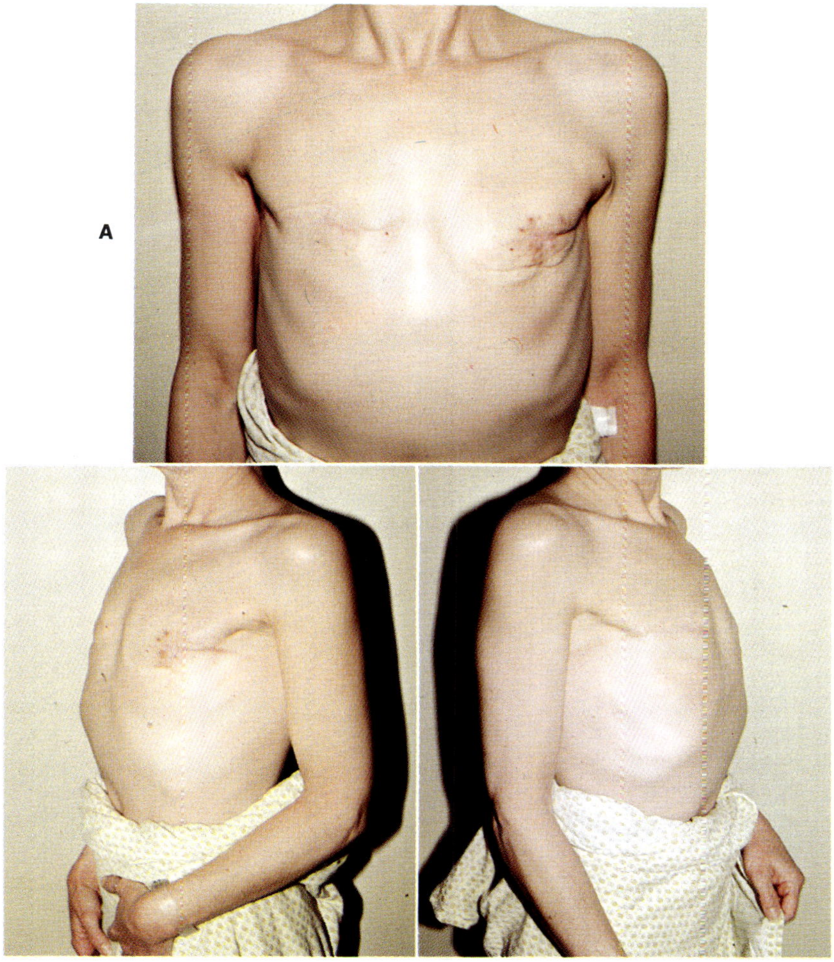

Fig. 13-4. Bilateral latissimus dorsi musculocutaneous flaps.
A to **C,** Bilateral total mastectomy.

Continued.

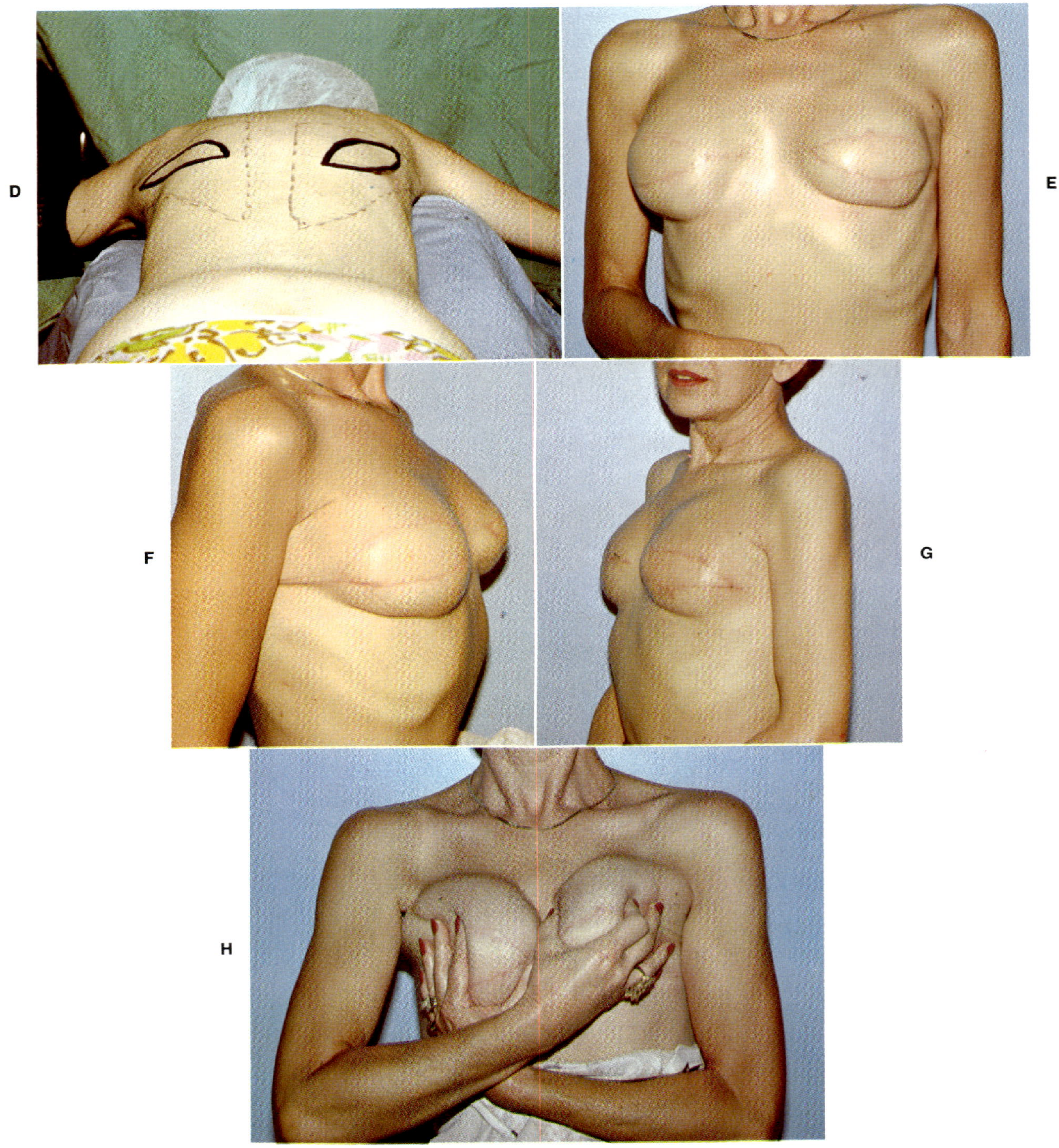

Fig. 13-4, cont'd
D, Bilateral latissimus dorsi flaps outlined.
E to **H,** One-year postoperative views.

MODIFIED RADICAL MASTECTOMY (Fig. 13-5)

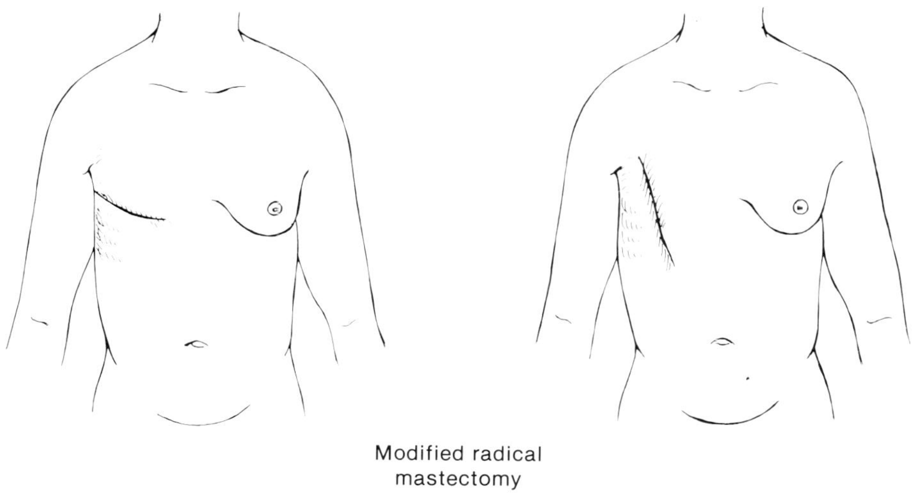

Fig. 13-5

Skin grafts

Skin grafts are not applicable in reconstruction of the breast after modified radical mastectomy

Local flaps

1. Pectoralis major muscle—the submuscular placement of implant
   a. This is the method of choice for immediate reconstruction
   b. Skin scar contracture and denervated inferior fibers of pectoralis major limit size, form, and position of resultant breast in late reconstruction (Fig. 13-6)
2. Serratus anterior—same as above

Distant flaps

1. Latissimus dorsi musculocutaneous flap is flap of choice for breast reconstruction
   a. Flap allows release of tight denervated fibers of inferior pectoralis major muscle with transposed latissimus dorsi muscle (Fig. 13-6)
   b. Flap allows release of tight skin scars with transposed skin island with latissimus dorsi muscle (Figs. 13-7 to 13-9)
2. Thoracoepigastric axial flap is flap of choice for immediate coverage of mastectomy defect when postoperative chest radiation therapy is planned
3. Rectus abdominis musculocutaneous flap is alternative flap for breast reconstruction (Figs. 13-10 and 13-11)
4. Free tissue transplantation is rarely indicated

A

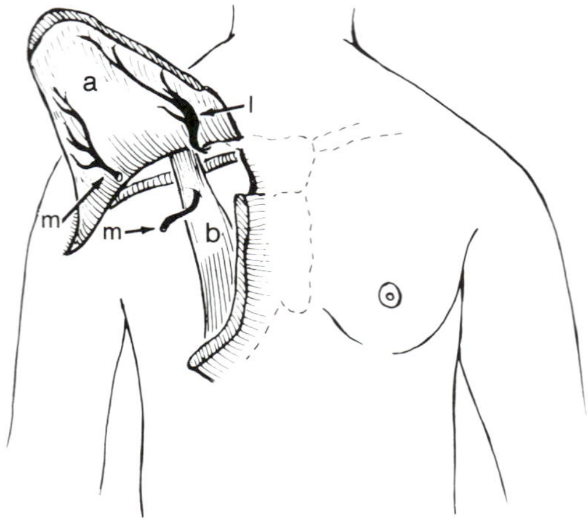

B

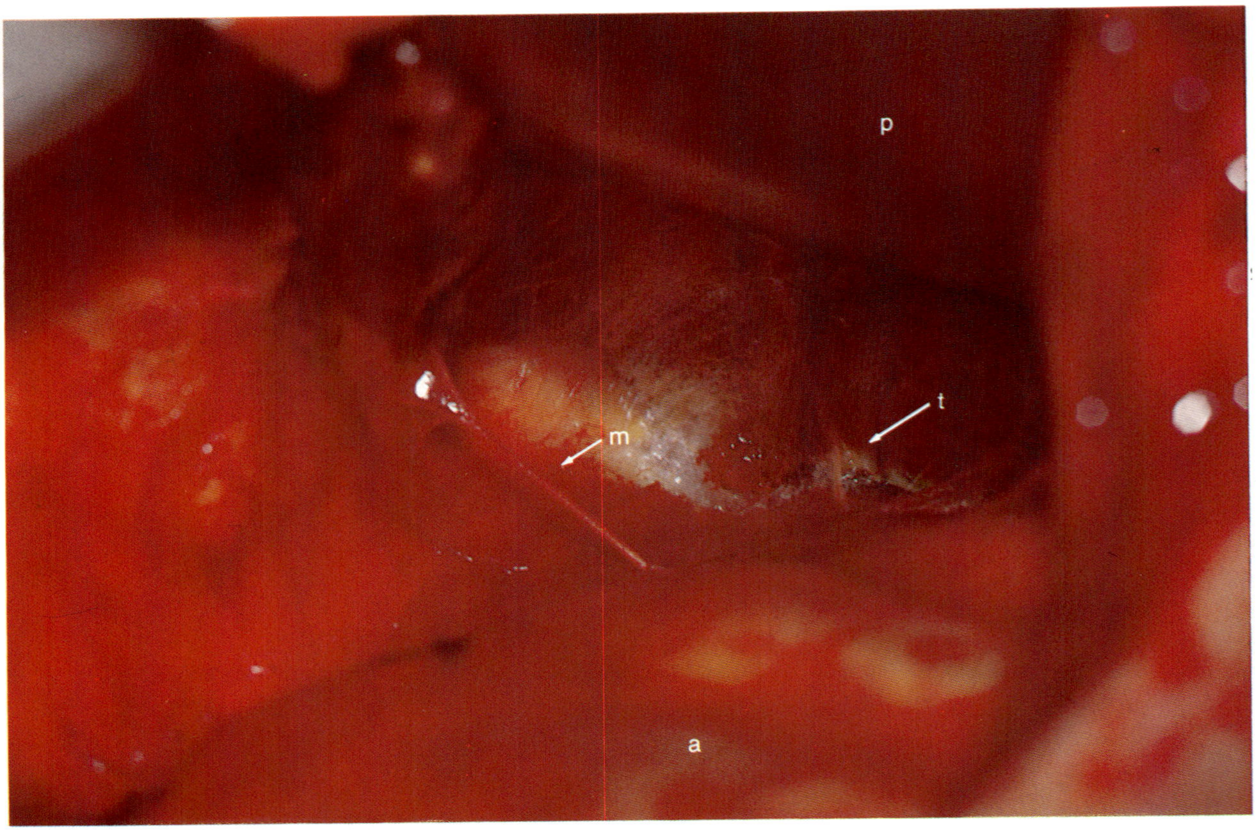

Fig. 13-6

A, During modified radical mastectomy division of medial pectoral nerve is common during the intermuscular dissection between pectoralis major and pectoralis minor. *a,* Pectoralis major elevated medially; *b,* pectoralis minor; *m,* medial pectoral nerve at site of frequent division during modified radical mastectomy; *l,* lateral pectoral nerve.

B, Intermuscular course of medial pectoral nerve results in frequent division during modified radical mastectomy. *p,* Retractor elevation of pectoralis major muscle; *a,* anterior chest wall; *m,* medial pectoral nerve coursing between pectoralis minor and lateral fibers of pectoralis major muscle; *t,* thoracoacromial vascular pedicle (location of lateral pectoral nerve).

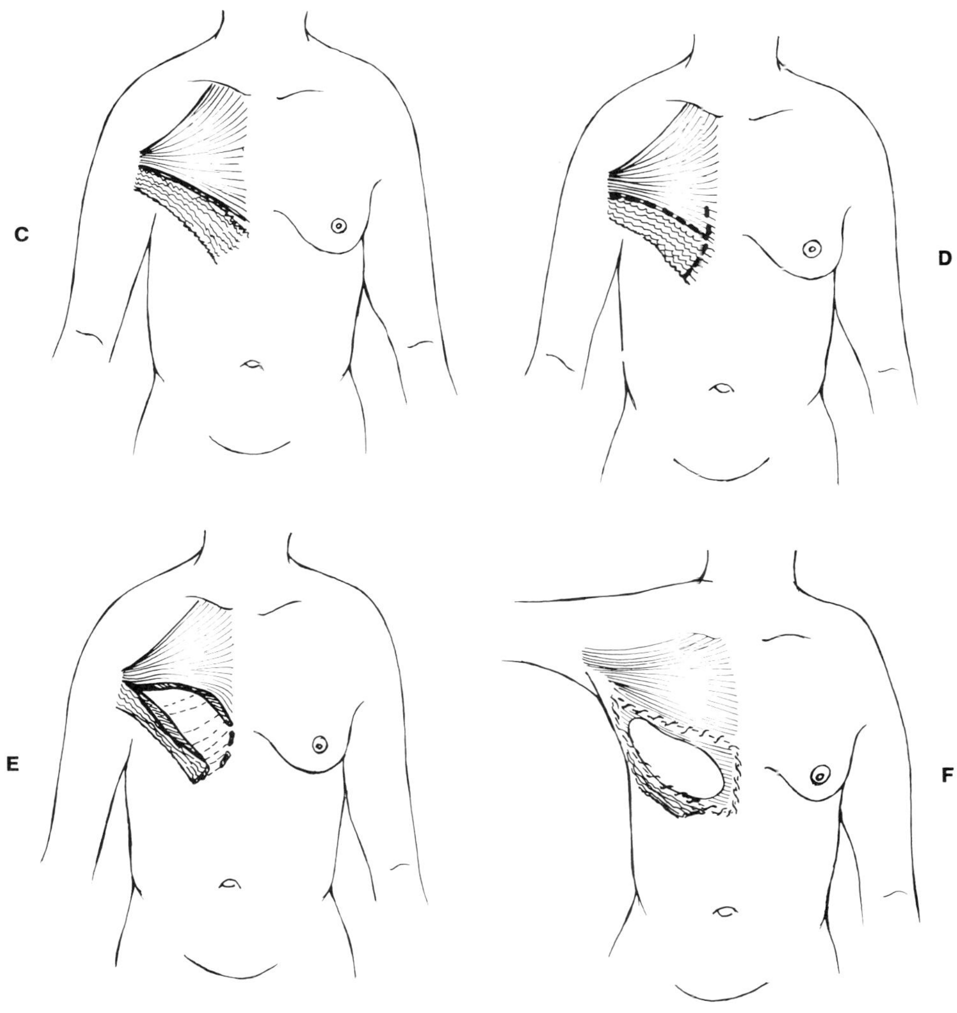

Fig. 13-6, cont'd

C, Postoperative deformity after modified radical mastectomy may include scar contracture of denervated lateral fibers of pectoralis major muscle.

D, Reconstruction of adequate breast projection with silicone implant requires release of tight denervated fibers.

E, Release of denervated lateral pectoral fibers allows restoration of breast projection.

F, Muscle coverage of implant and release of tight chest wall skin provided by transposition of latissimus dorsi musculocutaneous flap.

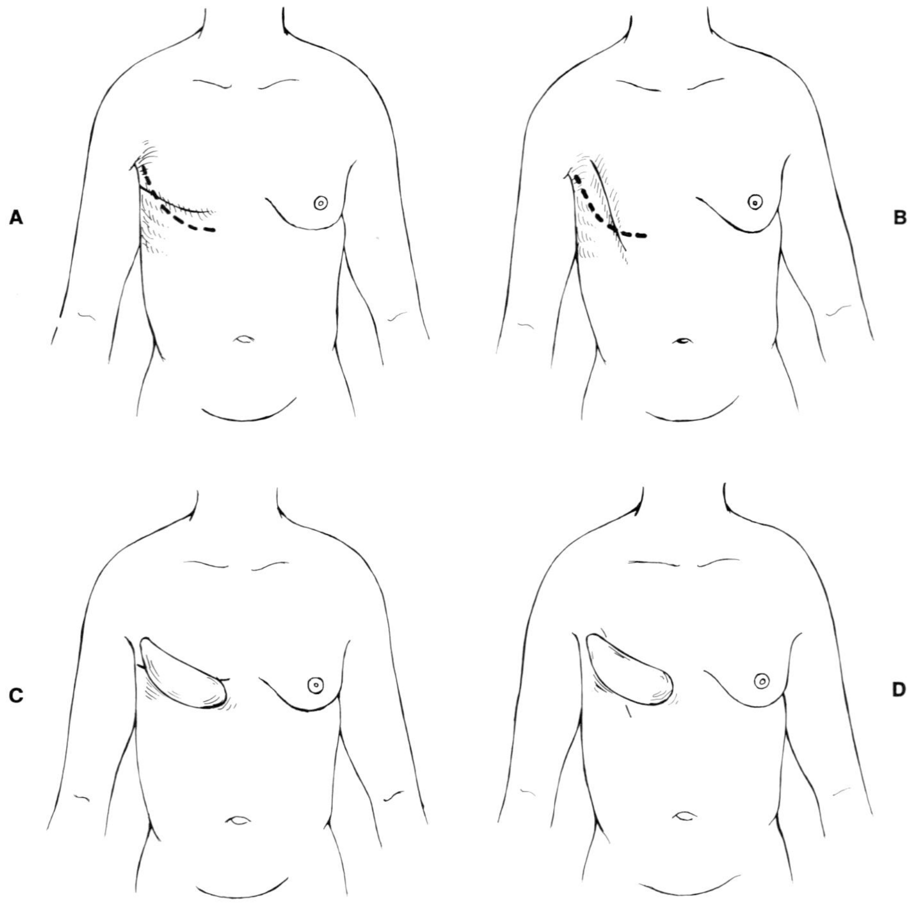

Fig. 13-7

A, Postoperative transverse scar after modified radical mastectomy. Dotted line represents site of planned flap inset.

B, Postoperative vertical scar after radical mastectomy. Dotted line represents site of planned flap inset.

C and **D,** Inset of skin island of latissimus dorsi musculocutaneous flap at inframammary line both releases transverse scar contracture and allows breast projection.

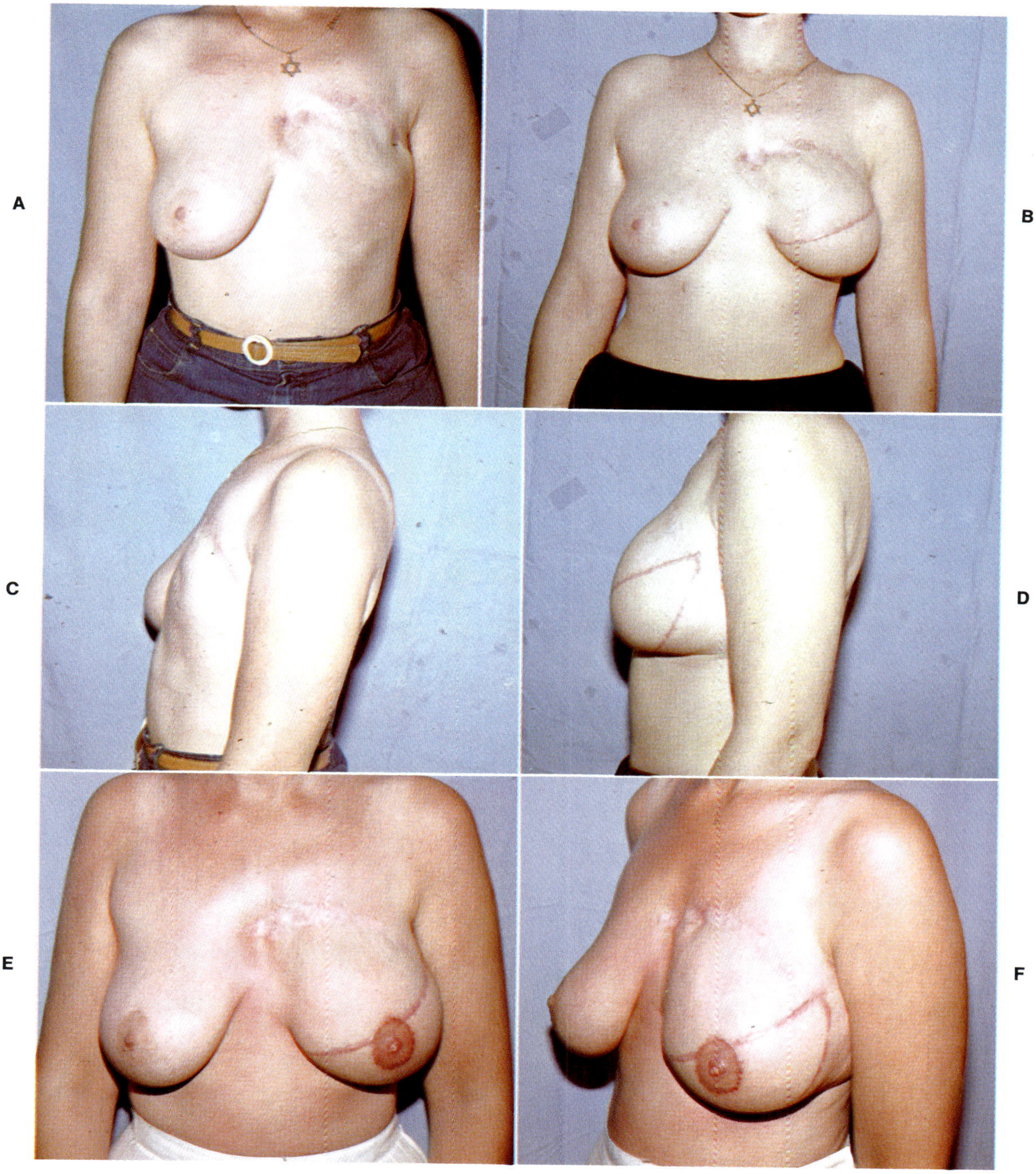

Fig. 13-8. Latissimus dorsi musculocutaneous flap.

A, Postoperative defect after left modified radical mastectomy. NOTE: High position of transverse scar.

B, Six-month postoperative view shows left breast reconstruction with use of latissimus dorsi musculocutaneous flap. NOTE: Flap inset at inframammary line to simulate contralateral breast ptosis.

C, Preoperative lateral view of left modified radical mastectomy defect. NOTE: At time of exploration inferior pectoralis major fibers which had been denervated resulted in tight scar contracture.

D, Six-month postoperative view of left modified radical mastectomy breast reconstruction. Natural symmetry provided despite superior placement of mastectomy scar.

E and **F,** One-year postoperative view of nipple-areola reconstruction completed.

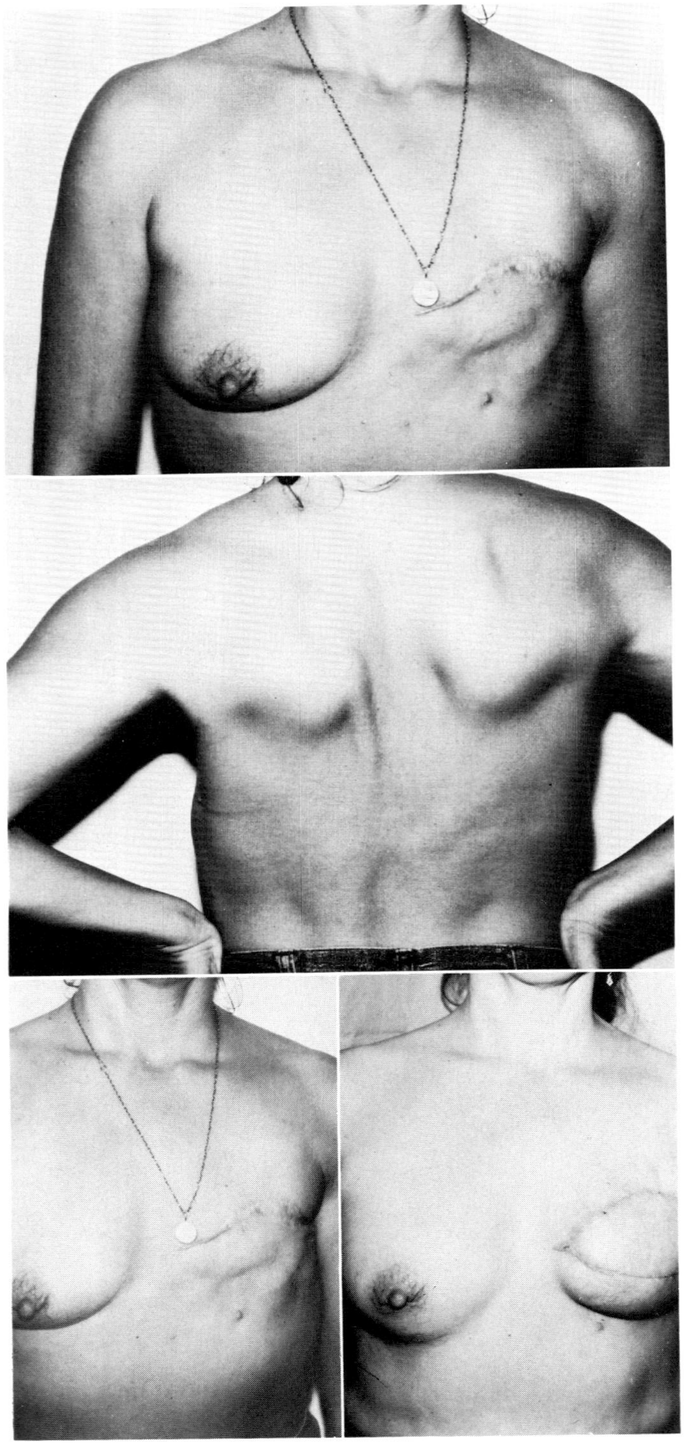

A

B

C

Fig. 13-9. Latissimus dorsi musculocutaneous flap.
A, Defect after modified radical mastectomy.
B, Preoperative evaluation demonstrates denervation of left latissimus dorsi muscle.
C, Preoperative and 6-month postoperative anterior views after transposition of denervated left latissimus dorsi musculocutaneous flap. Flap is based on vascular communications through the serratus branch of the thoracodorsal artery. Flap is inset at site of old incision.

D

E

F

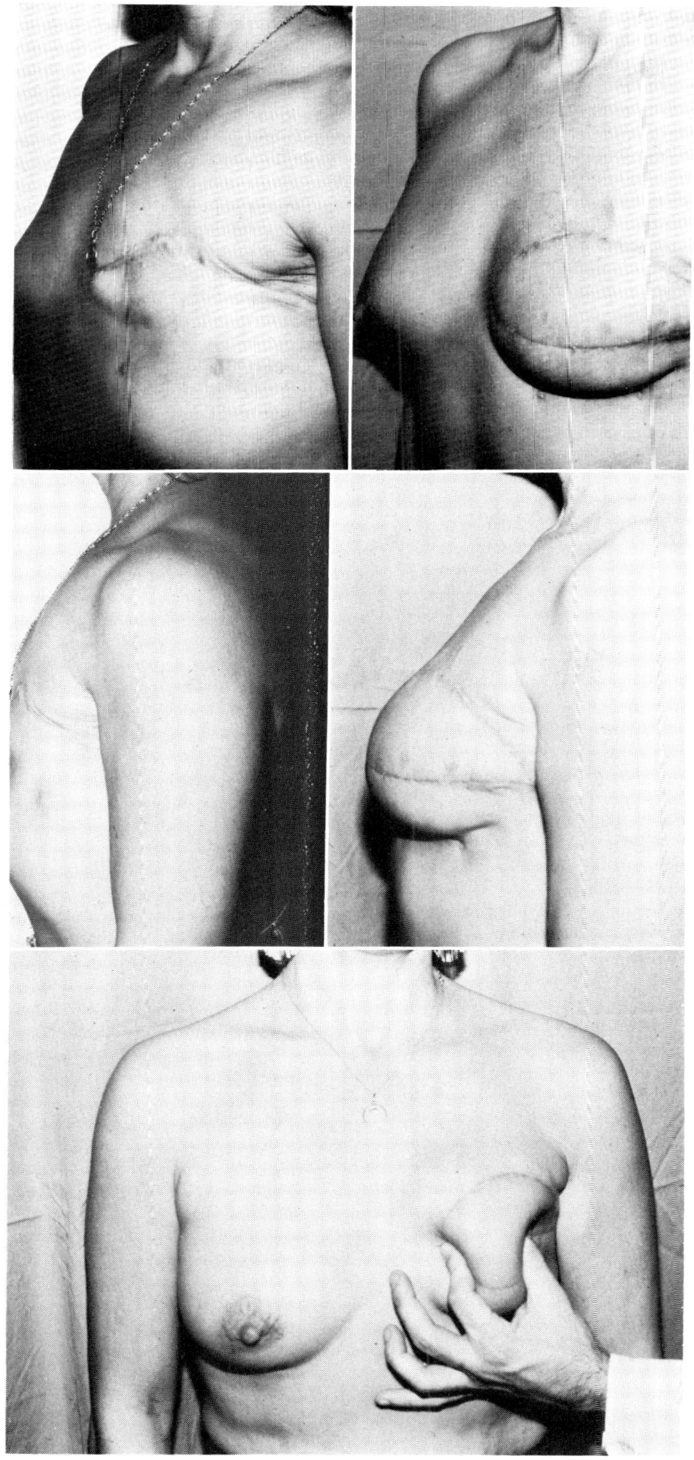

Fig. 13-9, cont'd

D, Oblique preoperative and 6-month postoperative views after left breast reconstruction.

E, Preoperative lateral and 6-month postoperative lateral views of left breast reconstruction with latissimus dorsi musculocutaneous flap. NOTE: Adequate breast projection is achieved with flap inset despite prior ligation of major vascular pedicle to latissimus dorsi musculocutaneous flap.

F, Six-month postoperative view demonstrates soft breast despite thin coverage of denervated latissimus dorsi muscle.

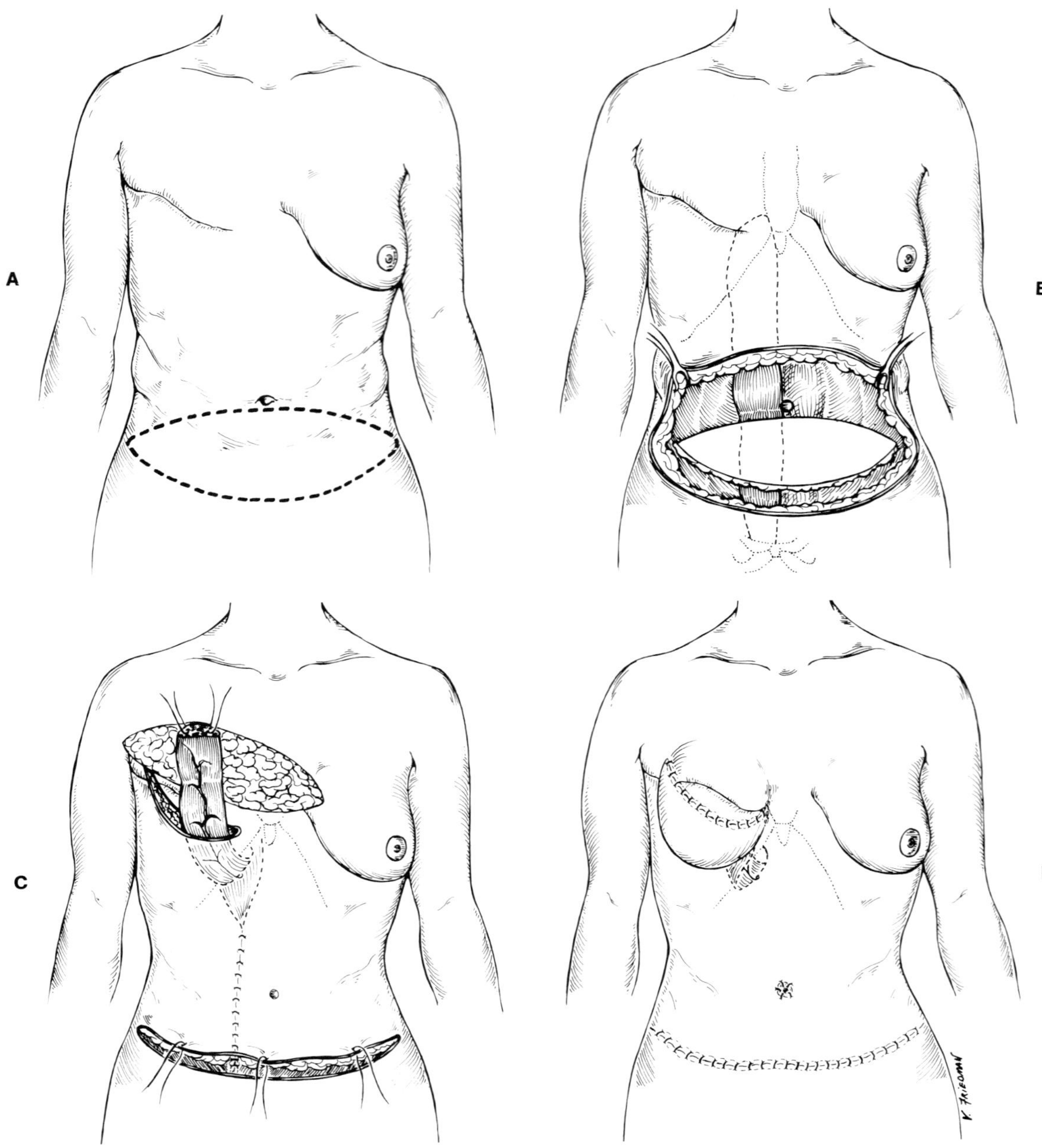

Fig. 13-10. Rectus abdominis musculocutaneous flap (transverse inferior skin island technique of Hartrampf and Scheflan).

A, Postoperative deformity after modified radical mastectomy. Design of inferior skin island for breast reconstruction in patient with redundant lower abdominal wall skin.

B, Inferior skin island elevated from abdominal fascia preserving vascular connections to rectus abdominis muscle on the side of the breast defect. Superior abdominal wall skin undermined in the supra-aponeurotic plane to level of costal margins.

C, Release of muscle origin at pubis allows transposition of flap beneath superior abdominal wall skin into inframammary incision. NOTE: Medial and lateral edges of anterior rectus sheath at flap donor site are preserved for closure to avoid hernia defect.

D, Flap is inset with the skin directly over the rectus muscle used to replace missing breast skin. Excessive skin from the random side of the flap is deepithelized and folded beneath the breast. Flap subcutaneous tissue is used to form the breast mound. The donor site is closed using the standard technique for abdominoplasty.

E

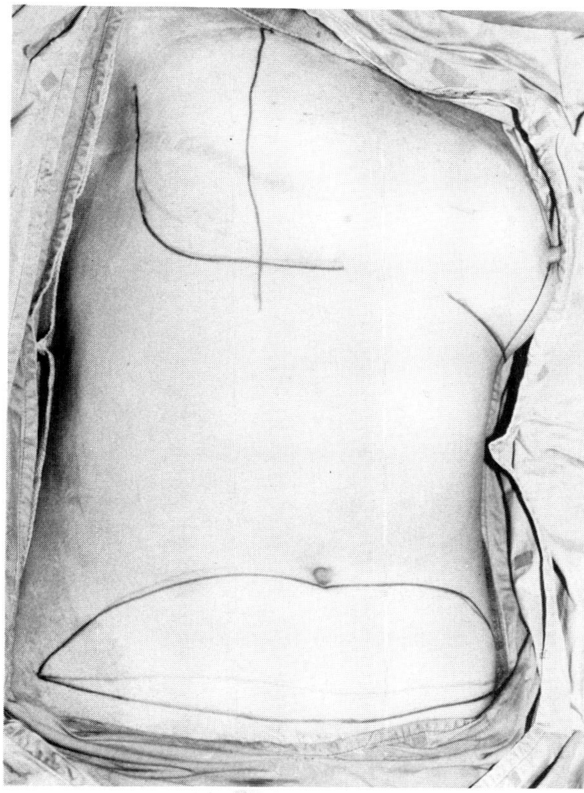

F

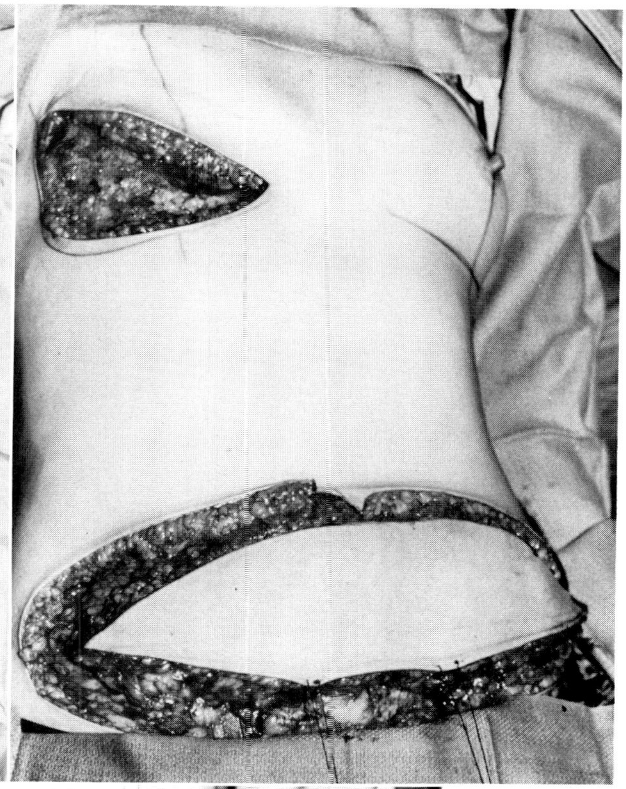

G

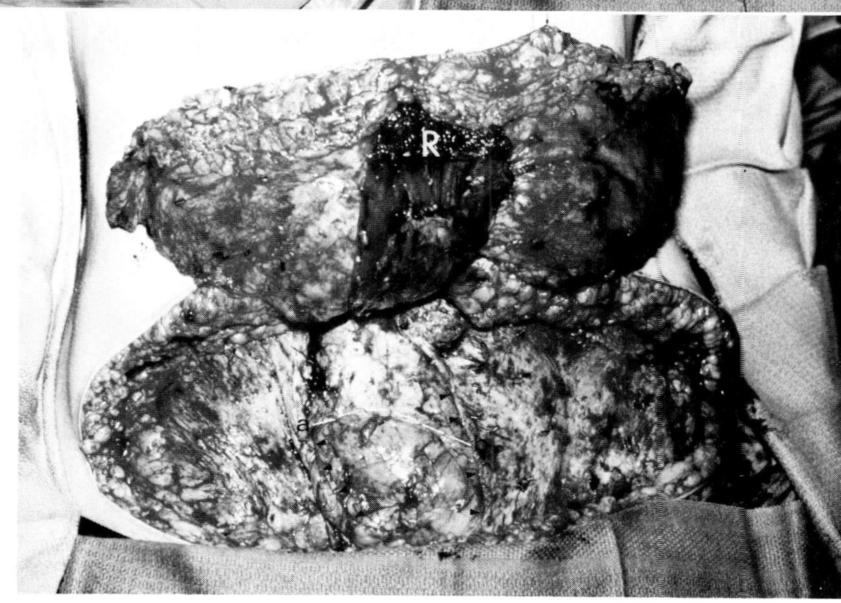

Fig. 13-10, cont'd

E, Postoperative right modified radical mastectomy. Design of inferior transverse skin island for breast reconstruction.

F, Superior abdominal wall skin elevated to costal margins for later advancement to pubis for abdominoplasty closure of donor defect. Inferior skin island vascular attachments to right rectus muscle carefully preserved.

G, Rectus abdominis musculocutaneous flap ready for transposition to chest wall beneath superior abdominal skin. *R,* Right rectus abdominis muscle; *line a-b,* semicircular line of Douglas; arrows denote preserved edge of anterior sheath allowing closure of fascial defect.

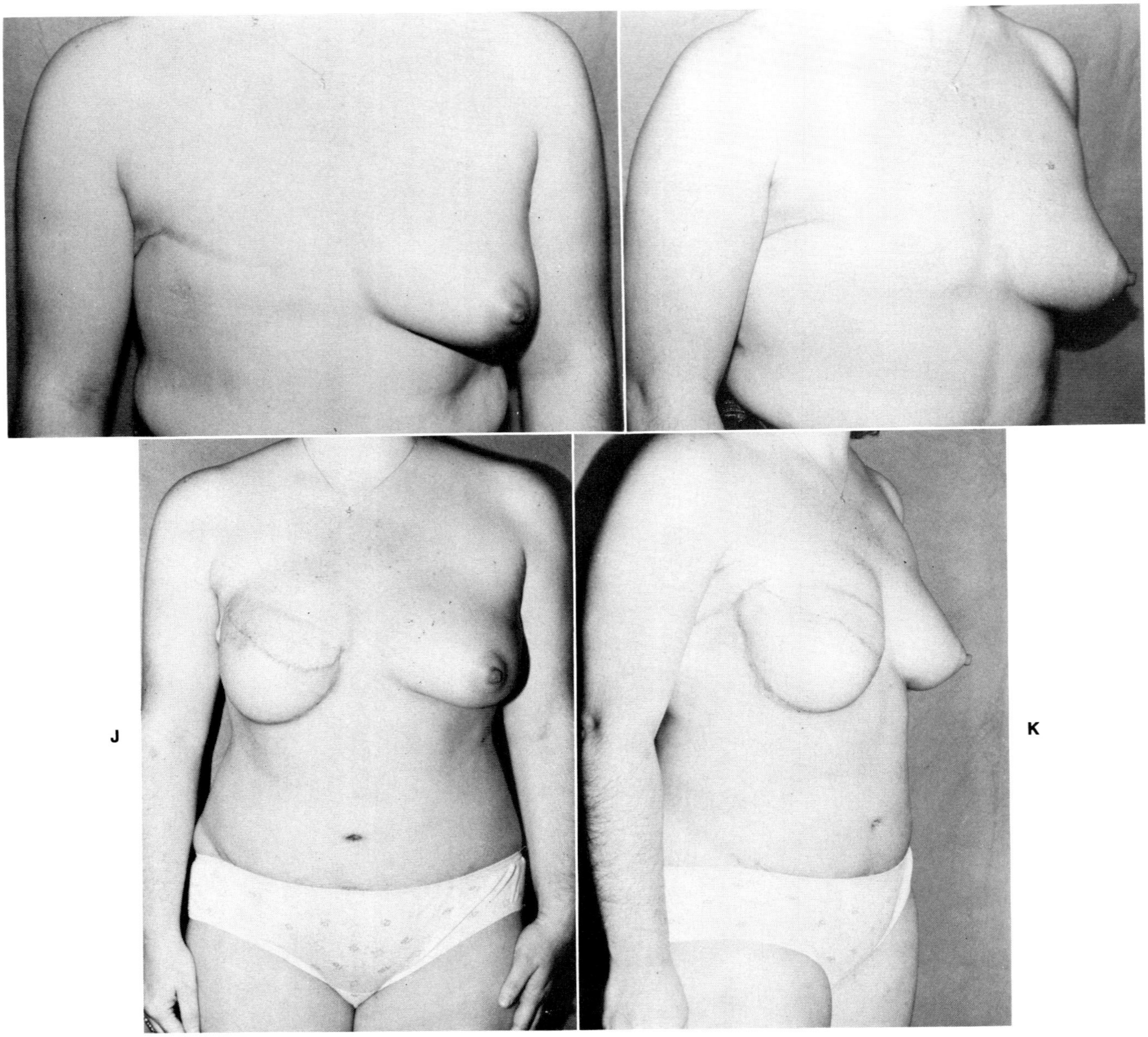

H I J K

Fig. 13-10, cont'd
H, Preoperative anterior view.
I, Preoperative oblique view.
J, Postoperative anterior view of right breast reconstruction. No implant required since fatty portion of skin island utilized to restore breast mound.
K, Oblique view demonstrates adequate fascial closure and abdominoplasty allows donor defect closure with improved abdominal wall contour.

A B C D

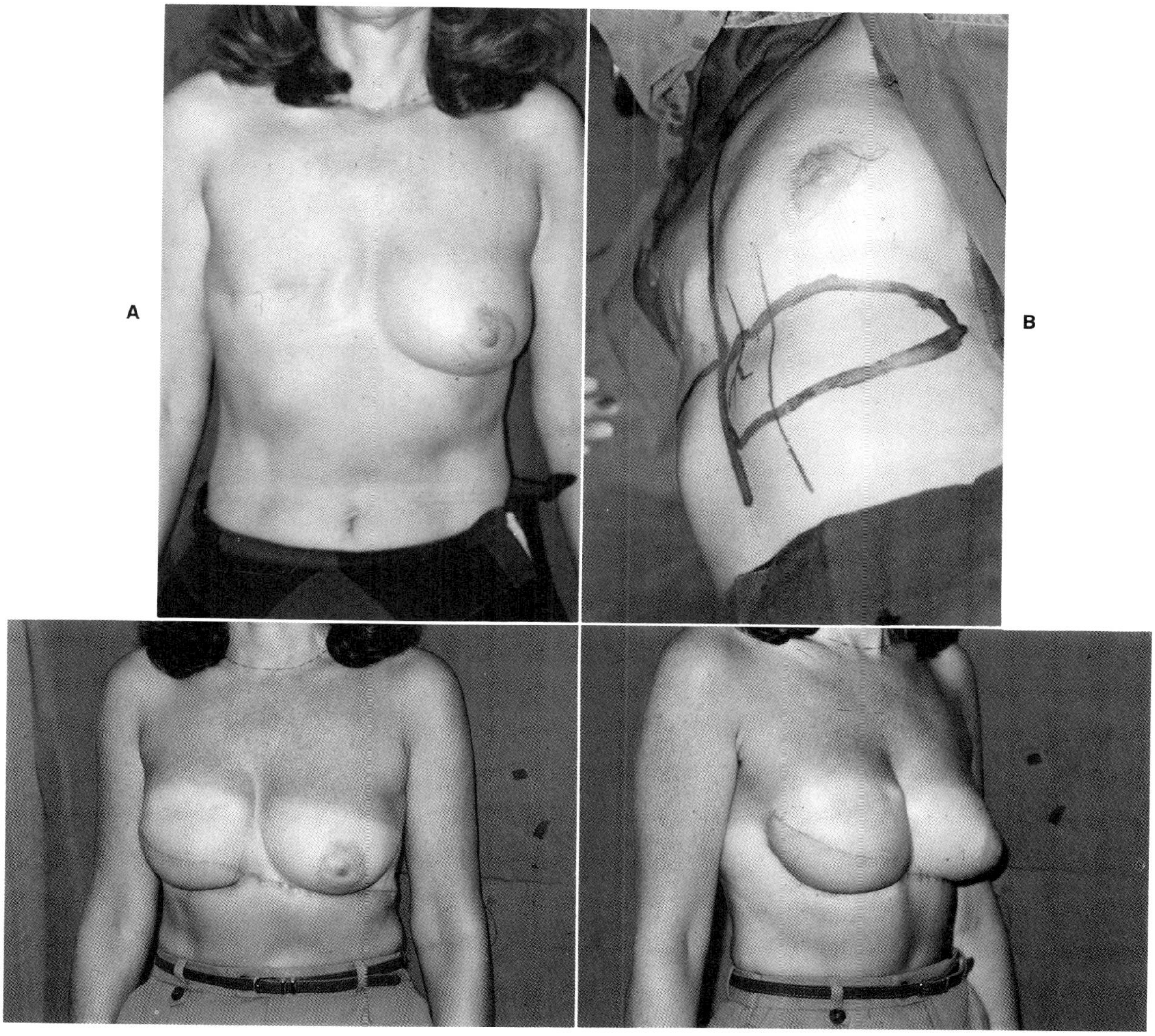

Fig. 13-11. Rectus abdominis musculocutaneous flap (superior transverse skin island).

A, Postoperative view of right modified radical mastectomy.

B, Design of horizontal skin island with lateral random extension based on contralateral superiorly based rectus abdominis musculocutaneous flap. NOTE: Use of contralateral upper abdominal skin allows direct closure of donor defect beneath breast.

C, Three-month postoperative view of breast reconstruction demonstrates closure of donor defect beneath inframammary fold.

D, Adequate breast form and projection are achieved with use of contralateral transverse rectus abdominis musculocutaneous flap.

(Courtesy Luis Vasconez, San Francisco.)

RADICAL MASTECTOMY (Fig. 13-12)

Skin grafts

Skin grafts are method of choice for immediate coverage of defect when radiation therapy is not planned

Local flaps

Local flaps are not available in reconstruction of the breast after radical mastectomy

Distant flaps (Fig. 13-12)

1. Latissimus dorsi musculocutaneous flap is flap of choice for immediate or late breast reconstruction (Figs. 13-13 and 13-14)
   a. Flap allows replacement of absent pectoralis major muscle with transposed latissimus dorsi muscle
   b. Flap allows replacement of absent skin with skin island of latissimus dorsi muscle
   c. Flap allows reconstruction of axillary fold with transposed latissimus dorsi muscle and deepithelialized skin island (Fig. 13-15)
2. Serratus anterior muscle is useful in conjunction with latissimus for coverage of implant
3. Rectus abdominis musculocutaneous flap is alternative flap
4. Thoracoepigastric axial flap is an alternative flap for immediate or late reconstruction
5. Free tissue transplantation is method of choice when latissimus dorsi musculocutaneous flap is not available; choices of free flap are opposite latissimus dorsi musculocutaneous flap and groin or inferior epigastric axial flap (Fig. 13-16)

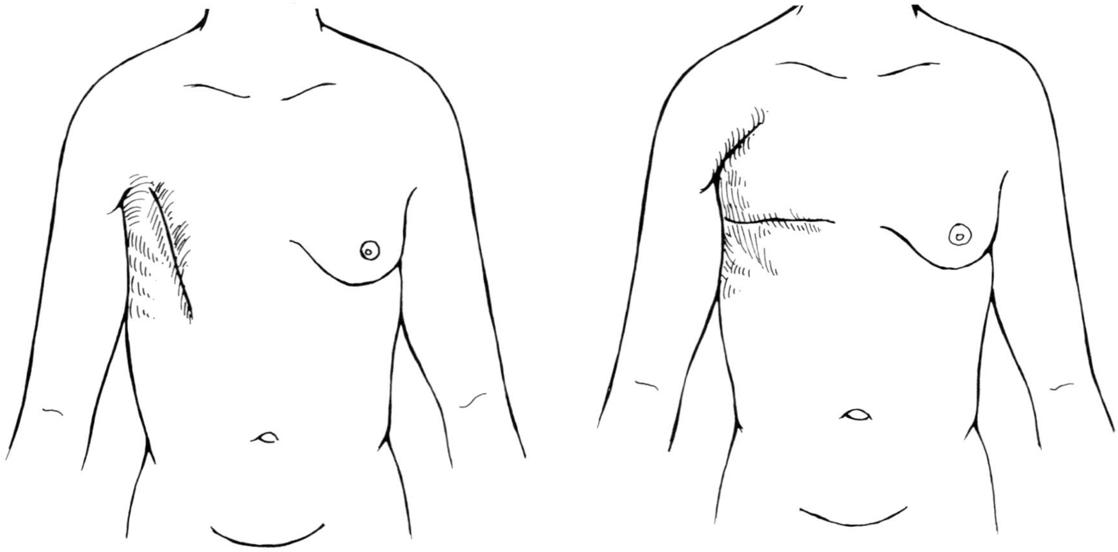

Radical mastectomy

Fig. 13-12. Bilateral latissimus dorsi musculocutaneous flap.

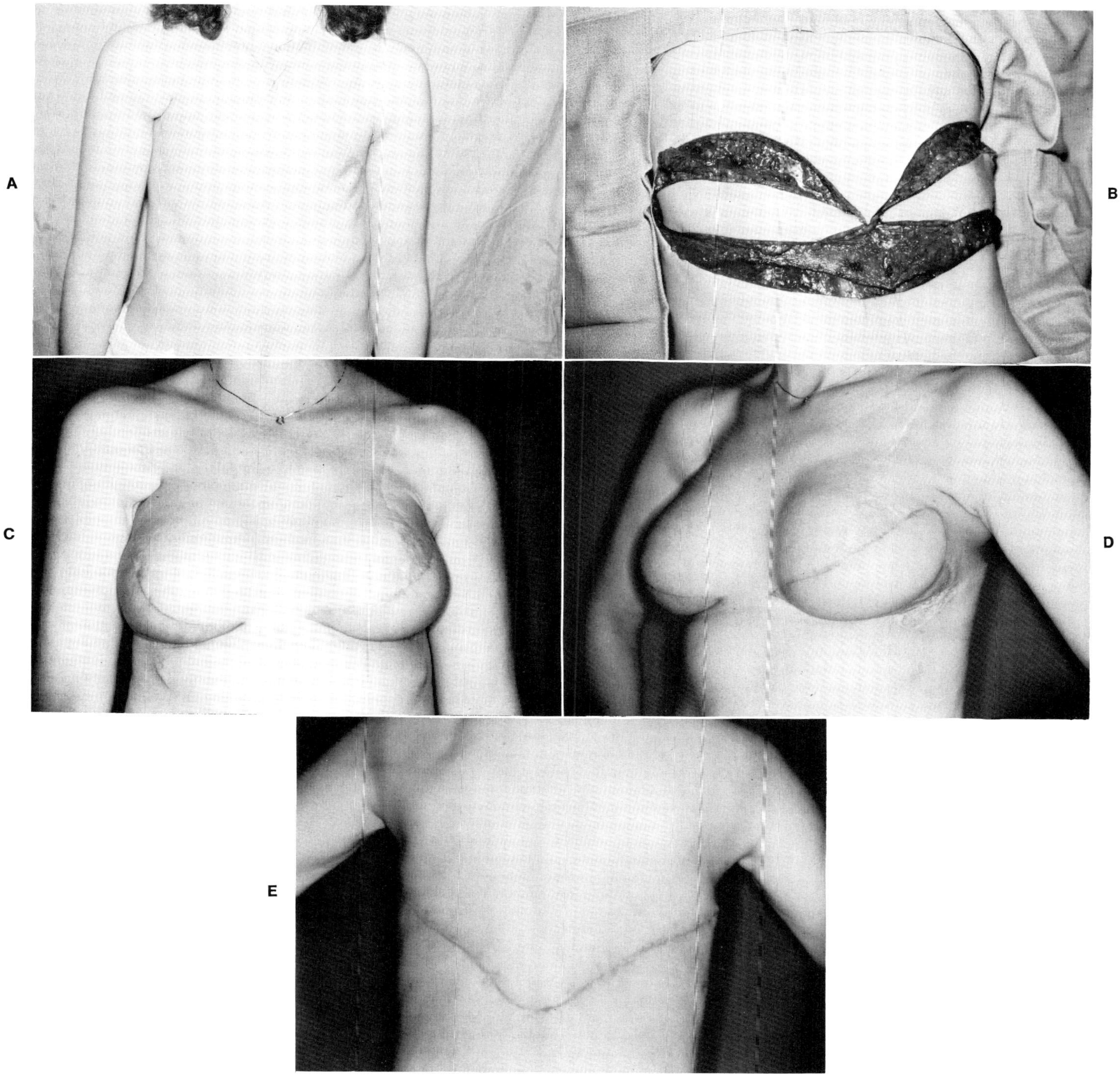

Fig. 13-13. Absence of pectoralis major muscle following radical mastectomy results in infraclavicular depression and loss of anterior axillary fold.

A, One-year postoperative view of right radical mastectomy and left modified radical mastectomy.

B, Bilateral latissimus dorsi musculocutaneous flaps elevated for anterior transposition.

C, Six-month postoperative view of bilateral breast reconstruction. NOTE: Flaps inset at inframammary region despite scar location and left anterior chest skin grafts.

D, Oblique view demonstrates adequate breast projection. Skin grafts survive based on vascular connections with pectoralis major muscle.

E, Direct donor site closure. No upper extremity disability noted by this nurse despite use of bilateral latissimus dorsi muscles.

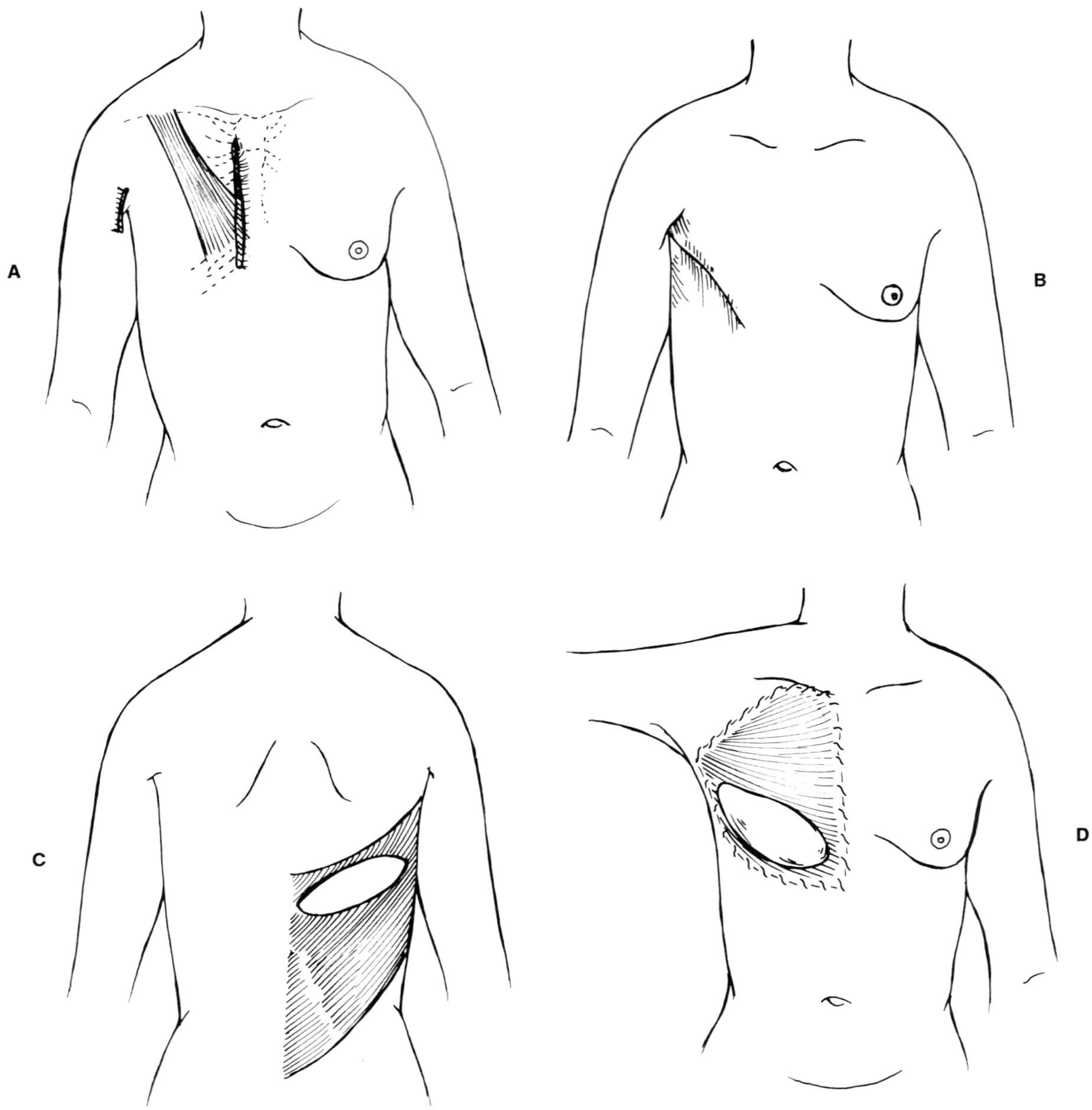

Fig. 13-14

A, Radical mastectomy defect includes loss of pectoralis major muscle.

B, Postoperative defect after radical mastectomy with loss of anterior axillary fold.

C, Design of skin island of latissimus dorsi musculocutaneous flap for reconstruction of radical mastectomy defect.

D, Transposition of muscle below skin island replaces absent pectoralis major muscle and fills intraclavicular depression.

E

F

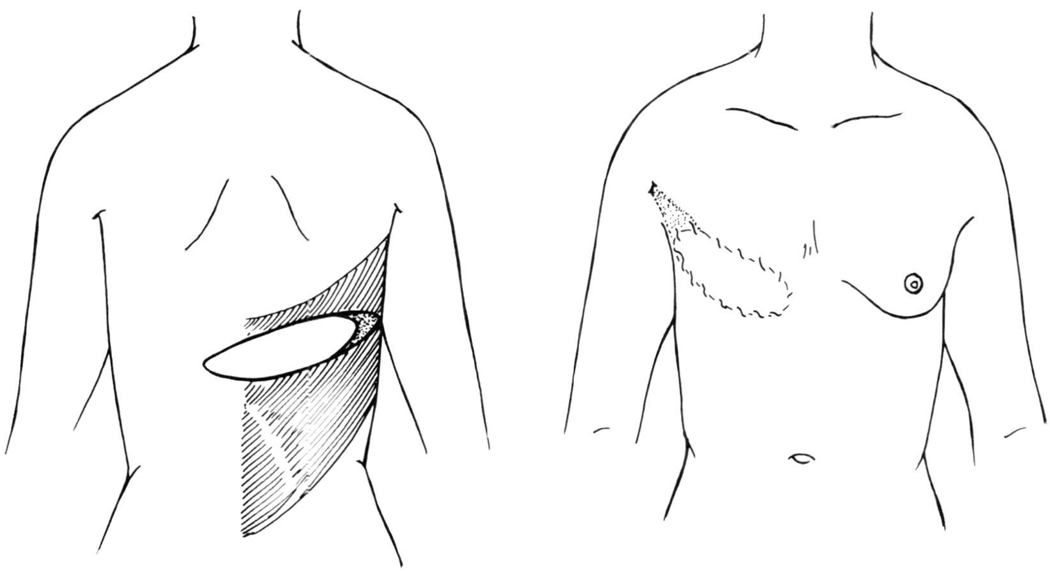

Fig. 13-14, cont'd

E, Design of skin island latissimus dorsi musculocutaneous flap for reconstruction after radical mastectomy. Lateral aspect of skin island is "deepithelialized."

F, Deepithelialized portion of skin island is sutured at desired location beneath axillary skin to restore anterior axillary fold.

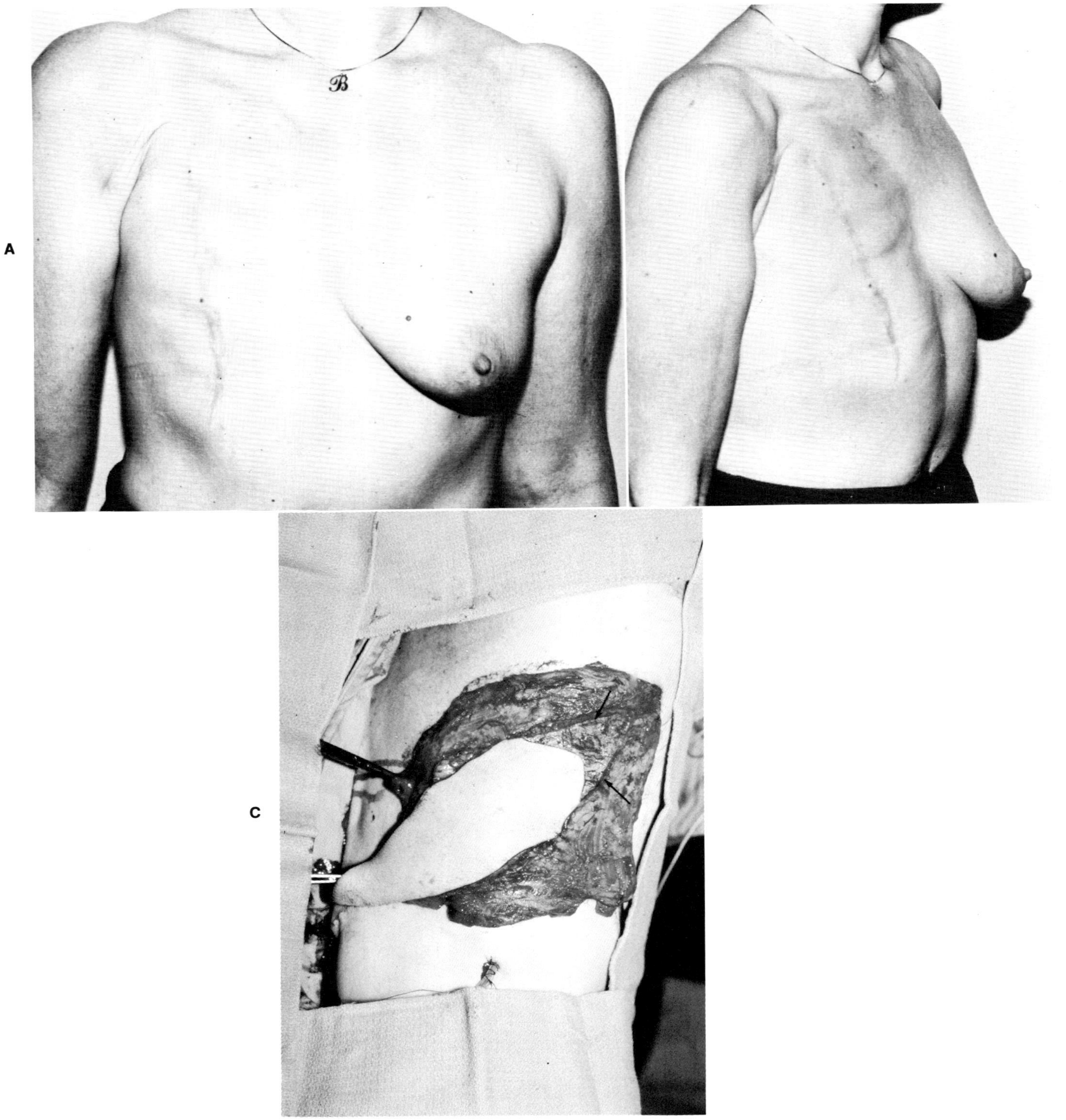

Fig. 13-15. Latissimus dorsi musculocutaneous flap.

A, Deformity after right radical mastectomy includes absent anterior chest skin, pectoralis major, breast mound, and nipple-areola complex.

B, Oblique view. Absence of anterior axillary fold and hollowness beneath clavicle is due to absent pectoralis major muscle.

C, Lateral aspect of skin island overlying latissimus dorsi muscle is deepithelialized before transposition of musculocutaneous flap to anterior chest.

D

E

F

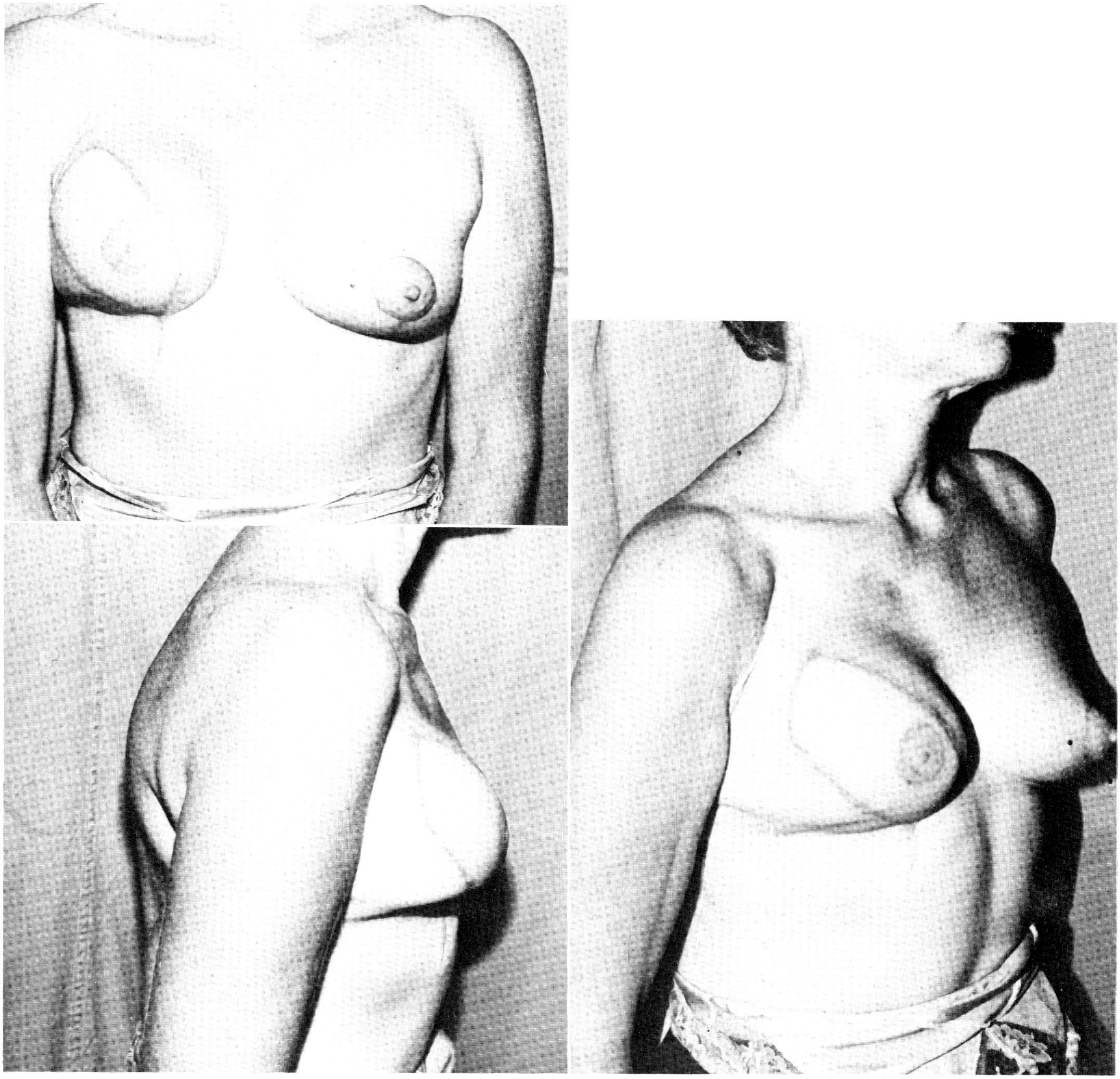

Fig. 13-15, cont'd

D, Six-month postoperative anterior view after right breast reconstruction, left mastopexy, and bilateral nipple-areola reconstruction.

E, Six-month postoperative lateral view demonstrates adequate breast projection.

F, Oblique view demonstrates restoration of axillary fold with deepithelialized proximal skin island and correction of intraclavicular hollowness with latissimus dorsi muscle.

A B C D

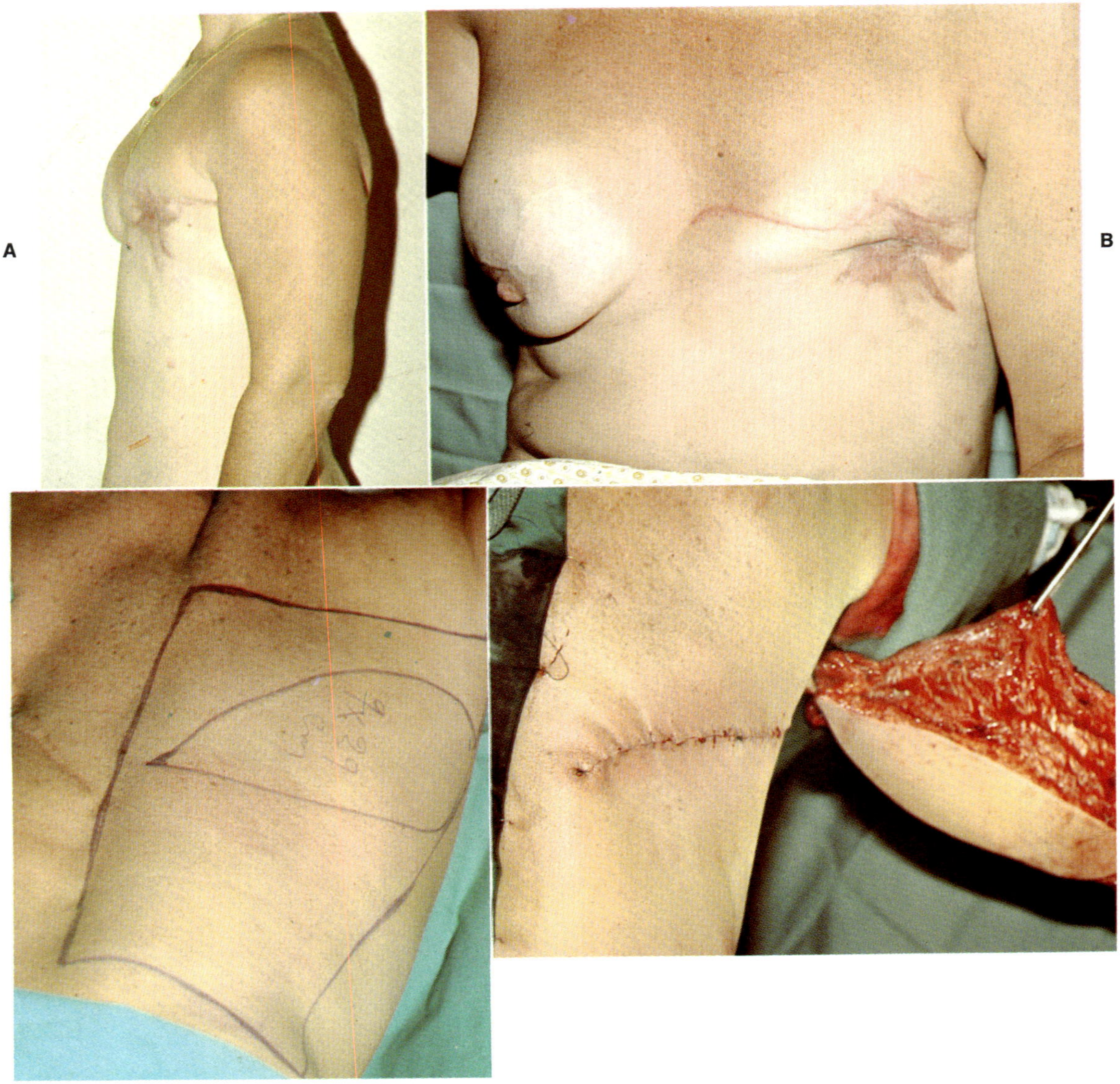

Fig. 13-16. Contralateral latissimus dorsi musculocutaneous transplantation flap.

A and **B,** Thirty-five-year-old woman with left radical mastectomy defect. Previous reconstruction with latissimus dorsi flap and abdominal flap had failed.

C, and **D,** Contralateral latissimus dorsi musculocutaneous flap is elevated as free flap. NOTE: Donor scar of ipsilateral latissimus flap.

E F G H

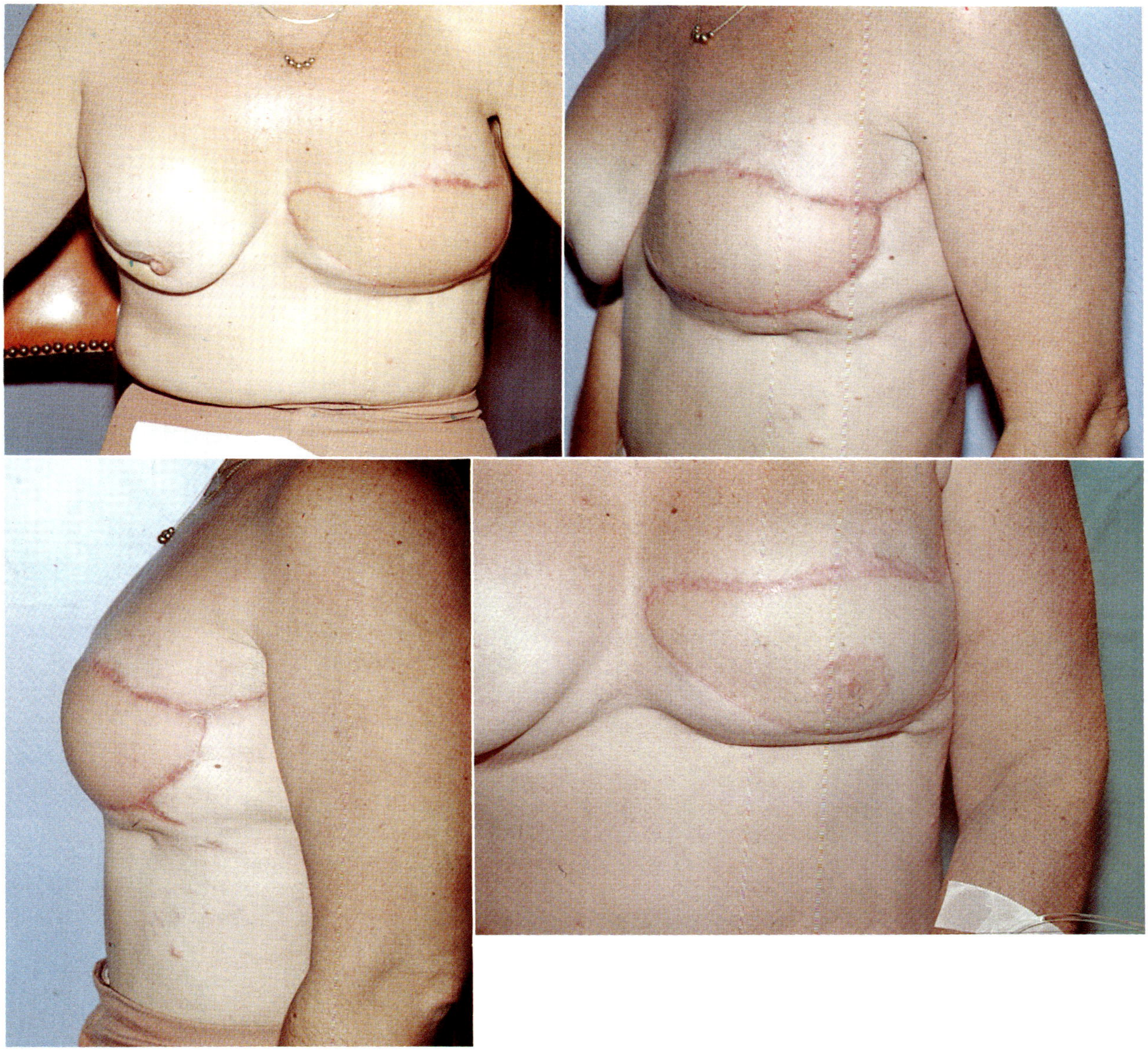

Fig. 13-16, cont'd
E to **G,** Free flap reconstruction of breast.
H, Nipple-areola reconstruction. NOTE: Capsular contraction of opposite breast secondary to prior subcutaneous mastectomy.

14

RECONSTRUCTION

John Bostwick III

A woman's breasts are a primary symbol of her femininity. The ideal breasts are youthful, full, and symmetric. The desired volume is usually 200 to 400 cc for each breast, depending on the patient's size and the general attitudes of her particular society.

For a variety of reasons, a woman may feel that her breasts do not fit the ideal image she has of herself. There may be failure of the breasts to develop (hypoplasia), or the breasts may develop then involute (breast ptosis). In other situations there may be overdevelopment of the breasts in relation to the patient's size. The breasts sometimes develop asymmetrically, or there may be asymmetry resulting from an operation.

Since the development of the speciality of plastic surgery, women have sought help to modify or reconstruct their breasts to obtain a more ideal shape. Patients with mammary hypertrophy sought a reduction of their breast volume. The ideal reduction mammoplasty of today reduces the breast volume, with inconspicuous scars along the lower aspect of the breast. Youthfully shaped, symmetric breasts with normal sensibility in the properly positioned nipple-areola are the goals of this procedure (Fig. 14-1).

The skin excision and nipple-areola positioning of the reduction mammoplasty were applied to the patient with ptotic breasts with good return to the ideal position.

BREAST AUGMENTATION

Women with small breasts often desire large breasts. A variety of materials were used during the first half of the 20th century for this purpose. Free fat flaps and local fat dermis flaps usually did not provide adequate and permanent fullness. Because of excessive scar tissue formation and associated firmness, Ivalon and Etheron sponges proved unacceptable for breast augmentation.

The development of the silicone gel prosthesis in 1964 by Cronin and Gerow provided a much less reactive material for enlargement of the small breast. This implant was placed deep to the breast tissue on the premuscular fascia. Although this is now the standard technique in breast augmentation, a large percentage (15% to 20%) of these patients develop fibrous capsules around the implant. The fibrous capsule results in a spherical, firm breast.

Attempts to minimize this problem include modification of the implant, use of corticosteroids, and submuscular placement of implants. Modifications of the implant include saline solution only and the double lumen implants. Corticosteroids placed in the outer lumen of the implant or in the dissected pockets result in softer breasts; however, the proper amount of steroid has not been determined. Steroids can result in atrophy and ptosis of the overlying breast tissue.

A

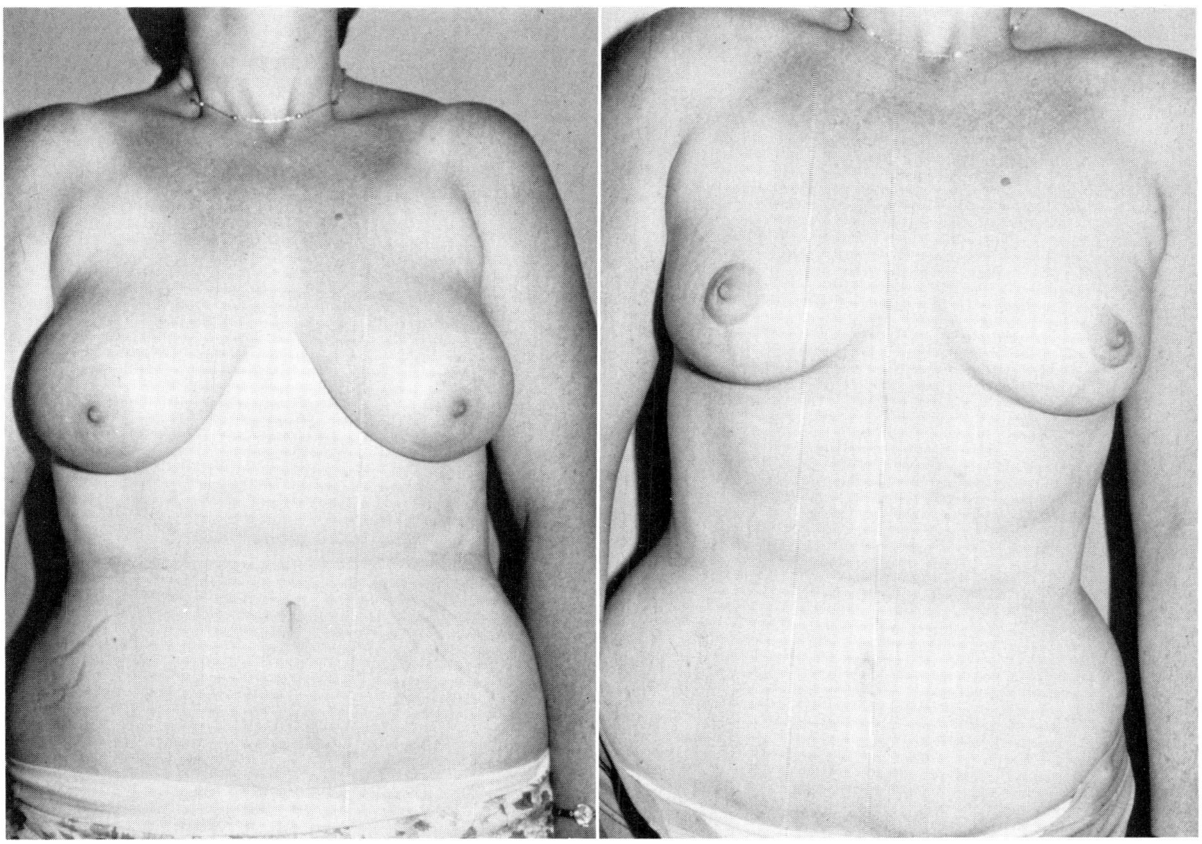

B

Fig. 14-1
A, Thirty-two-year-old women with mammary hypertrophy.
B, Two-year postoperative view after reduction mammoplasty of 800 gm.

Submuscular augmentation

Placement of the breast implant below the anterior chest wall musculature results in a lower incidence of capsular contracture. The development of fibrous capsule around a submuscular implant is less likely to deform the breast (Figs. 14-2 and 14-3).

Technique. Incisions for submuscular augmentation may be inframammary (Fig. 14-2), periareolar, or axillary. For the inframammary approach the usual 3.5-cm inframammary incision is made and the breast elevated off the muscle fascia to the lower margin of the pectoralis major. This fascia is opened, and with blunt dissection, the pectoralis major is elevated off the ribs up to the third rib. The pectoralis major origin is bluntly detached from the sternum at the level of the nipple-areola. Inferiorly it is also elevated off the ribs, and dissection is carried beneath the rectus fascia to the proposed inframammary crease. Dissection is then carried laterally, elevating the serratus anterior fascia. Care is taken to preserve the sensory nerves. With an adequate pocket developed, sutures are placed between the pectoralis major fascia and the serratus anterior fascia. The implant is inserted and these sutures tied.

When the periareolar incision is used, the dissection is carried subcutaneously down to the inframammary area, then up to the lower edge of the pectoralis major muscle. The procedure then is as described for the inframammary approach.

A

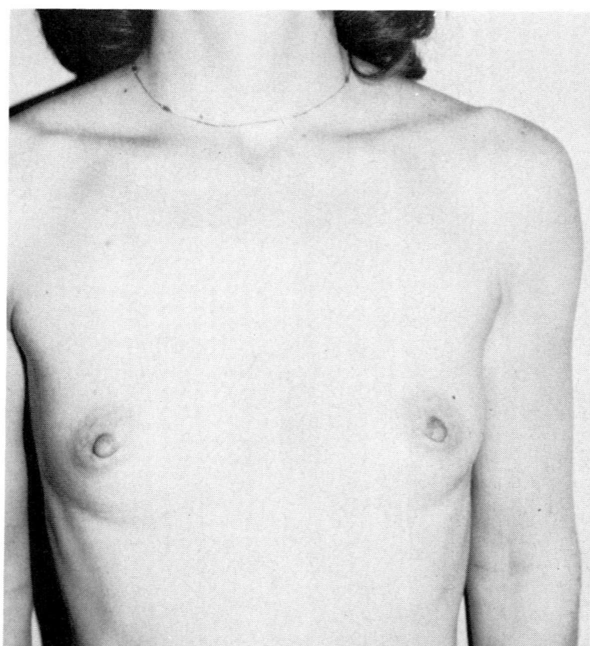

B

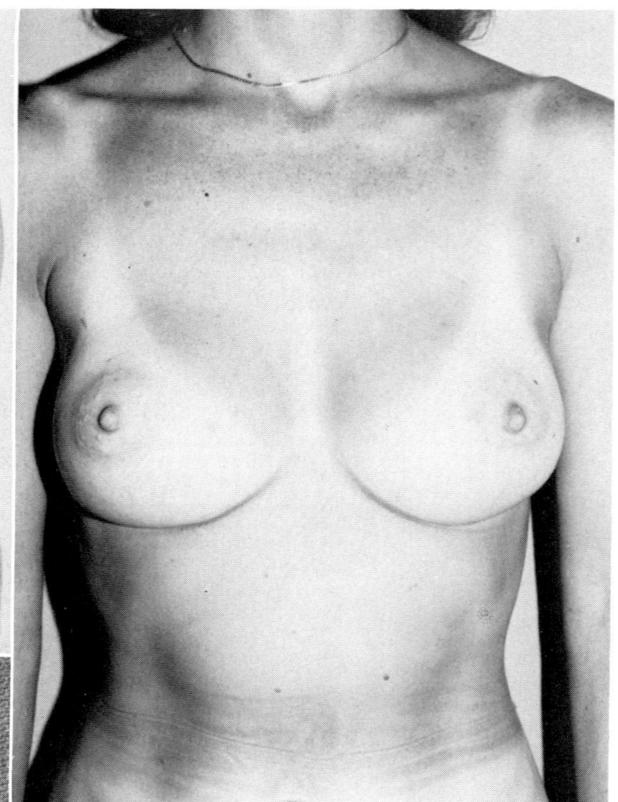

C

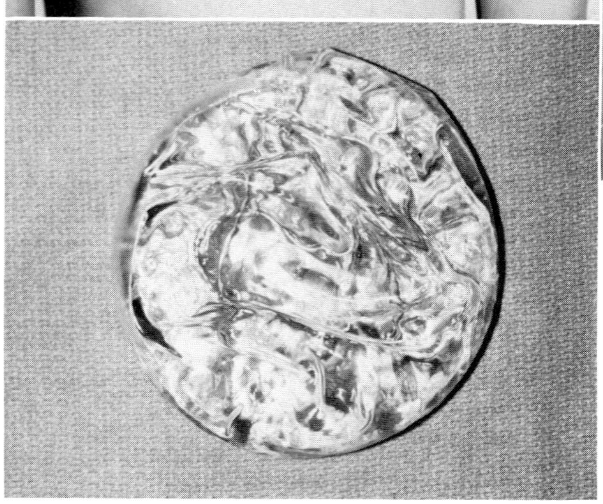

Fig. 14-2

A, Twenty-six-year-old woman with mammary hypoplasia.

B, Two years after augmentation mammoplasty (220 cc) in submuscular position with double lumen silicone breast prosthesis.

C, Double lumen silicone breast prosthesis.

The axillary incision for submuscular augmentation is best suited for the very small breast with no ptosis. A 4- to 5-cm incision is made transversely in the lower axillary area through the area of hair growth. The lateral margin of the pectoralis major muscle is identified and its fascia divided. The upper dissection beneath the muscle is performed bluntly. The lower dissection requires an instrument. We have found the largest urethral sound, a wide, firm ribbon or flat retractor, to be best for this dissection. After placement of the implant, the fascia is resutured and the skin closed. A light elastic support is placed across the upper part of the breast to keep the implant in the lowest portion of the dissected pocket.

There are some drawbacks to the submuscular positioning of the implant. The contraction of the muscle can give an unnatural, irregular contour to the breast. Occasionally some patients have discomfort associated with the implant in this position. Additional implant volume of 30 to 60 cc is necessary to achieve a similar improvement in size compared to the premuscular augmentation. It is also important to place the implant low; it is not only beneath the pectoralis major but also beneath the serratus anterior and upper rectus sheath.

When capsular contractures develop beneath the muscular layer, breasts become firm, but the contour is usually not as distorted. Displacement of the implant from capsular contracture is rare in the submuscular position.

A 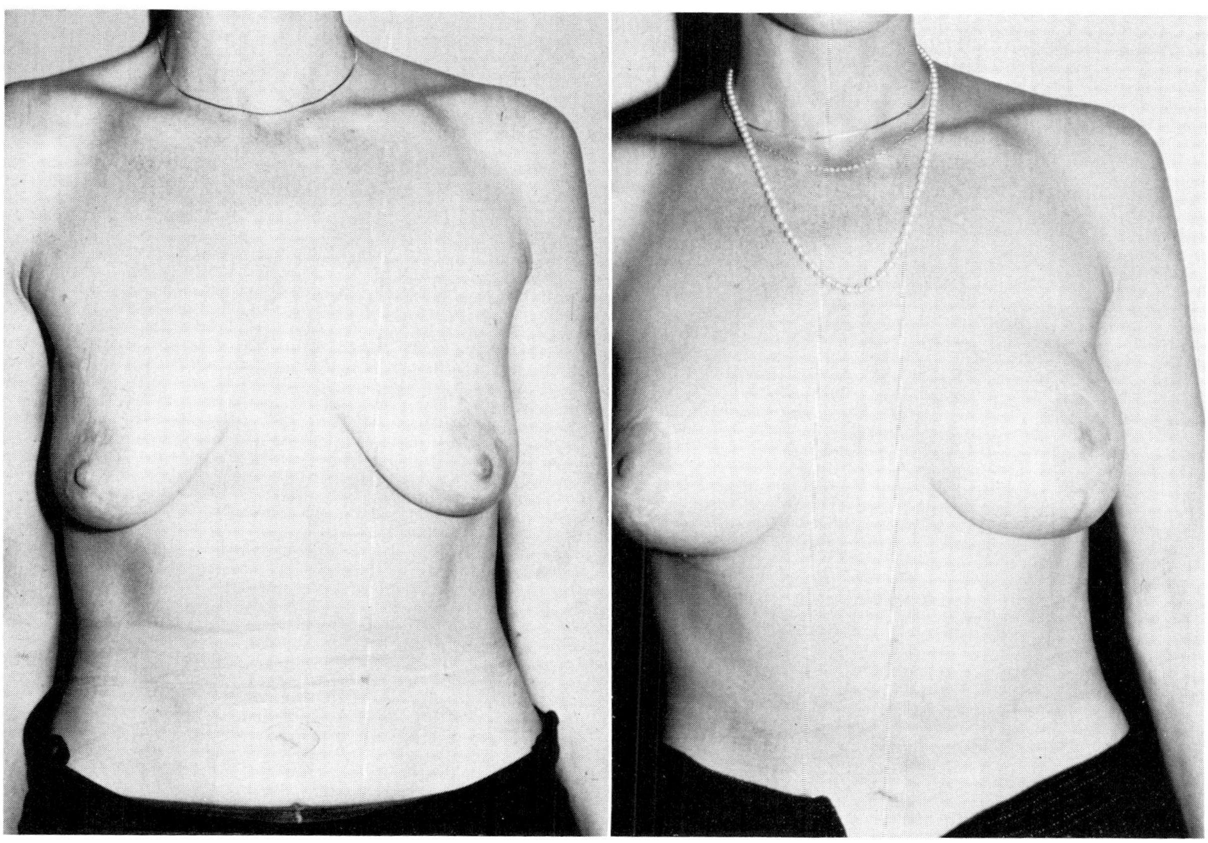 B

Fig. 14-3

A, Twenty-eight-year-old woman with postpartum involutional atrophy and ptosis of breasts.

B, One year after submuscular augmentation mammoplasty (180 cc) and mastopexy.

SUBCUTANEOUS MASTECTOMY

Prophylactic subcutaneous mastectomy is widely practiced as a technique to remove abnormal or premalignant breast tissue. Delayed or immediate reconstruction of the breasts has been accomplished with placement of silicone gel implant beneath the preserved breast skin. With the thin skin flaps after subcutaneous mastectomy, a fibrous capsule formation causes firmness and breast deformity. The incidence and severity of this deformity seem to be related to the thickness of the tissue overlying the prosthesis (Fig. 14-4). Other problems associated with this procedure include skin loss and exposure of the implant.

Submuscular reconstruction

In an attempt to avoid these problems in reconstruction after subcutaneous mastectomy, the silicone implant has been placed beneath the pectoralis major, serratus anterior, and upper aspect of the fascial sheath of the rectus abdominis. Although no controlled series are available, there is an overall impression that there is less breast deformity and implant exposure with submuscular placement (Fig. 14-5).

Technique. Before the description of the technique a review of the anatomy is appropriate. The pectoralis major muscle will cover the superior portion and medial lower portion of a properly positioned breast implant. This is a thick, triangular muscle, which originates from the fifth, sixth, and seventh ribs anteriorly as well as the lateral aspect of the sternum. A clavicular portion originates from the lower aspect of the clavicle and then converges into an insertion onto the lateral lip of the intertubercular sulcus of the humerus. The lower fibers toward the insertion define the anterior axillary fold. The blood supply to the pectoralis major is primarily from its thoracoacromial pedicle. Other significant supply is from the medial perforators from the internal mammary artery. Other intercostal perforators enter the deep aspect of the pectoralis major muscle. If the intercostal and sternal origin fibers are divided, the muscle remains innervated and well vascularized from the thoracoacromial pedicle. The medial and lateral pectoral nerves are the motor nerves. The serratus anterior muscle and fascia are used to cover the breast implant laterally.

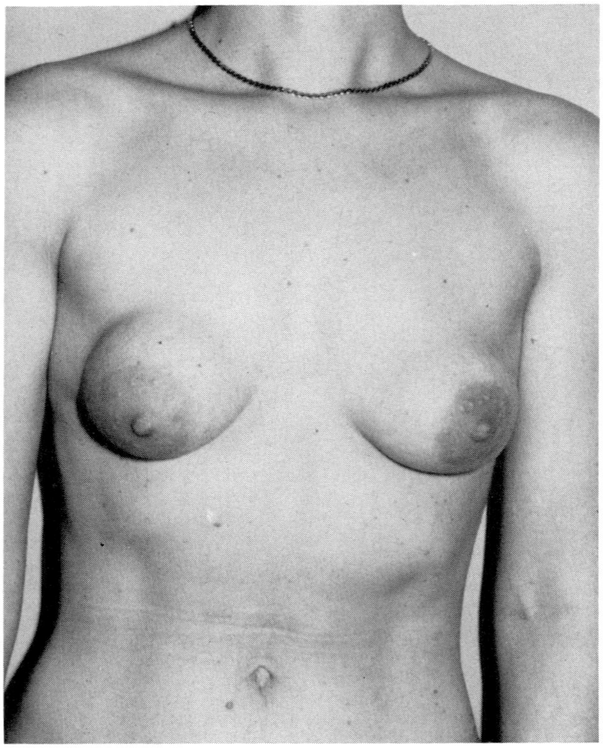

Fig. 14-4. Tight capsular contracture about 160 cc implants placed in subcutaneous position after prophylactic subcutaneous mastectomy.

The serratus anterior muscle inserts on the deep surface of the scapula from a broad origin from the upper eighth to tenth ribs anterolaterally. The blood supply to the serratus anterior is from the thoracodorsal and lateral thoracic arteries. These vessels and the motor nerve, the long thoracic nerve, enter the muscle superiorly and are not affected by mobilization for implant coverage.

The submuscular plane is entered by dividing the fascia between the pectoralis major and serratus anterior. Sharp dissection is performed to elevate the pectoralis major origin from the sternum at the sixth costal cartilage. This dissection is continued inferiorly and laterally. Inferiorly this dissection is continued beneath the anterior rectus sheath to 2 cm below the future inframammary crease. The serratus anterior fascia is then elevated to 2 cm lateral to the future lateral projection of the breast. Sutures are then placed between the pectoralis major and serratus fascia, and the implant is positioned and the sutures tied (Fig. 14-5).

If the breast is hypertrophic or ptotic, and excess skin must be excised, an inferior dermis flap may be developed. The submuscular space is approached through this inframammary incision. First, the rectus abdominis fascia is elevated, then the other muscles are elevated. The implant is placed beneath the muscle and dermis flap and the tissues closed.

In some patients, particularly those who require a prosthesis larger than 250 to 300 cc, there is insufficient cover for the lower aspect of the implant. A flap of latissimus dorsi muscle may be transposed to cover this lower portion. This is more fully discussed later.

It is essential that the submuscular pocket be adequately developed in all directions. The most frequent problem is a failure to dissect low enough, particularly failure to detach all the lower fibers of the origin of the pectoralis major.

Postoperative complications such as malpositioned implant, hematoma, and infection have been much lower with the submuscular technique.

A

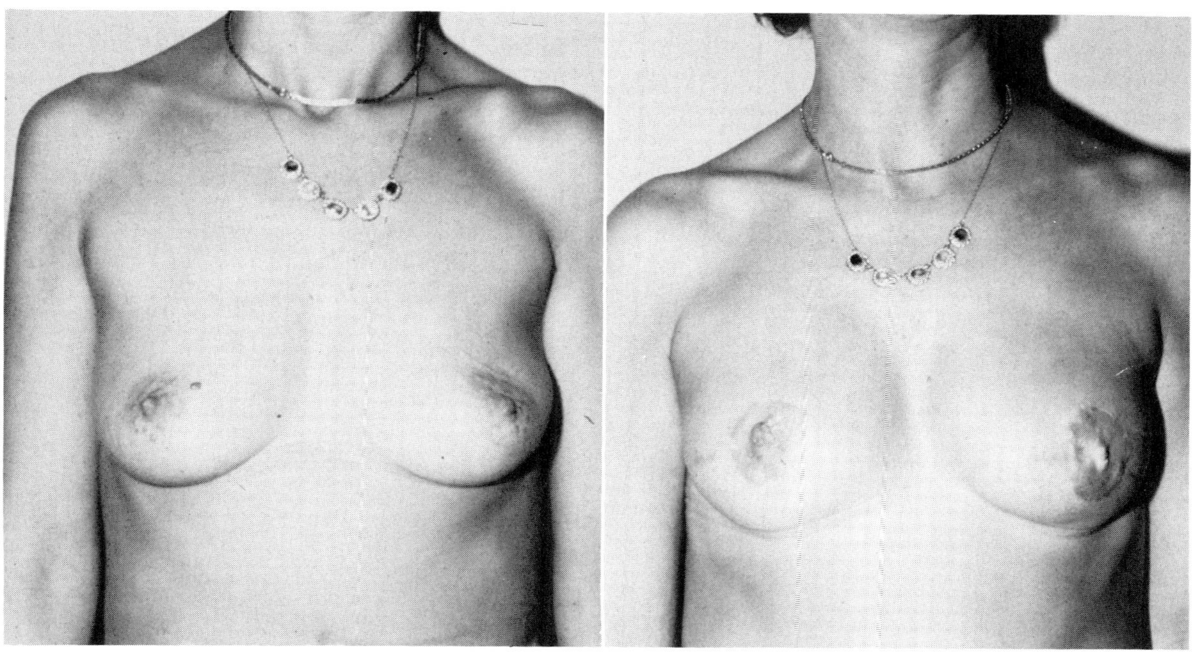

B

Fig. 14-5

A, Forty-two-year-old woman with biopsy-proven in situ lobular carcinoma of left breast.

B, Two years after bilateral total mastectomy and immediate submuscular reconstruction with 260 cc double lumen silicone breast prosthesis and immediate nipple-areola full-thickness grafts.

(From Bostwick, J., III: Breast reconstruction: a comprehensive approach, Clin. Plast. Surg. **6**[2]:143, 1979; reproduced with permission.)

BREAST RECONSTRUCTION FOLLOWING MASTECTOMY

The radical mastectomy evolved around 1900 as a primary treatment for breast carcinoma. Patients had large breast tumors necessitating wide skin excision and skin graft coverage. Axillary metastases were common, and lymphadenectomy was facilitated by removal of the pectoral muscles.

The deformity after radical mastectomy is a source of profound emotional distress with marked deleterious effects on the body image of a patient (Fig. 14-6). Early attempts to reconstruct this deformity were largely unsuccessful and were condemned by the general surgeons.

The defect after a radical mastectomy includes a missing pectoralis major muscle and the resultant flatness beneath the clavicle and hollowness of the axilla. The skin is tight and often replaced with a split-thickness skin graft. There is no breast mound or nipple-areola. The challenge is to replace these tissues and obtain a reconstruction that restores the impaired body image.

One of the earliest reconstructive efforts in 1900 was by an Italian, Tansini, who moved a transposition flap of latissimus dorsi muscle and its overlying skin into the mastectomy defect. This ameliorated the pectoral muscle defect and replaced resected skin; however, the operation did not become widely accepted.

Reconstructive surgeons then moved a portion of a large contralateral breast across to the defect after mastectomy. This operation placed a potentially malignant breast onto an area of previous cancer excision and was not acceptable to general surgeons. It also did not correct the pectoral and axillary defects associated with the radical mastectomy.

The abdominal tube pedicle was also used to move tissue onto the chest. It required many procedures and had the usual complications and restrictions that accompanied the use of this particular method of reconstruction. It was difficult to obtain a smooth infraclavicular and axillary reconstruction. The shape of the breast mound is not always predictable with this technique.

In 1970 Snyderman and Guthrie reported the creation of a breast mound by insertion of a silicone gel prosthesis beneath the skin following a modified radical mastectomy. With a nipple-areola graft and appropriate reshaping of the opposite breast, an acceptable breast reconstruction was obtained for patients who had modified radical mastectomies.

Cronin introduced the use of the transverse abdominal flap for placement of absent skin in the radical mastectomy defect. Radical mastectomy reconstruction was then accomplished with a combination of the abdominal flap and the use of the silicone breast implant. This technique also did not address the problem of the infraclavicular and axillary defect.

If a radical mastectomy has been performed, the infraclavicular area must be filled and the anterior axillary fold reconstructed. There should be minimal scars. Every effort should be made to achieve symmetry with the opposite breast if it has good shape, size, and contour. The use of the latissimus dorsi musculocutaneous flap is especially helpful in accomplishing these goals in the radical mastectomy patient.

Study of the latissimus dorsi muscle shows it to be

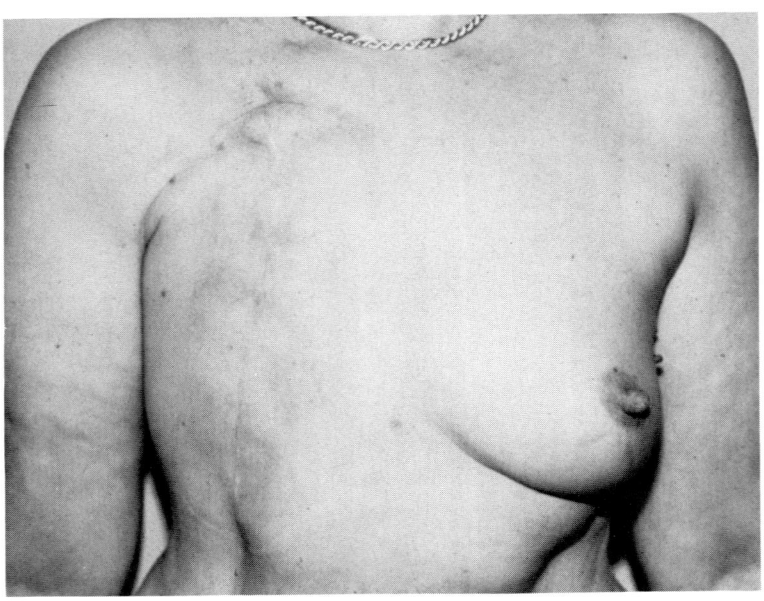

Fig. 14-6. Fifty-eight-year-old woman 10 years after standard radical mastectomy for 1-cm infiltrating ductal carcinoma.

analogous to the pectoralis major, only situated posteriorly. Both of their insertions and primary blood supply are in proximity. The thoracodorsal pedicle will support and nourish the entire latissimus dorsi muscle as well as its overlying skin. The latissimus dorsi muscle when transposed anteriorly will satisfactorily restore the contour after pectoralis major muscle resection. The versatility of the skin island placement over the latissimus dorsi is such that a compound skin island flap may be designed to simultaneously replace the resected skin and restore the pectoral muscle bulk after a radical mastectomy.

Experience with the pectoral muscle flap as an additional layer of cover for a silicone breast implant has led to the successful use of the latissimus dorsi muscle as a well-vascularized, functioning layer in situations without available pectoral muscle (e.g., after radical mastectomy [Fig. 14-6], Poland's syndrome, and subcutaneous mastectomy with threatened or actual implant exposure). The indications for the use of the latissimus dorsi muscle or musculocutaneous flap have broadened with increasing experience with this very reliable and versatile musculocutaneous unit.

It is now used whenever additional pectoral muscle bulk simulation is needed, such as after radical or modified radical mastectomies in which the pectoral nerves have been divided and the muscle atrophied. It is very useful in patients with Poland's syndrome, congenital absence of the pectoralis major muscle. It is also used when there is a skin graft or the skin is tight, extremely thin, or of poor quality and after radiation therapy.

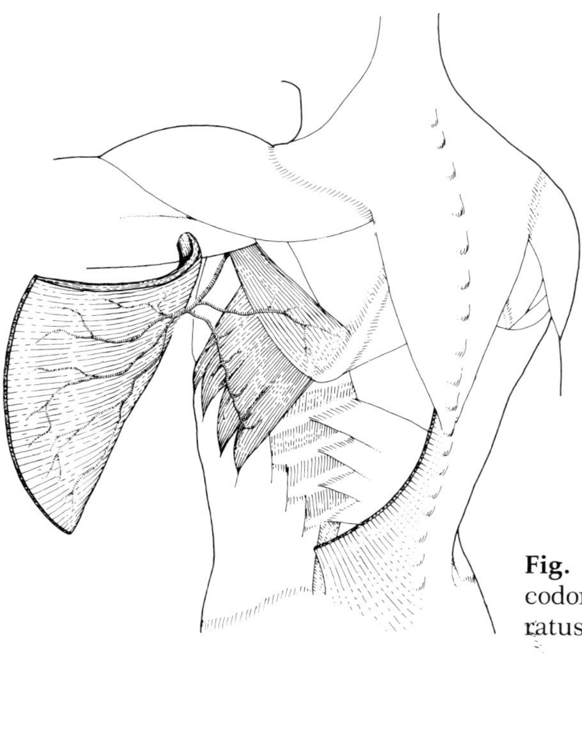

Fig. 14-7. Primary blood supply to latissimus dorsi is via thoracodorsal artery. Branch from thoracodorsal artery to adjacent serratus anterior muscle is also noted.

Anatomy

The latissimus dorsi muscle is a broad, flat muscle covering most of the lower portion of the back. It originates from the spinous processes from T6 to L5 and from the posterior iliac crest. Additional fibers of origin are deep near the scapula and the lateral ribs. The fibers converge toward the axilla and become closely associated with the teres major before inserting as a flat broad tendon into the bottom of the intertubercular groove of the humerus. This axillary portion along with the teres major forms the posterior axillary fold. The principal blood supply is from the thoracodorsal artery, a terminal branch of the subscapular artery. This thoracodorsal pedicle is 10 to 12 cm long and contains the 1.5- to 2.5-mm artery, motor nerve, and usually two veins (Fig. 14-7). Another significant blood supply comes from a crossing collateral vessel from the surface of the serratus anterior that enters near the entrance of the pedicle into the latissimus dorsi muscle. Preservation of this crossing collateral vessel is advisable if there is any question that the thoracodorsal vessel may have been previously divided. (Some surgeons divide the thoracodorsal artery and vein and preserve the nerve at the time of mastectomy.) Other significant blood supply comes from segmental paraspinous perforators. Other vessels enter on the deep surface via intercostal perforators in the midposterior thorax, and these will also nourish islands of skin and muscle. Intercostal perforating vessels also enter along the lateral margin of the muscle. Segmental sensory nerves accompany these intercostal perforating vessels. In addition, there are some direct vessels into the muscle from the axillary artery through the peritenon of the tendon and from the teres major muscle.

DEFECT RECONSTRUCTION AFTER RADICAL MASTECTOMY USING THE LATISSIMUS DORSI MUSCULOCUTANEOUS FLAP

The missing skin as well as the missing pectoralis major muscle bulk and the anterior axillary fold may be restored with a latissimus dorsi musculocutaneous island flap. The additional latissimus dorsi muscle is used to cover the silicone breast implant (Figs. 14-8 and 14-9).

To provide sufficient fill, the latissimus dorsi muscle must be innervated and functional. A discussion with the general surgeon, a check of the operative note, and a physical examination should confirm the functional status of the latissimus dorsi muscle. A nonfunctioning latissimus dorsi results in a winging of the scapular tip, and the latissimus dorsi cannot be seen or palpated during forced adduction against resistance (Fig. 14-10).

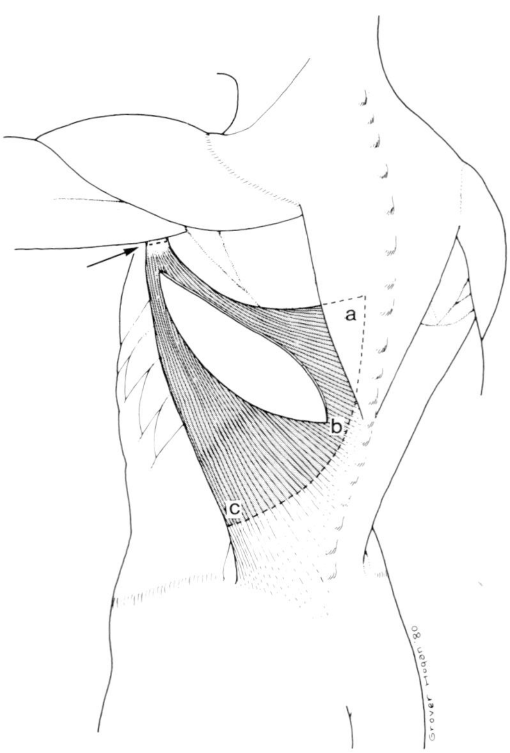

Fig. 14-8. Outline of skin island over latissimus dorsi to be transposed for reconstruction after radical mastectomy. *a,* Superior muscle will transpose to inframammary region on anterior chest. *b,* Central muscle covers breast implant. *c,* Inferior muscle will transpose to fill infraclavicular region of anterior chest.

The skin island of appropriate size is designed over the latissimus dorsi muscle on the back to replace the missing skin. The arc of rotation of the flap must be kept in mind (Figs. 14-11 to 14-13). For most radical mastectomy defects, the skin island may be placed beneath the brassiere in the back. Unless a large skin graft is present, sufficient skin may be obtained and the donor site closed directly. Most flaps have been 6 to 7 cm wide; however, defects 8 to 11 cm wide may be closed directly. The flap rotates so that the muscle tissue below the flap in the back moves up to the clavicle and medial-sternal area. The muscle tissue above the flap rotates so that it fits below the transposed flap when the incision is oblique or transverse.

The latissimus dorsi muscle is sutured to the former sternal and clavicular origin of the pectoralis major. If the clavicular portion has been spared, the latissimus dorsi is sutured to the lower edge of the remaining pectoralis major.

Reconstruction of the anterior axillary fold is accomplished by dividing the tendon of insertion of the latissimus dorsi and transposing it to the cut edge of the pectoralis major. This should only be done if a pulsatile artery in the pedicle is first identified.

Additional bulk for the axilla may be obtained by

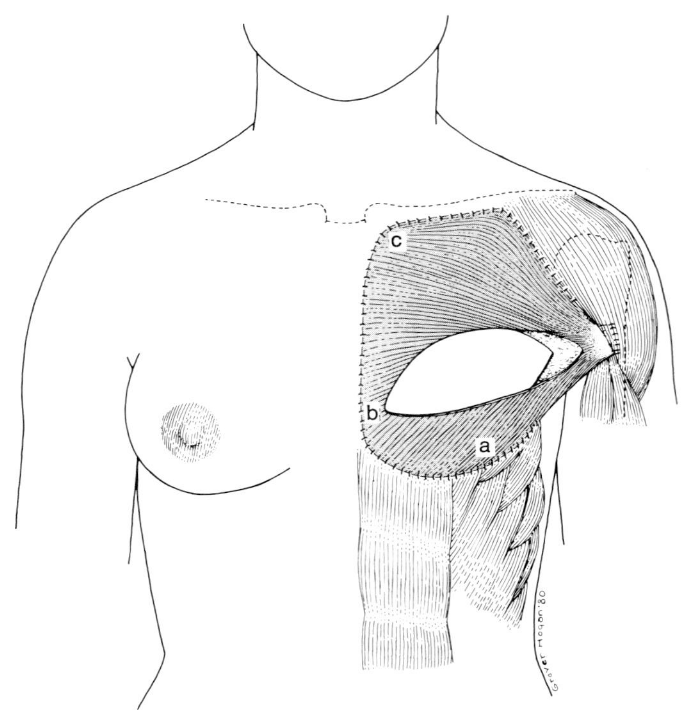

Fig. 14-9. Latissimus dorsi flap is transposed anteriorly along with its insertion. *a, b, c,* Same as in Fig. 14-8.

creating a longer skin flap and using its deepithelialized end for axillary fill. The mastectomy scar is excised over the anterior chest and the skin island sutured into its place. A small lateral opening is left for insertion of the breast prosthesis.

Elevation of the flap on the back is accomplished by incising around the skin island down to the latissimus dorsi muscle. The skin is elevated over the muscle to the extent of muscle required for reconstruction as determined during the preoperative planning. The patient is in the lateral position so that the anterior and posterior dissections can be made. Care is taken not to elevate the serratus anterior muscle with the latissimus dorsi. When the dissection reaches the serratus anterior muscle, the blood vessels on the surface of this muscle should be preserved. These lead to the crossing collateral vessels that go to the pedicle and supplement blood flow to the flap. As the latissimus dorsi is elevated, its axial blood vessels may be seen, and they also lead to the pedicle. The lateral edge of the latissimus dorsi is the most lateral aspect of the mastectomy dissection and is usually involved with scar. The latissimus dorsi must also be separated from the teres major to rotate it anteriorly.

To restore the anterior axillary fold, the flap must be tunneled high in the axilla. Further augmentation of the anterior axillary fold is possible by anterior transposition of the muscle insertion. When the latissimus dorsi insertion is to be transposed, it is necessary to ascertain if there is pulsatile flow in the pedicle, and the crossing vessel to the serratus anterior is left intact (Fig. 14-7).

Anteriorly the mastectomy scar is excised and the skin flaps are elevated so that the pectoralis major muscle defect can be restored and the implant positioned. The axillary vein is closely related to the overlying dermis, particularly after radiation. Careful dissection is essential to prevent a tear of this vessel. The flap is passed through the tunnel high in the axilla, and the back incision is closed after placing a suction drain. The patient is then placed in the supine position and redraped and the reconstruction completed.

The latissimus dorsi musculocutaneous flap is positioned so that the skin island will fit properly into the area of the excised scar. The muscle is then sutured to the clavicle and the sternum. If necessary, the divided insertion is sutured to the divided insertion of the pectoralis major insertion. The remainder of the muscle is sutured at a level just below the future inframammary crease. The skin island is closed except for a small opening laterally, and the prosthesis is inserted. The patient is placed in the sitting position so that final size determination and positioning are accomplished.

Text continued on p. 324.

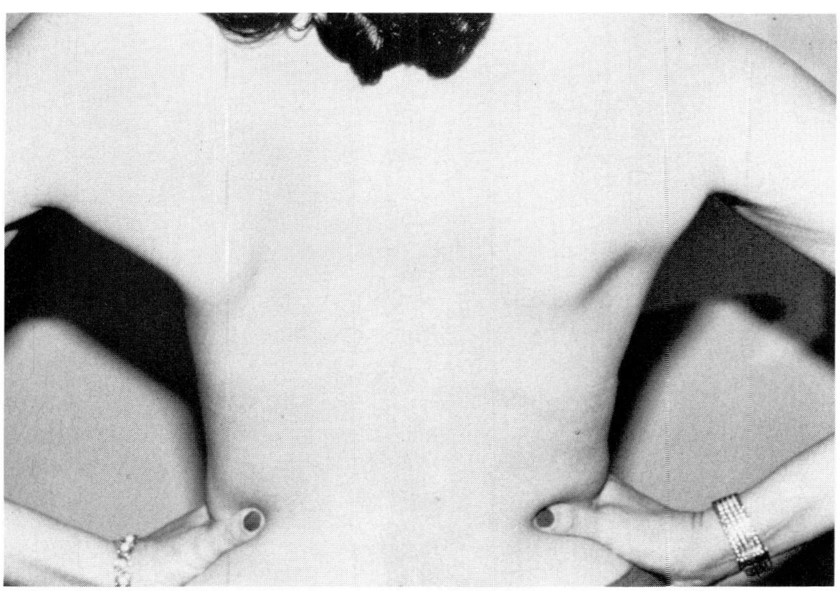

Fig. 14-10. Patient with denervated and atrophied right latissimus dorsi muscle after radical mastectomy. NOTE: Wing of scapula and lack of definition of latissimus dorsi laterally. Thoracodorsal vessels are usually divided in patients with denervated muscle.

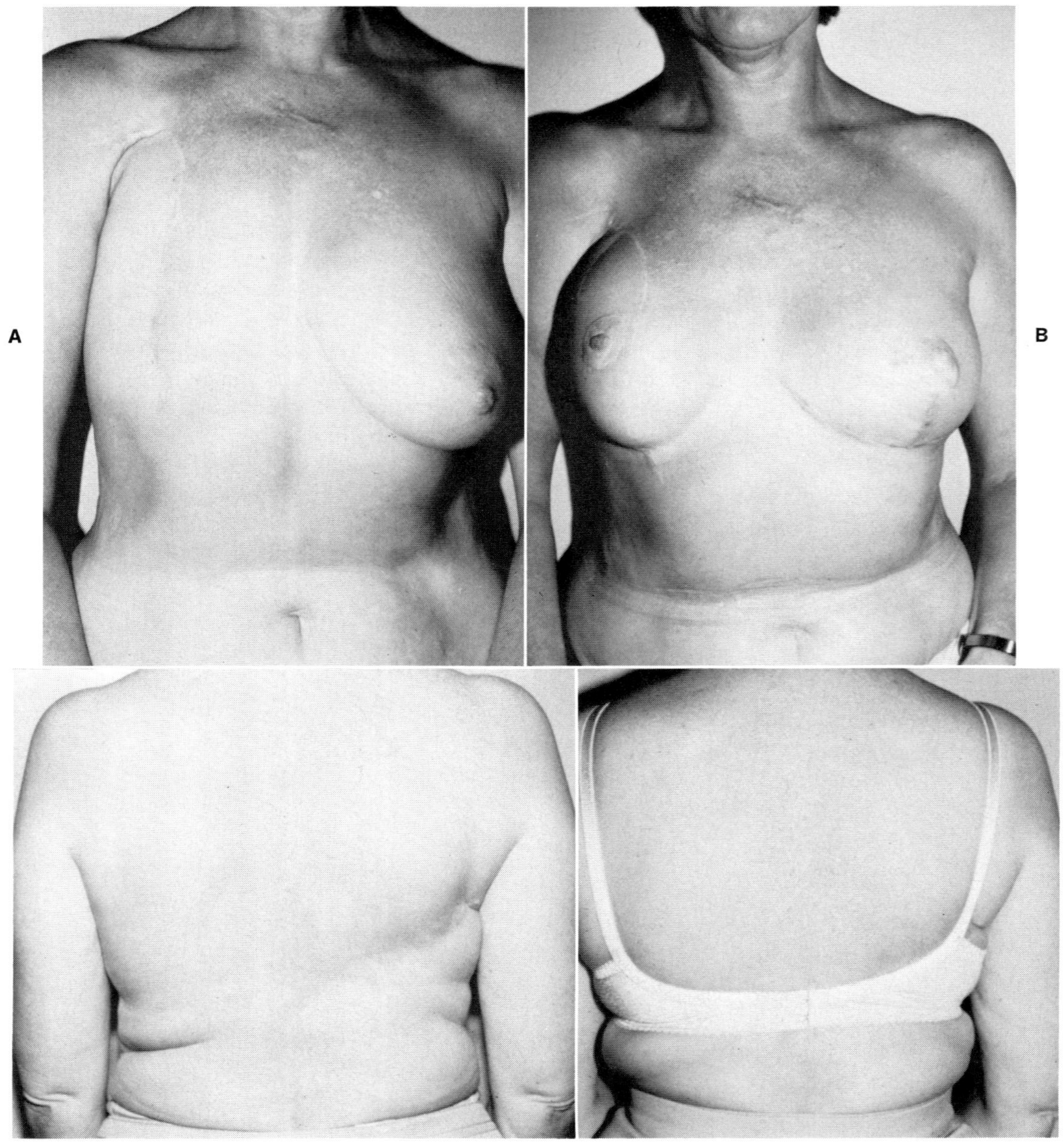

Fig. 14-11
A, Fifty-two-year-old woman 5 years after right radical mastectomy.
B, One year after reconstruction with right latissimus dorsi musculocutaneous island flap and 280 cc silicone breast prosthesis. Left reduction mammoplasty was done.
C, Patient's back 1 year after operation.
D, Brassiere covers donor site scar.

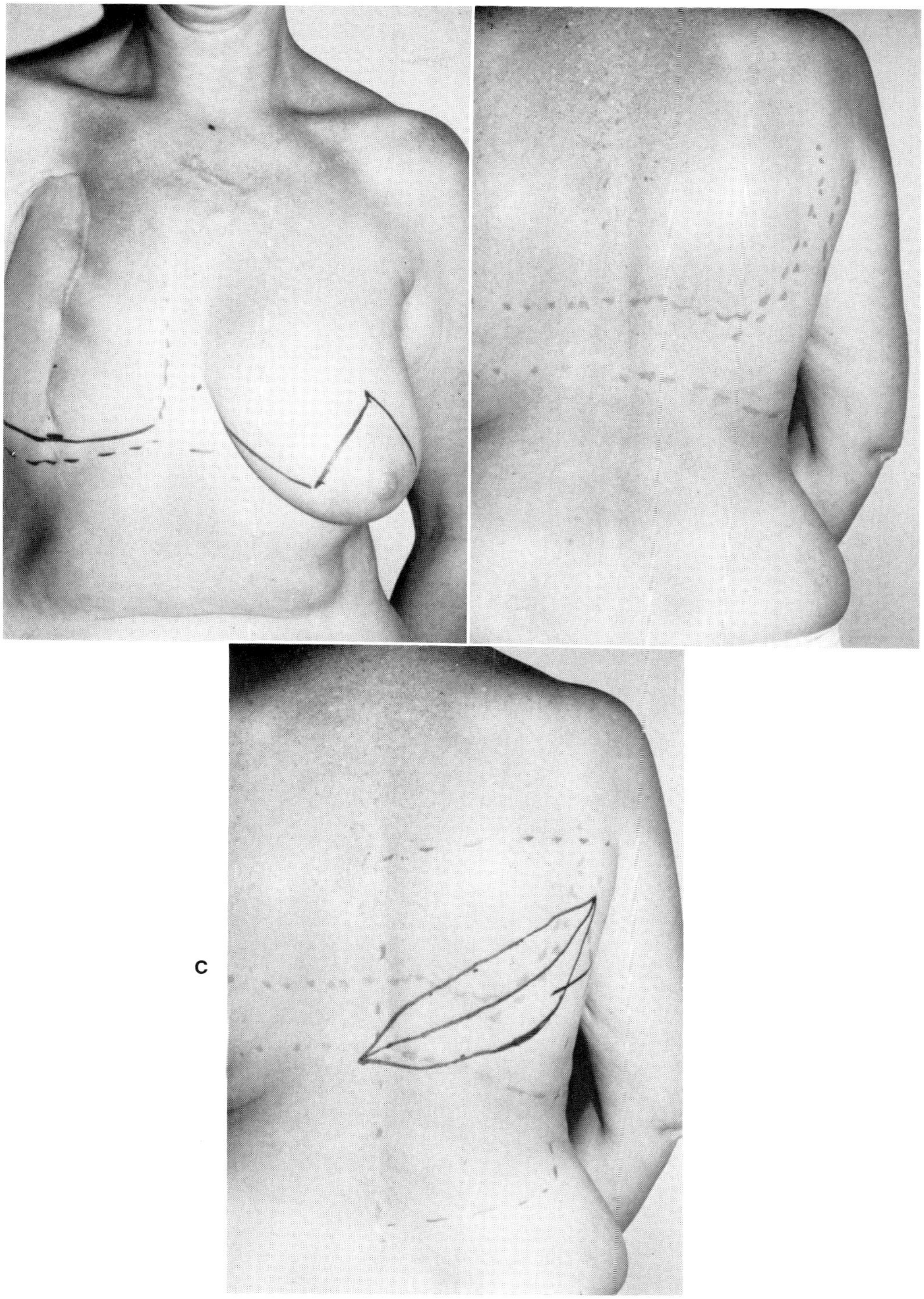

A B C

Fig. 14-12

A, Preoperative markings—anterior. Latissimus dorsi flap to be placed after excision of scars.

B, Outline of brassiere.

C, Flap is drawn to be covered by brassiere.

A 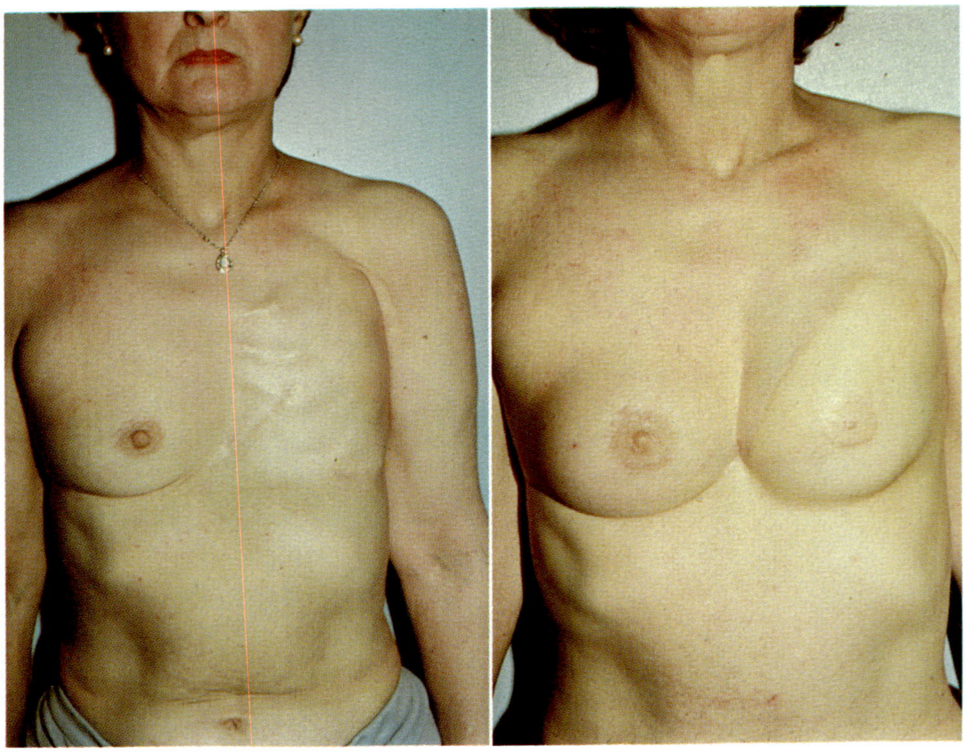 B

Fig. 14-13

A, Forty-seven-year-old woman 4 years after radical mastectomy.

B, One year after left breast reconstruction with latissimus dorsi musculocutaneous island flap. Insertion of latissimus dorsi was moved around to divided insertion of pectoralis major.

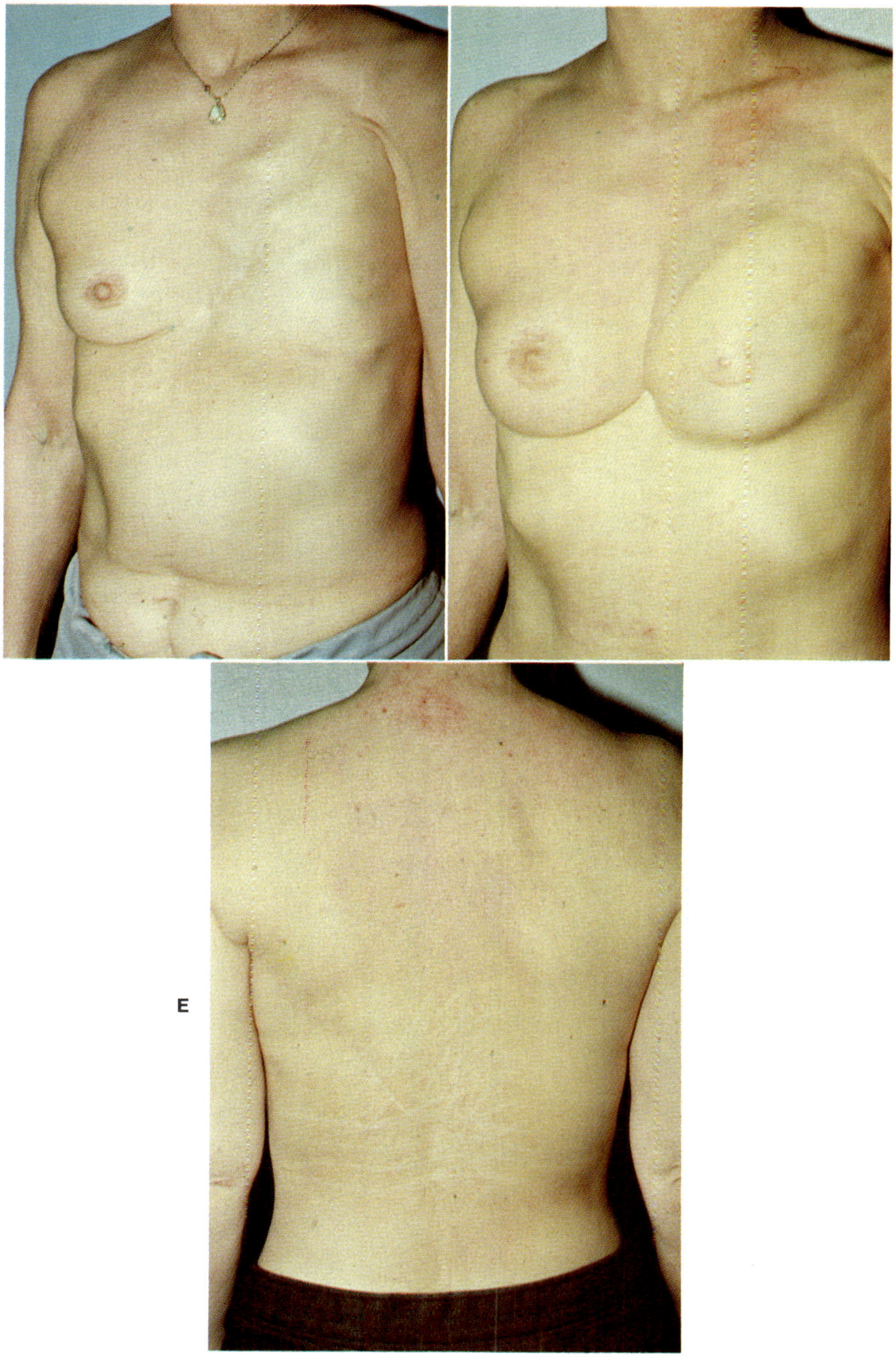

C

D

E

Fig. 14-13, cont'd
C and **D,** Oblique views.
E, Back donor site 1 year after operation.

BREAST RECONSTRUCTION AFTER MODIFIED RADICAL MASTECTOMY

Some general surgeons question the need to perform the usual radical mastectomy in each patient with breast cancer. Many tumors are detected while small and without axillary spread, particularly to the apex of the axilla. The radical mastectomy is modified to preserve the pectoralis major and often the pectoralis minor muscles (Fig. 14-14). Wide skin excision and extremely thin skin flaps are also modified, thus preserving more tissue and simplifying future breast reconstruction. Today the modified radical mastectomy is the accepted local treatment for most invasive breast cancers, although many surgeons continue to perform the radical mastectomy.

If the patient has had a modified radical mastectomy, and there is sufficient skin, no additional pedicled skin or muscle is usually necessary. A skin graft, prior radiation, or very thin skin of poor quality necessitates additional tissue (Fig. 14-15).

If the pectoral nerves, especially the medial pectoral nerve, have been divided, the pectoralis major muscle atrophies, and additional muscle tissue must be moved in to obtain adequate contour and symmetry in the infraclavicular area.

In most cases after modified radical mastectomy up to a 300-cc implant may be placed during the first operation. Because multiple problems are associated with subcutaneous placement and fibrous contracture, exposure, and positional change of the implant, the submuscular position is preferred.

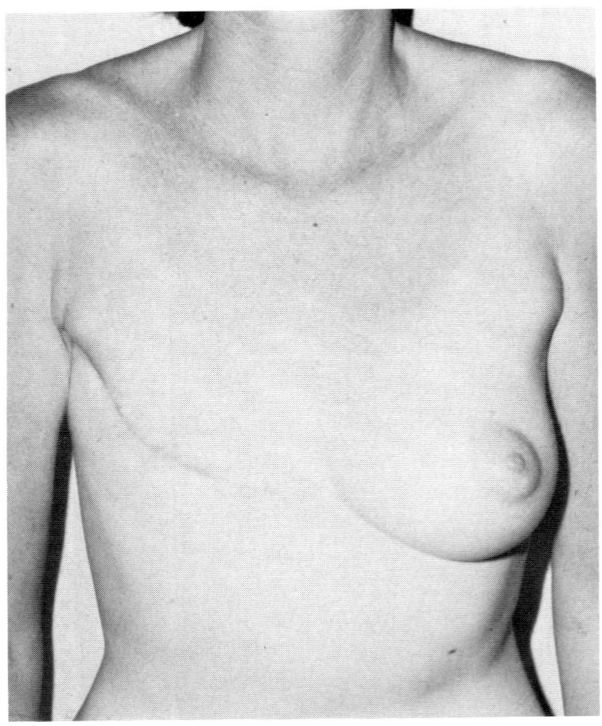

Fig. 14-14. Thirty-seven year-old woman 3 months after total mastectomy with axillary dissection (modified radical mastectomy), with preservation of pectoral nerves.

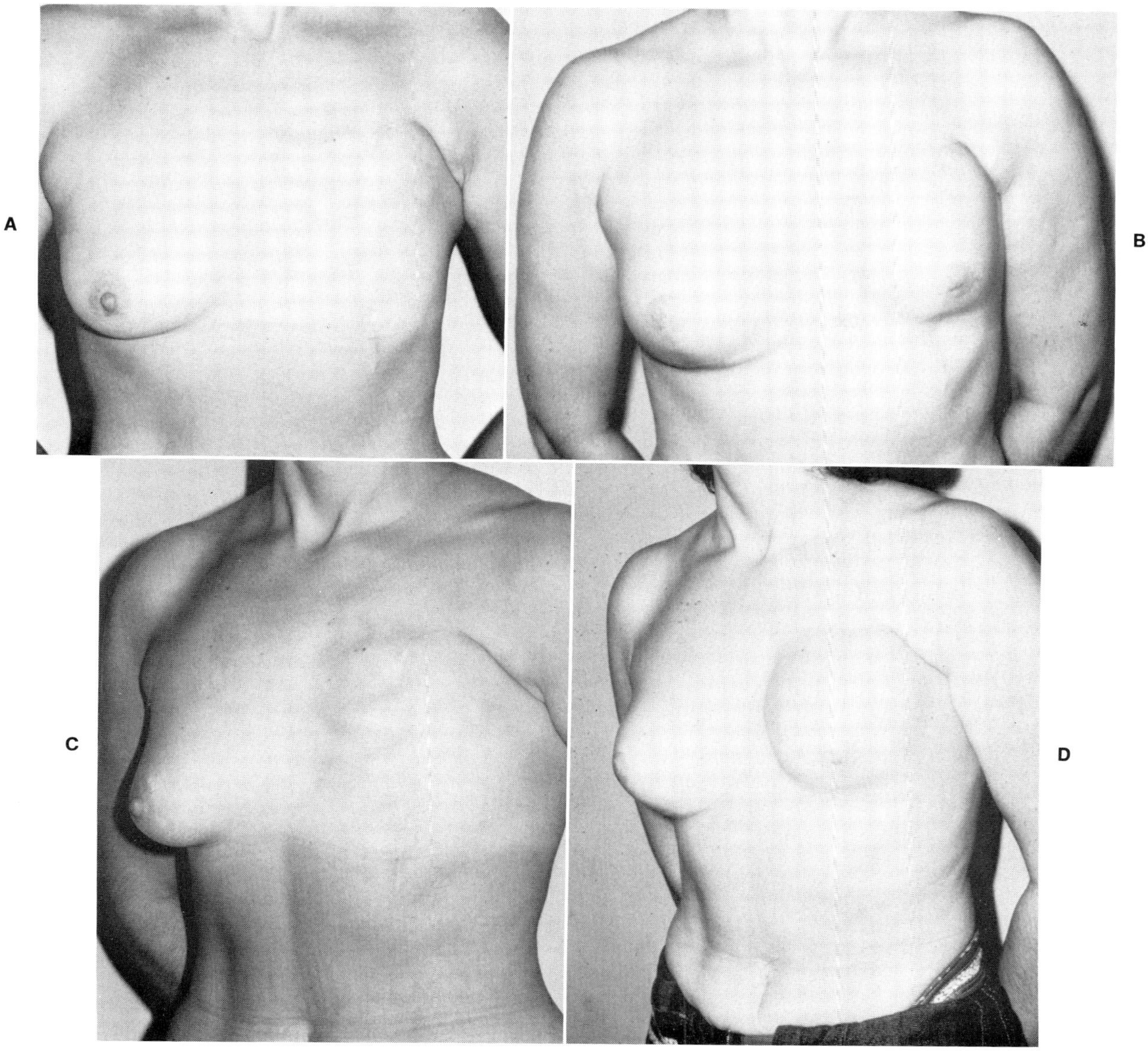

Fig. 14-15. Thirty-two-year-old woman 2 years after left radical mastectomy and postoperative irradiation. She was reconstructed with subcutaneous implant only, resulting in suboptimal reconstruction and tight capsular contracture.
A and **B,** Anterior views after mastectomy and 2 years postoperative reconstruction.
C and **D,** Oblique views after mastectomy and 2 years postoperative reconstruction.

Submuscular technique

The old incision is usually opened laterally over the junction of the serratus anterior and pectoralis major muscles. This fascial interval is opened and the pectoralis major muscle, serratus anterior fascia, and upper rectus sheath are elevated bluntly. The main vessels encountered are the intercostal perforators to the pectoralis major. The pectoralis major flaps are elevated from both the ribs as well as the lower sternum. The overlying fascia is left intact. The rectus abdominis fascia must be elevated 1.5 to 2 cm below the opposite inframammary crease, depending on the size of the implant to be placed. This avoids an unnaturally high inframammary crease on the reconstructed side (Figs. 14-16 and 14-17).

A 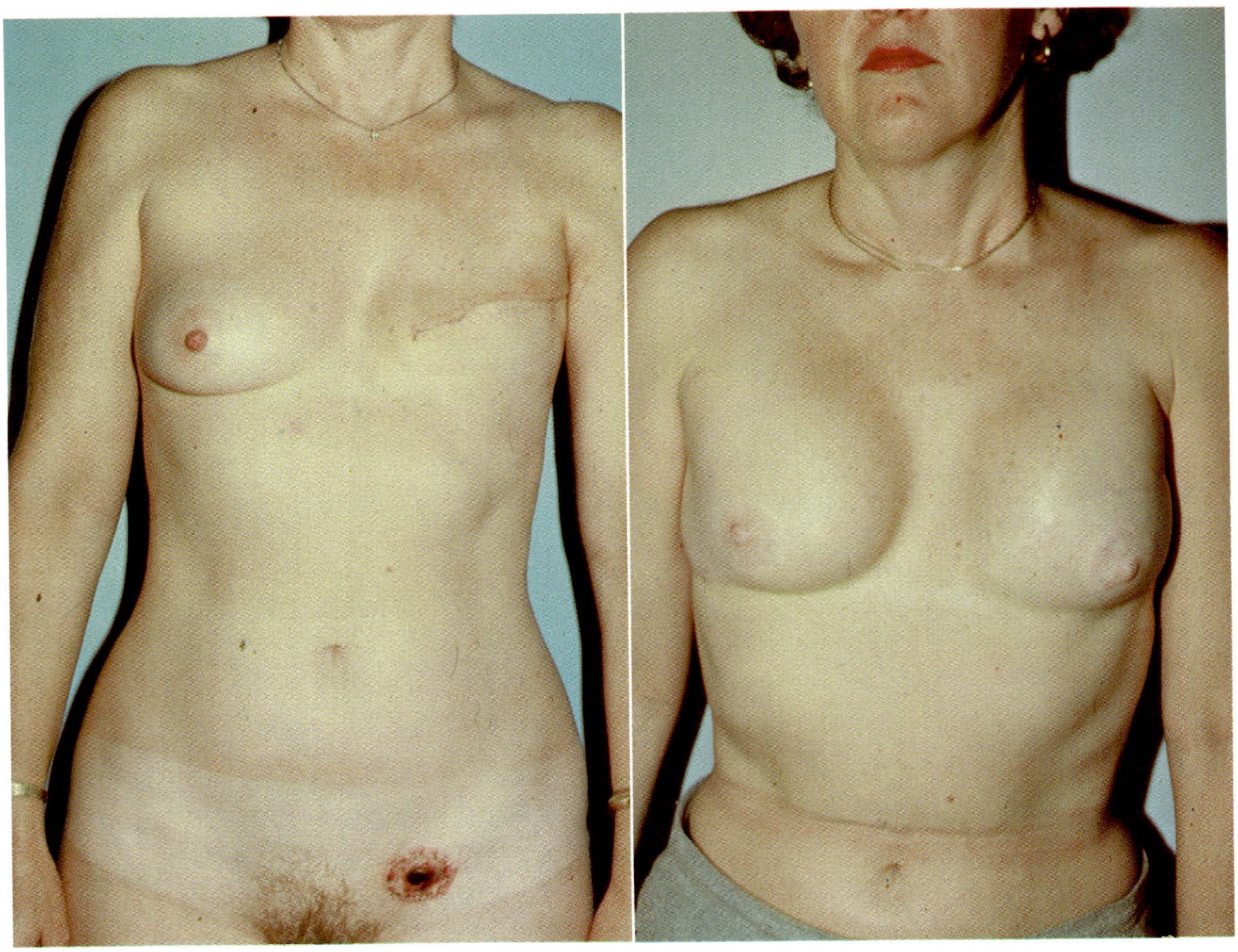 B

Fig. 14-16
A, Thirty-one-year-old woman 4 weeks after left modified radical mastectomy for 1-cm infiltrating ductal carcinoma. Nipple-areola was stored in groin.
B, Three years after left submuscular reconstruction and right total mastectomy and immediate nipple-areola reconstruction. We no longer bank nipple-areola in groin because of aesthetic and oncologic considerations.

The serratus anterior fascia is elevated laterally 1 to 2 cm beyond the lateral projection of the opposite breast. Sutures are placed in the divided fascia and the implant positioned and sutures tied. The patient should be in the sitting position to ascertain the final proper positioning. Placing the arms to the side reduces the tightness of the pectoralis major muscle. The implant will rarely go lower after the operation, but any fibrous contracture elevates the implant along with the inframammary crease.

The incidence of exposure and capsular contracture is much less when the implant is placed submuscularly. When a capsule develops in the submuscular position, it is usually firm; however, the breast does not appear as spherical as the contractures that develop in the subcutaneous position.

A

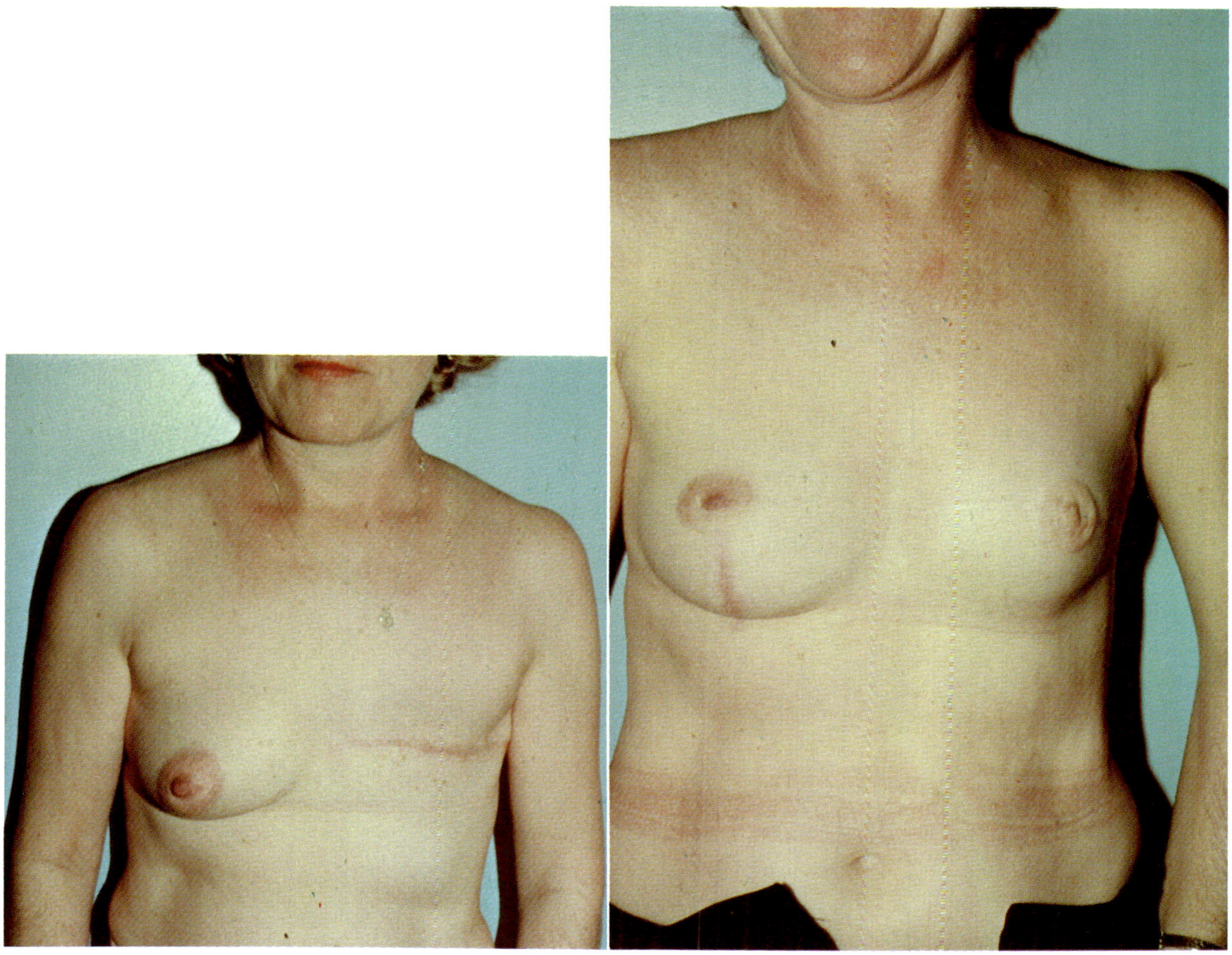

B

Fig. 14-17
A, Forty-four-year-old woman 1 year after left modified radical mastectomy.
B, View 9 months after left submuscular reconstruction and right mastopexy.

The latissimus dorsi musculocutaneous flap

After many modified radical mastectomies, if the medial pectoral nerve has been divided, there slowly develops an infraclavicular hollowness secondary to muscle atrophy. Other patients who have thicker subcutaneous tissue also have a notable asymmetry and flattening after the mastectomy. These patients benefit from the usual latissimus dorsi musculocutaneous island flap, as used in patients after radical mastectomy.

For patients who have a good pectoralis major muscle but with insufficient skin, the latissimus dorsi flap is modified so that the skin is replaced, and the latissimus dorsi muscle is sutured to the lower pectoralis major and to just below the inframammary crease (Figs. 14-18 and 14-19). This will allow an incision to be placed more laterally for a patient who prefers to wear clothes with an exposed back (Figs. 14-20 and 14-21). After the preoperative markings for the approximate skin and muscle tissue, the operation proceeds essentially the same as described for the reconstruction of the standard radical mastectomy (Fig. 14-22).

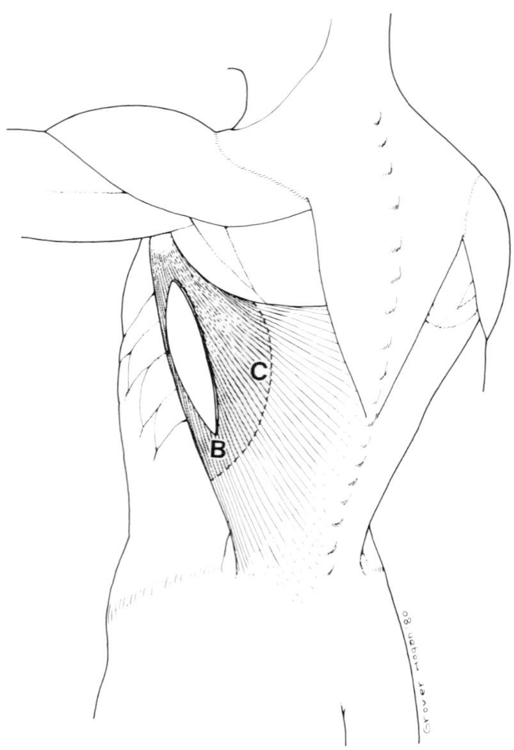

Fig. 14-18. Lateral design of latissimus dorsi flap with skin island for reconstruction after modified radical mastectomy. *B* and *C* show position of muscle during transposition.

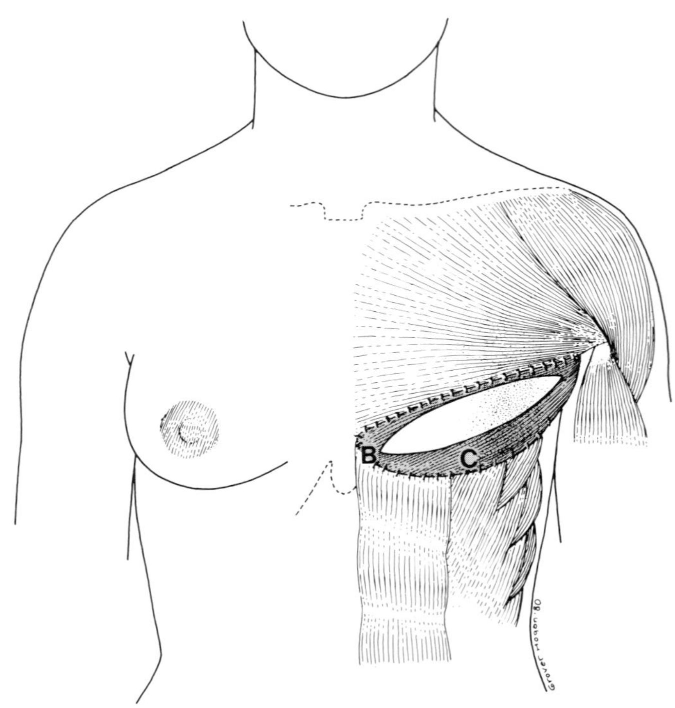

Fig. 14-19. Lateral latissimus flap is transposed and sutured to pectoralis major. *B* and *C* show position of muscle during transposition.

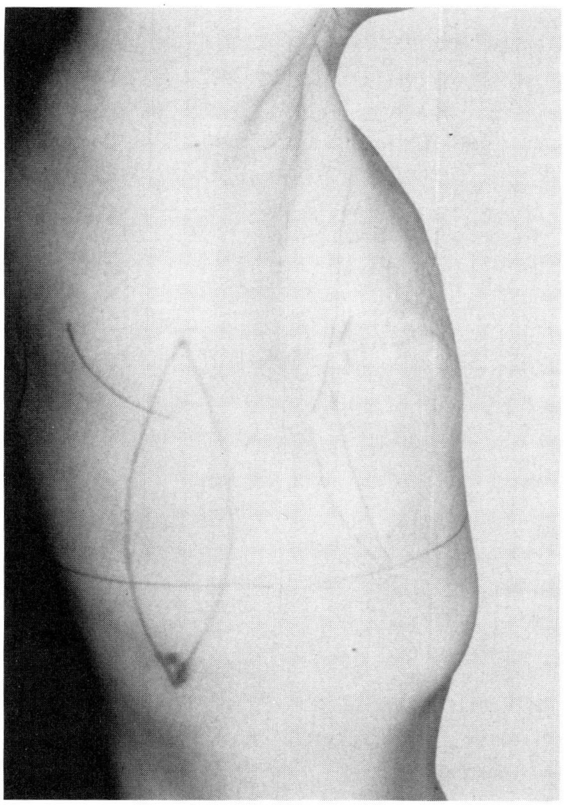

Fig. 14-20. In patients with modified radical mastectomy, skin island is designed laterally. Orientation is excellent when scar is oblique and pectoralis major muscle is preserved.

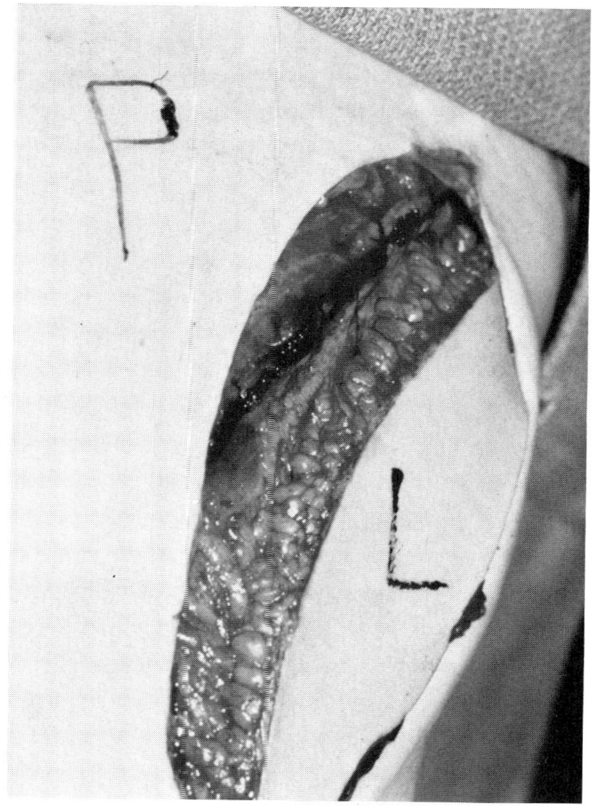

Fig. 14-21. In patients with modified radical mastectomy, the latissimus dorsi muscle is sutured to lower edge of the pectoralis major muscle. *P,* Pectoralis major; *L,* latissimus dorsi.

A

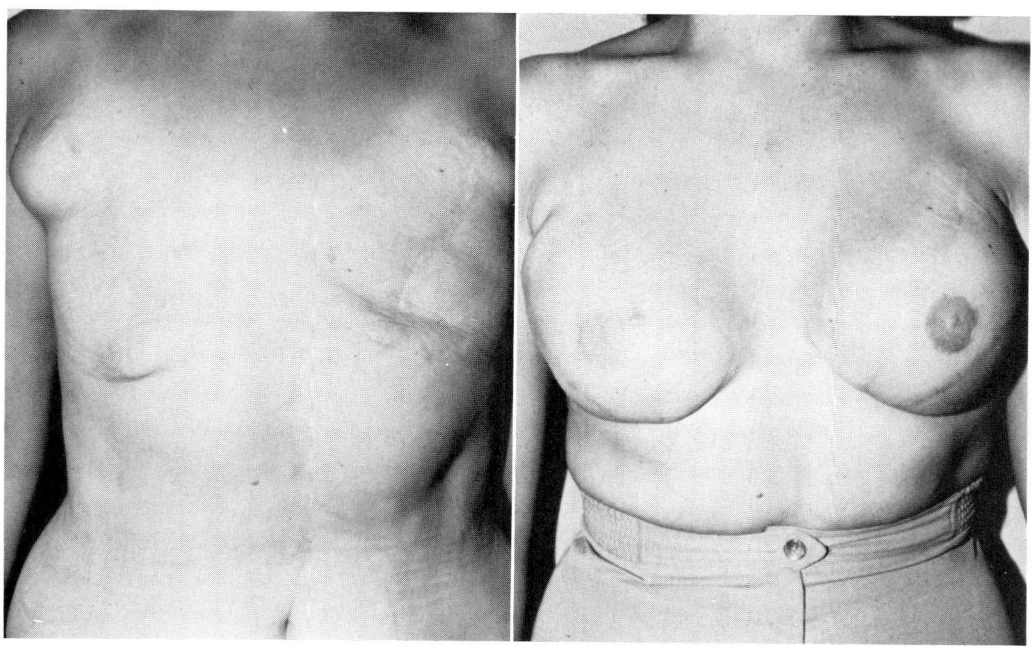

B

Fig. 14-22

A, Forty-three-year-old woman 5 years after bilateral modified radical mastectomies. She desired large breasts.

B, One year after one-stage bilateral latissimus dorsi musculocutaneous flaps and immediate nipple-areola reconstructions.

Poland's syndrome. Patients with Poland's syndrome have an absent pectoralis major muscle. They often have absence of the pectoralis minor and the serratus anterior. There is also malposition of the nipple-areola and underdevelopment of the breast. The latissimus dorsi muscle flap has been very useful in restoring the contour resulting from the absent pectoralis major muscle. It also allows for a submuscular positioning of a breast implant if this is necessary to achieve symmetry.

The procedure is delayed until after puberty. At this time the patient is fully aware of the defect and is cooperative. As in all patients after latissimus dorsi transposition, postoperative exercises and isometrics are necessary to maintain the bulk of the muscle and to contract over the silicone implant (Fig. 14-23).

A

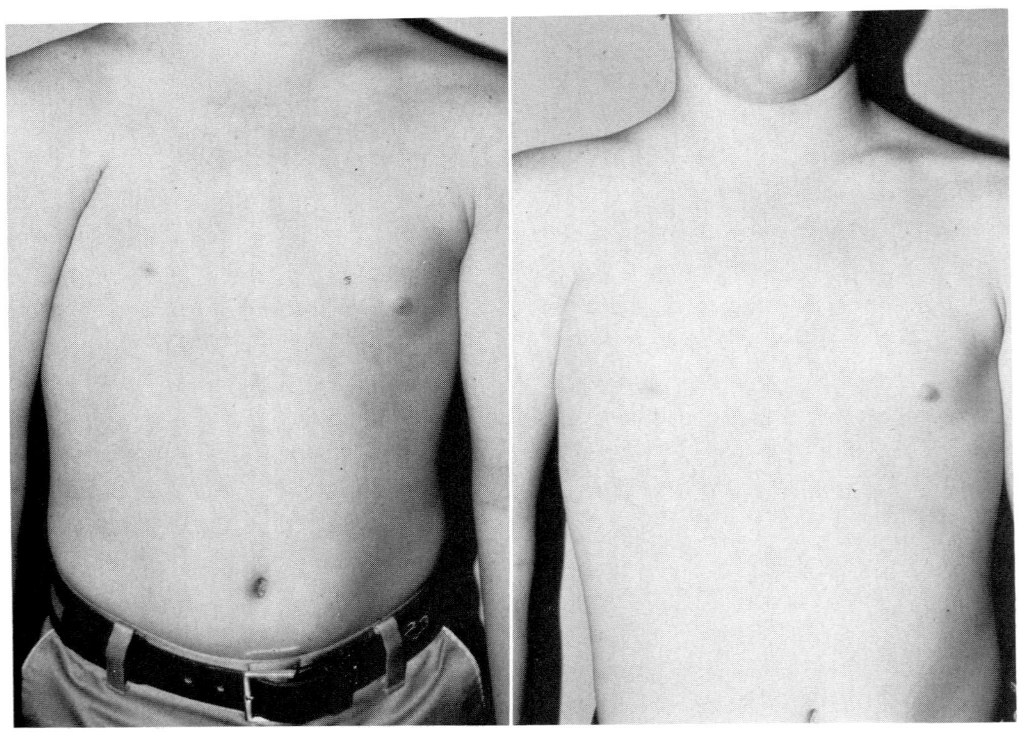

B

Fig. 14-23
A, Fourteen-year-old boy with Poland's syndrome.
B, Boy 1 year after transposition of latissimus dorsi muscle.

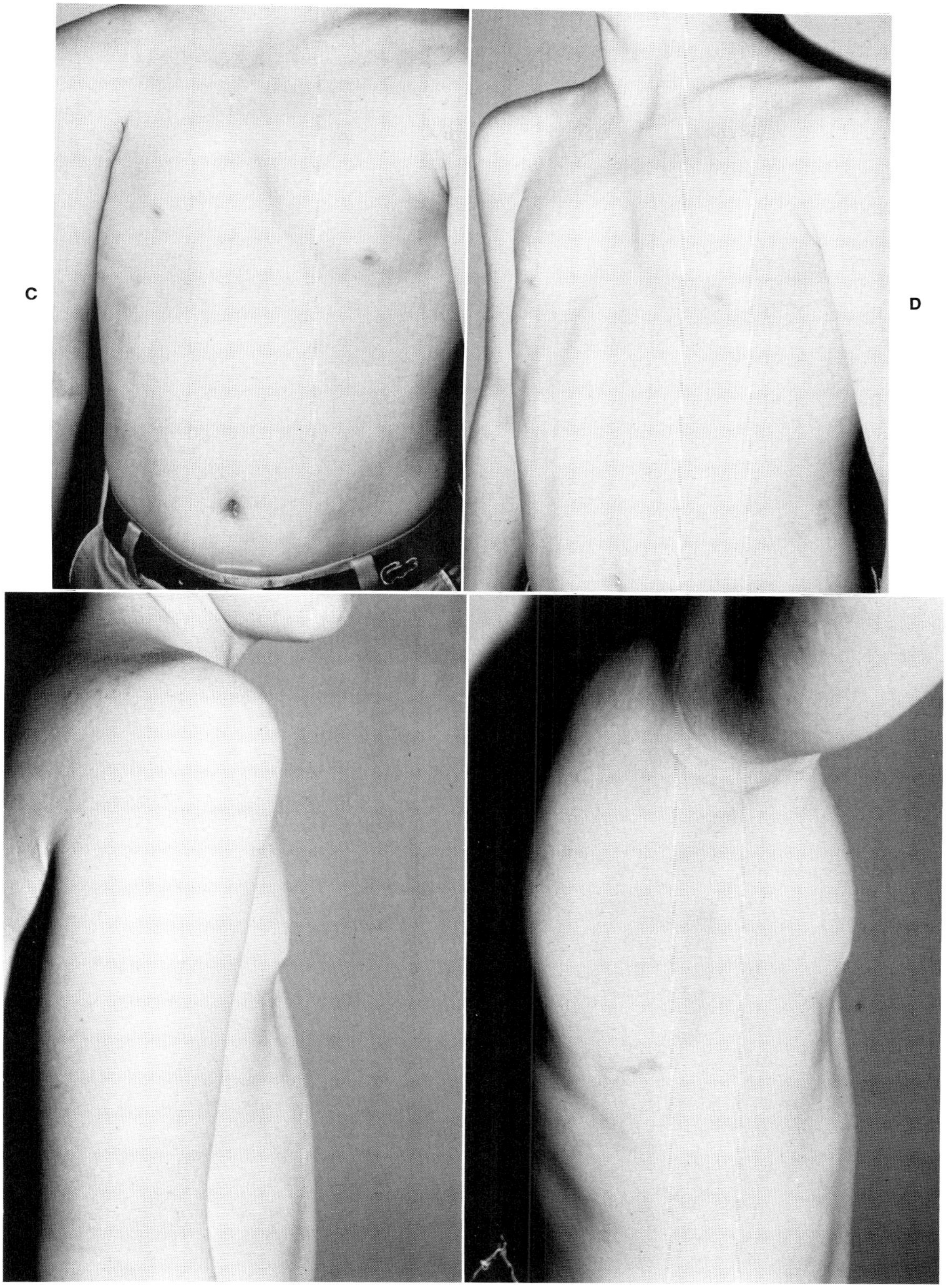

C D E F

Fig. 14-23, cont'd
C, Oblique view (preoperative).
D, Oblique view (pcstoperative).
E, Lateral view.
F, Lateral view showing donor scars.

USE OF THE LATISSIMUS DORSI FOR PATIENTS WITH COMPLICATIONS AFTER SUBCUTANEOUS MASTECTOMY AND AUGMENTATION MAMMOPLASTY

Some patients, after long-term implantation of a silicone gel breast implant beneath thin skin and who have undergone several procedures to relieve capsular contractures, develop impending or actual exposure of the breast prosthesis. Local measures and flaps of pectoralis major or serratus anterior have been suggested; however, what is so often needed is a small portion of skin and a layer of functioning, well-vascularized muscle between the implant and the skin.

A

B

C

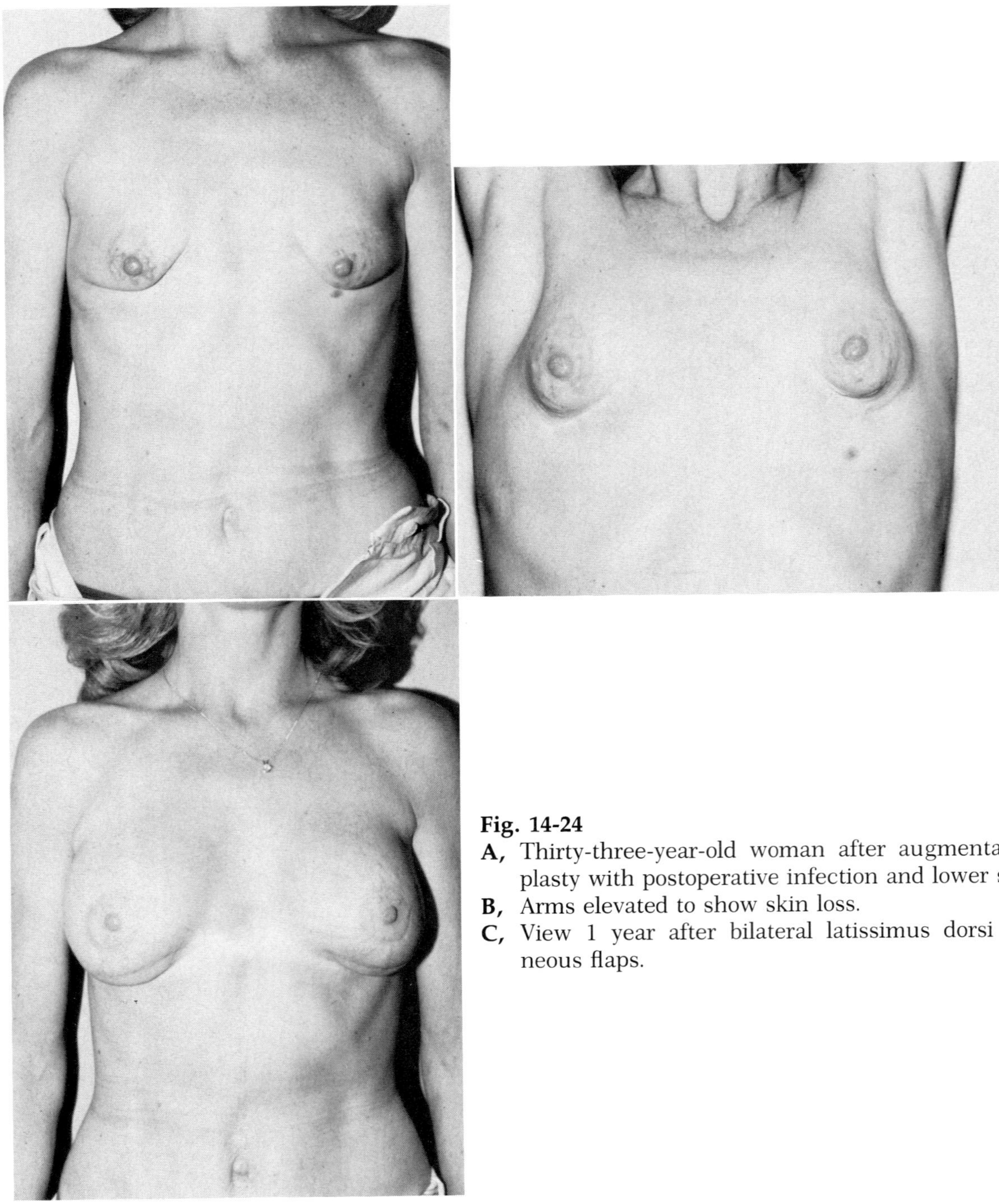

Fig. 14-24
A, Thirty-three-year-old woman after augmentation mammoplasty with postoperative infection and lower skin loss.
B, Arms elevated to show skin loss.
C, View 1 year after bilateral latissimus dorsi musculocutaneous flaps.

The latissimus dorsi muscle flap, and occasionally a small skin island obtained from the brassiere line in the back or a small laterally located skin island, provides this needed tissue. The muscle is elevated through the skin incision in the back. An axillary incision is necessary to complete the flap dissection. The breast incision is opened and the implant removed. The fibrous capsule is opened widely and a tunnel is made into the axilla. The latissimus dorsi muscle is then brought into the anterior breast pocket and sutured to the periphery of the pocket. The sutures are placed in position. The breast implant is positioned and the sutures tied. This gives a very satisfactory protective muscular covering to the implant, and most patients have continued to have soft breasts after the procedure (Figs. 14-24 and 14-25).

Custom-sized skin islands may be designed to replace skin that is thinned, particularly in the inframammary crease area. An alternative incision for obtaining the muscle is laterally over the anterior margin of the latissimus dorsi

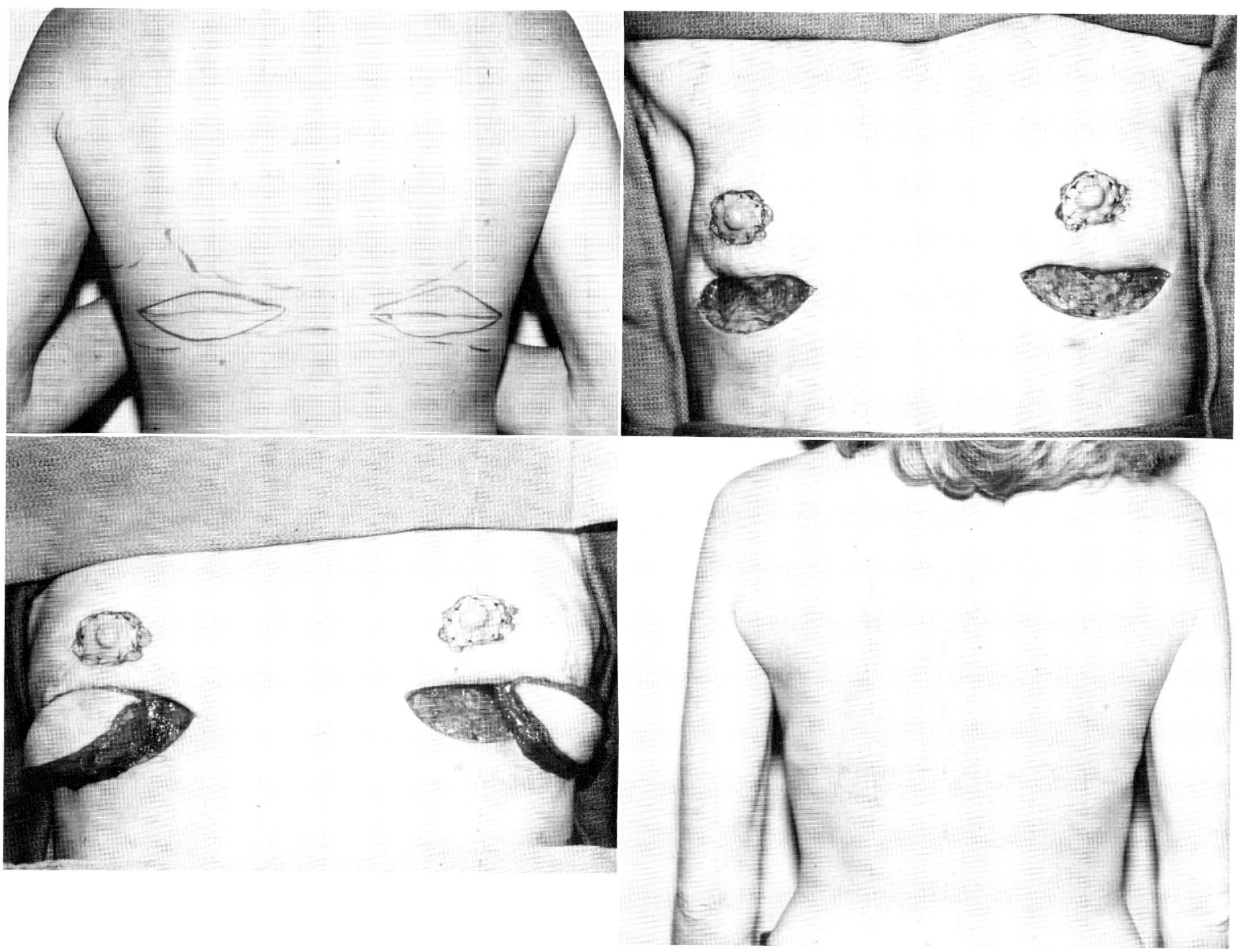

Fig. 14-25
A, Outline of flaps on back.
B, Defect made and periareolar reduction of nipple.
C, Flaps transposed anteriorly to be inset.
D, Donor site.

(From Pendergrast, W., Jr., Bostwick, J., III, and Jurkiewicz, M.J.: The subcutaneous mastectomy cripple, Plast. Reconstr. Surg. **66:**554, 1980.)

A

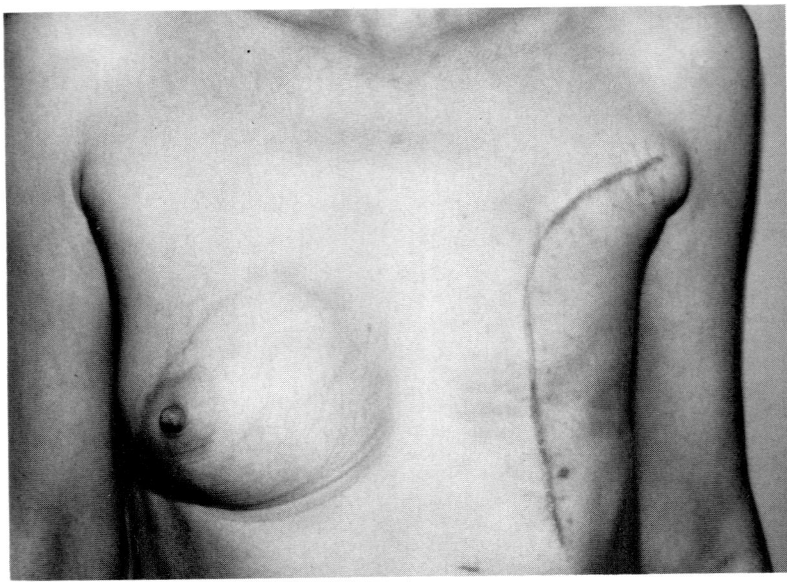

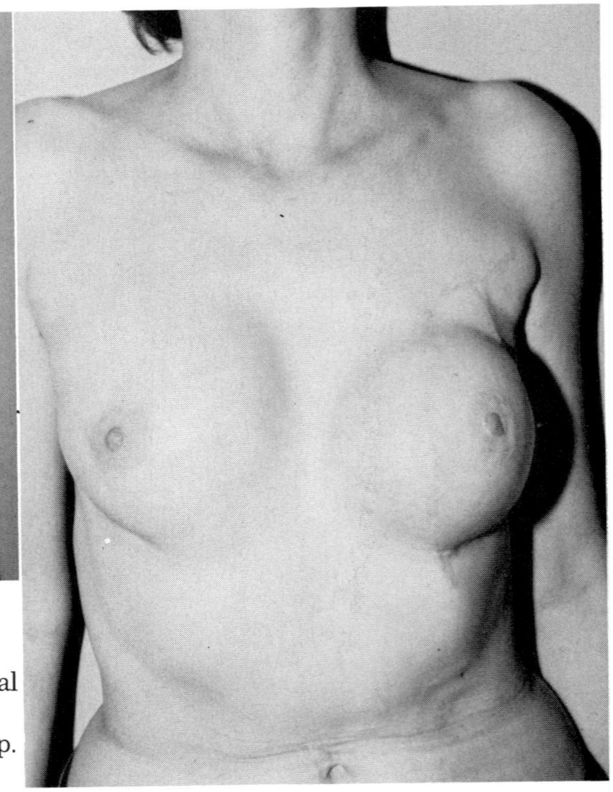

B

Fig. 14-26
A, Forty-four-year-old woman 6 months after modified radical mastectomy.
B, View 1 year after reconstruction with latissimus dorsi flap. Capsular contracture developed.

Complications of the latissimus dorsi flap

Experience with over 300 latissimus dorsi flaps have proved its versatility and safety. Seroma of the donor site is the most common complication and is seen in 30% of the patients. Use of suction drains and avoidance of unnecessary motion of the shoulder for the first few weeks seem to decrease the incidence. Capsular contraction has been seen in 30% of the patients 1 year after implantation of the breast prosthesis (Fig. 14-26). Color discrepancies have been noted (Fig. 14-27).

Hematoma was seen in 1% of the patients and is usually seen anteriorly; 1% of the implants developed a surrounding infection and required temporary removal. Partial skin island loss occurred in 2% of the flaps and was related to division of the pedicle at a previous mastectomy. It is necessary to preserve the crossing collateral vessel from the serratus anterior in these situations.

When the thoracodorsal artery, nerve, and vein have been divided at the time of mastectomy, a skin island may be successfully transferred; however, there is no muscle bulk. Infraclavicular fill for these patients is obtained either with a custom-designed prosthesis or dermis fat grafts taken from an abdominoplasty resection. As long as the serratus anterior crossing vessel is preserved, the denervated flap survives.

Nipple-areola reconstruction. The best tissue for nipple-areola reconstruction is from the opposite breast. Areola tissue may be obtained if a simultaneous total mastectomy or reduction mammoplasty is done.

If opposite areola tissue is not available, a thinned full-thickness graft from the upper inner thigh usually provides an excellent match. It does not develop hyperpigmentation as does vulva tissue.

The lower half of the opposite nipple is the best tissue for nipple reconstruction, and it is sutured first to the center of the deepithelialized proposed nipple-areola. We do not recommend banking of the nipple-areola because of loss of pigment, necessity for additional procedure, and possibility of transfer of tumor.

Auricular tissue is also useful at times, and the earlobe is an alternative source of tissue for a nipple. Retroauricular skin provides pink tissue when this is needed for areola reconstruction.

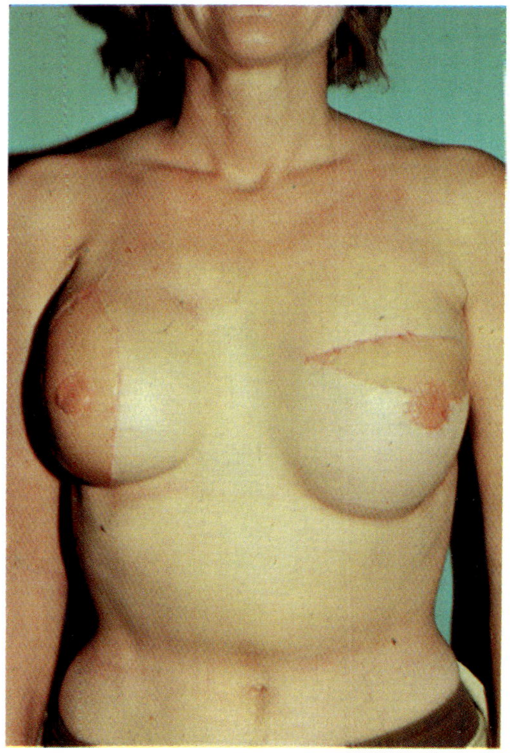

Fig. 14-27. Color discrepancy when tanned skin from back is moved anteriorly to untanned area.

ANNOTATED BIBLIOGRAPHY

1. Bostwick, J., III, Vasconez, L.O., and Jurkiewicz, M.J.: Breast reconstruction after radical mastectomy, Plast. Reconstr. Surg. **61:**682, 1978.

A description of a series of patients with breast reconstruction is presented. Techniques include transverse abdominal flap and latissimus dorsi musculocutaneous island flap.

2. Brent, B., and Bostwick, J., III: Nipple-areola reconstruction with auricular tissues, Plast. Reconstr. Surg. **60:**353, 1977.

This article describes the use of auricular tissues for nipple-areolar reconstruction—earlobe for nipple, reconstruction, auricular cartilage for nipple augmentation, and Montgomery's gland and retroauricular skin for areola reconstruction.

3. Cronin, T.D., and Gerow, F.J.: Augmentation mammoplasty: a new "natural feel" prosthesis. Trans. Third Int. Cong. Plast. Surg., Amsterdam, 1964, Excerpta Medica Foundation.

A description of the development and use of the silicone gel prosthesis is presented. This is the current material used for creating or augmenting the breast mound.

4. D'Este, S.: Le technique de l'amputation de la mamelle pour carcinome mammaire, Rev. Chir. Orthop. Paris **45:**164, 1912.

D'Este describes the technique of Tansini. This is a skin flap that contains the latissimus dorsi muscle and sometimes the teres major. The flap is used for closure of the defect after radical mastectomy. This is the first description of the latissimus dorsi musculocutaneous flap.

5. Gillies, H.: Surgical replacement of the breast, Proc. R. Soc. Med. **52:**597, 1959.

This describes the use of the abdominal tube pedicle flap to transfer tissue to the chest for use in breast reconstruction. It requires many operations, and the reliability of the tube pedicle is not as great as musculocutaneous flaps.

6. Halsted, W.S.: The results of radial operations for the cure of cancer of the breast, Ann. Surg. **46:**1, 1907.

This paper outlines Halsted's experience in the development of the radical mastectomy for the local treatment of breast cancer. This is the first effective local treatment described for breast cancer. All local surgical treatment today has evolved from Halsted's work.

7. Hutchins, E.H.: A method for the prevention of elephantiasis, Surg. Gynecol. Obstet. **69:**795, 1939.

   A description of the use of the latissimus dorsi muscle to cover the axillary vessels after radical mastectomy is presented. This was proposed to decrease the incidence of lymphedema after radical mastectomy.

8. McCraw, J.B., Dibbell, D.G., and Carraway, J.H.: Clinical definition of independent myocutaneous vascular territories, Plast. Reconstr. Surg. **60:**341, 1977.

   A description of evolution and definition of many musculocutaneous territories and the clinical applications of these flaps are presented.

9. Pendergrast, W., Jr., Bostwick, J., III, and Jurkiewicz, M.J.: The subcutaneous mastectomy cripple, Plast. Reconstr. Surg. **66:**554, 1980.

   This article describes patients with problems with skin loss and implant exposure following subcutaneous mastectomy and the use of the latissimus dorsi musculocutaneous flap for this problem.

10. Pontes, R.: Single stage reconstruction of the missing breast, Br. J. Plast. Surg. **27:**377, 1973.

    This is the best description of the technique of sharing the sound breast to reconstruct the breast after radical mastectomy. It requires a large contralateral breast. There is concern on oncologic grounds for transfering the breast to the opposite side.

11. Schlenker, J.D., and co-workers: Loss of silicone implants after subcutaneous mastectomy and reconstruction, Plast. Reconstr. Surg. **62:**853, 1978.

    This article outlines the complications seen after subcutaneous mastectomy. Schlenker and co-workers particularly note delayed infections, capsular contractures, and exposure and loss of implants.

12. Schneider, W.J., Hill, H.L., and Brown, R.G.: Latissimus dorsi myocutaneous flap for breast reconstruction, Br. J. Plast. Surg. **30:**277, 1977.

    A description of the latissimus dorsi musculocutaneous flap first used as an island to replace missing skin and muscle after a radical mastectomy is presented. This is the first description of a one-stage reconstruction of the breast, using a musculocutaneous flap.

13. Snyderman, R. K., and Guthrie, R.H.: Reconstruction of the female breast following radical mastectomy, Plast. Reconstr. Surg. **47:**565, 1971.

    This article describes the use of the silicone prosthesis for creating a breast mound beneath the flaps after radical mastectomy. This paper led to increasing interest in breast reconstruction.

15

COMPLICATIONS

Stephen J. Mathes ■ Foad Nahai

SUBMUSCULAR BREAST AUGMENTATION

Planning errors

Current techniques using the silicone implant are effective in correction of congenital hypomastia. Although the frequency of breast shape distortion as a result of capsule formation is reduced by submuscular placement of the implant, this complication may still occur.

Submuscular breast augmentation may accentuate existing breast ptosis.

Technical errors

Satisfactory local anesthesia is difficult to achieve for a submuscular muscle augmentation and often requires use of interoperative, intravenous hypnotics and analgesics.

Failure to release the pectoralis major fibers of insertion to the fifth and sixth ribs may result in an inadequate breast pocket. Unsatisfactory breast projection and shape result if the implant is held superiorly to the inframammary line.

Postoperative errors

Contractions of the pectoralis major muscle associated with strenuous exercise may cause distortion of the breast shape and discomfort.

TOTAL MASTECTOMY

Planning errors

The primary objective of total mastectomy is removal of breast tissue with increased potential for development of carcinoma. Optimal breast shape and projection are not uniformly achieved.

Residual breast tissue in the axilla and adjacent to breast skin are possible sources of neoplasia.

Total mastectomy may not relieve mastodynia.

Technical errors

Failure to maintain the continuity of the inferior pectoralis major muscle with the serratus anterior muscle may result in difficulty in providing complete muscular coverage of the silicone implant.

Failure to release pectoralis major fibers of insertion to the fifth and sixth ribs may result in an inadequate breast pocket. The lower fibers of the pectoralis major muscle may then push the implant toward the clavicle, resulting in unsatisfactory breast shape and projection.

Postoperative errors

The submuscular location of the implant generally avoids implant exposure after skin flap necrosis. However, wound infection occurring within the implant pocket requires temporary removal of the implant (Fig. 15-1).

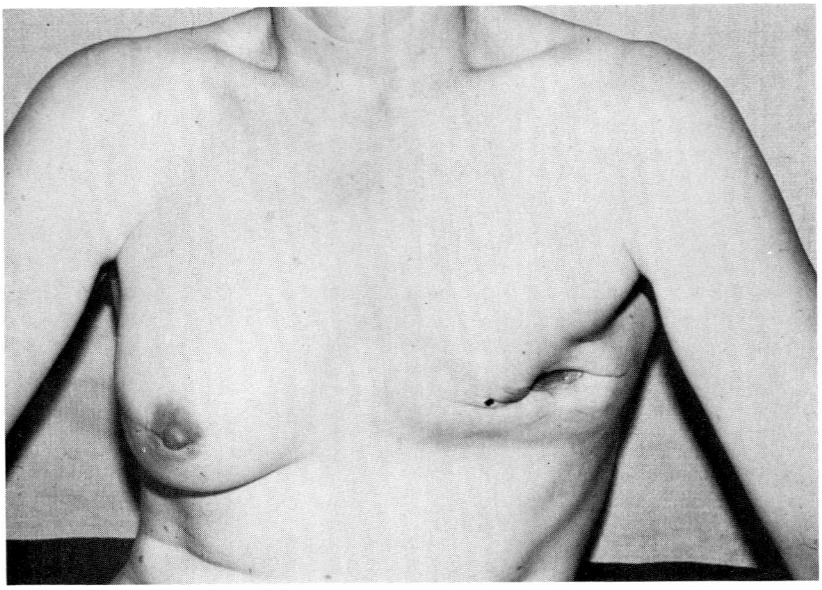

Fig. 15-1. Postoperative wound infection on left side after bilateral total mastectomy. Despite placement of implant beneath muscle, implant removal required. Reconstruction of left breast performed 6 months after wound healed.

MODIFIED RADICAL MASTECTOMY

Planning errors

Breast reconstruction provides a breast mound with acceptable shape, projection, and consistency but does not completely simulate the normal breast. Scars persist. Sensation is absent. Firmness may develop as a result of capsule formation. Patient expectations must not exceed the usual result of breast reconstruction observed by the reconstructive surgeon.

The reconstructive surgeon must interact with the general surgeon and medical and radiation oncologists in planning the timing of breast reconstruction. Breast reconstruction does not adversely affect the potential for cure in the patient after mastectomy. However, both local and systemic recurrent breast cancers are possible. Future treatment in patients with recurrence is not impaired by the presence of a reconstructed breast.

The inability to palpate the voluntary contractions of the latissimus dorsi muscle may indicate operative division of the thoracodorsal nerve during mastectomy. This indicates possible division of the adjacent thoracodorsal artery. However, transposition of the latissimus dorsi muscle is generally possible based on the reverse blood flow via the arterial branch from the thoracodorsal artery to the serratus anterior muscle.

Technical errors

Breast reconstruction by submuscular implant placement may not achieve optimal position and projection because of tight, denervated inferior fibers of the pectoralis major muscle and tight skin scars. Failure to release the pectoralis major fibers of insertion of the fifth and sixth ribs may result in an inadequate breast pocket pushing the silicone implant above the inframammary line.

Design of the skin island width of the latissimus dorsi musculocutaneous flap of more than 8 cm may not allow direct closure of the donor defect.

The denervated latissimus dorsi muscle is very thin and may be inadvertently divided during initial flap elevation.

During elevation of the latissimus dorsi flap, the arterial branch of the thoracodorsal artery to the serratus anterior muscle must be preserved. If the thoracodorsal artery has been divided during the mastectomy, a latissimus dorsi musculocutaneous flap may be safely transposed based on the vascular pedicle between the latissimus dorsi and serratus anterior muscle. Division of the thoracodorsal artery has been observed with normal muscle innervation. This vascular communication with the serratus anterior muscle is never divided unless an intact proximal thoracodorsal artery is visualized.

Release of axillary adhesions along the lateral edge of the preserved pectoralis major muscle allows the proximal portion of a transposed muscle to lie anteriorly to the midaxillary line. This prevents bending of the muscle in the axillary tunnel and avoids both muscle compression and a lateral bulge of a transposed muscle within the inferior axillae. The latissimus dorsi muscle is adherent to the serratus anterior muscle at the anterior axillary line. Prior separation

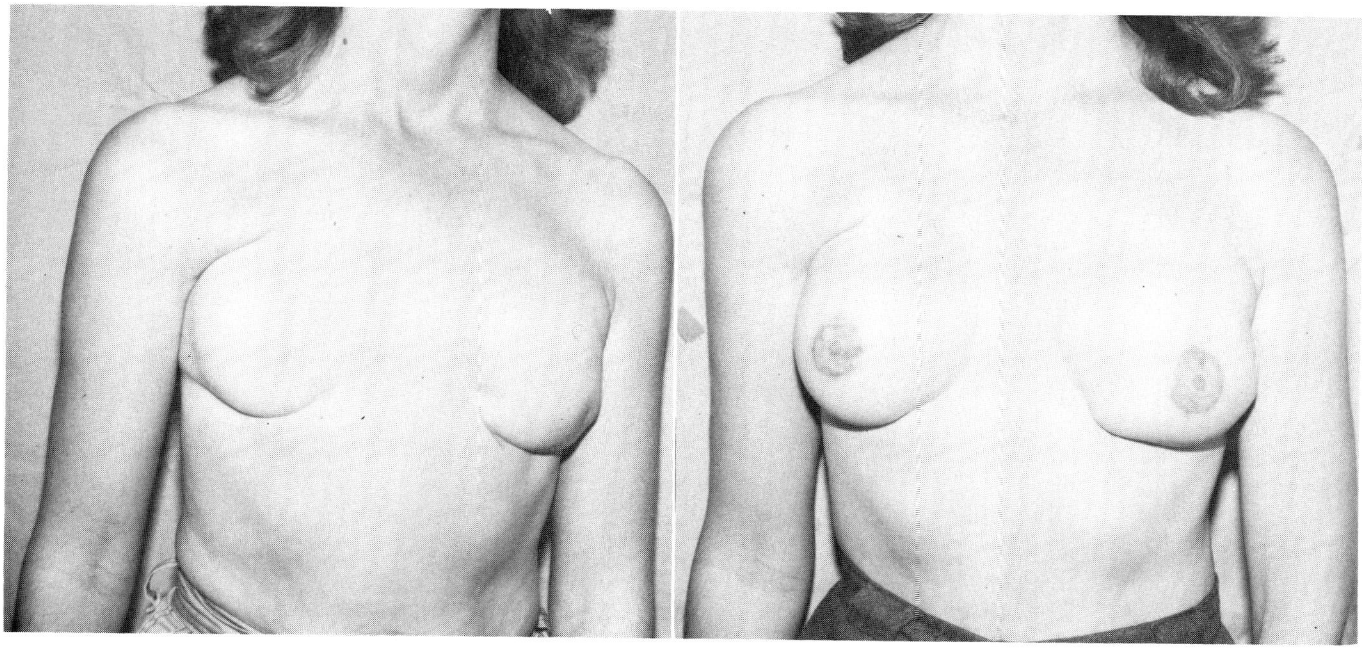

Fig. 15-2
A, Spontaneous deflation of bilateral saline implants 2 years after breast reconstruction.
B, Bilateral insertion of silicone gel implants and nipple-areola reconstruction.

of these muscles at the tip of the scapula avoids inadvertent elevation of both muscles during the dissection along the posterior axillary line.

Failure to close the communication of the breast pocket with the back wound beneath the transposed latissimus dorsi musculocutaneous flap may result in herniation of the implant into the axilla.

Tight, denervated fibers of the inferior pectoralis major muscle are divided to allow optimal breast projection.

The inferior border of the breast pocket on the anterior chest walls must be located at the level of the sixth rib slightly beneath the superior rectus sheath. An inadequate breast pocket at the level of the inframammary line results in unsatisfactory breast position.

The skin island of the latissimus dorsi musculocutaneous flap is inset obliquely at the inframammary line. Inset of this island transversely or vertically may result in inadequate release of inferior breast skin with impairment of breast projection. An abnormal fold may also result immediately above the inframammary line if the inferior border of the skin island is inset above the inframammary line.

Severe contralateral breast ptosis may require mastopexy. The use of implants on the breast reconstruction site in which volume adjustment (double lumen implant) is possible facilitates matching breast sizes (Fig. 15-2).

Postoperative errors

Postoperative closed suction drainage of the back donor area is recommended. Compressive chest dressings are avoided.

Intraoperative ligation of the secondary segmental blood supply of the latissimus dorsi muscle via the perforating intercostal lumbar arteries avoids postoperative hematoma formation. However, later seroma formation may occur, requiring needle aspiration.

Superficial wound infection may respond to appropriate antibiotic therapy. Infection located within the breast pocket requires implant removal (Fig. 15-3). Repeat operation for implant insertion when prior infection has occurred requires antibiotic coverage based on the prior bacterial sensitivity. Residual bacteria may reside in the breast pocket despite a long, asymptomatic interval between procedures.

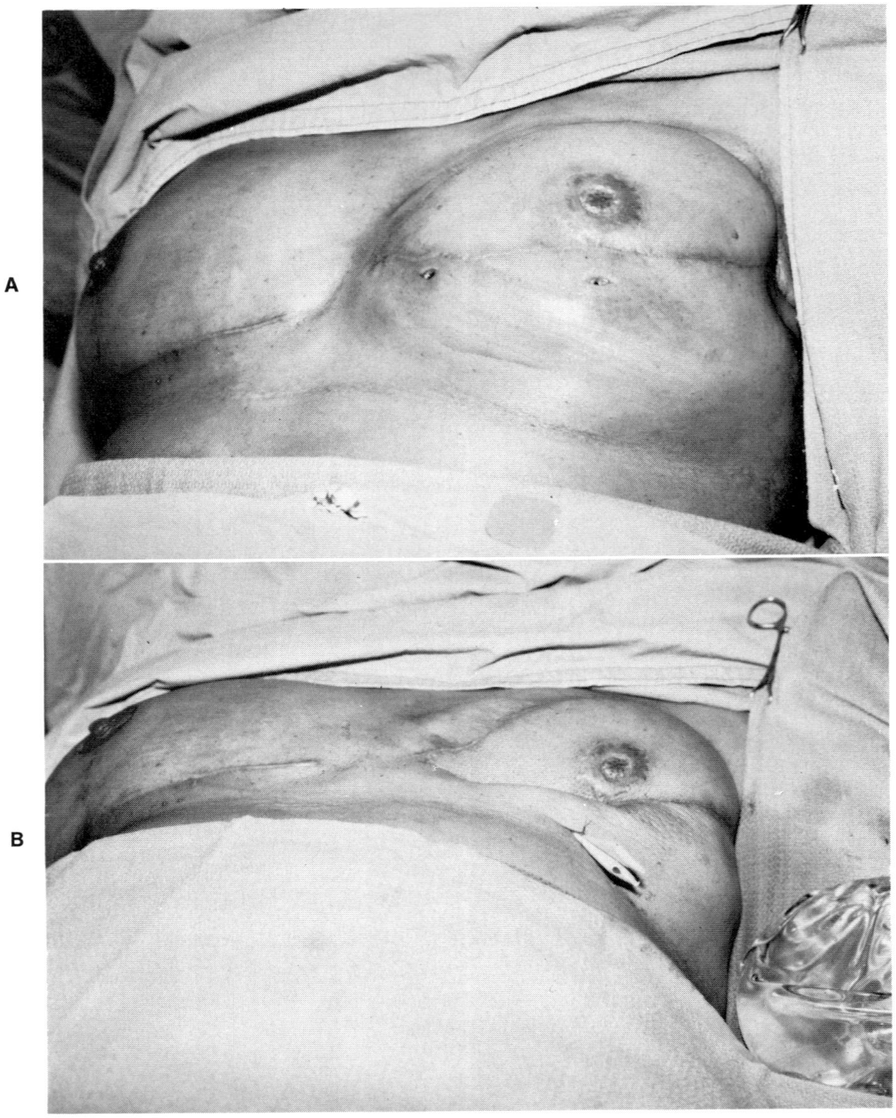

Fig. 15-3
A, Three-month postoperative view of breast reconstruction. Infection beneath latissimus dorsi musculocutaneous flap.
B, Implant removed and pocket drained. Gel implant reinserted 6 months after wound healed.

RADICAL MASTECTOMY

Planning errors

Planning errors in the radical mastectomy are the same as for the modified radical mastectomy. Design of the skin island obliquely may result in inadequate muscle for simulation of the pectoralis major muscle. The skin island located close to the superior muscle border allows a majority of the muscle that is located inferiorly to the skin island to transpose superiorly to the skin island on the anterior chest wall. Thus the latissimus dorsi muscle can be used to replace a missing pectoralis major muscle.

Technical errors

Technical errors in the radical mastectomy are the same as for the modified radical mastectomy. The anterior skin may have adhesions to the axillary and the subclavian vein as a result of the absence of the pectoralis major muscle. The dissection within the axilla and in proximity to the clavicle is performed under direct vision to avoid vascular injury.

Release of latissimus dorsi insertion and transposition to the anterior axilla may be used for restoration of the anterior axillary fold. This requires dissection adjacent to the thoracodorsal artery. Although this technique restores the anterior axillary fold, loss of the latissimus dorsi origin-insertion (length-width) ratios may accentuate late muscle atrophy.

Postoperative errors

Postoperative care in the radical mastectomy is the same as for the modified radical mastectomy.

16

A SYSTEMATIC APPROACH TO FLAP SELECTION

Stephen J. Mathes ■ Foad Nahai

Skin grafts

Skin grafts are method of choice provided vertebral column is not exposed

Local flaps (Figs. 16-1 and 16-2)

1. Posterior trapezius musculocutaneous flap is flap of choice for the upper third of the posterior trunk
2. Latissimus dorsi muscle or musculocutaneous flap is flap of choice for the middle third of the posterior trunk (Figs. 16-3 to 16-6)
3. Gluteus maximus muscle or musculocutaneous flap is flap of choice for the lower third of the posterior trunk (Fig. 16-7)
4. Transverse back flap is an alternative flap for coverage of midline back defects

Distant flaps

Distant flaps are rarely indicated in reconstruction of the posterior trunk

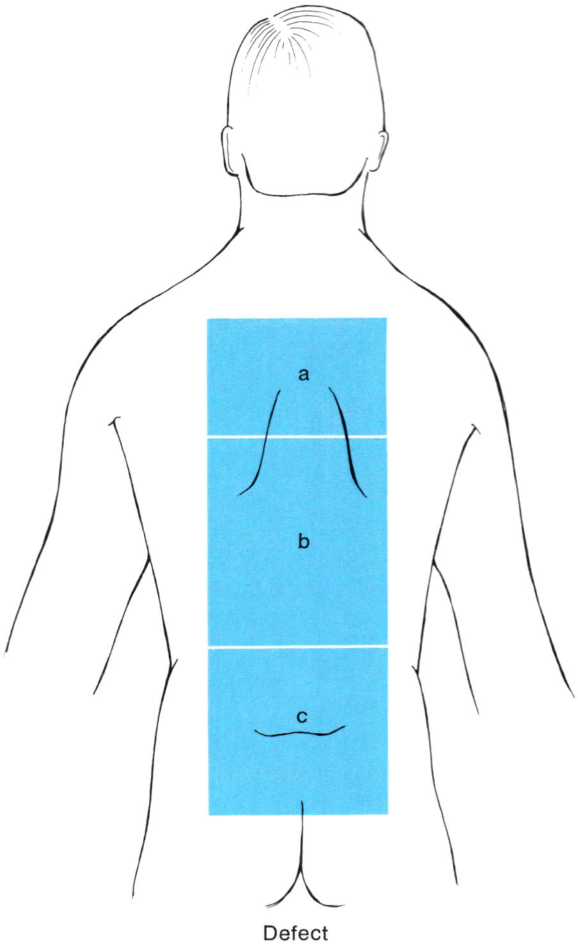

Fig. 16-1. *a*, Upper third; *b*, middle third; *c*, lower third.

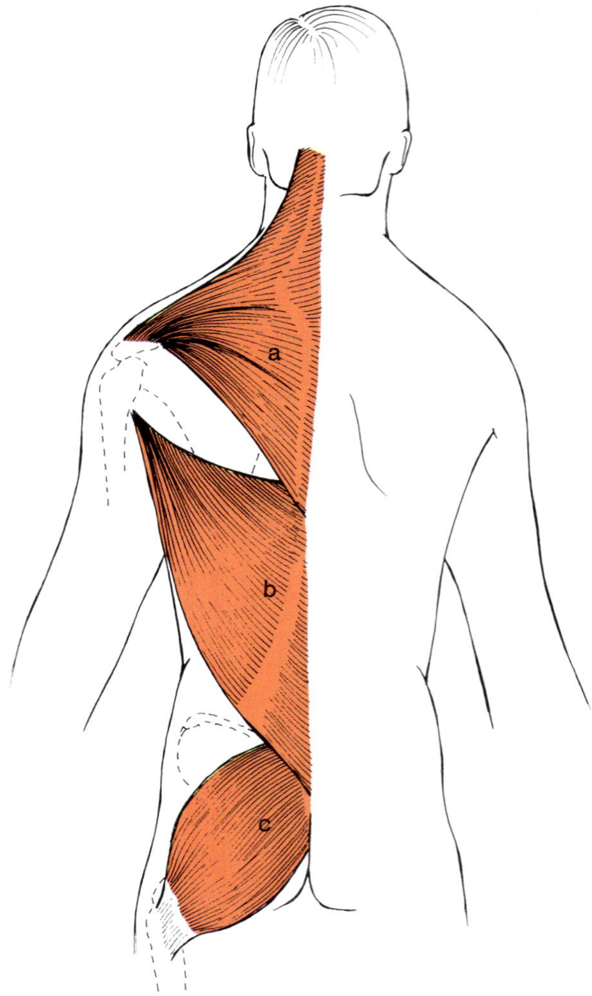

Fig. 16-2. *a*, Trapezius; *b*, latissimus dorsi; *c*, gluteus maximus.

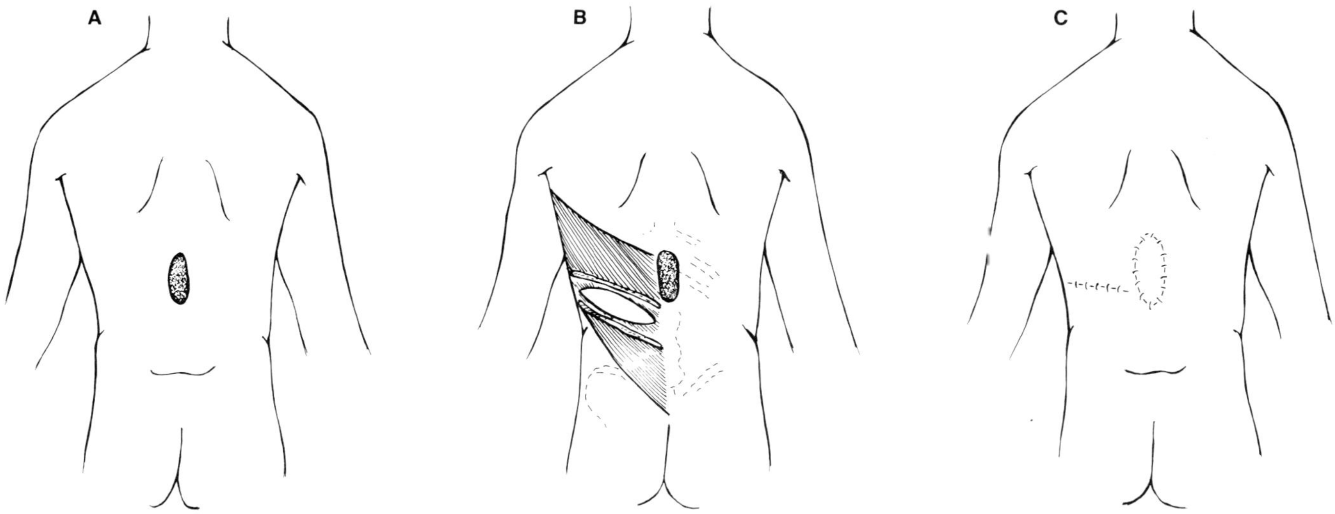

Fig. 16-3. Distally based latissimusa dorsi flap.
A, Defect in midposterior back.
B, Design and elevation of skin island on latissimus dorsi based on secondary segmental pedicles.
C, Transposition of musculocutaneous flap provides coverage of defect.

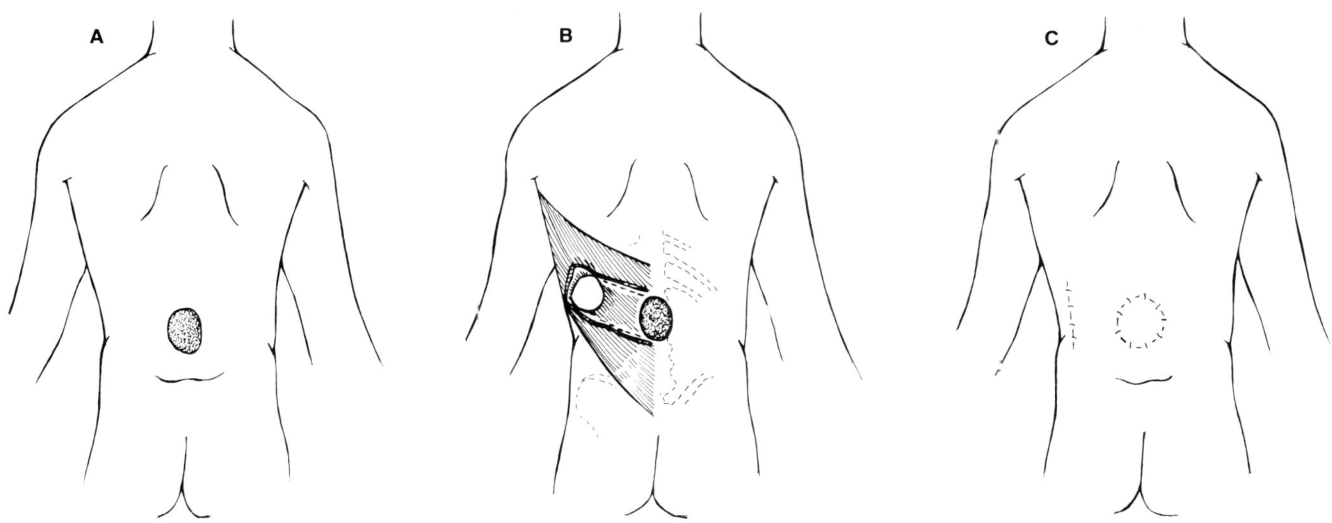

Fig. 16-4
A, Inferior back defect requiring flap coverage.
B, Design and elevation of skin island on latissimus dorsi muscle based on secondary segmental pedicles.
C, Muscle folded to allow skin to cover defect.

A

B

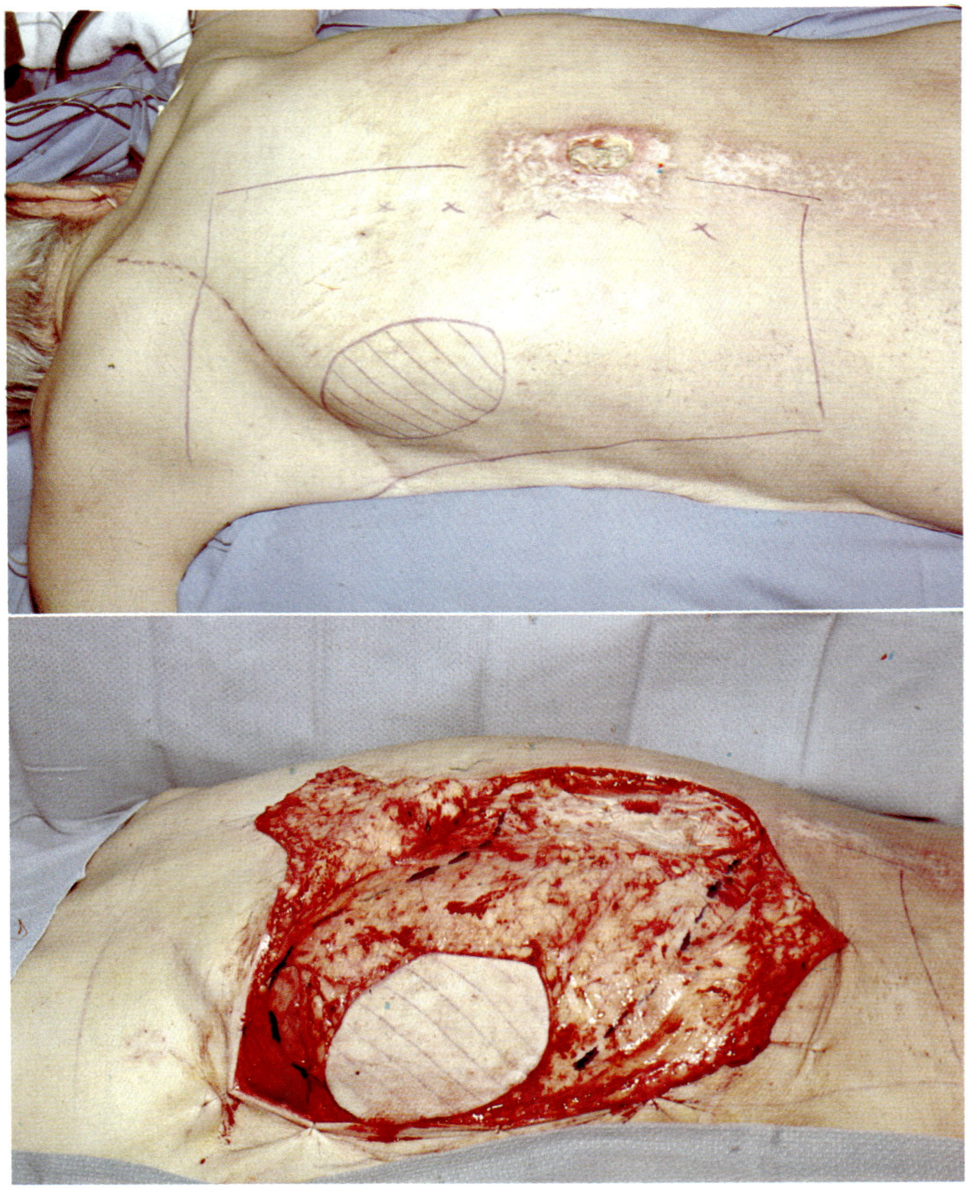

Fig. 16-5. Reverse latissimus dorsi flap.

A, Osteoradionecrosis of midback involving inferior thoracic vertebrae. NOTE: Scar overlying left posterior shoulder at site of prior thoracotomy. Despite prior division of proximal latissimus dorsi muscle fibers, posterior muscle is still vascularized via secondary segmental vascular pedicles from lumbar and thoracic perforating arteries and posterior intercostal arteries. Design of reverse latissimus dorsi musculocutaneous flap.

B, Dotted lines indicate site of planned division of latissimus dorsi muscle based on secondary segmental vascular pedicles.

C

D

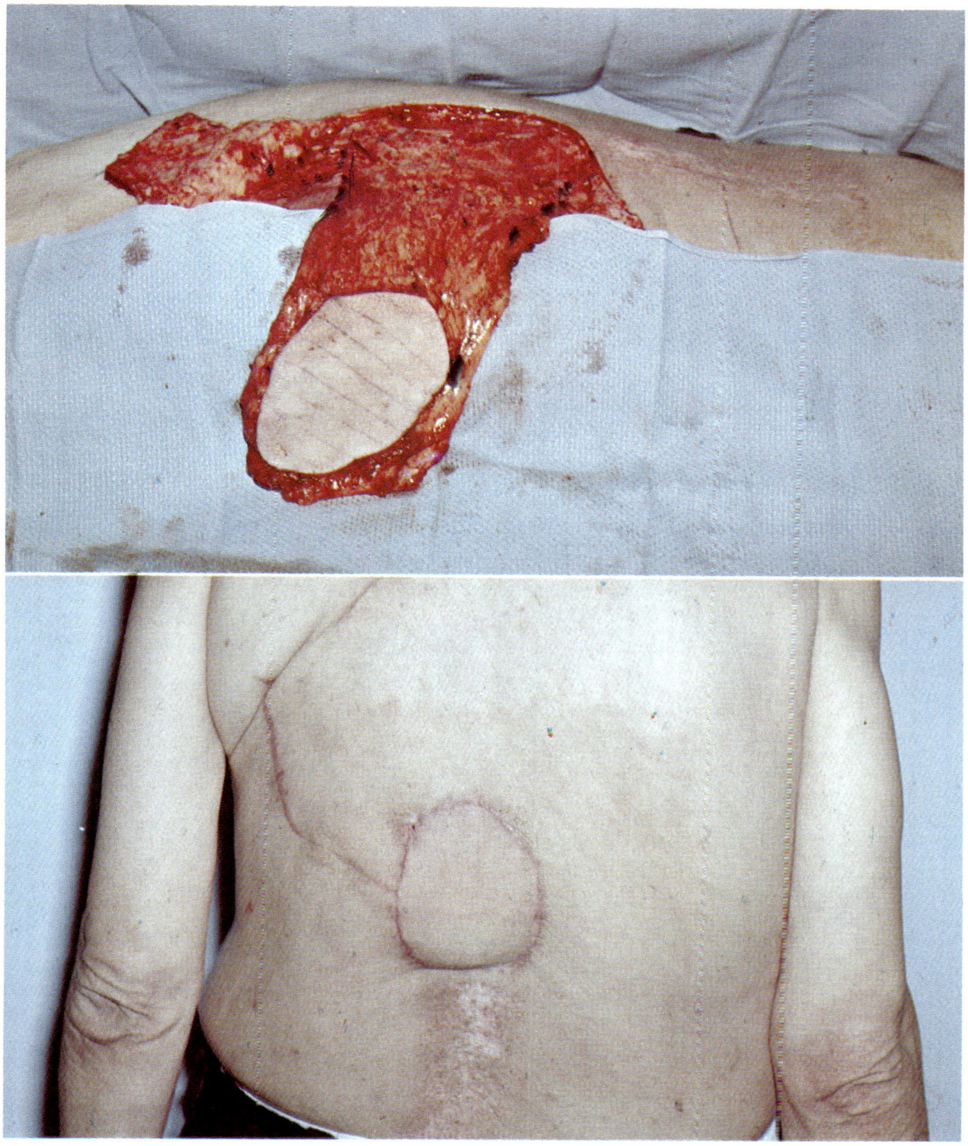

Fig. 16-5, cont'd

C, Reverse latissimus dorsi musculocutaneous flap is folded on itself to transpose skin island to back where coverage is required.

D, Stable coverage provided by reverse latissimus dorsi musculocutaneous flap. NOTE: Donor defect is closed directly.

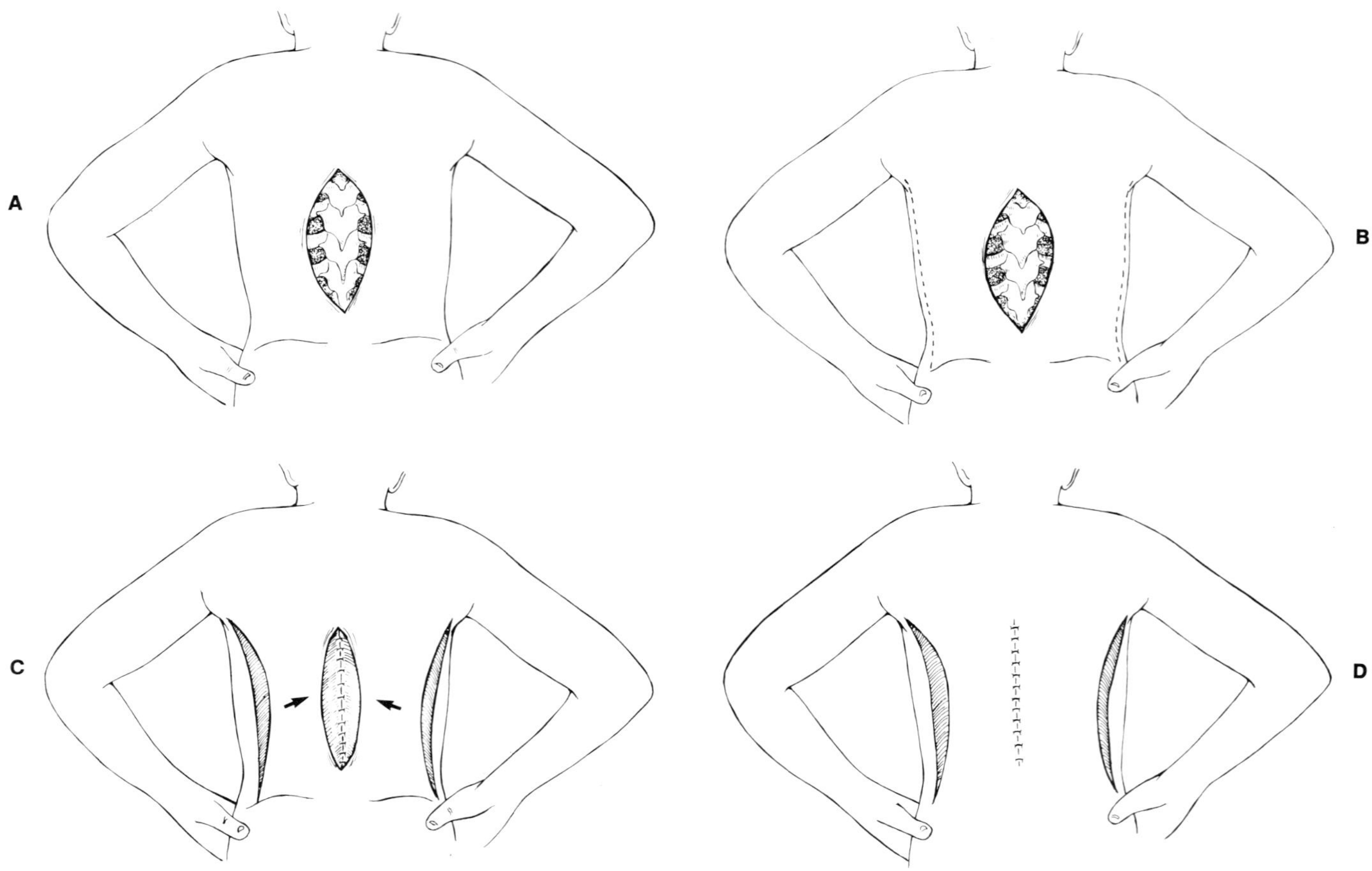

Fig. 16-6. Latissimus dorsi musculocutaneous flap.

A, Middle third defect of back in the midline.

B, Bilateral elevation of latissimus dorsi musculocutaneous flaps based on proximal major vascular pedicles. Dotted lines indicate site where connections between latissimus dorsi muscle and serratus anterior and abdominal fascia are divided. Technique is preferred for myelomeningocele closure.

C, Skin incisions bilaterally allow further transposition of musculocutaneous flaps to posterior midline for large defects.

D, Muscle and skin approximated in midline. Donor defects on lateral trunk require skin grafts for closure without tension.

A

B

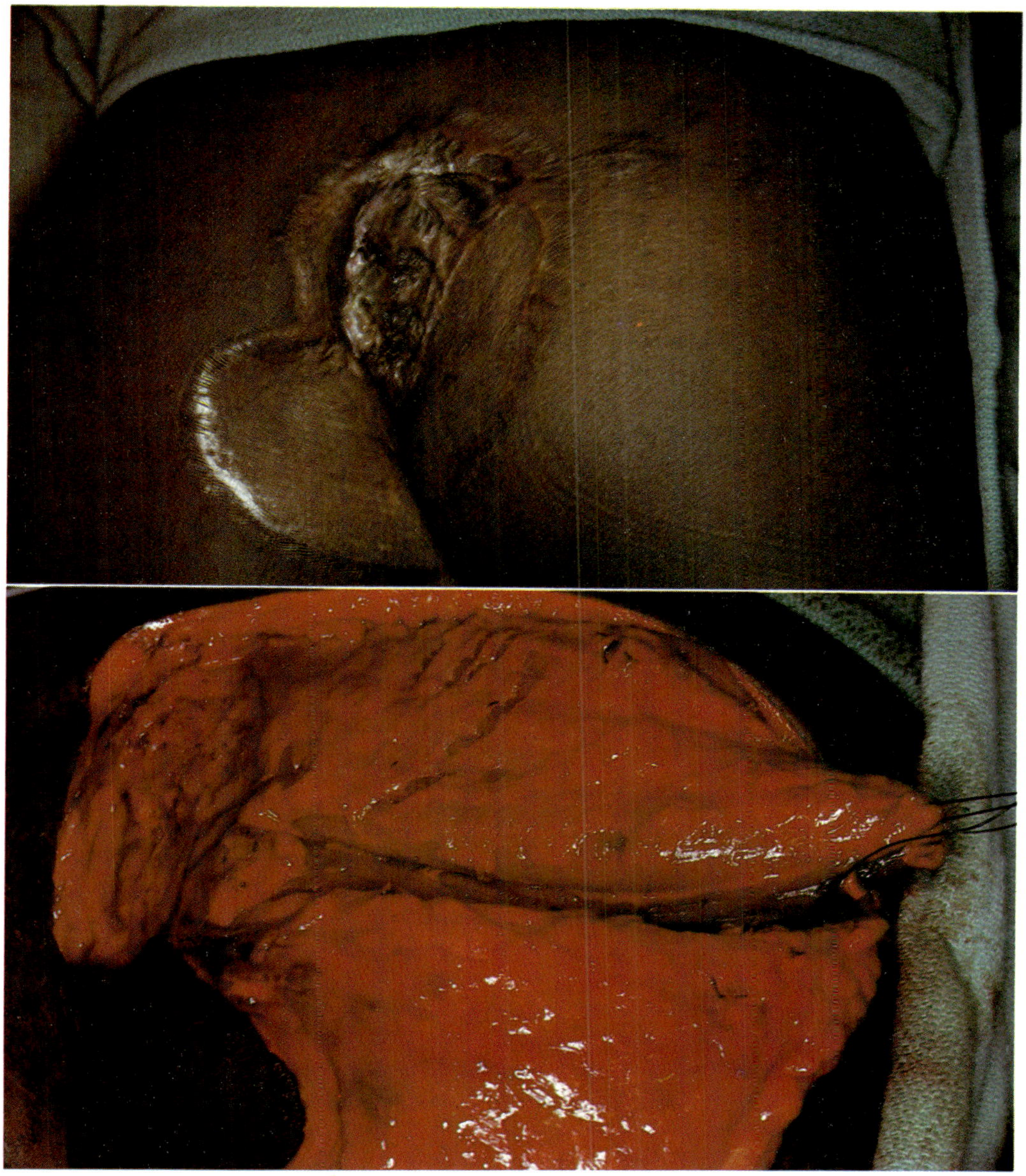

Fig. 16-7

A, Osteomyelitis of sacrum secondary to prior major blunt trauma.

B, Sequestrectomy includes remaining sacrum. NOTE: Rectum is located at base of wound. Since patient is ambulatory, only superior gluteus maximus muscle is transposed to provide coverage for defect.

(From Mathes, S.J., Vasconez, L.O., and Jurkiewicz, M.J.: Extensions and further applications of muscle flap transposition, Plast. Reconstr. Surg. **60:**6, 1977.)

Continued.

C

D

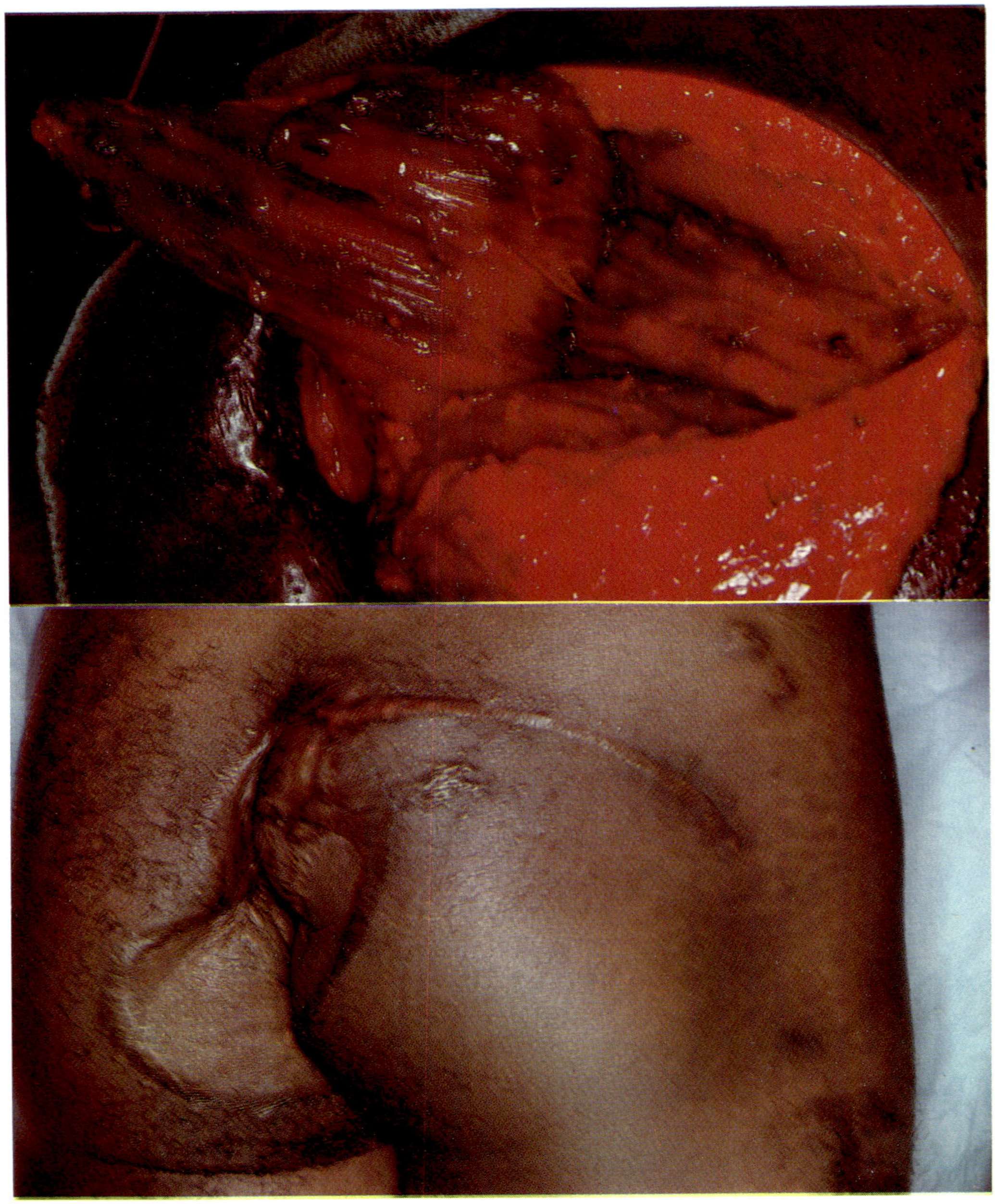

Fig. 16-7, cont'd

C, Gluteus maximus muscle is split, transposing superior gluteal muscle based on superior gluteal artery. Inferior half of gluteus maximus muscle, including origin, insertion, vascular supply, and innervation, is preserved.

D, One-year postoperative view demonstrates stable coverage provided by superior gluteus maximus muscle transposition with function preservation. Patient is ambulatory with no hip instability after reconstructive procedure because inferior half of muscle was preserved.

17

RECONSTRUCTION

Foad Nahai ■ Michael Scheflan ■ John Bostwick III

SOFT TISSUE COVERAGE OF THE POSTERIOR TRUNK

Muscle and musculocutaneous flaps have specific applications for reconstruction of the posterior trunk. Most of the defects are midline and are the result of congenital defects, previous surgery, or radiation therapy. The availability of these flaps has simplified the reconstruction of such defects.

The three most useful muscles are the trapezius, latissimus dorsi, and gluteus maximus. The trapezius will cover the upper third of the back, the latissimus the middle third, and the gluteus maximus the lower third. Other muscles with limited application for back coverage include the serratus anterior, serratus posterior, and the paraspinous muscles. The transverse back flap, although strictly not a musculocutaneous flap, is based on perforating musculocutaneous vessels through the paraspinous muscles (Figs. 17-1 and 17-2).

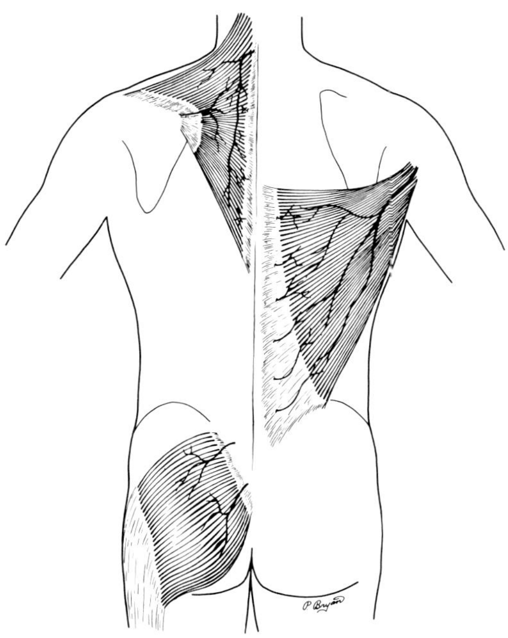

Fig. 17-1. Blood supply of flaps for coverage of posterior trunk.

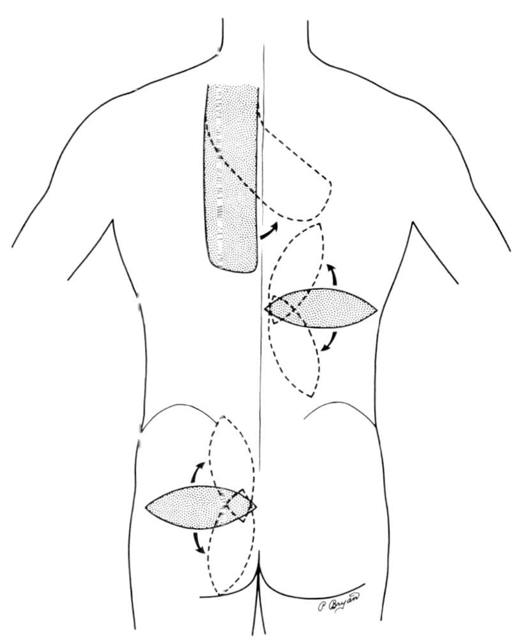

Fig. 17-2. Arc of rotation.

TRAPEZIUS

Anatomy

The trapezius is a large, flat, thick muscle that originates posteriorly from the base of the skull over the occipital protuberance, nuchal line, and spinous processes of the seventh cervical and all thoracic vertebrae. This muscle inserts into the lateral third of the clavicle, acromion, and along the spine of the scapula.

The trapezius muscle has a type II vascular pattern. The dominant vascular pedicle, the transverse cervical artery, enters the muscle in the base of the neck. The muscle has a segmental secondary vascular pedicle from the occipital artery and posterior intercostal arteries entering adjacent to the midline.

Flap elevation

The trapezius muscle may be transposed based either on its dominant vascular pedicle, the transverse cervical artery, or its secondary segmental pedicles. Since the majority of posterior truncal defects are located in the midline, injury or abnormality of the secondary segmental pedicles from the posterior intercostal arteries is common. For this reason the standard trapezius muscle or musculocutaneous transposition flap is based on its dominant vascular pedicle from the transverse cervical artery.

The muscle is elevated from its vertebral origin and scapular attachments. The overlying skin is generally included with the muscle. This skin is located between the posterior midline and scapula. The skin may be extended 3 to 4 cm inferiorly to the scapula. At the base of the flap located at the shoulder, the continuity of the skin is maintained for defects over the upper thoracic vertebrae. However, for defects over the shoulder or neck, a skin island may be designed over the posterior trapezius muscle (Fig. 17-3).

Flap elevation begins inferiorly with elevation of the extended portion of the skin flap over the upper edge of the latissimus dorsi. With the inferior edge of the trapezius muscle identified, the musculocutaneous flap is elevated by release of its vertebral origin. At the level of the scapula, the muscle plane between the trapezius and rhomboid muscles is identified. Caution is required to avoid elevation of the rhomboid muscles with the flap. At the superior edge of the scapula, this musculocutaneous unit will have adequate length for transposition over defects in the upper third of the posterior trunk.

Functional considerations

Elevation of the posterior trapezius muscle requires release of its vertebral origin and acromial insertion. Winging of the scapula is prevented by the underlying rhomboid muscle and the latissimus dorsi muscle. The continuity of the superior trapezius fibers between the base of the skull and clavicle is not disturbed. Both the motor nerve and the dominant vascular pedicle enter the flap at the base of the neck and are distant from the site of flap elevation. Use of posterior trapezius muscle or musculocutaneous flap

A

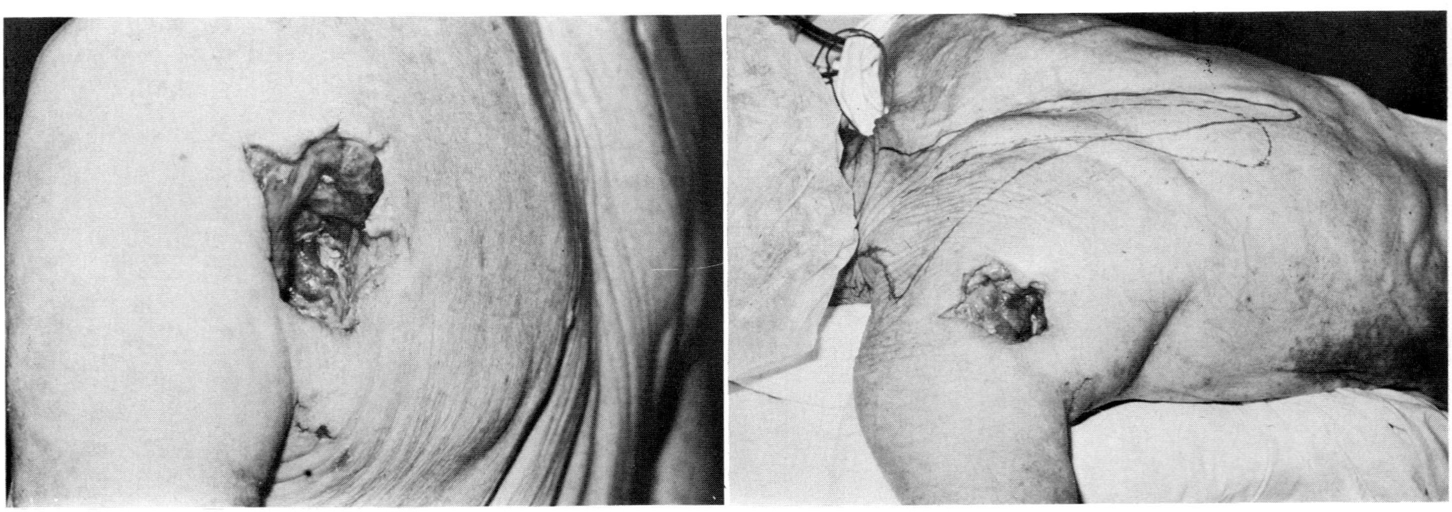

Fig. 17-3. Coverage of shoulder defect with posterior trapezius flap.
A, Shotgun blast of left shoulder posteriorly.
B, Deep defect extending to scapula and shoulder joint. Design of trapezius flap.

(Courtesy Stephen Gordon, Atlanta.)

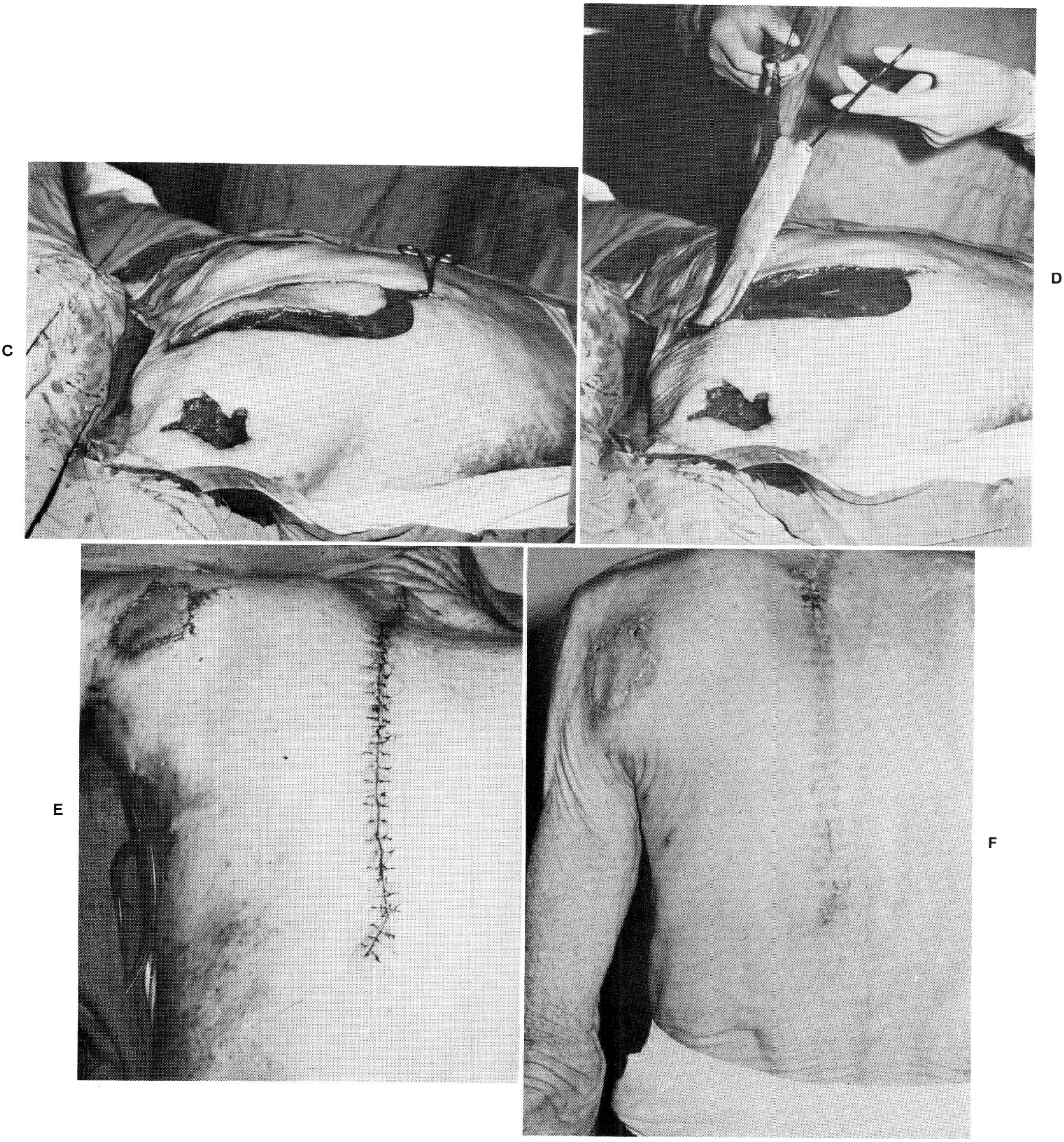

Fig. 17-3, cont'd
C, Flap dissected.
D, Flap elevated.
E, Flap tunneled through to defect over shoulder. Donor defect closed directly.
F, Early postoperative result.

is not associated with shoulder drop and its associated functional deformity.

The skin over the shoulder may be elevated with the superior trapezius muscle as a musculocutaneous flap. This musculocutaneous unit requires division of the dominant vascular pedicle, the decending branch of the transverse cervical artery, to the trapezius muscle. The muscle unit is based on a superior segmental pedicle, the occipital artery. This flap design may result in shoulder droop related to release of superior trapezius fibers and possible denervation of posterior fibers as a result of injury to the eleventh nerve.

LATISSIMUS DORSI

Anatomy

The latissimus dorsi is a large, flat muscle with origins from the lower six thoracic vertebrae, the lumbar and sacral vertebrae, and the posterior iliac crest. The fibers converge toward the insertion and constitute the posterior axillary fold. The tendon then inserts into the bicipital groove of the humerus.

The latissimus dorsi has a type V pattern of blood supply. The dominant pedicle, the thoracodorsal artery, enters the muscle over the deep surface—the axilla approximately 10 to 12 cm from the insertion. The secondary segmental blood supply is based on posterior perforating branches of the intercostal and lumbar arteries. The segmental vessels enter the muscle at its origin 3 to 4 cm from the midline posteriorly. The latissimus dorsi can be based on either blood supply for coverage of the back (Figs. 17-4 and 17-5). The origin may be completely divided, then a large flap is advanced toward the opposite side for midline coverage, or for smaller defects a portion of the muscle based on the secondary segmental blood supply may be elevated and transposed for coverage of smaller defects.

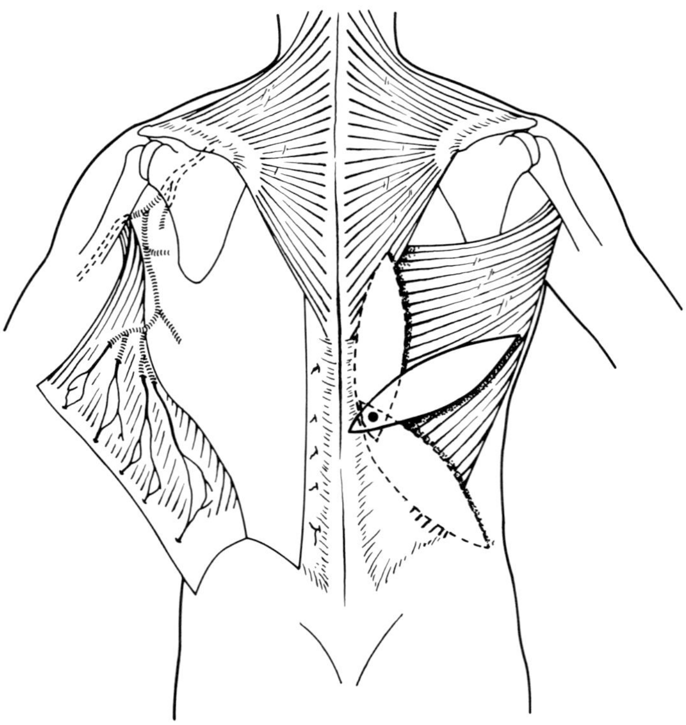

Fig. 17-4. Two arcs of rotation of latissimus dorsi flap are based on either the major pedicle or secondary segmental vascular pedicles.

Fig. 17-5. Reconstruction of shoulder defect with latissimus dorsi flap.
A, Fifty-four-year-old man 20 years after deep burns of posterior trunk with nonhealing wound at base of neck and two failed skin grafts.
B, Wide excision of the unstable scar did not reveal any neoplastic component in the specimen. An island of skin is designed over the superior border of the latissimus dorsi muscle.
C, The island musculocutaneous flap, based on the major thoracodorsal blood supply is shown in place and the donor area closed directly.
D, Two-year postoperative view of stable coverage of shoulder.

A B C D

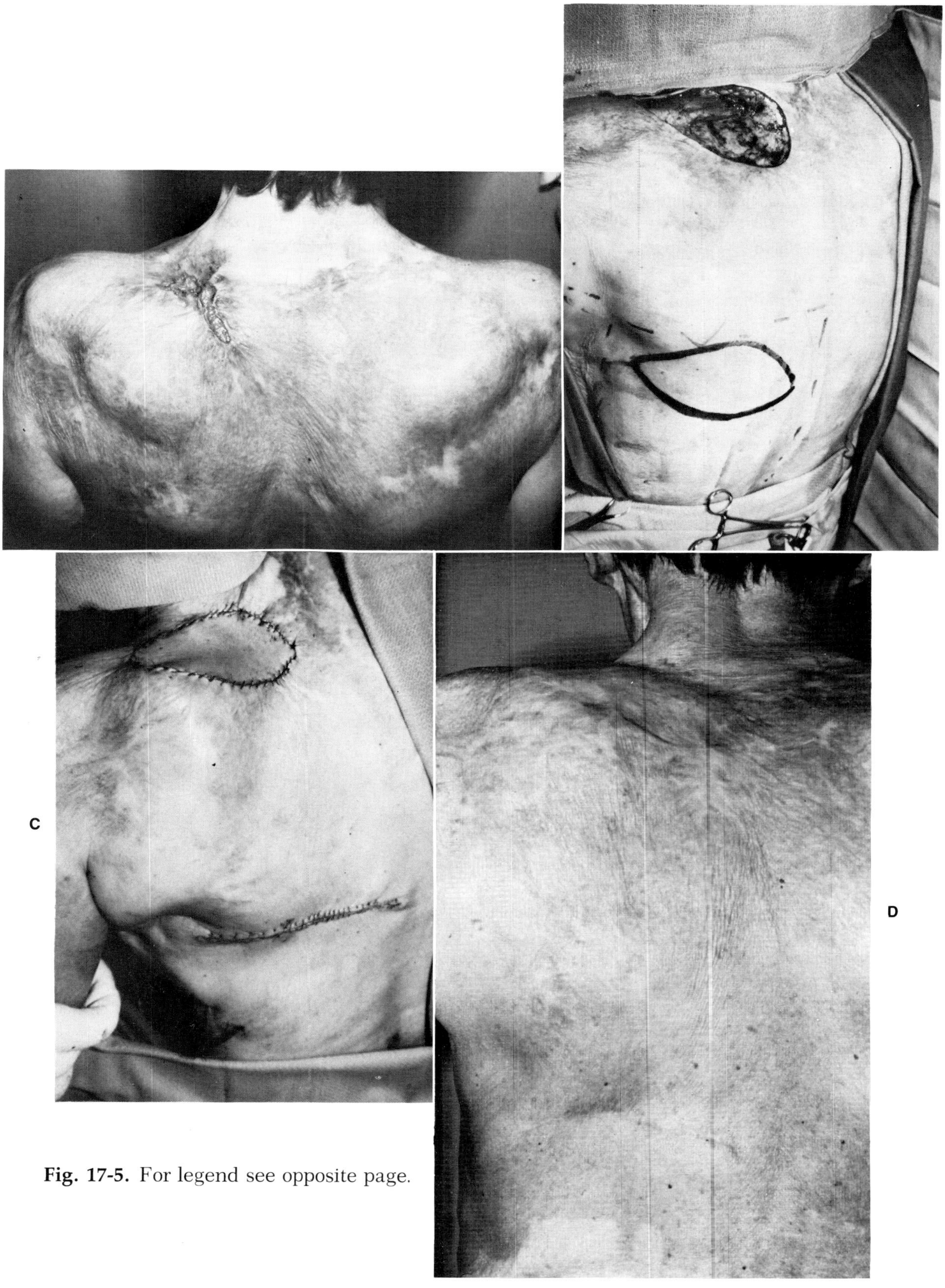

Fig. 17-5. For legend see opposite page.

Flap evaluation

Standard flap. For coverage of large defects such as a myelomeningocele, the entire origin of the muscle is divided, and the muscle is dissected free from the underlying tissues. This unit with or without the overlying skin will then advance toward the midline (Fig. 17-6). For extensive defects both latissimus dorsi flaps are elevated. Flap elevation is started through the existing midline defect; the plane below the latissimus dorsi is identified. It should be noted that inferiorly the latissimus dorsi has a fascial origin. This origin is divided, the segmental perforators through the paraspinous muscle are divided, and then the origin of the muscle from posterior iliac crest is sharply divided. Superiorly perforating vessels through the serratus anterior and serratus posterior muscles are divided. At this stage the entire muscle should be free, and it will then advance at least 6 to 8 cm. If necessary at this stage, the skin overlying the anterior free border of the latissimus dorsi muscle in the posterior axillary line may be incised from the axilla to the posterior iliac crest. This will allow further advancement of the skin.

A

B

C

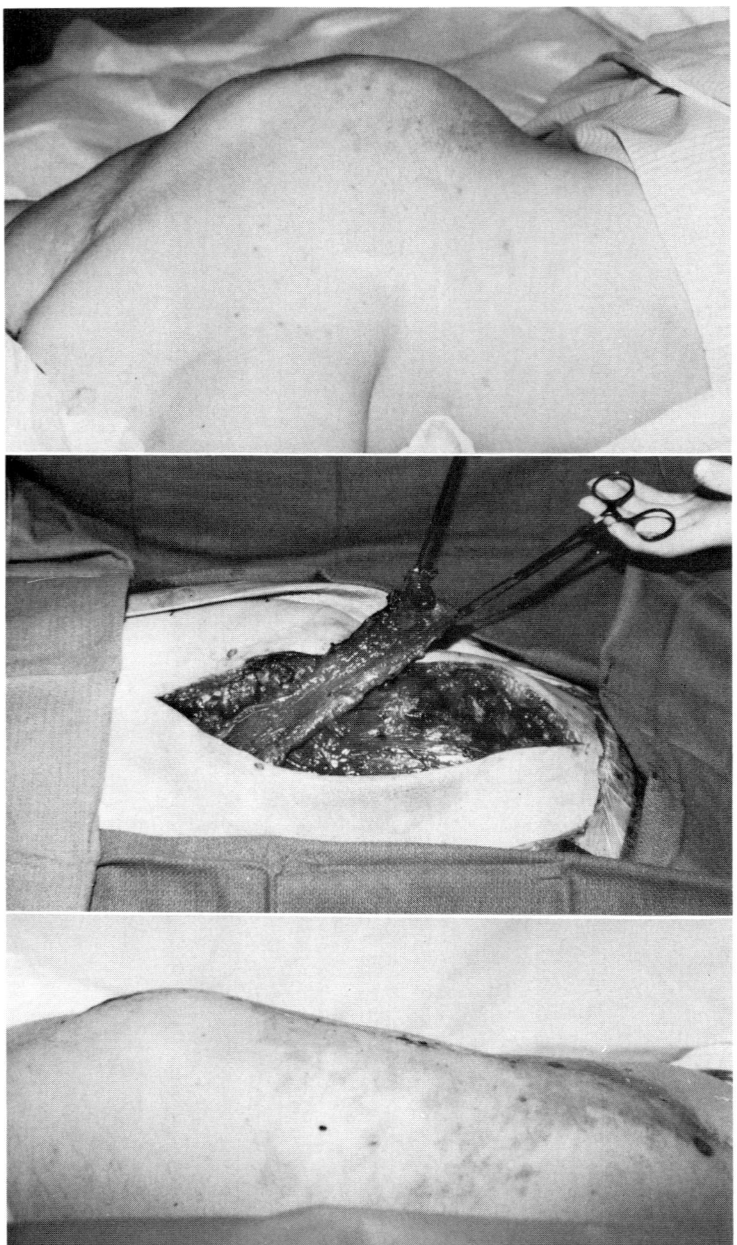

Fig. 17-6
A, Fourteen-year-old girl with severe gibbus deformity secondary to myelomeningocele.
B, Patient after partial resection of bony gibbus (in preparation for future insertion of Harrington rods). Latissimus dorsi muscle flap based on thoracodorsal artery has been detached distally, medially, and laterally.
C, Muscle has been placed over the vertebral defect and skin closed primarily.

Distally based flap. The distally or segmentally based latissimus dorsi flap is elevated as an island for closure of smaller defects. The island is outlined transversely across the back with the arc of rotation based 2 to 3 cm from the midline posteriorly (Fig. 17-4). The skin island is extended to the posterior axillary line. The island of skin is then incised and the underlying latissimus dorsi muscle identified. The muscle is then divided and the unit elevated laterally and dissected toward the midline. The perforating segmental vessels enter the flap approximately 2 to 3 cm from the midline. The vessels are 1 to 2 mm in size and need not be identified. The dissection is stopped at this point and the island transposed either superiorly or inferiorly (Fig. 17-7).

A

B

C

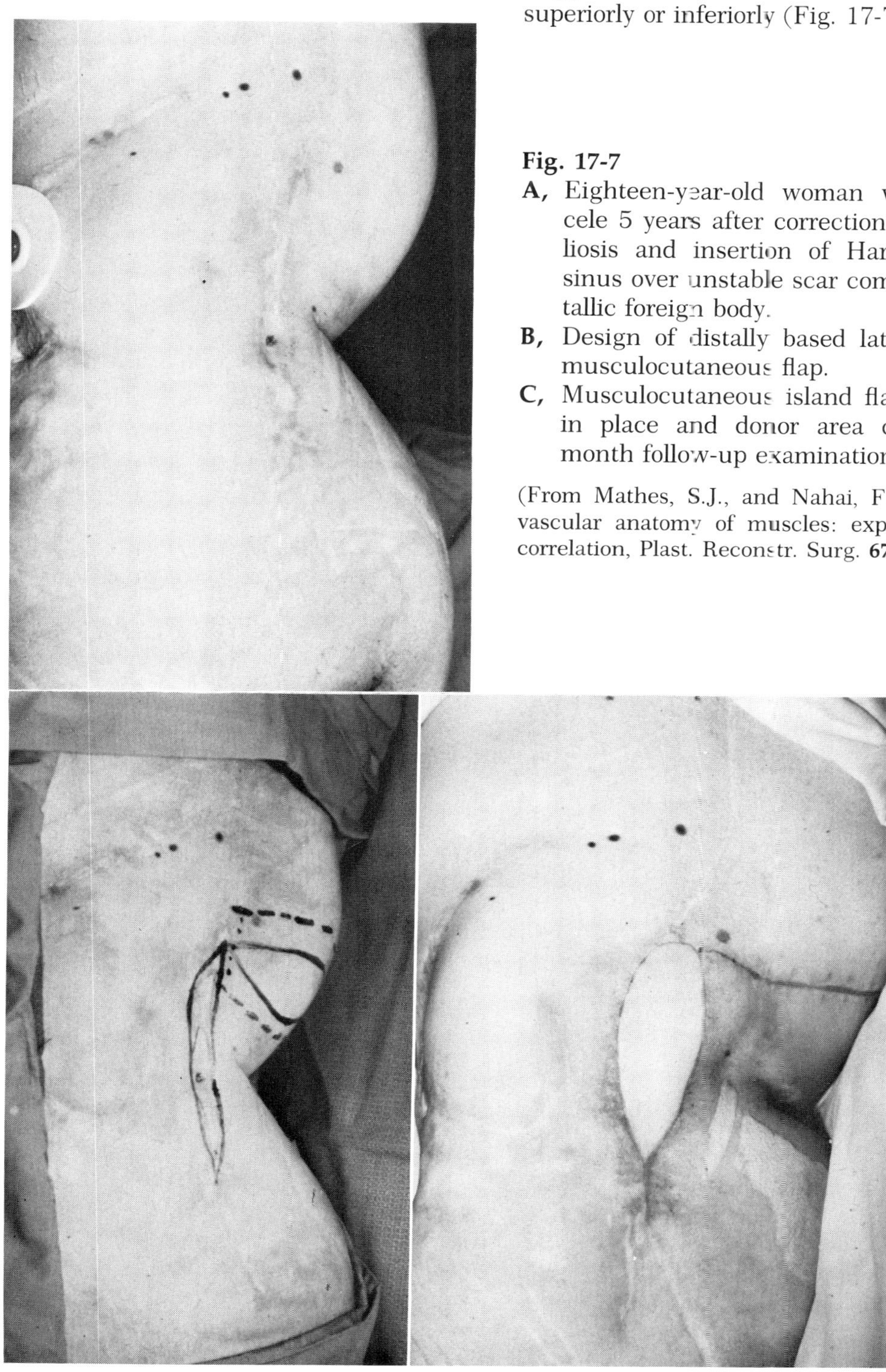

Fig. 17-7

A, Eighteen-year-old woman with myelomeningocele 5 years after correction of severe kyphoscoliosis and insertion of Harrington rods. Small sinus over unstable scar communicates with metallic foreign body.

B, Design of distally based latissimus dorsi island musculocutaneous flap.

C, Musculocutaneous island flap is shown sutured in place and donor area closed primarily (9-month follow-up examination).

(From Mathes, S.J., and Nahai, F.: Classification of the vascular anatomy of muscles: experimental and clinical correlation, Plast. Reconstr. Surg. **67**[2]:177, 1981.)

Functional considerations. Functional considerations are essential in selecting flaps for closure of defects in patients with myelomeningoceles. Schottstaedt and co-workers have pointed out the function of the latissimus dorsi in the polio victim. Similarly, in the paraplegic patient the upper limb girdle must not be weakened by indiscriminate use of the latissimus dorsi. The standard latissimus dorsi flap just described should not alter function. Although the origin of the muscle is divided, it is again resutured, usually to the other side. The reverse flap may, however, completely disrupt the muscle if it is designed in the middle of the muscle. However, a very low or a very high reverse flap would leave most of the muscle intact.

GLUTEUS MAXIMUS

Musculocutaneous flaps based on the gluteus maximus flap will cover low midline defects of the back (Fig. 17-8).

A

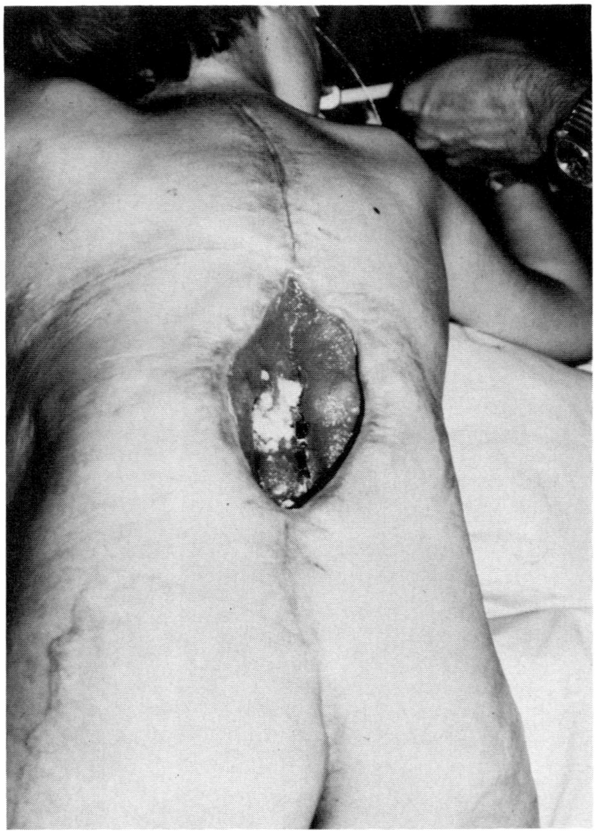

Fig. 17-8. Coverage of back with combined latissimus dorsi and gluteus maximus flaps.

A, Course of 12-year-old boy with myelomeningocele 6 weeks after Harrington rod insertion complicated by infection and resultant wound disruption.

B

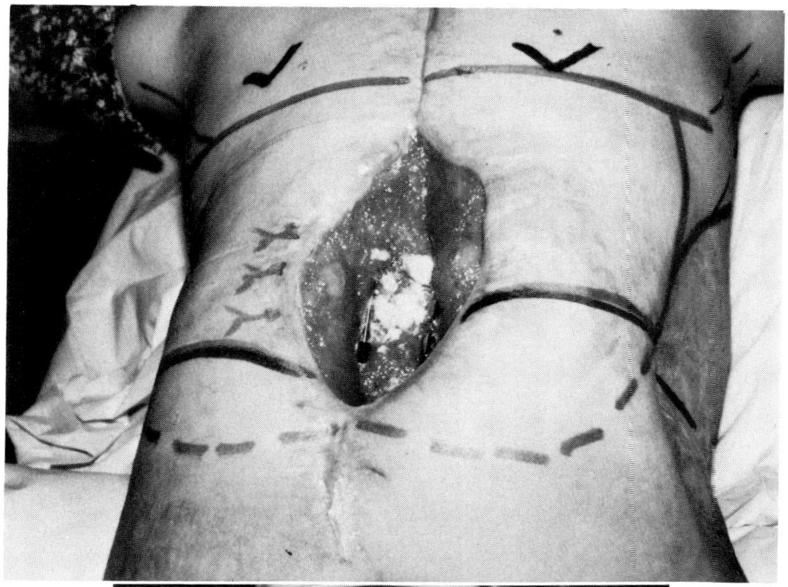

C

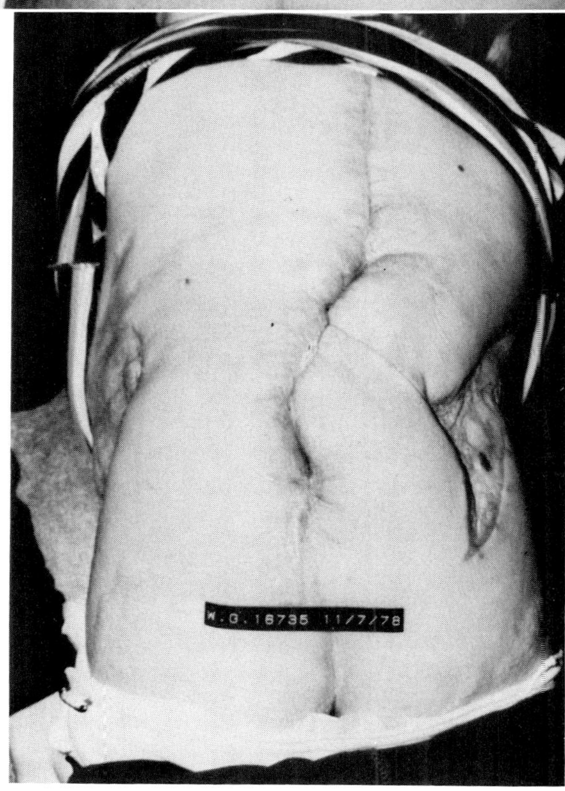

Fig. 17-8, cont'd

B, Latissimus dorsi territory and paraspinal perforators are marked; gluteus maximus territory is marked inferiorly.

C, One-year postoperative view after application of two musculocutaneous flaps. Superior portion is latissimus dorsi flap based on thoracodorsal artery, and inferior portion is a gluteus maximus flap based on superior gluteal artery. Continuity between two flaps was divided to facilitate transposition into midline. Donor area over right flank has been skin grafted.

Anatomy

The gluteus maximus is a thick, broad muscle with origin from the gluteal line of the ilium and sacrum and insertion into the greater trochanter and iliotibial tract. It has a type III pattern of blood supply with the superior gluteal artery supplying the upper half and the inferior gluteal the lower half of the muscle. Both these vessels are branches of the internal iliac artery. The superior gluteal artery enters the upper half of the muscle medially just above the piriformis muscle, and the inferior gluteal artery enters the muscle medially below the piriformis muscle. The inferior gluteal artery lies closer to the sciatic nerve.

Flap elevation

The gluteus maximus flap may be elevated as a turnover muscle flap or as a musculocutaneous flap for coverage of midline defects.

Turnover flap. An incision is made directly over the muscle; the insertion into the trochanter is divided and the flap elevated from the lateral border to the medial border. The plane between the gluteus maximus and the underlying gluteus medius and minimus is indistinct, therefore sharp dissection is necessary. The sciatic nerve is exposed deep to the lower half of the muscle.

Musculocutaneous island. The island of skin over the upper or lower half of the muscle is outlined. The skin incision is made, the muscle insertion divided, and the flap elevated from the lateral border to the medial border. The gluteal vessel is then identified and preserved. To develop an arc of rotation that will allow coverage of the midline, the origin of the muscle is divided, allowing the entire muscle to cover the sacral defect. This unit is then transposed as an island musculocutaneous flap (Fig. 17-9).

Functional considerations

Functional considerations are essential in gluteus maximus selection and design. This muscle is not dispensable, and the selection of the entire muscle in an ambulatory patient will lead to significant functional loss. However, one half can be elevated without significant functional impairment.

PARASPINOUS MUSCLES

Anatomy

The paraspinous group of muscles takes origin from the spinous processes of the lumbar vertebrae and inserts into the sacrum.

These muscles have a type IV pattern of blood supply with multiple vascular pedicles from the lumbar arteries. Based on its segmental vascular pedicles, this muscle has a limited arc of rotation. The muscle has been used to cover myelomeningocele defects. However, identification of muscle units with vascular patterns allowing safer and more reliable coverage (e.g., latissimus dorsi) has obviated the use of this muscle.

Transverse back flap. The transverse back flap is based over the paraspinous muscle and extends horizontally across the midline with a length-width relationship of $2\frac{1}{2}$:1. This flap represents a random skin extension of the paraspinous muscular perforating vessels of the lumbar arteries across the midline.

Flap elevations

The transverse back flap is useful for coverage of midline defects located in the lower third of the posterior trunk. The flap is designed immediately adjacent to the defect. The flap base is centered over the paraspinous muscle with a width of 8 cm and extends across the midline to the lateral aspect of the posterior trunk. Flap elevation begins distally with incision to the level of the lumbosacral fascia. The flap is elevated between this fascia and the subcutaneous tissue. Then this flap is elevated across the midline, the paraspinous fascia is elevated with the flap. Perforating branches of the lumbar artery are identified and carefully preserved. The base of the flap is located at the medial edge of the paraspinous muscle. The arc of rotation may be safely increased by a short back cut in the skin edge distant from the defect.

This flap may be designed at any level of the inferior trunk for a defect located adjacent to the midline. Since the circulation to the portion of the flap crossing the midline is located in the subdermal plexus, this flap is not reliable when its territory has been subjected to prior radiation therapy.

CONCLUSION

Defects of the posterior trunk are closed with the trapezius, latissimus dorsi, and gluteus maximus flaps. The trapezius is the flap for the upper third of the posterior trunk, the latissimus dorsi for the middle third, and the gluteus maximus for the lower third.

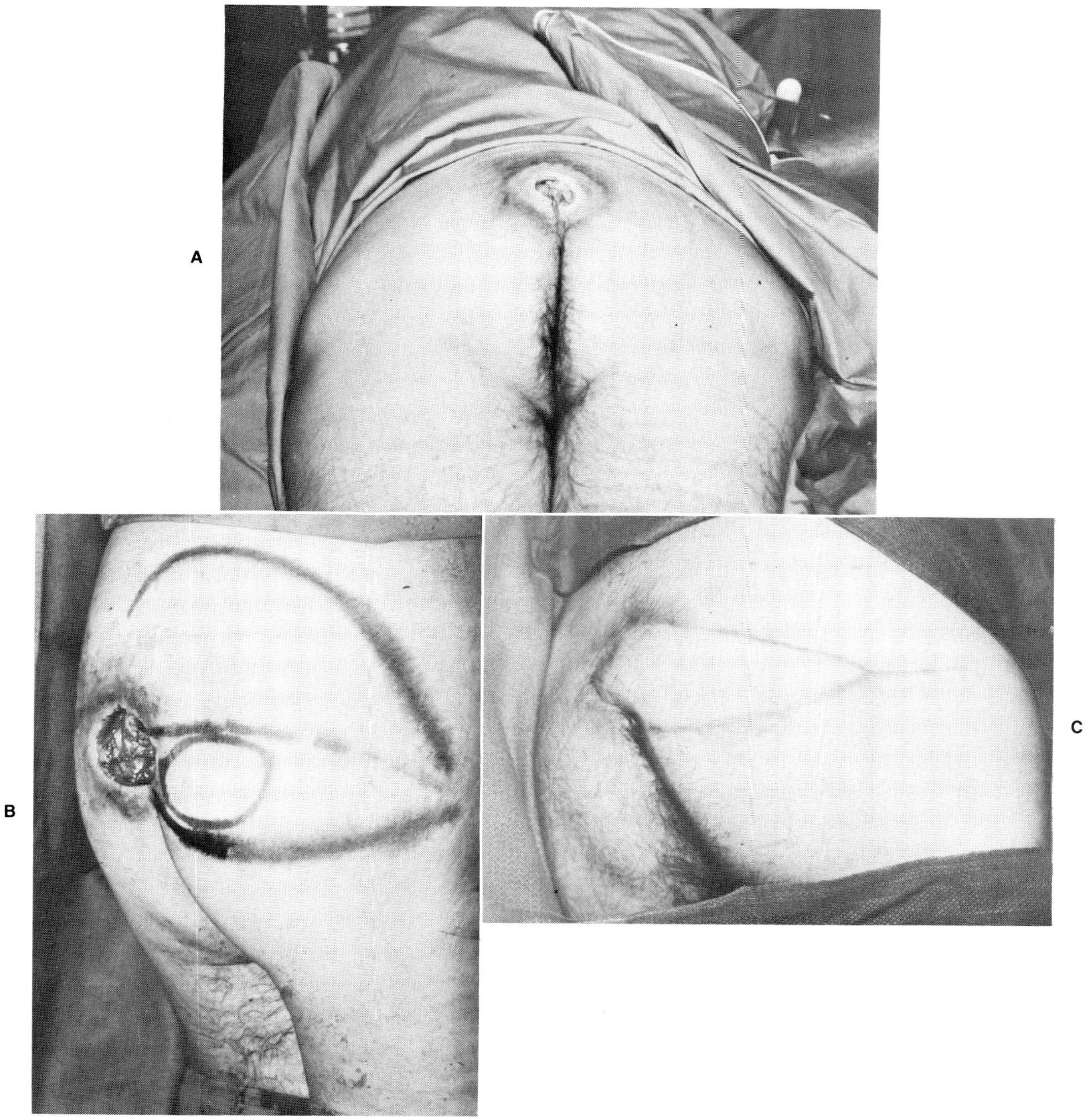

Fig. 17-9. Coverage of sacrum with gluteus maximus musculocutaneous island flap.
A, Pressure-induced defect over sacrum.
B, Wound excised and flap outlined.
C, Flap advanced into defect in V-Y fashion and view of healed flap.

ANNOTATED BIBLIOGRAPHY

1. Bostwick, J., III, and co-workers: Sixty latissimus dorsi flaps, Plast. Reconstr. Surg. **63:**31, 1979.

   The anatomy and various applications of the latissimus dorsi flap are described, including closure defects over the shoulder with an island flap.

2. Bostwick, J., III, and co-workers: The "reverse" latissimus dorsi muscle and musculocutaneous flap: anatomical and clinical considerations, Plast. Reconstr. Surg. **65:**395, 1980.

   Anatomic description and clinical application of the distal segmentally based latissimus dorsi flap for coverage of the posterior trunk are presented.

3. Daniel, R.K., Verrigan, C.L., and Gard, D.A.: The great potential of the intercostal flap for torso reconstruction, Plast. Reconstr. Surg. **61:**653, 1978.

   Experimental evaluation, anatomic description, and clinical application of the intercostal flap for closure of posterior trunk defects, including myelomeningocele and pressure sores, are presented. The advantages of this innervated neurosurgery flap are emphasized.

4. Desprez, J.D., Kiehn, C.L., and Echstein, W.: Closure of large meningomyelocele defects by composite skin muscle flaps, Plast. Reconstr. Surg. **47:**234, 1971.

   Several large myelomeningocele defects are closed with bilateral musculocutaneous flaps based on the latissimus dorsi, trapezius, and sacrospinalis muscles. Lateral incisions are made to allow greater advancement across the midline. The lateral defects are closed in V-Y fashion. A layered closure is achieved over the defect.

5. Ger, R., and Levine, S.A.: The management of decubitus ulcers by muscle transposition: an 8 year review, Plast. Reconstr. Surg. **58:**419, 1976.

   An 8-year review of 40 muscle flap transpositions for pressure sores, including 23 gluteus maximus muscle flaps, is presented. Transposition of the upper half for sacral coverage and lower half for ischial coverage is demonstrated.

6. Hill, H.L., Brown, R.G., and Jurkiewicz, M.J.: The transverse lumbosacral back flap, Plast. Reconstr. Surg. **63:**177, 1978.

   A description of the anatomic basis and clinical applications of the lumbosacral back flap for closure of sacral defects is presented.

7. McGraw, J.B., Penix, J.O., and Baker, J.W.: Repair of major defects of the chest wall and spine with latissimus dorsi myocutaneous flap, Plast. Reconstr. Surg. **62:**197, 1978.

   Five cases are presented, and the application of the latissimus dorsi musculocutaneous flap for closure of spinal defects in two infants is described. Bilateral latissimus dorsi musculocutaneous flaps are elevated on the dominant thoracodorsal pedicle and advanced across the midline defect. No lateral or relaxing incisions are made.

8. Mathes, S.J., and Nahai, F.: Clinical atlas of muscle and musculocutaneous flaps, St. Louis, 1979, The C.V. Mosby Co.

   A description of the posterior trapezius flap and its clinical application for coverage of the upper third of the spine is presented.

9. Mathes, S.J., Vasconez, L.O., and Jurkiewicz, M.J.: Extensions and further applications of muscle flap transposition, Plast. Reconstr. Surg. **60:**6, 1977.

A review of the first 94 muscle flaps performed at Emory University–affiliated hospitals between 1972 and 1976 is presented. The application of a gluteus maximus musculocutaneous flap for closure of a myelomeningocele is reported.

10. Mendleson, B.C., and Masson, J.K.: Treatment of chronic radiation injury over the shoulder with a latissimus dorsi myocutaneous flap, Plast. Reconstr. Surg. **60:**681, 1977.

Application of the latissimus dorsi flap based on the thoracodorsal artery for coverage of the shoulder is described.

11. Minami, R.T., Mills, R., and Pardoe, R.: Gluteus maximus myocutaneous flaps for repair of pressure sores, Plast. Reconstr. Surg. **60:**242, 1977.

The gluteus maximus musculocutaneous unit is described, and one example of its transposition to cover the sacrum is presented.

12. Mustarde, J.C.: Reconstruction of the spinal canal in severe spina bifida, Plast. Reconstr. Surg. **42:**109, 1968.

A new spinal canal is reconstructed with osteomuscular flaps. Part of the paraspinal muscles and the bifid spinous process of the involved vertebrae are transposed over the exposed cord, sutured to each other, and skin grafted.

13. Schottstaedt E.R., Larson, L.J., and Bost, F.C.: Complete muscle transposition, J. Bone Joint Surg. (Am) **37:**897, 1955.

An exhaustive review of functional muscle transposition, including the application of the latissimus dorsi muscle for elbow flexion and extension, is presented. The authors warn of the functional loss that may result after transposition of the latissimus dorsi muscle in polio victims.

14. Stallings, J.O., Delgado, J.P., and Converse, J.M.: Turnover island flap of gluteus maximus muscle for the repair of sacral decubitus ulcer, Plast. Reconstr. Surg. **54:**52, 1974.

The anatomy, blood supply, and application of turnover flaps of the gluteus maximus are described for closure of sacral defects.

18

COMPLICATIONS

Foad Nahai ■ Stephen J. Mathes

The trapezius, latissimus dorsi, and gluteus maximus muscles and the transverse back flap are available for reconstruction of posterior trunk defects. Specific complications related to use of these flaps are discussed to prevent flap failure in reconstruction of posterior trunk defects.

TRAPEZIUS

Planning errors

Prior radiation, trauma, or operative procedures on the posterior trunk are not likely to have affected the blood supply to the trapezius. The posterior trapezius is most useful for coverage for the upper third of the posterior trunk, and the flap should be designed to preserve the upper fibers and hence function, which avoids shoulder drop.

Technical errors

The underlying rhomboid muscles are intimately associated with the trapezius muscle. Dissection proceeds from the distal border to the proximal border, and the trapezius is first elevated off the latissimus dorsi, then off the rhomboids. Dissection deep to the rhomboids leads under the scapula.

As elevation of the posterior trapezius flap allows function and preservation, branches of the eleventh nerve to the anterior fibers of the muscle must be preserved.

Postoperative errors

Appropriate suction drainage and positioning of the patient are essential to avoid postoperative complications.

LATISSIMUS DORSI

Planning errors

In planning a latissimus dorsi flap for posterior trunk coverage, the blood supply and the muscle function preservation must be taken into account. It is unlikely that trauma, radiation, or surgical intervention in the posterior trunk would affect the blood supply through the thoracodorsal pedicle. However, the secondary segmental blood supply through the perforating branches of the intercostal and lumbar arteries may easily be compromised by radiation, trauma, or surgical intervention on the back.

The standard latissimus dorsi flap based on the thoracodorsal vascular pedicle that is advanced over the midline will not lead to functional problems, provided that after advancement the divided muscle origin is sutured to preserve muscle tension. However, reverse or distally based flaps designed over the midportion of the muscle may well affect function by dividing the muscle. However, a very high or low reverse flap would preserve function by leaving most of the muscle intact.

Technical errors

Standard flap. For the standard flap, elevation of the latissimus dorsi musculocutaneous flap is started from the midline usually through the defect, and the dissection proceeds from the medial border to the lateral border. Correct identification of the latissimus dorsi is essential. This is difficult after previous radiation and in children with myelomeningocele who have a large gibbus deformity. Once the latissimus dorsi is identified, it is dissected off the underlying paraspinous muscles. It is essential to take the fascial origin of the latissimus dorsi off the posterior iliac crest completely and to elevate the muscle all the way to the lateral border. This dissection should then allow adequate advancement of the flap to cover the midline. If further advancement is required, a skin incision may be made laterally on the anterolateral border of the latissimus dorsi.

Distally based flap. The segmental or distally based flap is elevated from the lateral border to the medial border. The vascular pedicle is located 2 to 3 cm from the midline, therefore medial dissection is stopped 3 to 4 cm from the midline.

Postoperative errors

Adequate suction and positioning of the patient to avoid pressure on the flaps are essential to prevent postoperative complications.

GLUTEUS MAXIMUS

Planning errors

Either half of the gluteus maximus muscle will cover the sacrum. The upper half of the muscle will reach the lower lumbar vertebrae. The entire muscle should not be used in ambulatory patients because this will lead to hip instability.

The skin island should be designed with dimensions that will allow direct skin closure. Skin grafting of the donor defect is difficult and may leave an unacceptable donor area.

Technical errors

Identification of the upper border of the gluteus maximus is difficult, because it lies deep to the thick lumbosacral fascia. This fascia has to be divided and then the muscle identified.

The piriformis muscle deep to the gluteus maximus muscle is an important landmark. It indicates the level of the muscle splitting, because the superior gluteal artery is located above the piriformis and the inferior gluteal artery below it.

Postoperative errors

Closed suction drainage and positioning so that pressure on the flap is avoided are essential to prevent postoperative complications.

TRANSVERSE BACK FLAP

Planning errors

This flap is based on the lumbar perforating arteries on one side and extends toward the opposite side across the midline. In length it should not be designed beyond the posterior superior iliac crest on the opposite side.

Technical errors

The perforators are located 2 to 3 cm laterally to the midline, therefore flap elevation should not extend beyond these vessels. Usually flap elevation proceeds rapidly toward the midline. Once the midline is crossed, flap elevation proceeds more slowly so these vessels can be identified and preserved.

Postoperative errors

Suction drainage and positioning to avoid pressure on the flap are essential to prevent postoperative complications.

19

A SYSTEMATIC APPROACH TO FLAP SELECTION

Stephen J. Mathes ■ Foad Nahai

UPPER ABDOMEN (Fig. 19-1)

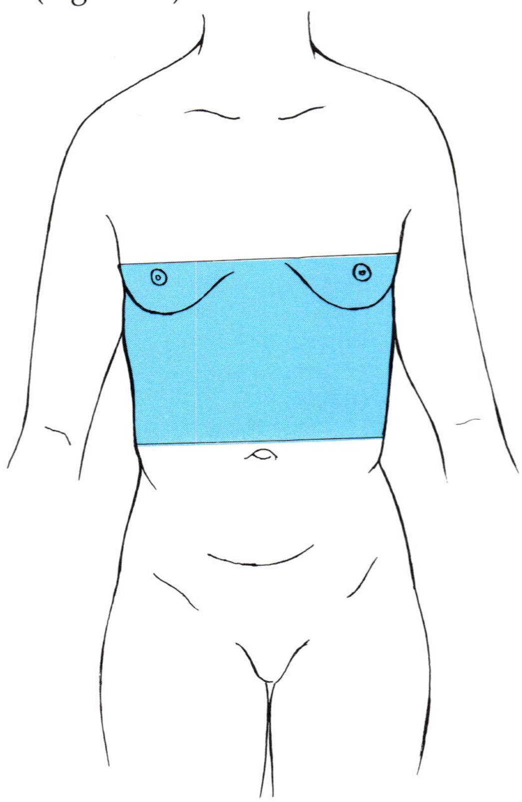

Fig. 19-1

Skin grafts

1. Skin grafts are method of choice for immediate coverage of defects
2. Grafts may be placed directly on abdominal contents
3. Grafts may be placed directly on synthetic mesh

Local flaps (Fig. 19-2)

1. Rectus abdominis flap, when available, is flap of choice for unilateral defects (Fig. 19-3)
2. External oblique musculocutaneous flap is alternative flap

Distant flaps

1. Latissimus dorsi musculocutaneous flap is alternative flap
2. Free tissue transplantation is rarely indicated

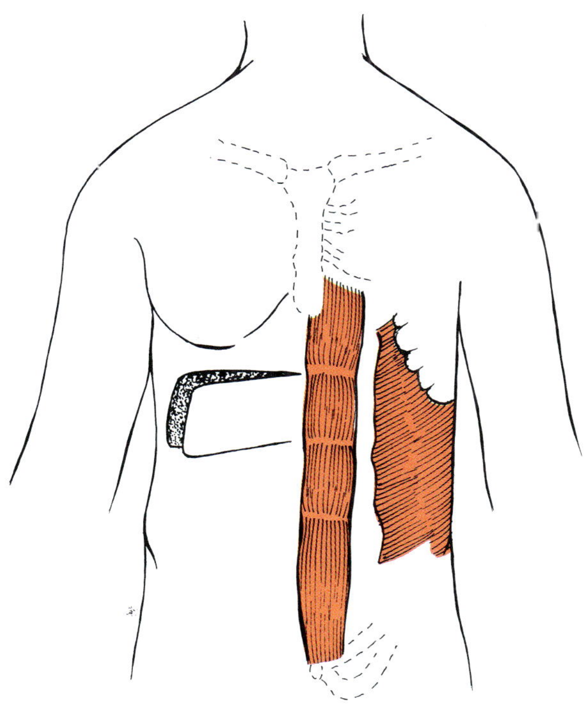

Rectus abdominis
Thoracoepigastric flap
External oblique

Fig. 19-2

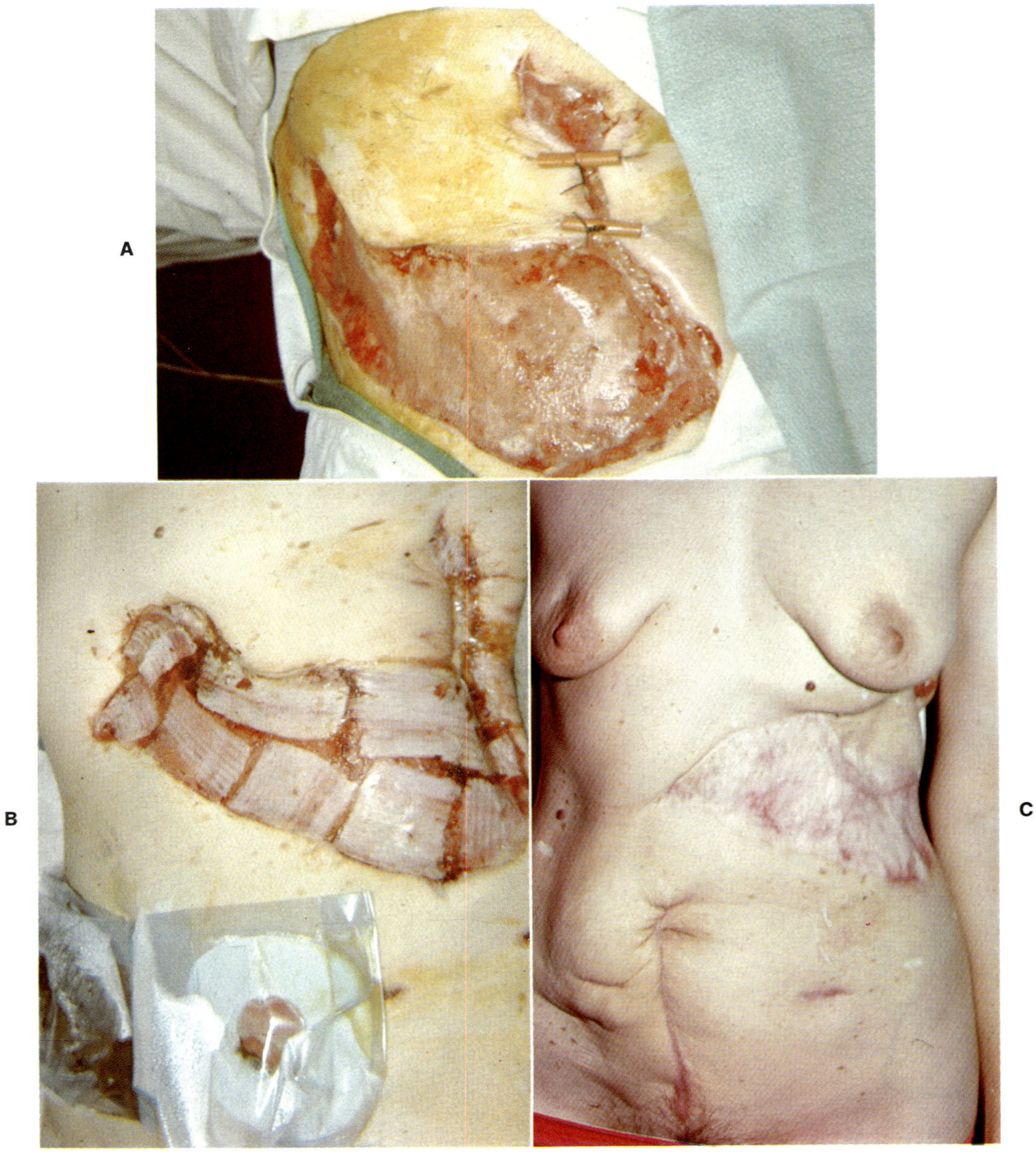

Fig. 19-3. Rectus abdominis musculocutaneous flap.

A, Immediate treatment of close-range shotgun blast includes ileostomy and restoration of abdominal wall continuity with prolene mesh.

B, Successful coverage provided by skin grafts placed directly over synthetic mesh.

C, Left anterior abdominal wound subject to frequent breakdown because of instability of skin graft over mesh.

D, Design of superiorly based rectus abdominis musculocutaneous flap. NOTE: Previous ileostomy placed through cutaneous territory of muscle unit. However, angiograms confirm patency of superior epigastric artery.

E, Anatomy of rectus abdominis muscle flap. *s,* Superior epigastric artery—major pedicle for superiorly based flap; *i,* inferior epigastric artery—major pedicle for inferiorly based flap. Elevation of superiorly based flap below semicircular line of Douglas results in potential site for abdominal hernia *(arrow)*.

(**C, D,** and **F** from Mathes, S.J., and Bostwick, J., III: Br. J. Plast. Surg. **30:**282, 1977.)

D E

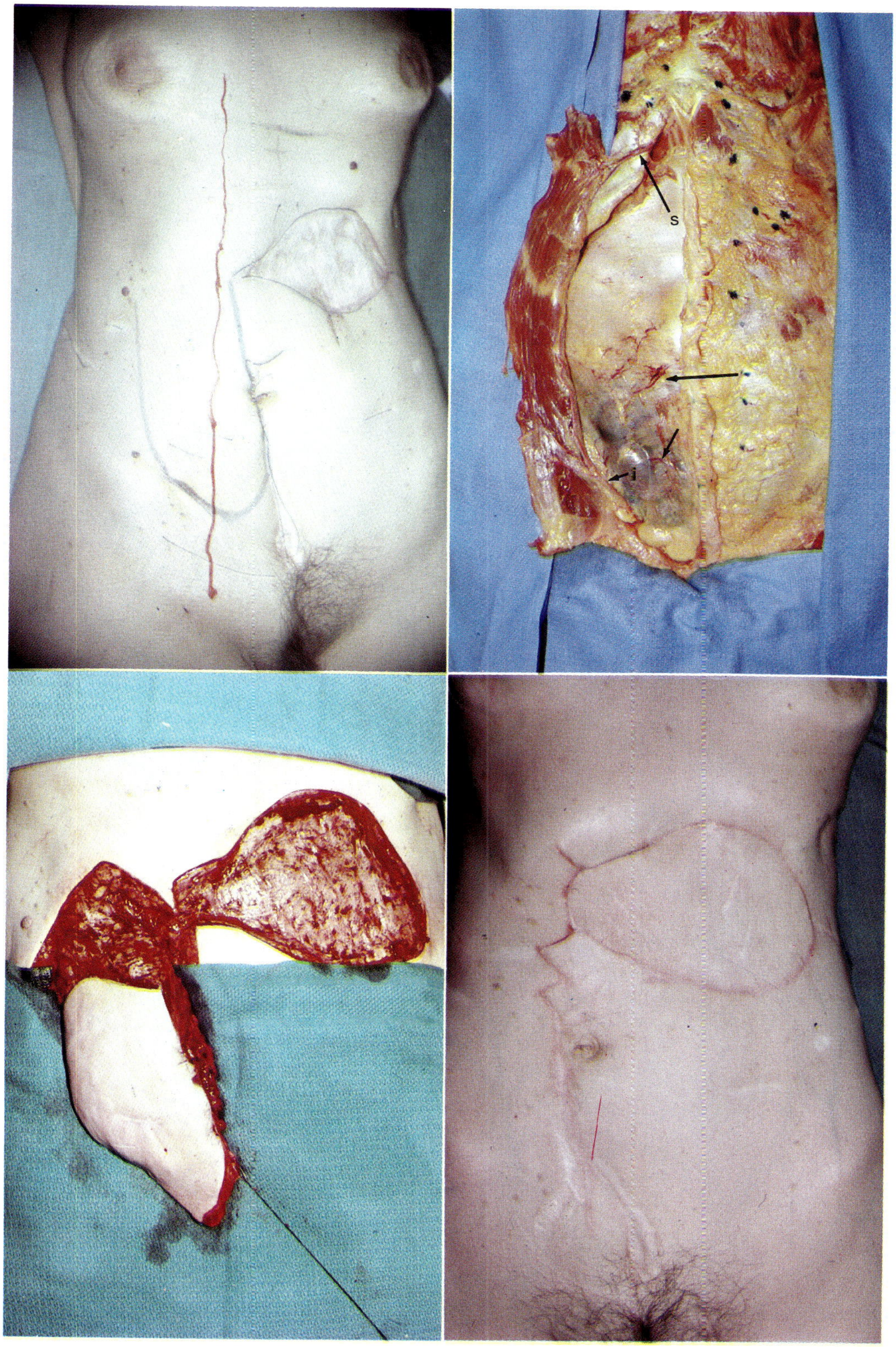

F G

Fig. 19-3, cont'd

F, Elevation of superiorly based rectus abdominis musculocutaneous flap.

G, Stable wound coverage provided over abdominal wall defect by transposition of rectus abdominis musculocutaneous flap. NOTE: Donor defect is closed directly with preservation of posterior rectus sheath without subsequent hernia formation. One-year postoperative view of anterior abdomen. (Recent Z-plasty performed on donor site scar.)

LOWER ABDOMEN (Fig. 19-4)

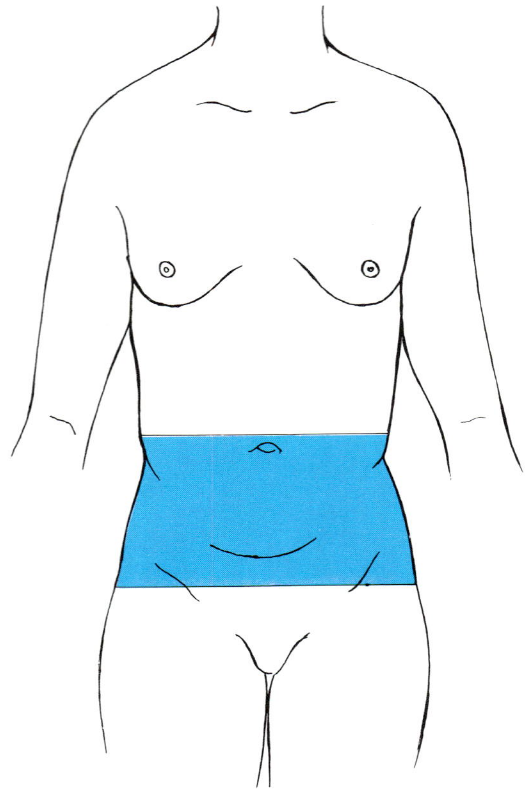

Fig. 19-4

Skin grafts

See upper abdomen

Local flaps (Fig. 19-5)

1. Rectus abdominis musculocutaneous flap, when available, is the flap of choice for unilateral defect
2. Thoracoepigastric axial flap and external oblique musculocutaneous flap are alternative flaps

Distant flaps (Fig. 19-5)

1. Tensor fascia lata musculofascial and musculocutaneous flaps are flaps of choice for reconstruction of the lower abdominal wall (in thin individuals this flap may well reach the upper abdomen) (Fig. 19-6)
2. Rectus femoris musculocutaneous flap is alternative flap for abdominal wall reconstruction and flap of choice for wounds with chronic infection or osteomyelitis of pelvis (Fig. 19-7)
3. Axial groin flap is the flap of choice for soft tissue defects that do not require fascial reconstruction (Fig. 19-8)

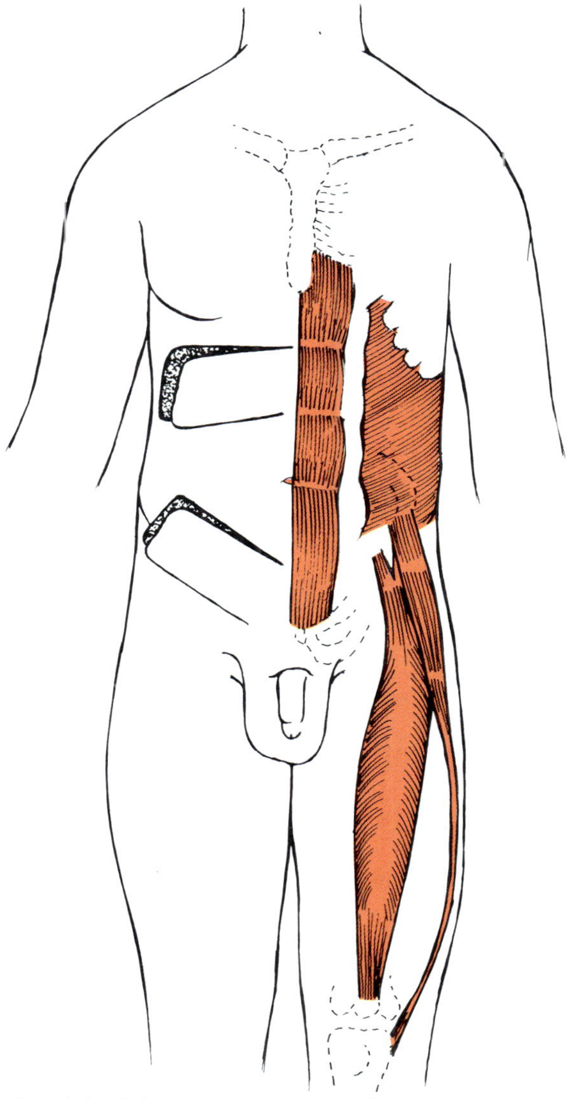

Rectus abdominis
External oblique
Thoracoepigastric flap
Groin flap

External oblique
Tensor fascia lata
Rectus femoris

Fig. 19-5

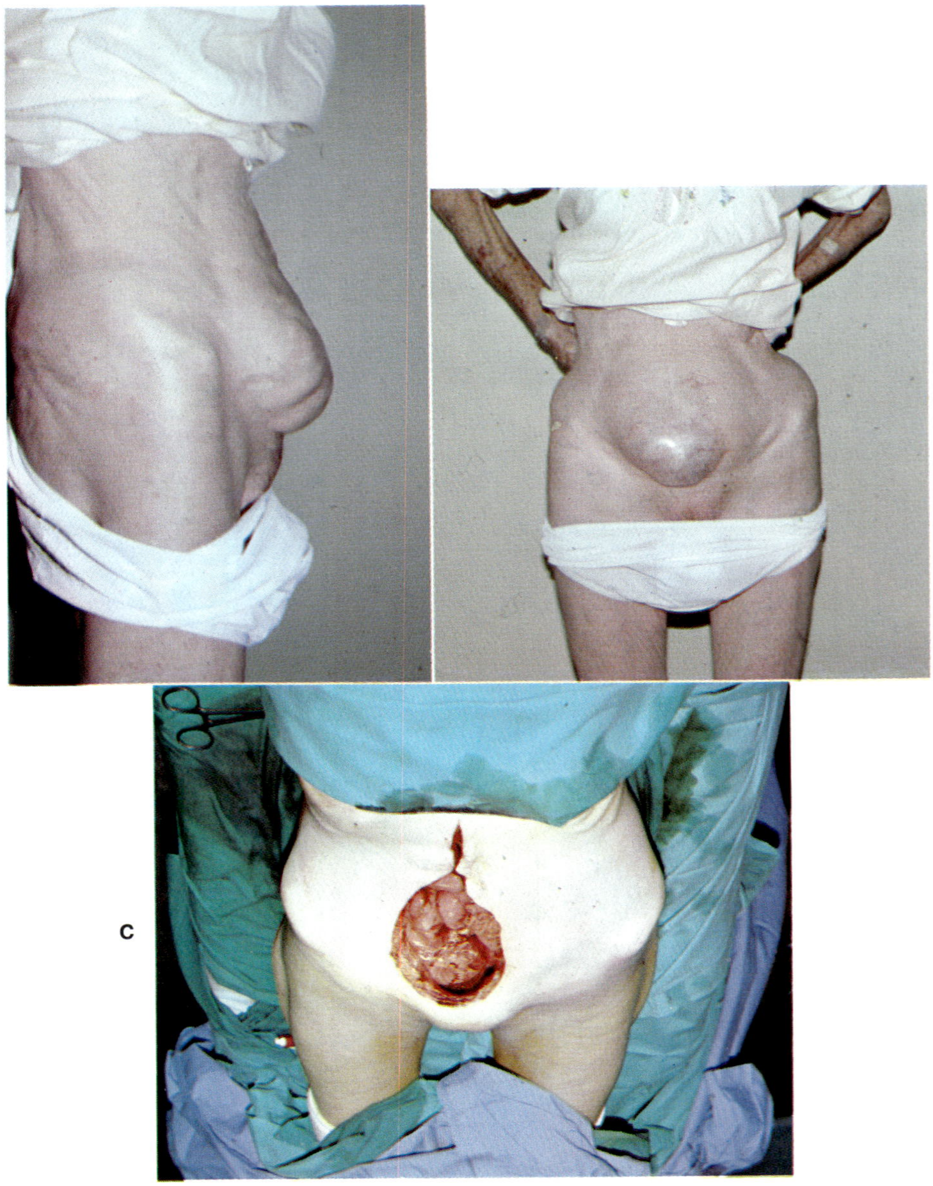

Fig. 19-6. TFL musculocutaneous flap.

A and **B,** Preoperative view of abdominal hernia in 60-year-old woman several years after resection and irradiation for carcinoma of cervix. She has small bowel fistula into vagina and abdominal wall hernia with previous skin grafts directly over bowel.

C, Resection of hernia and small bowel fistula.

D

E

F

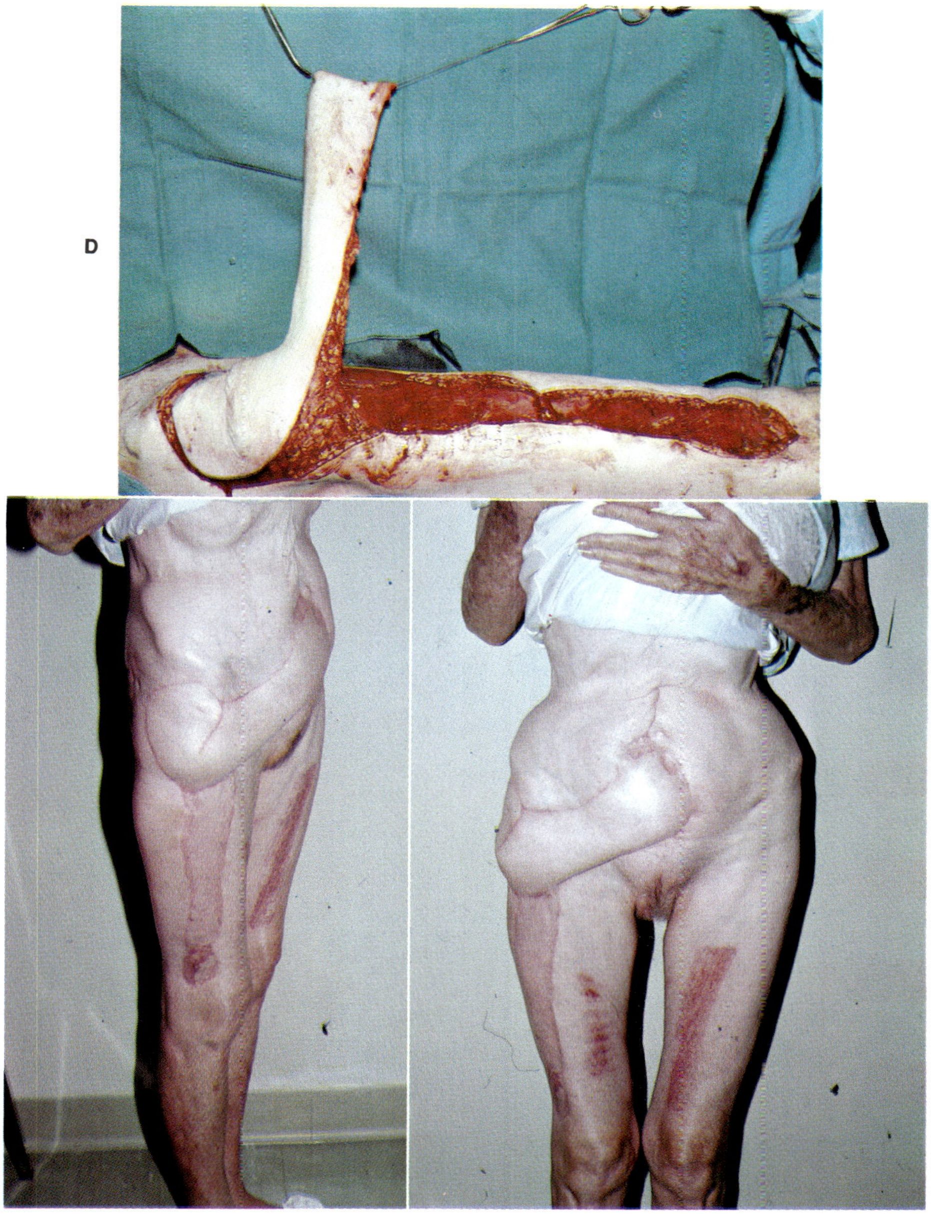

Fig. 19-6, cont'd

D, Extended TFL flap elevated and transposed onto abdominal wall defect.

E and **F,** Abdominal wall after operation reconstructed with TFL flap. Fascia lata replaces absent abdominal wall fascia and musculature.

A B C

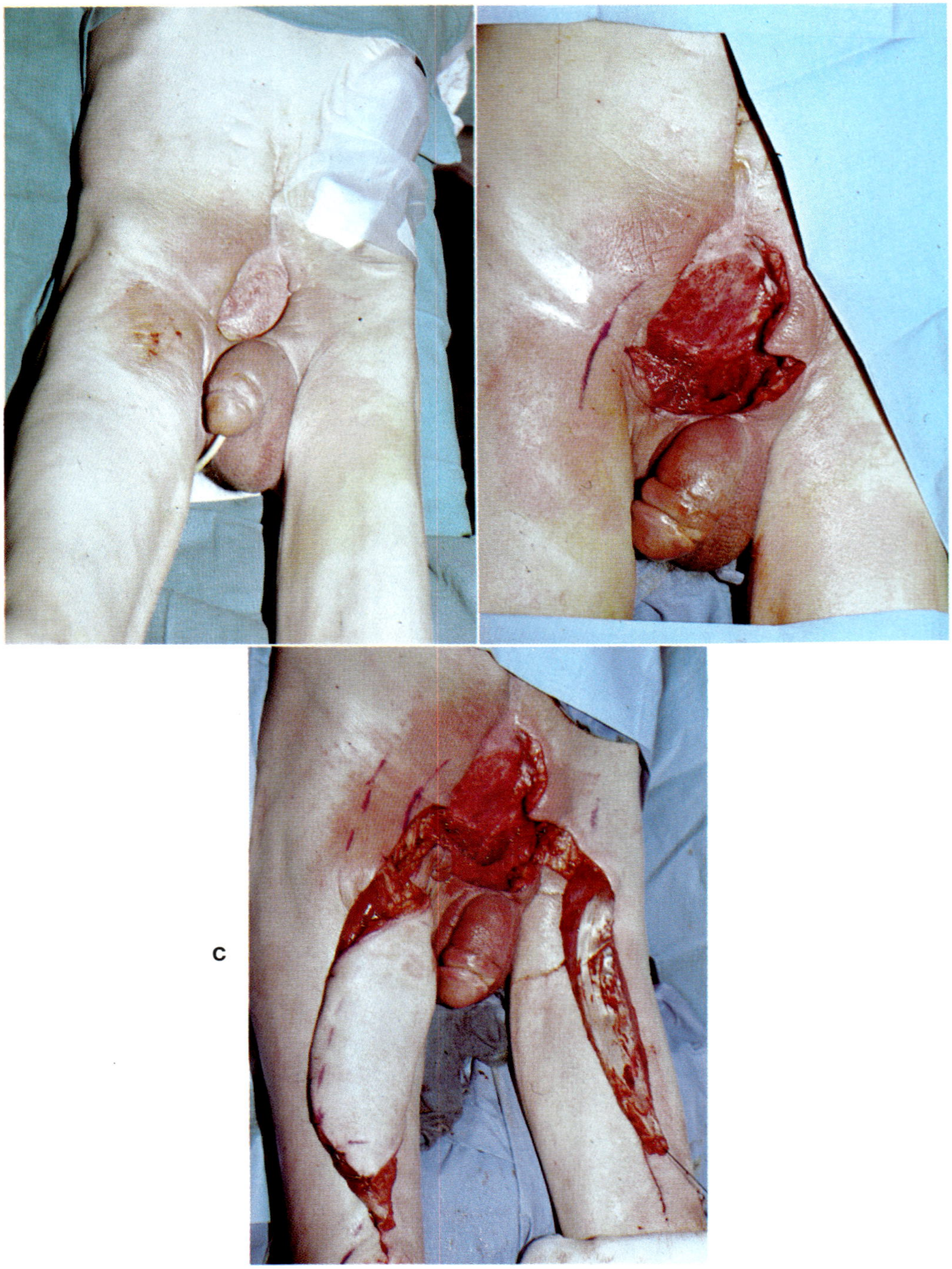

Fig. 19-7. Bilateral rectus femoris flaps.

A, Osteoradionecrosis of inferior abdominal wall and sacrum after recent abdominal peroneal resection and radiation therapy. Bilateral exposure of femoral vessels is noted.

B, Wound débridement includes pubis and portion of sacrum in retroperitoneal area.

C, Elevation of left rectus femoris muscle and right musculocutaneous flap.

D E

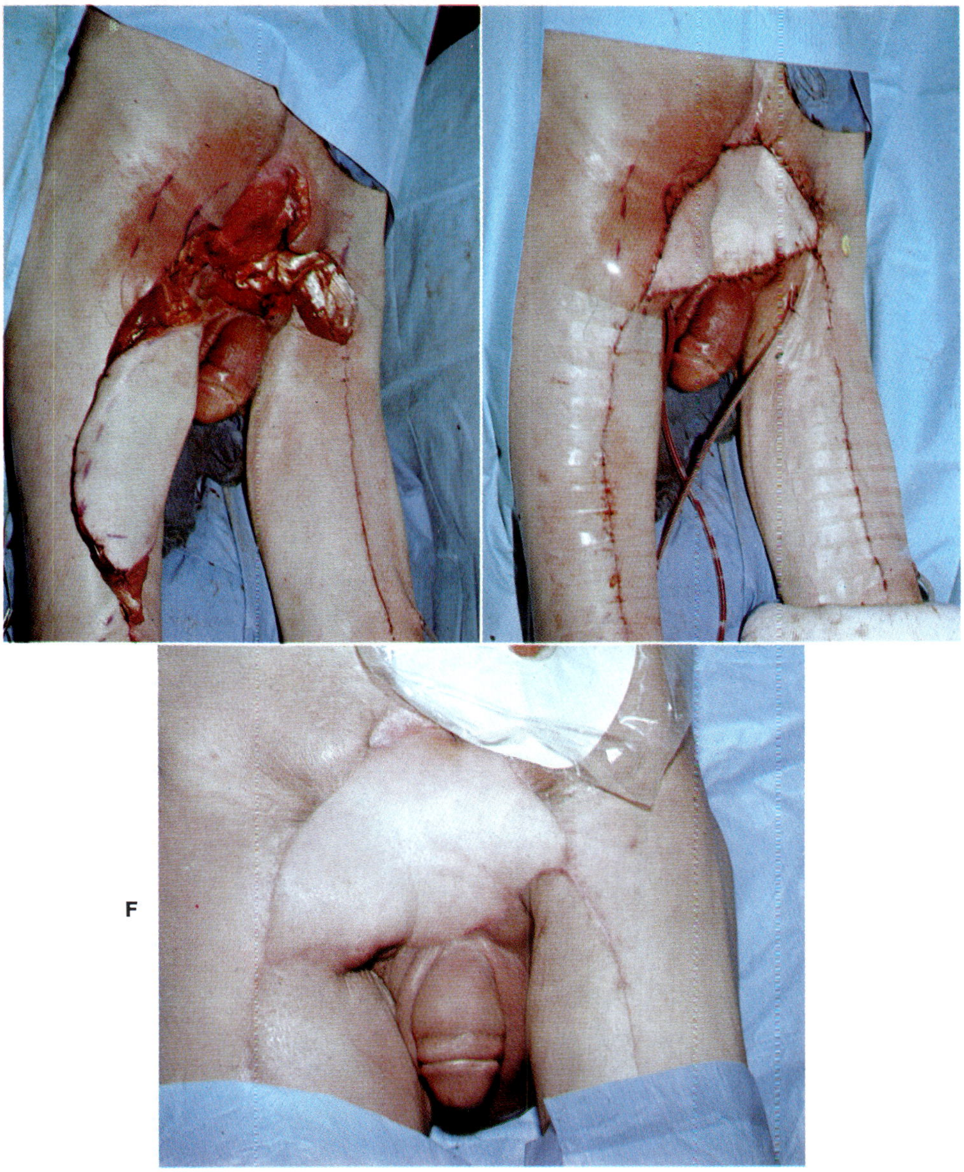

F

Fig. 19-7, cont'd

D, Left rectus femoris muscle flap transposed for coverage over left femoral vessels and muscle placed into sacral defect.

E, Left rectus femoris muscle flap transposed for coverage of left femoral vessels and for pelvic defect. Right rectus femoris musculocutaneous flap transposed for coverage of right femoral vessels and coverage of inferior abdominal wall defect. Both donor defects closed directly.

F, Three-month postoperative view demonstrates stable wound coverage provided by bilateral rectus femoris muscle and musculocutaneous flaps.

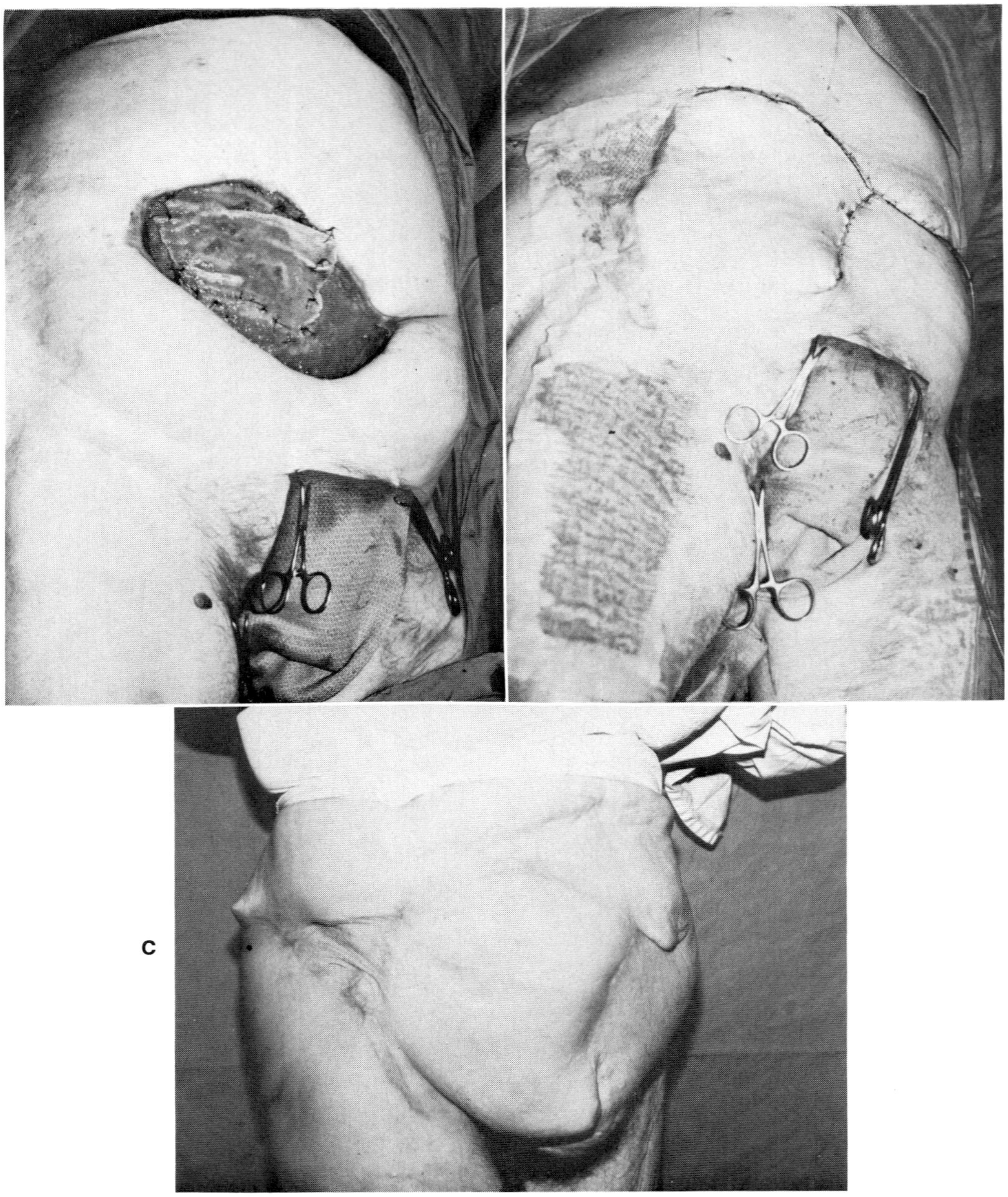

A

B

C

Fig. 19-8. Groin flap.

A, Exposed prolene mesh used to establish continuity of right anterior abdominal wall after emergency treatment for perforated colon carcinoma.

B, Right axial groin flap transposed over exposed mesh used for abdominal wall reconstruction.

C, One-year postoperative view demonstrates stable wound coverage provided by axial flap transposition. (Donor site covered with skin graft.)

20

RECONSTRUCTION

Foad Nahai ■ Michael Scheflan ■ Stephen J. Mathes

The goals of reconstructive surgery of the abdomen are two-fold and include reestablishment of the integrity of the abdominal wall and provision of soft tissue coverage. With full-thickness loss of the abdominal wall, both are required. However, in some instances only soft tissue coverage or only the reestablishment of the integrity of the abdominal wall is necessary. Large hernias where soft tissue coverage is adequate, for example, can be corrected by the repair or replacement of the weakened muscles only. Muscle and musculocutaneous flaps will provide soft tissue coverage, and some of the flaps can replace abdominal wall muscles and fascia to reestablish the integrity of the abdominal wall.

Synthetic mesh, Marlex or Prolene, may be used to reconstruct the abdominal wall musculature and then covered with a skin graft or flap. Mathes and Stone have shown the value and reliability of this method for the immediate reconstruction of large full-thickness traumatic wounds of the abdominal wall resulting from close-range shotgun wounds. The elevation of local or distant flaps is best avoided in acute injuries where bowel perforation and fecal contamination are common. Conservative management with synthetic mesh, followed by skin graft, is more appropriate initially. Later definitive reconstruction with flaps is undertaken if indicated. After tumor resection or elective repair of large hernias, flaps may be elevated primarily for reconstruction of full-thickness defects of the abdominal wall.

A variety of muscle and musculocutaneous flaps may be used for coverage or reconstruction of the abdominal wall. These include the rectus abdominis and external oblique muscles, which themselves are abdominal wall muscles; TFL; rectus femoris; gracilis from the thigh; and latissimus dorsi from the posterior trunk (Fig. 20-1).

TFL

The TFL flap is the most useful flap for reconstruction of the abdominal wall. Wangensteen describes the use of the fascia lata as a pedicled fascial flap for reconstruction of abdominal wall hernias. With the overlying skin as an extended TFL flap, this fascia can be used for reconstruction of full-thickness defects. Therefore this unit may be elevated either as an extended TFL flap or as a musculofascial flap for reconstruction of the abdominal wall.

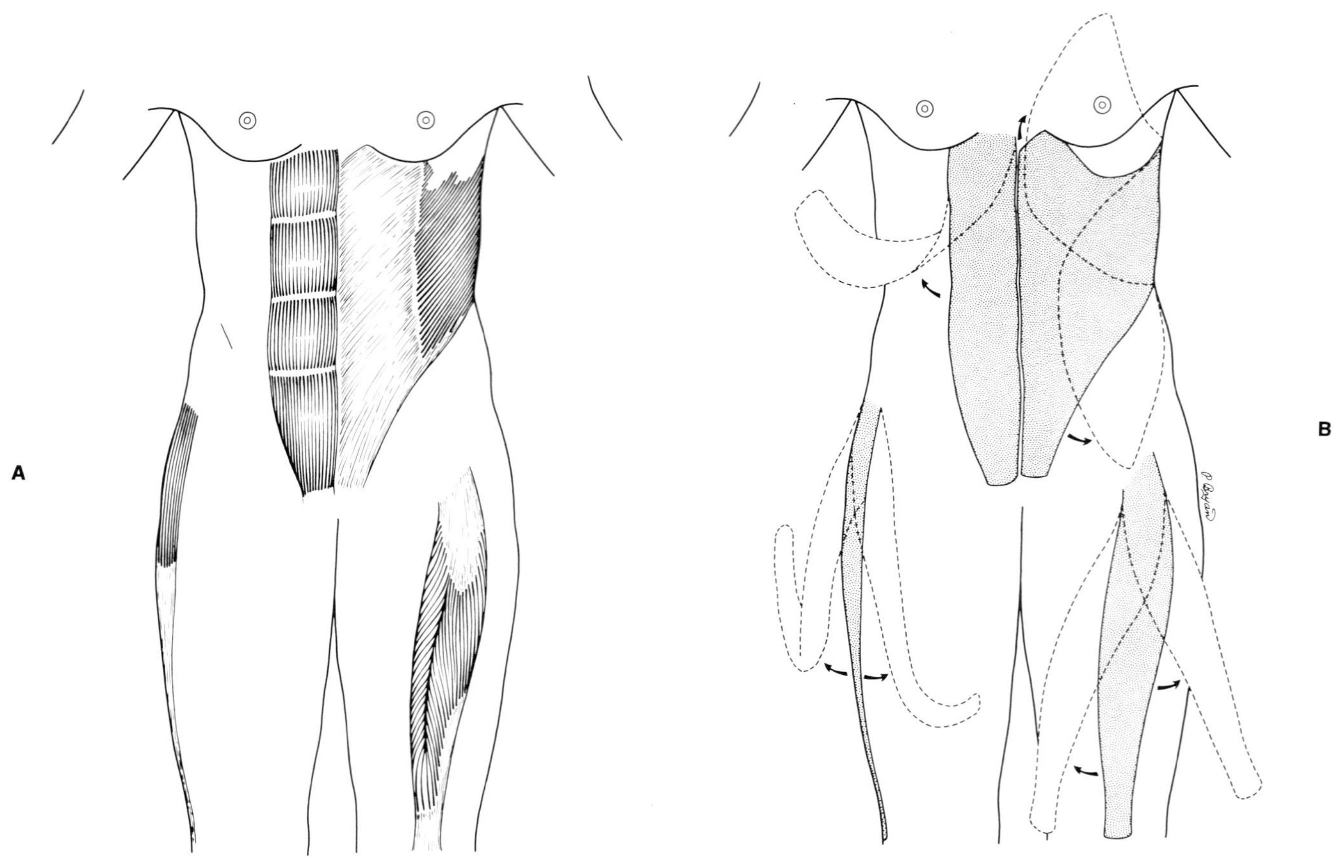

Fig. 20-1
A, Muscles and musculocutaneous flaps for reconstruction of abdominal wall.
B, Arc of rotation of flaps.

Flap elevation

Extended TFL musculocutaneous flap. The flap is designed on the anterolateral thigh. The point of rotation based on the dominant pedicle from the lateral circumflex femoral artery is approximately 10 cm below the anterior superior iliac spine. The length and width of the flap are designed according to the size of the defect to be reconstructed (Fig. 20-2). The anterior border is a line from the anterior superior iliac spine to the knee, and the posterior border is a straight line from the greater trochanter to the knee. The flap can be safely extended anteromedially to include skin over the rectus femoris if a wider flap is required (Fig. 20-3). The flap can also be extended posteriorly, but anteromedial extension is preferred, because the subcutaneous fat layer is thicker posteriorly.

Flap elevation proceeds from the distal border to the proximal border in the avascular plane between the fascia lata and the vastus lateralis. At a level approximately 15 cm below the anterior superior iliac spine, the deep surface of the TFL muscle is identified, and dissection then continues deep to the muscle. The pedicle is then identified entering the deep surface of the muscle medially. The lateral circumflex femoral artery emerges deep to the rectus femoris and superficial to the origin of the vastus later-

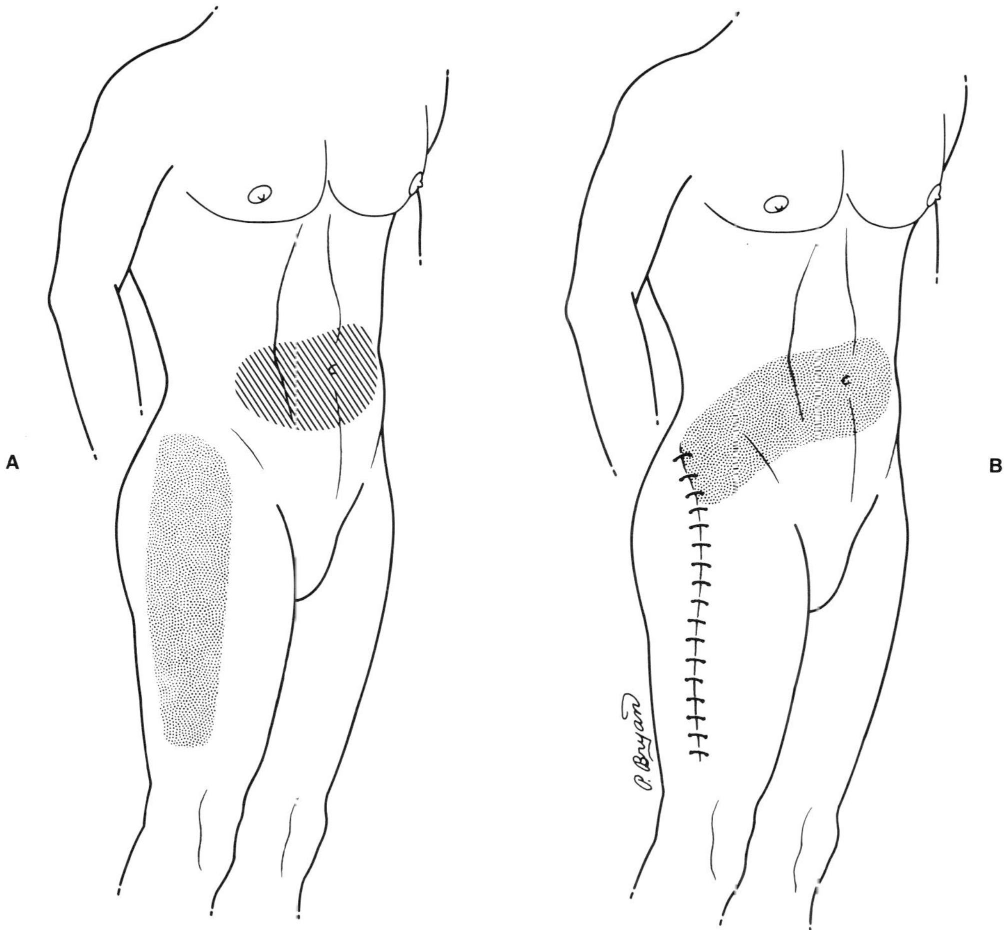

Fig. 20-2. Reconstruction of abdominal wall with TFL flaps.
A, Outline of flap and abdominal wall defect.
B, Flap transposed as island and donor area closed directly.

alis. The pedicle is dissected from the surrounding tissues to allow the 180° rotation of the flap for abdominal wall coverage. The skin incisions are then completed and the flap transposed. The flap is transposed as an island to avoid "dog ears" over the iliac crest. Tunneling of the flap should be avoided, even if this means excision of intact abdominal wall skin so that the flap can reach the defect. The fascia lata is then sutured into the muscle and fascial margins of the abdominal wall defect, while the skin of the flap is sutured to the skin margins of the defect. The fascia lata is strong enough to reestablish abdominal wall integrity, and synthetic mesh is not required. However, if, in the judgment of the reconstructive surgeon, mesh is preferred, then the TFL flap may be placed on the mesh to provide soft tissue coverage.

For flaps up to about 8 cm in width the donor area can be closed directly. Wider flaps will require skin grafting. However, this may vary with the body build of the patient. In general, the arc of rotation in a thin patient is greater, and the donor area is closed more readily.

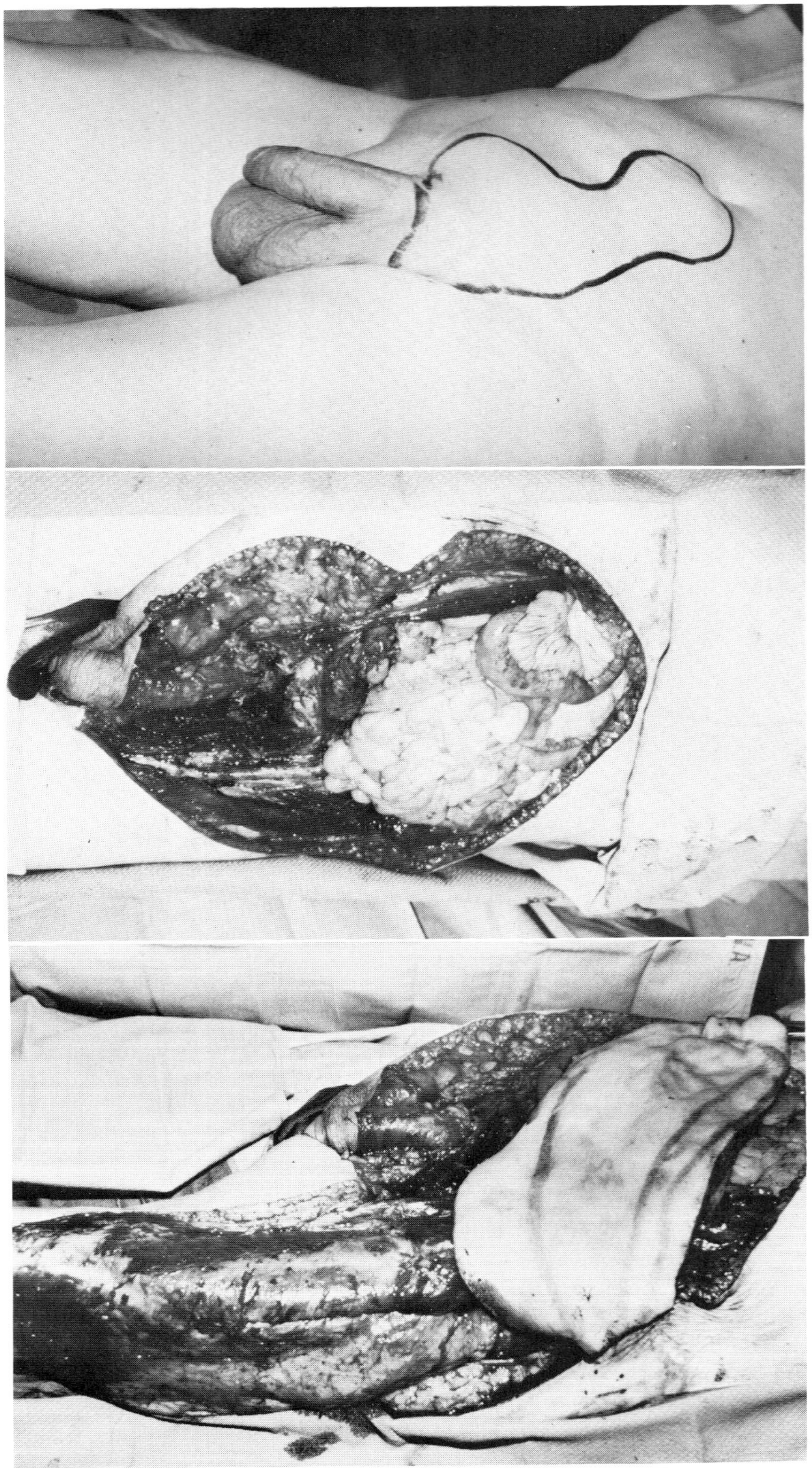
A

B

C

Fig. 20-3. Reconstruction of full-thickness defect of abdominal wall with extended TFL flap.

A, Sixty-five-year-old man with extensive liposarcoma of abdominal wall involving muscles.

B, Resultant defect after wide resection of tumor. NOTE: Exposed bowel and external iliac vessels.

C, Wide, extended TFL flap is elevated and transposed as island. NOTE: Flap extends well over rectus femoris muscle.

D

E

F

G

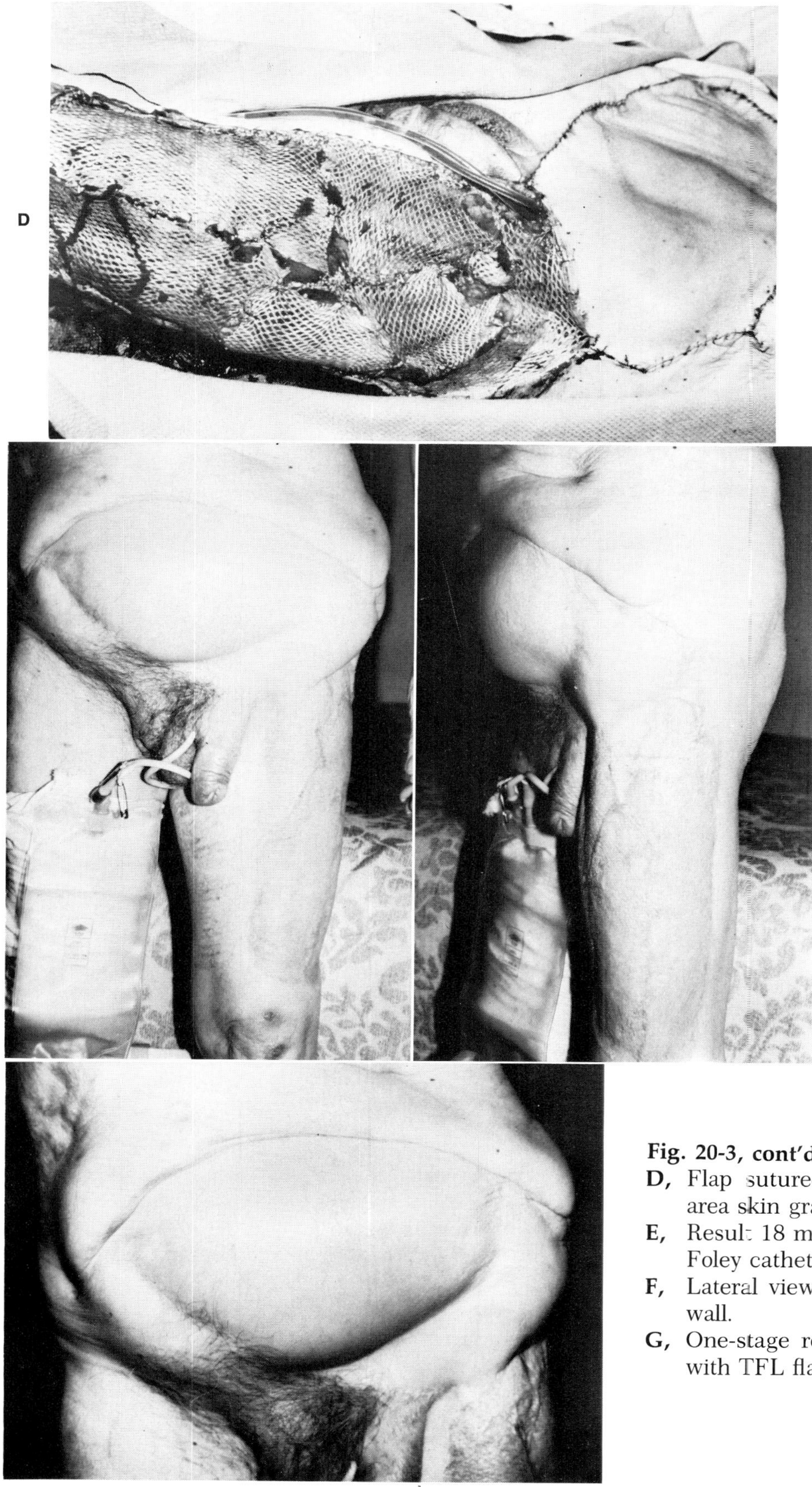

Fig. 20-3, cont'd

D, Flap sutured to abdominal wall and donor area skin grafted.

E, Result 18 months after operation. Indwelling Foley catheter for unrelated problem.

F, Lateral view to show integrity of abdominal wall.

G, One-stage reconstruction of abdominal wall with TFL flap.

Musculofascial flap. The TFL flap may be elevated as muscle and fascia lata without the overlying skin. This musculofascial flap is useful for reconstruction of abdominal wall defects where soft tissue coverage is adequate but muscle or fascia has been destroyed, as in large hernias (Fig. 20-4). The fascia lata is exposed through a longitudinal incision on the anterolateral thigh. The width and length of the fascia lata are designed to fit the defect. The overlying skin is dissected off the fascia lata and the fascia lata incised distally. The fascia is then dissected off the vastus lateralis from the distal border to the proximal border. The muscle and vascular pedicle are identified, as previously described. The muscle and fascia are then turned over, and if necessary, the fascia may be tunneled through into the defect where it is sutured into the margins of the abdominal wall. Provided a wide tunnel is created, the fascia may be tunneled safely. The fascia is sutured into the defect under adequate tension. The TFL muscle remains innervated so that abdominal wall tension can be maintained through it. The donor area is closed directly.

A

B

C

D

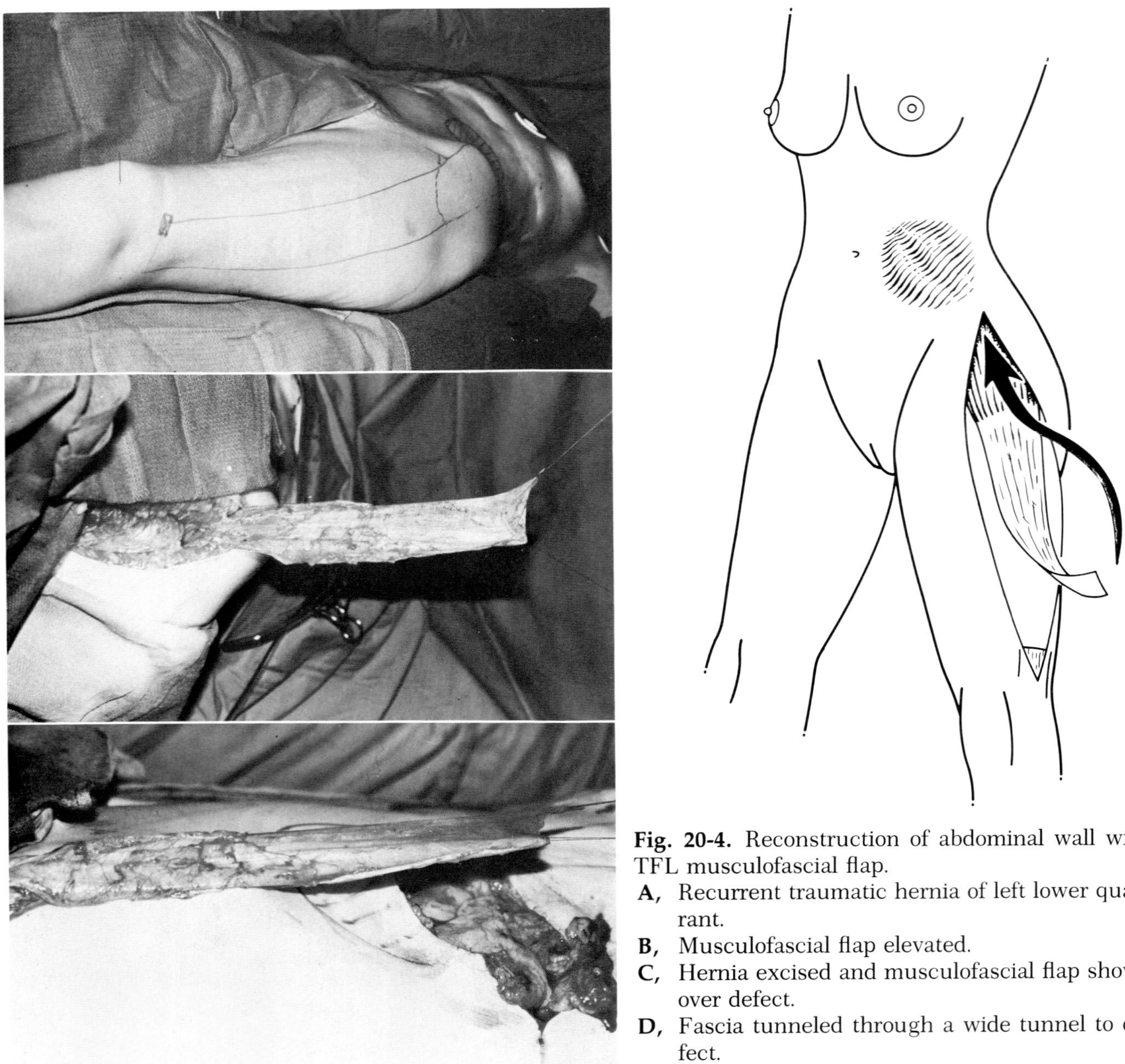

Fig. 20-4. Reconstruction of abdominal wall with TFL musculofascial flap.
A, Recurrent traumatic hernia of left lower quadrant.
B, Musculofascial flap elevated.
C, Hernia excised and musculofascial flap shown over defect.
D, Fascia tunneled through a wide tunnel to defect.

E

F

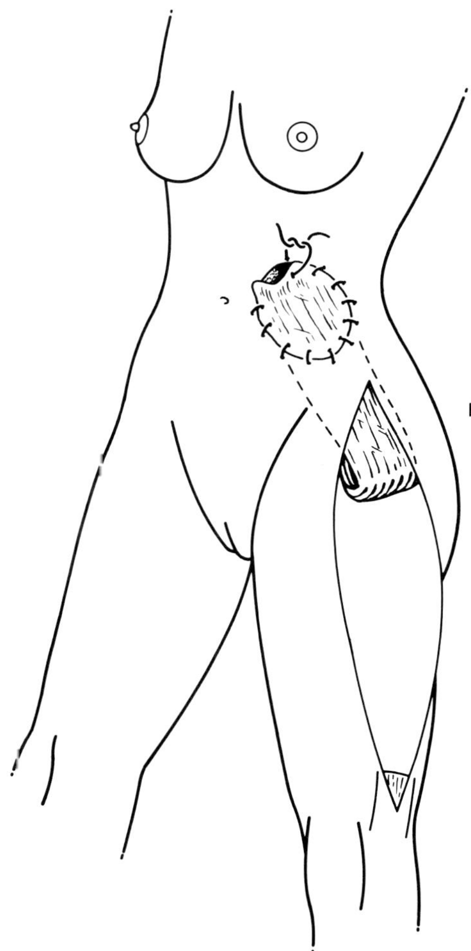

G

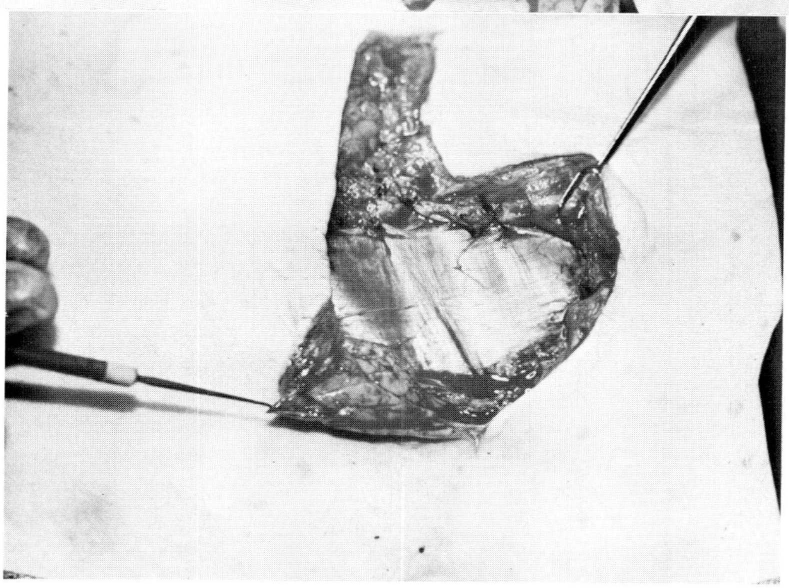

Fig. 20-4, cont'd
E, Fascia brought into defect.
F, Flap sutured in place.
G, Fascia lata sutured into defect.

Continued.

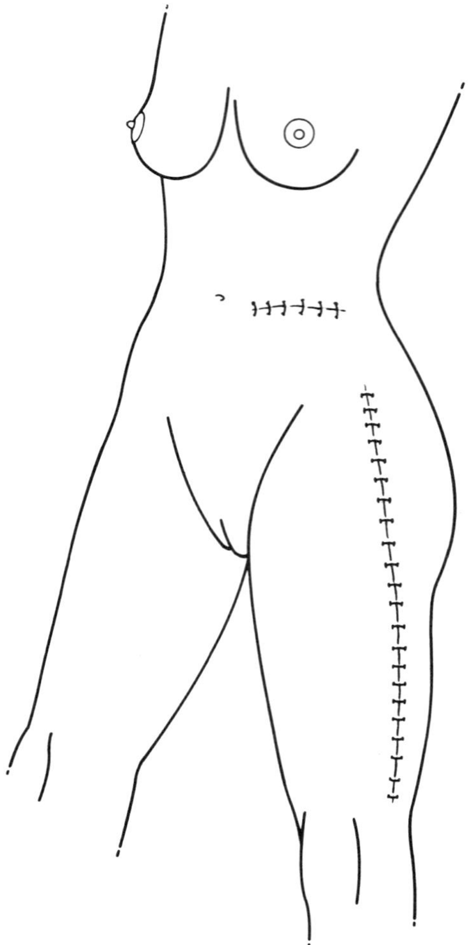

H

Fig. 20-4, cont'd
H and I, Final result donor site closed directly.

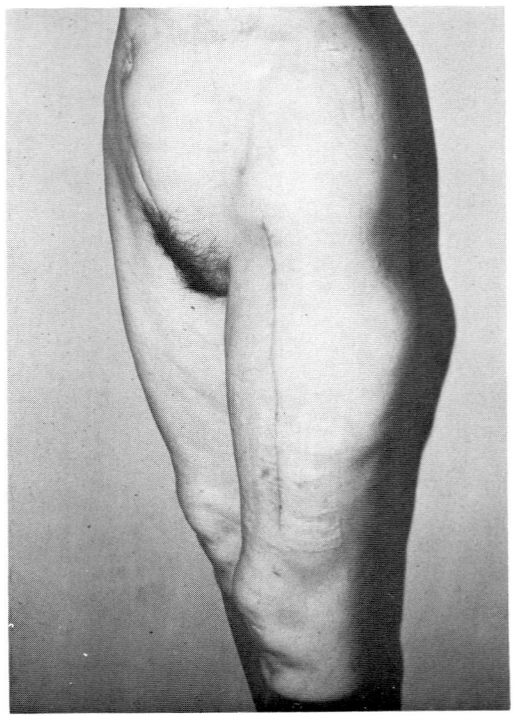

I

RECTUS FEMORIS

The rectus femoris is an excellent and reliable unit that will provide soft tissue coverage for the abdominal wall. It can be elevated with the fascia lata to reconstruct muscular and fascial defects. Its arc of rotation is similar to that of the TFL, but it is a flap of second choice if the TFL is available. The rectus femoris is a much thicker flap and is not suitable for repair of hernias where skin coverage is adequate. The potential functional loss after transposition of this flap should be considered before flap selection. In defects with chronic infection resulting from trauma or osteoradionecrosis, the rectus femoris muscle is preferred for coverage after débridement of infected bone (sequestrectomy) and soft tissue.

Flap elevation

The skin territory of the rectus femoris is located centrally on the anterior thigh. The skin island is placed over the upper two thirds of the muscle. The incision is made, and the distal muscle is identified. The muscle is divided at a point 8 cm above the knee. The distally cut tendinous portion of the muscle is then sutured to the underlying vastus intermedius to minimize functional loss. The flap is then elevated from the distal border to the proximal border of the vastus intermedius. Dissection is continued proximally until the sartorius muscle is identified, crossing in front of the rectus femoris. At this level the vascular pedicle, a branch of the lateral circumflex femoral artery, is identified. The lateral femoral circumflex artery runs deep to the rectus femoris, and the pedicle of the flap courses anteriorly into the muscle. Dissection beyond the vascular pedicle point of entrance into the muscle is rarely required. The donor site for islands of skin 8 cm or less in width can be closed directly, whereas wide islands of skin require skin grafting of the donor area.

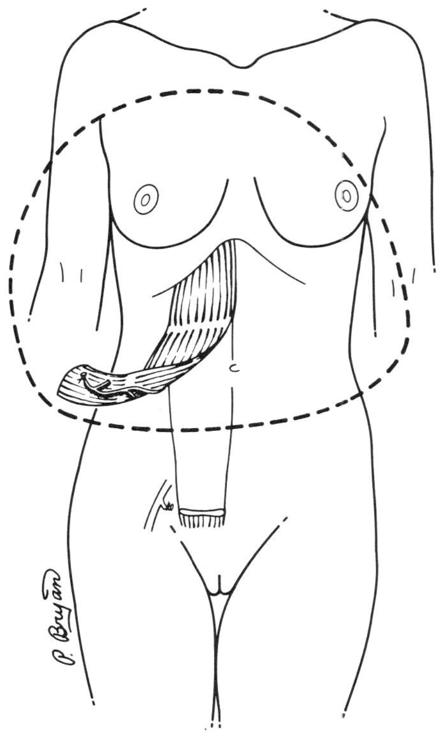

Fig. 20-5. Arc of rotation of superiorly based rectus abdominis flap *(superior arc)*.

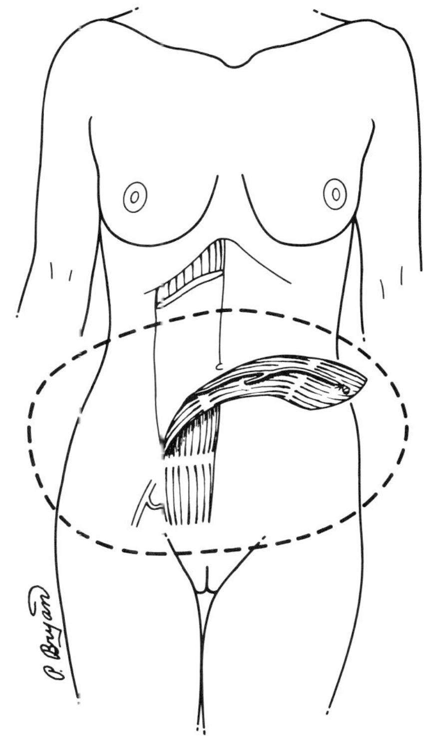

Fig. 20-6. Arc of rotation of inferiorly based rectus abdominis flap *(inferior arc)*.

RECTUS ABDOMINIS

The rectus abdominis flap may be elevated as a superiorly based flap on the superior epigastric artery (Fig. 20-5) or inferiorly based on the inferior epigastric artery (Fig. 20-6). This unit will only provide soft tissue coverage. Although the anterior rectus sheath is elevated with the musculocutaneous flap, the muscle that is segmentally innervated through its lateral border is denervated during flap elevation and cannot contribute to the maintenance of abdominal wall integrity. The foregoing, coupled with the fact that the rectus abdominis is a muscle on the abdominal wall and may well be damaged or resected when the defect is created, limits the usefulness of this flap for abdominal wall reconstruction.

Flap elevation

Superior arc. The skin island is outlined over the muscle and incised. The anterior rectus sheath is incised at the level of the umbilicus or a few centimeters below for longer flaps. The muscle is then divided and the flap elevated from the distal border to the proximal border of the posterior rectus sheath. The muscle dissects off easily with blunt dissection between the tendinous intersections. However, at the tendinous intersections careful, sharp dissection is essential. Laterally the segmental motor nerves are divided and the flap elevated to the level of the costal margin. The donor area is closed directly. The medial border of the external oblique is advanced and sutured to the linea alba and the skin closed directly. The use of the superior half of the rectus abdominis does not weaken the abdominal wall, because the strength of the anterior and posterior rectus sheath is the same at this level.

Inferior arc. The use of the inferiorly based rectus abdominis musculocutaneous flap is not recommended, because it may lead to weakening of the lower abdominal wall. Below the semicircular line of Douglas, the anterior rectus sheath is made up of the external and internal oblique fascia, but the posterior sheath is made up only of the transversalis fascia. For this reason alternative flaps are preferred to the inferior rectus abdominis flap.

EXTERNAL OBLIQUE

As an abdominal wall muscle with a segmental nerve and blood supply, the external oblique also has limited application for reconstruction of abdominal wall defects. However, under certain circumstances it will provide full-thickness tissue for reconstitution of abdominal wall integrity and soft tissue coverage. Hershey and Butcher report the use of compound fascia muscle and skin flaps based on the external oblique muscle for reconstruction of abdominal wall defects. The entire muscle and overlying skin are transposed superiorly or inferiorly (Fig. 20-1, *B*). The donor defect over the internal oblique muscle is then skin grafted.

LATISSIMUS DORSI

The latissimus dorsi based on the thoracodorsal pedicle will reach and cover the upper abdomen just below the costal margin. The lower portions of the origin of the latissimus dorsi muscle are fascial, and this fascia can be transposed anteriorly into the lateral abdominal wall for closure of full-thickness defects. This limited arc of rotation to the abdominal wall limits the usefulness of the latissimus dorsi in reconstruction of the abdominal wall.

GRACILIS

The gracilis muscle will reach the lower abdomen. However, this is a relatively narrow muscle, and its arc of rotation to the abdomen is limited. These two factors will therefore limit this unit's usefulness for reconstruction of the abdominal wall.

CONCLUSION

Reliable one-stage reconstruction of a variety of defects of the abdominal wall is possible with the application of muscle and musculocutaneous flaps. The TFL and rectus femoris are particularly useful, because the blood supply to these flaps is based on the lateral circumflex femoral artery and therefore is not compromised by radiation, trauma, or tumor resection of the abdominal wall. This, of course, is also true of the latissimus dorsi and gracilis, which have a more limited application in reconstruction of the abdominal wall.

Choice of flap

The TFL is the flap of first choice for abdominal wall reconstruction. As a musculocutaneous flap, it will provide one-stage reconstruction of full-thickness defects and as a musculofascial flap for correction of hernial defects. The TFL will easily reach up to the level of the umbilicus. In thin individuals it will reach up to the costal margin. In general, however, the use of the TFL is limited to the lower half or two thirds of the abdominal wall.

The arc of rotation and clinical applications of the rectus femoris in abdominal wall reconstruction are similar to that of the TFL musculocutaneous flap. However, this is a flap of second choice after the TFL. The rectus femoris is a much thicker flap and functional loss may result after its use.

The rectus abdominis will only provide soft tissue coverage. The upper half of the rectus abdominis based on the superior epigastric artery is preferred because elevation of the lower half may weaken the abdominal wall. This flap is most useful in coverage of synthetic mesh reconstruction of the opposite upper half of the abdominal wall.

The external oblique is useful for full-thickness reconstruction of small defects of the upper abdomen. For fascial defects the external oblique fascia can be turned over as a flap. In this situation its application is useful for correction of midline hernial defects.

The gracilis and latissimus dorsi have limited applications to the abdominal wall. The latissimus dorsi will reach the lateral abdominal wall, and the gracilis will reach the midline inferiorly.

Timing of reconstruction

Full-thickness defects resulting from tumor excision may be safely reconstructed in one stage at the time of resection. Similarly, large hernial defects can be reconstructed definitively in one operation. However, for traumatic loss of the abdominal wall, especially with perforation and fecal contamination, immediate flap coverage should be avoided. Reconstruction of abdominal wall integrity with synthetic mesh and later skin graft coverage is preferred. At a later stage definitive reconstruction with flaps may be undertaken.

ANNOTATED BIBLIOGRAPHY

1. Bostwick, J., III, and co-workers: Sixty latissimus dorsi flaps, Plast. Reconstr. Surg. **63:**31, 1979.

   The application of the latissimus dorsi muscle and fascia flap for reconstruction of defects of the lateral abdominal wall is described.

2. Bruck, H.: A method of reconstructing the whole abdominal wall, Br. J. Plast. Surg. **9:**108, 1956.

   Reconstruction of the abdominal wall in a 3½-year-old girl with a pedicled fascia lata flap, which was then covered with a tubed, pedicled skin flap from the other thigh, is discussed.

3. Farr, R.E.: Closure of large hernial defects in the upper abdomen, Surg. Gynecol. Obstet. **34:**264, 1922.

   Small defects of the upper abdomen are reconstructed with turnover flaps of fascia from the rectus sheath and the external oblique fascia.

4. Hershey, F.B., and Butcher, H.R.: Repair of defects after partial resection of the abdominal wall, Amer. J. Surg. **107:**586, 1964.

   Reconstruction of the abdominal wall with rotation flaps of the external oblique muscle, fascia, and overlying skin is described. Further applications of the pedicled fascia lata flap are also described.

5. Lesnick, G.J., and Davids, A.M.: Repair of surgical abdominal wall defect with a pedicled musculofascial flap, Ann. Surg. **137:**569, 1953.

   Case report of abdominal reconstruction with a turnover flap of external oblique fascia from the opposite side is discussed.

6. McPeal, C.J., and Miller, T.R.: Abdominal wall replacement, Surgery **47:**944, 1960.

   The application of free grafts of the fascia lata for correction of defects of the abdominal wall is described.

7. Mathes, S.J., and Bostwick, J., III: A rectus abdominis myocutaneous flap to reconstruct abdominal wall defects, Br. J. Plast. Surg. **30:**282, 1977.

   Case report of reconstruction of abdominal wall with synthetic mesh and rectus abdominis musculocutaneous flap.

8. Mathes, S.J., and Stone, H.H.: Acute traumatic losses of abdominal wall substance., J. Trauma **15:**386, 1975.

   Management of acute injuries of the abdominal wall using mesh and skin graft is described. The problems associated with closure of abdominal wall defects under tension are discussed.

9. Nahai, F.: The tensor fascia lata flap, Clin. Plast. Surg. **7:**51, 1980.

   The anatomy, variety of flaps, and clinical applications of the tensor fascia lata flap are described.

10. Nahai, F., Hill, H.L., and Hester, T.R.: Experiences with the tensor fascia lata flap, Plast. Reconstr. Surg. **63:**788, 1979.

    The fascia lata and overlying skin as an extended tensor fascia lata musculocutaneous flap are described for one-stage reconstruction of full-thickness defects of the abdominal wall. The application of the rectus femoris for abdominal wall reconstruction is also described.

11. Parkash, S., and Ramakrishnan, K.: A myocutaneous island flap in the treatment of a chronic radionecrotic ulcer of the abdominal wall, Br. J. Plast. Surg. **33:**138, 1980.

    Another case report using the rectus abdominis musculocutaneous island flap for soft tissue coverage of the abdominal wall is discussed.

12. Venugopalan, S.: Repair of midline abdominal incisional hernia by gracilis muscle transfer, Br. J. Plast. Surg. **33:**43, 1980.

    The application of the gracilis for repair of abdominal incisional hernias is described. Twenty patients with a 5-year follow-up are presented. Successful reconstruction with no recurrences are reported.

13. Wangensteen, O.H.: Repair of recurrent and difficult hernias and other large defects of the abdominal wall employing the iliotibial tract of fascia lata as a pedicle flap, Surg. Gynecol. Obstet. **59:**766, 1934.

    Use of tensor fascia lata as vascularized fascia is described for abdominal wall reconstruction.

14. Wangensteen, O.H.: Repair of large abdominal defects by pedicled fascial flaps, Surg. Gynecol. Obstet. **82:**144, 1946.

    The application of the fascia lata as a pedicled flap for closure of large defects of the abdominal wall and recurrent groin hernias is described.

ABDOMEN

21

COMPLICATIONS

Foad Nahai ■ Stephen J. Mathes

ABDOMINAL WALL

A number of flaps are available for reconstruction of the various sections of the abdominal wall. These flaps may be used alone or with synthetic mesh. Abdominal wall integrity and soft tissue coverage must be achieved to avoid complications. For large defects a combination of flaps may be necessary.

TFL

Planning errors

There is some margin for error in designing the skin territory over the TFL; however, the posterior extension of the skin is not reliable. The posterior thigh skin is supplied through the hamstrings and the inferior gluteal artery. The flap can be extended to within 5 to 8 cm of the knee. Beyond this length the skin may not be reliable. The arc of rotation of the TFL will vary greatly with the habitus of the patient. In obese patients, especially women, the TFL is a bulky flap, and the arc of rotation is limited.

In patients with significant arteriosclerotic vascular disease or vascular trauma, angiography should be considered to evaluate the profunda femoris artery. The donor defect from a wide TFL musculocutaneous flap in heavy patients will not close primarily, and a skin graft is necessary.

The deep surface of the TFL musculocutaneous unit is comprised of fascia. Although this is advantageous for abdominal wall reconstruction, this fascia appears less resistant to bacterial contamination. A musculocutaneous flap with muscle on the deep surface of the distal flap (e.g., rectus femoris) is preferable for wounds with poor vascularity and chronic infection.

Technical errors

Excessive bleeding between the fascia lata and the underlying vastus lateralis may indicate unsuspected damage to the pedicle of the TFL, the terminal branch of the lateral circumflex femoral artery. In this case an alternative flap should be elevated.

RECTUS ABDOMINIS

Planning errors

The rectus abdominis has limited use in abdominal wall reconstruction because of involvement of the muscle or its vascular pedicle in the area of injury. It is mandatory preoperatively to establish patency of the appropriate inferior or superior vascular pedicle, depending on the direction of flap transposition. Despite a normal internal mammary and superior epigastric artery, the distal rectus abdominis muscle and its cutaneous island may not survive or be consistently reliable.

Technical errors

Elevation of the inferior muscle, below the semicircular line of Douglas, weakens the lower abdominal wall. Hernia is avoided by closure of the inferior anterior rectus sheath or advancement of external oblique fascia to midline fascia. The superior epigastric artery running just below the rectus abdominis over the costal margin is very vulnerable to injury during this part of the flap elevation.

Postoperative errors

Nasogastric suction in the immediate postoperative period may prevent problems with gastric distention and increased intra-abdominal pressure.

EXTERNAL OBLIQUE

Planning errors

The external oblique muscle has limited application in abdominal wall reconstruction because of its short arc of rotation. Its use is primarily reserved for defects in the lateral abdominal wall. Use of the external oblique muscle may weaken the abdominal wall in the flap donor region.

Technical errors

The segmental nature of the blood supply to the external oblique muscle limits its useful arc of rotation. Extensive dissection to facilitate flap rotation may result in division of blood supply necessary for flap survival. A back cut to facilitate flap rotation may compromise the blood supply by division of segmental vascular pedicles.

Postoperative errors

Nasogastric tube decompression to reduce intraabdominal pressure may be necessary if abdominal distention is noticed.

LATISSIMUS DORSI

Planning errors

The latissimus dorsi will reach portions of the abdomen, but the arc of rotation may limit its use to upper lateral defects. If the muscle has been previously denervated, its bulk will be inadequate for use in abdominal wall reconstruction. The latissimus dorsi should never be used in the presence of a previous thoracotomy scar, because the distal muscle and the skin will not survive.

Technical errors

Use of the latissimus dorsi flap for the abdominal wall will require mobilization of the entire muscle. All proximal dissection must be performed under direct vision to avoid damage to the neurovascular pedicle or other structures in the axilla. The motor nerve to the latissimus dorsi should be preserved to retain muscle bulk. Tunneling should be avoided. The latissimus dorsi muscle must be separated from the underlying serratus anterior muscle during flap elevation. Inadvertent elevation of the serratus anterior muscle with the latissimus dorsi must be avoided.

Postoperative errors

Closed suction drainage to the donor deficit avoids serum collection in the large donor defect in the back. The patient should be positioned to avoid pressure on the flap.

THORACOEPIGASTRIC FLAP

Planning errors

The thoracoepigastric flap has limited arc of rotation that restricts its use to the lower abdominal wall. It is important to consider preoperatively possible damage to the perforating vessels that supply this flap from the internal mammary artery through the rectus abdominis muscle. The presence of a previous paramedian incision precludes the use of the flap. This flap alone will not reestablish abdominal wall integrity.

Technical errors

In elevating the thoracoepigastric flap, it is crucial to preserve the perforating vessels between the rectus abdominis and the flap.

RECTUS FEMORIS

Planning errors

The rectus femoris muscle is useful only for coverage of the lower abdomen. This is not an expendable muscle because of impairment of full knee extension, especially if other muscles in the quadriceps group are weak. In patients with significant peripheral vascular disease, preoperative angiography may be necessary to evaluate the blood supply to the flap. The fascia lata may be elevated with this flap to establish abdominal wall continuity.

Technical errors

The dissection plane between the rectus femoris and vastus medialis is not distinct and may be difficult to establish. This should be done precisely to avoid continuing proximally in an improper plane. The proximal dissection of this muscle must be performed under direct vision, because the branches of the femoral nerve are in proximity to the vascular pedicle.

Postoperative errors

The patient should be positioned to avoid pressure on the flap. Closed suction drainage to the donor defect is recommended.

22

A SYSTEMATIC APPROACH TO FLAP SELECTION

Stephen J. Mathes ■ Foad Nahai

GROIN (Figs. 22-1 and 22-2)

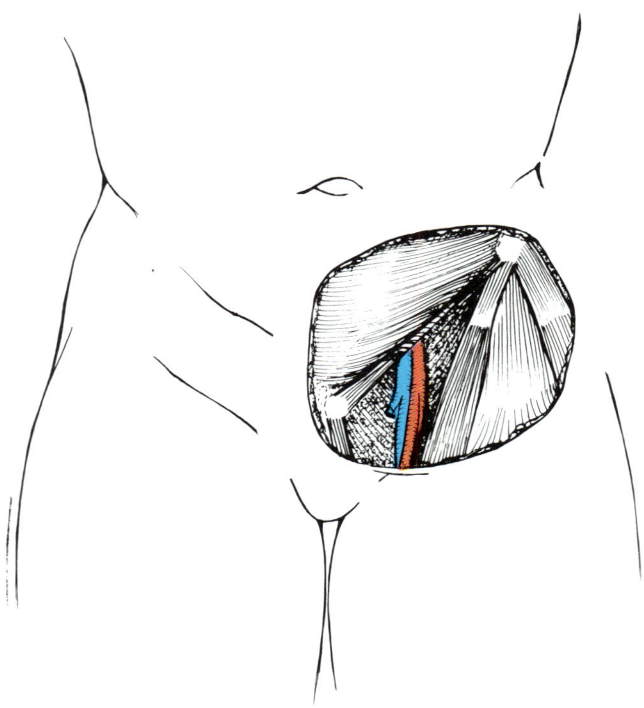

Fig. 22-1. Exposed femoral vessels.

Skin graft

Skin grafts are method of choice provided vital structures are not exposed (femoral vessels and nerve)

Local flaps

1. Sartorius muscle flap is flap of choice for routine coverage of vessels after radical groin lymphadenectomy (Fig. 22-3)
2. Tensor fascia lata flap is flap of choice for more extensive defects or after radiation
3. Rectus femoris and gracilis muscle or musculocutaneous flaps are alternative flaps (Fig. 22-4)
4. Vastus lateralis muscle flap and skin graft are alternative flap

Distant flaps

1. Omentum is rarely indicated
2. Free tissue transplantation is rarely indicated
3. Abdominal skin flaps are not indicated

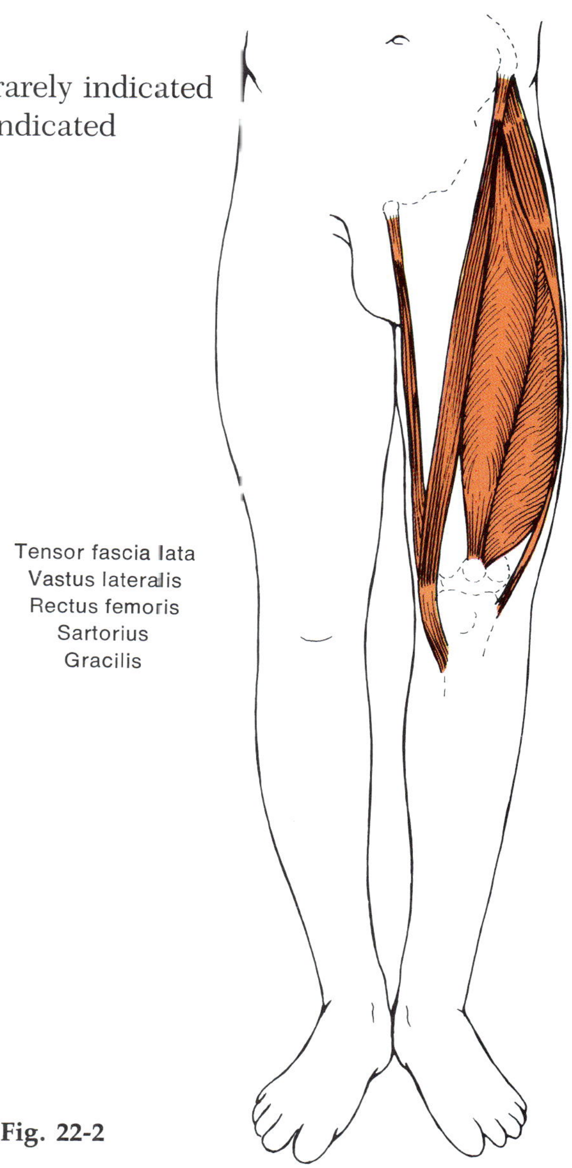

Fig. 22-2

A

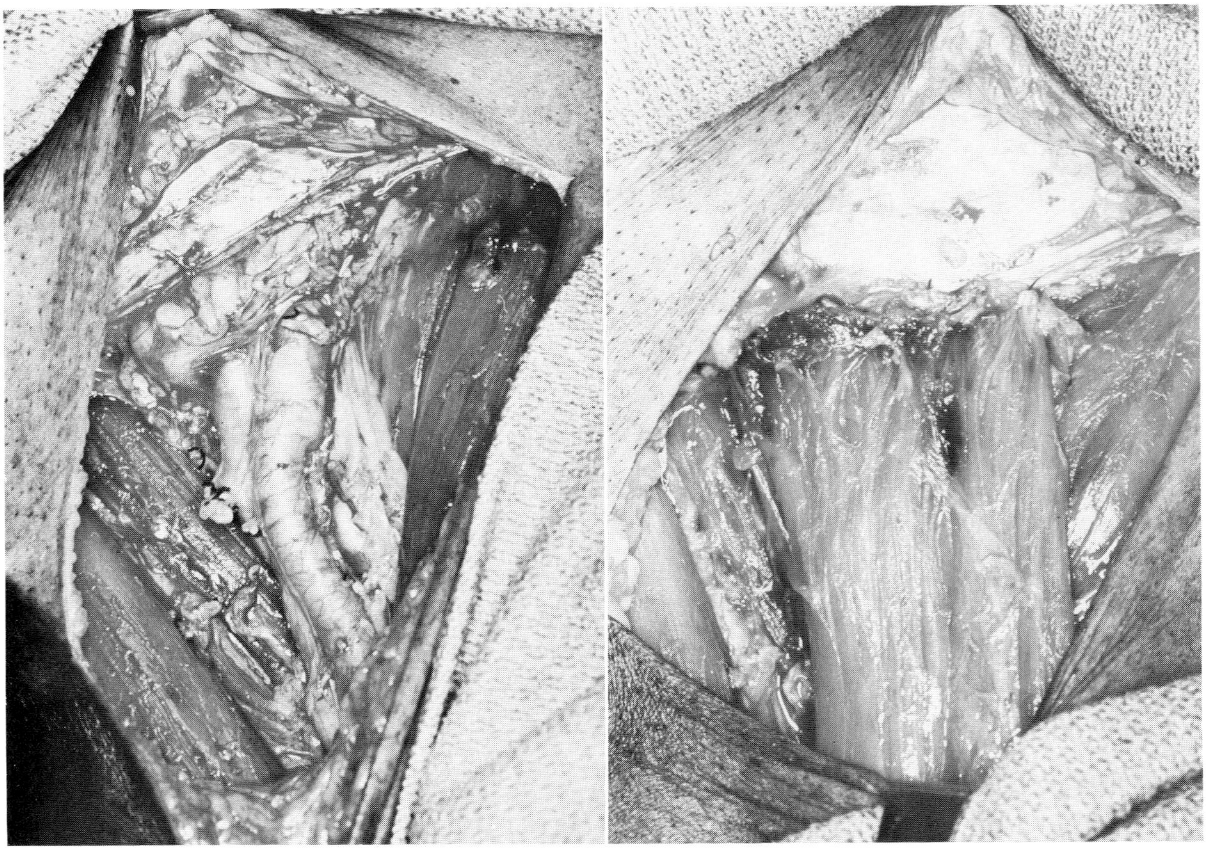

B

Fig. 22-3. Sartorius muscle for coverage of groin vessels.
A, Exposed femoral vessels in left groin after radical lymphadenectomy for level V melanoma of foot.
B, Routine transposition of proximal sartorius muscle provides coverage over exposed femoral vessels.

A

B

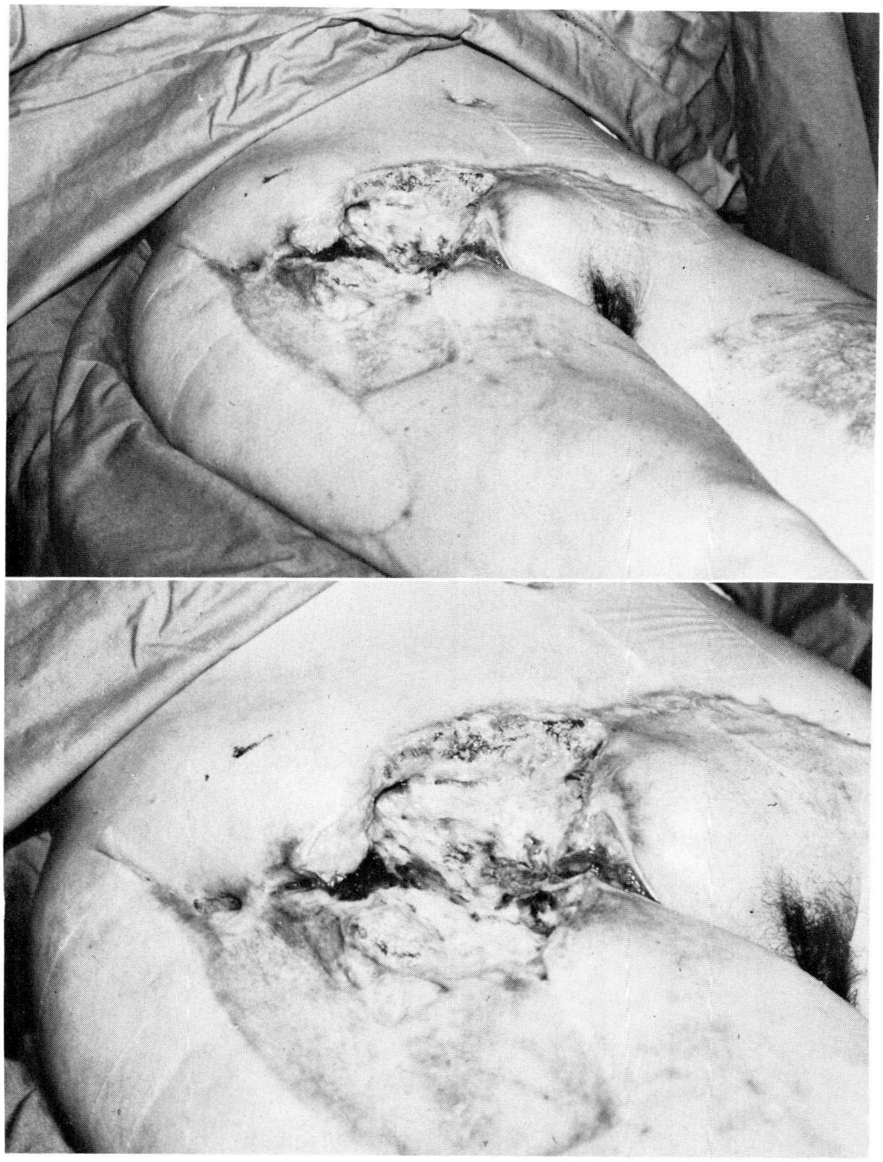

Fig. 22-4. Coverage of exposed irradiated femoral vessels in 42-year-old woman after resection and radiation of liposarcoma of lateral thigh.
A, Radiation ulcer on right groin with exposed femoral vessels.
B, Closeup view.

Continued.

C

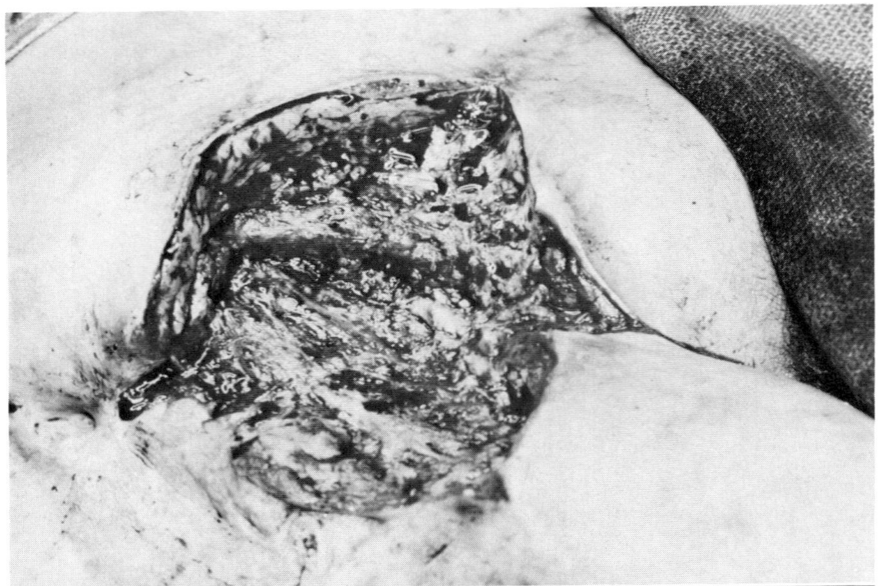

D

E

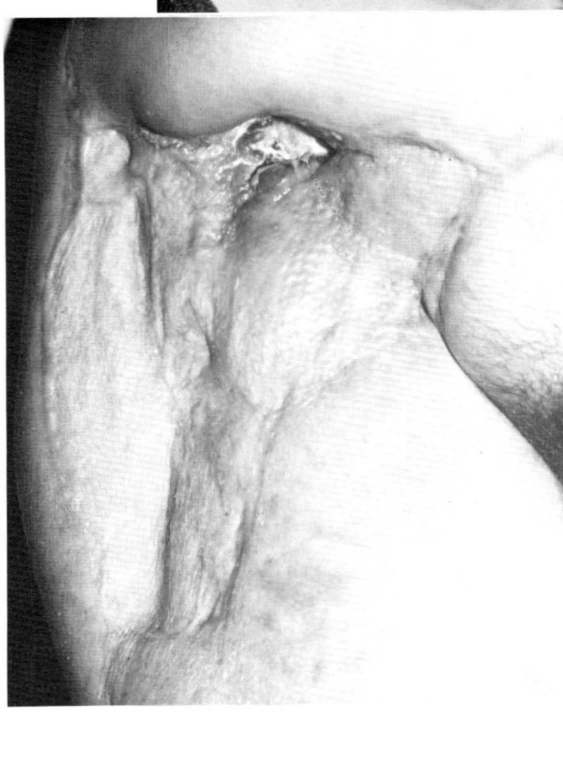

Fig. 22-4, cont'd

C, Wound debrided.

D, Rectus femoris muscle flap covers exposed vessels.

E, Postoperative view demonstrates stable coverage of vessels but recurrent tumor has invaded flap.

PERINEUM (Fig. 22-5)

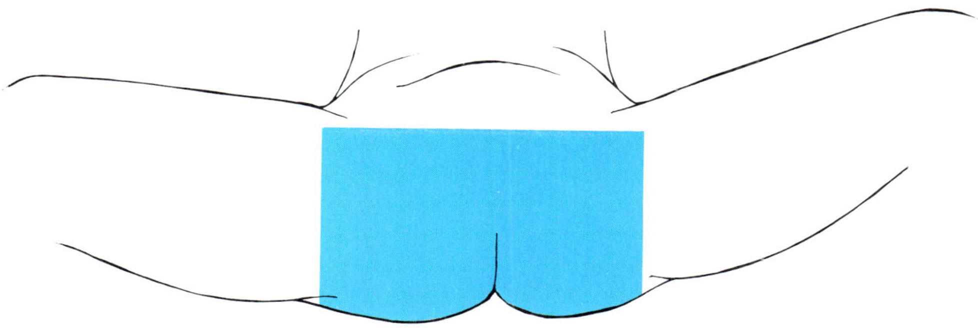

Fig. 22-5

Skin grafts

Skin grafts are method of choice for coverage of uncomplicated wounds and agenesis of the vagina

Local flaps

1. Random flaps are alternative method for penile reconstruction
2. Groin axial flap is alternative flap for penile reconstruction

Distant flaps (Figs. 22-6 and 22-7)

1. Gracilis muscle and musculocutaneous flaps
   a. Flap of choice for vaginal reconstruction after extirpative surgery or radiation (Fig. 22-8)
   b. Flap of choice for penile reconstruction
   c. Useful as functional muscle for rectal incontinence (Fig. 22-9)
   d. Flap of choice for chronic perineal wound after abdominal perineal resection (Fig. 22-10)
2. TFL and inferior gluteal thigh flap are alternative flaps
3. Rectus femoris musculocutaneous flap is alternative flap
4. Gluteal thigh flap is flap of choice for immediate reconstruction for extensive perineal defects after extirpative surgery (Fig. 22-11)

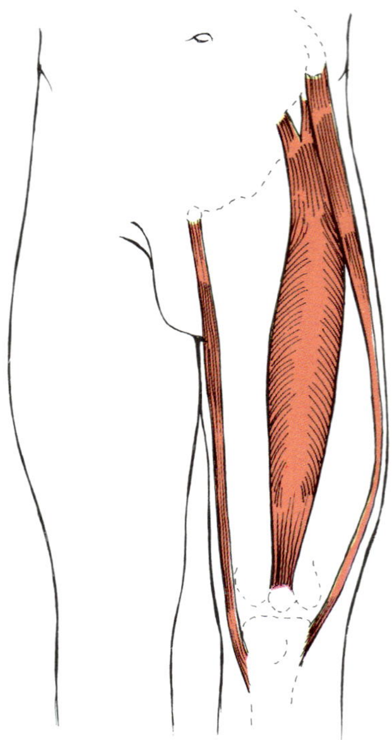

Fig. 22-6

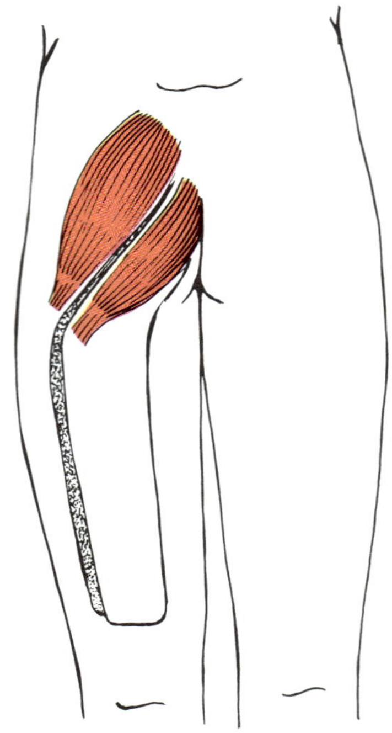

Fig. 22-7

Fig. 22-8. Reconstruction of vagina.

A, Absence of pelvic musculature and vagina after radical hysterectomy. Cutaneous territory marked for bilateral gracilis musculocutaneous flaps. Territory is inferior to line connecting pubis with medial condyle.

B, Bilateral flaps elevated.

C, Musculocutaneous flaps passed through subcutaneous tunnel into perineal region.

D, Bilateral cutaneous territories sutured to form reconstructed vagina. Muscle is then sutured to encircle neovagina.

E, Reconstructed vagina is placed into perineal defect. Skin edges are sutured to labial skin to complete single-stage vaginal reconstruction.

(From Mathes, S., and Nahai, F.: Clinical atlas of muscle and musculocutaneous flaps, St. Louis, 1979, The C.V. Mosby Co.)

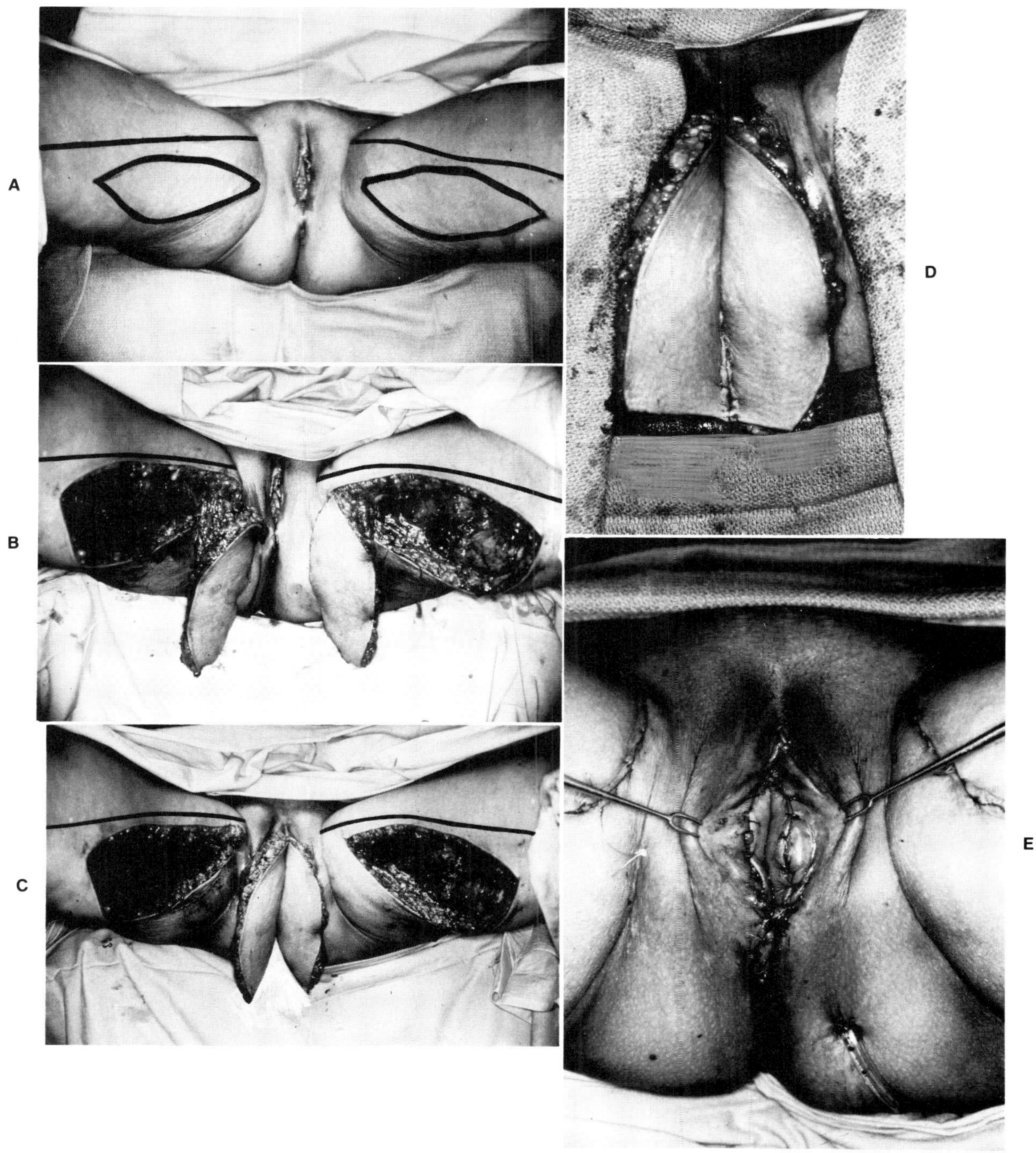

Fig. 22-8. For legend see opposite page.

A

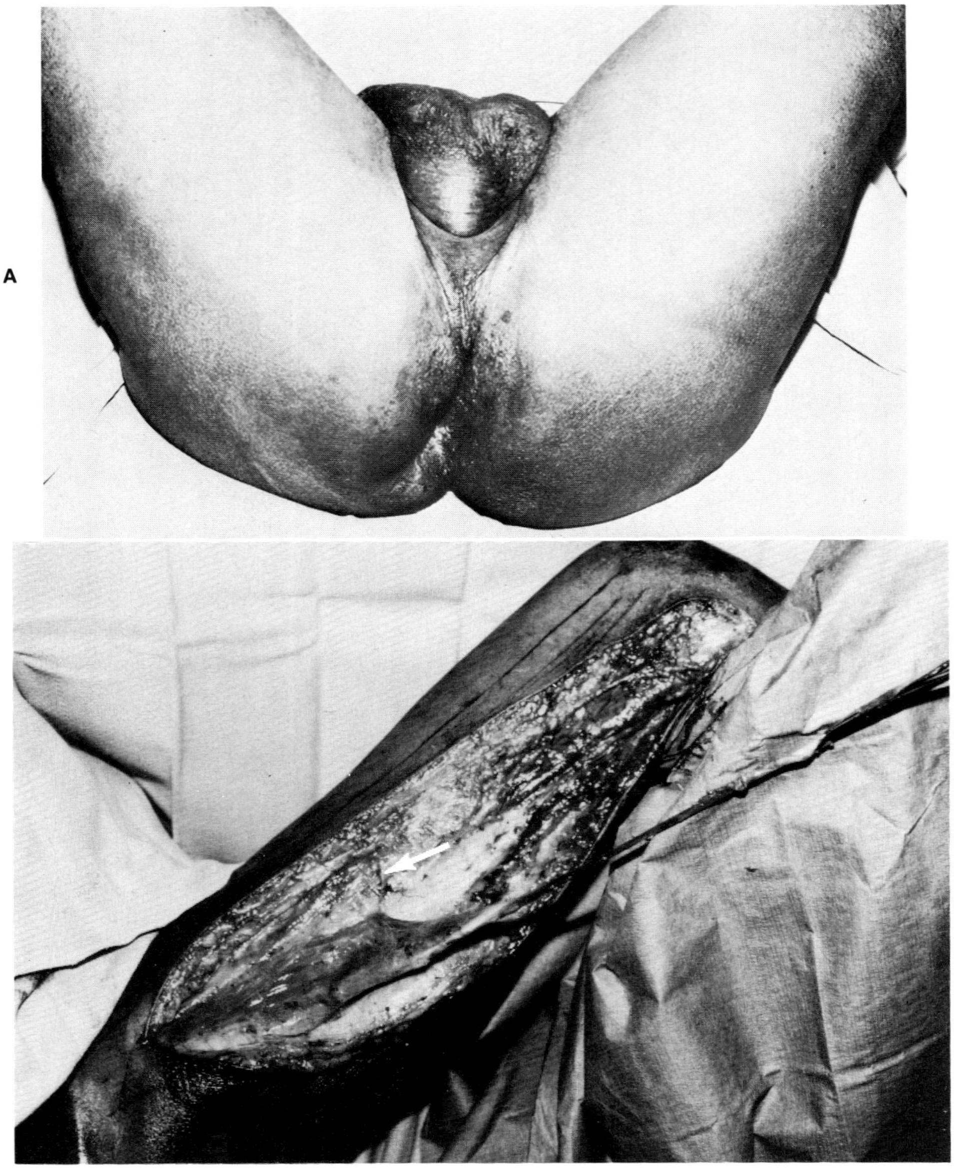

B

Fig. 22-9. Gracilis muscle functional transposition.

A, Absence of rectal sphincter after multiple fistulectomies and attempts at reconstruction of anal musculature.

B, Elevation of gracilis musculocutaneous flap. Arrow denotes minor pedicle before its division.

C

D

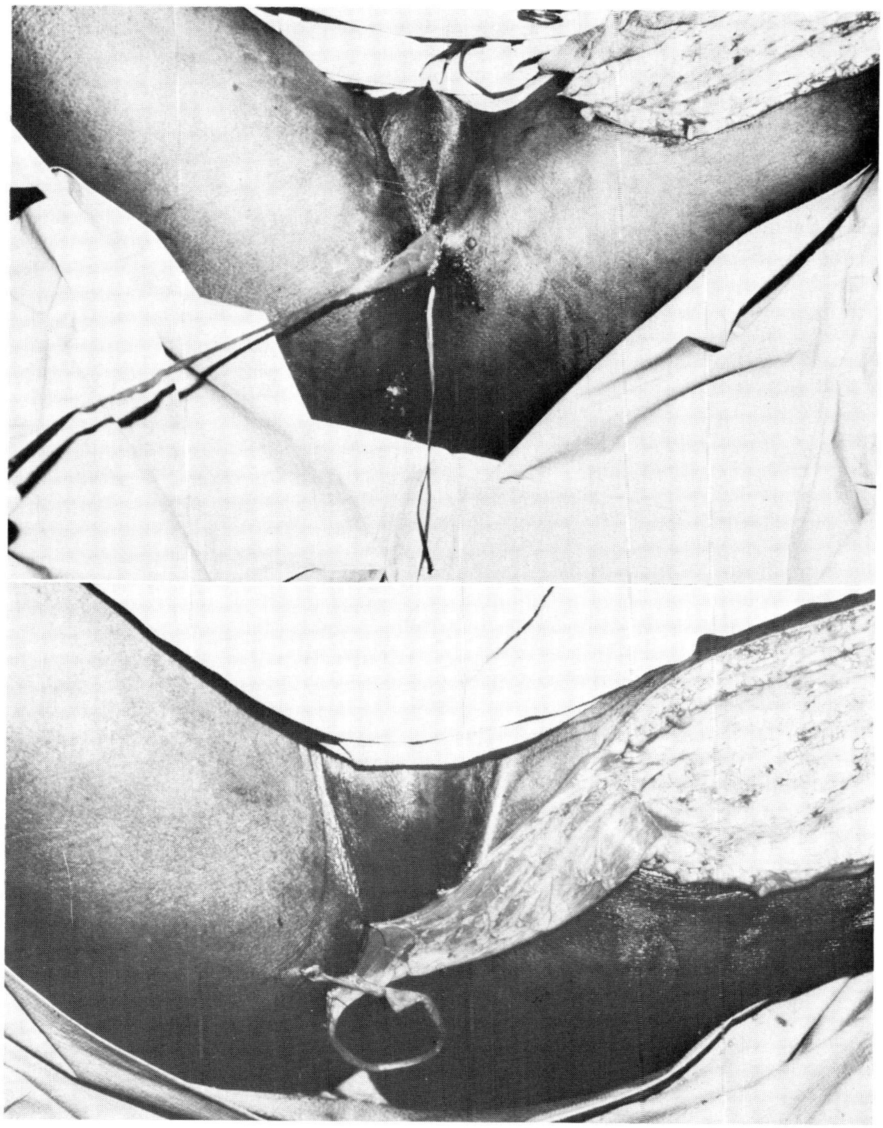

Fig. 22-9, cont'd

C, Muscle passed through subcutaneous tunnel into perianal region.

D, External position of muscle demonstrates its planned configuration for use as functional transfer for reconstitution of anal spincter.

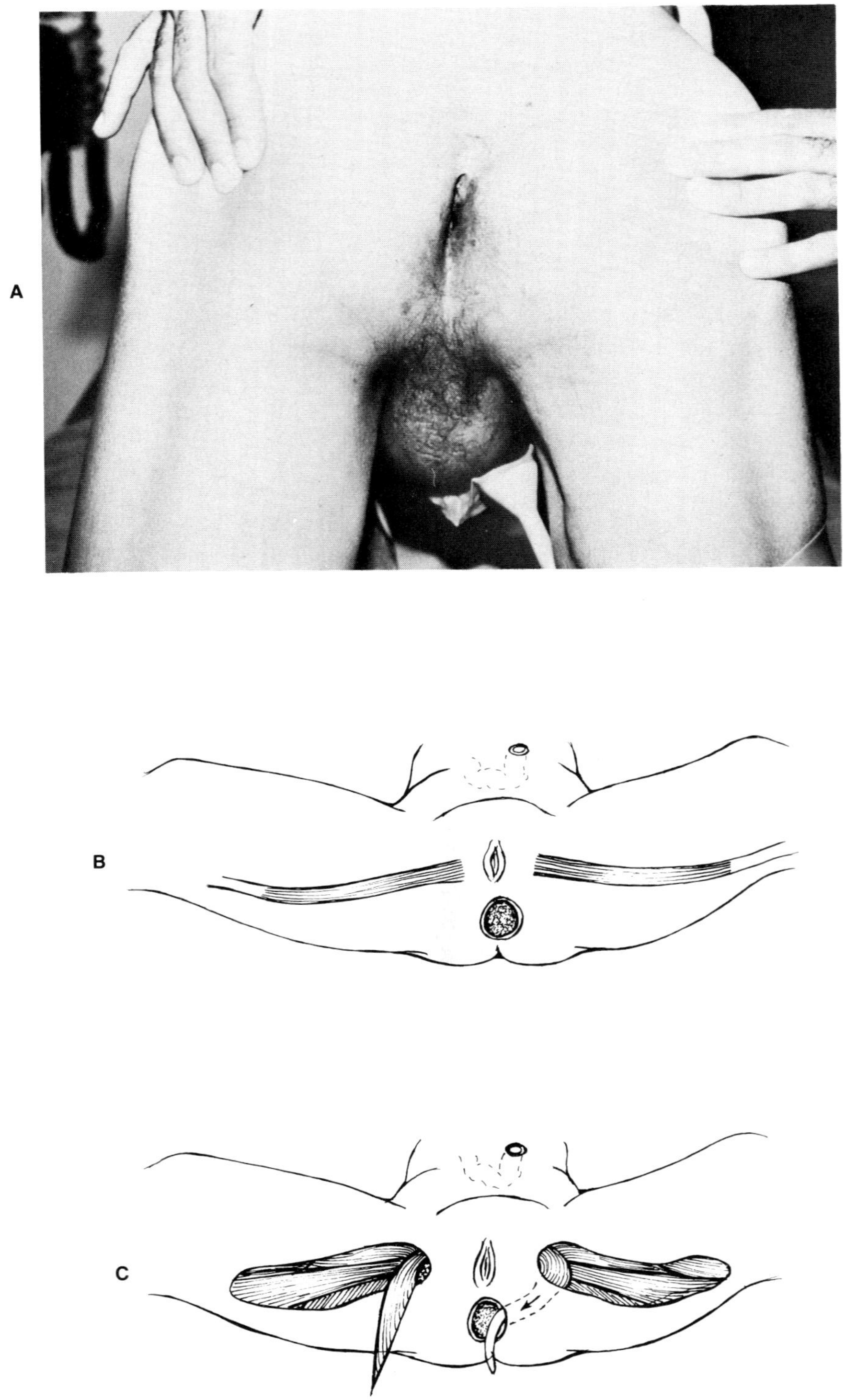

Fig. 22-10. Abdominal perineal reconstruction.
A, Chronic wound infection resulting from wound healing failure at site of previous abdominoperineal resection (2 years).
B, Wound débridement includes removal of chronic granulation tissue and nonviable bone in retroperitoneum.
C, One or both gracilis muscles are transposed into pelvic defect.

D

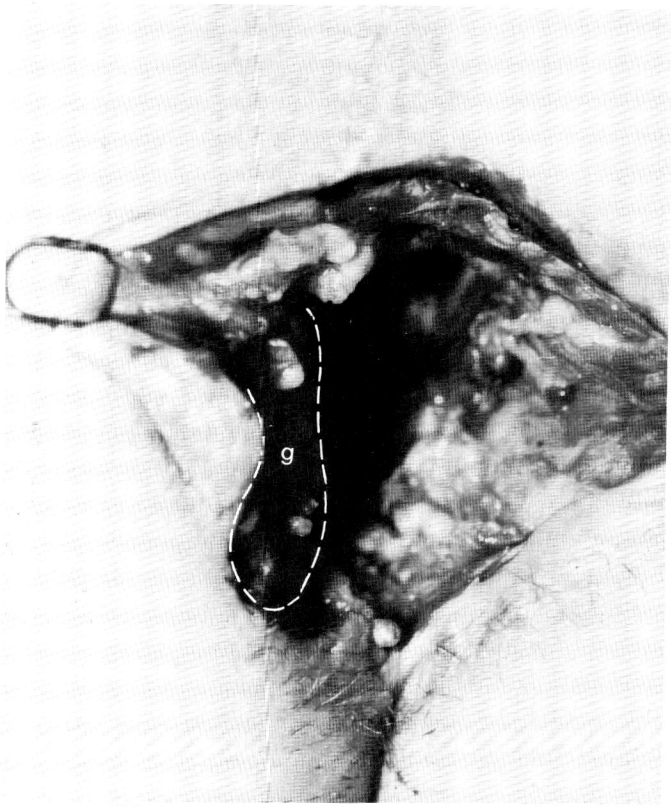

E

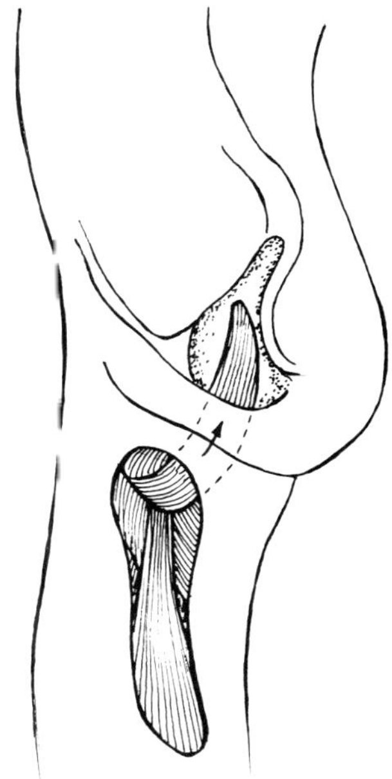

F

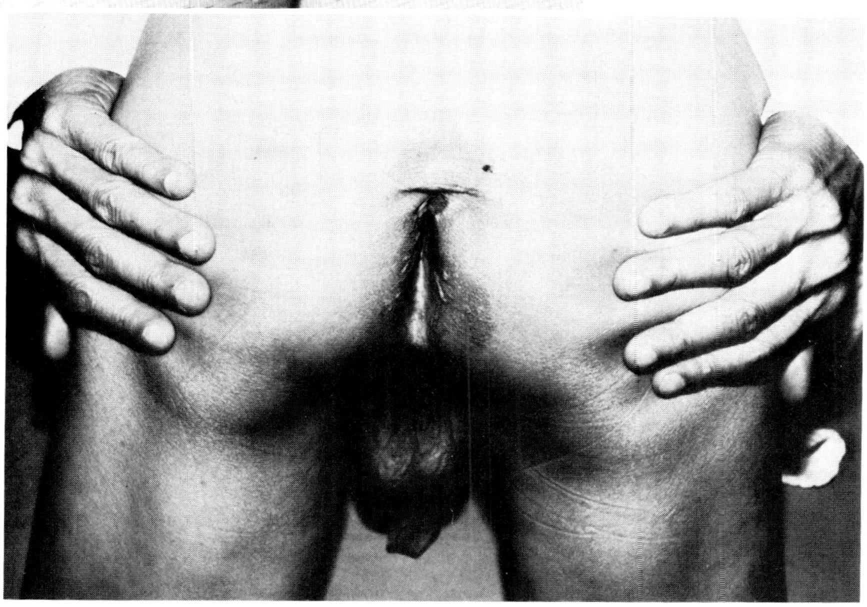

Fig. 22-10, cont'd

D, Débridement of sinus tract into retroperitoneum superficial to sacrum. NOTE: Position of gracilis muscle before inset into retroperitoneal cavity. *g*, Gracilis before insertion into pelvic defect.

E, Muscle is placed into retroperitoneum over exposed anterior sacral bone. Donor site is closed directly.

F, One-year postoperative view of gracilis muscle transposition demonstrates complete healing of chronic abdominal perineal wound.

A

B

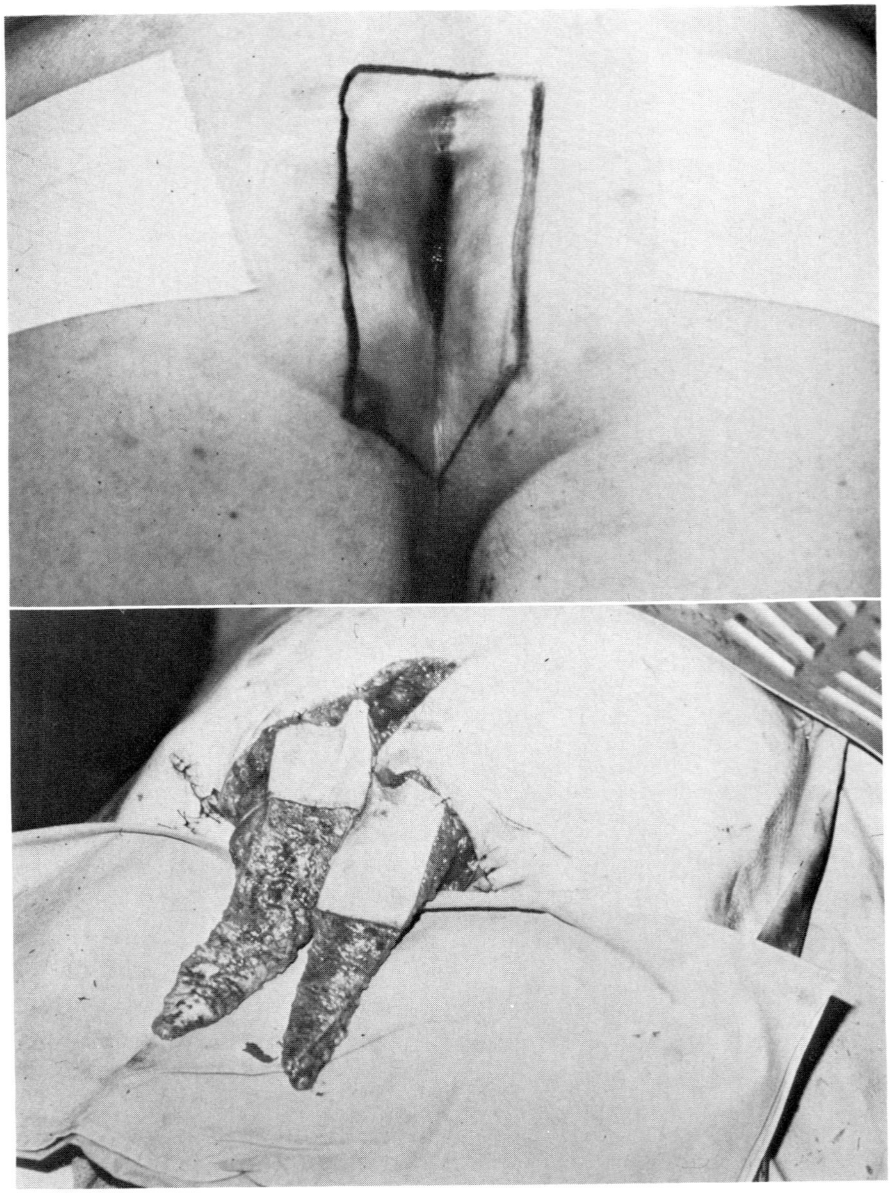

Fig. 22-11. Inferior gluteal thigh flap.

A, Preoperative view of squamous cell carcinoma of rectum before abdominal perineal resection.

B, Elevation of bilateral inferior gluteal thigh flaps with deepithelialization of inferior flaps for insertion into retroperitoneal defect. Skin territories used for cutaneous coverage.

(From Hurwitz, D. J., Schwartz, W. M., and Mathes, S. J.: The gluteal thigh flap: a reliable, sensate flap for closure of buttock and perineal wounds, Plast. Reconstr. Surg. **68**[4]:3, 1981.)

C

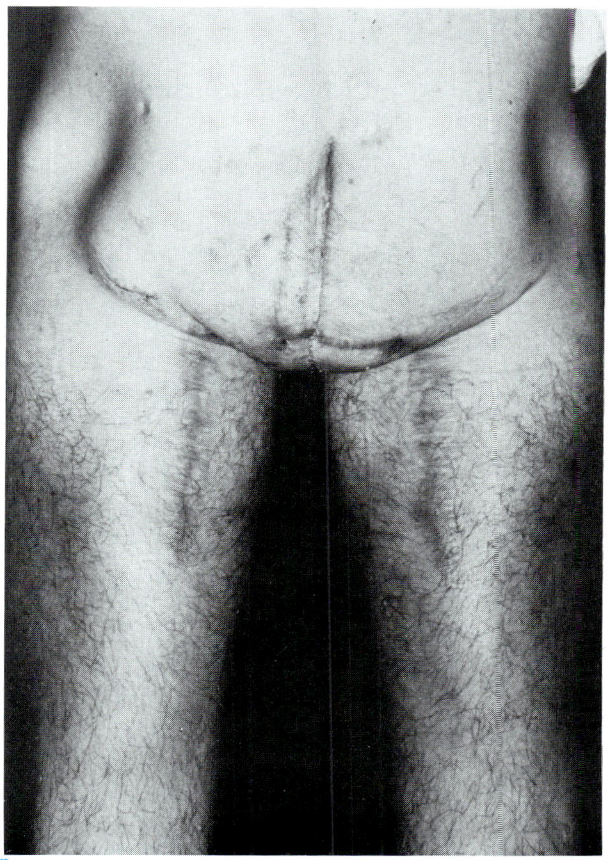

Fig. 22-11, cont'd

C, Six-month postoperative view demonstrates stable wound coverage after transposition of bilateral inferior gluteal thigh flaps for immediate reconstruction after abdominoperineal resection.

23

RECONSTRUCTION

H. Louis Hill, Jr. ■ T. Roderick Hester ■ Foad Nahai

The successful management of soft tissue defects in the groin and perineum has long been a challenging problem for the reconstructive surgeon. Multistage procedures were necessary for reconstruction in this region. The application of muscle and musculocutaneous flaps offers a simple, safe, one-stage method for reconstruction of such defects. Soft tissue defects of the groin and perineum are most frequently the result of trauma or ablative surgery for tumors of the genitalia or perianal regions. Tumor resection in this area is often complicated by radiation therapy. Moreover, this anatomic region is susceptible to the problems of skin maceration (especially in the obese), urinary and fecal contamination, fistulization, and infection, leading to delayed or incomplete wound closure. Vital anatomic structures such as the femoral vessels are exposed as a result of extirpative surgery in this region. Moreover, the total reconstruction of the penis, vaginia, and vulva makes reconstruction in this area even more challenging.

Traditional methods for tissue coverage and reconstruction in this area require multiple procedures often with incomplete or inadequate results leading to prolonged periods of recovery after trauma, tumor resection, or radiation therapy. The muscle and musculocutaneous flaps offer the following advantages:

1. A blood supply that is based out of the field of resection or radiation therapy
2. A blood supply precisely known as is the exact location of vascular pedicles
3. A one-stage procedure for reconstruction and coverage

Several flaps have been used in this area: TFL, gracilis, sartorious, rectus femoris, vastus lateralis, and biceps femoris. All have been reliable, however, four of these flaps have been particularly useful and far more commonly used: TFL, gracilis, sartorius, and rectus femoris. These four flaps and their applications will be discussed in detail.

TFL

The TFL takes its origin from the anterior 5 cm of the iliac crest. The small muscle belly blends with and inserts into the iliotibial tract of fascia lata in the upper thigh. It is an expendable muscle that is an accessory medial rotator of the thigh.

The single dominant vascular pedicle (type I), the terminal branch of the lateral circumflex femoral artery, enters the muscle on its deep surface at a point approximately 8 to 10 cm below the anterior superior iliac spine. The muscle is innervated by a branch of the superior gluteal nerve. The skin territory of this unit is innervated by the lateral femoral cutaneous nerve of the thigh (L2-3) and a lateral branch of T12.

The flap may be elevated as a standard or extended flap. The standard flap includes the muscle and overlying skin and may measure up to 10 to 15 cm. The extended flap includes muscle and fascia lata down to within 4 to 8 cm of the knee, with the overlying skin of the anterolateral thigh. The extended flap may measure 15 × 35 cm. With the vascular pedicle located close to a limb girdle, this unit has an anterior and a posterior arc of rotation (Fig. 23-1). The anterior arc will cover the groin, perineum, and abdominal wall, whereas the posterior arc will reach the trochanter and ischium.

It is not necessary to delay the extended flap, and either flap can safely be elevated as an island. The donor defect is closed directly for small flaps, however, for wider flaps the donor defect may require skin grafts.

Flap elevation

The anterior border of the flap is marked by drawing a line from the anterior superior iliac spine to the lateral condyle of the tibia. The greater trochanter marks the posterior boundary. The length is then marked according to the reconstructive needs. For vulval coverage and abdominal wall reconstruction a long flap is necessary; for groin coverage a shorter flap is necessary. The width may be safely extended

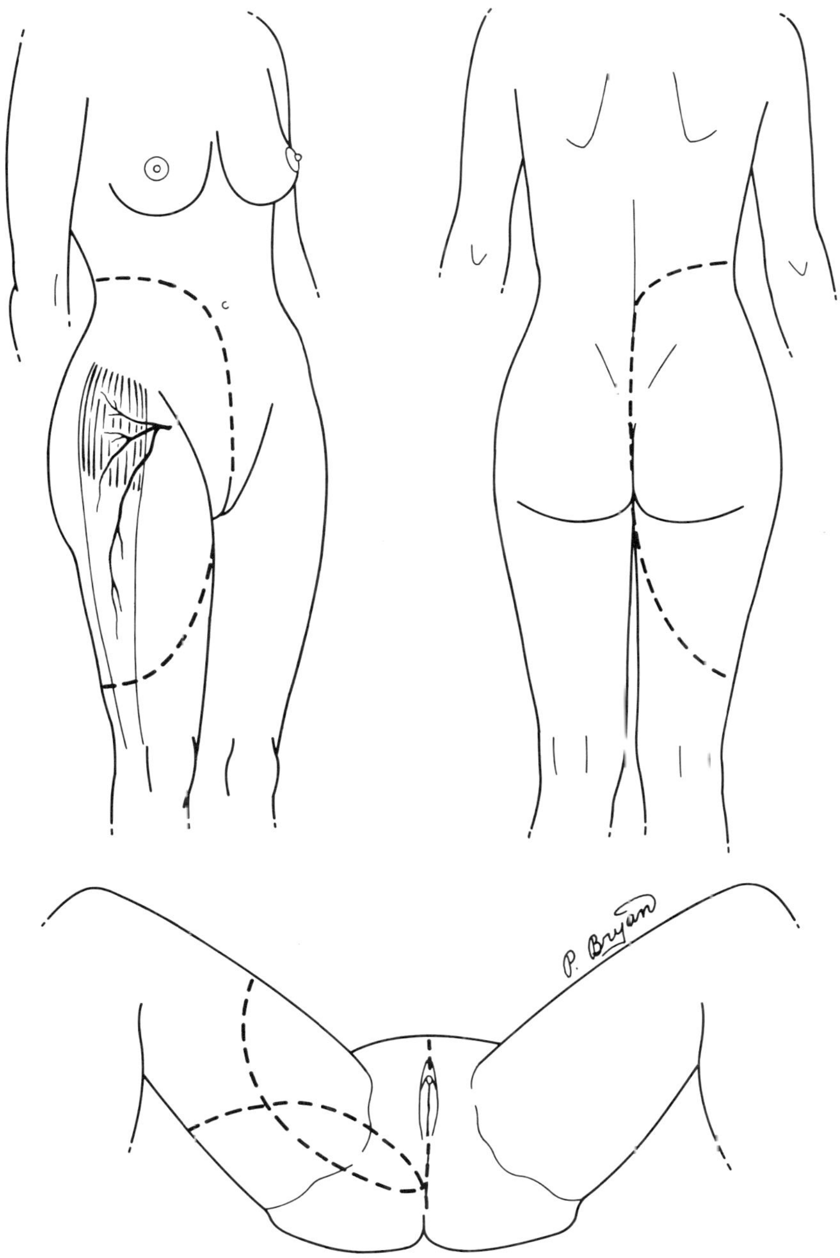

Fig. 23-1. Blood supply and arc of rotation of tensor fascia lata musculocutaneous flap.

beyond these markings for several centimeters if desired. Flap dissection is then started distally and continued from the distal border to the proximal border. The skin and fascia lata are incised, skin margins are temporarily sutured to the fascia lata, and the flap is then elevated off the vastus lateralis in a relatively avascular plane. At a level approximately 8 to 10 cm from the anterior superior iliac spine, the vascular pedicle is identified as it emerges deep to the rectus femoris muscle. At this stage the flap is ready for transposition. If the flap is to be an island, the necessary further incisions can then be made.

Clinical application

Groin coverage. The standard TFL flap has proved most useful for coverage of groin defects, particularly those after radical lymphadenectomy.

Vulval reconstruction. The extended TFL flap will not only cover the exposed vessels after radical groin lymphadenectomy but has been reliable for vulval reconstruction in patients undergoing radical vulvectomy and groin dissection for carcinoma. Postoperative recovery time, expecially in the irradiated patient, has been considerably shortened and postoperative morbidity dramatically reduced (Fig. 23-2).

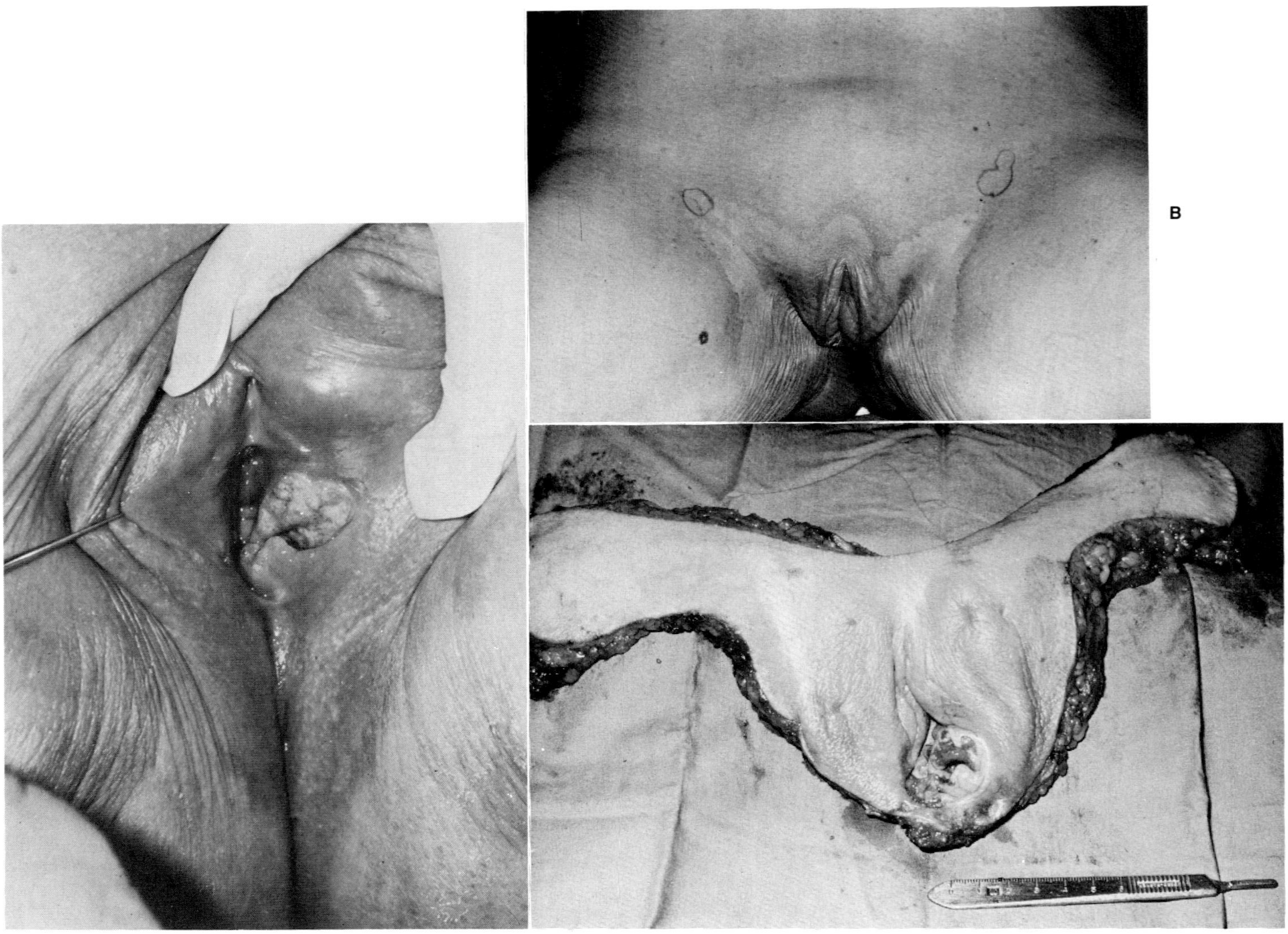

Fig. 23-2. Vulval reconstruction and groin coverage with TFL musculocutaneous flap after radiation and radical groin dissections.
A, Extensive vulval carcinoma.
B, Bilateral groin metastasis required bilateral lymphadenectomy in continuity with radical vulvectomy.
C, Surgical specimen.

(From Nahai, F., Hill, H.L., and Hester, T.R.: Experiences with the tensor fascia lata flap, Plast. Reconstr. Surg. **63:**788, 1979.)

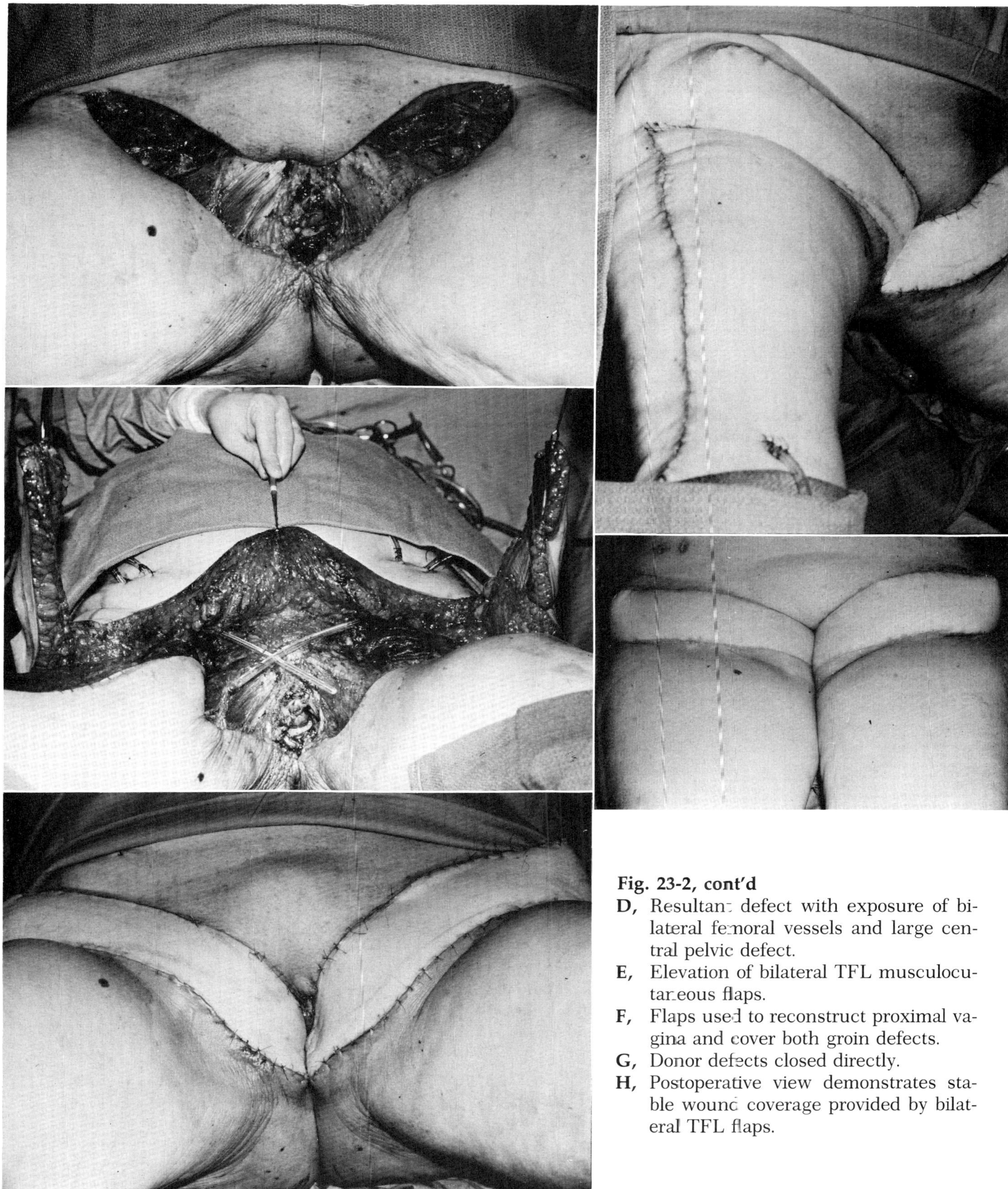

D

E

F

G

H

Fig. 23-2, cont'd

D, Resultant defect with exposure of bilateral femoral vessels and large central pelvic defect.

E, Elevation of bilateral TFL musculocutaneous flaps.

F, Flaps used to reconstruct proximal vagina and cover both groin defects.

G, Donor defects closed directly.

H, Postoperative view demonstrates stable wound coverage provided by bilateral TFL flaps.

Reconstruction of recurrent inguinal hernia. Wangensteen in 1934 reported his experience with the use of "the iliotibial tract of the fascia lata as a pedicle flap" for repair of difficult and recurrent hernias. This method has also proved most useful for reconstruction of recurrent inguinal hernias (Fig. 23-3). For use in reconstruction of hernias the flap is elevated as a musculofascial flap. Flap dissection proceeds as for the musculocutaneous flap, but the overlying skin is not elevated with the fascia lata. The muscle and fascia lata are isolated and transposed into the hernial defect where the fascia is sutured into position under tension.

A

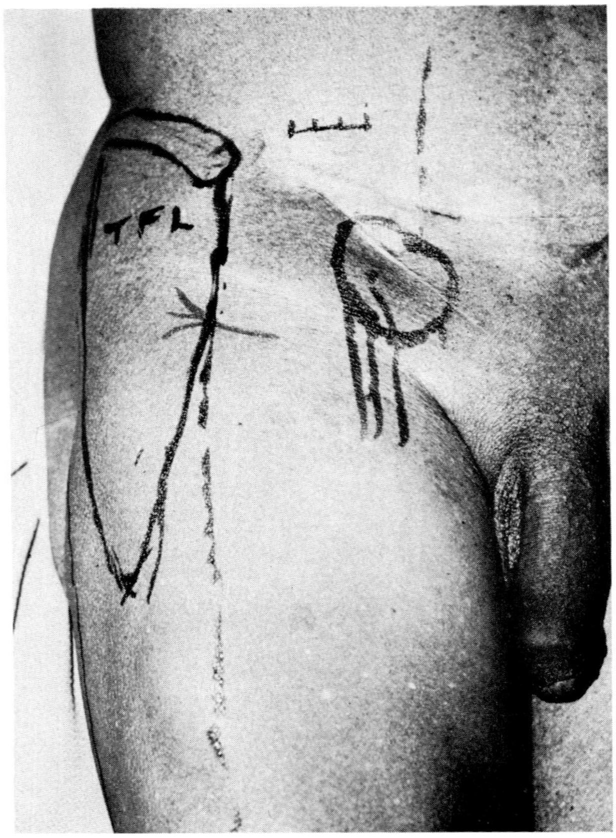

B

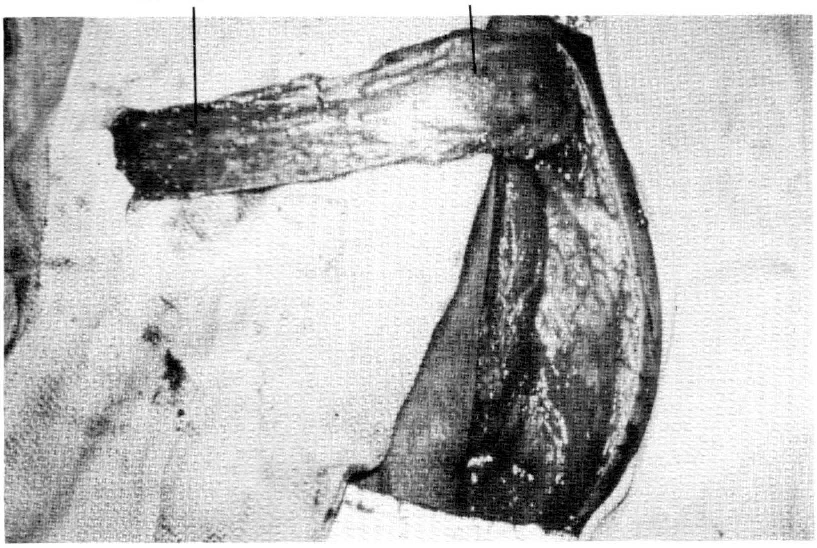

Fig. 23-3. Reconstruction of recurrent groin hernia with musculofascial TFL flap.
A, Fifty-six-year-old man with four times recurrent right groin hernia. Hernia sac and TFL flap outlined.
B, TFL muscle and fascia lata elevated.

C

D

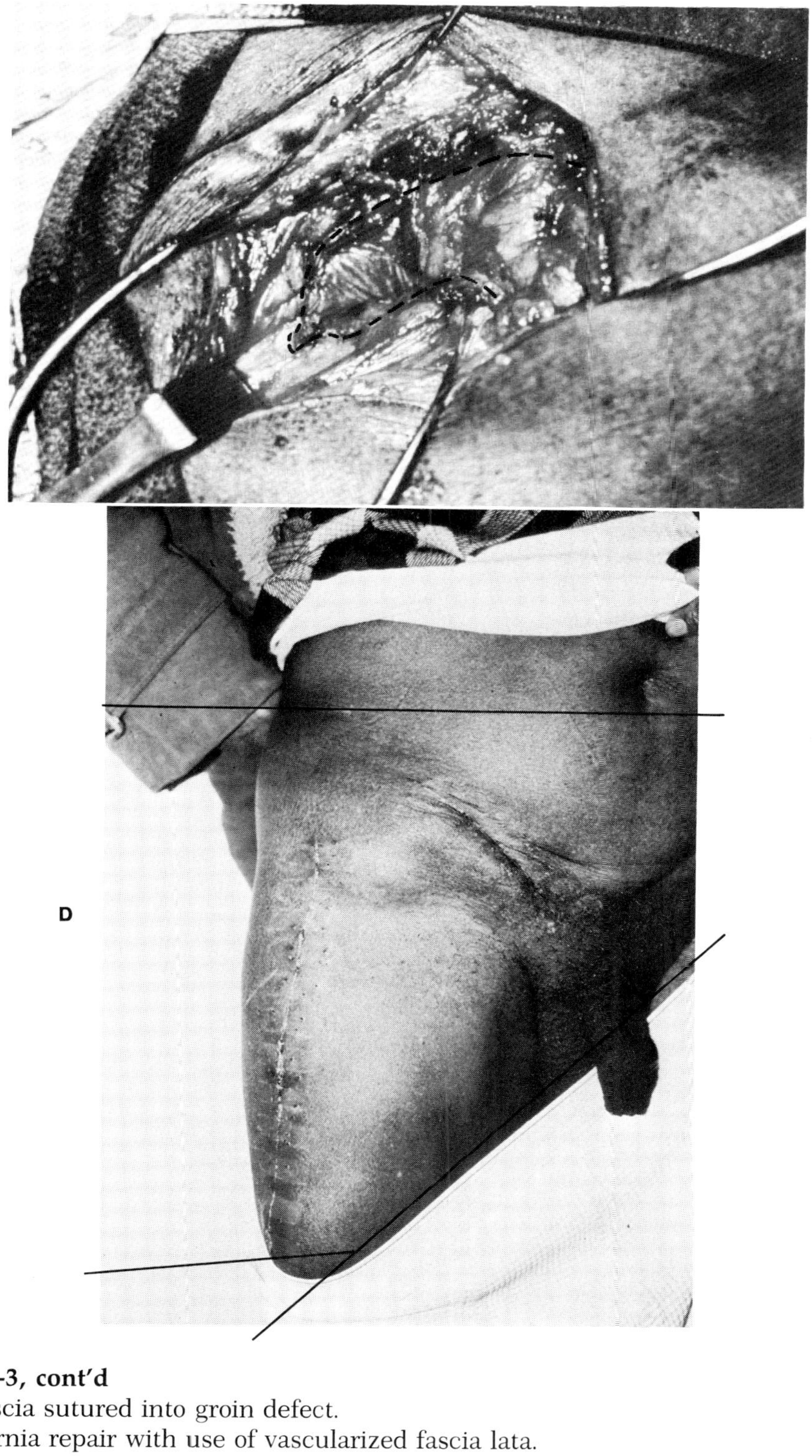

Fig. 23-3, cont'd
C, Fascia sutured into groin defect.
D, Hernia repair with use of vascularized fascia lata.

GRACILIS

The gracilis is a long, thin muscle on the medial aspect of the thigh. It takes origin from the pubic symphysis and inserts into the medial tibial condyle. This muscle has one proximal dominant and one or two distal minor pedicles (type II). The dominant pedicle is the medial circumflex femoral artery that enters the muscle and its deep surface at a point approximately 8 to 10 cm below the pubic symphysis. Based on this pedicle the muscle has an anterior and a posterior arc of rotation (Fig. 23-4). Anteriorly it will reach the groin and perineum, and posteriorly it will reach the ischial and perirectal areas. The flap may be raised as a musculutaneous unit with a skin island measuring up to 10 × 20 cm. The muscle is innervated by the obturator nerve and acts as an accessory adductor of the thigh. This muscle is dispensible.

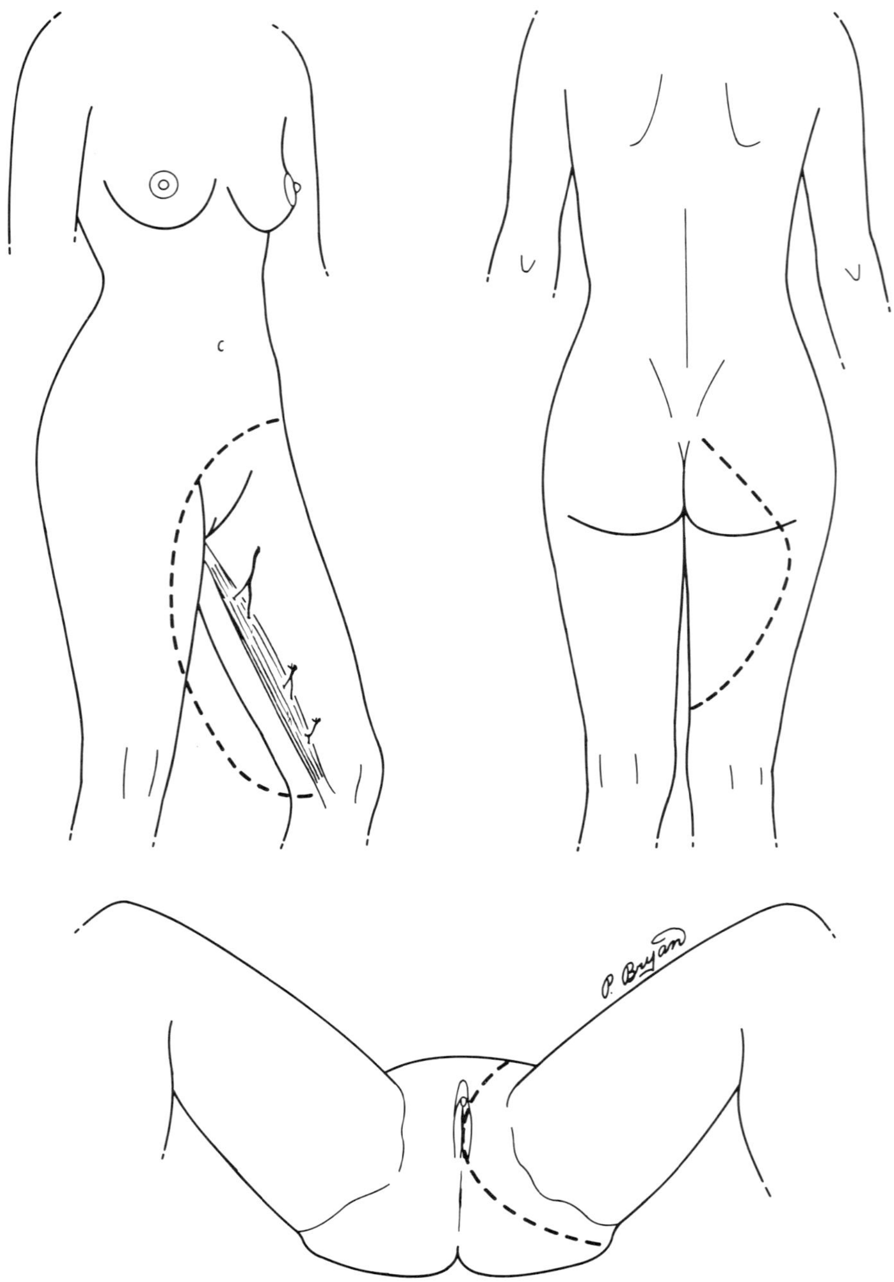

Fig. 23-4. Blood supply and arc of rotation of gracilis muscle and musculocutaneous flap.

Flap elevation

The key to safe elevation of this musculocutaneous flap is the accurate outlining of the skin island so that it is directly over the muscle. This is a thin muscle, measuring only 5 to 6 cm in width, and a skin island twice this width is usually elevated based on the muscle (Fig. 23-5). The accurate outlining of the island is more difficult with the patient in the lithotomy position. It is also more difficult in the obese patient. The correct orientation of the skin island over the muscle is facilitated by identification of the gracilis tendon just above the knee. A line is drawn from the pubic symphysis to the medial tibial condyle (Fig. 23-6, *A*). The gracilis muscle and its skin territory lie posterior to this line. A small incision is then made 5 to 6 cm above the knee and 2 cm posterior to this line. Through this incision the gracilis tendon is identified. The tendon of the gracilis is found between the muscular fibers of the sartorius anteriorly and the fascial expanse of the semimembranosus posteriorly (Fig. 23-6, *B*). The gracilis tendon is then firmly retracted. With traction on the tendon, the muscle is palpated and the skin island outlined directly over the muscle (Fig. 23-6, *C*). Because the skin of the medial thigh is so mobile, especially in the obese, the skin territory of this muscle cannot be outlined by external landmarks. Alternate methods for identifying skin territory include exposure and identification of the muscle origin.

Once the skin island is outlined, flap elevation proceeds from the distal border to the proximal border. The skin island is usually twice as wide as the muscle. It is advisable to suture the skin temporarily to the underlying muscle and bevel the incision in the subcutaneous tissue away from the muscle to preserve the musculocutaneous perforators. As the gracilis muscle is elevated off the adductor magnus muscle, the distal minor pedicles, one or two direct branches of the superficial femoral artery, are identified. These are clamped and divided. The dominant pedicle is identified at a point 8 to 10 cm below the pubic symphysis. The pedicle emerges deep to the medial border of the adductor longus muscle to enter the gracilis muscle on its deep surface medially. The muscle may then be divided at its origin and the flap transposed. For greater arc of rotation the major pedicle may be dissected or mobilized by retracting the adductor longus. The donor defect is always closed directly.

Because the gracilis is a type II muscle, with one dominant and one or two distal pedicles, the skin territory over the lower third of the muscle is not reliable. For this reason the skin island is usually based on the upper two thirds of the muscle.

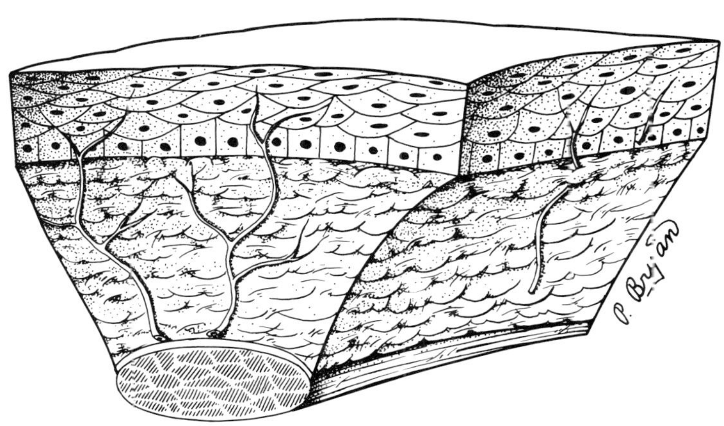

Fig. 23-5. Gracilis muscle (5 to 6 cm in width) is elevated with a much wider island of skin.

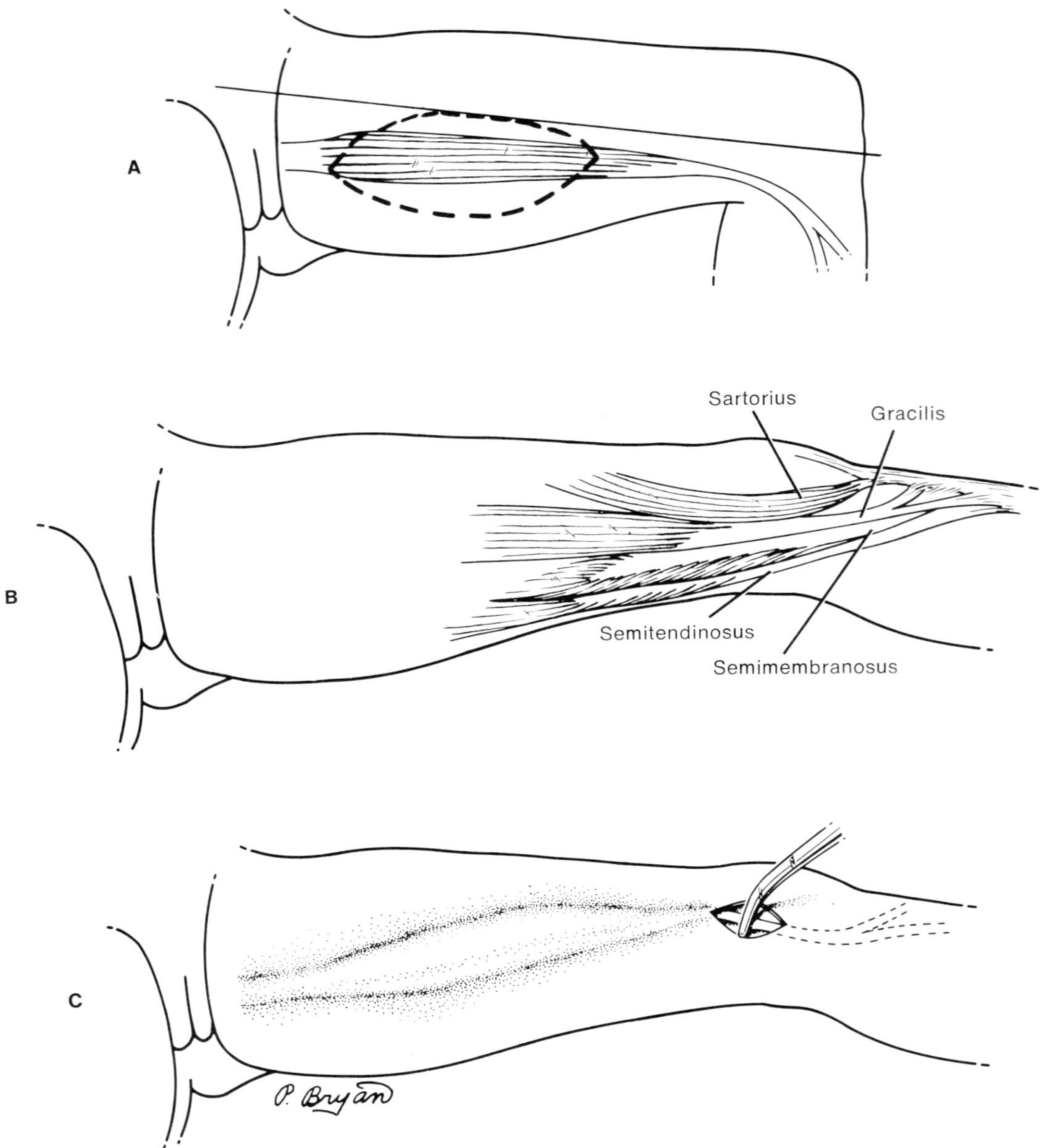

Fig. 23-6. Identification of gracilis muscle and skin island.
A, Skin markings.
B, Relationship of gracilis to sartorius and semimembranosus.
C, Gracilis tendon identified. Traction on tendon demonstrates skin overlying muscle.

Clinical applications

The gracilis has proved most useful for reconstruction of the vagina, penis, and anal sphincter and coverage of the groin, perineum and perirectal region.

Vaginal reconstruction. McCraw and co-workers demonstrate the use of bilateral gracilis flaps for vaginal reconstruction with good results. The operative details of this procedure are outlined in Fig. 23-7. For successful vaginal reconstruction with these flaps, it is essential to create a wide tunnel below the intact skin between the muscle and the vaginal defect. Excessive tension on the muscle and hence its pedicle is to be avoided. Dissection and mobilization of the pedicle are advised to avoid tension. The newly created vaginal pouch is sutured into the pelvis preferably through an abdominal incision to avoid postoperative prolapse.

The gracilis flap offers a one-stage method of total vaginal reconstruction. The detailed anatomy of the vascular supply is known, and the pedicle is out of

A

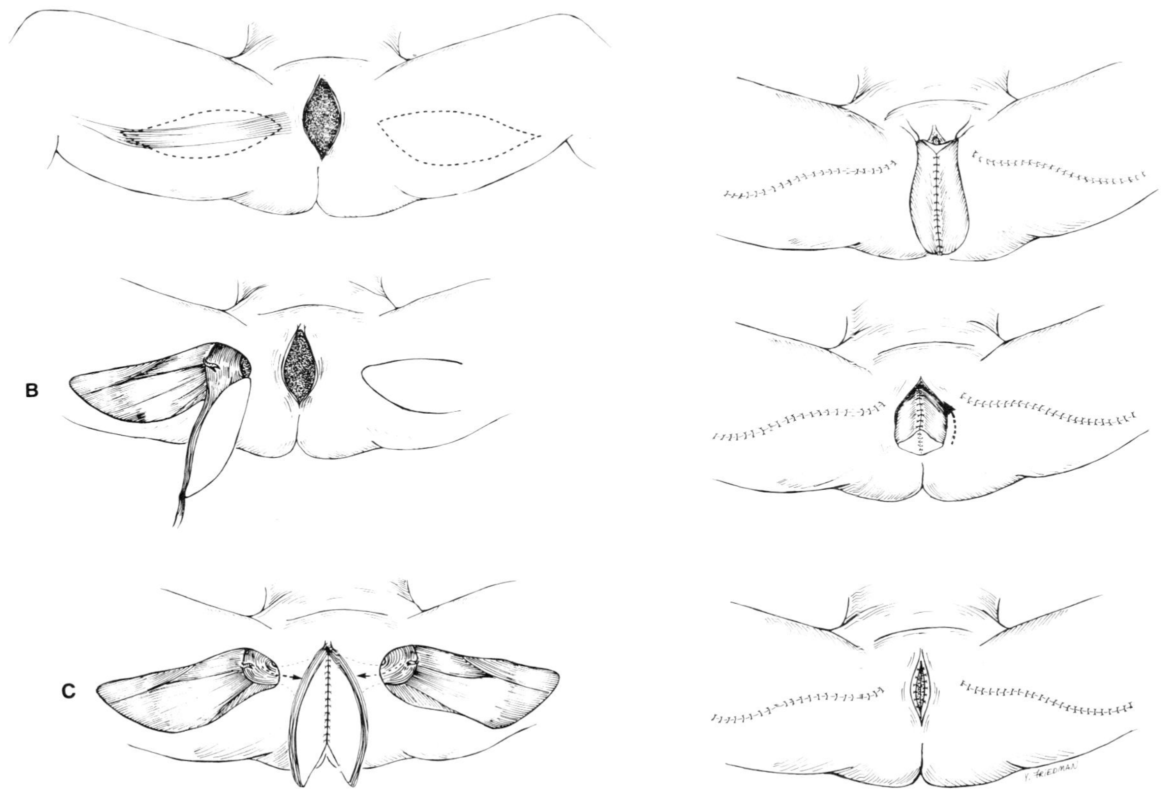

D

B

E

C

F

Fig. 23-7. Reconstruction of vagina. Absence of pelvic musculature and vagina after radical hysterectomy.

A, Cutaneous territory marked for bilateral gracilis musculocutaneous flaps. Territory is inferior to line connecting pubis with medial condyle.

B, Bilateral gracilis musculocutaneous island flap elevated with proximal pedicle intact and muscle origin intact.

C, Musculocutaneous flaps passed through subcutaneous tunnel into perineal region.

D, Bilateral cutaneous territories sutured to form reconstructed vagina. Muscle is then sutured to encircle neovagina.

E and **F,** Reconstructed vagina is placed into perineal defect. Skin edges are sutured to labial skin to complete single-stage vaginal reconstruction.

(From Mathes, S.J., and Nahai, F.: Clinical atlas of muscle and musculocutaneous flaps, St. Louis, 1979, The C.V. Mosby Co.)

the field of tumor resection and radiation. Although the muscle remains innervated, there is significant atrophy as a result of the division of the muscle. However, the bulk of this flap is not made up of muscle but of subcutaneous fat that will not atrophy. The flap has some limited pressure sensibility through the muscle but of course no skin sensibility. It is fair to conclude that although vaginal reconstruction with these flaps leaves a lot to be desired in terms of normal sexual function, the application of these flaps for reconstruction of pelvic defects after radical surgery with or without radiation has reduced postoperative morbidity and recovery time.

Unilateral gracilis flaps with or without the overlying skin have had some application for the reconstruction of vulvectomy defects. If the vulvectomy includes a radical groin dissection, our preference is to use the TFL flap for groin coverage and vulval reconstruction (Fig. 23-2).

Penile reconstruction. Orticochea describes the use of a unilateral gracilis musculocutaneous flap as a five-stage procedure for reconstruction of the penis. McCraw subsequently uses the flap as a one-stage procedure, and Hester demonstrates the use of a unilateral muscle-only flap covered with skin for reconstruction of a ventral penile defect in a young man with sickle cell disease. The gracilis musculocutaneous flap, which includes a thick subcutaneous layer, produces a rather bulky facsimile of a penis (Fig. 23-8). An excellent result using bilateral muscle-only flaps covered with split skin grafts has been reported by Hester and co-workers and is now our preferred method of penile reconstruction (Fig. 23-9). However, longer follow-up is necessary so that the effect of muscle atrophy on the reconstruction can be assessed. Muscle atrophy in a musculocutaneous flap would not lead to much shortening as similar atrophy in the muscle with skin graft. One advantage of the muscle is that it provides a vascularized bed for the urethral skin graft. The ideal penile reconstruction still eludes us. The problems of sensibility and sexual function have not been addressed. A combination of the gracilis procedure and the groin flap procedure may combine the best of both methods. The distal portion of the groin flap, for example, is innervated by the lateral branch of the T12 and may well form the basis of the neurosensory flap for penile reconstruction.

A

B

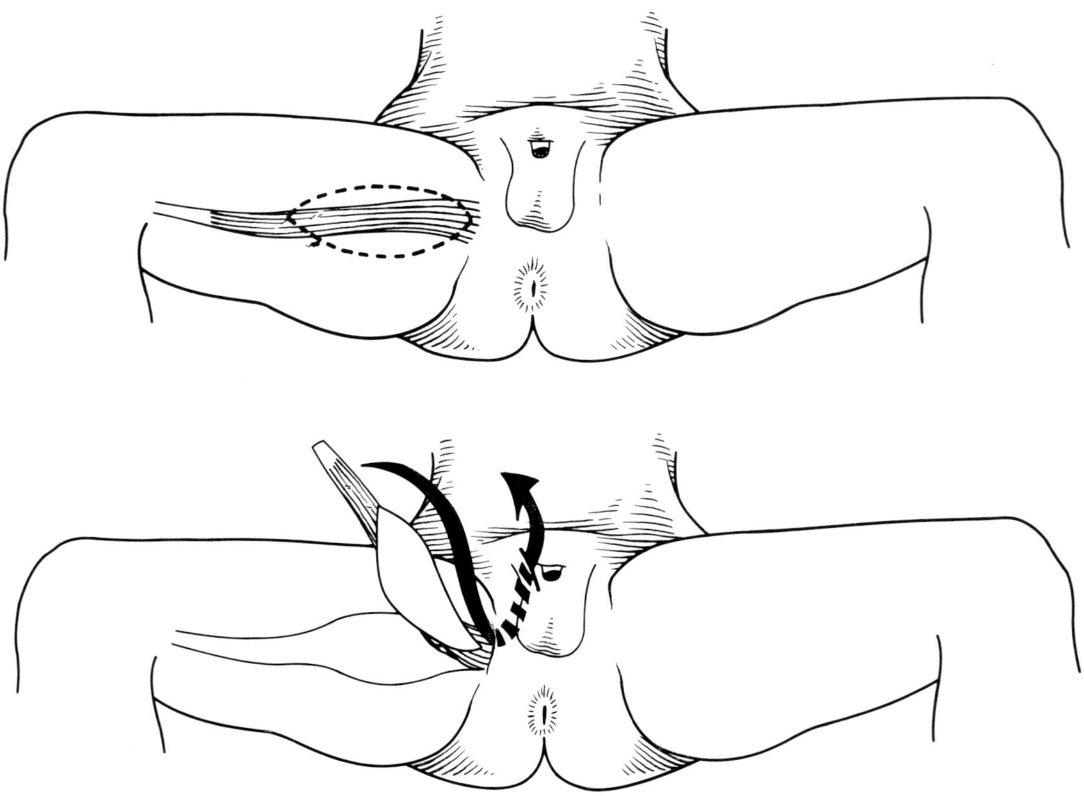

Fig. 23-8. Reconstruction of penis with unilateral gracilis musculocutaneous flap.
A, Musculocutaneous flap outlined.
B, Flap elevated.

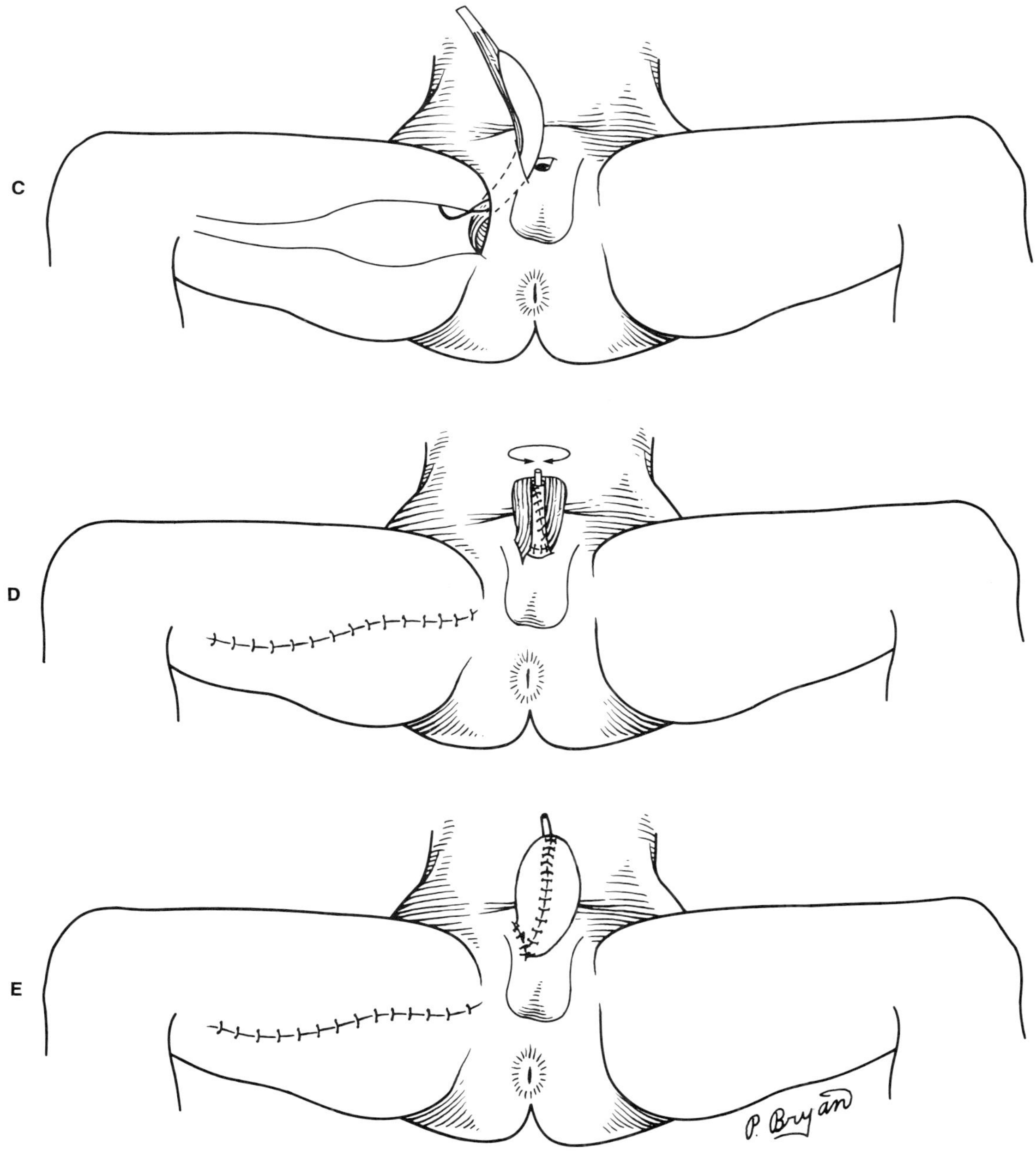

Fig. 23-8, cont'd
C, Flap tunneled through groin skin.
D, Muscle sutured around skin-grafted urethral reconstruction.
E, Skin island closed.

A B C D

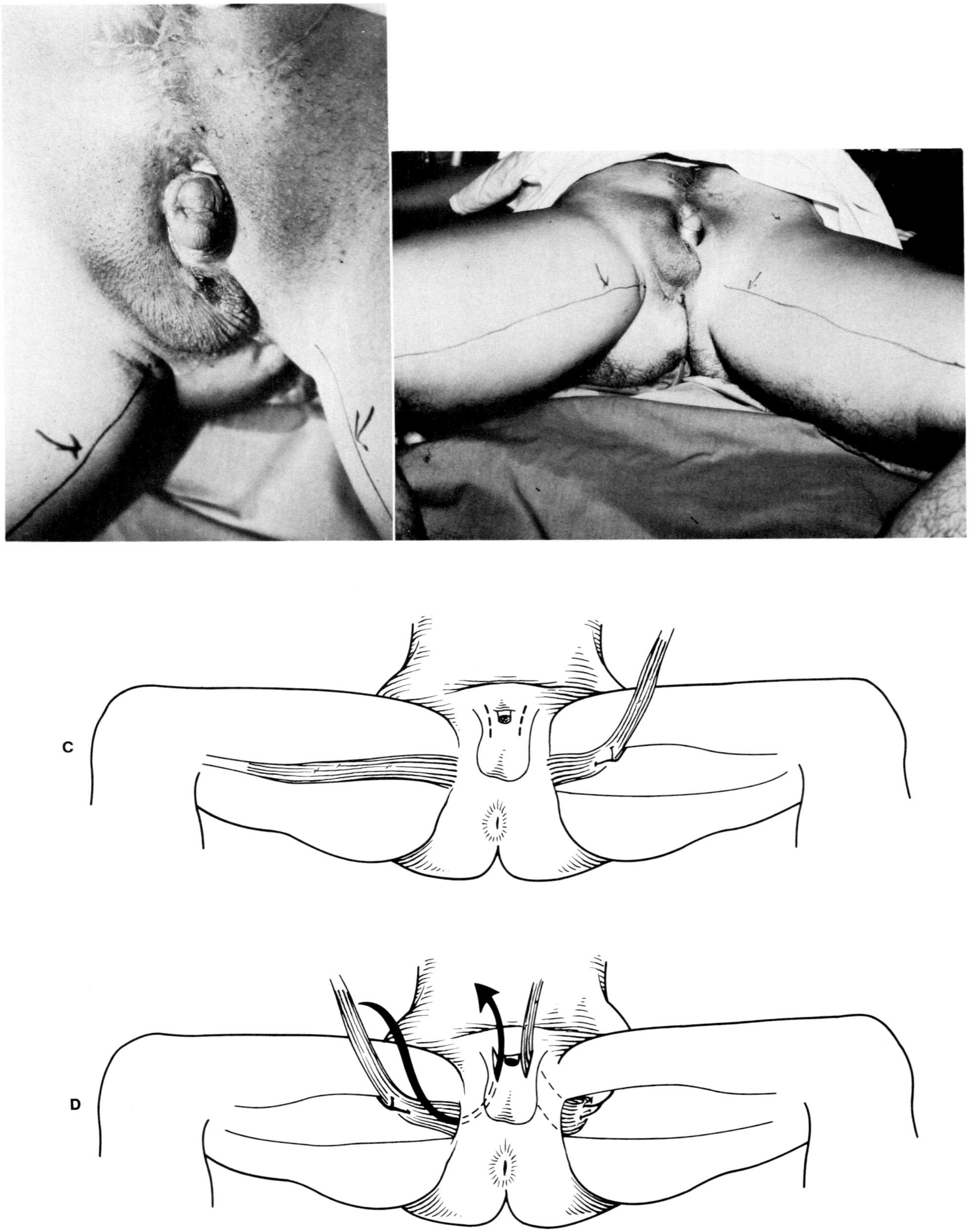

Fig. 23-9. Penis reconstruction with bilateral gracilis muscle flaps and skin graft.
A, Sixteen-year-old boy with epispadias.
B, Incisions outlined.
C, Muscles elevated on dominant vascular pedicle.
D, Muscles tunneled through groin skin.

E

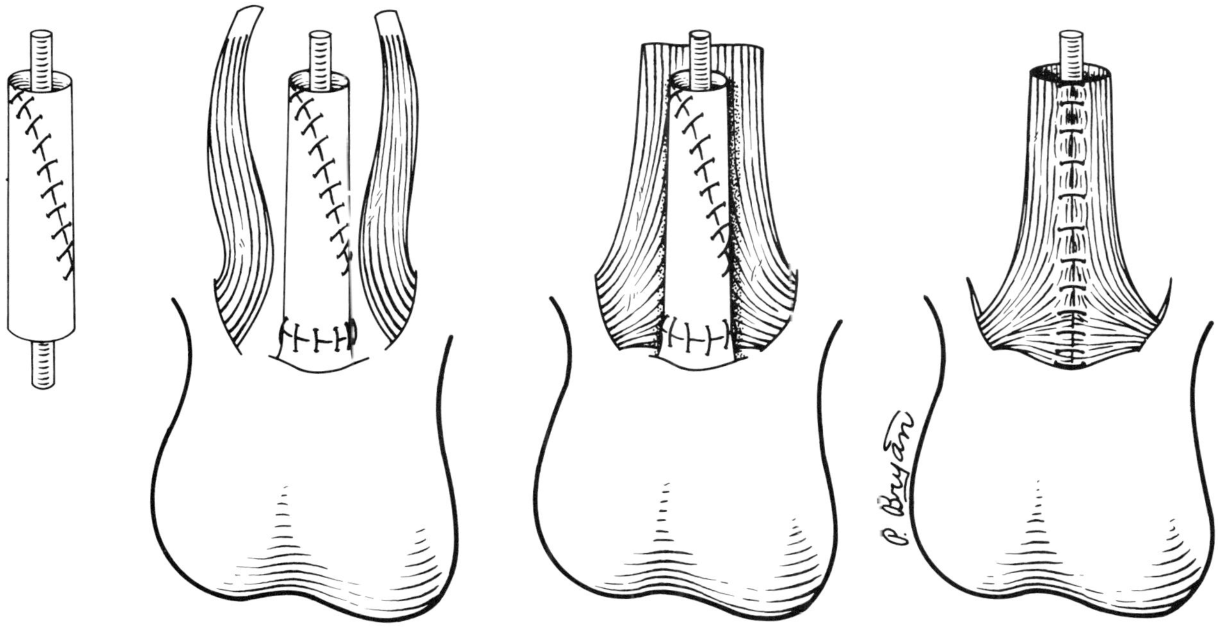

F

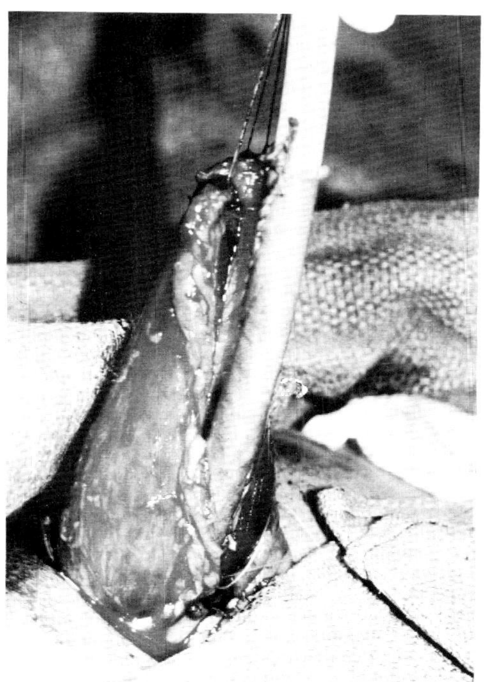

Fig. 23-9, cont'd
E and **F,** Urethral reconstruction with skin graft.

G

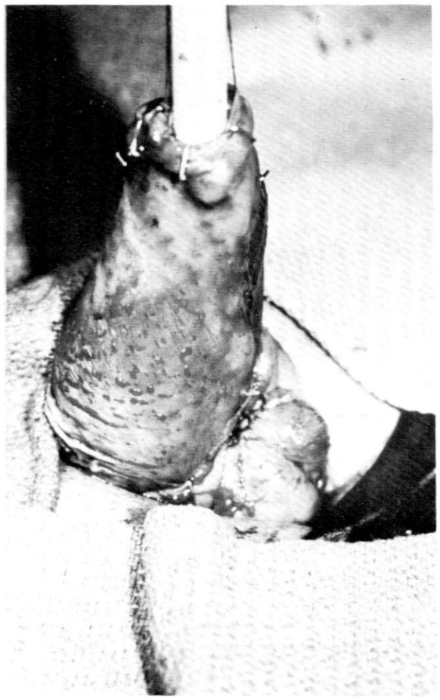

H

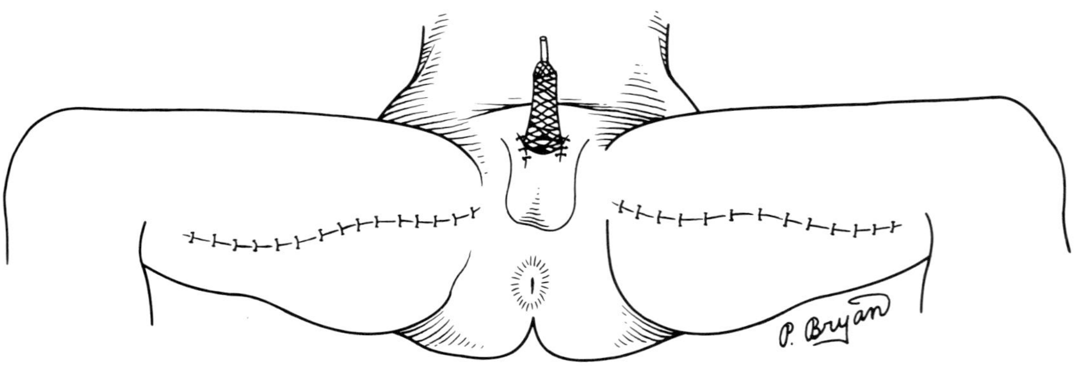

I

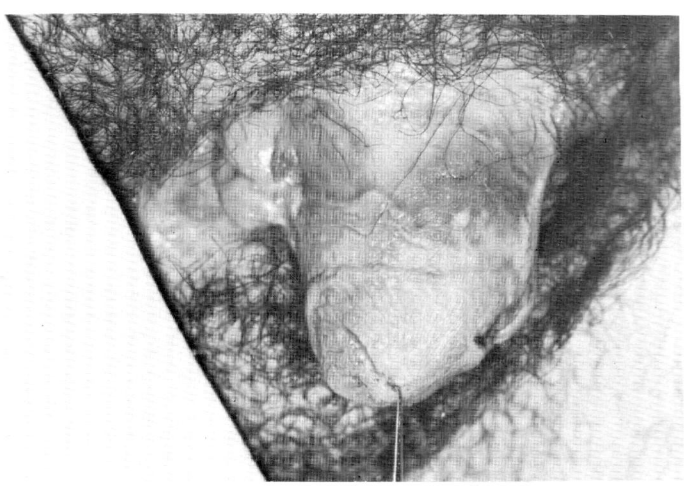

Fig. 23-9, cont'd
G and **H,** Muscles covered with skin graft.
I, Final result. NOTE: Muscle atrophy.

Anal sphincter reconstruction. The gracilis sling was described by Pickrell and co-workers for the correction of anal incontinence and later urinary incontinence with good results in children. However, Medgyesi and co-workers were unable to correct urinary continence with this method in seven men. We have had only limited but encouraging experience with this method for correction of anal sphincter incontinence in the adult.

Groin coverage. The gracilis unit will easily reach and cover the medial groin area (Fig. 23-10). It will not readily reach the femoral vessels, especially in obese patients.

Perineal coverage. Failure of primary healing of the abdominal perineal wound results in a defect resistant to local wound care. Failure of this wound to heal may relate to inadequate nutrition as a result of long-term illness. Correction of nutritional and vitamin deficiencies often results in secondary wound healing. Persistance of inflammatory bowel disease is sometimes associated with nonhealing, and a small bowel series is required during patient evaluation. Persistence of local tumor will result in nonhealing, and biopsy of suspicious areas is required in the postoperative cancer patient.

A

B

C

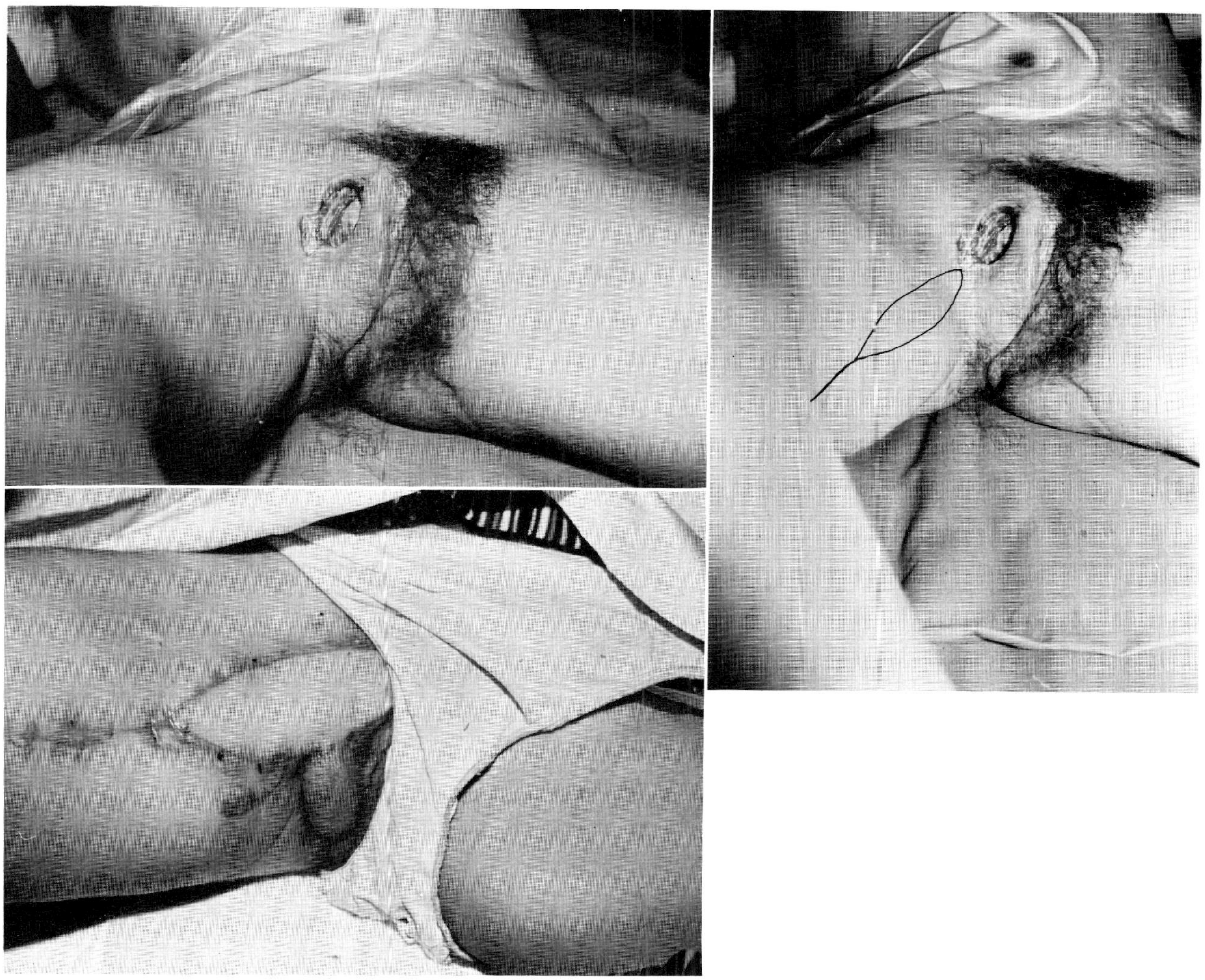

Fig. 23-10. Gracilis musculocutaneous flap coverage of groin.
A, Full-thickness defect over right pubis in young girl with exposed bone.
B, Flap outlined.
C, Stable wound coverage with gracilis musculocutaneous flap.

Perirectal coverage. Generally, failure of wound healing in the abdominal perirectal wound is related to absence of well-vascularized soft tissue. This is especially common if the patient receives postoperative radiation. The transposition of the gracilis muscle in association with radical débridement of the pelvic wound results in rapid wound healing. Since a chronic sinus tract extends deep into the retroperitoneum, a currette is helpful in removing the chronic granulation tissue where small bowel is often present at the wound base.

The muscle is generally elevated and transposed through a skin tunnel into the pelvic defects. The proximal or midthigh skin over the gracilis muscle will not reach the anal region after transposition of the gracilis muscle. Since distal gracilis skin is unreliable, a musculocutaneous flap is rarely used. Furthermore there is generally adequate skin in the perineum for wound coverage.

SARTORIUS

The sartorius is a long, thin, flat muscle that crosses the thigh from the lateral border to the medial border. It takes origin from the anterior superior iliac spine and inserts into the medial tibial condyle. The blood supply is segmental (type IV) with eight to eleven direct pedicles from the superficial femoral artery. These segmental vessels enter the muscle on its medial border. It is innervated by a branch of the femoral nerve. This is an expendable muscle that is a lateral rotator and flexor of the thigh. As a type IV muscle, it has a very limited arc of rotation. However, with ligation of the upper one or two pedicles the muscle can be divided from its origin and transposed medially into the groin (Fig. 23-11).

Flap elevation

Through a transverse groin incision, the muscle that is the most superficial is identified. The upper one or two pedicles are divided, and the muscle origin from the anterior superior iliac spine is divided. The muscle may then be transposed medially.

Clinical application

This flap has proved useful and reliable for coverage of exposed femoral vessels or vascular prosthesis in the inguinal region (Fig. 23-11).

RECTUS FEMORIS

The rectus femoris is a large, fusiform muscle that is located superficially in the central anterior thigh. It takes origin from the inferior iliac spine and blends with the other quadriceps muscles to insert into the patella. It has a dominant pedicle (type I), a branch of the lateral circumflex femoral artery, that enters the muscle on its deep surface at a point approximately 8 to 10 cm below the anterior superior iliac spine. It is innervated by the femoral nerve and is a strong extensor of the leg and the thigh flexor. The patella mechanism must be preserved, and the remaining quadriceps must be intact to minimize the functional loss resulting from the use of this muscle. The skin territory of this flap is innervated by the anterior cutaneous nerve of the thigh.

Based on its single dominant pedicle, this unit has an anterior and a posterior arc of rotation (Fig. 23-12). Anteriorly it will cover the groin perineum and abdominal wall, whereas posteriorly it will reach the trochanter and ischium.

Flap elevation

A skin island is planned over the proximal two thirds of the muscle. The proximal island is preferred so that the skin coverage over the patellar tendon is not disturbed. The island is incised over the muscle, and the muscle is then divided distally several centimeters above the knee, taking care to avoid injury to the patella tendon mechanism. The muscle is then elevated off the vastus intermedius and the dissection continued from the distal border to the proximal border. At the level where the sartorius muscle is visualized the vascular pedicle can be identified entering the rectus femoris on its deep surface. This point is usually 8 to 10 cm below the anterior superior iliac spine. At this stage the unit may be transposed. The unit can safely be made into an island. The donor defect is closed directly for standard size flaps. However, for wider flaps skin grafting of the donor defect may be necessary.

Clinical applications

The rectus femoris will cover the groin and perineum with a similar arc of rotation to the TFL. The extended TFL flap is fascial on its deep surface, whereas the rectus femoris is musculotendinous. Under certain circumstances muscle may be preferable to fascia despite the greater bulk. However, we favor the use of the TFL flap, because its use is less likely to produce any functional deficit in the thigh or leg.

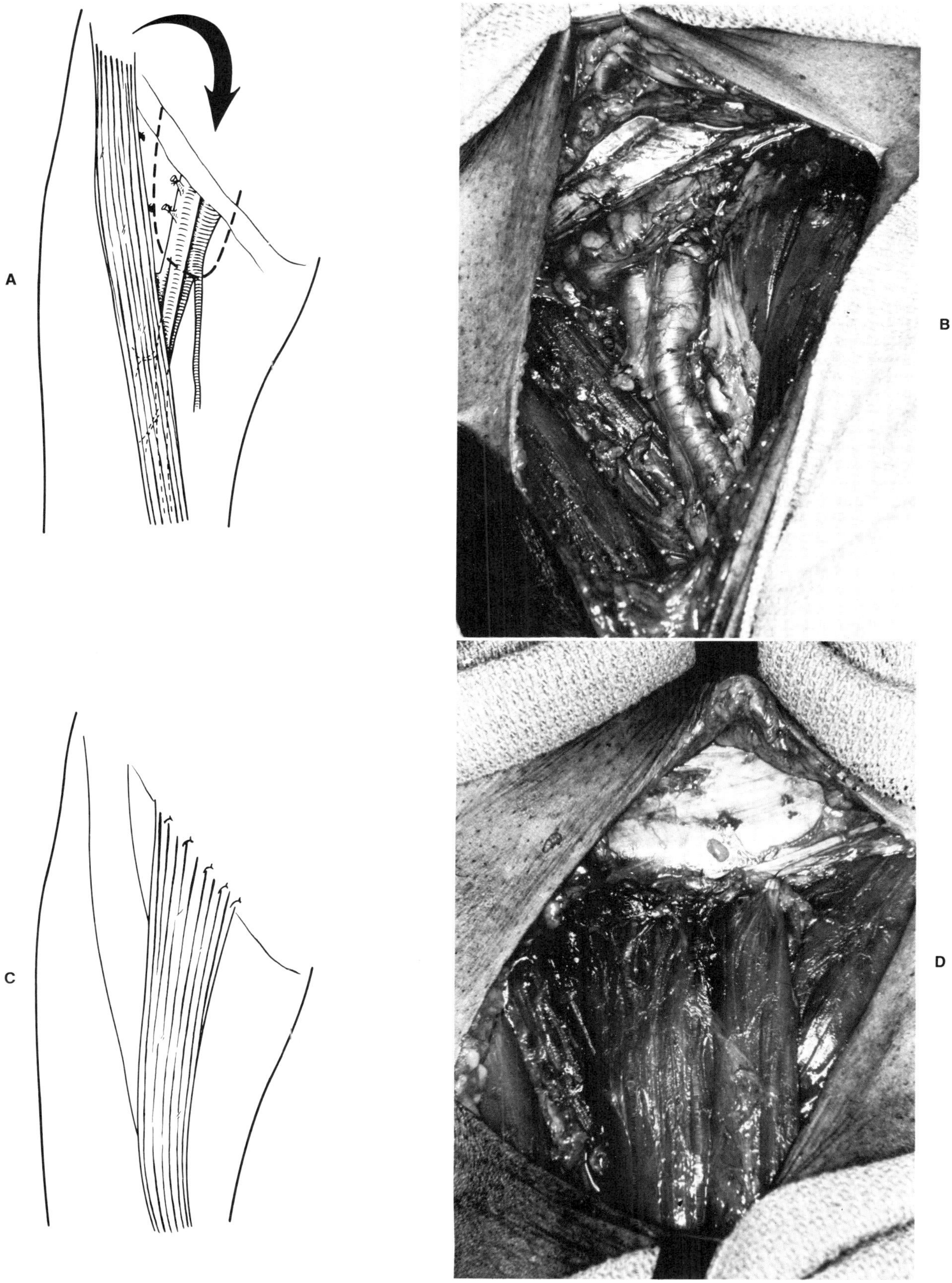

Fig. 23-11. Transportation of sartorius medially to cover exposed femoral vessels.
A and **B**, Exposed femoral vessels after groin lymphadenectomy.
C and **D**, Transposed sartorius covers vessels.

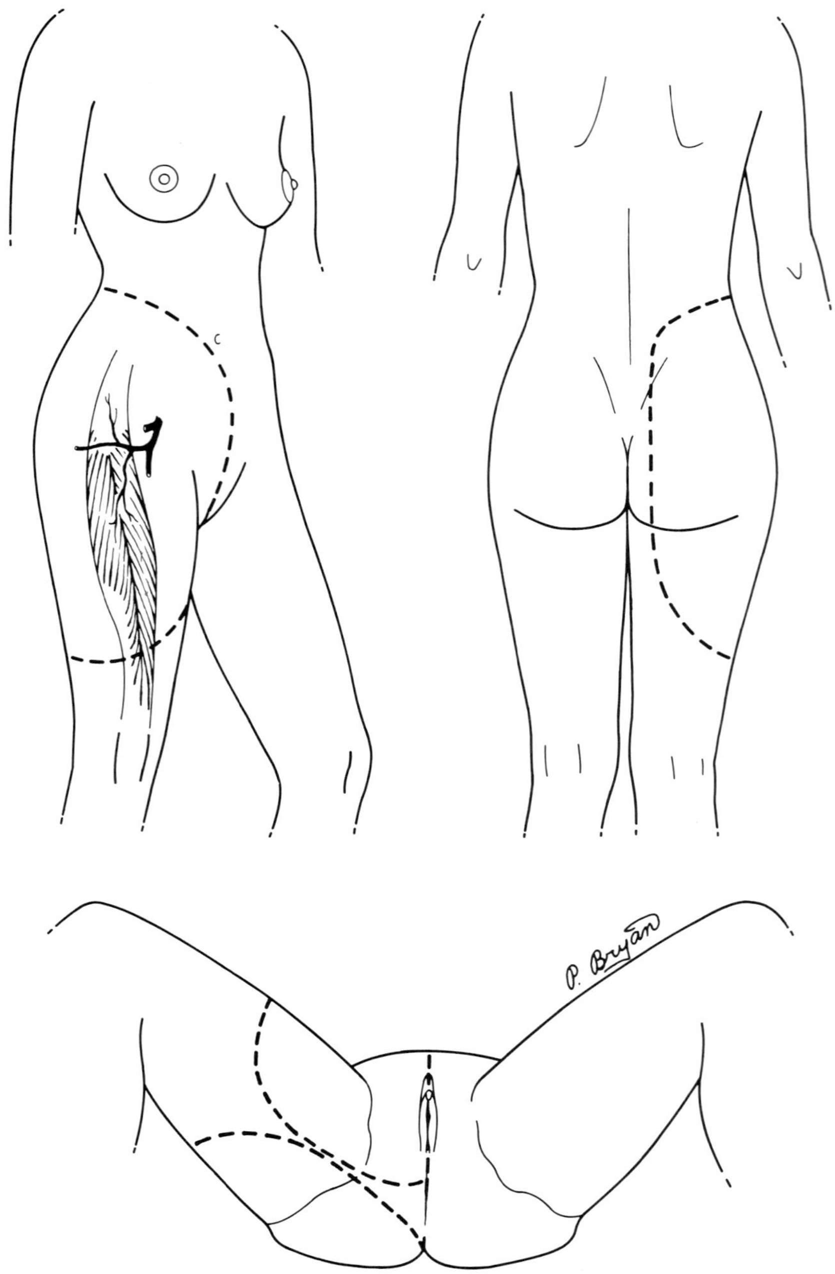

Fig. 23-12. Blood supply and arc of rotation of rectus femoris musculocutaneous flap.

CONCLUSION

Although all of the flaps described have been useful in reconstructive surgery in the groin and perineum, some flaps are more ideally suited to special problems. However, certain areas of overlap exist where each of the flaps could be used. Specifically, although the gracilis is most suitable for total vaginal and penile reconstruction, all of these flaps have some role in coverage of the groin. For groin coverage the TFL is our first choice. It is an expendable unit that is thin and will cover the entire groin, since it is located laterally. The rectus femoris is as reliable a unit but is more bulky, is located more medially, and is not an entirely dispensible muscle. The gracilis especially in an obese patient may not reach the lateral groin area.

Special mention must be made of the infected exposed vascular prosthesis. Although the TFL, rectus femoris, and sartorius will cover these, the TFL would not be our first choice. It would be preferable to cover the prosthesis with muscle such as the sartorius with a TFL flap over the sartorius, as opposed to placing the fascia directly over the prosthesis. If the sartorius is not adequate, then a rectus femoris flap or vastus lateralis would be preferable to the TFL. Exposed infected prostheses respond more favorably when covered by highly vascularized muscle rather than fascia, which is relatively less vascular. Under these circumstances where an exposed or infected vascular prosthesis threatens limb survival, the use of the rectus femoris or vastus lateralis, which may lead to functional disability, appears justified.

ANNOTATED BIBLIOGRAPHY

1. Baronofsky, I.D.: Technique of inguinal node dissection, Surgery **24:**555, 1948.

   The transposition of the sartorius for coverage of the exposed femoral vessels after radical groin dissection is described.

2. Bhagwat, B.M., Pearl, R.M., and Laub, D.R.: Uses of the rectus femoris myocutaneous flap, Plast. Reconstr. Surg. **62:**698, 1978.

   Four case reports using the rectus femoris for closure of complex wounds in the groin and buttock areas are presented. Anatomy and surgical techniques are detailed. The muscle is a reasonable alternative for wound coverage in these areas in the paraplegic patient. The question of functional loss in ambulatory patients after the application of this flap is raised.

3. Bostwick, J., III, Hill, H.L., and Nahai, F.: Repairs in the lower abdomen, groin, or perineum with myocutaneous or omental flaps, Plast. Reconstr. Surg. **63:**186, 1979.

   Thirty-one cases of complicated wound coverage with musculocutaneous flaps or omentum are discussed. Anatomy and arc of rotation of each flap are outlined along with recommendations for flap selection. One-stage reconstruction is again stressed.

4. Boxer, T.B.: Reconstruction of the male genitalia, Surg. Gynecol. Obstet. **141:**939, 1975.

   A review of the history of the reconstructin of male genitalia is presented. At the time of publication of this article, the application of muscle and musculocutaneous flaps to this region had not been fully described.

5. Gillier, H.D., and Harrison, R.J.: Congenital absence of the penis with embryological consideration, Br. J. Plast. Surg. **1:**8, 1948.

   A detailed discussion of this relatively rare deformity is presented.

6. Hester, T.R., Hill, H.L., and Jurkiewicz, M.J.: One-stage reconstruction of the penis, Br. J. Plast. Surg. **81:**279, 1978.

   One-stage successful reconstruction of the penis is demonstrated using bilateral gracilis muscles, covering a full-thickness neourethral graft. A split-thickness skin graft covers the muscles. Disadvantages and possible improvements are discussed.

7. Hill, H.L., Nahai, F., and Vasconez, L.O.: The tensor fascia lata myocutaneous free flap, Plast. Reconstr. Surg. **61:**517, 1978.

   An initial account of the tensor fascia lata used as a myocutaneous free flap. Anatomy, surgical technique, advantages, and disadvantages are outlined.

8. Hill, H.L., Hester, T.R., and Nahai, F.: Covering large groin defects with the tensor fascia lata myocutaneous flap, Br. J. Plast. Surg. **32:**12, 1979.

   A brief presentation of two cases of large groin defects covered with transposition flaps of the TFL is discussed. One-stage reconstructive procedure is stressed.

9. McCraw, J.B., and co-workers: Vaginal reconstruction with gracilis myocutaneous flaps, Plast. Reconstr. Surg. **58:**176 1976.

   The application of the gracilis musculocutaneous flap in 22 patients for reconstruction of the vagina after radical pelvic surgery is presented. There is a detailed discussion of anatomy, technique, and complications.

10. McCraw, J.B., Dibbell, D.G., and Carraway, J.H.: Clinical definition of independent myocutaneous vascular territories, Plast. Reconstr. Surg. **60:**212, 1977.

    An extensive review in the definition of the vascular territories of clinical musculocutaneous flaps is presented.

11. Mathes, S.J., Vasconez, L.O., and Jurkiewicz, M.J.: Extensions and further applications of muscle flap transpositions, Plast. Reconstr. Surg. **60:**6, 1977.

    Ninety-four cases of muscle flap transposition from 1973 to 1976 are presented. Muscles of the head and neck, chest, and upper extremities are included with particular emphasis placed on the perineal, sacral, and lower extremity regions. The precise knowledge of functional and vascular anatomy is stressed in this paper.

12. Mathes, S.J., and Nahai, F.: Clinical atlas of muscle and musculocutaneous flaps, St. Louis, 1979, The C.V. Mosby Co.

    An indepth clinical atlas and anatomic guide for reconstructive surgeons using muscle and musculocutaneous flaps. All useful muscles are included, and anatomic studies and clinical cases are presented with the advantages and disadvantages of each flap.

13. Medgyesi, S., Mortensen, S., and Nerstrom, B.: The failure of free muscle transplants in the treatment of urinary incontinence, Br. J. Plast. Surg. **32:**336, 1979.

    Free transplant of the extensor hallucis brevis muscle for urinary incontinence is reported in seven patients. The incontinence was unaffected in five patients. Incidentally, it is reported that only one out of seven patients with epispadias has a successful result from the gracilis sling procedure.

14. Nahai, F., and co-workers: The tensor fascia lata myocutaneous flap, Ann. Plast. Surg. **1:**372, 1978.

    The first in a series of articles on this reliable and versatile musculocutaneous flap is presented. Anatomy and technique are detailed. Early clinical cases are presented.

15. Nahai, F., Hill, H.L., and Hester, T.R.: Experiences with the tensor fascia lata flap, Plast. Reconstr. Surg. **63:**788, 1979.

    An extensive experience with 60 cases is presented. Anatomy and surgical techniques are stressed. The reliability and versatility of the flap are demonstrated as a rotation, island, or free flap. Coverage of the groin and reconstruction of the vulva are illustrated.

16. Orticochea, M.: A new method of total reconstruction of the penis, Br. J. Plast. Surg. **25:**347, 1972.

    A classic article that is partly responsible for the great interest in and applications of muscle and musculocutaneous flaps over the last 10 years is presented. A reasonable result is presented, although the procedure required five stages over a 2-year period.

17. Pickrell, K., and co-workers: Construction of a rectal sphincter and restoration of anal continence by transplanting the gracilis muscle, Ann. Surg. **135:**853, 1952.

    The gracilis sling operation is described for the correction of anal incontinence.

18. Pickrell, K., and co-workers: Gracilis muscle transplant for correction of urinary incontinence in male children, Ann. Surg. **143:**764, 1956.

    The gracilis sling procedure is successfully applied for the correction of urinary incontinence in male children.

19. Puckett, C.L., and Montle, J.E.: Construction of male genitalia in the transsexual using a tubed groin flap for the penis and a hydraulic inflation device, Plast. Reconstr. Surg. **61:**523, 1978.

    Although this article does not address muscle flap reconstruction or urinary reconstruction, one might consider a sensory groin flap surrounding bilateral gracilis flaps as a method of reconstructing a sensory organ.

20. Robinson, D.W.: Surgical problems in excision and repair of radiated tissue, Plast. Reconstr. Surg. **51:**41, 1975.

    The hazards of reconstructing radiated tissues are outlined. Wide excision of involved tissue is stressed. In most instances in this article flap coverage is provided through random tubed flaps. This article was written before the widespread application of muscle and musculocutaneous flaps to the groin region. However, it does emphasize the problems involved in reconstruction of irradiated tissues.

21. Wangensteen, O.H.: Repair of recurrent and difficult hernias and other large defects of the abdominal wall employing the iliotibial tract of fascia lata as a pedicled flap, Surg. Gynecol. Obstet. **59:**766, 1934.

    The application of the fascia lata as a pedicled fascial flap is described for the repair of groin hernias and abdominal wall defects.

24

COMPLICATIONS

Stephen J. Mathes ▪ Foad Nahai

The groin is the source of difficult coverage problems, because the exposed femoral vessel carries the blood supply to the flaps available for coverage. Adequate preoperative assessment of atherosclerotic vascular disease is particularly important in this region. Successful coverage of groin wounds through flap selection may avoid future vascular complications.

TFL

Planning errors

Preoperative assessment of the patient's habitus is important in considering use of the TFL for perineal coverage. The bulky TFL flap of an obese patient will not always reach the perineum.

When the flap is used to cover an exposed vascular prosthesis, the fascia of the TFL is directly over the prosthesis and may not adhere to the prosthesis, resulting in a space for abscess formation. For this reason alternative flaps may be more suitable for coverage of the exposed vascular prosthesis.

Technical errors

Excessive bleeding between the fascia lata and the underlying vastus lateralis muscle may indicate unsuspected damage to the pedicle of the TFL, the terminal branch of the lateral femoral circumflex artery. Under these circumstances the vastus lateralis is elevated with the TFL as an alternative flap. Posterior proximal dissection of this flap is difficult, because the TFL muscle is adherent to the gluteus minimus muscle. Adequate débridement of infected or necrotic tissue is essential, especially in the radiated wound.

Postoperative errors

Soft suction drainage should be used postoperatively to avoid injury to the femoral vessels. Prolonged antibiotic coverage and Betadine irrigation are recommended for the exposed infected vascular prosthesis. Despite adequate flap coverage, salvage of the infected prosthesis is not always possible because of late graft thrombosis or recurrent infection (Fig. 24-1).

A

B

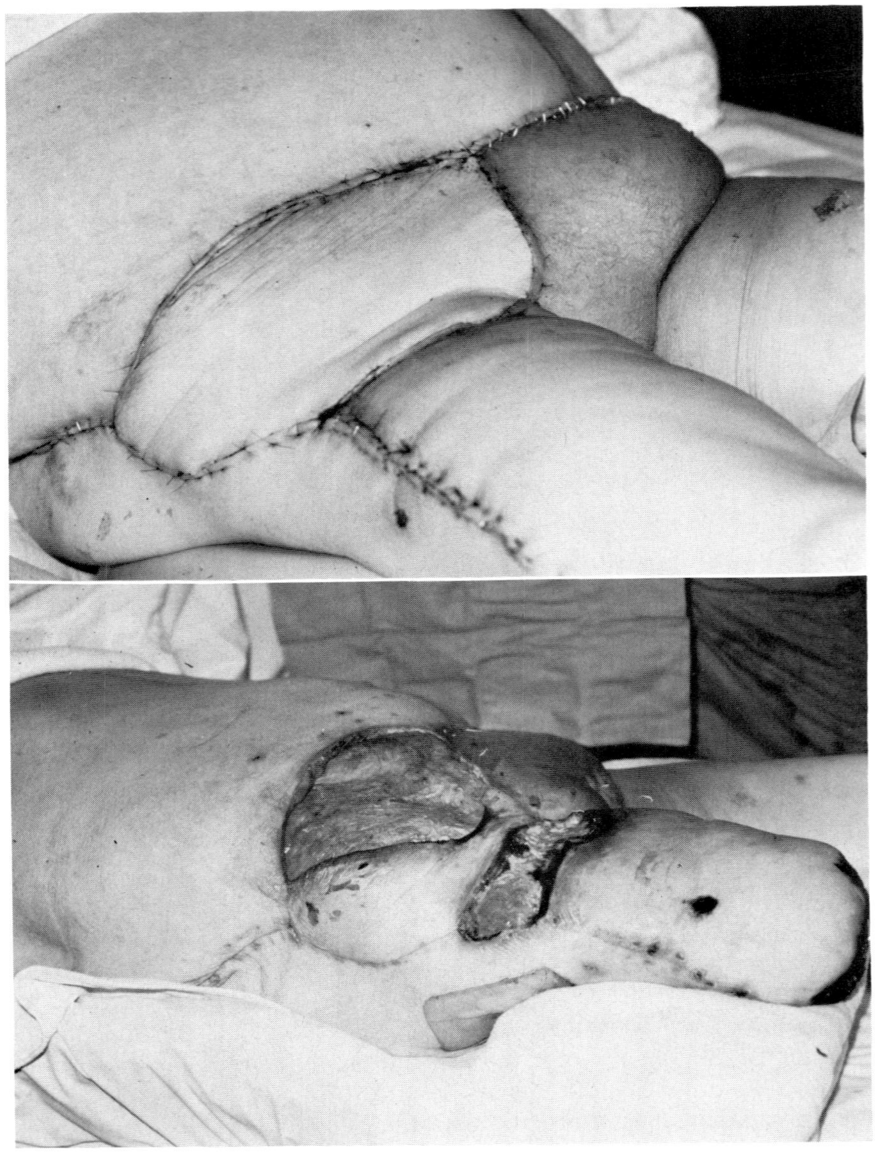

Fig. 24-1

A, Musculocutaneous flap transposed for coverage of exposed vascular graft to right lower extremity.

B, Thrombosis of vascular graft resulted in loss of arterial perfusion to musculocutaneous flap. Above-knee amputation required after flap loss and recurrent vascular graft exposure.

GRACILIS

Planning errors

The incorrect orientation of the skin island over the gracilis muscle is a common source of flap failure. The distal skin island is unreliable. However, the skin island must be located sufficiently distal to the point of flap rotation to reach beyond the tunnel. Since the skin island must be located over the distal third of the muscle to reach the perineum, transposition of muscle alone is preferred for this defect. In obese patients the musculocutaneous flap is bulky, which may reduce the arc of rotation.

Technical errors

Failure to localize the skin territory properly over the gracilis muscle is one of the most common technical errors leading to flap failure. This is avoided by either proximal or distal identification of the muscle before the skin island is designed. The cutaneous territory should be sutured to the muscle during flap dissection to avoid shearing of the musculocutaneous perforators. Excessive traction on the dominant vascular pedicle should be avoided during transposition.

Postoperative errors

The patient should be positioned to avoid pressure on the flap and venous congestion. Hip extension should be avoided, because this may put excessive tension on the flap and its vascular pedicle. Suction drainage of the donor area is recommended.

RECTUS FEMORIS

Planning errors

The rectus femoris is not an expendable muscle. Use of this muscle may result in weakness of knee extension, especially if the other quadriceps muscles are injured.

Technical errors

In dissecting the rectus femoris in a distal-to-proximal fashion, the plane between this muscle and the vastus medialis is not distinct, and the muscles may be difficult to separate. Proximal dissection must be under direct vision, because branches of the femoral nerve course close to the vascular pedicle of the rectus femoris.

Postoperative errors

Suction drainage is required when the donor region is closed directly.

SARTORIUS

Planning errors

The segmental pattern of blood supply to the sartorius limits its effective arc of rotation. Division of more than two adjacent pedicles may result in muscle devascularization. This flap will not reach the perineum. Use as a musculocutaneous unit is not recommended because of the limited arc of rotation and small skin territory size.

Technical errors

Correct identification of the sartorius, located between the TFL laterally and rectus femoris medially at its origin, is essential.

Postoperative errors

Suction drainage is required in the donor site after its direct closure.

25

A SYSTEMATIC APPROACH TO FLAP SELECTION

Stephen J. Mathes ■ Foad Nahai

SACRUM (Fig. 25-1)

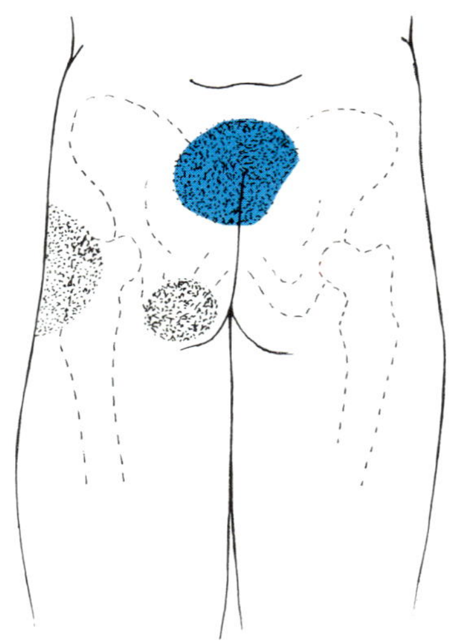

Fig. 25-1 Sacral defect.

Skin grafts

Skin grafts are method of choice for shallow defects without exposed bone in the patient with sensibility

Local flaps (Fig. 25-2)

1. Gluteus maximus muscle or musculocutaneous flap is flap of choice (Figs. 25-3 to 25-6)
2. Transverse back flap is alternative flap for sacral sores (Fig. 25-7)
3. Rotational flaps are rarely indicated in reconstruction of sacral pressure sores

Distant flaps

Intercostal flap is alternative flap that will provide sensibility

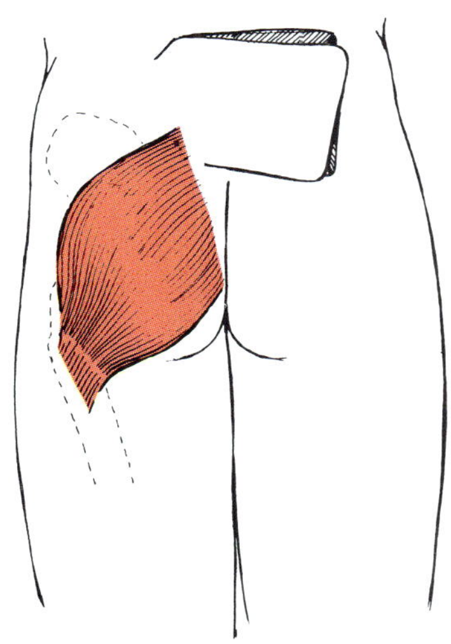

Gluteus maximus
Transverse back flap

Fig. 25-2

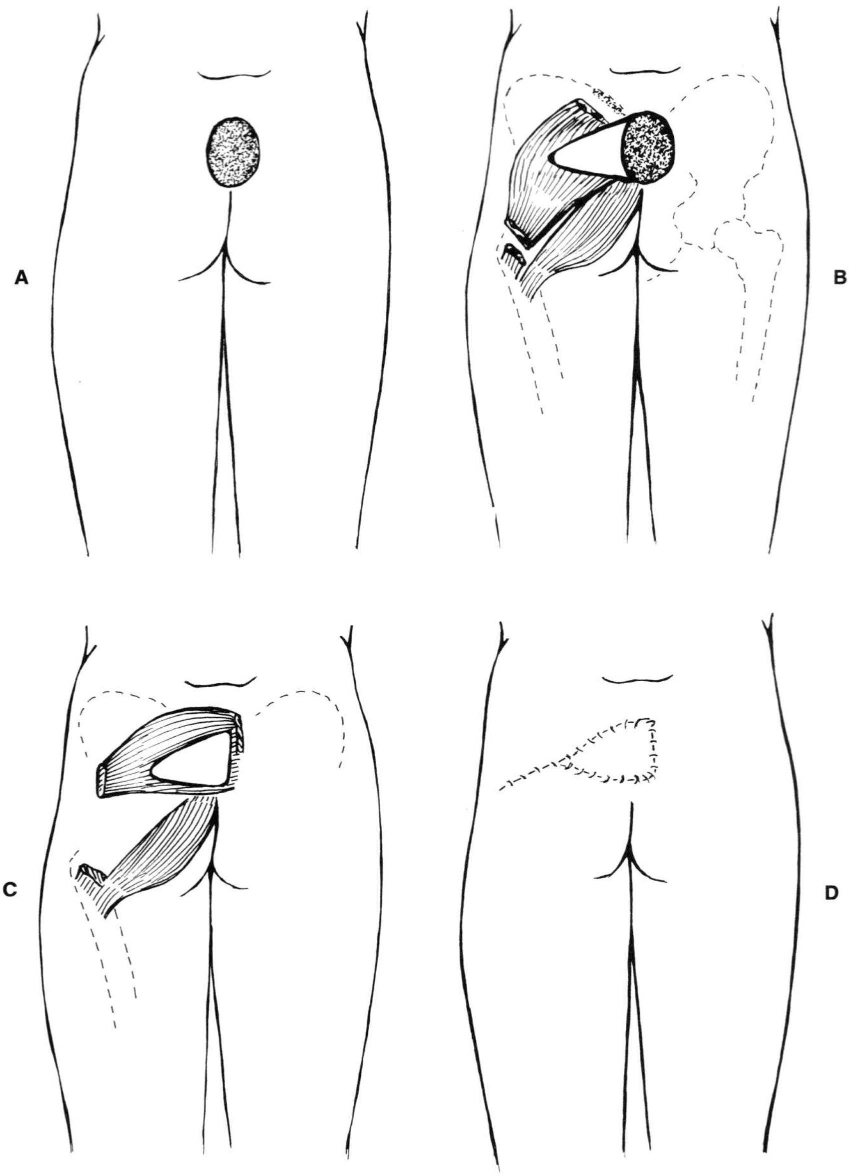

Fig. 25-3

A, Sacral pressure sore.

B, Design of proximal skin island on gluteus maximus.

C, Preservation of inferior half gluteus maximus muscle. NOTE: Sacral origin of muscle is preserved.

D, Transposition of musculocutaneous flap provides sacral coverage.

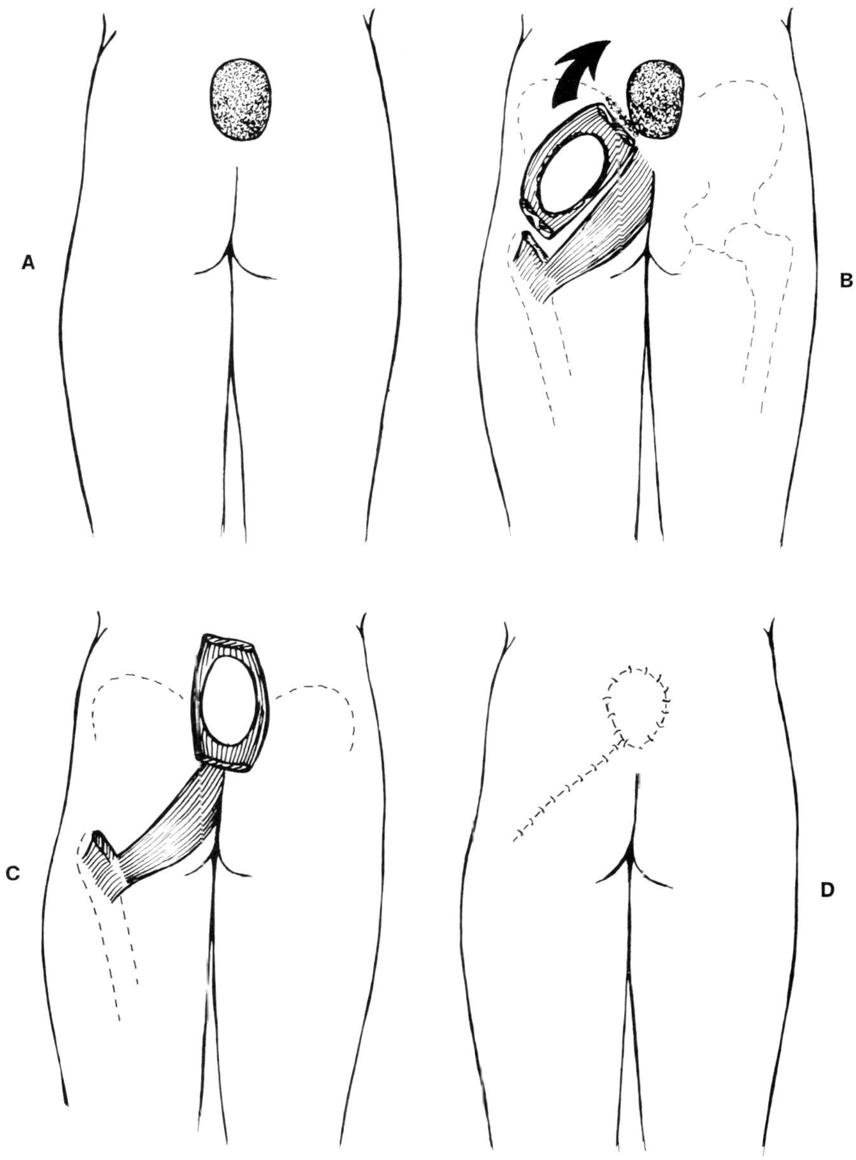

Fig. 25-4
A, Sacral pressure sore.
B, Design of skin island on superior gluteus maximus muscle.
C, Preservation of inferior half gluteus maximus muscle. NOTE: Detachment of muscle origin increases arc of rotation.
D, Transposition of musculocutaneous flap provides sacral coverage.

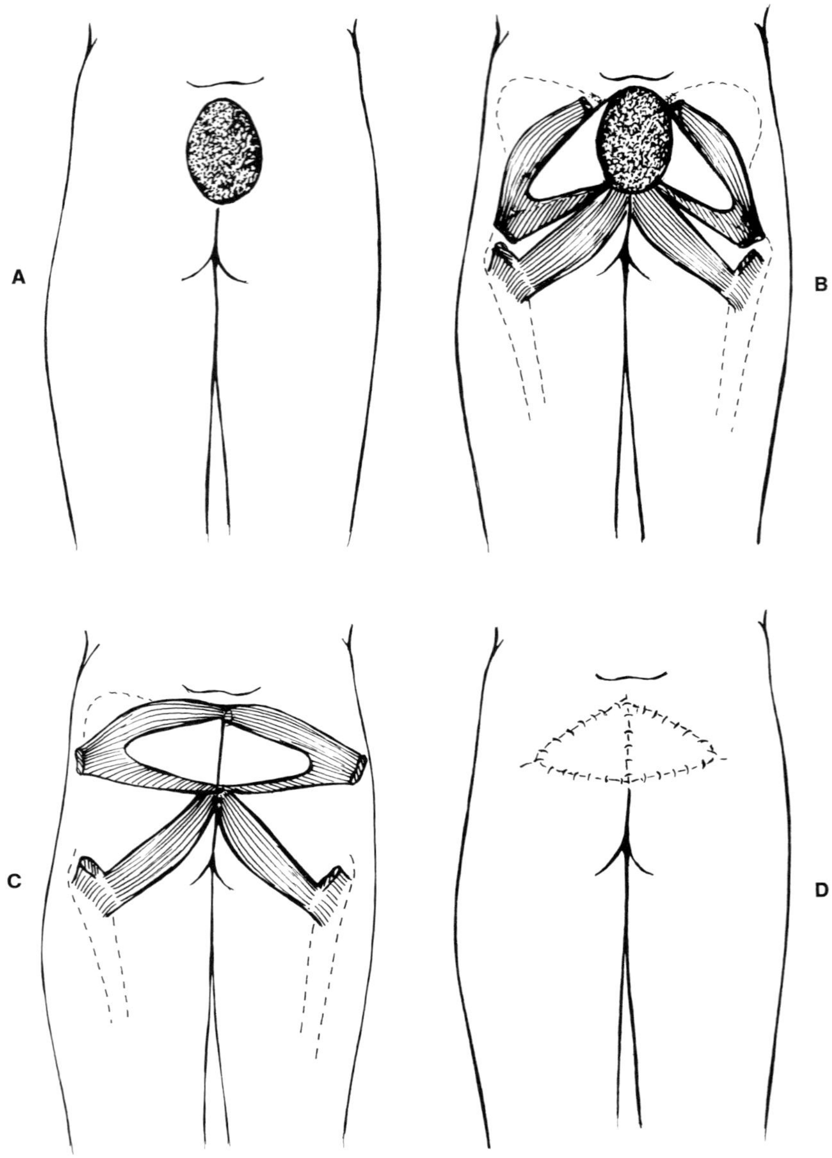

Fig. 25-5
A, Large sacral pressure sore.
B, Design of skin island on bilateral superior gluteus maximus musculocutaneous flap.
C, Preservation of inferior halves of gluteus maximus muscle.
D, Transposition of musculocutaneous flap provides coverage.

A

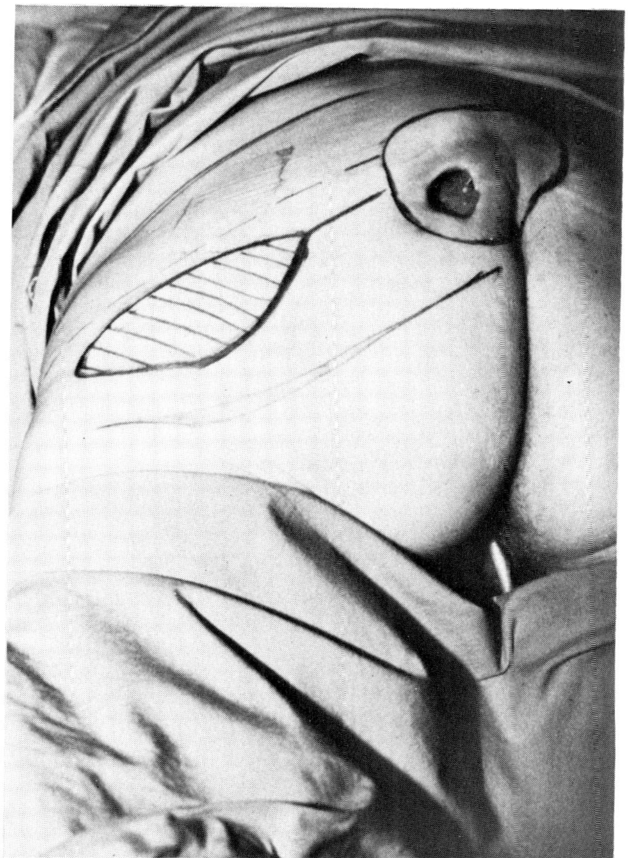

B

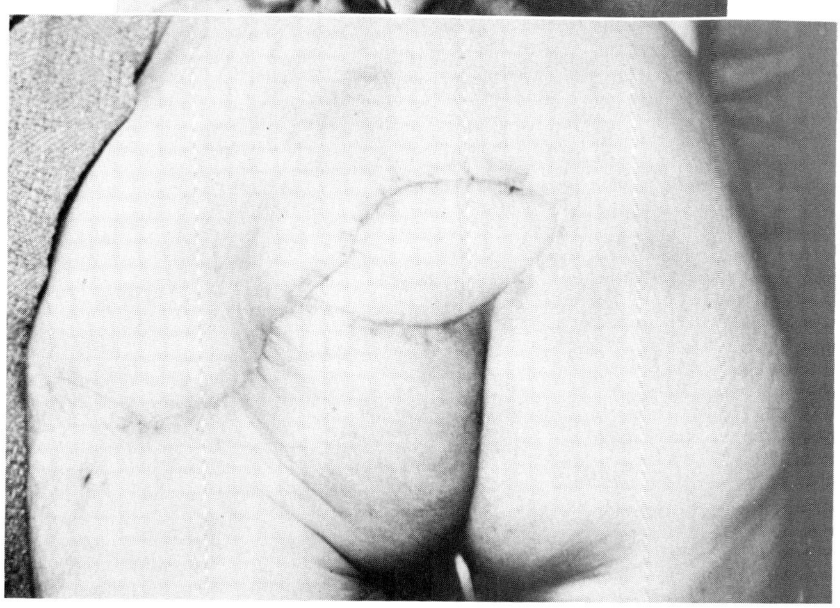

Fig. 25-6. Gluteus maximus musculocutaneous flap.
A, Chronic sacral pressure sore in paraplegic patient. Design of left superior gluteus maximus musculocutaneous flap.
B, Three-month postoperative view demonstrates stable coverage provided by transposition of superior gluteus maximus musculocutaneous flap. NOTE: Donor defect is closed directly. Normal skin is preserved over region of greater trochanter.

A

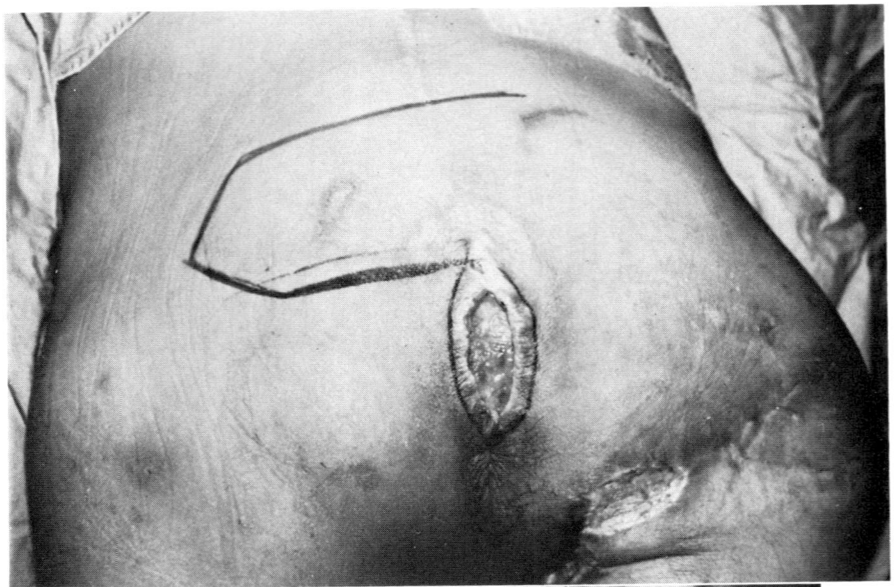

B

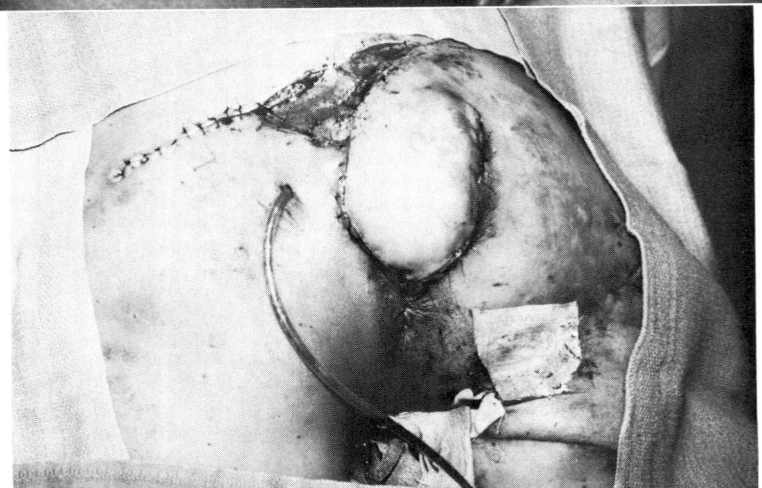

C

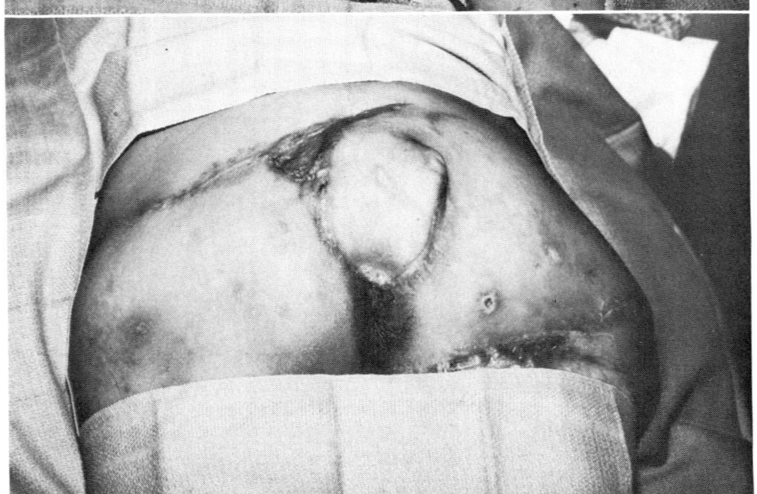

Fig. 25-7. Transverse back flap.

A, Chronic sacral pressure sore in paraplegic patient. Design of transverse back flap based on right lumbar perforating arteries.

B, Immediate postoperative view after transposition of transverse back flap. NOTE: Skin graft is required for midportion of donor defect.

C, Six-month postoperative view demonstrates stable coverage provided by transverse back flap.

ISCHIUM (Fig. 25-8)

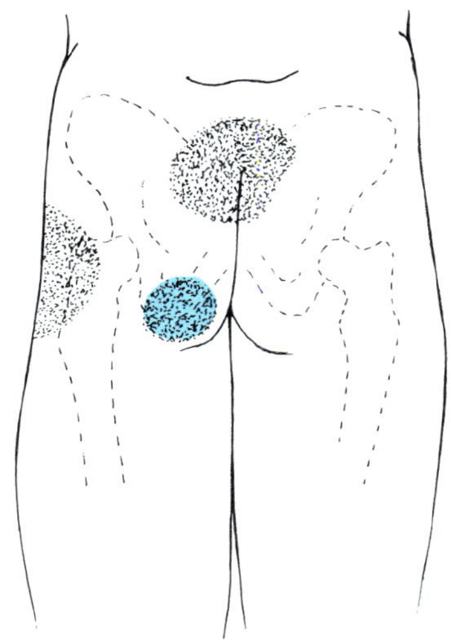

Fig. 25-8

Skin grafts

Skin grafts are not indicated in reconstruction of ischial pressure sores

Local flaps (Fig. 25-9)

Gluteus maximus muscle and skin graft or musculocutaneous flap is the flap of choice (Figs. 25-10 and 25-11)

Distant flaps (Fig. 25-9)

1. Gracilis musculocutaneous flap is the flap of choice (Fig. 25-12)
2. TFL and gluteal thigh flaps are alternative flaps with potential for providing sensibility and are useful alternative flaps for coverage (Fig. 25-13)
3. Hamstring muscles or V-Y advancement musculocutaneous flap is alternative flap (Fig. 25-14)
4. Posterior thigh advancement flap is rarely indicated in reconstruction of ischial pressure sores

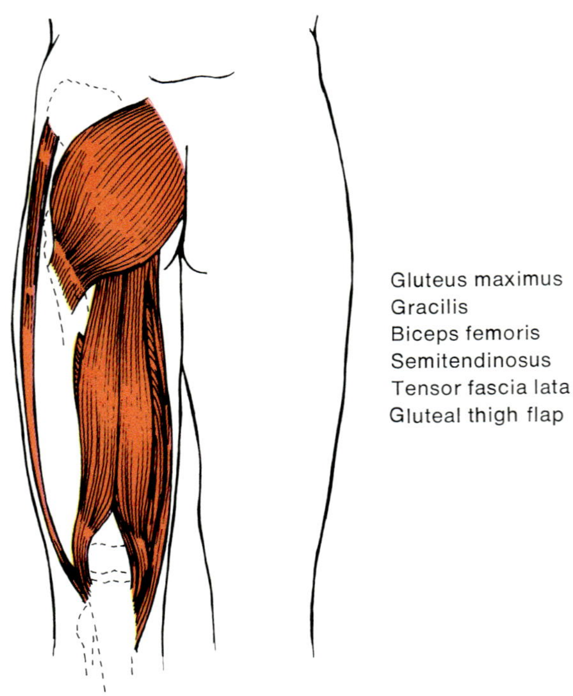

Fig. 25-9

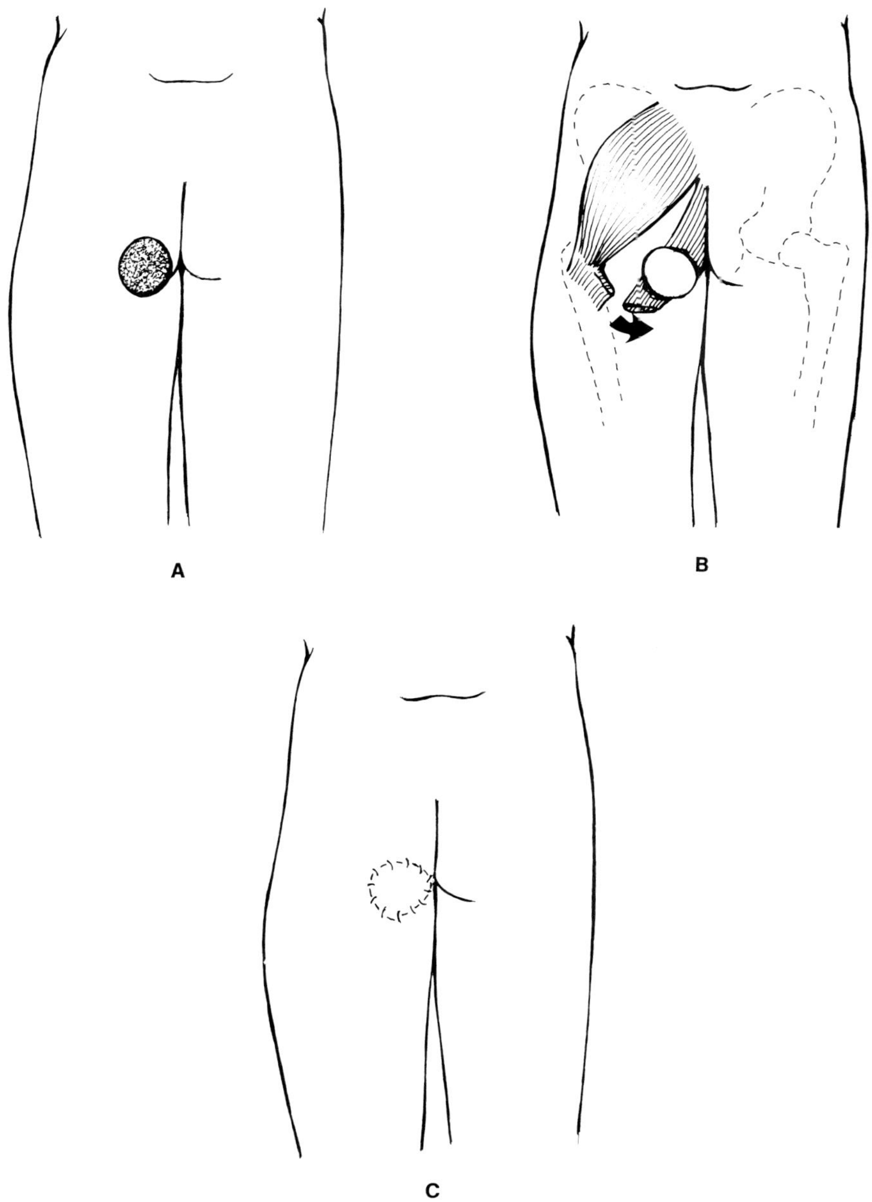

Fig. 25-10
A, Ischial pressure sore.
B, Design of skin island on inferior gluteus maximus. NOTE: Superior half of muscle is preserved.
C, Transposition of flap provides ischial coverage.

A

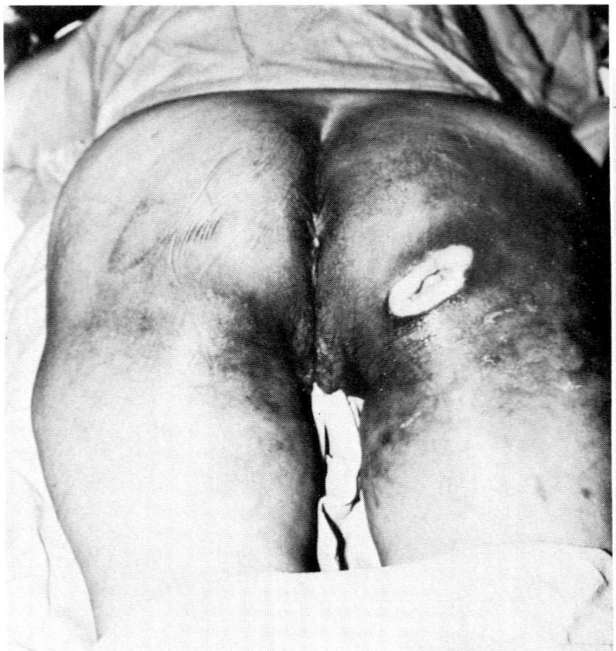

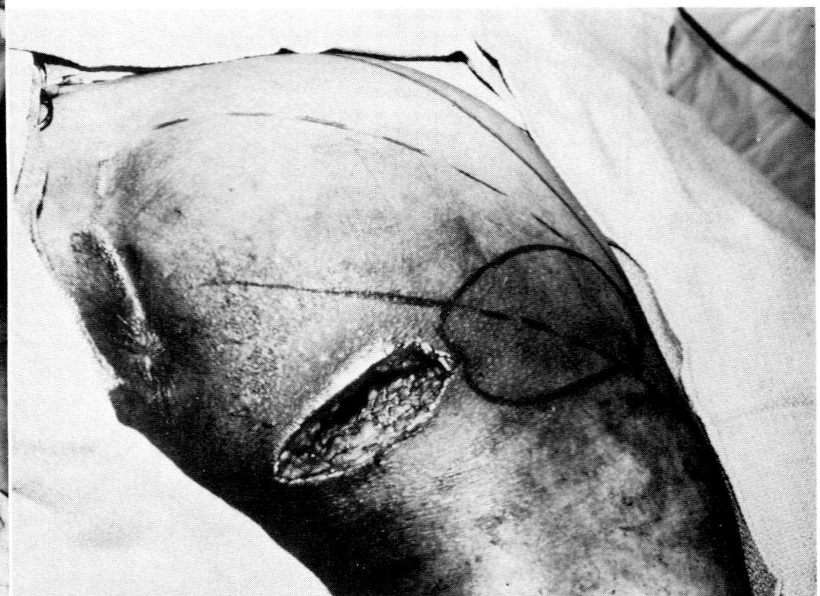

C

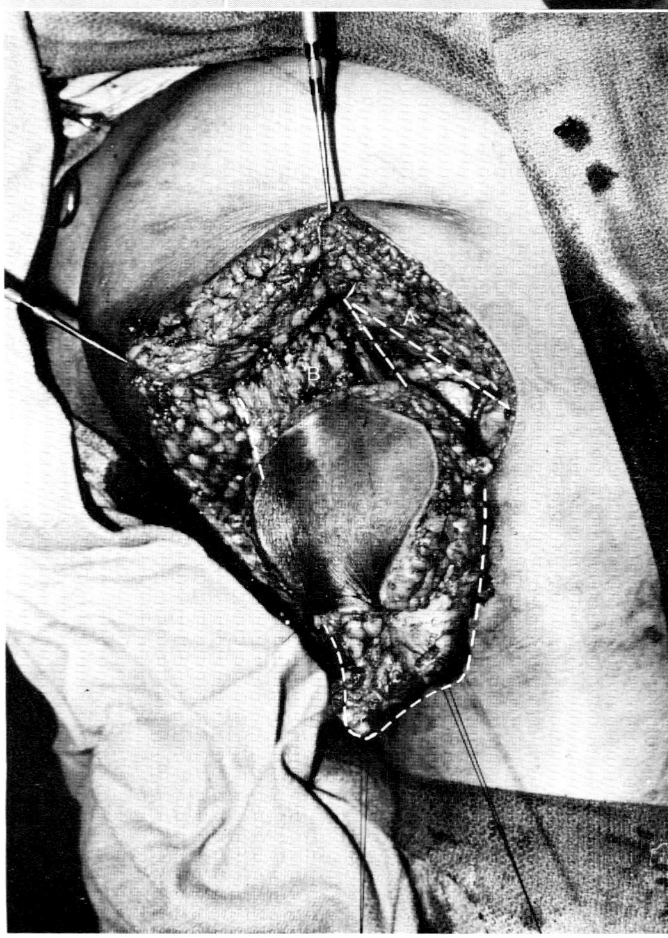

Fig. 25-11. Gluteus maximus musculocutaneous flap (inferior half of muscle).
A, Chronic right ischial pressure sore in paraplegic patient.
B, Design of skin island over right inferior half of gluteus maximus muscle.
C, Inferior arc of rotation of right gluteus maximus musculocutaneous flap over ischium. *B,* Inferior gluteus musculocutaneous flap; *A,* superior gluteus muscle. NOTE: Origin and insertion remain intact.

D

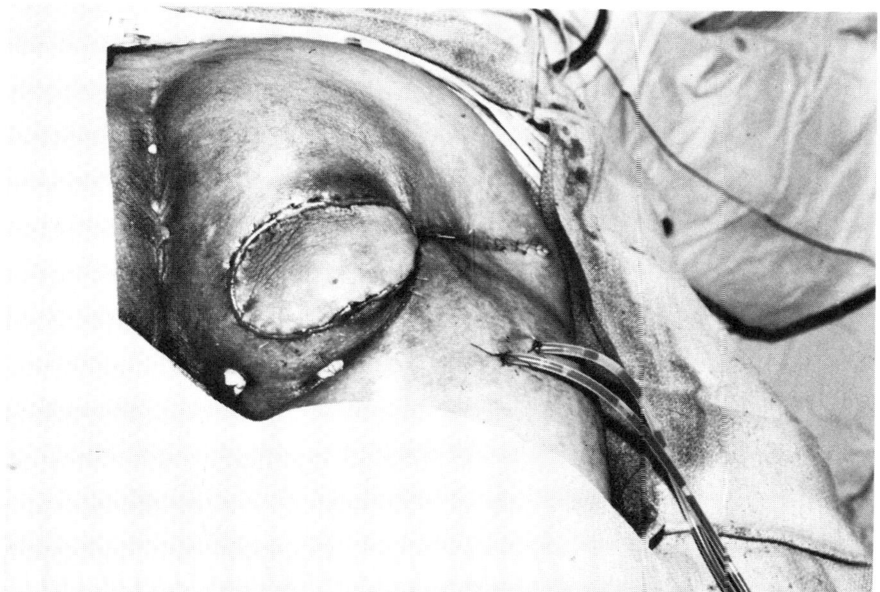

E

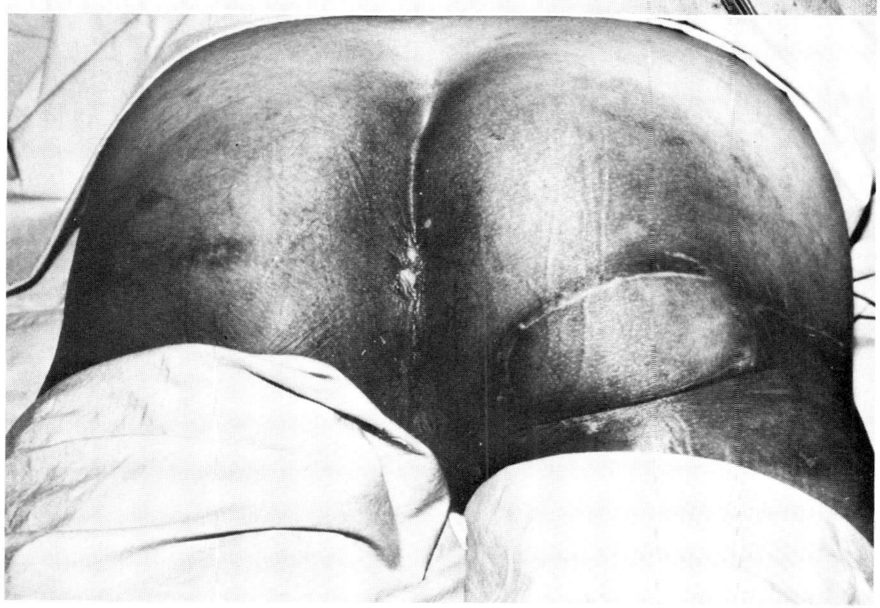

Fig. 25-11, cont'd

D, Immediate postoperative view demonstrates inset of skin island over ischial defect. NOTE: distal gluteus maximus muscle fibers beyond skin island now extend to posterior midline. Superior half of gluteus maximus is preserved.

E, Six-month postoperative view demonstrates stable coverage of right ischial defect provided by transposition of right inferior gluteus maximus musculocutaneous flap. NOTE: Donor defect is closed directly.

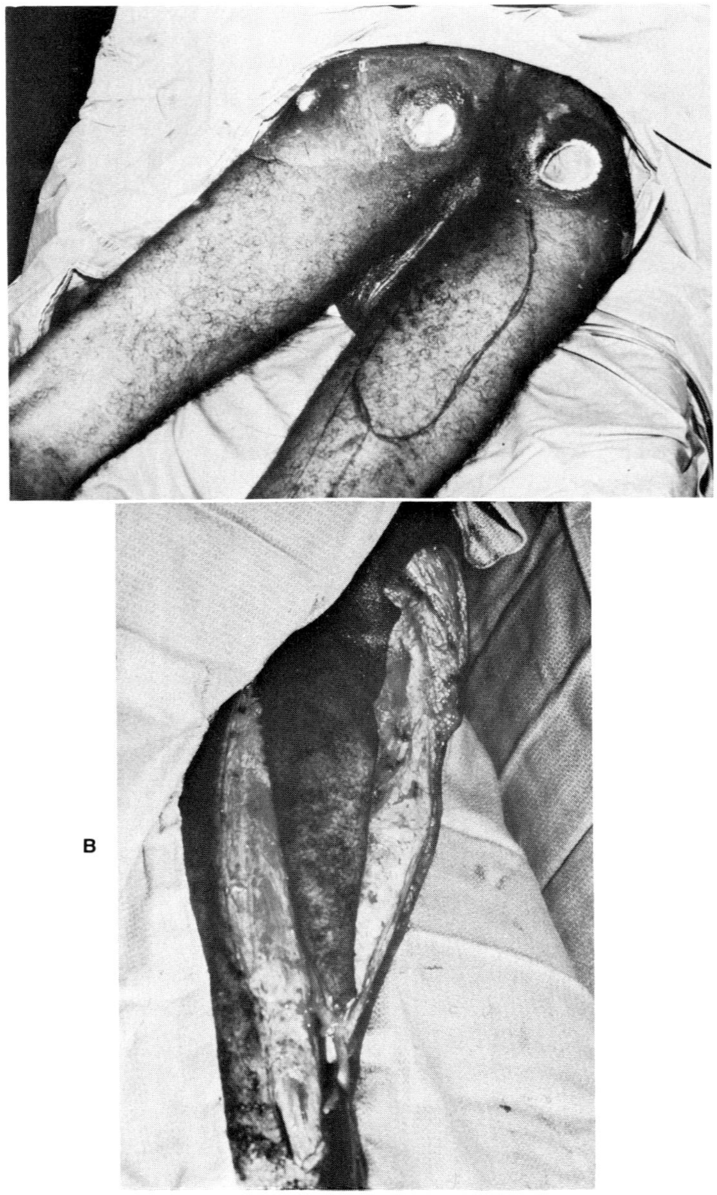

A

B

Fig. 25-12. Gracilis musculocutaneous flap.

A, Bilateral chronic ischial pressure sores. Design of right gracilis musculocutaneous flap.

B, Gracilis musculocutaneous flap is elevated from posterior approach for transposition to ischium. NOTE: Proximal skin continuity is maintained.

(From Mathes, S. J., and Alpert, B. S.: Advances in muscle and musculocutaneous flaps, Clin. Plast. Surg. **7**:15, 1980; reproduced with permission.)

C

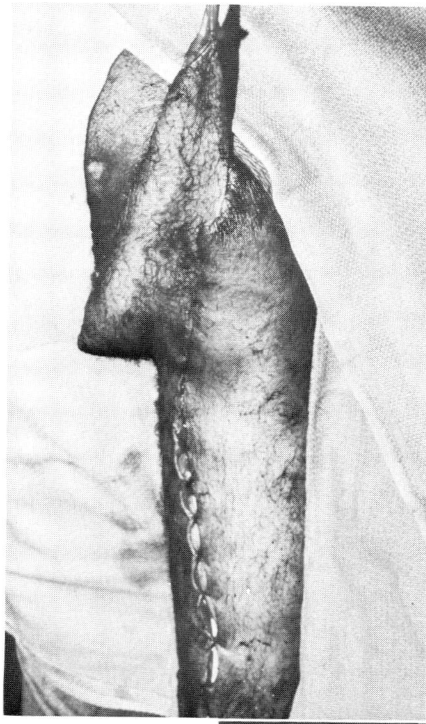

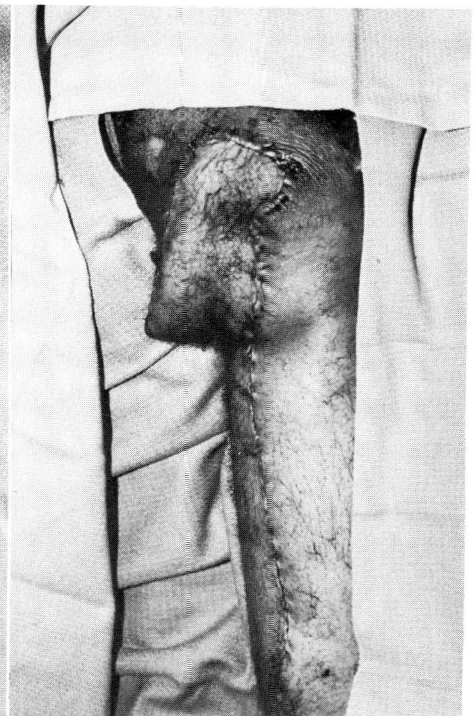

D

E

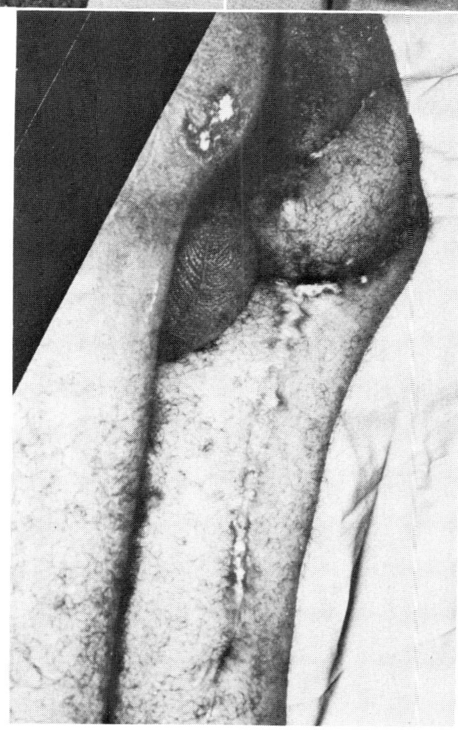

Fig. 25-12, cont'd

C, Demonstration of posterior superior arc of rotation of gluteus maximus musculocutaneous flap.

D, Inset of gracilis musculocutaneous flap over ischial pressure sore. NOTE: Distal skin is deepithelialized and folded with distal muscle into ischial defect.

E, Six-month postoperative view demonstrates stable wound coverage provided by gracilis musculocutaneous flap transposition. NOTE: Donor defect is closed directly.

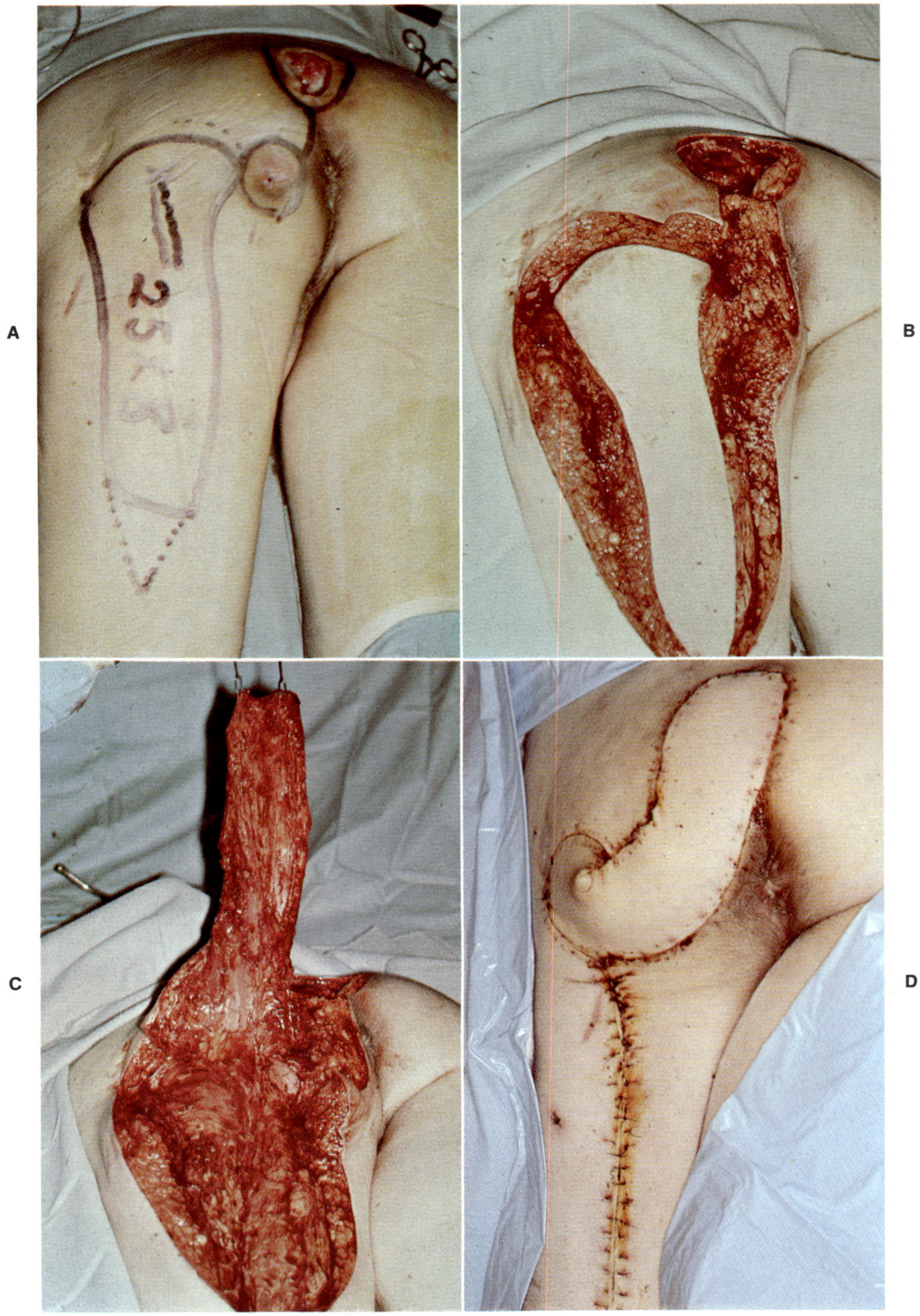

Fig. 25-13. Gluteal thigh flap.

A, Combined sacral and ischial pressure sore. Design of left gluteal thigh flap.

B, Elevation of island axial flap based on descending branch of inferior gluteal artery.

C, Posterior superior arc of rotation of gluteal thigh flap. NOTE: Posterior plane of elevation is deep to posterior thigh fascia, including posterior femoral cutaneous nerve and descending vascular pedicle from inferior gluteal artery.

D, Inset of gluteal thigh flap. NOTE: Direct closure of donor defect.

(From Hurwitz, D. J., Swartz, W. M., and Mathes, S. J.: The gluteal thigh flap: a reliable, sensate flap for closure of buttock and perineal wounds, Plast. Reconstr. Surg. **68**[4]:3, 1981.)

E

F

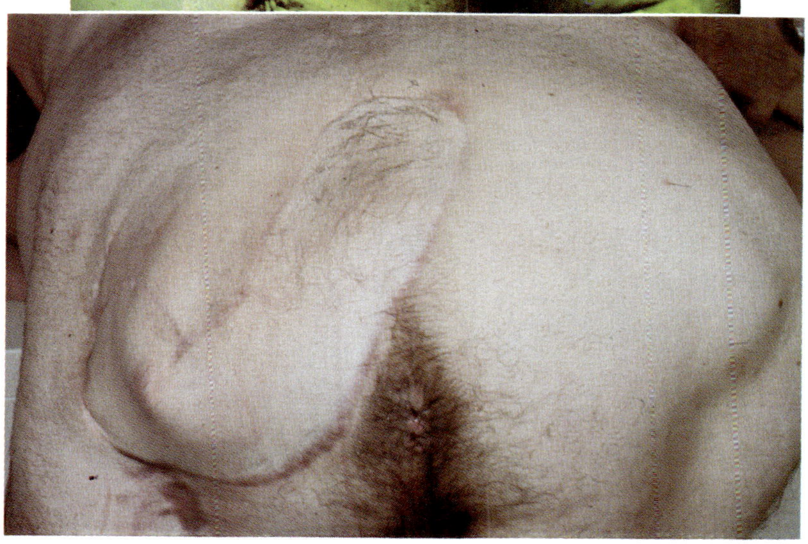

Fig. 25-13, cont'd
E, Fluorescein injection reveals adequate perfusion of entire flap.
F, Six-month postoperative view demonstrates stable coverage of both ischial and sacral pressure sores after use of gluteal thigh flap.

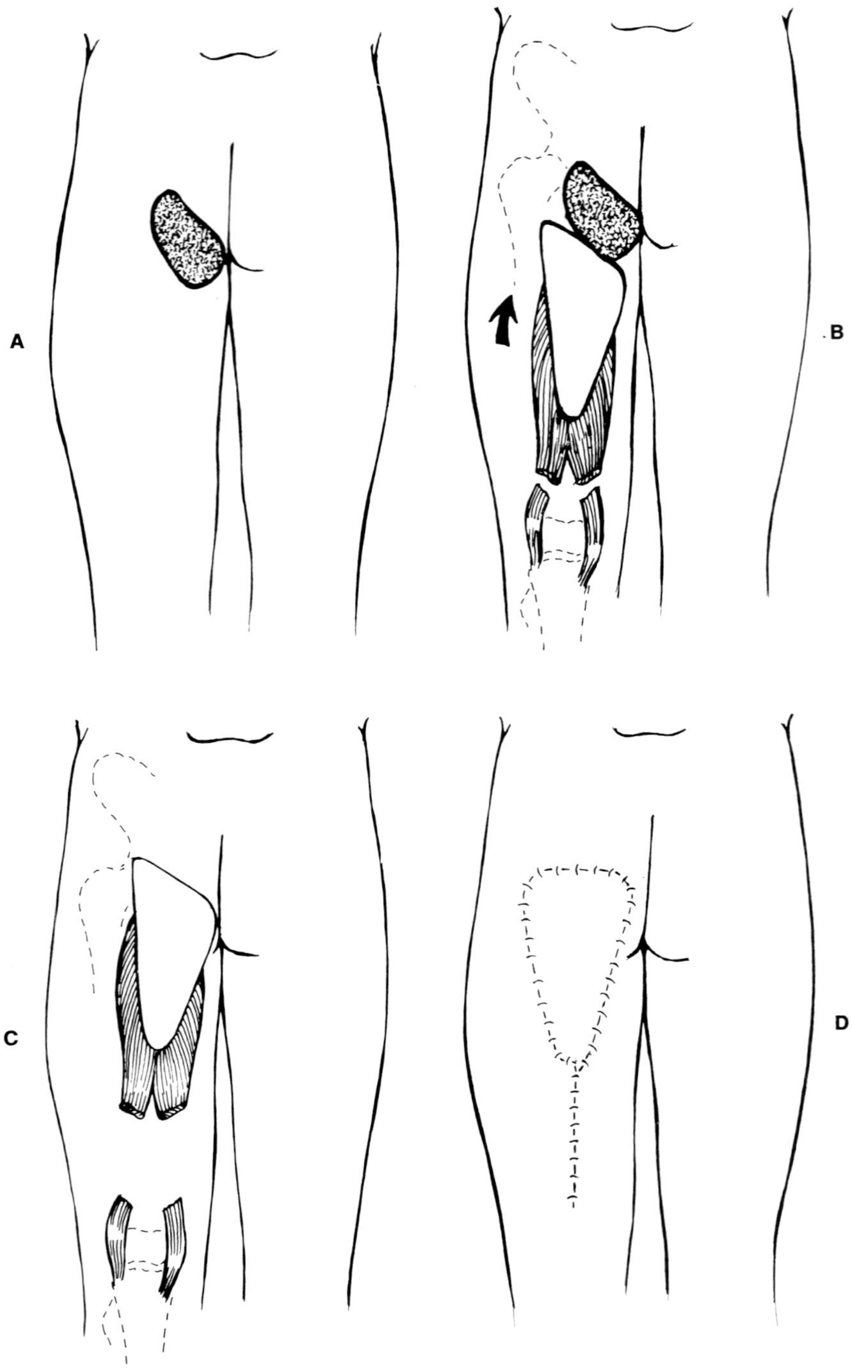

Fig. 25-14. Biceps femoris musculocutaneous flap (V-Y advancement).
A, Large ischial pressure sore.
B, Design of skin island on biceps femoris and semitendinosus muscles.
C, Advancement of musculocutaneous flap into defect after release of origin and insertion of muscle.
D, Coverage of ischium provided by musculocutaneous flap advancement. Donor defect closed directly.

GREATER TROCHANTER (Fig. 25-15)

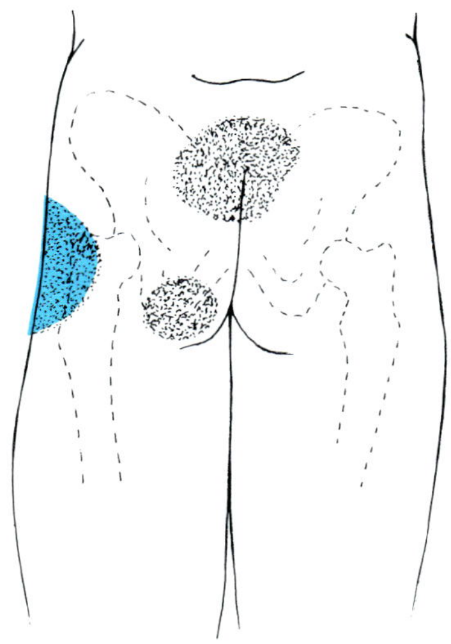

Fig. 25-15. Trochanteric defect.

Skin grafts

Skin grafts are not applicable in reconstruction of greater trochanteric pressure sores

Local flaps (Fig. 25-16)

1. TFL is flap of choice with potential for providing sensibility (Fig. 25-18)
2. Rectus femoris musculocutaneous flap is alternative flap
3. Vastus lateralis muscle flap with skin graft is the flap of choice for reconstruction after hip disarticulation for greater trochanteric pressure sore with septic hip joint (Fig. 25-18)

Distant flaps (Fig. 25-17)

Gluteal thigh flap is alternative flap (Fig. 25-19)

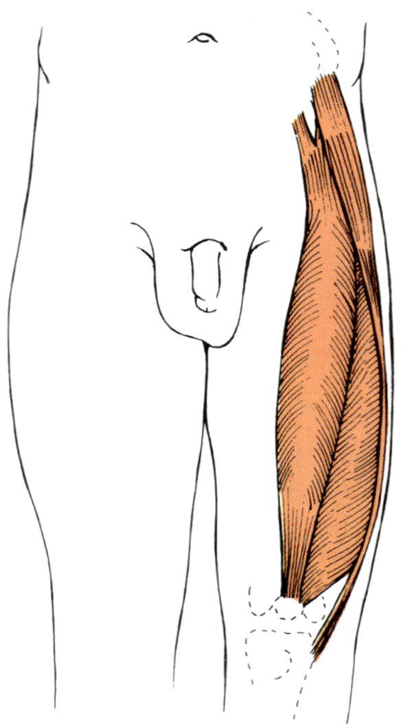

Fig. 25-16

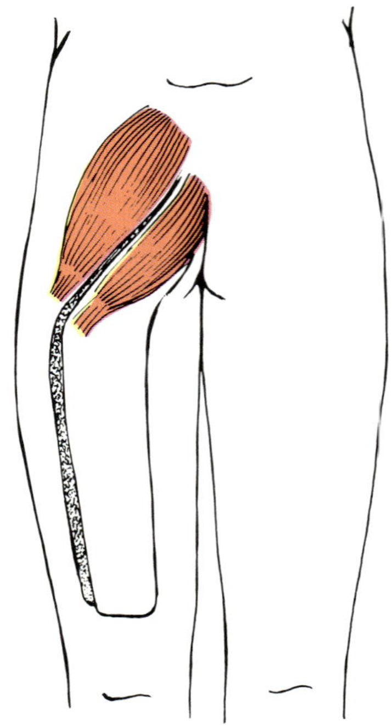

Fig. 25-17

A

B

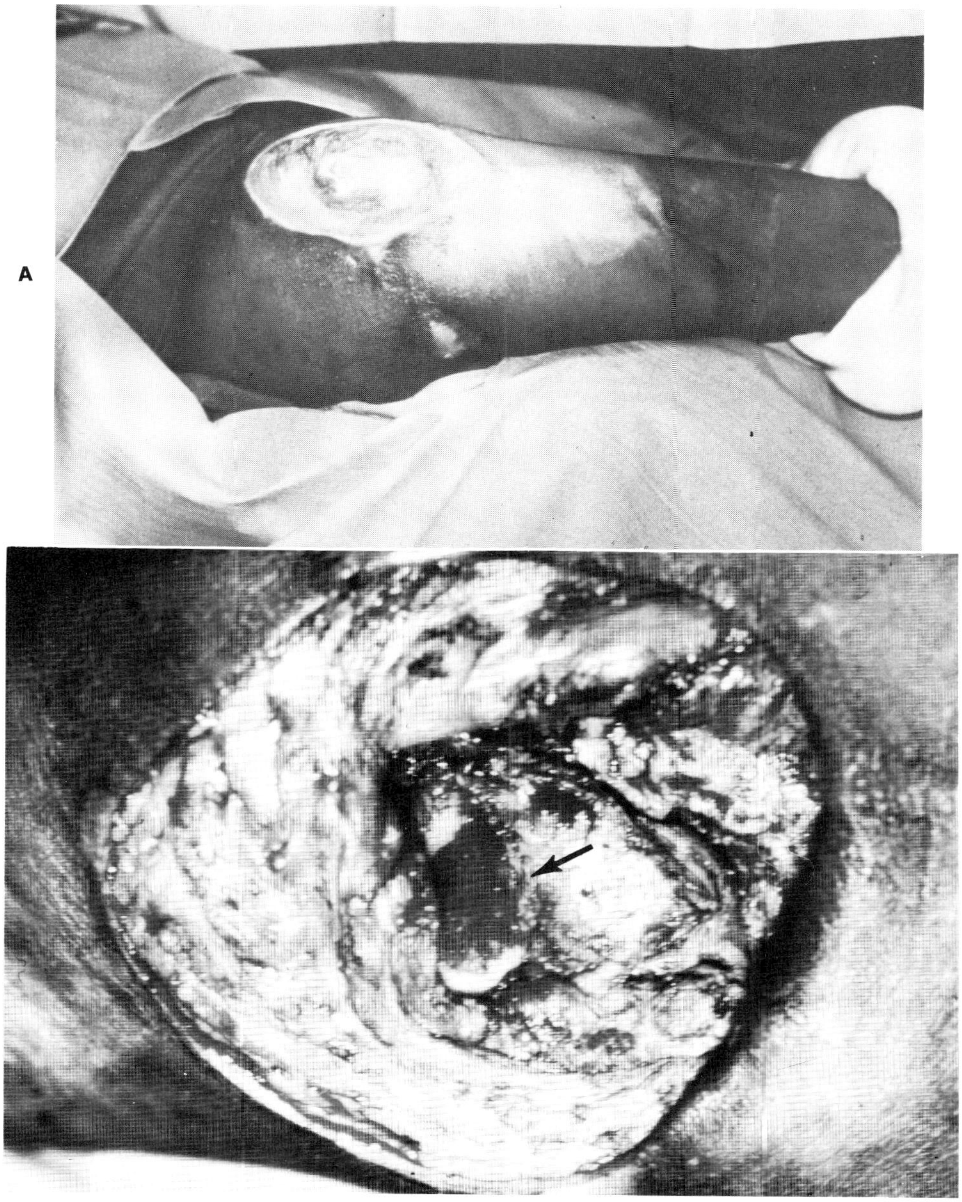

Fig. 25-18. Vastus lateralis muscle TFL and TFL musculocutaneous flaps.
A, Chronic pressure sore over greater trochanter with septic joint.
B, Femoral head ostectomy with excision of infected synovium from acetabulum. Arrow denotes ostectomy site below greater trochanter of femur.

(From Mathes, S. J., and Alpert, B. S.: Advances in muscle and musculocutaneous flaps, Clin. Plast. Surg. **7**:15, 1980; reproduced with permission.)

Continued.

C

D

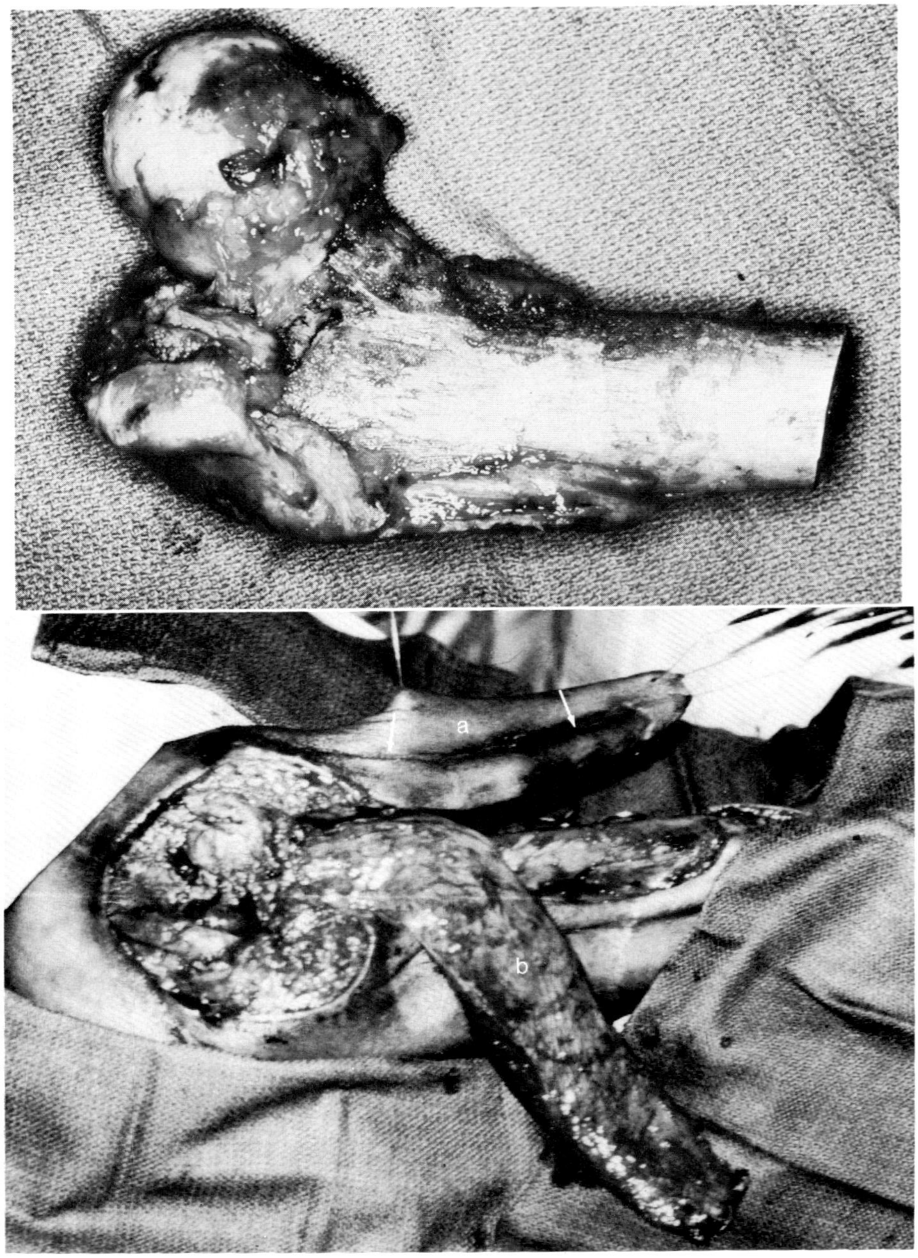

Fig. 25-18, cont'd
C, Femoral resection required for treatment of septic hip joint.
D, TFL musculocutaneous flap is elevated for cutaneous coverage of defect (*flap a*). Vastus lateralis muscle flap is elevated for coverage of exposed acetabulum (*flap b*). NOTE: Old horizontal scar in middle of cutaneous territory of TFL. Arrows denote prior incision.

E

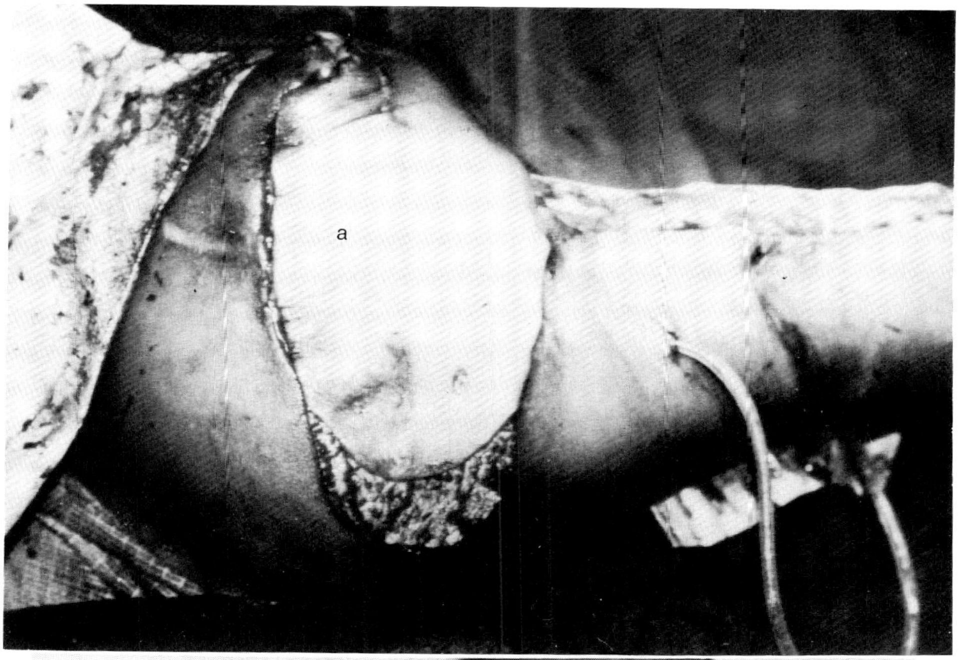

F

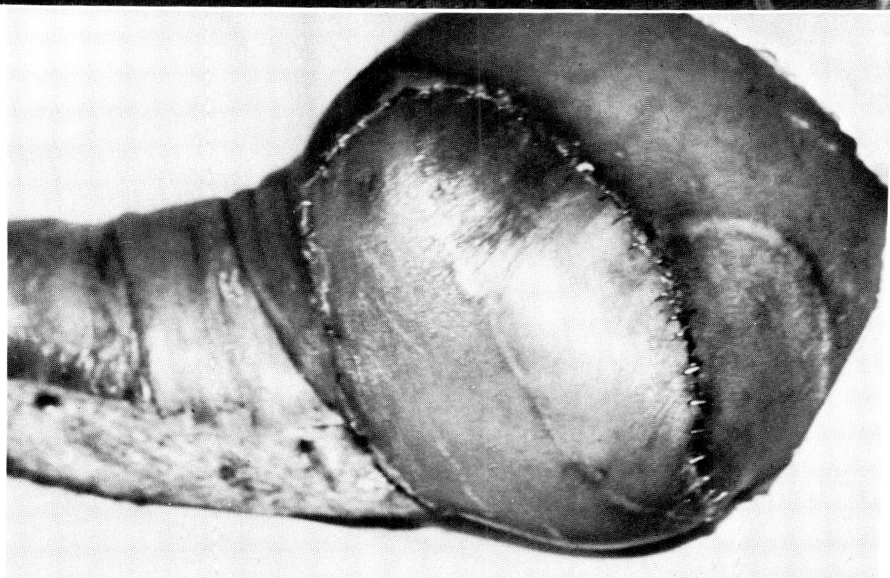

Fig. 25-18, cont'd

E, TFL musculocutaneous flap *(a)* used for skin coverage after hip disarticulation for trochanteric pressure sore. NOTE: Vastus lateralis muscle flap placed in exposed acetabulum.

F, Stable wound coverage provided by vastus lateralis muscle and TFL musculocutaneous flap for septic hip joint. Donor area closed directly.

A B C

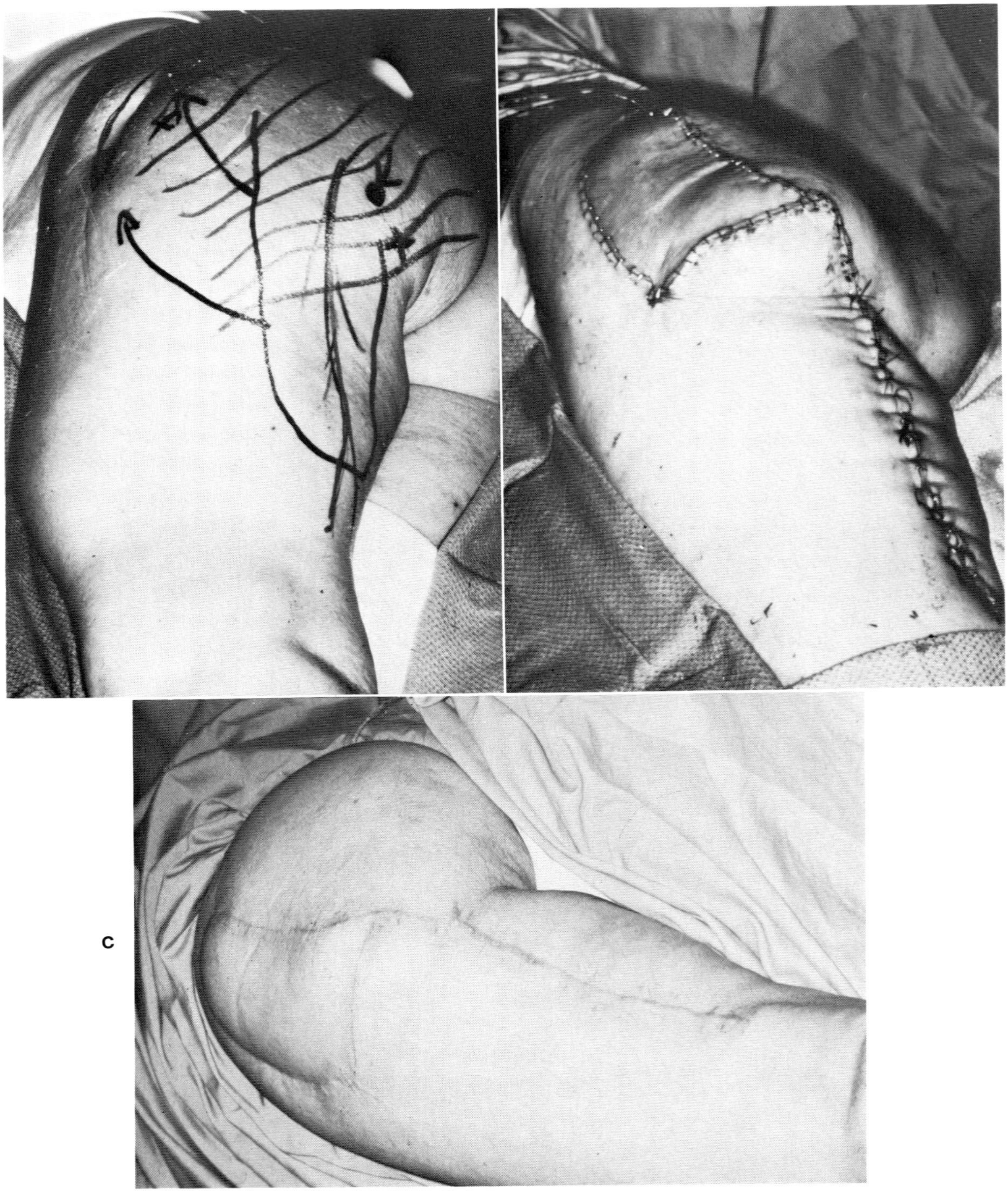

Fig. 25-19. Gluteal thigh flap.

A, Wound breakdown over greater trochanter after hip joint prosthesis insertion in ambulatory patient. Design of gluteal thigh flap.

B, Inset of gluteal thigh island flap over defect. Donor defect closed directly.

C, Three-month postoperative view demonstrates stable wound coverage provided by gluteal thigh flap transposition.

(From Hurwitz, D.J., Swartz, W.M., and Mathes, S.J.: The gluteal thigh flap: a reliable, sensate flap for closure of buttock and perineal wounds, Plast. Reconstr. Surg. **68**[4]:3, 1981.)

RECURRENT AND COMBINED GREATER TROCHANTERIC AND ISCHIAL PRESSURE SORES

Skin grafts

Skin grafts are not applicable in reconstruction of recurrent and combined pressure sores

Local flaps

1. Anterior thigh flap is flap of choice when hip disarticulation and amputation of the limb are necessary for resection of the pressure sore (Fig. 25-20)
2. TFL is flap of choice for coverage of combined greater trochanteric and ischial pressure sores (Fig. 25-21)

A

B

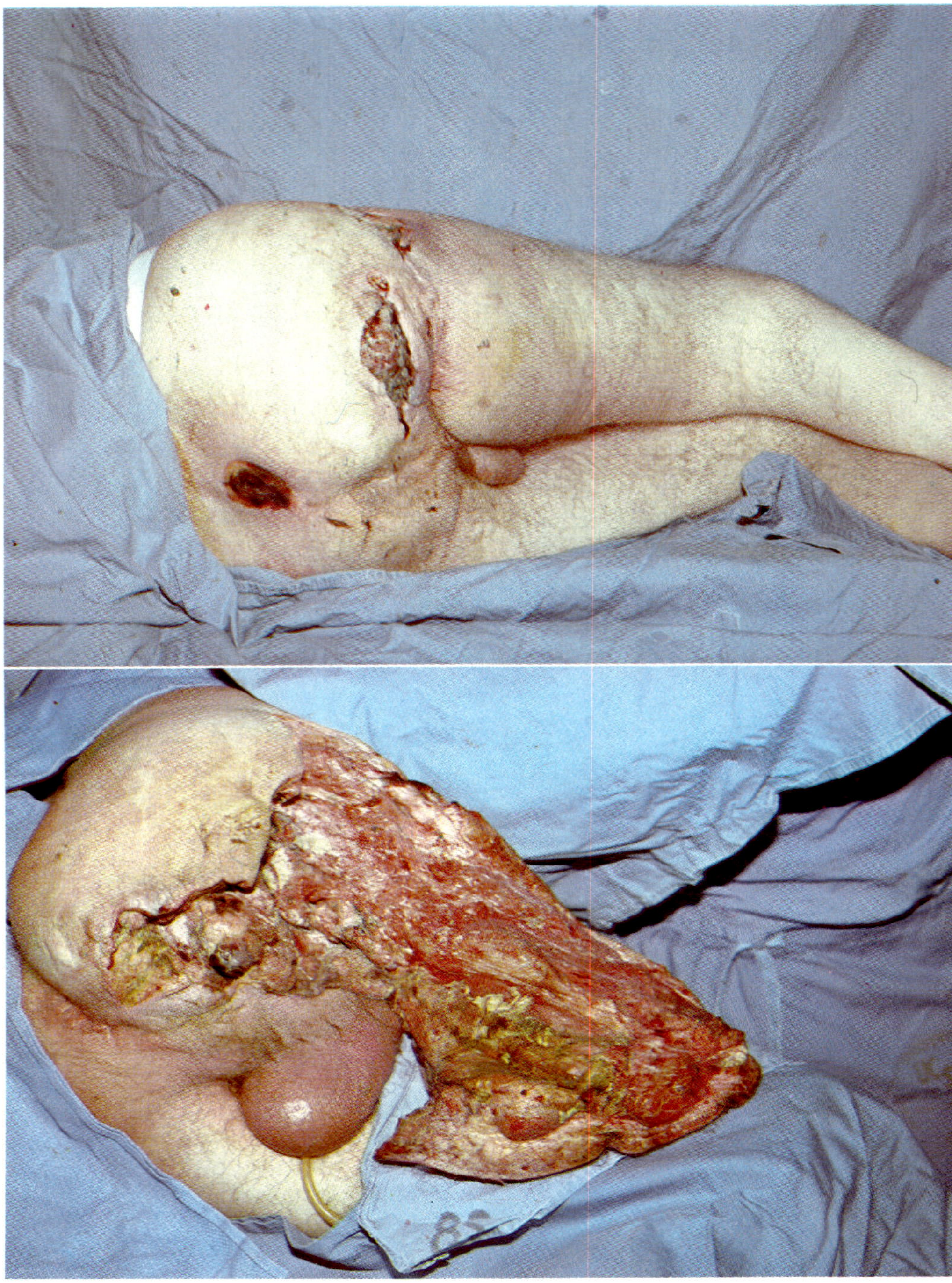

Fig. 25-20. Anterior thigh flap.

A, Synergistic gangrene of entire posterior thigh extending into retroperitoneum from infected ischial pressure sore in paraplegic patient.

B, Emergency wound débridement includes right lower extremity amputation with preservation of anterior thigh muscles and overlying skin.

C

D

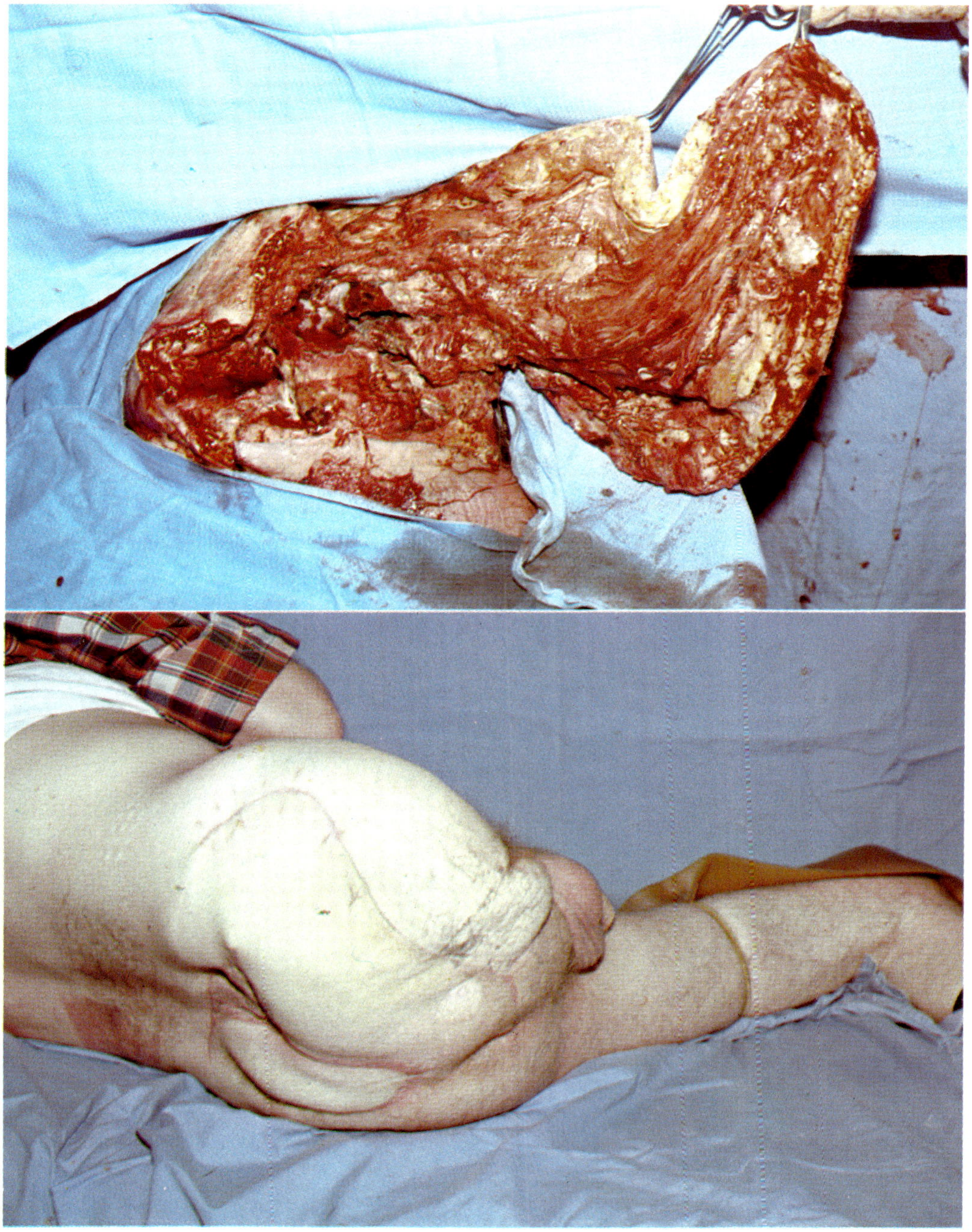

Fig. 25-20, cont'd

C, Management in this septic patient included emergency colostomy and multiple wound débridements, including excision of entire ischium. Anterior thigh flap is preserved. Arc of rotation of flap is demonstrated.

D, Six-month postoperative view. Stable wound coverage provided by anterior thigh flap. NOTE: Transverse back flap used for sacral pressure sore.

A

B

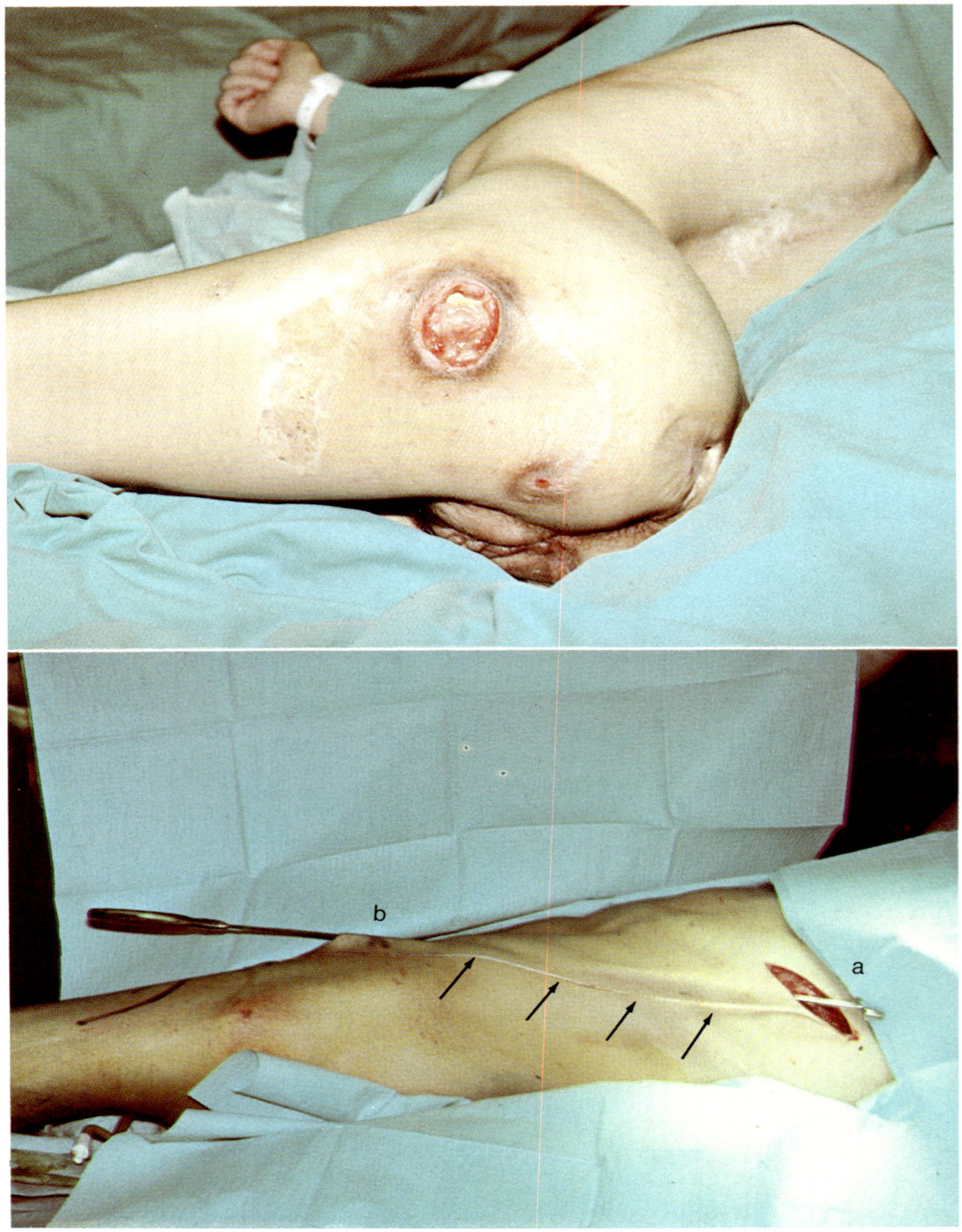

Fig. 25-21. For legend see opposite page.

C

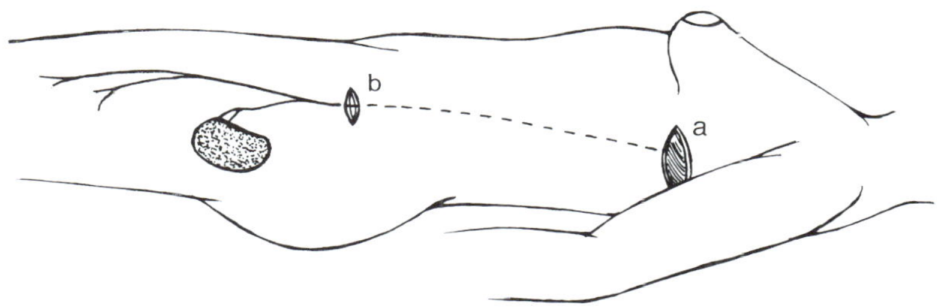

D

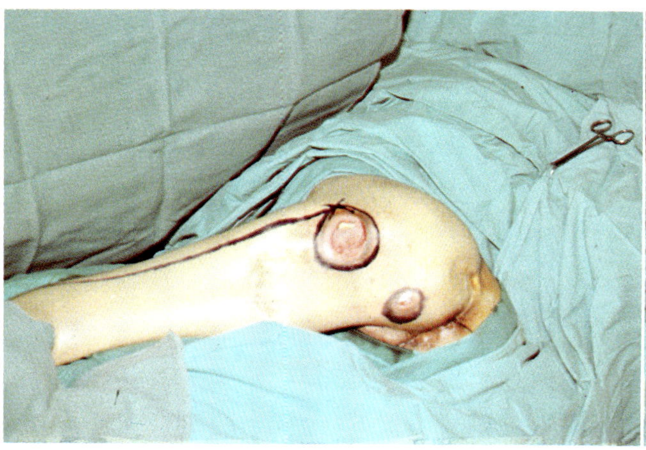

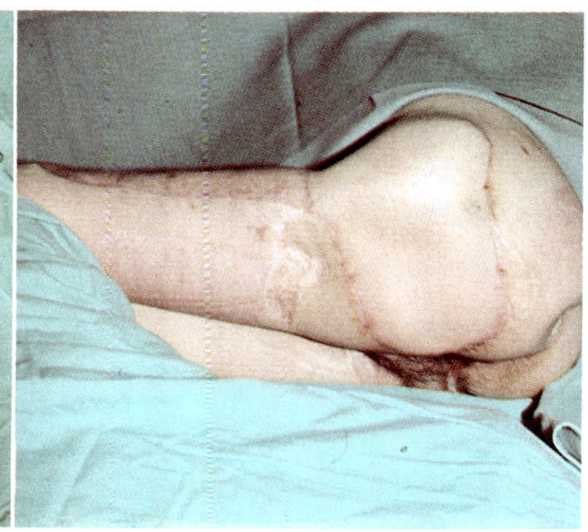

E

Fig. 25-21. Reconstruction of combined ischial and trochantric pressure sores with reinnervated TFL musculocutaneous flap.

A, Recurrent trochanteric and ischial pressure sores in 16-year-old paraplegic girl with T10 level. NOTE: Scars from previous flaps for pressure sore closure.

B and **C,** Lateral femoral cutaneous nerve of thigh is sutured to T8 intercostal nerve with sural nerve interposition graft. *a,* Site of nerve repair with T8 intercostal nerve; *b,* site of nerve repair with lateral femoral cutaneous nerve. Arrows denote sural nerve graft before placement in tunnel.

D, Several months later territory of TFL demonstrates return of protective sensation. The TFL musculocutaneous flap is outlined for closure of two defects.

E, TFL flap now provides stable wound coverage with protective sensation over ischial and trochanteric pressure sore areas.

PRESSURE SORES

26

RECONSTRUCTION

T. Roderick Hester ▪ William J. Schneider ▪ H. Louis Hill

The use of muscle flaps for coverage of pressure sores was suggested by Blackman and co-workers in 1949 with the description of the biceps femoris and gluteus maximus for coverage of ischial defects. A second report by Conway and Griffin in 1956 reported a large series using the biceps femoris for coverage of ischial sores. After Conway and Griffin's report little mention is made of the use of muscle flaps in closure of pressure sores until the concept was reintroduced by Ger in 1971, who later described the use of the gluteus maximus, rectus femoris, and sartorius in the closure of ischial, sacral, and trochanteric sores. Subsequently the muscle and musculocutaneous units have become the standard flap for coverage of pressure sores.

PRINCIPLES

Certain basic principles should be considered in the initial assessment of the patient with a pressure sore.

Preoperative management

SEPSIS: Initial control of sepsis may require use of preoperative incision, drainage, and débridement. Both topical and systemic antibiotics are started and continued through the operative period.

FLUID RESUSCITATION: Before the operation many patients will require fluid and electrolyte replacement and blood transfusion.

URINARY TRACT: Evaluation and drainage of the urinary tract are essential aspects of preoperative care.

NUTRITION: If present, the catabolic state must be reversed either through the alimentary tract or, if necessary, intravenously.

SPASMS: Control of flexor spasms may be advisable.

Operative management

PLANNING: The flap is planned and the operation performed so that donor defects and incisions are not placed over potential pressure areas. Alternative flaps for future use must be preserved.

EXCISION: Total excision of ulcer bursa and reduction of bony prominences are required.

HEMOSTASIS: Complete hemostasis is essential, because postoperative hematomas can lead to failure in an otherwise well-designed and executed procedure. Hematomas predispose to infection, suture line disruption, and bursa formation beneath the flap. Although adequate hemostasis is key, the use of a closed suction system is recommended to avoid postoperative seroma.

TENSION: Suture line tension must be avoided.

Postoperative management

POSITIONING: The patient is positioned so that there is no pressure on the flap or suture line for 2 to 3 weeks. Secondary pressure areas are protected during the postoperative period.

ANTIBIOTICS: Appropriate antibiotics selected on the basis of preoperative cultures are continued for at least 7 days after the operation.

DRAINS: The closed suction system is maintained for 5 days and then removed when serous drainage has stopped.

SUTURES: Sutures are left in place for a minimum of 21 days.

PATIENT EDUCATION: The patient and the family are counseled preoperatively and postoperatively in an attempt to reduce the likelihood of a future recurrence.

SACRAL SORES
General considerations

Sacral pressure sores most commonly occur in patients with cervical spine disease or trauma resulting in quadriplegia. These sores are less commonly seen in paraplegic patients. Sacral sores are occasionally a complication of prolonged illness, particularly in elderly patients. Comatose patients of any age group may develop sacral sores.

Some smaller sacral sores, especially in patients without permanent neurologic disease, can be closed with simple transposition flaps or skin grafts; however, many close spontaneously as the patient recovers from the underlying illness.

Sacral sores associated with neurologic defects are usually extensive with undermining of the surrounding tissues. In this instance surgical management with large flaps is required. The transverse back flap and variations of the gluteus maximus muscle or musculocutaneous flap are most useful and readily available.

Transverse back flap

The transverse back flap is an excellent choice in the primary care of many sacral sores. The flap is oriented transversely on the back with its base over the paraspinous muscle. The major blood supply is via lumbar and intercostal perforators (Fig. 26-1). This flap can be elevated with safety from the posterior axillary line. The flap is elevated medially toward the midline. The perforators on the side of flap elevation are divided (Fig. 26-1, *B*). However, after crossing the midline great care must be taken to avoid damage to the lumbar perforators on the opposite side. These vessels are the axial supply to this flap. The flap can be elevated 2:1 (length-width ratio) without a delay.

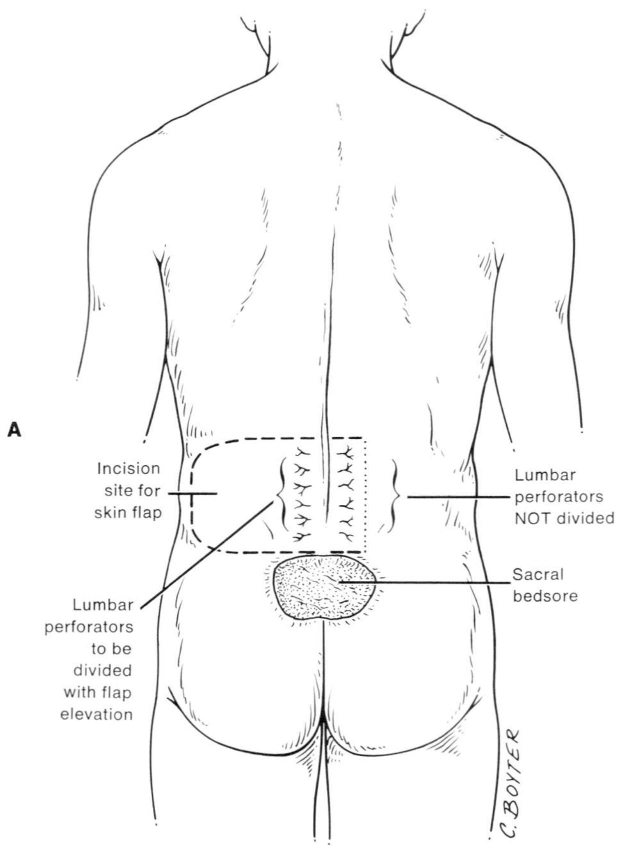

Fig. 26-1
A to **C,** Anatomy and technique of elevation of transverse back flap.

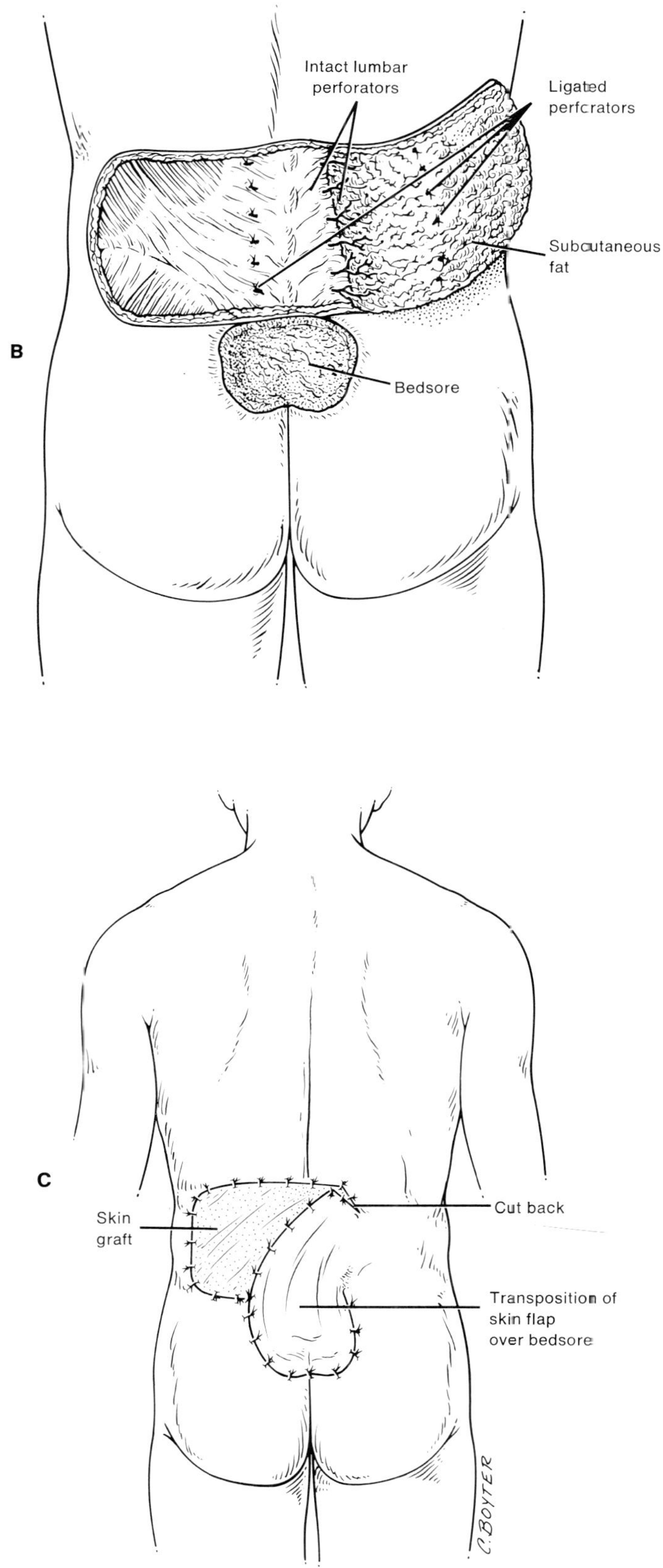

Fig. 26-1, cont'd. For legend see opposite page.

A

B

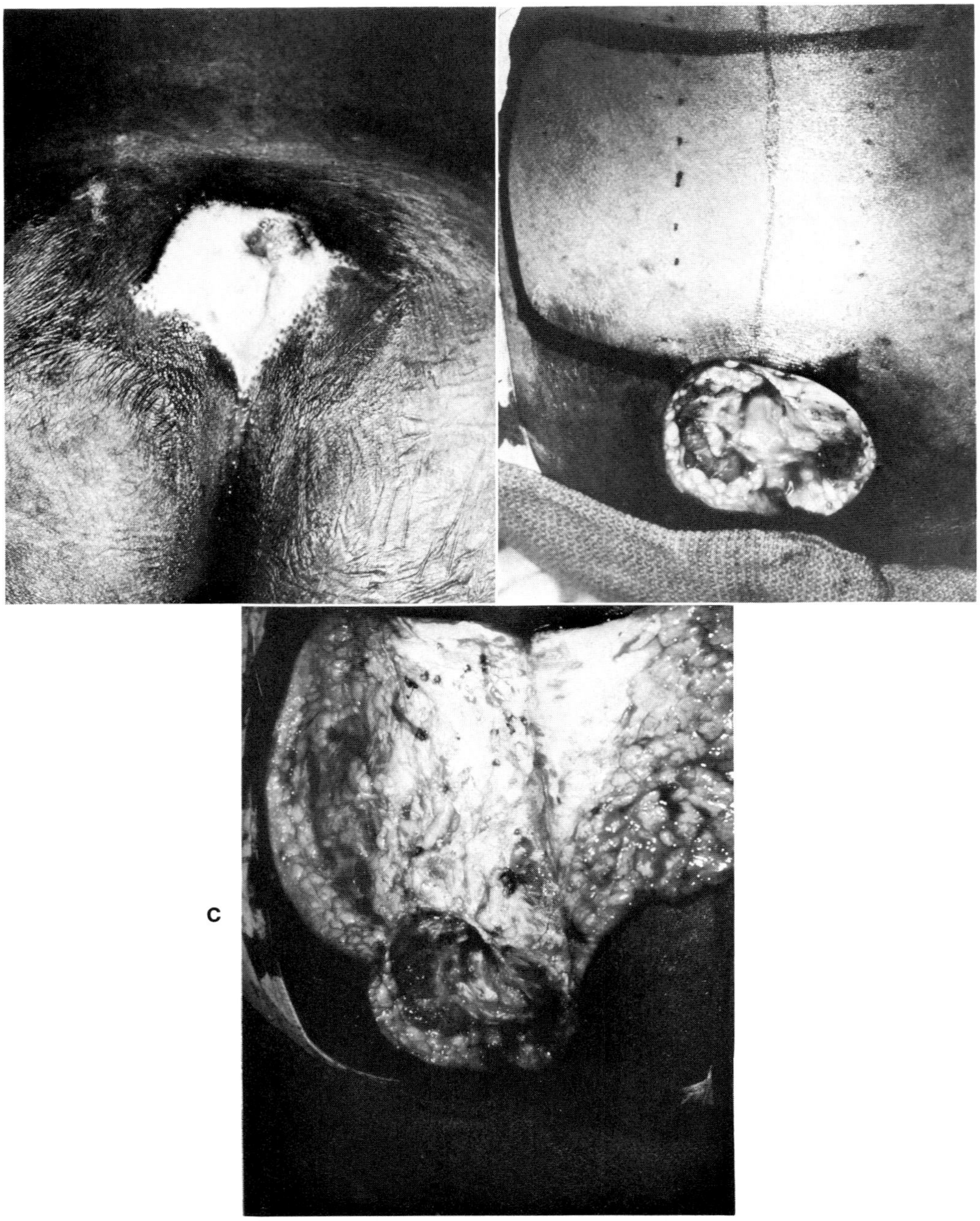

C

Fig. 26-2. Transverse back flap for sacral defect.
A, Sacral pressure sore.
B, Sore excised and flap outlined.
C, Flap elevated.

This flap has been used extensively with no flap loss (Fig. 26-2). It is easily elevated and transposed. The secondary defect requires skin grafting; however, the grafted site is not in a potential pressure area. It is especially useful in the nonparalyzed patient in that muscle function is not affected.

D

E

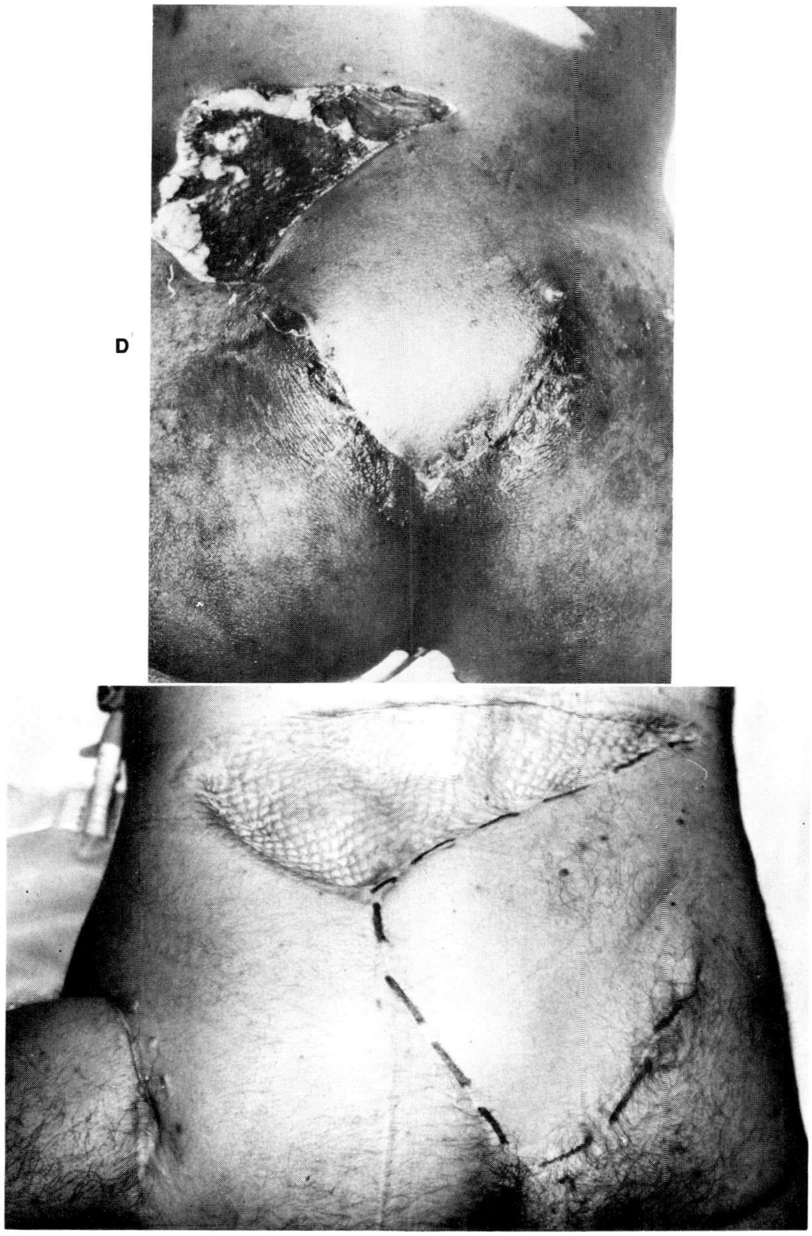

Fig. 26-2, cont'd
D, Flap transposed and donor defect skin grafted.
E, One-year postoperative view of similar patient. NOTE: TFL flap coverage of left greater trochanter.

Gluteus maximus flap

The gluteus maximus muscle and musculocutaneous unit is most useful as a flap for coverage of sacral sores. The anatomy and arc of rotation of this muscle have been described (Chapter 17). The primary blood supply (type III) enters the medial aspect of the muscle, via the superior and inferior gluteal arteries (Fig. 26-3). This dual blood supply can be used to design flaps based on either the upper or lower halves of the muscle.

Transposition of muscle. Depending on the size of the defect, either the entire muscle or the upper or lower segment of the muscle can be used as a transposition turnover muscle flap and skin grafted (Fig. 26-4).

The gluteus maximus can be rapidly and safely elevated through an incision over the muscle. The insertion is divided and the flap dissected from the lateral border to the medial border and turned over the sacrum. The muscle is then grafted and the donor defect closed directly. There have been no problems with skin graft durability when grafted over this thick muscle belly. In nonparalyzed patients, if this flap is to be used, an attempt should be made to preserve either the upper or lower segment of the muscle, thereby preserving function.

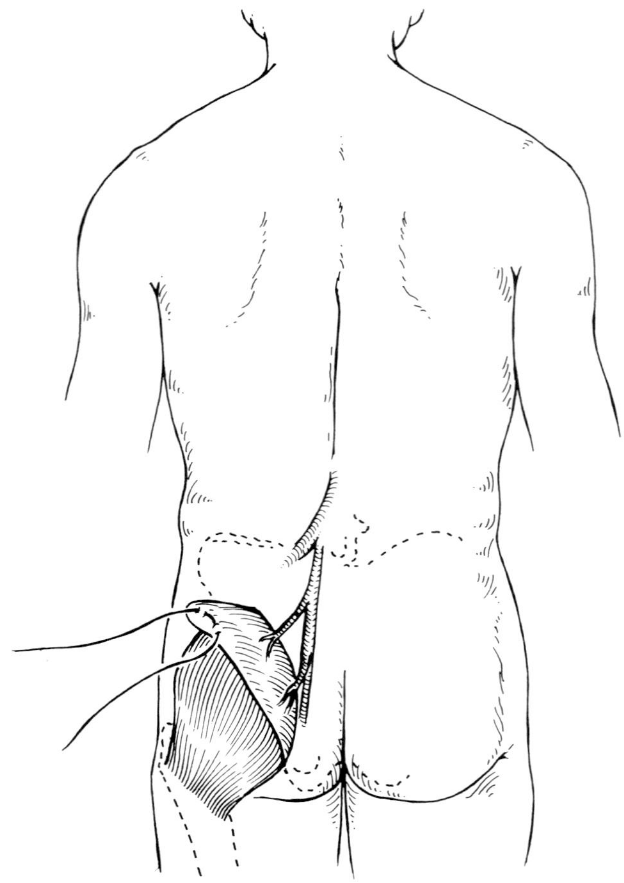

Fig. 26-3. Blood supply of gluteus maximus flap.

A

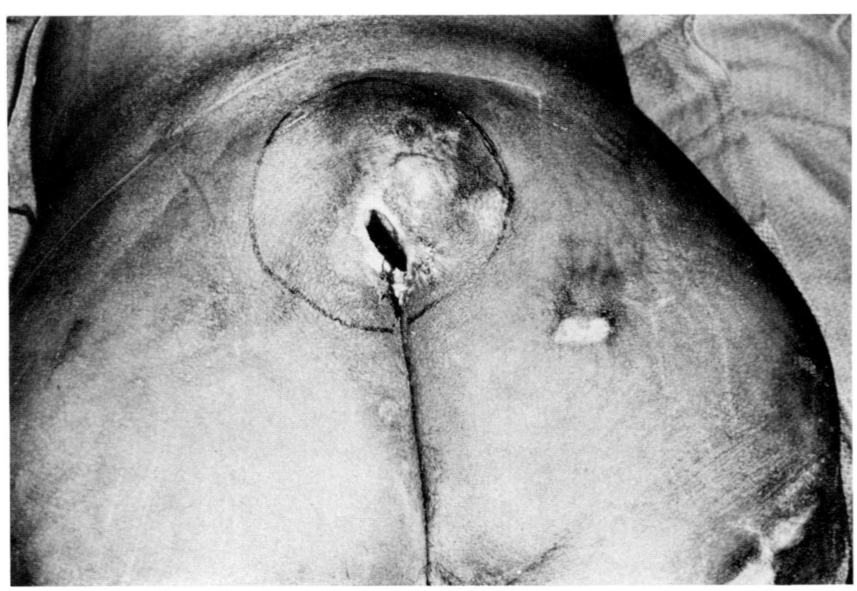

Fig. 26-4. Gluteus maximus turnover flap with skin graft.
A, Sacral defect.

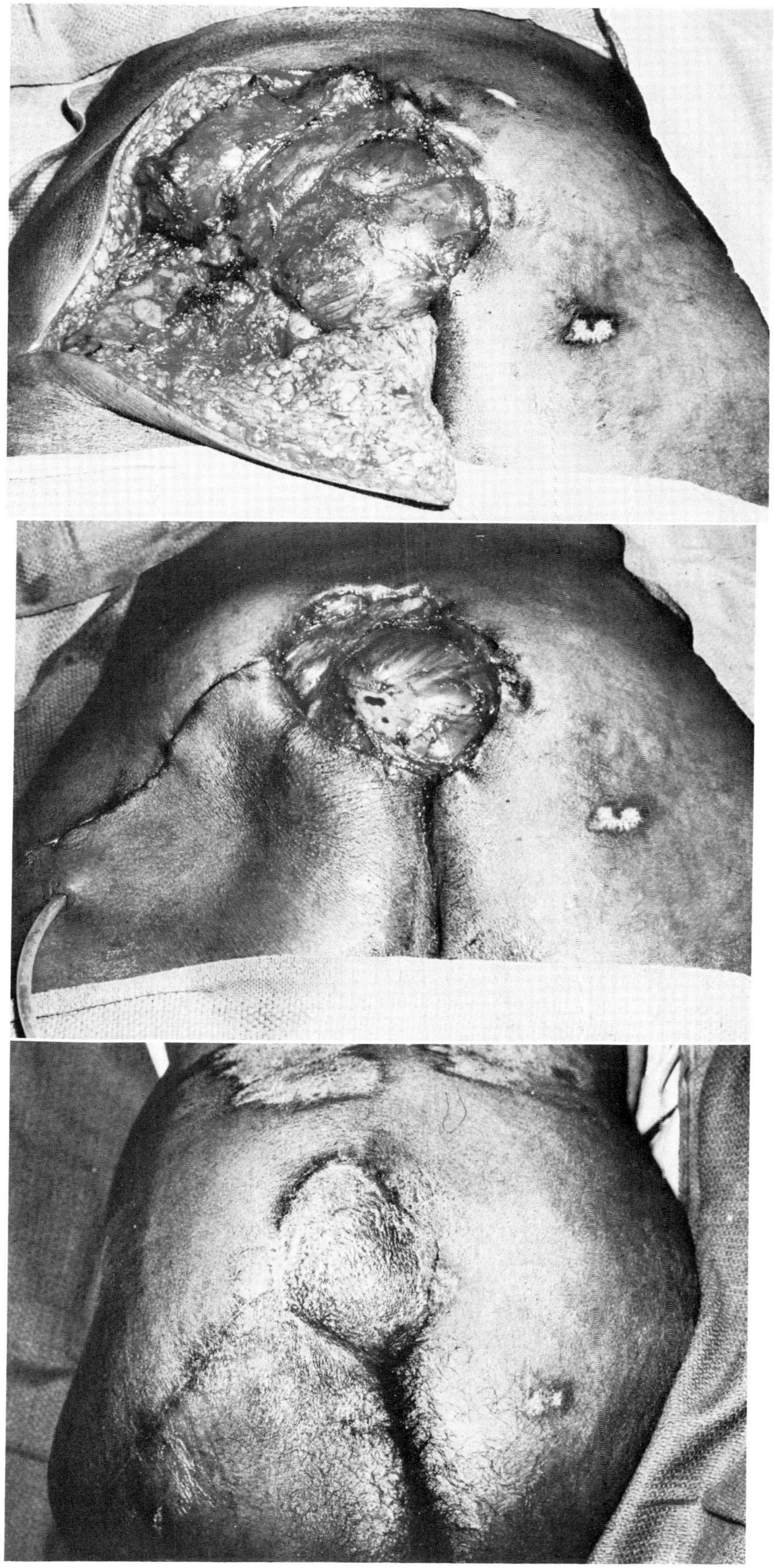

B

C

D

Fig. 26-4, cont'd

B, Defect excised, and upper half of gluteus maximus turned into defect.
C, Muscle in defect, and donor site closed directly.
D, One-year postoperative view of skin grafted muscle.

Transposition of musculocutaneous flap. The gluteus maximus can also be used as musculocutaneous unit. The entire muscle and skin unit is elevated by dividing the trochanteric attachments. The entire unit is transposed medially. Occasionally a skin graft is required for closure of the secondary defect (Fig. 26-5).

Because experience has been gained with the use of the gluteus maximus muscle, the use of a distal island musculocutaneous unit has evolved as a more appropriate method of transferring muscle and skin into sacral defects. The donor defect can be closed primarily. The musculocutaneous unit can be transposed or advanced in a V-Y manner (Fig. 26-6). Bilateral flaps may be required for larger defects (Fig. 26-7).

Flap design and elevation. This variation of the gluteus maximus musculocutaneous flap is medially based and can be designed on either the superior or inferior gluteal artery. The flap is designed in the following manner. The muscle is transected by lines drawn from the midsacrum to the greater trochanter. An ellipse of skin large enough to cover the defect is drawn over the distal muscle. The ellipse is incised, allowing exposure of greater trochanter, and the upper or lower one half of the insertion is divided.

A

B

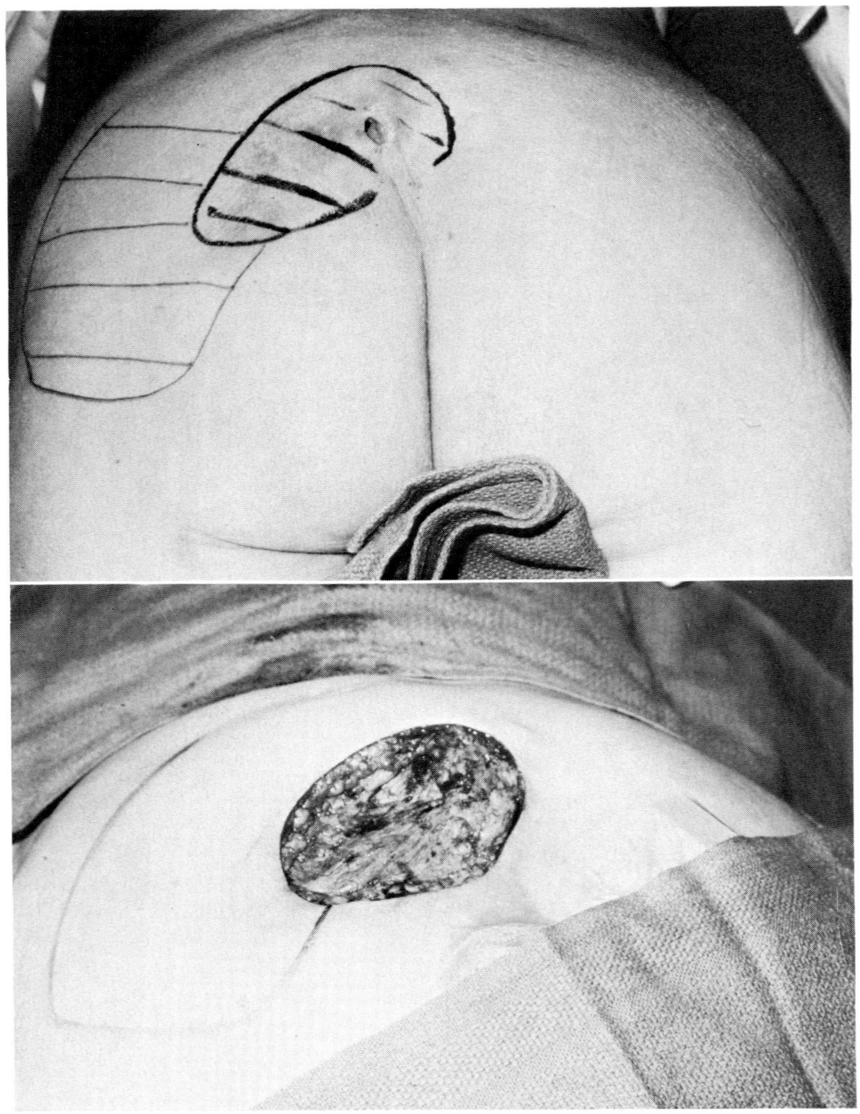

Fig. 26-5. Gluteus maximus musculocutaneous flap transposition for sacral defect.
A, Sacral defect. Area of underlying skin and flap are outlined.
B, Defect excised, and superior gluteal flap outlined.

C

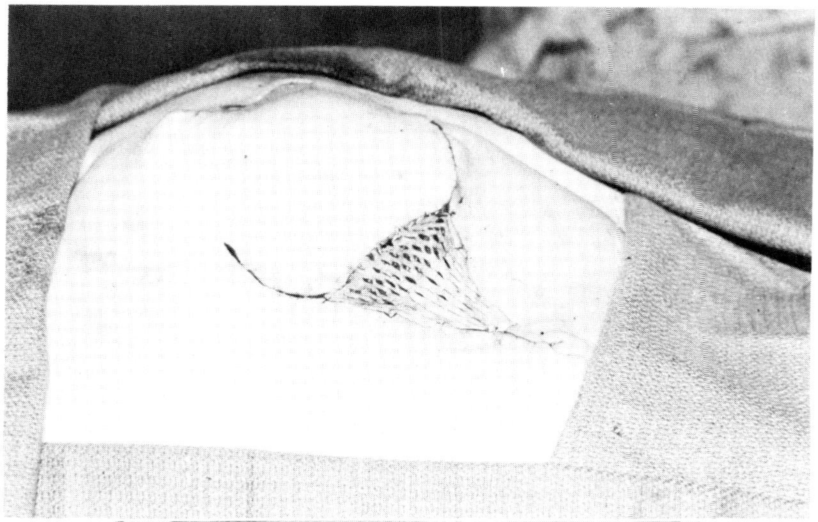

D

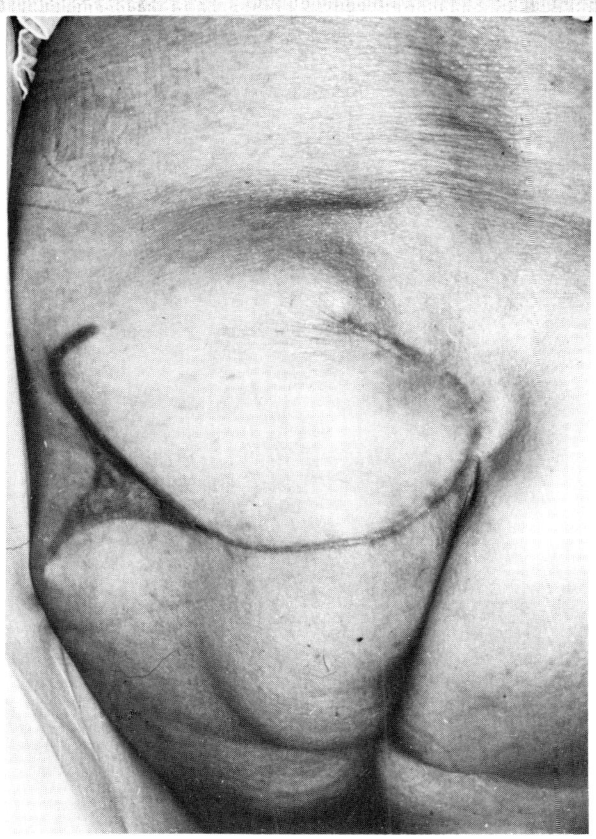

Fig. 26-5, cont'd
C, Flap transposed, and part of donor defect grafted.
D, One-year postoperative view of patient. Skin-grafted donor defect in non-weight-bearing area.

A B

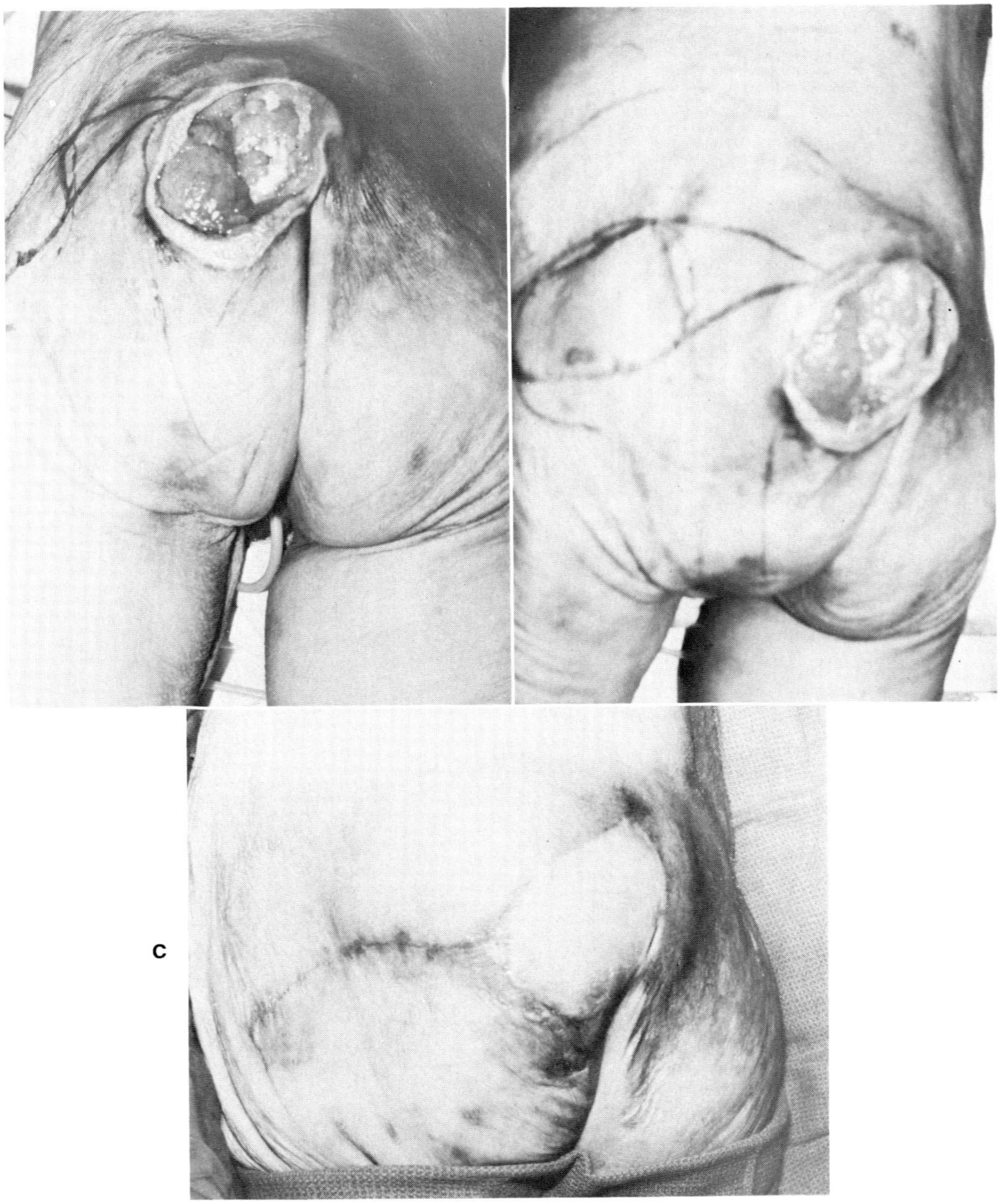

C

Fig. 26-6. Gluteus maximus island musculocutaneous flap for sacral sore.
A, Sacral defect.
B, Skin island based on upper half of gluteus maximus muscle is outlined.
C, One-year postoperative view.

The skin island and muscle are then dissected from the lateral border to the medial border toward the origin. The pedicle is identified and preserved. The origin is then divided as necessary to increase arc rotation.

Several technical points that are important in the use of this flap are the following:

1. The skin island should be over the distal muscle to allow rotation on the relatively short muscular pedicle. Division of the origin of the muscle will facilitate advancement.
2. There is usually enough buttock skin to allow direct closure of the donor defect. The gluteus maximus island can also be used in a V-Y advancement technique either unilaterally or bilaterally. This technique requires division of the muscle origin.

A

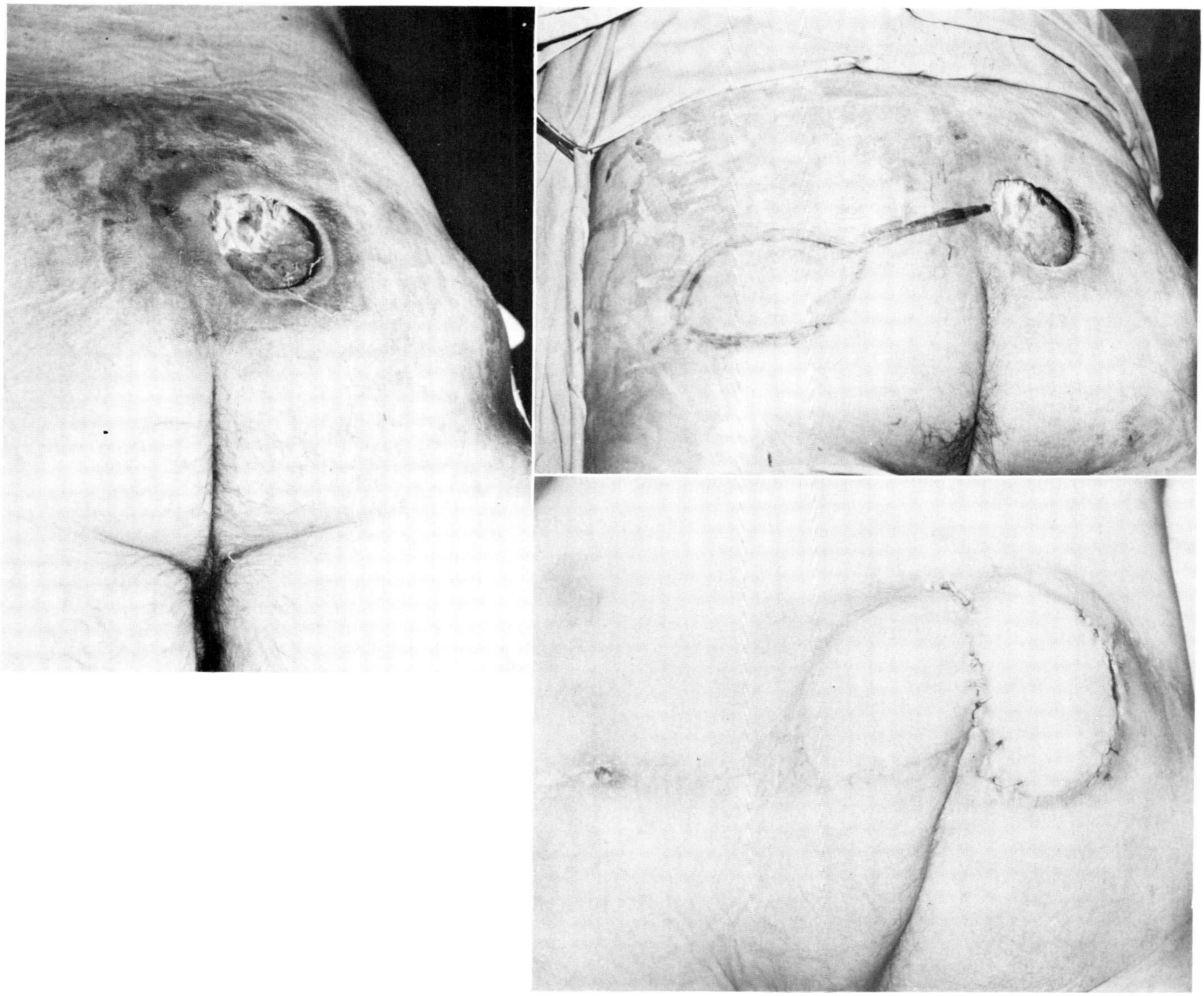

B

C

Fig. 26-7. Bilateral gluteus maximus island musculocutaneous flaps for extensive sacral defect.
A, Defect before excision.
B, Defect and one flap outlined.
C, Bilateral flaps over sacral defect. Donor areas closed directly.

Intercostal flaps

An alternative flap for sacral pressure sore coverage is the intercostal flap, which is designed as a neurosensory flap. Dibbell first described the use of an innervated intercostal island flap for coverage of sacral sores. The primary advantage of this flap is its usefulness in transferring innervated tissue to the sacrum in patients with lower spinal cord lesions. As originally described by Dibbell, the procedure was performed in multiple stages. Using a modification of the intercostal flap described by Daniel, a larger flap is transferred (Fig. 26-8). With this technique, the wound is closed with a neurosensory flap.

Flap design and elevation (Stage 1). With the patient awake, an intercostal block with 1% Xylocaine is performed in the midaxillary line of the selected intercostal muscle group. The area of anesthesia is then mapped out by response to pinprick. Alternatively the intercostal nerves above and below the selected unit can be blocked in the area of skin innervated marked out on the anterior abdominal wall. Within this area, the appropriate size skin island is selected and marked out. Anesthesia is induced, and the incision is carried through the skin and subcutaneous tissue to the rectus abdominis muscle, leaving the lateral border of the skin island intact. The rectus abdominis muscle is divided transversely down to the posterior rectus sheath cephalad and caudad, defining the skin island. In doing this the blood supply to this segment from the epigastric vessels is divided, basing the entire blood supply on the intercostals. A midline incision is made, connecting the two transverse incisions, and the rectus abdominis muscle is dissected free of its insertion at the linea alba. Using sharp dissection, the rectus abdominis is elevated from its posterior sheath. Special care must be used in dissecting laterally where the transversus abdominis muscle attenuates to the transversalis fascia of the posterior rectus sheath, for it is here that the intercostal bundle emerges from between the internal oblique muscle and the transversus abdominis muscle and runs on the deep surface and in the substance of the rectus abdominis muscle. The sensory nerve arborizes to supply the overlying skin of that dermatome via the anterior cutaneous bundles. The blood vessels communicate with the inferior and superior epigastric vessels. A piece of Silastic sheeting is placed between the rectus abdominis muscle and the posterior sheath, and the skin is resutured.

A

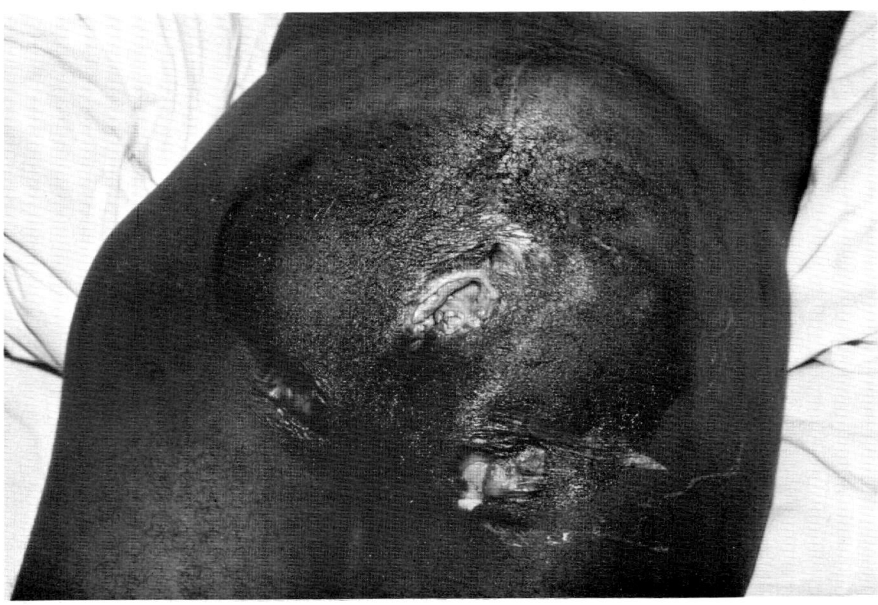

Fig. 26-8. Neurosensory intercostal flap for coverage of sacral defect. **A,** Extensive sacral defect.

(Courtesy John Coleman, Atlanta)

B

C

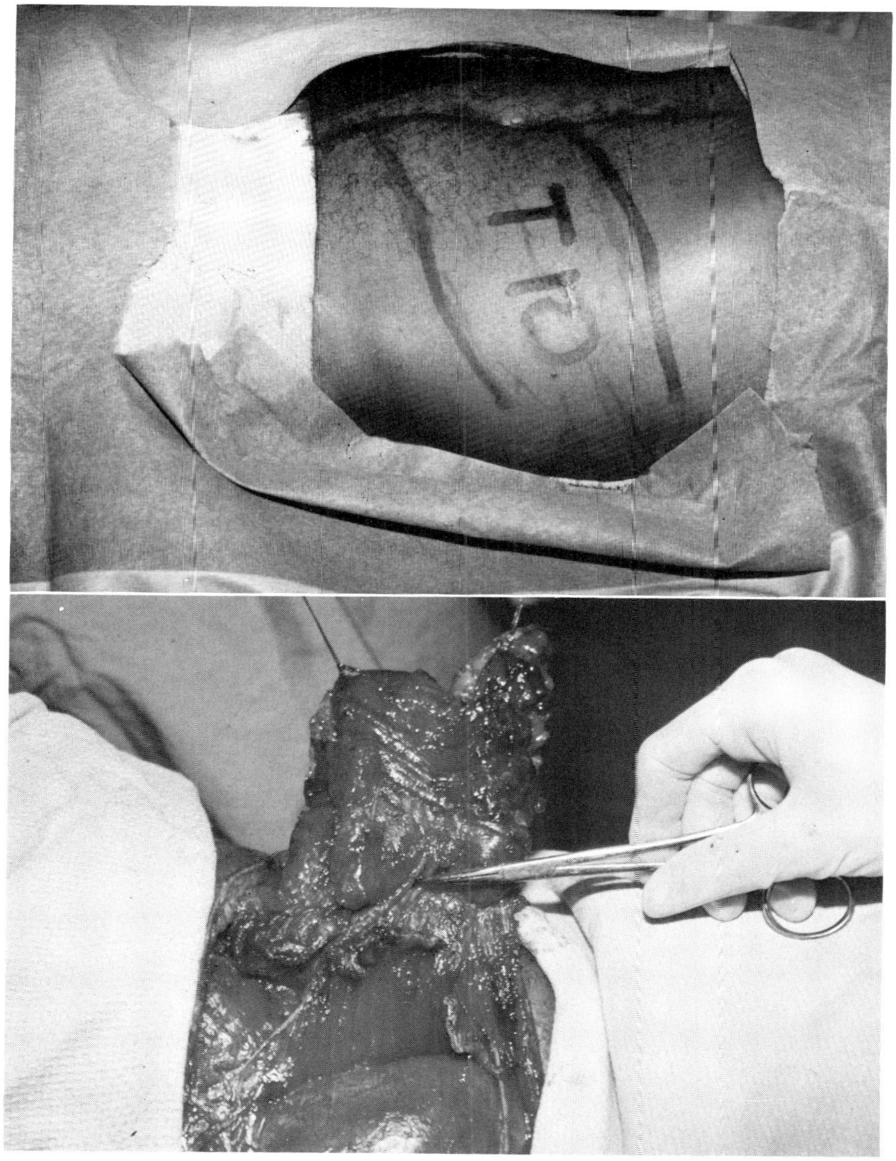

Fig. 26-8, cont'd

B, Neurosensory flap outlined and delayed once on anterior abdominal wall.

C, Stage 1—flap delay. Elevation of rectus abdominis muscle and overlying skin. NOTE: Clamp points to sensory branch of intercostal nerve coursing between the internal oblique and transversus abdominis muscles to arborize beneath the rectus abdominis muscle.

Continued.

Stage 2. After 10 days, the second stage of the procedure is begun. The patient is placed on the table in an oblique position with the flap side elevated on rolls. Incision is made directly over the rib overlying the selected intercostal bundle. The skin incision is carried forward to the lateral limit of the skin island and toward the midline of the back to the anterior border of the paraspinous muscles, dissecting directly on top of the rib. The periosteal elevator is used to free the rib subperiosteally. Extreme care must be taken not to perforate the periosteum on the internal side of the rib where the intercostal bundles are located. Dissection is carried posteriorly, and the latissimus dorsi and posterior serratus inferior are divided as necessary. Using a rib cutter, the rib is detached near its articulation with the transverse process of the vertebrae. To ensure viability of the skin island, the subjacent intercostal bundle may be included in the flap by dissecting the next rib free in a similar fashion. The bundle is situated just inside the rib near the midline of the back and more anteriorly runs between the innermost intercostal muscle just below the rib. Anteriorly it courses between the internal oblique and the transversus abdominis. The bundles are protected by dissecting the muscles free of the periosteum and pleura cephalad to the highest rib resected and cephalad to the next rib below the lowest rib resected.

The previously delayed portion of the flap is now elevated, and the transversus abdominis muscle is sharply dissected off the peritoneum. Posteriorly a small amount of the diaphragm may need to be excised if the T10 intercostal bundle is used. The anterior and posterior portions of the flap are connected by dividing the external oblique, internal oblique, and transversus abdominis in continuity with the previously divided intercostals. Since the bundles, like the ribs, run in a slightly inferior incline from back to front, care must be exercised in connecting the two incisions. The flap is now free to be rotated.

A sizable defect is left by this procedure. The diaphragm may be moved slightly cephalad and sutured to the free margin of the intercostals to close the posterior defect. Air is suctioned from the pleural space before closure, but no chest tube is necessary if the lung has not been injured. Anteriorly synthetic mesh is sutured to the fascia of the internal and external oblique and the anterior rectus fascia. Laterally based skin flaps superiorly and inferiorly are rotated to cover the mesh and are sutured to the midline incision. Suction drains are used to obliterate the dead space anteriorly. The patient is now turned prone to complete the reconstructive portion of the operation.

A skin incision connects the origin of the flap and the site where it is inset. Undermining on both sides and loose closure of this incision prevent pressure on the intercostal bundle.

D

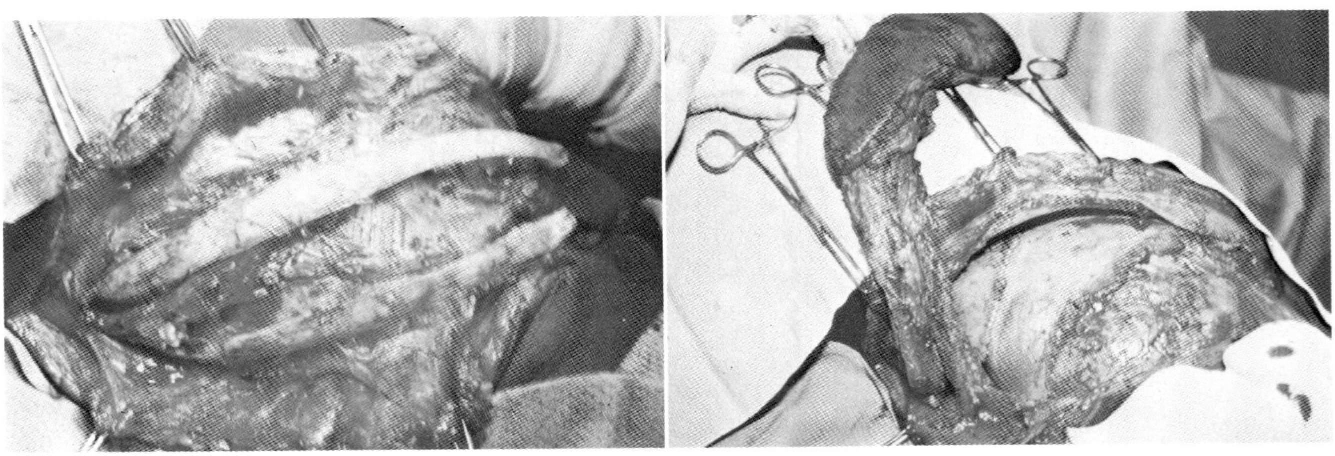

E

Fig. 26-8, cont'd

D, Stage 2—flap elevation. Separation of ribs from intercostal muscles and associated vascular pedicle and sensory nerve by subperiosteal dissection.

E, Stage 2. Intercostal flap is elevated on its neurovascular bundle and surrounding intercostal muscles with resultant donor site defect. NOTE: Exposed peritoneum and diaphragm.

F G H I

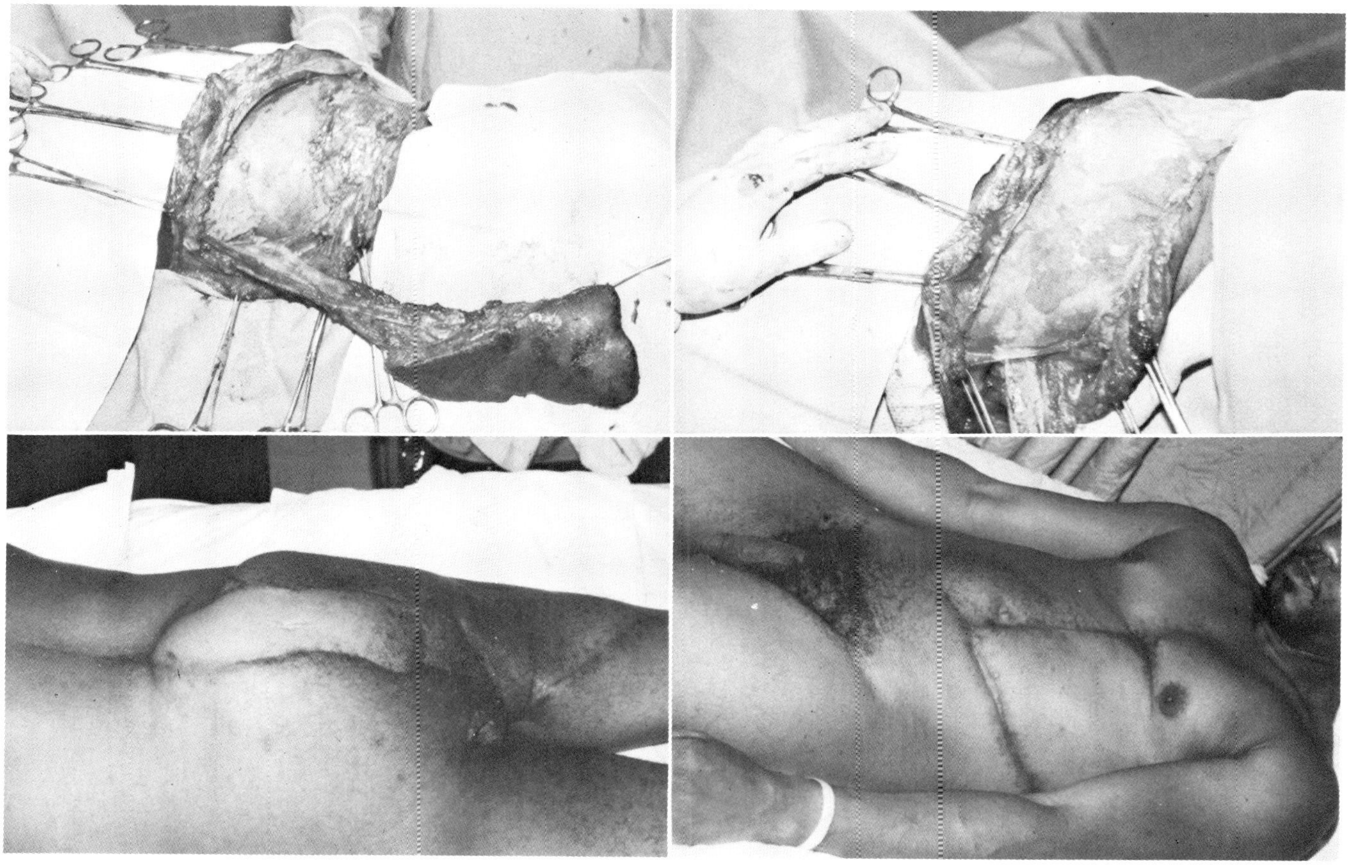

Fig. 26-8, cont'd

F, Stage 2—flap inset. Vascular pedicle has adequate length for flap to reach sacral region. NOTE: Tunnel beneath back skin allows placement of neurosensory flap in sacral defect.

G, Donor defect is closed with prolene mesh and local flap advancement.

H, Eighteen-month postoperative view. Flap demonstrates protective sensation.

I, Patient in sitting position demonstrates adequate donor site coverage and abdominal wall reconstruction.

Distally based latissimus dorsi flap

For high sacral and lower lumbar defects, the lower part of the latissimus dorsi muscle may be used based on its secondary segmental blood supply (Fig. 26-9). It should be emphasized that this flap will not always reach the lower sacral area and thus has limited usefulness in sacrococcygeal sores (Fig. 26-10).

Summary

The transverse back flap and variations of the gluteus maximus muscle flap are dependable and easily transferred in one stage. The intercostal flap has the advantage of transferring innervated skin in paraplegic patients but is a more difficult flap to dissect; its application is limited to selected patients.

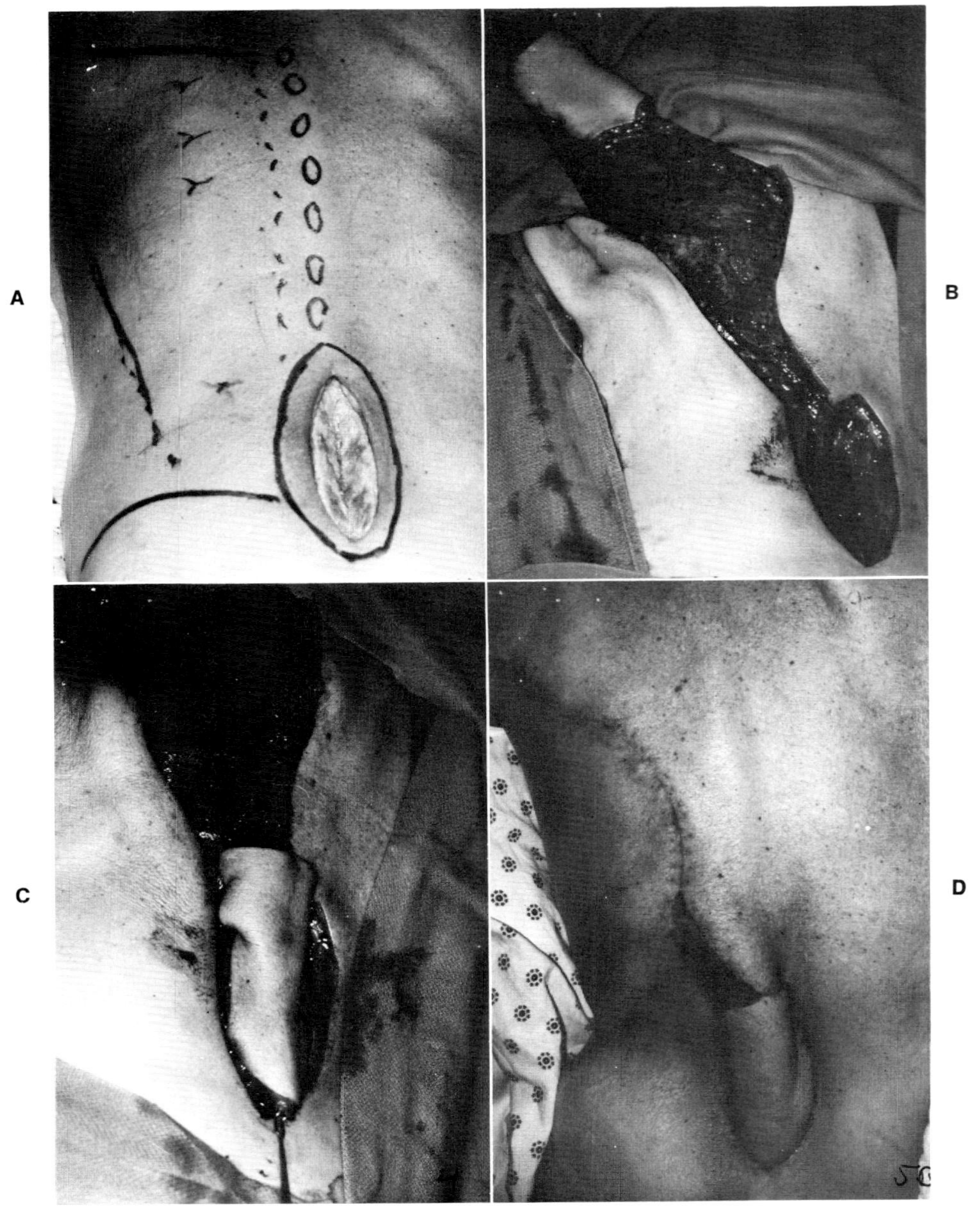

Fig. 26-9. Distally based latissimus dorsi musculocutaneous flap for sacral coverage.
A, Large sacral defect at site of tumor resection and irradiation.
B, Defect excised and skin island isolated on distally based muscle.
C, Latissimus dorsi muscle is turned over and island of skin with underlying muscle placed in defect.
D, View several months after operation. Small skin graft placed over muscle.

A

B

C

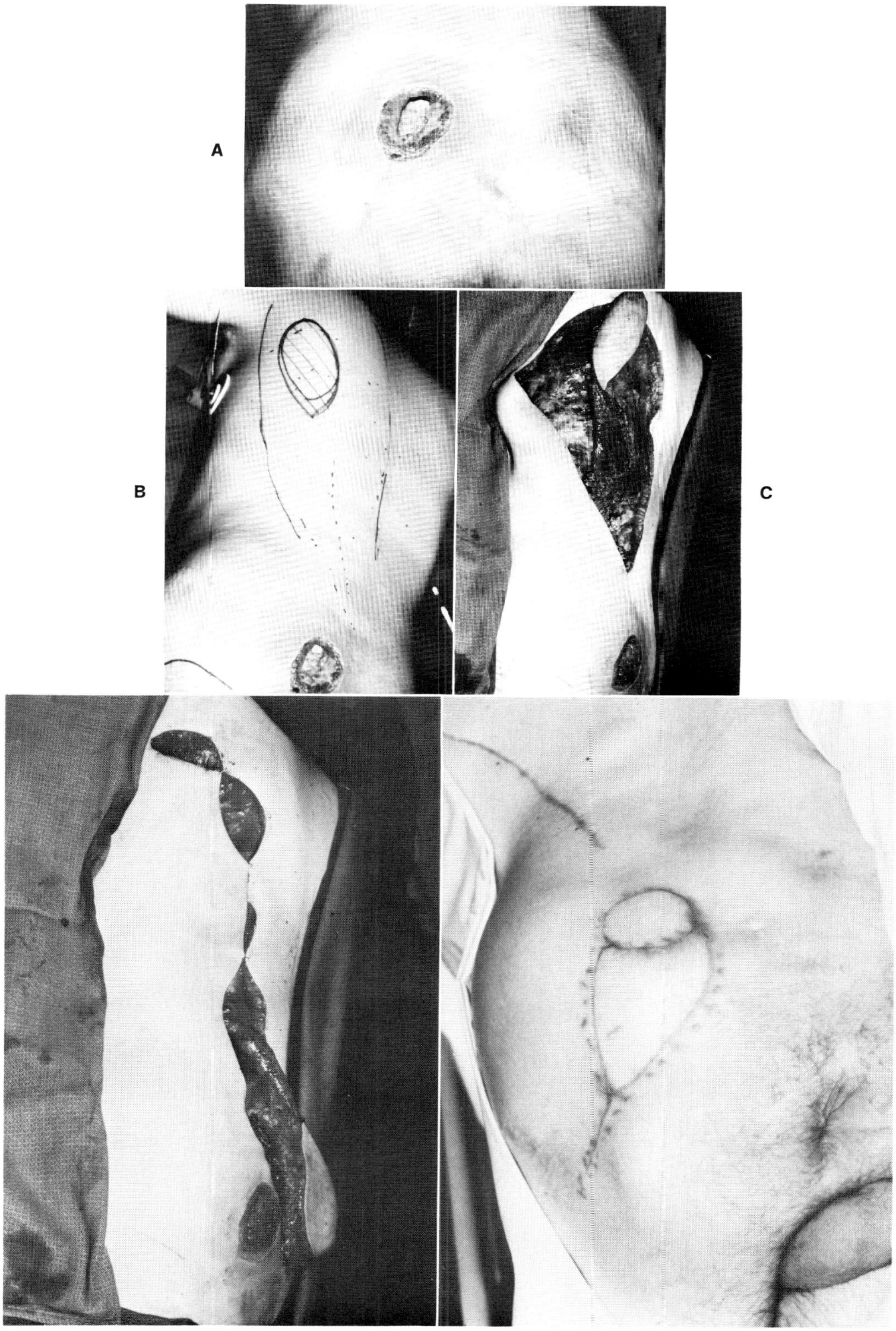

D

E

Fig. 26-10. Distally based latissimus dorsi musculocutaneous flap and V-Y advancement gluteus maximus flap for coverage of posterior iliac crest.

A, Large pressure area over left posterior iliac crest.

B, Distally based latissimus dorsi island flap outlined.

C, Island of skin isolated and muscle mobilized on secondary segmental pedicles.

D, Donor defect closed directly, muscle turned over, and island of skin placed over defect.

E, Defect closed by latissimus dorsi superiorly and V-Y advancement superior gluteus maximus musculocutaneous gluteal flap inferiorly.

ISCHIAL SORES

General considerations

The ischial sore is the "sitting sore" seen in paraplegic and quadriplegic patients. With its location between three major muscle groups (gluteal, hamstrings, and thigh adductors), a variety of muscle flaps are available for coverage of this defect.

The hamstrings

The muscles of the hamstring group include the biceps femoris, semimembranosus, and semitendinosus. The biceps femoris is the most powerful and laterally located of the three. The biceps femoris (type II) and the semimembranosus (type II) are supplied by perforating vessels from the profunda femoris. The semimembranosus has a type III pattern of blood supply with a proximal pedicle from the profunda femoris and secondary pedicle distally from the superior femoral artery.

Flap design and elevation

Muscle transposition and direct skin closure. The transposition of one or more of the hamstring group into an ischial defect is a very simple and easily accomplished procedure (Fig. 26-11). This was one of the first muscle flaps described for closure of pressure sores. An incision is made, extending from the defect inferiorly. The distal muscle (usually biceps femoris) is divided from its insertion, and dissection is carried proximally toward the pedicles from the profundus femoris. The most distal of these pedicles can be divided safely to allow for more extensive transposition. The muscle is sutured into the base of the defect after partial ischiectomy. The secondary defect is closed directly. The muscle is grafted or the skin closed directly over it. If the defect is larger, the biceps femoris and semitendinosus or semimembranosus can be used (Fig. 26-12).

A

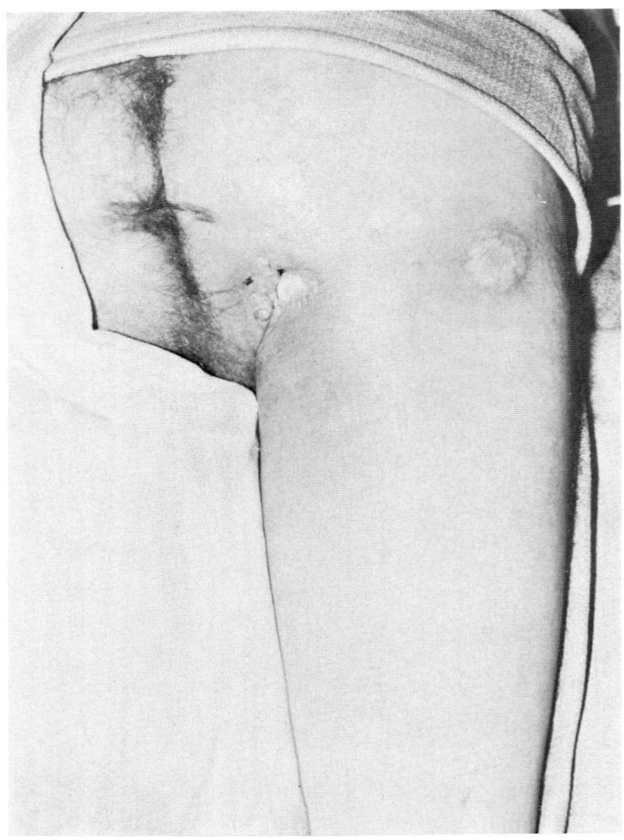

Fig. 26-11. Closure of ischial pressure sore with biceps femoris muscle transposition and skin flap.
A, Ischial pressure sore.

B

C

D

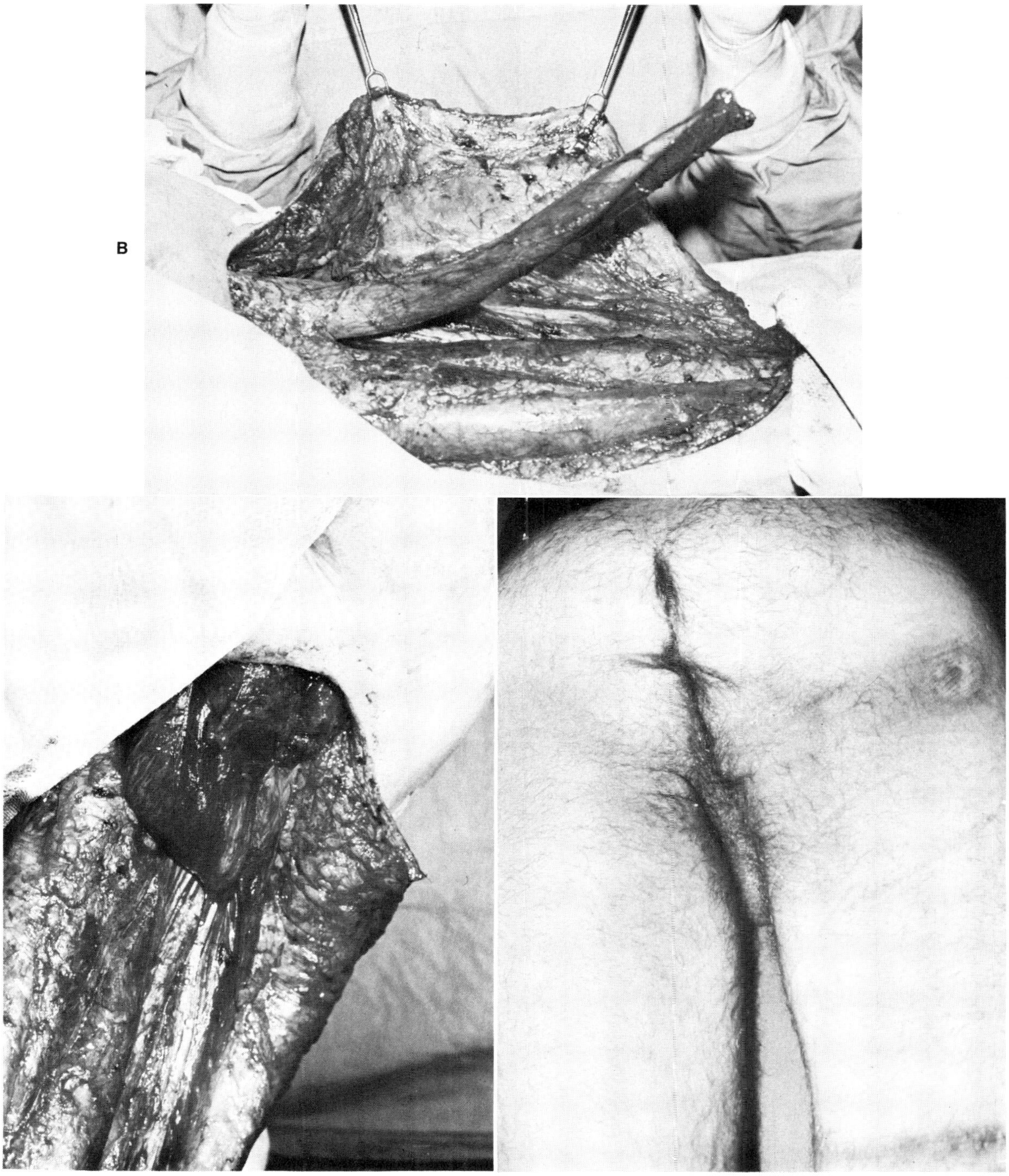

Fig. 26-11, cont'd

B, Biceps femoris muscle flap elevated with preservation of proximal vascular pedicles.

C, Muscle sutured into ischial defect.

D, Donor defect closed directly with advancement of posterior thigh skin over muscle flap.

A

B

C

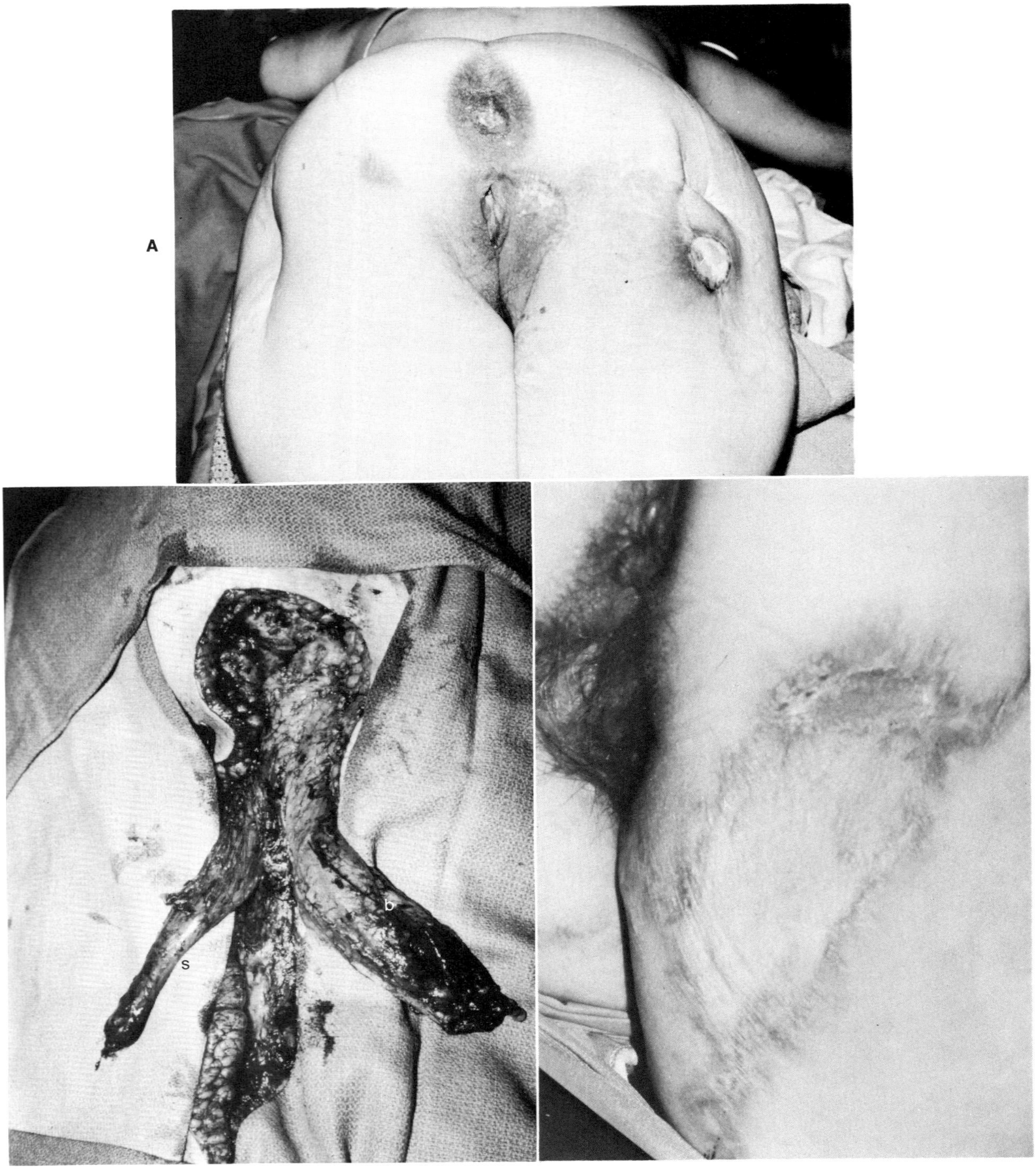

Fig. 26-12. Hamstring muscles and skin graft for ischial coverage.
A, Extensive recurrent ischial pressure ulcer.
B, Ulcer excised and hamstring muscles elevated. *b*, Biceps femoris; *s*, semitendinosus.
C, Grafted muscles several months after operation.

Musculocutaneous V-Y advancement. The use of a musculocutaneous island flap based on the entire hamstring compartment has recently been used in ischial sores. A triangular skin island is designed with the base of the triangle at the inferior margin of the defect. This triangle includes the cutaneous territory over the hamstrings. The entire muscle compartment is mobilized after division of muscle insertion. Distal muscle pedicles are divided if necessary, and the muscle group with the overlying skin triangle is then advanced proximally into the defect. Muscle origin is usually released at the time of wound excision (Fig. 26-13). This flap is easily elevated and advanced. Although the flap can be based on the biceps femoris alone, it is safer to elevate and advance the entire hamstring compartment. Greater muscle bulk is transferred to the ischial defect.

A

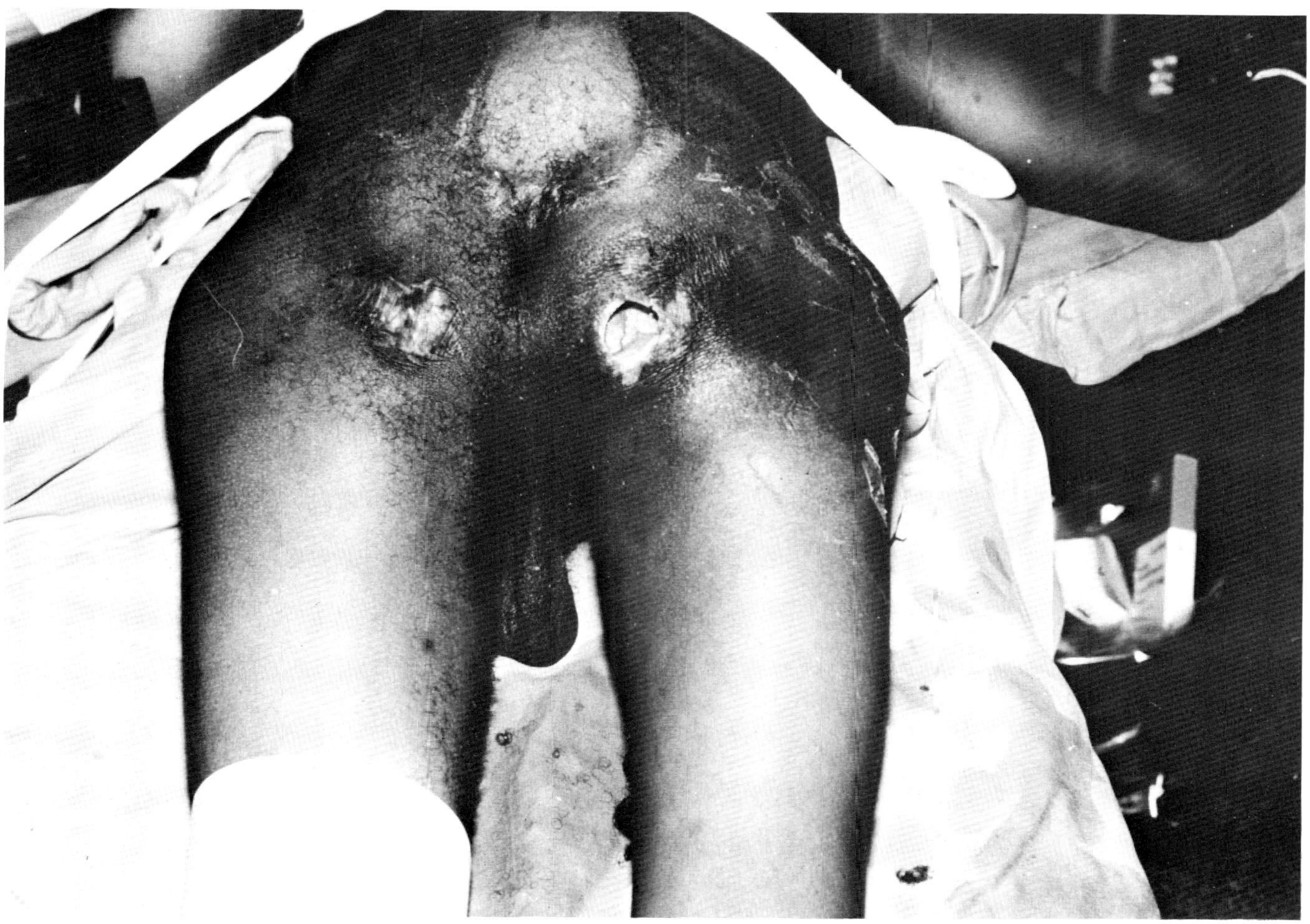

Fig. 26-13. Closure of ischial pressure sore with hamstring V-Y advancement island musculocutaneous flap.
A, Bilateral ischial pressure sores.

Continued.

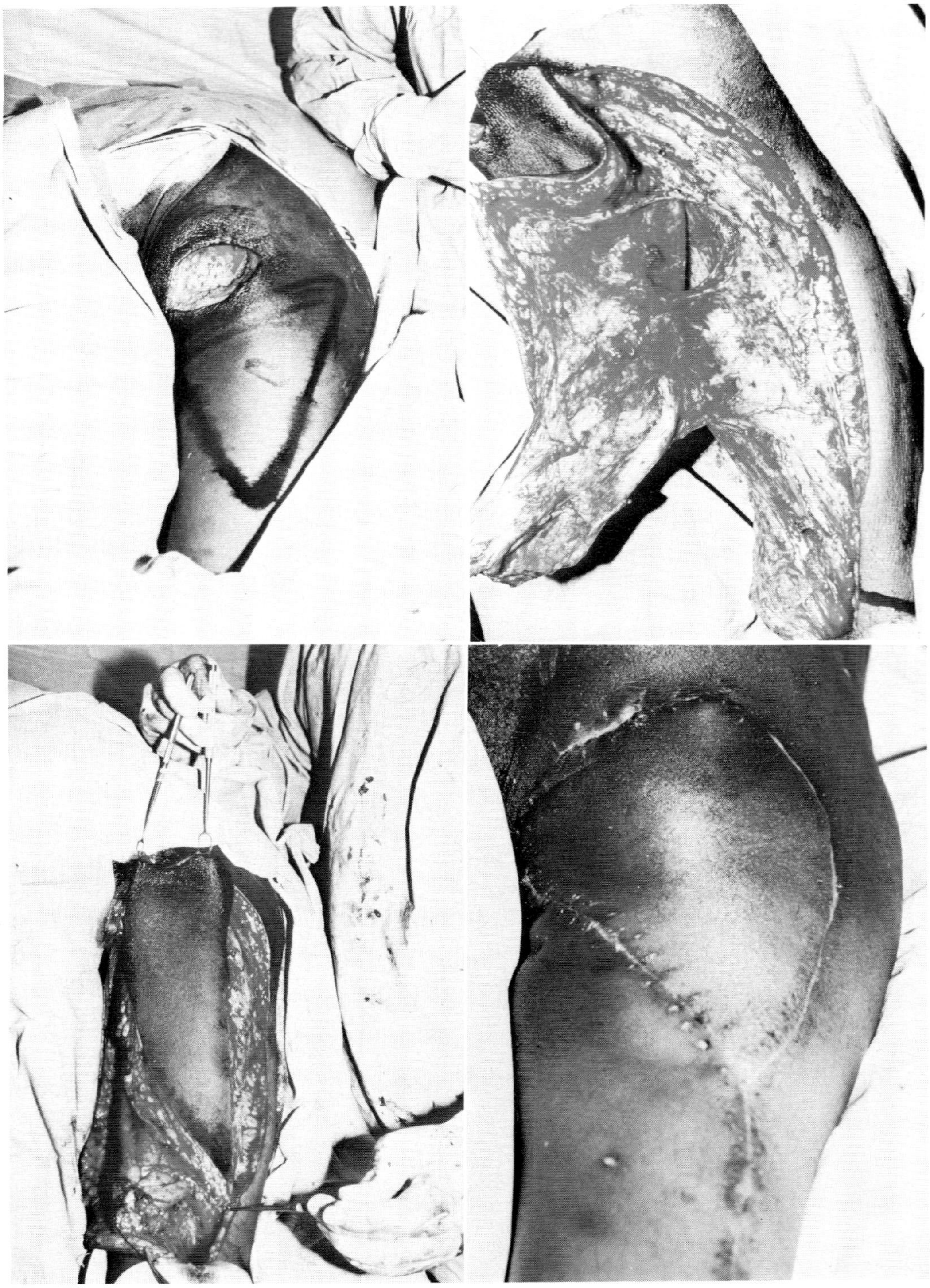

B C D E

Fig. 26-13, cont'd
B, Right ischial defect excised and flap outlined.
C, Flap elevated and muscles divided and isolated on vascular pedicles.
D, Flap advanced over defect.
E, Flap provides stable coverage with donor area closed directly in V-Y fashion.

Gluteus maximus

The gluteus maximus muscle may be transposed as a musculocutaneous flap for coverage of the ischial defect.

The inferior half of the gluteus maximus may be transposed for ischial defects. A skin island is designed inferiorly over the lateral aspect of muscle close to its insertion. The inferior muscle is then transposed with its distal skin island to close the ischial defect. The donor defect can generally be closed directly.

The use of the inferior gluteus maximus muscle is preferred, since the superior muscle is left intact. The superior muscle is preserved for future use as a transpositional flap. The inferior muscle with the overlying skin provides stable coverage for the ischial pressure sore (Fig. 26-14).

A

B

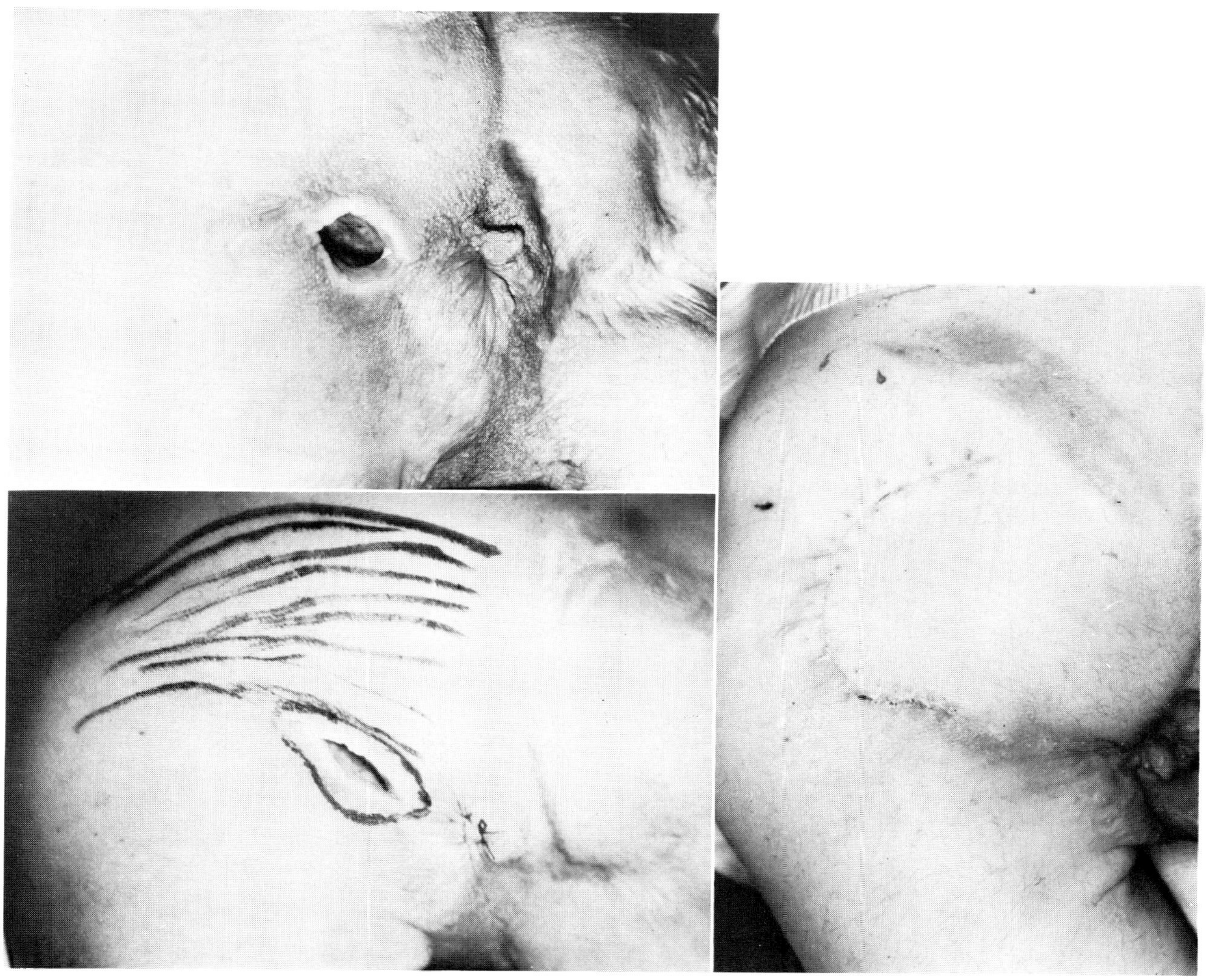

C

Fig. 26-14. Gluteus maximus flap for coverage of ischial defect.
A, Ischial defect.
B, Inferior half of gluteus maximus outlined.
C, Flap provides stable coverage of ischial defect. NOTE: Skin may be designed as island.

Gracilis

The gracilis musculocutaneous flap may be used for coverage of the ischial pressure sore. The gracilis muscle (type II) receives its dominant vascular pedicle from the medial femoral circumflex artery. After release of its minor pedicles, the muscle has a posterior arc of rotation that reaches the ischium. As a musculocutaneous unit, this flap provides adequate coverage for the ischial defect.

The cutaneous territory of the gracilis muscle is difficult to accurately design with the patient prone. However, an incision in the medial distal third of the thigh allows identification of the muscle. At this level, the gracilis muscle is half tendon and half muscle and is located between the fleshy sartorius muscle anteriorly and the fascial insertion of the semitendinous muscle posteriorly. With identification of the gracilis muscle distally, traction on the muscle allows accurate design of its overlying skin island. The skin is incised proximally and immediately sutured to the gracilis muscle to avoid disruption of musculocutaneous perforators. The distally located minor pedicles are divided and the muscle elevated until the proximal thigh is reached. At 10 cm inferior to the pubic tubercle, the major pedicle, the medial circumflex femoral artery, enters the muscle. From this posterior approach, the pedicle is noted between the adductor magnus and adductor longus muscles. Further proximal flap elevation is not generally required for an adequate flap arc to reach the ischial defect. The distal muscle is sutured to the ischiectomy site. The skin provides adequate coverage of the ischial defect.

The skin is not transposed as an island when the gracilis musculocutaneous flap is transposed for ischial coverage. The donor area is closed directly. When the cutaneous territory is correctly identified, the gracilis musculocutaneous unit is a reliable flap for ischial sore coverage.

Tensor fascia lata

The TFL flap is not the flap of choice for an isolated ischial sore. However, under certain circumstances this flap is used to cover the ischial area. It is the flap of choice for combined ischial and trochanteric sores, when one flap can cover two defects, and as a neurosensory flap for coverage of the sitting area. For ischial coverage the distal portion of the flap, which is only skin, subcutaneous tissue, and fascia lata, is often too thin for durable coverage. Under these circumstances the inferior gluteus maximus muscle or the vastus lateralis muscle may be used under the TFL flap for added bulk.

TROCHANTERIC SORES

The trochanteric sore occurs in both paraplegic and quadriplegic patients. These sores may occur when patients are turned side to side in an attempt to prevent sacral sores. As with all sores, total excision of the ulcer and reduction of the bony prominence is indicated before transposing tissue for coverage.

The TFL, rectus femoris, vastus lateralis, and gluteus maximus muscles all have applications in closure of this defect.

TFL

The standard TFL flap will cover the trochanteric area, and the extended flap will cover the trochanteric and ischial region. This flap has great potential as a neurosensory flap based on the lateral femoral cutaneous nerve of the thigh.

The TFL has a type I pattern of blood supply. The lateral circumflex femoral artery enters the medial aspect of the muscle at a point 8 to 10 cm below the anterior superior iliac spine. Based on this dominant pedicle, the flap has an arc of rotation posteriorly that will cover the trochanter and ischium.

Flap elevation. For coverage of the trochanter a standard flap is elevated. For ischial coverage an extended flap is required. The anterior marking for the flap is a line drawn from the anterior superior iliac spine to the lateral condyle of the knee. The greater trochanter represents the posterior extent of this flap. Flap dissection is begun distally. The skin, subcutaneous tissues, and underlying fascia lata are incised. The flap is dissected from the distal border to the proximal border. This dissection is in an avascular plane overlying the vastus lateralis. At a level of approximately 8 to 10 cm below the anterior superior iliac spine, the pedicle is identified and preserved. The flap is then transposed into the defect. The flap may be safely made into an island if desired (Fig. 26-15).

A

B

C

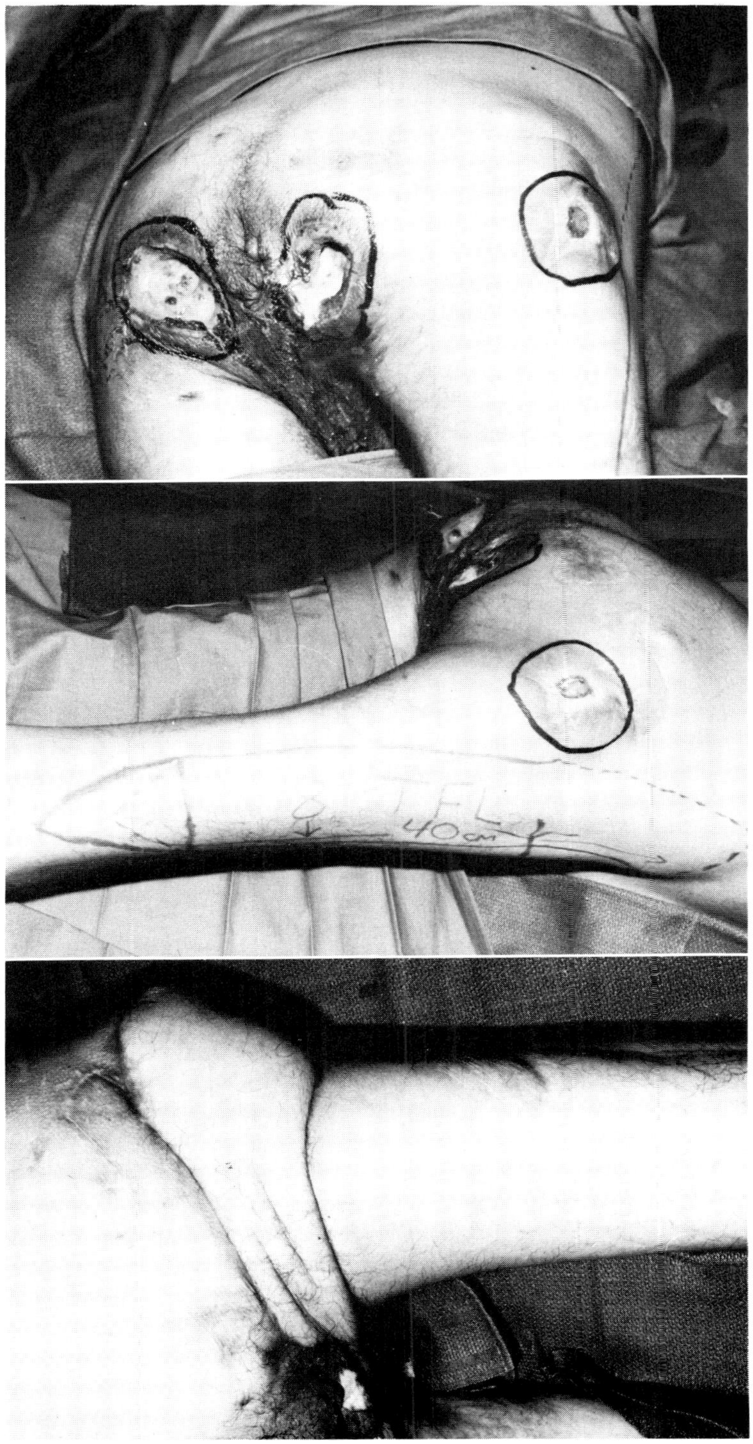

Fig. 26-15. Extended TFL flap for closure of combined ischial and trochanteric pressure sores.
A, Extensive bilateral ischial and right trochanteric sore.
B, Extended TFL flap outlined.
C, One-year postoperative view of transposed flap.

Rectus femoris

The rectus femoris unit may be used either as a muscle or musculocutaneous flap for the trochanteric pressure sore. This broad muscle is located on the anterior thigh, extending from its origin on the inferior iliac spine to its insertion to the patella. The muscle receives its blood supply from the lateral femoral circumflex artery (type I). This unit provides a reliable source of coverage for the trochanter.

A vertical incision is made between the rectus femoris and vastus lateralis to gain exposure to the rectus femoris muscle. After release of the muscle insertion, the muscle is elevated from the underlying vastus medialis muscle. The vascular pedicle, lateral femoral circumflex artery, enters the deep aspect of the muscle approximately 10 cm inferiorly to the anterior superior iliac crest. Exposure of the vascular pedicle is generally not required to gain adequate flap length for trochanteric coverage. The muscle is transposed over the trochanteric defect. The muscle is skin grafted and the donor defect closed directly.

The rectus femoris may also be transposed as a musculocutaneous flap. This skin may be designed either as a simple skin muscle flap or as a skin island over the distal muscle. Generally the donor defect can be closed directly (Fig. 26-16).

The rectus femoris muscle is a useful transposition flap for trochanteric coverage. It is generally reserved as an alternative flap for this purpose. Transposition of this muscle in the ambulatory patient will result in a functional deficit with loss of full leg extension.

A

B

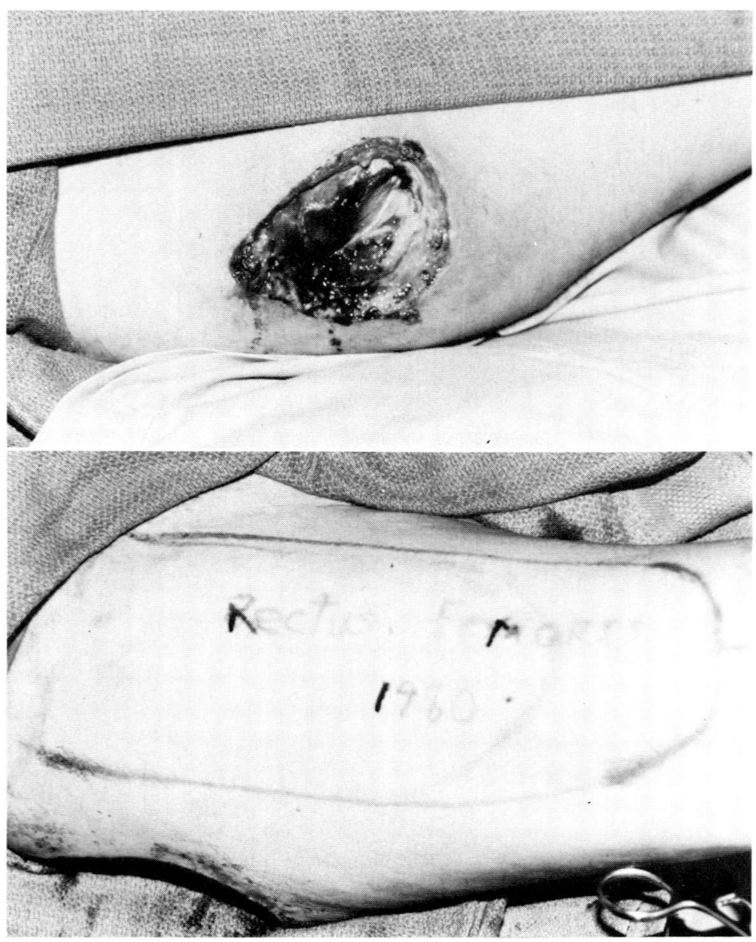

Fig. 26-16. Rectus femoris musculocutaneous flap for coverage of trochanter pressure sores.

A, Right trochanteric pressure sores.

B, Sore excised, and flap outlined.

C

D

E

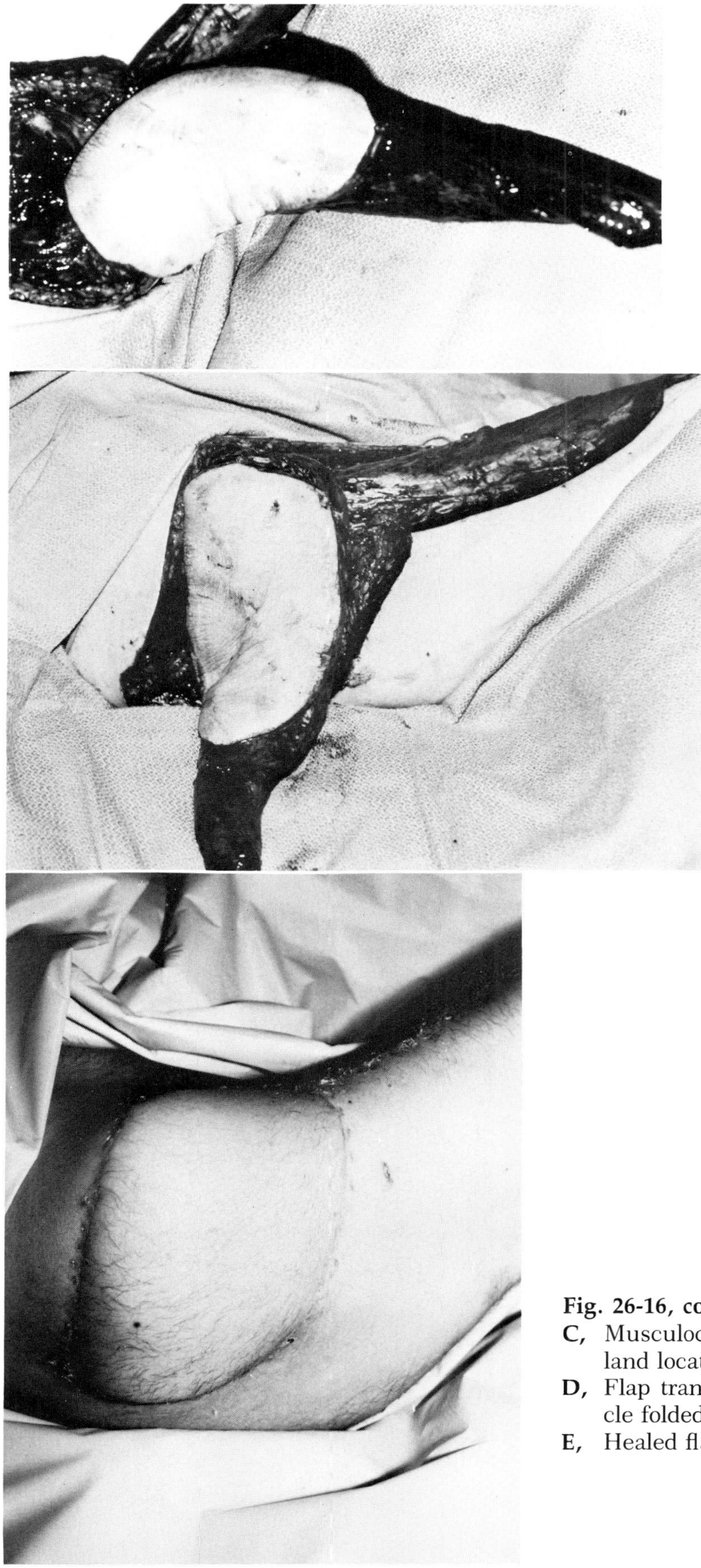

Fig. 26-16, cont'd

C, Musculocutaneous flap elevated. NOTE: Skin island located over proximal muscle.

D, Flap transposed over trochanter and distal muscle folded over bone for padding.

E, Healed flap with donor defect closed directly.

Vastus lateralis

The vastus lateralis is a large muscle that has proved most useful for obliteration of large trochanteric defects, especially when excision of the femoral head is required (Fig. 26-17). It has a type II pattern of blood supply with the dominant pedicle, a branch of the lateral circumflex femoral artery, entering the muscle at a level approximately 8 to 10 cm below the anterior superior iliac spine.

Flap elevation. After excision of the trochanteric sore and removal of the head of the femur, the vastus lateralis is identified through a lateral thigh incision located between the anterior superior iliac crest and the lateral condyle of the knee. The flap is dissected from the distal border to the proximal border and folded over into the defect. The muscle is then skin grafted and the donor defect closed directly. Alternately the muscle is advanced up into the defect rather than used as a transposition flap.

A combination flap of the TFL and vastus lateralis is useful for coverage of ischial defects; the vastus lateralis provides the bulk, and the TFL provides the skin, which has potential for a sensory flap.

A

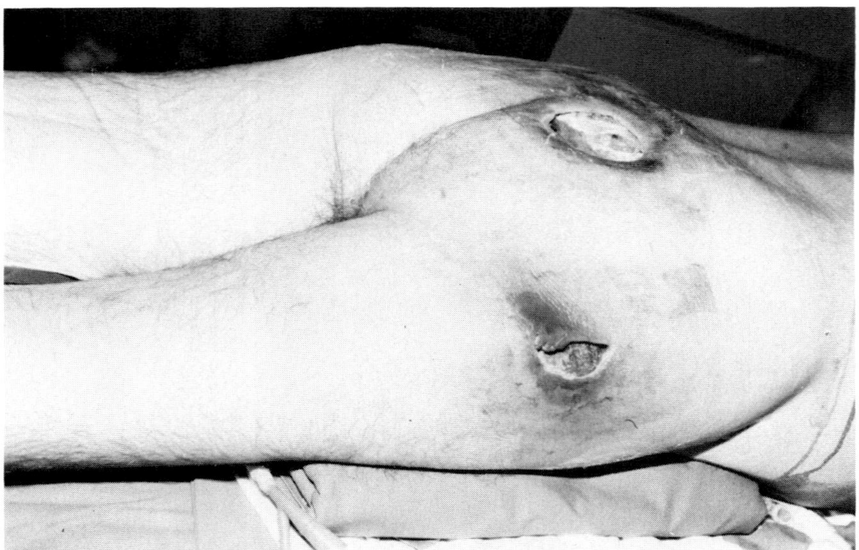

B

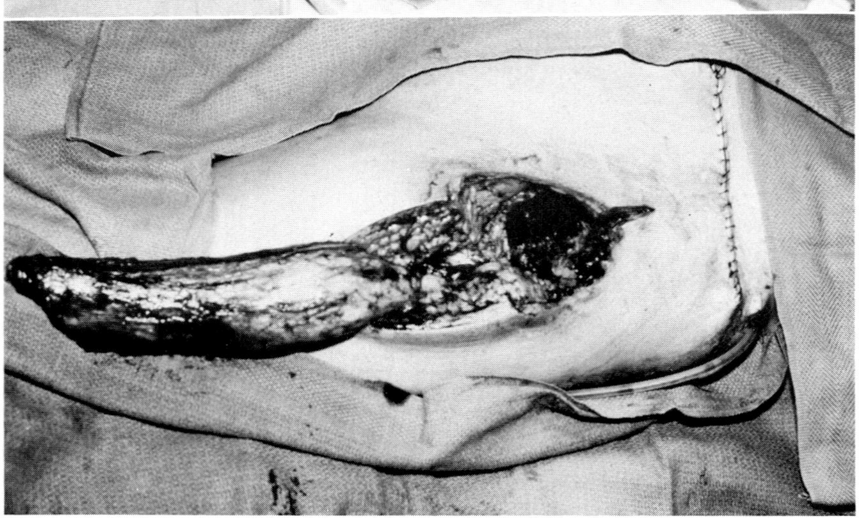

Fig. 26-17. Vastus lateralis muscle and skin graft for closure of trochanteric defect. **A,** Extensive sacral and right trochanteric pressure sores. **B,** Trochanteric sore excised and vastus lateralis elevated.

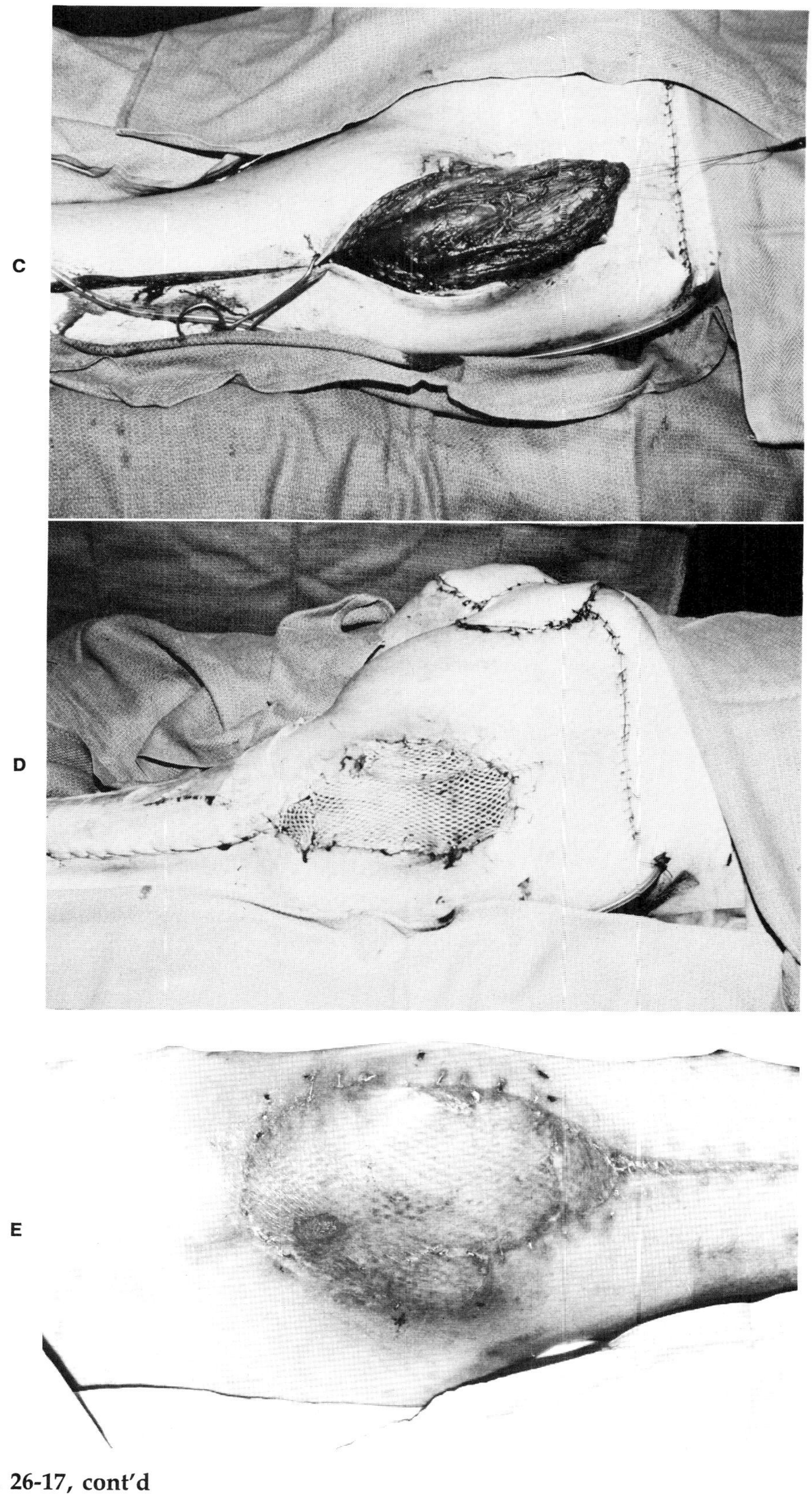

Fig. 26-17, cont'd
C, Vastus lateralis transposed into trochanteric defect.
D, Skin graft applied over muscle. Sacral sore closed with bilateral gluteus maximus island musculocutaneous flaps.
E, Healed skin graft over muscle.

MULTIPLE BED SORES

Patients are occasionally seen with any combination of the three classic bed sores. Special mention should be made of the use of the TFL when ischial and trochanteric sores occur on the same side. In this specific situation the trochanterectomy and débridement of the ulcer reduce the distance to the ischial defect from the point of rotation of the TFL flap, and in most cases this single flap will cover both sores without excess tension on the ischial suture line.

In patients with massive multiple or confluent sores, amputation and total thigh flaps may be necessary for coverage (Fig. 26-18).

A

B

C

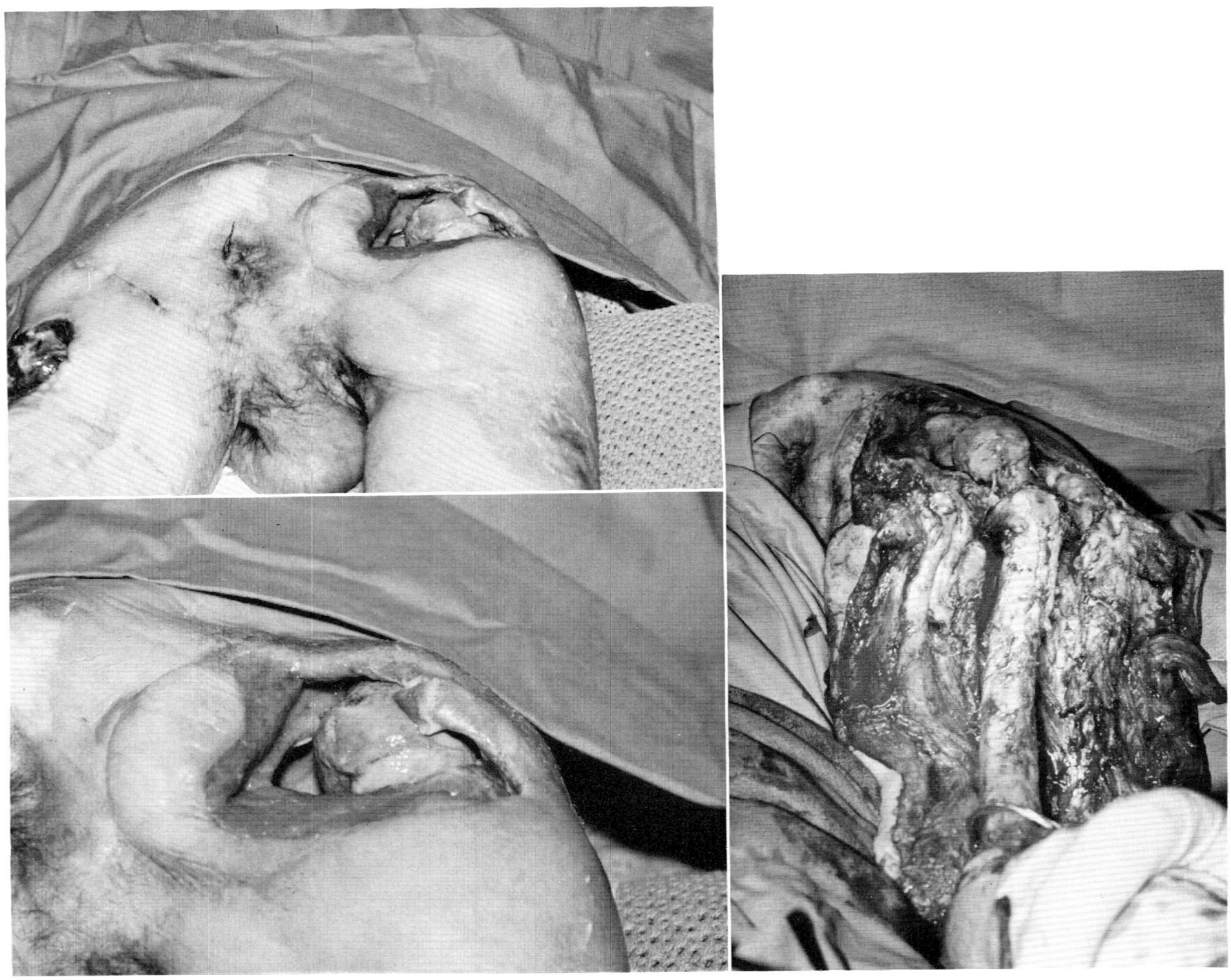

Fig. 26-18. Coverage of extensive right trochanteric pressure sore with total thigh flap and left infected hip with vastus lateralis and skin graft.

A, Extensive right trochanteric pressure sore with exposed necrotic femoral head. Left trochanteric pressure sore with septic hip joint.

B, Infected necrotic right femoral head.

C, Right femur "filleted out" through posterior approach.

D

E

F

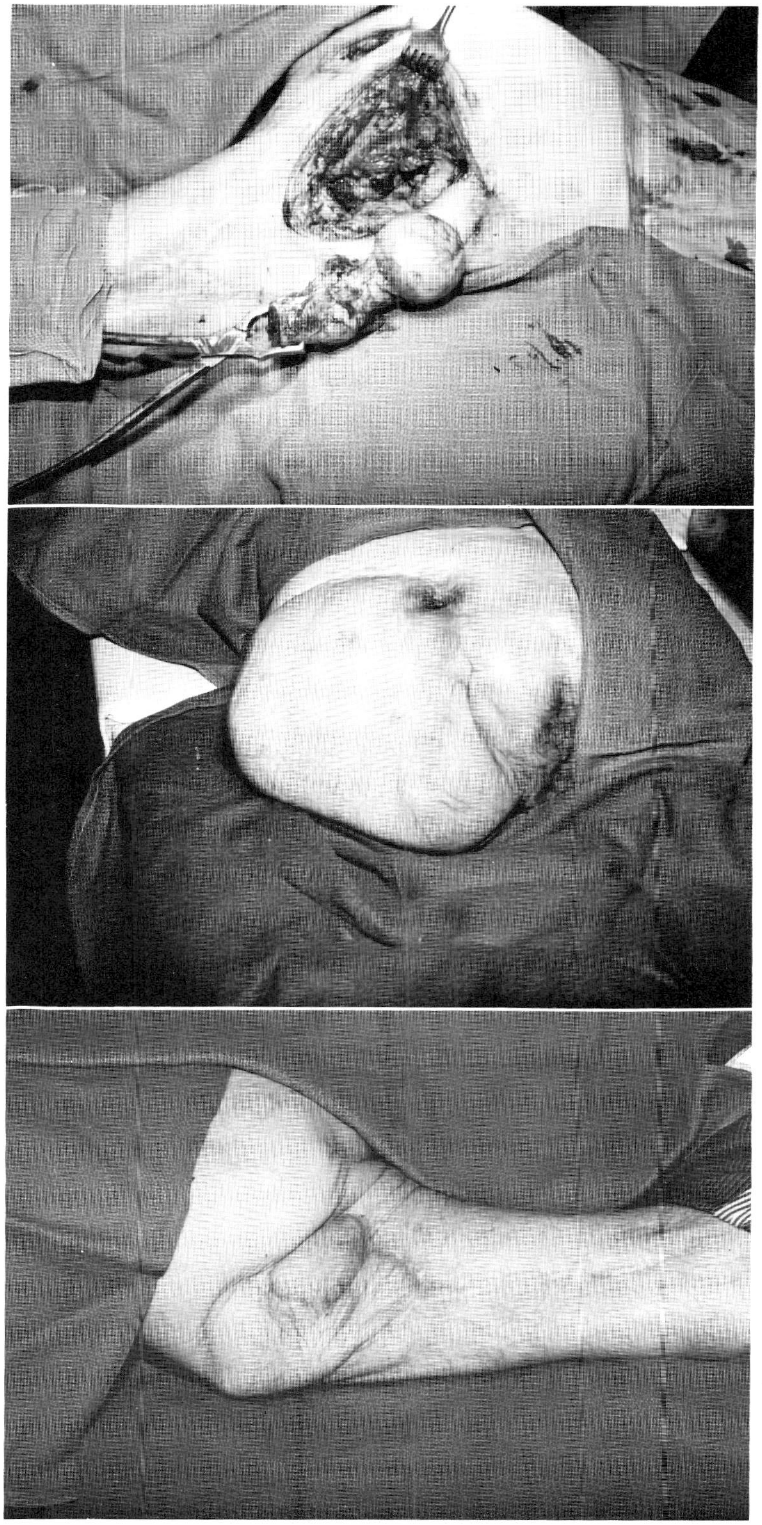

Fig. 26-18, cont'd
D, Left infected femoral head resected.
E, Coverage of right side with total thigh flap.
F, Coverage of left side with vastus lateralis muscle flap and skin graft.

The total thigh flap

In a few patients with large recurrent pressure sores (or multiple sores), hip disarticulation and use of a "thigh" flap may be indicated. The so-called "total thigh flap" was one of the first "musculocutaneous flaps" used. This flap contains multiple rather than single muscle units with an attended marked increase in skin perforators overlying the muscles used. For this reason these flaps are very dependable. In our experience an anterior and posterior variation of the thigh flap has been the most useful.

Anterior thigh flap (Fig. 26-19). In this flap the rectus femoris, sartorius, vastus lateralis, vastus medialis, and vastus intermedius are the muscles saved during the amputation. When using the muscles a large skin territory (anterior thigh to the knee) can be used. Care is taken during amputation to not injure muscular branches from the profunda femoris during removal of the femur. Also great care is taken to not dissect or avulse needed skin off the underlying muscle. The flap is tailored as needed and rotated into defect. Large recurrent sacral, ischial, or combination sores can be easily covered with this flap.

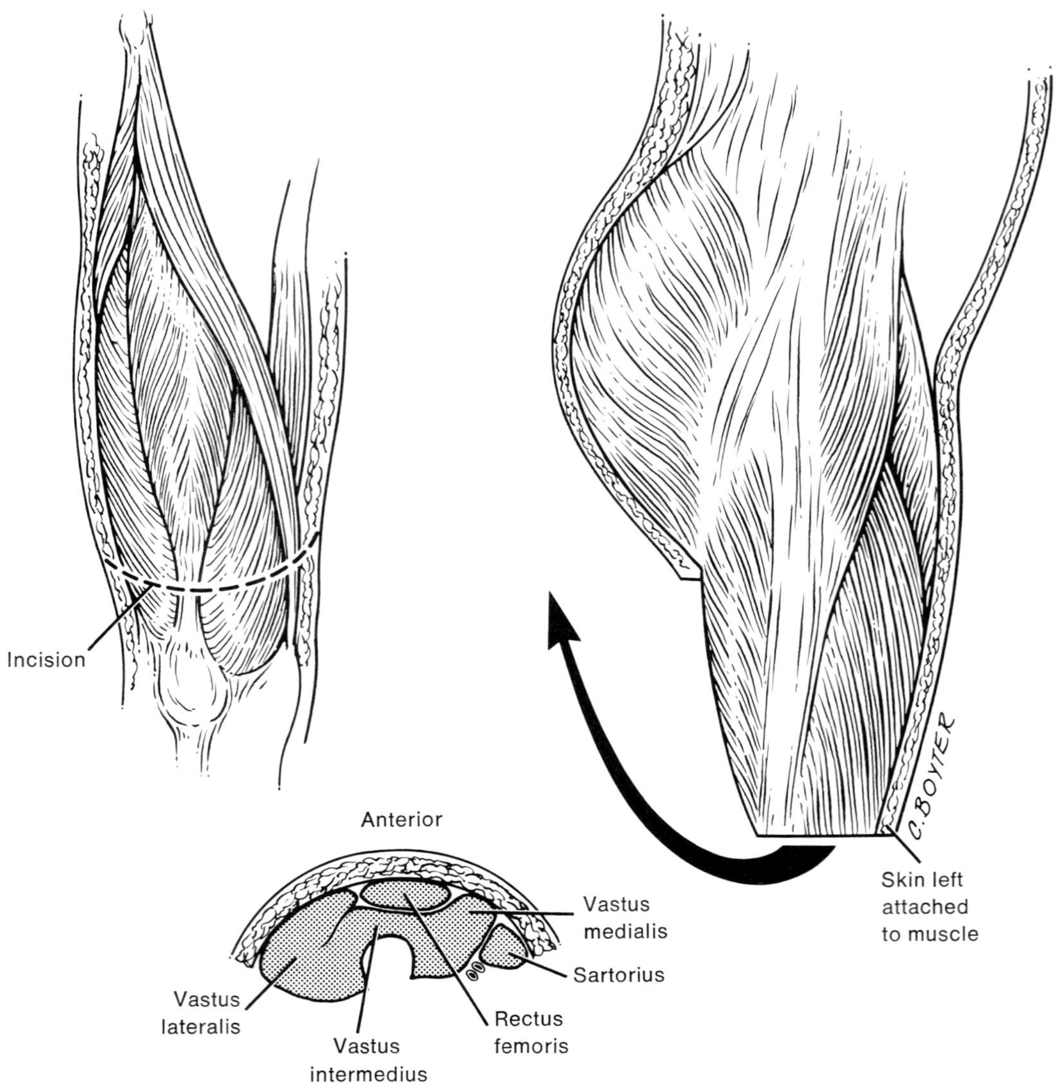

Fig. 26-19. Anatomy and arc of rotation of anterior thigh flap.

Posterior thigh flap (Fig. 26-20). The posterior variation of the flap usually includes the hamstrings (biceps femoris, semimembranous, and semitendinosus), adductor magnus, and gracilis. This group will carry all of the skin of the posterior thigh and can be used to cover large defects of anterior lower abdominal wall that result after hemipelvectomy.

Obviously, medial and lateral variations are also possible, but anterior and posterior variations are usually more applicable.

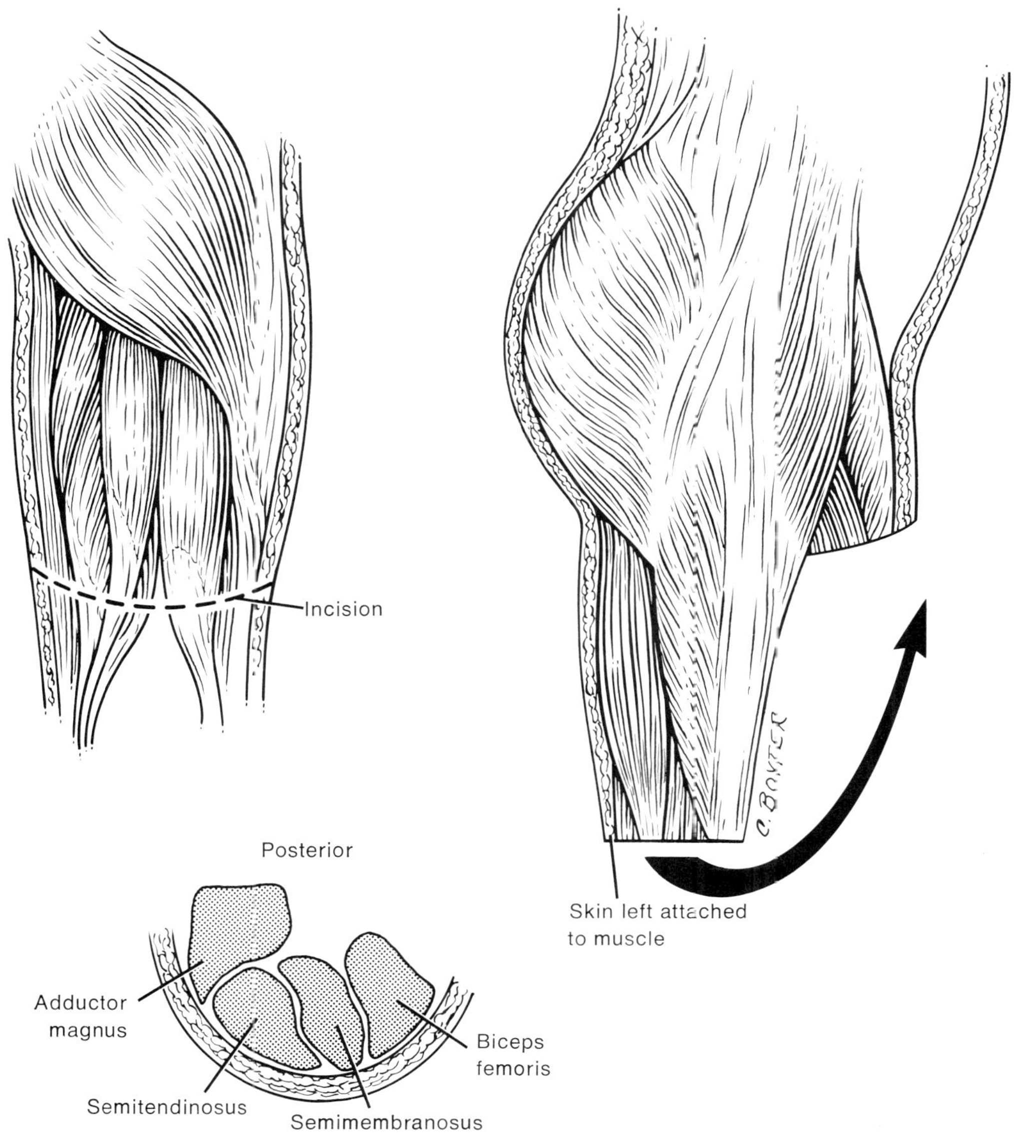

Fig. 26-20. Anatomy and arc of rotation of posterior thigh flap.

CONCLUSION

The management of the pressure sore depends on the sensory status of the involved region. When pressure sore development is related to temporary sensory loss such as associated with coma, successful wound management is usually possible with local wound débridement and skin graft coverage. However, when sensory loss in the region of the pressure sore is permanent as associated with spinal cord damage, management must consider both immediate wound coverage and the need for future coverage related to continuing pressure injury.

Unlike the random skin rotation flap, the muscle or musculocutaneous flap is designed on specific vascular pedicles to muscle. Only the skin required for sore coverage is transferred based on musculocutaneous perforating vessels. The remaining skin is undisturbed, providing coverage and a source of future flaps.

The transposition of muscle with either overlying skin grafts or muscle with skin (musculocutaneous units) provides a thick tissue with healthy circulation. The distal skin of the random transposition skin flap is generally that portion of the flap covering the critical pressure zone.

There is still much controversy concerning both the necessity for and extent of resection of the bony prominence beneath the pressure sore. Obviously, the infected outer cortex of bone must be resected. However, major resections as with the ischium may cause instability of the pelvis and merely shift pressure zones to new areas. With the increased bulk provided by transposition of muscle and musculocutaneous flaps, as compared to random flaps, there is less need for extensive resection of bone.

Restoration of sensibility remains the key to treatment of the pressure sore. The concept of transfer of skin with sensibility for pressure sores still unfortunately has limited application. When the level of spinal cord injury is low, both the intercostal flap and the innervated TFL flap represent the ideal methods of treatment. However, efforts to use long sural grafts between intact sensory nerves to the sensory nerves of pressure sore regions or to the sensory nerves of musculocutaneous flaps may have application in the management of the pressure sore.

Efforts to improve the design of muscle and musculocutaneous flaps for pressure sore coverage have assisted the reconstructive surgeon. Both flap reliability and the potential for future flaps have been improved. Hopefully, future refinements in the sensory flap will eliminate the problem of recurrence. Until then the reconstructive surgeon and the rehabilitation therapist must work together in the prevention and management of the pressure sore.

ANNOTATED BIBLIOGRAPHY

1. Baker, D.C., Barton, F.E., and Converse, J.M.: A combined biceps and semitendinosus muscle flap in repair of ischial sores, Br. J. Plast. Surg. **31**:26, 1978.

   The vascular anatomy of the hamstring muscle group and technique of transposition of biceps and semitendinosus to fill ischial defects are described.

2. Becker, H.: The distally based gluteus maximus muscle flap, Plast. Reconstr. Surg. **63**:653, 1979.

   Use of a distally based gluteus maximus muscle flap for coverage of trochanteric sores and the vascular anatomy of the so-called "cruciate" anastomosis that supplies this variation are described.

3. Blocksma, R., Kostrubala, J.G., and Greely, P.W.: The surgical repair of decubitus ulcer in paraplegics, Plast. Reconstr. Surg. **4**:123, 1949.

   Early description of use of ishiectomy and obturator internus transpotion for coverage of ischial sores is presented.

4. Conway, H., and Griffin, B.H.: Plastic surgery for closure of decubitius patients with paraplegia, Am. J. Surg. **91**:946, 1956.

   Successful use of biceps femoris muscle flap and medially based skin flap for coverage of large ischial sore is described. This article is one of the first descriptions of the use of muscle flaps for this problem.

5. Daniel, R.K., Terzis, J.K., and Cunningham, D.M.: Sensory skin flaps for coverage of pressure sores in paraplegic patients, Plast. Reconstr. Surg. **58**:317, 1976.

   Use of three intercostal neurovascular island flaps based at T10, T9, and T7 levels for two sacral sores and one lumbar defect is described, emphasizing the technical difficulty of raising the flap and high rate of complications (e.g., pneumothorax and loss of distal skin). Also discussed is the transplantation of the neurovascular island flap. The clinical usefulness of this technically difficult procedure remains to be proven.

6. Dibbell, D.G.: Use of a long island flap to bring sensation to the sacral area in young paraplegics, Plast. Reconstr. Surg. **54**:220, 1974.

   Four cases of sacral defects closed with innervated, intercostal island flaps are described, emphasizing potential benefits of transposing innervated skin. Mention is made of technical difficulties and multiple stages required to complete procedure.

7. Ger, R.: The surgical management of decubitus ulcers by muscle transposition, Surgery **69:**106, 1971.

Early description of 17 muscle transfers, including rectus femoris (8), gluteus maximus (6), and sartorius (3), for treatment of sacral and trochanteric ulcers is presented. Transposed muscle was skin grafted.

8. Ger, R., and Levine, S.A.: The management of decubitus ulcers by muscle transposition, Plast. Reconstr. Surg. **58:**419, 1976.

An excellent review of 40 muscle tranposition procedures for primarily sacral (22) and trochanteric (17) ulcers is presented. The gluteus maximus was used in 23 cases and rectus femoris in 10. The vastus lateralis and sartorius were also used. Follow-up was poor, but recurrent ulceration was related to general health of patient and quality of aftercare.

9. Hill L., Brown, R.G., and Jurkiewicz, M.J.: The transverse lumbosacral back flap, Plast. Reconstr. Surg. **62:**177, 1978.

Vascular anatomy of skin of lumbosacral area and the use of a transverse flap based on lumbar perforating arteries are described. This technique is used in closure of 20 sacral decubiti.

10. Labandter, H.P.: The gracilis muscle flap and musculocutaneous flap in the repair of perineal and ischial defects, **33:**95, 1980.

Successful use of gracilis island musculocutaneous flap for closure of ischial pressure sores is described. Also described is use of bilateral gracilis muscle flaps for successful closure of chronic perineal defects after total prostatectomy for granulomatous disease of colon.

11. Maruyama, Y., and co-workers: A gluteus maximus island flap for the repair of a sacral decubitus ulcer, Br. J. Plast. Surg. **33:**150, 1980.

Successful use of skin island over trochanteric portion of gluteus maximus transposed medially to close sacral sores is described. Donor defect was closed primarily.

12. Minami, R.T., Hentz, V.P., and Vistnes, L.M.: Use of vastus lateralis muscle flap for coverage of trochanteric pressure sores, Plast. Reconstr. Surg. **60:**364, 1977.

The successful use of vastus lateralis transposed without skin for closure of trochanteric ulcers is described. Skin grafts were placed over transposed muscle bellies in 18 patients.

13. Minami, R.T., Mills, R., and Pardoe, R.: Gluteus maximus musculocutaneous flaps for repair of pressure sores, Plast. Reconstr. Surg. **60:**242, 1977.

Vascular anatomy of medially based gluteus maximus musculocutaneous flap and the use of variations of this flap for closure of sacral, ischial, and trochanteric sores are described. As described for sacral and ischial sores, the flap is useful. Better flaps are now available for trochanteric ulcers.

14. Nahai, F.: The tensor fascia lata flap, Clin. Plast. Surg. **7:**51, 1980.

Excellent description of the use of the sensory TFL and sensory innervation of skin of TFL cutaneous territory by the twelfth thoracic nerve and lateral cutaneous nerve is presented.

15. Nahai, F. Hill, L., and Hester, T.R.: Experiences with the tensor fascia lata flap, Plast. Reconstr. Surg. **63:**788, 1979.

An excellent description of the neurovascular anatomy of the tensor fascia lata musculocutaneous flap is presented. Uses of this unit in perineal, groin, and abdominal reconstruction and in five cases of microsurgical transfer are described.

16. Stallings, J.O., Delgado, J.P., and Converse, J.M.: Turnover flap of gluteus maximus muscle for the repair of sacral decubitus ulcer, Plast. Reconstr. Surg. **54:**42, 1947.

Successful use of bilateral medially based gluteus maximus turnover flap with skin grafting for closure of sacral sores is described.

17. Vasconez, L.O., Schneider, W.J., and Jurkiewicz, M.J.: Pressure sores, Curr. Probl. Surg. **14:**1, 1977.

This monograph describes the management of pressure sores with emphasis on the use of the muscle and musculocutaneous flap for coverage.

18. Wingate, G.B., and Friedland, J.A.: Repair of ischial pressure ulcers with gracilis musculocutaneous island flaps, Plast. Reconstr. Surg. **62:**245, 1978.

Successful use of gracilis island musculocutaneous flaps in four cases of ischial pressure sores is described.

27

INFERIOR GLUTEAL THIGH FLAP

Robert L. Walton

The inferior gluteal flap is a skin-fascial flap based on an axial vascular pedicle, a descending branch of the inferior gluteal artery (Fig. 27-1). This flap extends beyond the inferior border of the gluteus maximus muscle, incorporating the posterior aspect of the thigh. It can be transposed with the gluteus maximus muscle as an axial extension of this flap or transferred alone as a skin-fascial flap. The descending branch of the inferior gluteal artery with its venae comitantes claims the cutaneous territory of the flap. Intimately associated with the vascular pedicle is the posterior cutaneous nerve of the thigh that provides sensibility within the confines of the flap's boundaries.

The advantages of the flap lie in its size, ease of elevation, wide arc of rotation, sensibility, and potential sparing of regional muscle tissue for coverage. The variable thickness of the subcutaneous tissue incorporated in the flap may well be its major disadvantage, particularly when considering coverage over major pressure points. Also the lack of muscle in the flap makes it less resistant to infection with increased chances for wound breakdown, pseudobursa formation, and slower healing time.

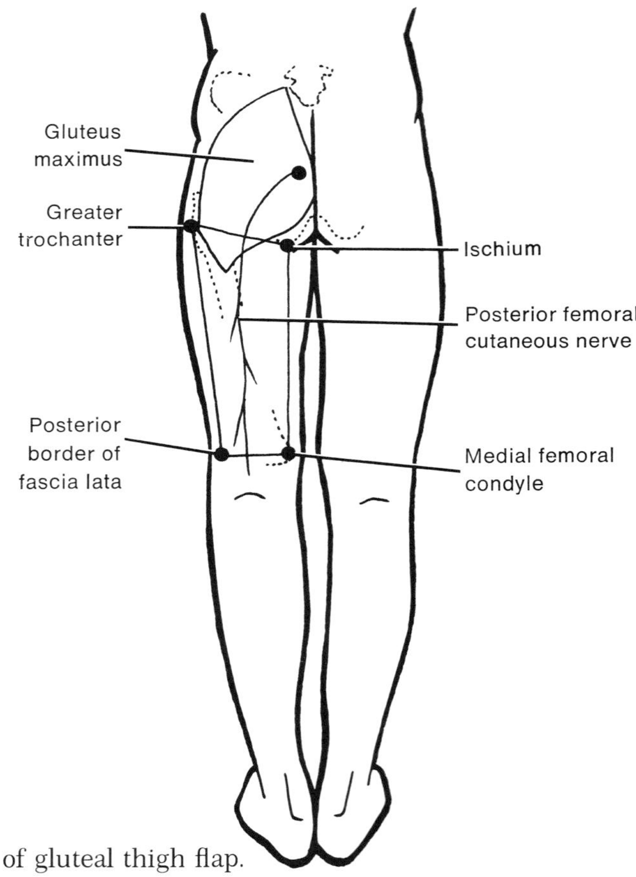

Fig. 27-1. Cutaneous territory of gluteal thigh flap.

NERVE SUPPLY

The flap receives its innervation via the posterior femoral cutaneous nerve (Fig. 27-2). This nerve courses through the infrapiriform foramen with the inferior gluteal vessels and enters the thigh at the lower border of the gluteus maximus muscle near a point midway between the ischium and the greater trochanter. It then extends axially down the posterior thigh slightly medial to the midline and terminates into multiple arborizing branches in the region of the popliteal fossa.

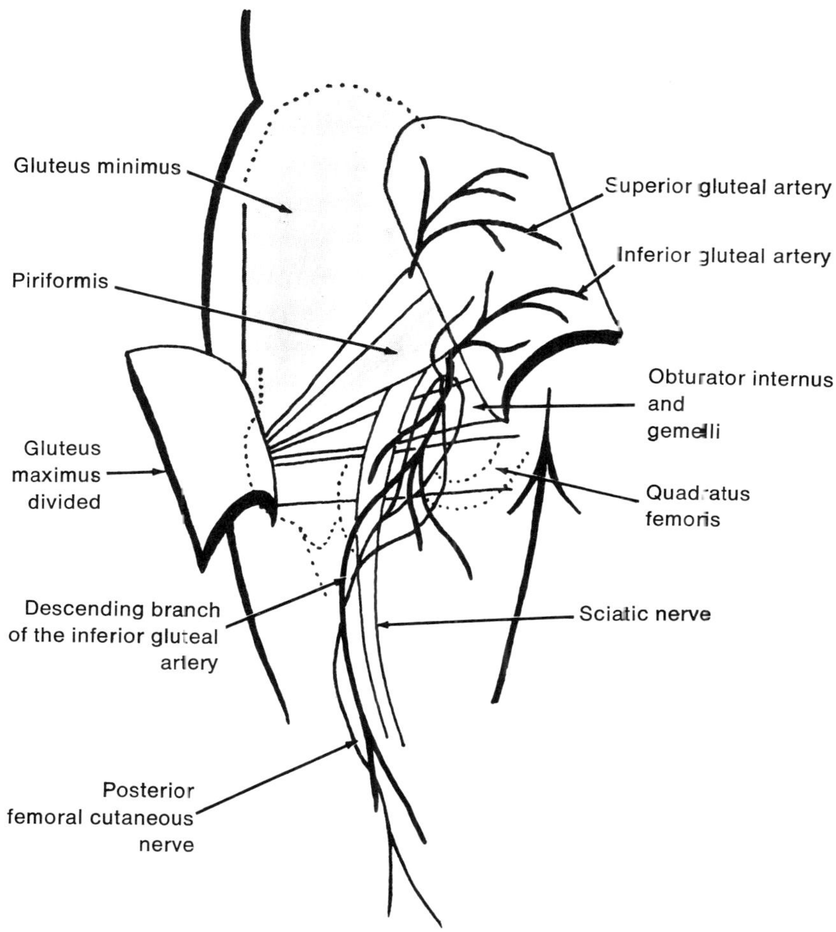

Fig. 27-2. Anatomy of gluteal region

The main body of the nerve lies deep to the femoral fascia, sending small perforating rami to the overlying skin. Its cutaneous territory is represented by the S2 dermatome and extends superiorly from a line joining the ischium and the posterior midpoint of the greater trochanter, laterally to the posterior border of the fascia lata, medially from a line joining the ischium to the medial femoral condyle, and inferiorly to the popliteal fossa (Fig. 27-3).

BLOOD SUPPLY

The descending branch of the inferior gluteal artery and its venae comitantes supply the cutaneous portion of the flap (Fig. 27-2). It enters the thigh at the lower border of the gluteus maximus muscle near a point midway between the ischium and the greater trochanter. Just before its emergence from beneath the gluteus maximus muscle it sends several deep branches to the hamstring muscles. It then extends axially down the thigh below the femoral fascia lateral to the posterior femoral cutaneous nerve. Proximally, the descending branch of the inferior gluteal artery is intimately associated with the nerve. Distally, the artery branches considerably, and this association becomes somewhat random.

The cutaneous territory of the artery is essentially that described for the posterior femoral cutaneous nerve. The venous drainage of the cutaneous territory is variable and consists of one or two venae comitantes that follow the artery in the subfascial plane. A superficial venous system arborizes extensively above the femoral fascia in the subcutaneous tissue. This system forms multiple anastomoses with surrounding muscle perforators superior, medial, and lateral to the cutaneous territory.

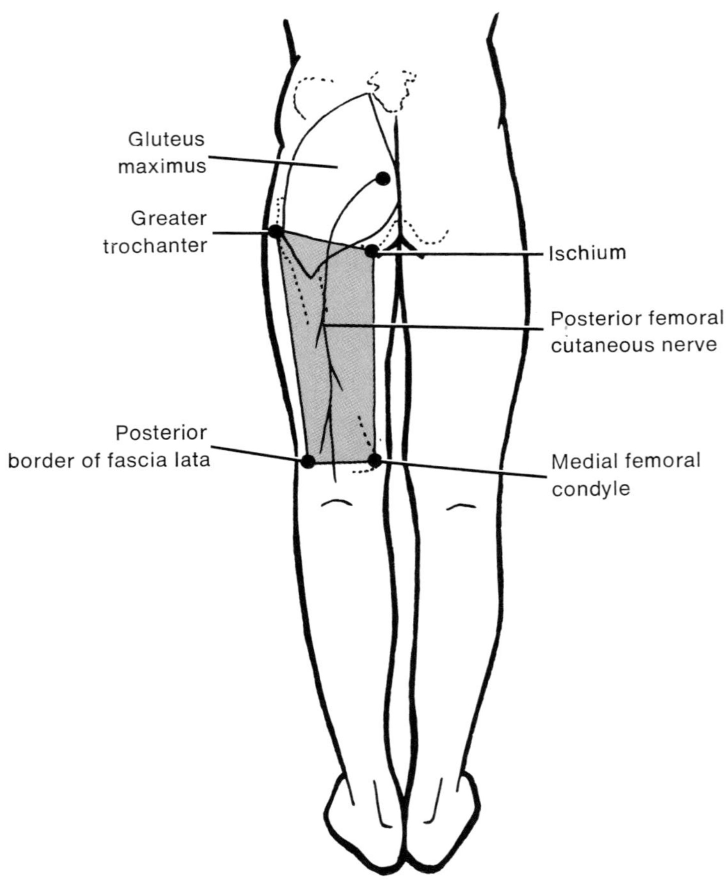

Fig. 27-3. Cutaneous territory of posterior femoral cutaneous nerve.

ARC OF ROTATION

Depending on the demands for coverage, the inferior gluteal thigh flap has two points of rotation. The first point is located at the lower border of the gluteus maximus muscle where the vascular pedicle enters the thigh (Fig. 27-4). This arc allows coverage of trochanteric, ischial, and gluteal defects. A wider arc of rotation can be achieved by splitting the gluteus maximus muscle and dissecting the vascular pedicle to its origin at the infrapiriform foramen (Fig. 27-5). This maneuver establishes the second more proximal point of rotation and allows extension of the flap for coverage of sacral, perineal, superior gluteal, and lateral hip defects.

Distant coverage can be achieved by free transfer of the flap and microneurovascular anastomosis. A potential disadvantage in this regard is the diameter of the flap vessels, which varies between 1 and 2 mm.

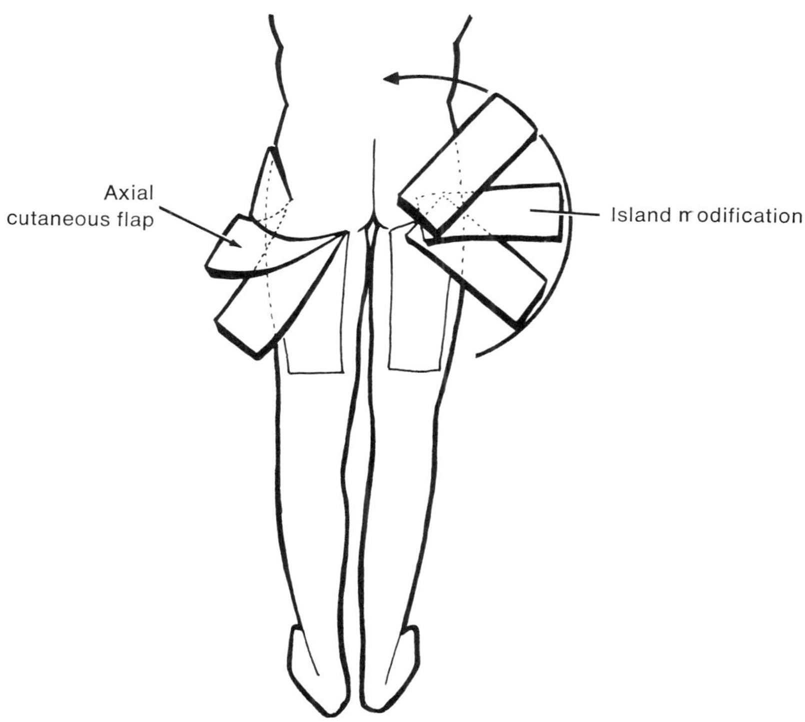

Fig. 27-4. Arc of rotation of inferior gluteal thigh flap. First point is the lower border of gluteus maximus muscle where vascular pedicle enters thigh. NOTE: Cutaneous flap has shorter arc of rotation compared to an island modification at same level of rotation.

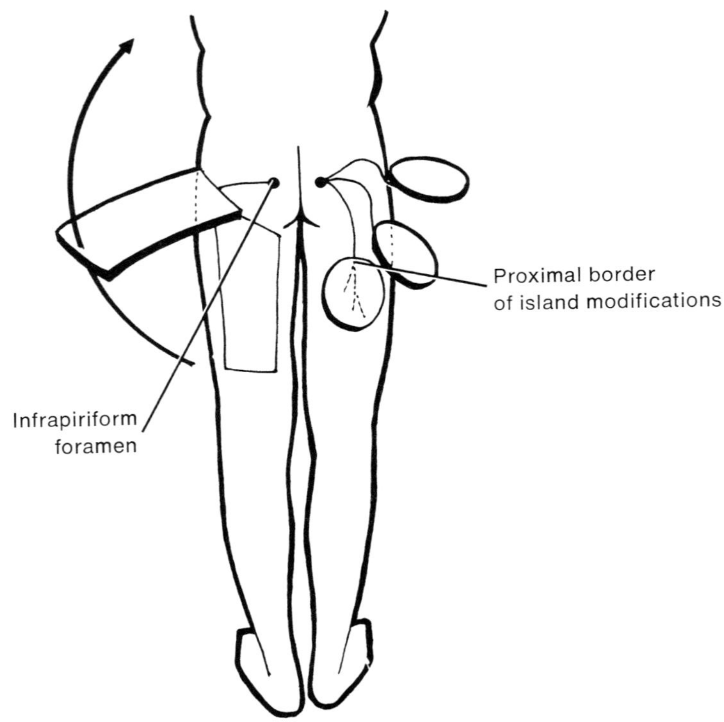

Fig. 27-5. Arc of rotation of inferior gluteal thigh flap. Second point of rotation is the infrapiriform foramen. By splitting gluteus maximus muscle and dissecting pedicle to its source, additional 8 to 12 cm increase in arc of flap is expected. This maneuver is helpful in coverage of sacral, lateral hip, and groin defects. Proximal border of island modification should not extend beyond midpoint of posterior thigh.

ELEVATION OF THE FLAP

The key to successful utilization of the inferior gluteal thigh flap lies in careful design regarding the demands for coverage. For ischial defects one may choose a simple axial transposition flap. In other situations an island pedicle flap may be advantageous. The size of the flap can be varied considerably. For trochanteric defects a longer flap is necessary. Oriented over the inferior gluteal vessels, the flap can be designed anywhere within its vascular territory, as previously outlined. Island pedicle flaps should not extend beyond the midpoint of the thigh because of the variability and branching of the axial vascular system beyond this point. For distant defects the range of coverage can be extended by dissecting the vascular pedicle to its origin at the infrapiriform foramen. This maneuver will extend the reach of the flap by 8 to 12 cm.

Elevation of the flap begins by first identifying the axial extension of the inferior gluteal artery. This is usually found near the midpoint of a line connecting the ischium with the greater trochanter (Fig. 27-6). Doppler assessment is oftentimes very helpful in precisely marking the course of the vascular pedicle. It is particularly important to confirm the patency of the inferior gluteal system if the patient has previously undergone gluteal flap surgery. In rare situations arteriography may be indicated.

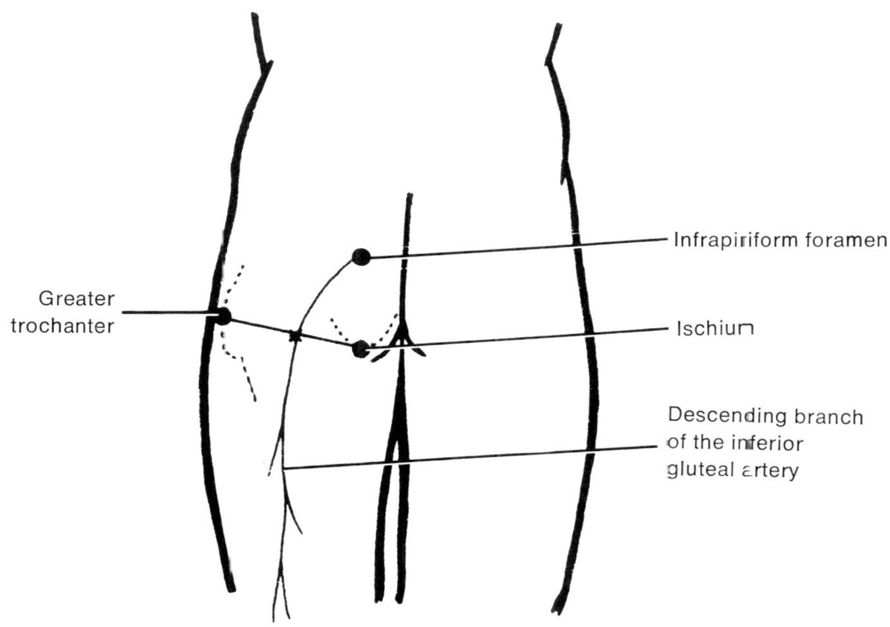

Fig. 27-6. Landmarks used to identify vascular supply of inferior gluteal thigh flap. Pedicle enters thigh near midpoint of a line, connecting ischium with greater trochanter.

After the appropriate design is chosen and oriented, an incision is made at the distal border of the flap (Fig. 27-7). The dissection is extended down through the femoral fascia. It is important to identify the posterior femoral cutaneous nerve, because this will define the level of dissection for flap elevation. Just lateral to the nerve is the axial vascular system. Blunt dissection will easily separate the flap from the underlying muscular bed. Laterally the fascia lata is confluent with the femoral fascia and must be divided sharply. Superiorly above this, one will encounter large vascular anastomoses between the inferior gluteal and femoral systems. These can be divided without compromising the flap's vascularity. At the inferior border of the gluteus maximus muscle the vascular pedicle can be easily identified beneath the flap. Care must be exercised to prevent damage to the pedicle at this point. The flap is then transposed to the appropriate recipient site. Back cuts can be made to facilitate transposition and insetting.

A

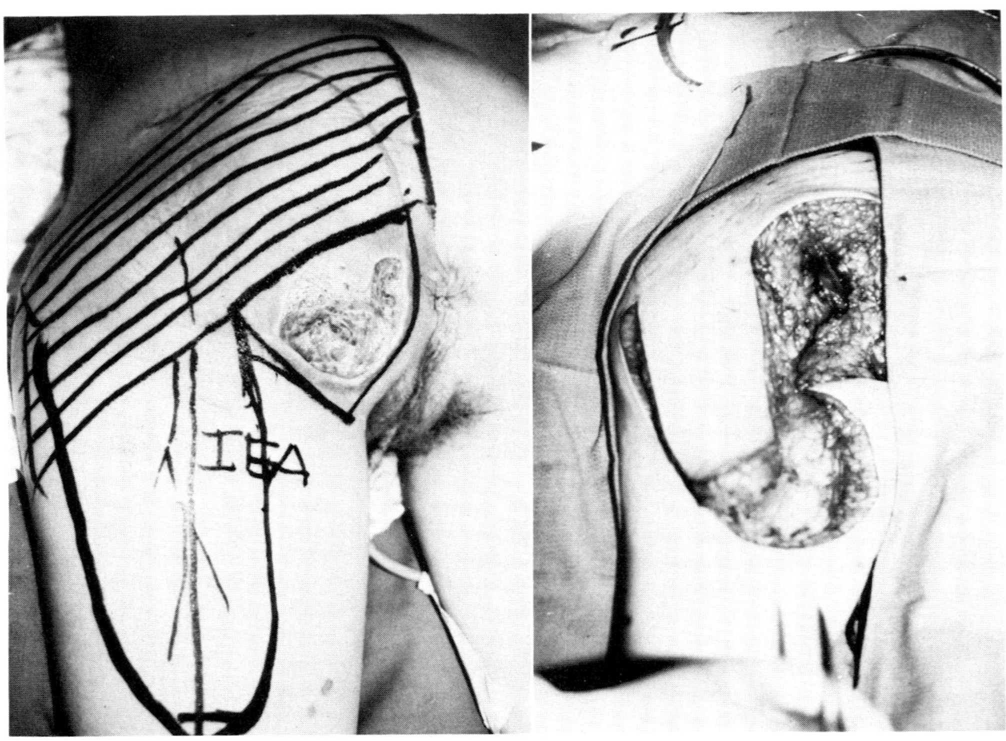

B

Fig. 27-7
A, Twenty-four-year-old paraplegic woman with large ischial pressure sore.
B, Flap has been elevated and dissection is extended to inferior border of gluteus maximus muscle.
C, Elevation of flap. Arrow points to deep branch of inferior gluteal artery.
D, Closer view of underside of flap. Clamp marks posterior femoral cutaneous nerve.
E, Flap is transposed into ischial defect.
F, Flap is inset, and donor site is closed primarily. NOTE: Cone of rotation at base of flap.

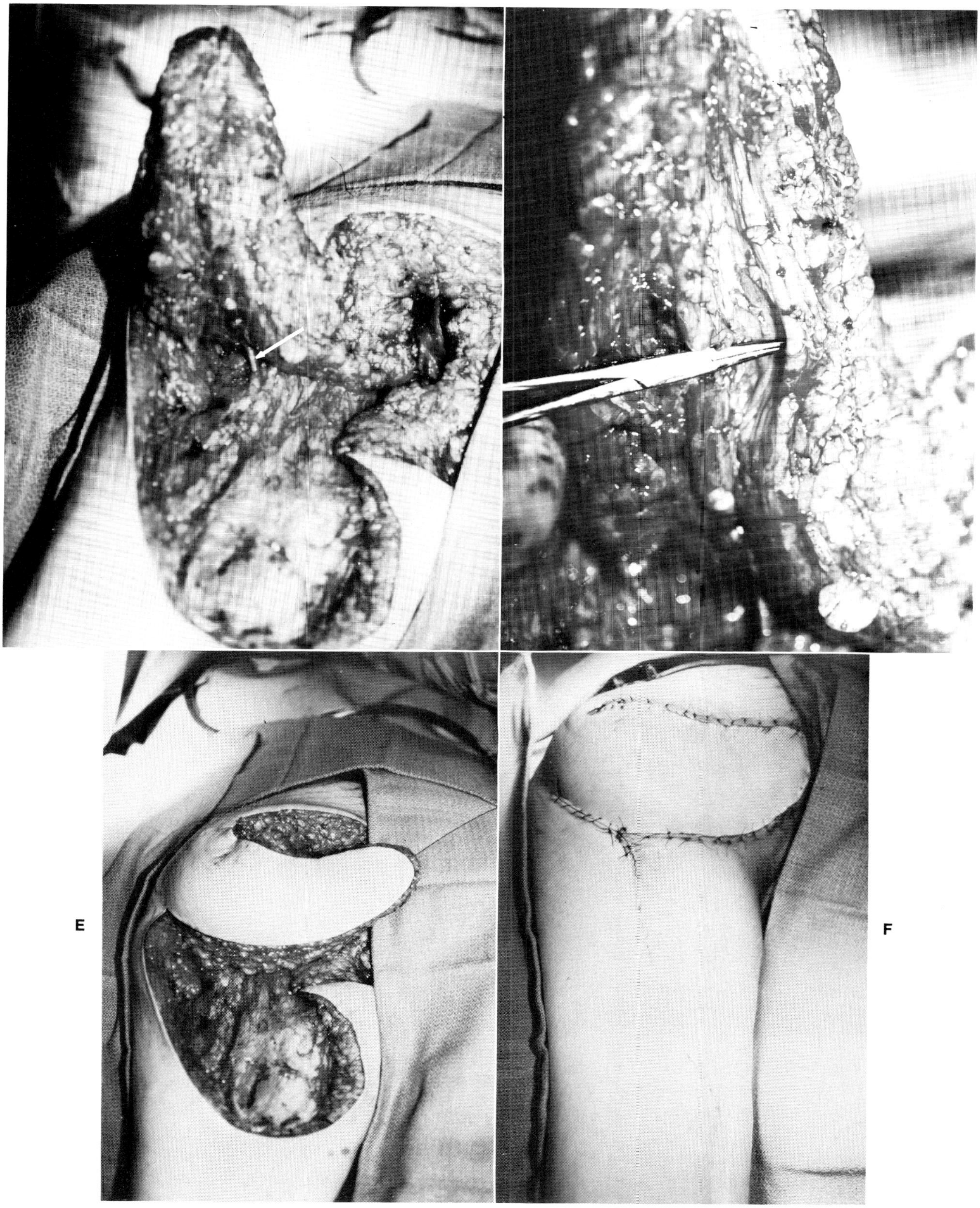

C D E F

Fig. 27-7, cont'd. For legend see opposite page.

If an island pedicle flap is chosen, the initial dissection is carried out, as just described (Fig. 27-8). At the desired level, the neurovascular pedicle is isolated and dissected proximally. Medial and lateral tributaries are individually isolated and divided.

The proximal border of the skin island is then incised and carried down through the femoral fascia. Several venous tributaries will be encountered in this dissection, and these should be ligated. At the level of the inferior border of the gluteus maximus muscle, the pedicle will give off several large branches. These branches are divided after precise visualization of the main vascular trunk. Gentle separation of the inferior gluteal pedicle from beneath the gluteus maximus muscle will allow the dissection to proceed to the origin of the pedicle. Numerous muscular branches will be encountered here, requiring some attentive care in the dissection. The island is then transferred to the recipient site. Frequently the overlying gluteus maximus muscle is divided to accommodate the mechanics of this transfer. Caution should be exercised to avoid kinking of the vascular pedicle.

The flap donor site can oftentimes be closed primarily. If the width of the flap exceeds 10 cm, however, primary closure may be tenuous. In this situation a split-thickness skin graft is often necessary.

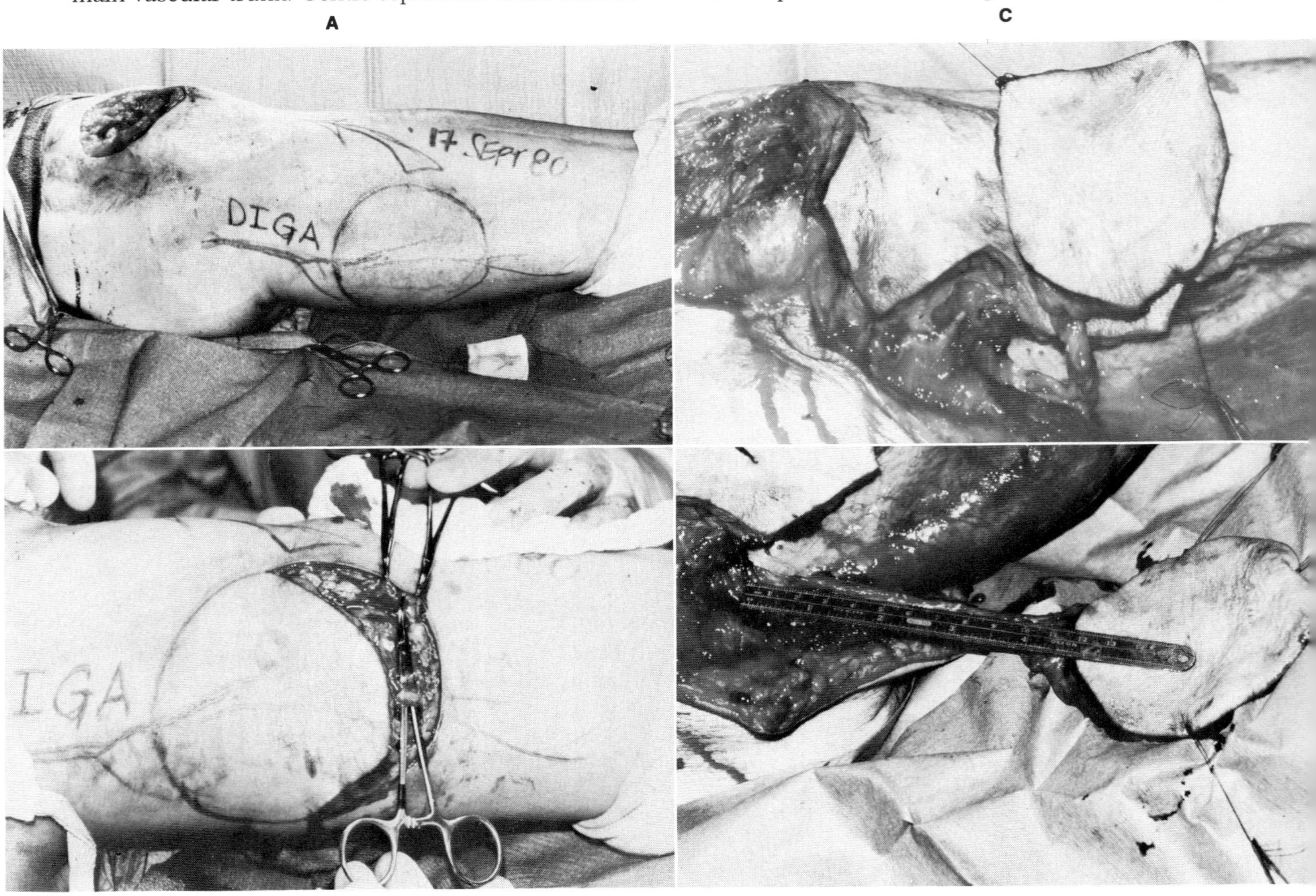

Fig. 27-8

A, Fifty-three-year-old ambulatory man with recurrent breakdown over right trochanteric region. Island pedicle modification of inferior gluteal thigh flap was planned with dissection to infrapiriform foramen.

B, Dissection starts with incision at inferior margin of flap. Clamp lies beneath posterior femoral cutaneous nerve. Dissection is performed deep to level of this nerve.

C, Cutaneous island is elevated after dissection to source of its vascular pedicle.

D, Neurovascular pedicle in this patient measured approximately 14 cm.

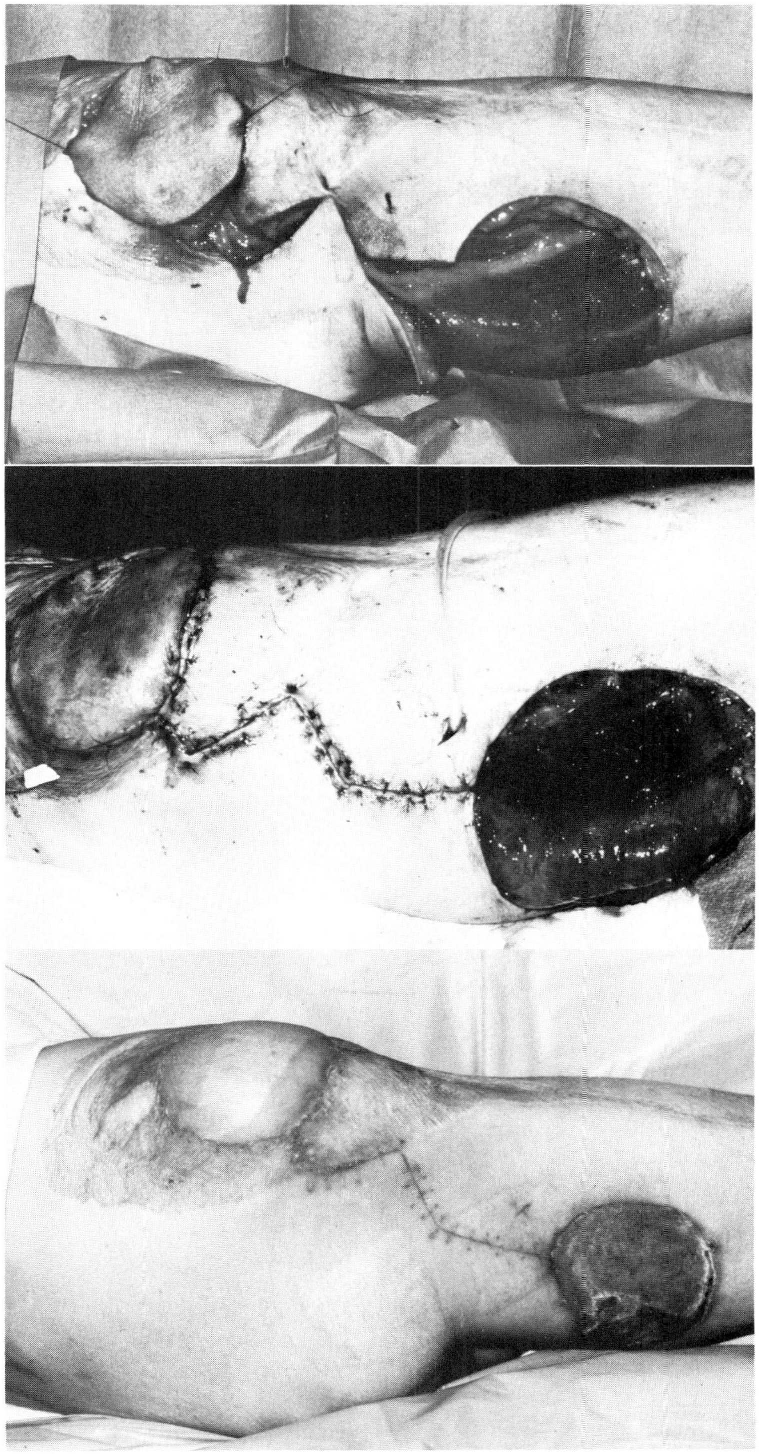

E

F

G

Fig. 27-8, cont'd

E, Transposition of island flap into wound defect.

F, Flap has been inset into trochanteric defect with primary closure of gluteal skin. Donor defect was covered with split-thickness skin graft.

G, One-month postoperative view. Wounds are well healed. Two-point discrimination in flap measured 3 to 4 cm. Patient could distinguish pinprick, light touch, and temperature variations.

The size of the flap should be designed to satisfy the needs for coverage. In most cases the point of rotation will be centered at the gluteal crease midway between the ischium and the greater trochanter. If one chooses to transpose the flap with an intact cutaneous pedicle, approximately one third of the planned length will be absorbed by the cone of rotation (Fig. 27-9). This fact must be considered to avoid inadequacies of reach. Also some patients may object to the "dog ear" deformity that inevitably results from this design.

Island modifications of this flap are inherently capable of further reach because of the absence of a cutaneous cone of rotation. However, the dissection is technically more difficult and carries a higher risk of injury to the vascular pedicle. In planning coverage of sacral and lateral hip defects, the arc of rotation can be extended approximately 10 cm by dissecting the pedicle to the infrapiriform foramen. This obviously requires an island modification of the flap as well as division of the inferior portion of the gluteus maximus muscle. Theoretically one might consider a "pull through" maneuver (i.e., bringing the island flap with its pedicle between the superior and inferior divisions of the gluteus maximus muscle). This may be of advantage in the ambulatory patient.

In all instances the flap should be centered over the course of the descending branch of the inferior gluteal artery. Failure to do so may result in injury to the vascular pedicle, or worse, necrosis of the flap. In island pedicle modifications of this flap, the proximal border should not be placed distal to the midlevel of the thigh because of variable vascular branching beyond this point.

The thickness of the inferior gluteal thigh flap is variable from patient to patient. For the most part, females tend to have thicker subcutaneous tissues in their posterior thigh regions. Extremes of flap thickness range from 2 to 12 mm. The distal portion of the flap is generally thinner than its proximal aspect. Flap thickness is grossly estimated by pinching the skin within the cutaneous territory. Thick, bulky flaps are cumbersome and difficult to transpose and must be designed somewhat larger to satisfy the demands for reach and coverage. A thick flap, however, may be advantageous, particularly when planning coverage over pressure areas. Conversely, this situation may well be the potential pitfall of a thin flap.

Operative management

It is important to identify the posterior femoral cutaneous nerve early in the dissection. Failure to do this may result in elevation of the flap in a plane superficial to its vascular pedicle.

Proximally, accurate identification of the main arterial trunk is essential when dissecting to the origin of the pedicle. Numerous branches are encountered at the inferior border of the gluteus maximus muscle. These should be meticulously isolated and ligated to avoid injury to the pedicle. Small muscle perforators will be encountered beneath the gluteus maximus muscle. These branches are very fragile and difficult to identify, because they embark from the fatty sheath of the pedicle. The use of fine vascular clips is often helpful at this level of the dissection.

The sciatic nerve lies just lateral to the origin of the pedicle and receives one or two large branches

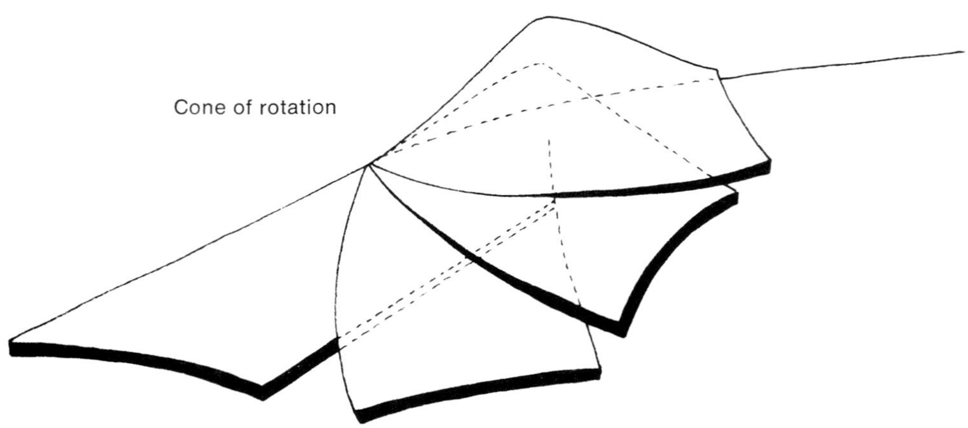

Fig. 27-9. Transposition of inferior gluteal thigh flap with intact cutaneous pedicle. This cone of rotation will absorb up to one third of flap's length.

from its proximal aspect. These branches should be preserved in the ambulatory patient. Care should also be exercised to avoid direct injury to the sciatic nerve at this level.

If the fatty stalk of the pedicle has not been violated, the integrity of the posterior femoral cutaneous nerve will be sound. Numerous nerve fascicles course through the fatty root of the pedicle and converge just proximal to the inferior border of the gluteus maximus muscle to form the main body of the nerve. This anatomic variance should be considered when planning free flap transfers for restoration of sensibility.

When transposing island pedicle modifications of the inferior gluteal thigh flap, one must avoid twisting, kinking, or traction of the neurovascular pedicle. This is especially important when the rotation exceeds 90°. Failure to appreciate this may result in flap necrosis, severe dysesthesia, or hyperalgesia (sensibility problems in the flap).

In the ambulatory patient there will be variable areas of numbness in the posterior thigh region as a result of the flap elevation. This is not a major concern for most patients and poses no functional problem. It is important, however, to inform the patient of this sensory deficit before operation.

Postoperative management

Routine nursing care is generally all that is required in the postoperative period. A certain amount of swelling in the inferior gluteal thigh flap is to be expected. This is particularly true for island pedicle modifications, because the lymphatic effluent is disrupted. Early venous congestion may also be encountered in the island flap. This usually improves over the course of 24 to 48 hours and is the result of venous spasm. If the flap becomes extremely tense, pale, or mottled, consideration should be given for reexploration to assess possible mechanical injury to the vascular pedicle.

Cellulitis of the flap is an infrequent complication that may appear within the first 48 hours after transposition. As with any cutaneous flap, the ability to handle bacterial inoculation is somewhat inferior to its muscular counterpart. For this reason prophylactic antibiotics are administered when the flap is transposed into a contaminated wound. Once infection has been established, appropriate antibiotic therapy is indicated. Flap necrosis is a rare complication of infection, providing these criteria are closely followed.

In the early postoperative period, direct pressure on the flap or its pedicle is to be avoided. Patients are allowed to ambulate after the first postoperative week. Weight bearing commences on the second week.

The inferior gluteal thigh flap is remarkably versatile and represents a major alternative to the muscle flap for coverage of sacral, ischial, perineal, trochanteric, and groin defects. If properly designed and executed the success rate is very high. In our initial series of 14 such flaps, there were no instances of flap necrosis. As our experience with the inferior gluteal thigh flap increases, indications for its use will become more apparent. As illustrated in Fig. 27-10, each patient represents a new and different challenge, the solution of which is met by a fundamental appreciation of basics tempered only by the imagination.

A

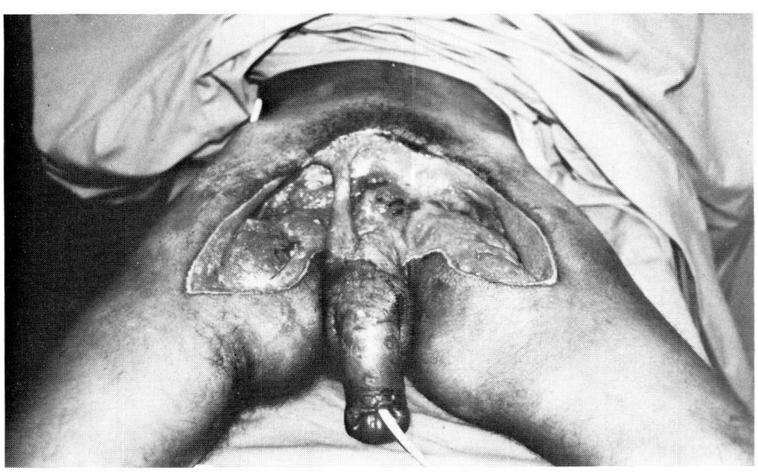

Fig. 27-10

A, Twenty-four-year-old quadriplegic patient had necrotizing infection of suprapubic region, extending to both hip joints. Wide débridement included extensive resection of adductor and quadriceps muscle groups with disarticulation of proximal one third of each femur. Resulting wound exposed both superficial femoral systems laterally and communicated with counterincisions over trochanteric regions. *Continued.*

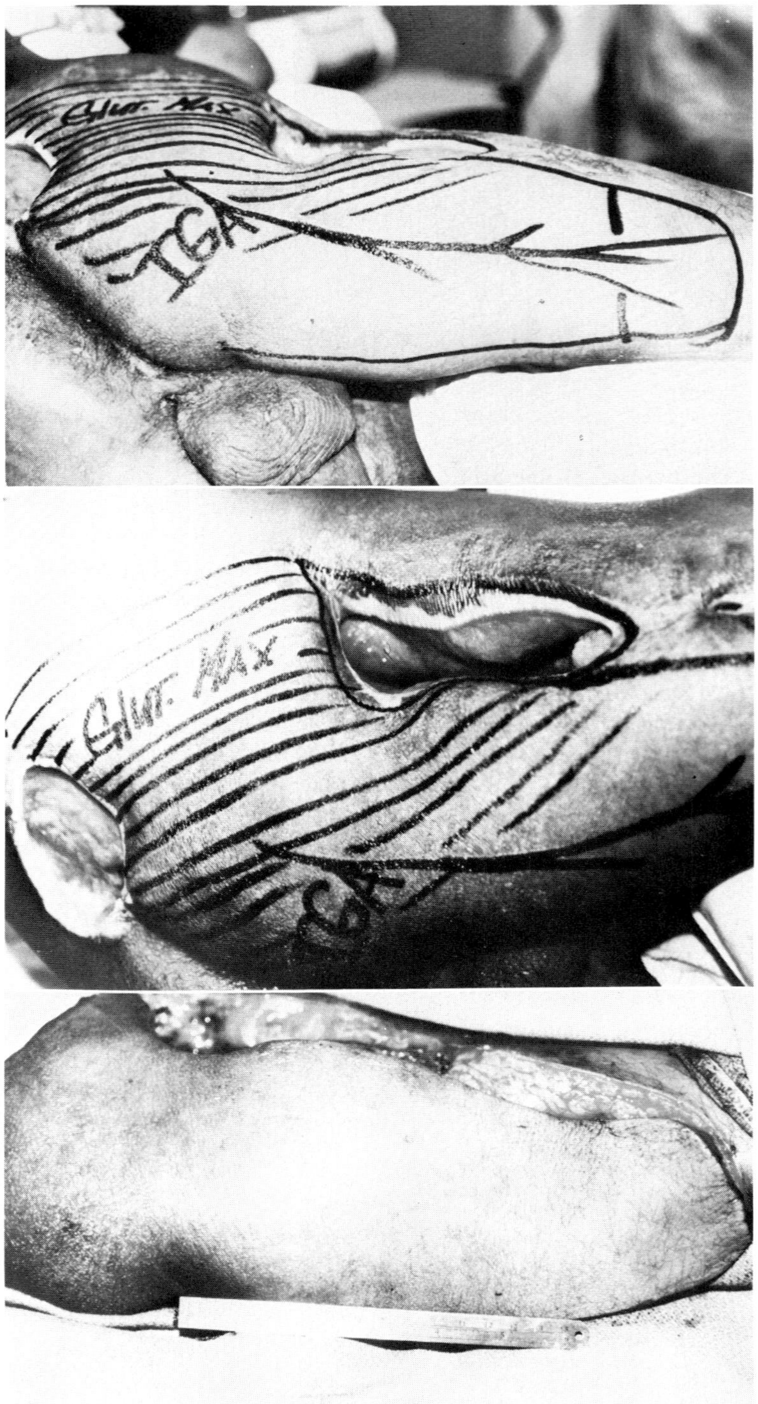

Fig. 27-10, cont'd

B, Bilateral inferior gluteal thigh flaps were designed, encompassing entire posterior aspects of both thighs. Each side was performed as staged procedures 1 week apart.

C, Lateral view of right thigh revealing trochanteric wound. Defect communicated with anterior groin wound.

D, Each flap was dissected to inferior border of gluteus maximus.

E
F
G

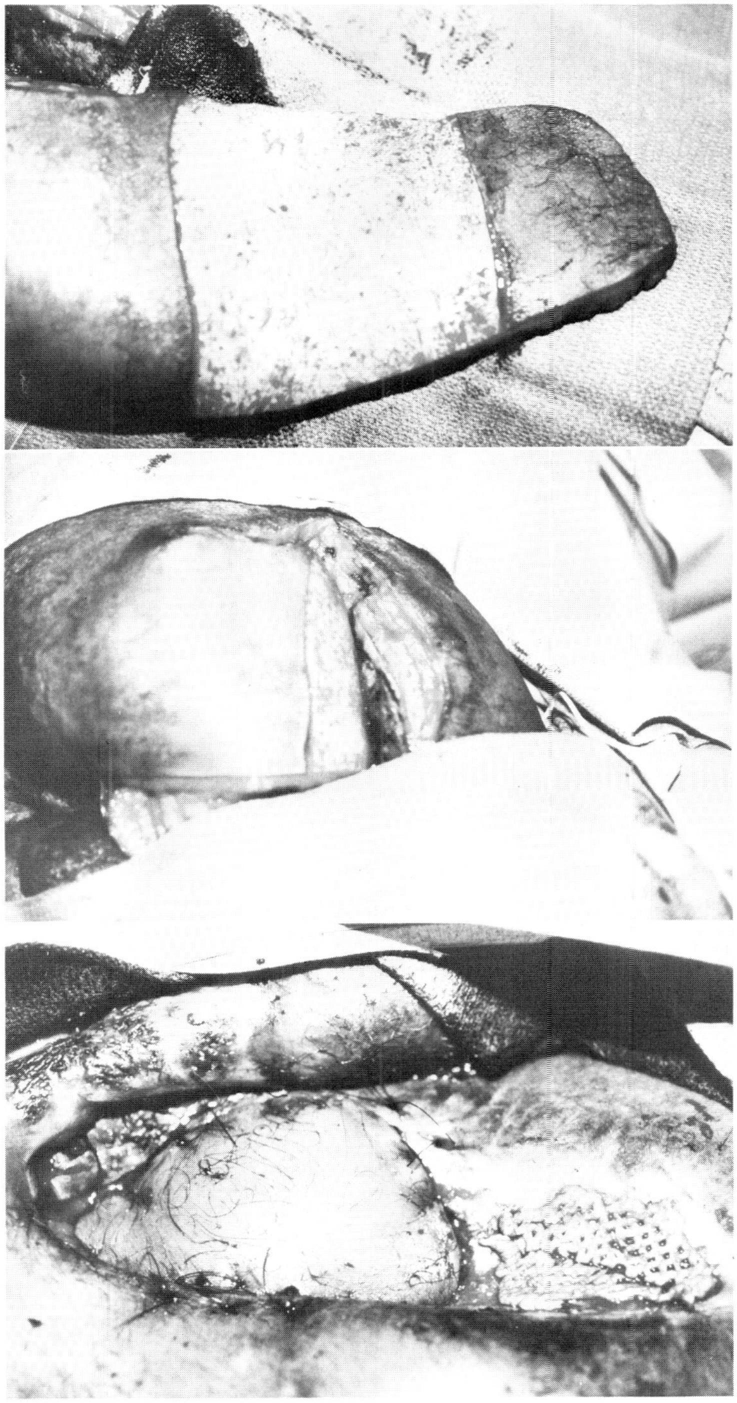

Fig. 27-10, cont'd

E, Flap was then "deepithelialized" in its midportion for transfer through thigh. Distal cutaneous paddle of skin was employed to provide anterior groin coverage.

F, Flap has been transposed into thigh wound. Flap enters thigh through trochanteric defect.

G, Anterior projection of flap is represented by distal skin paddle. This portion of the flap provided coverage of exposed femoral vessels in groin.

Continued.

H

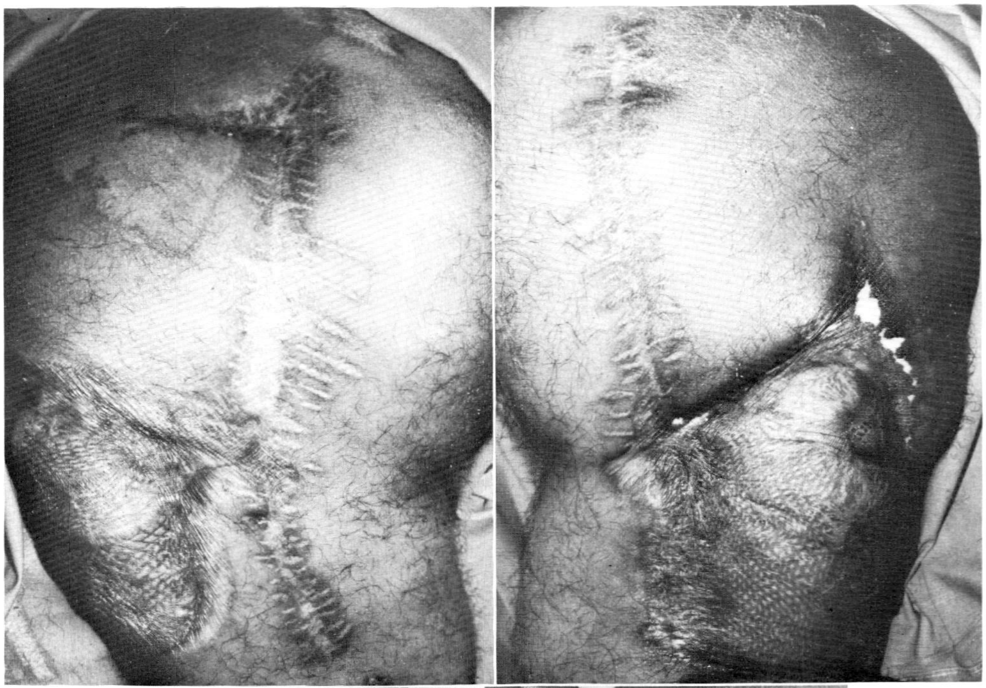

I

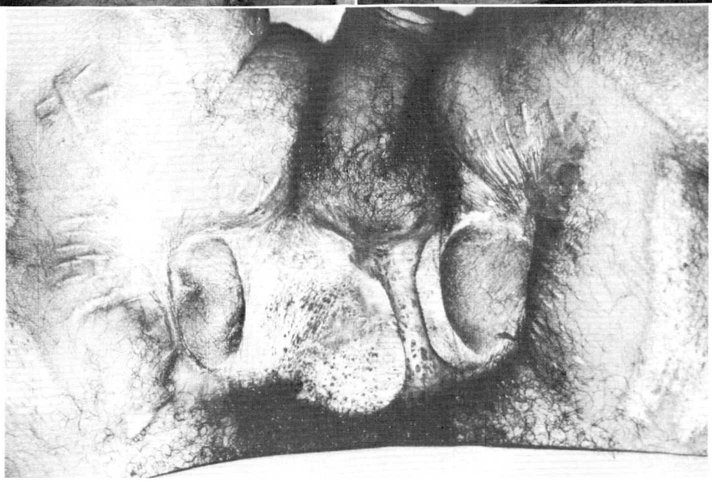

Fig. 27-10, cont'd

H, Inferior gluteal thigh flap was used to close both hip wounds (6-week postoperative lateral view of wound). Split-thickness skin grafts were used to close suprapubic region.

I, Distal flaps provide stable coverage of groin wounds.

J 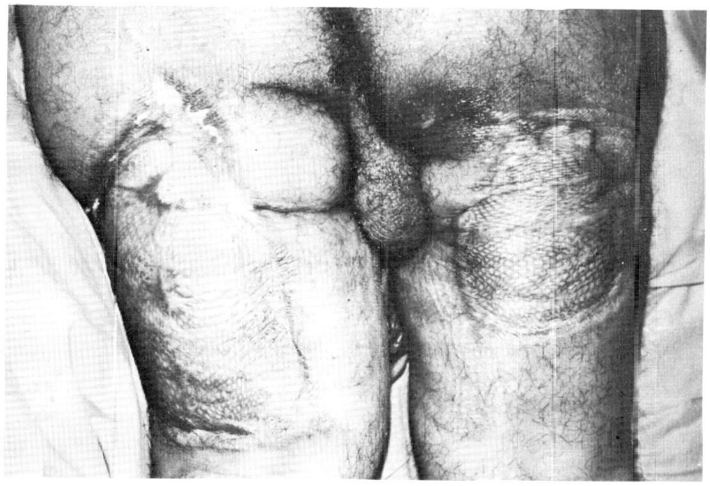

Fig. 27-10, cont'd
J, Posterior view reveals primary healing of both thigh wounds.

ANNOTATED BIBLIOGRAPHY

1. Hurwitz, D.J.: Closure of a large defect of the pelvic cavity by an extended compound flap based on the inferior gluteal artery, Br. J. Plast. Surg. **33:**256, 1980.

   The initial description of the vascular anatomy of the gluteal thigh flap and its application for reconstruction of a significant pelvic defect are presented.

2. Hurwitz, D.J., Schwartz, W.M., and Mathes, S.J.: A reliable sensate flap for the closure of buttock and perineal wounds, Plast. Reconstr. Surg. **68:**40, 1981.

   The vascular anatomy, arc of rotation, and sensory territory via the posterior femoral cutaneous nerve for this skin fascial flap is established. Case reports demonstrate the use of this flap for ischial, trochanteric, and pelvic defect coverage. The addition of a new neurosensory flap is a welcome addition to the flaps available for coverage of these areas.

28

COMPLICATIONS

Bernard S. Alpert ■ Luis O. Vasconez ■ Stephen J. Mathes

The treatment of pelvic pressure sores is greatly facilitated by use of axial, muscle, and musculocutaneous flaps. Despite expert assistance in rehabilitation, control of spasticity, and pressure distribution devices, recurrence is common unless sensation is restored to pressure zones. Unfortunately, sensory restoration is rarely feasible. Therefore flap choice and design must consider both coverage of the defect and future recurrent problems. A flap may successfully cover a defect but may create future complications because of loss of adjacent potential flaps.

SACRUM

Bony resection for the sacral pressure sore should be limited to infected bone. It is rarely necessary to resect beyond the outer cortex of this bone. Aggressive bone resection may result in injury to the underlying sacral venous plexus with resultant excessive bleeding and to the rectum with subsequent infection and fecal fistulae.

Gluteus maximus

Planning errors. Although either half of the gluteus maximus muscle will reach the sacrum, the inferior half of the gluteus maximus muscle is best preserved for future use in ischial coverage. The design of the skin island should not exceed the dimensions whereby direct donor site closure is not possible. Skin grafting of the donor defect is difficult because of the depth of the wound. Use of bilateral superior gluteal musculocutaneous flaps is preferable when the sacral defect is too large for a single musculocutaneous flap.

Technical errors. Identification of the superior border of the gluteus maximus muscle is difficult because of the thick lumbosacral fascia. Inadvertent incision into the muscle belly is possible, since the muscle is thick and often infiltrated with excessive fibrofatty tissue in the paralyzed patient. Identification of the piriformis muscle beneath the gluteus maximus muscle indicates the level of muscle splitting to stay between the superior and inferior gluteal vascular pedicles. Elevation of the superior border of the gluteal muscle and incision of the superior muscle fibers close to its insertion allow safe identification of the underlying piriformis muscle, vascular pedicles, and sciatic nerve.

Postoperative errors. Closed suction drainage of the donor site is recommended.

Transverse back flap

Planning errors. The transverse back flap is based on lumbar perforating vessels on one side and should not extend in length beyond the posterior superior iliac crest on the opposite side.

Technical errors. The transverse back flap should not be elevated beyond the midportion of the paraspinous muscle on the side of the flap base.

Postoperative errors. The patient is positioned prone postoperatively to avoid pressure on the flap base. Suction drainage is recommended.

ISCHIUM

Bony resection for ischial pressure sores should be limited to infected bone. Partial ischiectomy generally provides adequate débridement. Total ischiectomy shifts the pressure to the opposite ischium and may be associated with later development of periurethral diverticulae.

TFL

Planning errors. For ischial coverage the extended portion of the TFL flap is used. The distal flap is thin (fascia and skin only) and may be unreliable, especially if flap length extends closer than 5 to 8 cm to the knee (Fig. 28-1). Early recurrence may be a problem, because the flap is thin or sutured under tension. In obese patients the TFL is bulky proximally, resulting in inadequate arc of rotation to reach the ischium.

Some paraplegic patients with levels of injury below L3 have intact lateral femoral cutaneous nerves. The sensory innervated TFL is useful in these patients despite the fact that sensibility in the distal portion of the flap covering the ischium is not as marked as it is in the proximal portion.

A

B

C

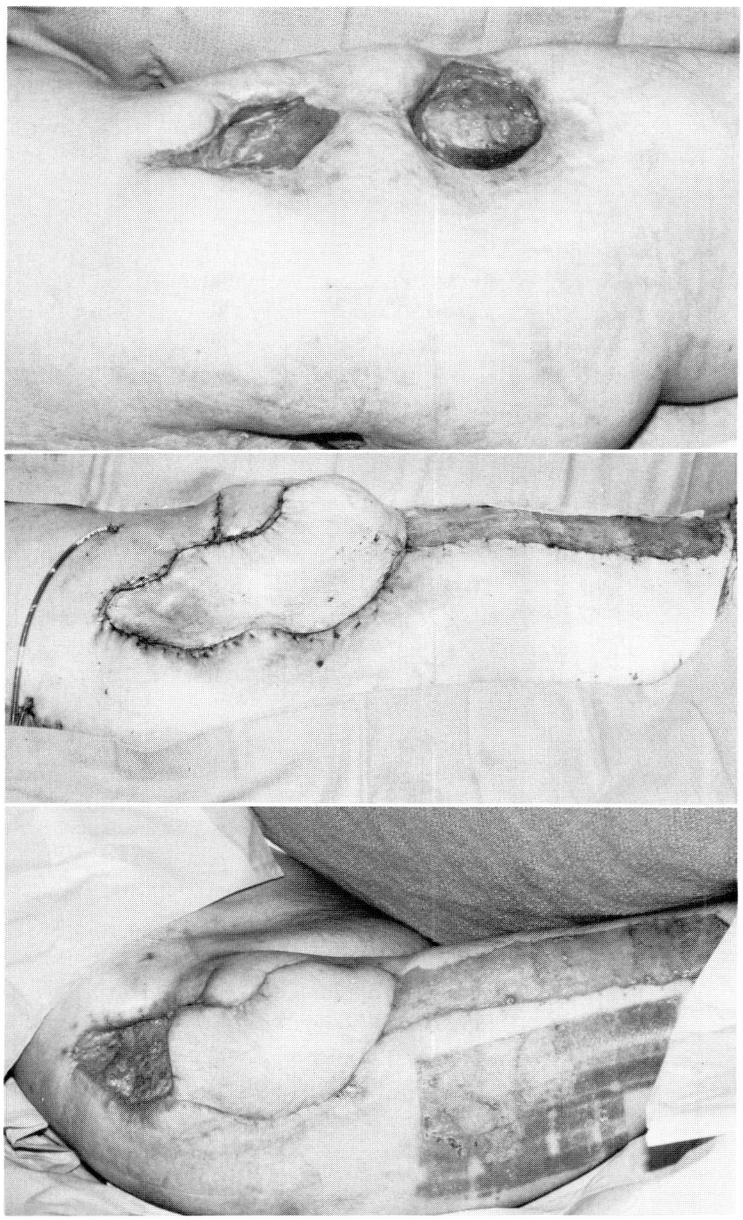

Fig. 28-1
A, Pressure sores located over trochanter and posterior iliac crest.
B, Extended TFL transposed for coverage of both defects.
C, Loss of distal 5 cm of flap may occur in extended TFL musculocutaneous flap.

Technical errors. An extended TFL will be required to reach the ischium. During flap elevation, if large perforating vessels are encountered between the vastus lateralis and the TFL, this indicates compromise of the TFL muscle. Under these circumstances the circulation of the extended portion of the flap may be more dependent on vascular communicators with the vastus lateralis muscle than on the TFL muscle. It is therefore advisable to elevate the vastus lateralis muscle with the overlying TFL. This will ensure survival of the overlying skin and provide the extra padding of the vastus lateralis muscle.

Tunneling of the flap to the ischium should be avoided. Intervening skin bridges should be incised or excised to avoid flap compression and flap compromise. Firm drainage catheters may cause vascular compromise if placed perpendicular to the vascular pedicle beneath the flap.

Postoperative errors. Pressure over the flap on the donor area is avoided. Suction drains are recommended.

Hamstrings

Planning errors. The hamstring muscles have multiple vascular pedicles (type II) that must be considered in flap design. Although this pattern of circulation limits safe transposition, division of origin and insertion allows advancement of this muscle group and its skin territory over the ischial area. Use of the hamstring muscles in the ambulatory patient results in loss of knee flexion.

Technical errors. In elevating the hamstring muscles, the most proximal and distal vascular pedicles may be divided for greater flap advancement. Tension should be avoided during flap advancement. The sciatic nerve courses between these muscles and should be identified and preserved during flap elevation.

Postoperative errors. The patient must be positioned prone or on the opposite side postoperatively. Closed suction drainage is required.

Gracilis

Planning errors. Incorrect orientation of the skin territory is the most common problem in designing a gracilis musculocutaneous flap. For ischial coverage the proximal skin continuity is maintained rather than a skin island created. With the patient positioned prone, the distal muscle is first identified before the skin territory is incised. This ensures correct orientation of the skin territory over the muscle.

Technical errors. The musculocutaneous portion of the gracilis is identified through an incision above the knee. This portion of the gracilis lies between the fleshy sartorius anteriorly and the fascial portion of the semimembranosus posteriorly. With traction on the gracilis tendon the correct skin territory is outlined. During flap elevation and transposition, the skin is temporarily sutured to the underlying muscle to avoid shearing of the musculocutaneous vessels. Excessive traction on the dominant vascular pedicle should be avoided during transposition.

Postoperative errors. The patient should be positioned prone. Suction drainage is required.

GREATER TROCHANTER

Trochanteric sores may extend into the hip joint. Débridement may include the greater trochanter or the entire joint. Preoperative assessment of the joint is necessary so that the reconstruction can be planned.

TFL

Planning errors. Previous incisions in the skin territory of the flap do not preclude the use of the flap. The posterior border of the flap is usually involved in the sore or the débridement. It is essential to design a flap of adequate length, because the point of rotation is at the level of the trochanter (Fig. 28-2).

Technical errors. The gluteus minimus is firmly adhered to the TFL posteriorly. The point of rotation of the flap is approximately 8 to 10 cm below the anterior superior iliac spine or at the level of the trochanter.

The lateral femoral cutaneous nerve is located along the medial border of the flap and should be identified as an initial step in flap elevation if a sensory flap is planned.

Postoperative errors. Suction drainage is recommended. The patient should be positioned on the opposite or prone.

Gluteal thigh flap

Planning errors. The lateral border of the gluteal thigh flap is usually involved in the sore or the débridement. If the wound extends to the posterior thigh, or a previous incision is located in the gluteal crease, the vascular pedicle may have been injured or divided. It is essential to design a flap of adequate length, because the point of rotation is at the level of the trochanter.

Technical errors. The dissection for this flap elevation includes the descending branch of the inferior gluteal artery and the posterior femoral cutaneous nerve. The sciatic nerve is identified deep to the flap and is preserved.

Postoperative errors. The patient must be positioned prone or on the opposite side postoperatively. Closed suction drainage is necessary.

Vastus lateralis

Planning errors. The skin overlying the vastus lateralis is supplied primarily by the TFL, and therefore the vastus lateralis is not usually elevated as a cutaneous flap without the TFL. This flap is usually used for coverage of the acetabulum after hip disarticulation and femoral head resection. Skin grafting of the muscle is necessary.

Technical errors. The vastus lateralis should be elevated through a skin incision that does not violate the territory of the TFL. An incision between the rectus femoris and TFL skin territories is recommended. The separation of the vastus lateralis from the other quadriceps distally is difficult. Failure to cover the donor area with adjacent muscles will leave the femur exposed.

Postoperative errors. Closed suction drainage is necessary.

A

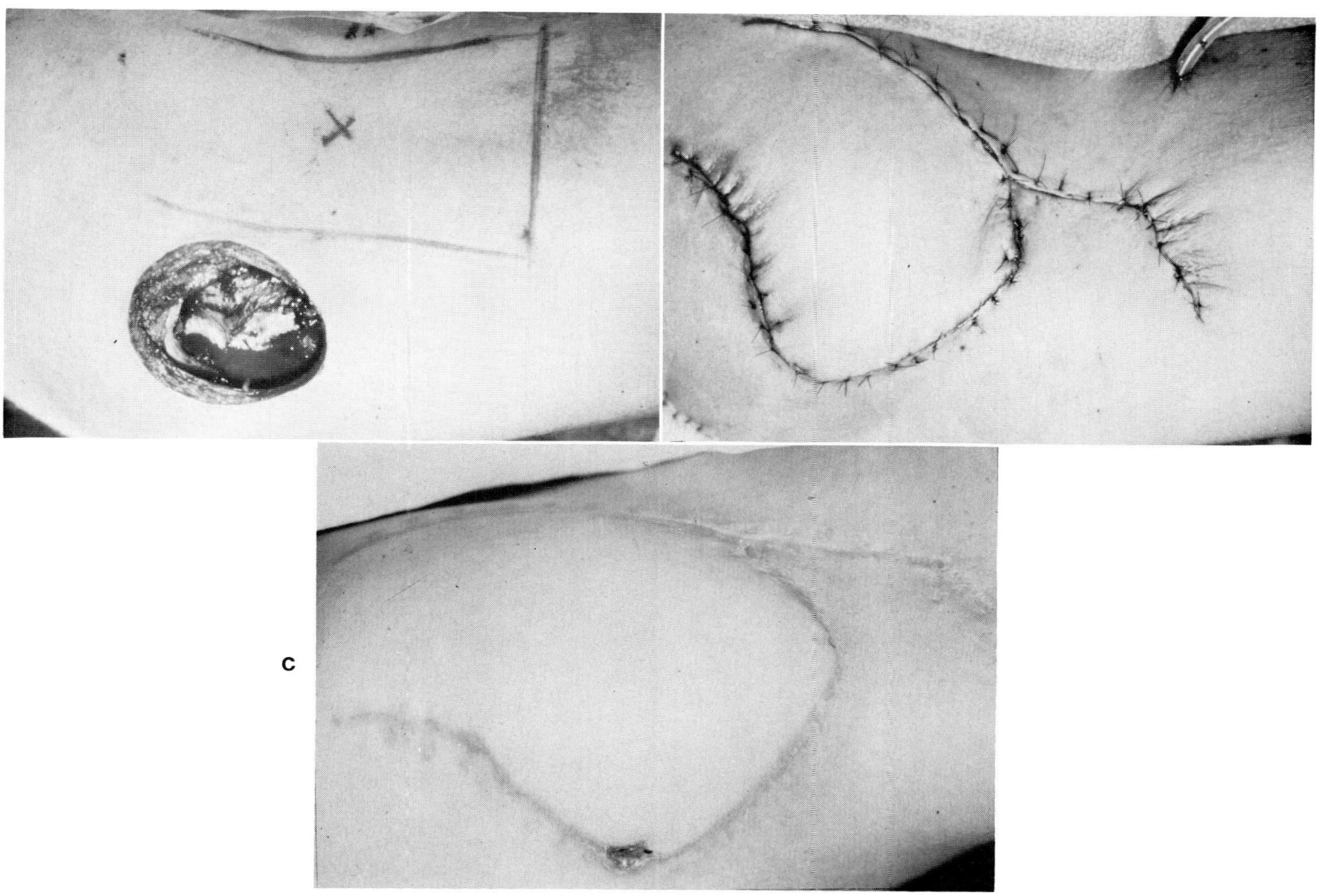

B

C

Fig. 28-2
A, Design of TFL for coverage of trochanteric defect. Flap does not extend adequate distance beyond point of rotation at site of vascular pedicle.
B, Inadequate length in musculocutaneous flap design may result in division of vascular pedicle for flap to reach defect.
C, Greater flap length results in less tension at site of flap closure.

29

A SYSTEMATIC APPROACH TO FLAP SELECTION

Stephen J. Mathes ▪ Foad Nahai

THIGH (Figs. 29-1 and 29-2)

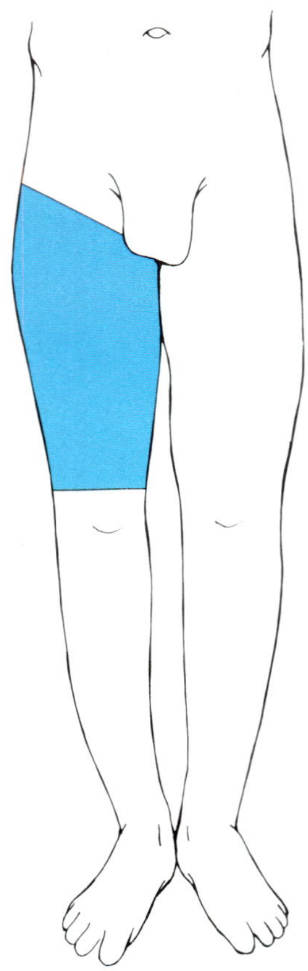

Fig. 29-1

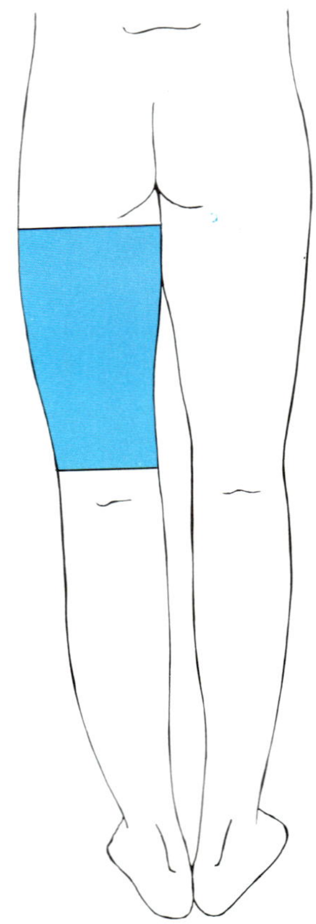

Fig. 29-2

Skin grafts

Skin grafts are method of choice provided no bone or vital structures are exposed

Local flaps (Fig. 29-3)

1. TFL musculocutaneous flap is flap of choice for posterior defects
2. Sartorius muscle flap is flap of choice for small defects superiorly and inferiorly
3. Vastus lateralis and medialis are useful for coverage of defects of middle and lower thigh
4. Distally based gracilis is alternative flap for small defects of the lower thigh

Distant flaps

1. Distant flaps are rarely indicated in reconstruction of the thigh
2. Muscle transplantation is method of choice for osteomyelitis with loss of adjacent local muscle flaps (Fig. 29-4)

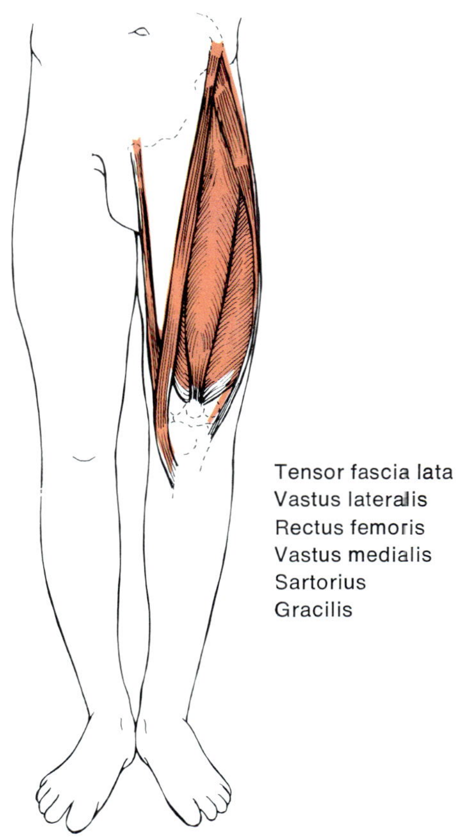

Fig. 29-3

A

B

C

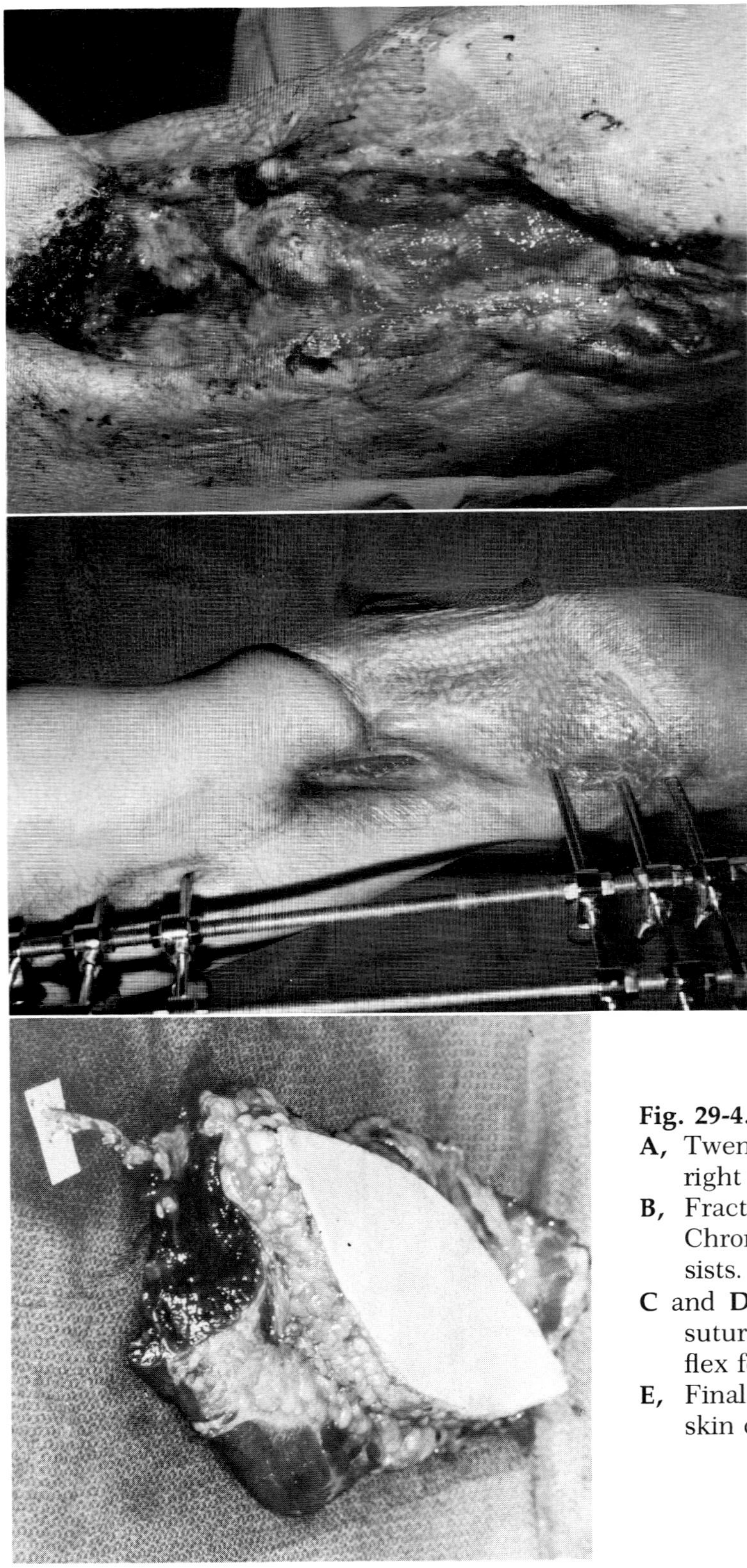

Fig. 29-4. Free flap coverage of infected right femur.

A, Twenty-six-year-old man with open fracture of right femur.

B, Fractures stabilized and soft tissues skin grafted. Chronic osteomyelitis with draining sinus persists. Multiple local flaps had failed.

C and **D,** Latissimus dorsi free flap elevated. Flap sutured into descending branch of lateral circumflex femoral artery.

E, Final result. Osteomyelitis resolved and stable skin coverage.

D
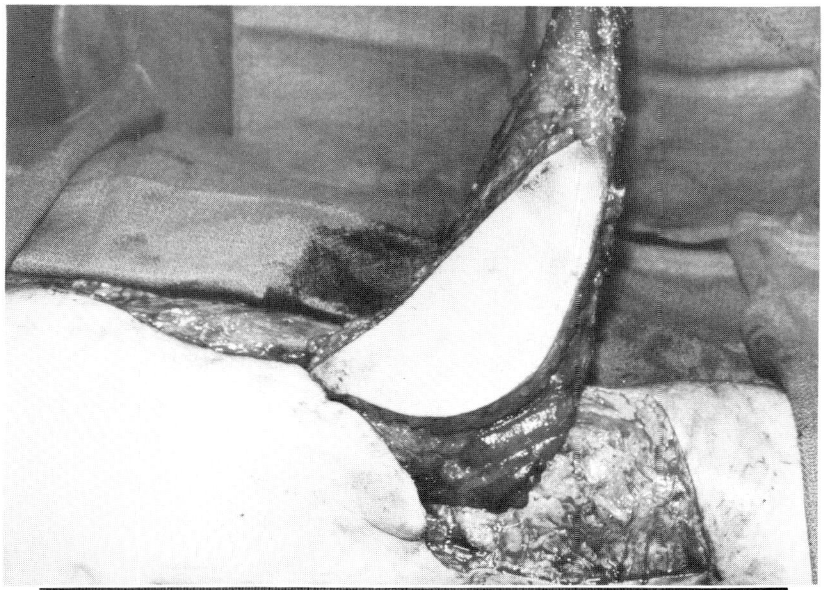

E
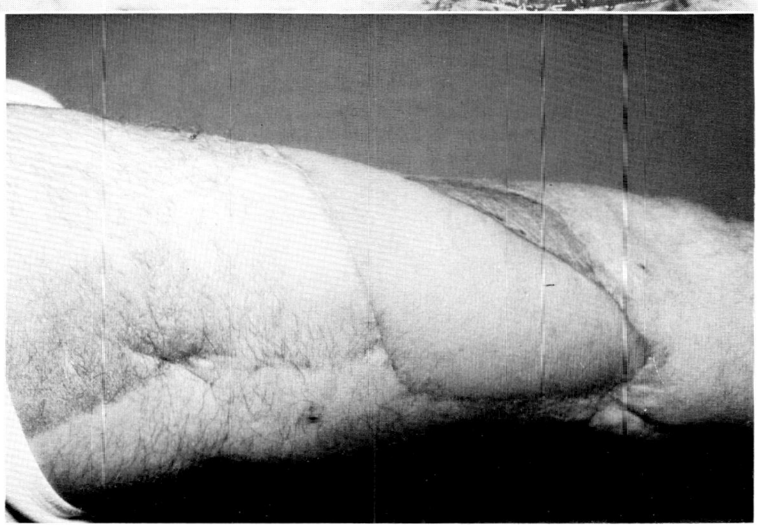

Fig. 29-4, cont'd. For legend see opposite page.

KNEE (Fig. 29-5)

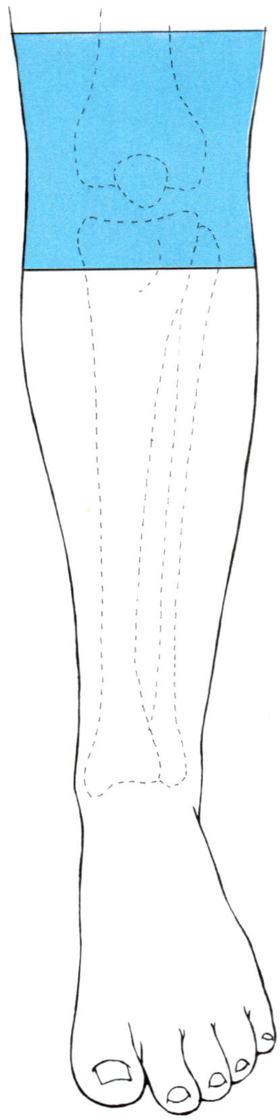

Defect

Fig. 29-5

Skin grafts

Skin grafts are method of choice for immediate coverage provided joint capsule is intact

Local flaps (Fig. 29-6)

1. Gastrocnemius muscle or musculocutaneous flap is flap of choice for coverage of the knee (Fig. 29-7)
2. Distally based gracilis flap
   a. Alternative flap
   b. Requires a strategic delay
3. Sartorius is alternative flap for small defects

Distant flaps (Fig. 29-6)

1. Free tissue transplantation is method of choice for extensive defects
2. Choice of free flap
   a. Latissimus dorsi for large defects
   b. Gracilis for small defects
3. Random cross-leg flap is alternative flap to free tissue transplantation
4. Musculocutaneous cross-leg flap is rarely indicated in reconstruction of the knee

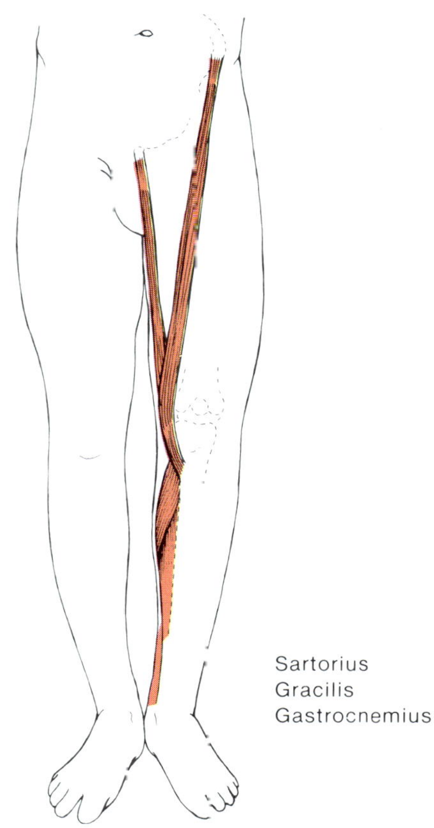

Fig. 29-6

A

B

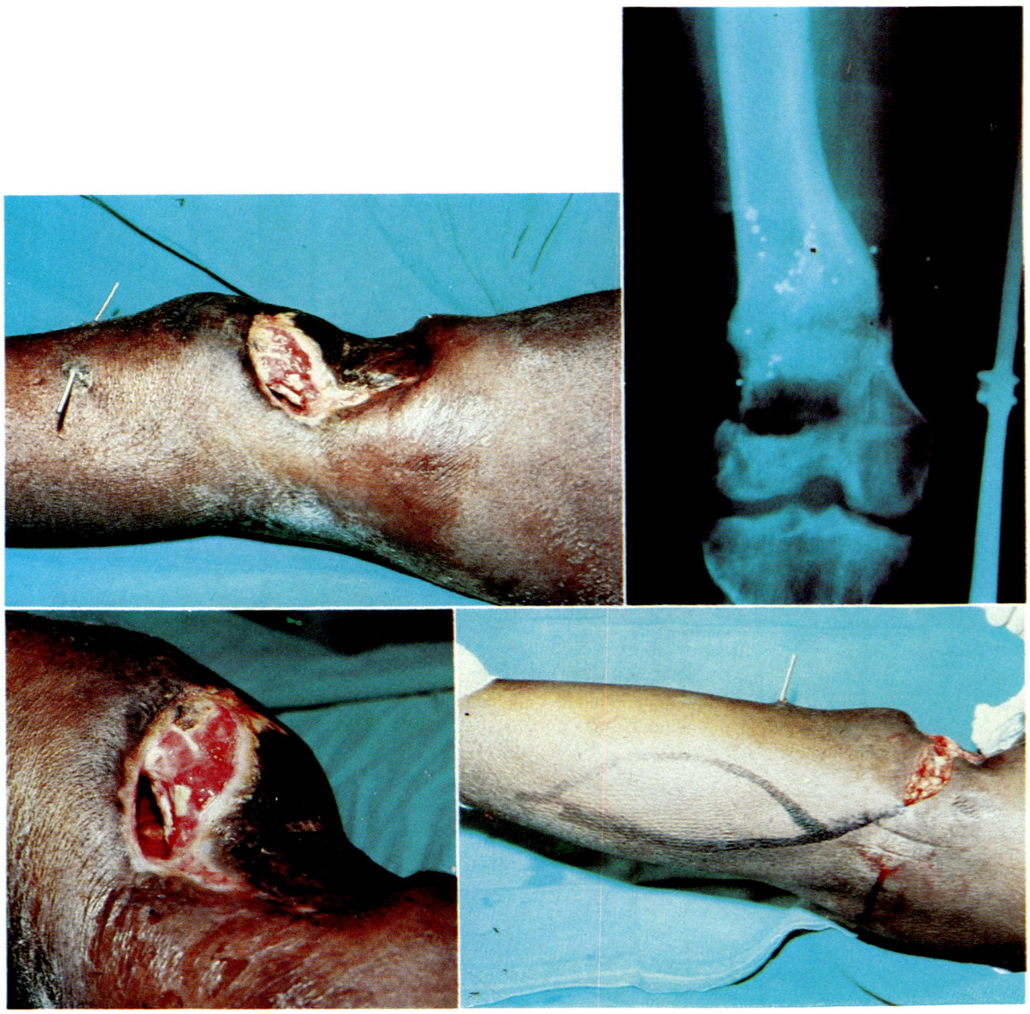

C

D

Fig. 29-7. Gastrocnemius musculocutaneous flap. Débridement of infected bony cavity and coverage of knee with lateral gastrocnemius musculocutaneous flap.
A, Soft tissue defect above knee as result of shotgun wound.
B, Radiograph showing bony defect in distal femur.
C, Closeup view of bony cavity.
D, Lateral gastrocnemius musculocutaneous flap outlined.

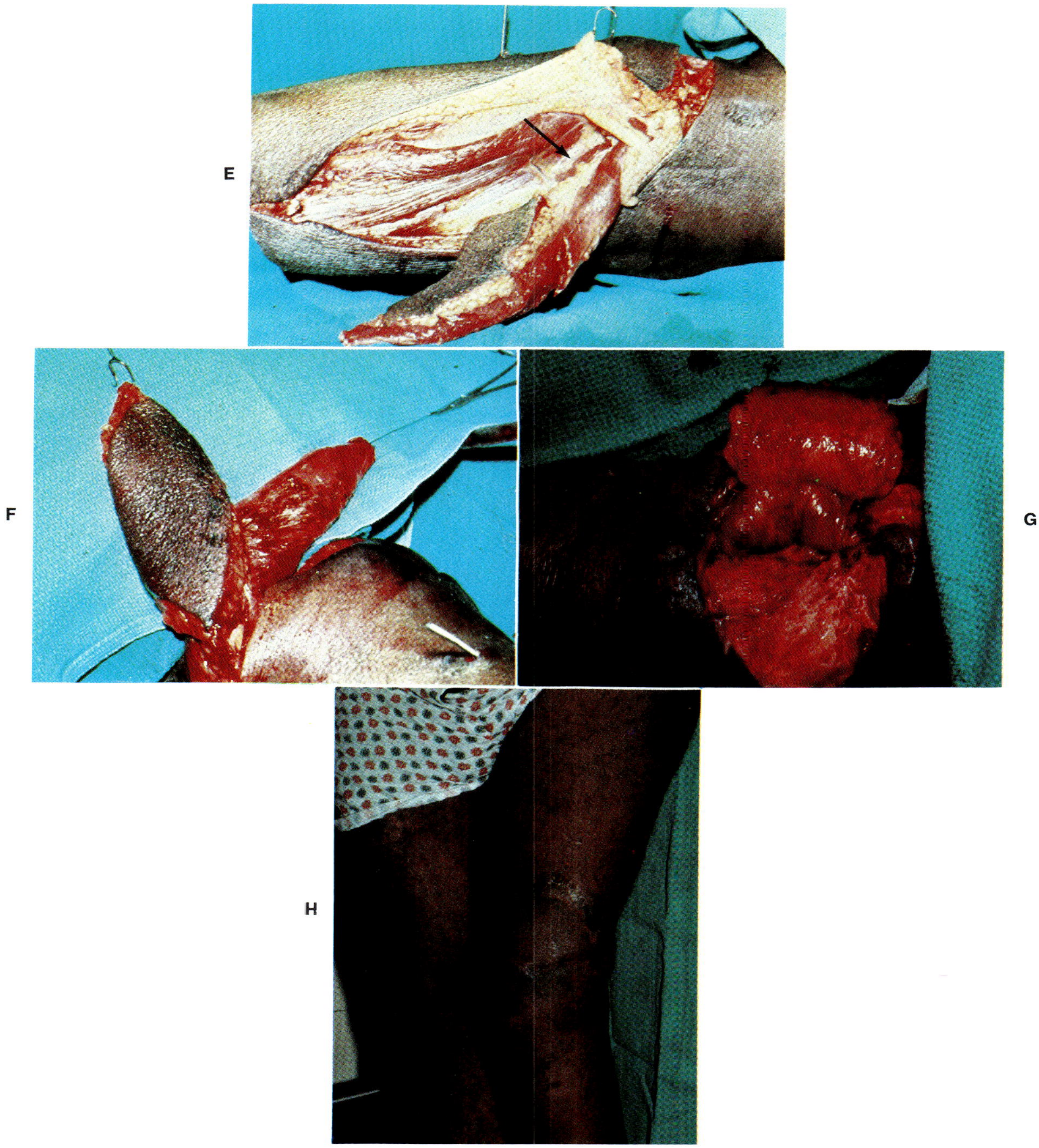

Fig. 29-7, cont'd

E, Flap elevated. NOTE: Arrows point to peroneal nerve.
F, Half of skin island separated from underlying muscle.
G, Muscle placed into bony defect after débridement, and skin island placed over defect.
H, Postoperative result. Flap provides stable wound coverage.

LOWER LEG
Upper third (Fig. 29-8)

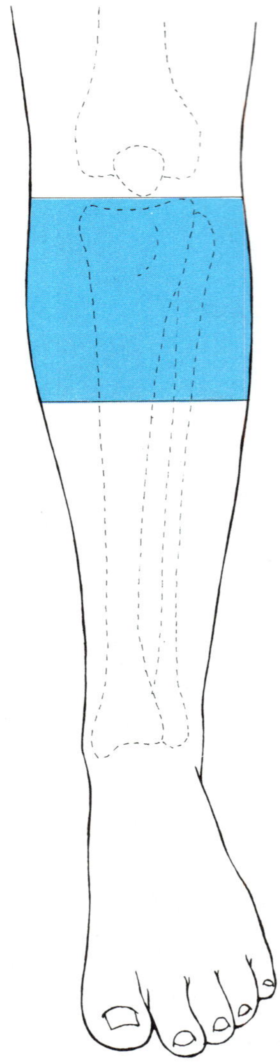

Fig. 29-8

Skin grafts

Skin grafts are applicable only if periosteum over bone is intact

Local flaps (Figs. 29-9 and 29-10)

1. Gastrocnemius muscle flap and skin graft are method of choice for coverage of exposed bone or fracture line
2. Soleus muscle will reach defect with extensive mobilization

Distant flaps

1. Free tissue transplantation is alternative technique for coverage, method of choice for extensive defects
2. Choice of free flap
   a. Gracilis (small defects)
   b. Latissimus dorsi (extensive defects)
3. Random cross-leg flap is alternative technique to free tissue transplantation
   a. Local flaps not available
   b. Method of choice if free tissue transplantation not possible
4. Musculocutaneous cross-leg flap is rarely indicated

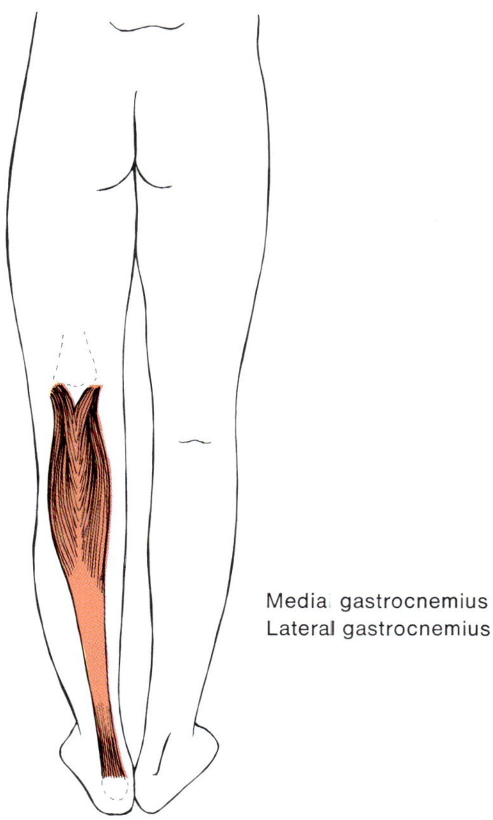

Fig. 29-9

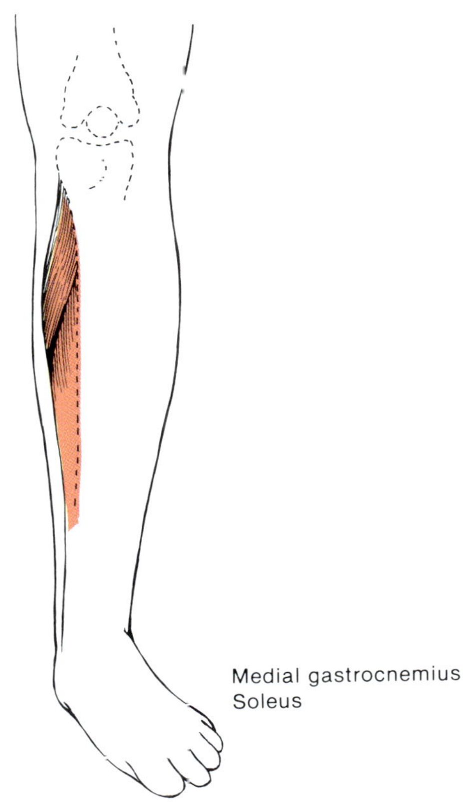

Fig. 29-10

Middle third (Fig. 29-11)

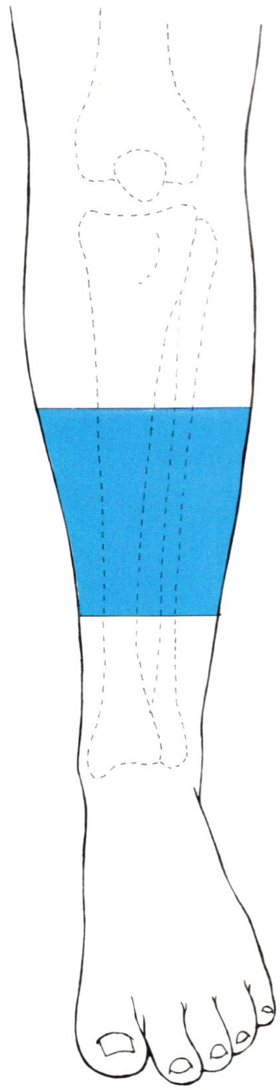

Fig. 29-11

Skin grafts

Skin grafts are applicable only if periosteum over bone is intact

Local flaps (Fig. 29-12)

1. Soleus muscle flap with skin graft is method of choice (Fig. 29-13)
2. Gastrocnemius musculocutaneous or skin-fascial flap is alternative flap
3. Flexor digitorum longus muscle flap is an adjunct to the soleus to cover a larger defect
4. Tibialis anterior muscle with skin graft is alternative flap for small lateral defects

Distant flaps

See upper third

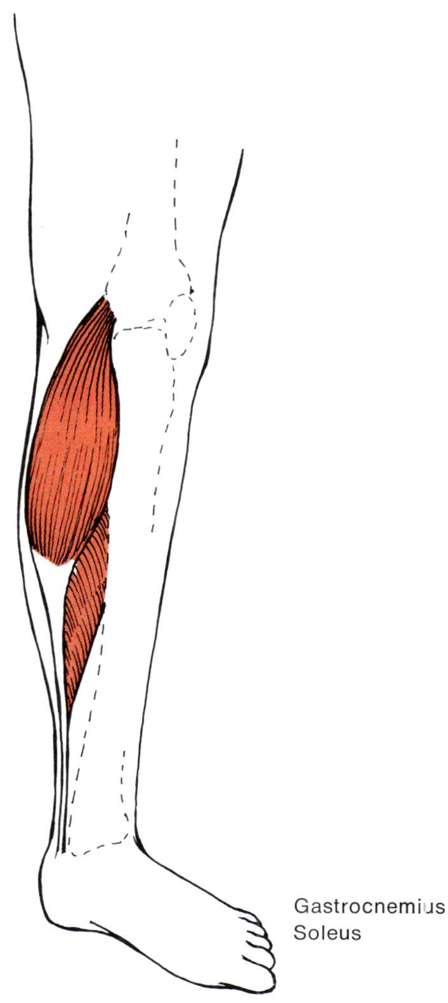

Fig. 29-12

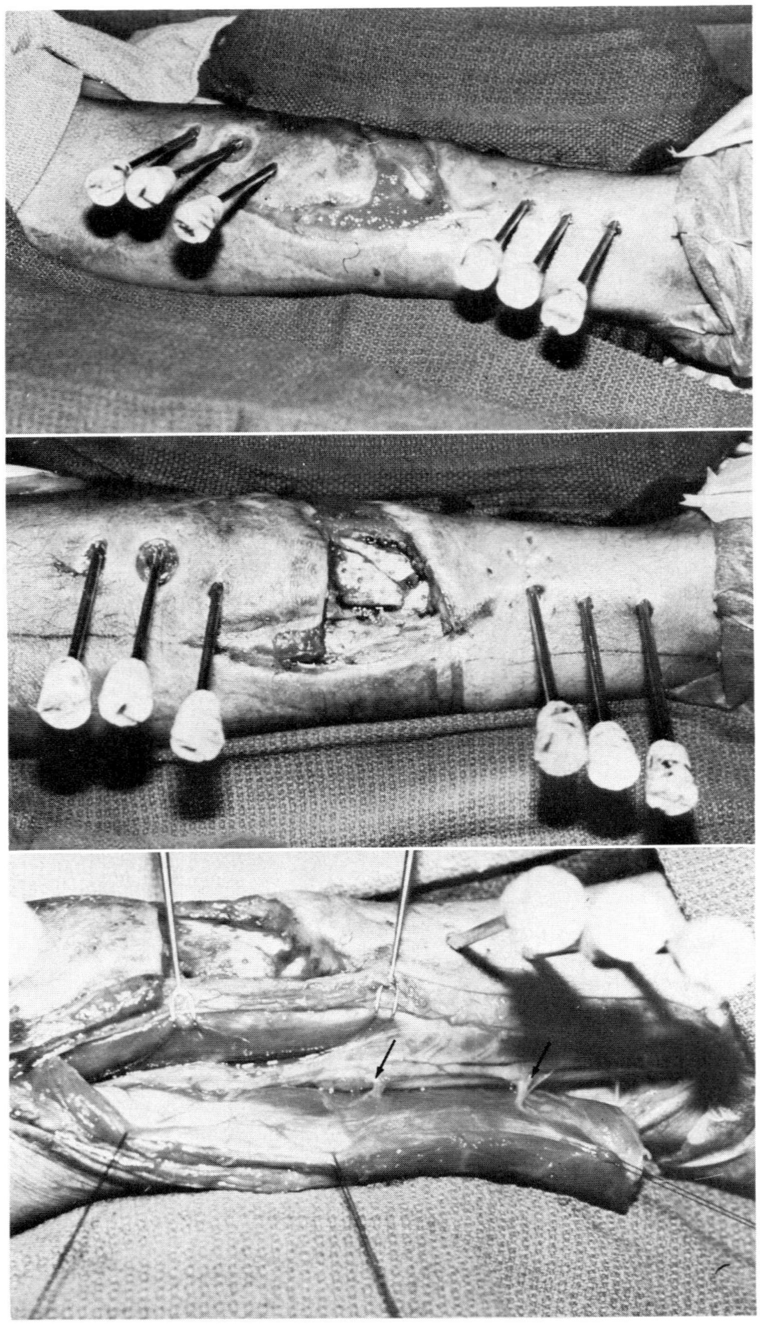

A

B

C

Fig. 29-13. Soleus muscle flap with skin graft.
A, Open fracture of middle third of tibia with soft tissue loss and exposed bone.
B, Wound debrided.
C, Soleus flap exposed. NOTE: Distal minor pedicles *(arrows)*.

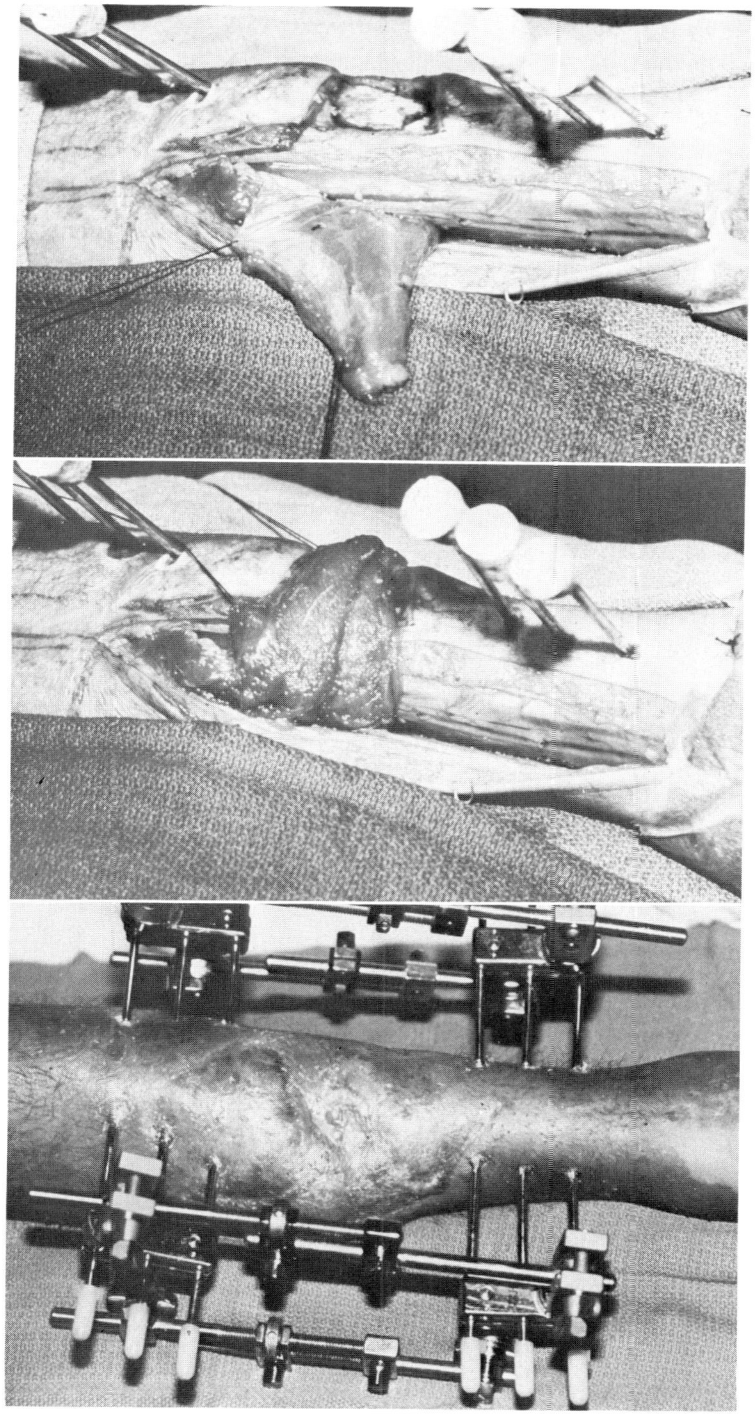

D

E

F

Fig. 29-13, cont'd
D, Soleus flap elevated.
E, Soleus flap transposed and skin grafted.
F, Stable wound coverage provided by muscle flap with skin grafts.

Lower third (Fig. 29-14)

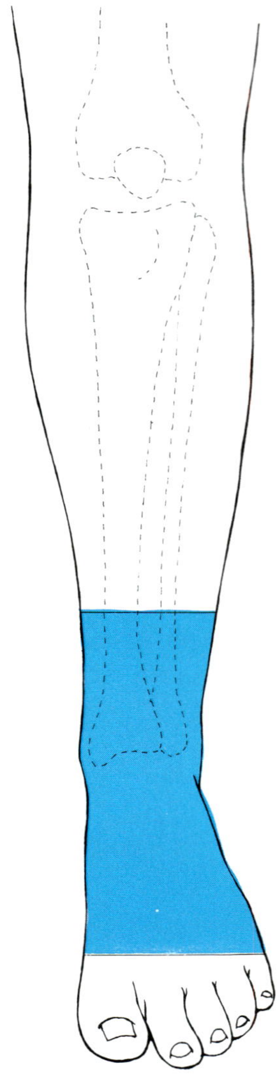

Fig. 29-14

Skin grafts

Skin grafts are applicable only if periosteum and paratenon are intact (Fig. 29-17)

Local flaps (Fig. 29-16)

1. Soleus muscle with skin graft is alternative flap for small defects
2. Peroneus brevis muscle with skin graft is alternative flap for small lateral defects
3. Soleus, extensor digitorum longus, and extensor hallucis longus are distally based muscle flaps and are alternative flaps for coverage when the defect is small, minor vascular pedicles are intact, and free tissue transplantation is not possible

Distant flaps (Fig. 29-15)

1. Free tissue transplantation is method of choice for large defects
2. Choice of free flap
   a. Gracilis (small shallow defect)
   b. Latissimus (larger defects) (Fig. 29-18)
   c. TFL (innervated) (Fig. 29-19)
   d. Deep groin flap (bone)
3. Random cross-leg flap (see upper third)
4. Musculocutaneous cross-leg flap (see upper third)

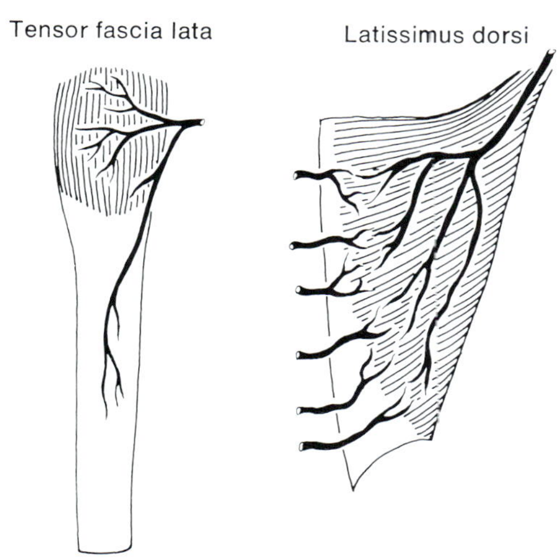

Fig. 29-15

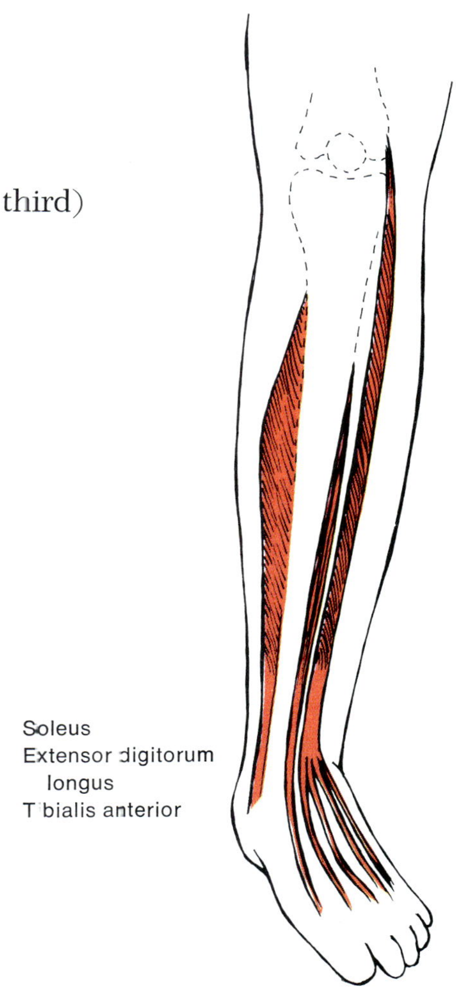

Fig. 29-16

A

B

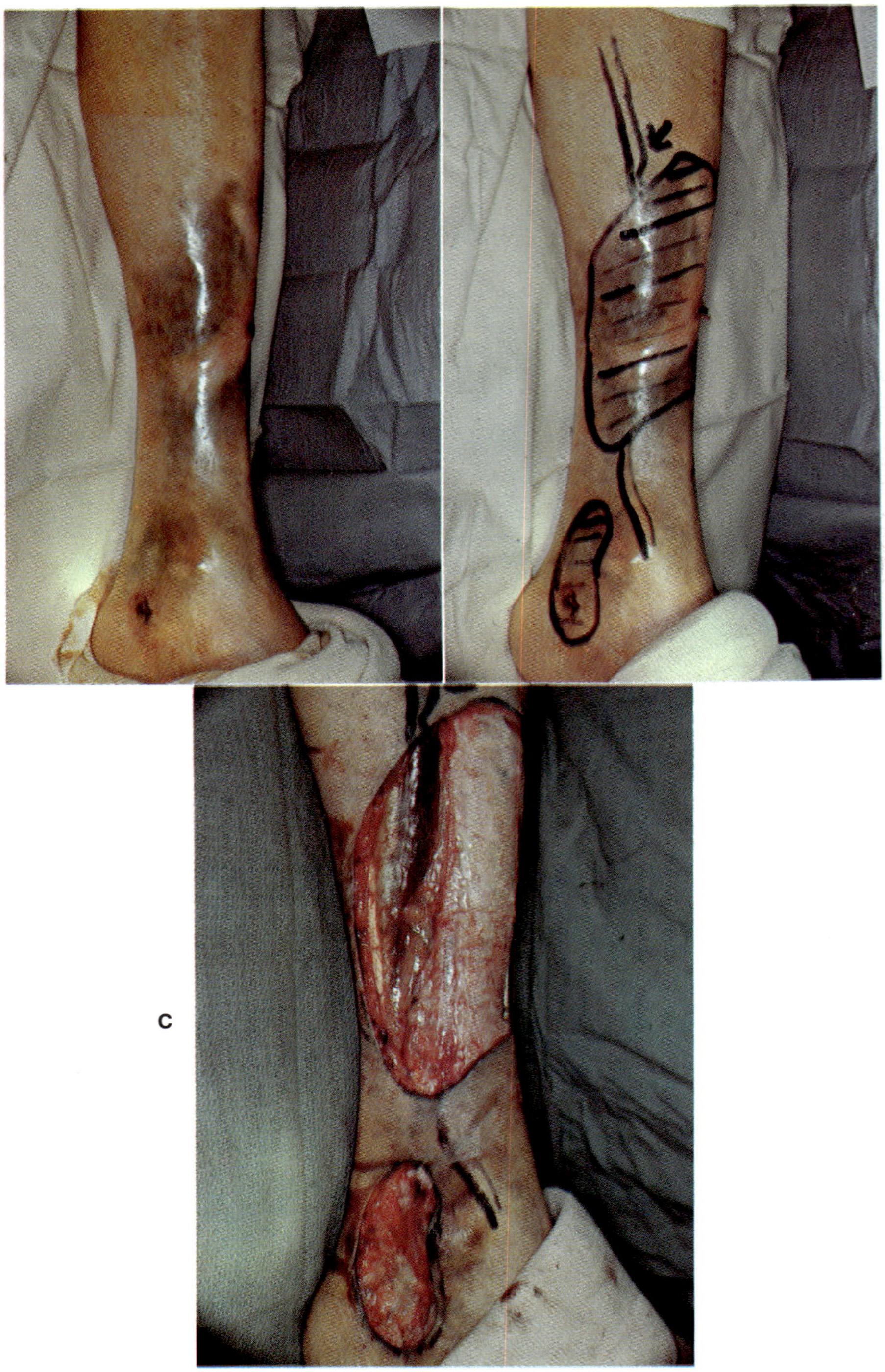

C

Fig. 29-17. Skin grafts.
A, Chronic venous ulceration secondary to lower extremity venous insufficiency.
B, Resection requires excision of all abnormal fibrotic skin, preserving underlying muscle and periosteum.
C, Defect after resection of skin with abnormal lymphatics and incompetent perforating veins.

D 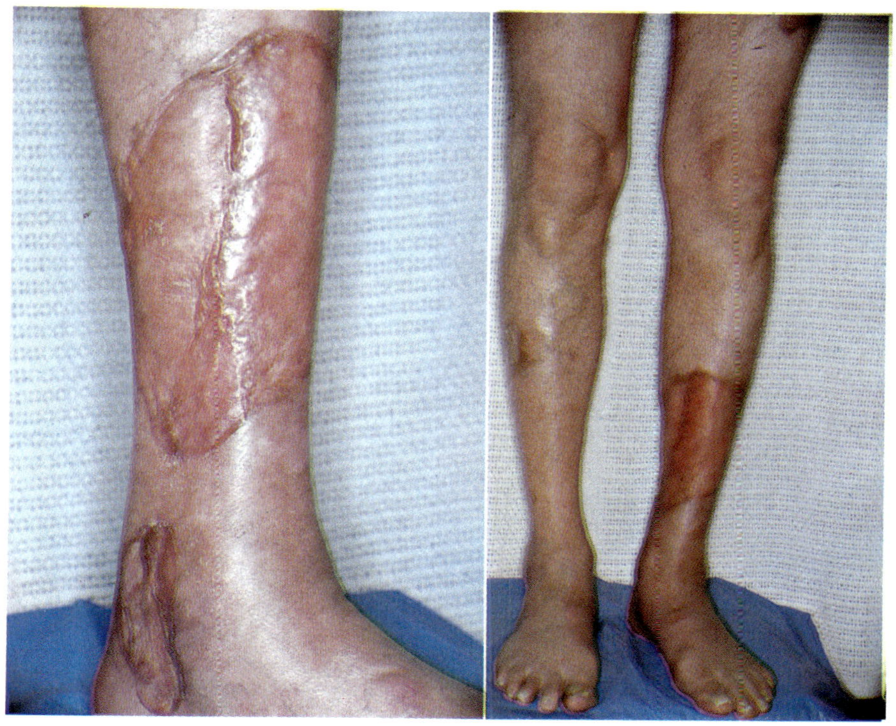 E

Fig. 29-17, cont'd

D, Six-month postoperative view demonstrates stable wound coverage at site of former venous ulcerations with use of split-thickness skin grafts.

E, Contour form and function preserved with use of skin grafts for lower leg defect.

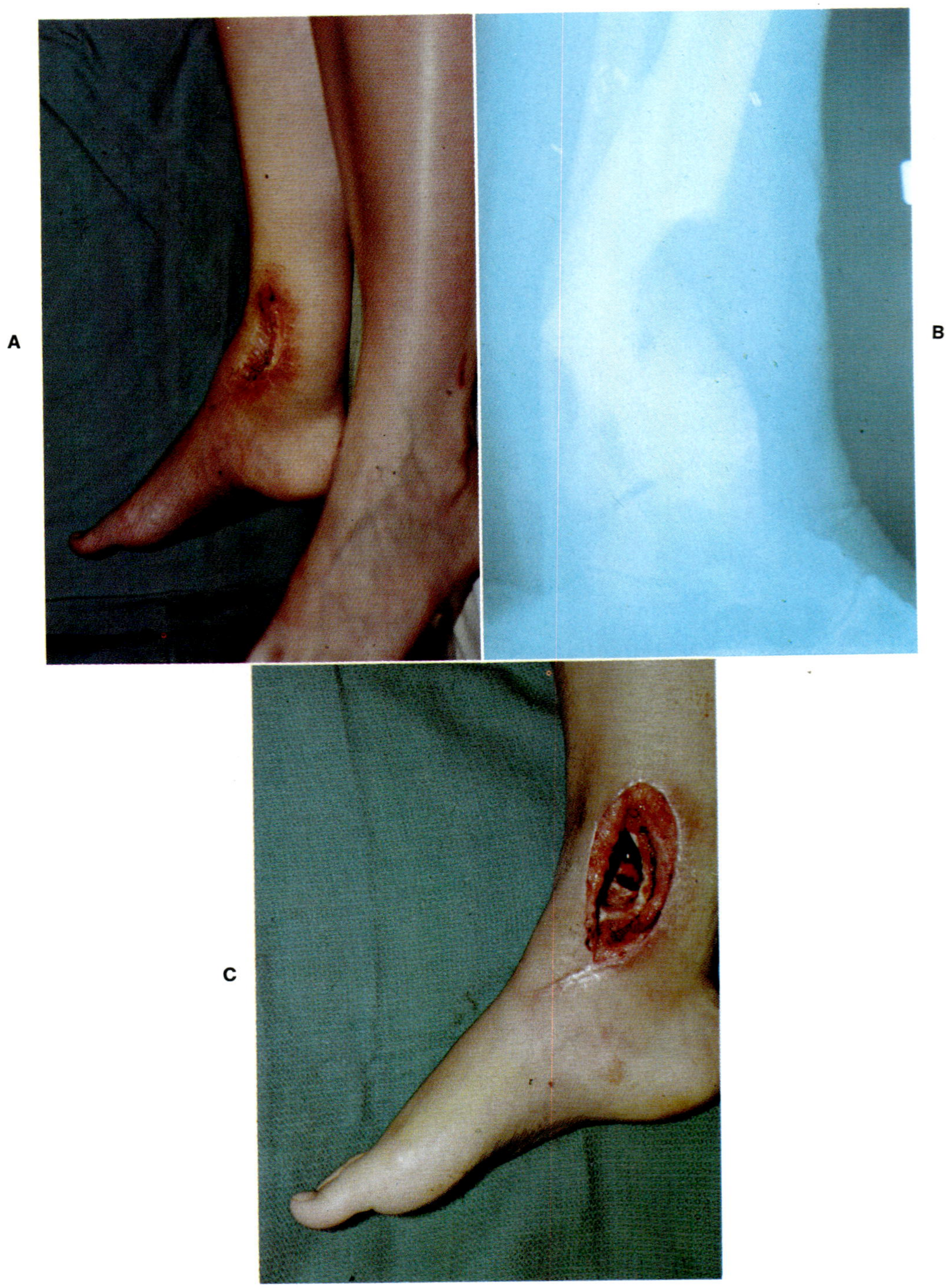

Fig. 29-18. Latissimus dorsi free flap.

A, Open wound and osteomyelitis of right distal tibia. Multiple previous local procedures had failed.

B, Radiograph showing defect in tibia.

C, Defect after débridement.

D

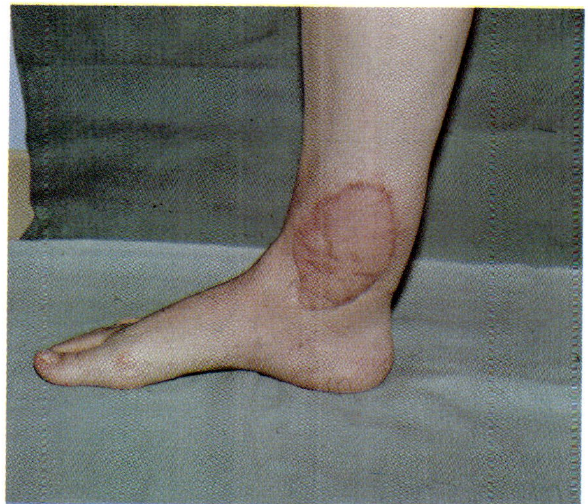

E

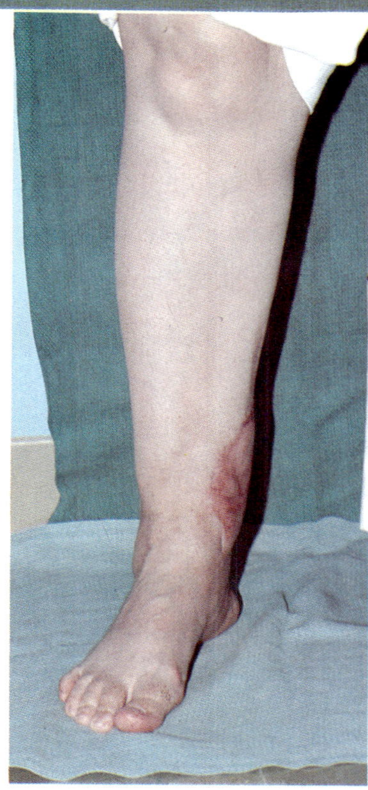

Fig. 29-18, cont'd
D, Transplanted latissimus dorsi muscle covered with skin grafts.
E, Eighteen-month postoperative anterior view demonstrates adequate contour. No recurrence of osteomyelitis.

A

B

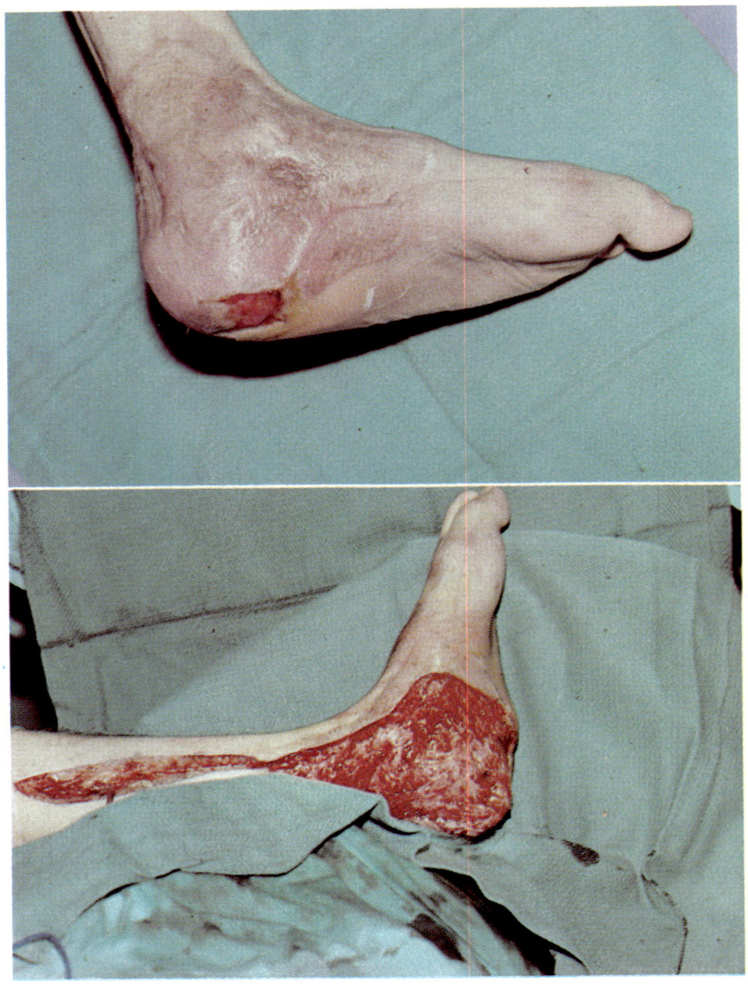

Fig. 29-19. Reconstruction of heel with neurosensory TFL musculocutaneous free flap.

A, Unstable wound of lower leg and heel in 18-year-old man after severe electrical burn injury. Previous skin grafts and local flaps have failed.

B, Wound resected.

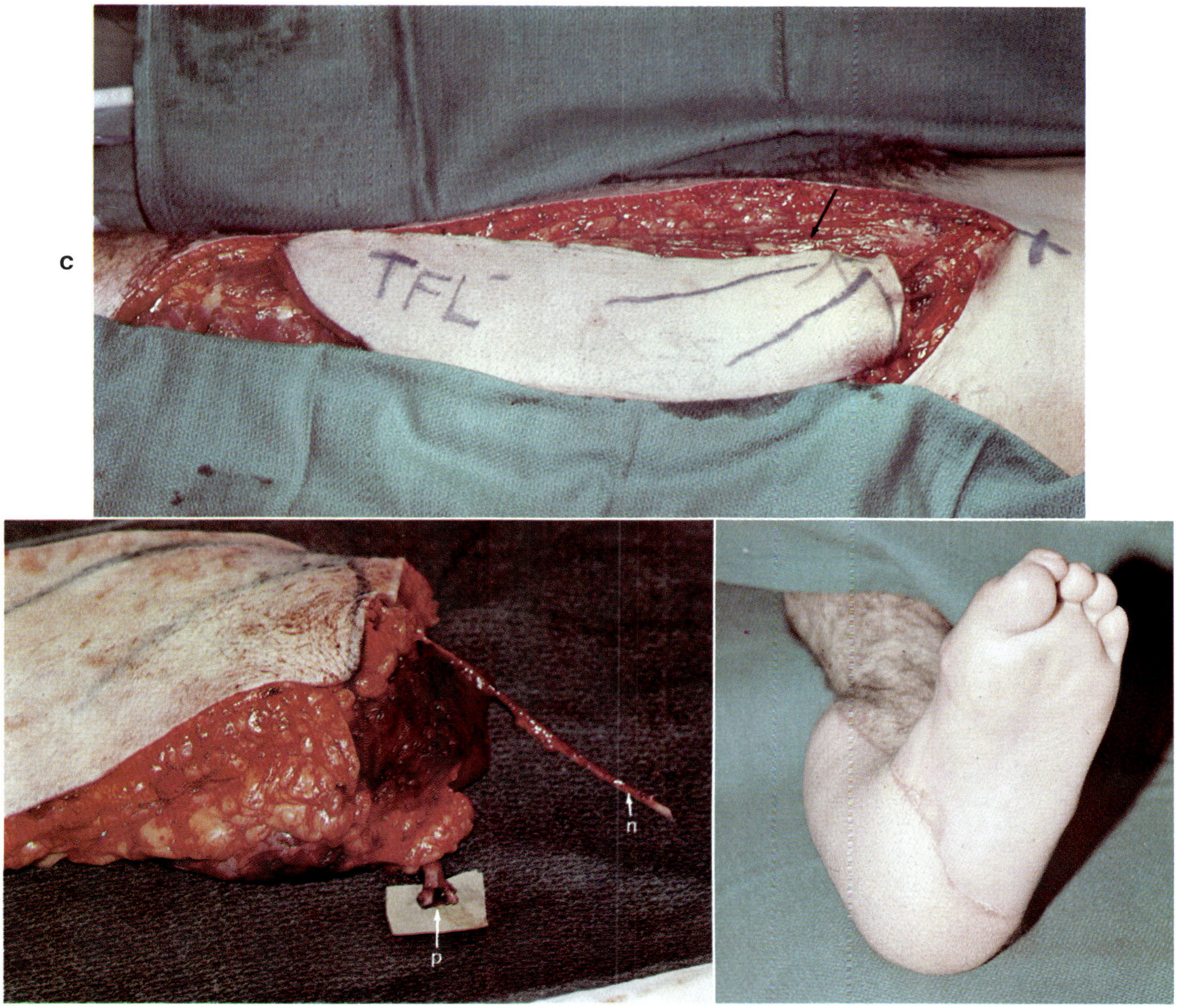

Fig. 29-19, cont'd

C, Neurosensory TFL flap elevated. Arrow denotes location of lateral femoral cutaneous nerve.

D, Vascular pedicle *(p)* and sensory nerve to flap *(n)*.

E, Final result. Coverage of lower leg and heel with neurosensory flap providing protective sensibility.

30

RECONSTRUCTION

Stephen J. Mathes ■ Luis O. Vasconez

A broad spectrum of methods is now available for lower extremity reconstruction. Over the past 10 years the reliability of the muscle and the musculocutaneous transposition flap has been well established. Furthermore with recent advances in microsurgical techniques, the more complex distant flap is now readily transplanted from varied distant body regions to the lower extremity for coverage. Therefore this spectrum of reconstructive methods now ranges from the simple skin graft to the more complex distant flap, whether transferred from the opposite leg in staged procedures or immediately transplanted as a free flap. The role of the muscle flap is placed in perspective within this spectrum of reconstructive methods.

EVALUATION OF RECONSTRUCTIVE REQUIREMENTS

Selection of a particular method for lower extremity reconstruction is based on the method's ability to best achieve the immediate requirements of coverage and to provide the long-term requirements of form and function. The process of method selection requires an analysis of defect etiology, location, and specific characteristics of each reconstructive method.

Etiology

Systemic disease. Coverage problems in the lower extremity have diverse causes stemming from both systemic and local problems. Systemic problems are likely to recur regardless of the method of coverage unless the systemic disease is simultaneously corrected. Certain systemic diseases associated with lower extremity ulceration require chronic medication and are associated with frequent recurrences of local ulceration. Examples of systemic diseases associated with such lower extremity ulceration include diabetes mellitus, lupus, and scleroderma. Impairment in wound healing associated with long-term maintenance on steroids often complicates the management of these patients. If possible, coverage by skin grafts avoids potential development of donor site wounds associated with both local and distant flaps.

Vascular disease. Small vessel disease associated with diabetes mellitus adversely affects muscle circulation, and donor site healing is often impaired. Although local muscle flaps may salvage lower extremity ulcerations resulting from diabetes, long-term prevention of amputation is uncommon. Use of skin grafts is generally preferred over random, axial, and distant flaps for coverage of the lower extremity wound in diabetic patients.

Although trauma commonly precedes the development of lower extremity ulceration, the difficult wound problem after minor trauma generally indicates underlying local disease. Evaluation of possible impairment of arterial or venous insufficiency is essential before selection of a method for coverage. Presence of claudication, rest pain, or absence of distal pulses requires specific evaluation for arterial vascular disease (i.e., arteriograms). Presence of the history of thrombophlebitis, chronic pain, and lower leg edema requires specific evaluation for venous insufficiency (i.e., phlebograms). When possible, correction of the underlying local vascular disease is an essential part of lower extremity reconstruction.

Débridement and skin grafts in conjunction with vascular surgery for arterial insufficiency generally provide adequate coverage. In venous insufficiency excision of the ulceration and adjacent abnormal skin and subcutaneous tissue with placement of skin grafts on muscle and periosteum provides stable coverage if support stockings are faithfully used by the patient.

The selection of the muscle flap in arterial vascular disease when skin grafts are not successful for coverage is not recommended without prior vascular reconstruction. However, successful vascular reconstruction should restore adequate profusion of lower leg muscles via their respective dominant vascular pedicles to allow safe muscle transposition. Use of the distant muscle or axial flap transplanted by microvascular techniques has been successfully performed when local muscle flaps are unavailable (i.e., distal third of leg) after successful arterial vascular reconstruction.

Distant flaps from the opposite leg transposed either as a cross-leg flap or free flap are avoided, since donor site healing problems may result from similar vascular disease.

The selection of the muscle transposition flap for venous disease when skin grafts are unsuccessful is a useful technique. However, recurrent ulceration and donor site healing problems are common, since the venous drainage of the transferred muscle is often into an incompetent venous system. If the phlebogram reveals complete thrombosis of the deep venous system, it is necessary to maintain proximal superficial venous connections to avoid muscle flap failure after its transposition for wound coverage. Distally based muscle flaps are not used in the presence of venous insufficiency. Distant flaps whether designed as cross-leg flaps or free flaps are rarely used unless the venous drainage of the flap is located proximally to the level of venous insufficiency.

Trauma. Coverage of the traumatic wound remains the most frequent problem encountered by the reconstructive surgeon. The technique selected depends on the timing and extent of the injury. A reconstructive ladder is again stressed for selection of the simplest method for closure. In multiple organ system injuries débridement, stabilization of fractures, and physiologic dressings may preclude any procedures for immediate coverage. However, exposure of major vessels, nerves, or tendons may require early attention for coverage. As soon as initial or subsequent wound débridement has adequately removed devitalized tissue, bone fractures are stabilized, and adequate circulation is confirmed, coverage should be accomplished. If skin grafts will not provide reliable coverage in the acute phase, the reconstructive surgeon must select either a local or distant flap. A lack of axial flaps and poor reliability of local random flaps result in the necessity to either use a local muscle transposition flap or a distant flap in management of a traumatic wound.

Location

Reconstructive requirements for lower extremity coverage vary with defect location. Factors such as flap availability and occurrence of trauma or pressure at the site of required coverage are considered. Although the etiology of the defect often demands selection of a particular method, long-term success is often based on the selection of the particular method that best establishes form and function at the site of reconstruction.

The lack of axial flaps and poor reliability of local random flaps in the lower extremity have resulted in the use of the distant random flap designed either as a tubed pedicle flap or a cross-leg flap to achieve defect coverage. With identification of the local muscle and musculocutaneous flap as a reliable single-stage technique, transposition of the muscle flap has become an accepted technique for coverage of lower extremity defects. Recently free flaps (muscle or axial) and the cross-leg musculocutaneous flap are used for coverage of lower extremity defects. The reconstructive surgeon now has several alternative techniques to establish coverage when standard closure or skin grafts will not cover the difficult lower extremity wound.

Distant flap vs muscle and musculocutaneous flap

Since both the distant flap (i.e., cross-leg and free flap) and the local muscle flap are well established as reliable techniques for lower extremity coverage, the reconstructive surgeon must consider the relative advantages of each method. In general, a muscle flap is preferred for lower extremity coverage.

Donor site proximity to wound. In the reconstructive ladder, a local flap is generally a more direct and simple method to achieve wound coverage. With exception of the distal third of the leg, muscles are available adjacent to lower extremity defects with vascular pedicles located proximally and distant from the zone of injury.

Reliability. Since a muscle or musculocutaneous flap is based on the single or dominant vascular pedicle(s) to muscle, flap survival is reliable without necessity of prior flap delay or reliance of a specific length-width ratio.

Accessibility of flap. Although elevation of a specific muscle requires knowledge of both the intended muscle for transposition and adjacent structure subject to injury (i.e., sensory and motor nerves and vessels), muscle flap elevation is performed between muscle planes with relative ease. The problems of maintaining the opposite leg in proximity to the cross-leg flap recipient site is avoided.

Single-stage procedure. Transposition of the muscle or musculocutaneous flap is generally completed in one procedure. With stable wound coverage completed, hospitalization is shortened, and emphasis is placed on rehabilitation and secondary reconstructive procedures when required.

Bacterial contamination. The difficult lower extremity wound after trauma is contaminated with bacteria and despite aggressive débridement may develop wound infection after coverage. The use of systemic and topical antibodies generally controls wound bacterial flora. Experimental data have demonstrated the superior resistance to bacterial inoculation of both the cutaneous and muscular surfaces of the musculocutaneous flap when compared to the random flap for coverage of such defects.

The muscle and musculocutaneous flap has certain disadvantages not inherent to the distant flap for coverage of lower extremity defects.

Loss of function. Transposition of a muscle generally requires release of the muscle insertion. This eliminates the locomotor function of the particular muscle. In general, the muscles selected for use have synergistic muscles that preserve function. Recent modifications in muscle flap design (e.g., segmental muscle transposition and skin-fascial flaps) allow maintenance of muscle-tendon continuity while providing coverage. However, in extensive injuries preservation of remaining muscles is vital to future leg function. When function-preserving techniques are unavailable, the use of a distant flap eliminates the necessity of disrupting remaining uninjured muscles.

Vascular injury. Extensive vascular injuries may have jeopardized the circulation of the injured extremity. Transposition of muscle may interrupt existing vascular communications from the dominant vascular pedicle through the muscle to the distal leg via the inferiorly located minor pedicles. The use of a distant flap eliminates the potential for lower extremity vascular compromise.

Distal leg. The transposition of muscle or musculocutaneous flaps is not reliable for coverage of lower third leg defects. The paucity of muscle fibers in the lower third of the leg allows coverage of only small defects as local transposition flap. The design of extended musculocutaneous flaps (i.e., gastrocnemius, soleus, and peroneus muscles) may cover such defects but often result in exposure of tendons in the distal leg, resulting in donor site coverage problems. The design of distally based muscle flaps (i.e., soleus and extensor digitorum longus) will also cover such defects but are not always reliable because of variability in number and location of minor pedicles. Injury of these pedicles resulting from their proximity to the wound is common. In this area, the distant flap is the procedure of choice for a difficult coverage problem.

Appearance. In the management of the difficult lower extremity wound, appearance is a secondary consideration. However, use of muscle as a transposition flap with skin graft coverage avoids both the bulky appearance at the recipient site and donor site deformity associated with the musculocutaneous flap.

In reconstruction of a contour defect or replacement of unstable skin, appearance is often a primary consideration. The tailored distant flap is generally the procedure of choice. If the cross-leg flap is selected, the donor site on the contralateral leg is often unacceptable. Selection of a free flap from a distant site where clothes will conceal the donor scar is the method of choice.

MUSCLE AND MUSCULOCUTANEOUS FLAPS FOR SPECIFIC REGIONS

If transposition of a particular muscle is selected as the method of choice for lower leg reconstruction, a review of regional and specific muscle anatomy is required. If the defect or zone of injury extends into the area of the vascular pedicles of the muscle, another muscle is considered as an alternative. Arteriograms are obtained to confirm the patency of the major or dominant vascular pedicles of the muscle if injury or vascular disease is suspected.

The timing of muscle flap transposition is important. If coverage of a defect associated with underlying fractures is required, stabilization of the fracture site is essential. Balanced skeletal traction does not generally provide adequate stabilization of bone fractures.

Muscle flap transposition is also delayed until débridement of devitalized tissue is complete. However, excessive delay is also avoided, since the lack of vascularized coverage often allows extension of the wound. Muscle flap transposition is performed at the same time as definitive wound débridement and not performed as a secondary procedure.

A tourniquet is used during both débridement of the defect and elevation of the muscle flap for knee and leg coverage. Identification of the appropriate muscle for transposition and avoidance of injury to adjacent structures during the muscle elevation is best achieved with a bloodless field. After flap elevation is completed, the tourniquet is released and hemostasis is obtained. A suction drainage system is placed within the donor defect to avoid seroma formation.

If a muscle flap is used, skin grafts are placed on exposed muscle fibers at the time muscle transposition is performed. The skin graft is meshed, since the graft contours better to the muscle surface. If the surface of the muscle contains thick fascia, this is excised so that the skin grafts are placed directly on muscle fibers. Delayed placement of skin grafts is not necessary, since a secondary procedure is routinely required, and the majority of grafts survive the initial application over the exposed muscle belly. However, occasional failure of initial graft survival may require a secondary graft procedure.

If a musculocutaneous flap is used, prior incisions into the cutaneous territory of the muscle do not preclude its use. However, if prior incisions or injuries have extended completely through the skin and muscle, the distal musculocutaneous flap will not generally survive transposition.

The technique of elevation of muscle units is described based on the ability of the muscle to cover defects within specific regions of the lower extremity. The flap design and surgical approach for flap elevation are determined by the specific reconstructive requirements of the defect and the muscle anatomy encountered during flap elevation. Although the selection of a specific muscle and the technique of elevation in each instance are always variable, guidelines are presented to assist the surgeon in muscle selection, design, and elevation for use as a transposition flap.

THIGH (Table 3)

Muscle flap transposition is occasionally useful in the thigh for coverage of defects associated with osteoradionecrosis, exposure of vascular prosthesis, or large tumor resection defects. The anterior thigh

TABLE 3 Thigh coverage

Muscle	Technique of elevation	Proximal third		Middle third		Distal third	
		Anterior	Posterior	Anterior	Posterior	Anterior	Posterior
Gracilis	Musculocutaneous	X	X				
	Distally based			X	X	X	
	Cross-thigh	X	X				
Gluteal thigh	Skin-fascia	X	X	X	X		
	Cross-thigh	X	X	X	X		
Medial gastrocnemius	Cross-leg skin-fascial					X	X
Rectus femoris	Muscle	X		X	X	X	
	Musculocutaneous	X	X	X	X	X	
	Cross-thigh	X		X			
Sartorius	Muscle	X				X	
	Advancement			X			
TFL	Musculocutaneous	X	X	X	X		
Vastus lateralis	Muscle	X	X	X	X	X	X
	Musculocutaneous	X	X	X	X		
Vastus medialis	Muscle			X	X	X	

muscles suitable for transposition either as muscle or musculocutaneous flaps include the following:

Sartorius
Gracilis
Vastus medialis
Vastus lateralis
Rectus femoris
TFL

Any of these muscle units are useful for proximally located thigh defects, although the arc of rotation is limited in the sartorius because of its segmental (type IV) pattern of circulation.

The TFL is not used for cover of exposed vascular prosthesis, since its fascial surface is not pliable, and contact of well-vascularized muscle with the vascular prosthesis is not obtained because of its thick fascial undersurface. Even with the use of optimal muscle units such as the rectus femoris or vastus lateralis, salvage of the infected vascular prosthesis is not always possible. For this reason bypass grafts around the site of the infection with removal of the infected graft are preferable if this alternative is available.

For small defects in the superior half of the thigh, transposition of the sartorius muscle or gracilis musculocutaneous flap is useful.

For small defects in the inferior half of the thigh, transposition of the vastus medialis or vastus lateralis muscles is useful. The fibers of insertion of either muscle are divided and the muscle mobilized proximally until defect coverage is achieved. Attachment of the vastus medialis to the joint capsule will restore leg extension for distal thigh central defects associated with loss of the rectus femoris muscle.

For coverage of large anterior or posterior thigh defects, transposition of the vastus medialis, vastus lateralis muscle, and TFL musculocutaneous units is useful. These muscles are expendable if the remaining quadriceps muscles are intact. However, significant loss of thigh muscles may require use of a distant flap to avoid further loss of leg extension after transposition of one of these remaining muscles. (Technique of elevation of the TFL musculocutaneous unit is described in Chapter 23.)

Sartorius

The defect is enlarged by a linear incision in the direction of the sartorius muscle belly. The sartorius muscle mobilization requires division of two or three segmental vascular pedicles to allow its transposition into the defect. (Further details regarding the anatomy and technique of elevation are available in Chapter 23.) If a wide arc of rotation is required, the muscle is not reliable, and another muscle unit is selected.

Gracilis

The gracilis musculocutaneous unit will cover both anterior and posterior thigh defects. Elevation of the flap in the prone position is possible. Careful identification of the distal third of the muscle is stressed before the skin portion of the musculocutaneous flap is designed. The distal third of the muscle will not safely carry its overlying skin as a musculocutaneous flap. (For detailed description of flap anatomy and technique of elevation see Chapter 23.)

Vastus medialis

The vastus medialis muscle is located on the medial thigh and is readily identified in the distal anterior thigh located medially to the sartorius muscle and laterally to the rectus femoris muscle. With a type II pattern of circulation, the muscle will survive transposition based on its dominant vascular pedicles entering the proximal third of the muscle from the profunda femoris artery and vein. The muscle has a small cutaneous territory located over the medial anterior thigh.

Generally this muscle is transposed as a muscle flap with skin graft coverage over exposed muscle. After division of its insertion into the quadriceps mechanism, medial border of the patella, and knee joint capsule, the muscle is elevated proximally by division of fibers of origin from the femur.

The muscle is safely elevated to the middle of the thigh. The femoral nerve is located medially and is preserved during flap elevation. The donor defect requires direct closure to avoid exposure of the distal femur.

Vastus lateralis

The vastus lateralis is located on the lateral thigh deep to the fascia lata located between the rectus femoris muscle medially and the biceps femoris muscle posteriorly. With a type II pattern of circulation, the muscle will survive transposition based on its dominant vascular pedicle from the lateral femoral circumflex artery and vein entering the proximal third of the muscle.

Generally this muscle is transposed without overlying skin, and the exposed muscle is skin grafted. However, the muscle has vascular connections to the distal lateral thigh skin (the distal portion of the territory of the TFL) and the skin may be included with the muscle unit as a musculocutaneous flap (Fig. 30-1).

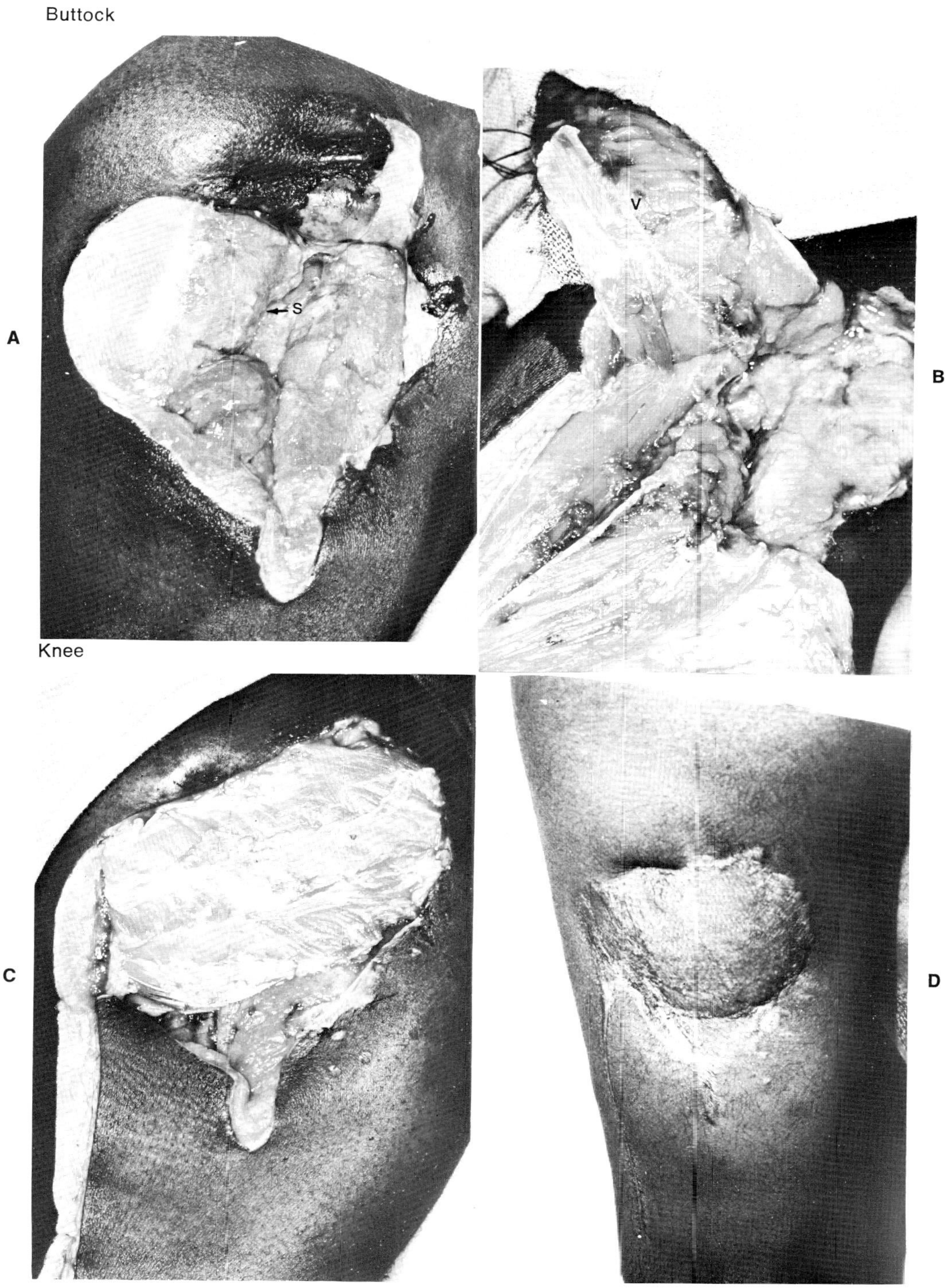

Fig. 30-1. Proximal thigh coverage. Use of vastus lateralis muscle with skin grafts for coverage of traumatic defect.

A, Close-range shotgun wound with exposure of sciatic nerve(s). Wound located on posterior superior thigh.

B, Elevation of vastus lateralis (*v*) muscle flap.

C, Excision of epimysium allows spreading of muscle for coverage of extensive defect.

D, One-year postoperative view demonstrates stable coverage provided by single-stage procedure.

(From Mathes, S.J., Vasconez, L.O., and Jurkiewicz, M.J.: Extensions and further applications of muscle flap transposition, Plast. Reconstr. Surg. **60:**6, 1977.)

The muscle is identified through a vertical thigh incision located over the lateral thigh between the lateral condyle of the knee and 10 cm inferior to the anterior superior iliac crest. After incising the fascia lata, the muscle is visualized. After division of its insertion to the lateral border of the patella and attachments to the quadriceps femoris tendon, the deep attachments of origin to the femur are released. Elevation of the muscle beyond the superior third of the thigh requires exposure of its vascular pedicle. Retraction of the rectus femoris muscle medially allows identification of the vascular pedicle, the descending branch of the lateral femoral circumflex artery and vein. With excision of the epimysium, the muscle expands and covers large anterior or posterior thigh defects. Direct donor site closure is required to avoid exposure of the femur.

Cross-thigh musculocutaneous flaps

Muscle units are not always available for large thigh defects resulting from vascular injury or absence of expendable muscle units. The transposition of musculocutaneous units from the opposite thigh is a useful alternative technique.

The gracilis and rectus femoris musculocutaneous units from the opposite leg will easily reach the contralateral anterior thigh. The gluteal thigh flap from the opposite leg will easily reach the contralateral posterior thigh. This method of coverage is only recommended for limb salvage. When used, division of the flap base is required 14 to 21 days after initial flap inset to avoid flap necrosis. A delay of the flap base is advisable before a flap inset. A description of the techniques of elevation of these flaps is included in the following chapters: gracilis and rectus femoris—Chapter 23, gluteal thigh flap—Chapter 27.

TABLE 4 Knee coverage

Muscle	Technique of elevation	Superior third		Middle third		Inferior third	
		Anterior	Posterior	Anterior	Posterior	Anterior	Posterior
Gracilis	Distally based	X					
Medial gastrocnemius	Muscle			X	X	X	X
	Musculocutaneous	X		X	X	X	X
	Cross-leg, skin-fascial	X	X	X	X	X	X
Vastus lateralis	Muscle	X					
Vastus medialis	Muscle	X					
Sartorius	Muscle	X					

KNEE (Table 4)

Muscle flap transposition is generally the method of choice for knee coverage. The lower extremity muscles suitable for transposition as muscle or musculocutaneous flaps include the following:

Vastus medialis
Vastus lateralis
Sartorius
Gracilis
Gastrocnemius

The gastrocnemius muscle is the most important muscle in this group for knee coverage. The other muscles are only useful for small defects and are less reliable.

Vastus medialis and vastus lateralis

As described in distal thigh coverage, the vastus medialis and vastus lateralis are occasionally used as transposition muscle flaps for coverage of defects located on the superior aspect of the knee joint. Skin grafts are placed on the exposed muscle and the donor site closed directly.

Sartorius

The distal sartorius inserts into the superior medial surface of the tibia. Since the muscle fibers of the sartorius extend inferiorly to its insertion, the muscle has a short arc of rotation to the knee joint. After division of the distal two to three segmental vascular pedicles, the muscle will cover small defects. Skin grafts are placed on the exposed muscle and the donor site closed directly.

Gracilis (distally based with delay)

The gracilis muscle has one or two minor vascular pedicles entering the distal third of the muscle. Ligation of the dominant vascular pedicle, the terminal branch of the medial femoral circumflex artery and vein, through an incision in the medial thigh is required 2 weeks before flap elevation (Fig. 30-2, *A* and *B*). (See Chapter 38 for technique to locate vascular pedicle.) The proximal muscle with its cutaneous territory designed as a skin island is elevated distally until the distal vascular pedicle entering the medial muscle is encountered. Since the position and number of minor vascular pedicles are variable, the point of rotation is occasionally too superior in the thigh for reliable knee coverage. Therefore this method is only used for superiorly located knee defects and then only as an alternative technique (Fig. 30-2, *C* to *F*).

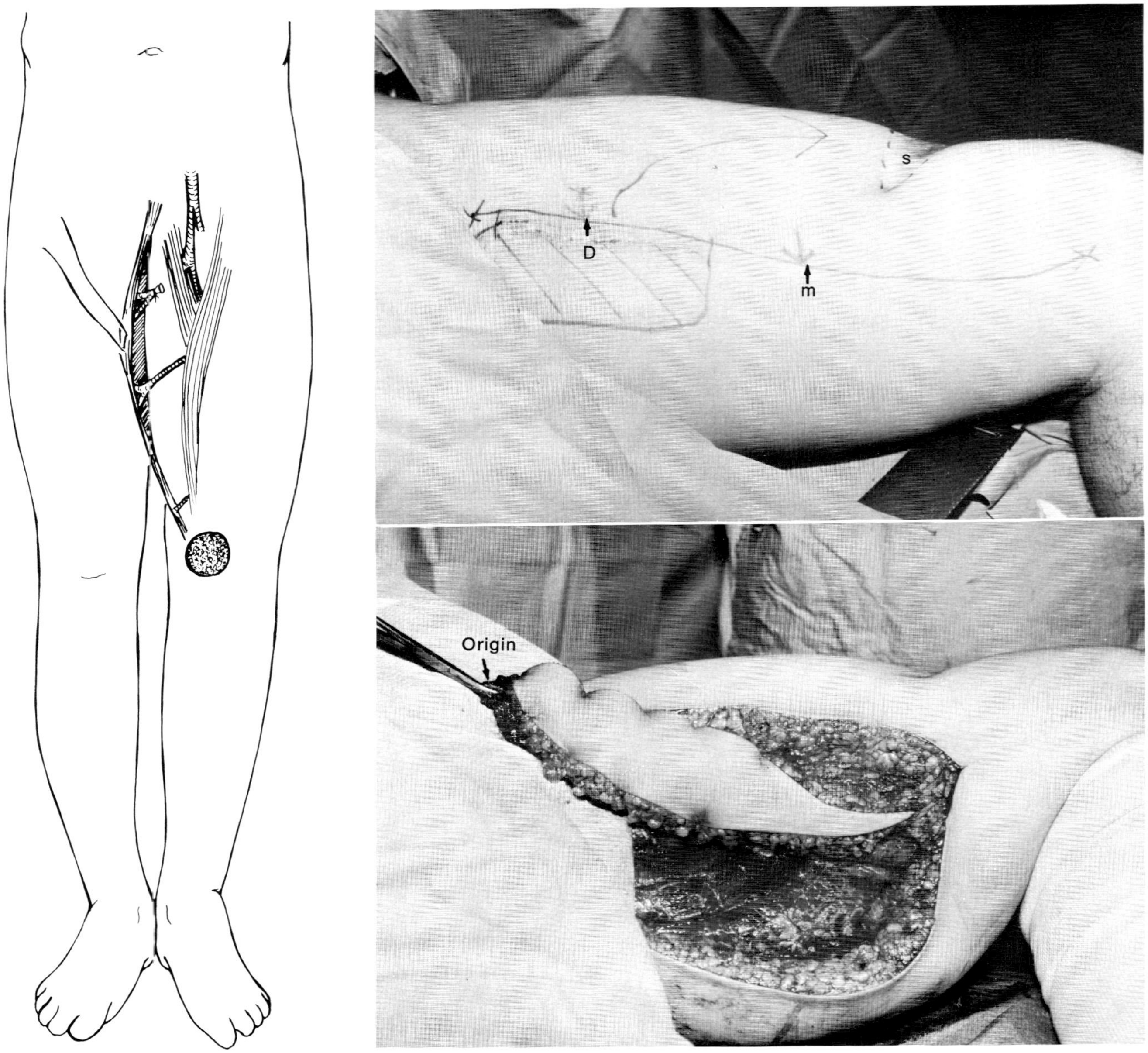

Fig. 30-2. Distal thigh and knee coverage. Use of distally based gracilis musculocutaneous flap for coverage of unstable wound at site of previous trauma.

A, Strategic delay performed 2 weeks before musculocutaneous flap transposition by ligation of proximal dominant pedicle (medial femoral circumflex artery).

B, Design of skin island over proximal gracilis muscle. Dotted line indicates site of prior incision for strategic delay. *D,* Location of divided dominant vascular pedicle; *m,* site of minor vascular pedicle; *s,* site of unstable skin over superior knee.

C, Origin of muscle divided with flap elevation.

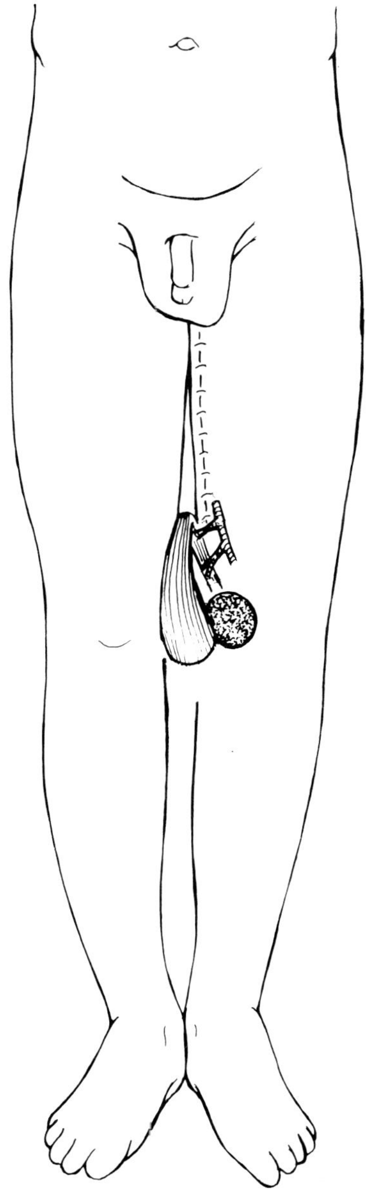

D

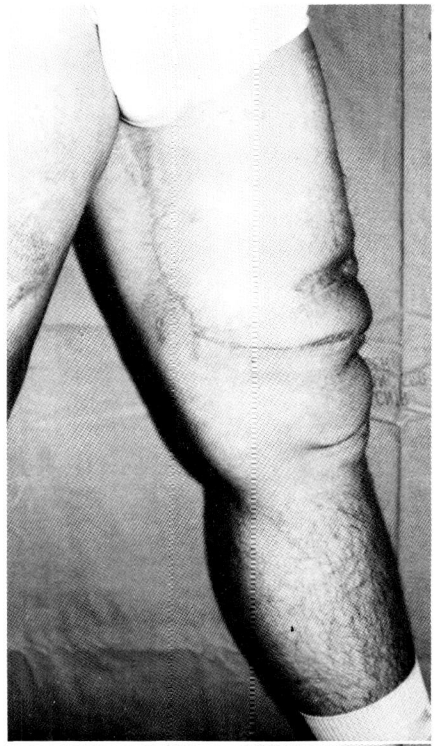

E

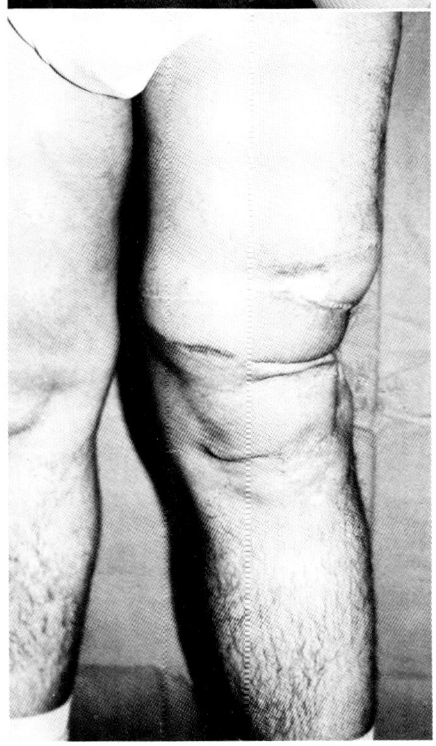

F

Fig. 30-2, cont'd

D, Extent of inferior arc of rotation determined by location of proximal minor vascular pedicle.

E and **F,** Six-month postoperative views demonstrate stable coverage at site of flap inset. NOTE: Flap will not reach middle knee region.

Gastrocnemius

The medial and lateral heads of the gastrocnemius are the primary muscle or musculocutaneous flaps used for knee coverage. With a type I pattern of circulation the medial and lateral heads of the gastrocnemius each receive a single vascular pedicle, the sural artery, entering the deep surface of the muscle within the popliteal fossa. Although both the medial and lateral gastrocnemius muscle or musculocutaneous units will reach the knee, the medial gastrocnemius muscle unit has a wider arc of rotation than the lateral gastrocnemius muscle unit (Fig. 30-3, *A*). The lateral gastrocnemius muscle unit is shorter, and its arc is also decreased because of the presence of the fibula.

The design of the gastrocnemius muscle or musculocutaneous flap depends on the location and size of the knee defect (Fig. 30-3, *B*).

Extensive knee defects. The use of the extended gastrocnemius musculocutaneous unit increases the arc of rotation and area of coverage of the flap, since the skin territory extends distally to the muscle (Fig. 30-3, *C*).

Inferior or medial knee defects. The medial gastrocnemius muscle or musculocutaneous flaps consisting of muscle with a skin island will cover the medial or inferior knee (Fig. 30-3, *D*). The muscle flap is preferable since the coverage is less bulky with muscle and skin grafts. Also the donor site is closed directly without the necessity of skin grafts. Thus, donor site deformity is decreased.

Lateral knee defects. The lateral gastrocnemius muscle or musculocutaneous flap consisting of muscle with a skin island will cover lateral knee defects. Again the muscle flap is preferable, since less bulk is noted at the recipient site, and donor site deformity is decreased. However, the muscle length is variable. For this reason if the lateral gastrocnemius appears short on physical examination, the extended musculocutaneous flap will provide a wider arc of rotation.

Posterior knee (popliteal fossa). Both medial and lateral gastrocnemius muscle or musculocutaneous flaps will cover the popliteal fossa. However, defects in this area may include loss of the vascular pedicle to the planned muscle flap. Unless arteriography or exploration confirms the patency of the sural artery, alternative flaps for coverage are used.

A

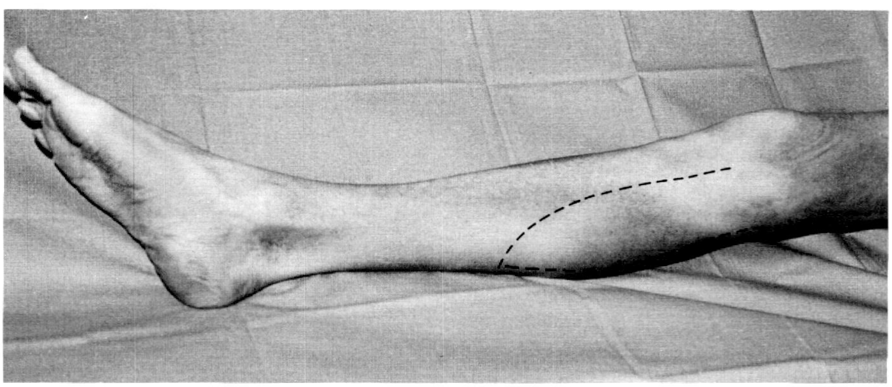

Fig. 30-3. Flap design of medial gastrocnemius.
A, Dotted lines indicate location of muscle.

B

C

D

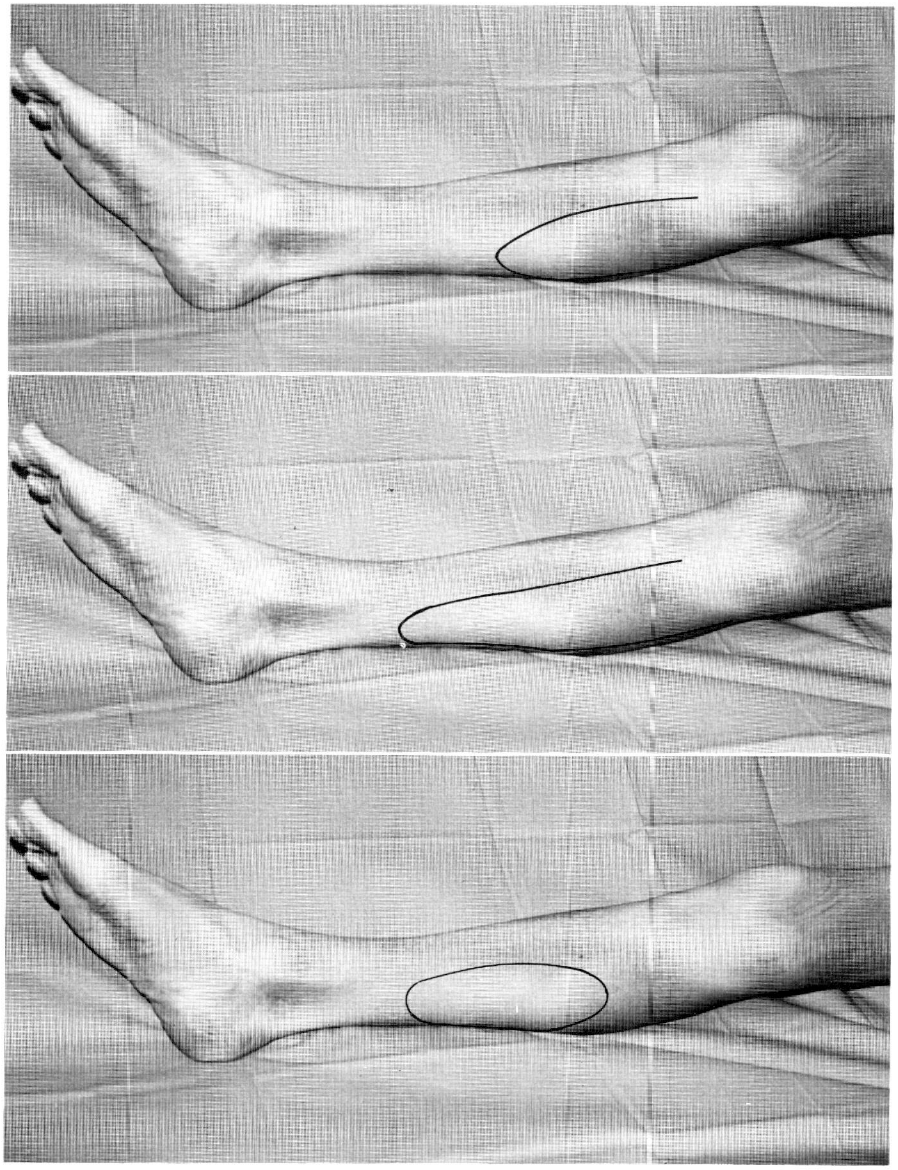

Fig. 30-3, cont'd
B, Standard design of musculocutaneous flap for coverage of middle or proximal third of leg.
C, Design of extended musculocutaneous flap for leg or knee coverage.
D, Design of skin island for gastrocnemius musculocutaneous flap for leg or knee coverage.

Medial gastrocnemius. The technique of elevation of the medial gastrocnemius muscle for transposition is similar whether designed as a muscle or musculocutaneous flap. The muscle is exposed through a vertical incision extending from the knee to the ankle and located 2 cm medially to the anterior edge of the tibia (Fig. 30-4, *A*). The saphenous vein is preserved when encountered beneath this incision (Fig. 30-4, *B*). After the soleus muscle is visualized deep to the gastrocnemius, the plane between these muscles is entered in the proximal third of the thigh. The plantaris tendon is visualized within this intermuscular plane and preserved (Fig. 30-4, *C*). After the soleus is completely separated from the gastrocnemius in the superior third of the leg, the deep surface of the gastrocnemius is examined. Generally a raphe is visualized between the medial and lateral heads of the gastrocnemius muscle. If this raphe is not clear, the medial head is separated from the lateral head at the midposterior leg. As the dissection progresses inferiorly, the medial half of the soleus is sharply separated from its junction with the gastrocnemius until the Achilles tendon is reached. At this level the medial half of the Achilles tendon is divided completely, releasing the distal medial gastrocnemius muscle from adjacent muscles and its tendon of insertion (Fig. 30-4, *D*).

If overlying skin is included with the muscle as a musculocutaneous flap, it is incised carefully, maintaining its vascular connection with the underlying muscle. If the muscle is elevated without overlying skin, the posterior leg skin is separated from the superficial (posterior) surface of the muscle to the site of prior separation of the raphe between the two heads of the gastrocnemius. The sural nerve and lesser saphenous veins are generally visualized at this level and are preserved.

A

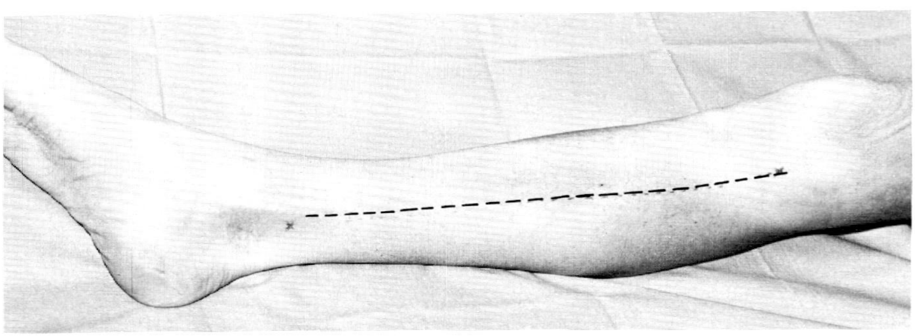

Fig. 30-4. Technique of elevation of medial gastrocnemius.
A, Location of initial anterior leg incision for elevation of gastrocnemius muscle or musculocutaneous flap.

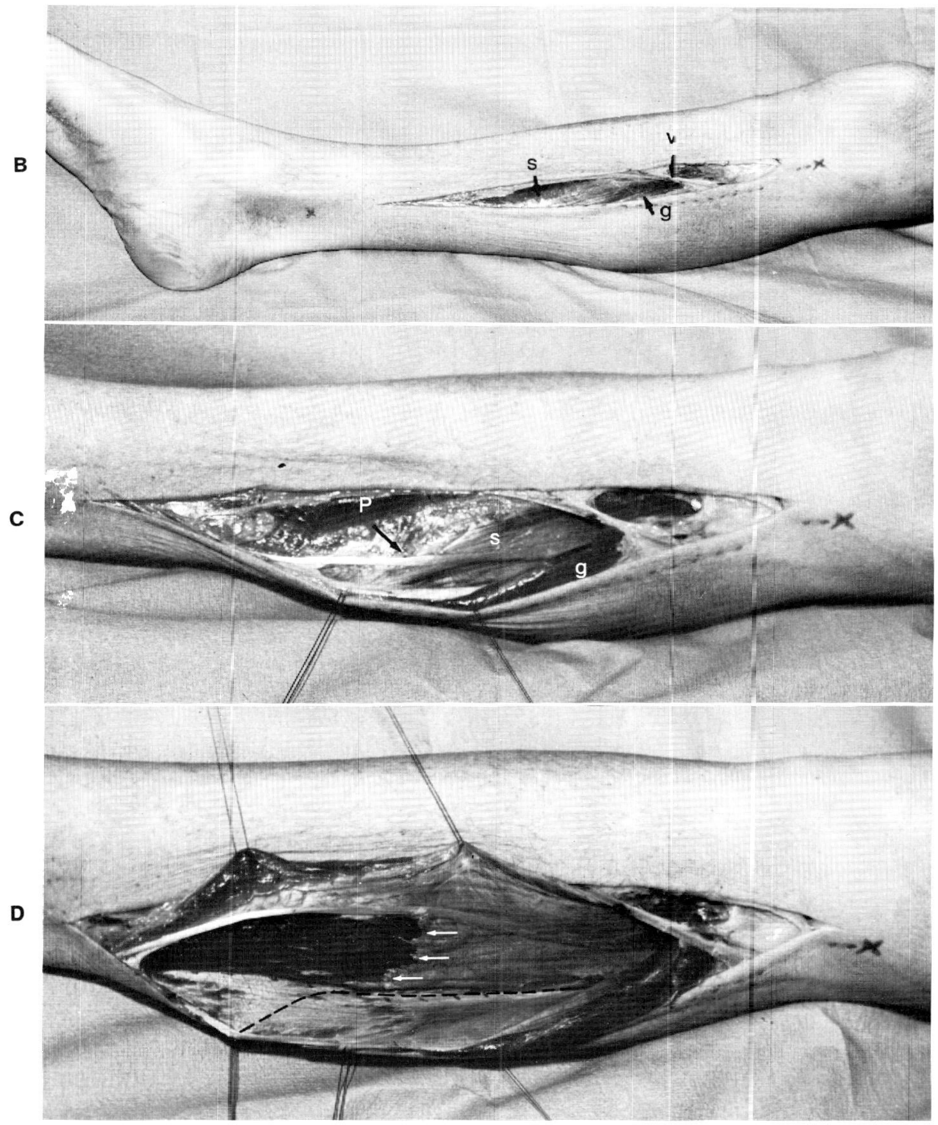

Fig. 30-4, cont'd

B, Identification of soleus muscle (*s*), medial gastrocnemius (*g*), and saphenous vein (*v*).

C, Intermuscular plane between soleus (*s*) and gastrocnemius (*g*) muscles is located in proximal third of leg. NOTE: Plantaris tendon (*p*) between muscles.

D, Arrows denote site of release of medial soleus muscle from medial gastrocnemius muscle at musculotendinous junction. Dashed line indicates site of release of fibers of insertion into Achilles tendon.

Continued.

The proximal dissection of the muscle is carried to the level of the medial condyle of the femur. Generally flap length is adequate at this time to reach the knee for defect coverage. (Fig 30-4, *E* and *F*). If length is still inadequate, dissection into the popliteal fossa will increase the arc of rotation, but the sural artery is subject to injury. Release of the muscle origin will still further increase the flap length, but the sural artery is identified before this maneuver is performed. Dissection of the medial gastrocnemius rarely extends superior to the medial condyle. Further dissection is avoided if possible, since the vascular pedicle and tibial nerve are subject to injury.

The thick posterior fascia of the gastrocnemius limits the expansion of the gastrocnemius muscle. Linear incisions or excision of this fascia is occasionally performed to increase the area of coverage, especially when the muscle is transferred without overlying skin (Fig. 30-4, *G* and *H*).

E

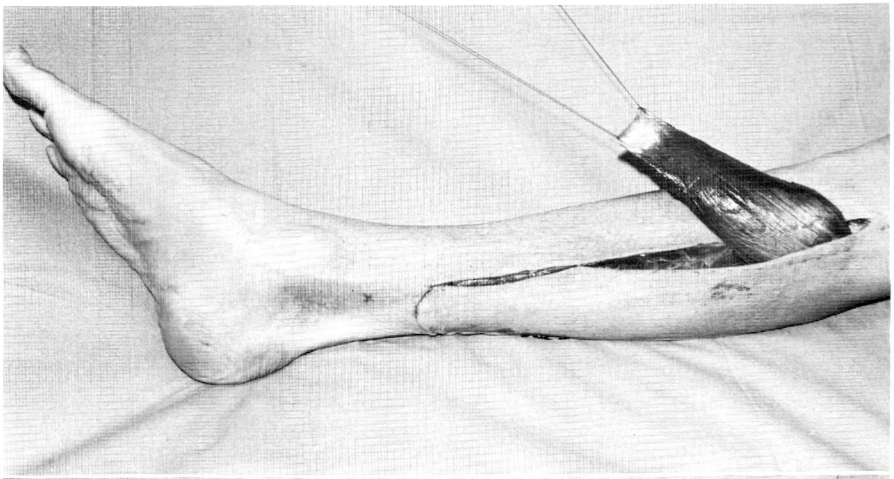

F

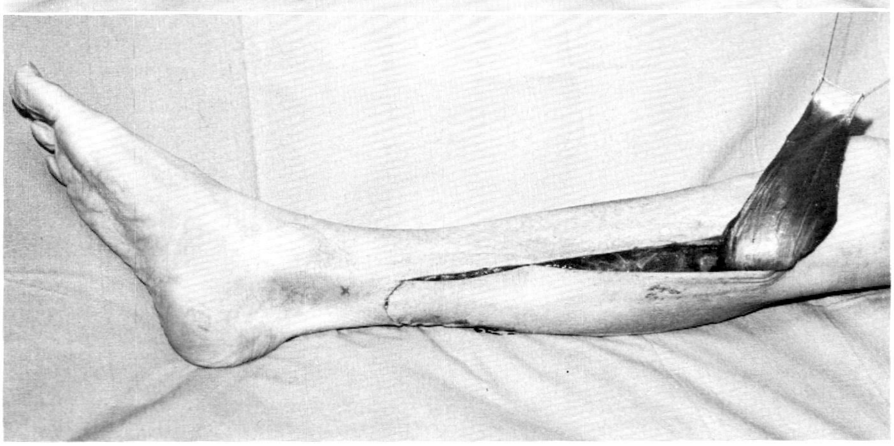

Fig. 30-4, cont'd

E, Transposition of medial gastrocnemius muscle to proximal third of leg. NOTE: Absence of severe contour deformity in mid calf with direct donor site closure.)

F, Transposition of muscle for knee coverage. NOTE: Proximal extent of deep dissection does not enter popliteal fossa.

G

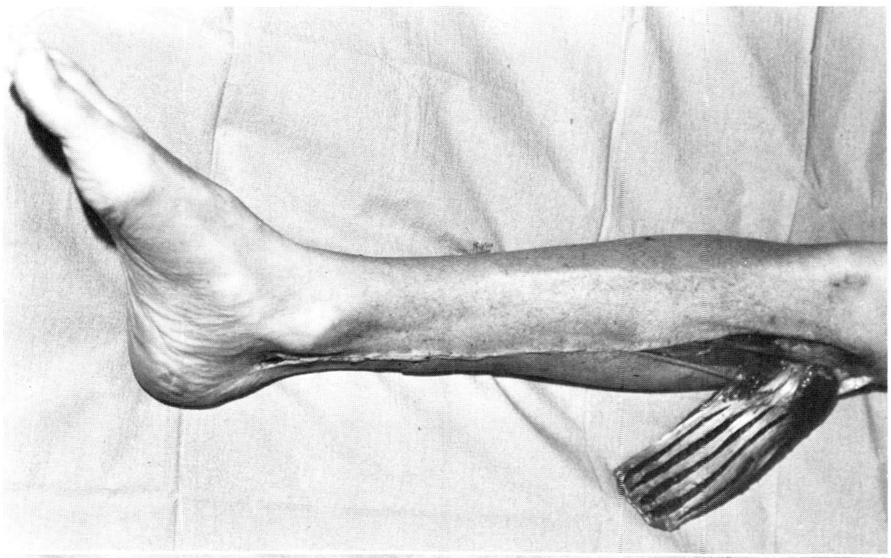

H

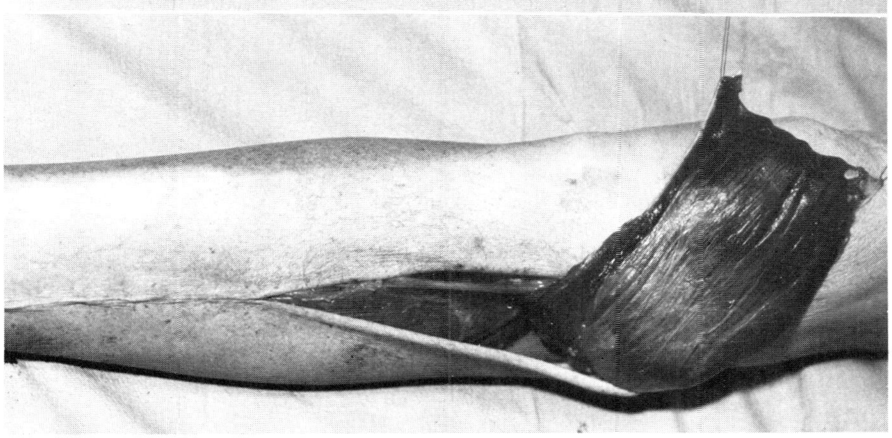

Fig. 30-4, cont'd

G, Linear incisions in thick deep fascia of muscle allow expansion of muscle width for greater coverage.

H, Transposition of expanded muscle flap for knee coverage.

Lateral gastrocnemius. The technique of elevation of the lateral gastrocnemius is similar, regardless of flap design as muscle or musculocutaneous flap. The muscle is exposed through a lateral vertical incision located 2 cm posteriorly to the fibula, extending from the knee to the ankle. The lateral edge of the soleus muscle is visualized. The intermuscular plane between the soleus and deep surface of the lateral gastrocnemius muscle is entered in the proximal third of the leg. The superficial peroneal nerve is encountered at the level of the head of the fibula and is preserved. When the raphe between the lateral and medial heads of the gastrocnemius muscle is encountered, the two heads of gastrocnemius muscle are separated. The soleus and lateral gastrocnemius muscles are separated at their junction with the Achilles tendon, and the attachment of the lateral gastrocnemius to the tendon of insertion is divided.

If overlying skin is included with the muscle as a musculocutaneous flap, it is incised carefully, maintaining its vascular connections with the underlying muscle. If the muscle is elevated without overlying skin, the posterior leg skin is separated from the superficial (posterior) surface of the muscle to the site of prior separation of the two heads of the gastrocnemius at the midposterior raphe. At this level the sural nerve and lesser saphenous veins are generally visualized and preserved.

The proximal dissection of the muscle is carried to the level of the lateral condyle of the femur. Generally flap length is adequate at this time to reach the knee. Further dissection into the popliteal fossa will increase the arc of rotation. The technique and precautions regarding this maneuver are similar to those described previously for the medial gastrocnemius muscle.

TABLE 5 Leg coverage

Muscle	Technique of elevation	Proximal third	Middle third	Distal third
Extensor digitorum longus	Muscle		X	
	Segmental		X	
	Distally based		X	X
Extensor hallucis longus	Muscle			X
	Distally based			X
Flexor digitorum longus	Muscle		X	
	Segmental		X	
Gastrocnemius	Musculocutaneous	X	X	
	Muscle	X	X	
	Skin-fascial	X	X	X
	Cross-leg	X	X	X
Peroneus brevis	Muscle			X
Peroneus longus	Muscle		X	
Soleus	Muscle	X	X	X
	Segmental		X	X
	Distally based			X
	Advancement		X	X
Tibialis anterior	Muscle		X	X
	Segmental		X	X
	Advancement		X	X

LEG (Table 5)

The muscle and musculocutaneous flap is the method of choice for defects in the proximal two thirds of the leg. In the distal third of the leg the lack of muscle fibers and superficial location of tendons and nerves prevents the reliable use of muscle and musculocutaneous flaps for coverage. However,the distal transposition of muscle will provide coverage of the distal leg. The skin-fascial flap based on the gastrocnemius musculocutaneous perforating vessels, cross-leg musculocutaneous flaps, and function-preserving techniques for lower extremity muscle transposition are recent modifications in the design of muscle flaps for lower leg coverage in an effort to improve coverage and provide optimal form and function.

The muscle and musculocutaneous flap is preferred for leg coverage when chronic infection is a component of the defect, usually involving bone (osteomyelitis). The muscle flap when placed within the wound after débridement demonstrates resistance to necrosis resulting from bacterial invasion. Both resolution of the chronic infection and coverage are generally achieved after use of muscle or musculocutaneous flap transposition to the debrided lower extremity wound.

The muscles useful for lower leg coverage and the region where their use is recommended include the following:

- Proximal third leg defects
  - Gastrocnemius
  - Skin-fascial gastrocnemius
- Medial third leg defects
  - Soleus
  - Skin-fascial gastrocnemius
  - Flexor digitorum longus
  - Peroneus longus
  - Tibialis anterior
  - Extensor digitorum longus
- Distal third leg defects
  - Soleus
  - Skin-fascial gastrocnemius
  - Extensor digitorum longus
  - Tibialis anterior
  - Extensor hallucis longus
  - Peroneus brevis
  - Cross-leg skin-fascial gastrocnemius

Proximal third leg defects

The gastrocnemius muscle is the primary muscle used in coverage of the proximal third of the lower leg. The soleus muscle transposed medially will reach this area, but the proximal dissection is difficult and only indicated in the absence of the medial or lateral heads of the gastrocnemius muscle.

Gastrocnemius. The design of the medial or lateral gastrocnemius for upper third leg coverage includes the following (Fig. 30-3):

- Muscle only
- Musculocutaneous flap
  - Standard
  - Island
- Skin-fascial

Muscle and musculocutaneous flap (Fig. 30-4). Transposition of muscle only with skin graft application to exposed muscle is preferred, since bulk is decreased at the site of reconstruction (Fig. 30-5). Also the donor site deformity noted with the musculocutaneous flap is avoided. The standard musculocutaneous flap maintaining proximal leg skin continuity at the flap base or the island musculocutaneous flap increases the area of flap coverage and is recommended for large defects. The use of the muscle to fill sequestrectomy defects is advocated when chronic osteomyelitis is present. The anatomy and technique of elevation of the medial and lateral gastrocnemius muscle are included in the section for the use of the gastrocnemius in knee coverage.

A

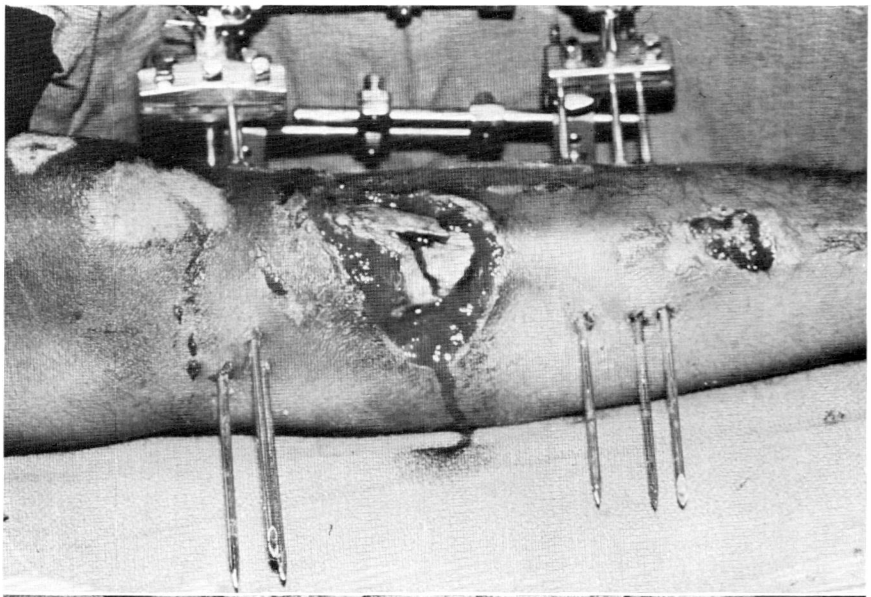

B

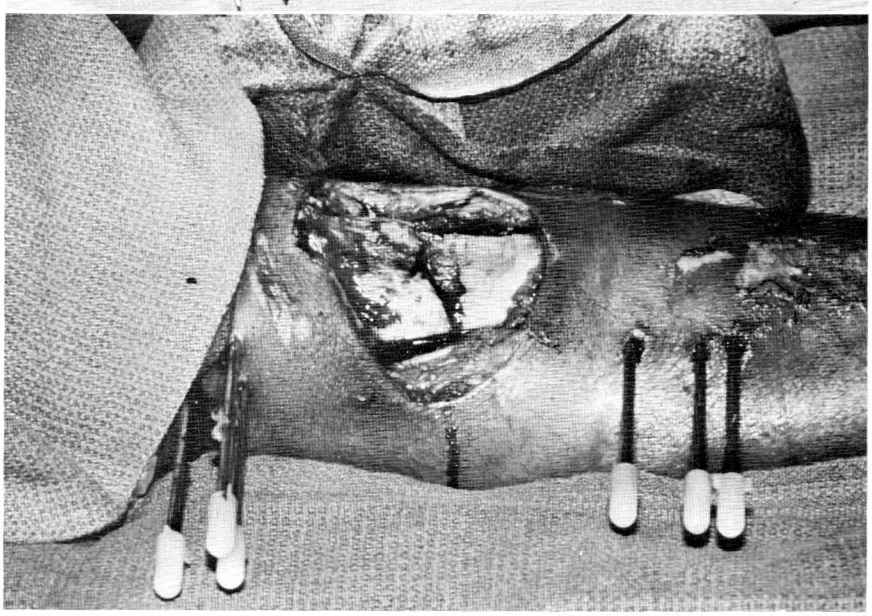

Fig. 30-5. For legend see opposite page.

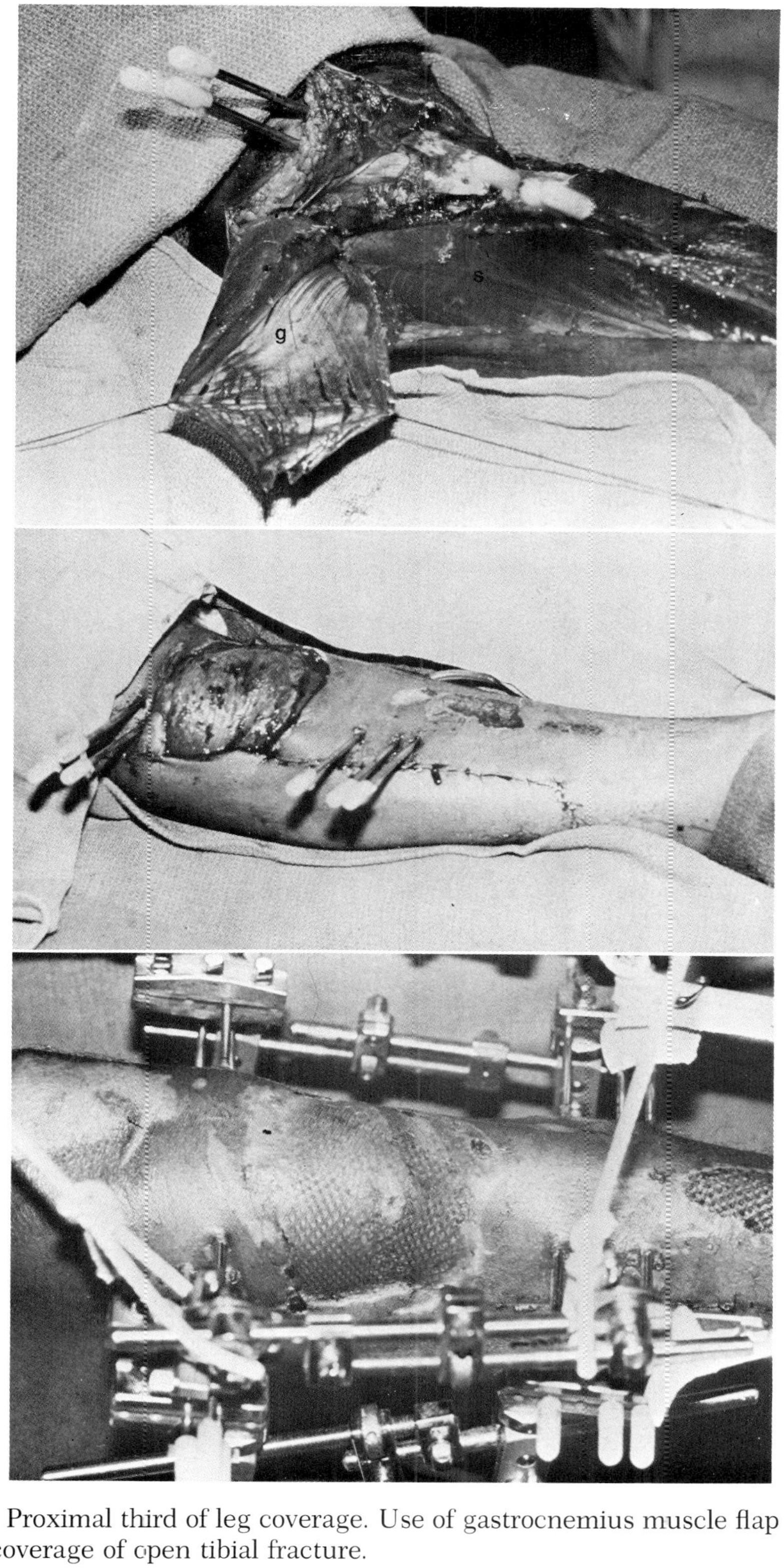

Fig. 30-5. Proximal third of leg coverage. Use of gastrocnemius muscle flap with skin graft for coverage of open tibial fracture.

A, Exposed proximal tibial fracture.

B, Fracture segments stabilized with external fixation apparatus.

C, Medial gastrocnemius muscle elevated (*g*). Linear incisions in deep fascia of muscle allow muscle expansion. *s*, Soleus muscle.

D, Transposed muscle covers exposed fracture site. Donor site closed directly.

E, Stable wound coverage in one operation.

Skin-fascial gastrocnemius flap. The use of the distal cutaneous territory of the medial and lateral gastrocnemius musculocutaneous flap with distal random extension is possible without elevation of the muscle. The large musculocutaneous perforators that enter the skin at the midportion of each head of the gastrocnemius muscle are the vascular pedicles for these flaps (Fig. 30-6). Both medial and lateral skin-fascial leg flaps will cover defects in the proximal and middle third of the leg.

The midportion of the medial or lateral gastrocnemius muscle is determined by preoperative inspection. This determines the probable base of the flap. The flap is then designed distal to this level, incorporating medial or lateral leg skin. The medial skin flap is located between the tibia and posterior midline and the lateral flap between the fibula and posterior midline (Fig. 30-7, *A*).

Flap elevation starts distally with elevation of skin and subcutaneous tissue until the junction of the gastrocnemius muscle with the Achilles tendon is visualized. At this level the epimysium is included with the flap. As the dissection approaches the midportion of the muscle belly, the musculocutaneous perforators are preserved (Fig. 30-7, *B*). If flap length is not adequate at this level, it is preferable to elevate the underlying medial or lateral gastrocnemius muscle with the flap. The paratenon is preserved on the Achilles tendon. Although the donor areas requires skin grafts, there is minimal contour deformity, and the continuity of the gastrocnemius muscle is preserved (Fig. 30-7, *C*).

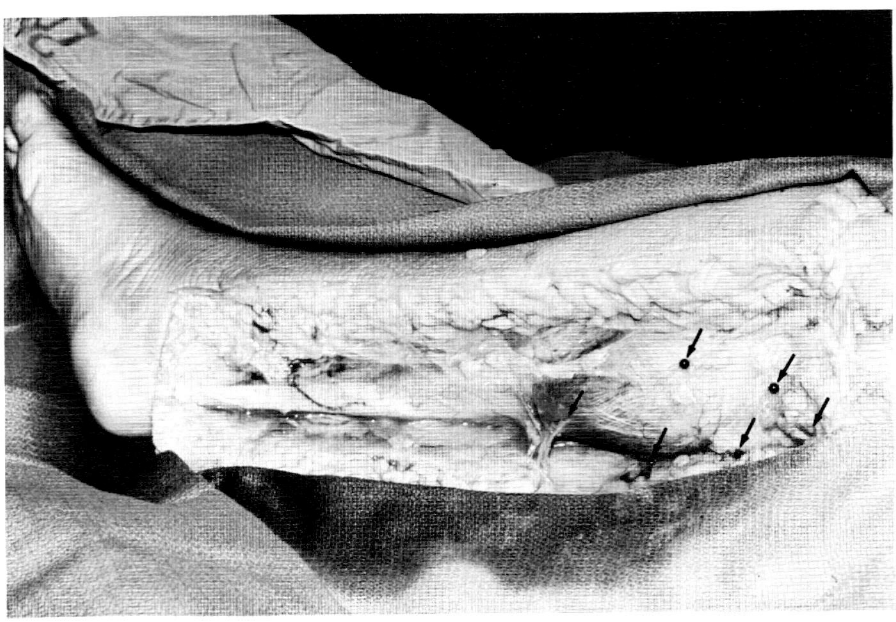

Fig. 30-6. Skin-fascial flap musculocutaneous perforators *(arrows)* from medial gastrocnemius muscle are vascular basis for skin-fascial flap. NOTE: Black pin indicates location of major musculocutaneous perforating vessel.

A

B

C

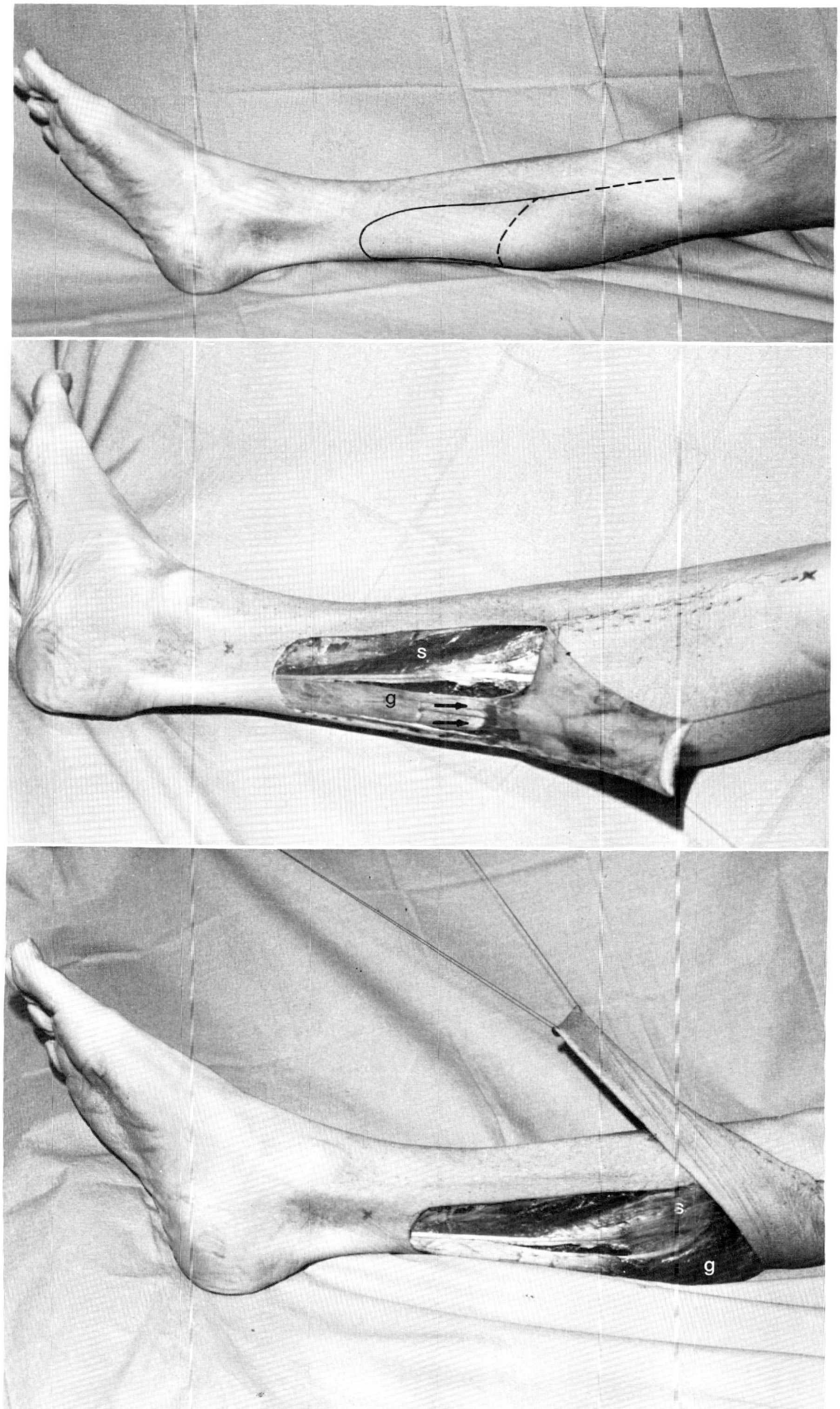

Fig. 30-7. Design and technique of elevation of skin-fascial flap.

A, Solid line denotes flap design. Dotted line denotes medial gastrocnemius muscle.

B, Flap base located at midportion of medial gastrocnemius muscle (*g*). Arrow denotes musculotendinous junction. NOTE: Musculocutaneous perforating vessels ligated only as required for adequate arc of rotation. *s*, Soleus muscle.

C, Transposition of skin-fascial flap will cover defects of proximal leg defects. *s*, Soleus muscle; *g*, gastrocnemius.

Middle third leg defects

The soleus is the primary muscle used for coverage of middle third defects. The soleus has adequate size to cover large defects and is occasionally used in combination with the medial gastrocnemius muscle flap if greater area of coverage is required (Figs. 30-8 and 30-9). The tibialis anterior, extensor digitorum longus, and flexor digitorum longus are not expendable muscles. However, segmental elevation of these muscles, maintaining tendon continuity with proximal muscle, is useful for coverage of small defects. The skin-fascial flap based on the gastrocnemius musculocutaneous perforating vessels will also cover middle third defects. This flap modification is discussed under gastrocnemius, proximal third leg coverage. The use of the soleus muscle to fill the sequestrectomy site is advocated when chronic osteomyelitis is present (Fig. 30-10). The muscles used for middle leg coverage are transposed as muscle without overlying skin. Skin graft coverage of exposed muscle is necessary.

A

B

C

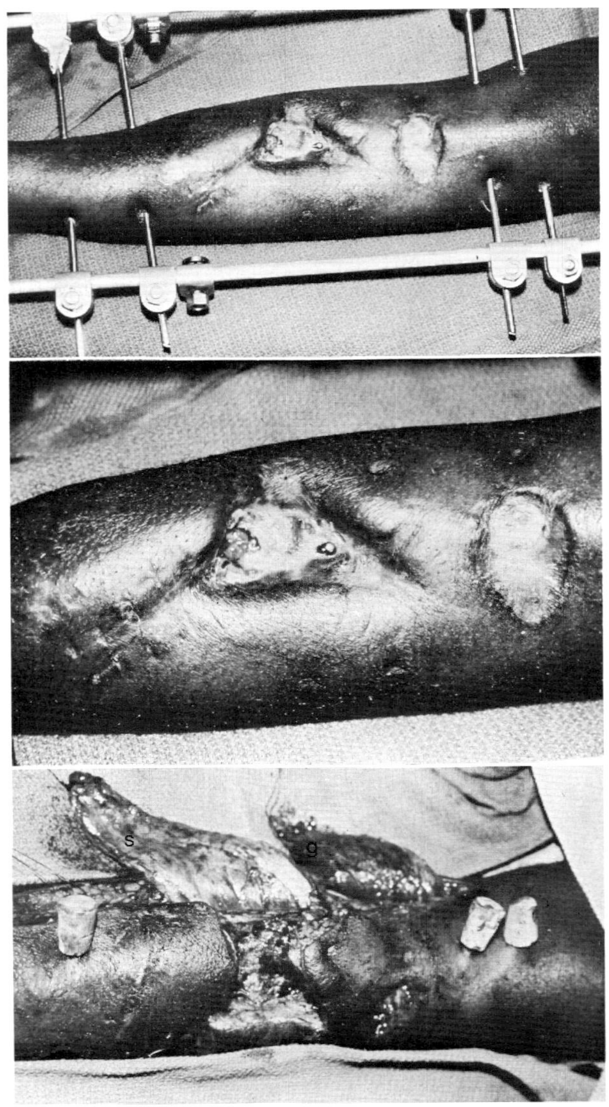

Fig. 30-8. Proximal and middle third leg coverage. Use of medial gastrocnemius muscle and soleus muscle flaps with skin graft for coverage of exposed tibial fracture.
A and **B,** Proximal tibia exposure and open tibial midshaft fracture. Fracture alignment maintained with external fixation apparatus.

D

E

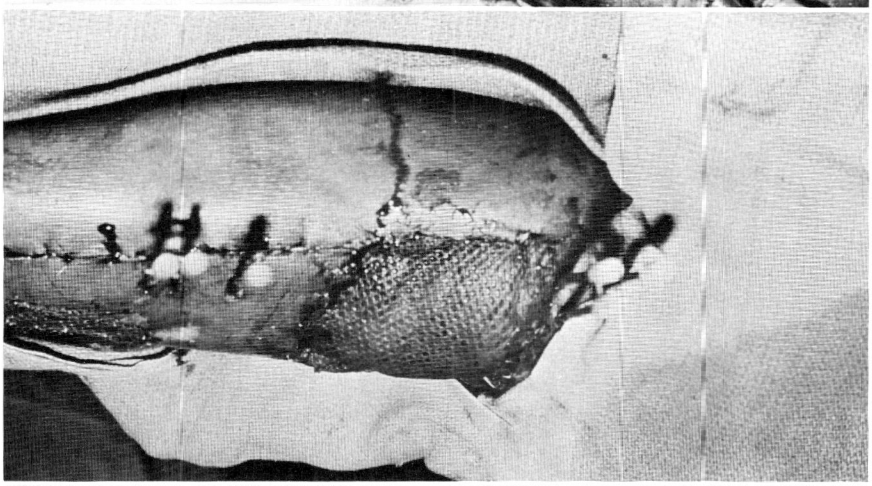

F

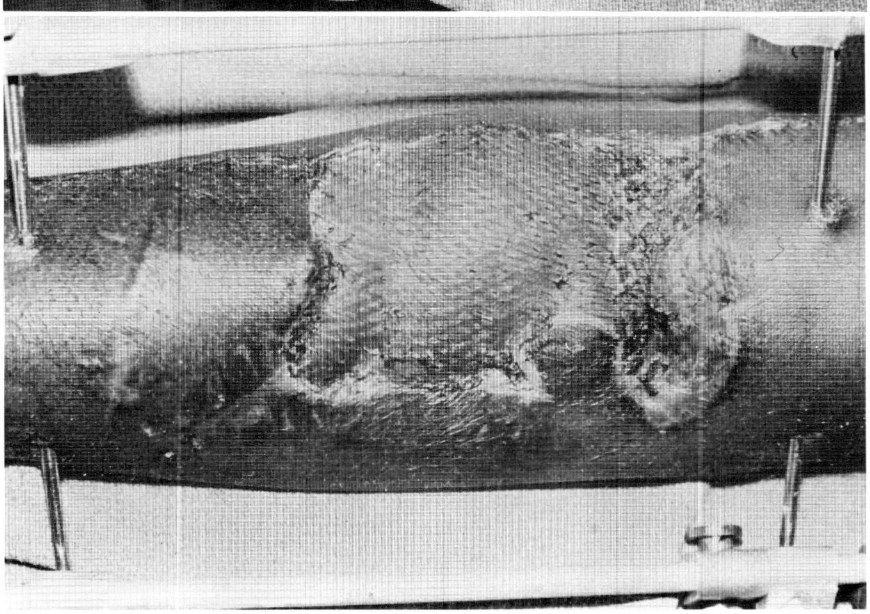

Fig. 30-8, cont'd
C and **D,** Medial gastrocnemius *(g)* and soleus *(s)* muscle elevation.
E, Meshed skin graft placed over exposed muscle. Donor site closed directly.
F, Stable wound coverage in one operation.

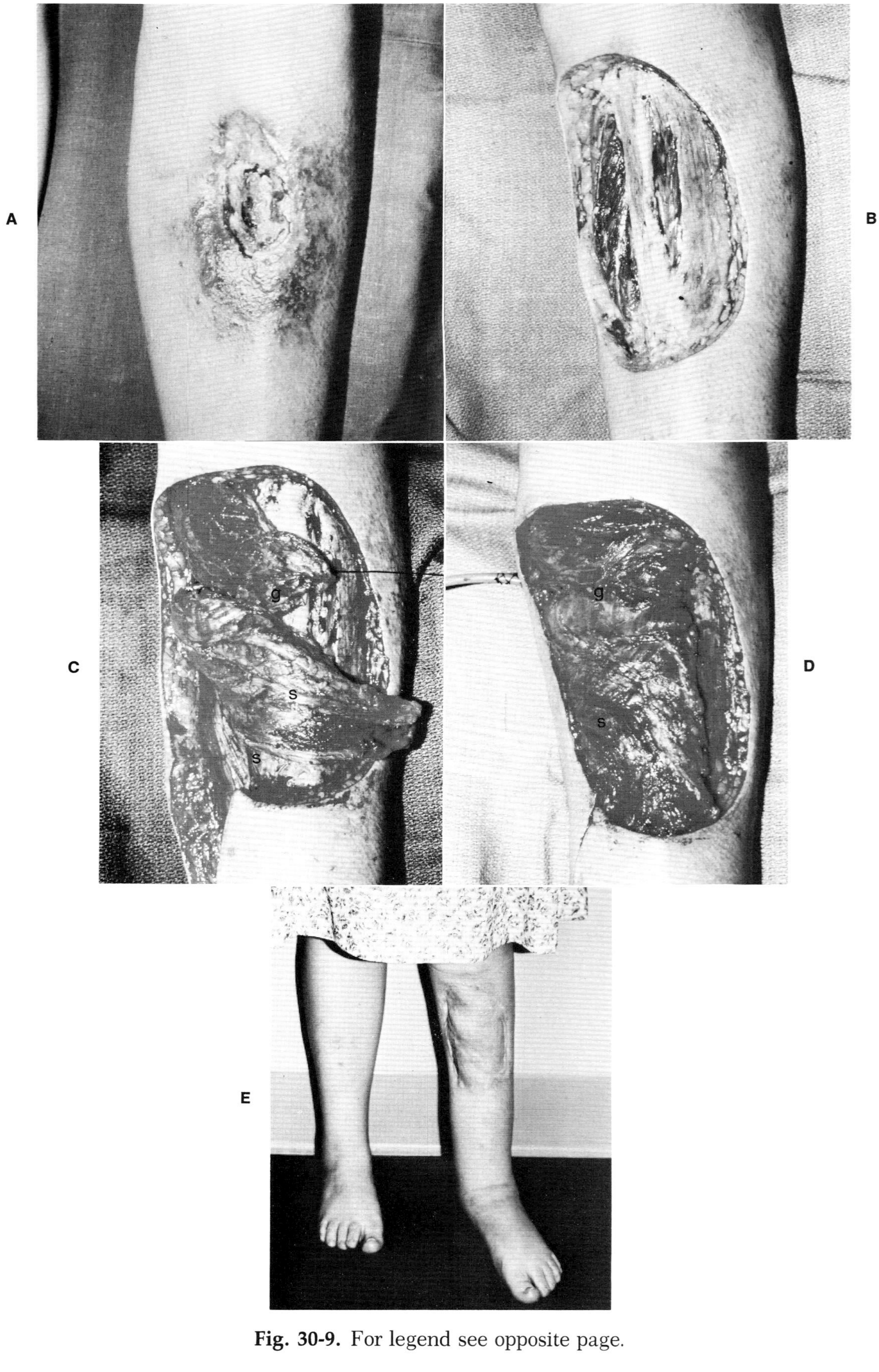

Fig. 30-9. For legend see opposite page.

Fig. 30-9. Proximal and middle third leg coverage. Use of medial gastrocnemius muscle and soleus muscle flaps with skin graft for coverage of osteoradionecrosis.
A, Osteoradionecrosis at site of prior radiation therapy for bone tumor.
B, Defect after resection of ulcer and adjacent scar tissue.
C and D, Medial gastrocnemius (*g*) and soleus (*s*) muscles transposed onto exposed anterior tibia.
E, Two-year postoperative view demonstrates stable coverage.

(Reproduced with permission from Brown, R.G., and Jurkiewicz, M.J.: Reconstructive Surgery in the Cancer Patient, in Hickey, R.C., et al. (eds.): CURRENT PROBLEMS IN CANCER. Copyright © 1977 by Year Book Medical Publishers, Inc., Chicago.)

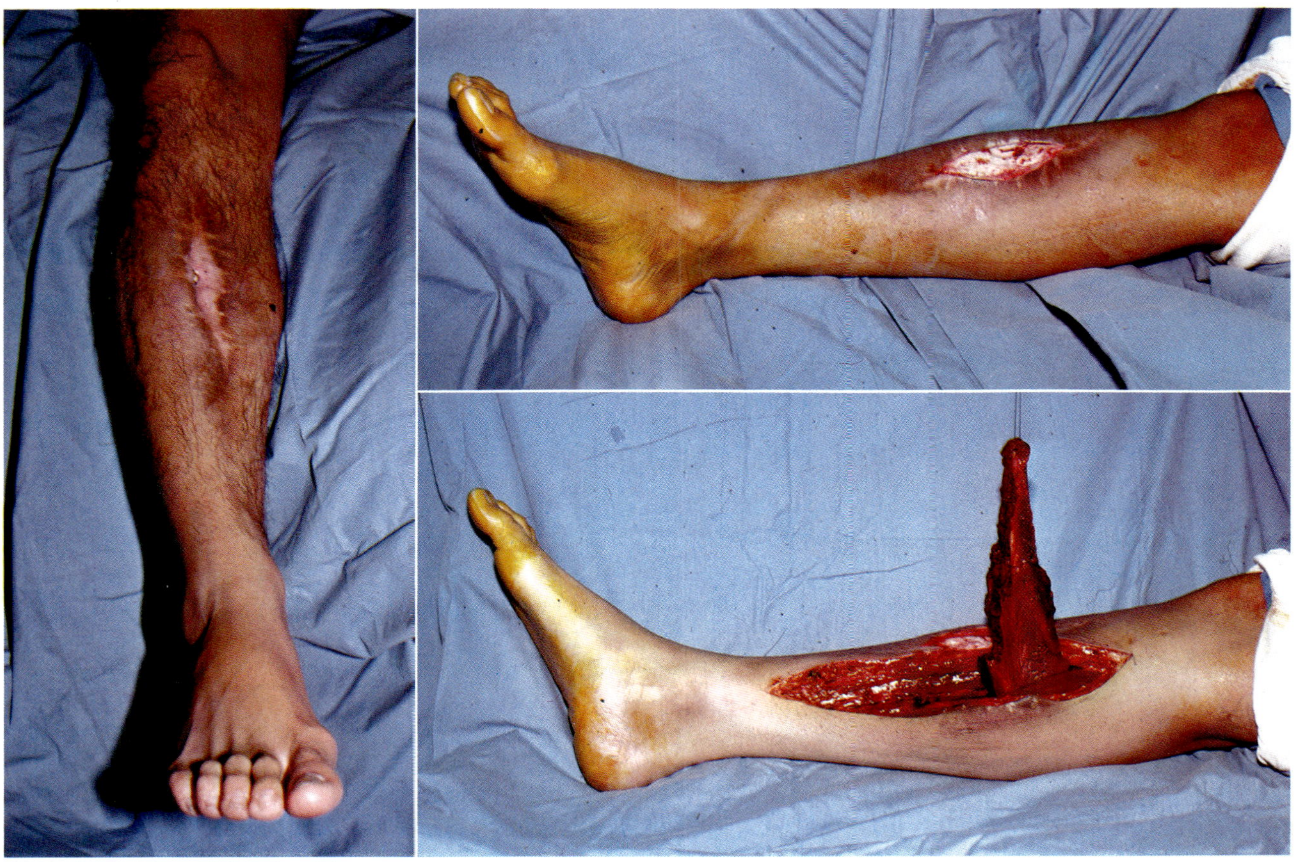
A
B
C

Fig. 30-10. Middle third leg coverage. Use of soleus muscle flap with skin graft for coverage of sequestrectomy defect of chronic osteomyelitis.
A, Four-year history of chronic osteomyelitis at site of prior open tibial fracture.
B, Sequestrectomy defect.
C, Soleus muscle elevated.

Continued.

D

E

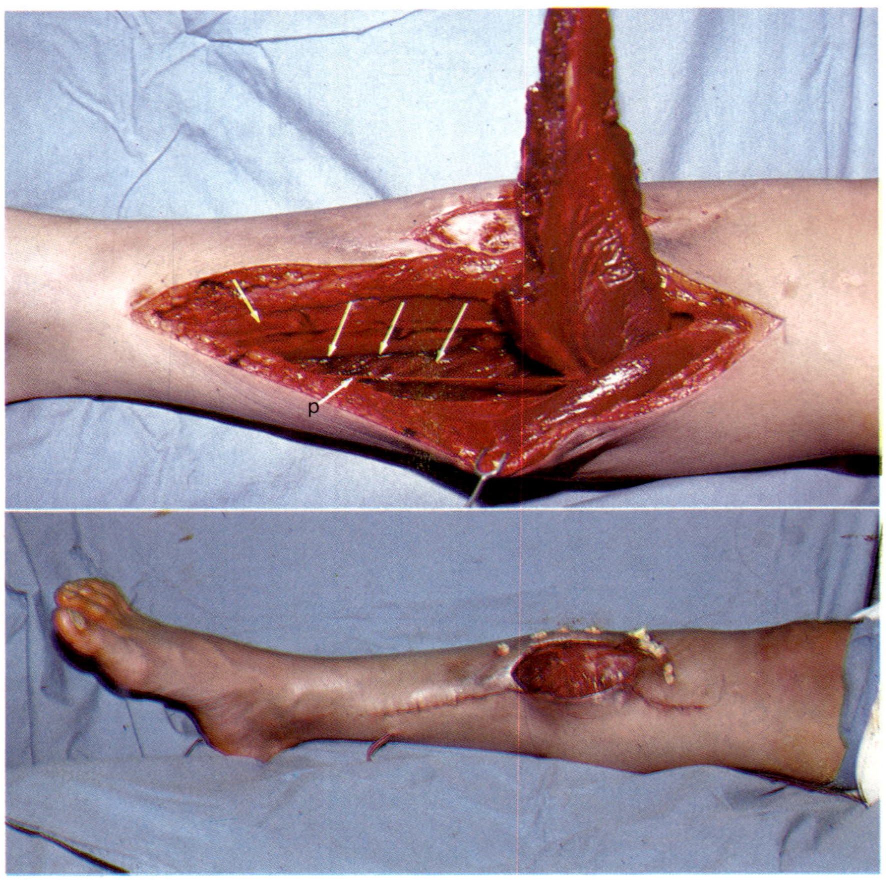

Fig. 30-10, cont'd

D, Segmental muscle flap elevation with preservation of lateral soleus muscle. Arrow denotes lateral half of soleus muscle. *p,* Plantaris tendon; *g,* medial gastrocnemius; *v,* posterior tibial vessels.

E, Soleus muscle placed into sequestrectomy defect. Donor defect closed directly.

F

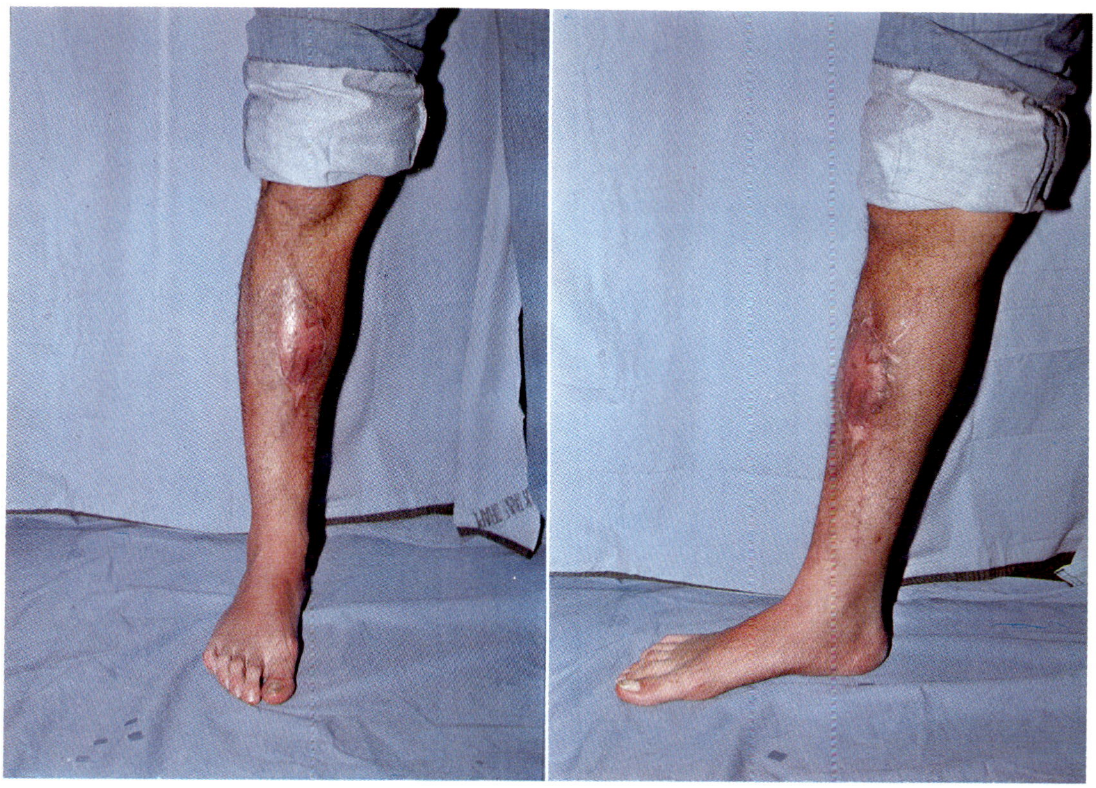

G

Fig. 30-10, cont'd
F, Stable wound coverage without recurrence of bone infection.
G, Use of muscle flap allows direct closure of donor site without contour deformity.

Soleus. The soleus is a posterior calf muscle located deep to the gastrocnemius muscle, extending the entire width and length of the lower leg. With a type II pattern of circulation, the muscle receives dominant vascular pedicles in its proximal third from the popliteal, posterior tibial, and peroneal arteries.

The muscle also receives minor pedicles distally that require division when muscle transposition is performed for middle leg coverage. The muscle is elevated from either the medial or lateral leg approach. However, the medial approach is technically easier and the transposition of the muscle is not limited by the fibula.

Medial approach. A vertical incision is made 2 cm medially to the tibia, extending from the medial condyle to the ankle (Fig. 30-11, *A*). The soleus muscle is identified beneath the medial head of the gastrocnemius muscle. The intermuscular plane between the soleus and gastrocnemius muscle is located in the proximal third of the leg. After separation of the two muscles proximally, the muscle is divided from its junction with the Achilles tendon from the midcalf level distally, as required for flap length (Fig. 30-11, *B*). The plantaris tendon located between the gastrocnemius and soleus muscle is preserved. The deep surface of the muscle is separated from the flexor digitorum longus at the middle third of the leg. When this plane is identified, the muscle is elevated proximally, dividing the minor vascular pedicles as they enter the medial deep aspect of the soleus muscle belly. Since these vascular pedicles are short, it is preferable to suture ligate rather than use the electrocautery to avoid injury to the posterior tibial vessels (Fig. 30-11, *C*). The dissection is carried proximally until adequate flap length and size are available for defect coverage. Once flap elevation reaches the midcalf level, no vascular pedicle is divided unless its release is required to increase flap length (Fig. 30-11, *D* to *G*).

If the defect does not involve the complete middle third of the leg, it is not necessary to elevate the entire muscle. Segmental flap elevation with function preservation is possible by splitting the muscle in the midposterior calf, leaving the lateral muscle in continuity with the Achilles tendon (Figs. 30-10, *D* and 30-11, *D*). This portion of the muscle is not devascularized because of laterally located dominant vascular pedicles from the peroneal artery. The medial muscle is transposed into defect, and the exposed muscle is skin grafted. The donor site is closed directly.

A

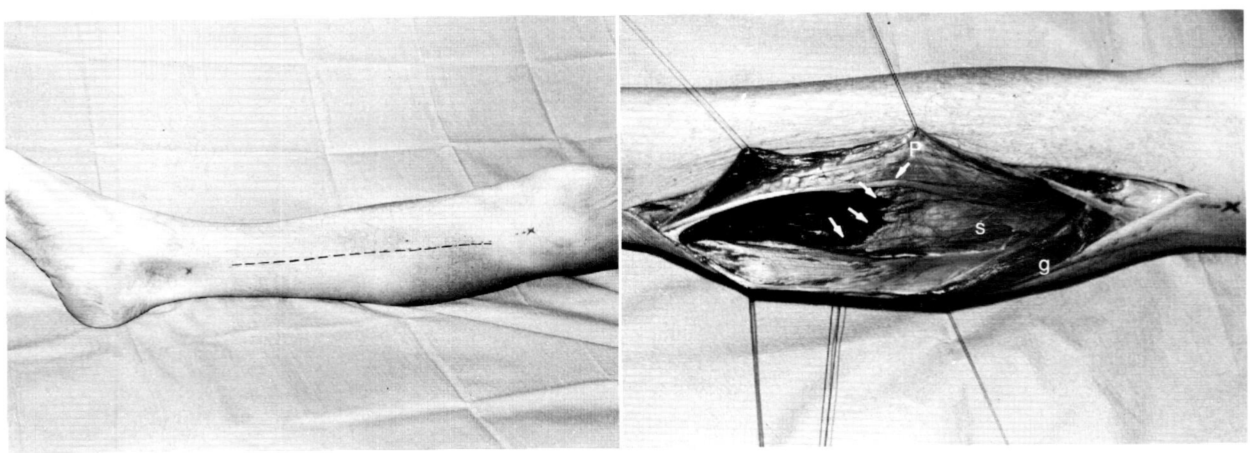

B

Fig. 30-11. Design and technique of elevation of soleus muscle.
A, Location of skin incision for muscle exposure.
B, Intermuscular plane in proximal third of leg allows separation of medial gastrocnemius (*g*) and soleus (*s*) muscles. NOTE: Location of plantaris tendon (*p*) on superficial surface of soleus muscle. Arrows denote site of division of soleus muscle from Achilles tendon in middle third of leg.

C

D

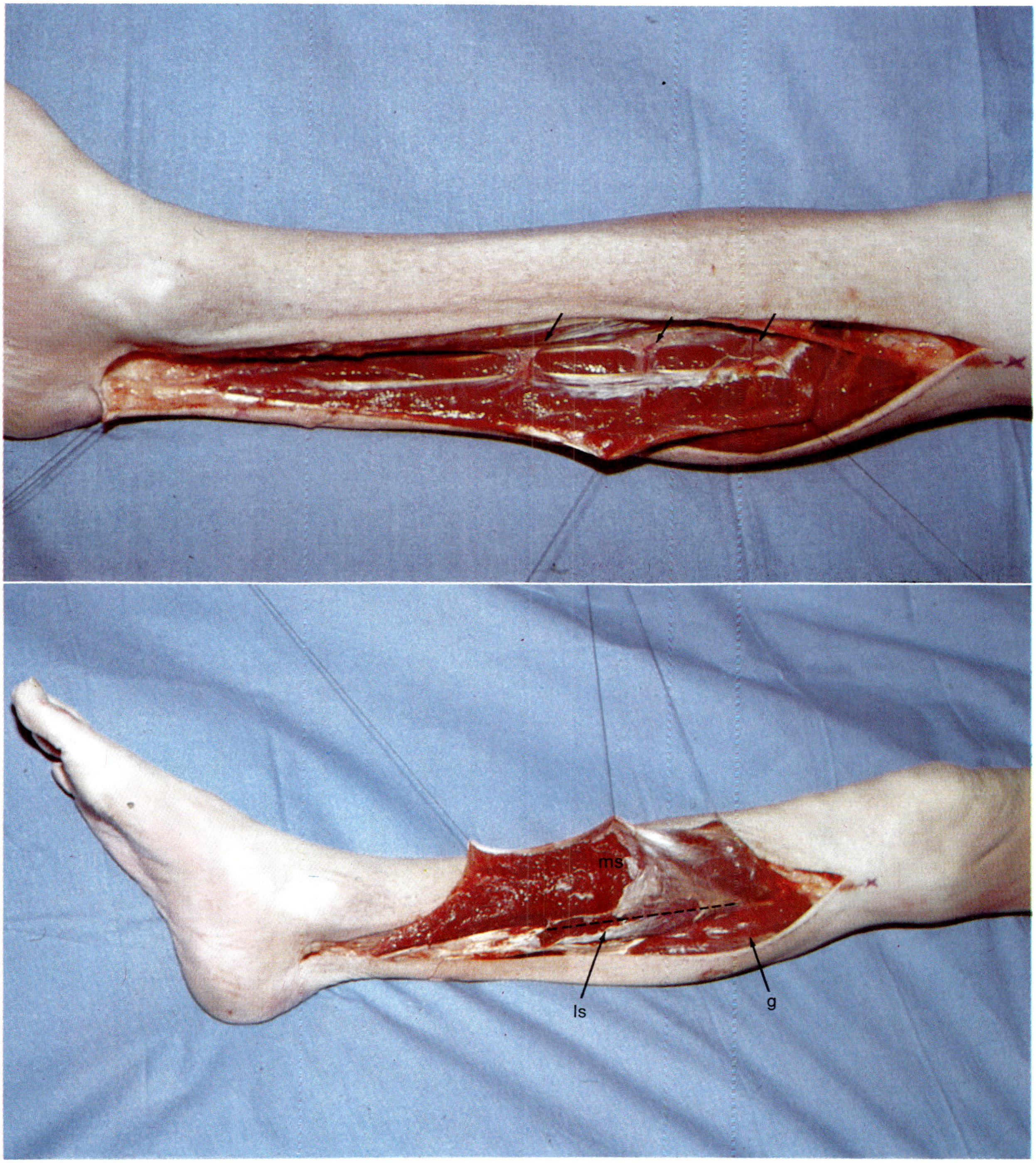

Fig. 30-11, cont'd

C, Mobilization of deep surface of soleus muscle requires identification of posterior tibial artery and veins with identification of inferior minor vascular pedicles to muscle *(arrows)*.

D, Muscle split at midposterior calf (dotted line denotes segmental transposition) or entire muscle elevated for transposition. Pedicles from the posterior tibial artery are only divided as required for adequate muscle arc of rotation. *ms,* Medial soleus; *ls,* lateral soleus; *g,* gastrocnemius.

Continued.

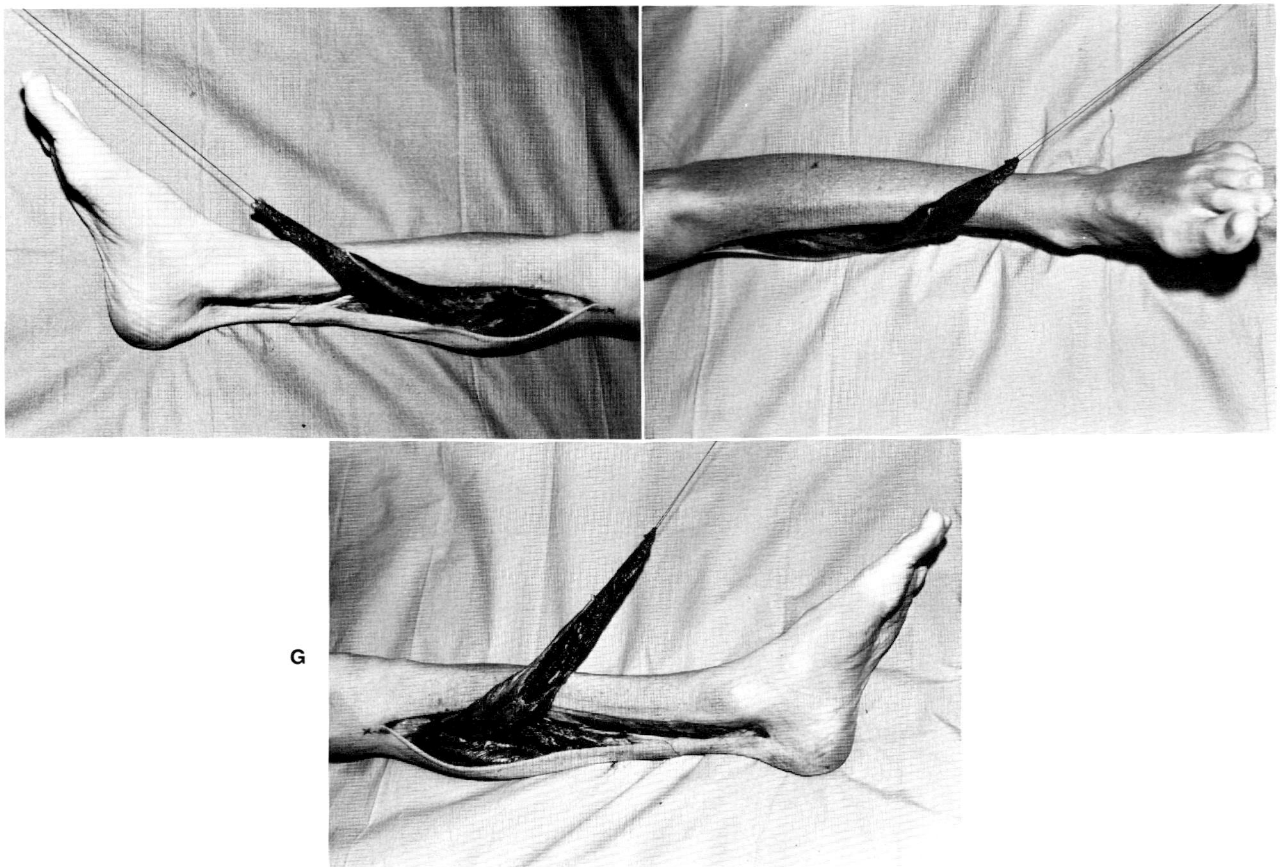

Fig. 30-11, cont'd
E and **F,** Medial soleus muscle transposed for coverage of distal leg defects.
G, Medial soleus muscle transposed for coverage of middle third leg defects.

Lateral approach. The lateral approach for soleus elevation is very difficult and is only used when (1) the defect is located lateral to the tibia; (2) the medial half of the soleus muscle is involved in the traumatic defect; and (3) the fibula is excised and soleus exposed during tumor extirpation or wound débridement.

A vertical incision is made 2 cm laterally to the fibula. The soleus muscle is visualized between the peroneus longus and gastrocnemius muscles. The intermuscular plane between the soleus and gastrocnemius muscles in the proximal third of the leg is identified. After the muscles are separated, the soleus is divided from its attachments to the Achilles tendon distally until an adequate length of muscle is mobilized. Proximal elevation of the muscle is difficult, since vision of the medial side of the muscle is obstructed by the fibula. However, ligation of medially located minor vascular pedicles to the soleus is required to avoid later hematoma formation. The muscle is elevated proximally to the middle third of the leg. At this time further muscle elevation is only performed until adequate muscle is available to reach the defect. The exposed muscle is skin grafted and the donor site closed directly.

Flexor digitorum longus. The flexor digitorum longus is a posterior calf muscle rarely used as a transposition flap. With a small muscle belly and a type IV pattern of circulation, the muscle is only used in conjunction with the soleus for small tibial defects.

Segmental transposition is possible by separation of muscle from its tendon distally while maintaining proximal continuity with the tendon insertion (Fig. 30-12).

A vertical incision is made 2 cm medial to the tibia, extending from the knee to the ankle. The soleus muscle is identified and elevated to visualize its deep surface. The flexor digitorum longus muscle is identified deep to the posterior tibial vessels. The tendon of insertion is divided distally, or distal muscle fibers are separated from the tendon (segmental muscle transposition). The distal segmental vascular pedicles from the posterior tibial artery are divided. Flap elevation continues proximally until the muscle reaches the tibial defect. Elevation of this muscle beyond the midcalf level may result in loss of distal muscle because of excessive division of segmental vascular pedicles. The exposed muscle belly is skin grafted and the donor defect closed directly.

A

B

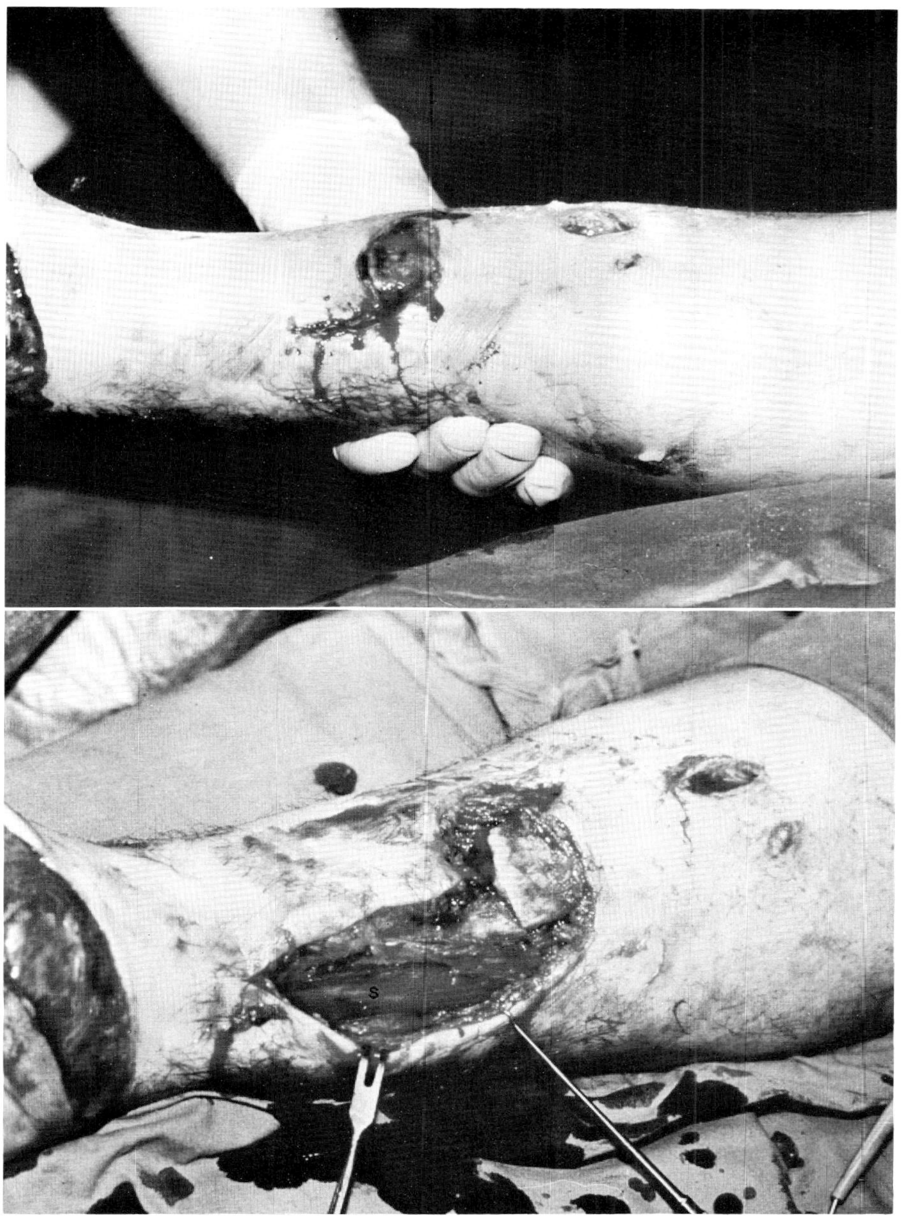

Fig. 30-12. Middle third leg coverage. Use of segmental transposition of soleus and flexor digitorum longus muscles for coverage of midtibial shaft fracture.

A, Open wound tibial fracture.

B, Wound débridement and identification of soleus muscle (*s*).

(**A, C,** and **D** from Mathes, S.J., and Nahai, F.: Muscle flap transposition with function preservation: technical and clinical considerations, Plast. Reconstr. Surg. **66:**242, 1980.) *Continued.*

C

D

E

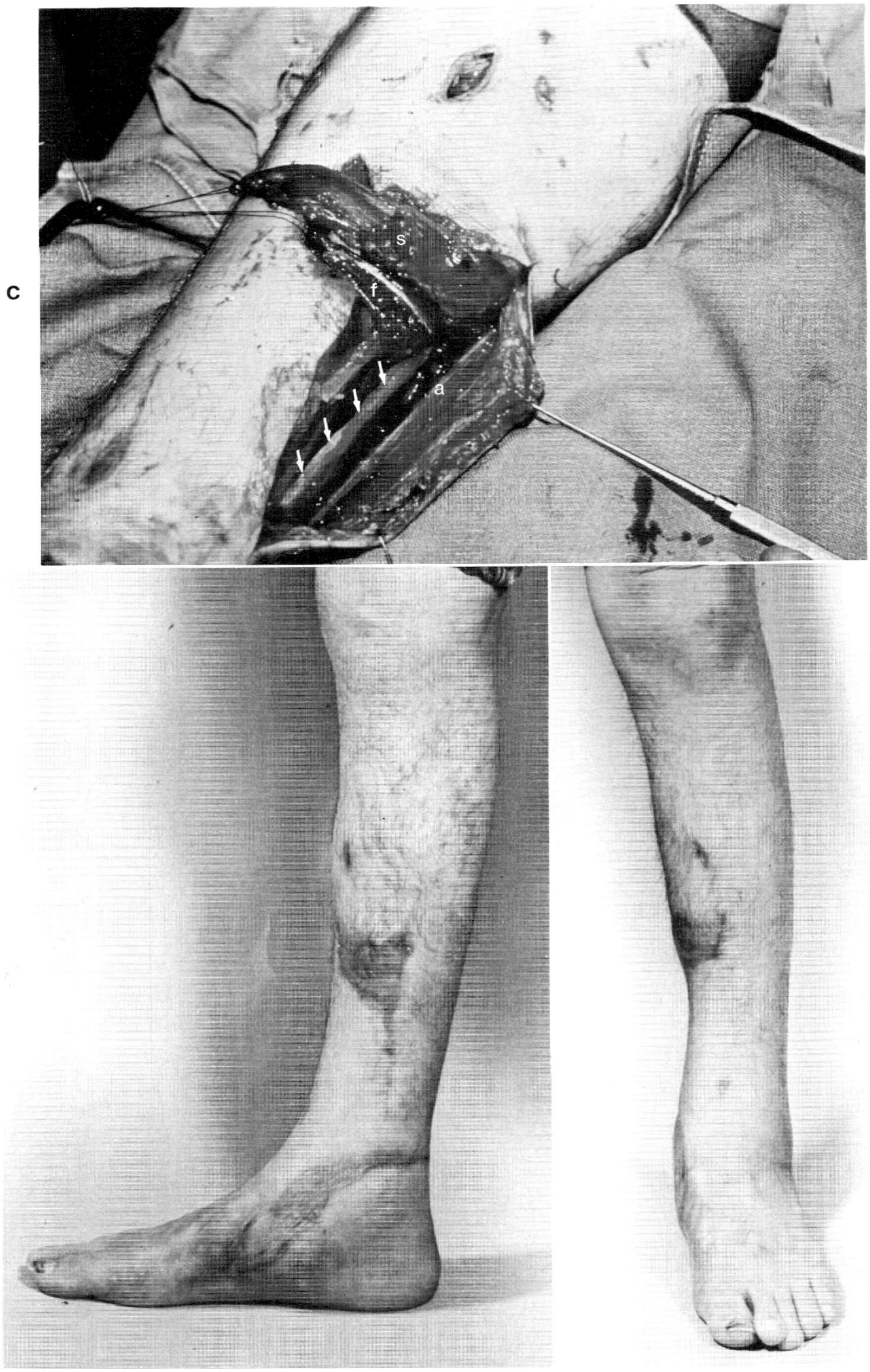

Fig. 30-12, cont'd

C, Segmental transposition of medial soleus (*s*) and flexor digitorum longus (*f*) muscles. Arrows denote preserved tendon of insertion of flexor digitorum longus. *a,* Achilles tendon.

D and **E,** Stable wound coverage in one operation with function preservation techniques.

Peroneus longus. Peroneus longus is a lateral leg muscle occasionally used as a transposition flap for coverage of small defects in the middle third of the leg. The muscle is transposed without its overlying cutaneous territory because of resultant exposure of the fibula.

A lateral leg vertical incision extending between the head of the fibula to the lateral condyle allows muscle exposure. The muscle is located between the tibialis anterior medially and the soleus muscle posteriorly. The superficial peroneal nerve is preserved during the proximal dissection. After division of its tendon of insertion, the muscle is elevated proximally to the midcalf level. Further elevation is avoided, since its dominant vascular pedicle enters the muscle belly in this area. The exposed muscle is skin grafted and the donor site closed directly.

Tibialis anterior. The tibialis anterior is an anterior leg muscle and is the primary foot extensor. Transposition of this muscle results in an unacceptable functional result. However, use of segmental transposition techniques has demonstrated reliable use of the muscle as a flap for small defects while maintaining continuity of the proximal muscle belly with its tendon of insertion (Fig. 30-13). Since the muscle has a type IV pattern of circulation, mobilization of the muscle with division of several distal vascular pedicles is possible without loss of flap circulation. With function-preserving techniques the muscle is useful for coverage of small defects in the middle and distal third of the leg.

A vertical leg incision 1 cm laterally to the tibia, extending from the level of the head of the fibula to the lateral malleolus, allows muscle exposure. The superficial muscle is located between the tibia and the extensor digitorum longus. The tendon of insertion is identified in the distal third of the leg. Proximal dissection separates the muscle from the tibia medially and from the adjacent extensor digitorum longus laterally. The segmented vascular pedicles entering the deep surface of the muscle from the anterior tibial artery and veins are preserved. The muscle is dissected from the tendon starting distally and is elevated proximally until the tendon starts to widen to form a broad fascial attachment with muscle. This elevation requires division of the most distal three or four segmental vascular pedicles. The muscle is then transposed over the defect. If muscle length is inadequate, another local muscle flap is selected, since further dissection proximally will result in vascular compromise to the muscle flap and loss of continuity between proximal muscle and tendon.

An alternative method for use of the tibialis anterior muscle requires separating the distal two thirds of the muscle belly from its tibial attachments while preserving its vascular pedicles and continuity with its tendon of insertion. The muscle is then advanced medially over the tibia. Use of this technique of muscle advancement allows coverage of the middle third of the lateral aspect of the tibia.

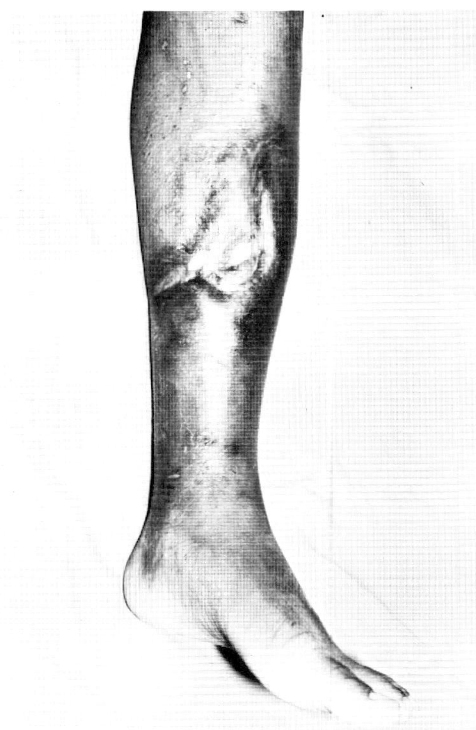

A

Fig. 30-13. Middle third leg coverage. Use of tibialis anterior muscle (segmental transposition) with skin graft for coverage of exposed tibia.

A, Exposed tibia at site of prior open fracture.

(**A** to **C** and **E** from Mathes, S., and Nahai, F.: Muscle flap transposition with function preservation: technical and clinical considerations, Plast. Reconstr. Surg. **66:**242, 1980.)

Continued.

B

C

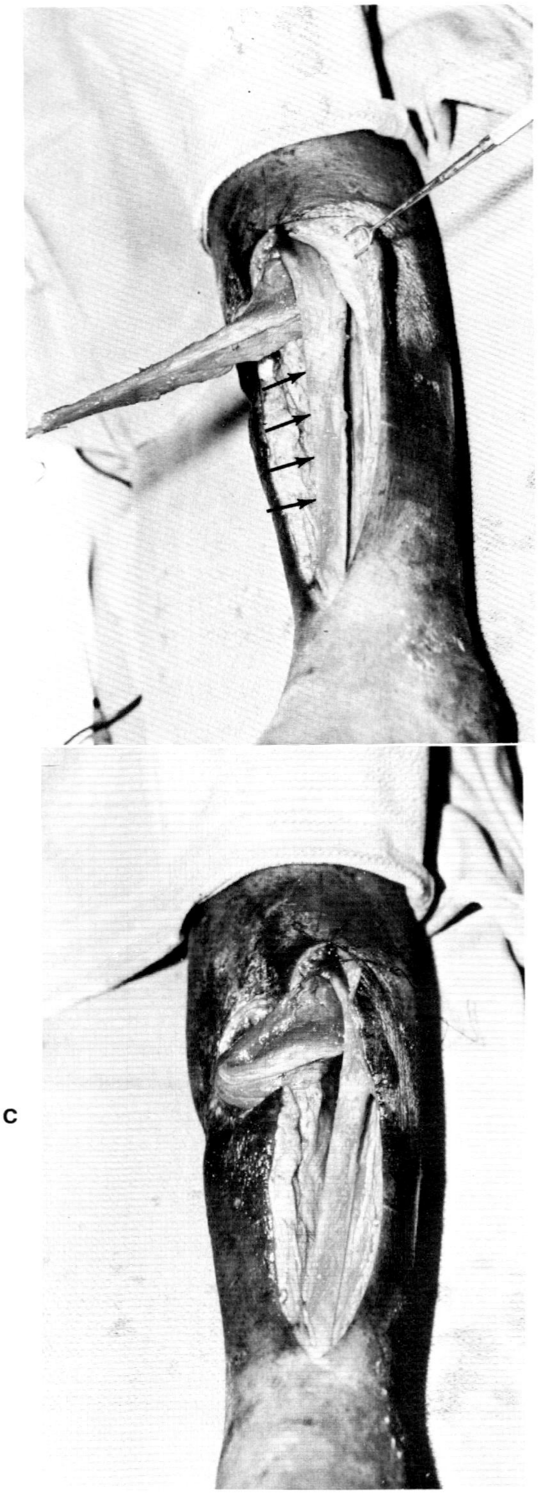

Fig. 30-13, cont'd. For legend see opposite page.

D E

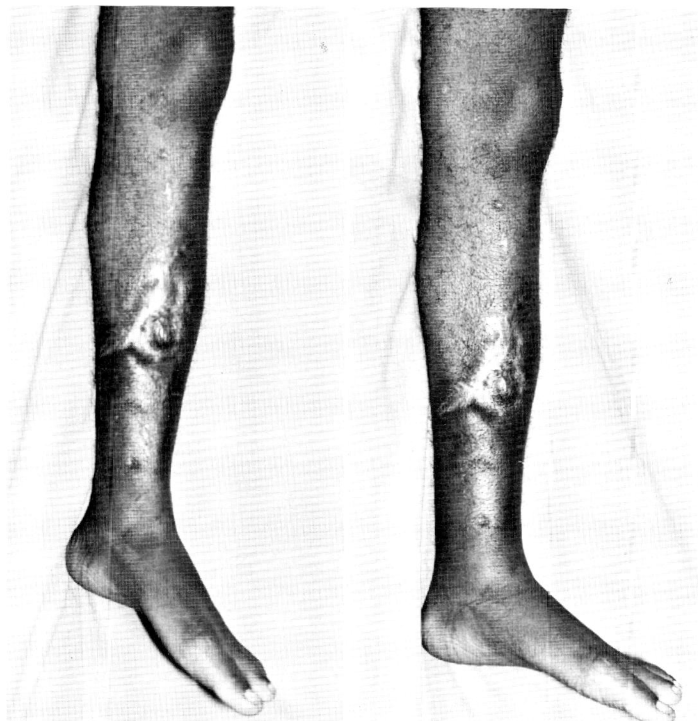

Fig. 30-13, cont'd

B, Elevation of tibialis anterior muscle with preserved continuity of proximal muscle with tendon of insertion. Arrows denote tendon of insertion.

C, Segmental muscle transposition covers small defects of mid tibia level.

D, Six-month postoperative view demonstrates stable coverage.

E, Active foot extension demonstrates function preservation of tibialis anterior muscle.

Distal third leg defects

The transposition of muscle or musculocutaneous flap is not a uniformly reliable method for extremity reconstruction in this area. Those muscles in which the muscle fibers extend into the distal leg are occasionally useful as transposition flaps (e.g., soleus, peroneus brevis, and extensor hallucis longus). However, the level that the muscular fibers extend distally and their size is variable. At best, only small defects are covered with muscle transposition based on proximal vascular pedicles.

The use of the distally located minor or segmental vascular pedicles as the vascular basis for distal transposition of muscle is an alternative method for coverage of this area. Unfortunately, this technique is not reliable because of variability in the location of the minor pedicles. Furthermore the success of muscle flaps in the lower extremity is based on the location of the vascular pedicles beyond the defect and its zone of injury. The distally based flap is, by necessity, based on vascular pedicles adjacent to the defect. These factors explain the high rate of failure reported in use of distally based muscle flaps for coverage. The distally based muscle flap (e.g., soleus and extensor digitorum longus) is used for small defects only when the distal vascular pedicles are identified by preoperative arteriography and intraoperative identification during flap elevation.

A 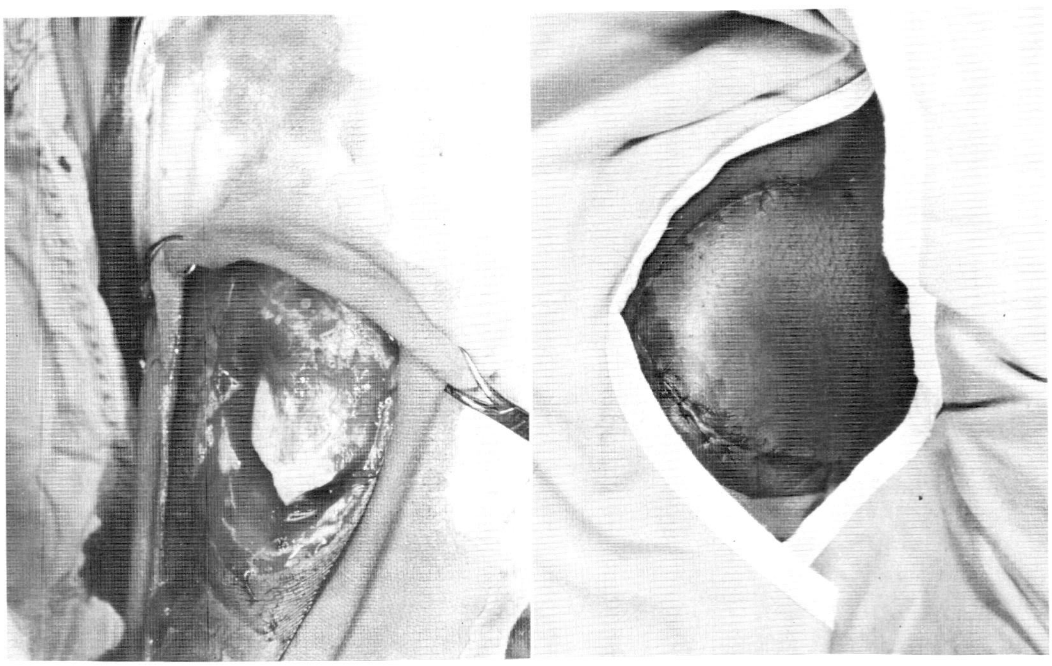 B

Fig. 30-14. Distal third of tibia coverage. Use of random cross-leg flap for coverage of exposed tibial fracture.
A, Exposed tibial fracture.
B, Inset of random cross-leg flap.

(Courtesy Gilbert Gradinger, San Mateo, Calif.)

The use of distant flaps remains the technique of choice for coverage of lower third leg defects. During the last 3 years no flap loss was observed after microvascular transplantation of distant flaps for coverage in this region (Chapter 38). The use of the cross-leg flaps (random or skin-fascial gastrocnemius) has been reserved for defects in which suitable receptor vessels are unavailable for composite tissue transplantation.

The standard use of the cross-leg random (Fig. 30-14) or skin-fascial gastrocnemius flap (Fig. 30-15) remains a more reliable technique for coverage of distal third leg defects than the use of local muscle flaps. The thin random flap generally provides a more superior result in terms of form and function than the bulky local musculocutaneous flap. The cross-leg musculocutaneous flap is not recommended for leg coverage, since the resultant contour deformity is objectionable both in donor and recipient sites. When a long flap is required from the contralateral leg, the use of the skin-fascial flap based on the gastrocnemius musculocutaneous perforators is reliable with no loss of muscle function and a more acceptable appearance in both the donor and recipient sites.

C

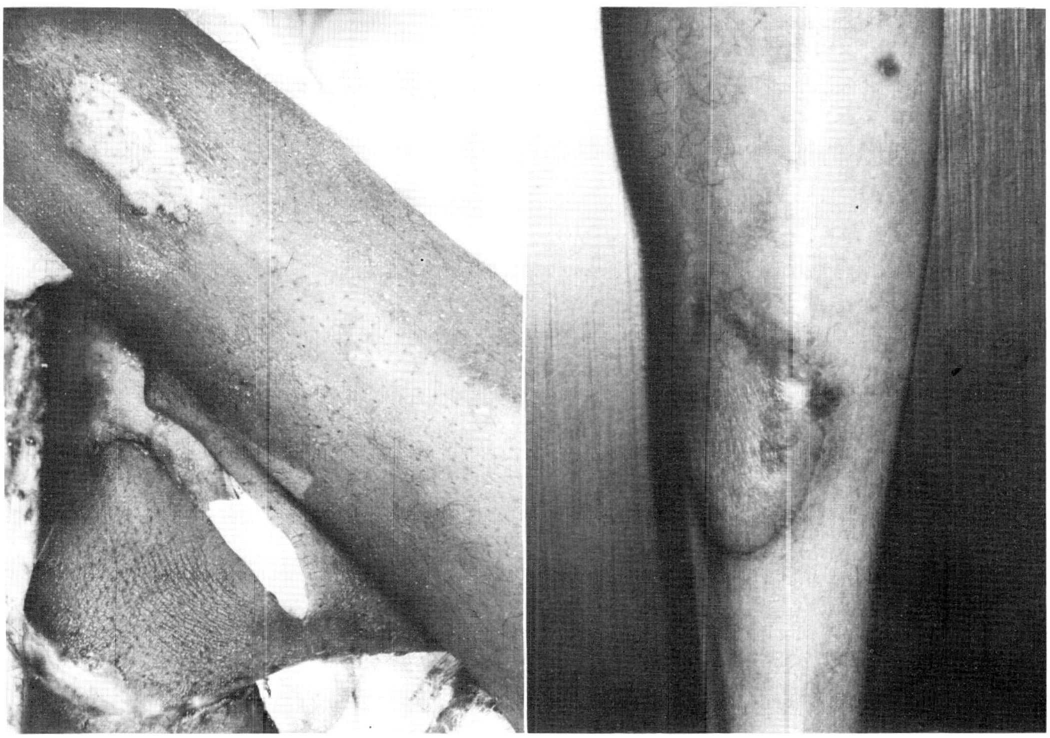

D

Fig. 30-14, cont'd
C, Staged delay of flap with partial division and inset of flap base.
D, Seventeen-year postoperative view demonstrates stable wound coverage.

A

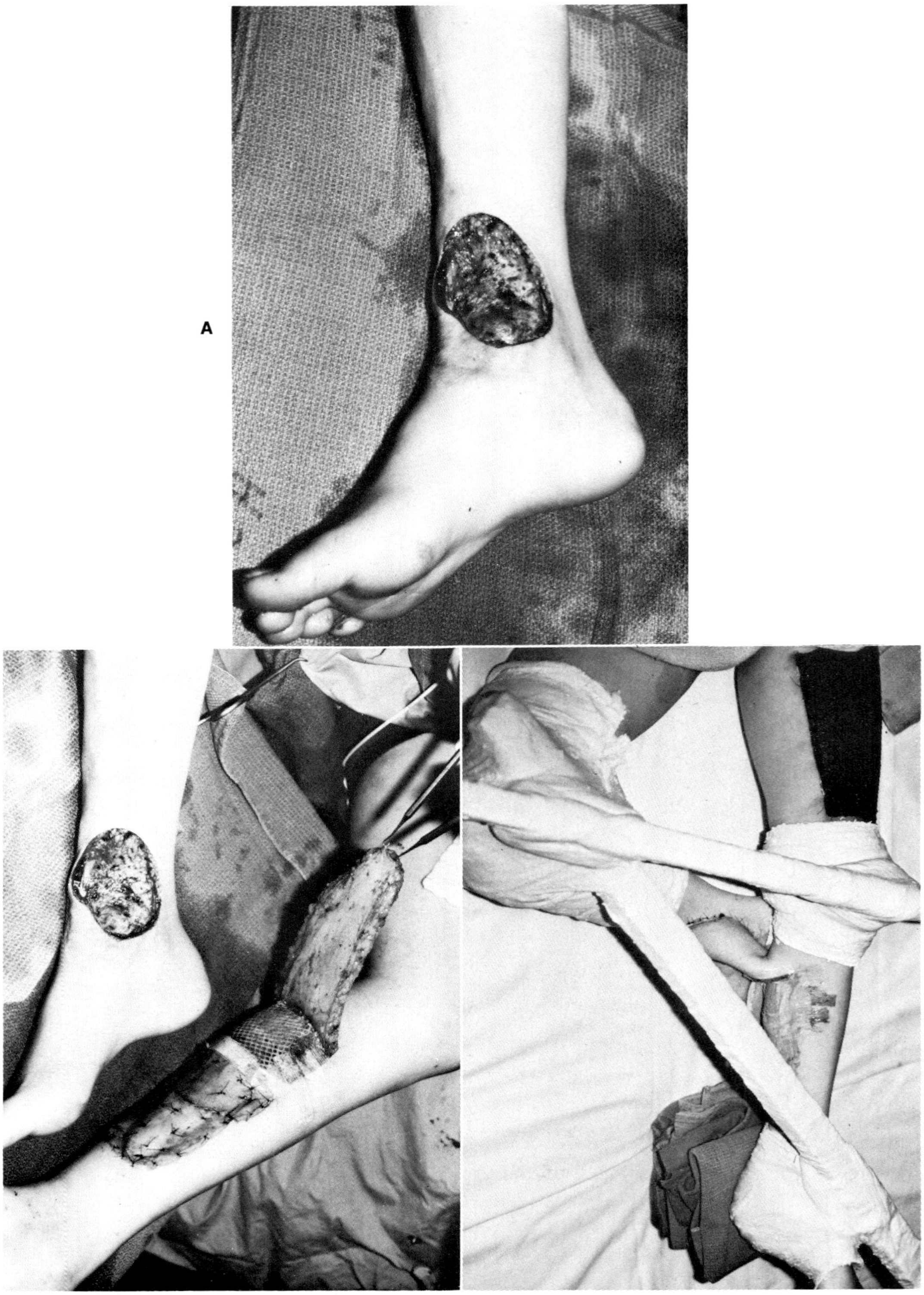

B

C

Fig. 30-15. Distal third leg coverage. Use of cross-leg skin-fascial gastrocnemius flap for coverage of exposed tibia.

A, Site of excision of skin defect.

B, Elevation of skin-fascial flap. NOTE: Flap elevated to level of gastrocnemius musculocutaneous perforating vessels. Skin graft applied to donor defect.

C, Flap inset.

(Courtesy William Schneider, Knoxville, Tenn.)

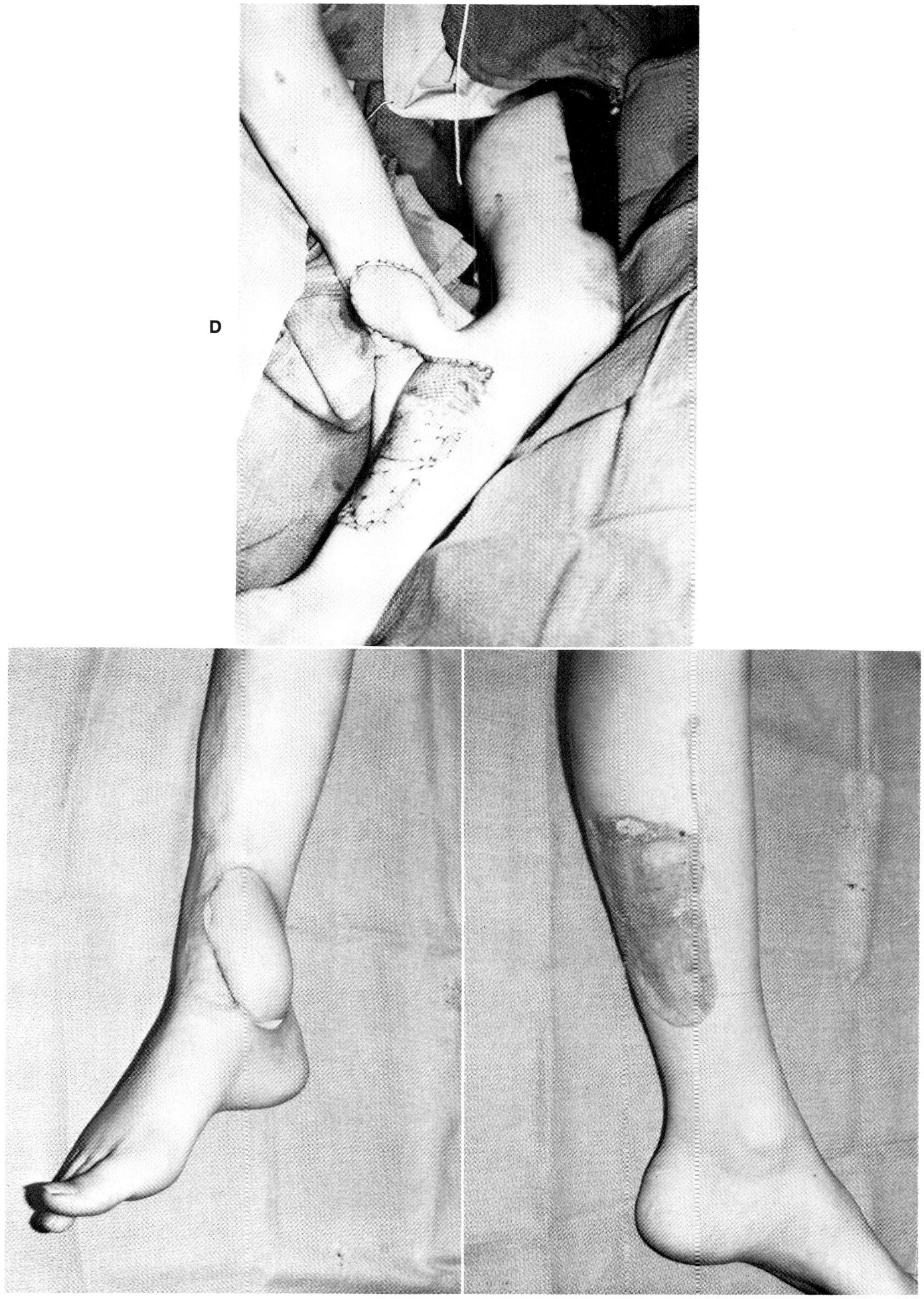

Fig. 30-15, cont'd

D, Flap base divided 2 weeks after initial flap inset.

E, Stable wound coverage.

F, Skin grafts on donor site with no contour deformity.

Soleus. The muscle fibers of the soleus extend to the level of the medial malleolus. The distal muscle fibers, when released from the tendon of insertion, will cover small defects in the distal third of the leg. Although this muscle contributes musculocutaneous perforating vessels to the distal medial leg skin, the donor defect after transposition of the musculocutaneous flap results in exposure of the tendons of insertion of the great toe and foot flexors. For this reason the muscle alone is preferred for transposition as a flap for defect coverage (Fig. 30-16). The technique of elevation of this muscle for transposition based proximally is described for coverage of middle third leg defects.

The distal transposition of this muscle based on minor vascular pedicles is possible. The muscle is exposed through a medial or lateral vertical incision, depending on defect location. It is preferable to explore the muscle on the same side as the defect. After the muscle is identified and separated from adjacent muscles, the minor vascular pedicles are identified. In the medial approach the vascular pedicles are located on its medial surface close to the muscle edge. Initial blunt dissection is avoided, since these pedicles are small and easily damaged. In the lateral approach the vessels are seen entering the deep muscle on the opposite side, so initial injury is less likely. Only a portion of proximal muscle is designed as a flap based on the distal two to three minor pedicles (Fig. 30-17). The small muscle flap is transposed into the defect based on these identified minor pedicles. The exposed muscle belly is skin grafted and the donor site closed directly.

Transposition of the entire muscle on the minor pedicles requires extensive proximal muscle dissection, and complete muscle survival is unlikely. If the defect is large and requires the entire muscle, the use of the distally based soleus flap is not a useful muscle flap for coverage.

A

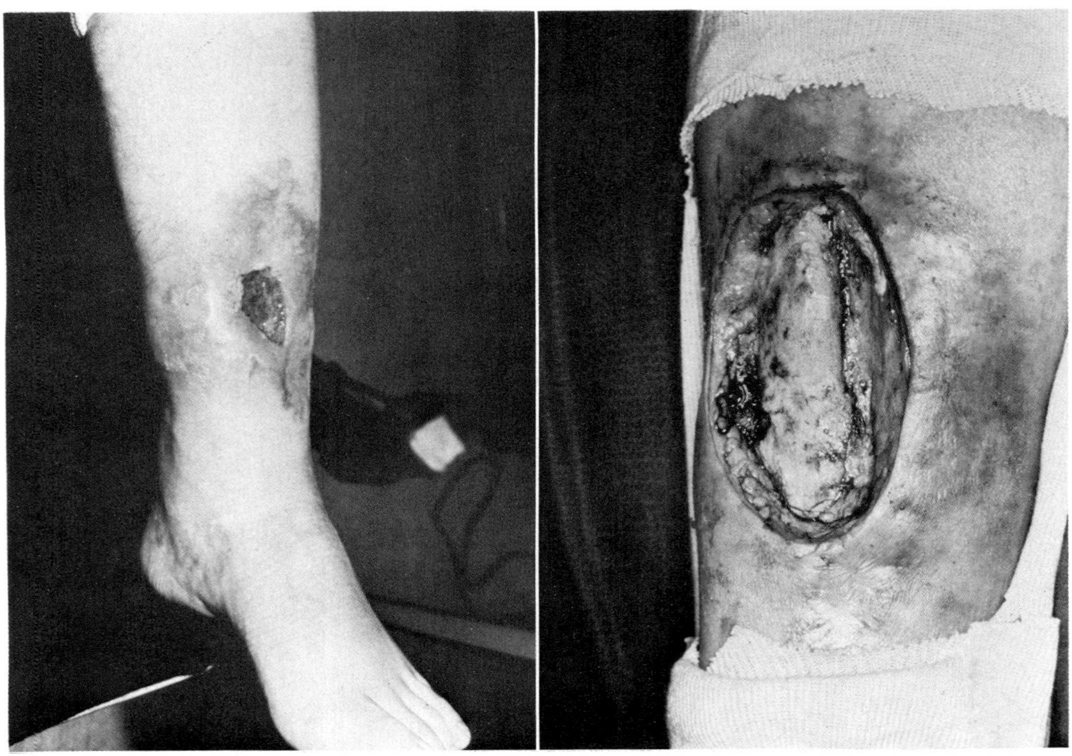

B

Fig. 30-16. Distal third leg coverage. Use of soleus muscle with skin graft for coverage of exposed tibia.
A, Exposed tibia.
B, Defect after wound débridement.

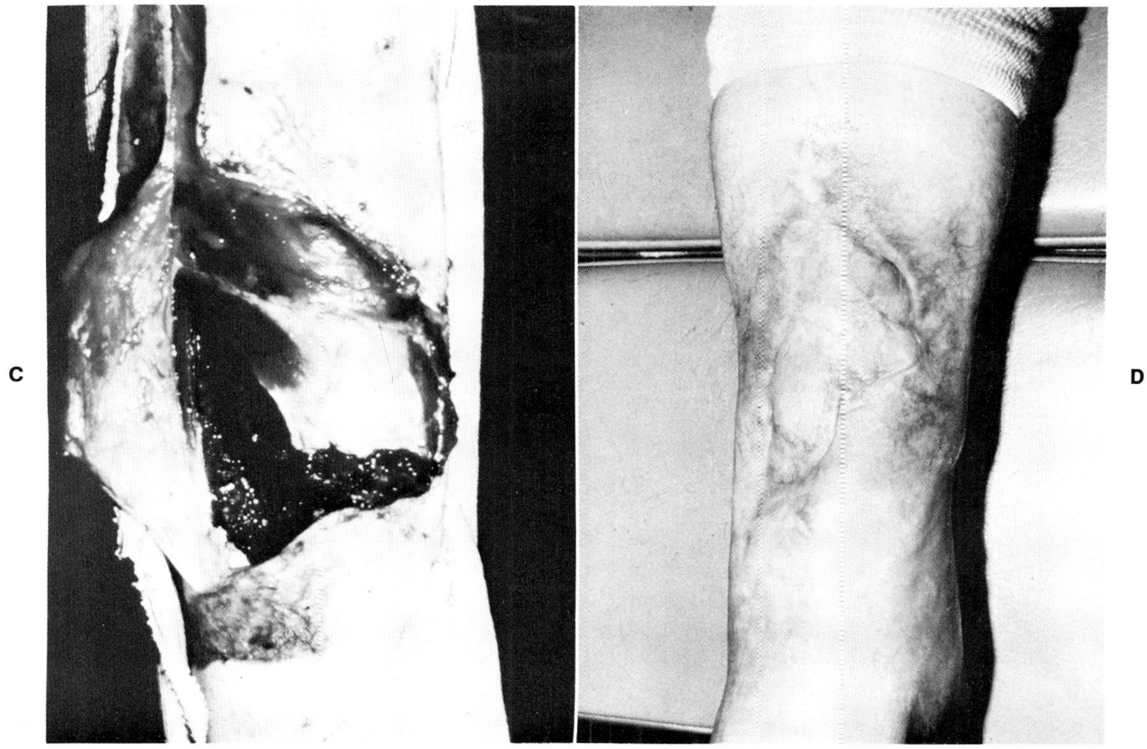

Fig. 30-16, cont'd
C, Soleus muscle transposed over defect.
D, Stable wound coverage in one operation.

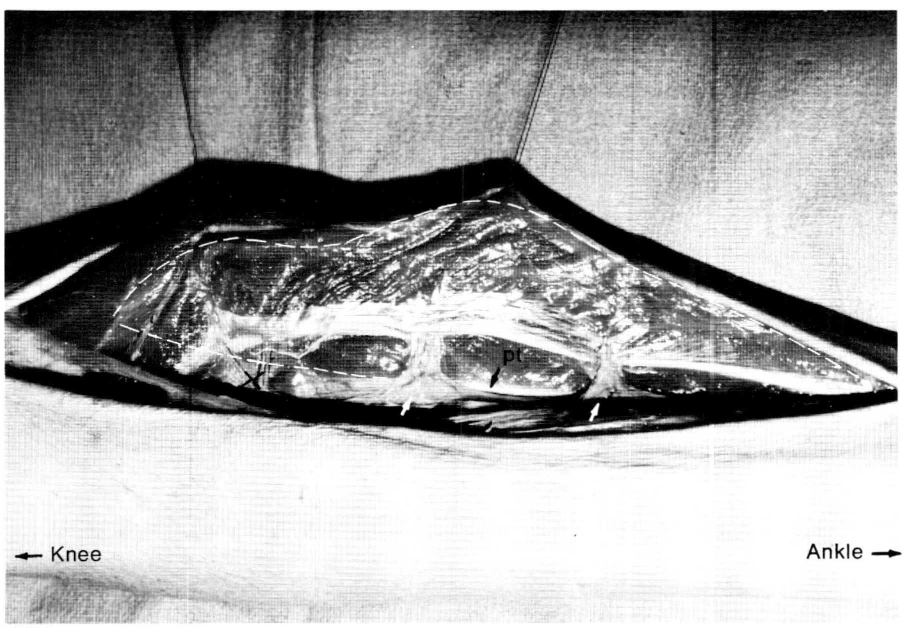

Fig. 30-17. Technique of elevation of distally based soleus (medial approach). Dotted lines denote muscle flap design based on inferior two minor pedicles *(arrows)*. *x,* Site of ligation of third most proximal minor pedicle; *pt,* posterior tibial artery.

Extensor digitorum longus. The extensor digitorum longus is a lateral leg muscle located between the tibialis anterior and peroneus longus. Its muscle fibers extend into the middle lower third of the leg, and its tendon of insertion is adjacent to the deeper extensor hallucis longus. This muscle has a type IV vascular pattern, receiving segmental vascular pedicles from the tibialis anterior. Transposition of the muscle based proximally is possible with function preservation (segmental transposition), but the muscle is small, and minimal coverage is possible. For the proximal portion of the distal third of the leg, the tibialis anterior with function preservation is preferred.

Transposition of the muscle based on the distally located segmental pedicles is another potential method to achieve distal third leg coverage (Fig. 30-18). Careful patient selection is required before this technique is undertaken. Natural adhesions between the tibialis anterior, extensor digitorum, and extensor hallucis longus are common in the noninjured extremity. In the injured extremity the addition of inflammatory adhesions often complicates muscle elevation with preservation of the distal segmental vascular pedicles.

The muscle is exposed through a vertical incision 2 cm laterally to the tibia, extending from the head of the fibula to the lateral malleolus (Fig. 30-18, *A*). The superficial peroneal nerve is preserved during the proximal dissection. An incision through the superficial fascia in the distal leg allows identification of the tibialis anterior and extensor digitorum longus musculotendinous junctions. The muscles are then separated proximally (Fig. 30-18, *B*). Between the muscles at the medial edge of the extensor digitorum longus, the segmental vascular pedicles are observed. In the distal leg the muscular fibers of the extensor hallucis longus are located deep to the extensor digitorum longus (Fig. 30-18, *C*). For distally based transposition no dissection is required in this area. The tendon of the peroneus longus is identified laterally to the extensor digitorum longus, and then the peroneus longus and extensor digitorium longus muscles are separated proximally. Immediately inferior to the fibula head, the origin of the extensor digitorum longus is released and the muscle elevated distally (Fig. 30-18, *D*). The two to three most distal segmented vascular pedicles are preserved (Fig. 30-18, *E*). If the muscle does not reach the defect at this level of dissection, the muscle may survive division of one additional pedicle, although fluorescein testing is recommended to establish adequate circulation (Chapter 40). The muscle is then transposed into the defect, and exposed muscle is skin grafted (Fig. 30-18, *F*). The donor site is closed directly.

A

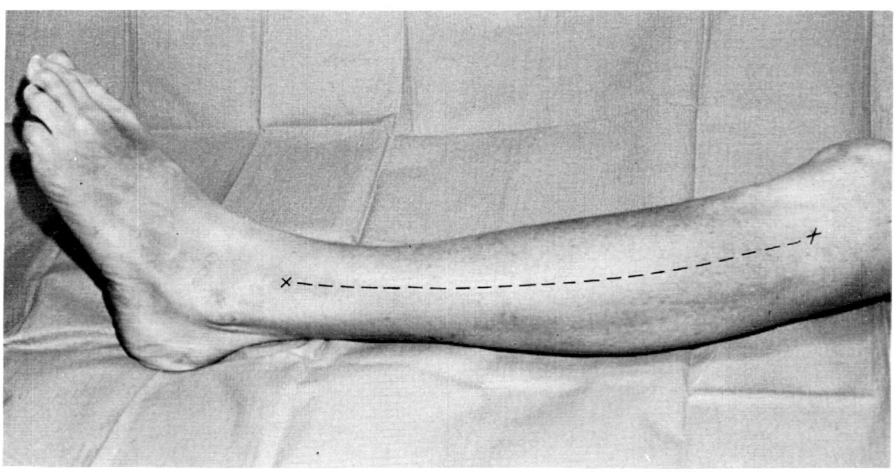

Fig. 30-18. Technique of elevation of distally based extensor digitorum longus muscle flap.
A, Location of leg incision for muscle exposure.

B

C

D

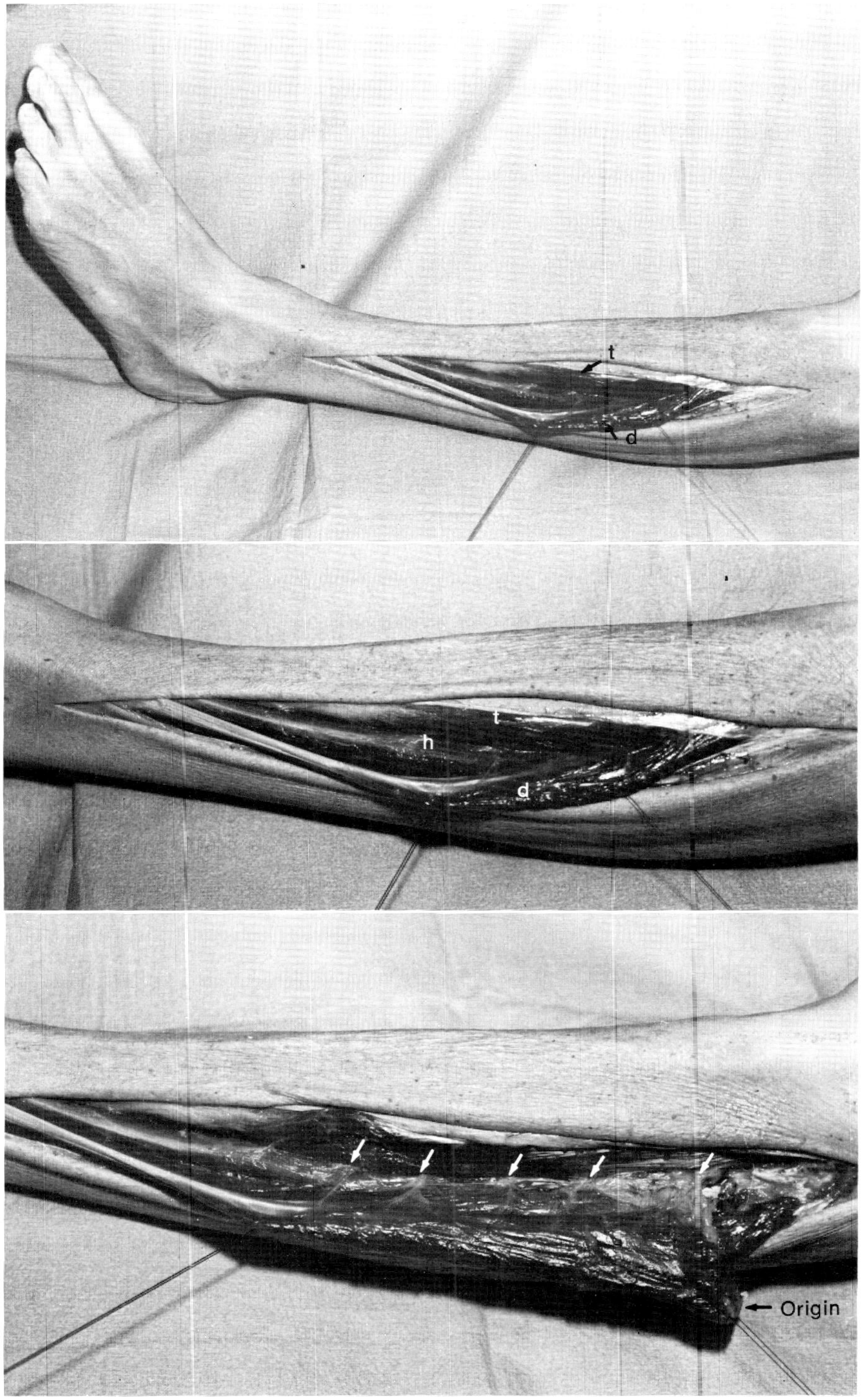

Fig. 30-18, cont'd

B, In middle third of leg musculotendinous junction of tibialis anterior (*t*) and extensor digitorum longus (*d*) are identified and muscles separated proximally.

C, Extensor hallucis longus muscle (*h*) is located deep to extensor digitorum longus muscle and is also transposed based distally if greater coverage is required. *t*, Tibialis anterior; *d*, extensor digitorum longus.

D, Muscle origin is released. NOTE: Arrows denote segmental vascular pedicles to muscle.

Continued.

E

F

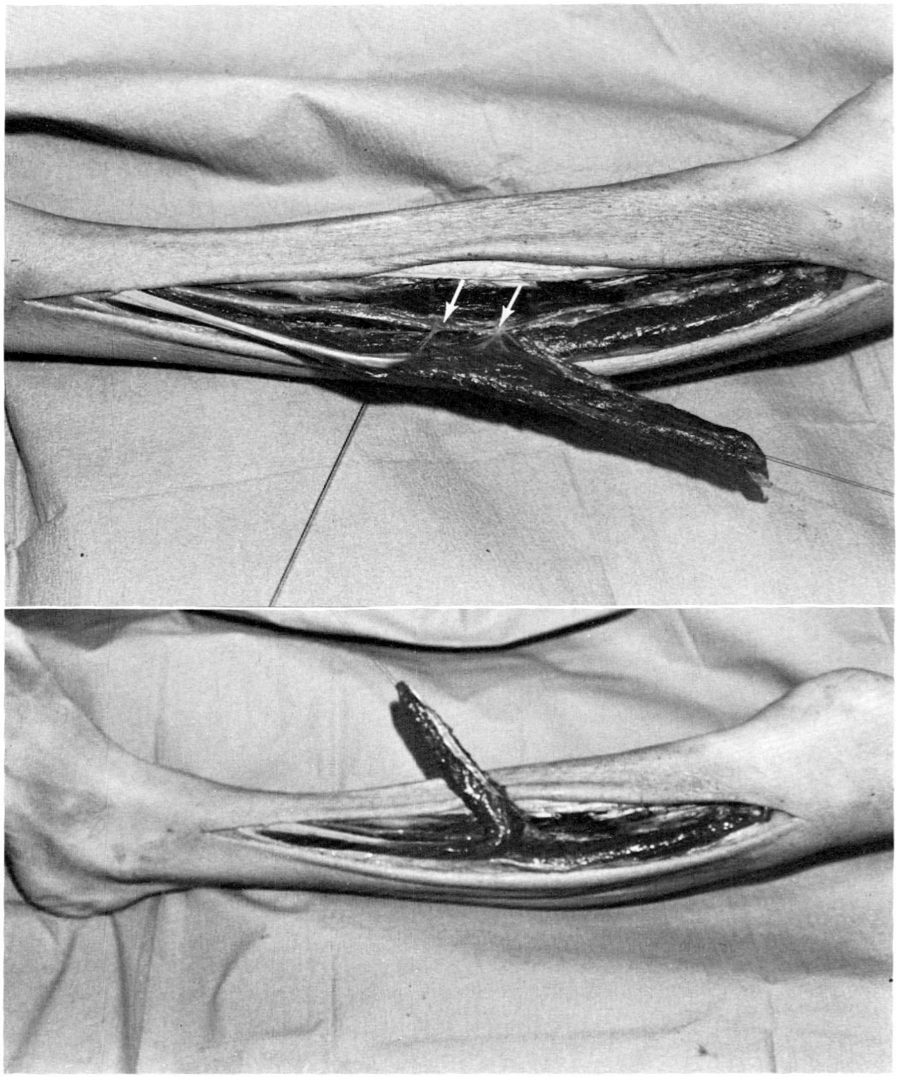

Fig. 30-18, cont'd
E, Distal two vascular pedicles *(arrows)* are preserved to muscle flap.
F, Muscle flap transposed to distal leg for small defects.

Tibialis anterior. The tibialis anterior is a lateral leg muscle that is occasionally useful for coverage of distal third small defects. The muscle fibers extend inferiorly to the musculotendinous junction. These fibers are either elevated (based proximally) or advanced for coverage of small defects. The technique of elevation of this muscle unit is described for middle third leg defects.

Extensor hallucis longus. The extensor hallucis longus is a lateral leg muscle that is occasionally useful for coverage of small defects in the distal leg. The muscle fibers are located deep to the medially located tibialis anterior and extensor digitorum longus muscles. The proximal muscle fibers are closely associated with the tendon of the extensor digitorum longus. With a type IV pattern of circulation, the muscle receives segmental vascular pedicles from the posterior and anterior tibial vessels. Based on the proximal segmental vascular pedicles, the distal muscle is transposed for small defects. Also a distally based flap is possible by release of the origin of the muscle with transposition based on preserved distally segmented vascular pedicles. The deep location and small size of this muscle make this unit impractical for coverage of most lower third defects.

Peroneus brevis. The peroneus brevis is a small lateral leg muscle occasionally useful as a transposition flap for small defects. With a type II pattern of circulation, standard transposition of the muscle is reliable based on its proximal dominant vascular pedicles from the peroneal artery. The muscle is not transposed as a musculocutaneous flap because of resultant exposure of the fibula.

The muscle is exposed through a vertical incision over the fibula, extending from the junction of the proximal and middle thirds of the leg to the lateral malleolus. The muscle is identified beneath the tendon of the peroneus longus and located between the extensor digitorum longus and soleus muscles. The tendon of insertion is divided and the muscle elevated proximally. The peroneus tertius has congenital adhesions to this muscle and is generally included with the peroneus brevis muscle. The distal two thirds of the muscle are elevated and the muscle transposed into the defect. If inadequate defect coverage is noted, further dissection proximally is avoided, since disruption of the dominant vascular pedicles is possible. If the tendon of insertion of the peroneus longus muscle is also divided, transposition of this muscle with the peroneus brevis will improve defect coverage in the superior part of the distal third of the leg. The exposed muscle belly is skin grafted and the donor site closed directly.

SUMMARY

The medial gastrocnemius and soleus muscles are the basis for most muscle or musculocutaneous flaps used in lower leg reconstruction. The thigh is generally covered with skin grafts. The distal third of the lower leg is generally covered with distant flaps. With precise knowledge of the anatomy of lower extremity muscles and careful assessment of the etiology and location of the defect, the majority of difficult lower leg wounds are safely covered in one operation. Certain defects will require the use of microsurgery for free flap transplantation or traditional methods of random cross-thigh, cross-leg, and skin-fascial flap transfer. Emphasis is placed on achieving wound coverage with optimal form and function at the site of reconstruction, using the entire spectrum of the reconstructive ladder, ranging from the skin graft and local muscle flap to the complex distant flap.

ANNOTATED BIBLIOGRAPHY

1. Arnold, P.G., and Prunes-Carrillo, F.: Vastus medialis muscle flap for functional closure of the exposed knee joint, Plast. Reconstr. Surg. **68:**69, 1981.

   The use of the vastus medialis muscle with skin graft is presented for coverage of the knee. In the two patients with exposure of the knee joint and loss of the rectus femoris muscle, the vastus medialis muscle is used both for knee coverage as a transposition flap and for restoration of knee extension by attachment of the distal muscle to the extensor mechanism of the knee.

2. Barfod, B., and Pers, M.: Gastrocnemius-plasty for primary closure of compound injuries of the knee, J. Bone Joint Surg. **52:**124, 1970.

   The use of the lateral gastrocnemius muscle flap for coverage of the exposed knee joint is demonstrated. External fixation device (Hoffmann apparatus) is used for bone stabilization in the management of lower leg defects with open fractures.

3. Byrd, H.S., Cierny, G., III., and Tebbetts, J.B.: The management of open tibial fractures with associated soft-tissue loss: external pin fixation with early flap coverage, Plast. Reconstr. Surg. **68:**73, 1981.

   A prospective study of 20 open tibial fractures with associated soft tissue injury demonstrates rapid union without infection after aggressive wound débridement, fasciotomy, external bone fixation, and flap coverage. Emphasis on flap coverage using muscle and musculocutaneous flaps is based on restoration of vascularized coverage of bones with injury-related loss of both periosteal and endosteal circulation. A useful classification of lower extremity open fractures segregates wounds and their management according to a progressive loss of vascularity.

4. Feldman, J.J., Cohen, B.E., and May, J.W., Jr.: Medial gastrocnemius myocutaneous flap, Plast. Reconstr. Surg. **61:**531, 1978.

The successful use of the medial gastrocnemius musculocutaneous flap in 11 patients for knee and leg coverage is presented. The extended medial gastrocnemius musculocutaneous flap is used when a longer flap with a wide arc of rotation is required.

5. Ger, R.: The operative treatment of the advanced stasis ulcer: a preliminary communication, Am. J. Surg. **111:**659, 1966.

The use of muscle transposition for coverage of the advanced stasis ulcer is introduced. The soleus and flexor digitorum longus muscles are used for coverage of unstable skin in the distal leg. The principle of muscle transposition is also applied to difficult wounds secondary to wound lymphedema and unstable scars over prior open fracture sites.

6. Ger, R.: The management of pretibial skin loss, Surgery **63:**757, 1968.

A description of the use of muscle transposition for coverage of lower leg defects is presented. Based on a comparison of available techniques for coverage of acute skin loss in the lower leg, the muscle flap demonstrated the following advantages: (1) prevention of skin adherence to bone, (2) consistent acceptance of skin grafts to the underlying muscle flap, (3) diminished incidence of infection because of the viable muscle tissue interposed between skin and bone, and (4) increase in the overall vascularity to the underlying wound. These clinical observations are based on 4-year follow-up for coverage for leg trauma and 3-year follow-up for coverage for venous disease. The clinical experience includes use of the soleus muscle flap in 29 patients, the flexor digitorum longus muscle in 14 patients, and the abductor hallucis in 5 patients. A review of these patients indicates maintenance of the healed wound with minimal loss of muscle bulk and no siginificant functional disability.

7. Ger, R.: Operative treatment of the advanced stasis ulcer using muscle transposition: a follow-up study, Am. J. Surg. **120:**376, 1970.

A retrospective study of 39 patients with advanced stasis ulcer in which muscle flap transposition was used for coverage. Muscle flap necrosis occurs in 11 patients, and 2 patients develop skin loss requiring secondary skin grafts. Analysis of causes of muscle flap failure provides excellent data regarding pitfalls in the use of muscle flaps as a technique for reconstructive surgery. Evaluation of long-term results after use of muscle flap transposition for venous ulcers reveals stable coverage in those patients in which the muscle survives its initial transposition. The soleus, flexor digitorum longus, tibialis anterior, abductor hallucis, peroneus brevis, and peroneus tertius are used as transposition flaps in this study.

8. Ger, R.: The technique of muscle transposition and the operative treatment of traumatic and ulcerative lesions, J. Trauma **2:**502, 1971.

The suitability and technique of elevation of muscles for transposition for coverage of the tibia are presented based on location of the defect: upper third defects—gastrocnemius, middle third defects—soleus, and lower third defects—flexor digitorum longus and abductor hallucis. Muscles used for distal fibula coverage include the peroneus longus and peroneus brevis.

9. Ger, R.: Surgical management of ulcerative lesions of the leg, Curr. Probl. Surg., March, 1972.

This review article summarizes the early pioneering work of Ger in the use of the muscle flap for coverage of lower leg defects. A comparison of techniques for coverage of leg defects related to trauma and venous disease demonstrates the reliability of the muscle transposition flap.

10. Ger, R.: The management of stasis ulcers of the leg where the tendo calcaneous is exposed, Plast. Reconstr. Surg. **60:**337, 1977.

The use of the flexor digitorum longus, soleus, and peroneus brevis muscles is described for coverage of ulcers with exposure of the tendo calcaneus. Since the muscular fibers of these muscles extend into the distal leg, these muscles are used as transposition flaps for coverage of such defects in the distal lower leg.

11. Kojima, T., Kohno, T., and Ito, T.: Muscle flap with simultaneous skin graft for skin defects of the lower leg, J. Trauma **19:**724, 1979.

Successful use of muscle flap transposition with meshed skin graft coverage in 12 patients with skin defects in the lower leg is presented. The soleus muscle is used in most cases. Muscle flap transposition is performed as early as possible in conjunction with reduction of the fracture.

12. McCraw, J.B., Dibbell, D.G., and Carraway, J.H.: Clinical definition of independent myocutaneous vascular territories, Plast. Reconstr. Surg. **60:**341, 1977.

The cutaneous territories of superficial muscles, including the biceps femoris and gastrocnemius, in the lower extremity are defined based on musculocutaneous perforating vessels. The incorporation of the skin overlying muscle based on these vascular connections is now the accepted method for flap design for lower extremity coverage.

13. McCraw, J.B., Fishman, J.H., and Sharzer, L.A.: The versatile gastrocnemius myocutaneous flap, Plast. Reconstr. Surg. **62:**15, 1978.

The successful use of the gastrocnemius musculocutaneous flap in 26 patients for coverage of lower leg defects is presented. The anatomy and pertinent arc of rotation are described both for the standard gastrocnemius musculocutaneous flap and the extended flap, which use skin distal to the muscle.

14. Mathes, S.J., and Buchanan, R.T.: Tensor fascia lata: neurosensory musculocutaneous free flap, Br. J. Plast. Surg. **32:**184, 1979.

The TFL musculocutaneous flap is transplanted by microvascular technique for coverage of lower extremity defects. The primary sensory nerve to the cutaneous territory of the TFL is defined as the lateral femoral cutaneous nerve. This nerve is included with the TFL musculocutaneous free flap to provide both coverage and sensory innervation to a lower extremity and foot defect.

15. Mathes, S.J., McCraw, J.B., and Vasconez, L.O.: Muscle transposition flaps for coverage of lower extremity defects: anatomical considerations, Surg. Clin. No Amer. **54:**1337, 1974.

The principles of muscle flap transposition are presented, emphasizing the vascular anatomy of muscles useful in coverage of lower extremity defects. The use of a latex injection technique is used to study the vascular anatomy of leg muscles, including the gastrocnemius, soleus, peroneus longus, and peroneus brevis.

16. Mathes, S.J., and Nahai, F.: Muscle flap transposition with function preservation: technical and clinical considerations, Plast. Reconstruct. Surg. **66:**242, 1980.

The concept of segmental muscle transposition, allowing use of muscle as a flap for coverage while maintaining the continuity of the muscle with its origin and insertion, is introduced. This technique is pertinent to lower leg reconstruction where the leg function is often diminished because of muscle injury. Techniques of function preservation and their anatomic basis are presented for the flexor digitorum longus, tibialis anterior, and soleus muscles.

17. Mathes, S.J., Nahai, F., and Vasconez, L.O.: Myocutaneous free flap transfer, anatomical and experimental considerations, Plast. Reconstr. Surg. **62:**162, 1978.

The gracilis musculocutaneous flap is transplanted by microvascular techniques for coverage of distal third lower extremity defects. The concept of strategic delay is introduced for transposition of the musculocutaneous flap based on its minor pedicles. Experimental data support this technique of distal muscle transposition.

18. Mathes, S.J., Vasconez, L.O., and Jurkiewicz, M.J.: Extension and further application of muscle flap transposition, Plast. Reconstr. Surg. **60:**6, 1977.

Muscle flap transposition is successfully used in 28 patients for coverage of lower extremity defects. Safe guidelines for use of the muscle flap transposition are presented.

19. Maxwell, G.P., Manson, P.N., and Hoopes, J.E.: Experience with 13 latissimus dorsi myocutaneous free flaps, Plast. Reconstr. Surg. **64:**1, 1979.

The latissimus dorsi musculocutaneous flap is transplanted by microvascular techniques for coverage of lower extremity defects. The anatomy and use of the latissimus dorsi muscle as an excellent donor flap for microvascular transplantation is presented.

20. Nahai, F., Hill, H.L., and Hester, T.R.: Experiences with the tensor fascia lata flap, Plast. Reconstr. Surg. **63:**788, 1979.

The TFL musculocutaneous flap is transplanted by microvascular techniques for coverage of lower extremity defects. The osseous territory of the TFL is introduced for transfer of vascularized bone. Sensory innervation of the proximal portion of the TFL is identified as the sensory branch of T12. This nerve is used with the flap to restore coverage with sensation.

21. Orticochea, M.: The musculocutaneous flap method: an immediate and heroic substitute for the method of delay, Br. J. Plast. Surg. **25:**106, 1972.

The concept of the musculocutaneous flap for coverage of a lower leg defect as a cross-leg flap is introduced. The author points out the importance of the perforating vessels between muscle and skin as a source of skin circulation.

22. Pers, M., and Medgyesi, S.: Pedicle muscle flaps and their application in the surgery of repair, Br. J. Plast. Surg. **26:**313, 1973

The use of the medial and lateral gastrocnemius, soleus, and extensor hallucis longus muscles for coverage of lower extremity defects in 11 patients is described. The muscle flap is recommended for coverage of avascular defects.

23. Petty, C., and Hogue, R., Jr.: Closure of an exposed knee joint by use of a sartorius muscle flap: case report, Plast. Reconstr. Surg. **62:**458, 1978.

The use of the sartorius muscle for knee coverage is demonstrated. The middle of the muscle is divided, and the inferior half of the muscle is transposed distally to cover the exposed knee joint based on the distal segmental vascular pedicles. This distally based sartorius muscle flap with a local rotation flap over it provides joint coverage and knee function restoration.

24. Robbins, T.H.: Use of fascio-muscle flaps to repair defects in the lower leg, Plast. Reconstr. Surg. **57:**460, 1976.

The use of the tibialis anterior muscle is described for coverage of middle lower leg defects. The muscle is used as a transposition flap to cover exposed tibial cortex.

25. Serafin, D., and co-workers: Reconstruction of the lower extremity with vascularized composite tissue: improved tissue survival and specific indications, Plast. Reconstr. Surg. **66:**230, 1979.

The use of musculocutaneous flap transplantation by microvascular techniques is compared to the use of the free groin flap in 48 patients requiring 50 vascularized flaps for reconstruction of difficult defects. Factors for improved survival with the musculocutaneous flap when free composite tissue transplantation techniques are used for lower extremity defects are presented.

26. Stark, W.J.: The use of pedicled muscle flaps in the surgical treatment of chronic osteomyelitis resulting from compound fracture, J. Bone Joint Surg. **28:**343, 1946.

The use of muscle flaps for coverage of the sequestrectomy wound of chronic osteomyelitis is compared with identical surgical débridement and delayed wound closure. Of the 32 wounds treated with muscle transposition, 84% were successfully treated, whereas only 43% of the wound débridement in the delayed closure patients remained healed. Of the 16 patients with osteomyelitis involving the tibia, the soleus and tibialis anterior muscles were transferred for coverage.

27. Townsend, P.L.: An inferiorly based soleus flap, Br. J. Plast. Surg. **31:**210, 1978.

The use of the distally based soleus flap for lower leg defects is described in four patients. It is stressed that the lower segmental blood supply to the soleus may be impaired by the injury, thereby eliminating the use of this muscle flap modification for coverage of distal leg defects.

28. Vasconez, L.O., Bostwick, J., III, and McCraw, J.: Coverage of exposed bone by muscle transplantation and skin grafting, Plast. Reconstr. Surg. **53:**526, 1974.

The use of muscle flap transposition for lower leg coverage is described in 20 patients. Muscles used in this series for leg coverage include the gastrocnemius, soleus, flexor digitorum longus, and peroneus longus.

31

COMPLICATIONS

Bernard S. Alpert ■ Foad Nahai ■ Luis O. Vasconez

THIGH

The abundance of muscles and soft tissue in the thigh generally protects vital neurovascular structures and bone. Most defects are adequately covered with skin grafts.

KNEE

Traumatic defects of the knee may have associated vascular injuries. Vascular evaluation is a prerequisite to flap design. This evaluation may include a preoperative angiogram to assess the vessels to the leg and vascular pedicles to the available flaps.

Gastrocnemius

Planning errors. The sural arteries are located within the popliteal fossa and may have been injured. Evaluation of these vessels by angiography is undertaken before flap selection when injury or vascular disease is suspected. The longer medial head of the gastrocnemius has a wider arc of rotation and is most useful for medial and anterior knee defects requiring coverage. The arc of rotation of the lateral gastrocnemius is limited by its shorter length and by the presence of the fibula. However, the lateral gastrocnemius is useful for defects located on the lateral aspect of the knee.

The design of the gastrocnemius as a musculocutaneous flap may result in a significant donor defect (Fig. 31-1). Transposition of the gastrocnemius as a muscle flap with skin grafts is preferred. Flap elevation under tourniquet control is recommended to avoid injury to adjacent structures.

Technical errors. During proximal elevation of the gastrocnemius, the popliteal artery and tibial nerve are identified and protected in the popliteal fossa. Dissection into the popliteal fossa is not always necessary to achieve adequate arc of rotation. During elevation of the lateral head of the gastrocnemius, the peroneal nerve is routinely identified at the anterior superior border of the flap and preserved. As the two heads are separated through their raphe in the posterior midline, the sural nerve is identified and preserved during the posterior dissection. The saphenous vein is identified and preserved anteriorly during elevation of the medial gastrocnemius. When the gastrocnemius is elevated as a musculocutaneous flap, the paratenon over the Achilles tendon must be preserved for successful skin grafting of the donor site.

Postoperative errors. Closed suction drainage should be placed in the donor area. Elevation of the leg for 2 to 3 weeks is recommended.

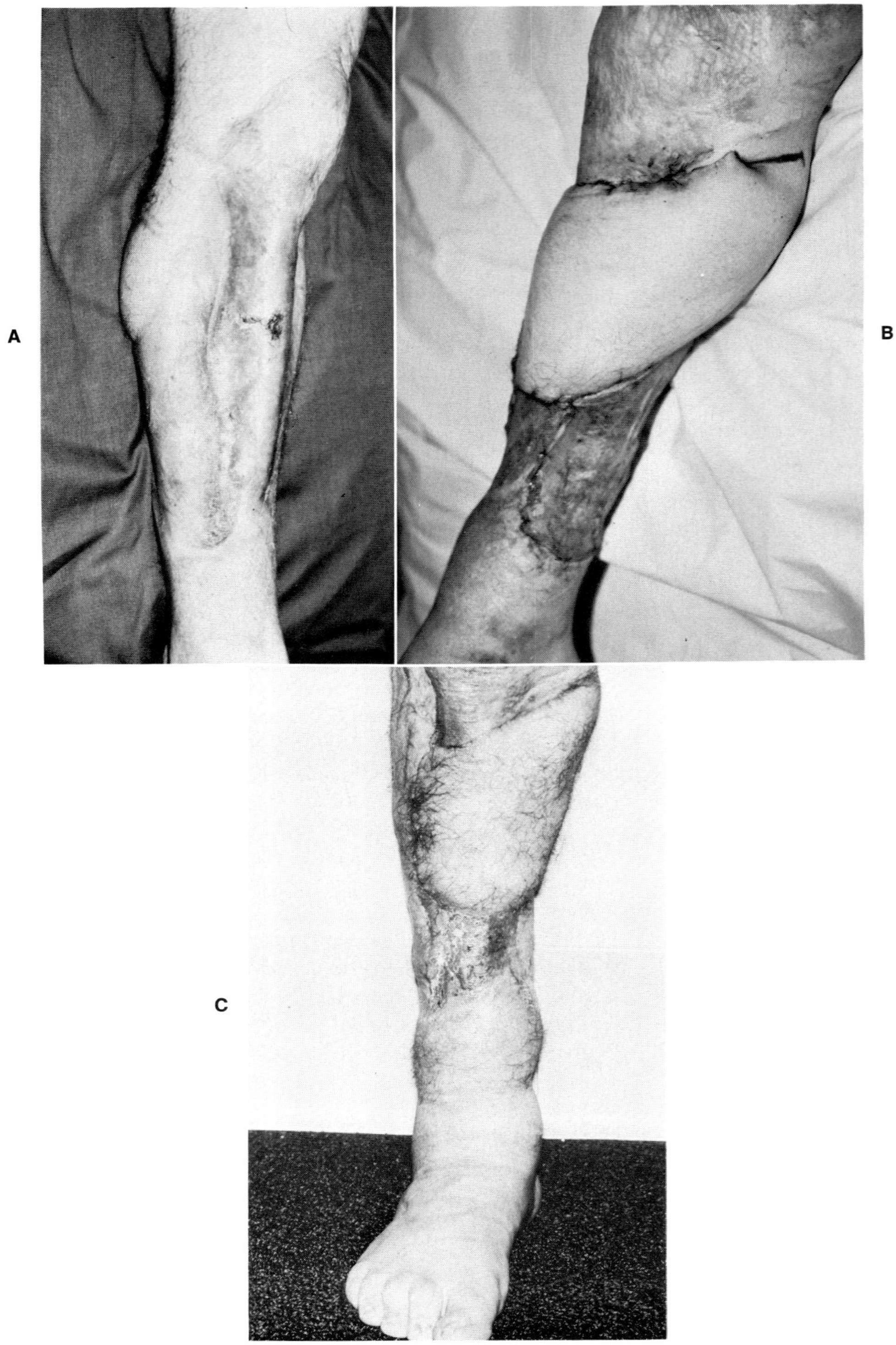

Fig. 31-1
A, Unstable skin grafts over tibia at site of prior injury.
B, Transposition of medial gastrocnemius musculocutaneous flap for defect coverage.
C, Stable wound coverage provided by flap. Donor defect is significant; free flap or cross-leg flap would provide better coverage and appearance.

Distally based gracilis

Planning errors. The distally based gracilis flap may reach the knee but requires a prior strategic delay. If the minor pedicle is not located in the distal third of the muscle, the flap will not have an adequate arc of rotation for knee coverage. Since the minor pedicle cannot be safely visualized at the time of a strategic delay, the status of the minor pedicles is not known until the second procedure. For this reason this flap is rarely indicated for knee coverage.

Technical errors. During this strategic delay, the proximal musculocutaneous perforators to the overlying skin must be preserved. Flap elevation is delayed for 2 weeks. Correct orientation of the skin island over the muscle is required. The cutaneous territory is temporarily sutured to the muscle borders during flap elevation to avoid disruption of musculocutaneous perforating vessels. Tension on the minor pedicle must be avoided during flap transposition.

Postoperative errors. Closed suction drainage in the donor defect is necessary.

Sartorius

Planning errors. The sartorius will cover only small medial defects. Its use as a musculocutaneous unit is limited, because the skin territory is small.

Technical errors. The segmental pattern of blood supply to the sartorius limits its effective arc of rotation. Division of more than two adjacent pedicles may result in muscle devascularization.

Postoperative errors. Closed suction drainage in the donor site is necessary.

LOWER LEG

Defects of the lower leg requiring flap coverage are generally related to underlying bony or vascular damage. Evaluation and successful management of bony problems and vascular injury are prerequisites for successful flap coverage.

Gastrocnemius

Planning errors. Use of the gastrocnemius muscle flap is preferred to avoid (1) bulky coverage associated with musculocutaneous flaps and (2) a donor defect that requires skin grafts when a musculocutaneous flap is used.

Technical errors. Flap elevation beyond the medial or lateral condyle is rarely required. The peroneal nerve is visualized and preserved during elevation of the lateral gastrocnemius muscle.

Postoperative errors. Closed suction drainage in the donor defect is necessary. Leg elevation is recommended for 2 to 3 weeks.

Soleus

Planning errors. Integrity of the popliteal and either posterior tibial or peroneal arteries must be established before a soleus flap is planned. The soleus muscle has a type II pattern of blood supply, and one or more of the minor pedicles will have to be divided, depending on the extent of rotation required.

Technical errors. Flap dissection under tourniquet control is recommended. The correct identification of the soleus muscle is the key to successful flap elevation. The soleus lies deep to the gastrocnemius with the plantaris tendon between the two. Distally the muscles are firmly adherent, and sharp dissection is used to separate the soleus and gastrocnemius. Minor pedicles are identified and ligated as necessary to increase the arc of rotation. The posterior tibial artery, vein, and nerve are located between the soleus and flexor hallucis longus.

Postoperative errors. Closed suction drainage should be placed in the donor area. Elevation of the leg for 2 to 3 weeks is recommended.

Flexor digitorum longus

Planning errors. The flexor digitorum longus muscle is small and only useful for small defects of the middle and lower tibia. The blood supply is segmental, so there is a limit to the number of pedicles that may be divided. This muscle is best used when additional inferior coverage is required for soleus transposition.

Technical errors. The posterior tibial artery and tibial nerve are located immediately lateral to this muscle and are vulnerable to injuries in its dissection.

Postoperative errors. Suction drainage should be established and the patient positioned with the leg elevated to facilitate venous drainage.

Tibialis anterior

Planning errors. The tibialis anterior is not an expendable muscle, and its use should only be considered with function preservation technique. The origin and insertion of the muscle are left intact, and only a portion of the muscle belly is transposed.

Technical errors. The posterior muscle belly is closely related to the anterior tibial artery and deep peroneal nerve, and injury to these structures must be avoided. Since the pattern of blood supply is segmental (type IV), proximal vascular pedicles to the muscle are preserved.

Postoperative errors. The leg should be elevated postoperatively and suction drainage provided.

Peroneus longus

Planning errors. The peroneus longus is a small muscle that will only cover a small part of the middle third of the leg and should only be considered if the soleus is not available. A secondary problem with the donor defect may arise if the fibula is left exposed. If both peroneus muscles are used, foot elevation is impaired.

Technical errors. The peroneus longus may be difficult to separate from the peroneus brevis in the dissection. The blood supply proximally from the peroneal artery must be preserved. The muscle will only cover a very small area.

Postoperative errors. The leg must be elevated postoperatively.

Peroneus brevis

Planning errors. The peroneus brevis is a small muscle that is useful occasionally for small defects of the distal third of the lateral leg. If both peroneus muscles are taken, foot eversion will be lost.

Technical errors. It is difficult to separate the peroneus longus from the peroneus brevis. The proximal blood supply from the peroneal artery must be preserved. The muscle will cover only a very small area.

Postoperative errors. The leg must be elevated postoperatively. Closed suction catheters should be placed in the donor area.

DISTALLY BASED FLAPS

The distally based muscle flaps depend on the minor pedicle entering the muscle in the distal leg. This technique is possible for muscles with type II or III vascular patterns, including the soleus, extensor digitorum longus, and extensor hallucis longus. Unfortunately, distal leg defects are commonly associated with trauma or vascular disease. In either situation the inferior pedicles may no longer exist because of associated vascular disease. When circulation to the lower leg is confirmed clinically or with an angiogram, muscles with distal minor pedicles may be transposed as distally based flaps into small defects within the area.

Soleus

Planning errors. Defects extending below the malleoli cannot be covered by the distally based soleus.

Technical errors. The dissection is performed under tourniquet control. Minor pedicles located inferiorly to the point of rotation must be identified and preserved.

Postoperative errors. Suction drainage should be established and the patient positioned with the leg elevated to facilitate venous drainage.

Extensor digitorum longus

Planning errors. Defects extending below the lateral malleolus cannot be covered by the distally based extensor digitorum longus.

Technical errors. The dissection is performed under tourniquet control. Minor pedicles located inferiorly to the point of rotation must be identified and preserved. Separation of the extensor digitorum longus and extensor hallucis longus is difficult and not always necessary.

Postoperative errors. Suction drainage should be established and the patient positioned with the leg elevated to facilitate venous drainage.

Extensor hallucis longus

Planning errors. Defects extending below the lateral malleolus cannot be covered by the distally based extensor hallucis longus. This muscle belly is small and rarely useful.

Technical errors. The dissection is performed under tourniquet control. Minor pedicles located inferiorly to the point of rotation must be identified and preserved.

Postoperative errors. Suction drainage should be established and the patient positioned with the leg elevated to facilitate venous drainage.

32

A SYSTEMATIC APPROACH TO FLAP SELECTION

Stephen J. Mathes ■ Foad Nahai

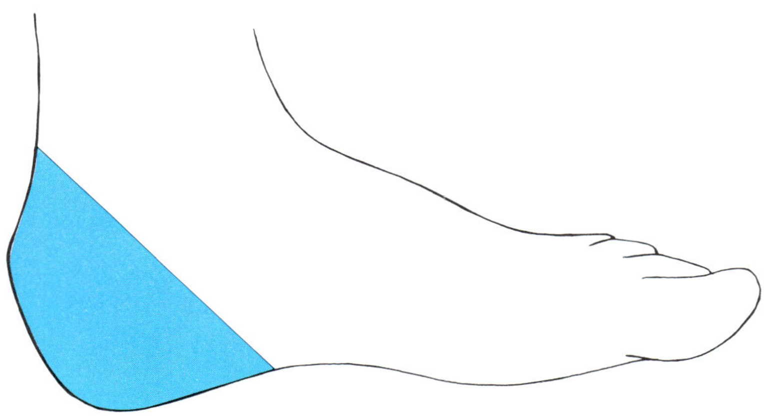

Fig. 32-1

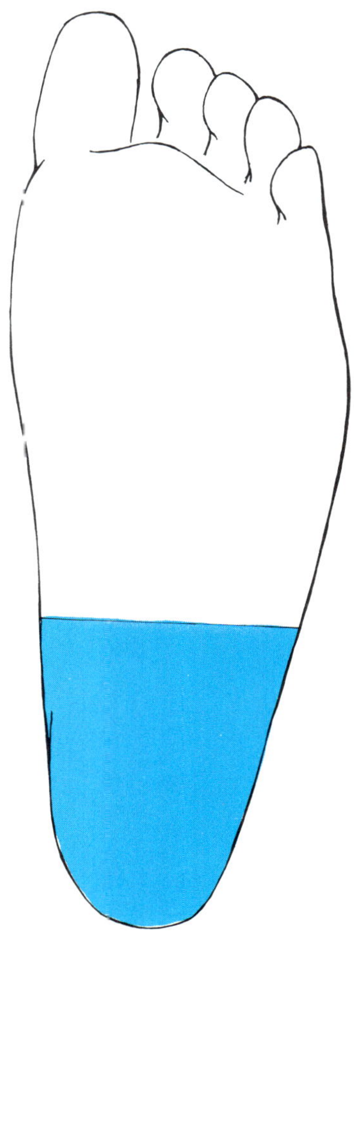

Fig. 32-2

SKIN GRAFTS

Skin grafts are method of choice when adequate soft tissue padding is present

LOCAL FLAPS (Fig. 32-4)

1. Axial innervated skin flaps are method of choice for weight-bearing areas (Fig. 32-3)
2. Flexor digitorum brevis muscle and skin graft is alternative flap for heel coverage (Fig. 32-5)
3. Abductor hallucis muscle with skin graft is alternative flap for defect below medial malleolus (Fig. 32-6)
4. Abductor digiti minimi muscle with skin graft is alternative flap for defects below lateral malleolus
5. Dorsalis pedis is alternative flap for medial and lateral ankle defects

DISTANT FLAPS (Fig. 32-7)

1. Free tissue transplantation is method of choice for extensive defects of foot when a local innervated flap is not available
2. Flaps of choice
   a. Gracilis (small dorsal defects) (Fig. 32-8)
   b. TFL (innervated for posterior defects and weight-bearing areas)
   c. Dorsalis pedis axial flap (contralateral innervated)
   d. Deltoid axial neurosensory flap
3. Random cross-leg flap is alternative technique to free tissue transplantation when local flap is not available, or free tissue transplantation is not possible
4. Musculocutaneous cross-leg flap is rarely indicated in reconstruction of the foot

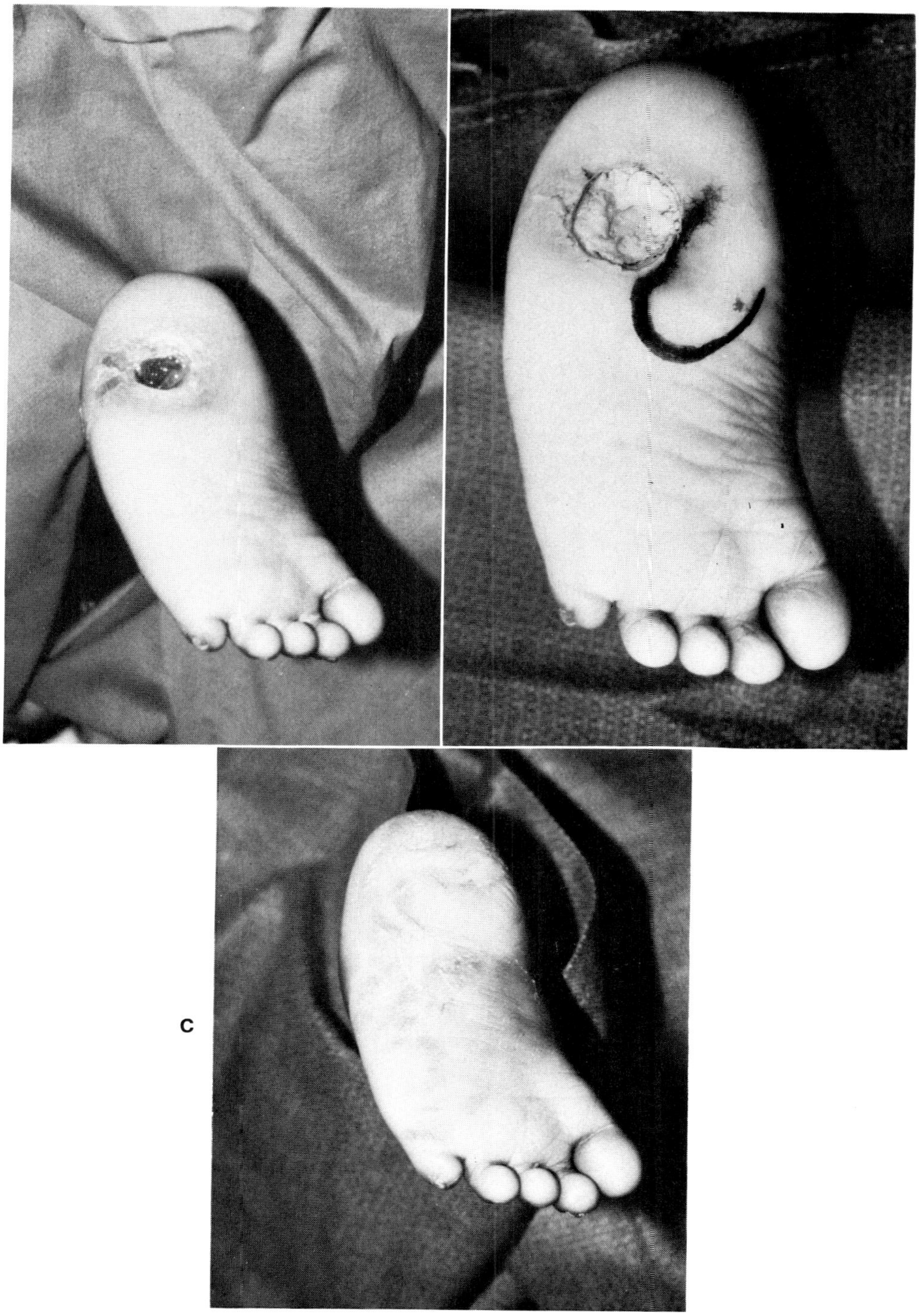

Fig. 32-3. Local flap.
A, Chronic wound of heel in 4-year-old boy with myelomeningocele. Previous skin grafts had failed.
B, Wound excised and flap outlined.
C, Postoperative result demonstrates stable wound coverage. Donor area closed directly.

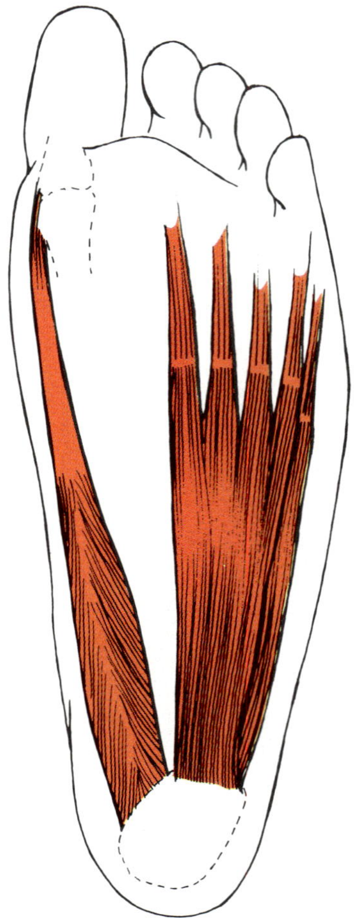

Abductor digiti minimi
Abductor hallucis

Fig. 32-4

Fig. 32-5. Flexor digitorum brevis.

A, Level V melanoma of heel.

B, After radical resection of heel and calcaneus, flexor digitorum brevis muscle elevation performed.

C, Flexor digitorum muscle transposition over exposed calcaneus.

D, Stable wound coverage provided in this ambulatory patient with flexor digitorum muscle transposition and skin grafts.

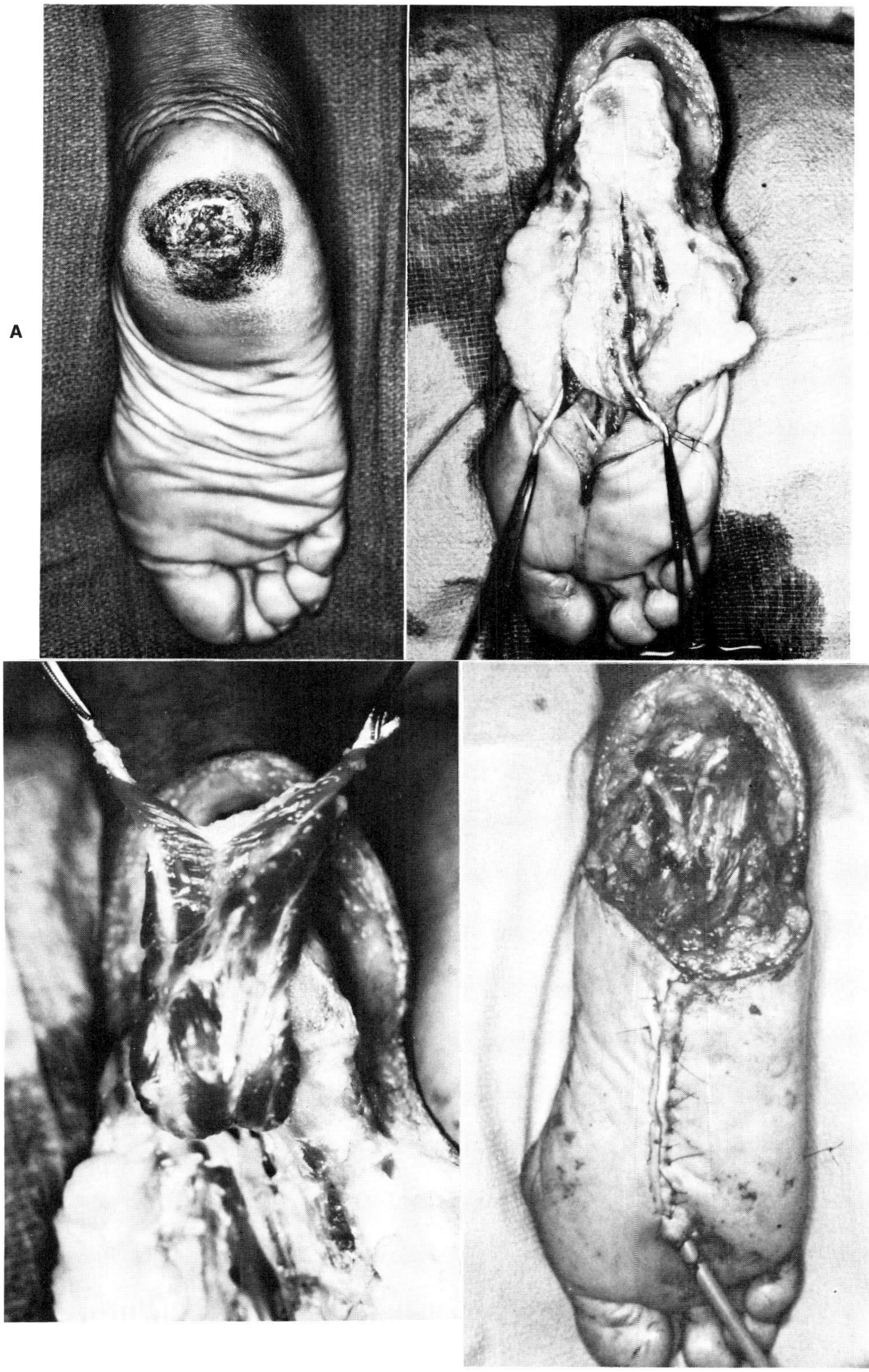

A B C D

Fig. 32-5. For legend see opposite page.

A B

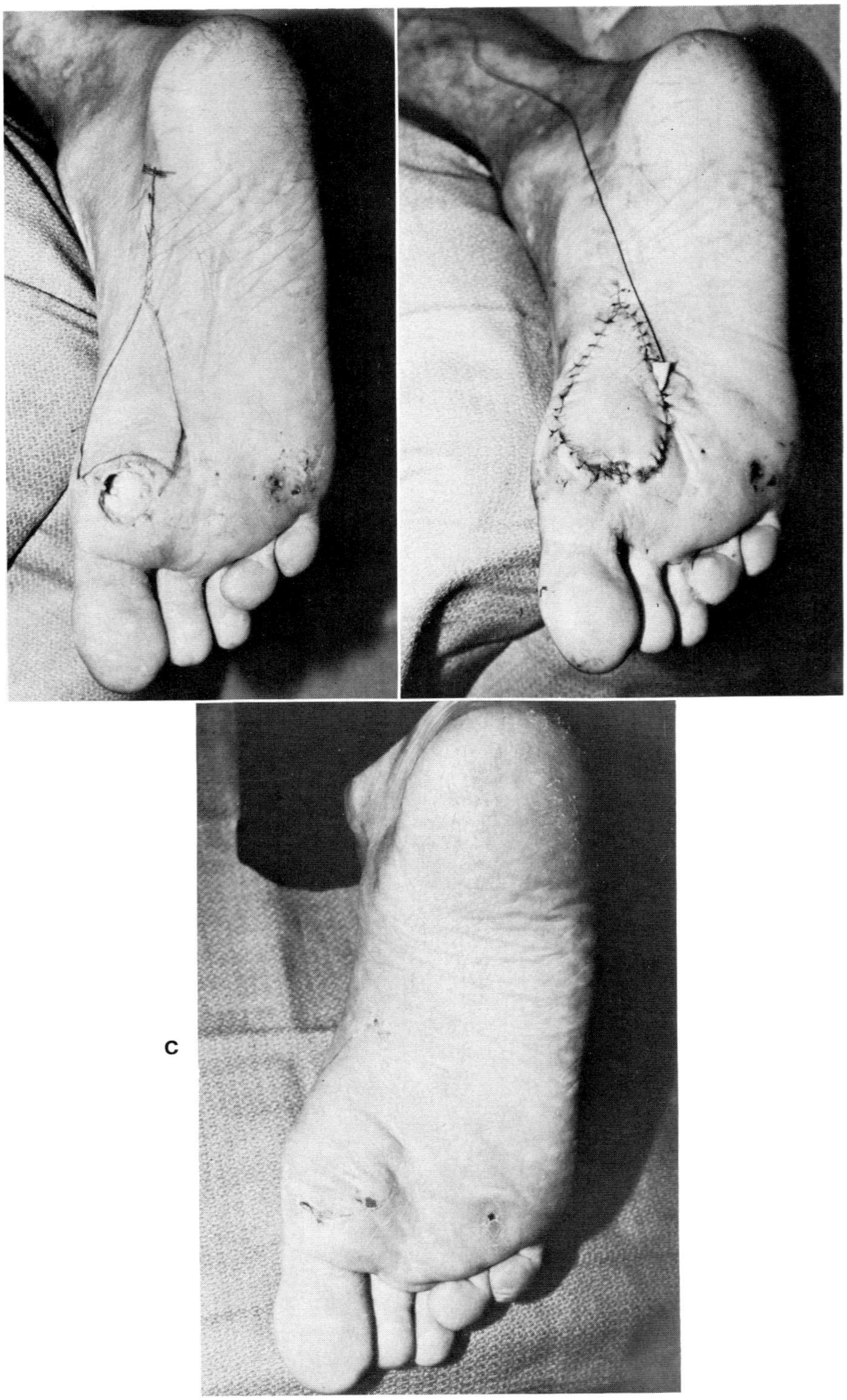

C

Fig. 32-6. Local axial flap based on skin territory of abductor hallucis muscle.
A, Chronic plantar foot ulcer in diabetic patient.
B, Advancement of axial flap based on medial plantar artery with preservation of medial plantar nerve branches to skin island.
C, Postoperative view demonstrates stable wound coverage provided by axial flap based on medial plantar artery.

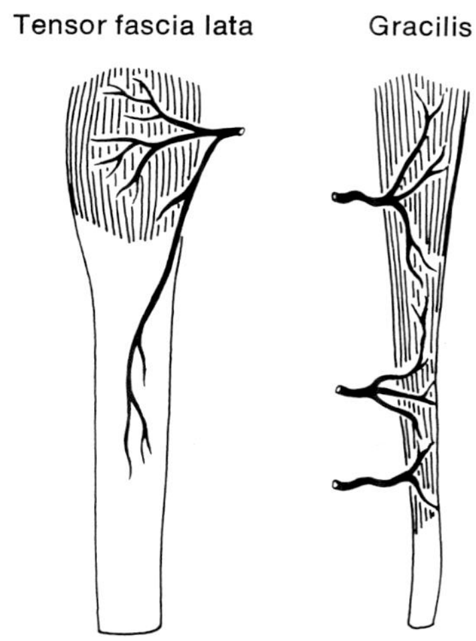

Fig. 32-7

A B C D

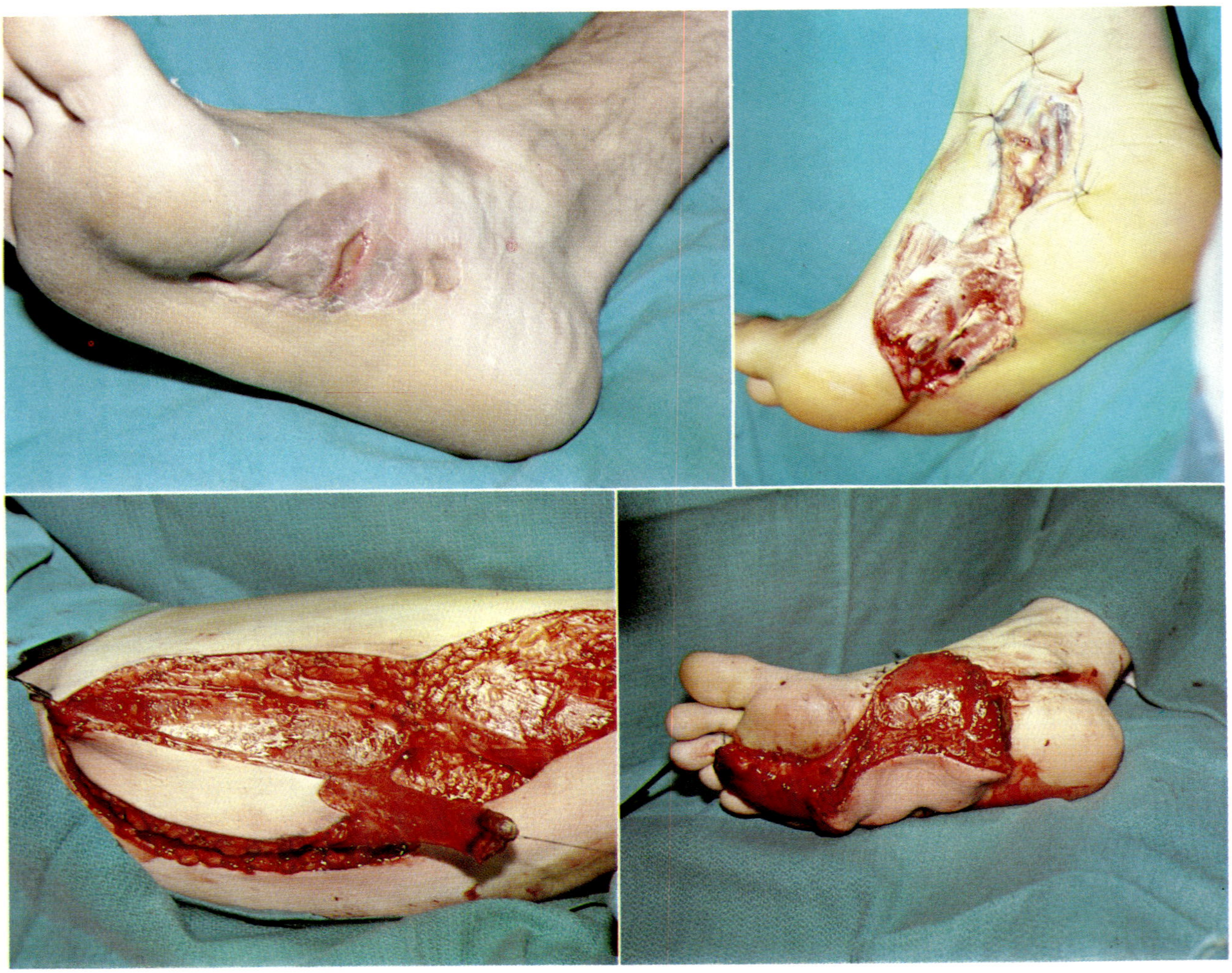

Fig. 32-8. Gracilis free flap with skin grafts.

A, Twenty-two-year-old man with chronic infected wound of right foot. Multiple skin grafts and local flaps had failed.

B, Wound excised and recipient vessels (posterior tibial artery and vein) exposed for microsurgical transplantation.

C, Gracilis musculocutaneous flap elevated.

D, Flap revascularized in foot. NOTE: Bulky flap.

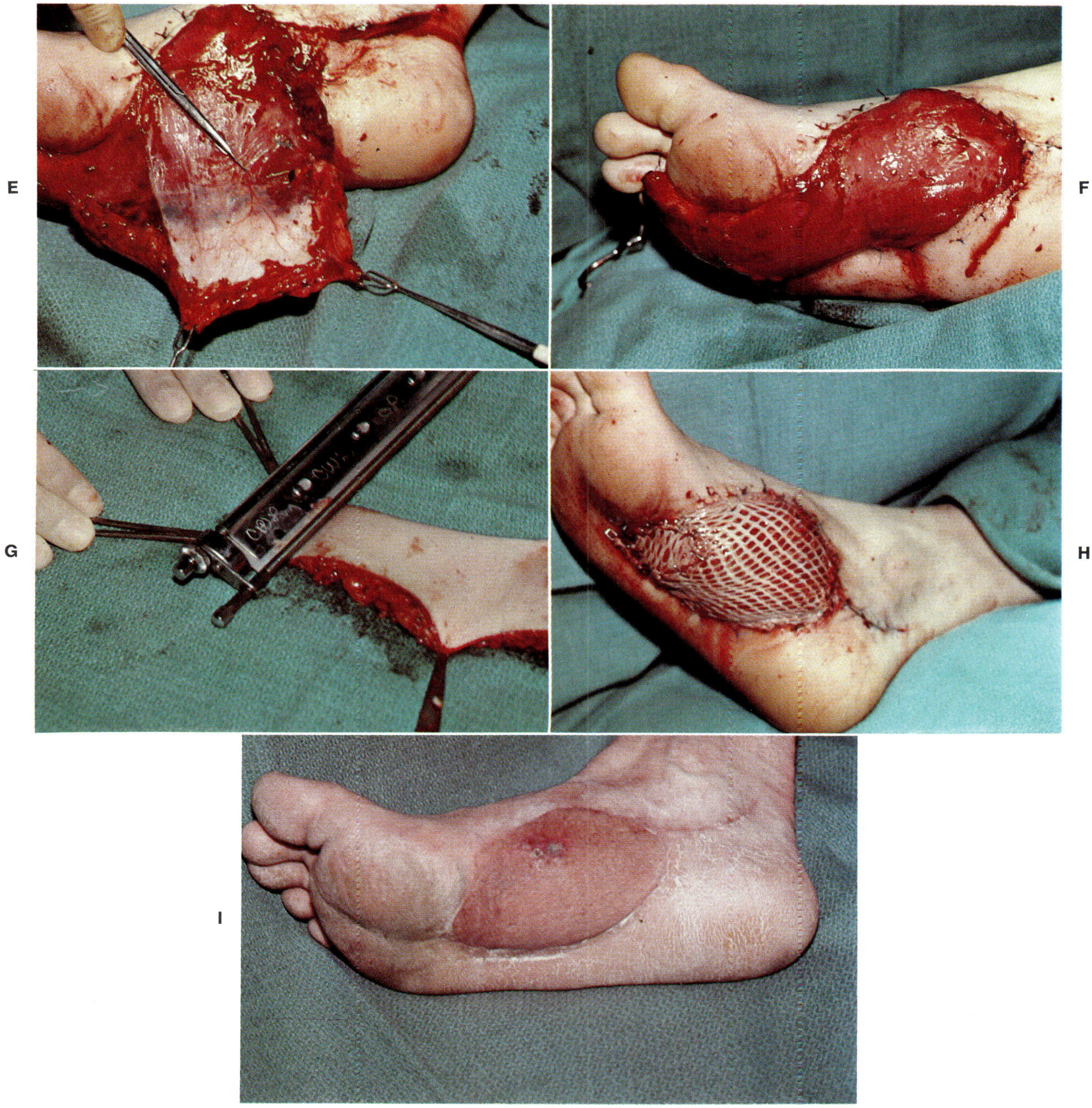

Fig. 32-8, cont'd

E, Skin and subcutaneous tissue excised from muscle after transplantation. Pointer denotes musculocutaneous perforating artery.
F, Gracilis muscle after excision of subcutaneous fat and skin island.
G, Skin harvested from resected skin island.
H, Muscle trimmed and covered with meshed skin graft.
I, Postoperative result demonstrates stable wound and excellent contour of foot, after muscle atrophy.

33

RECONSTRUCTION

Michael Scheflan ▪ Foad Nahai

The advent of muscle and musculocutaneous flap surgery has provided new flaps for soft tissue reconstruction of the foot. The foot is a difficult area to reconstruct. The determining factors in this often complex and difficult problem consist of (1) the lack of available nearby donor sites for soft tissue transposition; (2) the unusual durability and resistant soft tissue that is required for weight bearing and locomotion; and (3) the nature of the underlying pathologic process (e.g., diabetes mellitus and atherosclerotic peripheral vascular disease), which often precludes the application of a certain or all surgical procedures. Furthermore soft tissue defects on a weight-bearing surface represent a different problem from those that occur on non-weight-bearing surfaces and require a different surgical approach. Present knowledge of dependable musculocutaneous territories has opened new possibilities in reconstructive foot surgery. The superficial muscle group of the foot consists of the abductor hallucis, flexor digitorum brevis, and abductor digiti minimi and is the mainstream of local soft tissue donor sites available for reconstruction.

It should be mentioned that local arterialized subfascial skin flaps are sometimes possible around the heel pad for reconstruction of small soft tissue defects and will be discussed next. Free tissue transfer to the foot is discussed in Chapter 38 and is therefore omitted from this chapter.

ABDUCTOR HALLUCIS MUSCLE

Anatomy

The abductor hallucis muscle pads the medial aspect of the arch of the foot (Fig. 33-1). It originates in the calcaneus and inserts on the base of the proximal phalanx of the first toe. Its innervation and blood supply enter proximally from the medial plantar neurovascular bundle. About 1 inch distally to its bifurcation from the posterior tibial artery, the medial plantar artery gives off two or three small proximal pedicles to the posterior aspect of the abductor hallucis muscle belly. The medial plantar artery then courses between the abductor hallucis and flexor digitorum brevis muscles to which it gives off distal pedicles as well. A deep branch to the tarsal bones is usually given at that point and the medial plantar artery continues distally and terminates as the most medial digital plantar artery.

Flap elevation

A medial foot incision on the non-weight-bearing surface is used to approach the muscle. The tendon is divided distally and separated from the fibers of the flexor hallucis brevis. (In some cases these fibers can be left attached to the tendon and elevated together with the abductor hallucis flap to augment its reach.) The muscle is then turned over on itself and elevated up to the pedicle on the posterior aspect of the proximal third. With this elevation the muscle will reach the proximal medial aspect of the dorsum of the foot and defects just below the medial malleolus (Fig. 33-2). Defects at or just above the malleolus as well as defects on the posterior heel will require division of the medial plantar artery distally and elevation of the artery with the flap up to the bifurcation of the posterior tibial artery. The overlying skin receives sensory innervation from the saphenous nerve. Although usually unnecessary, the flap may be elevated as a musculocutaneous flap. It is important, however, that the donor site over this sensitive area is closed primarily and not skin grafted.

The amount of skin that can be elevated with this flap is therefore limited. Since dissection of this muscle flap for further reach often requires the distal ligation of the medial plantar artery, patency of both dorsalis pedis and posterial tibial arteries must be established. In patients that have one of these two vessels occluded or traumatized, this operative procedure should not be undertaken (Fig. 33-3).

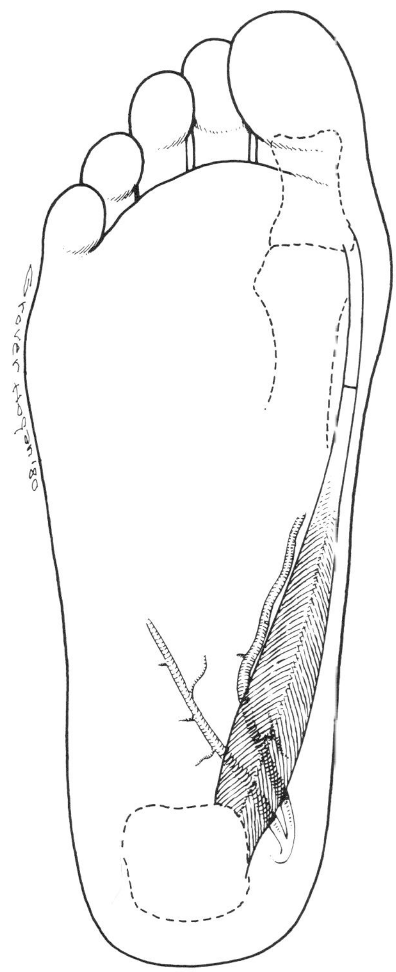

Fig. 33-1. Abductor hallucis muscle pads medial aspect of foot and receives its proximal pedicle from medial plantar artery.

(From Scheflan, M., Nahai, F., and Hartrampf, C.R., Jr.: Ann. Plast. Surg. **7**:385, 1981.)

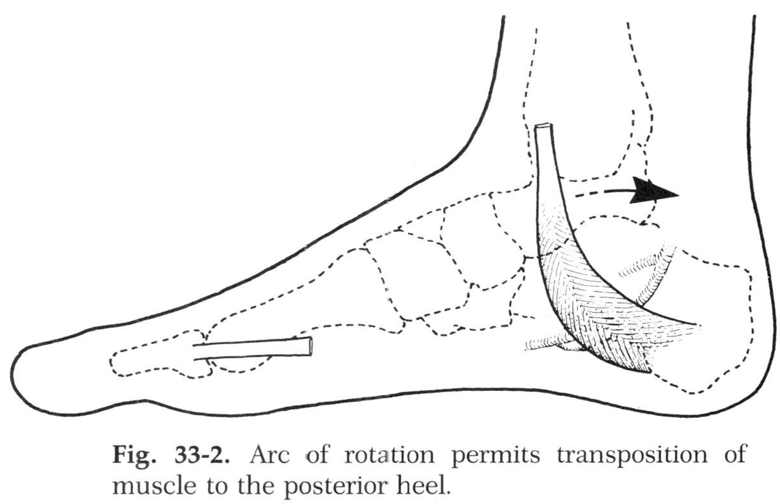

Fig. 33-2. Arc of rotation permits transposition of muscle to the posterior heel.

(From Scheflan, M., Nahai, F., and Hartrampf, C.R., Jr.: Ann. Plast. Surg. **7**:385, 1981.)

A 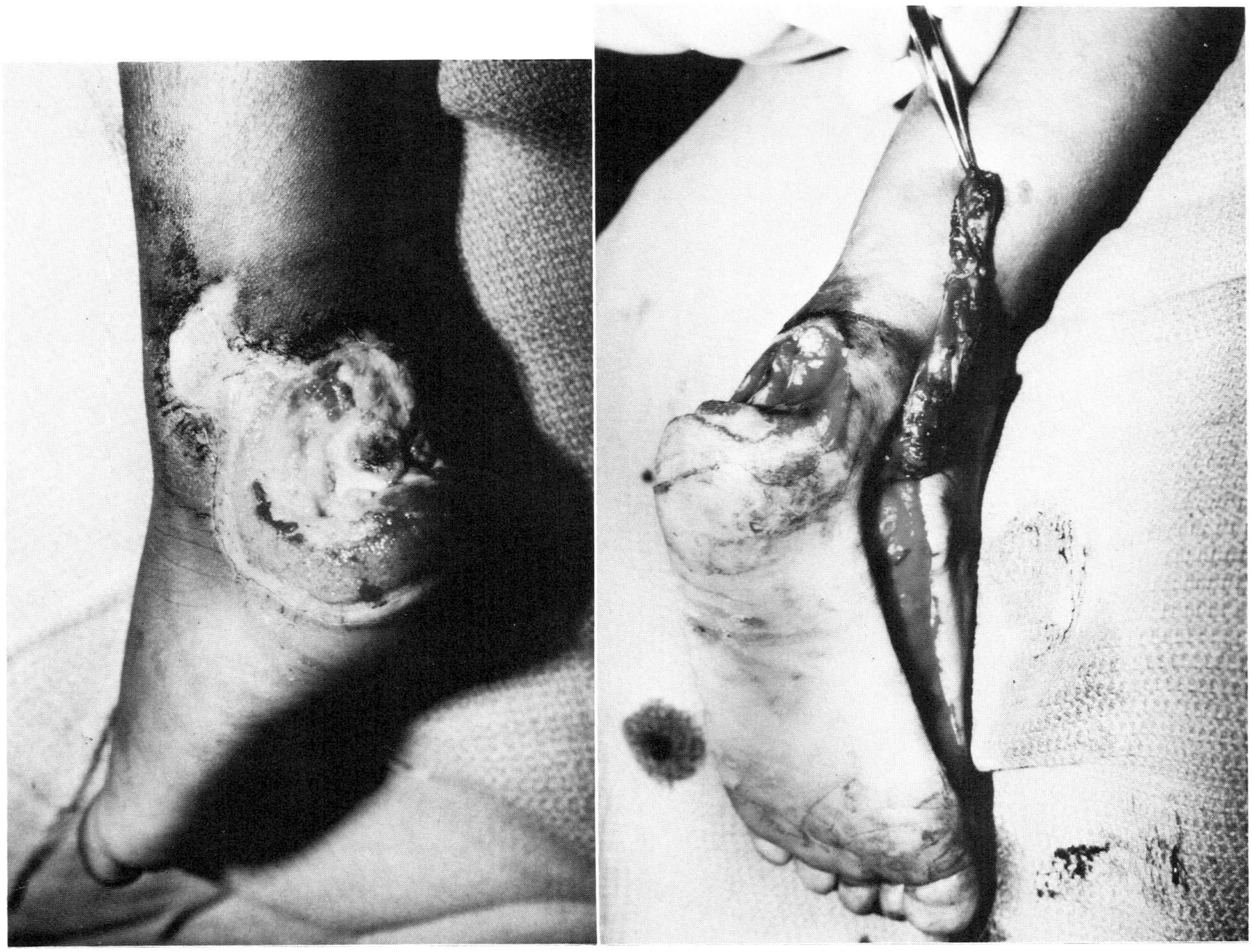 B

Fig. 33-3. Abductor hallucis muscle.
A, Nine-year-old patient with traumatic avulsion of posterior heel.
B, Abductor hallicus muscle has been isolated on its vascular pedicle and transposed posteriorly.

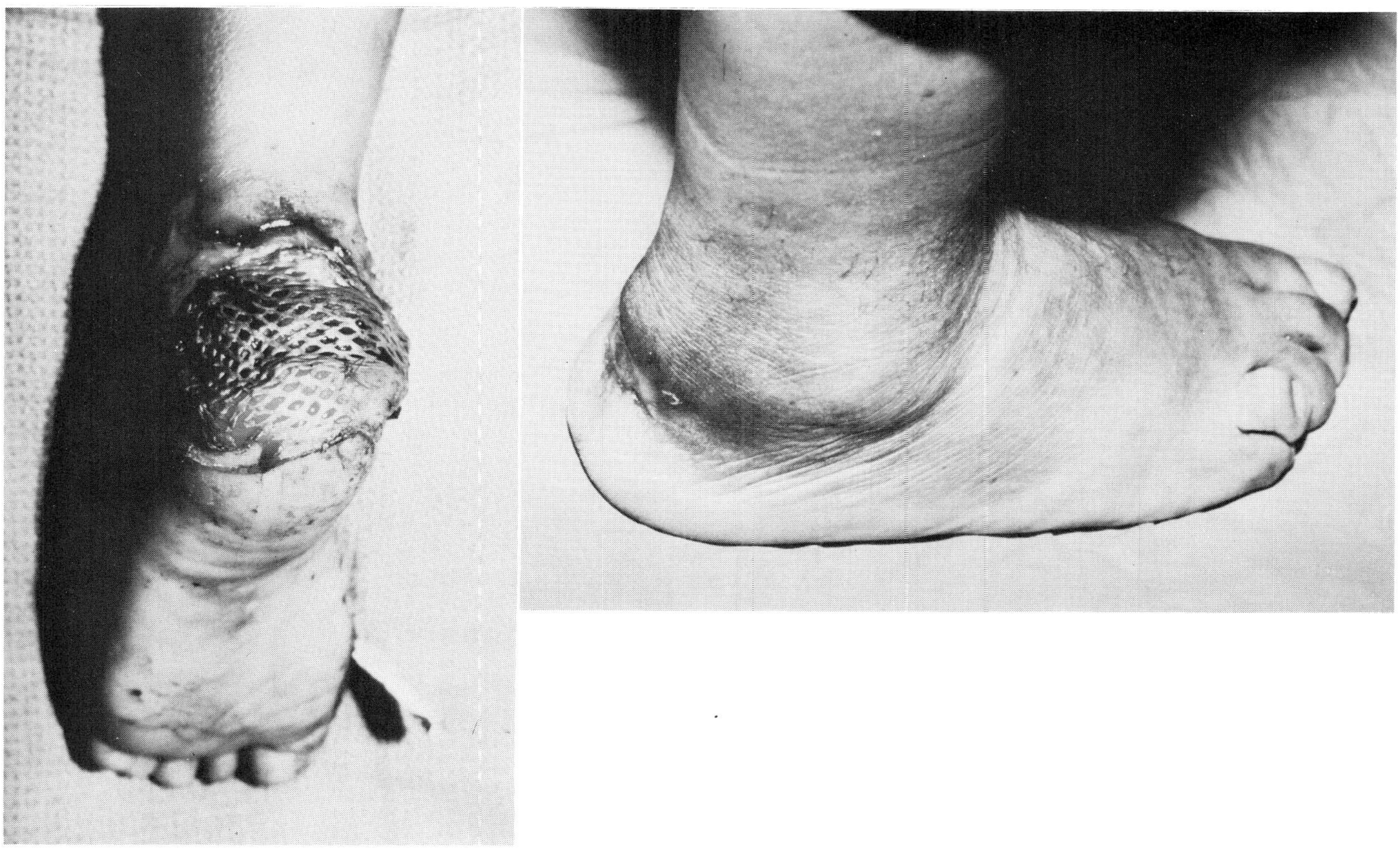

C

D

Fig. 33-3, cont'd
C, Abductor hallicus muscle is in place and covered with split-thickness mesh graft.
D, One-year postoperative view of patient without walking difficulties or wound problems.

FLEXOR DIGITORUM BREVIS MUSCLE
Anatomy

The flexor digitorum brevis originates from the medial process of the calcaneus and inserts into the middle phalanges of the second to fifth toes (Fig. 33-4). It is intimately adhered to but easily separated from the plantar aponeurosis and the intramuscular septi.

The blood supply comes from pedicles from both the medial and lateral plantar arteries, with the latter being the dominant one. The lateral plantar artery branches off the posterior tibial artery just below the medial malleolus. It then courses under the origin of the abductor hallucis muscle and crosses from medial to lateral under the proximal third of the flexor digitorum brevis. Under the muscle it gives off a dominant pedicle and then proceeds laterally. On the lateral aspect of the foot, it travels in a groove between the abductor digiti minimi and the flexor digitorum brevis muscles.

The lateral plantar artery divides at the level of the base of the fifth metatarsal into a deep branch and a superficial branch. The superficial branch ends as the most lateral digital plantar artery. The deep branch courses medially to form the deep plantar arch and give off the four plantar metatarsal arteries to the third, fourth, and medial half of the fifth toes.

The plantar and dorsal metatarsal arteries communicate through vessels that course through the first interosseous space, connecting the dorsalis pedis and posterior tibial arterial systems. The skin overlying the flexor digitorum brevis muscle gets its blood supply through perforating branches from the muscle. Motor innervation to the muscle is through the lateral plantar nerve that courses with the vascular bundle.

Skin sensibility to the medial plantar aspect to the foot is through the saphenous nerve. The lateral plantar aspect gets its sensory innervation through the sural nerve and the territory in between is innervated through a sensory branch from the posterior tibial nerve. This is important, since it allows elevation of a flexor digitorum brevis musculocutaneous flap as a neurosensory unit (Fig. 33-5).

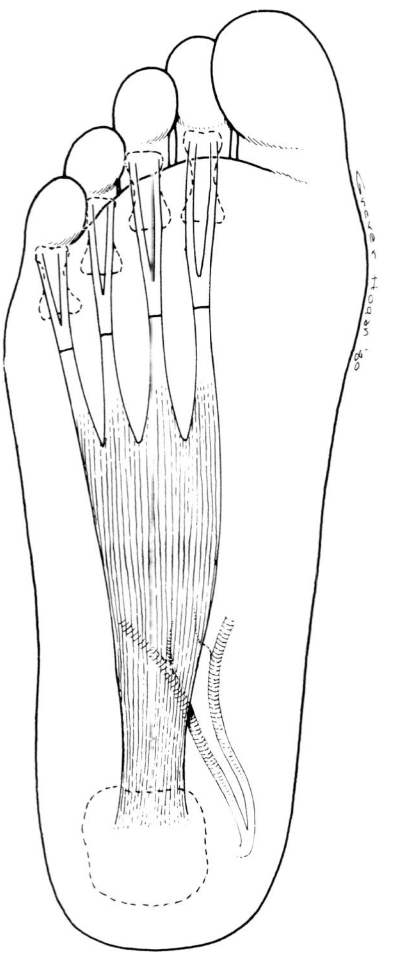

Fig. 33-4. Flexor digitorum brevis muscle occupies center portion of plantar aspect of foot and receives its blood and nerve supply from lateral plantar neurovascular bundle.

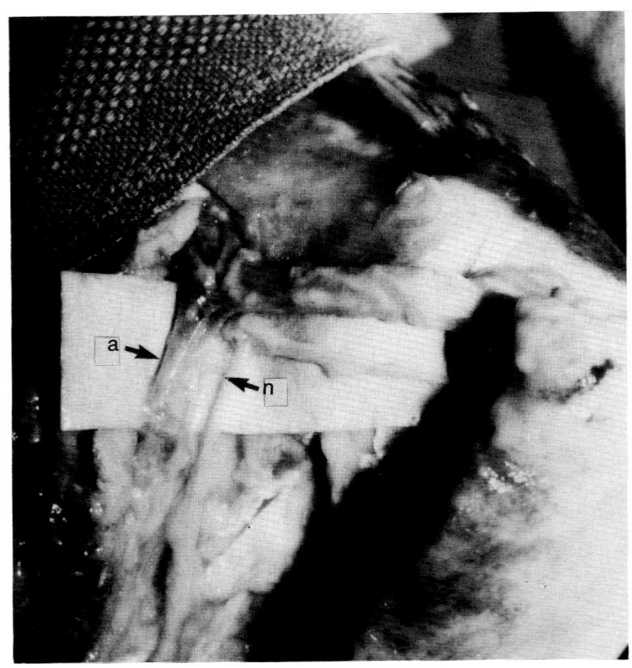

Fig. 33-5. Cadaver dissection demonstrates sensory nerve to skin overlying flexor digitorum brevis. Nerve is branch of lateral plantar nerve. *n,* Nerve; *a,* lateral plantar artery.

Flap elevation

Muscle flap. A midline foot incision is used and carried down to the plantar fascia. The fascia can be elevated with the muscle so that the fibers hold together better. The medial and lateral skin flaps are elevated over the plantar fascia. The four tendons are divided distally and the muscle turned over on itself and traced back in a loose areolar tissue plane between the flexor digitorum brevis and the quadratus plantae muscles. Small or moderately large plantar aspect heel defects may be closed in this fashion. For more proximal (higher) defects, the origin of the muscle must be divided off the calcaneus. This will permit another 1 cm of coverage (Figs. 33-6 to 33-8).

Defects on the high posterior heel or at the malleolus level require distal division of the lateral plantar artery. The flap is then elevated with its vascular pedicle and traced back to the bifurcation of the posterior tibial artery. To achieve a larger pedicle with a wider arc of rotation, an incision is best carried out over the distal posterior tibial artery just behind and below the medial malleolus. The bifurcation is identified and the vessels traced down under the origin of the abductor hallucis muscle, which is divided. The island of muscle isolated on its vascular pedicle can then be easily transposed further posteriorly or superiorly. After suturing the muscle to the wound edges, a split-thickness skin graft is applied to the wound. Patency of the dorsalis pedis and posterior tibial arteries must be assessed before ligation of the distal lateral plantar artery.

Musculocutaneous flap. When both dorsalis pedis and posterior tibial arteries are intact, and there is a large soft tissue defect over the weight-bearing surface of the heel, then moving non-weight-bearing muscle and skin to the weight-bearing area is desirable. Several cadaver dissections demonstrate that it is possible to transfer the flexor digitorum brevis and its overlying skin as a musculocutaneous unit that receives sensory innervation through a branch of the posterior tibial nerve.

To elevate this flap the skin island is incised within the non-weight-bearing surface, and the unit is raised as a musculocutaneous flap as just described. The secondary donor defect is covered with a thick split-thickness skin graft. It is important to design the skin island area within the non-weight-bearing surface of the foot to avoid skin grafts and scars over weight-bearing areas. The lateral plantar artery must be divided distally for elevation of this flap, and the presence of pulses in the dorsalis pedis and tibial posterior arterial systems must be established preoperatively.

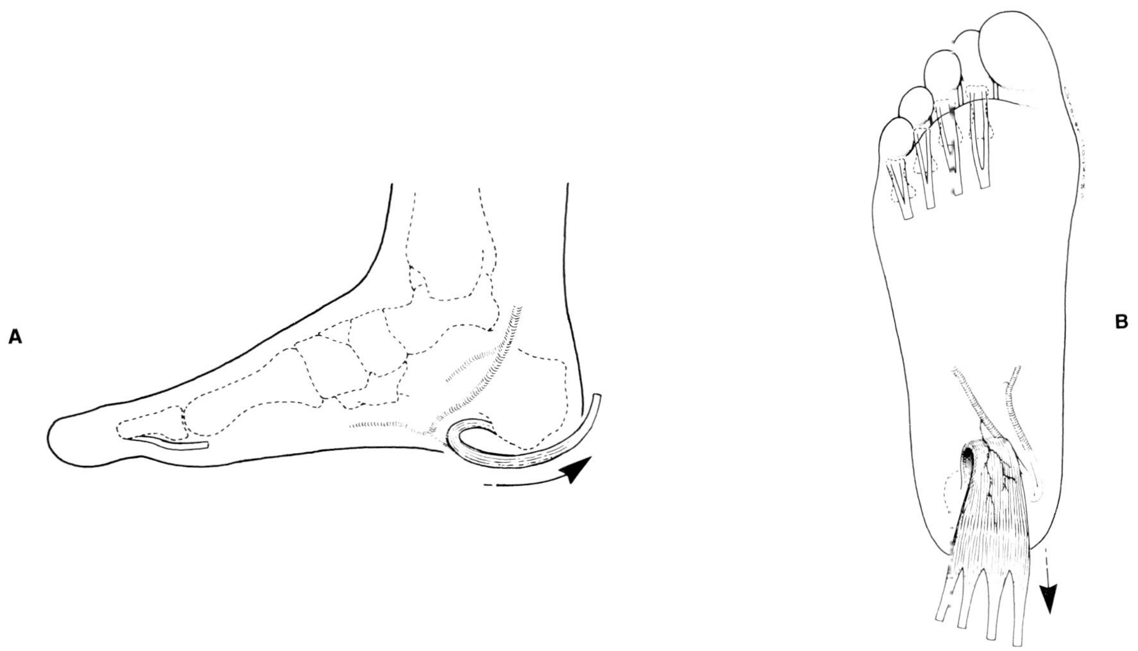

Fig. 33-6
A and **B,** Standard elevation of flexor digitorum brevis flap with limited reach over head.

(**A** to **D** from Scheflan, M., Nahai, F., and Hartrampf, C.R., Jr.: Ann. Plast. Surg. **7**:385, 1981.)

Continued.

C

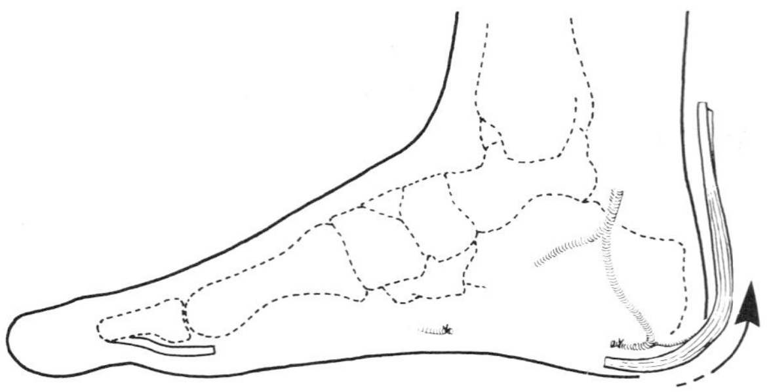

D

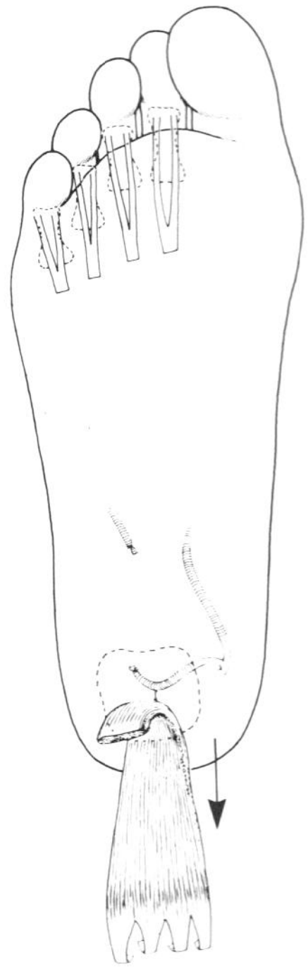

Fig. 33-6, cont'd

C and **D,** Extended flexor digitorum brevis flap with division of lateral plantar artery allows for transposition of flap posteriorly over calcaneus and Achilles tendons. This division should only be undertaken after establishing that both anterior tibial and posterior tibial vessels are patent.

A

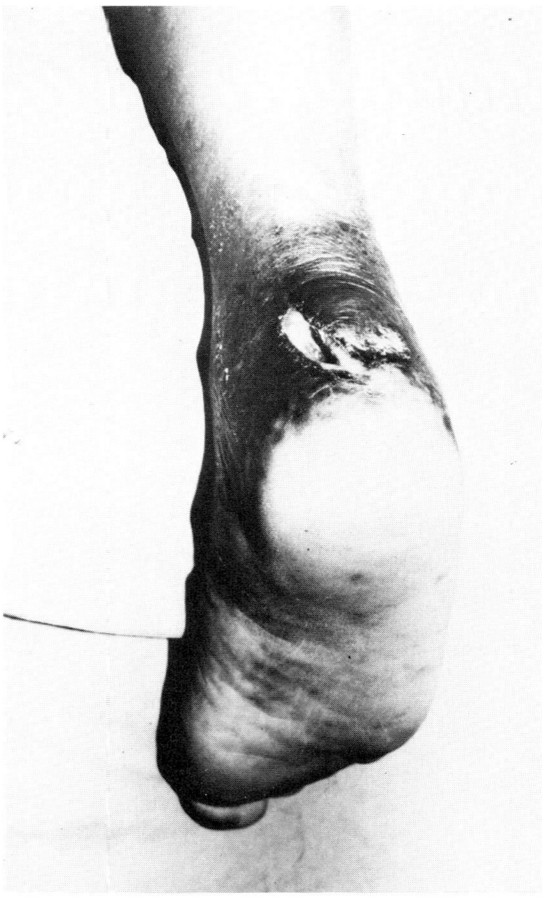

Fig. 33-7

A, Fifty-five-year-old patient with 2-year history of draining ulcer in posterior heel after calcaneal fracture.

(**A, C** to **G** from Scheflan, M., Nahai, F., and Hartrampf, C.R., Jr.: Ann. Plast. Surg. **7:**385, 1981.)

B

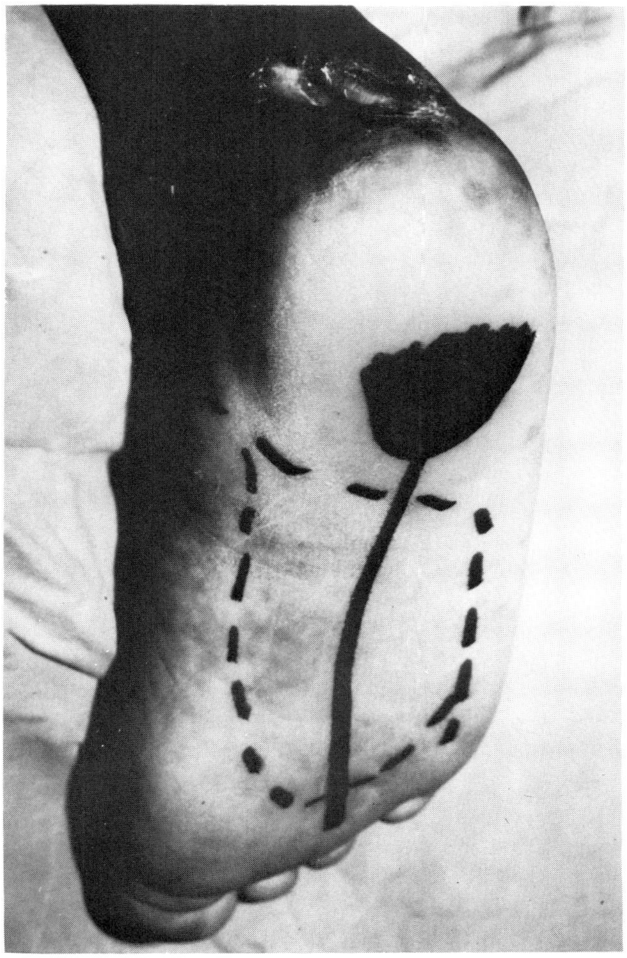

Fig. 33-7, cont'd

B, Midline plantar incision is demonstrated, extending from medial calcaneal tuberosity to metatarsal heads. Non-weight-bearing surface is shown by broken line.

C

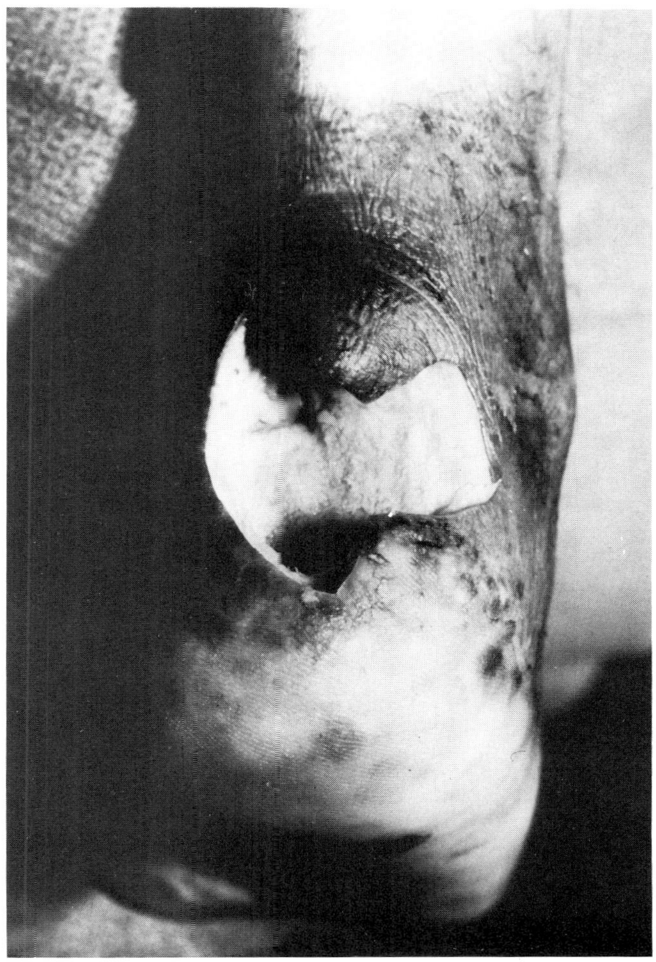

C, Posterior heel ulcer after surgical débridement that included outer table of posterior calcaneus.

Continued.

D

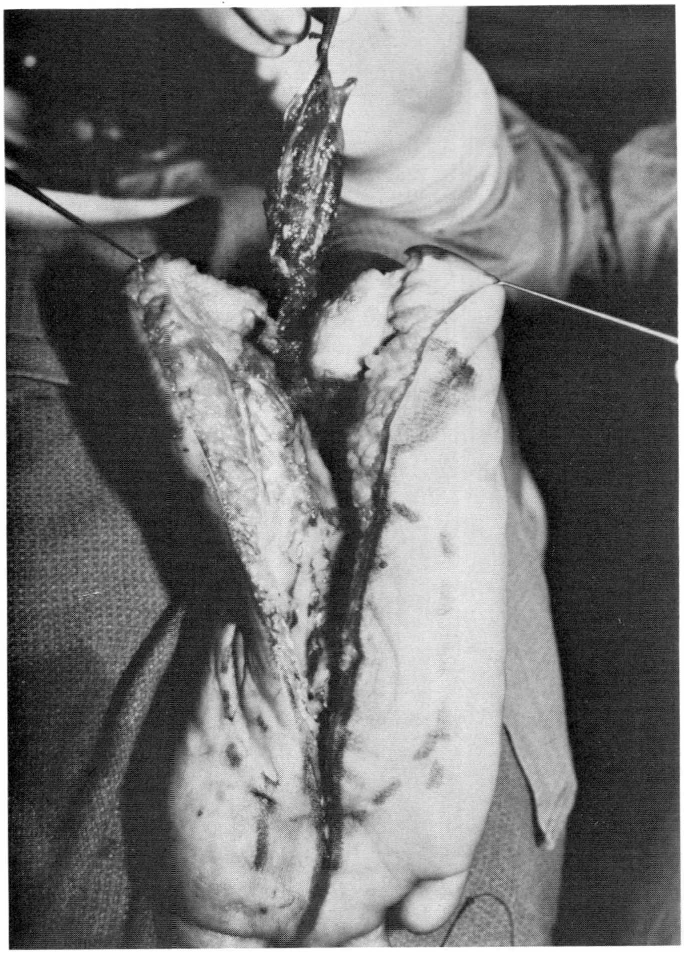

Fig. 33-7, cont'd

D, After division of tendons of insertion and distal lateral plantar artery, island muscle flap isolated on its vascular pedicle. The lateral plantar artery is dissected posteriorly back to bifurcation of posterior tibial artery through heel-splitting incision.

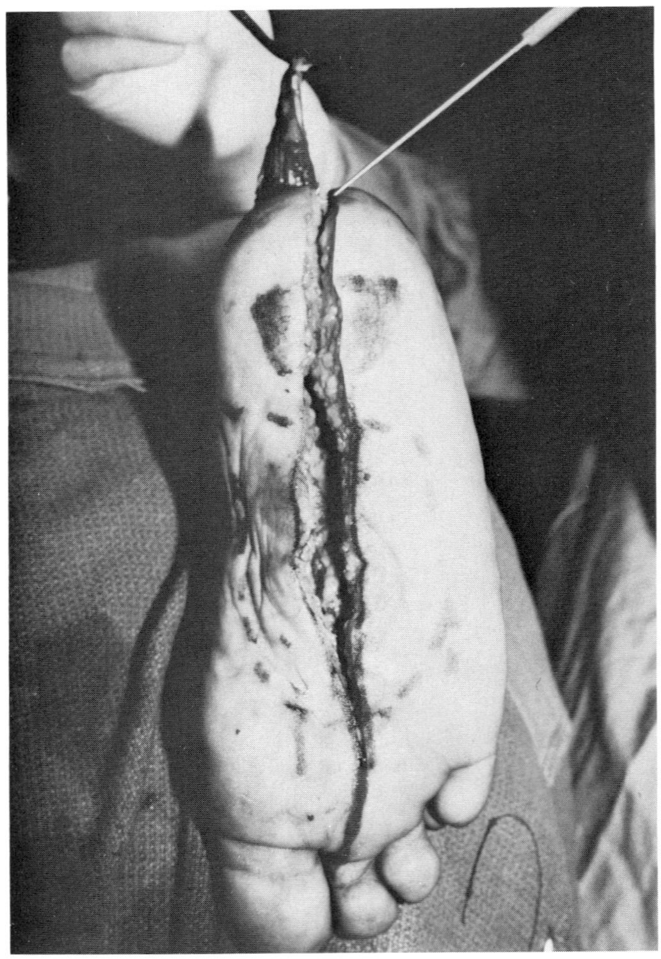

E

E, Muscle is placed in the posterior heel cavity, and plantar incision is approximated.

F

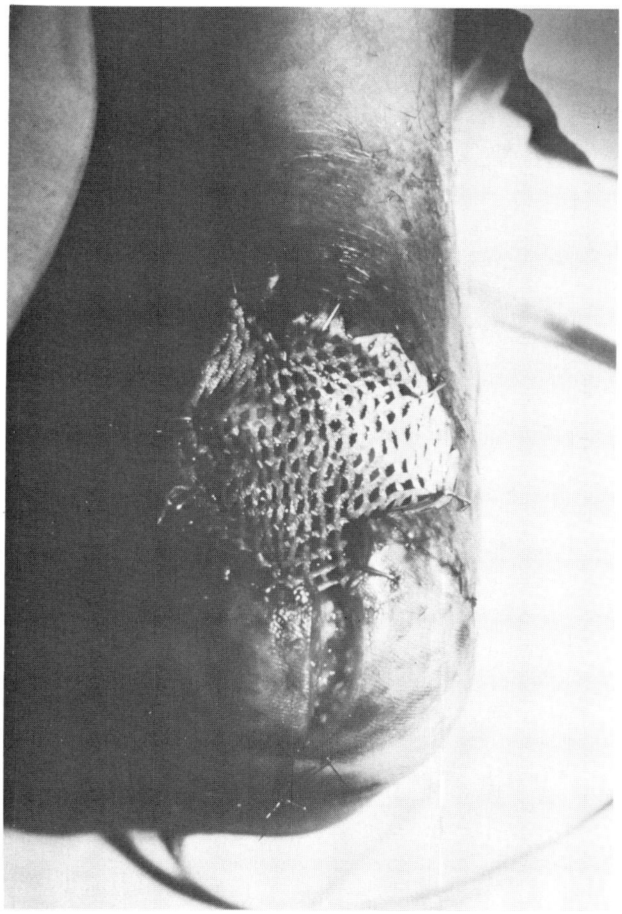

Fig. 33-7, cont'd
F, A mesh split-thickness skin graft is placed on muscle.

G

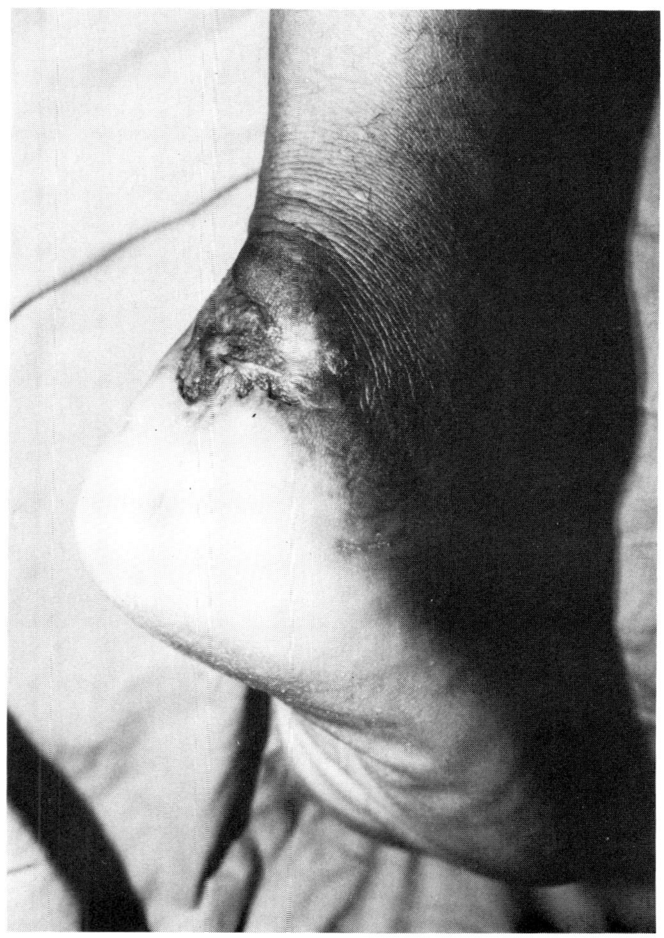

G, Two-year postoperative view demonstrates wound is healed. Plantar scar on the weight-bearing surface has caused patient no difficulties.

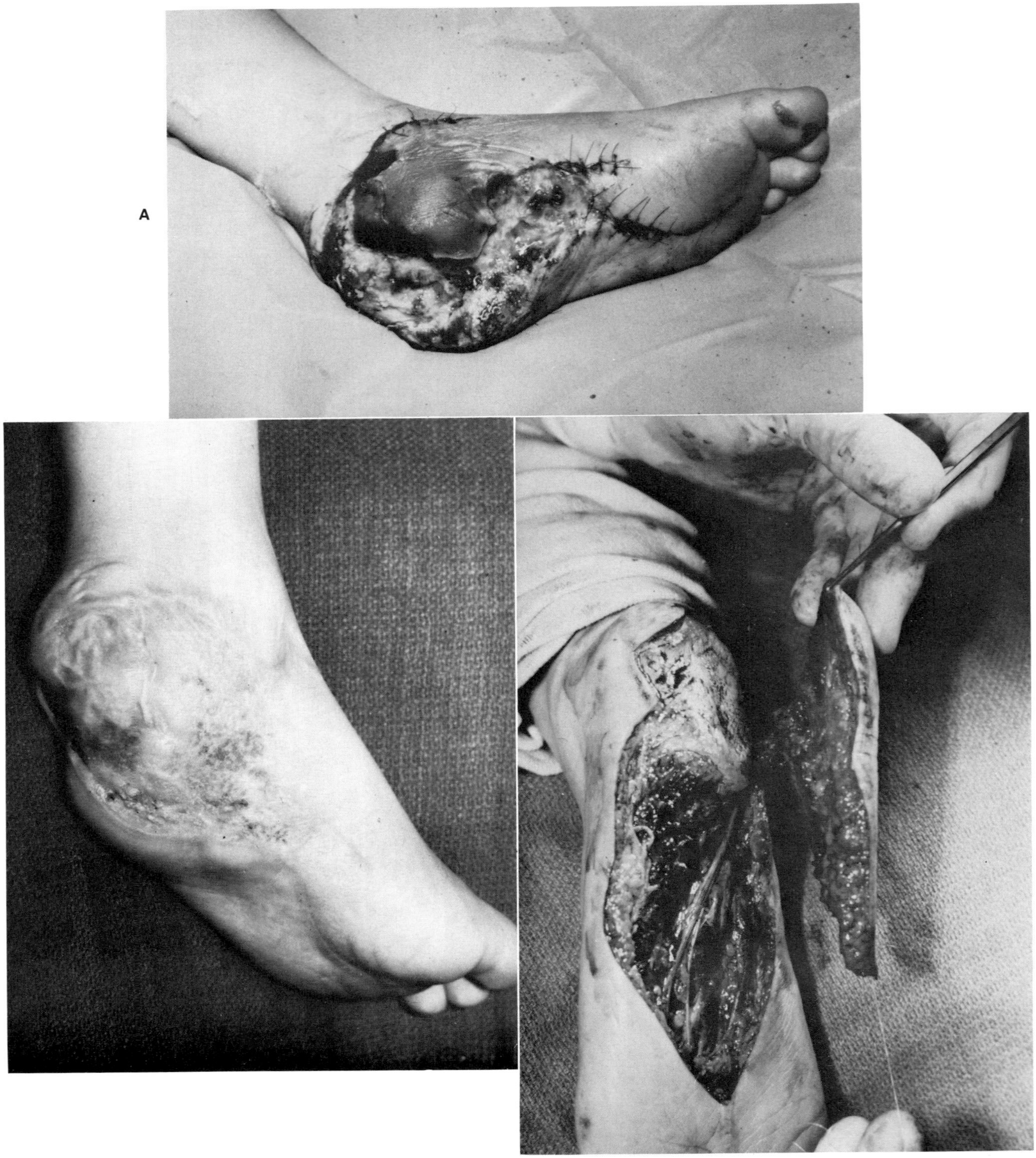

Fig. 33-8

A, Fourteen-year-old boy several days after lawn mower injury that amputated entire soft tissue pad of weight-bearing heel.

B, Same patient 6 months later after application of thick split-thickness skin graft. Patient complains of tight, sensitive scars and inability to walk.

C, Design and elevation of flexor digitorum brevis musculocutaneous sensory flap.

(**A** and **B** from Scheflan, M., Nahai, F., and Hartrampf, C.R., Jr.: Ann. Plast. Surg. **7**:385, 1981.)

D

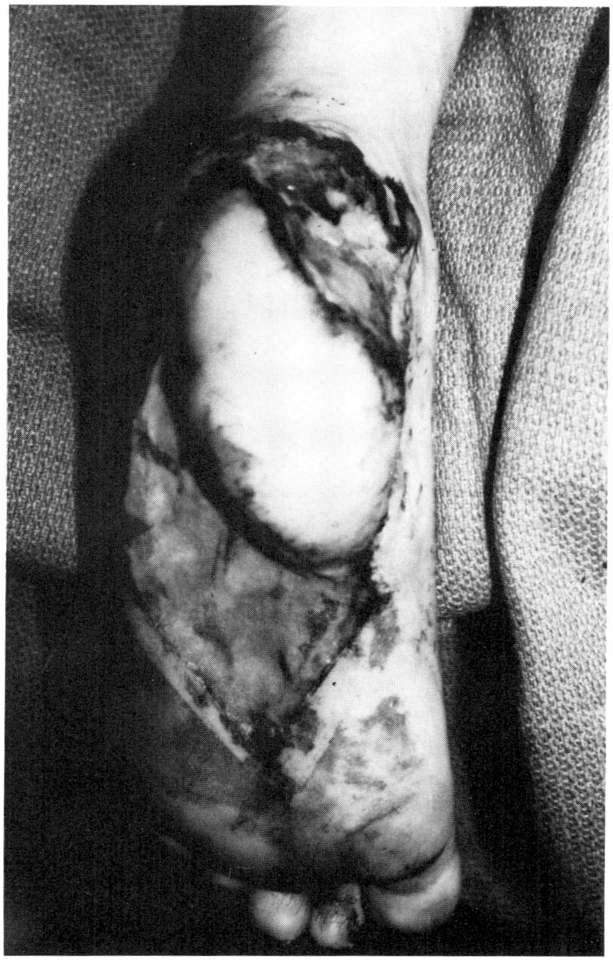

Fig. 33-8, cont'd
D, Flexor digitorum brevis musculocutaneous unit is transposed posteriorly, and donor site is grafted.

E

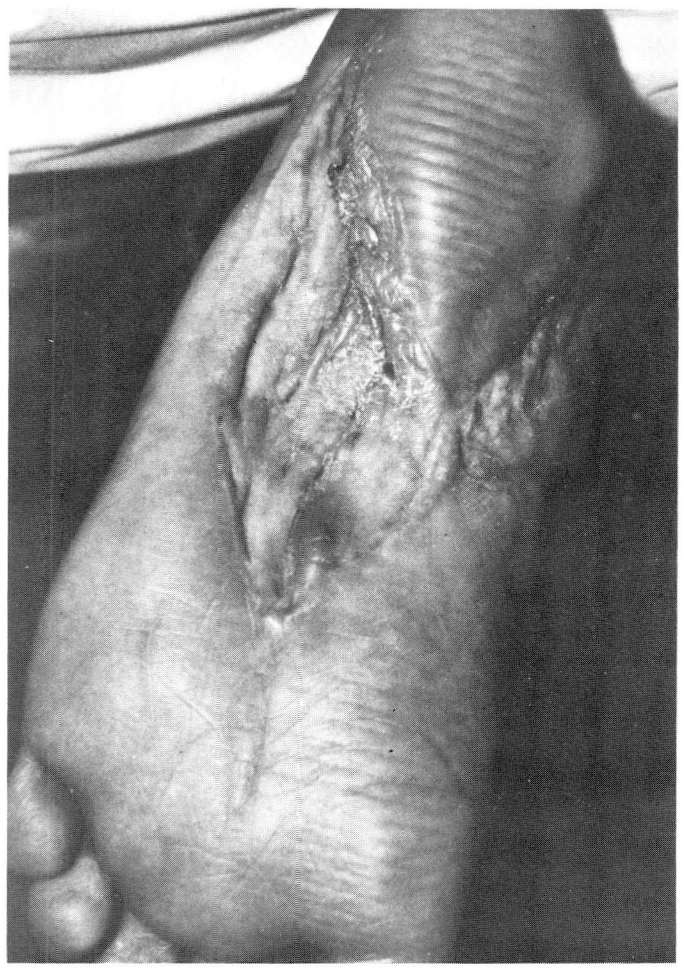

E, Same patient 2 years later with durable sensitive flexor digitorum brevis musculocutaneous unit on which he bears weight.

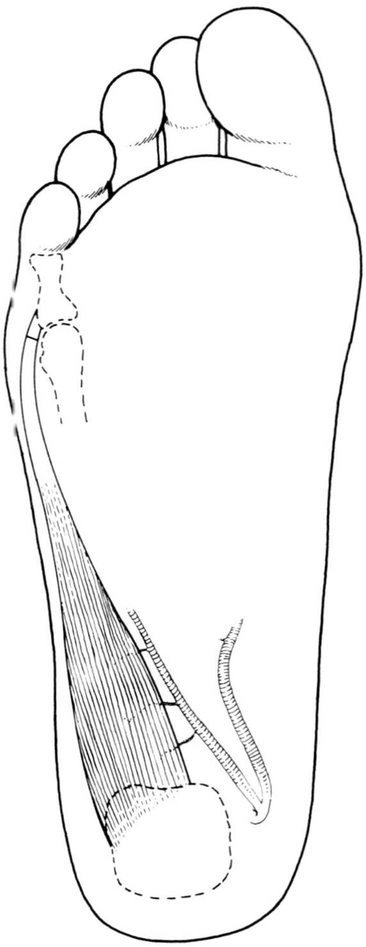

Fig. 33-9. Abductor digiti minimi muscle pads lateral aspect of plantar foot and receives its blood supply from lateral plantar artery.

(From Scheflan, M., Nahai, F., and Hartrampf, C.R., Jr.: Ann. Plast. Surg. **7**:385, 1981.)

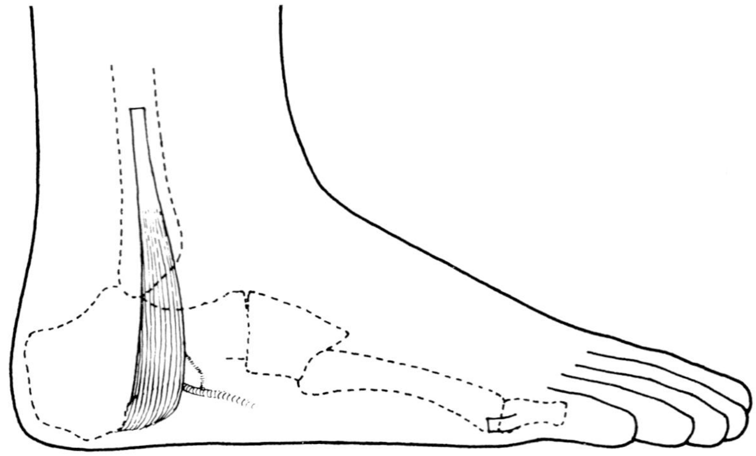

Fig. 33-10. Arc of rotation of abductor digiti minimi muscle allows it to cover soft tissue defects in lateral proximal foot, posterior heel, and areas below lateral malleolus.

(From Scheflan, M., Nahai, F., and Hartrampf, C.R., Jr.: Ann. Plast. Surg. **7**:385, 1981.)

ABDUCTOR DIGITI MINIMI MUSCLE

Anatomy

The abductor digiti minimi muscle is located on the lateral plantar aspect of the foot. It originates on the lateral and medial processes of the calcaneal tuberosity and inserts into the lateral aspect of the base of the proximal phalanx of the fifth toe (Fig. 33-9).

Both motor innervation and blood supply are derived from the lateral plantar neurovascular bundle. The overlying skin is innervated through branches of the sural nerve. The lateral plantar artery runs in a groove between the abductor digiti minimi and the flexor digitorum brevis and gives off two small proximal pedicles to the former.

Flap elevation

A lateral foot incision is used on the non-weight-bearing skin of the foot. The incision is carried down to the muscle, and the skin flaps are elevated. The muscle is detached from its attachments to the fifth metatarsal bone, and the tendinous insertion is cut. The muscle is then reflected over on itself for proximal reach. Small distal pedicles to the muscle are divided. Without further dissection this unit would reach to just below the lateral malleolus (Fig. 33-10). Division of the lateral plantar artery distal to the pedicle to this muscle will provide a greater arc of rotation.

Local arterialized heel pad flaps

Recent recognition of the microcirculation to the heel enables the surgeon to use whatever is left of a damaged heel for the reconstruction. The blood supply of the heel pad itself is based on small segmental branches that originate from the lateral plantar artery. These anastomose inside the thick heel pad with small arterial branches of the posterior tibial and peroneal arteries. In small to moderate defects subfascial heel pad flaps have been used to reconstruct soft tissue defects of the weight-bearing portion of the heel. These are local, rotational flaps based on the segmental arteries from the lateral plantar vessels. The donor defect is closed by local tissue advancement (Fig. 33-11).

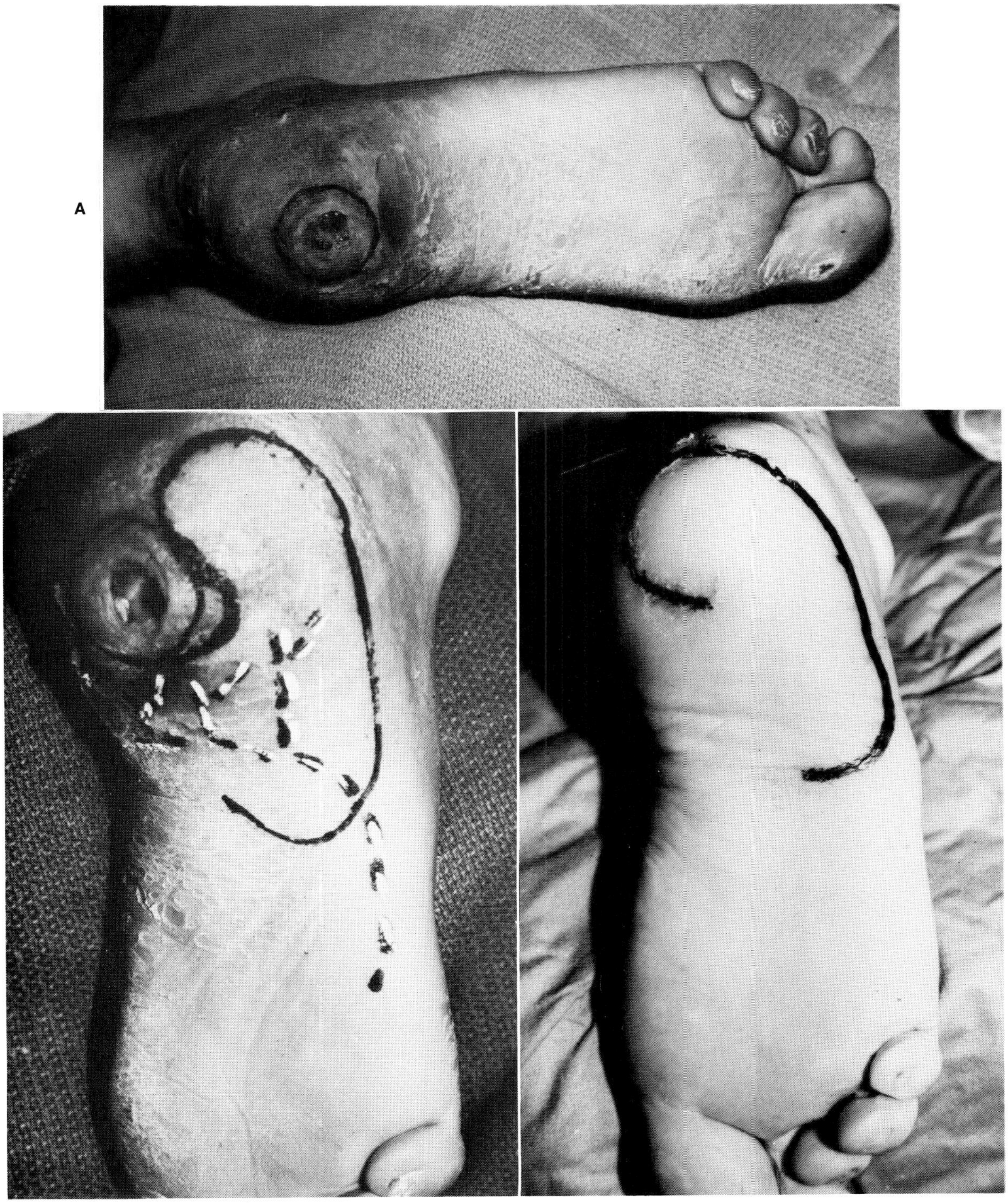

A

B

C

Fig. 33-11

A, Thirty-five-year-old man with low meningocele and altered sensibility in left foot with long-standing, deep ulcer of plantar surface of heel.

B, Proposed local, arterialized, subfascial skin flap. Blood supply is based on segmental perforators from lateral plantar artery.

C, Patient 3 months after operation has healed wound and no walking difficulties, although sensory status has not been altered.

(From Scheflan, M., Nahai, F., and Hartrampf, C.R., Jr.: Ann. Plast. Surg **7**:385, 1981.)

COMPLICATIONS

Muscle flap necrosis

Inaccurate or blunt dissection may lead to avulsion of any of these tiny proximal pedicles and cause muscle flap necrosis. Meticulous dissection, using loop magnification in a dry field under tourniquet control, should effectively prevent this problem.

Vascular ischemia of the foot

If distal division of either the lateral or medial plantar pedicle is undertaken in a patient with peripheral vascular compromise, it may lead to foot claudication or gangrene. Careful preoperative clinical assessment aided with Doppler flow studies or microangiography is indicated in doubtful cases.

Unstable and painful scars

Skin grafts and scars should be placed away from weight-bearing surfaces whenever possible. Interestingly, heel-splitting incisions have caused no difficulties. A painful scar on a weight-bearing area may alter gait characteristics and should be avoided. Hammer toe has not been a problem in any of our patients.

CONCLUSION

Soft tissue and bony ailments of the foot are associated with pain and disability. An average period of 17 months associated with disability has been recently reported. Clear definitions of dependable muscle and musculocutaneous territories along with increased knowledge of the microcirculation of the foot have all provided techniques for reconstruction of the foot.

Transposition of small muscles in the foot may be the only alternative to microvascular tissue transplantation or cross-leg flaps. In some patients the large sizes of the defects exclude the possibility of small local muscle flaps. In others the very nature of the disease process (e.g., diabetes mellitus and pe-

TABLE 6 Classification of heel ulcers

Location	Size	Therapy
Non-weight-bearing surfaces		
Posterior heel	Superficial	Skin graft
	Deep (down to bone)	Skin graft
	Very deep (bone involved)	Extended flexor digitorum brevis muscle or abductor hallucis muscle flap
	Very large	Microvascular tissue transplant
Foot instep		Skin graft
Weight-bearing surfaces		
Plantar heel	Superficial	Skin graft
	Deep (down to bone; up to 1/2 the circumference or width of the heel pad)	Local arterialized skin flaps
	Very deep (bone involved)	Flexor digitorum brevis or abductor hallucis muscle flap and skin graft or local arterialized skin flap for coverage
	Very large	Flexor digitorum brevis musculocutaneous flap* or microvascular tissue transplant
Metatarsal heads pad	Small	"Reverse flow" muscle flap or local arterialized skin flap†
	Large	Transmetatarsal amputation

*Transfer of either the extended flexor digitorum brevis muscle flap or the flexor digitorum brevis musculocutaneous flap requires the distal division of the lateral plantar artery. Both the dorsalis pedis and tibialis posterior arteries must therefore be intact.

†Transposition of the "reverse flow" flexor digitorum brevis and abductor hallucis muscles and musculocutaneous flaps requires proximal interruption of the lateral and medial plantar arteries respectively. Both the dorsalis pedis and tibialis posterior arteries must therefore be intact. These flaps survive on their minor distal pedicles and the perforating vessels between the dorsalis pedis and lateral plantar arcade.

ripheral vascular disease) precludes this mode of surgical reconstruction.

Although the application of skin grafts to weight-bearing surfaces is feasible, it often severely alters gait to produce new weight-bearing areas and awkward limping. Skin grafts and local transposition flaps are usually adequate on the weight-bearing aspect of the foot. Coverage of weight-bearing defects requires durable skin, thick padding, and protective sensibility. The best skin to reconstruct the sole of the foot is skin from the sole of the foot. Thus use of these techniques, using small muscle and skin flaps from the plantar aspect of the foot, provides coverage and avoids postoperative foot trauma and gait deformity (Table 6).

ANNOTATED BIBLIOGRAPHY

1. Bostwick, J., III: Reconstruction of the heel pad by muscle transposition and split-skin graft, Surg. Gynecol. Obstet. **143:**972, 1976.

   This article presents a description of the use of the flexor digitorum brevis muscle for heel coverage.

2. Ger, R.: The management of chronic ulcers of the dorsum of the foot by muscle transposition and free skin grafting, Br. J. Plast. Surg. **29:**199, 1976.

   The coverage of ulcers or scar tissue of the dorsum of the foot is often difficult to accomplish with use of skin grafts. The use of the abductor digiti minimi and abductor hallucis as transposition flaps is presented as a technique to provide such coverage. However, these muscles are small with a limited arc of rotation and will only cover defects located close to the medial or lateral aspect of the dorsum of the foot.

3. Ger, R.: The surgical management of ulcers of the heel, Surg. Gynecol. Obstet. **140:**909, 1975.

   The author describes the surgical anatomy of the abductor hallucis and abductor digiti minimi muscles. The use of both muscles for coverage of resistant ulcers in the posterior heel is described. Successful operations in six instances are noted.

4. Hartrampf, C.R., Jr., Scheflan, M., and Bostwick, J., III: The flexor digitorum brevis muscle island pedicle flap: a new dimension in heel reconstruction, Plast. Reconstr. Surg. **66:**264, 1980.

   The technique for elevation of the flexor digitorum brevis muscle, its overlying skin, sensory nerve, and vascular pedicle is presented for transposition to the heel for both coverage and sensation.

5. Mathes, S.J., McCraw, J.B., and Vasconez, L.O.: Muscle transposition flaps for coverage of lower extremity defects: anatomic consideration, Surg. Clin. North Am. **54:**1337, 1974.

   A description of the vascular anatomy of intrinsic foot muscles is presented. The arc of rotation of these muscles is demonstrated for coverage of heel and ankle.

6. Reiffel, R.S., and McCarthy, J.G.: Coverage of heel and sole defects: a new subfascial arterialized flap, Plast. Reconstr. Surg. **66**(2):250, 1980.

   The authors describe a subfascial arterialized flap of skin and subcutaneous tissues based on segmental perforators from the lateral plantar artery. This flap is superior for coverage of small to moderate soft tissue defects of the weight-bearing aspect of the heel. The secondary donor defect created by the transposition of this flap can be closed primarily by mobilizing the skin overlying the Achilles tendon.

7. Scheflan, M., Nahai, F., and Hartrampf, C.R., Jr.: Surgical management of heel ulcers—a comprehensive approach, Ann. Plast. Surg. **7:**385, 1981.

   A critical review of all the modalities available for reconstruction of the sole of the foot is presented. A classification of heel ulcers is offered based on their etiology, size, location, and status of the peripheral vascular tree.

8. Sommerlad, B.C., and McGrouther, D.A.: Resurfacing the sole: long-term follow-up and comparison of techniques, Br. J. Plast Surg. **31:**107, 1978.

   The authors present 51 patients observed for an average of 5.8 years after resurfacing of the sole of the foot. Regardless of the mode of reconstruction, whether skin graft, local flaps, cross-leg flaps, or free flaps, all patients alter their gait characteristics so that weight-bearing on all resurfaced areas is avoided whenever possible.

9. Woltering, E.A., and co-workers: Split-thickness skin grafting of the plantar surface of the foot after wide excision of neoplasms of the skin, Surg. Gynecol. Obstet. **149:**229, 1979.

   The authors present 13 patients observed from 10 months to 18 years after resurfacing of the foot with split-thickness skin grafts. Ten of the patients have the grafts placed directly over the periosteum on weight-bearing surfaces. All patients do well, and there is an insignificant incidence of graft breakdown. All patients, however, change their gait patterns to protect the graft.

34

COMPLICATIONS

Foad Nahai ▪ Stephen J. Mathes

The muscles available for local coverage are small, have limited arcs of rotation, and are subject to ischemia from any proximal vascular problem.

FLEXOR DIGITORUM BREVIS

Planning errors

Patency of the posterior tibial artery must be confirmed preoperatively. Angiograms may be indicated when injury is suspected. The flexor digitorum brevis will not reach the posterior heel unless the calcaneus is resected.

Technical errors

The tendons of the flexor digitorum brevis must be distinguished from the tendons of the flexor digitorum longus during flap elevation. The medial and lateral plantar nerves deep to the muscle belly must be preserved.

Postoperative errors

Foot elevation is required for a minimum of 2 weeks.

ABDUCTOR HALLUCIS

Planning errors

The abductor hallucis is a small muscle with a limited arc of rotation. Adequate blood flow through the posterior tibial and medial plantar arteries must be ensured preoperatively. The extensive muscle mobilization required for use of this muscle to reach the ankle may injure the lateral plantar artery. Flap tension must be avoided.

Technical errors

The sensory branch of the medial plantar nerve to the great toe is in close relationship to the tendon of the abductor hallucis muscle and should be preserved.

Postoperative errors

Suction drainage should be established and the patient positioned with the leg elevated to facilitate venous drainage.

ABDUCTOR DIGITI MINIMI

Planning errors

The abductor digiti minimi muscle is very small and has a limited arc of rotation. Patency of the posterior tibial and lateral plantar arteries must be confirmed preoperatively. The musculocutaneous flap will leave an unacceptable donor defect.

Technical errors

The muscle is very thin in the region of the head of the fifth metatarsal, and the distal muscle circulation can be damaged during the dissection in this region. The lateral plantar artery, as it crosses the posterior aspect of the flexor digitorum brevis, is the point of rotation of this flap. Tension must be avoided.

Postoperative errors

Suction drainage should be established and the patient positioned with the leg elevated to facilitate venous drainage.

35

A SYSTEMATIC APPROACH TO FLAP SELECTION

Foad Nahai ■ Stephen J. Mathes

UPPER ARM

Skin grafts

Skin grafts are method of choice provided bone and vital structures are not exposed

Local flaps

1. Local flaps have limited indications in reconstruction of the upper arm
2. Biceps brachii has limited application for coverage of vascular access for dialysis (segmental transposition with function preservation)

Distant flaps

1. Distant flaps are method of choice if bone or vital structures are exposed
2. Latissimus dorsi flap is flap of choice for this area
3. Transverse thoracic flap is alternative flap for coverage of upper arm (Fig. 35-1)
4. Free tissue transplantation is rarely necessary in reconstruction of the upper arm

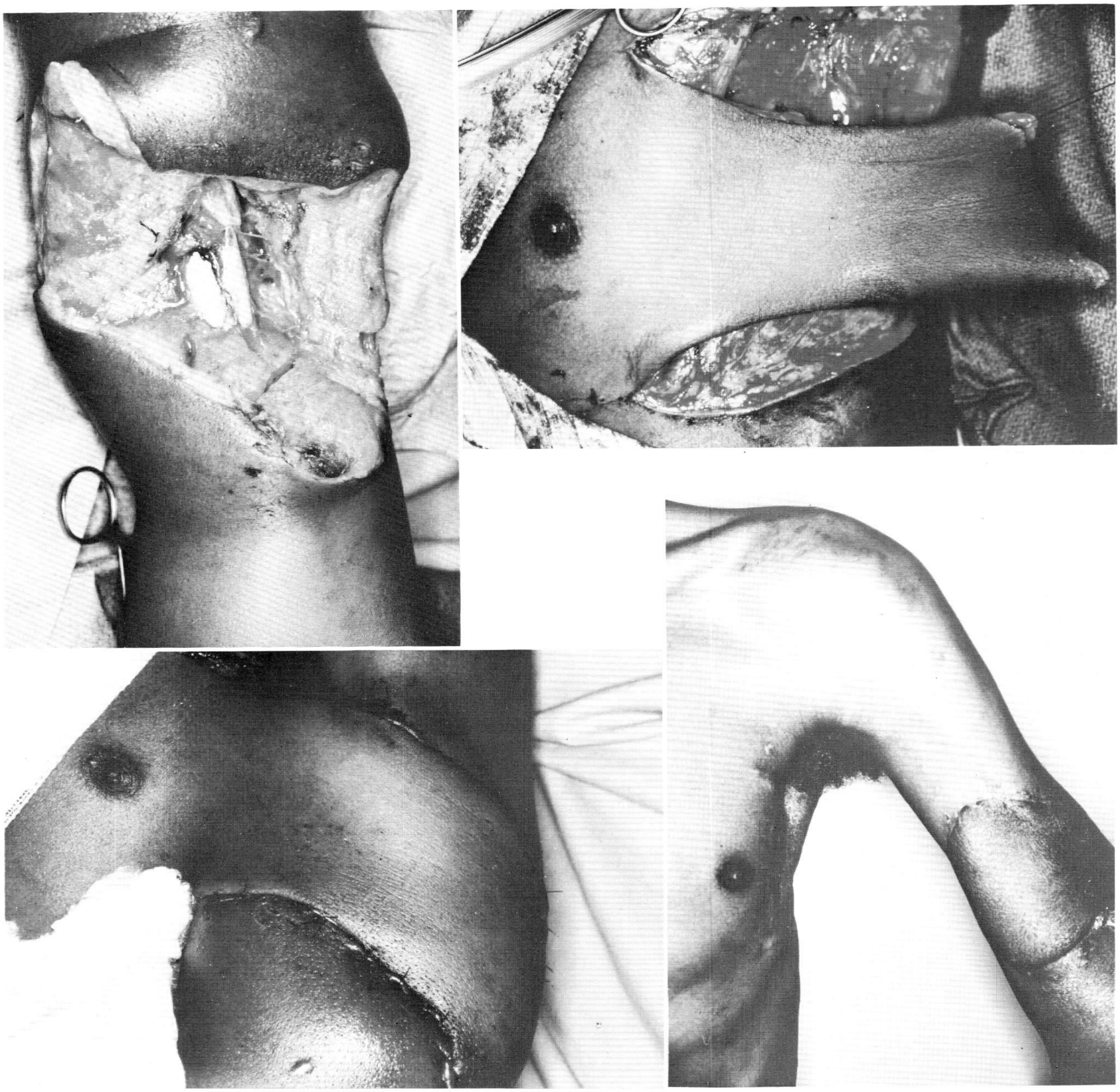

A

B

C

D

Fig. 35-1. Transverse thoracic flap for upper arm coverage.
A, Shotgun wound to upper arm requires coverage of exposed vein graft for arterial reconstruction and exposed humeral fracture.
B, Transverse thoracic flap based on musculocutaneous perforators from pectoralis major and rectus abdominis muscle.
C, Flap is inset over defect after wound débridement, and fracture stabilization with medullary rod is completed.
D, Flap provides stable coverage over defect in upper arm.

ELBOW

Skin grafts

Skin grafts are method of choice only if joint and vital structures are not exposed

Local flaps

1. Local skin transposition flaps based on perforators through flexor carpi ulnaris
2. Anconeus flap is excellent for coverage of the posterior elbow and is flap of choice in reconstruction of the elbow
3. Flexor carpi ulnaris will cover elbow and is useful for osteomyelitis; functional loss is not generally acceptable (Fig. 35-2)

Distant flaps

1. Distant flaps are method of choice if bone and joints or vital structures are exposed
2. Abdominal flap is flap of choice for coverage of elbow
3. Free tissue transplantation is alternative method for coverage but rarely is indicated in reconstruction of the elbow
4. Distally based latissimus dorsi flap (staged procedure) is rarely indicated in reconstruction of the elbow

FOREARM

Skin grafts

Skin grafts are method of choice if bone and vital structures are not exposed

Local flaps

All forearm muscles are potential flaps, but the functional deficit resulting from such transfer does not generally justify their use

Distant flaps

1. Distant flaps are method of choice if bones or vital structures are exposed
2. Transverse abdominal flap is flap of choice
3. Groin flap is alternative flap
4. Pectoralis major alternative flap for distal forearm
5. TFL and free tissue transplantation are rarely indicated in reconstruction of the forearm

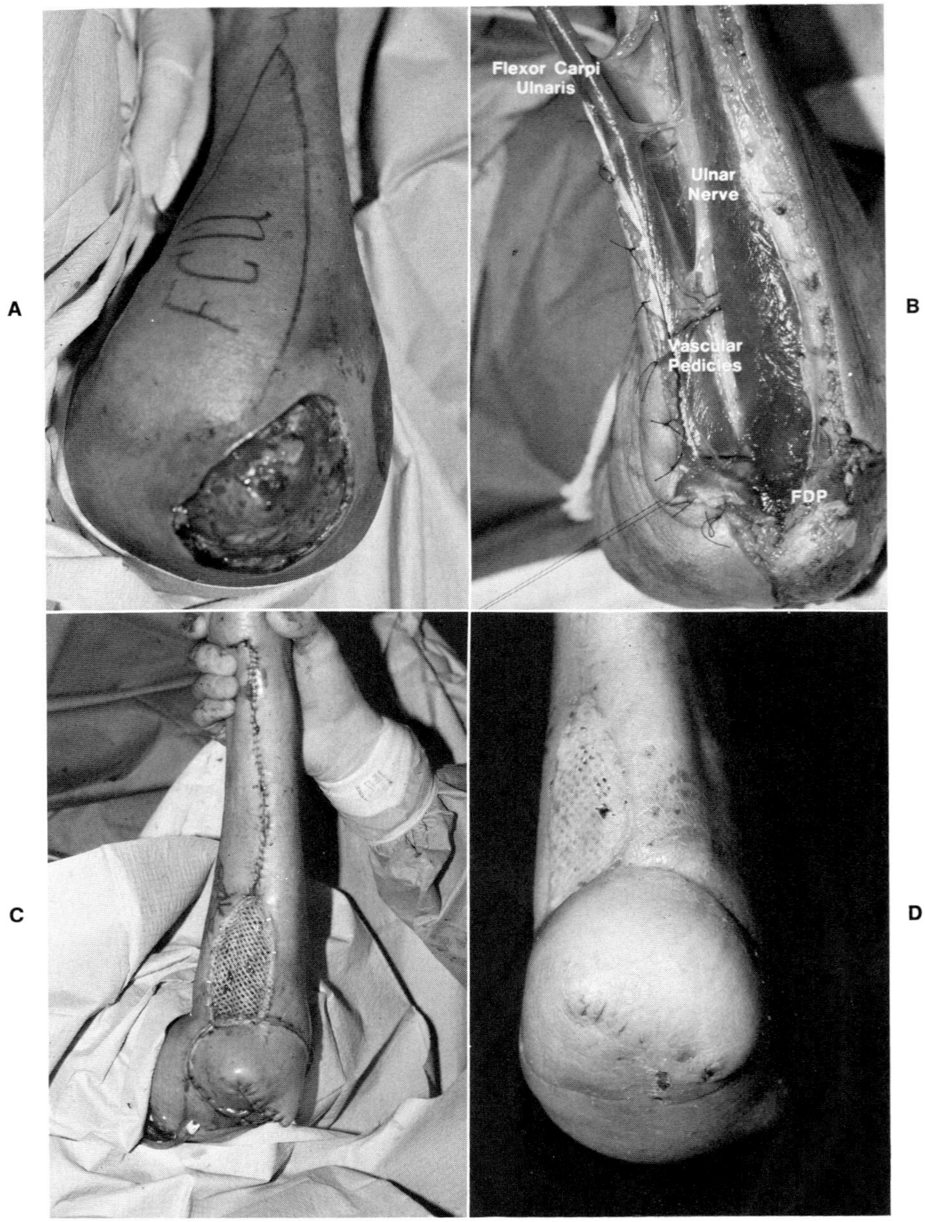

Fig. 35-2

A, Sequestrectomy required for chronic osteomyelitis of proximal ulna at elbow (design of flexor carpi ulnaris musculocutaneous flap).

B, Flap elevation demonstrates vascular pedicles to muscle.

C, Muscle placed within bone defect. Donor site closure requires skin grafts.

D, Stable wound coverage without evidence of recurrent infection provided by flexor carpi ulnaris musculocutaneous flap.

HAND

Skin graft

Skin grafts are method of choice if vital structures or bone are not exposed

Local flaps

1. Skin flaps are method of choice for reconstruction of the hand, especially in the palm and fingers
2. Muscle flaps are rarely indicated in reconstruction of the hand, but under certain circumstances functional loss is justified to obtain required coverage (Figs. 35-3 and 35-4)

Distant flaps

1. Distant tissue transfer is method of choice if local flap is not available, or the defect is too large
2. Groin flap is flap of choice (Fig. 35-5)
3. Abdominal flap is alternative flap
4. Free tissue transplantation is alternative flap but may well be the flap of choice if innervated (Fig. 35-6)

A

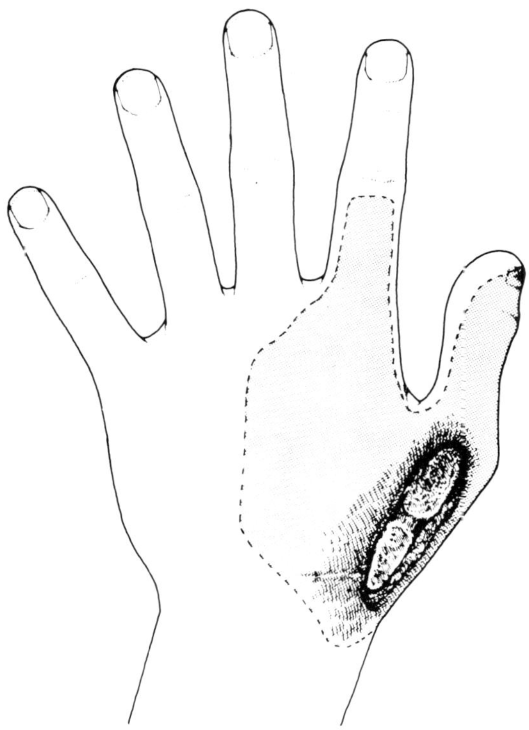

Fig. 35-3. First dorsal interosseous muscle for hand coverage.
A, Chronic osteomyelitis in thumb metacarpal bone. Patient recovering from 80% body surface burn with lack of donor flap sites. *Continued.*

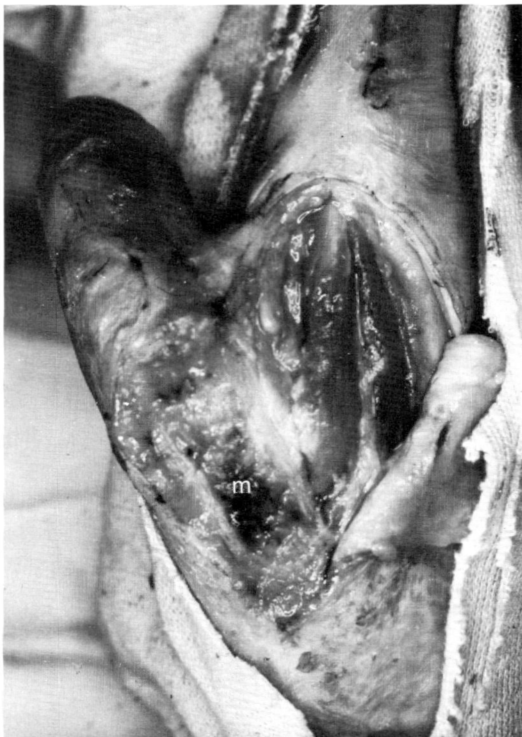

B

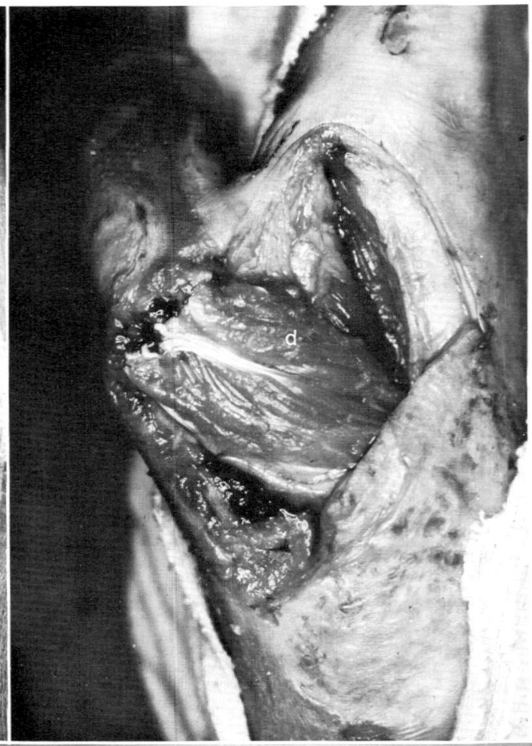

C

D

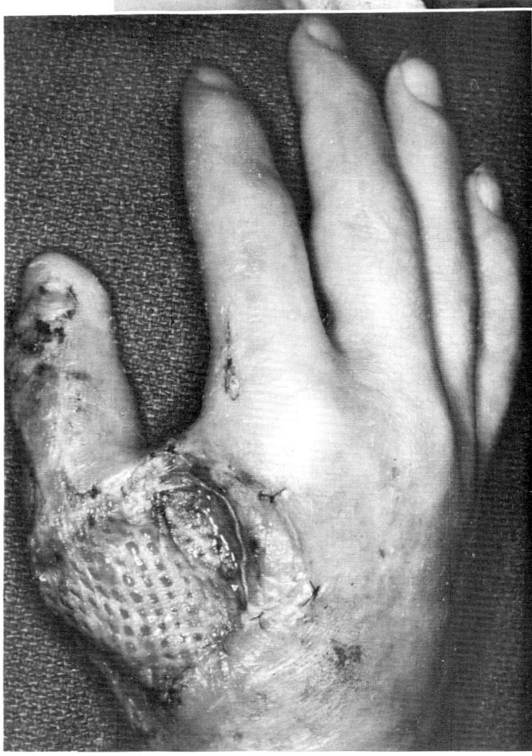

E

Fig. 35-3, cont'd

B, Wound débridement and sequestrectomy of thumb metacarpal (*m*). First dorsal interosseous muscle is available for transposition.

C, First dorsal interosseous muscle (*d*) transposed over bone defect.

D, Skin grafts over muscle and donor site.

E, Two-year postoperative view demonstrates stable coverage with no recurrence of bone infection after muscle coverage.

(From Mathes, S. J., Vasconez, L. O., and Jurkiewicz, M. J.: Extensions and further applications of muscle flap transposition, Plast. Reconstr. Surg. **60:**6, 1977.)

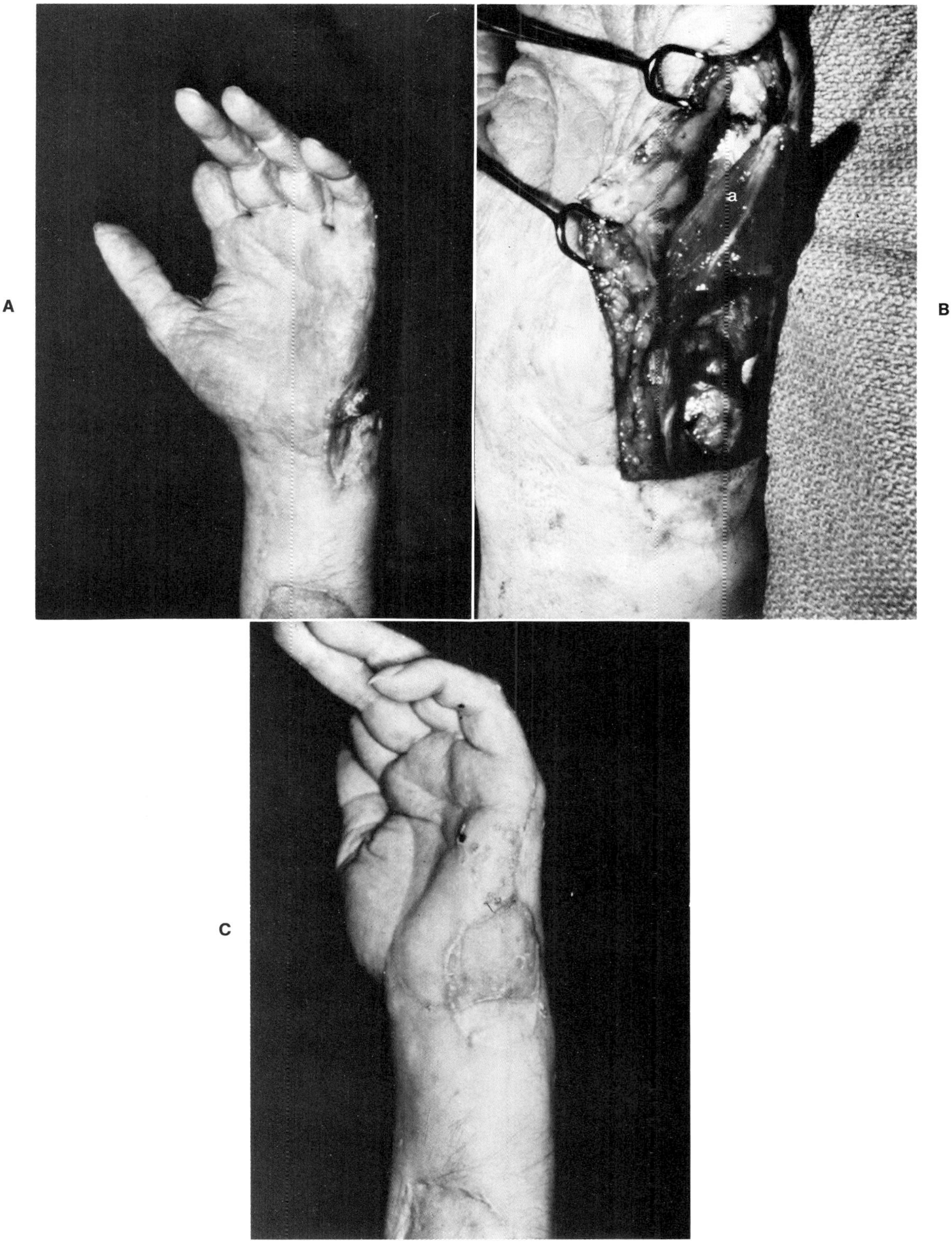

Fig. 35-4. Abductor digiti minimi muscle from hand coverage.
A, Recurrent squamous cell carcinoma at site of prior radiation injury.
B, Elevation of abductor digiti minimi (*a*) for coverage of exposed wrist joint.
C, One-year postoperative view demonstrates stable coverage at site of muscle transposition and skin graft coverage.

(From Mathes, S. J., Vasconez, L. O., and Jurkiewicz, M. J.: Extensions and further applications of muscle flap transposition, Plast. Reconstr. Surg. **60:**6, 1977.)

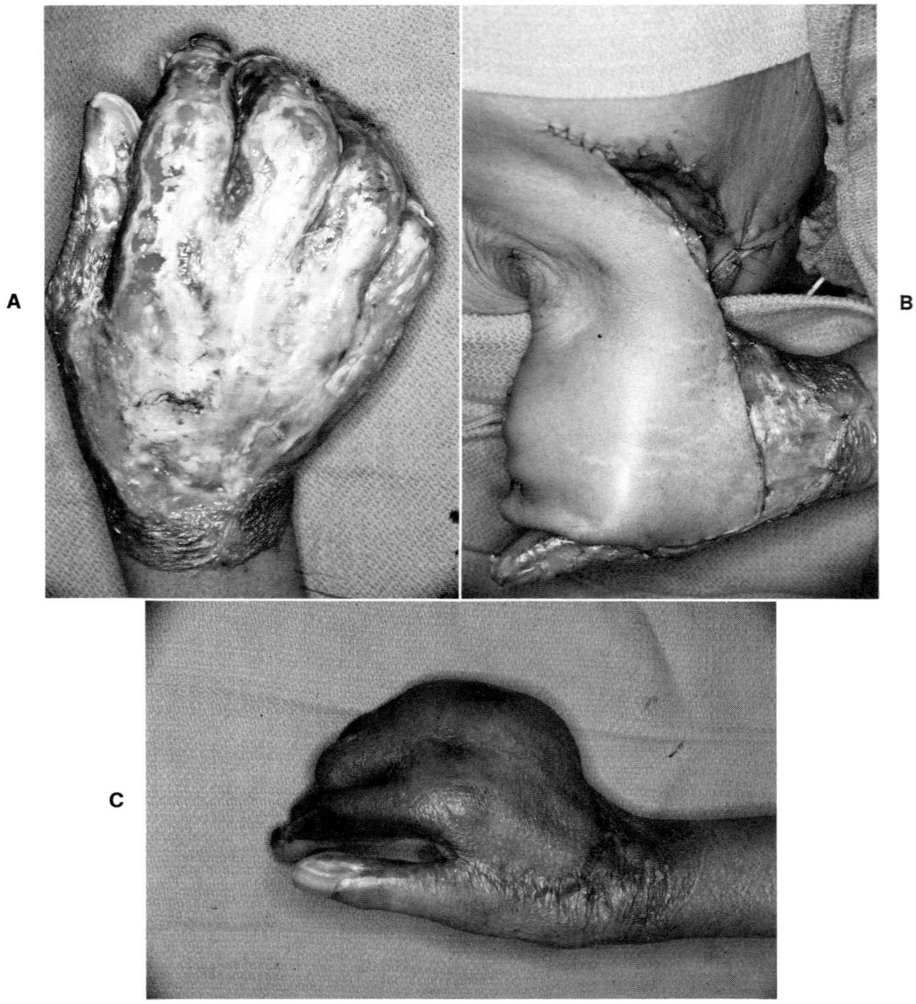

Fig. 35-5
A, Full-thickness burn of dorsum of hand resulting from steam injury.
B, Groin flap provides coverage of exposed extensor tendons.
C, Flap divided and inset in 2 weeks. Stable coverage provided by groin flap.

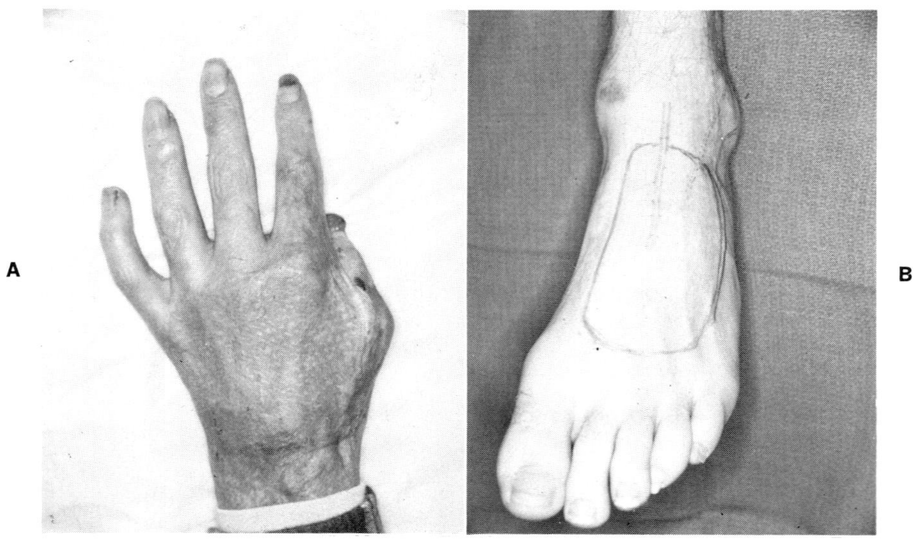

Fig. 35-6
A, Full-thickness burn to hand with loss of stable skin coverage to radial hand and absence of thumb-index finger web space and long extensor to thumb.
B, Design of dorsalis pedis flap.

C

E

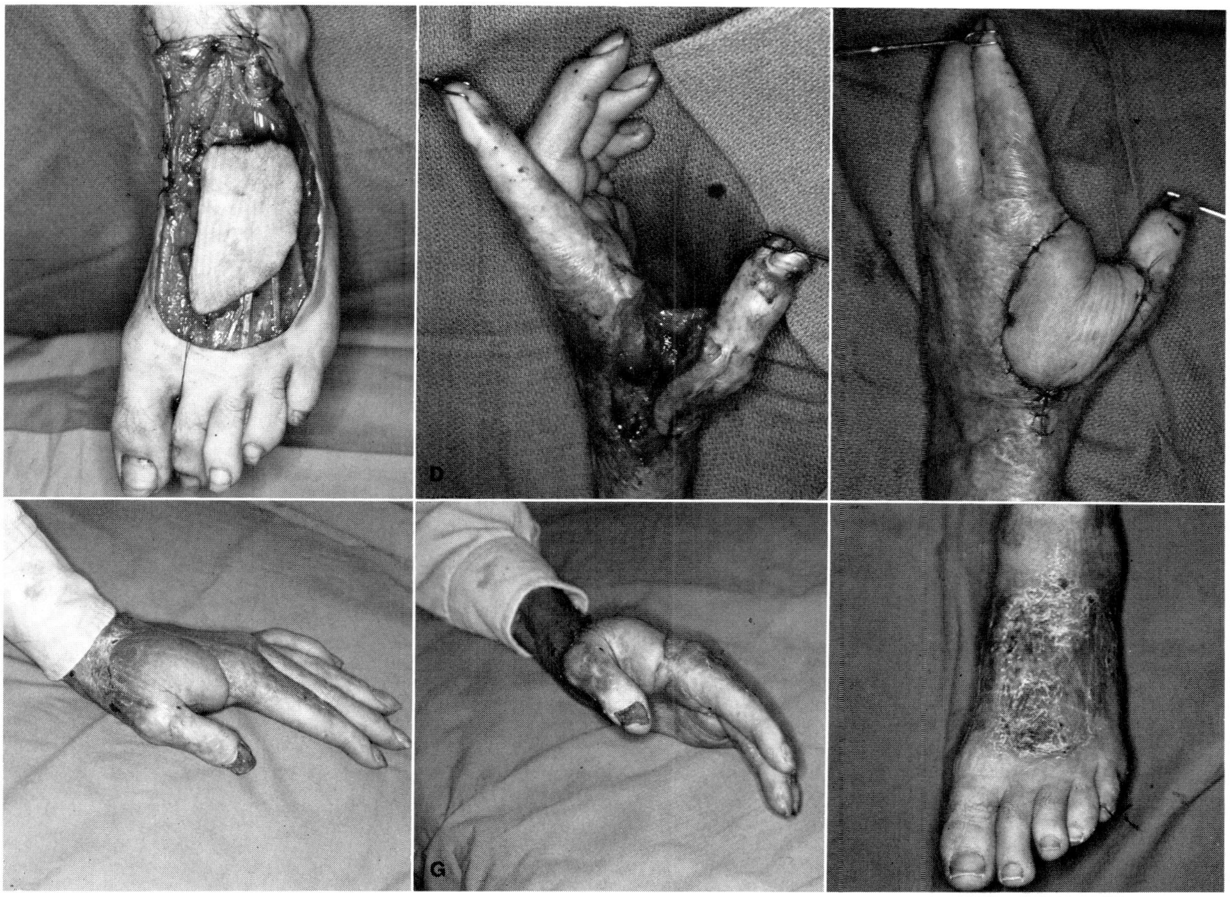

F

H

Fig. 35-6, cont'd

C, Flap includes extensor tendon to second toe for reconstruction of thumb extensor and superficial peroneal nerve for necessary innervation.

D, Release of thumb-index finger web space at flap recipient site.

E, Dorsalis pedis flap vascularized by end-to-end anastomosis between the dorsalis pedis artery and associated venae comitantes with the princeps pollicis artery and dorsal veins.

F and **G,** Free flap provides stable wound coverage 1 year after transplantation with sensation provided by sensory nerve repair and vascularized extensor tendon graft for thumb.

H, Stable coverage of donor defect provided by skin grafts.

(**A** to **H** from Mathes, S.J.: The role of microsurgery in acute hand injuries. In Weeks, P.M., and Wray, R.C.: Management of acute hand injuries: a biological approach, ed. 2, St. Louis, 1978, The C.V. Mosby Co., p. 164.)

36

RECONSTRUCTION

Foad Nahai ■ Stephen J. Mathes

Muscle and musculocutaneous flaps have had a dramatic impact on reconstructive surgery of the lower extremity. The flaps are widely used as the method of choice in reconstruction of the lower extremity. However, muscle and musculocutaneous flaps have only a limited application in reconstructive surgery of the upper extremity and are rarely the method of choice. Two major factors account for this difference between the upper and lower extremity: the relative nonexpendability of the muscles of the upper extremity and the proximity of the upper extremity to the trunk. Therefore distant flaps from the trunk or free tissue transfer is selected in preference to local muscle flaps. Under certain circumstances where free flap transfer is not possible, and a truncal flap is not available, then local muscle flap transposition is justified and indicated.

It is with this realization that this brief chapter on the application of muscle flaps to the upper extremity is included. Only the most useful of the flaps are described, and the functional deficit from the use of these flaps is emphasized.

The most useful muscle and musculocutaneous flaps for coverage of the upper extremity are on the trunk and are not local muscles within the extremity. The trunk muscles include the latissimus dorsi, rectus abdominis, and pectoralis major.

LATISSIMUS DORSI

The latissimus dorsi is the single most useful musculocutaneous flap and the distant flap of choice for coverage of the upper arm. It will easily reach and cover the anterior or posterior aspect of the arm (Fig. 36-1) from the shoulder to the elbow (Fig. 36-2). The functional deficit from this transfer is minimal, and in fact the latissimus dorsi is an ideal unit for functional transfer to restore elbow flexion or extension.

Flap elevation

The flap is based on the thoracodorsal pedicle for coverage of the arm. The distally based flap may be used to cover the posterior aspect of the elbow as a staged procedure. This application of the latissimus dorsi would rarely be indicated.

The flap is elevated from the distal border to the proximal border, as for breast reconstruction, but the dissection is continued further proximally and the muscle dissected free within the axilla. This proximal dissection is essential so that the flap can be transposed onto the arm. An adequate anterior or posterior tunnel is then made in the axilla and the flap pulled through onto the arm. It will easily reach the elbow.

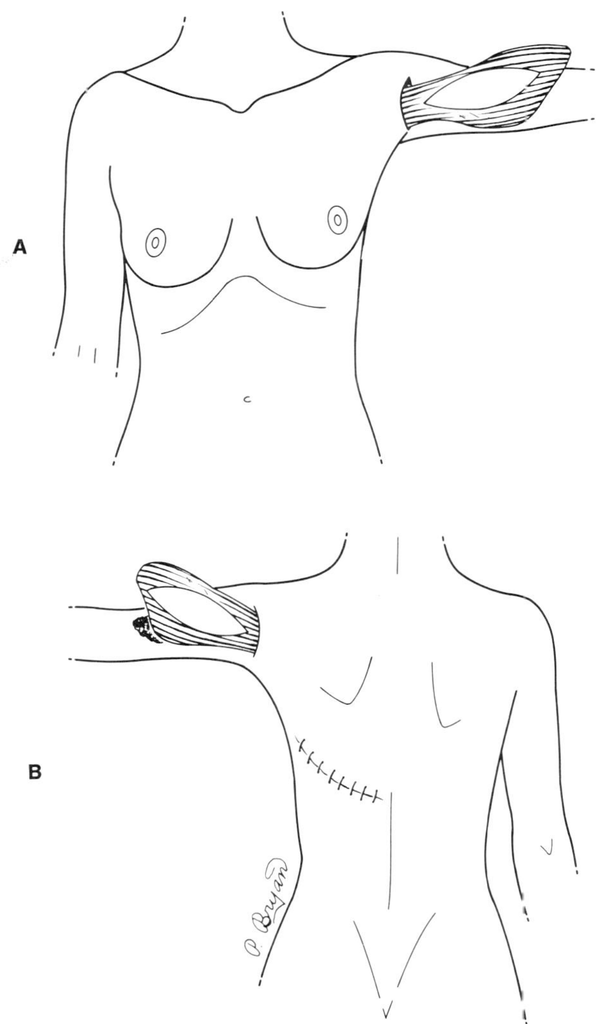

Fig. 36-1. Coverage of arm with latissimus dorsi flap.
A, Anterior aspect of arm.
B, Posterior aspect of arm.

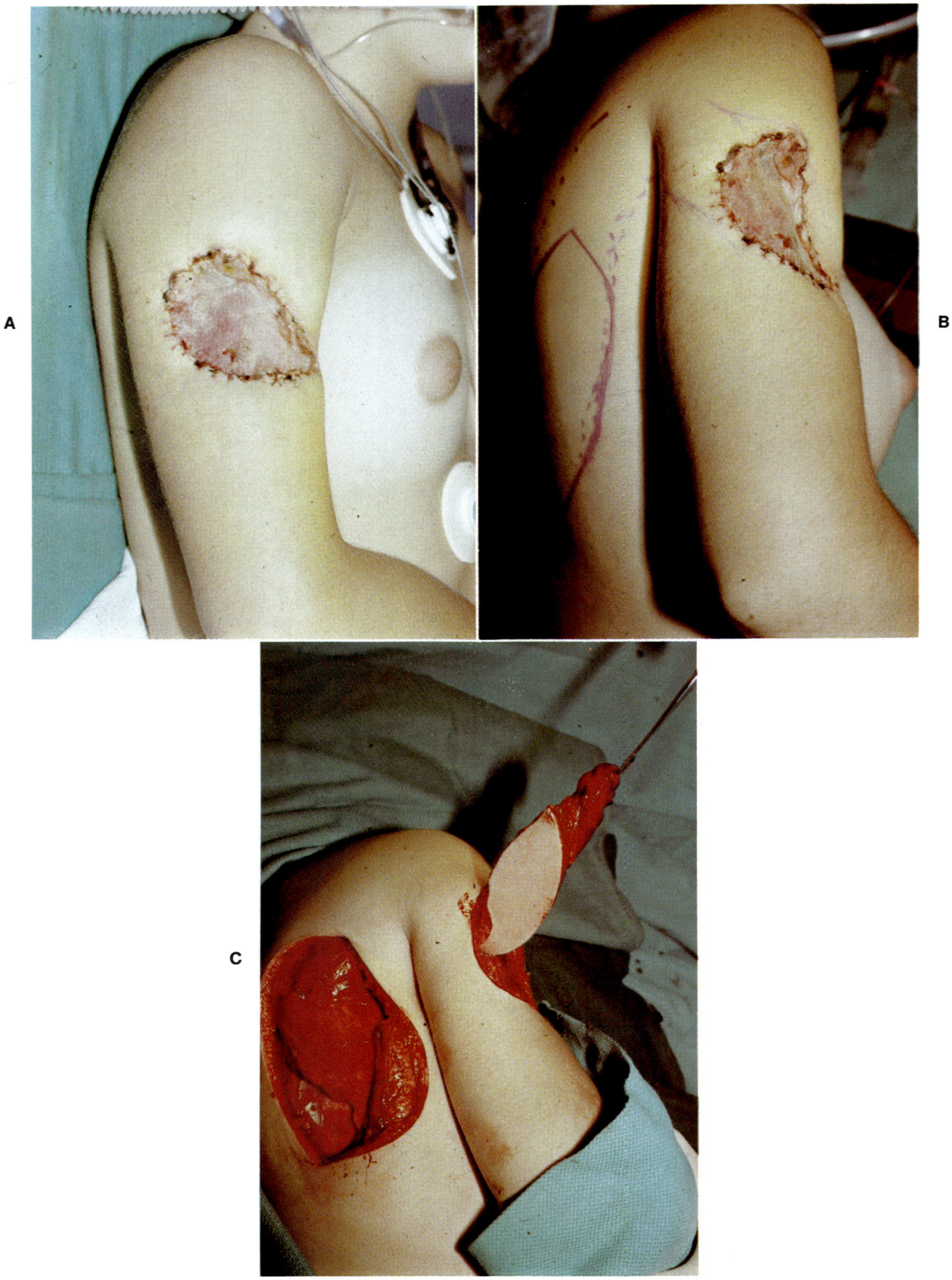

Fig. 36-2. For legend see opposite page.

D

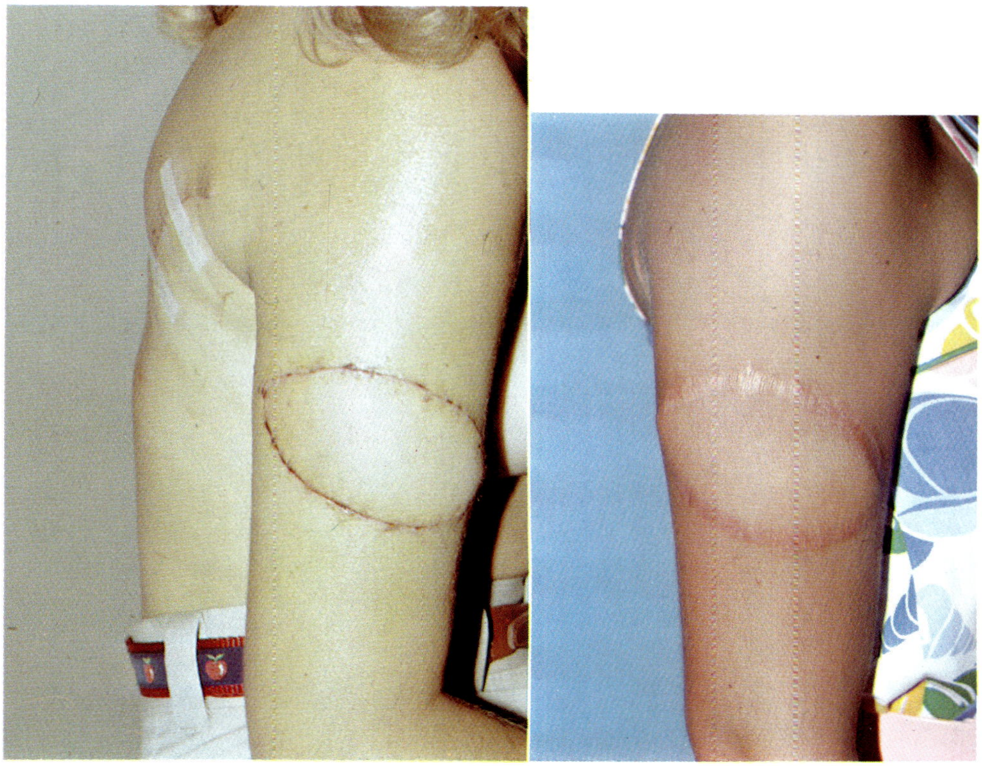

E

Fig. 36-2. Latissimus dorsi musculocutaneous flap for coverage of upper arm.
A, Defect of lateral aspect of upper arm after excision of malignant histiocytoma in 18-year-old girl.
B, Flap outlined. NOTE: Contour deformity of arm.
C, Flap elevated and tunneled through axilla onto upper arm.
D, Postoperative result demonstrates stable coverage of arm and restoration of contour.
E, One-year postoperative view.

(Courtesy T. Roderick Hester, Atlanta)

RECTUS ABDOMINIS

Flaps based on the rectus abdominis are the distant flaps of choice for coverage of the forearm. A variety of flaps may be based on the rectus abdominis. The flap may be elevated as a musculocutaneous unit (the vertical flap) or as a skin flap that is based on the perforating musculocutaneous arteries through the rectus abdominis (the transverse flap).

Vertical rectus abdominis flap

The vertical abdominal flap may be elevated either as a musculocutaneous unit or as a skin flap based on the perforating vessels. The flap may be based superiorly or inferiorly (Figs. 36-3 and 36-4).

Flap elevation. The vertical flap is designed so that the defect is covered and the forearm held in satisfactory position, usually with the forearm level with or just above the umbilicus (Figs. 36-3 and 36-4).

A

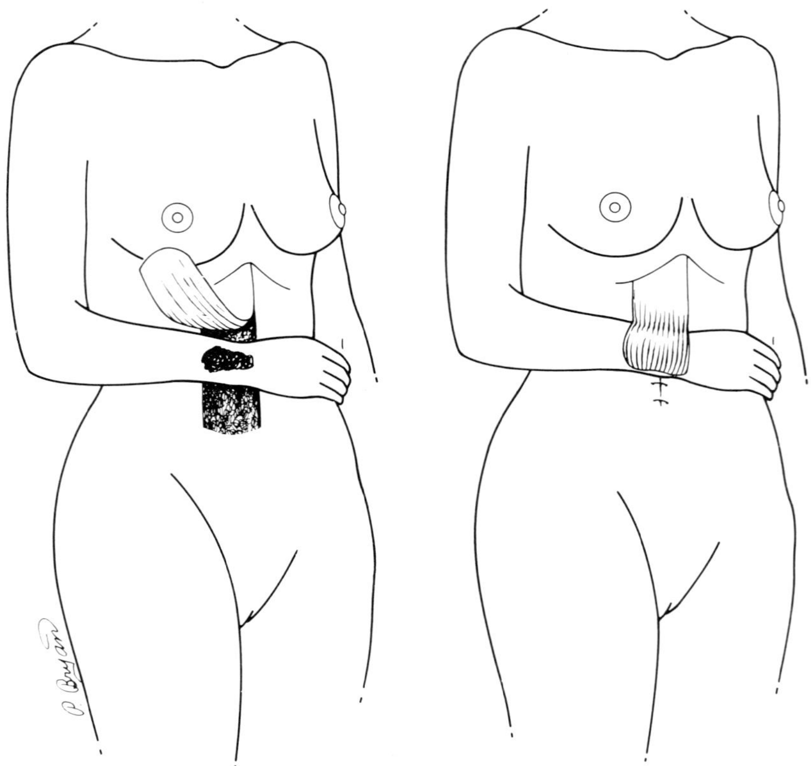

B

Fig. 36-3. Superiorly based vertical rectus abdominis flap for forearm coverage.
A, Flap elevated.
B, Defect covered and donor area closed directly.

The superiorly based flap is preferred for defects located centrally on the radial side of the forearm, and the inferiorly based flap is preferred for defects located on the ulnar side (Fig. 36-5).

The distal margin of the flap is then incised and elevated superiorly or inferiorly toward the base of the flap. If the flap is elevated as a musculocutaneous unit, it may safely be dissected up or down to the vascular pedicle. However, if the flap is elevated as a skin flap based on the musculocutaneous perforators, the length of the flap should then not exceed 12 to 15 cm.

The donor defect is usually closed directly. The posterior rectus sheath above the line of Douglas is substantial and will preserve the integrity of the abdominal wall; however, below this level the posterior sheath may not be adequate, and hence elevation of the inferior flap as a musculocutaneous unit may lead to weakening of the lower abdominal wall.

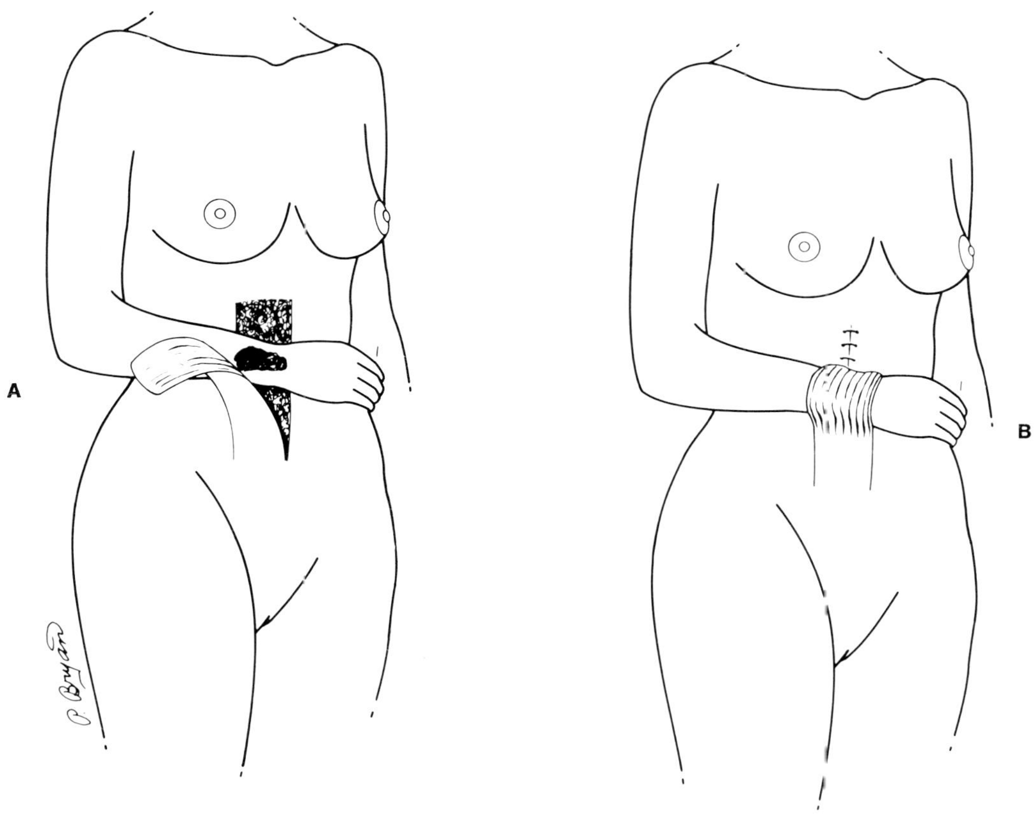

Fig. 36-4. Inferiorly based vertical rectus abdominis flap for forearm coverage.
A, Flap elevated.
B, Defect covered and donor area closed directly.

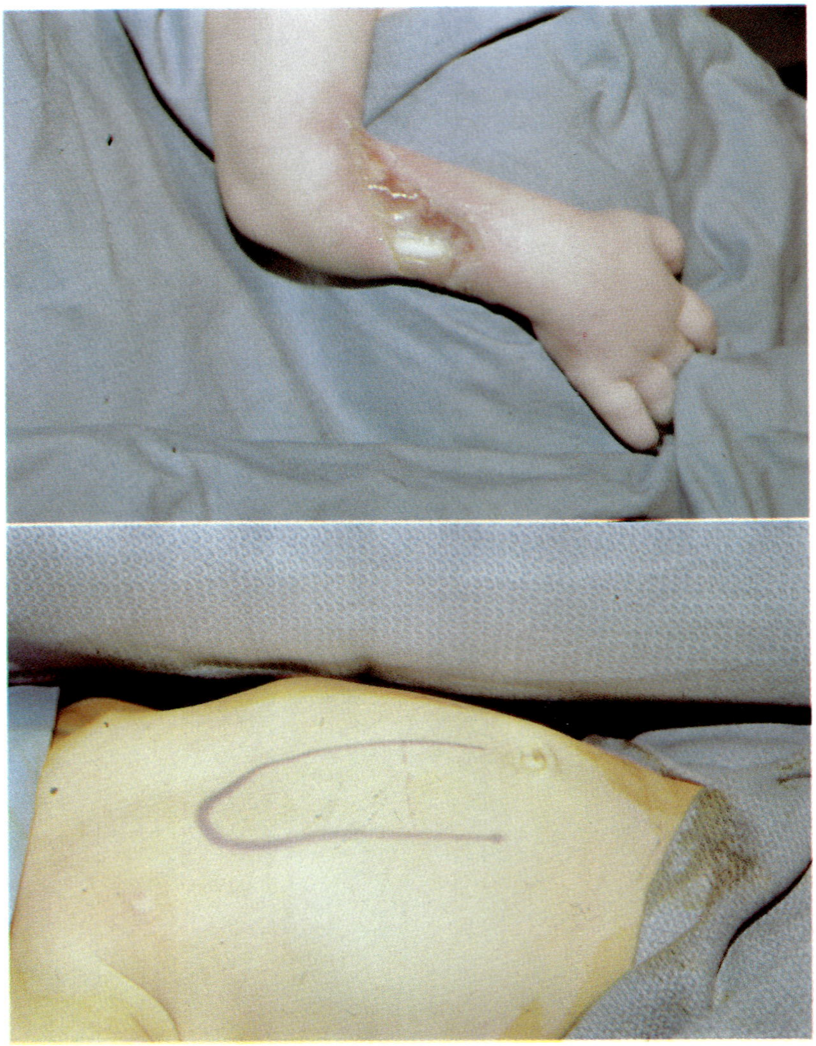

A

B

Fig. 36-5. Coverage of ulnar side of forearm with inferiorly based skin flap based on perforators through rectus abdominis.
A, Newborn has congenital defect of ulnar border of forearm with exposed ulna.
B, Design of inferiorly based skin flap.

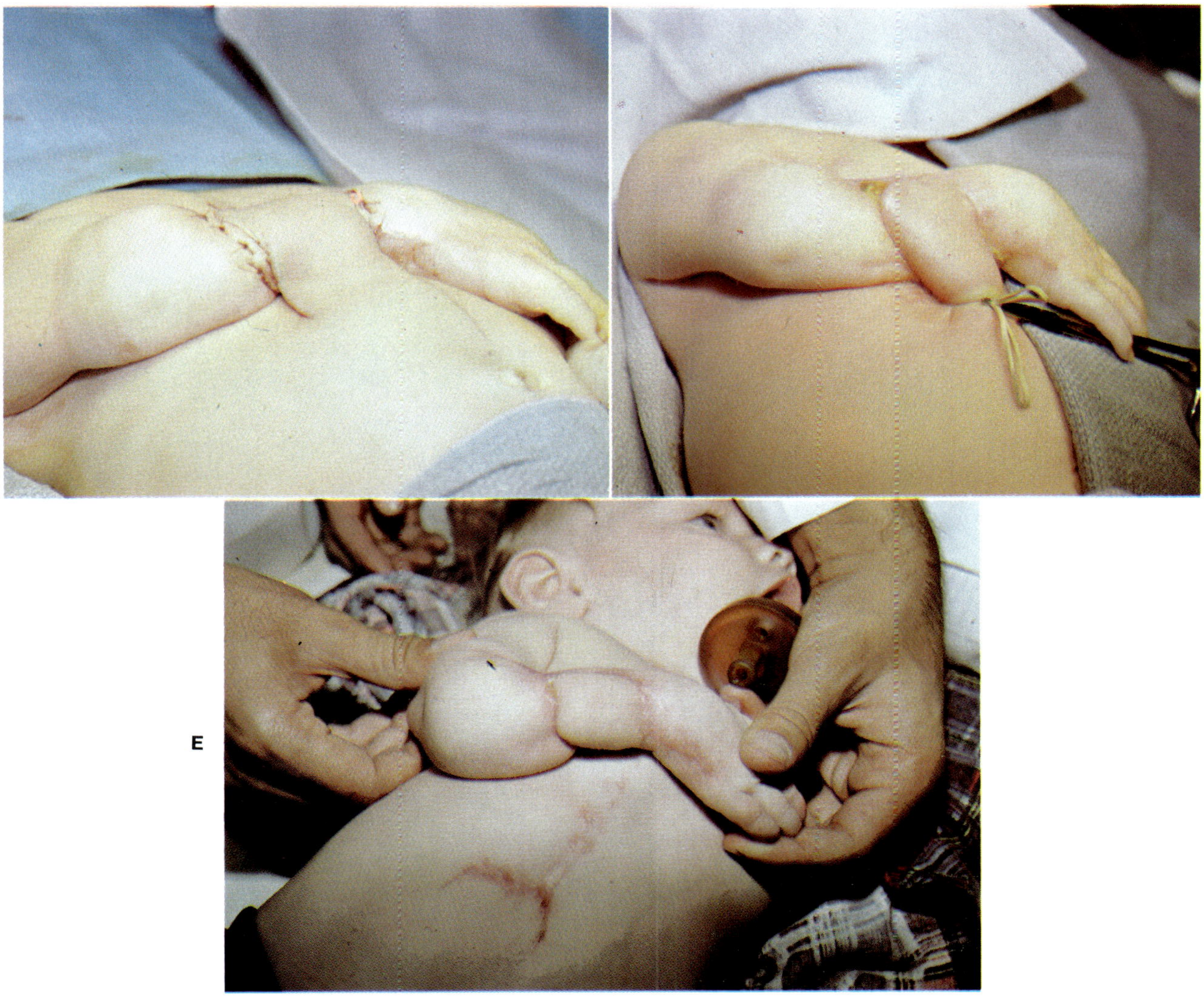

Fig. 36-5, cont'd
C, Flap elevated and sutured into defect.
D, Tourniquet test at 3 weeks just before flap division.
E, Postoperative result demonstrates stable coverage of bone and donor defect closed directly.

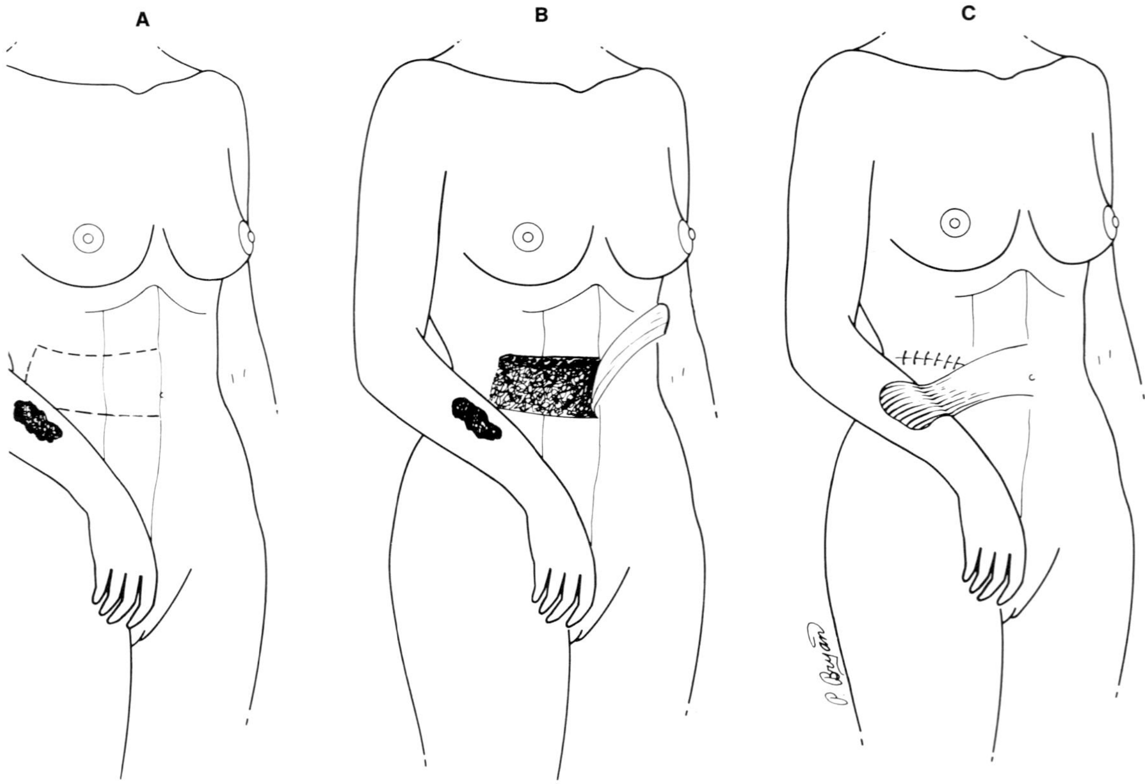

Fig. 36-6. Transverse rectus abdominis flap for coverage of forearm.
A, Flap outlined.
B, Flap elevated.
C, Forearm defect covered and donor area closed directly. NOTE: Dependent position of hand.

Transverse rectus abdominis flap

The transverse abdominal flap based medially on the perforating musculocutaneous vessels will cover defects of the forearm. However, the vertical flap is preferred, because it allows elevation of the hand, whereas the hand remains dependent on the transverse flap (Fig. 36-6).

Flap elevation. The transverse abdominal flap is based anywhere along the length of the rectus abdominis muscle. Laterally it may be safely extended to the anterior axillary line. The width is designed to suit the defect. The distal flap margin is incised and the flap elevated from the lateral border to the medial border across the external oblique into the anterior rectus sheath. It is elevated across the rectus abdominis to within 2 to 3 cm of the midline where the perforating vessels are located. The donor defect is usually closed directly unless a wide flap is elevated.

PECTORALIS MAJOR

The pectoralis major based on the dominant pedicle may be transposed into the axilla or onto the proximal part of the upper arm to provide coverage. The pectoralis major may also be transferred for elbow flexion.

Vertical and transverse skin flaps similar to those based on the rectus abdominis may also be elevated on the medial row of musculocutaneous perforating arteries through the pectoralis major. These flaps based on the pectoralis major that are higher on the trunk have the advantage of maintaining hand elevation. However, the donor defect may be less acceptable on the chest than on the abdomen. In women the breast would preclude the use of these flaps.

Tensor fascia lata

The tubed TFL flap will reach and provide coverage of the dorsum of the hand and wrist. This is not a flap of choice, but if abdominal and groin flaps are not available, and free flap coverage is not possible, as in some burn patients, the TFL flap may be useful (Fig. 36-7).

Flap elevation

The TFL flap is elevated as an extended flap and then tubed and rotated toward the lower abdomen, where it is sutured into the defect. The donor area is usually skin grafted. The tube is divided and the flap inset at 2 to 3 weeks postoperatively.

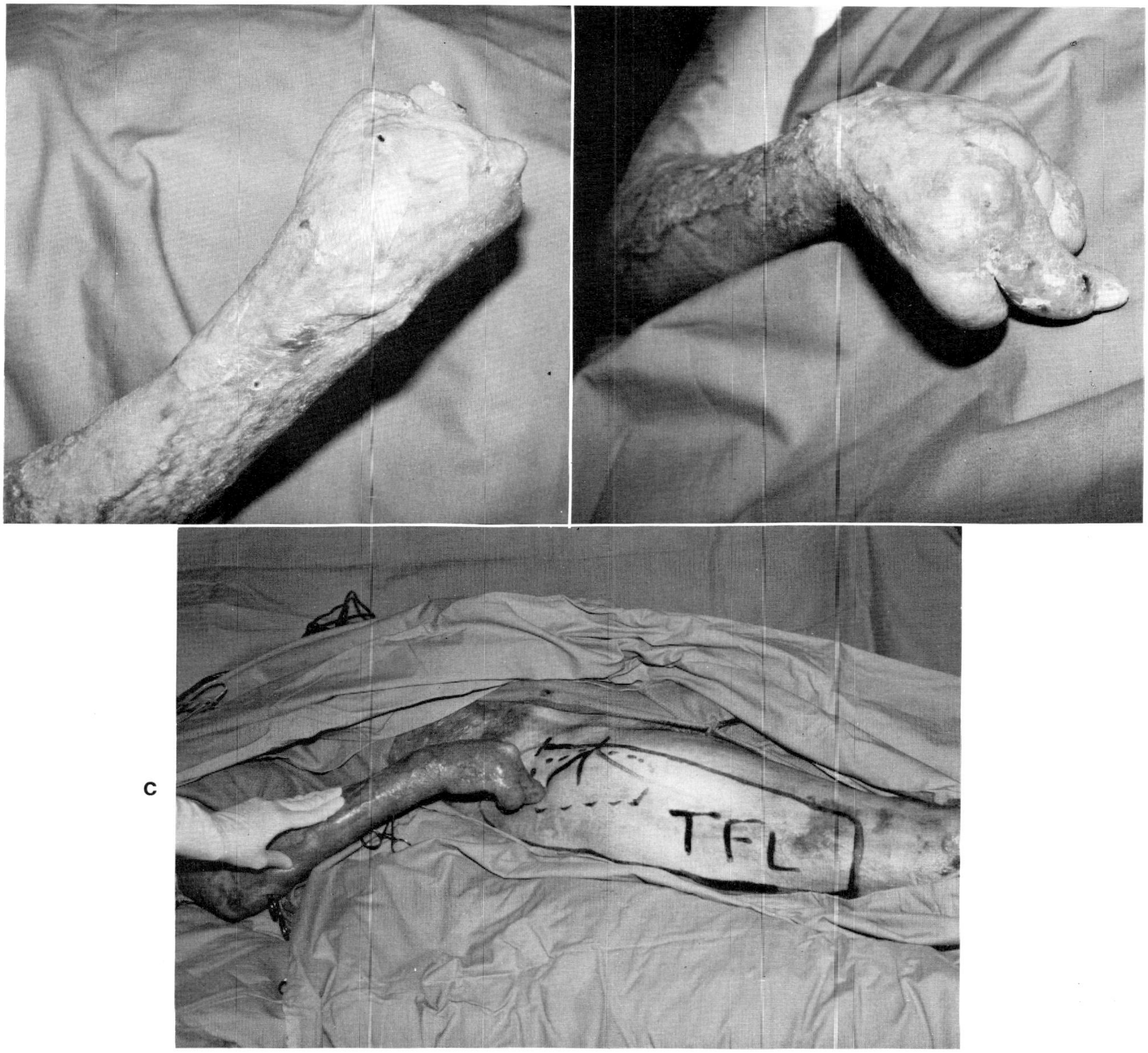

Fig. 36-7. Tubed TFL flap for coverage of burned hand.
A and **B,** Severely burned and deformed right hand with destruction of metacarpophalangeal joints and extensor tendons.
C, Burns of groin and abdominal skin preclude use as distant flaps.

Continued.

D

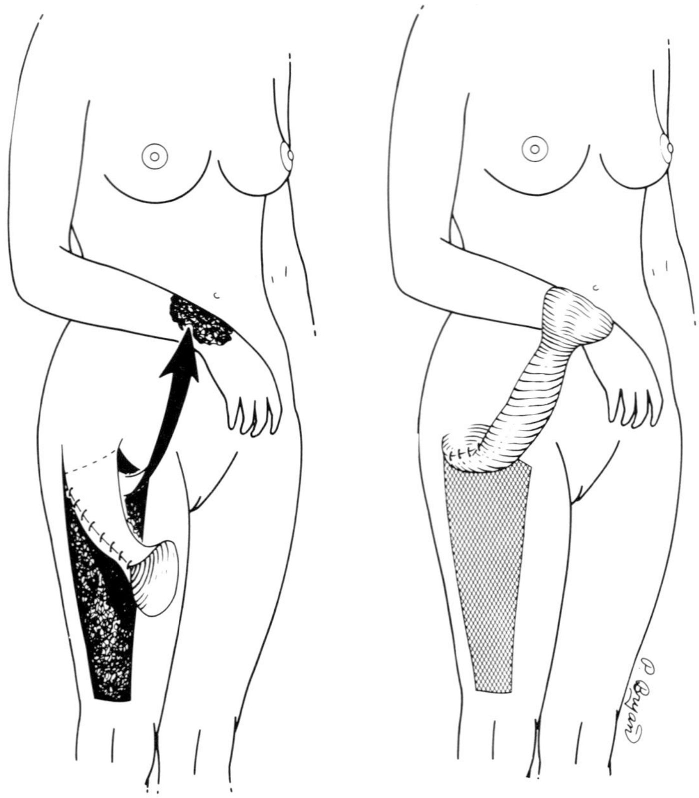

E

F

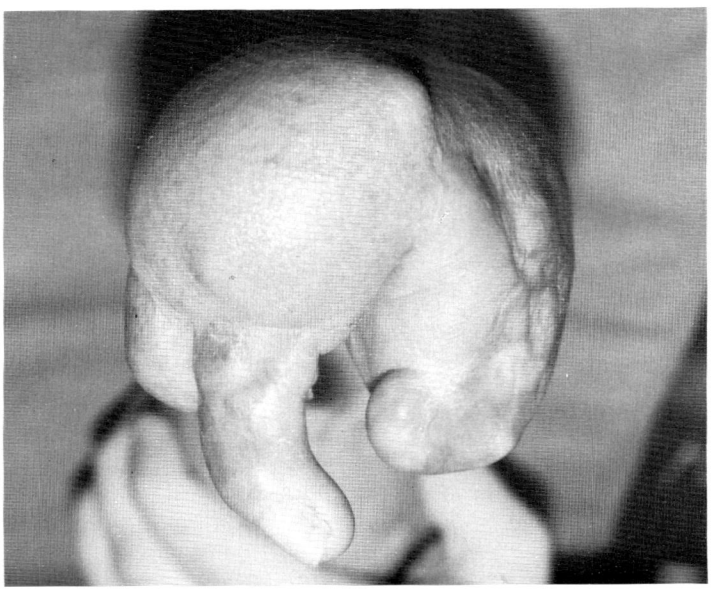

Fig. 36-7, cont'd

D, TFL flap elevated and tubed.

E, Defect closed and donor area skin grafted.

F, Satisfactory coverage of wrist and dorsum of hand in two stages.

BICEPS BRACHII

The biceps brachii is an essential muscle for flexion and supination of the forearm. Its total use as a muscle flap is rarely justified. Transposition of part of the muscle with functional preservation can provide excellent coverage for exposed dialysis access shunts. The medial fibers of the muscle are dissected, leaving the tendon functionally intact. The fibers are then transposed over the shunt.

BRACHIORADIALIS

The brachioradialis is a flexor of the forearm, and in comparison to other forearm muscles it is relatively expendable. However, its application as a flap should only be undertaken if alternative flaps are not available for coverage of the antecubital fossa.

The brachioradialis is the most superficial of the muscles on the radial side of the forearm. It takes origin from the lateral supracondylar ridge of the humerus and inserts into the styloid process of the radius. The blood supply is through a dominant proximal pedicle that is a branch of the radial recurrent artery. Based on this pedicle the flap has an arc of rotation that will cover the antecubital fossa and reach the lower part of the arm.

Flap elevation

The brachioradialis flap may be elevated as muscle only or as a musculocutaneous flap. The flap is usually transposed as muscle only. As a musculocutaneous flap, the donor defect must leave the radial artery and superficial radial nerve exposed.

The muscle is approached through an incision along the radial border of the forearm. The muscle is retracted medially, and the superficial radial nerve and radial artery are exposed. The muscle is then divided close to its insertion and dissection continues from the distal border to the proximal border. The pedicle is identified proximally and the flap transposed. The donor defect is closed directly.

FLEXOR CARPI ULNARIS

The flexor carpi ulnaris is a powerful wrist flexor and should only be transposed as a flap under special circumstances when alternatives are not available, or osteomyelitis of the elbow is evident.

The flexor carpi ulnaris is the most superficial of the muscle on the ulnar side of the forearm. It has two heads of origin, one from the medial condyle of the humerus and one from the posterior border of the ulna. The ulnar nerve runs between these two heads of origin. This muscle inserts into the pisiform, hamate, and fifth metacarpal. The pattern of blood supply is type II through a major pedicle based on the posterior ulnar recurrent branch of the ulnar artery. The muscle also has one or two distal minor pedicles that are direct branches of the ulnar artery. Based on the dominant proximal pedicle the muscle has an arc of rotation that will cover the antecubital fossa, elbow, and lower part of the arm.

Flap elevation

Although the flap may be elevated as a muscle or musculocutaneous flap, the musculocutaneous flap would leave the ulnar artery and nerve exposed. The muscle is identified through an incision on the ulnar border of the forearm. The muscle is retracted outward and the ulnar artery and nerve identified. The muscle is divided close to its insertion and dissected from the distal border to the proximal border. The distal pedicle is identified and ligated. The proximal pedicle is identified approximately 5 to 6 cm below the antecubital fossa. The muscle is then transposed, and the donor defect is closed directly.

Skin flap based on flexor carpi ulnaris

There are numerous small perforating vessels from the proximal part of the flexor carpi ulnaris muscle into the overlying skin (Fig. 36-8). A skin flap useful for coverage of the elbow may be elevated based on these perforating vessels. The flap is based on the ulnar side of the antecubital fossa and extends distally to the midforearm over the flexor carpi ulnaris. The donor defect requires skin grafting.

It should be noted that for osteomyelitis involving the olecranon or elbow joint, it is preferable to use the flexor carpi ulnaris and cover the sequestrectomy defect with muscle, as opposed to using a skin flap.

ANCONEUS

Cardany and co-workers have described the anconeus flap for coverage of the elbow posteriorly.

The anconeus is a small, triangular muscle that takes origin from the lateral humeral condyle and inserts into the posterior surface of the ulna and the lateral side of the olecranon. It is an expendable muscle that acts as an extensor of the forearm. It has a proximal dominant pedicle that is a branch of the profunda brachii artery. It is elevated with the overlying skin as a musculocutaneous unit for coverage of the posterior aspect of the elbow joint.

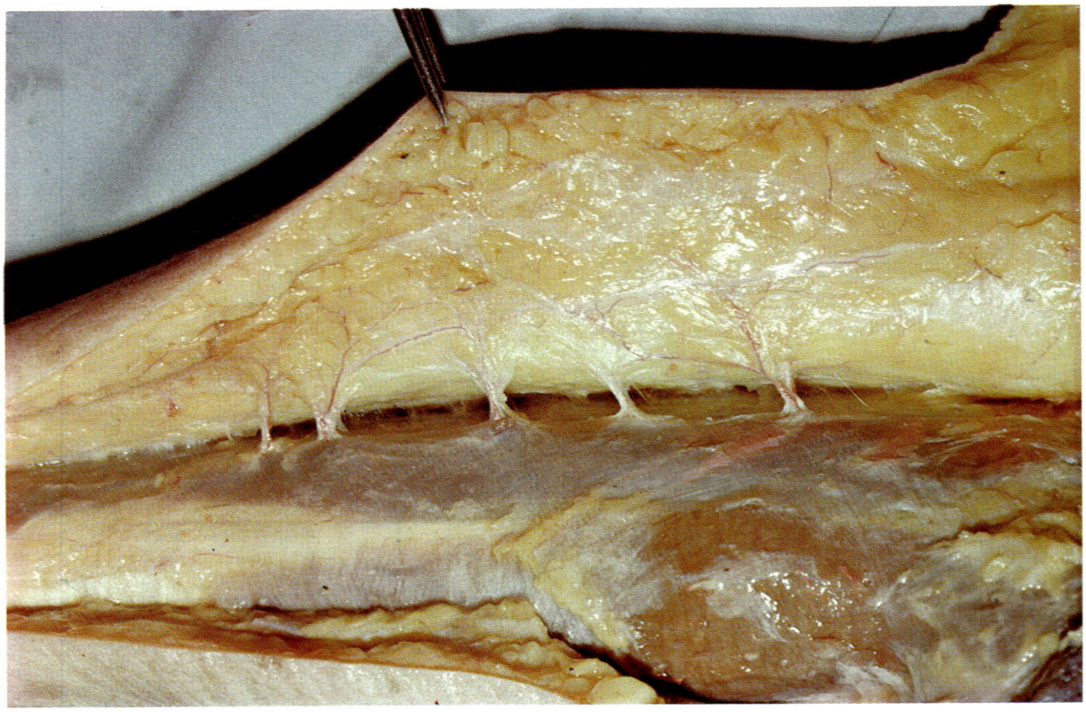

Fig. 36-8. Perforating musculocutaneous arteries from the flexor carpi ulnaris muscle into the overlying skin.

INTRINSIC HAND MUSCLES

The abductor digiti minimi and dorsal interosseous muscles have been transposed as flaps for coverage of small defects in the hand. The intrinsic muscles of the hand are not expendable and should only be transposed as flaps if other methods are not available.

Abductor digiti minimi

This is a small, relatively expendable hypothenar muscle with limited application for coverage of the wrist.

The abductor digiti minimi is the most superficial of the hypothenar muscles. It takes origin from the pisiform bone and inserts on the ulnar side of the first phalanx of the little finger. The blood supply is through a major proximal pedicle that is a branch of the ulnar artery. The pedicle enters the muscle just beyond Guyon's canal. Based on this vascular pedicle at the level of the pisiform bone the flap has an arc of rotation that will cover a small area on the volar or dorsal aspect of the wrist.

Flap elevation. The abductor digiti minimi muscle is identified through an incision on the ulnar border of the hand. Although there are musculocutaneous perforating arteries between the muscle and the overlying skin, the flap is not elevated as a musculocutaneous flap, because the secondary defect may be unacceptable. The muscle is dissected from the distal border to the proximal border, pedicle identified, and flap transposed. The donor defect is closed directly.

CONCLUSION

To preserve form and function within the extremity, flap coverage of upper extremity defects are best achieved through distant tissue transfer. Flaps based on the latissimus dorsi, rectus abdominis, and pectoralis major muscles and the axial groin flap are preferred to local muscle or musculocutaneous flaps based on muscles within the extremity. If distant flap transfer is not possible, then the use of local muscles as flaps may be justified.

ANNOTATED BIBLIOGRAPHY

1. Bashir, A.H.: Transposition of the sterno-costal head of the pectoralis major in extensive burn contractures of the axilla, Br. J. Plast. Surg. **29:**377, 1976.

This report describes the transposition of the sterno-costal head of the pectoralis major muscle into the axilla for release of axillary burn contractures. The muscle is transposed and then skin grafted.

2. Bostwick, J., and co-workers: Sixty latissimus dorsi flaps, Plast. Reconstr. Surg. **63:**31, 1979.

A review of 60 latissimus dorsi flaps includes use of latissimus dorsi in one patient as a functional muscle transfer to restore elbow flexion.

3. Cardany, C., Maxwell, P., and Gilbert, D.: Personal communication.

The original description of the anconeus myocutaneous flap for coverage of the extensive surface of the elbow is presented.

4. Ger, R.: The coverage of vascular repairs by muscle transposition, J. Trauma **16:**974, 1976.

The application of the brachioradialis as a muscle flap for coverage of exposed vascular repairs in the antecubital fossa is described. Advantage of muscle flap coverage of vascular repairs is discussed.

5. Landra, A.P.: The latissimus dorsi musculocutaneous flap used to resurface a defect on the upper arm and restore extension of the elbow, Br. J. Plast. Surg. **32:**275, 1979.

A case report describes the use of a delayed latissimus dorsi musculocutaneous flap for coverage of the posterior arm and restoration of extension to the elbow. Donor defect is skin grafted.

6. Luce, E., and Gottleib, S.: Pectoralis major flap in upper extremity coverage, J. Hand Surg. 1981.

Original description of the various applications of the pectoralis major musculocutaneous flap for coverage of the upper extremity is presented.

7. Mathes, S.J., Vasconez, L.O., and Jurkiewicz, M.J.: Extensions and further applications of muscle flap transposition, Plast. Reconstr. Surg. **60:**6, 1977.

A review of 94 muscle flaps is presented. Included are two flaps in the upper extremity: the first dorsal interosseous muscle flap for coverage of osteomyelitis of the first metacarpal in a burned hand and an abductor digiti minimi muscle flap to cover exposed ulnar and carpal bones after tumor resection.

8. Millard, T.M., Stott, W.G., and Kleinert, H.E.: The abductor digiti minimi muscle flap, Hand **9:**82, 1977.

A case report where the abductor digiti minimi muscle is transposed to cover both the median and ulnar nerves at the wrist in an attempt to provide coverage and separate the nerve from overlying skin scars is presented.

9. Nahai, F., Hill, H.L., and Hester, T.R.: Experience with the tensor fascia lata flap, Plast. Reconstr. Surg. **63:**788, 1979.

A review of 60 TFL flaps is presented. Included is one flap for coverage of the hand in a burn patient. The flap is transposed as a tubed flap. Abdominal and groin flaps are not available.

10. Schottstaedt, E.R., Larson, L.J., and Bost, F.C.: Complete muscle transposition, J. Bone Joint Surg. (Am) **37:**397, 1955.

An exhaustive review of functional muscle transfers is presented. A detailed description of latissimus dorsi transfer for elbow flexion and extension is included.

11. Silverton, J.S., Nahai, F., and Jurkiewicz, M.J.: The latissimus dorsi myocutaneous flap to replace a defect on the upper arm, Br. J. Plast. Surg. **31:**29, 1978.

A case report describes the application of the latissimus dorsi musculocutaneous flap for reconstruction of a defect on the anterior surface of the upper arm.

37

COMPLICATIONS

Foad Nahai ■ Stephen J. Mathes

GENERAL CONSIDERATIONS

Complications of flap coverage of the upper extremity not only include flap and donor site problems but also may lead to significant functional problems. Mere tissue coverage is not the goal. Rather, tissue coverage with preservation or even improvement of hand function is the end point.

Planning errors

Flap selection for a given defect is the single most important preoperative step in avoiding postoperative complications. Problems related to hand position and postoperative immobilization may have far-reaching consequences on hand function. Injudicious selection of local flaps may also alter function.

Technical errors

Precise operative technique is essential in handling tissue in the upper extremity, particularly in the hand. The tourniquet is used to allow surgical débridement and flap elevation with optimal visualization of vital structures.

Postoperative errors

Postoperative elevation of the extremity is essential. If the extremity is immobilized for a prolonged period, as in staged flap transfers, then passive and active motion of the fingers are recommended to minimize joint stiffness.

SPECIFIC FLAPS

Latissimus dorsi

Planning errors. Correct orientation of the latissimus dorsi skin island for transposition to the upper arm is essential. The flap will only reach as far as the elbow and antecubital fossa.

Technical errors. Separation of the flap from the underlying serratus muscle is an important early step. Dissection of the muscle in the axilla is necessary for transposition to the arm. The tunnel should be made large enough to avoid flap constriction.

Postoperative errors. Suction drainage of the donor defect is required.

Rectus abdominis

Planning errors. The rectus abdominis flap is planned to minimize dependency of the hand during the period of immobilization. The vertical flap will permit more elbow flexion and hence elevation of the hand. Elevation of the lower half of the rectus abdominis of a musculocutaneous flap may lead to weakening of the abdominal wall below the semicircular line of Douglas.

Technical errors. The rectus abdominis vertical and transverse flaps are reliable flaps. The vertical skin flap that is based on the perforators, however, cannot be made too long, and in general is limited in length of 15 cm or less. Flap division is planned at 2 to 3 weeks postoperatively, depending on the condition of the recipient wound. A tourniquet test before flap division is recommended.

Postoperative errors. Positional kinking and twisting of the flap are avoided through proper dressings and splinting.

Pectoralis major

Planning errors. The pectoralis major scars are unacceptable in women, and the skin territory may be too hairy in men. The flap, however, will allow satisfactory elevation of the hand.

Technical errors. Exposure of the costochondral junction should be avoided. Excessive medial dissection will divide perforating vessels. A tourniquet test before flap division is recommended.

Postoperative errors. Satisfactory position is maintained through dressing and splinting.

TFL

Planning errors. The TFL flap should only be used if the preferred alternative distant flaps are not available. The donor defect usually requires skin graft. The hand is in a dependent position.

Technical errors. The flap must be made wide enough to facilitate tubing. A tourniquet test is recommended before flap division.

Postoperative errors. Flap and extremity position are maintained through splinting and dressing techniques.

Biceps brachii

Planning errors. The biceps branchii should only be used as a flap with function preservation.

Technical errors. The brachial artery and median nerve are related to the muscle distally. The ulnar and musculocutaneous nerves are related to the muscle more proximally.

Postoperative errors. Immobilization of the elbow for 3 to 4 days postoperatively is recommended.

Brachioradialis

Planning errors. Use of the brachioradialis musculocutaneous flap leaves an unacceptable donor defect.

Technical errors. The radial artery and superficial radial nerve lie deep to the muscle and must be preserved during dissection.

Postoperative errors. Immobilization of the elbow postoperatively is recommended.

Flexor carpi ulnaris

Planning errors. Use of the flexor carpi ulnaris as a musculocutaneous flap results in an unacceptable donor defect. Use of this muscle should be limited, because this is a very useful muscle preserved for functional transfer.

Technical errors. The ulnar nerve and ulnar artery lie deep to this muscle and must be preserved during dissection.

Postoperative errors. Immobilization of the elbow is recommended postoperatively.

Intrinsic hand muscles

Planning errors. The small hand muscles should only be used if alternative flaps are not available.

Technical errors. Careful dissection of these muscles with magnification is recommended to prevent damage to nerves and tendons within the hand.

Postoperative errors. Appropriate mobilization of the wrist is necessary postoperatively.

III

TRANSPLANTATION

38

TRANSPLANTATION OF MUSCLE AND MUSCULOCUTANEOUS FLAPS

Stephen J. Mathes ■ Foad Nahai

ANATOMIC CONSIDERATIONS

With the development of microsurgical techniques, the muscle and musculocutaneous flap is no longer limited to an arc of rotation based on the site of entrance of the dominant vascular pedicle. If the circulation to the dominant vascular pedicle is established by microsurgical anastomoses to suitable recipient vessels, the muscle or musculocutaneous unit may be transplanted to other body regions. When local muscle units are not available for coverage of difficult wounds or for replacement of needed muscle function, the free transplantation of muscle or musculocutaneous units represents a useful technique in reconstructive surgery.

Vascular anatomy

Like standard transposition techniques, the limiting factor regarding free transplantation of the muscle or musculocutaneous unit is the vascular anatomy. Muscles with a single vascular pedicle (type I and type V) or a dominant vascular pedicle (type II) are best suited for transplantation, since the muscle or musculocutaneous unit will generally survive microvascular transplantation. Muscles with segmental vascular pedicles (type IV) would require multiple vascular anastomoses between their pedicles and recipient vessels to ensure survival. Muscles with two vascular pedicles of equal importance (type III) generally require two vascular anastomoses for successful transplantation. If the entire vascular system is excised with the muscle (e.g., anterior tibial artery with the extensor digitorum longus), the segmental vascular circulation may be established with a single microvascular repair. However, the required extensive dissection and sacrifice of a major lower extremity vessel are rarely justified. Knowledge of the internal vascular anatomy of muscles allows transplantation of portions of muscle, and this modification is discussed under pertinent muscle units.

The arterial pedicles to muscles have an external lumen diameter generally in the range of 1.5 to 2.0 mm. With proximal dissection of the pedicle to the junction with its proximal arterial source, this lumen diameter increases in the range of 2 to 2.5 mm. This dissection is performed between muscle planes and generally requires division of vascular pedicles to adjacent muscles. Mobilization of the vascular pedicle increases the pedicle length. The long pedicle length associated with muscle flaps generally eliminates the need for vein grafts. Anomalies in the configuration and location of the dominant vascular pedicle to muscle are rare. However, the number and variability in the location of minor pedicles in types II and V muscles are more common.

The venous circulation of muscles parallels the arterial pedicle as paired venae comitantes. The paired venae comitantes form a single vein immediately before joining the main venous drainage of the region of the muscle. At this level the external lumen diameter of the vessel is 2 to 3 mm. The veins associated with the dominant vascular pedicles of muscle provide adequate venous drainage for successful free transplantation of muscle units. If the vascular pedicle is not dissected to the level of the junction with its confluence with the regional veins, both venae comitantes should be repaired with receptor veins in proximity to the transplantation site

Neural (motor) anatomy

When functional transplantation of muscle is required, the motor nerve must be preserved for transplantation with the muscle unit. The motor nerve enters the proximal muscle and generally is located in proximity to the point of entrance of the dominant vascular pedicle. The nerve can be mobilized proximally to gain length so that nerve grafts are not required for repair with recipient nerves in proximity to the area of transplantation.

Neural (sensory) anatomy

When coverage is required in weight-bearing regions, a flap with sensation is desirable. Sensory nerves entering the cutaneous territory of muscles can be incorporated with the musculocutaneous flap. With repair of the sensory nerve with receptor sensory nerves in proximity to the site of transplantation, a sensory musculocutaneous flap is possible.

TRANSPLANTATION VS TRANSPOSITION

The versatility of axial, muscle, and musculocutaneous flaps for coverage of difficult wounds as a transposition flap is well recognized by reconstructive surgeons. The transposition of muscles for functional purposes is also a well-established procedure, especially in upper extremity reconstructive surgery. Free transplantation of muscle either for coverage or function requires the complex techniques associated with microvascular surgery. Therefore free transplantation of muscle or musculocutaneous flaps is a technique in reconstructive surgery reserved for problems not readily solved by the simpler, more direct methods of flap transposition.

Absent or inadequate flap

When local flaps are either absent, damaged, or inadequate in size, free tissue transplantation may be required. The head and distal leg are body regions where this situation is commonly encountered.

Number of procedures

The multiple procedures required for distant transfer of pedicle flaps may be avoided by free transplantation of composite tissue. When local tissue is unavailable, and a reconstruction can be accomplished in one operation with microvascular transplantation of muscle units, rehabilitation may be more rapid with earlier, successful wound healing.

Flap durability

When a distant flap is inset into an area of poor vascularity (e.g., cross-leg flap in lower leg defect), the flap may not survive or may not provide long-term stable coverage. Successful free flap transplantation requires vascularization from undamaged recipient vasculature. Since the muscle unit transplanted by microvascular techniques is not dependent on local revascularization from underlying tissue, the coverage is more durable when the defect is associated with an area of poor vascularity.

Specialized requirements

Certain musculocutaneous flaps have specialized features such as cutaneous nerves (e.g., TFL) or vascular connections to bone. Free microvascular transplantation of these specialized units allows use of these composite tissue in body regions where local flaps do not possess these special requirements.

Function

After transposition of a muscle unit for coverage of a local defect, the muscle can no longer perform its locomotor function. When muscle units adjacent to a defect are lost either because of trauma or tumor resection, the remaining intact muscle units may require preservation to avoid severe functional deficits. In this instance the free transplantation of muscle or musculocutaneous flaps from other body regions may be indicated (e.g., defects associated with severe lower extremity trauma). Free transplantation of muscle units from other body regions may avoid further impairment in function of the injured extremity. With repair of the motor nerve of the transplanted muscle or musculocutaneous units to recipient motor nerves, both coverage and function may be provided by one composite tissue transplantation.

SELECTION OF DONOR FREE FLAP

When free tissue transplantation is chosen as a technique for coverage, the reconstructive surgeon has the option of transplantation of either axial or muscle units. Although both techniques remain useful for microvascular surgeons, the muscle or musculocutaneous unit as a donor flap has certain advantages over currently defined axial flaps.

Size

For small defects requiring microvascular flap transplantation, both muscle and axial flaps will generally cover the defect. Flap selection is then generally based on thickness or special requirements such as the need for sensation or vascularized bone. Flap thickness is generally greater in the musculocutaneous flap when compared to the axial flap. Use of the muscle only with application of skin grafts eliminates the excessive thickness associated with the majority of musculocutaneous units. When the axial donor sites are associated with thick adipose tissue in the obese patient, the muscle unit with skin grafts is often preferable for defect coverage.

Large axial flaps such as the groin or deltopectoral flaps have been successfully transplanted for extensive defects. However, the flap size does not compare with the size of the latissimus dorsi muscle or extended TFL musculocutaneous units. For this reason selection of the muscle unit either as a muscle with skin grafts or musculocutaneous unit is preferable for coverage of defects larger than 15 × 8 cm.

Flap elevation

The vascular pedicles entering muscles are located deep to the subcutaneous tissue. The pedicles are therefore exposed by retraction of adjacent muscles. Since the entire dissection for muscle or musculocutaneous flap elevation is carried out between muscle planes, flap elevation is generally not difficult. Most axial flaps require dissection within the subcutaneous plane, thus increasing the risk of injury to the small vascular pedicles and the difficulty of flap elevation.

Vascular anomalies

The dominant vascular pedicles to muscles are consistent in their location and size. Although location and size of minor pedicles to muscles are variable, these vessels are ligated for free flap transplantation (e.g., gracilis). Variability in the vasculature of axial flaps is a common operative finding and occasionally does not allow the use of the flap for microsurgical transplantation.

Donor deformities

The muscles currently useful as donor flaps for transplantation are expendable muscles. Other synergistic muscles preserve function. However, the donor site scar is often objectionable, unlike the groin flap in which the scar is readily concealed.

Transplantation of the muscle without its overlying skin allows direct closure of the donor site without tension. This technique eliminates undesirable scars, especially if the muscle is located in the back (e.g., latissimus dorsi) or medial thigh (e.g., gracilis).

SPECIFIC INDICATIONS FOR MUSCLE TRANSPLANTATION

Clinical applications for coverage

The reconstructive surgeon who is knowledgeable in three areas—muscle anatomy, defect anatomy, and techniques of microsurgery—can perform free composite tissue transplantations with safety. The last 50 microsurgical transplantations of muscle and musculocutaneous flaps have been accomplished with no failures. In areas where local, axial, or muscle flaps are unavailable, inadvisable, or associated with a high failure rate, the procedure of choice is composite tissue transplantation. Such areas include the skull and distal lower leg. These areas are discussed in regard to specific indications for use of microsurgical transplantation of muscle and musculocutaneous flaps in reconstructive surgery.

Skull

Numerous axial flaps based on the scalp circulation are both useful in scalp reconstruction and represent the techniques of choice. However, these flaps may not be available after neurosurgical procedures and intracranial radiation therapy. Distant axial and musculocutaneous flaps based on vascular pedicles in the base of the neck or chest generally will not adequately reach the parietal and temporal skull, especially when the defect extends across the anterior-posterior midline. Under these circumstances, when flap coverage is required because of exposed irradiated skull, free microsurgical composite tissue transplantation represents the procedure of choice for scalp coverage (Fig. 38-1 and Table 7).

TABLE 7 Free flaps for skull coverage

Flap	Technique	Defect
Latissimus dorsi	Segmental	1/3-1/2 of skull
	Entire muscle	
	With skin island	1/2 of skull
	Without skin island	Entire skull
Omentum	Flap with skin grafts	Entire skull
Groin flap	Superficial	1/3-1/2 of skull
	Deep with iliac bone	1/3-1/2 of skull

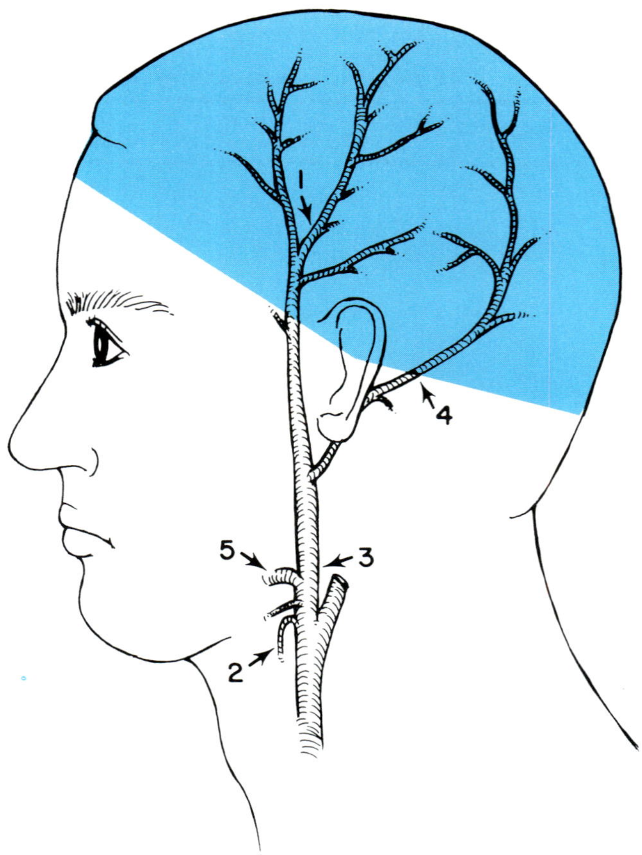

Fig. 38-1. Skull coverage of defects involving one third to entire skull. Composite tissue transplantation is method of choice when local scalp flaps are unavailable. Recipient vessels in order of preference are superficial temporal artery and vein *(1);* superior thyroid artery and vein *(2);* external carotid artery and external jugular vein *(3);* occipital artery and vein *(4);* and facial artery and vein *(5).*

Distal lower leg

The inferior half of the lower leg, ankle, and dorsum of the foot are areas not easily covered by local, axial, or muscle flaps. When coverage is required in this area, coverage may be achieved from cross-leg flaps or distally based muscle flaps. Although these flaps are discussed under extremity coverage, these techniques are not completely reliable. The cross-leg flap must achieve its blood supply from the underlying defect before inset is possible. This may require a lengthy period before flap inset and still may not achieve stable coverage. The distally based muscle flap requires transposition of muscle flap with type II or IV vascular patterns. The transposition has a great risk of flap failure (25%) and may impair distal leg circulation because of manipulation of proximal vascular pedicles and a loss of muscle function. Both these techniques have provided coverage for difficult lower leg defects and should remain part of the armamentarium of the reconstructive surgeon.

The microvascular transplantation of muscle and musculocutaneous flaps provides coverage of lower leg defects in one procedure. Intact musculature is left undisturbed, and flap design is based on the precise requirements for the defect. Using techniques of end-to-side microanastomoses, neither distal arterial inflow nor venous return is impaired. However, the most significant aspect of this technique is the vascularity of the flap. The successfully transplanted free muscle or musculocutaneous flap does not rely on impaired lower leg blood flow but contributes to the blood flow and venous return of the defect. Thus microsurgical composite tissue transplantation for lower leg defects is associated with successful coverage and contributes to improved healing of underlying bony injury and control of infection (Fig. 38-2 and Table 8).

TABLE 8 Free flaps for lower extremity coverage

Flap	Technique	Defect
Latissimus dorsi	Segmental	1/4 of lower leg, foot; osteomyelitis, contour deformity
	Entire muscle	
	With skin island	1/3 of lower leg
	With skin grafts	1/2 of lower leg; osteomyelitis
TFL	Musculocutaneous	1/3-1/2 lower leg
	Neurosensory	1/3 distal leg and foot
	Osseous	Leg or foot
Gracilis	Musculocutaneous	1/4 of lower leg
	Entire muscle with skin grafts	1/4 leg; contour deformity, osteomyelitis
Groin	Superficial	1/4 of lower leg; contour deformity
	Deep (with iliac bone)	Leg or foot

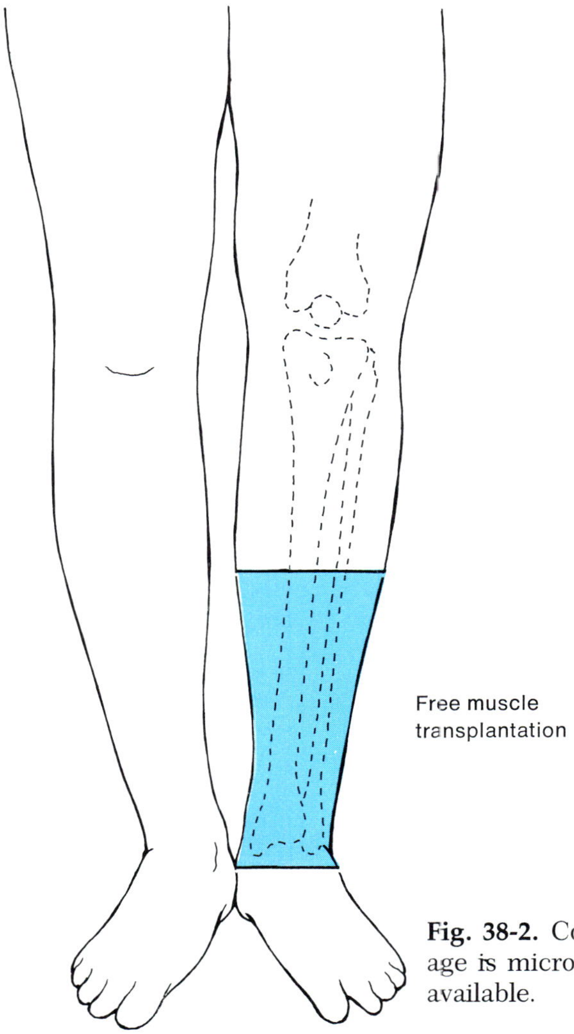

Fig. 38-2. Coverage of lower leg and foot. Method of choice for defect coverage is microvascular flap transplantation when suitable receptor vessels are available.

Special defects

Axial, muscle, and musculocutaneous flaps have been identified as the procedures of choice in reconstruction in most body regions. In areas where these flaps are either unavailable or fail, microsurgical transplantation may be indicated as a secondary procedure. Such areas include the following:

1. Head and neck—failure of flap coverage or mandibular reconstruction, extensive upper and middle third facial defects, contour defects (e.g., hemifacial atrophy)
2. Breast—failure of breast reconstruction techniques such as loss or absence of latissimus dorsi muscle flap
3. Back—unavailability of local flaps
4. Perineum—unavailability or failure of gracilis and TFL for standard muscle transposition procedures

If suitable receptor vessels in proximity to the defect are available, microsurgical composite tissue transplantation is often the procedure of choice. However, the use of vein grafts often becomes essential under these circumstances, since prior dissections and injury may require use of receptor vessels some distance from the actual defect.

Clinical application for function

The well-established techniques of muscle transfer are the procedures of choice after functional muscle loss in the face and upper and lower extremities. However, when the injury has destroyed multiple muscle groups, then local muscle transfer may not be indicated because (1) lack of muscles suitable for transfer and (2) risk of further loss of already impaired extremity function. In this situation functional transplantation of muscle represents an alternative approach to provide the required muscle unit. Muscle transplantation has been successfully applied in face and upper and lower extremity reconstructive surgery when local muscles could not be transferred to replace lost motor function.

The transplantation of muscle is associated with significant loss of muscle strength. For this reason the muscles selected for transplantation must be stronger than the unit they are replacing. Furthermore since the origin-insertion length relationship must be maintained, the donor muscle length must be similar to the recipient defect. Finally, its vascular pattern must be suitable for microsurgical transplantation (preferably Types I, II, and V). With these specific limitations, the muscle for transplantation must be carefully selected and used only when local muscle cannot be transferred. With future sophistication of techniques of motor nerve repair, the use of free muscle transplantation for function may become routine in reconstructive surgery.

ANATOMIC AND TECHNICAL CONSIDERATIONS FOR SPECIFIC MUSCLE AND MUSCULOCUTANEOUS UNITS

The following three muscle units are the most suitable for transplantation based on the parameters of size, minimal thickness, potential for function, and specific features:

1. Latissimus dorsi
2. Gracilis
3. TFL

The anatomy and techniques of elevation for these units are described. The usefulness of these flaps for specific reconstructive problems is presented.

LATISSIMUS DORSI

The latissimus dorsi, a posterior back muscle, may be transplanted either as a muscle or musculocutaneous unit for coverage or functional muscle. Since this muscle covers the entire lower back from the spine of the scapula to the iliac crest in the vertical axis and posterior axillary line to posterior midline in the horizontal axis, a flap of great dimensions is possible (35 × 20 cm). This unit, when transplanted as a musculocutaneous unit flap, may include skin of similar dimensions, or the skin island may be designed over any portion of the muscle surface, depending on specific reconstructive requirements.

Vascular anatomy

The latissimus dorsi muscle has a type V vascular pattern with its major vascular pedicle entering the muscle within the axilla at the posterior axillary line. This vascular pedicle, the thoracodorsal artery, arises from the subscapular artery, a branch of the axillary artery. After the scapular circumflex artery branches from the main artery, this pedicle is designated as the thoracodorsal artery. The thoracodorsal artery enters the latissimus dorsi muscle 10 cm inferiorly to its insertion into the humerus (Figs. 38-3 and 38-4). This vascular pedicle, the thoracodorsal artery and its associated vein, has a pedicle length of 8 cm between the vessel of origin, the axillary artery and vein, and its junction with the muscle belly (Fig. 38-5). The pedicle has branches to the adjacent serratus anterior and teres major muscles. At the level of junction of the vascular pedicle with the axillary vessels, the external lumen diameter of the subscapular artery and vein measures 2 to 3 mm. At the level of the thoracodorsal artery and vein (now paired venae comitantes), the external lumen diameter is 1.5 to 2 mm. On entering the anterior surface of the muscle, the thoracodorsal artery and its venae comitantes course between the muscle and the epimysium. The pedicle divides into several branches that course along the long axis of the muscle (axillae to posterior midline). The most anterior of these branches courses 3 to 4 cm from the anterior muscle edge (Fig. 38-5). This arterial pedicle with its paired venae comitantes has an external lumen diameter of 1 to 1.5 mm at its junction with the thoracodorsal artery immediately distal to its branch to the serratus anterior muscle.

The secondary segmental vascular pedicles enter the muscles adjacent to the posterior midline near its origin. These pedicles may be safely divided, when the entire muscle is transplanted based on the thoracodorsal vascular pedicle, without adverse effects either on muscle or skin circulation (Fig. 38-3).

The venous drainage of the latissimus dorsi muscle parallels the arterial pedicles. At the level of the arterial branch of the thoracodorsal artery to the serratus anterior muscle, this vein is formed by paired venae comitantes draining both the serratus anterior and the latissimus dorsi muscles. Occasionally, the thoracodorsal vein persists as two venae comitantes with its junction with the axillary vein. The venous pedicle dimensions are similar to those described for the thoracodorsal artery (Fig. 38-5).

Neural (motor) anatomy

The motor nerve of the latissimus dorsi muscle is the thoracodorsal nerve, a branch of the posterior cord of the brachial plexus. This nerve courses adjacent to the thoracodorsal artery and vein, entering the muscle 10 cm inferiorly to its insertion into the intertubercular groove of the humerus. This motor nerve also courses between the epimysium and the anterior surface of the muscle adjacent to the branches of the vascular pedicle. The motor nerve divides into three or four branches that course with the vascular pedicles from the posterior axilla to anterior, lateral, posterior, and superior muscle segments (Fig. 38-5).

Neural (sensory) anatomy

The sensory nerves to the posterior cutaneous territory of the latissimus dorsi muscle enter the muscle with the secondary segmental vascular pedicles. These nerves are branches of the intercostal nerves coursing with the perforating branches of the posterior intercostal arteries. These nerves are small, with one or two fascicles, and short, with length of 2 cm, depending on the thickness of the posterior latissimus dorsi muscle and subcutaneous tissue.

Flap characteristics

SIZE: The latissimus dorsi unit as a free flap may include the entire skin of the posterior inferior back overlying the muscle. Anterior skin over the serratus anterior muscle may be included as random skin for a distance of 3 to 4 cm, as long as 4 to 6 cm of skin over the anterior latissimus dorsi muscle are included with the flap.

THICKNESS: Thickness of this unit depends on flap design. A distal anterior skin island will have a thickness of 3 to 4 cm, except in the very obese patient. Proximal muscle close to the axilla is much thicker, and this bulk may not be desirable in certain instances.

PLIABILITY: The muscle or musculocutaneous flap is pliable and folds into irregular defects easily.

DONOR SITE: When the skin island's greatest width is 8 cm, the back skin can generally be closed directly. Closure in the obese patient may be difficult even at this dimension. The function preservation technique that maintains a posterior muscle avoids complete loss of muscle function. Despite the transplantation of the entire muscle, the donor disability has been minimal to none.

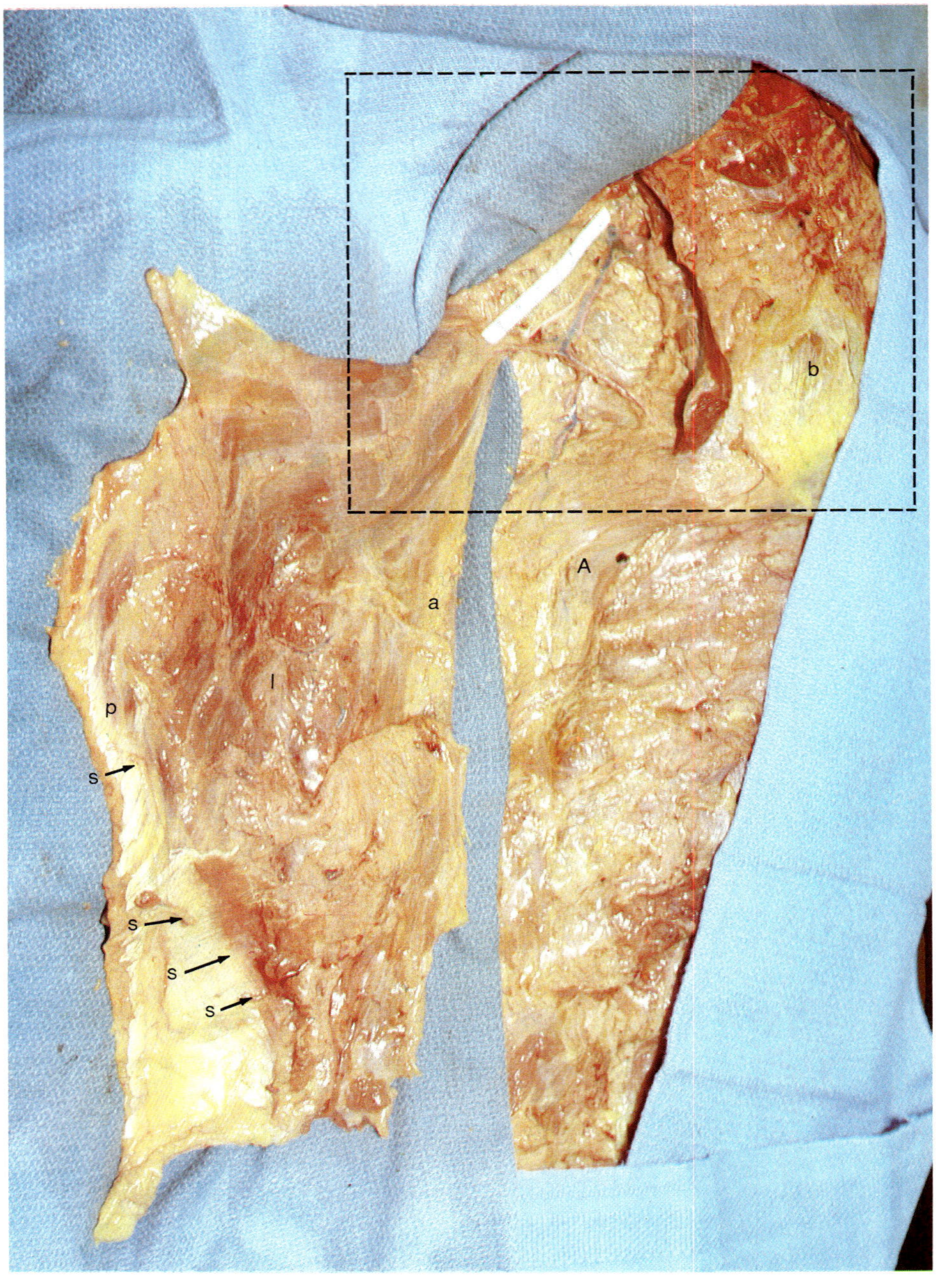

Fig. 38-3. Vascular anatomy of latissimus dorsi. Inset denotes appearance of vascular anatomy of latissimus dorsi (*l,* deep surface) with patient prone. Muscle origin (*p*) is released, and anterior free muscle edge (*a*) is separated from serratus anterior muscle (*A*). *s,* Divided posterior secondary segmental pedicles; *b,* bony prominence of scapula.

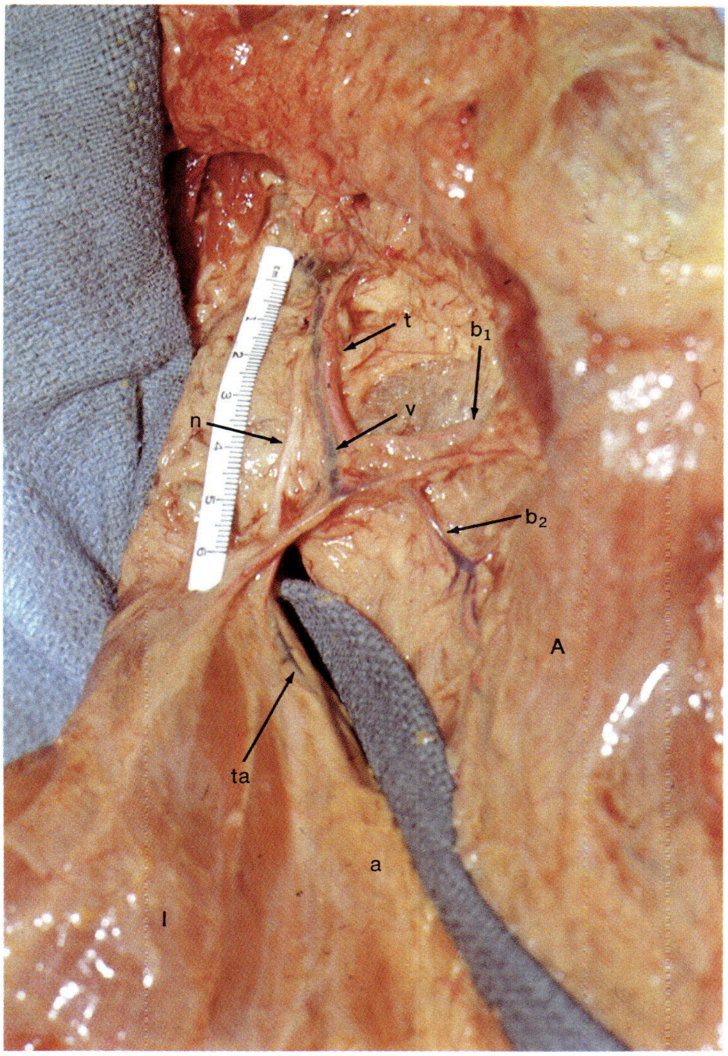

Fig. 38-4. Vascular anatomy of latissimus dorsi. Inset of Fig. 38-3 demonstrates major vascular pedicle of latissimus dorsi muscle. *l,* Deep surface; *a,* anterior muscle edge; *A,* serratus anterior. Vascular pedicle consists of thoracodorsal artery *(t),* nerve *(n),* and vein *(v).* NOTE: Branches to teres major (b_1) and serratus anterior (b_2) muscles. *ta,* Anterior branch of thoracodorsal artery to anterior muscle *(a).*

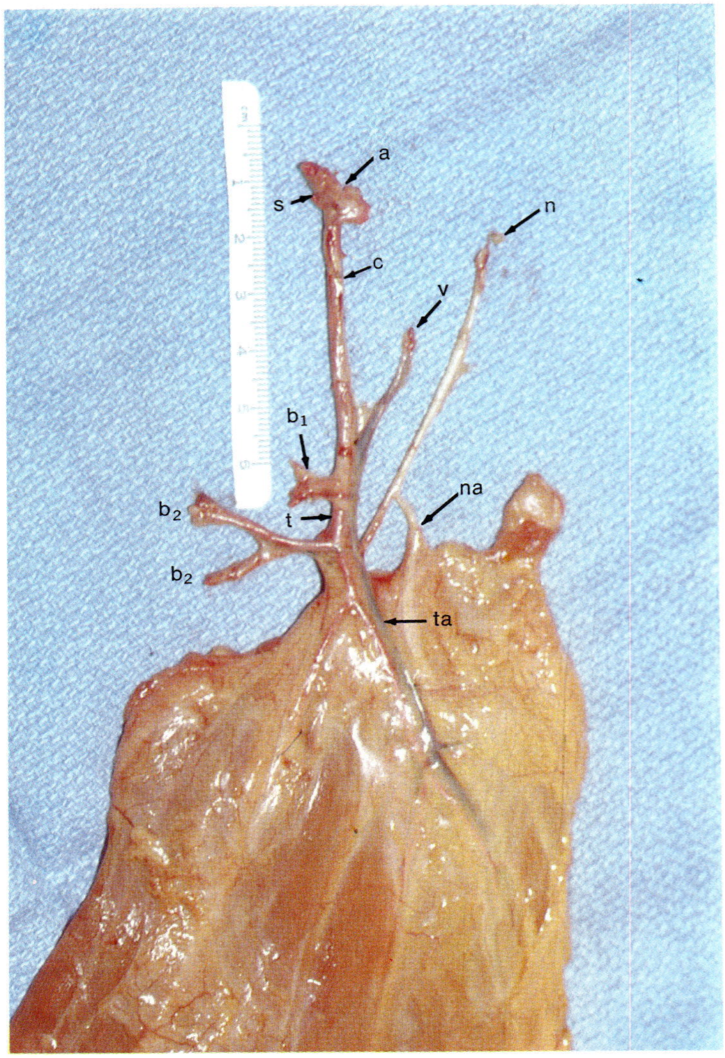

Fig. 38-5. Vascular anatomy of latissimus dorsi. *a,* Axillary artery; *s,* subscapular artery; *c,* circumflex scapular artery; *t,* thoracodorsal artery; b_1, branch to teres major; b_2, branch to serratus anterior; *ta,* branch to anterior latissimus dorsi muscle; *v,* subscapular vein; *n,* thoracodorsal nerve; *na,* branch to anterior latissimus dorsi.

Distant coverage

The latissimus dorsi unit may be transplanted either as a muscle or musculocutaneous flap for distant coverage. By orientation of the muscle or muscle and overlying skin on the anterior portion of the muscle, both circulation and motor innervation to the posterior muscle may be preserved. This flap design is designated as the anterior latissimus dorsi flap with function preservation (Fig. 38-6). When the entire muscle is elevated for coverage, skin grafts are placed over the transplanted muscle. Since elevation of the entire muscle and its overlying skin would leave a significant posterior defect, this flap design is rarely indicated.

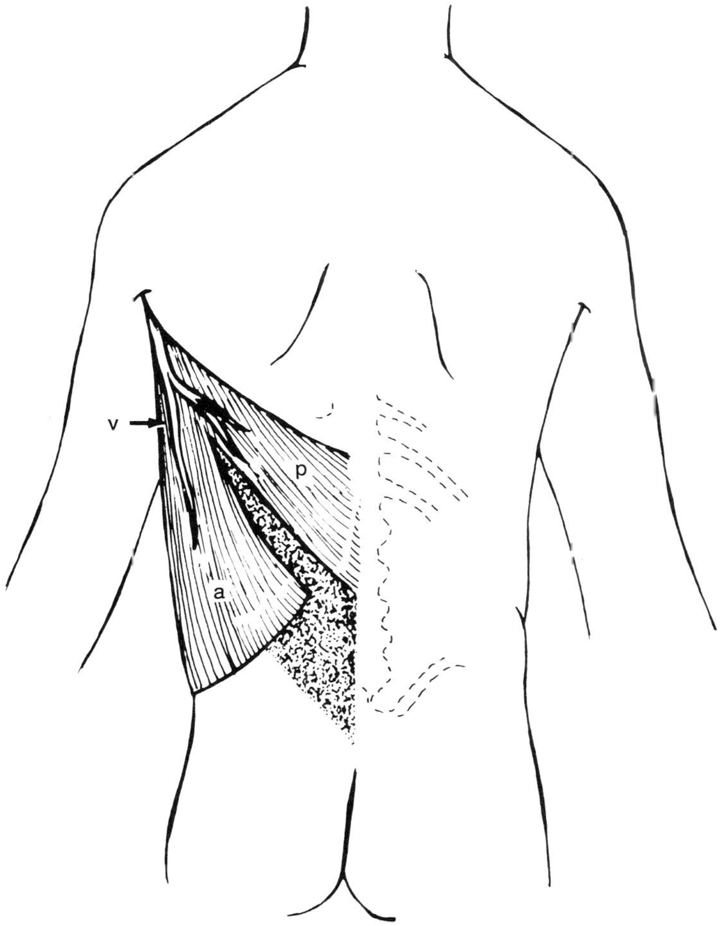

Fig. 38-6. Technique of elevation for latissimus dorsi. Segmental transplantation of anterior muscle *(a)* based on anterior vascular pedicle *(v)* allows preservation of posterior latissimus dorsi muscle *(p)* for function preservation.

Anterior latissimus dorsi flap with function preservation

Technique of elevation. The anterior margin of the latissimus dorsi muscle is located along the posterior axillary line. The muscle border is defined by the patient's voluntary contraction of the muscle while in the standing position (Fig. 38-7). If anterior muscle only is required, this line defines the skin incision for exposure of the muscle. However, if muscle with skin transposition is planned, then the anterior skin incision extends anteriorly to the muscle border by 3 to 4 cm (Fig. 38-8). This random skin is included with the skin over the anterior muscle, since this decreases flap bulk without adverse affect on flap circulation. The skin island is oriented with the long axis vertically. The skin length may range from 6 to 30 cm, depending on reconstructive requirements. The skin width generally ranges from 6 to 8 cm. At least 3 to 4 cm of skin must be designed actually over the anterior muscle. This ensures the presence of

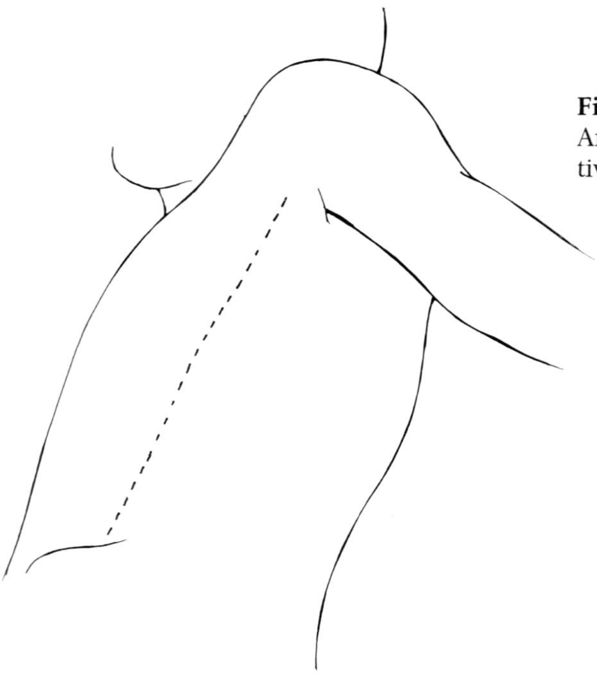

Fig. 38-7. Technique of elevation for segmental latissimus dorsi. Anterior edge of latissimus dorsi muscle is located preoperatively by palpation of muscle contraction.

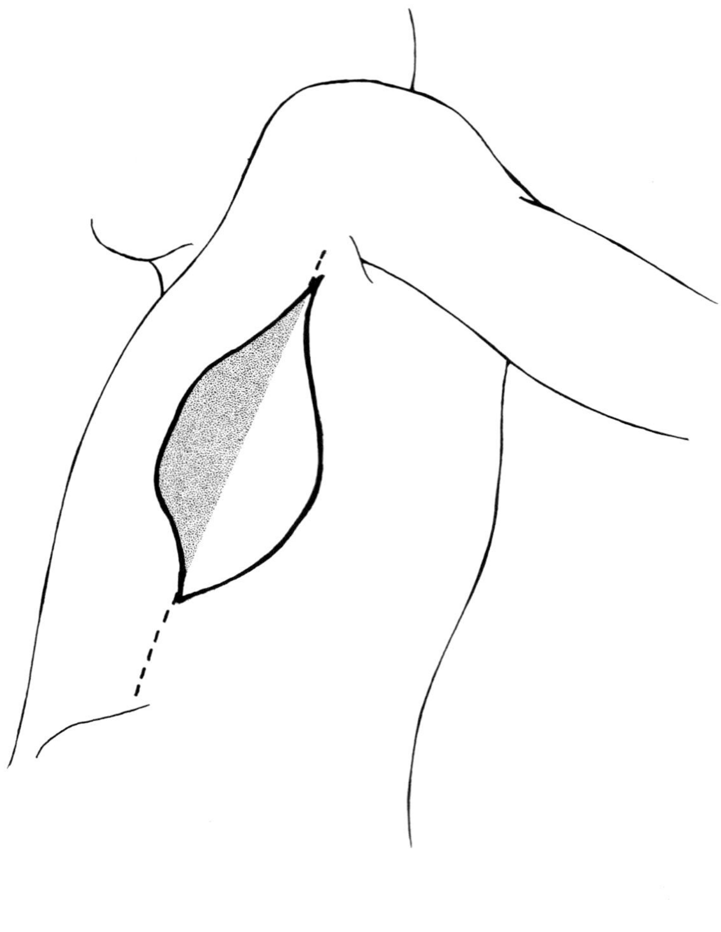

Fig. 38-8. Technique of elevation for segmental latissimus dorsi. Skin island is designed with part of skin located anterior to muscle edge *(stippled area)*. This results in thinner flap design.

musculocutaneous perforators to support the overlying skin. When the width is greater than 8 cm, skin grafts will probably be required for donor site closure. Obviously, the skin territory should be carefully planned. The ability to precisely define skin to fit the reconstructive defect is the unique advantage of microsurgical composite tissue transplantation. Since the skin territory of the latissimus dorsi is large and reliable, the surgeon should be able to design an adequate flap for coverage of the patient's defect.

After preoperative markings are completed, the patient is placed in the lateral decubitus position for flap elevation. The arm should be included in the operative field, since manipulation of the arm will assist in proximal axillary dissection of the thoracodorsal vascular pedicle. The anterior incision is extended to the serratus anterior muscle anteriorly and the posterior and superior incisions to the underlying latissimus dorsi muscle. The inferior latissimus dorsi muscle is divided and the skin muscle island reflected posteriorly so that the muscle is folded on itself (Fig. 38-9). The anterior branch of the thoracodorsal artery is identified superiorly to the proposed skin muscle segment for transplantation (Fig. 38-10).

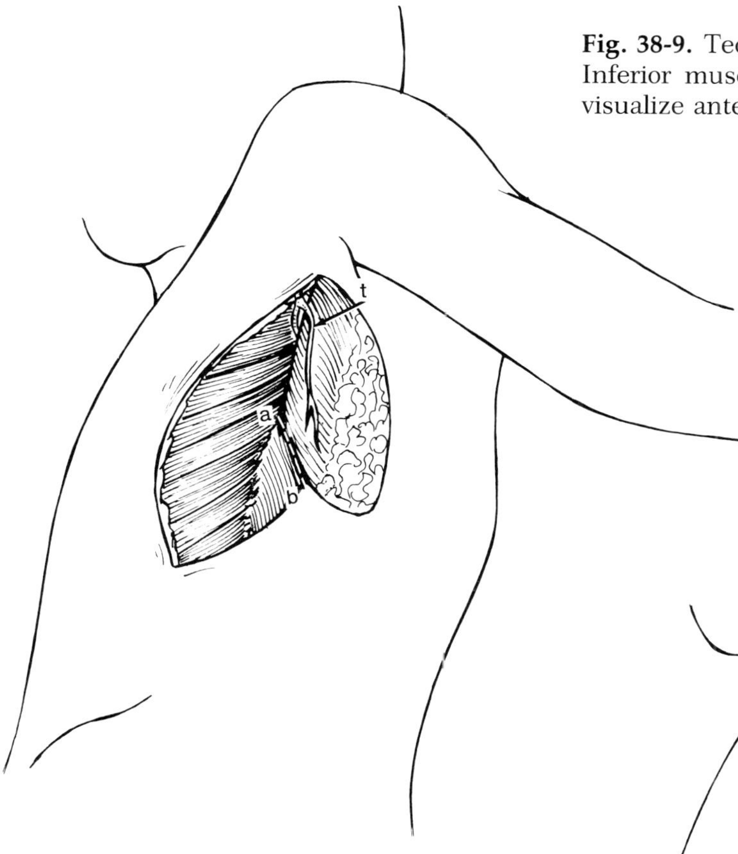

Fig. 38-9. Technique of elevation for segmental latissimus dorsi. Inferior muscle is divided *(a-b)* and flap folded posteriorly to visualize anterior branch of thoracodorsal pedicle *(t)*.

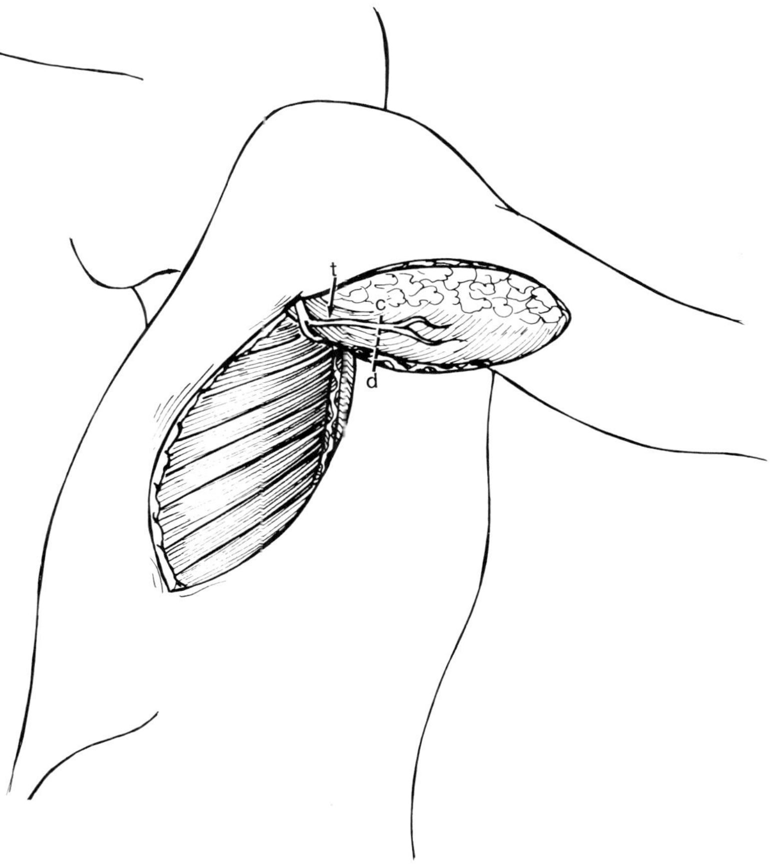

Fig. 38-10. Technique of elevation for segmental latissimus dorsi. Muscle is divided immediately proximal to bifurcation of anterior branch of thoracodorsal artery *(c-d)*. Vascular pedicle is divided distal to junction with posterior branches of thoracodorsal vascular pedicle *(t)*.

This vascular pedicle, the anterior branch of the thoracodorsal artery and its paired venae comitantes, and the associated anterior branch of the motor nerve are dissected proximally to the respective junction of these vessels and nerve to the main vascular-nerve pedicle. This dissection requires division of the epimysium and use of bipolar electrocoagulation to divide the multiple small branches into proximal muscle. This portion of the dissection is generally performed under loupe magnification.

If the flap has been designed over the inferior anterior muscle, generally with the superior apex of skin island at least 10 cm below the axillae, a pedicle length of 6 to 8 cm is achieved before reaching the junction of the anterior artery and paired venae comitantes with the posterior branches of the thoracodorsal vessels. Likewise, the anterior motor branch of the thoracodorsal nerve joins its posterior branches at this level. If this pedicle length is adequate, further proximal dissection is not necessary. Thus the continuity of the vascular pedicle and nerve to the posterior muscle remains undisturbed. More proximal dissection requires division of the nerve and vascular connections to the posterior latissimus dorsi, teres major, and serratus anterior muscles. Proximal dissection may increase pedicle length and lumen diameter, however, function preservation is no longer possible for the posterior muscle fibers. When pedicle dissection is completed, the remaining proximal muscle is divided, using care to preserve the vascular pedicle. After division of the vascular pedicle and motor nerve, the flap is ready for transplantation.

Clinical application

1. Coverage of moderate scalp defects (Fig. 38-11): method of choice

   ADVANTAGES

   a. The muscle contours to skull or dura.
   b. Donor site closure is possible.
   c. Function preservation is important if defect is associated with prior intracranial tumor resection with loss of contralateral motor function.
   d. Simultaneous donor and recipient site dissection are possible with patient either in prone or lateral decubitus position.

2. Coverage of small defects of lower leg and foot: alternative method (Fig. 38-12)

   ADVANTAGES

   a. The distal design of skin island decreases flap bulk.
   b. Function preservation is possible.

   DISADVANTAGES

   a. Simultaneous dissection of donor and recipient site is generally not possible.
   b. Flap bulk is often excessive.

3. Facial defects (e.g., orbital exenteration with maxillectomy): alternative method

   ADVANTAGES

   a. Coverage is accomplished in one operation.
   b. Simultaneous coverage of palate and cutaneous defects is possible.

   DISADVANTAGES

   a. Simultaneous donor and recipient site dissection is not always possible.
   b. Flap bulk is excessive.

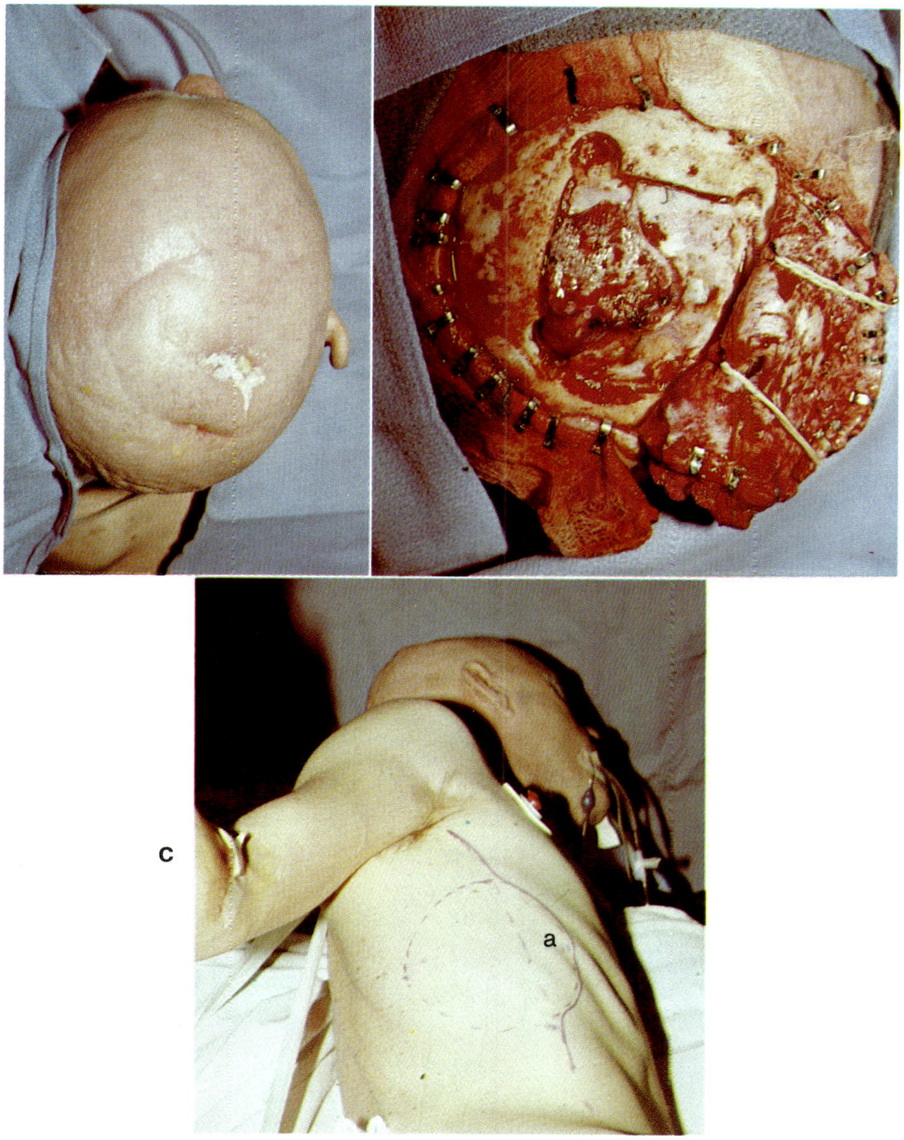

Fig. 38-11. Latissimus dorsi transplantation for skull coverage: segmental muscle with skin island and function preservation.

A, Osteoradionecrosis of skull at site of prior craniotomy and radiation therapy for intracranial tumor.

B, Site of débridement of nonviable infected bone.

C, Design of skin island on ipsilateral anterior latissimus dorsi muscle. *a,* Random portion of skin island.

(From Mathes, S.J., Vasconez, L.O. and Rosenblum, M.L.: Management of the difficult scalp and intracranial wound, Clin. Plast. Surg. **8:**331, 1981; reproduced with permission.)

Continued.

D

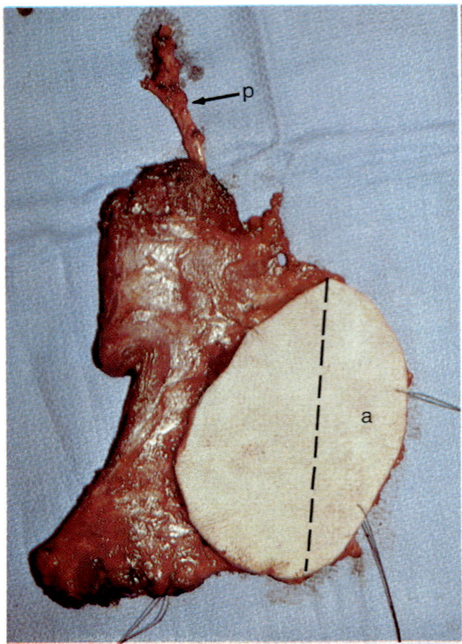

E

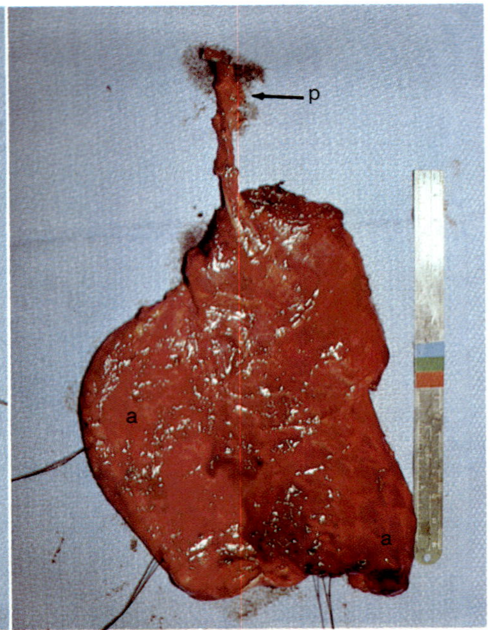

F

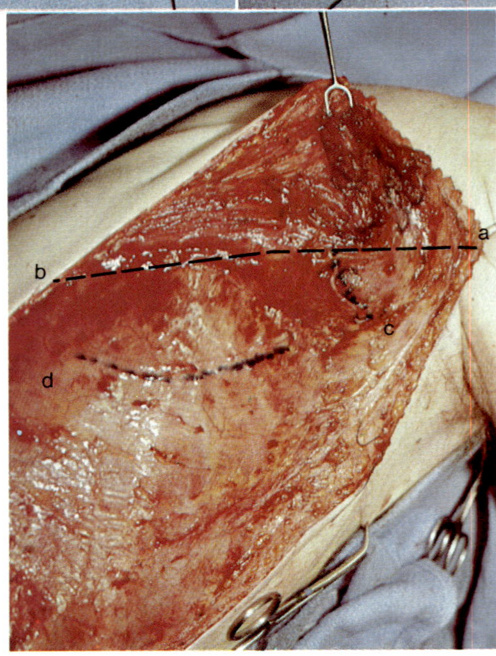

Fig. 38-11, cont'd

D, Anterior latissimus dorsi musculocutaneous flap. *a,* Random portion of skin island; *p,* vascular pedicle to flap (consists of anterior branch of thoracodorsal artery and vein).

E, Deep surface of free flap demonstrates random portion *(a)* of skin island. *p,* Vascular pedicle.

F, Segmental transplantation of latissimus dorsi preserves posterior muscle. *(a-b),* Edge of posterior latissimus dorsi with intact vascular pedicle and motor nerve. *c-d,* Former location of anterior edge of latissimus dorsi muscle.

G

H

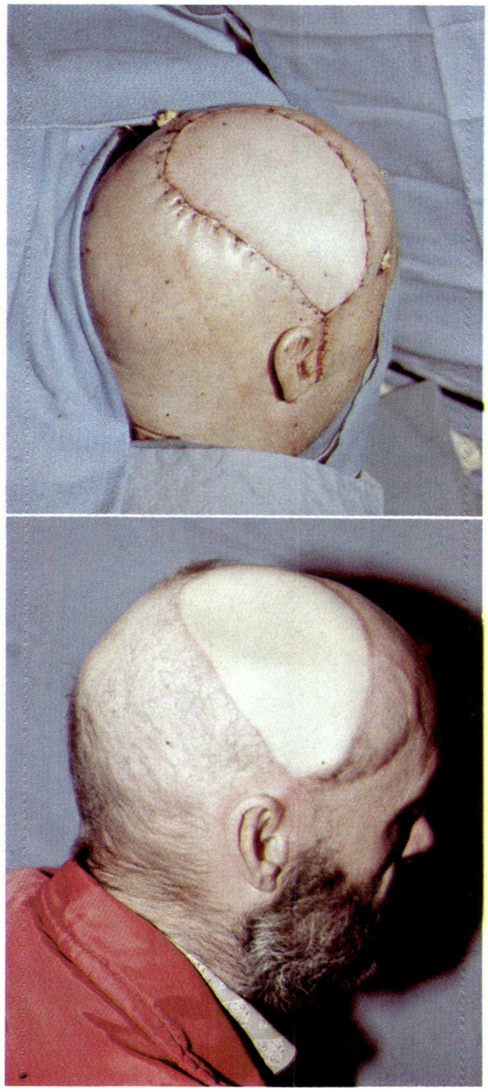

Fig. 38-11, cont'd

G, Free flap revascularized by end-to-side anastomoses between anterior branch of thoracodorsal artery and venae comitantes to superficial temporal artery and vein. Donor site closed directly.

H, Three-month postoperative view demonstrates stable wound coverage with preserved latissimus dorsi muscle function. Patient resumed chemotherapy 2 weeks after surgical reconstruction.

A B C D

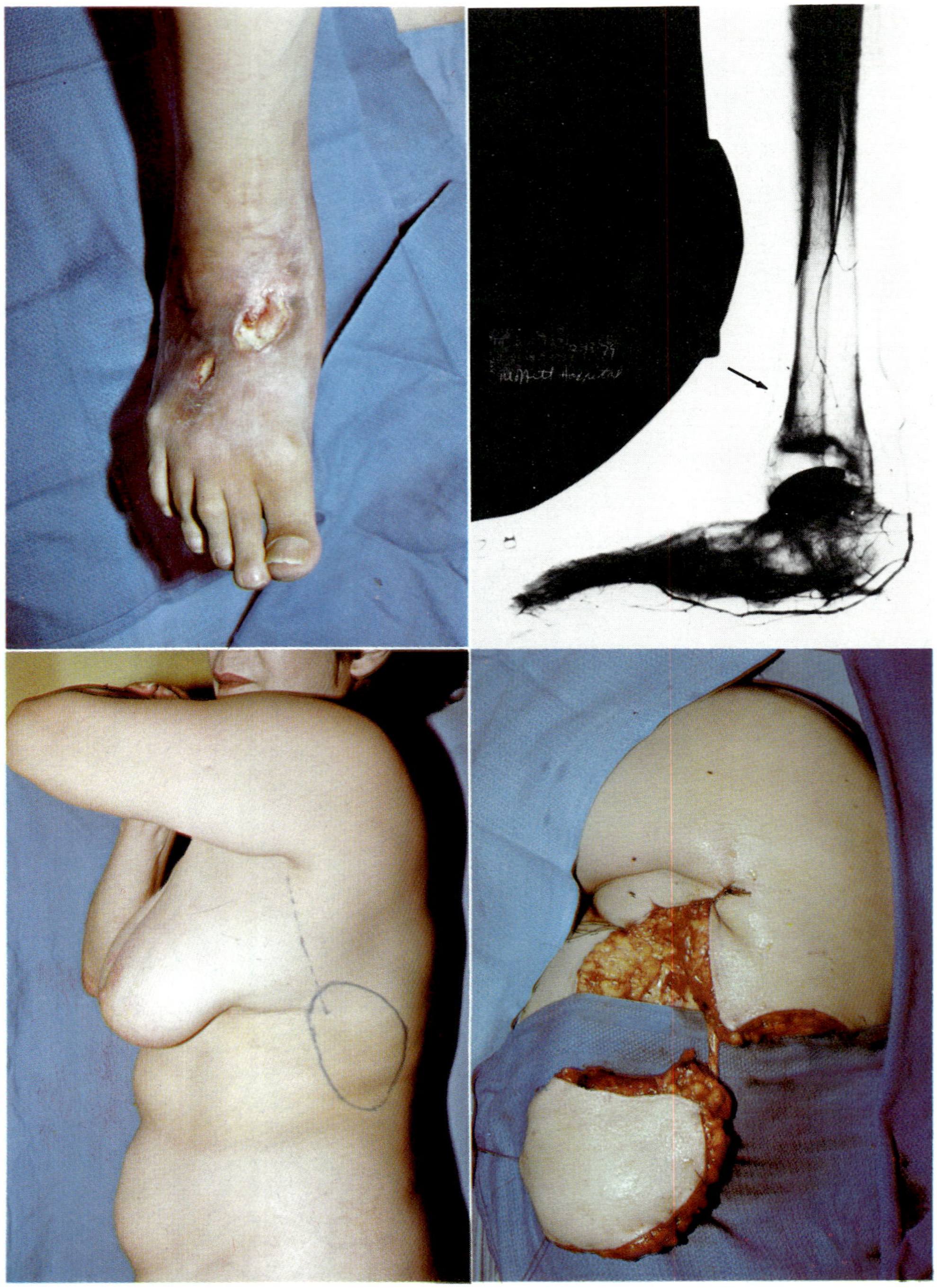

Fig. 38-12. Latissimus dorsi transplantation for lower extremity coverage: segmental muscle with skin island and function preservation.

A, Exposed ankle prosthesis in patient with rheumatoid arthritis.

B, Radiograph demonstrates normal posterior tibial artery and absence of dorsalis pedis artery. Arrow denotes site of patent anterior tibial artery used as recipient site for flap revascularization.

C, Skin island located over inferior anterior muscle to obtain thin flap and preserve posterior half of latissimus dorsi muscle.

D, Anterior latissimus dorsi musculocutaneous flap isolated on vascular pedicle.

(From Mathes, S. J., and Nahai, F.: Classification of the vascular anatomy of muscles: experimental and clinical correlation, Plast. Reconstr. Surg. **67**:177, 1981.)

E

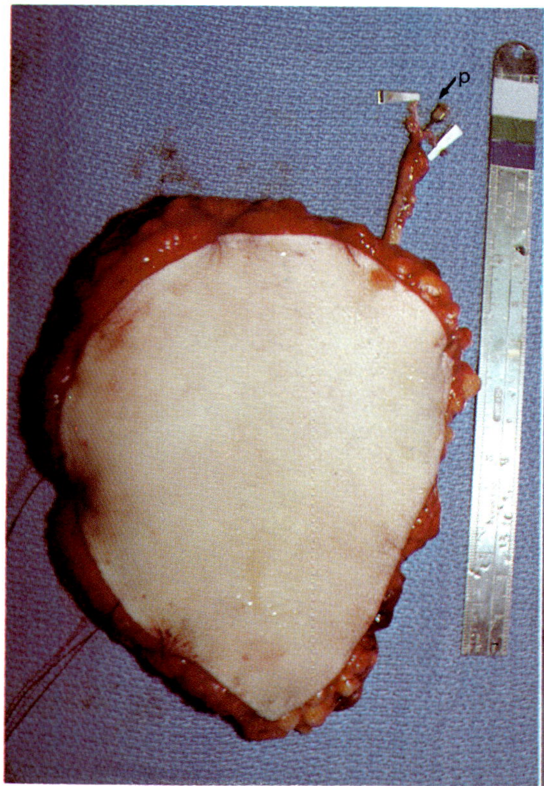

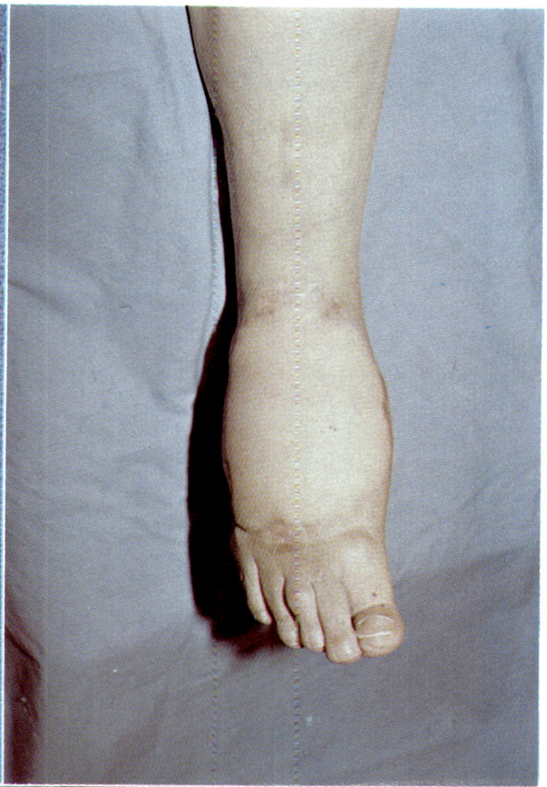

F

G

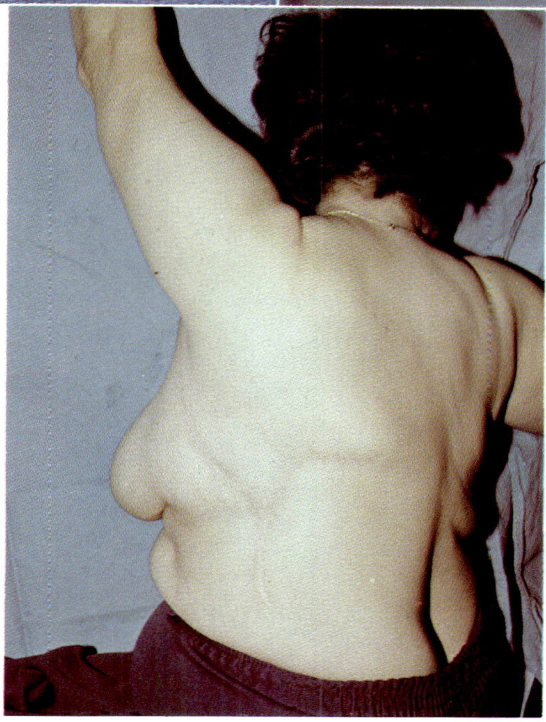

Fig. 38-12, cont'd

E, Anterior segmental latissimus dorsi musculocutaneous flap ready for transplantation. *p*, Vascular pedicle, consisting of anterior branch of thoracodorsal artery and paired venae comitantes.

F, One-year postoperative view of lower extremity. Stable coverage of wound provided by free flap.

G, Donor site closed directly. Posterior half of latissimus dorsi muscle preserves muscle function.

Latissimus dorsi muscle transplantation

Technique of elevation. The entire muscle may be transplanted for coverage of large defects or muscle function. The muscle either includes a skin island or requires skin grafts. The incision is made at the anterior border of the muscle along the posterior axillary line. The posterior muscle is exposed by elevation of the back skin between the scapula and iliac crest. The muscle origin is divided from the iliac crest, lumbosacral fascia, and posterior vertebral column. Before the origin is entirely released, clips or sutures are placed at designated intervals along the muscle edge so that proper muscle length is restored if the unit is transplanted for functional purposes.

The muscle unit is now reflected onto the axilla where the anterior muscle surface is visualized. At this level, 10 cm distally to the muscle insertion, the thoracodorsal artery, vein, and nerve are easily visualized entering the anterior surface of the muscle belly. The vascular pedicles to the serratus anterior and teres major are divided. Proximal dissection of the nerve, thoracodorsal artery, and vein is performed, depending on required pedicle lengths previously determined by measurement of the distance of recipient vessels to the donor defect.

After division of the muscle insertion, nerve, artery, and vein, the flap is ready for transplantation.

Clinical application

1. Large scalp defects (Fig. 38-13): method of choice

   ADVANTAGES

   a. The muscle will cover the entire skull.
   b. Skin grafts on muscle surface avoid bulky reconstruction.
   c. Donor site closure is possible.

2. Upper extremity coverage: alternative method

   ADVANTAGES

   a. The muscle will cover extensive extremity defects.
   b. Single-stage reconstruction allows early attention to rehabilitation.

   DISADVANTAGES

   a. Simultaneous donor and recipient site dissection are generally not possible.
   b. Muscle bulk is often excessive.

3. Lower extremity coverage: method of choice for large defects (Fig. 38-14)

   ADVANTAGES

   a. Muscle will cover extensive defects.
   b. Skin island is large enough to cover one third to one half of lower leg defects.

   DISADVANTAGES

   a. Simultaneous donor and recipient site dissection is not possible.
   b. Donor site closure is not always possible.
   c. Musculocutaneous flap is often too bulky in the obese patient.

A

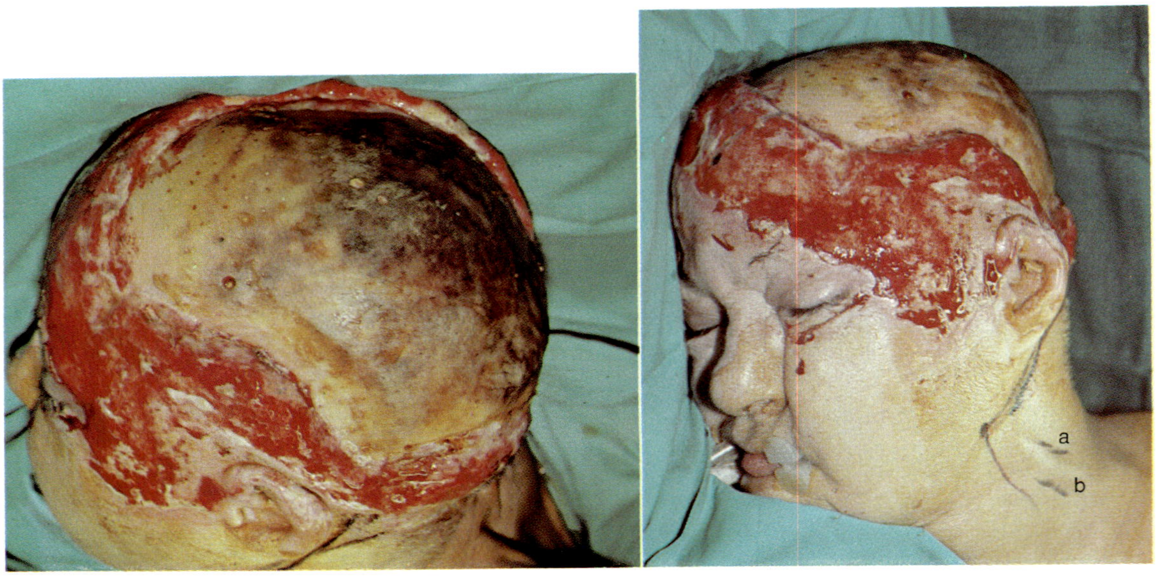

B

Fig. 38-13. Latissimus dorsi transplantation for skull coverage (entire muscle with skin island).

A, Full-thickness burn of entire scalp and half of outer table of skull.

B, Lateral decubitus position allows simultaneous exposure of receptor vessels in neck *(a-b)* and elevation of latissimus dorsi musculocutaneous flap.

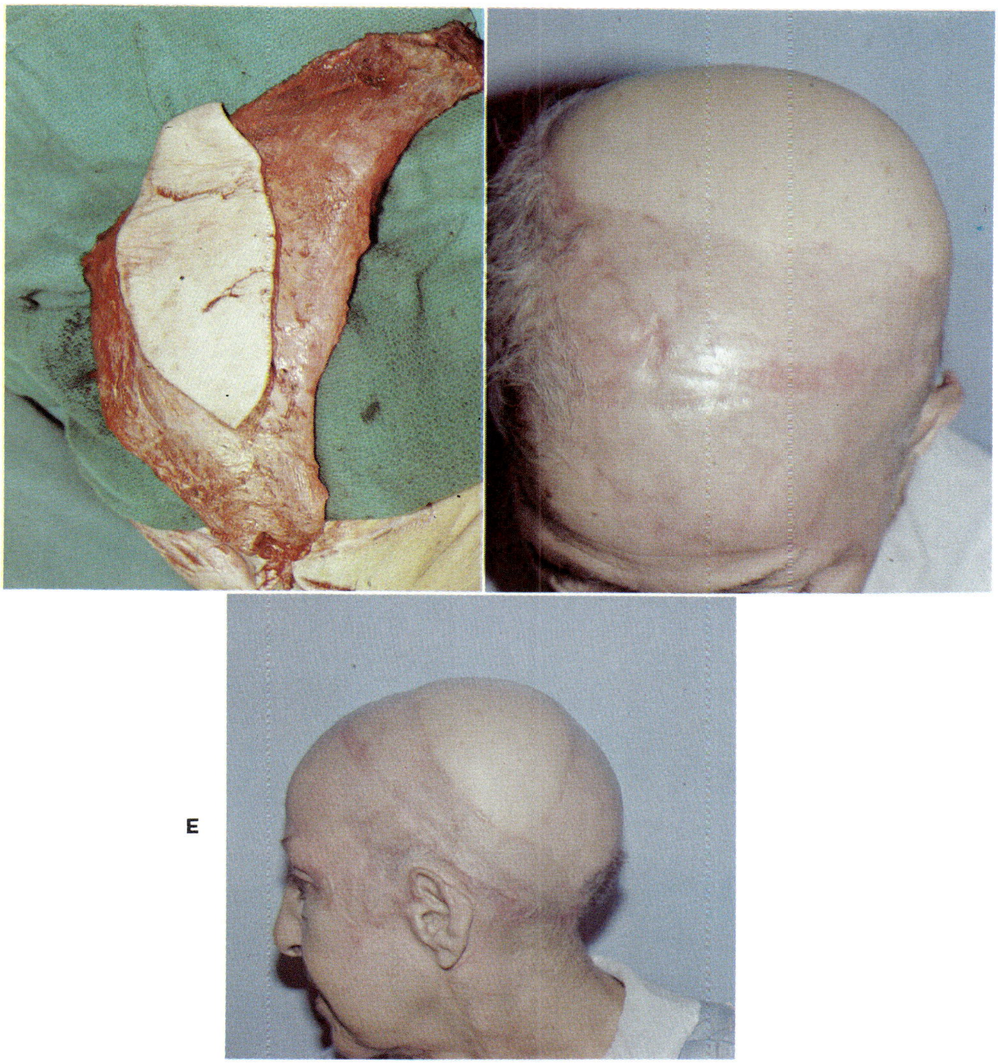

Fig. 38-13, cont'd

C, Entire latissimus dorsi muscle with large skin island revascularized by end-to-end anastomoses between thoracodorsal artery and vein and superior thyroid artery and external jugular vein.

D, Revascularized musculocutaneous flap placed directly over skull. Exposed muscle is covered with skin grafts.

E, Postoperative view demonstrates stable coverage provided by free flap.

A B C D

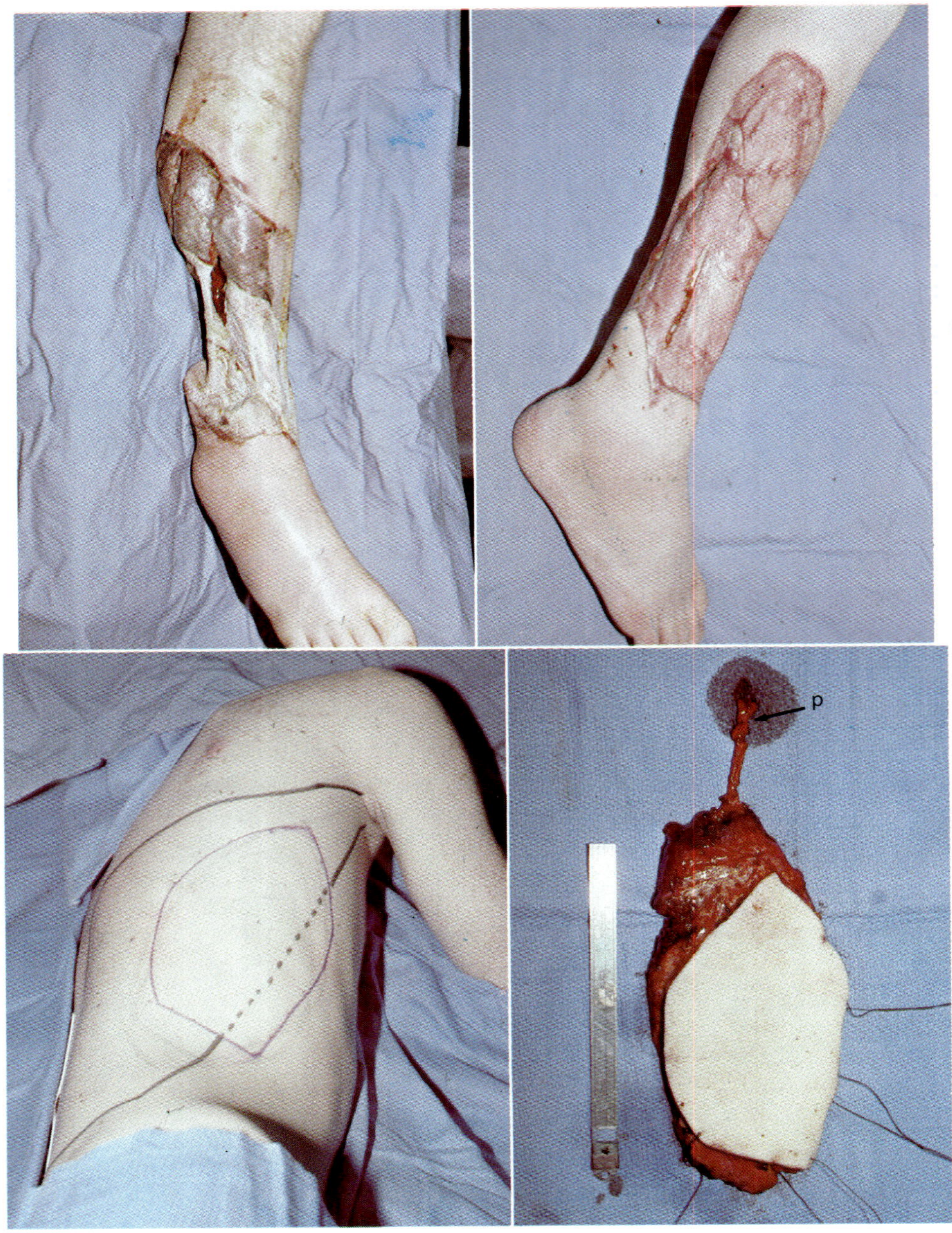

Fig. 38-14. Latissimus dorsi transplantation for lower extremity coverage: muscle with anterior skin island.

A, Porcine heterografts provide temporary coverage for extensive avulsion injury of lower half of leg.

B, Skin grafts provide immediate cover. Skin unstable over areas of exposed bone.

C, Design of latissimus dorsi musculocutaneous flap for coverage of lower third of leg.

D, Latissimus dorsi musculocutaneous flap ready for transplantation. *p,* Vascular pedicle consisting of thoracodorsal artery and vein.

E F G H

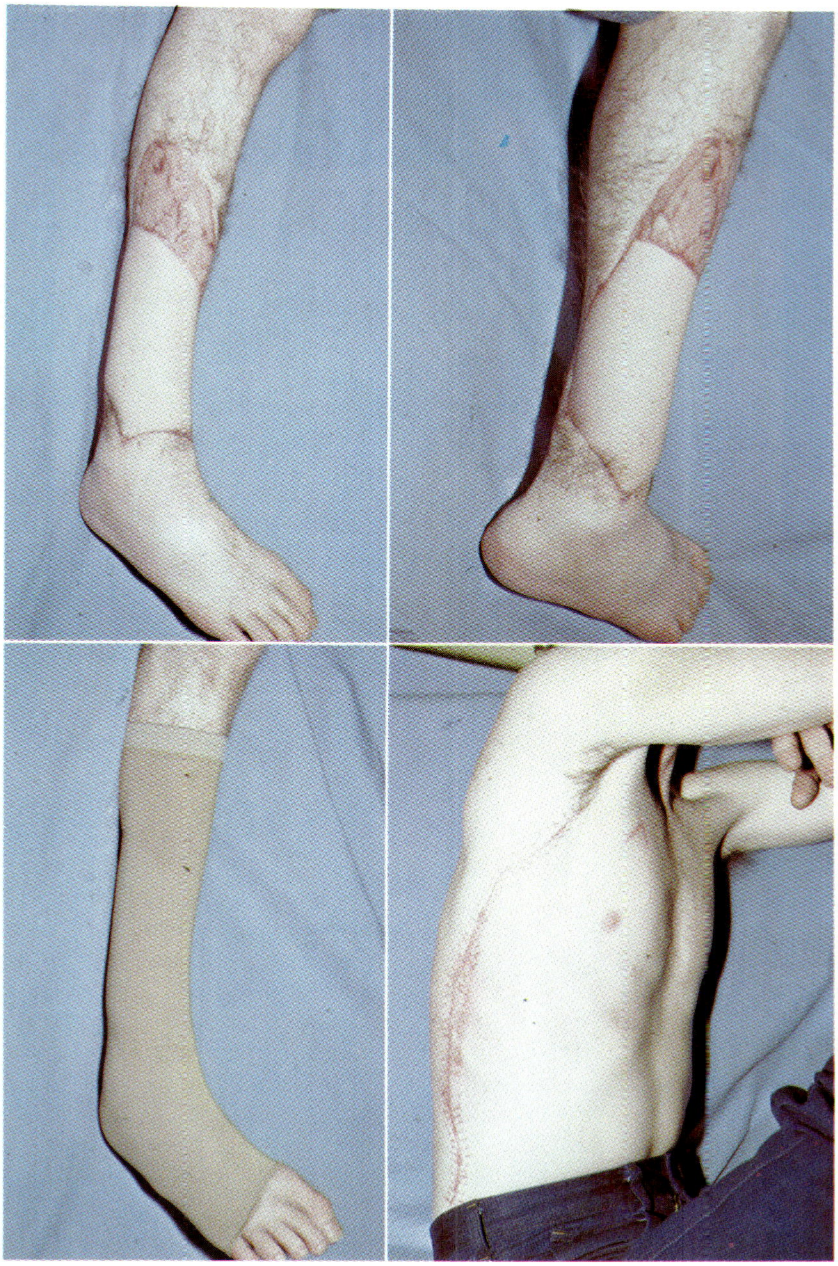

Fig. 38-14, cont'd

E, Flap vascularized by end-to-side anastomoses of thoracodorsal artery and vein to anterior tibial artery and vein.

F, Three-month postoperative view demonstrates stable coverage of wound by free flap. NOTE: Skin grafts on exposed gastrocnemius muscle belly do not require flap coverage.

G, Jobst compressive stocking recommended for initial 6 months after lower extremity flap transplantation to avoid flap edema.

H, Donor site is closed directly. Patient notes no functional disability. Motor nerve is preserved to remaining posterior fibers of latissimus dorsi muscle.

Functional transplantation

Technique of elevation. Flap design and elevation are similar to method described for complete muscle transplantation. The motor nerve, the thoracodorsal nerve, is located adjacent to the vascular pedicle (Figs. 38-3 to 38-5). The nerve has a length of 8 cm. Both tendon of insertion and the lumbosacral fascia of origin are preserved for use for attachment of muscle at its site of transplantation. Sutures at specific intervals before division of the flap from the donor site ensure maintenance of appropriate length at transplantation site.

Clinical application

1. Upper extremity muscle function (long flexors): alternative technique

   ADVANTAGES
   a. Muscle provides both coverage and function.
   b. The nerve has long length, eliminating need for nerve grafts to reach suitable recipient nerve.

   DISADVANTAGES
   a. The muscle has excessive power.
   b. The muscle has excessive size and bulk.
   c. Simultaneous donor and recipient site dissections are not generally possible.

2. Lower extremity muscle function: alternative technique

   ADVANTAGES
   a. Muscle has adequate power to replace long flexor or extensor muscle function.
   b. Simultaneous defect coverage is accomplished.

   DISADVANTAGES
   a. Simultaneous dissection of donor flap and recipient site is generally not possible.

Neurosensory latissimus dorsi flap

Technique of elevation. If the skin island of the latissimus dorsi unit is located over the posterior muscle, the sensory branches of the posterior intercostal nerve may be preserved during flap elevation. Unfortunately, these nerves are short and segmentally distributed and are rarely useful as donor sensory nerves. The sensory branches are usually too short to reach recipient sensory nerves adjacent to a defect requiring both coverage and sensation. However, in certain instances these nerves should be considered where coverage and sensation are required and appropriate recipient vessels are available immediately adjacent to the defect.

GRACILIS

Gracilis, an anterior thigh muscle, may be transplanted either as a muscle or musculocutaneous unit for coverage or functional muscle. The donor defect is minimal, since the closure scar is located on the medial thigh. Its function as an adductor is preserved by the stronger adductor longus and adductor magnus muscles. The muscle width is 4 to 6 cm, and it extends the entire length of the medial thigh. The cutaneous territory of this muscle may include the skin over the proximal half of the muscle with a width of 8 cm.

Vascular anatomy

The gracilis muscle has a type II vascular pattern with a dominant arterial pedicle, the terminal branches of the medial femoral circumflex artery entering the proximal muscle, and one or two minor pedicles, branches of the superficial femoral artery entering the distal muscle. The proximal dominant pedicle enters the muscle approximately 10 cm inferiorly to the pubic tubercle (Fig. 38-15). This pedicle originates from the profunda femoris artery as the medial femoral circumflex artery and courses beneath the adductor longus muscle and superficial to the adductor magnus muscle. After providing muscular branches to these muscles, the artery terminates as two or three branches entering the medial deep surface of the gracilis muscle. The pedicle length between the profunda femoris vessels and gracilis muscle is generally 6 cm (Figs. 38-16 and 38-17). This pedicle has an external lumen diameter of 2 mm. Division of the one or two minor pedicles from the superficial femoral artery may be accomplished without adverse effect on muscle survival. However, the survival of skin overlying the distal muscle is not reliable after the minor pedicles are divided. Although the point of entrance of the major pedicle into the gracilis muscle may vary slightly, major vascular anomalies in regard to the major pedicle have not been observed.

The veins of the gracilis muscle are located with the arterial pedicle as paired venae comitantes (Fig. 38-17). The entire muscle and proximal half of its cutaneous territory are adequately drained by the paired venae comitantes associated with the medial femoral circumflex artery. These veins course with the artery and either unite as a single vein or continue as paired veins, eventually draining into the deep veins associated with the profunda femoris artery. The proximal veins of the dominant pedicle have an external lumen diameter in the range of 1.5 to 2.5 mm and a pedicle length of 6 cm.

Neural (motor) anatomy

The motor nerve of the gracilis muscle is the anterior branch of the obturator nerve. This nerve enters the muscle approximately 2 to 3 cm superiorly to the point of entrance of the major vascular pedicle (Figs. 38-16 and 38-17). The motor nerve may be dissected proximally between the adductor longus and adductor magnus muscles to increase donor nerve length.

Neural (sensory) anatomy

The obturator nerve has small sensory branches entering the medial thigh skin. There is no specific sensory branch that can be located consistently with the proximal cutaneous territory as a donor sensory nerve.

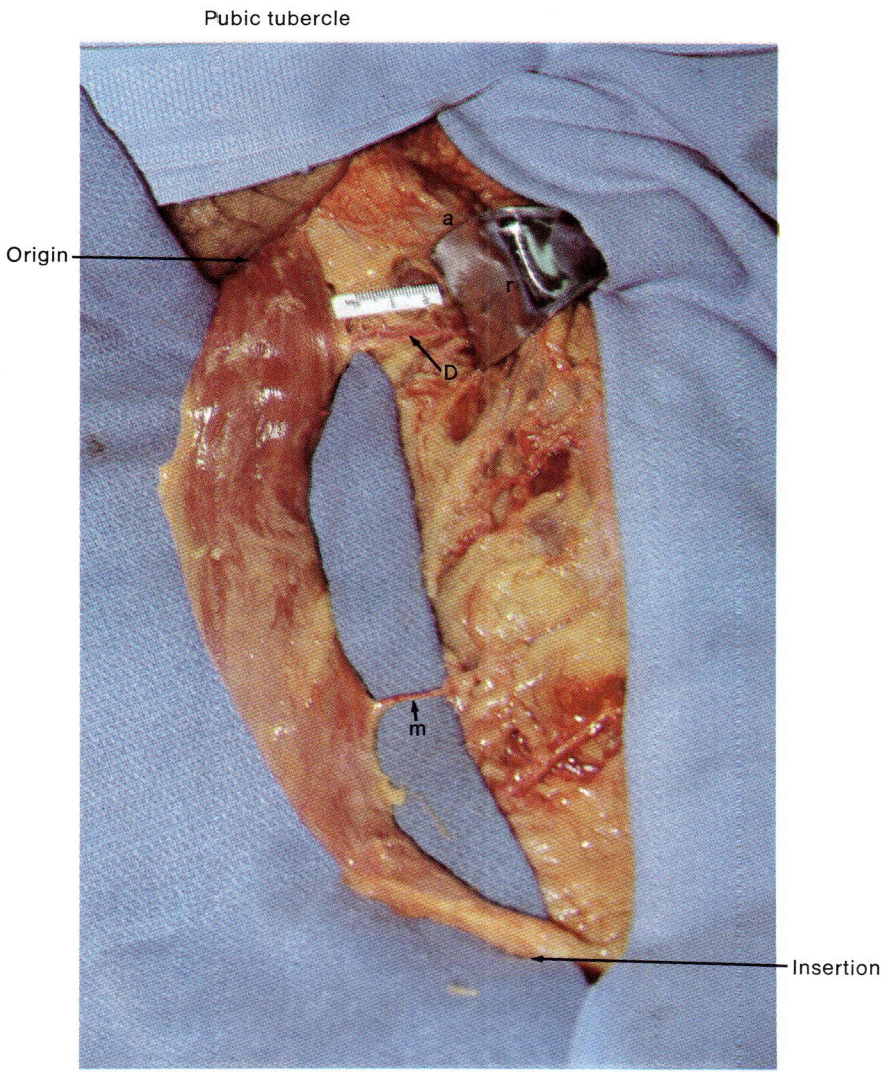

Fig. 38-15. Vascular anatomy of gracilis. Anterior view of medial thigh with exposed gracilis muscle. Retraction *(r)* of adductor longus muscle *(a)* exposes dominant vascular pedicle *(D)* (medial femoral circumflex artery and vein). Minor pedicle *(m)*, branch of superficial femoral artery and vein, is demonstrated.

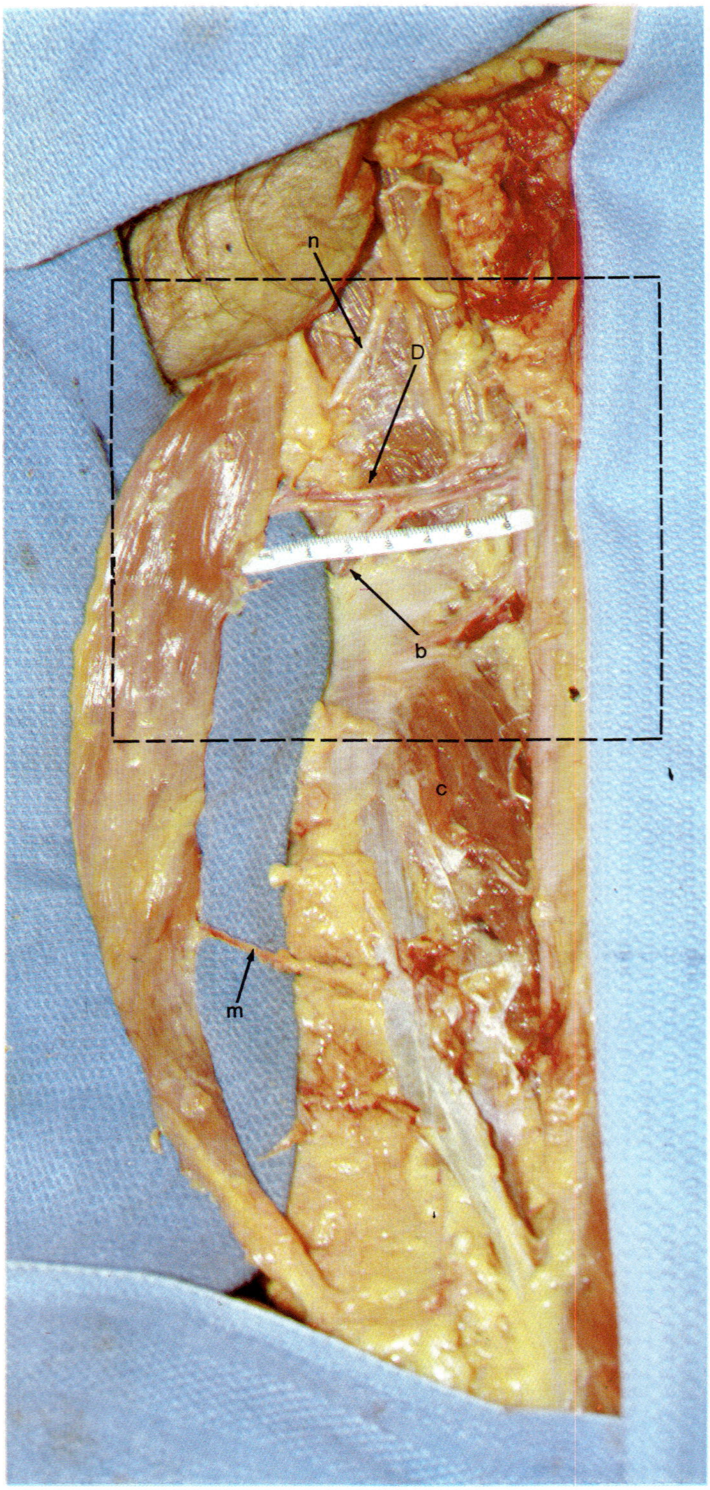

Fig. 38-16. Vascular anatomy of gracilis. Resection of adductor longus muscle in cadaver allows demonstration of complete course of dominant vascular pedicle *(D)*, medial femoral circumflex artery, and paired venae comitantes. *b*, Branch of vascular pedicle to adductor magnus muscle *(c)*; *m*, minor pedicle to muscle; *n*, obturator nerve.

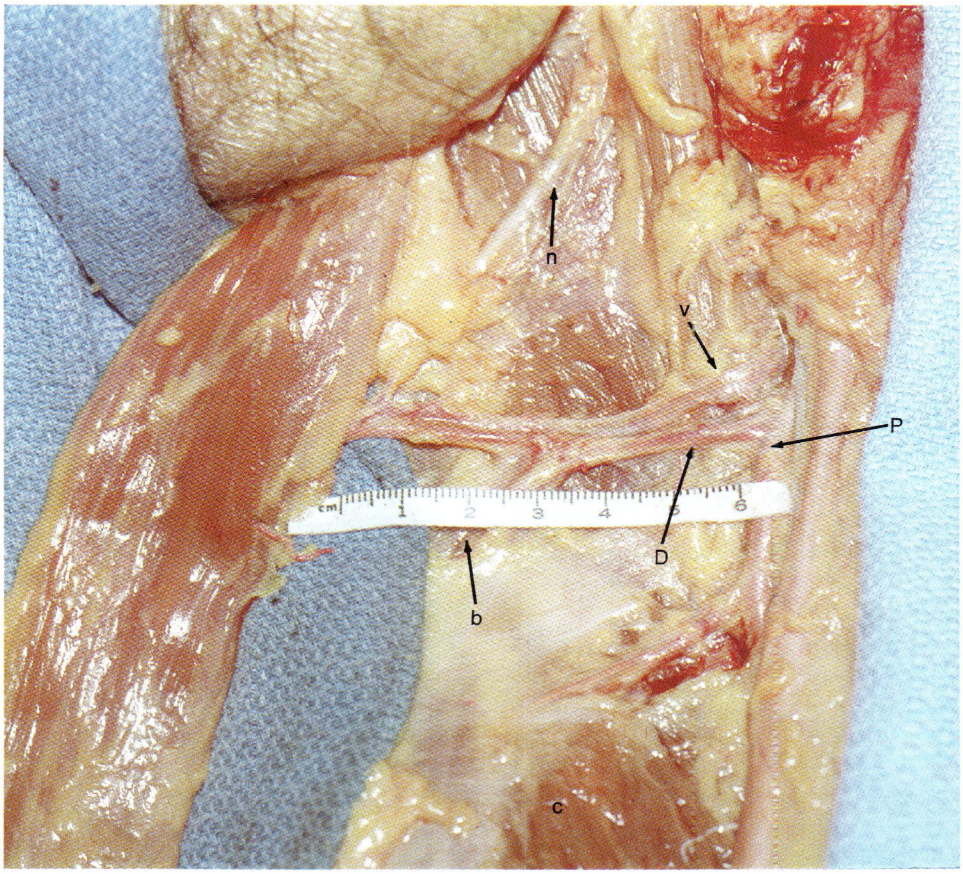

Fig. 38-17. Vascular anatomy of gracilis. Closeup view of vascular anatomy of dominant vascular pedicle to gracilis muscle. *p,* Profunda femoris artery; *D,* medial femoral circumflex artery; *b,* branch to adductor magnus muscle *(c); v,* medial femoral circumflex vein; *n,* obturator nerve.

Flap characteristics

SIZE: The gracilis muscle measures 5 to 6 cm in width over the proximal 15 cm. Its distal portion gradually tapers as it approaches the knee. When transplanted as free muscle flap for coverage, the proximal useful portion measures 6 × 15 cm. For functional transplantation the muscle unit measures 30 cm, although its length may be shortened by excision of muscle proximal to the dominant vascular pedicle.

Only the proximal half of the cutaneous territory of the gracilis muscle is reliable when this unit is transplanted on its proximal vascular pedicle. This skin territory measures 6 to 8 cm in width and 15 cm in length. Survival of this proximal skin territory requires accurate localization of the skin territory over the muscle and careful preservation of the musculocutaneous perforating vessels.

THICKNESS: The gracilis muscle is ideal for coverage of small defects. The muscle is thin, measuring 2 to 3 cm in thickness, and can be tailored as required to fit small bone or soft tissue defects.

The musculocutaneous unit is generally thick because of excessive subcutaneous tissue in the medial thigh. In the male patient this skin-muscle unit may be thin enough to serve as a musculocutaneous unit, especially when skin coverage is required for functional transplantation of the gracilis unit in upper extremity reconstruction.

PLIABILITY: The gracilis muscle unit is pliable and can be folded as required to fill irregular defects. Its small size is ideally suited to fill bone defects (e.g., sequestrectomy site in treatment of osteomyelitis).

DONOR SITE: When only muscle is transplanted for use in distant coverage, a vertical incision overlying the proximal two thirds of the muscle belly is required. A closed suction system is placed in the donor region and skin closed directly. The donor scar is well accepted by patients. The remaining strong adductor muscles allow functional preservation. Direct closure of skin is also possible after transplantation of the gracilis musculocutaneous unit, although the donor site scar now extends inferiorly to the knee. There also may be slight contour deformity in the medial thigh.

Distant coverage

Technique of elevation. The gracilis unit may be transplanted either as a muscle or musculocutaneous flap for distant coverage. An incision directly over the muscle allows rapid identification of the gracilis. However, the gracilis musculocutaneous flap requires careful planning, since the muscle is narrow, and the overlying skin is not adherent to the muscle.

The design of the cutaneous territory of the gracilis muscle begins with identification of the distal gracilis muscle adjacent to its insertion (Fig. 38-18). The flap elevation is performed with the patient in the supine position. A line is drawn between the pubic tubercle and the medial condyle of the knee. The muscle is located slightly posterior to the skin marking. A preliminary incision is made over the distal one third of the gracilis muscle. Through this incision the saphenous vein and fleshy muscle belly of the sartorius muscle are encountered. The gracilis muscle is posterior to these structures and may be recognized by its muscle-tendon configuration. Slightly posterior to the distal gracilis muscle is the tendinous insertion of the semimembranosus muscle. After the distal gracilis muscle is isolated and encircled with a Penrose drain, traction is applied to the muscle. The proximal muscle can now be palpated beneath the medial thigh skin and the cutaneous territory outline centered over the muscle. When the skin island is placed directly over the muscle, the skin dimensions may include the proximal half of the medial thigh skin with a maximum width of 8 cm. Since the dominant vascular pedicle enters the medial muscle belly approximately 10 cm inferiorly to the pubic tubercle, the exact design of the skin island on the muscle belly must be based on the relationship of the recipient defect and the planned receptor vessels.

The skin is incised around the designed skin island with immediate placement of temporary sutures approximating the skin edge to the underlying gracilis muscle. Minimal traction of the skin is necessary until these sutures are placed to avoid disruption of musculocutaneous perforating vessels. The one or two minor pedicles are ligated as they enter the distal half of the muscle.

The medial adductor longus is retracted anteriorly at 10 cm inferior to the pubic tubercle. Immediately beneath the adductor longus muscle, the large muscle belly of the adductor magnus muscle is visualized. The terminal branches of the medial circumflex

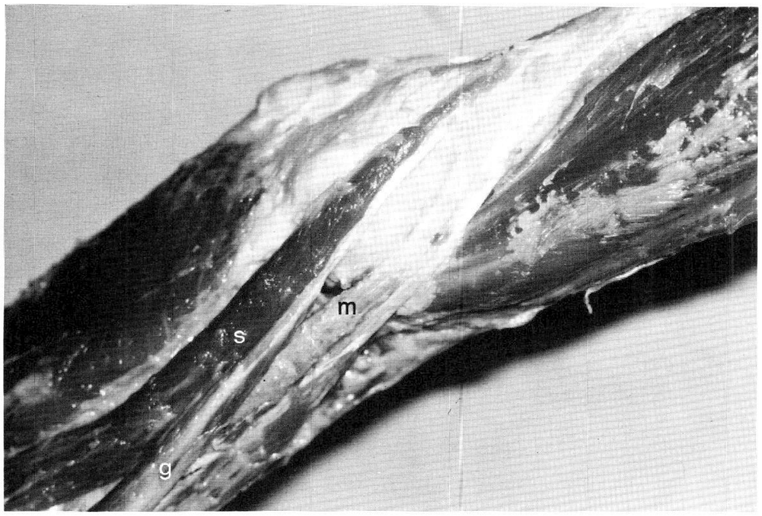

Fig. 38-18. Technique of elevation for gracilis musculocutaneous flap. Distal location of gracilis tendon allows correct design of skin island on gracilis muscle. Round tendon of gracilis *(a)* lies between muscular sartorius *(s)* anteriorly and fascial tendon of semimembranosus *(m)* posteriorly.

artery with its paired venae comitantes course between the adductor longus and adductor magnus muscles, entering the deep surface of the gracilis muscle, and are easily visualized (Fig. 38-15). After the vascular pedicle is identified, the pedicle is mobilized by division of muscular branches to the adductor longus and adductor magnus muscles until its junction with the profunda femoris artery and vein is reached. The pedicle is not divided after its dissection is completed until the remainder of the donor and recipient site dissections are completed.

The motor nerve is divided as it enters the proximal muscle, and this portion of the muscle is mobilized. Generally, the muscle continuity is also preserved until the recipient site dissection is completed. This prevents accidental traction injury to the vascular pedicle until it is ready for transplantation. When the proximal and distal muscle and vascular pedicles are divided, the flap is ready for microvascular transplantation to the recipient site.

Clinical application

1. Facial defects (e.g., orbital exenteration and maxillectomy) and contour deformities (e.g., hemifacial atrophy): alternative method
   ADVANTAGES
   a. Muscle has minimal bulk and is tailored to exact dimensions to fill defect.
   b. Simultaneous dissection of donor and recipient sites is possible.
   DISADVANTAGES
   a. Muscle is covered with skin grafts or beneath existing skin since musculocutaneous flap is generally too bulky.
2. Lower extremity and foot (Fig. 38-19) small defect: method of choice
   ADVANTAGES
   a. Muscle has minimal bulk and is tailored for exact dimensions to fill defect.
   b. Exposed muscle is covered with skin grafts.
   c. Simultaneous dissection of donor and recipient sites is possible.
3. Lower extremity osteomyelitis (Figs. 38-20 to 38-23): method of choice
   ADVANTAGES
   a. Muscle fits into sequestrectomy site.
   b. Extensive skin coverage is rarely required.
   c. Muscle dissection is performed on opposite medial thigh and is rapidly accomplished, since skin island is not required.
   d. Skin grafts are applied to exposed muscle surface.
   e. Simultaneous dissection of donor and recipient sites is always possible.
   f. The muscle can be turned in any direction, depending on local requirements and site of recipient vessels.

A B

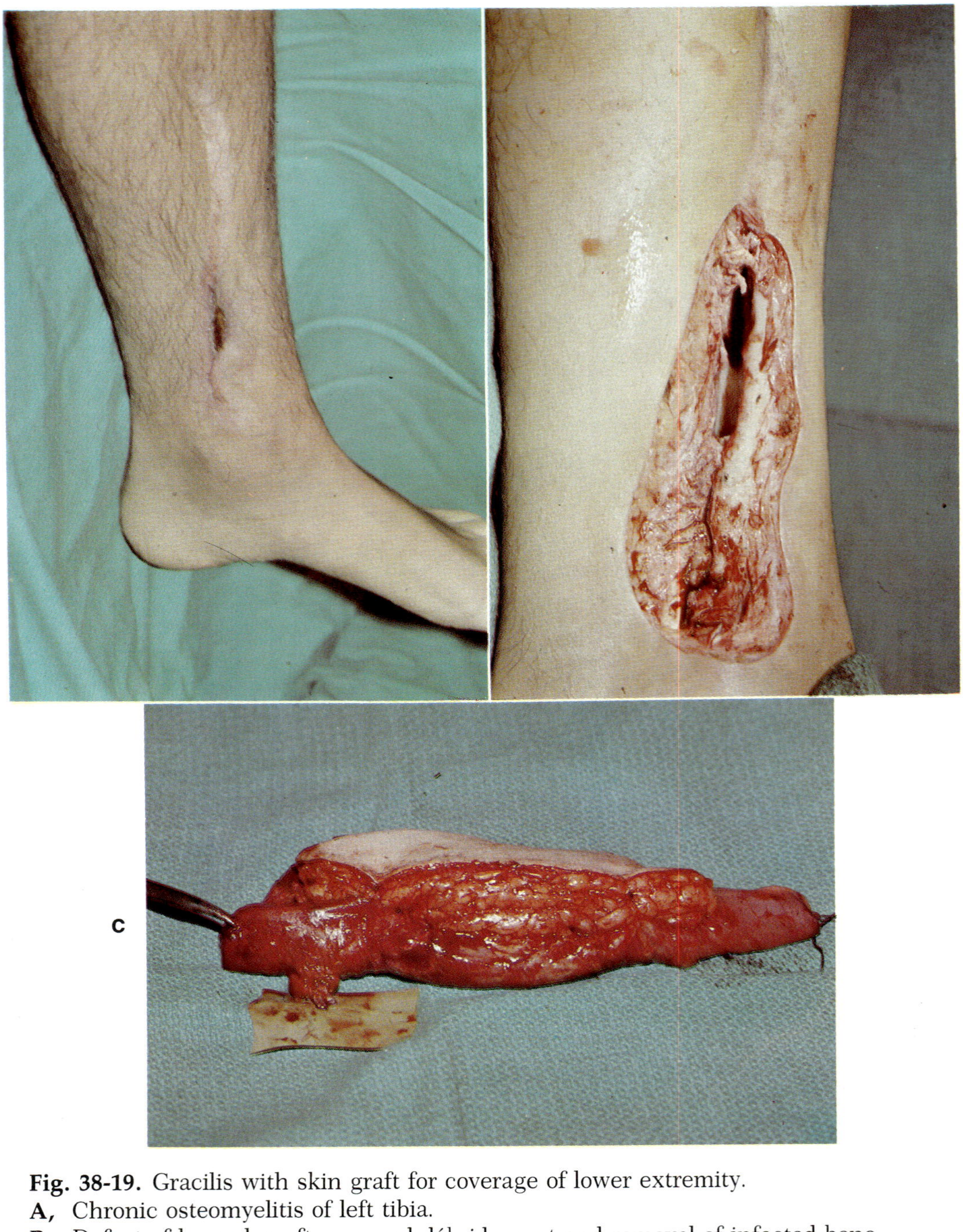

C

Fig. 38-19. Gracilis with skin graft for coverage of lower extremity.
A, Chronic osteomyelitis of left tibia.
B, Defect of lower leg after wound débridement and removal of infected bone.
C, Gracilis musculocutaneous flap isolated. NOTE: Thick subcutaneous layer.

(**A** to **D** from Nahai, F., and Mathes, S.J.: Aesthetic aspects of reconstructive microsurgery of the lower extremity, Clin. Plast. Surg. **8**:369, 1981; reproduced with permission.)

D

E

F

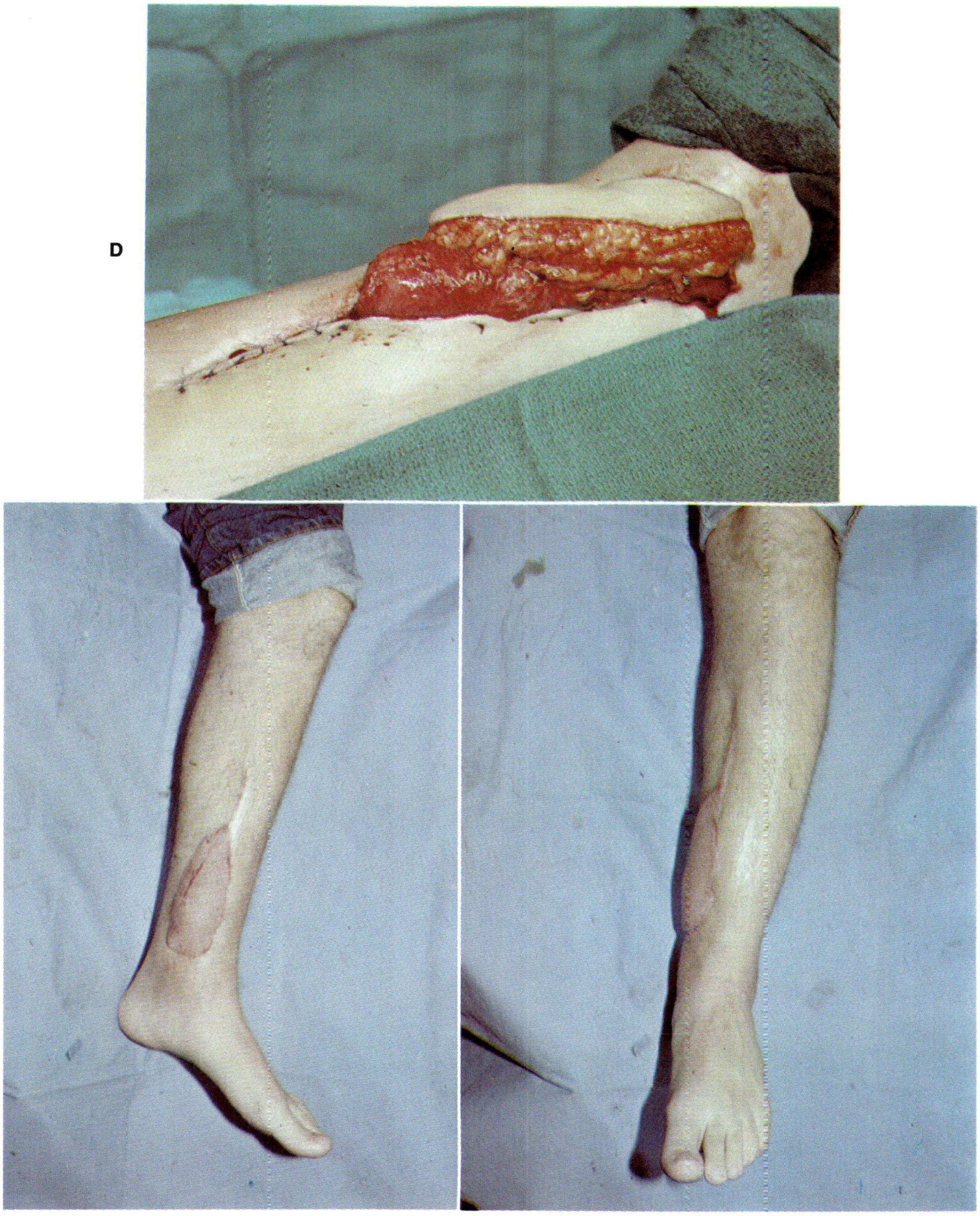

Fig. 38-19, cont'd

D, Flap revascularized and muscle inset. Skin island and subcutaneous fat were immediately excised and the skin harvested as split skin to cover the muscle.

E and **F,** Postoperative result. Stable coverage with resolution of osteomyelitis. Satisfactory contour of lower leg.

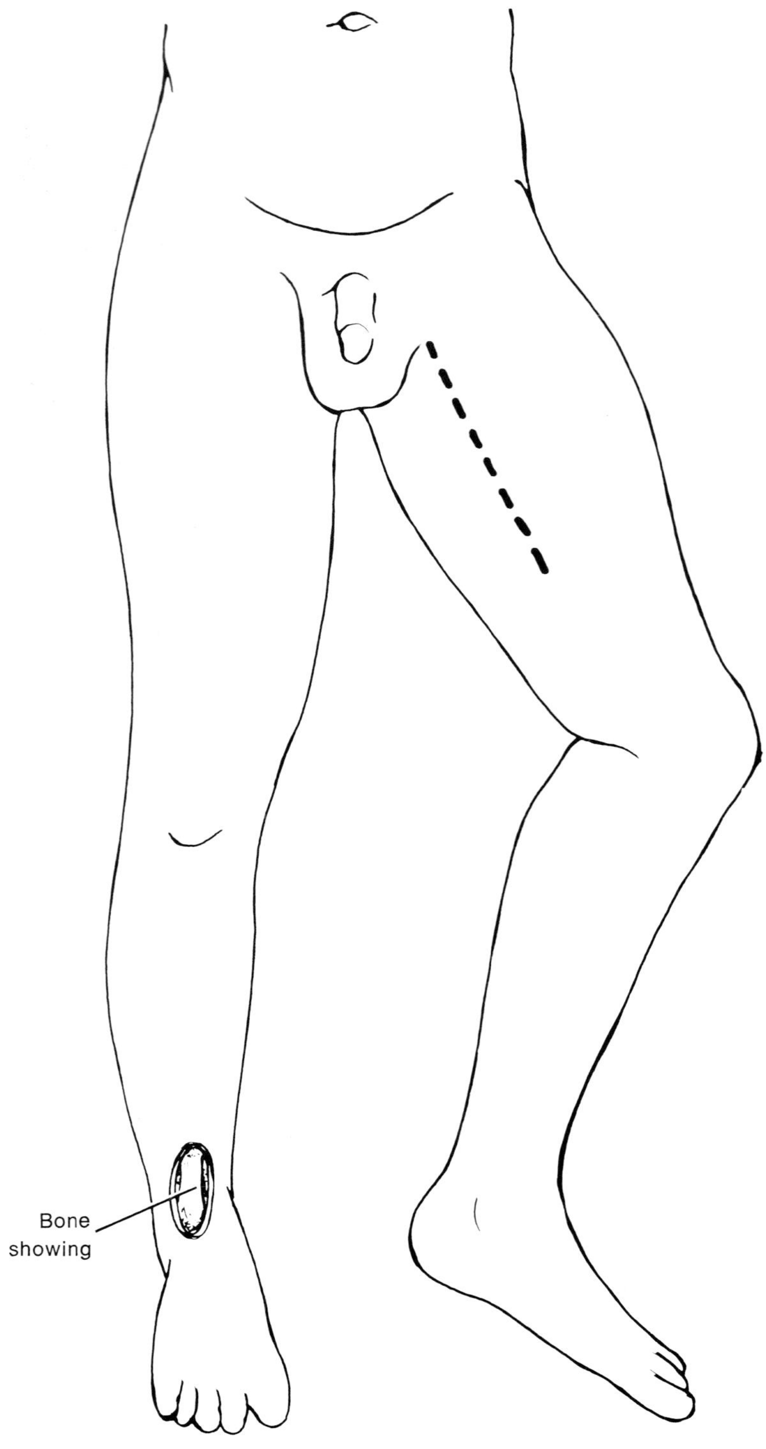

Fig. 38-20. Technique of elevation: gracilis muscle for coverage of osteomyelitis defects. Sequestrectomy is performed at site of osteomyelitis. Donor site incision (opposite leg) over muscle is determined preoperatively by palpation of voluntary muscle contraction of adductor longus muscle. Incision is immediately posterior to adductor longus muscle.

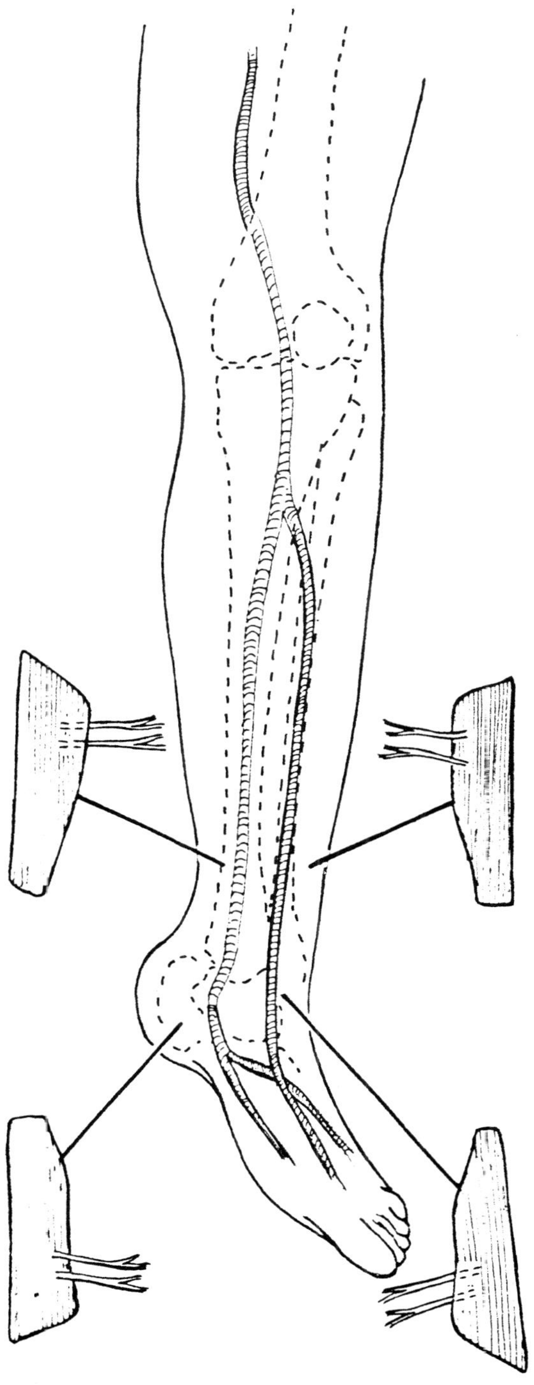

Fig. 38-21. Technique of elevation: gracilis muscle for coverage of osteomyelitis defects. Preoperative arteriogram determines suitable recipient vessels for gracilis muscle. Since muscle without skin is required, muscle is placed in four potential positions, depending on location of recipient vessels for gracilis muscle and shape of bone defects. End-to-side arterial and venous anastomoses are preferred. Skin grafts are applied to exposed muscle.

A

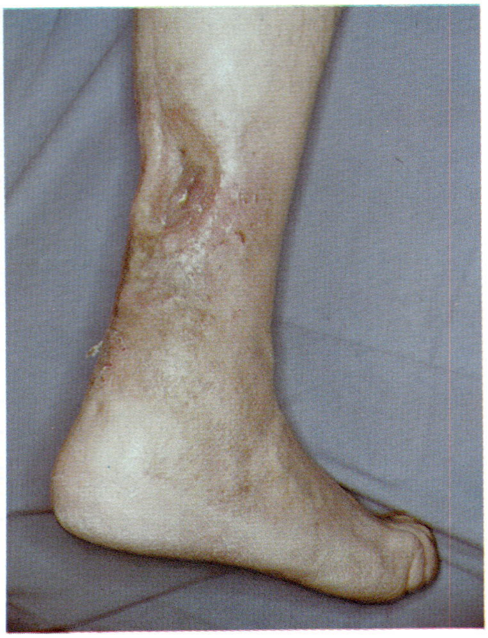

B

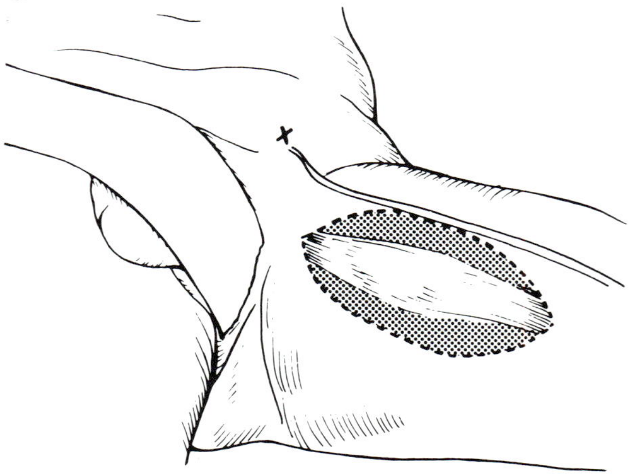

Fig. 38-22. Clinical application for gracilis muscle transplantation for lower extremity osteomyelitis: sequestrectomy and musculocutaneous flap.

A, Chronic osteomyelitis of distal tibia at site of bone injury.

B, Gracilis musculocutaneous flap is elevated from contralateral leg for coverage of sequestrectomy site.

(**A** to **C** from Mathes, S.J., Nahai, F., and Vasconez, L.O.: Myocutaneous free flap transfer: anatomical and experimental considerations, Plast. Reconstr. Surg. **62:**162, 1978.)

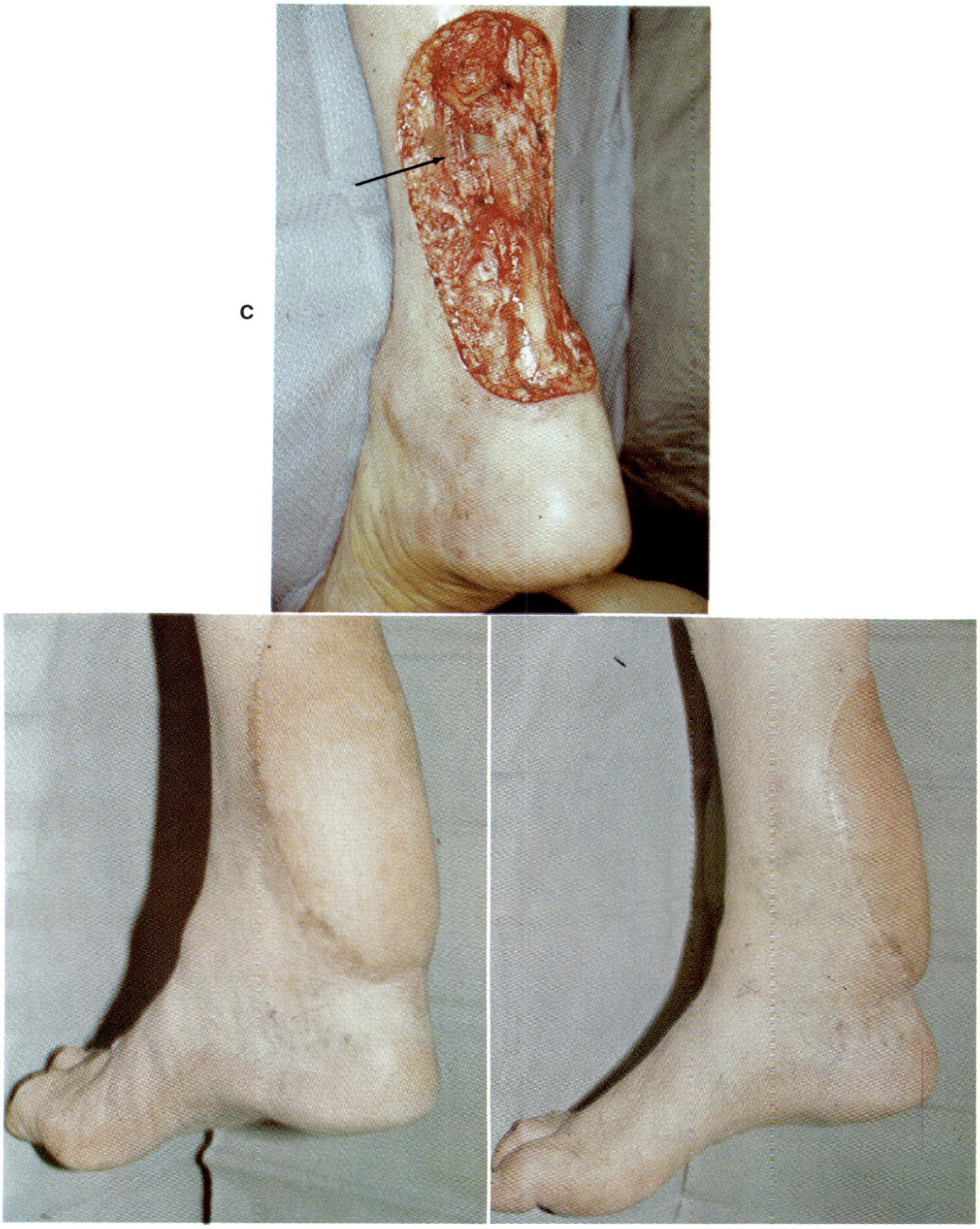

Fig. 38-22, cont'd

C, Débridement of infected soft tissue and bone. Arrow denotes site of posterior tibial vessels.

D, Gracilis musculocutaneous flap revascularized by end-to-end anastomoses between medial femoral circumflex artery and vein to posterior tibial artery and vein.

E, Five-and-one-half-year follow-up examination demonstrates stable wound coverage without recurrent infection.

(**D** and **E** from Mathes, S.J., Alpert, B.S., and Chang, N.: Use of the muscle flap in chronic osteomyelitis: experimental and clinical correlation, Plast. Reconstr. Surg., vol. 69, 1982.)

A

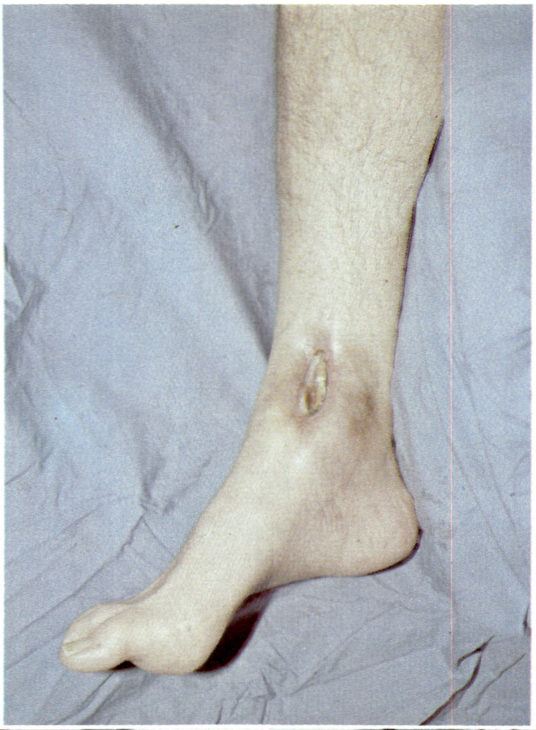

B

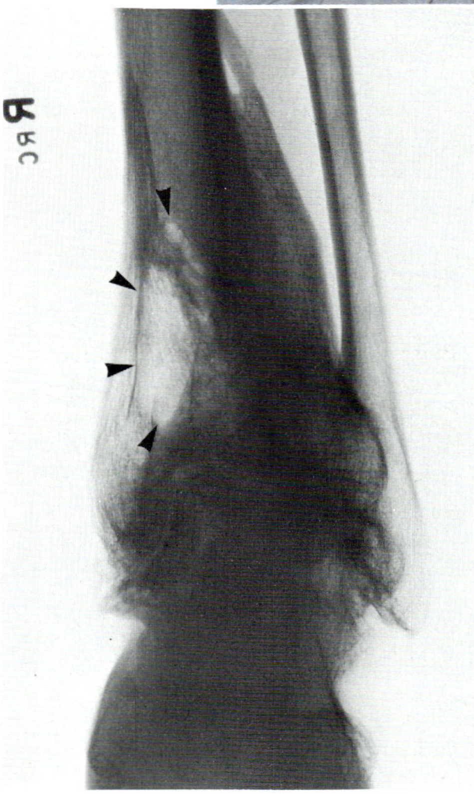

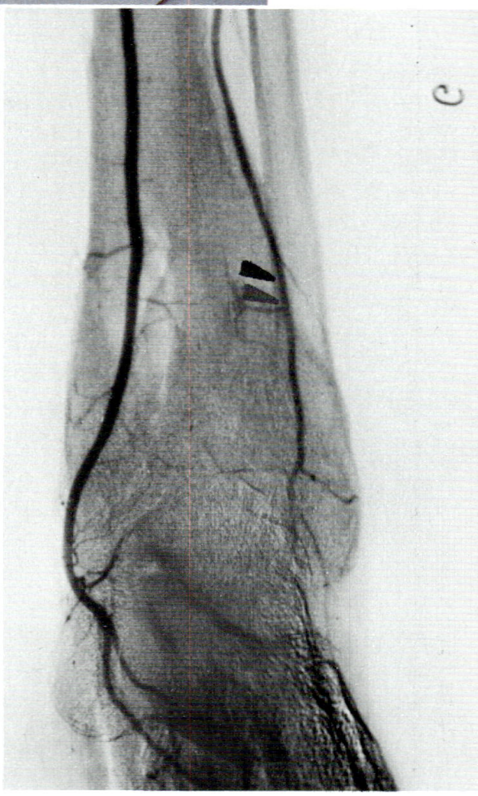

C

Fig. 38-23. Clinical application for gracilis transplantation for lower extremity osteomyelitis. Sequestrectomy and muscle flap with skin grafts.

A, Distal third lower leg. Site of chronic osteomyelitis at site of open fracture.

B, Radiograph of distal tibia demonstrates radiolucency *(arrows)* typical of bone defect associated with chronic osteomyelitis.

C, Arteriogram demonstrates patency of both anterior and posterior tibial arteries at site of osteomyelitis. Anterior tibial artery and associated veins *(arrows)* are selected as recipient vessels for muscle flap transplantation.

(**E** and **I** from Mathes, S.J., Alpert, B.S., and Chang, N.: Use of the muscle flap in chronic osteomyelitis: experimental and clinical correlation, Plast. Reconstr. Surg., vol. 69, 1982.)

D E F

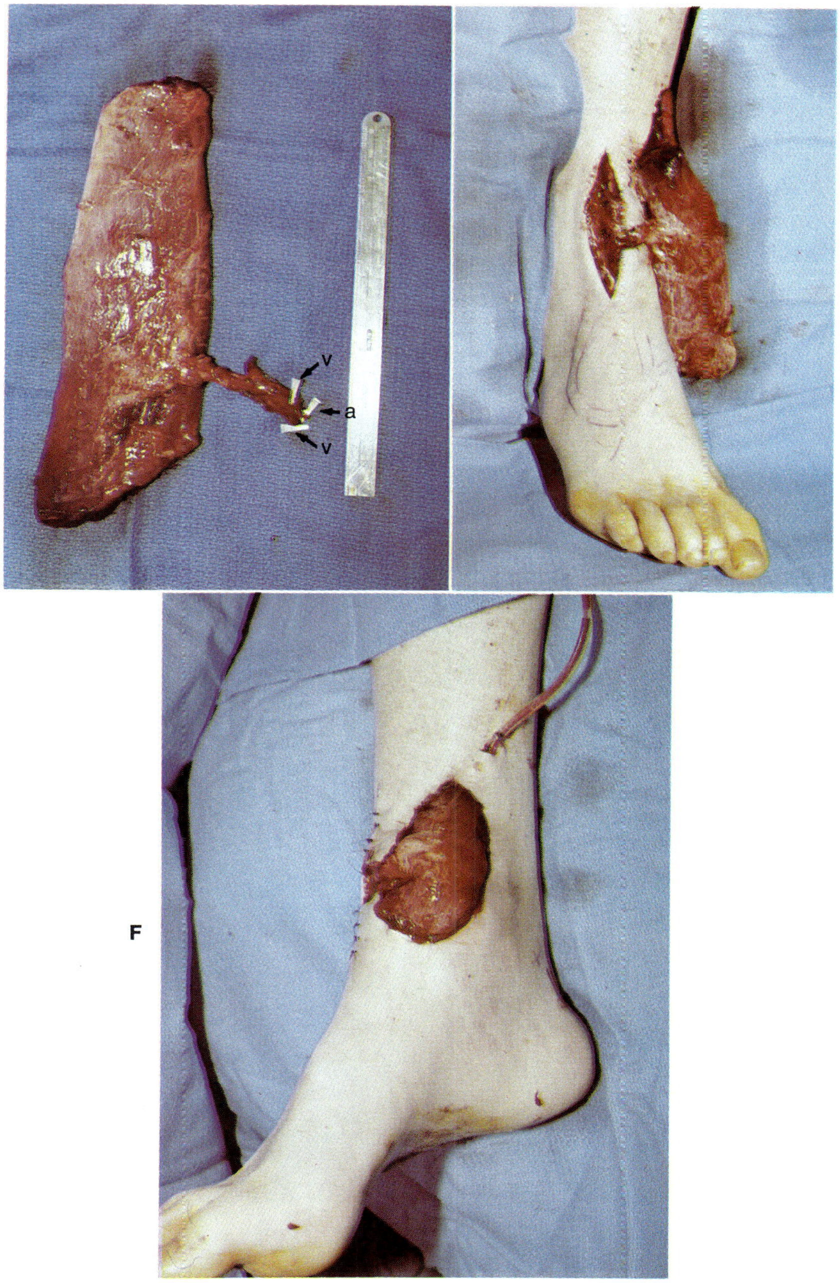

Fig. 38-23, cont'd

D, Gracilis muscle ready for transplantation. *a,* Medial femoral circumflex artery; *v,* venae comitantes.

E, Gracilis vascularized by end-to-side anastomosis of medial femoral circumflex artery and venae comitantes with anterior tibial artery and vein.

F, Muscle fits into sequestrectomy site. Exposed muscle is covered with skin grafts.

Continued.

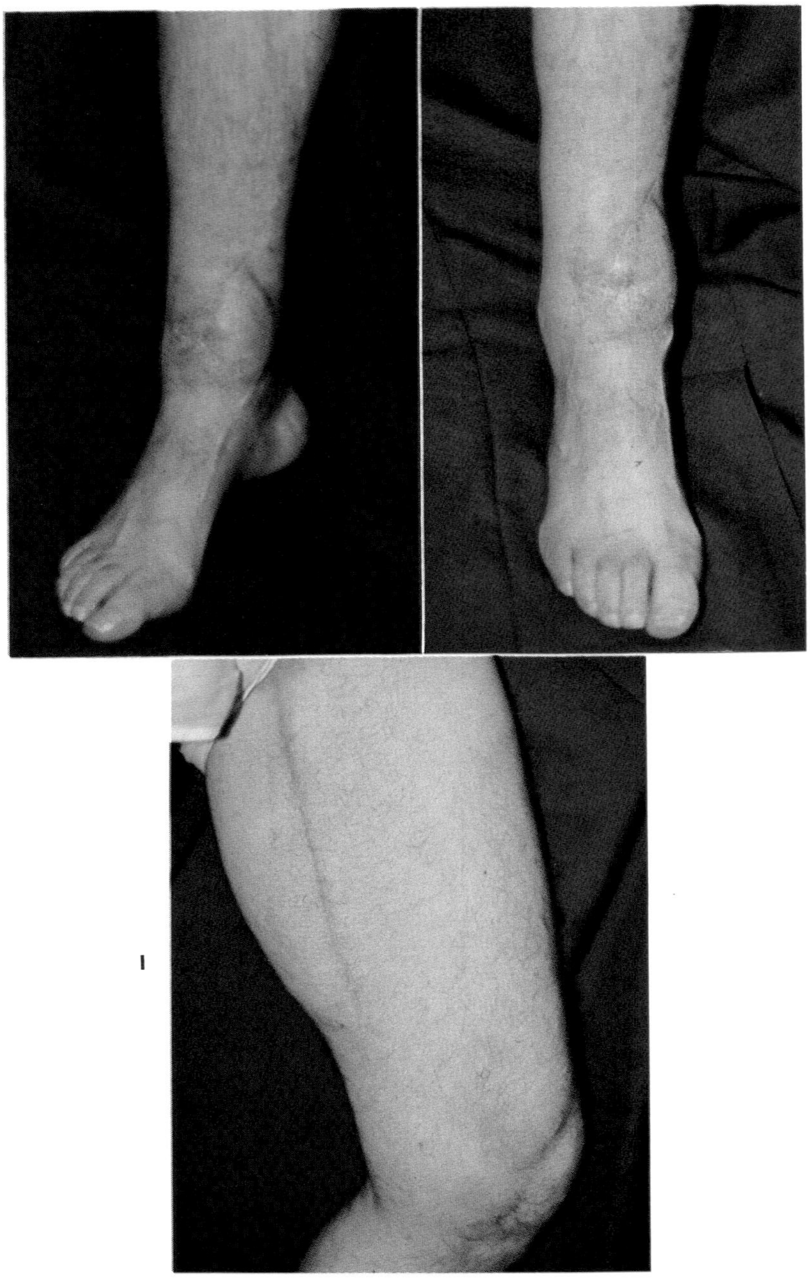

G H I

Fig. 38-23, cont'd

G and **H**, Anterior and lateral views of lower leg demonstrate stable wound coverage without recurrent infection 2 years after reconstructive surgery.

I, Donor site closes directly with minimal scarring and no functional disability.

Functional transplantation

The gracilis may be transplanted with its motor nerve as a functional motor unit. The muscle has adequate strength for use as a flexor for the upper extremity. This unit may be transplanted only as a muscle unit or with its overlying skin to provide both muscle function and skin coverage.

Flap design and elevation are similar to the techniques described for distant coverage. After the vascular pedicle is identified and mobilized, the motor branch of the obturator nerve is identified as it enters the proximal deep muscle belly. This motor nerve enters the muscle belly 2 to 3 cm superiorly to the vascular pedicle. The nerve is dissected proximally between the muscle bellies of the adductor longus and adductor magnus muscles. A nerve length of 5 to 6 cm may be mobilized before branches to adjacent adductor muscles are reached. The nerve is divided at this level and included with the gracilis unit for transplantation.

Before the muscle origin and insertion are divided, sutures or clips are placed at intervals along the muscle belly. These markers allow maintenance of normal muscle length when the muscle is transplanted as a functional unit. The vascular pedicle to the muscle is not divided until recipient site and a suitable recipient motor nerve are prepared to receive the muscle for microvascular transplantation.

Clinical application

1. Facial reanimation: alternate method (Fig. 38-24)

   ADVANTAGES

   a. Muscle is split as required for facial muscle replacement.
   b. Muscle motor nerve is repaired to one of the following (order of preference):
      (1) Proximal ipsilateral facial nerve
      (2) Contralateral facial nerve via cross nerve facial grafts
      (3) Ipsilateral hypoglossal nerve

   DISADVANTAGES

   a. Muscle power and bulk may be excessive.

2. Upper extremity long flexor replacement: alternative method

   ADVANTAGES

   a. Excessive muscle bulk in the forearm is avoided.
   b. Simultaneous defect coverage is provided.
   c. Simultaneous donor and recipient site dissection are possible.

   DISADVANTAGES

   Muscle may have inadequate power as flexor unit.

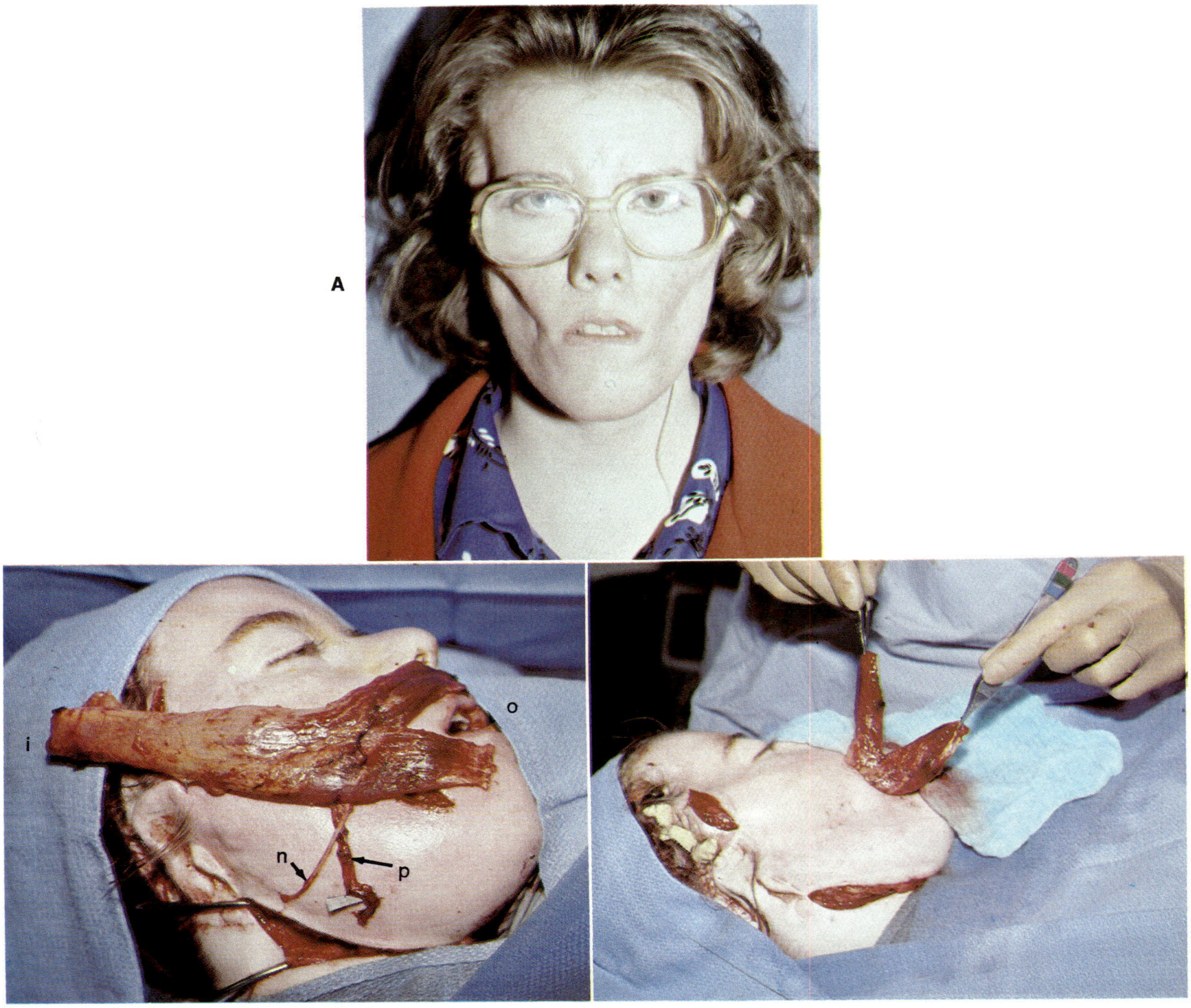

Fig. 38-24. Gracilis transplantation for facial muscle function: bilateral muscle flaps with motor nerve repair.

A, Bilateral congenital absence of facial nerves. Severe contour deformity resulting from lack of muscles of facial animation. Prior temporalis muscle transfers provide protective muscles for eye closure.

B, Staged transplantation of gracilis muscles to face. Muscles vascularized by anastomoses of medial femoral circumflex artery and vein *(p)* end-to-side to facial artery and vein. Obturator nerve repaired to half of fascicles of hypoglossal nerve. *i,* Muscle insertion; *o,* muscle origin.

C, Origin is split to provide orbicularis oris function.

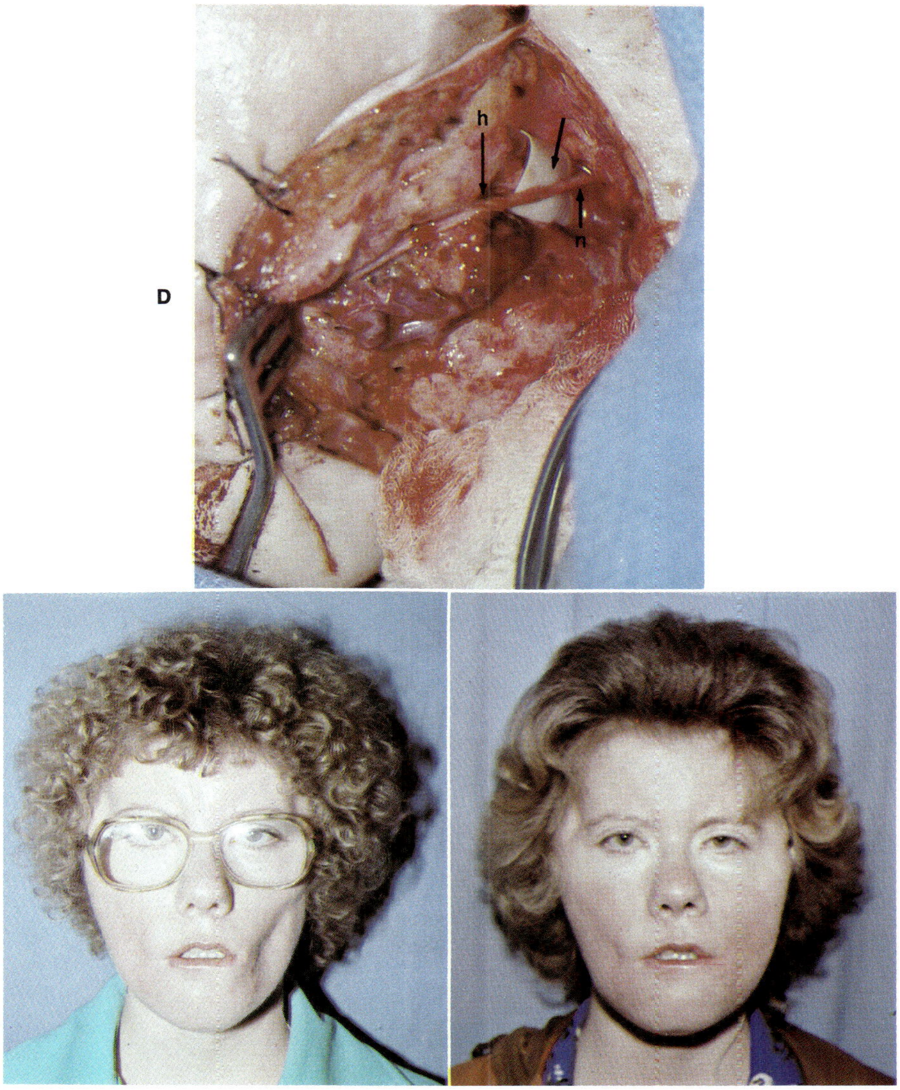

Fig. 38-24, cont'd

D, Arrow denotes site of group fascicles repair of obturator nerve *(n)* with half of fascicles of hypoglossal nerve *(h)*.

E, Six months after right muscle transplantation, left gracilis muscle transplantation is performed.

F, Three months after second muscle transplantation, contour deformity is corrected. Both muscles demonstrate voluntary muscle contraction.

Tensor fascia lata

The TFL is a small anterior thigh muscle. Based on its vascular anatomy, flap design, and special features, this musculocutaneous unit is a useful flap for microvascular transplantation. This versatile musculocutaneous unit may be used for the following: distant coverage, neurosensory flap, osseous-musculocutaneous flap, neurosensory osseous-musculocutaneous flap, and functional muscle.

The cutaneous territory of this muscle unit is located on the lateral thigh. A line connecting the anterior superior iliac crest and the lateral condyle of the knee denotes the anterior border. The posterior border is located at the level of the greater trochanter. The cutaneous territory extends between the iliac crest superiorly and the knee inferiorly. The cutaneous territory may be extended as random skin medially over the rectus femoris, laterally over the biceps femoris, and superiorly above the iliac crest when a larger flap is required by the defect size.

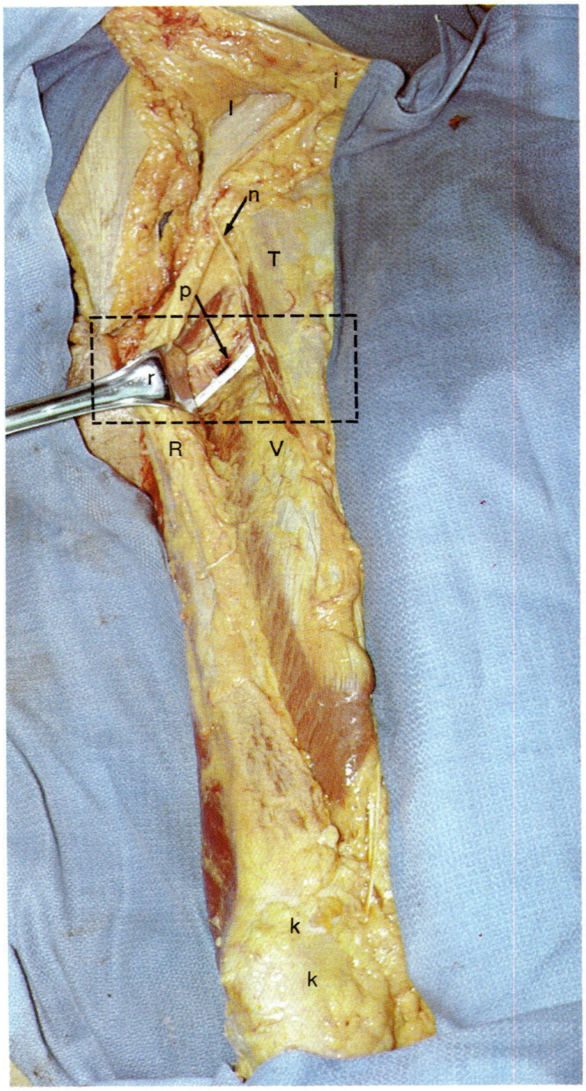

Fig. 38-25. Vascular anatomy of TFL. TFL muscle *(T)* is demonstrated with cadaver supine. Superiorly iliac spine *(i)* and inguinal ligament *(l)* and inferiorly patella *(k)* are visualized. Relationship of lateral femoral cutaneous nerve *(n)* with medial border of TFL and inguinal ligament is noted. Retraction *(r)* of the rectus femoris *(R)* medially at 10 cm inferiorly to iliac crest demonstrates vascular pedicle *(p)*, terminal branch of lateral femoral circumflex artery, and vein to TFL, coursing superficial to vastus lateralis muscle *(V)*.

Vascular anatomy

The TFL muscle unit receives a single arterial vascular pedicle, entering the medial muscle belly approximately 10 cm inferiorly to the medial border of the cutaneous territory (Fig. 38-25). The vascular pedicle, the transverse branch of the lateral femoral circumflex artery, originates from the profunda femoris artery. The artery courses beneath the rectus femoris, providing branches to the rectus femoris, vastus lateralis, and gluteus minimus muscles (Fig. 38-26). The lateral femoral circumflex artery enters the muscle as two or three terminal branches.

The venous drainage of the muscle parallels the arterial pedicle as paired venae comitantes. The venae comitantes form a single vein proximally and drain into the deep vein adjacent to the profunda femoris artery.

Both artery and vein have a proximal external lumen diameter of 2 to 2.5 cm. A pedicle of 6 to 8 cm is possible with dissection of the pedicles to their respective junction with the profunda femoris artery and vein.

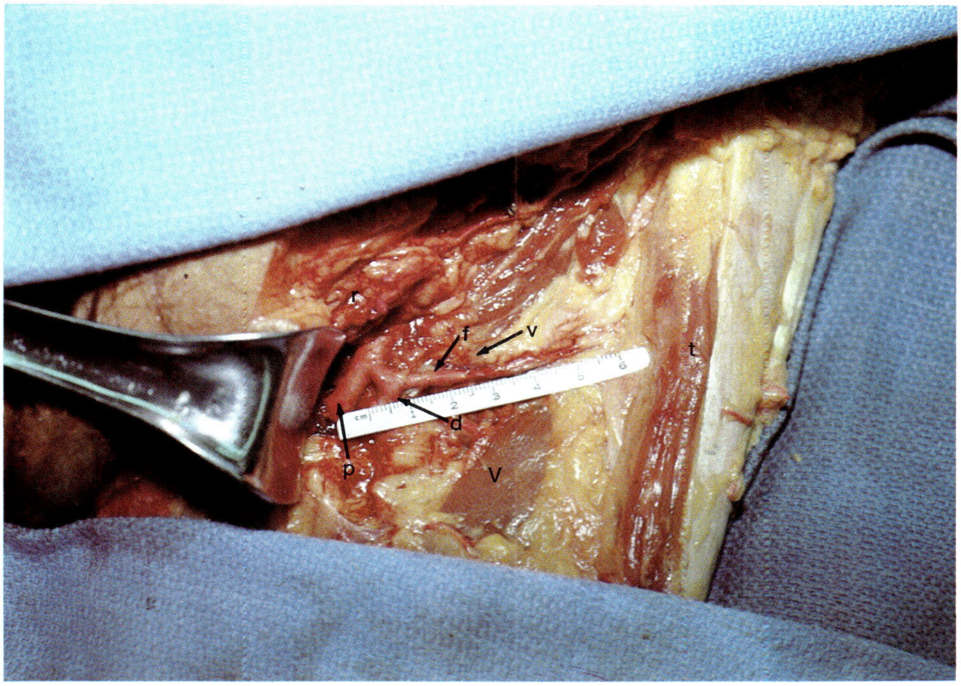

Fig. 38-26. Vascular anatomy of TFL. Closeup view of vascular pedicle to TFL. With medial retraction of rectus femoris muscle *(r)*, lateral femoral circumflex artery *(f)*, and vein *(v)*, branches of profunda femoris artery *(p)* and vein are visualized. Division of descending branch of lateral femoral circumflex artery to vastus lateralis *(V)* increases pedicle length.

Neural (motor) anatomy

The motor nerve to the TFL muscle is a branch of the superior gluteal nerve. The nerve enters the deep surface of the muscle slightly proximal to the site of entrance of the vascular pedicle.

Neural (sensory) anatomy

The superior portion of the cutaneous territory of the TFL muscle receives the cutaneous branch of T12. This sensory nerve is located in the subcutaneous tissue at the superior posterior aspect of the flap.

The middle and inferior portions of the cutaneous territory of this muscle receive their innervation via the lateral femoral cutaneous nerve: This sensory nerve enters the medial edge of the flap within the subcutaneous tissue 8 to 10 cm inferiorly to the anterior superior iliac crest (Figs. 38-27 and 38-28).

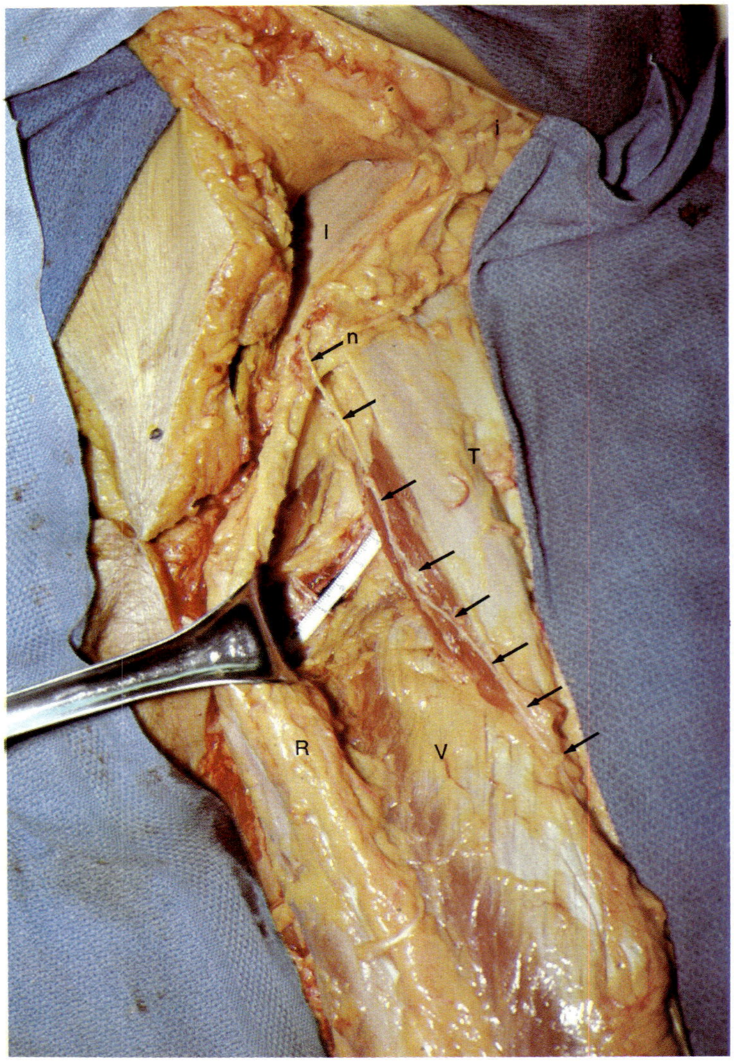

Fig. 38-27. Sensory nerve of TFL. Arrows denote course of lateral femoral cutaneous nerve from beneath inguinal ligament *(l)* to cutaneous territory of TFL muscle unit *(T)*. *i,* Iliac spine; *R,* rectus femoris muscle; *V,* vastus lateralis muscle.

Flap characteristics

SIZE: The TFL unit as a free flap may include the entire skin of the lateral thigh. The distal 4 cm of the flap immediately above the lateral condyle are less reliable in survival. However, flap dimensions of 8 × 24 cm are safely transplanted in this musculocutaneous unit.

THICKNESS: Thickness of the TFL unit is variable. In the male patient the flap is generally fairly thin, an advantage of this donor area. In the female patient the adipose tissue in this area may be thick, an undesirable feature especially for distal leg and foot defects.

PLIABILITY: The fascia lata comprises the undersurface of the distal half of the TFL flap. This portion of the flap is stiff and does not fold well over concave surfaces. Also, the fascia lata does not adhere well to the underlying defect. In a contaminated defect the potential dead space between the flap and the defect may result in secondary abscess formation.

DONOR SITE: Generally, the defect after transplantation of the TFL musculocutaneous unit cannot be closed directly. Since the vastus lateralis lies at the base of the defect, the donor site is skin grafted. In the thin patient the subsequent donor deformity is minimal.

When the entire TFL muscle or the iliac crest is included with flap transplantation, there is a noticeable depression at the superior edge of the defect. However, in this area the skin can usually be closed directly.

There has been no problem with parasthesias after division of the lateral femoral cutaneous nerve or the sensory branch of T12. Since the nerve is generally dissected proximally to the site of its penetration of the deep fascia, proximal neuromas are located deep in the lower extremity and are seldom symptomatic.

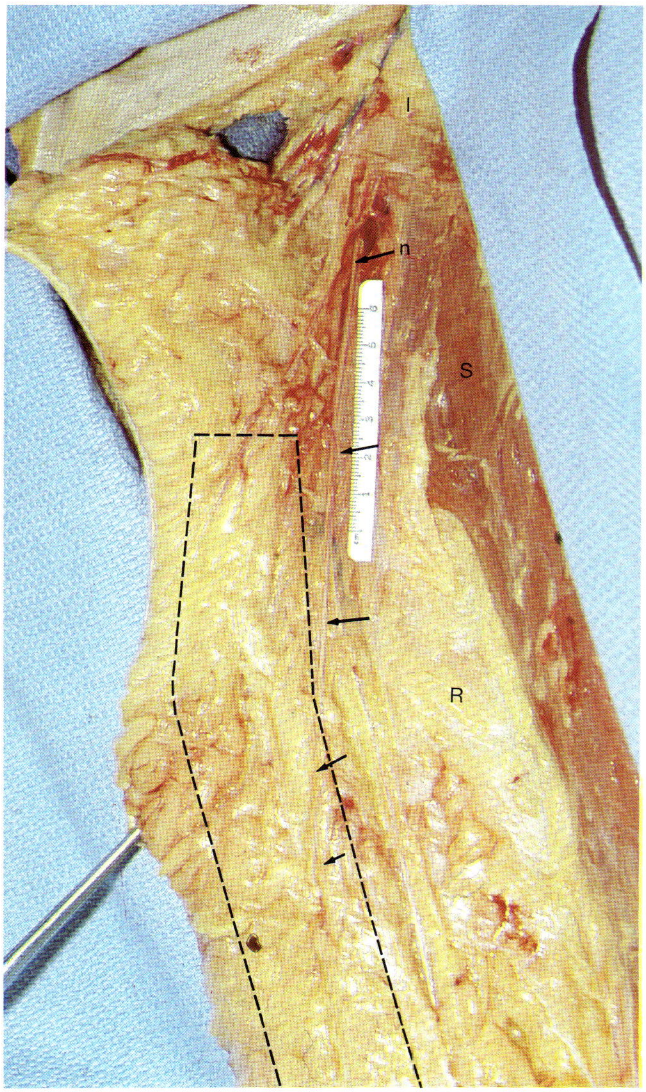

Fig. 38-28. Sensory nerve of TFL. Lateral thigh skin is reflected laterally off anterior thigh. Dotted lines demonstrate undersurface of TFL cutaneous territory. Medial edge of TFL muscle is visualized. Arrows denote course of lateral femoral cutaneous nerve into cutaneous territory of TFL unit. *l*, Inguinal ligament; *S*, sartorius muscle; *R*, rectus femoris muscle.

Musculocutaneous flap

Technique of elevation. Proper planning of the flap design is necessary before flap elevation is started. The borders of the cutaneous territory of the TFL muscle are outlined on the lateral thigh (Fig. 38-29). The site of projected entrance of the vascular pedicle (10 cm inferiorly to the iliac crest) is denoted. The required dimensions of skin are outlined within the confines of the cutaneous territory. The required location of the vascular pedicle in relation to both the flap design and the location of the recipient vessels adjacent to the defect must be considered in flap planning. Generally, the flap design is located slightly inferior to the point of entrance of the vascular pedicle. The pedicle then extends above the flap, reaching predetermined recipient vessels adjacent to the defect.

The lateral, medial, and inferior borders of the flap are incised through subcutaneous tissue and underlying fascia lata. Temporary sutures are placed between the muscle superiorly and fascia lata inferiorly and the skin to avoid disruption of the musculocutaneous perforating vessels during flap elevation. Starting at the inferior flap edge, the deep surface of the fascia lata is elevated from the underlying vastus lateralis muscle.

When the junction of muscle and fascia lata is noted on the deep surface of the flap, caution is required until the vascular pedicle (lateral circumflex

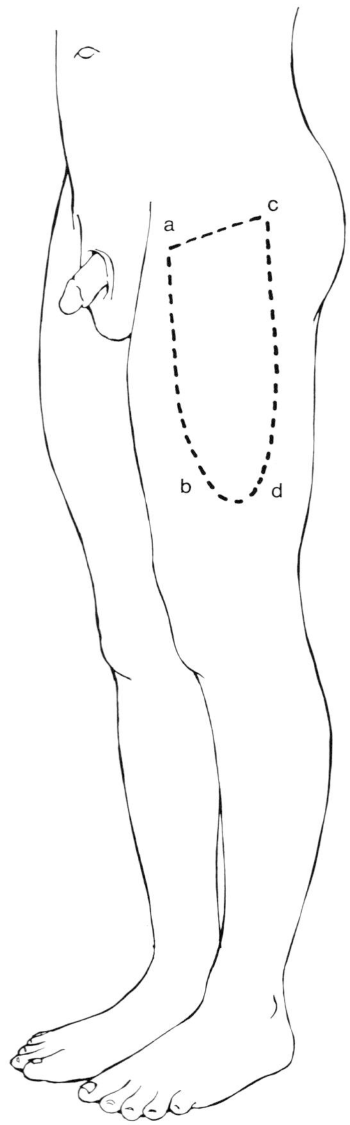

Fig. 38-29. Technique of elevation for TFL musculocutaneous flap for coverage. Design of skin island over TFL muscle and distal fascia lata. Line between anterior iliac spine and knee *(a-b)* determines anterior edge. Posterior edge *(c-d)* located at prominence of greater trochanter. Superior *(a-c)* and inferior *(b-d)* edges determined by specifications of defect and location of recipient vessels.

femoral artery) is located at the medial surface of the muscle. With medial retraction of the rectus femoris muscle at this level, the underlying transverse branch of the lateral circumflex femoral artery and its paired venae comitantes are easily visualized (Fig. 38-27). The vascular pedicle is now dissected medially to its junction with the profunda femoris vessels. Branches to adjacent musculature are divided, using the bipolar coagulator. While protecting the pedicle from excessive traction, the superior portion of the flap is elevated. The superior margin of the flap is incised and proximal muscle divided immediately superior to the site of entrance of the vascular pedicle to the muscle. The muscle is divided with the knife, since the muscular contraction associated with electric cautery may distort the muscle, resulting in inadvertent pedicle injury.

Since the muscular portion of the flap is often bulky, minimal muscle in the superior flap is preferred. Only the muscle associated with the inferior terminal branch of the lateral circumflex artery is required for successful microvascular transplantation for distant coverage (Fig. 38-30).

The TFL musculocutaneous unit is now ready for microvascular transplantation. The pedicle is not divided until the recipient area is completely prepared, including isolation of appropriate receptor artery and veins in proximity to the defect.

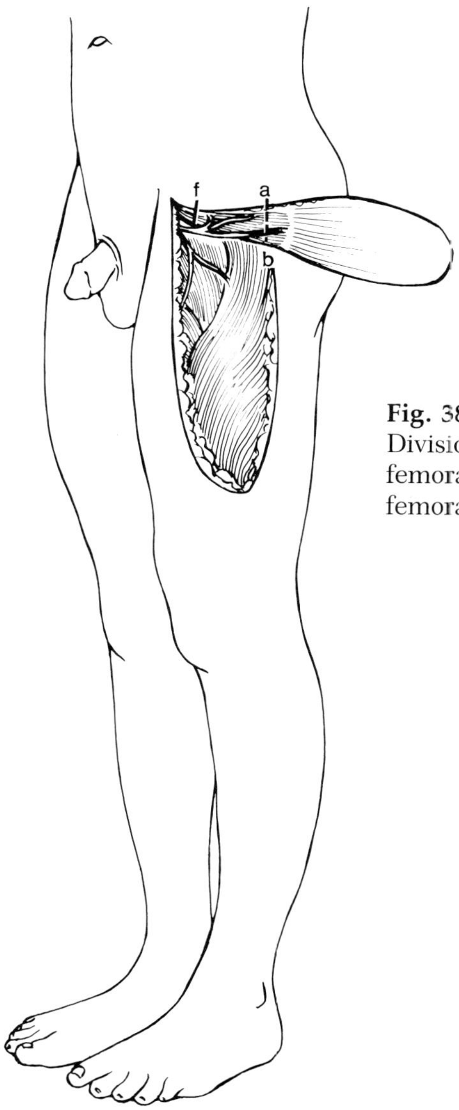

Fig. 38-30. Technique of elevation for TFL musculocutaneous flap for coverage. Division of muscle immediately proximal to inferior terminal branches of lateral femoral circumflex pedicle into muscle decreases flap bulk (line *a-b*). *f*, Lateral femoral circumflex artery and venae comitantes.

Neurosensory TFL musculocutaneous flap

Technique of elevation. Preoperative planning includes precise flap design and the location and distance of a receptor sensory nerve from the defect requiring both coverage and protective sensation. A neurosensory TFL musculocutaneous flap requires identification and preservation of the lateral femoral cutaneous nerve as it enters the medial flap border (Figs. 38-27 and 38-28). The nerve enters the cutaneous territory of the flap between 5 to 10 cm inferiorly to the anterior superior iliac crest. After the flap is designed on the lateral thigh, the medial flap incision is made through only the skin. The subcutaneous tissue is carefully incised until the nerve is identified. Once identified, the nerve is dissected proximally. Since the nerve must be divided before the vascular pedicle can be identified because of its deeper position beneath the rectus femoris muscle, adequate proximal nerve length should be obtained at this phase of flap elevation. The preserved lateral femoral cutaneous nerve is divided, and flap elevation proceeds, as described for the standard free TFL flap for distant coverage (Figs. 38-31 and 38-32).

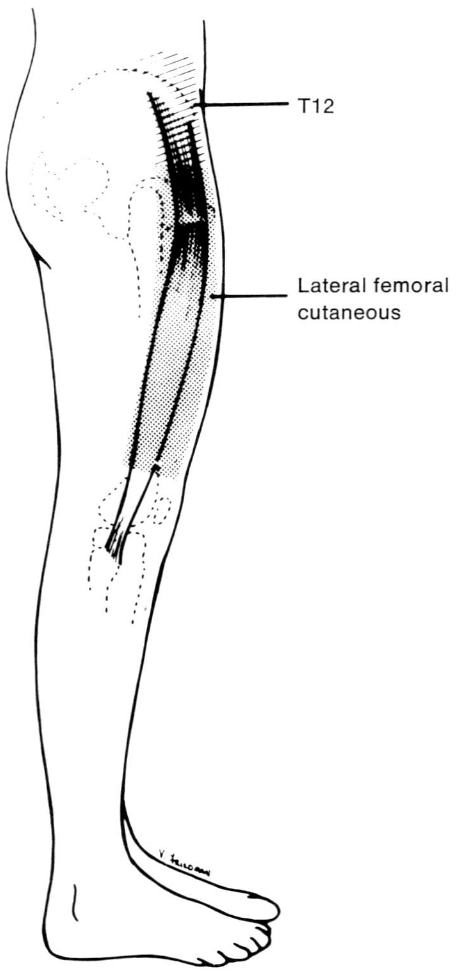

Fig. 38-31. Technique of elevation for neurosurgery flap for coverage. Location of sensory nerves to cutaneous territory of TFL unit.

(From Mathes, S.J., and Buchanan, R.T.: Br. J. Plast. Surg. **32:**184, 1979.)

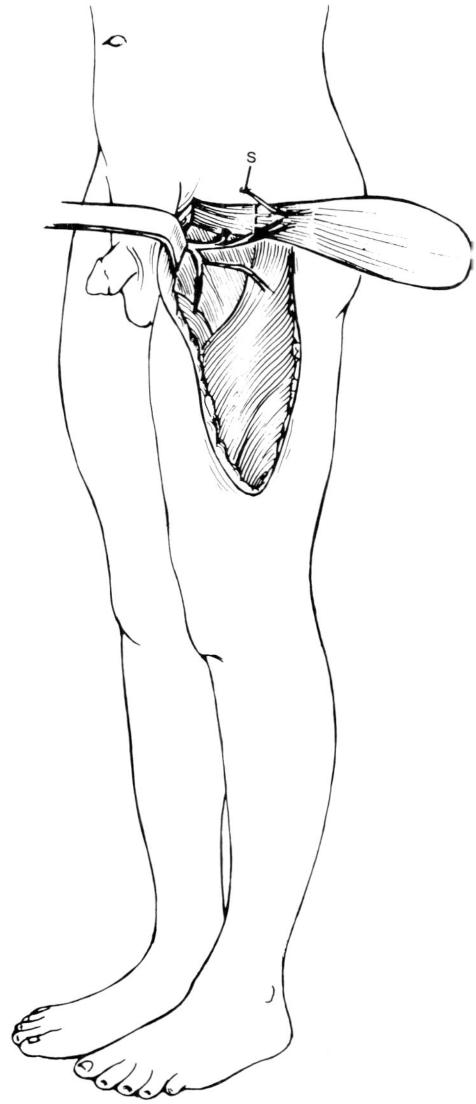

Fig. 38-32. Technique of elevation for neurosensory flap for coverage. Preservation of lateral femoral cutaneous nerve allows transplantation of neurosensory flap. *s*, Sensory nerve.

Clinical application

Coverage of weight-bearing surfaces of lower extremity (e.g., distal leg and foot) (Fig. 38-33): method of choice

ADVANTAGES

a. Repair of lateral femoral cutaneous nerve provides protective sensation to cutaneous territory of flap.
b. Large dimensions of flap will cover distal leg and foot.
c. Simultaneous dissection of donor and recipient site is possible.

DISADVANTAGES

a. Flap is thick in obese patients.
b. Fascial deep surface of flap does not adhere well to recipient site.

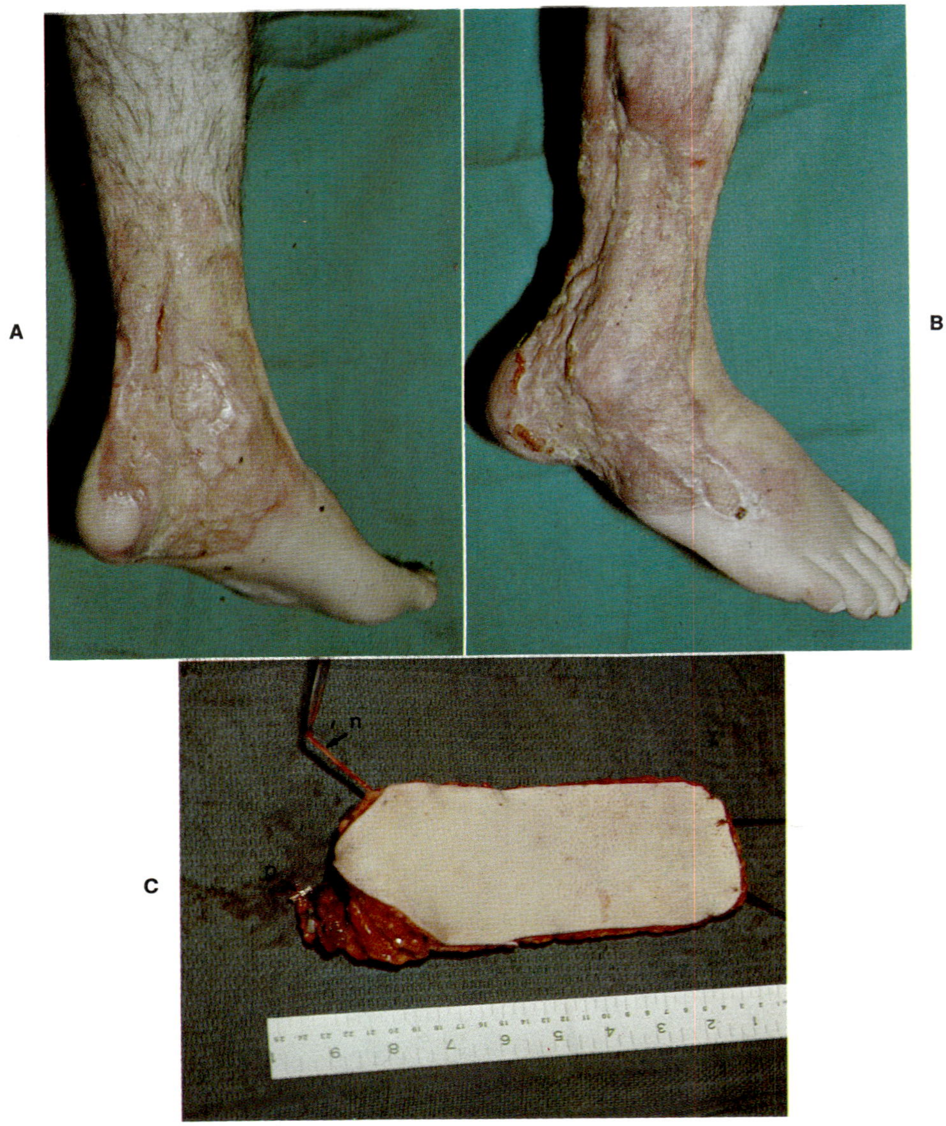

Fig. 38-33. Clinical application of neurosensory flap for coverage of lower extremity.
A and **B,** Bilateral electrical burn injuries treated with initial wound débridement and skin grafts. Skin grafts unstable over distal tibia of left leg *(A)* and fibula of right leg *(B)*.
C, TFL musculocutaneous flap ready for transplantation as neurosensory flap. *n*, Lateral femoral cutaneous nerve; *p*, lateral femoral circumflex artery and vein.

(**A** and **C** from Mathes, S.J., and Buchanan, R.T.: Br. J. Plast. Surg. **32:**184, 1979.)

D

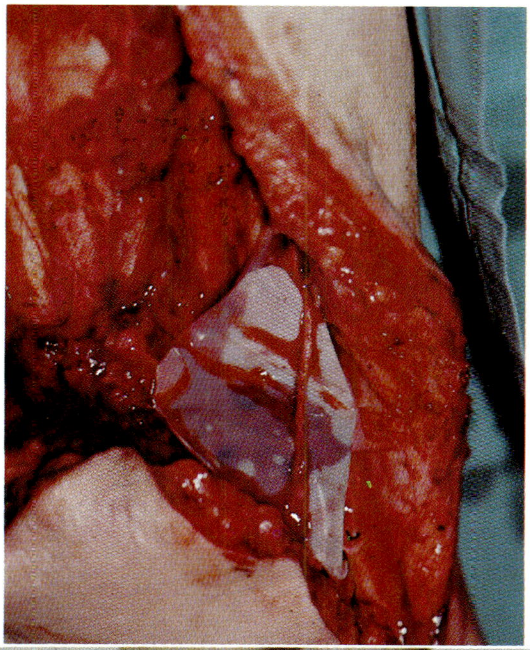

E

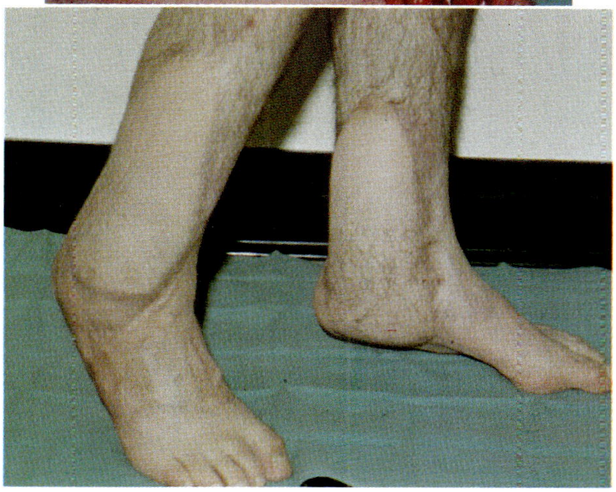

Fig. 38-33, cont'd

D, Site of repair of lateral femoral cutaneous nerve with saphenous nerve on left lower extremity. Sural nerve used as recipient nerve for right lower extremity.

E, Three-year postoperative view demonstrates bilateral stable wound coverage. Both flaps have protective sensation.

Osseous-musculocutaneous TFL flap

Technique of elevation. The TFL muscle originates from the anterior 5 to 10 cm of the outer lip of the iliac crest. The vascular connections between the muscle and its bony origin allow transplantation of this unit as an osseous-musculocutaneous flap.

Preoperative planning includes determining the relationship of the bone, skin, and vascular pedicle of the flap to the location of the bony defect, the required amount of skin coverage, and location of the receptor vessels. The flap design is centered over the proximal cutaneous territory of this musculocutaneous unit. Up to 4 cm of random skin superior to the iliac crest may be included with the flap. Generally, skin superior to the bone is required to provide complete coverage of the bone in the defect requiring reconstruction.

Flap elevation begins inferiorly, as discussed in techniques for standard TFL flap transplantation. After the muscle is elevated and the pedicle isolated, the dissecton proceeds superiorly to the bony origin. The muscular attachments between the iliac crest and the proximal TFL muscle are preserved. Care also must be taken to separate the gluteus minimus posteriorly from the TFL muscle.

The superior skin is incised and elevated to the level of the iliac crest. Using wide osteotomies, the bone is transected deep to the TFL origin. Before complete release of the bone, drill holes are made and wire passed for use in later bony fixation to the recipient site. The bone is then completely released. The flap is now completely elevated with the vascular pedicle still intact. The vascular pedicle is divided only after the defect is completely prepared and ready for free flap transplantation.

Neurosensory osseous-musculocutaneous flap

Technique of elevation. The sensory branch of T12 provides sensory innervation to the superior cutaneous territory of the TFL unit. When the defect requires a combination of bone and skin coverage with protective sensation, this can be provided by the TFL unit (Fig. 38-34).

The flap elevation is similar to that described for the osseous-musculocutaneous flap. When flap elevation is completed inferiorly to the iliac crest, the superior skin incision is made. The sensory branch of T12 is located as it enters the posterior superior flap edge. Once identified, the nerve is dissected proximally to gain adequate length before it is divided. After completion of the sensory nerve dissection, the remainder of the flap elevation is performed as previously described.

A 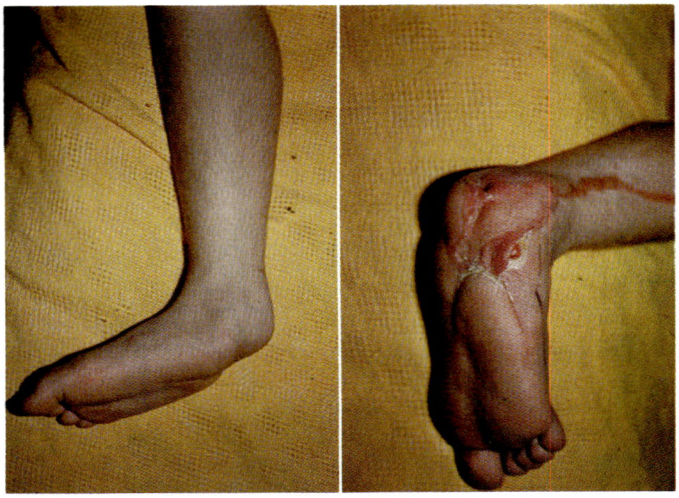 B

Fig. 38-34. Reconstruction of the heel with a TFL neurosensory osseous musculocutaneous free flap.

A and **B,** Ten-year-old boy lost skin soft tissue and calcaneal bone of left heel in a lawn mower accident. Skin graft over exposed bone did not provide stable coverage.

(From Nahai, F., Hill, L., and Hester, T.R.: Experiences with the tensor fascia lata flap, Plast. Reconstr. Surg. **63:**788, 1979.)

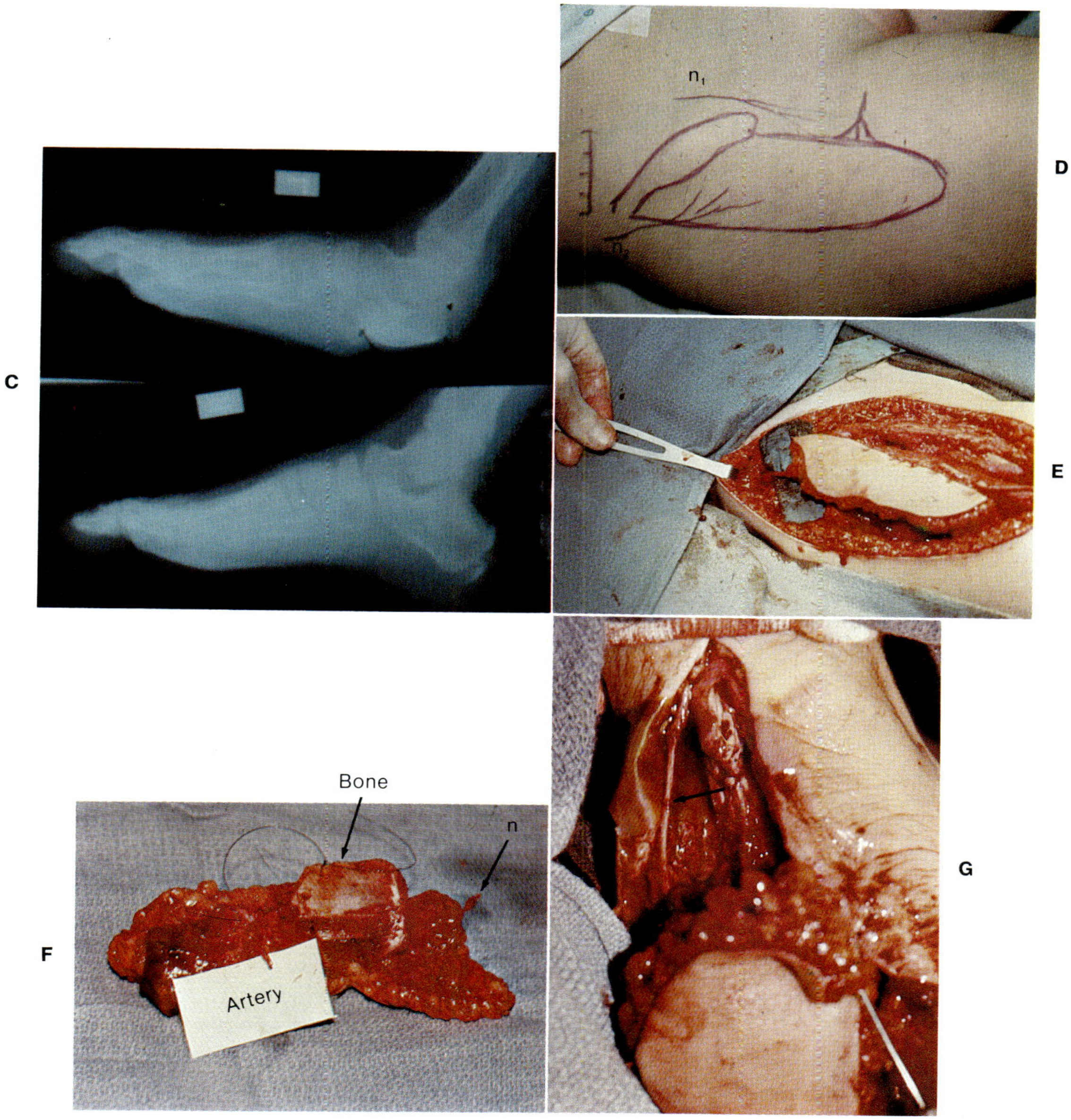

Fig. 38-34, cont'd

C, Radiograph of both feet. Injured left foot above and reversed radiograph of normal foot below for comparison.

D, TFL flap designed and bone (iliac crest) and two sensory nerves outlined. n_1, Lateral femoral cutaneous nerve; n_2, T12.

E, Flap elevated and lateral branch of T12 isolated.

F, Isolated free flap. Note bone.

G, Flap revascularized and sural nerve sutured to lateral branch of T12 *(arrow).*

Continued.

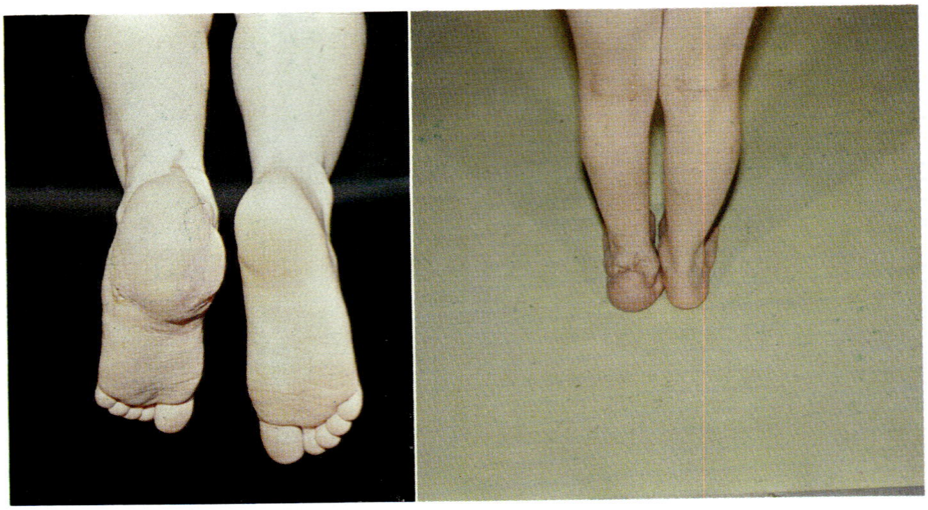

H I

Fig. 38-34, cont'd
H and I, Result after 9 months reveals stable coverage with protective sensibility.

Functional transplantation

Technique of elevation. When coverage and function is desired, the TFL muscle unit may be transplanted with its motor nerve. With its small muscle belly, the TFL is not useful for replacement of strong flexors but may be considered for functional transplantation as an extensor unit. Preoperative flap design may include only the muscle-fascial unit or its overlying cutaneous territory, depending on recipient site requirements.

Preoperative planning must include localization of a suitable motor nerve in proximity to the recipient defect. If wound coverage is required, the skin is designed within the cutaneous territory of the TFL. If skin coverage is adequate in the receptor site (e.g., dorsum of the hand), the proximal muscle may be covered with skin grafts. In most instances where functonal transposition is required because of absence of local muscle units, skin coverage is an associated problem.

When transplantation of the musculocutaneous unit is elected, the TFL flap is elevated exactly as described for its use in distant coverage.

After the borders of the cutaneous territory are incised to the level of the underlying TFL unit, the distal fascia lata is not divided. After isolation of the vascular pedicle beneath the rectus femoris muscle, the motor nerve is located. The motor nerve, a branch of the superior gluteal nerve, enters the muscle on its deep surface immediately superior to the point of entrance of the vascular pedicle. Proximal dissecton between the gluteus medius and gluteus maximus muscles may be required for adequate motor nerve length to reach a motor nerve in the recipient defect.

Before the muscle is released from its origin and insertion, temporary sutures or clips are placed on the muscle belly at specific intervals so that normal muscle length is restored in its transplanted position.

The distal fascia lata is not divided at the same level as the overlying skin island. Rather, it is extended inferiorly 5 to 6 cm beyond the distal margins of the skin. This fascial extension will be used for repair with distal tendon stumps. It may be divided into multiple fascial strips for repair with multiple tendons (e.g., restoration of common extensors to hand). After release of the muscle insertion and origin, an isolation of both motor nerve and vascular pedicle, the TFL unit is ready for transplantation.

Clinical application

1. Upper extremity functional transplantation for long extensor function: alternative method

   ADVANTAGES

   a. Minimal muscle bulk is present.
   b. Fascial extension of distal flap will reach fingers with minimal bulk.
   c. Simultaneous donor and recipient site dissection are possible.

   DISADVANTAGES

   a. Bulk of the musculocutaneous flap can be excessive.
   b. TFL muscle may have inadequate motor strength.

ANNOTATED BIBLIOGRAPHY

1. Bostwick, J., III, and co-workers: Sixty latissimus dorsi flaps, Plast. Reconstr. Sur. **63:**31, 1979.

   A review of the various applications of the latissimus dorsi flap is presented. This report includes our first free latissimus dorsi musculocutaneous flap for coverage of a defect in the lower extremity that was performed in October 1977.

2. Godina, M.: Preferential use of end-to-side arterial anastomosis in free flap transfers, Plast. Reconstr. Surg. **64:**673, 1979.

   The end-to-side anastomosis is described and compared to the end-to-end anastomosis. This paper includes an excellent review of 32 latissimus dorsi flaps, most of which were performed by Godina before any latissimus dorsi free flaps had been performed in the United States.

3. Harii, K., Ohmari, K., and Sekiguchi, J.: The free musculocutaneous flap, Plast. Reconstr. Surg. **57:**294, 1976.

   This is the first experimental and clinical article on transfer of the gracilis musculocutaneous unit. The anatomy of the muscle is well described, and three cases illustrate the utilization of this musculocutaneous flap for coverage of defects of the lower extremity and the head and neck.

4. Harii, K., Ohmari, K., and Torii, S.: Free gracilis muscle transplantation with microvascular anastomosis for treatment of facial paralysis, Plast. Reconstr. Surg. **57:**133, 1976.

   This article is the first clinical report of successful transfer of the gracilis muscle for correction of unilateral facial paralysis. The anatomy of the gracilis muscle and the surgical technique are well described. Excellent clinical evidence of muscle reinnervation and function is presented.

5. Hill, H.L., Nahai, F., and Vasconez, L.O.: The tensor fascia lata myocutaneous free flap, Plast. Reconstr. Surg. **61:**517, 1978.

   A case report of the first TFL free flap that was used for coverage of a defect of the lower extremity is presented. The anatomy of the flap and its advantages and disadvantages are outlined.

6. Ikuta, Y., Kubo, T., and Tsuge, K.: Free muscle transplantation by microsurgical technique to treat severe Volkmann's contracture, Plast. Reconstr. Surg. **58:**407, 1976.

   The first clinical report of free muscle transfer by microsurgical technique for correction of Volkmann's contracture is presented. The pectoralis major muscle is used with a good result.

7. Manktelow, R.T., and McKee, N.H.: Free muscle transplantation to provide active finger flexion, J. Hand Surg. **3:**416, 1978.

   A description of the use of the pectoralis major and gracilis muscles as free flaps with microneurovascular anastomosis to replace destroyed forearm flexor muscles is presented. Excellent results are shown. The anatomy and operative technique are well described.

8. Mathes, S.J., Albert, B., and Chang, N.: Use of the muscle flap in chronic osteomyelitis: experimental and clinical correlation, Plast. Reconstr. Surg. (In press.)

   The use of the gracilis muscle with skin graft for correction of chronic osteomyelitis of the lower extremity is described.

9. Mathes, S.J., and Buchanan, R.T.: Tensor fascia lata neurosensory musculocutaneous free flap, Br. J. Plast. Surg. **32:**184, 1979.

   The sensory innervation to the TFL territory is introduced and used for reconstruction of an electrical burn deformity of the lower leg.

10. Mathes, S.J., Nahai, F., and Vasconez, L.O.: Myocutaneous free flap transfer: anatomical and experimental considerations, Plast. Reconstr. Surg. **62:**162, 1978.

    Applications of the gracilis musculocutaneous free flap for coverage of the lower extremity are presented. The vascular anatomy and feasibility for free flap transfer of other muscle units are discussed. The concept of a strategic delay of muscle flaps is also explored.

11. Maxwell, G.P., Manson, P.N., and Hoopes, J.E.: Experiences with 13 latissimus dorsi myocutaneous free flaps, Plast. Reconstr. Surg. **64:**1, 1979.

    A review of 13 latissimus dorsi free flaps is presented. Applications for head and neck and lower extremity reconstruction are discussed. Of the 13 flaps, 11 are successful. The two failures are attributed to compression of the vascular pedicles within a tunnel.

12. Maxwell, G.P., Stueber, K., and Hoopes, J.E.: A free latissimus dorsi myocutaneous flap, Plast. Reconst. Surg. **62:**462, 1978.

    The first published report of the latissimus dorsi musculocutaneous free flap is presented. The flap was transferred on December 5, 1977 for reconstruction of an extensive defect of the scalp.

13. Nahai, F., Hill, H.L., and Hester, T.R.: Experiences with the tensor fascia lata flap, Plast. Reconstr. Surg. **63:**788, 1979.

A review of an extensive experience with the TFL flap is presented. Five free flaps are included in this paper. One of these was a neurosensory flap including bone. The lateral femoral cutaneous nerve of the thigh and the lateral branch of T12 were used as sensory nerves in reconstruction of a heel defect.

14. Nahai, F., and Mathes, S.J.: Esthetic aspects of reconstructive surgery of the lower extremity, Clin. Plast. Surg. **8:**369, 1981.

The use of the gracilis muscle with skin graft as opposed to a gracilis musculocutaneous flap is described. The skin-grafted muscle provides better contouring than the musculocutaneous flap with a rather thick subcutaneous layer. Experiences with the latissimus dorsi free muscle flap and skin graft are also described.

15. O'Brien, B.M., Franklin, J.D., and Morrison, W.A.: Cross-facial nerve grafts and microneurovascular free muscle transfer for long established facial palsy, Br. J. Plast. Surg. **33:**202, 1980.

This article is a review of a 5-year experience with 20 patients undergoing treatment of unilateral facial nerve palsy in two stages: a cross facial nerve graft and a free muscle transfer with microneurovascular anastomosis. Excellent results are reported with the gracilis (12 patients). Of eight patients with extensor digitorum brevis transfers, only four had muscle reinnervation. The gracilis muscle is preferred. The technique is demonstrated in great detail, and excellent clinical results are shown.

16. Pennington, D.G., Lai, M.F., and Pelly, A.D.: The rectus abdominis myocutaneous free flap, Br. J. Plast. Surg. **33:**277, 1980.

Anatomic description and one case report of a free flap based on the inferior half of the rectus abdominis muscle are presented. The inferior epigastric vessels are used. The advantages and disadvantages are discussed. Specifically, the need for reconstruction of the lower abdominal wall is pointed out.

17. Schenk, R.: Rectus femoris muscle and composite skin transplantation by microneurovascular anastomosis for avulsion of forearm muscles, J. Hand Surg. **3:**60, 1978.

A case report of a rectus femoris musculocutaneous free flap to restore finger flexion after destruction of the forearm flexors is presented.

18. Serafin, D., and co-workers: Reconstruction of the lower extremity with vascularized composite tissue: improved tissue survival and specific indications, Plast. Reconstr. Surg. **66:**230, 1980.

A review of 50 free flaps in 48 patients for reconstruction of defects of the lower extremity is presented. The advantages of the latissimus dorsi musculocutaneous free flap are outlined. An improved rate of flap survival is attributed in part to the application of the latissimus dorsi for reconstruction of the lower extremity.

19. Serafin, D., Georgiade, N.G., and Given, K.S.: Transfer of free flaps to provide well-vascularized thick cover for breast reconstruction after radical mastectomy, Plast. Reconstr. Surg. **62:**527, 1978.

This article discusses 12 breast reconstructions with free flaps. Ten of these are free groin flaps, and two are free latissimus dorsi musculocutaneous flaps from the contralateral side. The indications for free flap reconstruction are discussed.

20. Tamai, S., and co-workers: Free muscle transplants in dogs with microsurgical, neurovascular anastomosis, Plast. Reconstr. Surg. **46:**219, 1970.

This is a classic article in which the rectus femoris muscle of the dog is transferred orthotopically and heterotopically. This article proves the feasibility of free muscle transfer by microsurgical techniques. This essentially paves the way for later work using free muscle transfer for the correction of Volkmann's contracture of the forearm.

21. Withers, E., and co-workers: Further experiences with the tensor fascia lata musculocutaneous flap, Ann. Plast. Surg. **4:**31, 1980.

A review of 11 TFL flaps in 9 patients is presented. Two of these are musculocutaneous free flaps. One out of the two flaps does not survive.

39

COMPLICATIONS

Foad Nahai ■ Stephen J. Mathes

Early in our experience with free tissue transplantation, success was measured only in terms of flap survival. Today mere flap survival, although essential, is no longer the end point, and we view successful free tissue transplantation in terms of restoration of form and function to the defect and preservation of form and function in the donor area. A discussion of complications is therefore aimed not only at ensuring flap survival but also at considering the donor defect and the final reconstructive result in terms of form and function.

The key to success in free tissue transplantation is *selection:* selection of the patient, selection of the flap, and selection of recipient vessels. Careful selection of all three will greatly reduce or even eliminate complications. Of the last 50 muscle and musculocutaneous flaps transferred as free flaps at Emory University in Atlanta and the University of California at San Francisco, all have survived with minimal associated complications. The associated complications have been limited to the donor areas.

The technical skills required for microsurgical anastomoses are essential for successful tissue transplantation. Once these skills are mastered, selection of patient, flap, and recipient vessels assumes a greater role in avoiding complications.

GENERAL FACTORS

Preoperative planning

Careful preoperative selection of the patient and the flap is essential.

Patient selection. Age per se is not a factor in patient selection. We have successfully transferred muscle and musculocutaneous flaps by microsurgical technique in children age 3 and in adults up to the age of 83. The patient's general state of health and ability to tolerate a 5- to 7-hour procedure, or longer, must be considered. Patients with diabetes or peripheral or coronary vascular disease must be carefully evaluated before microsurgical tissue transfer is considered. Preoperative angiography, especially after lower extremity trauma, is an essential part of the preoperative planning. Angiography of the donor flap is not normally undertaken.

Flap selection. Flap choice is made after a thorough evaluation of the defect and assessment of the reconstructive needs. Size and depth of the defect are considered. On the whole, the gracilis is our choice for small, shallow defects and the latissimus dorsi for large, deep defects. For shallow defects muscle without the overlying skin and fat is transferred and covered with skin grafts.

Technical (intraoperative) planning

Adequate débridement and excision of the defect are essential. Inadequate excision of necrotic, infected, or neoplastic tissue will lead to postoperative complications.

Selection of recipient vessels is based on preoperative evaluation of pulses and, if indicated, angiography. The selection of suitable recipient vessels for microsurgical anastomosis and their careful isolation is probably the single most important intraoperative step in successful microsurgery. We prefer end-to-side arterial and venous anastomoses wherever possible in the lower extremity to preserve all arterial and venous channels. End-to-side anastomoses also eliminate the problem of vessel size discrepancy.

It should be emphasized that only *normal* vessels should be selected. The vessels should be free from atherosclerotic disease and out of the field of previous radiation. Pulsatile flow in the vessel should be confirmed immediately before the anastomosis is performed.

The donor flap should be elevated with as long a pedicle as possible. The longer pedicle facilitates the microsurgical anastomosis by avoiding tension, and through proximal dissection on the vascular pedicle, the lumen diameter of the vessels will be larger, facilitating the microanastomosis.

Postoperative planning

Careful monitoring of the flap postoperatively is essential. Early recognition of vascular complications and immediate reexploration of the anastomoses will prevent flap loss.

Our postoperative regime includes suction drainage of the donor site and under the flap. Our patients receive low molecular weight dextran postoperatively for 5 days. They are not allowed to smoke postoperatively for 3 weeks. We have seen bleeding complications in two patients related to the dextran therapy.

SPECIFIC FLAPS

Latissimus dorsi

Planning errors. The latissimus dorsi muscle is a very versatile and reliable flap. Preoperative planning is aimed at minimizing the donor defect. If large islands of skin or a posterior island is elevated (Fig. 39-1), the donor area requires skin grafting. This is not a reliable sensory flap and will not provide adequate sensibility or protection over weight-bearing areas.

As a musculocutaneous flap, it may be too thick for some defects (Fig. 39-2). It is the best transferred as muscle with skin graft in these instances.

Technical errors. The correct plane between the latissimus dorsi and serratus anterior must be identified. The flap elevation should proceed from the distal border to the proximal border, because this facilitates the identification of the vascular pedicle on the deep surface of the muscle. Once the pedicle is seen, dissection should then continue under loupe magnification. The pedicle is isolated up to the axillary vessels for maximum vessel length and diameter. This dissection is facilitated if the upper extremity is prepared and draped in the field. However, excessive traction on the arm should be avoided. Suction drains are always left in the donor area.

Postoperative errors. Despite the use of suction drains, seroma formation in the donor area is a common problem. It may be necessary to aspirate serous fluid from the back once or twice after the drains are removed. If the flap is transferred to the lower extremity, elevation of the lower extremity postoperatively is essential.

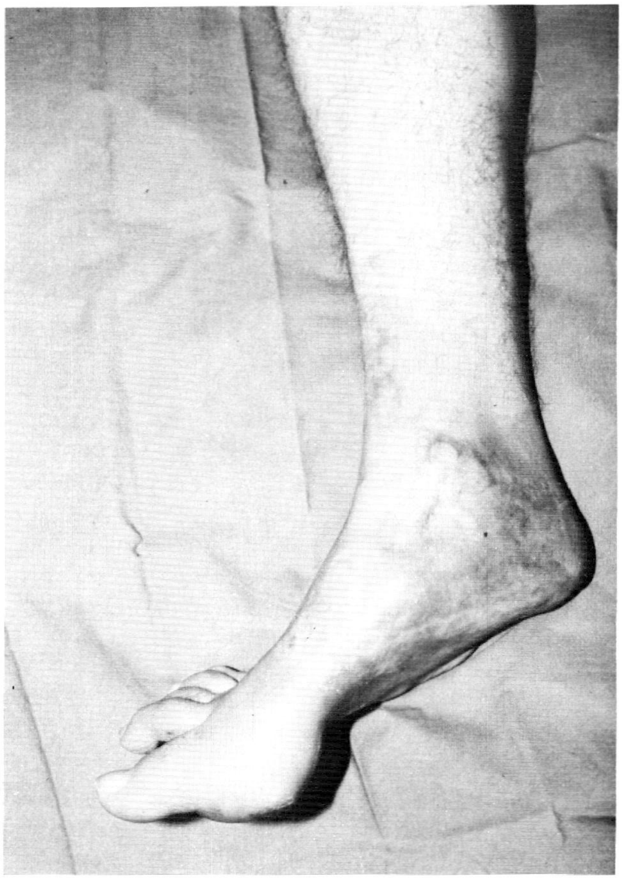

A

Fig. 39-1. Latissimus dorsi musculocutaneous neurosensory flap.

A, Unstable skin graft coverage of entire plantar surface of foot and lack of plantar sensation.

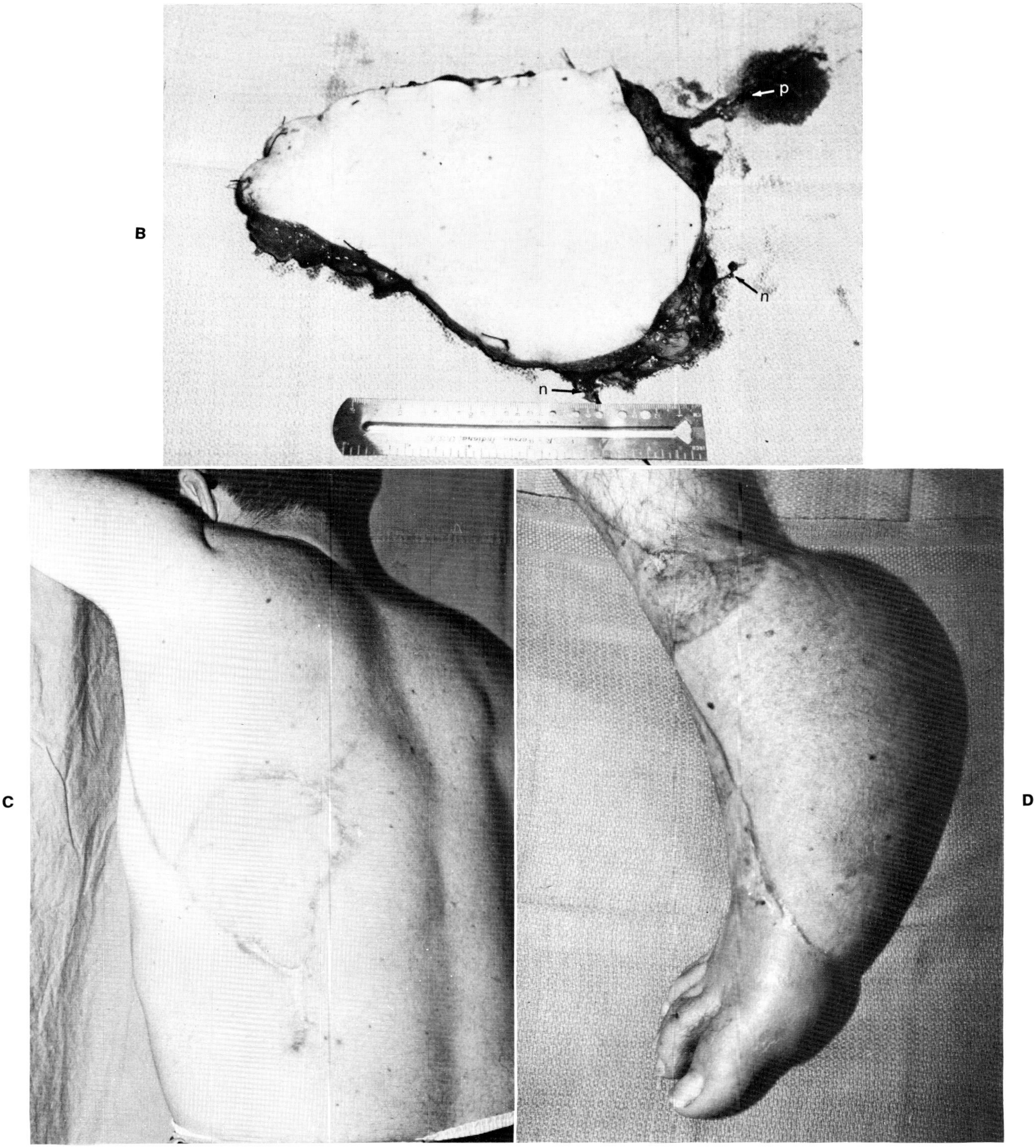

Fig. 39-1, cont'd

B, Design of flap, including posterior skin island, to include posterior intercostal sensory nerves (*n*) and thoracodorsal artery, vein, and nerve (*p*).

C, Donor site closure requires skin graft when all skin island dimensions exceed 15 cm. Posterior location of skin island over territory of the posterior intercostal sensory nerves.

D, Postoperative view ($1\frac{1}{2}$ years). Large skin island has provided coverage over plantar surface of foot. Flap sensory nerve repair with saphenous nerve has failed to provide protective sensation in flap. Frequent episodes of cellulitis result at site of pressure sores on distal plantar surface of skin and flap.

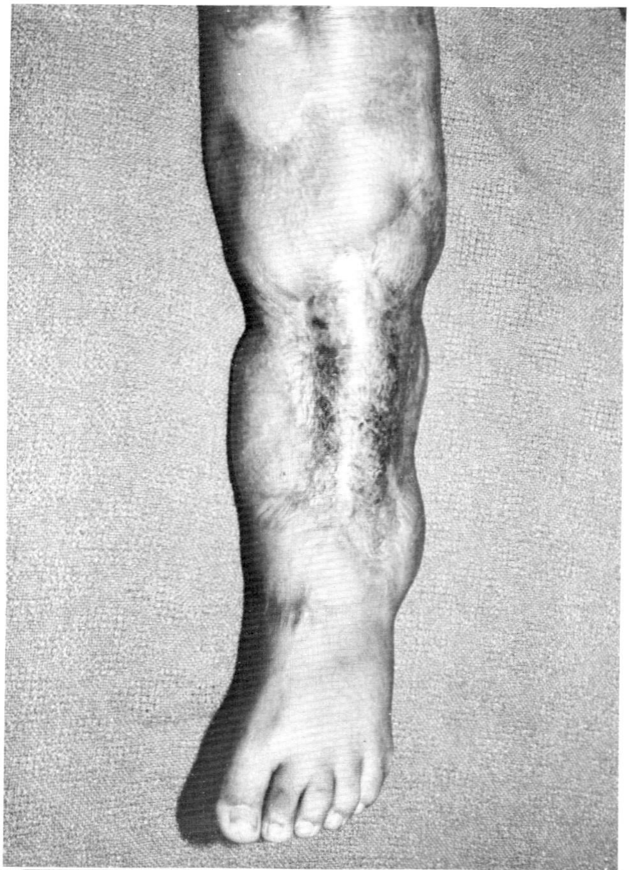

A

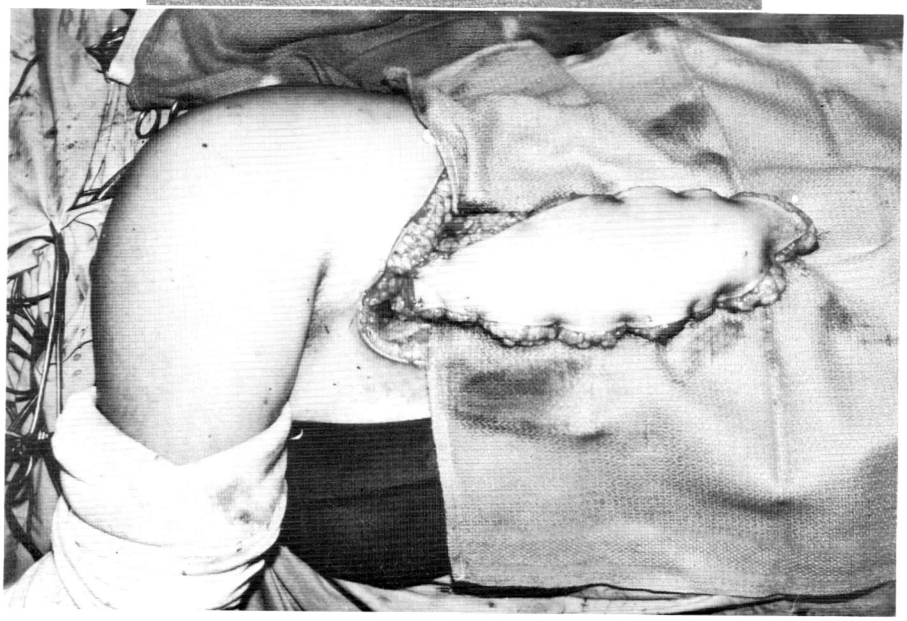

B

Fig. 39-2. Segmental latissimus dorsi musculocutaneous flap.

A, Unstable skin grafts over middle third of leg. Multiple prior injuries preclude use of local muscle flaps.

B, Distal latissimus dorsi muscle with anterior random skin allows preservation of posterior latissimus dorsi muscle for function and represents thinnest portion of cutaneous territory of this muscle.

C

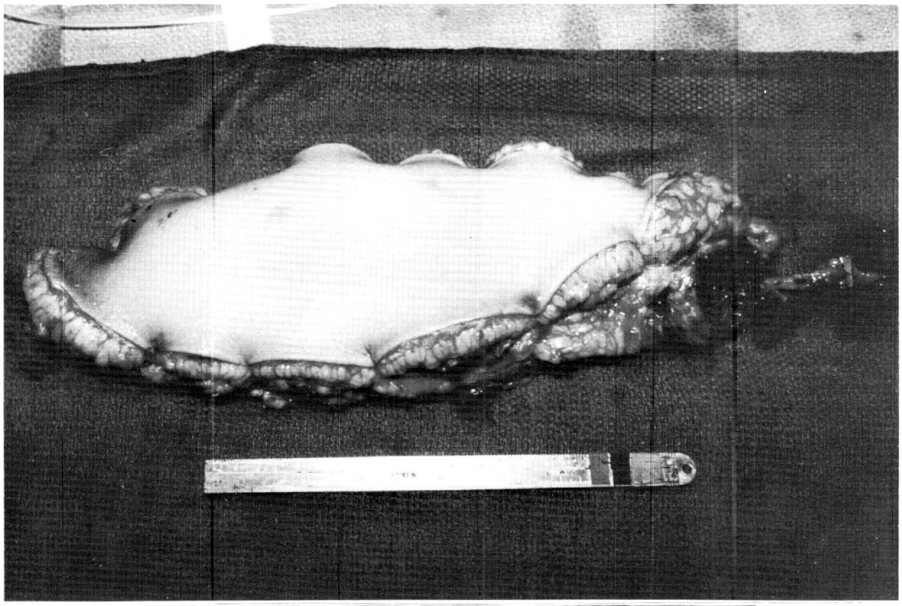

D

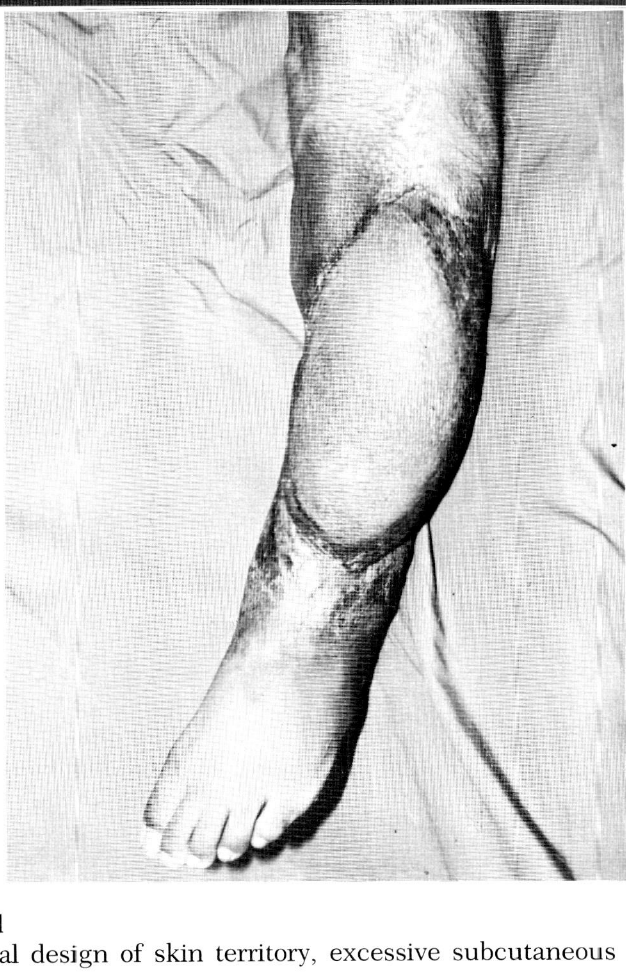

Fig. 39-2, cont'd

C, Despite distal design of skin territory, excessive subcutaneous tissue results in thick flap.

D, Three-month postoperative view of free flap coverage. Although stable wound coverage is provided by flap, excessive thickness of subcutaneous tissue is undesirable. Later thinning of flap required.

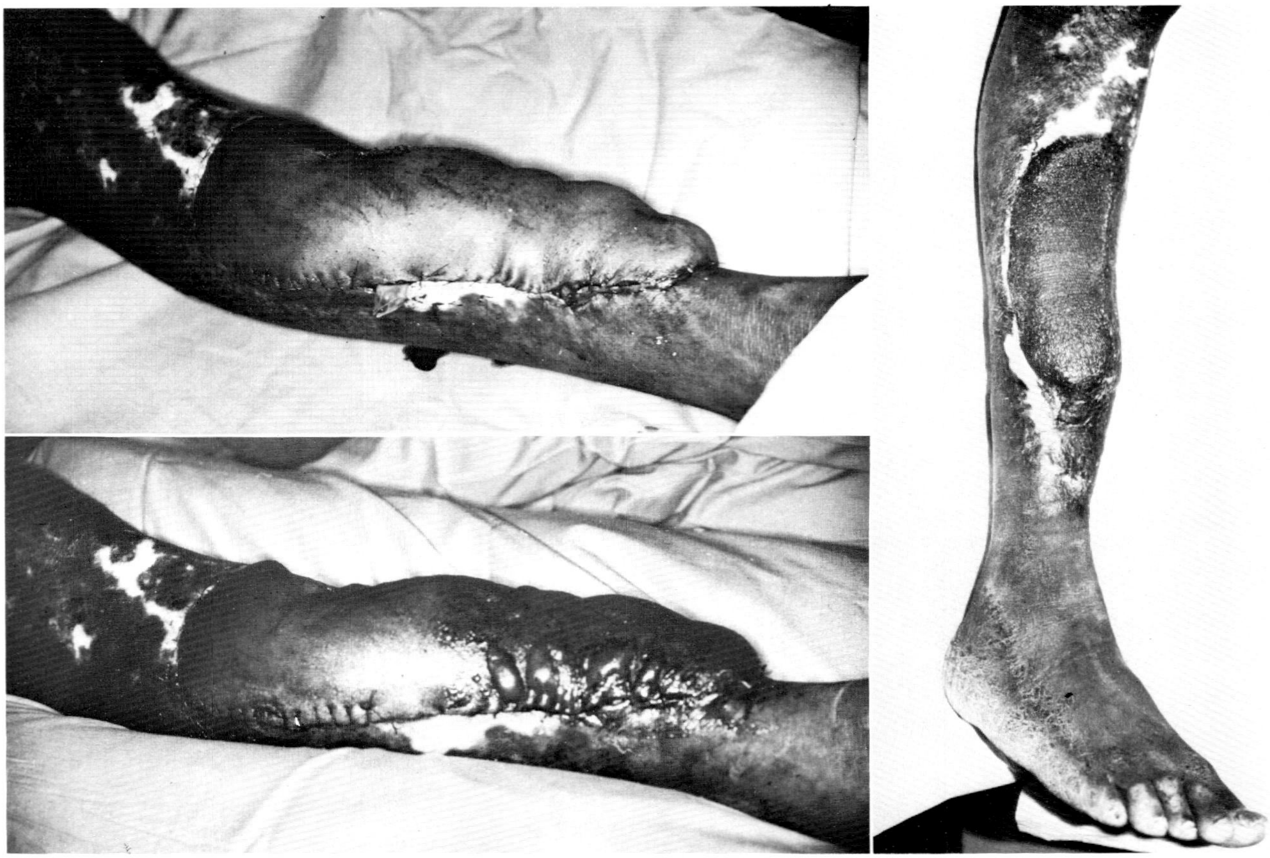

A

B

C

Fig. 39-3. Gracilis musculocutaneous flap with loss of distal skin.
A, Gracilis with entire overlying skin is transferred as free flap to lower extremity for coverage of unstable burn scar and exposed tibia.
B, Distal one third of skin island is necrotic. This was excised, and viable muscle underneath was skin grafted.
C, Nine-month postoperative result demonstrates stable wound coverage. NOTE: Muscle atrophy.

(From Mathes, S.J., Nahai, F., and Vasconez, L.O.: Myocutaneous free flap transfer: anatomical and experimental considerations, Plast. Reconstr. Surg. **62:**162, 1978.)

Gracilis

Planning errors. The skin over the distal one third of the gracilis is unreliable when elevated as a musculocutaneous island flap (Fig. 39-3). The musculocutaneous flap may be too bulky, and under these circumstances it is best to use the flap as muscle covered with split-skin graft.

Technical errors. The correct placement of the skin island over the gracilis muscle is essential. The muscle is first identified near its insertion. Then the skin island is designed over the proximal two thirds of the muscle. The skin should be sutured to the underlying muscle during the flap elevation.

Dissection of the pedicle is performed under loupe magnification. By retraction on the adductor longus muscle, a long pedicle is developed by dissection to the origin of the vascular pedicle with the profunda femoris artery and vein.

Postoperative errors. Suction drains are placed in the donor defect. If the flap has been transferred to the lower extremity, elevation of the lower extremity postoperatively is essential.

TFL

Planning errors. The proximal portion of the TFL flap may be too bulky in obese patients. The distal portion of the flap, made up of fascia and skin, is not suitable for transplantation into contaminated wounds. The donor defect usually requires skin grafting and must be taken into account during flap selection.

Technical errors. Separation of the TFL muscle from the gluteus minimus muscle may be difficult. During proximal dissection on the pedicle, the branches of the femoral nerve to the vastus lateralis are encountered. These nerves must be preserved.

Postoperative errors. If the TFL flap is transferred to the lower extremity, elevation of the extremity postoperatively is essential.

IV

EVALUATION

40

ASSESSMENT OF FLAP VIABILITY WITH FLUORESCEIN

Bernard S. Alpert

The value of intravenous fluorescein as an intraoperative indicator of tissue viability has been well demonstrated. Used and interpreted correctly, it is reliable in predicting survival of skin, muscle, and musculocutaneous flaps. Correct interpretation of the test is gained only through experience.

ADMINISTRATION

Fluorescein is available as fluorescein sodium 25% (Fluorescite) in 2-ml vials at (250 mg/ml). The safe clinical dose is 10 to 20 mg/kg body weight, although 10 mg/kg is sufficient for most patients. In black patients 20 mg/kg may be necessary for skin evaluation. The median lethal dose (LD_{50}) in experimental animals is 1000 mg/kg.

The drug is not administered until all dissection and flap placement is complete. After a 1-ml test dose, the remainder of the drug is given by intravenous drip for 2 to 5 minutes. Fluorescein rapidly distributes itself in the extracellular space in those areas with intact microcirculation. It has been used accurately in the past to determine circulation times in adults, consistently fluorescing through the lip mucosa in 15 to 20 seconds after injection. Evaluation therefore may safely be done 10 to 20 minutes after injection. Distribution of the drug does not reflect or correlate with blood flow or intravascular concentration. It is, however, a reliable predictor of the integrity of the microcirculation in a given area.

INTERPRETATION OF RESULTS

Fluorescein dye emits a strong fluorescence when exposed to a light source in the ultraviolet (3600 to 4000 Angstrom) wavelength. All lights in the operating room must be turned off, and a source of ultraviolet light such as a strong Wood's lamp is directed onto the area to be examined. Viable skin fluoresces brightly with a yellow or greenish hue (Fig. 40-1). Nonviable tissue appears dark or purple. Areas with spotty fluorescence are difficult to interpret but usually survive if there is a relatively uniform distribution of the spots. Additional manipulation of a flap after fluorescein injection invalidates the original interpretation. Increasing flap tension may result in necrosis in an area originally predicted to survive.

Vascular spasm may result in a false negative test (i.e., an area that proves viable later that does not fluoresce on testing). False positive tests do not occur as long as there is no further dissection or manipulation after the test.

PRECAUTIONS

Intravenous fluorescein is a safe drug. It stains the skin yellow for up to 24 hours after injection, and the patient and nursing persons should be warned. It is excreted unchanged in the urine within 36 hours of administration. In a compiled series of over 10,000 intravenous injections for ophthalmic angiography, the mean rate of reactions was 0.6%. Nausea, vomiting, and urticaria were the most common reactions. The incidence of severe reactions was 0.41%. Serious reactions included anaphylaxis, respiratory obstruction and arrest, hypotension with loss of consciousness, and one case of myocardial infarction and cardiac arrest. There was one death in over 10,000 cases.

The drug should not be administered without an emergency tray and oxygen available. A preoperative history of allergy should be taken but is not a contraindication to use. A test dose may be administered in these cases.

A

B

C

D

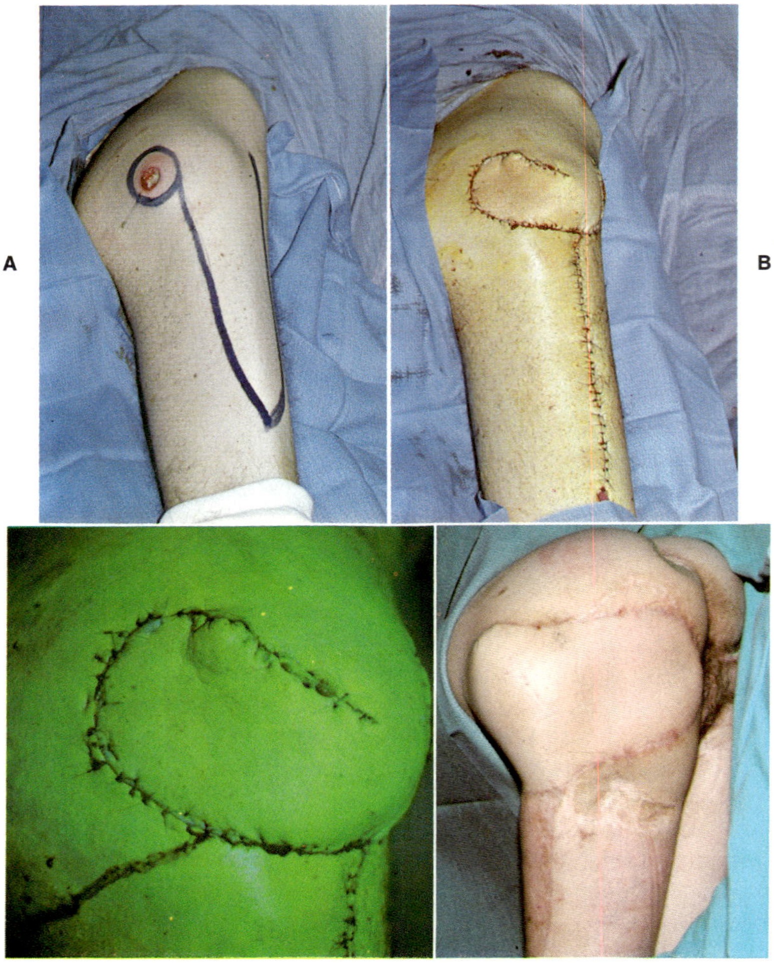

Fig. 40-1

A, Trochanteric pressure sore. Design of standard TFL musculocutaneous flap.

B, Flap inset and donor site closed. Intravenous fluorescein injected to assess flap viability.

C, Fluorescence noted throughout flap, indicating adequate capillary perfusion at time of flap inset.

D, Postoperative view demonstrates complete flap survival.

ANNOTATED BIBLIOGRAPHY

1. Lange, K., and Boyd, L.J.: The use of fluorescein to determine the adequacy of circulation, Med. Clin. North Am. **26:**943, 1942.

This article is an excellent review of the biophysical properties of fluorescein. Multiple clinical and experimental observations made, including evaluation of intestinal blood supply after incarceration, use in peripheral vascular obstruction, determination of circulation times in experimental animals and humans with cardiac failure, hyperthyroidism, hypothyroidism, and normal metabolic rates.

2. McCraw, J.B., Myers, B., and Shanklin, K.D.: The value of fluorescein in predicting the viability of arterialized flaps, Plast. Reconstr. Surg. **60:**710, 1977.

A clinical series demonstrating the predictive reliability of fluorescein in 159 cutaneous and 126 musculocutaneous flaps is presented. The importance of experience with interpretation is emphasized.

3. Myers, M.B.: Prediction of skin sloughs at the time of operation with the use of fluorescein dye, Surgery **51:**158, 1962.

A clinical study demonstrating the value of fluorescein in predicting skin flap viability at radical mastectomy is presented. Use in lower limb amputations is also discussed, demonstrating that the wound must be closed before the fluorescein test is performed. The additional tension of wound closure may produce false positive results.

4. Stein, M.R., and Parker, C.W.: Reactions following intravenous fluorescein, Am. J. Ophthalmol. **72:**861, 1971.

A compilation of adverse reactions in over 10,000 cases of use of intravenous fluorescein is presented. Overall reaction rate is 0.6%. Mild and serious reactions, allergic and nonallergic, are discussed. There is one death in the series.

41

LONG-TERM EVALUATION OF MUSCLE AND MUSCULOCUTANEOUS FLAPS

John J. Coleman, III ■ Foad Nahai ■ Stephen J. Mathes

. . . . Do they develop a sufficient secondary blood supply to every millimeter from the adjacent and subjacent tissues and, if so, how long does this process take? Can thinning (and carving and trimming be done as quickly, easily, and safely as in ordinary pedicle skin flaps? or more so?), or does it entail more time, more danger and more difficulty? Can it be done at all in some instances? When, if ever, can it not be done?

Frank McDowell, 1979

In his editorial *Logs vs. Harpsichords, blobby flaps vs. finished results,* McDowell raises these questions about free tissue transplantation and muscle and musculocutaneous flaps. The answers to these and many other questions may only be given after an analysis of long-term results, ideally through a prospective study with close follow-up. In this chapter we present conclusions based on long-term follow-up of patients who have undergone muscle and musculocutaneous flap procedures during the last 10 years. Many questions have now been answered. However, the evaluation of a reconstructive method or a given flap is a dynamic process. Progress in the field continues at a rapid pace. Thus the design and selection of flaps are constantly changing in an effort to improve the reconstructive process.

In the reconstructive ladder the muscle and musculocutaneous flaps are represented on the top two steps of the ladder (i.e., they are available as both local and distant flaps for reconstructive surgery) (Fig. 41-1). This evaluation of these flaps therefore includes a comparison with the alternate tissues available as local or distant flaps. In addition, an analysis of results in terms of donor site deformity, especially functional changes and long-term durability of the reconstruction, is presented.

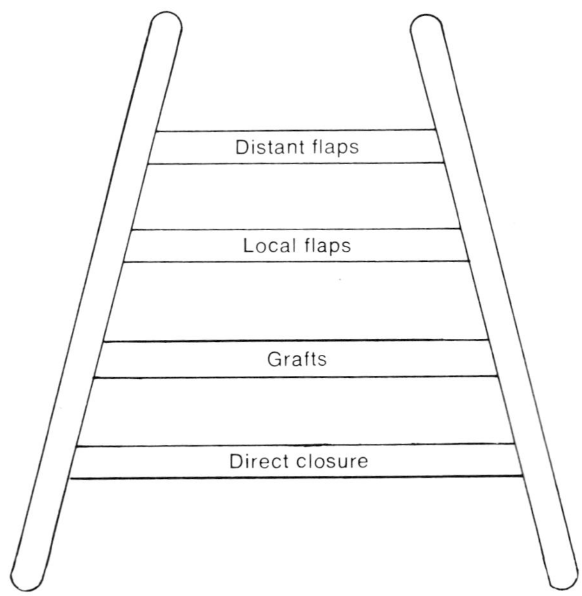

Fig. 41-1. Reconstructive ladder from simple to more complex reconstructive methods.

GENERAL EVALUATION

In comparing muscle and musculocutaneous flaps with axial or random skin flaps, three general factors must be discussed: (1) blood supply and hence flap reliability, (2) flap size and thickness, and (3) functional and aesthetic aspects of donor regions. Comparison of results in these terms will assist the reconstructive surgeon in selection of a method of reconstruction for specific regions.

Blood supply and flap reliability

The location, size, and number of vascular pedicles to each muscle and musculocutaneous flap are precisely defined and the intramuscular course of the vessels clearly delineated. Preservation of these pedicles during flap elevation will preserve circulation to muscle and overlying skin. This is in sharp contrast to the random skin flap in which skin circulation is adversely altered the moment the flap is elevated. Flap reliability has been a constant feature when the muscle and musculocutaneous flap is used in reconstructive surgery.

For tissue transplantation both the axial and the muscle and musculocutaneous flaps are useful. However, anatomic variations in the axial flap vasculature are far more common. Also, the vascular pedicle lumen diameter is much smaller in size, as compared to the dominant or major vascular pedicle to muscle. Both greater selection in donor flap availability and the ease of pedicle location have shifted the emphasis to the muscle and musculocutaneous flap for use in free tissue transplantation. The axial skin flaps also have well defined and precisely delineated vascular pedicles. However, anatomic variations are far more commonly seen with the axial skin flaps, and the arteries are much smaller in size (rarely over 1 to 1.5 mm in diameter), factors that make muscle and musculocutaneous flaps far more attractive for tissue transplantation.

The strong vascular supply to muscle and musculocutaneous flaps and to the axial skin flaps raises several questions concerning division of the pedicle after flap transposition. Provided the flaps are transposed into well-vascularized tissues, pedicle division is safe after 2 weeks. Experimental data have shown that the muscle or musculocutaneous flaps are revascularized by the surrounding tissues primarily through the underlying muscle. In our experimental model this revascularization is sufficient to routinely support the flap as early as 6 days after transposition into a well-vascularized bed. However, clinically pedicle division is recommended in 2 or 3 weeks after transposition, depending on the vascular status of the defect. Staged pedicle division is preferred when the flap is transposed over bone or into an irradiated field. An excellent example is the use of the gastrocnemius cross-leg flap when transposed over infected or exposed bone. A delay of the pedicle division is recommended to avoid muscle flap necrosis. The vascular pedicle is isolated and divided at 2 or 3 weeks after transposition. The muscle is then divided a week later. The tourniquet test is recommended to assess flap viability before flap pedicle division. A Penrose drain is tied tightly around the flap base and the perfusion of the area is evaluated with fluorescein before flap division and inset are completed.

A clinical impression that has now been demonstrated by experimental studies is the superior resistance of the skin and muscle of the musculocutaneous flap to bacterial inoculation when compared to the skin and subcutaneous tissue of the random flap. The resistance of the muscle and musculocutaneous flap to infection appears related to the vascularity of the flap.

Flap size and thickness

Flaps based on the vascular pedicles to muscle or on direct axial cutaneous vessels may safely be designed as large flaps. Random skin flaps are limited in size. The largest flaps in terms of surface area are those based on the latissimus dorsi, rectus abdominis, and TFL muscles.

In terms of the aesthetics of the final result, the thickness of the flap is a most important consideration. Skin flaps are thinner than musculocutaneous flaps and may require less secondary procedures for thinning to restore form and contour. However, a muscle flap with skin graft is often thin enough to not require any secondary procedures.

Prediction of the extent of muscle atrophy after transposition of transplantation of the muscle or musculocutaneous flap is still not always possible. Two factors affect muscle atrophy: (1) division of the motor nerve and (2) alteration of muscle tension through division of origin or insertion. After transposition as a flap, one or both of these factors will affect the muscle and hence flap bulk. Under certain circumstances (e.g., breast reconstruction or correction of Poland's syndrome) the motor nerve is spared, and the muscle after transposition is sutured under adequate tension so that the muscle remains functional. In this manner, significant atrophy is avoided.

Finally, if a bulky musculocutaneous flap is transposed, the muscle will atrophy, but the thick subcutaneous layer of fat will not. Therefore it is important to recognize how much of the thickness of a musculocutaneous flap is the subcutaneous fat and how much is the muscle.

Thinning of a thick flap in terms of excision of subcutaneous fat or trimming of muscle may be done safely at the time of flap elevation if the pedicle, the intramuscular vessels, and musculocutaneous perforators are not disturbed. If the process of thinning should involve or endanger these vessels, it is essential to wait at least 2 weeks before the thinning is undertaken.

Functional and aesthetic aspects of the donor deformity

The functional and aesthetic aspects of the donor defect must be considered in selecting flaps. The donor defects of most axial flaps are closed directly with easily concealed scars. The donor defects of random flaps often require skin grafts. Although the donor defects of muscle flaps are always closed directly, the donor defects of musculocutaneous flaps may often require skin grafts. The larger the skin component of a musculocutaneous flap, the less the likelihood of direct closure.

Some musculocutaneous flaps leave such unacceptable donor defects (e.g., gastrocnemius) that their use as musculocutaneous flaps is avoided when possible.

The functional deficit after transposition of a muscle as a flap is minimized by the action of synergistic muscles. Muscles whose function can be adequately preserved by synergistic muscles have been termed *dispensable muscles*. Our clinical observations have shown no observable deficit in function after transposition of these muscles as flaps. Because the transposition of muscle flaps with functional preservation is relatively recent, long-term evaluation of these flaps is not as yet possible. However, segmental transposition of muscle while maintaining the continuity of muscle origin and insertion appears to preserve adequate muscle function and still provide reliable coverage.

SPECIFIC EVALUATION

Head and neck

Reconstruction after extirpative head and neck surgery demonstrates the recent rapid progression of flap selection and design. The skin of the anterior chest is now transposed as part of a musculocutaneous flap, the pectoralis major, and less commonly as an axial flap, the deltopectoral flap for replacement of missing facial skin and oral lining. Likewise the skin of the shoulder and back is now designed as a composite of skin and muscle, the anterior and posterior trapezius musculocutaneous flap, and rarely as an axial skin flap, the cervical-humeral flap (Fig. 41-2). These changes in flap selection and design will continue in an effort to improve the reliability, appearance, and function of head and neck reconstruction.

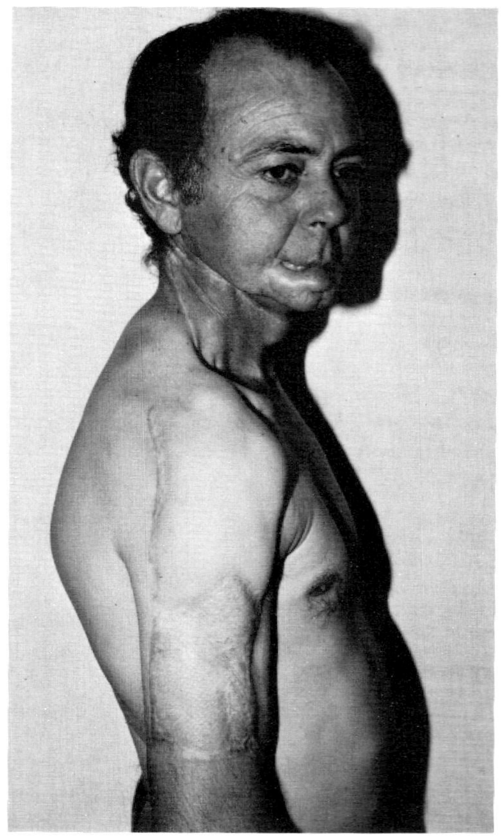

Fig. 41-2. Wide local excision of squamous cell carcinoma of lip and radical neck dissection were performed on this patient, followed by immediate reconstruction with cervicohumeral flap. Five-year long-term follow-up result shows no evidence of recurrent disease. NOTE: Grafted donor area on right arm.

The pectoralis major flap is now the flap most often used for reconstruction of the head and neck. The pectoralis major has certain advantages. When compared to the deltopectoral flap, one-stage reconstruction is the rule rather than the exception. The muscle carrier for the skin island tunnelled under the neck flap has provided excellent vascularized coverage for the carotid artery.

The donor defect for the pectoralis major has proved aesthetically more acceptable than that of the deltopectoral flap. The donor defect of the pectoralis major is usually closed directly, and if is placed in the inframammary fold in women, it is easily concealed.

The functional deficit from the pectoralis major has been minimal. Muscle and musculocutaneous flaps have tolerated postoperative irradiation without any problems (Fig. 41-3).

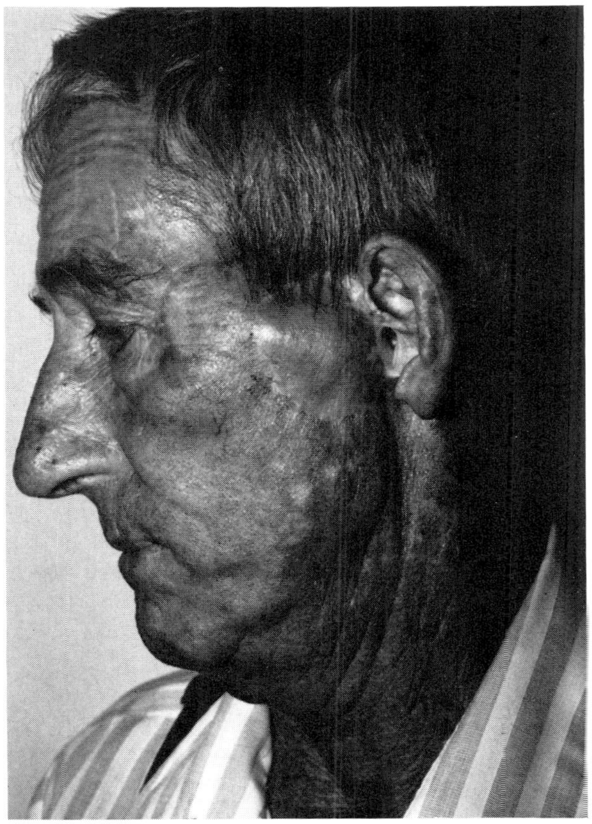

Fig. 41-3. Squamous cell carcinoma of lip was resected in this patient. Superficial parotidectomy and resection of metastatic carcinoma were required 22 years later. Wound was covered with temporalis muscle flap and skin graft, followed by postoperative irradiation to 5100 rads. Six-year long-term follow-up result shows stable coverage with skin graft intact.

Trunk

Reconstruction of the trunk has included the pectoralis major, latissimus dorsi, rectus abdominis, trapezius, and external oblique muscles. The application of the pectoralis major and rectus abdominis for coverage of the infected median sternotomy wound has had a dramatic impact on the management of this difficult problem. All the flaps have survived with no recurrence of the infection at 3 years after reconstruction. The only alternative flap for this problem is the greater omentum, which has been equally effective. However, the omentum carries the added risk of a laparotomy and should be a flap of second choice if the pectoralis major or the rectus abdominis muscles are not available.

For coverage of the abdominal wall, the TFL has proved excellent with no breakdown or abdominal wall hernias at 4 years after reconstruction. However, the rectus abdominis as a musculocutaneous flap will provide coverage for the abdominal wall, but as the muscle is denervated, it will not reestablish the integrity of the abdominal wall, and late herniation has been seen (Fig. 41-4).

A

B

C

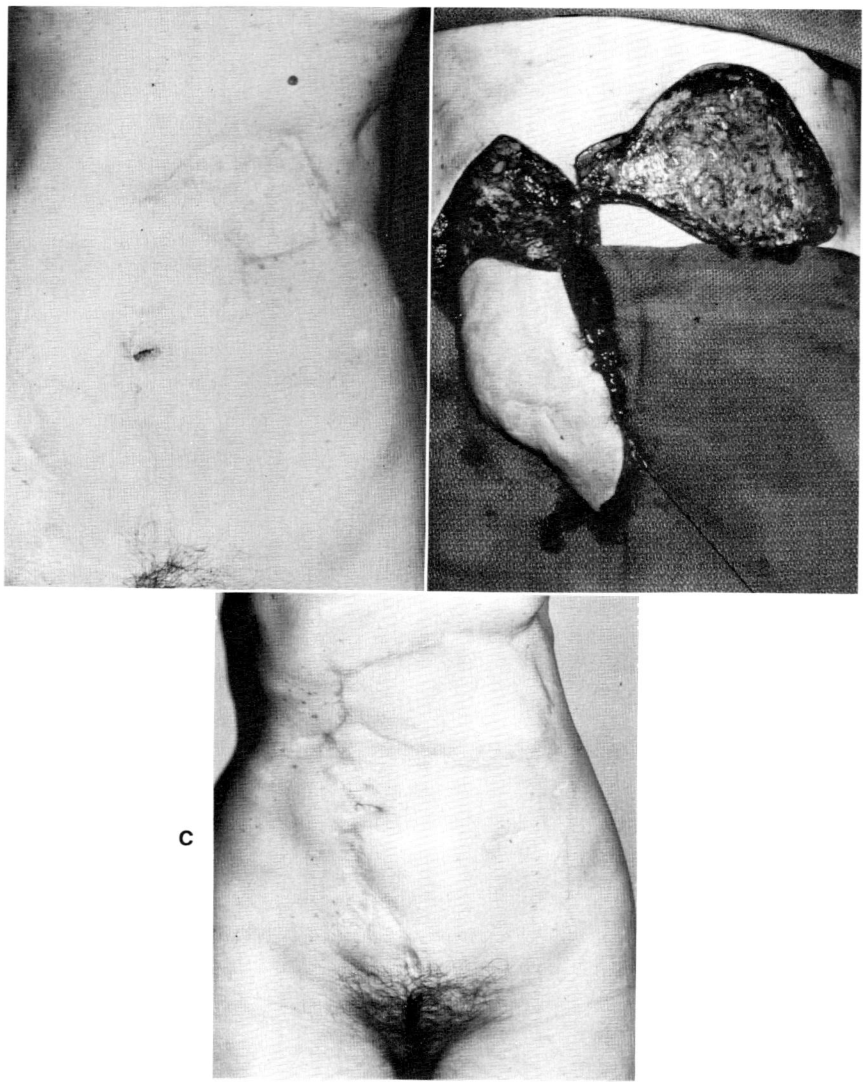

Fig. 41-4. For legend see opposite page.

Breast

The advantages of placing an implant under muscle in breast reconstruction have already been emphasized. Implant exposure has almost been eliminated by submuscular placement. The question of capsular contracture remains unanswered. Capsules do form around implants under muscle, but these appear to be less perceptible.

The use of the latissimus dorsi musculocutaneous flap has provided a reliable method for single-stage breast reconstruction. There has been no significant functional deficit after transposition of the latissimus dorsi muscle. Although the donor scar is not desirable, it is well concealed and accepted by the patient. Earlier diagnosis of breast cancer, less radical extirpative procedures (e.g., preservation of the pectoralis major muscle), and improvements in the technique of flap inset (e.g., low placement of the skin island regardless of location of the mastectomy scar) have improved the results of breast reconstruction.

All problems of breast reconstruction are not answered. Restoration of the anterior axillary fold and avoidance of muscle atrophy are not fully solved in breast reconstruction after radical mastectomy. The search for new techniques for breast reconstruction has not stopped. Recent innovations in breast reconstruction using the rectus abdominis musculocutaneous flap show promise. The superior transverse skin island based on the contralateral rectus abdominis muscle allows closure of the donor defect in the inframammary region of the opposite breast, while providing missing skin for the breast reconstruction without the sequelae of the back scar associated with the latissimus dorsi musculocutaneous flap. The use of the inferior abdominal wall skin based on the rectus abdominis muscle provides enough skin and subcutaneous tissue when transposed for breast reconstruction to eliminate the need for a synthetic breast implant with its inherent potential for capsular contracture. Naturally these new approaches raise new questions, such as the potential for abdominal hernia formation, but also indicate the required search for improvements in current techniques and design of new ones.

Recurrence of the underlying breast carcinoma has been seen after breast reconstruction. However, detection has been possible and treatment not adversely effected. Overall, the knowledge that breast reconstruction is available has encouraged the patient with breast cancer to accept the required extirpative therapy.

Fig. 41-4. Shotgun wound to left lower chest and upper abdomen. Immediate reconstruction was accomplished with Marlex mesh closure and skin graft. However, skin grafts over mesh were unstable and flap coverage was required.

A, Unstable skin over mesh in left upper abdominal wall.

B, Reconstruction of abdominal wall with rectus abdominis island musculocutaneous flap.

C, Hernia noted 2 years later at junction of mesh with abdominal wall fascia and was repaired with latissimus dorsi musculocutaneous flap. Hernia recurred for the third time 2 years later at junction of rectus abdominis and latissimus dorsi flaps and was repaired with TFL musculofascial flap.

(**A** and **B** from Mathes, S. J., and Bostwick, J., III: Br. J. Plast. Surg. **30:**282, 1977.)

Pressure sores

With the exception of sensory flaps, the extensive application of muscle and musculocutaneous flaps for closure of pressure sores has not altered the rates of recurrence (Fig. 41-5). However, the immediate problems related to flap failure and postoperative infection seen with the random flaps have been greatly reduced. Therefore the muscle and musculocutaneous flaps have not prevented long-term recurrence but have greatly reduced postoperative morbidity and hence postoperative recovery time. Skin grafts placed directly over muscle have also proved as durable as musculocutaneous flaps or skin flaps.

The sensory TFL flap has significantly altered the course of patients in whom it has been available for transposition. There has been no recurrence at 3 years after transposition on the side of the flap, and in patients with bilateral pressure sores a unilateral sensory flap has allowed spontaneous closure of the opposite pressure defect and has prevented further recurrence.

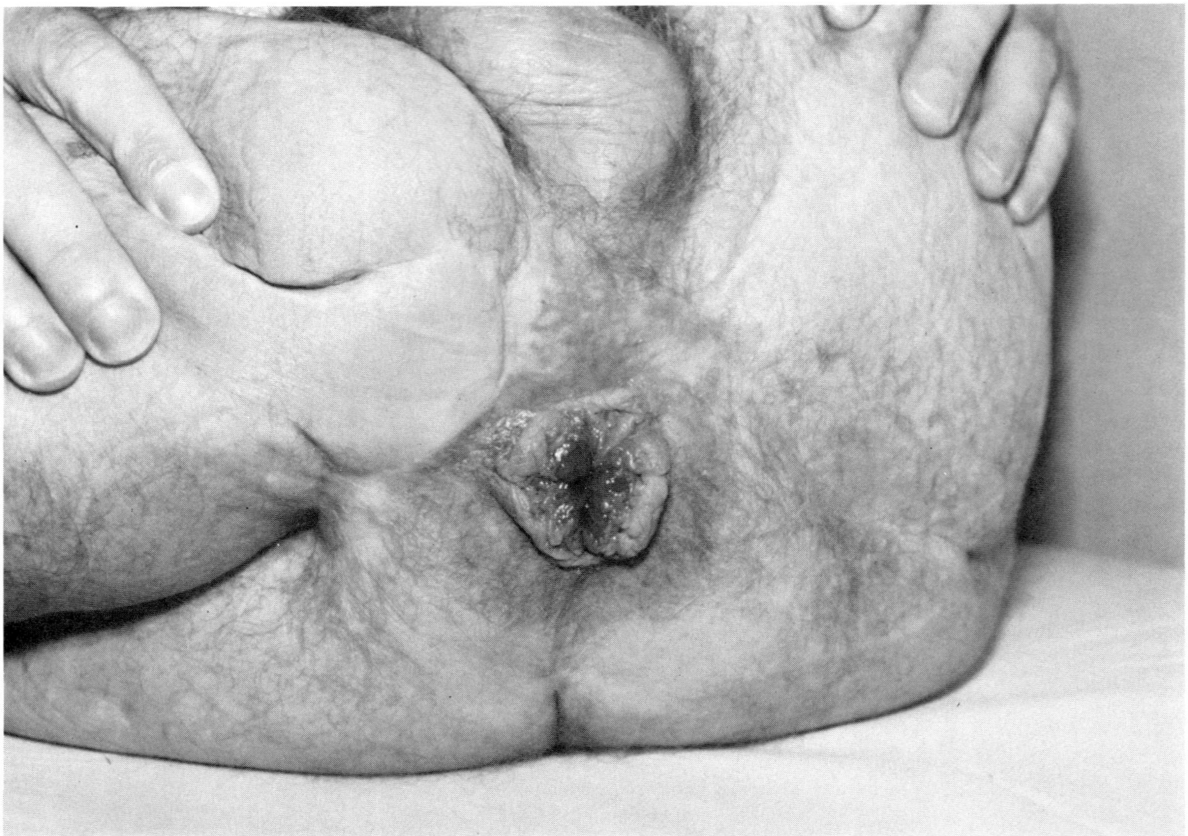

Fig. 41-5. Bilateral hip disarticulations in young patient with T11 paraplegia, as result of automobile accident, and recurrent pressure ulceration. Surgical procedures required in 11-year period after accident include ischiectomy and closure of right ischial pressure sore with posterior thigh flap, coccygectomy and primary closure of sacral pressure sore, closure of lateral malleolar pressure sore with skin graft, excision of right trochanteric pressure sore and closure with TFL flap, excision of recurrent right ischial pressure sore and direct closure, right hip disarticulation and closure with thigh flap, gracilis musculocutaneous flap for closure of recurrent left ischial pressure sore, left TFL flap for recurrent trochanteric pressure sore, and left hip disarticulation and total thigh flap. Despite numerous procedures, including multiple muscle and musculocutaneous flaps, this patient developed recurrent ulceration and eventually underwent bilateral hip disarticulation with thigh flaps for closure.

Extremities

Our longest follow-up has been with flaps to the lower extremity. In the patients available for long-term follow-up, these flaps have proved durable with no breakdown. Where flaps have been transposed into infected bone, there has been no recurrence of the infection. In patients where the flaps had failed to control the infection, the failure was evident at an early stage. There has been no clinically detectable functional deficit or alteration of gait as a result of transposition of the soleus or one head of the gastrocnemius (Figs. 41-6 and 41-7).

The donor defect of the medial gastrocnemius musculocutaneous flap has presented some long-term problems (Fig. 41-8). It has proved unacceptable to our patients. (Two have asked for further reconstructive procedures to improve on the donor deformity.) We no longer use the gastrocnemius as a musculocutaneous flap if alternative methods are feasible for defect coverage.

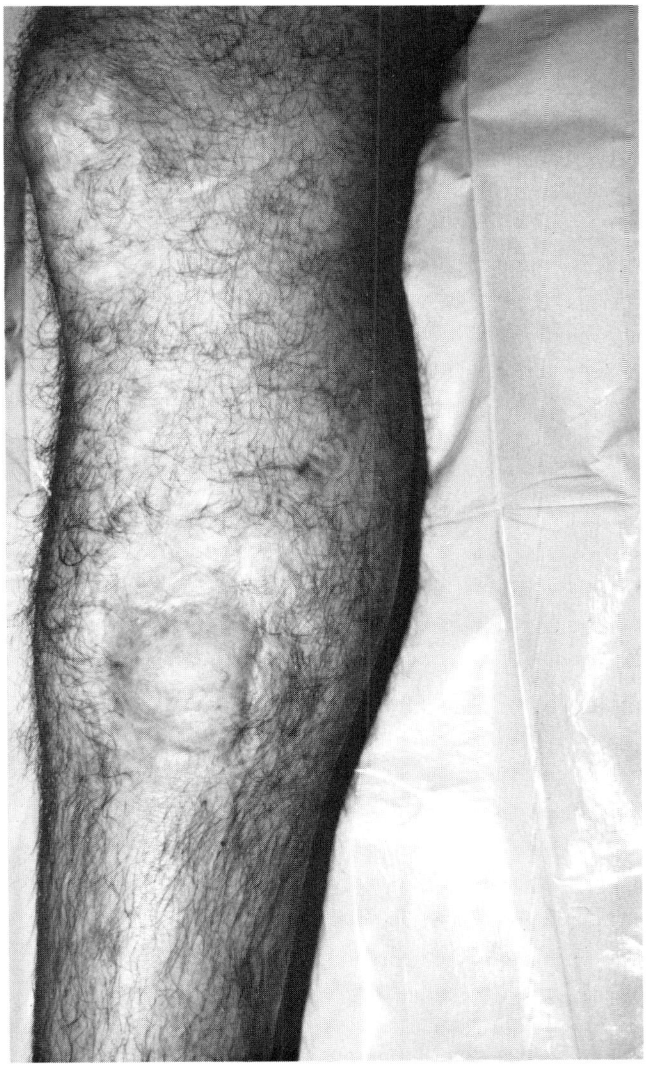

Fig. 41-6. Patient had open fracture of middle third of tibia as result of motorcycle accident. Fracture was primarily treated by débridement and casting, however, patient developed chronic nonhealing wound over tibia. Wound was debrided 1 year after accident and closed with soleus muscle flap and delayed skin graft. Follow-up at $8^1/_2$ years shows skin graft over muscle is stable and patient, who works as truck driver, has no functional deficit.

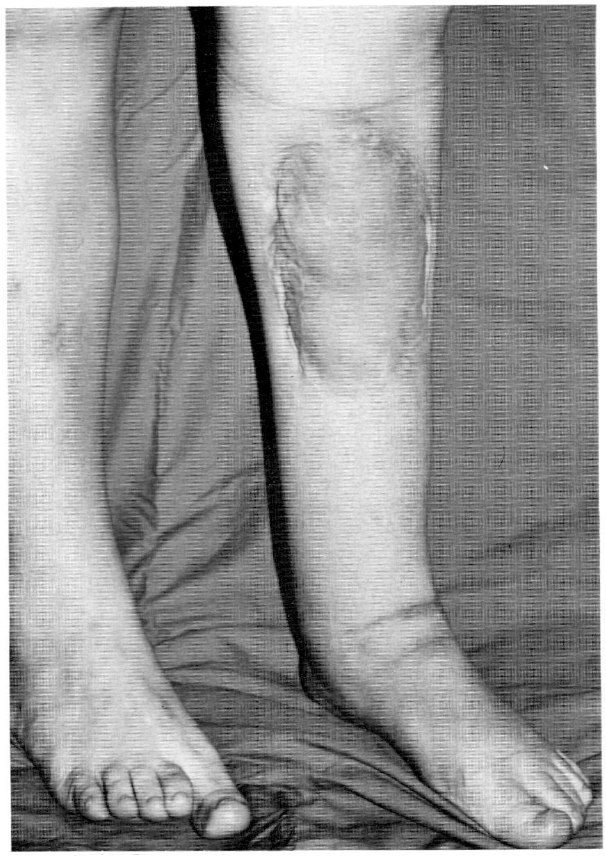

Fig. 41-7. 4-year-follow up of medial gastrocnemius and soleus flaps with skin graft for coverage of irradiated defect of the upper half of the tibia as a result of soft tissue malignancy of leg. Wound was covered with gastrocnemius and soleus muscle flaps. Four-year follow-up shows stable wound, no evidence of recurrence, and no functional deficit.

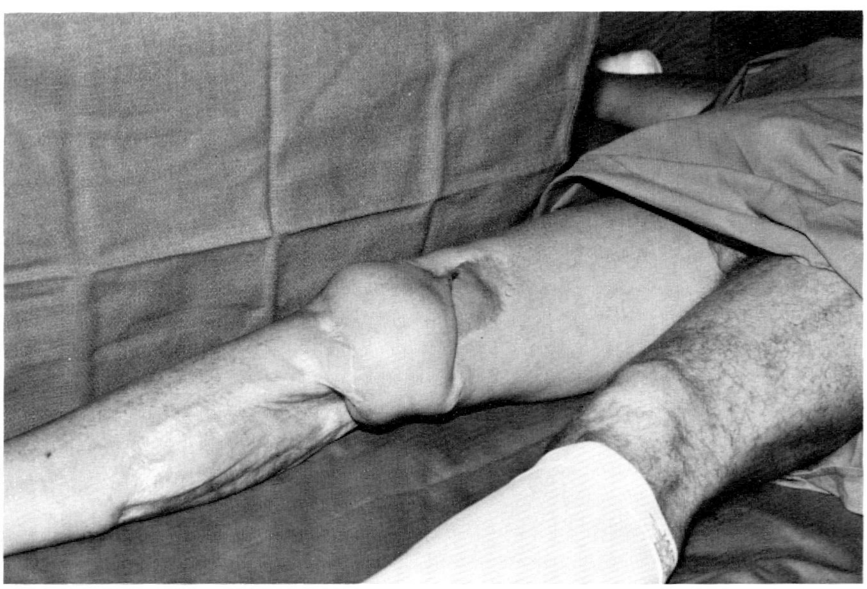

Fig. 41-8. Unacceptable donor defect of medial gastrocnemius musculocutaneous flap.

Feet

The role of sensibility in the foot cannot be over emphasized (Figs. 41-9 and 41-10). Muscle flaps in innervated feet have performed well and have provided excellent coverage for small defects. There has been no noticeable functional deficit, and the skin grafts over these muscles have held up well.

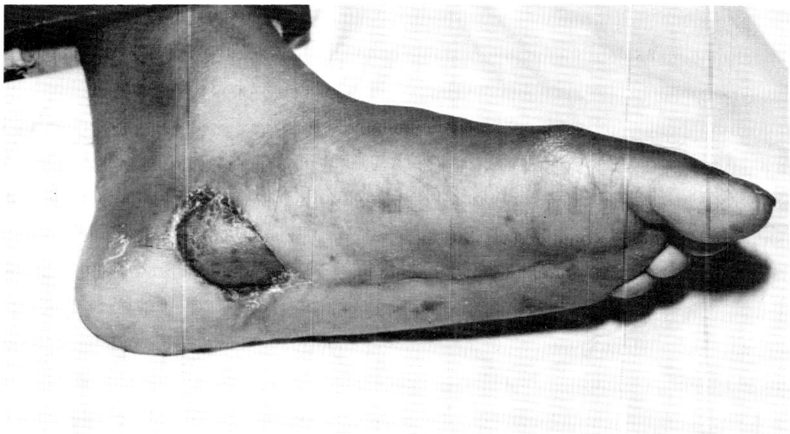

Fig. 41-9. Shotgun wound of foot in 25-year-old man with loss of soft tissue and bone of heel. Wound was covered with flexor digitorum brevis flap and skin graft 3 weeks after trauma. Three years after trauma patient has stable wound and protective sensibility with no functional deficit of the foot.

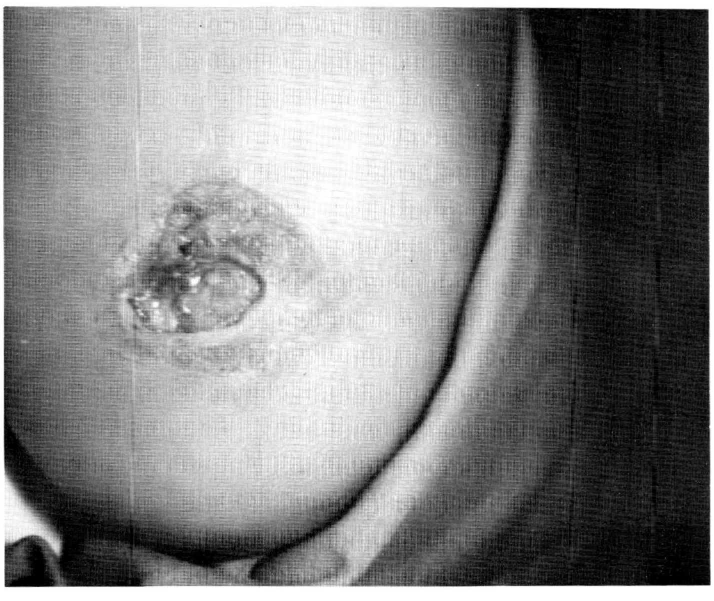

Fig. 41-10. Long-term follow-up of flexor digitorum brevis and split-thickness skin graft in patient with myelomeningocele and heel ulcer. Within 5-year period after pressure ulcer developed, patient underwent six skin grafting procedures with recurrence of breakdown of graft. Persistent ulcer covered with flexor digitorum brevis muscle flap with skin graft. Four-year follow-up view shows breakdown of skin graft over muscle, but muscle was intact and was successfully regrafted. In contrast to patient in Fig. 41-9, this patient had no sensation in his foot, and graft over flap broke down after 4 years. However, despite breakdown, skin graft over flap had lasted 4 years, in contrast to frequent skin graft procedures performed during prior 5-year period.

Free tissue transplantation

Muscle and musculocutaneous flaps have been transplanted by microvascular techniques with increasing frequency over the last 5 years. The predictable anatomy of the vascular pedicles and the large size of the vessels together with the length of the pedicles make the muscle and musculocutaneous flaps easy to elevate and transplant as free flaps. Our success rate in terms of flap survival over the last 5 years with axial skin flaps has been approximately 80%, but it has been over 90% with muscle and musculocutaneous flaps, with the last 50 muscle and musculocutaneous flaps transplanted as free flaps without a loss (Fig. 41-11).

By far the majority of the flaps were transferred to the lower extremity into contaminated wounds, over open fractures, or into cavities resulting from osteomyelitis. In the majority of patients the infection was brought under control, the fractures healed, and the chronic osteomyelitis resolved. In two patients, however, the flap failed to resolve the chronic osteomyelitis. These failures were obvious at a very early stage.

A

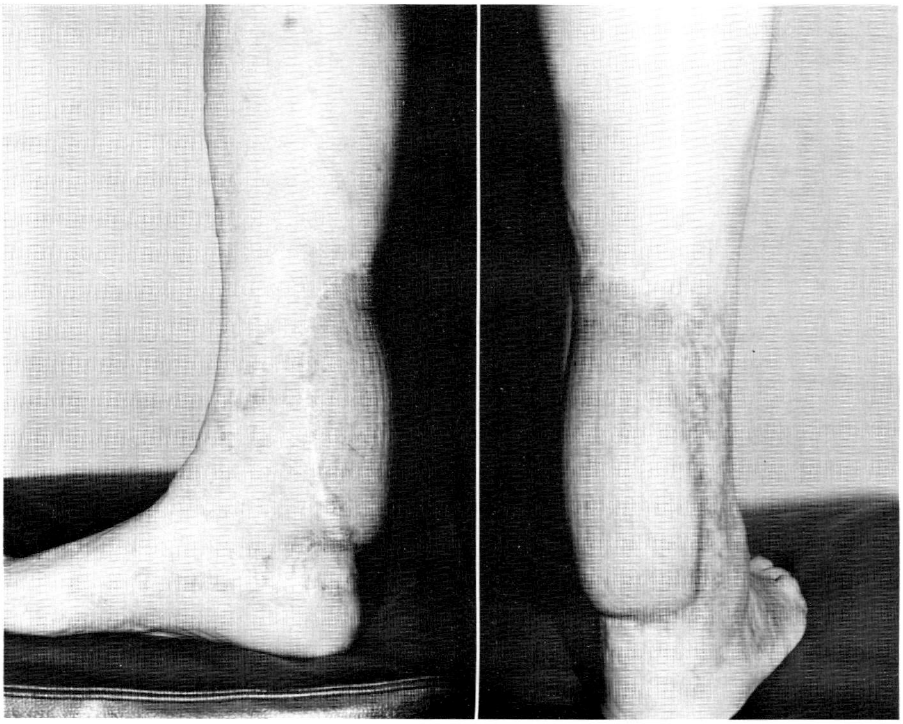

B

Fig. 41-11
A and **B,** Long-term follow-up on gracilis musculocutaneous free flap transplanted to the lower leg of patient with development of chronic osteomyelitis at site of traumatic wound. Multiple débridements with skin graft coverage failed. Attempted soleus flap transfer was unsuccessful, and defect was again skin grafted. Gracilis musculocutaneous free flap transfer was performed 20 years after accident as result of recurrent breakdown of wound. Follow-up ($5^1/_2$ years) after gracilis musculocutaneous transplantation shows wound is stable with no evidence of breakdown. However, flap remains bulky because of thick layer of subcutaneous fat. Patient has no functional deficit and is active as county sheriff.

A limited number of free flaps have been transposed into patients with chronic venous ulceration. The numbers are too small for any conclusions, however, Fig. 41-12 shows a latissimus dorsi flap with the island of skin transferred with the muscle has remained intact at 4 years after transposition, but the skin grafted portion of the free flap has broken down.

The sensory TFL flap has proved durable after transfer to weight-bearing areas in younger patients (Fig. 41-13). In older patients, however, areas of breakdown have been seen despite the return of excellent sensibility.

A

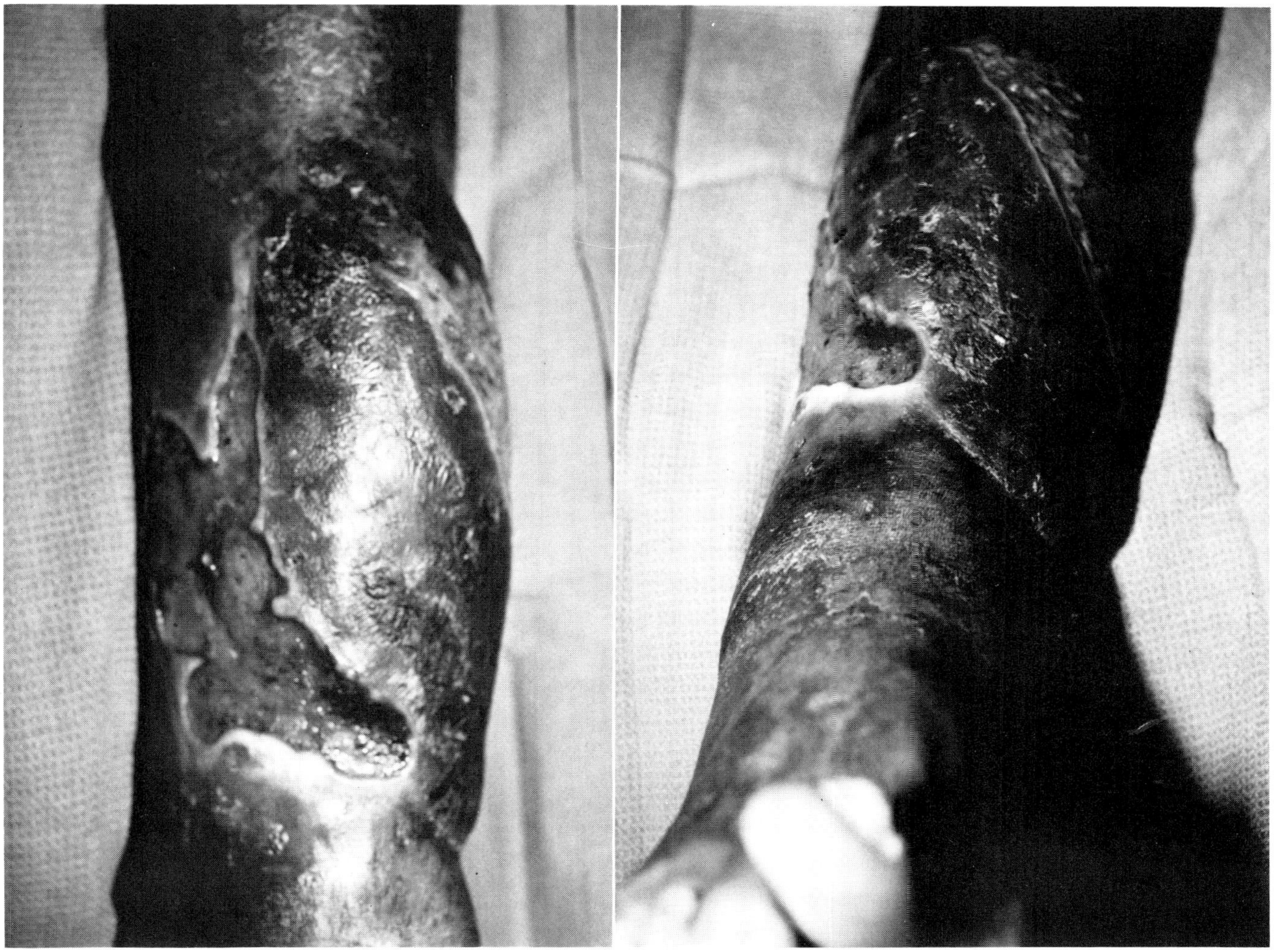

B

Fig. 41-12
A and **B,** Four-year follow-up of latissimus dorsi musculocutaneous free flap transplantation and skin graft for chronic venous stasis ulcer of leg. Multiple skin grafts failed. Small island of skin with latissimus dorsi muscle was taken so that donor defect in back could be closed directly. Part of transferred muscle was therefore skin grafted. Four-year follow-up shows skin island has remained intact, but skin graft over muscle has again ulcerated. Patient is laborer and has been working over past 4 years with no deficit from latissimus dorsi muscle donor area.

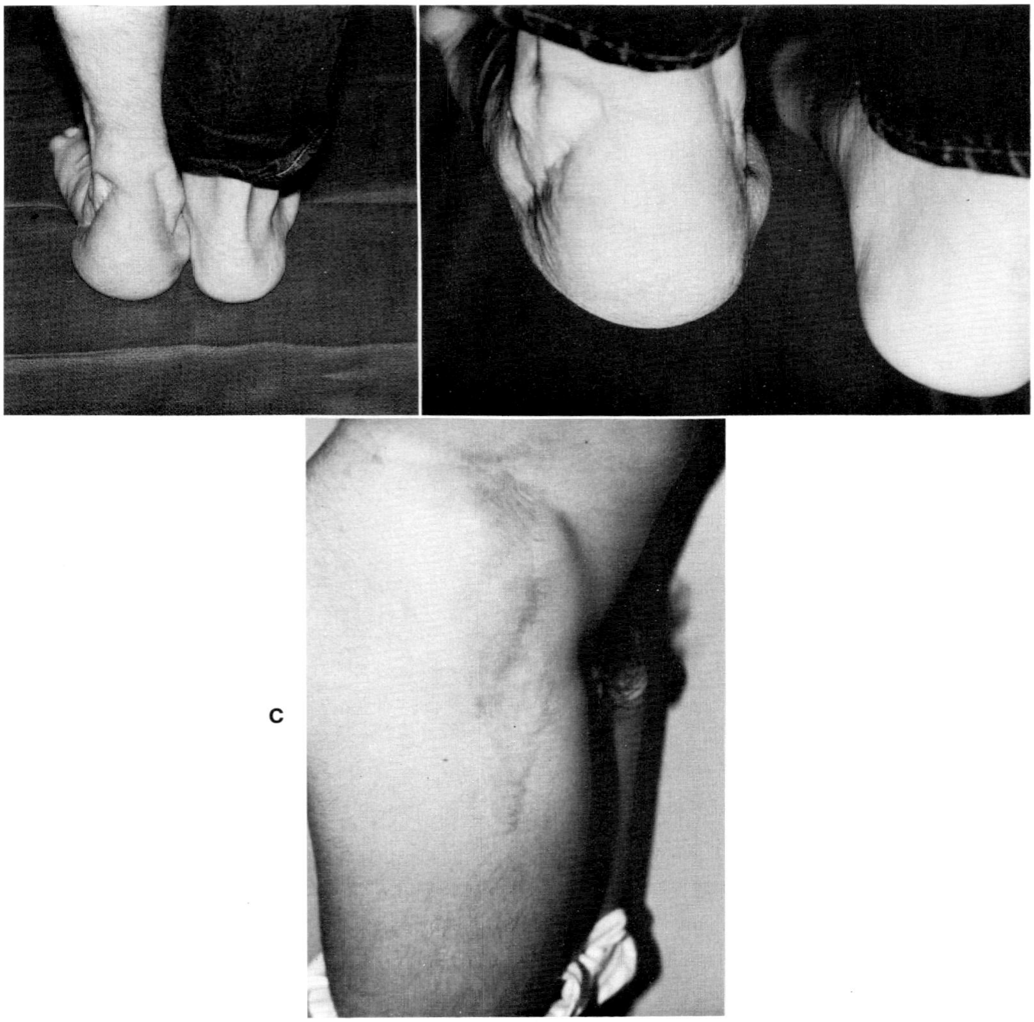

Fig. 41-13. Free TFL neurosensory flap with iliac crest bone was transferred to left heel of patient with posttraumatic amputation as result of lawn mower accident.

A and **B,** Follow-up at $3^1/_2$ years after composite flap transplantation shows that flap remains bulky. (Flap bulk is made up now of subcutaneous tissue.) This is a neurosensory flap with no breakdown in this active young man.

C, Donor area that was closed directly has not caused problems, and patient has no functional deficit related to flap donor site.

ANNOTATED BIBLIOGRAPHY

1. Ger, R., and Levine, S.R.: The management of decubitus ulcers by muscle transposition, Plast. Reconstr. Surg. **58:**419, 1976.

   An 8-year review of pressure sore closure with muscle flaps and skin grafts with 7-year follow-up is presented. Skin graft over muscle is shown to be stable.

2. McDowell, F.: Logs vs. harpsichords, blobby flaps vs finished results (editorial), Plast. Reconstr. Surg. **64:**249, 1979.

   An eloquent plea for finished products is presented. Questions concerning free tissue transfer and muscle and musculocutaneous flaps are raised.

3. Mathes, S.J., Alpert, B., and Chang, N.: Use of the muscle flap in chronic osteomyelitis: experimental and clinical correlation, Plast. Reconstr. Surg., vol. 69, 1982.

   Experimental data comparing the muscle and musculocutaneous flap with the random pattern flap is presented, demonstrating increased resistance of the muscle to bacterial inoculation and increased oxygen tension in the distal muscle. Successful management of 11 consecutive patients with chronic osteomyelitis utilizing sequestrectomy and gracilis muscle transplantation is presented.

4. Mathes, S.J., and Nahai, F.: Classification of the vascular anatomy of muscles: experimental and clinical correlation, Plast. Reconstr. Surg. **67:**177, 1981.

   A systematic classification of the blood supply and the known clinical useful musculocutaneous flaps is presented. Vascular anatomy reliably predicts bulk of tissue and clinical usefulness in these flaps.

5. Mathes, S.J., and Nahai, F. Muscle flap transposition with function preservation: technical and clinical considerations, Plast. Reconstr. Surg. **66:**242, 1980.

   This article is a graphic demonstration that portions of indispensable muscles can be rotated as adequate carriers for a large unit of muscle and skin and that function of the muscle can be preserved.

INDEX

N